The Finite Element Method

Linear Static and Dynamic Finite Element Analysis

The Finite Element Method

Linear Static and Dynamic Finite Element Analysis

Thomas J. R. Hughes

Professor of Mechanical Engineering
Chairman of the Division of Applied Mechanics
Stanford University

PRENTICE-HALL, INC., Englewood Cliffs, New Jersey 07632

Library of Congress Cataloging-in-Publication Data

HUGHES, THOMAS J. R.
The finite element method.

Bibliography.
Includes index.
1. Finite element method. 2. Boundary value problems. I. Title.
TA347.F5H84 1987 620'.0042 86-22558
ISBN 0-13-317025-X

Editorial/production supervision and interior design: Joan McCulley
Cover sketch: M. Igarashi and K. Shimomaki
Suzuki Motor Co., Ltd.
Cover design: 20/20 Services, Inc.
Manufacturing buyer: Rhett Conklin

Printed in the United States of America

10 9 8 7 6 5 4 3 2 1

ISBN 0-13-317025-X 025

PRENTICE-HALL INTERNATIONAL (UK) LIMITED, *London*
PRENTICE-HALL OF AUSTRALIA PTY. LIMITED, *Sydney*
PRENTICE-HALL CANADA INC., *Toronto*
PRENTICE-HALL HISPANOAMERICANA, S. A., *Mexico*
PRENTICE-HALL OF INDIA PRIVATE LIMITED, *New Delhi*
PRENTICE-HALL OF JAPAN, INC., *Tokyo*
PRENTICE-HALL OF SOUTHEAST ASIA PTE. LTD., *Singapore*
EDITORA PRENTICE-HALL DO BRASIL, LTDA., *Rio de Janeiro*

To my wife,
my mother,
and my children

Contents

Part Two Linear Dynamic Analysis

Preface

This book is based on courses that I have taught at the California Institute of Technology and Stanford University during the last 10 years. It deals with the finite element method in linear static and dynamic analysis. It is intended primarily for engineering and physical science students who wish to develop comprehensive skills in finite element methodology, from fundamental concepts to practical computer implementations.

Some sections of this book touch upon the frontiers of research and have been used as lecture notes in a number of more advanced short courses I have taught in Europe, Japan, and the United States. Consequently, I believe it will also be of interest to more experienced analysts and researchers working in the finite element field.

SUBJECTS COVERED

The first chapter of the book introduces the finite element method in the context of simple one-dimensional model problems. Chapter 2 deals with variational formulations of two- and three-dimensional boundary-value problems in heat conduction (in fact, all problems governed by Laplace/Poisson equations, such as electrostatics, potential flow, elastic membranes, and flow in porous media) and elasticity theory. These serve as the basis for finite element discretization and the techniques developed illustrate the relationship between "strong," or classical, statements of boundary-value problems and their "weak," or variational, counterparts. The Galerkin method of approximate solution is emphasized in Chapter 2 and throughout the book, rather than "variational principles" due to the significantly greater generality of the former approach. In Chapter 3 a variety of finite element interpolatory schemes are developed.

These apply to triangles, quadrilaterals, tetrahedra, hexahedra, wedges, etc. The isoparametric concept is emphasized, and special-purpose interpolatory strategies are also developed (e.g., "singular elements"). Programming techniques are introduced for numerically integrated finite elements. Chapter 4 deals with basic error estimates for standard "displacement" finite element methods and introduces mixed methods for constrained media applications such as incompressible elasticity and Stokes flow. A variety of "variational crimes" are described: for example, incompatible elements, reduced and selective integration, strain projection (i.e., $\bar{\boldsymbol{B}}$-) methods, etc. (Most of these have been decriminalized in recent years.) The mathematical analysis of finite element methods for incompressible media is rather complex. David Malkus, an authority on this subject, explains some of the subtleties in an appendix to Chapter 4. Chapter 5 is concerned with finite element methods for Reissner-Mindlin plates and elastic frameworks composed of straight beam elements accounting for axial, torsional, bending, and transverse shear deformations (i.e., Timoshenko beams). Chapter 6 deals with three-dimensional curved shell elements and two-dimensional special cases such as rings, tubes, and shells of revolution. The problem classes discussed in Chapters 1–6 give rise to associated eigenvalue problems and initial-value problems. The formulations of problems of these types are the subjects of Chapter 7. Chapter 8 presents time-stepping algorithms for first-order ordinary differential equation systems such as those arising from unsteady heat conduction ("parabolic case"). Chapter 9 deals with algorithms for second-order ordinary differential equation systems such as those emanating from elastodynamics and structural dynamics ("hyperbolic case"). Chapter 10 presents basic algorithmic strategies for symmetric elliptic eigenvalue problems such as those encountered in free vibration and structural stability. A very efficient major software package for matrix eigenvalue and eigenvector calculations based on the Lanczos method is also presented in Chapter 10. The documentation of the Lanczos algorithm and software were written by Bahram Nour-Omid, an expert on procedures of this type. Chapter 11 presents a linear static and dynamic analysis computer program based on the methods developed in the book. My student, Robert Ferencz, and colleague, Arthur Raefsky, collaborated with me in the writing of the program. The program is named DLEARN and it contains a very complete library of finite element software tools. This program is suitable for homework assignments, projects (e.g., programming additional elements), and research studies. DLEARN is highly structured for readability, maintainability, and extendability and has been written specifically to complement and enhance the procedures described in the remainder of the book.

LEVEL AND BACKGROUND

This book is primarily intended for the graduate level, although advanced undergraduates will find much of it accessible. An undergraduate degree in engineering, mathematics, computer science, or any of the physical sciences constitutes essential background. Courses in applied linear algebra and elementary ordinary and partial differential equations are desirable prerequisites. A working knowledge of FORTRAN (the dominant language used in finite element programming) is also necessary for understanding the software presented in Chapters 10 and 11.

The book emphasizes heat conduction and elasticity as primary vehicles for developing finite element methods because of the widespread interest in these theories in the applied sciences. By virtue of the fact that the partial differential equation of heat conduction, namely, the Laplace/Poisson equation, appears under different names in virtually all branches of engineering and physics, most students have had some familiarity with it. Some exposure to the theory of elasticity is also desirable. This can be obtained, for example, in a good advanced course on strength of materials. Students at Caltech and Stanford frequently have taken courses in elasticity simultaneously with finite elements. Background in structural mechanics (i.e., the theory of beams, plates, and shells) is certainly an asset when it comes to studying this book but is not essential. Only Chapters 5 and 6 deal exclusively with this subject, and these chapters may be ignored by students whose primary interests lie elsewhere. It is worth noting that students who have taken this finite element course at Stanford and Caltech have had very diverse backgrounds (e.g., geophysics, chemical engineering, planetary sciences, coastal engineering, electrical engineering, computer science, mathematics, bioengineering, aeronautics and astronautics, material science, physics, earth sciences, environmental engineering, biomechanics, thermosciences, engineering design, applied mechanics, earthquake engineering, and, of course, mechanical and civil engineering). In this spirit the book emphasizes fundamental finite element concepts and techniques applicable to a very broad range of problems and thus constitutes a suitable text for most students in the physical sciences.

It is my experience that most finite element methodology can and should be taught initially within the confines of linear behavior before introducing nonlinear effects. A solid understanding of finite elements in linear analysis is, of course, very useful in its own right. In addition, it makes the subject of nonlinear finite element analysis much more accessible.

WHAT IS UNIQUE ABOUT THIS BOOK

Although this book deals with many standard aspects of the finite element method, there are many unique features, some of which are enumerated below.

- A comprehensive presentation and analysis of algorithms for time-dependent phenomena.
- Beam, plate, and shell theories are derived directly from three-dimensional elasticity theory.
- An extensive static and dynamic finite element analysis program, DLEARN, was specially prepared based on the text material. It is written using structured programming concepts and contains many advanced procedures.
- Although written for students without serious mathematical training, the book contains introductory material on the mathematical theory of finite elements and

many important mathematical results. It thus serves as a primer for more advanced works on this subject.

- A systematic treatment of "weak," or variational, formulations for various classes of initial- and boundary-value problems.
- Many of the procedures described are presented in book form for the first time, for example, strain projection (i.e., $\bar{\boldsymbol{B}}$-) methods, implicit-explicit finite element mesh partitions in transient analysis, element-by-element iterative solvers, complete computer implementation of predictor/multicorrector implicit-explicit time-stepping algorithms based upon the Hilber-Hughes-Taylor α-method, etc.
- A Lanczos software package for eigenvalue/eigenvector extraction.

NOTATIONAL CONVENTIONS

As far as possible, we have adoped fairly standard notations. Vectors and matrices are denoted by boldface italic characters. Scalars and components of vectors, tensors, and matrices are denoted by lightface italic characters. Cartesian tensor notation is used in the presentation of heat conduction, elasticity, plates, etc. The symbol ■ is used to denote the end of a proof. Unavoidably, in a book of this size and scope there are some minor conflicts of notation. In instances when this occurs, it is hoped that surrounding explanation and context will make the intended meaning clear. All notations are defined when they are introduced. Principal notations are summarized in a brief glossary following the Preface.

NUMBERING CONVENTIONS

The book is divided into two parts. Chapters 1–6 deal with static analysis, and Chapters 7–11 deal with dynamic analysis. The chapters are subdivided into sections and several of the chapters include appendices. Some of the sections are further subdivided into subsections. Equations, tables, and figures are numbered consecutively within each section, with each number consisting of chapter, section, and item number. Exercises are numbered consecutively beginning with 1 in each section. References are numbered consecutively at the end of each chapter. Equation numbers are enclosed in parentheses, and reference numbers appearing in the text are enclosed in square brackets. If, for example, we wish to refer to Eq. 8 of Section 3, Chapter 4, then we would write this as (4.3.8).

If we wish to draw the reader's attention to a subsection, then the section number is followed by a decimal point and the subsection number. For example, subsection 2 of Section 3, Chapter 5, is written as Sec. 5.3.2.

EXERCISES

Exercises appear throughout the text. In some cases answers or complete solutions are provided. Many of the exercises are placed in the proximity of the pertinent text material rather than at the ends of sections or chapters.

INTERDEPENDENCE OF CHAPTERS

Prior to consulting particular chapters it may be worthwhile to peruse the "reading paths," which indicate the interdependence of the chapters. Of course, if one has sufficient background, the suggested reading paths may be bypassed. Knowledge of Chapters 1–3 is assumed for all subsequent chapters with the exception of Chapter 10, which is essentially self-contained.

Chapter Reading Paths

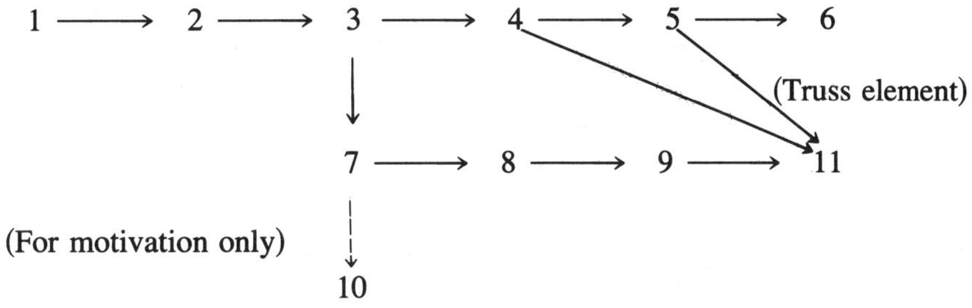

A NOTE FOR INSTRUCTORS REGARDING THE ORGANIZATION OF COURSES

Several courses can be developed based on the material in this book. Here are a few examples.

Almost the entire book can be covered in a one-year course:

Title: Linear Static Finite Element Analysis (1 semester)
Syllabus: Chapters 1–3, selected topics from Chapters 4–6, and computing assignments using DLEARN (Chapter 11).

Title: Linear Dynamic Finite Element Analysis (1 semester)
Syllabus: Chapters 7–10 and computing assignments using DLEARN (Chapter 11).

An abbreviated version of the above covering two quarters can be organized as follows:

Title: Linear Static Finite Element Analysis (1 quarter)

Syllabus: Chapters 1–3, selected topics from Chapter 4, and computing assignments using DLEARN (Chapter 11).

Title: Linear Dynamic Finite Element Analysis (1 quarter)

Syllabus: Sections 1–3 of Chapter 7; Sections 1, 2, and 5 of Chapter 8; Sections 1–4 of Chapter 9; Sections 1–5 of Chapter 10; and computing assignments using DLEARN (Chapter 11).

A course on finite element programming can be taught in one quarter or one semester:

Title: Finite Element Programming

Syllabus: Chapter 11 and selected sections from earlier chapters as background material. Lectures should describe overall program architecture (subroutines DLEARN and DRIVER), the structure of element routines (e.g., QUADC and subroutines called by QUADC), equation-solving (subroutines FACTOR and BACK), assembly routines (ADDLHS and ADDRHS), and important finite element utility routines (BC, COLHT, DCTNRY, DIAG, EQSET, FORMLM, LOCAL, and MPOINT). Programming assignments, such as adding new elements and extending capabilities, are essential.

ACKNOWLEDGMENTS

I would like to thank many former and current students, colleagues, and friends for their constructive comments and criticisms, especially Ted Belytschko, Alec Brooks, Marty Cohen, Leo Franca, Hans Hilber, Greg Hulbert, Itzhak Levit, Wing Kam Liu, Michel Mallet, Jerry Marsden, Isidoro Miranda, Arthur Muller, Juan Simo, Gary Stanley, Tayfun Tezduyar, Jim Winget, and Thomas Zimmermann.

I would also like to express my appreciation to those individuals who have made contributions to the writing of this book: Bob Ferencz, David Malkus, Bahram Nour-Omid, and Arthur Raefsky.

In addition, I would like to thank Alex Tessler who prepared notes for me on some recent developments in plate elements, which I used as the basis of comments included in Chapter 5. Wanda Gooden typed most of the manuscript. I wish to thank her for her patience, dedication, and good-naturedness in the face of a seemingly infinite number of corrections.

I would like to acknowledge the following organizations for permission to reprint various materials: the American Society of Mechanical Engineers for Figs. 5.3.21–5.3.31; John Wiley and Sons, Ltd., for Figs. 3.II.1, 3.II.2, 9.3.1–9.3.3, 9.3.5, and Table 3.I.1; Pergamon Journals, Ltd., for Figs. 5.3.11, 5.3.13–5.3.15, 5.3.18–5.3.20; and North-Holland Publishing Co. for portions of Chapters 8 and 9, which originally appeared in T. J. R. Hughes, "Analysis of Transient Algorithms with Particular Reference to Stability," pp. 67–155, *Computational Methods for Transient Analysis* (eds. T. J. R. Hughes and T. Belytschko), 1983.

Finally, I would like to thank M. Igarashi and K. Shimomaki of the Suzuki Motor Co., Ltd., for providing the mesh for the cover, and Dave Benson and John Hallquist of Lawrence Livermore National Laboratory for the accompanying color graphics.

NOTE: The software in this book is available on diskette and magnetic tape for several computers. Further information and an order form may be obtained by writing to the author: Professor Thomas J. R. Hughes, Division of Applied Mechanics, Durand Building, Stanford University, Stanford, California 94305.

A Brief Glossary of Notations

Sets

$\mathbb{R}$	real numbers
$\mathbb{C}$	complex numbers
$\cup$	set union
$\cap$	set intersection
$\emptyset$	empty set
$\in$	is a member of; belongs to
$\not\in$	is not a member of; does not belong to
$\subset$	is a subset of
$\not\subset$	is not a subset of
$\forall$	for all; for each

Various Integers

n_{np}	number of nodal points
n_{en}	number of element nodes
n_{eq}	number of equations
n_{ee}	number of element equations
n_{dof}	number of degrees of freedom
n_{ed}	number of element degrees of freedom
n_{el}	number of elements
A, B	global node numbers $(1 \leq A, B \leq n_{np})$
a, b	element node numbers $(1 \leq a, b \leq n_{en})$
P, Q	global equation numbers $(1 \leq P, Q \leq n_{eq})$
p, q	element equation numbers $(1 \leq p, q \leq n_{ee})$
e	element number $(1 \leq e \leq n_{el})$

Element Arrays

$\boldsymbol{m}^e$	mass matrix of the eth element
$\boldsymbol{c}^e$	damping matrix of the eth element

$\boldsymbol{k}^e$	stiffness matrix of the eth element
$\boldsymbol{f}^e$	force vector of the eth element
$\boldsymbol{a}^e$	acceleration vector of the eth element
$\boldsymbol{v}^e$	velocity vector of the eth element
$\boldsymbol{d}^e$	displacement vector of the eth element
Γ_h, Γ_{h_i}	part of the boundary where Neumann conditions are specified
δ_{ij}	Kronecker delta
η	set of node numbers
η_g, η_{g_i}	set of node numbers at which Dirichlet conditions are specified

Global Arrays

$\boldsymbol{M}$	mass matrix
$\boldsymbol{C}$	damping matrix
$\boldsymbol{K}$	stiffness matrix
$\boldsymbol{F}$	force vector
$\boldsymbol{a}$	acceleration vector
$\boldsymbol{v}$	velocity vector
$\boldsymbol{d}$	displacement vector
$\mathbf{A}$	finite element assembly operator $\left(\text{e.g., } \boldsymbol{K} = \mathop{\mathbf{A}}_{e=1}^{n_{el}} \boldsymbol{k}^e\right)$

Boundary-Value Problems

n_{sd}	number of space dimensions
i, j, k, l	spatial indices $(1 \leq i, j, k, l \leq n_{sd})$
$\mathbb{R}^{n_{sd}}$	Euclidean n_{sd}-space
Ω	domain in $\mathbb{R}^{n_{sd}}$
$\overline{\Omega}$	closure of Ω
x_i, $\boldsymbol{x}$	point in $\overline{\Omega}$
Γ	boundary of Ω
n_i, $\boldsymbol{n}$	unit outward normal vector to Γ
Γ_g, Γ_{g_i}	part of the boundary where Dirichlet conditions are specified

Variational Methods

h	mesh parameter
$\mathcal{S}$, $\mathcal{S}_i$, $\boldsymbol{\mathcal{S}}$	collections of trial solutions
$\mathcal{V}$, $\mathcal{V}_i$, $\boldsymbol{\mathcal{V}}$	collections of weighting functions
$\mathcal{S}^h$, $\mathcal{S}_i^h$, $\boldsymbol{\mathcal{S}}^h$	finite-dimensional collections of trial solutions
$\mathcal{V}^h$, $\mathcal{V}_i^h$, $\boldsymbol{\mathcal{V}}^h$	finite-dimensional collections of weighting functions
u, u_i, $\boldsymbol{u}$	trial solutions
w, w_i, $\boldsymbol{w}$	weighting functions
u^h, u_i^h, $\boldsymbol{u}^h$	finite-dimensional trial solutions
w^h, w_i^h, $\boldsymbol{w}^h$	finite-dimensional weighting functions
f, f_i, $\boldsymbol{f}$	source terms
g, g_i, $\boldsymbol{g}$	Dirichlet boundary data
h, h_i, $\boldsymbol{h}$	Neumann boundary data
g^h, g_i^h, $\boldsymbol{g}^h$	extensions of g, g_i, $\boldsymbol{g}$, respectively, to Ω
e, e_i, $\boldsymbol{e}$	errors
$L_2(\Omega)$	Hilbert space of square-integrable functions

$H^s(\Omega)$	Sobolev space of functions with s square-integrable generalized derivatives
$\lVert\cdot\rVert$	L_2 norm on Ω
$\lVert\cdot\rVert_s$	H^s norm on Ω
$a(\cdot,\cdot)$	symmetric bilinear form; strain-energy inner product
$(\cdot,\cdot)$	symmetric bilinear form; L_2 inner product on Ω
$(\cdot,\cdot)_\Gamma$	symmetric bilinear form; L_2 inner product on Γ

Data Processing Arrays

IEN (a, e)	element nodes matrix
ID (i, A)	destination matrix
LM (i, a, e)	location matrix

Heat Conduction

q_i	heat flux vector
κ_{ij}	conductivities
u	temperature
$u_{,i}$	temperature gradient
ℓ	prescribed heat supply
g	prescribed boundary temperature
h	prescribed boundary heat flux
d_A, d_a^e	nodal temperature
g_A, g_a^e	prescribed nodal temperature

Elasticity

ϵ_{ij}	infinitesimal strain tensor
σ_{ij}	Cauchy stress tensor
c_{ijkl}	elastic coefficients
u_i	displacement vector
$u_{(i,j)}$	symmetric part of the displacement gradients (equals ϵ_{ij})
ℓ_i	prescribed body force vector
g_i	prescribed boundary displacement vector
h_i	prescribed boundary traction vector
d_{iA}, d_{ia}^e	nodal displacements
g_{iA}, g_{ia}^e	prescribed nodal displacements
B	bulk modulus
E	Young's modulus
ν	Poisson's ratio
G, μ	shear modulus
λ	Lamé modulus
$\boldsymbol{B}$, $\boldsymbol{B}_A$, $\boldsymbol{B}_a$	strain-displacement matrices
$\boldsymbol{D}$, $\widetilde{\boldsymbol{D}}$	material moduli matrices

Isoparametric Elements

Ω^e	element domain $\subset \Omega$
Γ^e	boundary of Ω^e
x_1, x_2, x_3, x, y, z	Cartesian coordinates
$\square$	parent domain of a quadrilateral or hexahedral element
$\boldsymbol{x}^e\colon \square \to \overline{\Omega}^e$	isoparametric mapping
ξ, η	element natural coordinates for quadrilaterals
ξ, η, ζ	element natural coordinates for hexahedra
$\boldsymbol{\xi}$	point in $\square$
j	Jacobian determinant of the mapping $\boldsymbol{x}^e(\boldsymbol{\xi})$

$C^0(\Omega)$	the class of continuous functions
$C^k(\Omega)$	the class of continuous functions possessing k continuous derivatives
N_A, N_a^e	shape functions
$\boldsymbol{x}_A$, $\boldsymbol{x}_a^e$	nodal points
r, s, t	triangular coordinates; area coordinates
r, s, t, u	tetrahedral coordinates; volume coordinates
$l_a(\xi)$	Lagrange polynomial associated with the ath element node
$\tilde{\boldsymbol{\xi}}_l$	location of the lth integration point
W_l	weight assigned to lth integration point
$U1$, $U2$, $U3$	uniform integration elements
$S1$, $S2$, $S3$	selective integration elements

Incompressible and Nearly-Incompressible Elasticity

$\mathcal{P}$	pressure trial solution and weighting function space
$\mathcal{P}^h$	finite-dimensional pressure trial solution and weighting function space
p, $q \in \mathcal{P}$	trial pressure and weighting function
p^h, $q^h \in \mathcal{P}^h$	finite-dimensional trial pressure and weighting function
$\boldsymbol{g}^e$, $\boldsymbol{G}$	element and global gradient matrices
$(\boldsymbol{g}^e)^T$, $\boldsymbol{G}^T$	element and global divergence matrices
$\boldsymbol{p}$	vector of nodal pressures
$\boldsymbol{B}_a^{\text{dev}}$	deviatoric strain-displacement matrix
$\boldsymbol{B}_a^{\text{dil}}$	dilatational strain-displacement matrix
$\bar{\boldsymbol{B}}_a^{\text{dil}}$	improved dilatational strain-displacement matrix
$\bar{\boldsymbol{B}}_a$	improved strain-displacement matrix

Plates

α, β, γ, δ	tensor indices ($1 \leq \alpha, \beta, \gamma, \delta \leq 2$)
$e_{\alpha\beta}$	alternator tensor
w	transverse displacement
θ_α	rotation vector
$\kappa_{\alpha\beta} = \theta_{(\alpha,\beta)}$	curvature tensor
γ_α	shear strain vector
$c_{\alpha\beta\gamma\delta}$, $c_{\alpha\beta}$	elastic coefficients
$m_{\alpha\beta}$	moment tensor
q_α	shear force vector
W	prescribed boundary displacement
Θ_α	prescribed boundary rotations
F	total applied transverse force
C_α	total applied couple
M_α	prescribed boundary moments
Q	prescribed boundary shear force
s	arc-length parameter of plate boundary
$\hat{\theta}_\alpha = e_{\alpha\beta}\theta_\beta$, $\hat{\theta}_s$, $\hat{\theta}_n$	right-hand-rule rotations
$\boldsymbol{k}_b^e$, $\boldsymbol{k}_s^e$	eth element bending and shear stiffness

$\boldsymbol{D}^b$, $\boldsymbol{D}^s$	matrices of bending and shear moduli
$\boldsymbol{B}^b$, $\boldsymbol{B}^s$	bending and shear strain-displacement matrices

Beams

w_i	displacements
θ_i	rotations
κ_α	curvatures
γ_α	shear strains
ϵ	axial strain
ψ	twist
m_i	moments
q_i	shear forces
W_i	prescribed end displacements
Θ_i	prescribed end rotations
F_i	total applied force per unit length
C_i	total applied couple per unit length
M_i	prescribed end moments
Q_i	prescribed end shears
A	cross-sectional area
A_α	shear areas
I_α	moments of inertia
J	polar moment of inertia
$\boldsymbol{k}_a^e$, $\boldsymbol{k}_b^e$, $\boldsymbol{k}_s^e$, $\boldsymbol{k}_t^e$	*e*th element axial, bending, shear, and torsional stiffness, respectively

Shells

$\boldsymbol{x}$	position vector
$\bar{\boldsymbol{x}}$, $\bar{\boldsymbol{x}}_a$	position vectors to references surface
$\boldsymbol{X}$, $\hat{\boldsymbol{X}}_a$	fiber vectors
$z_a(\zeta)$	thickness function
$\boldsymbol{e}_1^l$, $\boldsymbol{e}_2^l$, $\boldsymbol{e}_3^l$	lamina basis vectors
$\boldsymbol{e}_1^f$, $\boldsymbol{e}_2^f$, $\boldsymbol{e}_3^f$	fiber basis vectors
$\boldsymbol{u}$	displacement vector
$\bar{\boldsymbol{u}}$, $\bar{\boldsymbol{u}}_a$	displacement of reference surface
$\boldsymbol{U}$, $\boldsymbol{U}_a$, $\hat{\boldsymbol{U}}_a$	fiber displacements
$\tilde{\boldsymbol{\sigma}}^l = \tilde{\boldsymbol{D}}^l \tilde{\boldsymbol{\epsilon}}^l$	reduced constitutive equation in the lamina basis
$m_{\alpha\beta}$	moments
n_α	membrane forces
q_α	transverse shear forces

Dynamics

t	time
Δt	time step
n	time step number
$\boldsymbol{a}_n$, $\boldsymbol{v}_n$, $\boldsymbol{d}_n$	approximations of $\boldsymbol{a}(t_n)$, $\boldsymbol{v}(t_n)$, and $\boldsymbol{d}(t_n)$, respectively
$\tilde{\boldsymbol{a}}_n$, $\tilde{\boldsymbol{v}}_n$, $\tilde{\boldsymbol{d}}_n$	predictor values of acceleration, velocity, and displacement, respectively
Δt_{crit}	critical time step
A	amplification factor
$\boldsymbol{A}$	amplification matrix
$\rho(\boldsymbol{A})$	spectral radius of amplification matrix
τ, $\boldsymbol{\tau}$	local truncation errors
λ, λ^h, $\bar{\lambda}^h$	eigenvalues
$\boldsymbol{\psi}$, $\boldsymbol{\phi}$	eigenvectors
ω, ω^h, $\bar{\omega}^h$	frequencies
$\bar{\xi}$	algorithmic damping ratio
$(\bar{T} - T)/T$	relative period error
Ω_{crit}	critical sampling frequency

$\boldsymbol{M}^I$, $\boldsymbol{C}^I$, $\boldsymbol{K}^I$, $\boldsymbol{F}^I$	implicit mesh partition of mass, damping, stiffness, and force
$\boldsymbol{M}^E$, $\boldsymbol{C}^E$, $\boldsymbol{K}^E$, $\boldsymbol{F}^E$	explicit mesh partition of mass, damping, stiffness, and force
u_{0i}, $\dot{u}_{0i}$	initial displacements and velocities
u_0	initial temperature
ρ	density
c	capacity

1

Fundamental Concepts; a Simple One-Dimensional Boundary-Value Problem

1.1 INTRODUCTORY REMARKS AND PRELIMINARIES

The main constituents of a finite element method for the solution of a boundary-value problem are

i. The variational or weak statement of the problem; and
ii. The approximate solution of the variational equations through the use of "finite element functions."

To clarify concepts we shall begin with a simple example.

Suppose we want to solve the following differential equation for u:

$$u_{,xx} + \ell = 0 \tag{1.1.1}$$

where a comma stands for differentiation (i.e., $u_{,xx} = d^2u/dx^2$). We assume ℓ is a given smooth, scalar-valued function defined on the unit interval. We write

$$\ell : [0, 1] \rightarrow \mathbb{R} \tag{1.1.2}$$

where $[0, 1]$ stands for the unit interval (i.e., the set of points x such that $0 \leq x \leq 1$) and $\mathbb{R}$ stands for the real numbers. In words, (1.1.2) states that for a given x in $[0, 1]$, $\ell(x)$ is a real number. (Often we will use the notation $\in$ to mean "in" or "a member of." Thus for each $x \in [0, 1]$, $\ell(x) \in \mathbb{R}$.). Also, $[0, 1]$ is said to be the ***domain*** of ℓ, and $\mathbb{R}$ is its ***range.***

We have described the given function ℓ as being smooth. Intuitively, you probably know what this means. Roughly speaking, if we sketch the graph of the function ℓ, we want it to be a smooth curve without discontinuities or kinks. We do this to avoid technical difficulties. Right now we do not wish to elaborate further as this would divert us from the main theme. At some point prior to moving on to the next chapter, the reader may wish to consult Appendix 1.I, "An Elementary Discussion of Continuity, Differentiability and Smoothness," for further remarks on this important aspect of finite element work. The exercise in Sec. 1.16 already uses a little of the language described in Appendix 1.I. The terminology may be somewhat unfamiliar to engineering and physical science students, but it is now widely used in the finite element literature and therefore it is worthwhile to become accustomed to it.

Equation (1.1.1) is known to govern the transverse displacement of a string in tension and also the longitudinal displacement of an elastic rod. In these cases, physical parameters, such as the magnitude of tension in the string, or elastic modulus in the case of the rod, appear in (1.1.1). We have omitted these parameters to simplify subsequent developments.

Before going on, we introduce a few additional notations and terminologies. Let $]0, 1[$ denote the unit interval without end points (i.e., the set of points x such that $0 < x < 1$). $]0, 1[$ and $[0, 1]$ are referred to as ***open and closed unit intervals,*** respectively. To simplify subsequent writing and tie in with notation employed later on in multidimensional situations, we shall adopt the definitions

$$\Omega =]0, 1[\qquad \text{(open)} \tag{1.1.3}$$

$$\overline{\Omega} = [0, 1] \qquad \text{(closed)} \tag{1.1.4}$$

See Fig. 1.1.1.

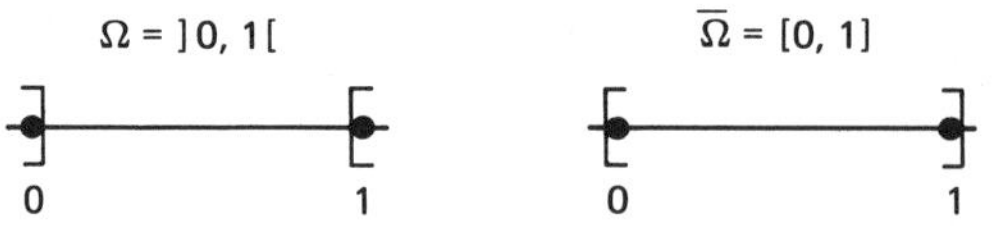

Figure 1.1.1

At this point, considerations such as these may seem pedantic. Our purpose, however, is to develop a language for the precise articulation of boundary-value problems, which is necessary for good finite element work.

1.2 STRONG, OR CLASSICAL, FORM OF THE PROBLEM

A boundary-value problem for (1.1.1) involves imposing ***boundary conditions*** on the function u. There are a variety of possibilities. We shall assume u is required to satisfy

$$u(1) = g \tag{1.2.1}$$

$$-u_{,x}(0) = h \tag{1.2.2}$$

where g and h are given constants. Equations (1.2.1) and (1.2.2) require that u take on the value g at $x = 1$ and the derivative of u (i.e., slope) take on the value $-h$ at $x = 0$, respectively. This set of boundary conditions will later enable us to illustrate certain key features of variational formulations. For obvious reasons, boundary conditions of the type (1.2.1) and (1.2.2) lead to so-called ***two-point boundary-value problems***.

The strong form of the boundary-value problem, (S), is stated as follows:

$$(S) \begin{cases} \text{Given } \ell : \overline{\Omega} \to \mathbb{R} \text{ and constants } g \text{ and } h, \text{ find } u : \overline{\Omega} \to \mathbb{R}, \text{ such that} \\ \qquad u_{,xx} + \ell = 0 \quad \text{on } \Omega \\ \qquad u(1) = g \\ \qquad -u_{,x}(0) = h \end{cases}$$

When we write $u_{,xx} + \ell = 0$ on Ω we mean $u_{,xx}(x) + \ell(x) = 0$ for all $x \in \Omega$. Of course, the exact solution of (S) is trivial to obtain, namely,

$$u(x) = g + (1 - x)h + \int_x^1 \left\{ \int_0^y \ell(z)\, dz \right\} dy \qquad (1.2.3)$$

where y and z are used to denote dummy variables. However, this is not the main concern here. We are interested in developing schemes for obtaining approximate solutions to (S) that will be applicable to much more complex situations in which exact solutions are not possible.

Some methods of approximation begin directly with the strong statement of the problem. The most notable example is the finite difference method (e.g., see [1]). The finite element method requires a different formulation, which is treated in the next section.

1.3 WEAK, OR VARIATIONAL, FORM OF THE PROBLEM

To define the weak, or variational, counterpart of (S), we need to characterize two classes of functions. The first is to be composed of candidate, or trial, solutions. From the outset, we shall require these possible solutions to satisfy the boundary condition $u(1) = g$. The other boundary condition will not be required in the definition. Furthermore, so that certain expressions to be employed make sense, we shall require that the derivatives of the trial solutions be square-integrable. That is, if u is a trial solution, then

$$\int_0^1 (u_{,x})^2\, dx < \infty \qquad (1.3.1)$$

Functions that satisfy (1.3.1) are called ***H^1-functions;*** we write $u \in H^1$. Sometimes the domain is explicitly included, i.e., $u \in H^1([0, 1])$.

Thus the collection of ***trial solutions,*** denoted by $\mathcal{S}$, consists of all functions which have square-integrable derivatives and take on the value g at $x = 1$. This is written as follows:

$$\mathcal{S} = \{u \mid u \in H^1,\ u(1) = g\} \qquad \text{(trial solutions)} \qquad (1.3.2)$$

The fact that $\mathcal{S}$ is a collection, or set, of objects is indicated by the curly brackets (called braces) in (1.3.2). The notation for the typical member of the set, in this case u, comes first inside the left-hand curly bracket. Following the vertical line (|) are the properties satisfied by members of the set.

The second collection of functions is called the ***weighting functions,*** or ***variations.*** This collection is very similar to the trial solutions except we require the homogeneous counterpart of the g-boundary condition. That is, we require weighting functions, w, to satisfy $w(1) = 0$. The collection is denoted by $\mathcal{V}$ and defined by

$$\mathcal{V} = \{w \mid w \in H^1, w(1) = 0\} \qquad \text{(weighting functions)} \tag{1.3.3}$$

It simplifies matters somewhat to continue to think of $f : \Omega \rightarrow \mathbb{R}$ as being smooth. (However, what follows holds for a considerably larger class of f's.)

In terms of the preceding definitions, we may now state a suitable weak form, (W), of the boundary-value problem.

(W) Given f, g, and h, as before. Find $u \in \mathcal{S}$ such that for all $w \in \mathcal{V}$

$$\boxed{\int_0^1 w_{,x} u_{,x} \, dx = \int_0^1 w f \, dx + w(0)h} \tag{1.3.4}$$

Formulations of this type are often called ***virtual work,*** or ***virtual displacement, principles*** in mechanics. The w's are the ***virtual displacements***.

Equation (1.3.4) is called the ***variational equation,*** or (especially in mechanics) the ***equation of virtual work***.

The solution of (W) is called the ***weak,*** or ***generalized, solution***. The definition given of a weak formulation is not the only one possible, but it is the most natural one for the problems we wish to consider.

1.4 EQUIVALENCE OF STRONG AND WEAK FORMS; NATURAL BOUNDARY CONDITIONS

Clearly, there must be some relationship between the strong and weak versions of the problem, or else there would be no point in introducing the weak form. It turns out that the weak and strong solutions are identical. We shall establish this assuming all functions are smooth. This will allow us to proceed expeditiously without invoking technical conditions with which the reader is assumed to be unfamiliar. "Proofs" of this kind are sometimes euphemistically referred to as "formal proofs." The intent is not to be completely rigorous but rather to make plausible the truth of the proposition. With this philosophy in mind, we shall "prove" the following.

Proposition

a. Let u be a solution of (S). Then u is also a solution of (W).
b. Let u be a solution of (W). Then u is also a solution of (S).

Another result, which we shall not bother to verify but is in fact easily established, is that both (S) and (W) possess unique solutions. Thus, by (a) and (b), the strong and weak solutions are one and the same. Consequently, (W) is equivalent to (S).

Formal Proof

a. Since u is assumed to be a solution of (S), we may write

$$0 = -\int_0^1 w(u_{,xx} + \ell)dx \tag{1.4.1}$$

for any $w \in \mathcal{V}$. Integrating (1.4.1) by parts results in

$$0 = \int_0^1 w_{,x}u_{,x}\,dx - \int_0^1 w\ell\,dx - wu_{,x}\Big|_0^1 \tag{1.4.2}$$

Rearranging and making use of the fact that $-u_{,x}(0) = h$ and $w(1) = 0$ results in

$$\int_0^1 w_{,x}u_{,x}\,dx = \int_0^1 w\ell\,dx + w(0)h \tag{1.4.3}$$

Furthermore, since u is a solution of (S), it satisfies $u(1) = q$ and therefore is in $\mathcal{S}$. Finally, since u also satisfies (1.4.3) for all $w \in \mathcal{V}$, u satisfies the definition of a weak solution given by (W).

b. Now u is assumed to be a weak solution. Thus $u \in \mathcal{S}$; consequently $u(1) = q$, and

$$\int_0^1 w_{,x}u_{,x}\,dx = \int_0^1 w\ell\,dx + w(0)h$$

for all $w \in \mathcal{V}$. Integrating by parts and making use of the fact $w(1) = 0$ results in

$$0 = \int_0^1 w(u_{,xx} + \ell)\,dx + w(0)[u_{,x}(0) + h] \tag{1.4.4}$$

To prove u is a solution of (S) it suffices to show that (1.4.4) implies[1]

i. $u_{,xx} + \ell = 0$ on Ω; and
ii. $u_{,x}(0) + h = 0$

First we shall prove (i). Define w in (1.4.4) by

$$w = \phi(u_{,xx} + \ell) \tag{1.4.5}$$

where ϕ is smooth; $\phi(x) > 0$ for all $x \in \Omega = \,]0, 1[$; and $\phi(0) = \phi(1) = 0$. For example, we can take $\phi(x) = x(1 - x)$, which satisfies all the stipulated requirements (see Figure 1.4.1). It follows that $w(1) = 0$ and thus $w \in \mathcal{V}$, so (1.4.5) defines a

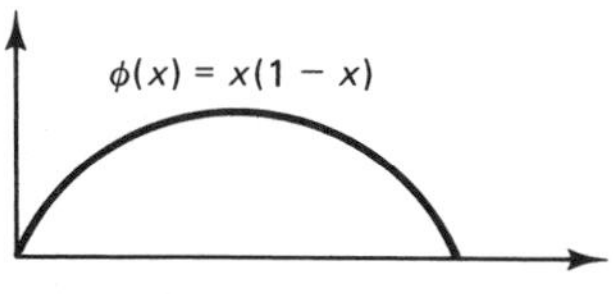

Figure 1.4.1

[1] These equations are sometimes called the ***Euler-Lagrange equations*** of the weak formulation.

legitimate member of $\mathcal{V}$. Substituting (1.4.5) into (1.4.4) results in

$$0 = \int_0^1 \underbrace{\phi(u_{,xx} + \ell)^2}_{\geq 0} dx + 0 \tag{1.4.6}$$

Since $\phi > 0$ on Ω, it follows from (1.4.6) that (i) must be satisfied.

Now that we have established (i), we may use it in (1.4.4) to prove (ii), namely,

$$0 = w(0)[u_{,x}(0) + h] \tag{1.4.7}$$

That $w \in \mathcal{V}$ puts no restriction whatsoever on its value at $x = 0$. Therefore, we may assume that the w in (1.4.7) is such that $w(0) \neq 0$. Thus (ii) is also shown to hold, which completes the proof of the proposition. ■

Remarks

1. The boundary condition $-u_{,x}(0) = h$ is not explicitly mentioned in the statement of (W). From the preceding proof, we see that this boundary condition is, however, implied by the satisfaction of the variational equation. Boundary conditions of this type are referred to as ***natural boundary conditions***. On the other hand, trial solutions are explicitly required to satisfy the boundary condition $u(1) = q$. Boundary conditions of this type are called ***essential boundary conditions***. The fact that solutions of the variational equation satisfy natural boundary conditions is extremely important in more complicated situations which we will consider later on.

2. The method used to prove part (b) of the proposition goes under the name of the ***fundamental lemma*** in the literature of the calculus of variations. In essence, it is the methodology that enables us to deduce the differential equations and boundary conditions implied by the weak formulation. To develop correct weak forms for complex, multidimensional problems, it is essential to have a thorough understanding of these procedures.

Now we see that to obtain approximate solutions to the original boundary-value problem we have alternative starting points, i.e., the strong or weak statements of the problem. *Finite element methods are based upon the latter*. Roughly speaking, the basic idea is to approximate $\mathcal{S}$ and $\mathcal{V}$ by convenient, finite-dimensional collections of functions. (Clearly, $\mathcal{S}$ and $\mathcal{V}$ contain infinitely many functions.) The variational equations are then solved in this finite-dimensional context. An explicit example of how to go about this is the subject of the next section. However, we first introduce some additional notations to simplify subsequent writing.

Let

$$a(w, u) = \int_0^1 w_{,x} u_{,x}\, dx \tag{1.4.8}$$

$$(w, \ell) = \int_0^1 w \ell\, dx \tag{1.4.9}$$

In terms of (1.4.8) and (1.4.9), the variational equation takes the form

$$\boxed{a(w, u) = (w, \ell) + w(0)h} \tag{1.4.10}$$

Here, $a(\cdot, \cdot)$ and $(\cdot, \cdot)$ are examples of ***symmetric, bilinear forms.*** What this means is as follows: Let c_1 and c_2 be constants and let u, v, and w be functions. Then the ***symmetry*** property is

$$a(u, v) = a(v, u) \tag{1.4.11}$$

$$(u, v) = (v, u) \tag{1.4.12}$$

Bilinearity means linearity in each "slot"; for example,

$$a(c_1 u + c_2 v, w) = c_1 a(u, w) + c_2 a(v, w) \tag{1.4.13}$$

$$(c_1 u + c_2 v, w) = c_1(u, w) + c_2(v, w) \tag{1.4.14}$$

Exercise 1. Use the definitions of $a(\cdot; \cdot)$ and $(\cdot, \cdot)$ to verify the properties of symmetry and bilinearity.

The above notations are very concise; at the same time they capture essential mathematical features and thus are conducive to a mathematical understanding of variational and finite element methods. Diverse classes of physical problems can be written in essentially similar fashion to (1.4.10). Thus ideas developed and results obtained are seen at once to have very broad applicability.

1.5 GALERKIN'S APPROXIMATION METHOD

We shall now describe a method of obtaining approximate solutions to boundary-value problems based upon weak formulations. Our introduction to this subject is somewhat of an abstract treatment. However, the meaning should be significantly reinforced by the remaining sections of the chapter. It may be worthwhile for the reader to consult this section again after completing the rest of the chapter to make sure a full comprehension of the material is attained.

The first step in developing the method is to construct finite-dimensional approximations of $\mathcal{S}$ and $\mathcal{V}$. These collections of functions are denoted by $\mathcal{S}^h$ and $\mathcal{V}^h$, respectively. The superscript refers to the association of $\mathcal{S}^h$ and $\mathcal{V}^h$ with a ***mesh,*** or ***discretization,*** of the domain Ω, which is parameterized by a characteristic length scale h. We wish to think of $\mathcal{S}^h$ and $\mathcal{V}^h$ as being subsets of $\mathcal{S}$ and $\mathcal{V}$, respectively. This is written as

$$\mathcal{S}^h \subset \mathcal{S} \qquad (\text{i.e., if } u^h \in \mathcal{S}^h, \text{ then } u^h \in \mathcal{S}) \tag{1.5.1}$$

$$\mathcal{V}^h \subset \mathcal{V} \qquad (\text{i.e., if } w^h \in \mathcal{V}^h, \text{ then } w^h \in \mathcal{V}) \tag{1.5.2}$$

where the precise meaning is given in parentheses.[2] Consequences of (1.5.1) and (1.5.2) are (respectively) that if $u^h \in \mathcal{S}^h$ and $w^h \in \mathcal{V}^h$, then

$$u^h(1) = g \tag{1.5.3}$$

$$w^h(1) = 0 \tag{1.5.4}$$

The collections, $\mathcal{S}$, $\mathcal{V}$, $\mathcal{S}^h$, and $\mathcal{V}^h$, are often referred to as ***function spaces***. The terminology *space* in mathematics usually connotes a linear structure. This has the following meaning: If c_1 and c_2 are constants and v and w are in $\mathcal{V}$, then $c_1 v + c_2 w$ is also in $\mathcal{V}$. Both $\mathcal{V}$ and $\mathcal{V}^h$ are thus seen to possess the property of a linear space. However, this property is clearly not shared by $\mathcal{S}$ and $\mathcal{S}^h$ due to the inhomogeneous boundary condition. For example, if u_1 and u_2 are members of $\mathcal{S}$, then $u_1 + u_2 \not\in \mathcal{S}$, since $u_1(1) + u_2(1) = g + g = 2g$ in violation of the definition of $\mathcal{S}$. Nevertheless, the terminology function space is still (loosely) applied to $\mathcal{S}$ and $\mathcal{S}^h$.

(Bubnov-) Galerkin Method

Assume the collection $\mathcal{V}^h$ is given. Then, to each member $v^h \in \mathcal{V}^h$, we construct a function $u^h \in \mathcal{S}^h$ by

$$u^h = v^h + g^h \tag{1.5.5}$$

where g^h is a *given* function satisfying the essential boundary condition, i.e.,

$$g^h(1) = g \tag{1.5.6}$$

Note that (1.5.5) satisfies the requisite boundary condition also:

$$\begin{aligned} u^h(1) &= v^h(1) + g^h(1) \\ &= 0 + g \end{aligned} \tag{1.5.7}$$

Thus (1.5.5) constitutes a definition of $\mathcal{S}^h$; that is, $\mathcal{S}^h$ is all functions of the form (1.5.5). The key point to observe is that, up to the function g^h, $\mathcal{S}^h$ and $\mathcal{V}^h$ are composed of *identical* collections of functions. This property will be shown later on to have significant consequences for certain classes of problems.

We now write a variational equation, of the form of (1.4.10), in terms of $w^h \in \mathcal{V}^h$ and $u^h \in \mathcal{S}^h$:

$$\boxed{a(w^h, u^h) = (w^h, f) + w^h(0)h} \tag{1.5.8}$$

This equation is to be thought of as defining an approximate (weak) solution, u^h.

[2] This condition may be considered standard. However, it is often violated in practice. Strang [2] coined the terminology "variational crimes" to apply to this, and other, situations in which the classical rules of variational methods are violated. Many "variational crimes" have been given a rigorous mathematical basis (e.g., see [2]). We shall have more to say about this subject in subsequent chapters.

Substitution of (1.5.5) into (1.5.8), and the bilinearity of $a(\cdot, \cdot)$ enables us to write

$$\boxed{a(w^h, v^h) = (w^h, \ell) + w^h(0)h - a(w^h, g^h)} \tag{1.5.9}$$

The right-hand side consists of the totality of terms associated with given data (i.e., ℓ, g, and h). Equation (1.5.9) is to be used to define v^h, the unknown part of u^h.

The (Bubnov-) Galerkin form of the problem, denoted by (G), is stated as follows:

$$(G)\begin{cases} \text{Given } \ell,\ g, \text{ and } h, \text{ as before, find } u^h = v^h + g^h, \text{ where } v^h \in \mathcal{V}^h, \\ \text{such that for all } w^h \in \mathcal{V}^h \\ \qquad a(w^h, v^h) = (w^h, \ell) + w^h(0)h - a(w^h, g^h) \end{cases}$$

Note that (G) is just a version of (W) posed in terms of a finite-dimensional collection of functions, namely, $\mathcal{V}^h$.

To make matters more specific, g^h and $\mathcal{V}^h$ have to be explicitly defined. Before doing this, it is worthwhile to mention a larger class of approximation methods, called ***Petrov-Galerkin methods,*** in which v^h is contained in a collection of functions other than $\mathcal{V}^h$. Recent attention has been paid to methods of this type, especially in the context of fluid mechanics. For the time being, we will be exclusively concerned with the Bubnov-Galerkin method. The Bubnov-Galerkin method is commonly referred to as simply the Galerkin method, terminology we shall adopt henceforth. Equation (1.5.9) is sometimes referred to as the ***Galerkin equation***.

Approximation methods of the type considered are examples of so-called ***weighted residual methods***. The standard reference dealing with this subject is Finlayson [3]. For a more succinct presentation containing an interesting historical account, see Finlayson and Scriven [4].

1.6 MATRIX EQUATIONS; STIFFNESS MATRIX K

The Galerkin method leads to a coupled system of linear algebraic equations. To see this we need to give further structure to the definition of $\mathcal{V}^h$. Let $\mathcal{V}^h$ consist of all linear combinations of given functions denoted by $N_A : \overline{\Omega} \to \mathbb{R}$, where $A = 1, 2, \ldots, n$. By this we mean that if $w^h \in \mathcal{V}^h$, then there exist constants c_A, $A = 1, 2, \ldots, n$, such that

$$\begin{aligned} w^h &= \sum_{A=1}^{n} c_A N_A \\ &= c_1 N_1 + c_2 N_2 + \cdots + c_n N_n \end{aligned} \tag{1.6.1}$$

The N_A's are referred to as ***shape, basis,*** or ***interpolation*** functions. We require that each N_A satisfies

$$N_A(1) = 0, \qquad A = 1, 2, \ldots, n \tag{1.6.2}$$

from which it follows by (1.6.1) that $w^h(1) = 0$, as is necessary. $\mathcal{V}^h$ is said to have dimension n, for obvious reasons.

To define members of $\mathcal{S}^h$ we need to specify g^h. To this end, we introduce another shape function, $N_{n+1} : \overline{\Omega} \rightarrow \mathbb{R}$, which has the property

$$N_{n+1}(1) = 1 \tag{1.6.3}$$

(Note $N_{n+1} \notin \mathcal{V}^h$.) Then g^h is given by

$$g^h = gN_{n+1} \tag{1.6.4}$$

and thus

$$g^h(1) = g \tag{1.6.5}$$

With these definitions, a typical $u^h \in \mathcal{S}^h$ may be written as

$$\begin{aligned} u^h &= v^h + g^h \\ &= \sum_{A=1}^{n} d_A N_A + gN_{n+1} \end{aligned} \tag{1.6.6}$$

where the d_A's are constants and from which it is apparent that $u^h(1) = g$.

Substitution of (1.6.1) and (1.6.6) into the Galerkin equation yields

$$\begin{aligned} a\left(\sum_{A=1}^{n} c_A N_A, \sum_{B=1}^{n} d_B N_B\right) = \left(\sum_{A=1}^{n} c_A N_A, \ell\right) + \left[\sum_{A=1}^{n} c_A N_A(0)\right]h \\ - a\left(\sum_{A=1}^{n} c_A N_A, gN_{n+1}\right) \end{aligned} \tag{1.6.7}$$

By using the bilinearity of $a(\cdot, \cdot)$ and $(\cdot, \cdot)$, (1.6.7) becomes

$$0 = \sum_{A=1}^{n} c_A G_A \tag{1.6.8}$$

where

$$G_A = \sum_{B=1}^{n} a(N_A, N_B)d_B - (N_A, \ell) - N_A(0)h + a(N_A, N_{n+1})g \tag{1.6.9}$$

Now the Galerkin equation is to hold for all $w^h \in \mathcal{V}^h$. By (1.6.1), this means for all c_A's, $A = 1, 2, \ldots, n$. Since the c_A's are arbitrary in (1.6.8), it necessarily follows that each G_A, $A = 1, 2, \ldots, n$, must be identically zero, i.e., from (1.6.9)

$$\boxed{\sum_{B=1}^{n} a(N_A, N_B)d_B = (N_A, \ell) + N_A(0)h - a(N_A, N_{n+1})g} \tag{1.6.10}$$

Note that everything is known in (1.6.10) except the d_B's. Thus (1.6.10) constitutes a system of n equations in n unknowns. This can be written in a more concise form as follows:

Let

$$K_{AB} = a(N_A, N_B) \tag{1.6.11}$$

$$F_A = (N_A, \ell) + N_A(0)h - a(N_A, N_{n+1})g \tag{1.6.12}$$

Then (1.6.10) becomes

$$\sum_{B=1}^{n} K_{AB} d_B = F_A, \qquad A = 1, 2, \ldots, n \tag{1.6.13}$$

Further simplicity is gained by adopting a matrix notation. Let

$$\boldsymbol{K} = [K_{AB}] = \begin{bmatrix} K_{11} & K_{12} & \cdots & K_{1n} \\ K_{21} & K_{22} & \cdots & K_{2n} \\ \vdots & \vdots & & \vdots \\ K_{n1} & K_{n2} & \cdots & K_{nn} \end{bmatrix} \tag{1.6.14}$$

$$\boldsymbol{F} = \{F_A\} = \begin{Bmatrix} F_1 \\ F_2 \\ \vdots \\ F_n \end{Bmatrix} \tag{1.6.15}$$

and

$$\boldsymbol{d} = \{d_B\} = \begin{Bmatrix} d_1 \\ d_2 \\ \vdots \\ d_n \end{Bmatrix} \tag{1.6.16}$$

Now (1.6.13) may be written as

$$\boldsymbol{Kd} = \boldsymbol{F} \tag{1.6.17}$$

The following terminologies are frequently applied, especially when the problem under consideration pertains to a mechanical system:

$\boldsymbol{K}$ = stiffness matrix

$\boldsymbol{F}$ = force vector

$\boldsymbol{d}$ = displacement vector

A variety of physical interpretations are of course possible.

At this point, we may state the matrix equivalent, (M), of the Galerkin problem.

$$(M)\left\{\begin{array}{l}\text{Given the coefficient matrix } \boldsymbol{K} \text{ and vector } \boldsymbol{F}, \text{ find } \boldsymbol{d} \text{ such that} \\ \qquad\qquad\qquad \boldsymbol{Kd} = \boldsymbol{F}\end{array}\right.$$

The solution of (M) is, of course, just $\boldsymbol{d} = \boldsymbol{K}^{-1}\boldsymbol{F}$ (assuming the inverse of $\boldsymbol{K}$, $\boldsymbol{K}^{-1}$, exists). Once $\boldsymbol{d}$ is known, the solution of (G) may be obtained at any point $x \in \overline{\Omega}$ by employing (1.6.6), viz.,

$$u^h(x) = \sum_{A=1}^{n} d_A N_A(x) + g N_{n+1}(x) \tag{1.6.18}$$

Likewise, derivatives of u^h, if required, may be obtained by term-by-term differentiation. It should be emphasized, that the solution of (G) is an *approximate* solution of (W). Consequently, the differential equation and natural boundary condition are only approximately satisfied. The quality of the approximation depends upon the specific choice of N_A's and the number n.

Remarks

1. The matrix $\boldsymbol{K}$ is symmetric. This follows from the symmetry of $a(\cdot, \cdot)$ and use of Galerkin's method (i.e., the same shape functions are used for the variations and trial solutions):

$$\begin{aligned} K_{AB} &= a(N_A, N_B) \\ &= a(N_B, N_A) \\ &= K_{BA} \end{aligned} \tag{1.6.19}$$

In matrix notation

$$\boldsymbol{K} = \boldsymbol{K}^T \tag{1.6.20}$$

where the superscript T denotes transpose. The symmetry of $\boldsymbol{K}$ has important computational consequences.

2. Let us schematically retrace the steps leading to the matrix problem, as they are typical of the process one must go through in developing a finite element method for any given problem:

$$\boxed{(S) \Leftrightarrow (W) \approx (G) \Leftrightarrow (M)} \tag{1.6.21}$$

The only apparent approximation made thus far is in approximately solving (W) via (G). In more complicated situations, encountered in practice, the number of approximations increases. For example, the data ℓ, g, and h may be approximated, as well as the domain Ω, calculation of integrals, and so on. Convergence proofs and error analyses involve consideration of each approximation.

3. It is sometimes convenient to write

$$u^h(x) = \sum_{A=1}^{n+1} N_A(x)d_A \tag{1.6.22}$$

where $d_{n+1} = g$.

1.7 EXAMPLES: 1 AND 2 DEGREES OF FREEDOM

In this section we will carry out the detailed calculations involved in formulating and solving the Galerkin problem. The functions employed are extremely simple, thus expediting computations, but they are also primitive examples of typical finite element functions.

Example 1 (1 degree of freedom)

In this case $n = 1$. Thus $w^h = c_1 N_1$ and $u^h = v^h + g^h = d_1 N_1 + g N_2$. The only unknown is d_1. The shape functions must satisfy $N_1(1) = 0$ and $N_2(1) = 1$ (see (1.6.2) and (1.6.3)). Let us take $N_1(x) = 1 - x$ and $N_2(x) = x$. These are illustrated in Fig. 1.7.1 and clearly satisfy the required conditions. Since we are dealing with only 1 degree of freedom, the matrix paraphernalia collapses as follows:

$$\boldsymbol{K} = [K_{11}] = K_{11} \tag{1.7.1}$$

$$\boldsymbol{F} = \{F_1\} = F_1 \tag{1.7.2}$$

$$\boldsymbol{d} = \{d_1\} = d_1 \tag{1.7.3}$$

$$K_{11} = a(N_1, N_1) = \int_0^1 \underbrace{N_{1,x}}_{-1}\underbrace{N_{1,x}}_{-1}\, dx = 1 \tag{1.7.4}$$

$$\begin{aligned} F_1 &= (N_1, \ell) + N_1(0)h - a(N_1, N_2)g \\ &= \int_0^1 (1 - x)\ell(x)\, dx + h - \int_0^1 \underbrace{N_{1,x}}_{-1}\underbrace{N_{2,x}}_{+1}\, dx\, g \\ &= \int_0^1 (1 - x)\ell(x)\, dx + h + g \end{aligned} \tag{1.7.5}$$

$$d_1 = K_{11}^{-1} F_1 = F_1 \tag{1.7.6}$$

Consequently

$$u^h(x) = \underbrace{\left[\int_0^1 (1 - y)\ell(y)\, dy + h + g\right]}_{d_1}(1 - x) + gx \tag{1.7.7}$$

In (1.7.7), y plays the role of a dummy variable. An illustration of (1.7.7) appears in Fig. 1.7.2. To get a feel for the nature of the approximation, let us compare (1.7.7) with the exact solution (see (1.2.3)). It is helpful to consider specific forms for ℓ.

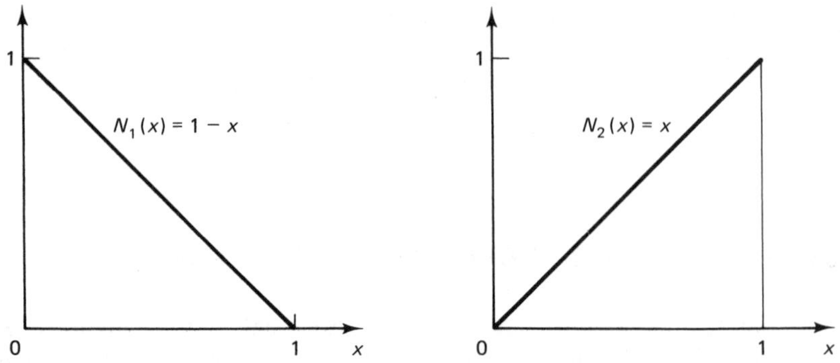

Figure 1.7.1 Functions for the 1 degree of freedom examples. (These functions are secretly the simplest finite element interpolation functions in a one-element context.)

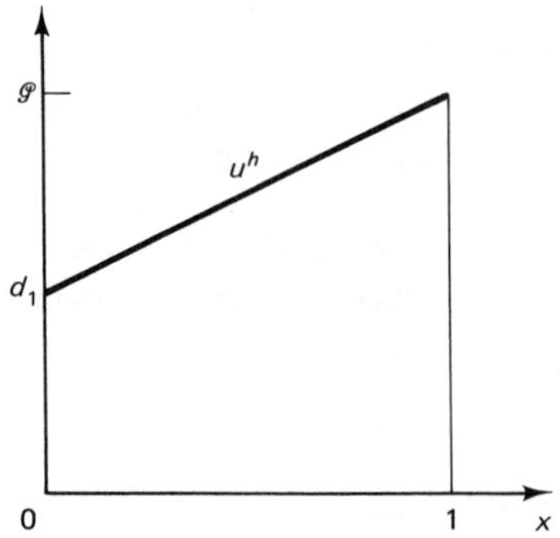

Figure 1.7.2 The Galerkin solution for the 1 degree of freedom example.

i. Let $f = 0$. Then

$$u^h(x) = u(x) = g + (1 - x)h \tag{1.7.8}$$

That is, the approximate solution is exact. In fact, it is clear by inspecting (1.7.7) and (1.2.3) that the homogeneous solution (i.e., the part of the solution corresponding to $f = 0$) is always exactly represented. The only approximation pertains to the particular solution (i.e., the part of the solution corresponding to $f \neq 0$).

ii. Now let us introduce a nonzero f. Assume $f(x) = p$, a constant. Then the particular solutions take the form

$$u_{\text{part}}(x) = \frac{p(1 - x^2)}{2} \tag{1.7.9}$$

and

$$u^h_{\text{part}}(x) = \frac{p(1 - x)}{2} \tag{1.7.10}$$

Equations (1.7.9) and (1.7.10) are compared in Fig. 1.7.3. Note that u^h_{part} is exact at $x = 0$ and $x = 1$ and that $u^h_{\text{part},x}$ is exact at $x = \frac{1}{2}$. (It should be clear that it is impossible for u^h_{part} to be exact at all x in the present circumstances. The exact solution, (1.7.9), contains a quadratic term in x, whereas the approximate solution is restricted to linear variation in x by the definitions of N_1 and N_2.)

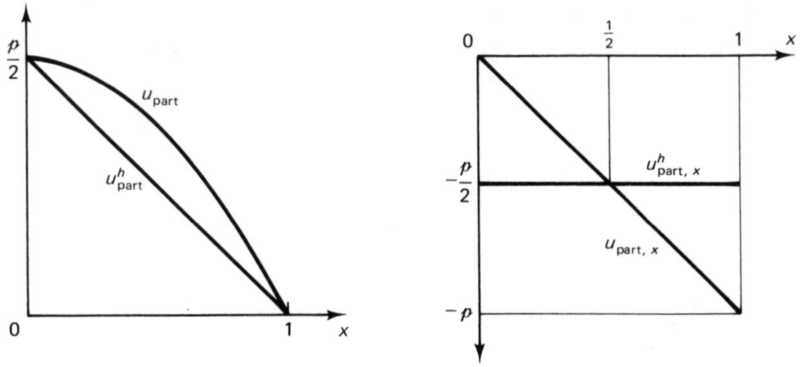

Figure 1.7.3 Comparison of exact and Galerkin particular solutions, Example 1, case (ii).

iii. This time let $\ell(x) = qx$, where q is a constant. This choice for ℓ leads to

$$u_{\text{part}}(x) = \frac{q(1 - x^3)}{6} \tag{1.7.11}$$

and

$$u^h_{\text{part}}(x) = \frac{q(1 - x)}{6} \tag{1.7.12}$$

which are compared in Fig. 1.7.4. Again we note that the u^h_{part} is exact at $x = 0$ and $x = 1$. There is one point, $x = 1/\sqrt{3}$, at which $u^h_{\text{part},x}$ is exact.

Let us summarize what we have observed in this example:

a. The homogeneous part of u^h is exact in all cases.
b. In the presence of nonzero ℓ, u^h is exact at $x = 0$ and $x = 1$.
c. For each case, there is at least one point at which $u^h_{,x}$ is exact.

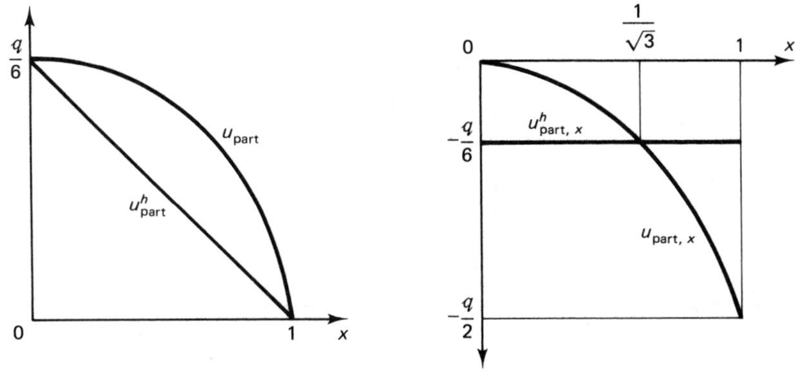

Figure 1.7.4 Comparison of exact and Galerkin particular solutions, Example 1, case (iii).

Example 2 (**2 degrees of freedom**)

In this case $n = 2$. Thus $w^h = c_1 N_1 + c_2 N_2$, where $N_1(1) = N_2(1) = 0$, and $u^h = d_1 N_1 + d_2 N_2 + g N_3$, where $N_3(1) = 1$. Let us define the N_A's as follows

$$N_1(x) = \begin{cases} 1 - 2x & 0 \le x \le \frac{1}{2} \\ 0 & \frac{1}{2} \le x \le 1 \end{cases} \tag{1.7.6}$$

$$N_2(x) = \begin{cases} 2x & 0 \le x \le \frac{1}{2} \\ 2(1 - x) & \frac{1}{2} \le x \le 1 \end{cases} \tag{1.7.7}$$

$$N_3(x) = \begin{cases} 0 & 0 \le x \le \frac{1}{2} \\ 2x - 1 & \frac{1}{2} \le x \le 1 \end{cases} \tag{1.7.8}$$

The shape functions are illustrated in Fig. 1.7.5. Typical $w^h \in \mathcal{V}^h$ and $u^h \in \mathcal{S}^h$ and their derivatives are shown in Fig. 1.7.6. Since $n = 2$, the matrix paraphernalia takes the following form:

$$\boldsymbol{K} = \begin{bmatrix} K_{11} & K_{12} \\ K_{21} & K_{22} \end{bmatrix} \tag{1.7.9}$$

$$\boldsymbol{F} = \begin{Bmatrix} F_1 \\ F_2 \end{Bmatrix} \tag{1.7.10}$$

$$\boldsymbol{d} = \begin{Bmatrix} d_1 \\ d_2 \end{Bmatrix} \tag{1.7.11}$$

$$K_{AB} = a(N_A, N_B) = \int_0^1 N_{A,x} N_{B,x}\, dx = \int_0^{1/2} N_{A,x} N_{B,x}\, dx + \int_{1/2}^1 N_{A,x} N_{B,x}\, dx \tag{1.7.12}$$

$$K_{11} = 2, \qquad K_{12} = K_{21} = -2, \qquad K_{22} = 4 \tag{1.7.13}$$

$$\boldsymbol{K} = 2\begin{bmatrix} 1 & -1 \\ -1 & 2 \end{bmatrix} \tag{1.7.14}$$

$$\begin{aligned} F_A &= (N_A, \ell) + N_A(0)h - a(N_A, N_3)g \\ &= \int_0^1 N_A \ell\, dx + N_A(0)h - \int_{1/2}^1 N_{A,x} N_{3,x}\, dx\, g \end{aligned} \tag{1.7.15}$$

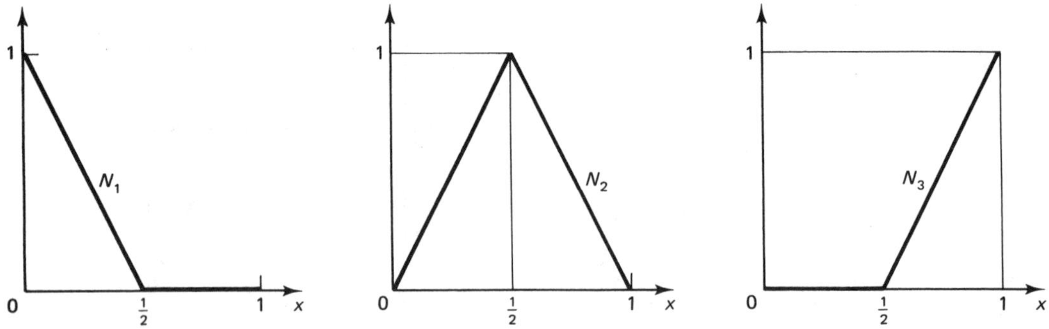

Figure 1.7.5 Functions for the 2 degree of freedom examples. (These functions are secretly the simplest finite element functions in a two-element context.)

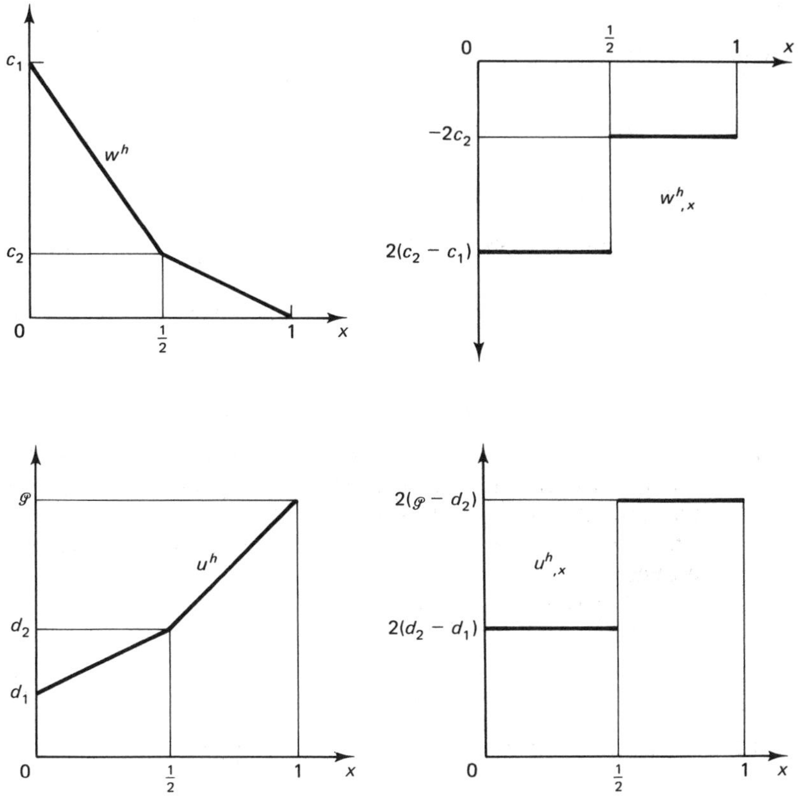

Figure 1.7.6 Typical weighting function and trial solution for the 2 degree of freedom example.

$$F_1 = \int_0^{1/2} (1 - 2x)\ell(x)\, dx + h \tag{1.7.16}$$

$$F_2 = 2\int_0^{1/2} x\ell(x)\, dx + 2\int_{1/2}^{1} (1 - x)\ell(x)\, dx + 2\mathcal{g} \tag{1.7.17}$$

Note that due to the shape functions' discontinuities in slope at $x = \frac{1}{2}$, it is convenient to express integrals over the subintervals $[0, \frac{1}{2}]$ and $[\frac{1}{2}, 1]$ (e.g., see (1.7.12) and (1.7.15)). We need not worry about the value of the derivative of N_A at $x = \frac{1}{2}$ (it suffers a discontinuity there and thus is not well-defined classically) since it has no effect on the integrals in (1.7.12). This amounts to employing the notion of a ***generalized derivative***.

We shall again analyze the three cases considered in Example 1.

i. $\ell = 0$.

$$\boldsymbol{F} = \begin{Bmatrix} h \\ 2\mathcal{g} \end{Bmatrix} \tag{1.7.18}$$

$$\boldsymbol{d} = \boldsymbol{K}^{-1}\boldsymbol{F}$$

$$= \begin{bmatrix} 1 & \frac{1}{2} \\ \frac{1}{2} & \frac{1}{2} \end{bmatrix} \begin{Bmatrix} h \\ 2g \end{Bmatrix}$$

$$= \begin{Bmatrix} g + h \\ g + \dfrac{h}{2} \end{Bmatrix} \tag{1.7.19}$$

This results in

$$u^h = (g + h)N_1 + \left(g + \frac{h}{2}\right)N_2 + gN_3$$

$$= g(N_1 + N_2 + N_3) + h\left(N_1 + \frac{N_2}{2}\right) \tag{1.7.20}$$

$$u^h(x) = g + h(1 - x) \tag{1.7.21}$$

Again, the exact homogeneous solution is obtained. (The reason for this is that the exact solution is linear, and our trial solution is capable of exactly representing any linear function. Galerkin's method will give the exact answer whenever possible—that is, whenever the collection of trial solutions contains the exact solution among its members.)

ii. $f(x) = p =$ constant.

$$F_1 = \frac{p}{4} + h \tag{1.7.22}$$

$$F_2 = \frac{p}{2} + 2g \tag{1.7.23}$$

$$\boldsymbol{d} = \begin{bmatrix} 1 & \dfrac{1}{2} \\ \dfrac{1}{2} & \dfrac{1}{2} \end{bmatrix} \begin{Bmatrix} \dfrac{p}{4} + h \\ \dfrac{p}{2} + 2g \end{Bmatrix} = \begin{Bmatrix} \dfrac{p}{2} + g + h \\ \dfrac{3p}{8} + g + \dfrac{h}{2} \end{Bmatrix} \tag{1.7.24}$$

The solution takes the form

$$u^h(x) = g + h(1 - x) + u^h_{\text{part}}(x) \tag{1.7.25}$$

$$u^h_{\text{part}} = \frac{p}{2}N_1 + \frac{3p}{8}N_2 \tag{1.7.26}$$

The approximate particular solution is compared with the exact in Fig. 1.7.7, from which we see that agreement is achieved at $x = 0, \frac{1}{2}$ and 1, and derivatives coincide at $x = \frac{1}{4}$ and $\frac{3}{4}$.

iii. $f(x) = qx$, $q =$ constant.

$$F_1 = \frac{q}{24} + h \tag{1.7.27}$$

$$F_2 = \frac{q}{4} + 2g \tag{1.7.28}$$

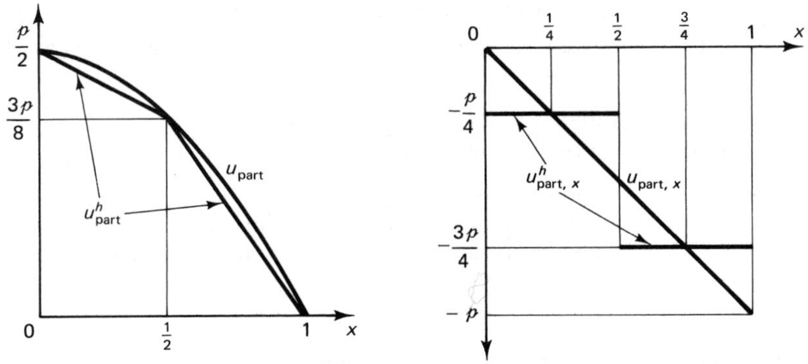

Figure 1.7.7 Comparison of exact and Galerkin particular solutions, Example 2, case (ii).

$$\boldsymbol{d} = \begin{Bmatrix} \dfrac{q}{6} + g + h \\ \dfrac{7q}{48} + g + \dfrac{h}{2} \end{Bmatrix} \tag{1.7.29}$$

Again u^h may be expressed in the form (1.7.25), where

$$u^h_{\text{part}} = \frac{q}{6}N_1 + \frac{7q}{48}N_2 \tag{1.7.30}$$

A comparison is presented in Fig. 1.7.8. The Galerkin solution is seen to be exact once again at $x = 0, \frac{1}{2}$, and 1, and the derivative is exact at two points.

Let us summarize the salient observations of Example 2:

a. The homogeneous part of u^h is exact in all cases, as in Example 1. (A rationale for this is given after Equation (1.7.21).)

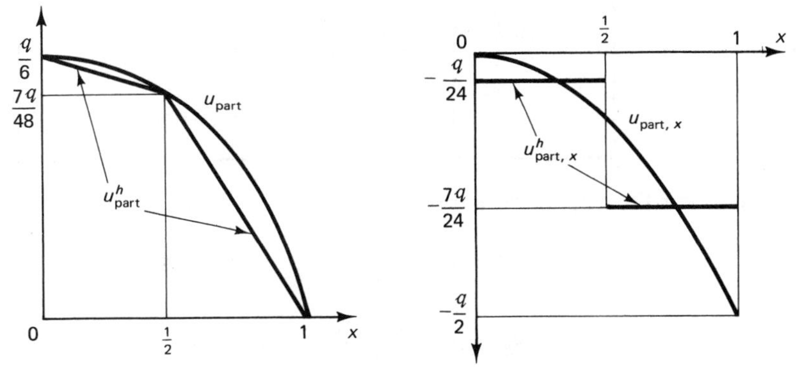

Figure 1.7.8 Comparison of exact and Galerkin particular solutions, Example 2, case (iii).

b. The Galerkin solution is exact at the endpoints of each subinterval for all cases.
c. In each case, there is at least one point in each subinterval at which $u^h_{,x}$ is exact.

After generalizing to the case of n subintervals in the following section, we shall show in Sec. 1.10 that the above observations are not accidental.

Exercise 1. If the reader has not had experience with calculations of the type presented in this section, it would be worthwhile to reproduce all results, providing all omitted details.

1.8 PIECEWISE LINEAR FINITE ELEMENT SPACE

The examples of the preceding section employed definitions of $\mathcal{V}^h$ and $\mathcal{S}^h$ which were special cases of the so-called piecewise linear finite element space. To define the general case in which $\mathcal{V}^h$ is n-dimensional, we partition the domain $[0, 1]$ into n nonoverlapping subintervals. The typical subinterval is denoted by $[x_A, x_{A+1}]$, where $x_A < x_{A+1}$ and $A = 1, 2, \ldots, n$. We also require $x_1 = 0$ and $x_{n+1} = 1$. The x_A's are called ***nodal points,*** or simply ***nodes.*** (The terminologies ***joints*** and ***knots*** are also used.) The subintervals are sometimes referred to as the ***finite element domains,*** or simply ***elements.*** Notice that the lengths of the elements, $h_A = x_{A+1} - x_A$, are *not* required to be equal. The mesh parameter, h, is generally taken to be the length of the maximum subinterval (i.e., $h = \max h_A$, $A = 1, 2, \ldots, n$). The smaller h, the more "refined" is the partition, or mesh. If the subinterval lengths are equal, then $h = 1/n$.

The shape functions are defined as follows: Associated to a typical internal node (i.e., $2 \leq A \leq n$)

$$N_A(x) = \begin{cases} \dfrac{(x - x_{A-1})}{h_{A-1}}, & x_{A-1} \leq x \leq x_A \\ \dfrac{(x_{A+1} - x)}{h_A}, & x_A \leq x \leq x_{A+1} \\ 0, & \text{elsewhere} \end{cases} \tag{1.8.1}$$

whereas for the boundary nodes we have

$$N_1(x) = \frac{x_2 - x}{h_1}, \qquad x_1 \leq x \leq x_2 \tag{1.8.2}$$

$$N_{n+1}(x) = \frac{x - x_n}{h_n}, \qquad x_n \leq x \leq x_{n+1} \tag{1.8.3}$$

The shape functions are sketched in Fig. 1.8.1. For obvious reasons, they are referred to variously as "hat," "chapeau," and "roof" functions. Note that $N_A(x_B) = \delta_{AB}$, where

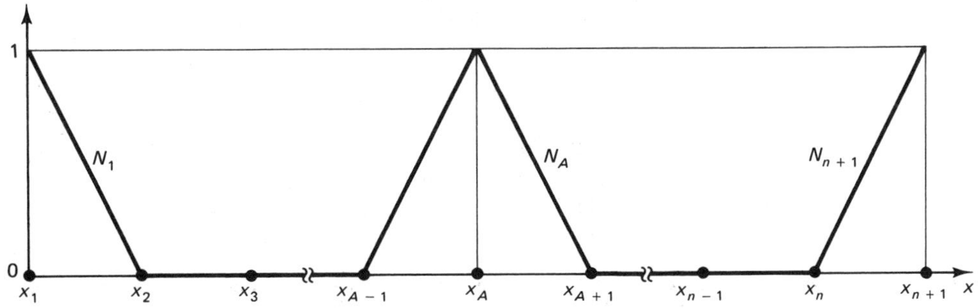

Figure 1.8.1 Basis functions for the piecewise linear finite element space.

δ_{AB} is the Kronecker delta (i.e., $\delta_{AB} = 1$ if $A = B$, whereas $\delta_{AB} = 0$ if $A \neq B$). In words, N_A takes on the value 1 at node A and is 0 at all other nodes. Furthermore, N_A is nonzero only in the subintervals that contain x_A.

A typical member $w^h \in \mathcal{V}^h$ has the form $\Sigma_{A=1}^{n} c_A N_A$ and appears as in Fig. 1.8.2. Note that w^h is continuous but has discontinuous slope across each element boundary. For this reason, $w^h_{,x}$, the generalized derivative of w^h, will be piecewise constant, experiencing discontinuities across element boundaries. (Such a function is sometimes called a ***generalized step function.***) Restricted to each element domain, w^h is a linear polynomial in x. In respect to the homogeneous essential boundary condition, $w^h(1) = 0$. Clearly, w^h is identically zero if and only if each $c_A = 0$, $A = 1, 2, \ldots, n$.

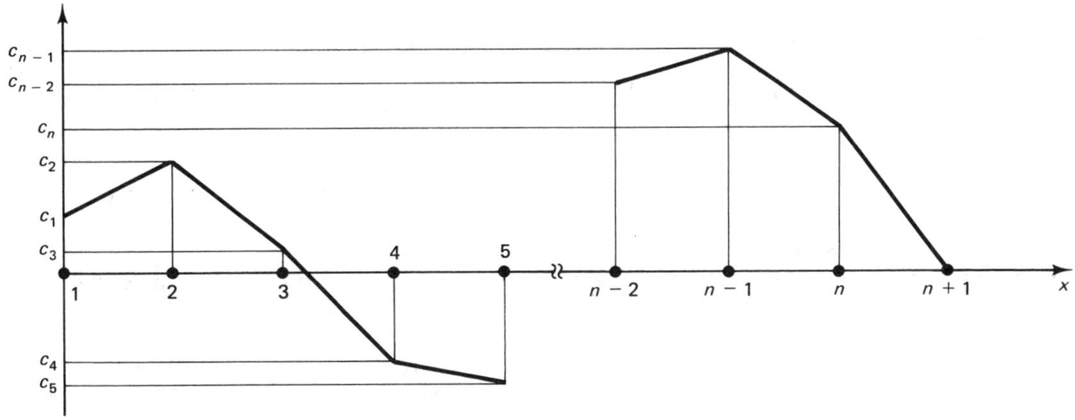

Figure 1.8.2 A typical member $w^h \in \mathcal{V}^h$.

Typical members of $\mathcal{S}^h$ are obtained by adding $g^h = gN_{n+1}$ to typical members of $\mathcal{V}^h$. This ensures that $u^h(1) = g$.

The piecewise linear finite element functions are the simplest and most widely used finite element functions for one-dimensional problems.

Exercise 1. Consider the weak formulation of the one-dimensional model problem:

$$\int_0^1 w_{,x} u_{,x} \, dx = \int_0^1 w \ell \, dx + w(0) h \tag{1.8.4}$$

where $w \in \mathcal{V}$ and $u \in \mathcal{S}$ are assumed to be smooth on element interiors (i.e., on $]x_A, x_{A+1}[$, $A = 1, 2, \ldots, n$), but may suffer slope discontinuities across element boundaries. (Functions of this class contain the piecewise linear finite element space described earlier.) From (1.8.4) and the assumed continuity of the functions, show that:

$$0 = \sum_{A=1}^{n} \int_{x_A}^{x_{A+1}} w(u_{,xx} + \ell) \, dx + w(0)[u_{,x}(0^+) + h] + \sum_{A=2}^{n} w(x_A)[u_{,x}(x_A^+) - u_{,x}(x_A^-)] \tag{1.8.5}$$

Arguing as in Sec. 1.4, it may be concluded that the Euler-Lagrange conditions of (1.8.5) are

i. $u_{,xx}(x) + \ell(x) = 0$, where $x \in]x_A, x_{A+1}[$ and $A = 1, 2, \ldots, n$,
ii. $-u_{,x}(0^+) = h$; and
iii. $u_{,x}(x_A^-) = u_{,x}(x_A^+)$, where $A = 2, 3, \ldots, n$.

Observe that (i) is the differential equation *restricted to element interiors*, and (iii) is a continuity condition across element boundaries. This may be contrasted with the case in which the solution is assumed smooth. In this case the continuity condition is identically satisfied and the summation of integrals over element interiors may be replaced by an integral over the entire domain (see Sec. 1.4).

In the Galerkin finite element formulation, an *approximate* solution of (i)–(iii) is obtained.

1.9 PROPERTIES OF K

The shape functions N_A, $A = 1, 2, \ldots, n + 1$, are zero outside a neighborhood of node A. As a result, many of the entries of $\boldsymbol{K}$ are zero. This can be seen as follows. Let $B > A + 1$. Then (see Fig. 1.9.1)

$$K_{AB} = \int_0^1 \underbrace{N_{A,x} N_{B,x}}_{0} \, dx = 0 \tag{1.9.1}$$

The symmetry of $\boldsymbol{K}$ implies, in addition, that (1.9.1) holds for $A > B + 1$. One says that $\boldsymbol{K}$ is ***banded*** (i.e., its nonzero entries are located in a band about the main diagonal). Figure 1.9.2 depicts this property. Banded matrices have significant advantages in that the zero elements outside the band neither have to be stored nor operated

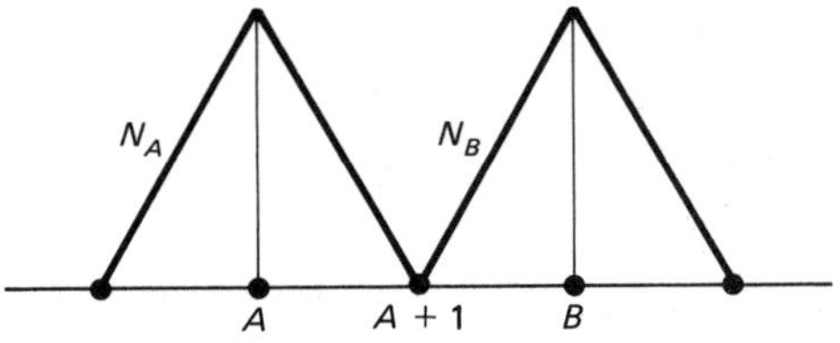

Figure 1.9.1 If $B > A + 1$, the nonzero portions of N_B and N_A do not overlap.

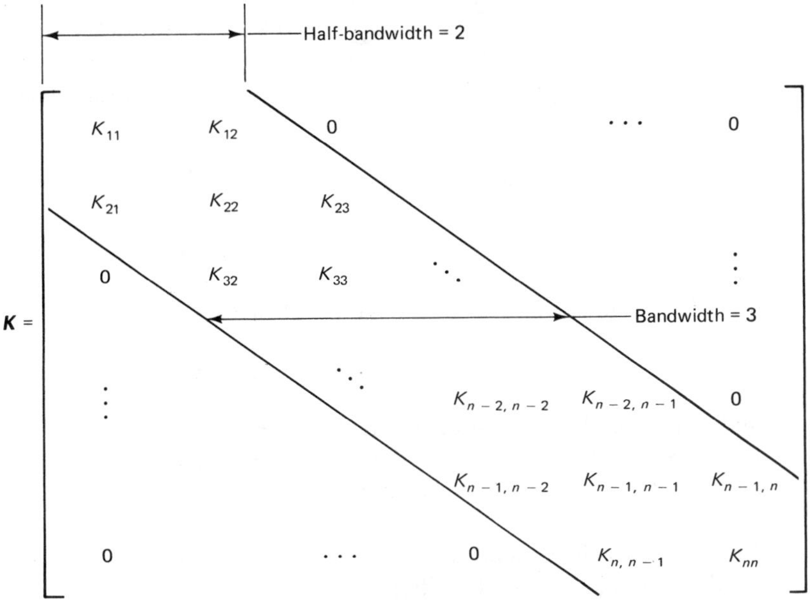

Figure 1.9.2 Band structure of $\boldsymbol{K}$.

upon in the computer. The stiffness matrix arising in finite element analysis is, in general, narrowly banded, lending itself to economical formation and solution.

Definition. An $n \times n$ matrix $\boldsymbol{A}$ is said to be ***positive definite*** if

i. $\boldsymbol{c}^T\boldsymbol{A}\boldsymbol{c} \geq 0$ for all n-vectors $\boldsymbol{c}$; and
ii. $\boldsymbol{c}^T\boldsymbol{A}\boldsymbol{c} = 0$ implies $\boldsymbol{c} = \boldsymbol{0}$.

Remarks

1. A symmetric positive-definite matrix posesses a unique inverse.
2. The eigenvalues of a positive-definite matrix are real and positive.

Theorem. The $n \times n$ matrix $\boldsymbol{K}$ defined by (1.6.11) is positive definite.

Proof

i. Let c_A, $A = 1, 2, \ldots, n$, be the components of $\boldsymbol{c}$ (i.e., $\boldsymbol{c} = \{c_A\}$), an arbitrary vector. Use these c_A's to construct a member of $\mathcal{V}^h$, $w^h = \sum_{A=1}^{n} c_A N_A$, where the N_A's are the basis functions for $\mathcal{V}^h$. Then

$$\begin{aligned}
\boldsymbol{c}^T\boldsymbol{K}\boldsymbol{c} &= \sum_{A,B=1}^{n} c_A K_{AB} c_B \\
&= \sum_{A,B=1}^{n} c_A a(N_A, N_B) c_B && \text{(definition of } K_{AB}\text{)} \\
&= a\left(\sum_{A=1}^{n} c_A N_A, \sum_{B=1}^{n} c_B N_B\right) && \text{(bilinearity of } a(\cdot\,,\cdot)\text{)} \\
&= a(w^h, w^h) && \text{(definition of } w^h\text{)} \\
&= \int_0^1 \underbrace{(w^h_{,x})^2}_{\geq 0} dx && \text{(by (1.4.8)} \\
&\geq 0
\end{aligned}$$

ii. Assume $\boldsymbol{c}^T\boldsymbol{K}\boldsymbol{c} = 0$. By the proof of part (i),

$$\int_0^1 (w^h_{,x})^2\, dx = 0$$

and consequently w^h must be constant. Since $w^h \in \mathcal{V}^h$, $w^h(1) = 0$. Combining these facts, we conclude that $w^h(x) = 0$ for all $x \in [0, 1]$, which is possible only if each $c_A = 0$, $A = 1, 2, \ldots, n$. Thus $\boldsymbol{c} = \boldsymbol{0}$. ■

Note that part (ii) depended upon the definition of $\boldsymbol{K}$ and the zero essential boundary condition built into the definition of $\mathcal{V}^h$.

Summary. $\boldsymbol{K}$, defined by (1.6.11), is

i. Symmetric
ii. Banded
iii. Positive-definite

The practical consequence of the above properties is that a very efficient computer solution of $\boldsymbol{Kd} = \boldsymbol{F}$ may be performed.

1.10 MATHEMATICAL ANALYSIS

In this section we will show that the observations made with reference to the example problems of Sec. 1.7 are, in fact, general results. To establish these facts rigorously requires only elementary mathematical techniques.

Our first objective is to establish that the Galerkin finite element solution u^h is exact at the nodes. To do this we must introduce the notion of a Green's function.

Let $\delta_y(x) = \delta(x - y)$ denote the ***Dirac delta function.*** The Dirac function is not a function in the classical sense but rather an operator defined by its action on (continuous) functions. Let w be continuous on [0, 1]; then we write

$$\begin{aligned}(w, \delta_y) &= \int_0^1 w(x)\delta(x-y)\,dx \\ &= w(y)\end{aligned} \tag{1.10.1}$$

By (1.10.1), we see why attention is restricted to continuous functions—δ_y sifts out the value of w at y. If w were discontinuous at y, its value would be ambiguous. In mechanics, we think of δ_y visually as representing a concentrated force of unit amplitude located at point y.

The Green's function problem corresponding to (S) may be stated as follows: Find a function g (i.e., the ***Green's function***) such that

$$g_{,xx} + \delta_y = 0 \qquad \text{on } \Omega \tag{1.10.2}$$

$$g(1) = 0 \tag{1.10.3}$$

$$g_{,x}(0) = 0 \tag{1.10.4}$$

Note that (1.10.2)–(1.10.4) are simply (S) in which ℓ is replaced by δ_y and q and h are taken to be zero.

This problem may be solved by way of formal calculations with ***distributions***, or ***generalized functions***, such as δ_y. (The theory of distributions is dealt with in Stakgold [5]. A good elementary account of formal calculations with distributions is presented in Popov [9]. This latter reference is recommended to readers having had no previous experience with this topic.) To this end we note that the (formal) integral of δ_y is the ***Heaviside, or unit step, function:***

$$H_y(x) = H(x-y) = \begin{cases} 0, & x < y \\ 1, & x > y \end{cases} \tag{1.10.5}$$

The integral of H_y is the ***Macaulay bracket:***

$$\langle x - y \rangle = \begin{cases} 0, & x \le y \\ x - y, & x > y \end{cases} \tag{1.10.6}$$

The preceding functions are depicted in Fig. 1.10.1.

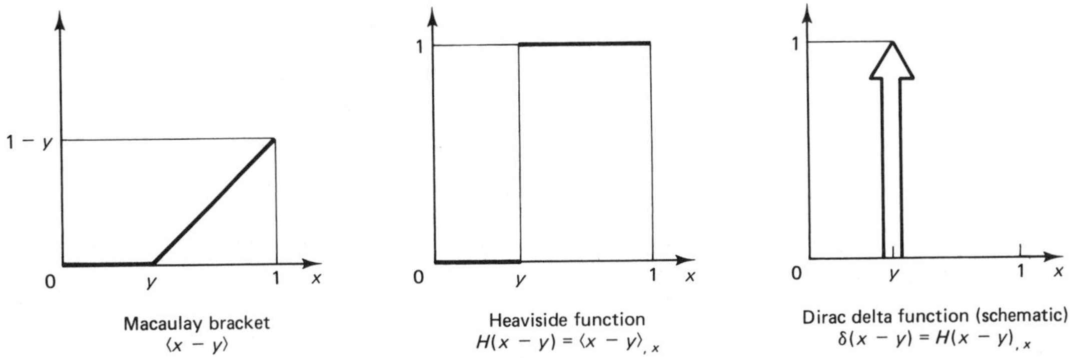

Figure 1.10.1 Elementary generalized functions (distributions).

To solve the Green's function problem, (1.10.2) is integrated, making use of (1.10.5), to obtain:

$$g_{,x} + H_y = c_1 \tag{1.10.7}$$

where c_1 is a constant of integration. A second integration and use of (1.10.6) yields

$$g(x) + \langle x - y \rangle = c_1 x + c_2 \tag{1.10.8}$$

where c_2 is another constant of integration. Evaluation of c_1 and c_2 is performed by requiring (1.10.7) and (1.10.8) to satisfy the boundary conditions. This results in (see Fig. 1.10.2)

$$g(x) = (1 - y) - \langle x - y \rangle \tag{1.10.9}$$

Observe that g is piecewise linear. Thus if $y = x_A$ (i.e., if y is a node), $g \in \mathcal{V}^h$.

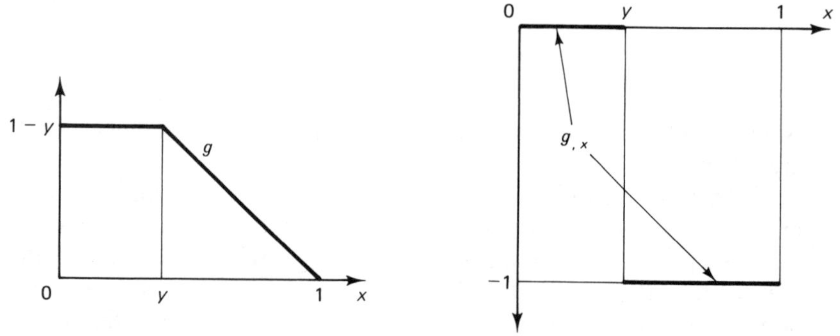

Figure 1.10.2 Green's function.

In the ensuing analysis we will need the variational equation corresponding to the Green's function problem. This can be deduced from (W) by replacing u by g, f by δ_y, and q and h by 0, viz.,

$$a(w, g) = (w, \delta_y) = w(y) \tag{1.10.10}$$

Equation (1.10.10) holds for all continuous $w \in \mathcal{V}$. The square-integrability of derivatives of functions $w \in \mathcal{V}$ actually implies the continuity of all $w \in \mathcal{V}$ by a well-known theorem in analysis due to Sobolev. (This result is true only in one dimension. The square-integrability of second derivatives is also required to ensure the continuity of functions defined on two- and three-dimensional domains.)

Theorem. $u^h(x_A) = u(x_A)$, $A = 1, 2, \ldots, n + 1$ (i.e., u^h is exact at the nodes). To prove the theorem, we need to establish two preliminary results.

Lemma 1. $a(u - u^h, w^h) = 0$ for all $w^h \in \mathcal{V}^h$.

Proof. We have observed previously that $\mathcal{V}^h \subset \mathcal{V}$, so we may replace w by w^h in the variational equation:

$$a(w^h, u) = (w^h, f) + w^h(0)h \tag{1.10.11}$$

Equation (1.10.11) holds for all $w^h \in \mathcal{V}^h$. Recall that the Galerkin equation is identical to (1.10.11) except that u^h appears instead of u. Subtracting the Galerkin equation from (1.10.11) and using the bilinearity and symmetry of $a(\cdot, \cdot)$ yields the required result. ■

Lemma 2. $u(y) - u^h(y) = a(u - u^h, g)$, where g is the Green's function.

Proof

$$\begin{aligned} u(y) - u^h(y) &= (u - u^h, \delta_y) && \text{(definition of } \delta_y\text{)} \\ &= a(u - u^h, g) && \text{(by (1.10.10))} \end{aligned}$$

Note that line 2 is true since $u - u^h$ is in $\mathcal{V}$. ■

Proof of Theorem. As we have remarked previously, if $y = x_A$, a node, $g \in \mathcal{V}^h$. Let us take this to be the case. Then

$$\begin{aligned} u(x_A) - u^h(x_A) &= a(u - u^h, g) && \text{(Lemma 2)} \\ &= 0 && \text{(Lemma 1)} \end{aligned}$$ ■

The theorem is valid for $A = 1, 2, \ldots, n + 1$. Strang and Fix [6] attribute this argument to Douglas and Dupont. Results of this kind, embodying exceptional accuracy characteristics, are often referred to as ***superconvergence*** phenomena. However, the reader should appreciate that, in more complicated situations, we will not be able, in practice, to guarantee nodal exactness. Nevertheless, as we shall see later on, weighted residual procedures provide a framework within which optimal accuracy properties of some sort may often be guaranteed.

Accuracy of the Derivatives

In considering the convergence properties of the derivatives, certain elementary notions of numerical analysis arise. The reader should make sure that he or she has a complete understanding of these ideas as they subsequently arise in other contexts. We begin by introducing some preliminary mathematical results.

Taylor's Formula with Remainder

Let $f: [0, 1] \rightarrow \mathbb{R}$ possess k continuous derivatives and let y and z be two points in $[0, 1]$. Then there is a point c between y and z such that

$$\begin{aligned} f(z) = f(y) &+ (z - y)f_{,x}(y) + \frac{1}{2}(z - y)^2 f_{,xx}(y) \\ &+ \frac{1}{3!}(z - y)^3 f_{,xxx}(y) + \cdots + \\ &+ \frac{1}{k!}(z - y)^k f_{,\underbrace{x\ldots x}_{k \text{ times}}}(c) \end{aligned} \tag{1.10.12}$$

The proof of this formula may be found in [7]. Equation (1.10.12) is sometimes called a ***finite Taylor expansion***.

Mean-Value Theorem

The mean-value theorem is a special case of (1.10.12) which is valid as long as $k \geq 1$ (i.e., f is continuously differentiable):

$$\boxed{f(z) = f(y) + (z - y)f_{,x}(c)} \tag{1.10.13}$$

Consider a typical subinterval $[x_A, x_{A+1}]$. We have already shown that u^h is exact at the endpoints (see Fig. 1.10.3). The derivative of u^h in $]x_A, x_{A+1}[$ is constant:

$$u^h_{,x}(x) = \frac{u^h(x_{A+1}) - u^h(x_A)}{h_A}, \qquad x \in \,]x_A, x_{A+1}[\tag{1.10.14}$$

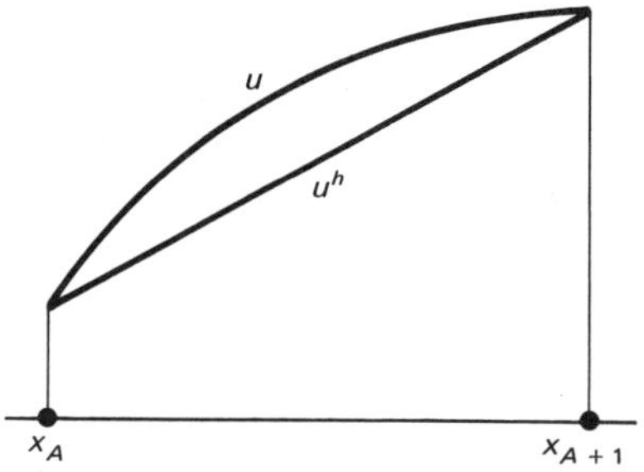

Figure 1.10.3

Theorem. Assume u is continuously differentiable. Then there exists at least one point in $]x_A, x_{A+1}[$ at which (1.10.14) is exact.

Proof. By the mean value theorem, there exists a point $c \in \,]x_A, x_{A+1}[$ such that

$$\frac{u(x_{A+1}) - u(x_A)}{h_A} = u_{,x}(c) \tag{1.10.15}$$

(We have used (1.10.13) with u, x_A, and x_{A+1}, in place of f, y, and z, respectively.) Since $u(x_A) = u^h(x_A)$ and $u(x_{A+1}) = u^h(x_{A+1})$, we may rewrite (1.10.15) as

$$\frac{u^h(x_{A+1}) - u^h(x_A)}{h_A} = u_{,x}(c) \tag{1.10.16}$$

Comparison of (1.10.16) with (1.10.14) yields the desired result. ■

Remarks

1. This result means that the constant value of $u^h_{,x}$ must coincide with $u_{,x}$ somewhere on $]x_A, x_{A+1}[$; see Fig. 1.10.4.

2. Without knowledge of u we have no way of determining the locations at which the derivatives are exact. The following results are more useful in that they tell us that the midpoints are, in a sense, optimally accurate, independent of u.

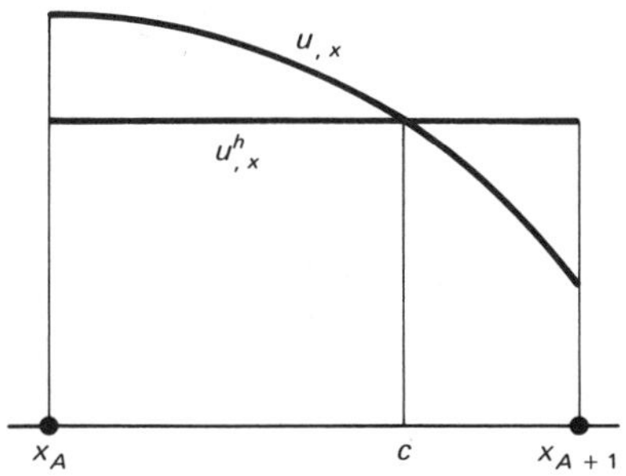

Figure 1.10.4

Let

$$e_{,x}(\alpha) \overset{\text{def.}}{=} u^h_{,x}(\alpha) - u_{,x}(\alpha) = \frac{u^h(x_{A+1}) - u^h(x_A)}{h_A} - u_{,x}(\alpha)$$

the ***error in the derivative*** at $\alpha \in [x_A, x_{A+1}]$. To establish the superiority of the midpoints in evaluating the derivatives, we need a preliminary result.

Lemma. Assume u is three times continuously differentiable. Then

$$\begin{aligned} e_{,x}(\alpha) &= \left(\frac{x_{A+1} + x_A}{2} - \alpha\right)u_{,xx}(\alpha) \\ &\quad + \frac{1}{3!\, h_A}[(x_{A+1} - \alpha)^3 u_{,xxx}(c_1) - (x_A - \alpha)^3 u_{,xxx}(c_2)] \end{aligned} \tag{1.10.17}$$

where c_1 and c_2 are in $[x_A, x_{A+1}]$.

Proof. Expand $u(x_{A+1})$ and $u(x_A)$ in finite Taylor expansions about $\alpha \in [x_A, x_{A+1}]$, viz.,

$$\begin{aligned} u(x_{A+1}) &= u(\alpha) + (x_{A+1} - \alpha)u_{,x}(\alpha) + \frac{1}{2}(x_{A+1} - \alpha)^2 u_{,xx}(\alpha) \\ &\quad + \frac{1}{3!}(x_{A+1} - \alpha)^3 u_{,xxx}(c_1), \qquad c_1 \in [\alpha, x_{A+1}] \end{aligned}$$

$$\begin{aligned} u(x_A) &= u(\alpha) + (x_A - \alpha)u_{,x}(\alpha) + \frac{1}{2}(x_A - \alpha)^2 u_{,xx}(\alpha) \\ &\quad + \frac{1}{3!}(x_A - \alpha)^3 u_{,xxx}(c_2), \qquad c_2 \in [x_A, \alpha] \end{aligned}$$

Subtracting and dividing through by h_A yields

$$\begin{aligned} \frac{u(x_{A+1}) - u(x_A)}{h_A} &= u_{,x}(\alpha) + \left(\frac{x_{A+1} + x_A}{2} - \alpha\right)u_{,xx}(\alpha) \\ &\quad + \frac{1}{3!\, h_A}[(x_{A+1} - \alpha)^3 u_{,xxx}(c_1) - (x_A - \alpha)^3 u_{,xxx}(c_2)] \end{aligned}$$

Replacing $u(x_{A+1})$ by $u^h(x_{A+1})$ and $u(x_A)$ by $u^h(x_A)$ in the left-hand side and rearranging terms completes the proof. ■

Discussion

To determine what (1.10.17) tells us about the accuracy of the derivatives, we wish to think of the situation in which the mesh is being systematically refined (i.e., we let h_A approach zero). In this case h_A^2 will be much smaller than h_A. Thus, for a given u, if the right-hand side of (1.10.17) *is* $O(h_A^2)$,[3] the error in the derivatives will be much smaller than if the right-hand side is only $O(h_A)$. The exponent of h_A is called the ***order of convergence*** or ***order of accuracy.*** In the former case we would have second-order convergence of the derivative, whereas in the latter case we would have only first-order convergence.

As an example, assume $\alpha \to x_A$. Then

$$e_{,x}(x_A) = \frac{h_A}{2} u_{,xx}(x_A) + \frac{h_A^2}{3!} u_{,xxx}(c_1) = O(h_A)$$

As $h_A \to 0$, the first term dominates. (We have seen from the example calculations in Sec. 1.8 that the endpoints of the subintervals are not very accurate for the derivatives.)

Clearly any point $\alpha \in [x_A, x_{A+1}]$ achieves first-order accuracy. We are thus naturally led to asking the question, are there any values of α at which higher-order accuracy is achieved?

Corollary. Let $x_{A+1/2} \equiv (x_A + x_{A+1})/2$ (i.e., the midpoint). Then

$$\begin{aligned} e_{,x}(x_{A+1/2}) &= \frac{h_A^2}{24} u_{,xxx}(c), \qquad c \in [x_A, x_{A+1}] \\ &= O(h_A^2) \end{aligned}$$

Proof. By (1.10.17)

$$e_{,x}(x_{A+1/2}) = \frac{h_A^2}{48}[u_{,xxx}(c_1) + u_{,xxx}(c_2)]$$

By the continuity of $u_{,xxx}$, there is at least one point c between c_1 and c_2 such that

$$u_{,xxx}(c) = \frac{1}{2}[u_{,xxx}(c_1) + u_{,xxx}(c_2)]$$

Combining these facts completes the proof. ■

Remarks

1. From the corollary we see that the derivatives are second-order accurate at the midpoints.

[3]A function $f(x)$ is said to be $O(x^k)$ (i.e., order x^k) if $f(x)/x^k \to$ a constant as $x \to 0$. For example, $f(x) = x^k$ is $O(x^k)$, as is $f(x) = \Sigma_{j=k}^{k+l} x^j$, $l \geq 0$. But neither is $O(x^{k+1})$. (Verify.)

2. If the exact solution is quadratic (i.e., consists of a linear combination of monomials $1, x, x^2$), then $u_{,xxx} = 0$ and—by (1.10.17)—the derivative is exact at the midpoints. This is the case when $f(x) = p =$ constant.

3. In linear elastic rod theory, the derivatives are proportional to the stresses. The midpoints of linear "elements" are sometimes called the ***Barlow stress points,*** after Barlow [8], who first noted that points of optimal accuracy existed within elements.

Exercise 1. Assume the mesh length is constant (i.e., $h_A = h, A = 1, 2, \ldots, n$). Consider the standard finite difference "stencil" for $u_{,xx} + f = 0$ at a typical internal node, namely,

$$\frac{u_{A+1} - 2u_A + u_{A-1}}{h^2} + f_A = 0 \tag{1.10.18}$$

Assuming f varies in piecewise linear fashion and so can be expanded as

$$f = \sum_{A=1}^{n+1} f_A N_A \tag{1.10.19}$$

where the f_A's are the nodal values of f, set up the finite element equation associated with node A and contrast it with (1.10.18). Deduce when (1.10.18) will also be capable of exhibiting superconvergence phenomena. (That is, what is the restriction on f?) Set up the finite element equation associated with node 1, accounting for nonzero h. Discuss this equation from the point of view of finite differences. (For further comparisons along these lines, the interested reader is urged to consult [6], Chapter 1.)

Summary. The Galerkin finite element solution u^h, of the problem (S), possesses the following properties:

i. It is exact at the nodes.
ii. There exists at least one point in each element at which the derivative is exact.
iii. The derivative is second-order accurate at the midpoints of the elements.

1.11 INTERLUDE: GAUSS ELIMINATION; HAND-CALCULATION VERSION

It is important for anyone who wishes to do finite element analysis to become familiar with the efficient and sophisticated computer schemes that arise in the finite element method. It is felt that the best way to do this is to begin with the simplest scheme, perform some hand calculations, and gradually increase the sophistication as time goes on.

To do some of the problems we will need a fairly efficient method of solving matrix equations by hand. The following scheme is applicable to systems of equations which *no pivoting* (i.e., reordering) is necessary. For example, symmetric, finite coefficient matrices never require pivoting. The procedure is as

ination

- Solve the first equation for d_1 and elminate d_1 from the remaining $n - 1$ equations.
- Solve the second equation for d_2 and eliminate d_2 from the remaining $n - 2$ equations.

  .
  .
  .

- Solve the $n - 1$st equation for d_{n-1} and eliminate d_{n-1} from the nth equation.
- Solve the n-th equation for d_n.

The preceding steps are called ***forward reduction***. The original matrix is reduced to upper triangular form. For example, suppose we began with a system of four equations as follows:

$$\begin{bmatrix} K_{11} & K_{12} & K_{13} & K_{14} \\ K_{21} & K_{22} & K_{23} & K_{24} \\ K_{31} & K_{32} & K_{33} & K_{34} \\ K_{41} & K_{42} & K_{43} & K_{44} \end{bmatrix} \begin{Bmatrix} d_1 \\ d_2 \\ d_3 \\ d_4 \end{Bmatrix} = \begin{Bmatrix} F_1 \\ F_2 \\ F_3 \\ F_4 \end{Bmatrix}$$

The ***augmented matrix*** corresponding to this system is

$$\Big[\underbrace{\begin{matrix} K_{11} & K_{12} & K_{13} & K_{14} \\ K_{21} & K_{22} & K_{23} & K_{24} \\ K_{31} & K_{32} & K_{33} & K_{34} \\ K_{41} & K_{42} & K_{43} & K_{44} \end{matrix}}_{\boldsymbol{K}} \;\Big|\; \underbrace{\begin{matrix} F_1 \\ F_2 \\ F_3 \\ F_4 \end{matrix}}_{\boldsymbol{F}}\Big]$$

After the forward reduction, the augmented matrix becomes

$$\Big[\underbrace{\begin{matrix} 1 & K'_{12} & K'_{13} & K'_{14} \\ 0 & 1 & K'_{23} & K'_{24} \\ 0 & 0 & 1 & K'_{34} \\ 0 & 0 & 0 & 1 \end{matrix}}_{\boldsymbol{U}} \;\Big|\; \underbrace{\begin{matrix} F'_1 \\ F'_2 \\ F'_3 \\ d_4 \end{matrix}}_{\boldsymbol{F}'}\Big] \tag{1.11.1}$$

corresponding to the upper triangular system $\boldsymbol{Ud} = \boldsymbol{F}'$.[4] It is a simply verified fact that if $\boldsymbol{K}$ is banded, then $\boldsymbol{U}$ will be also.

Employing the reduced augmented matrix, proceed as follows:

- Eliminate d_n from equations $n - 1, n - 2, \ldots, 1$.

[4]Primes will be used to denote intermediate quantities throughout this section.

- Eliminate d_{n-1} from equations $n-2, n-3, \ldots, 1$.

 .
 .
 .

- Eliminate d_2 from the first equation.

This procedure is called ***back substitution***. For example, in the example just given, after back substitution we obtain

$$\left[\underbrace{\begin{matrix} 1 & 0 & 0 & 0 \\ 0 & 1 & 0 & 0 \\ 0 & 0 & 1 & 0 \\ 0 & 0 & 0 & 1 \end{matrix}}_{\boldsymbol{I}} \left| \underbrace{\begin{matrix} d_1 \\ d_2 \\ d_3 \\ d_4 \end{matrix}}_{\boldsymbol{d}} \right.\right] \tag{1.11.2}$$

corresponding to the identity $\boldsymbol{Id} = \boldsymbol{d}$. The solution winds up in the last column.

Hand-Calculation Algorithm

In a hand calculation, Gauss elimination can be performed on the augmented matrix as follows.

Forward reduction

- Divide row 1 by K_{11}.
- Subtract $K_{21} \times$ row 1 from row 2.
- Subtract $K_{31} \times$ row 1 from row 3.

 .
 .
 .

- Subtract $K_{n1} \times$ row 1 from row n.

Consider the example of four equations. The preceding steps reduce the first column to the form

$$\left[\begin{matrix} 1 & K'_{12} & K'_{13} & K'_{14} \\ 0 & K''_{22} & K''_{23} & K''_{24} \\ 0 & K''_{32} & K''_{33} & K''_{34} \\ 0 & K''_{42} & K''_{43} & K''_{44} \end{matrix} \left| \begin{matrix} F'_1 \\ F''_2 \\ F''_3 \\ F''_4 \end{matrix} \right.\right]$$

Note that if $K_{A1} = 0$, then the computation for the Ath row can be ignored. Now reduce the second column

- Divide row 2 by K''_{22}.
- Subtract $K''_{32} \times$ row 2 from row 3.
- Subtract $K''_{42} \times$ row 2 from row 4.

.
.
.

- Subtract $K''_{n2} \times$ row 2 from row n.

The result for the example will look like

$$\begin{bmatrix} 1 & K'_{12} & K'_{13} & K'_{14} & F'_1 \\ 0 & 1 & K'''_{23} & K'''_{24} & F'''_2 \\ 0 & 0 & K'''_{33} & K'''_{34} & F'''_3 \\ 0 & 0 & K'''_{43} & K'''_{44} & F'''_4 \end{bmatrix}$$

Note that only the submatrix enclosed in dashed lines is affected in this procedure.

Repeat until columns 3 to n are reduced and the upper triangular form (1.11.1) is obtained.

Back substitution

- Subtract $K'_{n-1,n} \times$ row n from row $n - 1$.
- Subtract $K'_{n-2,n} \times$ row n from row $n - 2$.

.
.
.

- Subtract $K'_{1,n} \times$ row n from row 1.

After these steps the augmented matrix, for this example, will look like

$$\begin{bmatrix} 1 & K'_{12} & K'_{13} & 0 & F_1'''' \\ 0 & 1 & K'_{23} & 0 & F_2'''' \\ 0 & 0 & 1 & 0 & d_3 \\ 0 & 0 & 0 & 1 & d_4 \end{bmatrix}$$

Note that the submatrix enclosed in dashed lines is unaffected by these steps, and, aside from zeroing the appropriate elements of the last column of the coefficient matrix, only the vector $\boldsymbol{F}'$ is altered.

Now clear the second-to-last column in the coefficient matrix:

- Subtract $K'_{n-2,n-1} \times$ row $n - 1$ from row $n - 2$.
- Subtract $K'_{n-3,n-1} \times$ row $n - 1$ from row $n - 3$.

.
.
.

- Subtract $K'_{1,n-1} \times$ row $n - 1$ from row 1.

Again we mention that the only nontrivial calculations are being performed on the last column (i.e., on $\boldsymbol{F}$).

Repeat as above until columns $n-2, n-3, \ldots, 2$ are cleared. The result is (1.11.2).

Remarks

1. In passing we note that the above procedure is *not* the same as the way one would implement Gauss elimination on a computer, which we shall treat later. In a computer program for Gauss elimination of symmetric matrices we would want all intermediate results to retain symmetry and thus save storage. This can be done by a small change in the procedure. However, it is felt that the given scheme is the clearest for hand calculations.

2. The numerical example with which we close this section illustrates the preceding elimination scheme. Note that the band is maintained (i.e., the zeros in the upper right-hand corner of the coefficient matrix remain zero throughout the calculations). The reader is urged to perform the calculations.

Example of Gauss elimination

$$\begin{bmatrix} 1 & -1 & 0 & 0 \\ -1 & 2 & -1 & 0 \\ 0 & -1 & 2 & -1 \\ 0 & 0 & -1 & 2 \end{bmatrix} \begin{Bmatrix} d_1 \\ d_2 \\ d_3 \\ d_4 \end{Bmatrix} = \begin{Bmatrix} 1 \\ 0 \\ 0 \\ 0 \end{Bmatrix}$$

Augmented matrix

$$\left[\begin{array}{cccc|c} 1 & -1 & 0 & 0 & 1 \\ -1 & 2 & -1 & 0 & 0 \\ 0 & -1 & 2 & -1 & 0 \\ 0 & 0 & -1 & 2 & 0 \end{array}\right]$$

Forward reduction

$$\left[\begin{array}{cccc|c} 1 & -1 & 0 & 0 & 1 \\ 0 & 1 & -1 & 0 & 1 \\ 0 & -1 & 2 & -1 & 0 \\ 0 & 0 & -1 & 2 & 0 \end{array}\right]$$

$$\left[\begin{array}{cccc|c} 1 & -1 & 0 & 0 & 1 \\ 0 & 1 & -1 & 0 & 1 \\ 0 & 0 & 1 & -1 & 1 \\ 0 & 0 & -1 & 2 & 0 \end{array}\right]$$

$$\left[\begin{array}{cccc|c} 1 & -1 & 0 & 0 & 1 \\ 0 & 1 & -1 & 0 & 1 \\ 0 & 0 & 1 & -1 & 1 \\ 0 & 0 & 0 & 1 & 1 \end{array}\right]$$

Back substitution

$$\left[\begin{array}{cccc|c} 1 & -1 & 0 & 0 & 1 \\ 0 & 1 & -1 & 0 & 1 \\ 0 & 0 & 1 & 0 & 2 \\ 0 & 0 & 0 & 1 & 1 \end{array}\right]$$

$$\left[\begin{array}{cccc|c} 1 & -1 & 0 & 0 & 1 \\ 0 & 1 & 0 & 0 & 3 \\ 0 & 0 & 1 & 0 & 2 \\ 0 & 0 & 0 & 1 & 1 \end{array}\right]$$

$$\left[\begin{array}{cccc|c} 1 & 0 & 0 & 0 & 4 \\ 0 & 1 & 0 & 0 & 3 \\ 0 & 0 & 1 & 0 & 2 \\ 0 & 0 & 0 & 1 & 1 \end{array}\right]$$

$$\begin{Bmatrix} d_1 \\ d_2 \\ d_3 \\ d_4 \end{Bmatrix} = \begin{Bmatrix} 4 \\ 3 \\ 2 \\ 1 \end{Bmatrix}$$

Exercise 1. Consider the boundary-value problem discussed in the previous sections:

$$u_{,xx}(x) + \ell(x) = 0 \qquad x \in]0, 1[$$
$$u(1) = g$$
$$-u_{,x}(0) = h$$

Assume $\ell = qx$, where q is constant, and $g = h = 0$.

a. Employing the linear finite element space with equally spaced nodes, set up and solve the Galerkin finite element equations for $n = 4 (h = \text{mesh parameter} = \frac{1}{4})$. Recall that in Sec. 1.7 this was carried out for $n = 1$ and $n = 2$ ($h = 1$ and $h = \frac{1}{2}$, respectively). Do *not* invert the stiffness matrix $\boldsymbol{K}$; use Gauss elimination to solve $\boldsymbol{Kd} = \boldsymbol{F}$ or a more sophisticated direct factorization scheme if you know one. You can check your answers since they must be exact at the nodes.
b. Let $re_{,x} = |u^h_{,x} - u_{,x}|/(q/2)$, the ***relative error*** in $u_{,x}$. Compute $re_{,x}$ at the midpoints of the four elements. They should all be equal. (This was also the case for $n = 2$.)
c. Employing the data for $h = 1, \frac{1}{2}$, and $\frac{1}{4}$, plot ln $re_{,x}$ versus ln h.
d. Using the error analysis for $re_{,x}$ at the midpoints presented in Sec. 1.10, answer the following questions:
   i. What is the significance of the slope of the graph in part (c)?
   ii. What is the significance of the y-intercept?

1.12 THE ELEMENT POINT OF VIEW

So far we have viewed the finite element method simply as a particular Galerkin approximation procedure applied to the weak statement of the problem in question. What makes what we have done a finite element procedure is the character of the selected basis functions; particularly their piecewise smoothness and "local support" (i.e., $N_A \equiv 0$ outside a neighborhood of node A). This is the mathematical point of view; it is a ***global*** point of view in that the basis functions are considered to be defined everywhere on the domain of the boundary-value problem. The global viewpoint is useful in establishing the mathematical properties of the finite element method. This can be seen in Sec. 1.10 and will be made more apparent later on.

Now we wish to discuss another point of view called the ***local***, or ***element***, point of view. This viewpoint is the traditional one in engineering and is useful in the computer implementation of the finite element method and in the development of finite elements.

We begin our treatment of the local point of view with a question: What is a finite element?

We shall attempt to give the answer in terms of the piecewise linear finite element space that we defined previously. An individual element consists of the following quantities.

Linear finite element (global description)

$(g1)$	Domain:	$[x_A, x_{A+1}]$
$(g2)$	Nodes:	$\{x_A, x_{A+1}\}$
$(g3)$	Degrees of freedom:	$\{d_A, d_{A+1}\}$
$(g4)$	Shape functions:[5]	$\{N_A, N_{A+1}\}$
$(g5)$	Interpolation function:	

$$u^h(x) = N_A(x)d_A + N_{A+1}(x)d_{A+1}, \qquad x \in [x_A, x_{A+1}]$$

(Recall $d_A = u^h(x_A)$.) In words, a linear finite element is just the totality of paraphernalia associated with the globally defined function u^h *restricted* to the *element* domain. The above quantities are in terms of *global* parameters—namely, the global coordinates, global shape functions, global node ordering, and so on. It is fruitful to introduce a *local* set of quantities, corresponding to the global ones, so that calculations for a typical element may be standardized. These are given as follows:

Linear finite element (local description)

$(l1)$	Domain:	$[\xi_1, \xi_2]$

[5] In weighted residual methods in which $\mathcal{S}^h$ and $\mathcal{V}^h$ are built up from different classes of functions (i.e., Petrov-Galerkin methods), we would also have to specify a set of ***weighting functions***, say $\{\widetilde{N}_A, \widetilde{N}_{A+1}\}$; the entire set of $\widetilde{N}_A$'s would then constitute a basis for $\mathcal{V}^h$. In Galerkin's method $\widetilde{N}_A = N_A$.

(*l*2) Nodes: $\{\xi_1, \xi_2\}$

(*l*3) Degrees of freedom: $\{d_1, d_2\}$

(*l*4) Shape functions: $\{N_1, N_2\}$

(*l*5) Interpolation function:

$$\boxed{u^h(\xi) = N_1(\xi)d_1 + N_2(\xi)d_2}$$

Note that in the local description, the nodal numbering begins with 1.

We shall relate the domains of the global and local descriptions by an "affine" transformation $\xi : [x_A, x_{A+1}] \to [\xi_1, \xi_2]$, such that $\xi(x_A) = \xi_1$ and $\xi(x_{A+1}) = \xi_2$. It is standard practice to take $\xi_1 = -1$ and $\xi_2 = +1$. Thus ξ may be represented by the expression

$$\xi(x) = c_1 + c_2 x \tag{1.12.1}$$

where c_1 and c_2 are constants which are determined by

$$\left.\begin{aligned} -1 &= c_1 + x_A c_2 \\ 1 &= c_1 + x_{A+1} c_2 \end{aligned}\right\} \tag{1.12.2}$$

Solving this system yields

$$\xi(x) = \frac{2x - x_A - x_{A+1}}{h_A} \tag{1.12.3}$$

(Recall $h_A = x_{A+1} - x_A$.) The inverse of ξ is obtained by solving for x:

$$x(\xi) = \frac{h_A \xi + x_A + x_{A+1}}{2} \tag{1.12.4}$$

In (1.12.1), ξ is a mapping and x is a point, whereas in (1.12.4), x is a mapping and ξ is a point.

In the sequel, we adopt the notational convention that subscripts $a, b, c, \ldots$ pertain to the local numbering system. The subscripts $A, B, C, \ldots$ will always pertain to the global numbering system. To control the proliferation of notations, we will frequently use the same notation for the local and global systems (e.g., d_a and d_A or N_a and N_A). This generally should not cause confusion as the context will make clear which point of view is being adopted. If there is danger of confusion, a superscript e will be introduced to denote a quantity in the local description associated with element number e (e.g., $d_a^e = d_A$, $N_a^e(\xi) = N_A(x^e(\xi))$, where $x^e : [\xi_1, \xi_2] \to [x_1^e, x_2^e] = [x_A, x_{A+1}]$, etc.).

In terms of ξ, the shape functions in the local description take on a standard form

$$\boxed{N_a(\xi) = \tfrac{1}{2}(1 + \xi_a \xi), \qquad a = 1, 2} \tag{1.12.5}$$

Note also that (1.12.4) may be written in terms of (1.12.5):

$$\boxed{x^e(\xi) = \sum_{a=1}^{2} N_a(\xi)x_a^e.} \tag{1.12.6}$$

This has the same form as the interpolation function (cf. *15*).

For future reference, we note the following results:

$$N_{a,\xi} = \frac{\xi_a}{2} = \frac{(-1)^a}{2} \tag{1.12.7}$$

$$x^e_{,\xi} = \frac{h^e}{2} \tag{1.12.8}$$

where $h^e = x_2^e - x_1^e$ and

$$\xi^e_{,x} = (x^e_{,\xi})^{-1} = \frac{2}{h^e} \tag{1.12.9}$$

The local and global descriptions of the eth element are depicted in Fig. 1.12.1.

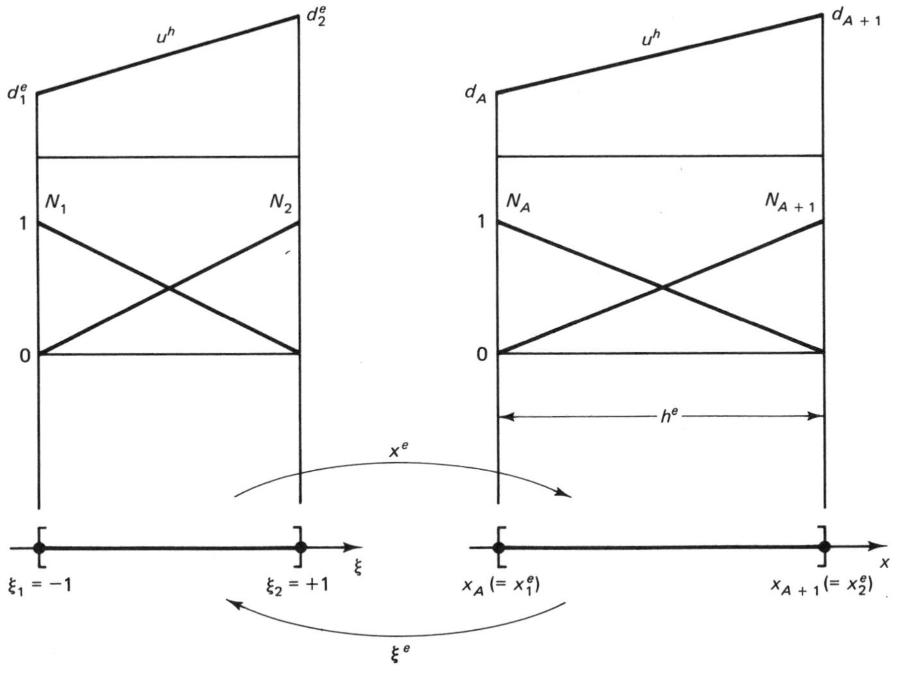

Figure 1.12.1 Local and global descriptions of the eth element.

NT STIFFNESS MATRIX AND FORCE VECTOR

elop the element point of view further, let us assume that our model consists lements, numbered as shown in Figure 1.13.1. Clearly $n_{el} = n$ for this case. take e to be the variable index for the elements; thus $1 \le e \le n_{el}$.

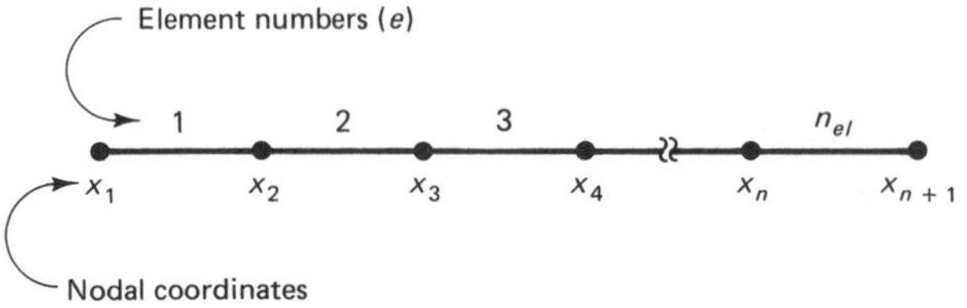

Figure 1.13.1

Now recall the definitions of the (global) stiffness matrix and force vector

$$\boldsymbol{K} = \underbrace{[K_{AB}]}_{n \times n}, \qquad \boldsymbol{F} = \underbrace{\{F_A\}}_{n \times 1} \tag{1.13.1}$$

where

$$K_{AB} = a(N_A, N_B) = \int_0^1 N_{A,x} N_{B,x}\, dx \tag{1.13.2}$$

$$\begin{aligned} F_A &= (N_A, \ell) + \delta_{A1} h - a(N_A, N_{n+1}) g \\ &= \int_0^1 N_A \ell \, dx + \delta_{A1} h - \int_0^1 N_{A,x} N_{n+1,x}\, dx\, g \end{aligned} \tag{1.13.3}$$

(In (1.13.3) we have assumed $N_A(x_1) = \delta_{A1}$, as for the piecewise linear finite element space.) The integrals over $[0, 1]$ may be written as sums of integrals over the element domains. Thus

$$\boldsymbol{K} = \sum_{e=1}^{n_{el}} \boldsymbol{K}^e, \qquad \boldsymbol{K}^e = [K^e_{AB}] \tag{1.13.4}$$

$$\boldsymbol{F} = \sum_{e=1}^{n_{el}} \boldsymbol{F}^e, \qquad \boldsymbol{F}^e = \{F^e_A\} \tag{1.13.5}$$

where

$$K^e_{AB} = a(N_A, N_B)^e = \int_{\Omega^e} N_{A,x} N_{B,x}\, dx \tag{1.13.6}$$

$$\begin{aligned} F^e_A &= (N_A, \ell)^e + \delta_{e1}\, \delta_{A1} h - a(N_A, N_{n+1})^e g \\ &= \int_{\Omega^e} N_A \ell\, dx + \delta_{e1}\, \delta_{A1} h - \int_{\Omega^e} N_{A,x} N_{n+1,x}\, dx\, g \end{aligned} \tag{1.13.7}$$

and $\Omega^e = [x^e_1, x^e_2]$, ***the domain of the eth element.***

The important observation to make is that $\boldsymbol{K}$ and $\boldsymbol{F}$ can be *constructed by summing the contributions of elemental matrices and vectors, respectively. In the literature, this procedure is sometimes called the* ***direct stiffness method*** [10].

By the definitions of the N_A's, we have that

$$K_{AB}^e = 0, \qquad \text{if } A \neq e \text{ or } e+1 \text{ or } B \neq e \text{ or } e+1 \tag{1.13.8}$$

and

$$F_A^e = 0, \qquad \text{if } A \neq e \text{ or } e+1 \tag{1.13.9}$$

The situation for a typical element, e, is shown in Fig. 1.13.2. In practice we would not, of course, add in the zeros but merely add in the nonzero terms to the appropriate locations. For this purpose it is useful to define the ***eth element stiffness matrix*** $\boldsymbol{k}^e$ and ***element force vector*** $\boldsymbol{f}^e$ as follows:

$$\boldsymbol{k}^e = \underbrace{[k_{ab}^e]}_{2\times 2}, \qquad \boldsymbol{f}^e = \underbrace{\{f_a^e\}}_{2\times 1} \tag{1.13.10}$$

$$k_{ab}^e = a(N_a, N_b)^e = \int_{\Omega^e} N_{a,x} N_{b,x}\, dx \tag{1.13.11}$$

$$f_a^e = \int_{\Omega^e} N_a \ell\, dx + \begin{cases} \delta_{a1} h & e = 1 \\ 0 & e = 2, 3, \ldots, n_{el} - 1 \\ -k_{a2}^e g & e = n_{el} \end{cases} \tag{1.13.12}$$

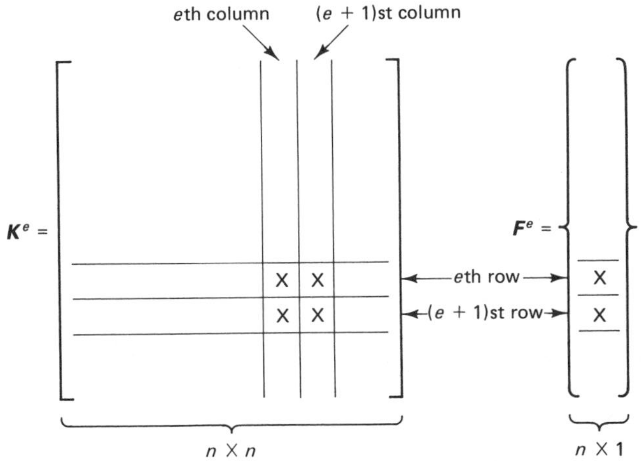

Figure 1.13.2 X's indicate nonzero terms; all other terms are zero.

Here $\boldsymbol{k}^e$ and $\boldsymbol{f}^e$ are defined with respect to the *local* ordering, whereas $\boldsymbol{K}^e$ and $\boldsymbol{F}^e$ are defined with respect to the global ordering. To determine where the components of $\boldsymbol{k}^e$ and $\boldsymbol{f}^e$ "go" in $\boldsymbol{K}$ and $\boldsymbol{F}$, respectively, requires keeping additional information. This is discussed in the following section.

1.14 ASSEMBLY OF GLOBAL STIFFNESS MATRIX AND FORCE VECTOR; LM ARRAY

In a finite element computer program, it is the task of a "finite element subroutine" to produce $\boldsymbol{k}^e$ and $\boldsymbol{f}^e$, $e = 1, 2, \ldots, n_{el}$, from given data and to provide an "assembly subroutine" enough information so that the terms in $\boldsymbol{k}^e$ and $\boldsymbol{f}^e$ can be added to the appropriate locations in $\boldsymbol{K}$ and $\boldsymbol{F}$, respectively. This assembly information is stored in an array named LM, the ***location matrix.***

Let us construct the LM array for the problem under consideration. The dimensions of LM are n_{en}, ***the number of element nodes,*** by the number of elements; in the present case, the numbers are 2 and n_{el}, respectively. Given a particular degree of freedom number and an element number (say a and e, respectively), the value returned by the LM array is the corresponding global equation number, A, viz.,

$$A = \mathrm{LM}(a, e) = \begin{cases} e & \text{if } a = 1 \\ e + 1 & \text{if } a = 2 \end{cases} \tag{1.14.1}$$

The complete LM array is depicted in Fig. 1.14.1. This is the way we envision it stored in the computer. Note that $\mathrm{LM}(2, n_{el}) = 0$. This indicates that degree of freedom 2 of element number n_{el} is prescribed and is not an unknown in the global matrix equation. Hence the terms $k_{12}^{n_{el}}$, $k_{21}^{n_{el}}$, $k_{22}^{n_{el}}$, and $f_2^{n_{el}}$ are *not* assembled into $\boldsymbol{K}$ and $\boldsymbol{F}$, respectively. (There are no places for them to go!)

Element numbers $1 \leq e \leq n_{el}$

Local node number ($n_{en} = 2$)	1	2	3	. . .	e	. . .	n_{el-1}	n_{el}
1	1	2	3	. . .	e	. . .	$n - 1$	n
2	2	3	4	. . .	$e + 1$	. . .	n	0

$n_{en} \times n_{el}$

Figure 1.14.1 LM array for example problem.

As an example, assume we want to add the eth elemental contributions, where $1 \leq e \leq n_{el-1}$, to the partially assembled $\boldsymbol{K}$ and $\boldsymbol{F}$. From the LM array, we deduce the following assembly procedure:

$$K_{ee} \leftarrow K_{ee} + k_{11}^e \tag{1.14.2}$$

$$K_{e,e+1} \leftarrow K_{e,e+1} + k_{12}^e \tag{1.14.3}$$

$$K_{e+1,e} \leftarrow K_{e+1,e} + k_{21}^e \text{ [6]} \tag{1.14.4}$$

$$K_{e+1,e+1} \leftarrow K_{e+1,e+1} + k_{22}^e \tag{1.14.5}$$

[6] Due to symmetry k_{21}^e would not actually be assembled in practice.

$$F_e \leftarrow F_e + f_1^e \qquad (1.14$$

$$F_{e+1} \leftarrow F_{e+1} + f_2^e \qquad (1.14.7$$

where the arrow ($\leftarrow$) is read "is replaced by."

For element n_{el} we have only that

$$K_{nn} \leftarrow K_{nn} + k_{11}^{n_{el}} \qquad (1.14.8)$$

$$F_n \leftarrow F_n + f_1^{n_{el}} \qquad (1.14.9)$$

With these ideas, we may construct, in sketchy fashion, an algorithm for the assembly of $\boldsymbol{K}$ and $\boldsymbol{F}$; see Fig. 1.14.2.

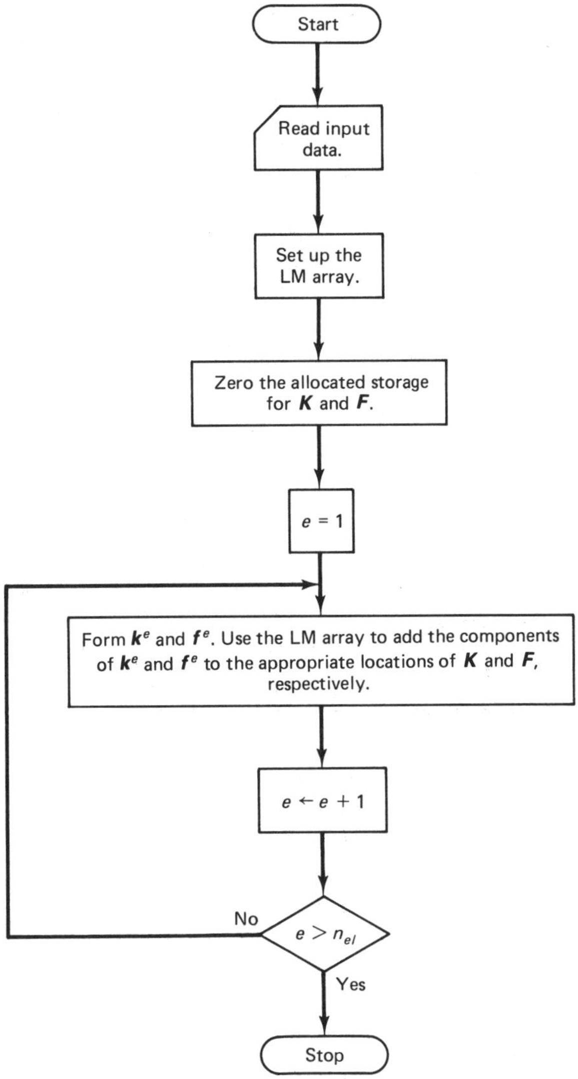

Figure 1.14.2 Flowchart of a finite element assembly algorithm.

ion of the assembly algorithm is denoted throughout by **A**, the ***assembly***

$$K = \mathop{\mathbf{A}}_{e=1}^{n_{el}} (\boldsymbol{k}^e), \qquad F = \mathop{\mathbf{A}}_{e=1}^{n_{el}} (\boldsymbol{f}^e) \tag{1.14.10}$$

1.15 EXPLICIT COMPUTATION OF ELEMENT STIFFNESS MATRIX AND FORCE VECTOR

The explicit computation of $\boldsymbol{k}^e$ and $\boldsymbol{f}^e$, for the problem under consideration, provides some preliminary insight into the type of calculations that must be performed in a finite element subroutine. Some preliminary results are required.

Change of Variables Formula (One-Dimensional Version)

Let $f:[x_1, x_2] \to \mathbb{R}$ be an integrable function and let $x:[\xi_1, \xi_2] \to [x_1, x_2]$ be continuously differentiable, with $x(\xi_1) = x_1$ and $x(\xi_2) = x_2$. Then

$$\int_{x_1}^{x_2} f(x)\, dx = \int_{\xi_1}^{\xi_2} f(x(\xi))x_{,\xi}(\xi)\, d\xi \tag{1.15.1}$$

Chain Rule

Let f and x be as above, and, in addition, assume f is differentiable. Then

$$\frac{\partial}{\partial \xi} f(x(\xi)) = f_{,x}(x(\xi))x_{,\xi}(\xi) \tag{1.15.2}$$

Proofs of these results may be found in [11].

The computation of $\boldsymbol{k}^e$ proceeds as follows:

$$k^e_{ab} = \int_{\Omega^e} N_{a,x}(x)N_{b,x}(x)\, dx \qquad \text{(by definition)}$$

$$= \int_{-1}^{+1} N_{a,x}(x(\xi))N_{b,x}(x(\xi))x_{,\xi}(\xi)\, d\xi$$

(Change of variables, where $x(\xi)$ is defined by (1.12.6))

$$= \int_{-1}^{+1} N_{a,\xi}(\xi)N_{b,\xi}(\xi)(x_{,\xi}(\xi))^{-1}\, d\xi$$

(Chain rule; $N_{a,\xi}(\xi) = (\partial/\partial\xi)N_a(x(\xi)) = N_{a,x}(x(\xi))x_{,\xi}(\xi)$)

$= (-1)^{a+b}/h^e$ (by (1.12.7)–(1.12.9))

Thus

$$\boldsymbol{k}^e = \frac{1}{h^e}\begin{bmatrix} 1 & -1 \\ -1 & 1 \end{bmatrix} \tag{1.15.3}$$

Observe that $N_{a,\xi}$ (see (1.12.7)) does not depend upon the particular element data, as $N_a = N_a(\xi)$. We shall see that this is generally true, and hence these computations may be done once and for all.

The derivatives $x_{,\xi}$ and $\xi_{,x}$ do depend on the particular element data (in the present case h^e), and subroutines will be necessary to compute the analogs of these quantities in more general cases.

Now we wish to compute $\boldsymbol{f}^e$. However, this cannot be done without explicitly knowing what $\ell = \ell(x)$ is. In practice, it would be inconvenient to reprogram every time we wanted to solve a problem involving a different function ℓ. Generally a convenient approximation is made. For example, we might replace ℓ by its linear interpolate over each element, namely,

$$\ell^h = \sum_{a=1}^{2} \ell_a N_a \tag{1.15.4}$$

where $\ell_a = \ell(x(\xi_a))$; see Fig. 1.15.1. The notation ℓ^h is used to indicate that the approximation depends upon the mesh. This represents an approximation that is sufficient for most practical applications. (It is, of course, exact for constant or linear "loading" of the element.) Now standardization of input to the program may be facilitated; that is, the nodal values of ℓ are the required data. Let us employ this approximation in the explicit calculation of an element force vector:

$$\int_{\Omega^e} N_a(x)\ell^h(x)\,dx = \int_{-1}^{+1} N_a(x(\xi))\ell^h(x(\xi))x_{,\xi}(\xi)\,d\xi \qquad \text{(change of variables)}$$

$$= \frac{h^e}{2}\sum_{b=1}^{2}\int_{-1}^{+1} N_a(\xi)N_b(\xi)\,d\xi\,\ell_b \qquad \text{(by (1.12.8))} \tag{1.15.5}$$

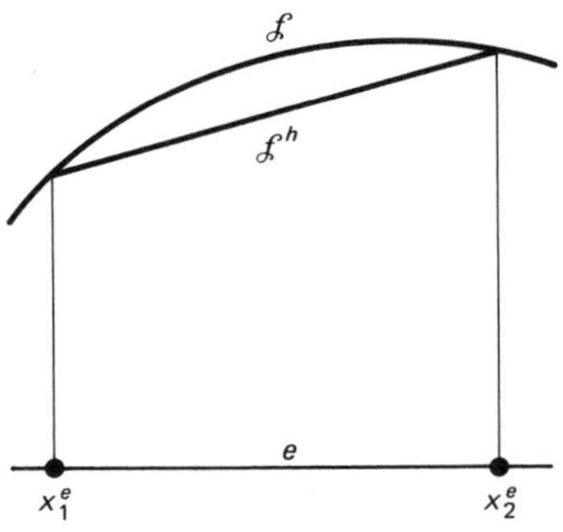

Figure 1.15.1 Approximation of ℓ by piecewise linear interpolation of nodal values.

Carrying out the integrations ($\int_{-1}^{+1} N_a N_b \, d\xi = (1 + \delta_{ab})/3$) yields

$$f^e = \frac{h^e}{6}\begin{bmatrix} 2 & 1 \\ 1 & 2 \end{bmatrix}\begin{Bmatrix} \ell_1 \\ \ell_2 \end{Bmatrix} \qquad \text{(+ boundary terms, cf. (1.13.12))}$$
$$= \frac{h^e}{6}\begin{Bmatrix} 2\ell_1 + \ell_2 \\ \ell_1 + 2\ell_2 \end{Bmatrix} \qquad \text{(+ boundary terms)} \tag{1.15.6}$$

Remark. It can be shown that, under suitable hypotheses, piecewise linear nodal interpolation produces $O(h^2)$ errors in the data; in this case, ℓ. (See [12], pp. 56–57, for basic estimates of interpolation errors.) It can be shown that, in appropriate measures of the error, this produces at worst $O(h^2)$ errors in u^h and $u^h_{,x}$.

The following exercise indicates that there may be better ways to approximate given data.

Exercise 1. Suppose $\ell(x)$ is quadratic (i.e., consists of a linear combination of the monomials $1, x, x^2$). Determine a piecewise linear approximation—not necessarily continuous—to ℓ over each element which results in exact nodal values. *Hint:* The analysis may be performed with respect to one element.

Exercise 2. The equation of a string on an elastic foundation is given by:

$$u_{,xx} - \lambda u + \ell = 0 \qquad \text{on } \Omega =]0, 1[$$

where λ, a positive constant, is a measure of the foundation stiffness. Assuming the same boundary conditions as for the problem discussed previously in this chapter, it can be shown that an equivalent weak formulation is:

$$\int_\Omega (w_{,x}u_{,x} + w\lambda u)\, dx = \int_\Omega w\ell \, dx + w(0)h$$

where $u \in \mathcal{S}$, $w \in \mathcal{V}$, and so on. This can also be written as

$$a(w, u) + (w, \lambda u) = (w, \ell) + w(0)h$$

i. Let $u^h = v^h + g^h$. Write the Galerkin counterpart of the weak formulation:

$$a(w^h, v^h) + \boxed{} =$$

$$(w^h, \ell) + w^h(0)h - a(w^h, g^h)$$

$$- \boxed{}$$

ii. Define $K_{AB} = a(N_A, N_B) + \boxed{}$

and

$$k^e_{ab} = a(N_a, N_b)^e + \boxed{}$$

iii. Determine $\boldsymbol{k}^e$ explicitly:

$$\boldsymbol{k}^e = [k^e_{ab}] = \begin{bmatrix} & \\ & \end{bmatrix}$$

iv. Show that $\boldsymbol{K}$ is symmetric.

v. Show that $\boldsymbol{K}$ is positive definite. Is it necessary to employ the boundary condition $w^h(1) = 0$? Why?

vi. The Green's function for this problem satisfies

$$g_{,xx} - \lambda g + \delta_y = 0$$

and can be written as

$$g(x) = \begin{cases} c_1 e^{px} + c_2 e^{-px}, & 0 \le x \le y \\ c_3 e^{px} + c_4 e^{-px}, & y \le x \le 1 \end{cases}$$

where $p = \lambda^{1/2}$ and the c's are determined from the following four boundary and continuity conditions:

$$g(1) = 0$$

$$g_{,x}(0) = 0$$

$$g(y^+) = g(y^-)$$

$$g_{,x}(y^+) = g_{,x}(y^-) - 1$$

Why is the piecewise linear finite element space incapable of attaining nodally exact solutions in this case?

vii. Construct *exponential* element shape functions $N_1(x)$ and $N_2(x)$ such that

$$u^h(x) = d^e_1 N_1(x) + d^e_2 N_2(x), \qquad x \in \Omega^e$$

where

$$u^h(x) = c_1 e^{px} + c_2 e^{-px}$$

and the c's are determined from

$$d^e_a = u^h(x^e_a), \qquad a = 1, 2$$

What is the attribute which this choice of functions attains?

1.16 EXERCISE: BERNOULLI-EULER BEAM THEORY AND HERMITE CUBICS

This problem develops basic finite element results for Bernoulli-Euler beam theory. The strong form of a boundary-value problem for a thin beam (Bernoulli-Euler theory) fixed at one end and subjected to a shear force and moment at the other end, may be stated as follows:

Let the beam occupy the unit interval (i.e., $\Omega =]0, 1[$, $\overline{\Omega} = [0, 1]$).

(S) Given $\ell: \Omega \to \mathbb{R}$ and constants M and Q, find $u: \overline{\Omega} \to \mathbb{R}$ such that

$$EI\, u_{,xxxx} = \ell \text{ on } \Omega \qquad \text{(transverse equilibrium)}$$

$$u(1) = 0 \qquad \text{(zero transverse displacement)}$$

$$u_{,x}(1) = 0 \qquad \text{(zero slope)}$$

$$EI\, u_{,xx}(0) = M \qquad \text{(prescribed moment)}$$

$$EI\, u_{,xxx}(0) = Q \qquad \text{(prescribed shear)}$$

where E is Young's modulus and I is the moment of inertia, both of which are assumed to be constant.

The setup is shown in Fig. 1.16.1.

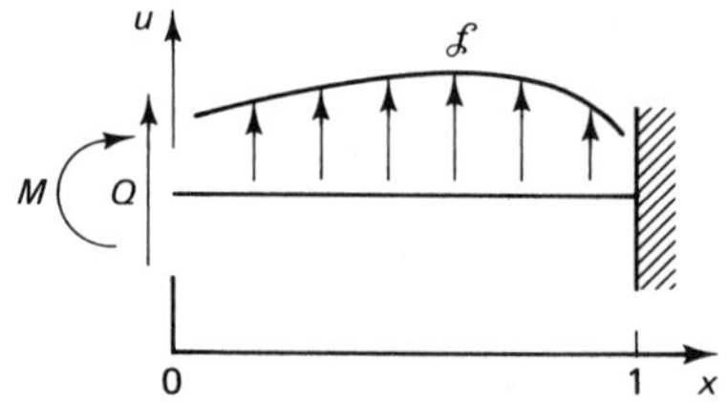

Figure 1.16.1

Let $\mathcal{S} = \mathcal{V} = \{w \mid w \in H^2(\Omega),\ w(1) = w_{,x}(1) = 0\}$[7]. Then a corresponding weak form of the problem is:

(W) Given ℓ, M, and Q, find $u \in \mathcal{S}$ such that for all $w \in \mathcal{V}$

$$\boxed{a(w, u) = (w, \ell) - w_{,x}(0)M + w(0)Q}$$

where

$$a(w, u) = \int_0^1 w_{,xx} EI\, u_{,xx}\, dx$$

$$(w, \ell) = \int_0^1 w \ell\, dx$$

[7] $w \in H^2(\Omega)$ essentially means that $w_{,xx}$ is square-integrable (i.e., $\int_0^1 (w_{,xx})^2\, dx < \infty$).

The collection of functions, $\mathcal{V}$, may be thought of as the space of finite strain-energy configurations of the beam, satisfying the kinematic (essential) boundary conditions at $x = 1$. It is a consequence of Sobolev's theorem that each $w \in \mathcal{V}$ is continuously differentiable. For reasonable ℓ, these problems possess unique solutions.

Let $\mathcal{S}^h = \mathcal{V}^h$ be a finite-dimensional approximation of $\mathcal{S}$. In particular, we assume $w^h \in \mathcal{V}^h$ satisfies $w^h(1) = w^h_{,x}(1) = 0$.

The Galerkin statement of the problem goes as follows:

$$(G)\begin{cases}\text{Given } \ell,\ M, \text{ and } Q, \text{ find } u^h \in \mathcal{S}^h \text{ such that for all } w^h \in \mathcal{V}^h \\ \boxed{a(w^h, u^h) = (w^h, \ell) - w^h_{,x}(0)M + w^h(0)Q}\end{cases}$$

a. Assuming all functions are smooth and bounded, show that the solutions of (S) and (W) are identical. What are the natural boundary conditions?

b. Assume $0 = x_1 < x_2 < \cdots < x_{n+1} = 1$ and $\mathcal{V}^h = \{w^h \mid w^h \in C^1(\overline{\Omega}),$ $w^h(1) = w^h_{,x}(1) = 0$, and w^h restricted to $[x_A, x_{A+1}]$ is a cubic polynomial (i.e., consists of a linear combination of $1, x, x^2, x^3$)$\}$[8]. This is a space of ***piecewise cubic Hermite shape functions***. Observe that $w^h \in \mathcal{V}^h$ need *not* have continuous second derivatives at the nodes.

On each subinterval, show that w^h may be written as

$$w^h(x) = N_1(x)w^h(x_1) + N_3(x)w^h(x_2) + N_2(x)w^h_{,x}(x_1) + N_4(x)w^h_{,x}(x_2)$$

where

$$N_1(x) = \frac{-(x - x_2)^2[-h + 2(x_1 - x)]}{h^3}$$

$$N_2(x) = \frac{(x - x_1)(x - x_2)^2}{h^2}$$

$$N_3(x) = \frac{(x - x_1)^2[h + 2(x_2 - x)]}{h^3}$$

$$N_4(x) = \frac{(x - x_1)^2(x - x_2)}{h^2}$$

Hint: Let $w^h(x) = c_1 + c_2x + c_3x^2 + c_4x^3$, where the c's are constants. Determine them by requiring the following four conditions hold:

$$w^h(x_1) = c_1 + c_2x_1 + c_3x_1^2 + c_4x_1^3$$

$$w^h(x_2) = c_1 + c_2x_2 + c_3x_2^2 + c_4x_2^3$$

$$w^h_{,x}(x_1) = c_2 + 2c_3x_1 + 3c_4x_1^2$$

$$w^h_{,x}(x_2) = c_2 + 2c_3x_2 + 3c_4x_2^2$$

[8] The notation $w^h \in C^1$ means w^h is continuously differentiable.

Sketch the element functions N_1, N_2, N_3, and N_4, and their typical global counterparts.

The finite element space described in part (b) results in *exact* nodal displacements and slopes (first derivatives), analogous to the case presented in Sec. 1.10. In part (g), you are asked to prove this. In problems of beam bending we are generally interested in curvatures (second derivatives) for bending moment calculations.

c. Locate the optimal curvature points in the sense of Barlow. *Warning:* The algebraic manipulations can be tiresome unless certain simplifications are observed. If we work in the ξ-element coordinate system introduced in Sec. 1.12 (recall $\xi = (2x - x_A - x_{A+1})/h_A$), the location of the Barlow curvature points may be expressed as $\xi = \pm 1/\sqrt{3}$. That is, there are two symmetrically spaced optimal locations to compute curvature.

d. What is the rate of convergence of curvature at these points? (*Ans.* $O(h^3)$).

e. If the segment of the beam $[x_A, x_{A+1}]$ is unloaded (i.e., $u_{,xxxx} = 0$, where u is the exact solution), which points are optimal?

f. Assume $n_{el} = 1$ (i.e., one element) and $\ell(x) = c =$ constant. Set up and solve the Galerkin-finite element equations. Plot u^h and u; $u^h_{,x}$ and $u_{,x}$; and $u^h_{,xx}$ and $u_{,xx}$. Indicate the locations of the Barlow curvature points.

g. Prove that

$$u^h(x_A) = u(x_A)$$

$$u^h_{,x}(x_A) = u_{,x}(x_A)$$

where x_A is a typical node (i.e., prove ***the displacements and slopes are exact at the nodes***). To do the second part you will have to be familiar with the ***dipole,*** $\delta_{,x}(x - x_A)$, which is the generalized derivative of the delta function.

h. Show that the Barlow curvature points are exact when $\ell(x) = c =$ constant.

i. Why do we require that the functions in $\mathcal{V}^h$ have continuous first derivatives?

j. Calculate the 4×4 element stiffness matrix,

$$k^e_{pq} = \int_{x^e_1}^{x^e_2} N_{p,xx} EI \, N_{q,xx} \, dx \qquad 1 \le p, q \le 4$$

where $h^e = x^e_2 - x^e_1$.

k. (See the exercise in Sec. 1.8.) Consider the weak formulation. Assume $w \in \mathcal{V}$ and $u \in \mathcal{S}$ are smooth on element interiors (i.e., on $]x_A, x_{A+1}[$) but may exhibit discontinuities in second, and higher, derivatives across element boundaries. (Functions of this type contain the piecewise-cubic Hermite functions.) Show that

$$
\begin{aligned}
0 = &\sum_{A=1}^{n} \int_{x_A}^{x_{A+1}} w(EI\, u_{,xxxx} - \ell)\, dx \\
&- w_{,x}(0)(EI\, u_{,xx}(0^+) - M) \\
&+ w(0)(EI\, u_{,xxx}(0^+) - Q) \\
&- \sum_{A=2}^{n} w_{,x}(x_A)\, EI\, (u_{,xx}(x_A^+) - u_{,xx}(x_A^-)) \\
&+ \sum_{A=2}^{n} w(x_A)\, EI\, (u_{,xxx}(x_A^+) - u_{,xxx}(x_A^-))
\end{aligned}
$$

from which it may be concluded that the Euler-Lagrange conditions are

i. $EI\, u_{,xxxx}(x) = \ell(x)$, where $x \in \,]x_A, x_{A+1}[$ and $A = 1, 2, \ldots, n$
ii. $EI\, u_{,xx}(0^+) = M$
iii. $EI\, u_{,xxx}(0^+) = Q$
iv. $EI\, u_{,xx}(x_A^+) = EI\, u_{,xx}(x_A^-)$, where $A = 2, 3, \ldots, n$
v. $EI\, u_{,xxx}(x_A^+) = EI\, u_{,xxx}(x_A^-)$, where $A = 2, 3, \ldots, n$

Note that (i) is the equilibrium equation *restricted to the element interiors,* and (iv) and (v) are continuity conditions across element boundaries of moment and shear, respectively. Contrast these results with those obtained for functions w and u, which are *globally* smooth.

The Galerkin finite element formulation yields a solution that *approximates* (i) through (v).

Appendix 1.1

An Elementary Discussion of Continuity, Differentiability, and Smoothness

Throughout Chapter 1 we have introduced mathematical terminologies and ideas in a gradual, as-needed format. Many of these ideas had to do with the continuity and differentiability of functions. The presentation was, admittedly, somewhat vague on these points in order that the main ideas would not be overencumbered. Careful characterization of the properties of functions is an essential ingredient in the development and analysis of finite element methods. However, to pursue this subject deeply would take us into the realm of serious mathematical analysis, which is outside the scope of this book. Nevertheless, we feel compelled to say a few additional words on the subject to round out the presentation in Chapter 1 and to expose the reader to notations and ideas that will probably be encountered if he or she attempts to read published papers on finite elements.

The discussion here will be restricted to one dimension. In Chapter 1 we spoke of continuously differentiable functions. If we have a grasp of the notion of a continuous function, then continuously differentiable functions pose no problem.

Definition: A function $f\colon \Omega \to \mathbb{R}$ (recall $\Omega =]0, 1[$) is said to be ***k-times continuously differentiable,*** or ***of class*** $C^k = C^k(\Omega)$, if its derivatives of order j, where $0 \leq j \leq k$, exist and are continuous functions.

A C^0 function is simply a continuous function. A C^∞ function is one that possesses a continuous derivative of any order (i.e., $j = 0, 1, \ldots, \infty$).

Definition: A function f is said to be ***of class*** C_b^k if it is C^k and bounded (i.e., $|f(x)| < c$, where c is a constant, for all $x \in \Omega$).

Example 1

The functions defined by monomials (i.e., $f(x) = 1, x, x^2$, etc.) are C_b^∞.

Example 2

The function $f(x) = 1/x$ is continuous on Ω, as are all its derivatives; hence it is C^∞, but it is not bounded (i.e., there does not exist a constant c such that $|1/x| < c$ for all $x \in \Omega$; see Fig. 1.I.1). Consequently this function is not of class C_b^k for any $k \geq 0$.

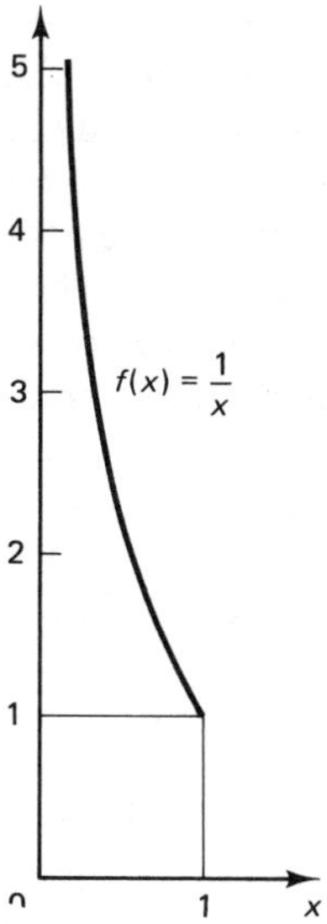

Figure 1.I.1 A continuous function that is not bounded.

Example 3

The function

$$f(x) = \begin{cases} x, & x \leq \frac{1}{2} \\ 1/2, & x > \frac{1}{2} \end{cases} \tag{1.I.1}$$

is continuous but not continuously differentiable (i.e., it is C_b^0 but not C_b^1). Functions in C_b^k, $k \geq 1$, but not in C_b^{k+1} may be constructed by integrating (1.I.1) k times. For example,

$$f(x) = \begin{cases} \dfrac{x^2}{2}, & x \leq \frac{1}{2} \\ \dfrac{(x - \frac{1}{4})}{2}, & x > \frac{1}{2} \end{cases} \tag{1.I.2}$$

is in C_b^1 but not C_b^2. (The reader may wish to verify this.)

There is no universally accepted definition of what is meant by a "smooth" function. However, it is generally taken to mean that at least one derivative exists and is continuous (i.e., either C^1 or C_b^1) and sometimes means $k > 1$, even ∞.

The C^k and C_b^k functions employ the classical notion of a derivative in their definitions. If we employ the closed unit interval, $\overline{\Omega} = [0, 1]$, instead of $\Omega =]0, 1[$, the difference between C^k and C_b^k disappears. This is because if f is $C^k([0, 1])$, $f(0)$ and $f(1)$ are real numbers and are not allowed to be ∞. Thus unboundedness, as in the example above, is precluded. Very often, we think of C^k functions in this light. However, in some situations the differences between $C^k(\Omega)$ and $C_b^k(\Omega)$ must be kept in mind.

Generally, finite element functions are smooth on element interiors (there are exceptions, however) but possess only low-order continuity across element boundaries. One might be tempted to characterize them as locally smooth but globally "rough." The piecewise linear finite element functions discussed in Sec. 1.8 are of class C_b^0. The Hermite cubics employed in Sec. 1.16 are C_b^1. To calculate derivatives of such functions we need to employ the notion of a "generalized derivative," as was used in solving the Green's function problem of Sec. 1.10. For example, the first derivative of a piecewise linear finite element function is a generalized step function; the second derivative is a generalized Dirac delta function (i.e., delta functions, of various amplitudes, acting at the nodes). In the case of the Hermite cubics, the first derivative is continuous, the second a generalized step function, and so on. We have seen in Sec. 1.16 that other generalized functions also arise in the analysis of finite element behavior (namely, the dipole). The useful examples of generalized functions are by no means exhausted by what we have seen thus far. However, the ones we have introduced are perhaps the most basic.

In the mathematical analysis of boundary-value problems, and consequently in finite element analysis, we need to introduce classes of functions that possess generalized derivatives and, in addition, certain integrability properties. We have encountered such functions in the statements of weak formulations in Sec. 1.3 and 1.16. These are particular examples of ***Sobolev spaces of functions*** defined as follows:

$$H^k = H^k(\Omega) = \{w \mid w \in L_2;\ w_{,x} \in L_2;\ \ldots\ ;\ w_{,\underbrace{x\ldots x}_{k \text{ times}}} \in L_2\} \tag{1.I.3}$$

where

$$L_2 = L_2(\Omega) = \{w \mid \int_0^1 w^2\, dx < \infty\} \tag{1.I.4}$$

In words, the Sobolev space of degree k, denoted by H^k, consists of functions that possess square-integrable generalized derivatives through order k. A square-integrable function is called an L_2-function, by virtue of (1.I.4). From (1.I.3), we see that $H^0 = L_2$ and that $H^{k+1} \subset H^k$. The Sobolev spaces are the most important for studying elliptic boundary-value problems.

The question naturally arises as to the relation between Sobolev spaces and the classical spaces of differentiable functions introduced previously. In particular, when is an H^k-function smooth in the classical sense? The answer is provided by ***Sobolev's theorem,*** which states that, in one dimension, $H^{k+1} \subset C_b^k$. That is, if a function is of class H^{k+1}, then it is actually a C_b^k function. For example, in Sec. 1.3 we required H^1 functions. By Sobolev's theorem, such functions are, additionally, continuous and bounded. In Sec 1.16, we employed H^2 functions. These are C_b^1 by Sobolev's theorem and thus possess bounded, continuous, classical derivatives.

Certain "singularities" are precluded by square-integrability. For example, $x^{-1/4}$ is in L_2, but $x^{-1/2}$ is not. (Verify!) Such considerations become important in many physical circumstances (e.g., in fracture mechanics).

The number of other types of function spaces that arise in mathematical analysis is large, and many are difficult to comprehend without serious training in "functional analysis." These topics are outside the scope of this book. The reader who wishes to delve further may consult [13, 14, 15] and references therein.

REFERENCES

Section 1.2

1. A. R. Mitchell and D. F. Griffiths, *The Finite Difference Method in Partial Differential Equations*. New York: John Wiley, 1980.

Section 1.5

2. G. Strang and G. J. Fix, *An Analysis of the Finite Element Method*. Englewood Cliffs, N. J.: Prentice-Hall, 1973.

3. B. A. Finlayson, *The Method of Weighted Residuals and Variational Principles*. New York: Academic Press, 1972.

4. B. A. Finlayson and L. E. Scriven, "The Method of Weighted Residuals—A Review," *Applied Mechanics Reviews,* 19, (1966), 735–738.

Section 1.10

5. I. Stakgold, *Boundary-Value Problems of Mathematical Physics*. Vols. I and II, New York: Macmillan, 1968.

6. G. Strang and G. J. Fix, *An Analysis of the Finite Element Method*. Englewood Cliffs, N. J.: Prentice-Hall, 1973.

7. J. E. Marsden, *Elementary Classical Analysis*. San Francisco: W. H. Freeman, 1974.

8. J. Barlow, "Optimal Stress Locations in Finite Element Models," *International Journal for Numerical Methods in Engineering,* 10 (1976), 243–251.

9. E. Popov, *Introduction to Mechanics of Solids*. Englewood Cliffs, N. J.: Prentice-Hall, 1968.

Section 1.13

10. M. J. Turner, R. W. Clough, H. C. Martin, and L. J. Topp, "Stiffness and deflection analysis of complex structures," *Journal of Aeronautical Sciences,* 23 (1956), 805–823.

Section 1.15

11. J. E. Marsden, *Elementary Classical Analysis*. San Francisco: W. H. Freeman, 1974.

12. P. J. Davis, *Interpolation and Approximation*. New York: Blaisdell, 1963.

Appendix 1.I

13. P. G. Ciarlet, *The Finite Element Method for Elliptic Problems*. New York: North-Holland, 1978.

14. J. T. Oden and J. N. Reddy, *An Introduction to the Mathematical Theory of Finite Elements*. New York: Academic Press, 1978.

15. J. T. Oden, *Applied Functional Analysis*. Englewood Cliffs, N. J.: Prentice-Hall, 1979.

2

Formulation of Two- and Three-dimensional Boundary-value Problems

2.1 INTRODUCTORY REMARKS

It makes no sense to attempt to "solve" a boundary-value problem without a precise knowledge of what the problem is. The truth of this statement seems self-evident. Unfortunately, attempts are often made to solve vaguely defined problems, creating considerable confusion and, sometimes, totally erroneous results. In this chapter we present precise statements of multidimensional boundary-value problems in classical linear heat conduction and elastostatics. The presentation is similar in many respects to that for the one-dimensional model problem of Chapter 1. In particular, we discuss strong and weak forms, their equivalence, corresponding Galerkin formulations, the definitions of element arrays, and pertinent data processing concepts. In multidimensions, the data processing ideas necessarily become more involved. The reader is urged to study them carefully as they are necessary in order to understand the computer implementation of finite element techniques.

2.2 PRELIMINARIES

Let $n_{sd}(= 2 \text{ or } 3)$ denote the ***number of space dimensions*** of the problem under consideration. Let $\Omega \subset \mathbb{R}^{n_{sd}}$ be an open set[1] with piecewise smooth boundary Γ. A general point in $\mathbb{R}^{n_{sd}}$ is denoted by $\boldsymbol{x}$. We will identify the point $\boldsymbol{x}$ with its position vector emanating from the origin of $\mathbb{R}^{n_{sd}}$. The ***unit outward normal vector to*** Γ is denoted by $\boldsymbol{n}$.

[1] For our purposes, it is sufficient to think of an open set as one without its boundary.

We shall employ the following alternative representations for $\boldsymbol{x}$ and $\boldsymbol{n}$:

$$(n_{sd} = 2): \quad \boldsymbol{x} = \{x_i\} = \begin{Bmatrix} x_1 \\ x_2 \end{Bmatrix} = \begin{Bmatrix} x \\ y \end{Bmatrix} \qquad \boldsymbol{n} = \{n_i\} = \begin{Bmatrix} n_1 \\ n_2 \end{Bmatrix} = \begin{Bmatrix} n_x \\ n_y \end{Bmatrix} \tag{2.2.1}$$

$$(n_{sd} = 3): \quad \boldsymbol{x} = \{x_i\} = \begin{Bmatrix} x_1 \\ x_2 \\ x_3 \end{Bmatrix} = \begin{Bmatrix} x \\ y \\ z \end{Bmatrix} \qquad \boldsymbol{n} = \{n_i\} = \begin{Bmatrix} n_1 \\ n_2 \\ n_3 \end{Bmatrix} = \begin{Bmatrix} n_x \\ n_y \\ n_z \end{Bmatrix} \tag{2.2.2}$$

where x_i and n_i, $1 \leq i \leq n_{sd}$, are the Cartesian components of $\boldsymbol{x}$ and $\boldsymbol{n}$, respectively; see Figure 2.2.1. Unless otherwise specified we shall work in terms of Cartesian components of vectors and tensors.

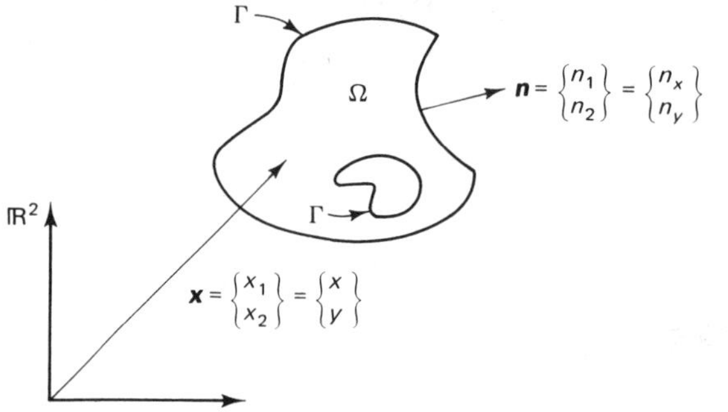

Figure 2.2.1

We assume that Γ admits the decomposition

$$\Gamma = \overline{\Gamma_g \cup \Gamma_h} \tag{2.2.3}$$

where

$$\Gamma_g \cap \Gamma_h = \varnothing \tag{2.2.4}$$

and Γ_g and Γ_h are open sets in Γ. The notations are defined as follows: $\cup$ is the ***set union*** symbol. Thus $\Gamma_g \cup \Gamma_h$ means the set of all points $\boldsymbol{x}$ contained in either Γ_g or Γ_h. Also, $\cap$ is the ***set intersection*** symbol. Thus $\Gamma_g \cap \Gamma_h$ means the set of all points contained in both Γ_g and Γ_h. The ***empty set*** is denoted by $\varnothing$. Thus (2.2.4) means that there is no point $\boldsymbol{x}$ contained in both Γ_g and Γ_h (i.e., Γ_g and Γ_h do not intersect or overlap). A bar above a set means ***set closure,*** i.e., the union of the set with its boundary. Thus

$$\overline{\Omega} = \Omega \cup \Gamma \tag{2.2.5}$$

To understand the meaning of $\overline{\Gamma_g \cup \Gamma_h}$, we must define the boundaries of Γ_g and Γ_h in Γ. We shall do this with the aid of an example.

Example 1

Let $\Omega = \{\boldsymbol{x} \in \mathbb{R}^2 \mid x^2 + y^2 < 1$, i.e., the interior of the unit disc$\}$. The boundary of Ω is $\Gamma = \{\boldsymbol{x} \in \mathbb{R}^2 \mid x^2 + y^2 = 1$, i.e., the unit circle$\}$. Let

$$\Gamma_g = \Gamma \cap \{\boldsymbol{x} \in \mathbb{R}^2 \mid y > 0\} \tag{2.2.6}$$

The "boundary of Γ_g" consists of the endpoints of the upper semicircle, i.e., $\{\boldsymbol{x} \in \mathbb{R}^2 \mid x = -1, y = 0, \text{ and } x = +1, y = 0\}$. Thus

$$\overline{\Gamma_g} = \Gamma \cap \{\boldsymbol{x} \in \mathbb{R}^2 \mid y \geq 0\} \tag{2.2.7}$$

Similarly, let

$$\Gamma_h = \Gamma \cap \{\boldsymbol{x} \in \mathbb{R}^2 \mid y < 0\} \tag{2.2.8}$$

Thus

$$\overline{\Gamma_h} = \Gamma \cap \{\boldsymbol{x} \in \mathbb{R}^2 \mid y \leq 0\} \tag{2.2.9}$$

Clearly

$$\overline{\Gamma_g \cup \Gamma_h} = \overline{\Gamma_g} \cup \overline{\Gamma_h} = \overline{\Gamma} \tag{2.2.10}$$

These sets are depicted in Fig. 2.2.2.

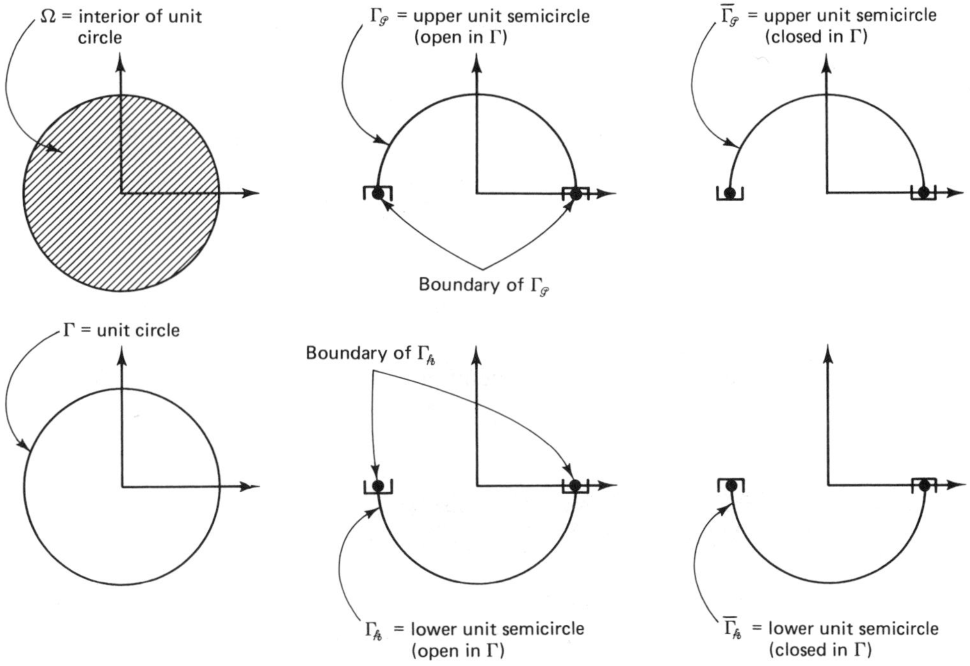

Figure 2.2.2

We shall assume throughout that $\Gamma_g \neq \varnothing$ but allow for the case $\Gamma_h = \varnothing$.

Let the indices i, j, k, l, run over the values $1, \ldots, n_{sd}$. Differentiation is denoted by a comma (e.g., $u_{,i} = u_{,x_i} = \partial u/\partial x_i$) and repeated indices imply summation (e.g., in $\mathbb{R}^3$, $u_{,ii} = u_{,11} + u_{,22} + u_{,33} = \partial^2 u/\partial x^2 + \partial^2 u/\partial y^2 + \partial^2 u/\partial z^2$). The summation convention only applies to the indices i, j, k, and l and only to two repeated indices. If there are three, or more, repeated indices in an expression, then the summation convention is *not* in effect. (This is in keeping with the usual convention.)

Divergence theorem. Let $f : \overline{\Omega} \to \mathbb{R}$ be C^1. Then

$$\int_\Omega f_{,i}\, d\Omega = \int_\Gamma f n_i\, d\Gamma \tag{2.2.11}$$

The proof may be found in [1].

Integration by parts. Let f be as above and also let $g : \overline{\Omega} \to \mathbb{R}$ be C^1. Then

$$\int_\Omega f_{,i} g\, d\Omega = -\int_\Omega f g_{,i}\, d\Omega + \int_\Gamma f g n_i\, d\Gamma \tag{2.2.12}$$

Proof. We integrate the identity (i.e., "product rules of differentiation," see [1])

$$(fg)_{,i} = f_{,i} g + f g_{,i}$$

to get

$$\int_\Omega (fg)_{,i}\, d\Omega = \int_\Omega f_{,i} g\, d\Omega + \int_\Omega f g_{,i}\, d\Omega$$

and then use the divergence theorem to convert the left-hand side into a boundary integral. ■

2.3 CLASSICAL LINEAR HEAT CONDUCTION: STRONG AND WEAK FORMS; EQUIVALENCE

Let q_i denote (Cartesian components of) the ***heat flux vector,*** let u be the ***temperature,*** and let ℓ be the ***heat supply per unit volume.*** Assume the heat flux vector is defined in terms of the temperature gradient by the ***generalized Fourier law***[2]:

$$q_i = -\kappa_{ij} u_{,j}, \qquad \kappa_{ij} = \kappa_{ji} \qquad \text{(symmetry)} \tag{2.3.1}$$

where the ***conductivities,*** κ_{ij}'s, are given functions of $\boldsymbol{x}$. (If the κ_{ij}'s are constant throughout Ω, the body is said to be ***homogeneous.***) The ***conductivity matrix,*** $\boldsymbol{\kappa} = [\kappa_{ij}]$, is assumed positive definite (see the definition in Sec. 1.9). The most common situation in practice is the ***isotropic case*** in which $\kappa_{ij}(\boldsymbol{x}) = \kappa(\boldsymbol{x})\delta_{ij}$, where δ_{ij} is the Kronecker delta.

[2]The generalized fourier law is a ***constitutive equation***, or ***equation of state***, which reflects the heat conduction properties of the body (i.e., Ω) under consideration.

A formal statement[3] of the strong form of the boundary-value problem is as follows:

$$(S)\begin{cases}\text{Given } \ell : \Omega \to \mathbb{R},\ g : \Gamma_g \to \mathbb{R} \text{ and } h : \Gamma_h \to \mathbb{R}, \text{ find } u : \overline{\Omega} \to \mathbb{R} \text{ such that} \\ q_{i,i} = \ell \quad \text{in } \Omega \quad (\textit{heat equation}) & (2.3.2) \\ u = g \quad \text{on } \Gamma_g & (2.3.3)^4 \\ -q_i n_i = h \quad \text{on } \Gamma_h & (2.3.4) \\ \text{where } q_i \text{ is defined by (2.3.1).}\end{cases}$$

The functions g and h are the ***prescribed boundary temperature and heat flux,*** respectively. This problem possesses a unique solution for appropriate restrictions on the given data.

In more mathematical terminology, (2.3.2) is a generalized Poisson equation, (2.3.3) is a Dirichlet boundary condition, and (2.3.4) is a Neumann boundary condition.

We shall now construct a weak formulation of the boundary-value problem analogous to that for the one-dimensional problem of Chapter 1. In particular, (2.3.3) and (2.3.4) will be treated as essential and natural boundary conditions, respectively. As before, let $\mathcal{S}$ denote the trial solution space and $\mathcal{V}$ the variation space. This time $\mathcal{S}$ and $\mathcal{V}$ consist of real-valued functions defined on $\overline{\Omega}$ satisfying certain smoothness requirements, such that all members of $\mathcal{S}$ satisfy (2.3.3), whereas if $w \in \mathcal{V}$, then

$$w = 0 \qquad \text{on } \Gamma_g \tag{2.3.5}$$

The weak formulation of the problem goes as follows:

$$(W)\begin{cases}\text{Given } \ell : \Omega \to \mathbb{R},\ g : \Gamma_g \to \mathbb{R} \text{ and } h : \Gamma_h \to \mathbb{R}, \text{ find } u \in \mathcal{S} \text{ such that for all } w \in \mathcal{V} \\ \boxed{-\int_\Omega w_{,i} q_i \, d\Omega = \int_\Omega w \ell \, d\Omega + \int_{\Gamma_h} w h \, d\Gamma} & (2.3.6) \\ \text{where } q_i \text{ is defined by (2.3.1).}\end{cases}$$

Theorem. Assume all functions involved are smooth enough to justify the manipulations. Then a solution of (S) is a solution of (W) and vice versa.

Proof. **1.** Assume u is the solution of (S). By virtue of (2.3.3), $u \in \mathcal{S}$. Pick any $w \in \mathcal{V}$ and proceed as follows:

[3] By a formal statement, we mean one in which we do not precisely delineate the spaces to which the functions involved belong.

[4] The statement $u = g$ on Γ_g means $u(\boldsymbol{x}) = g(\boldsymbol{x})$ for all $\boldsymbol{x} \in \Gamma_g$, and so on.

$$0 = \int_\Omega w(\underbrace{q_{i,i} - \ell}_{0})\, d\Omega \qquad \text{(heat equation, i.e., (2.3.2))}$$

$$= -\int_\Omega w_{,i} q_i \, d\Omega + \int_\Gamma w q_i n_i \, d\Gamma - \int_\Omega w \ell \, d\Omega \qquad \text{(integration by parts)}$$

$$= -\int_\Omega w_{,i} q_i \, d\Omega - \int_{\Gamma_h} w h \, d\Gamma - \int_\Omega w \ell \, d\Omega$$

(generalized Fourier law, $w = 0$ on Γ_g, and heat flux boundary condition, i.e. (2.3.4))

Therefore (2.3.6) is satisfied and so u is a solution of (W).

2. Assume u is the solution of (W). Then $u = g$ on Γ_g, and for all $w \in \mathcal{V}$

$$0 = \int_\Omega w_{,i} q_i \, d\Omega + \int_\Omega w \ell \, d\Omega + \int_{\Gamma_h} w h \, d\Gamma \qquad \text{(by (2.3.6))}$$

$$= \int_\Omega w(-q_{i,i} + \ell)\, d\Omega + \int_{\Gamma_h} w(n_i q_i + h)\, d\Gamma \tag{2.3.7}$$

Let

$$\alpha = -q_{i,i} + \ell$$
$$\beta = q_i n_i + h$$

To show (2.3.2) and (2.3.4) are satisfied and thus complete the proof, we must prove that

$$\alpha = 0 \quad \text{on } \Omega$$
$$\beta = 0 \quad \text{on } \Gamma_h$$

First pick $w = \alpha\phi$ where

i. $\phi > 0$ on Ω;
ii. $\phi = 0$ on Γ; and
iii. ϕ is smooth.

(These conditions insure that $w \in \mathcal{V}$.) With this choice for w, (2.3.7) becomes

$$0 = \int_\Omega \alpha^2 \phi \, d\Omega$$

which implies $\alpha = 0$ on Ω.

Now pick $w = \beta\psi$, where

i′. $\psi > 0$ on Γ_h;
ii′. $\psi = 0$ on Γ_g; and
iii′. ψ is smooth.

(These conditions insure that $w \in \mathcal{V}$.) With this choice for w, (2.3.7) becomes (making use of $\alpha = 0$):

$$0 = \int_{\Gamma_h} \beta^2 \psi \, d\Gamma$$

from which it follows that $\beta = 0$ on Γ_h. Thus u is a solution of (S). ■

It is convenient to introduce an abstract version of (2.3.6). Let

$$a(w, u) = \int_{\Omega} w_{,i} \kappa_{ij} u_{,j} \, d\Omega \tag{2.3.8}$$

$$(w, \ell) = \int_{\Omega} w \ell \, d\Omega \tag{2.3.9}$$

$$(w, h)_{\Gamma} = \int_{\Gamma_h} w h \, d\Gamma \tag{2.3.10}$$

Then (2.3.6) may be written as

$$a(w, u) = (w, \ell) + (w, h)_{\Gamma} \tag{2.3.11}$$

Exercise 1. Verify that $a(\cdot\,,\,\cdot)$, $(\cdot\,,\,\cdot)$ and $(\cdot\,,\,\cdot)_{\Gamma}$, as just defined, are symmetric bilinear forms. (Note that the symmetry of $a(\cdot\,,\,\cdot)$ follows from the symmetry of the conductivities.)

In manipulating terms in theories involving vector and tensor quantities, the indicial notation used is very explicit and convenient. However, when we come to the Galerkin formulation analogous to (2.3.6), additional indices necessarily appear. The situation becomes very complicated due to the greater number of indices involved and due to the ranges of the various indices being different. When we come to elasticity theory, the situation is even worse as the corresponding terms have an even greater number of indices. For these reasons it is useful at this point to adopt an *index-free* notation for (2.3.6). Aside from stemming the proliferation of indices, we shall find later on that this formulation is conducive to the computer implementation of the element arrays, especially in more complicated situations such as elasticity.

In introducing our index-free notation we shall assume for definiteness that $n_{sd} = 2$. Let ∇ denote the gradient operator; thus

$$\nabla u = \{u_{,i}\} = \begin{Bmatrix} u_{,1} \\ u_{,2} \end{Bmatrix} \tag{2.3.12}$$

$$\nabla w = \{w_{,i}\} = \begin{Bmatrix} w_{,1} \\ w_{,2} \end{Bmatrix} \tag{2.3.13}$$

In the case of two space dimensions, the conductivity matrix may be written as

$$\boldsymbol{\kappa} = [\kappa_{ij}] = \begin{bmatrix} \kappa_{11} & \kappa_{12} \\ \kappa_{21} & \kappa_{22} \end{bmatrix} \qquad \text{(symmetric)} \tag{2.3.14}$$

In the isotropic case, (2.3.14) simplifies to

$$\boldsymbol{\kappa} = \kappa[\delta_{ij}] = \kappa\begin{bmatrix} 1 & 0 \\ 0 & 1 \end{bmatrix} \tag{2.3.15}$$

In terms of the above expressions, the integrand of (2.3.8) may be written in index-free fashion:

$$\boxed{w_{,i}\kappa_{ij}u_{,j} = (\boldsymbol{\nabla} w)^T\boldsymbol{\kappa}(\boldsymbol{\nabla} u)} \tag{2.3.16}$$

Thus in place of (2.3.8) we may write

$$\boxed{a(w, u) = \int_\Omega (\boldsymbol{\nabla} w)^T\boldsymbol{\kappa}(\boldsymbol{\nabla} u)\, d\Omega} \tag{2.3.17}$$

Exercise 2. Verify (2.3.16) for the cases $n_{sd} = 2$ and 3.

2.4 HEAT CONDUCTION: GALERKIN FORMULATION; SYMMETRY AND POSITIVE-DEFINITENESS OF K

Let $\mathcal{S}^h$ and $\mathcal{V}^h$ be finite-dimensional approximations to $\mathcal{S}$ and $\mathcal{V}$, respectively. We assume all members of $\mathcal{V}^h$ vanish, or vanish approximately, on Γ_g and that each member of $\mathcal{S}^h$ admits the representation

$$u^h = v^h + g^h \tag{2.4.1}$$

where $v^h \in \mathcal{V}^h$ and g^h results in satisfaction, or at least approximate satisfaction, of the boundary condition $u = g$ on Γ_g.

The Galerkin formulation is given as follows:

(G) Given ℓ, g, and h [as in (W)], find $u^h = v^h + g^h \in \mathcal{S}^h$ such that for all $w^h \in \mathcal{V}^h$ (cf. Sec. 1.5):

$$\boxed{a(w^h, v^h) = (w^h, \ell) + (w^h, h)_\Gamma - a(w^h, g^h)} \tag{2.4.2}$$

We now view our domain as "discretized" into element domains Ω^e, $1 \leq e \leq n_{el}$. In two dimensions the element domains might be simply triangles and quadrilaterals; see Fig. 2.4.1. Nodal points may exist anywhere on the domain but most frequently appear at the element vertices and interelement boundaries and less often in the interiors.

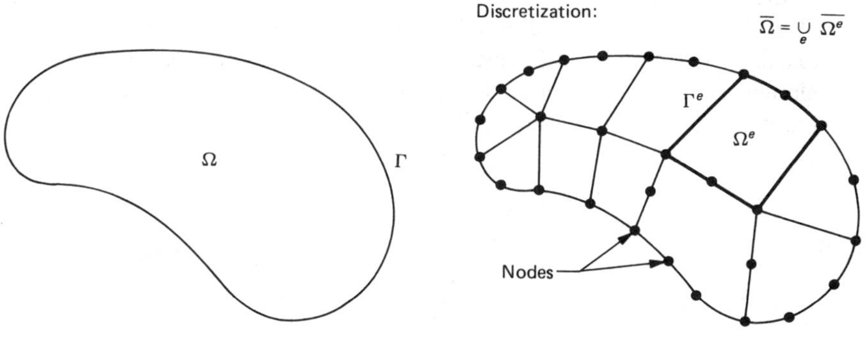

Figure 2.4.1

In Sec. 1.9 the global nodal ordering and ordering of equations in the matrix system coincided. In multidimensional applications this would prove to be an inconvenient restriction with regard to data preparation.

In what follows, a more flexible scheme is described. Let $\eta = \{1, 2, \ldots, n_{np}\}$, the ***set of global node numbers*** where n_{np} is the ***number of nodal points.*** By the terminology g-node we shall mean a node, A, at which it is prescribed that $u^h = g$. Let $\eta_g \subset \eta$ be the set of "g-nodes." The ***complement of*** η_g ***in*** η, denoted by $\eta - \eta_g$, is the set of nodes at which u^h is to be determined. The number of nodes in $\eta - \eta_g$ equals n_{eq}, the ***number of equations.***

A typical member of $\mathcal{V}^h$ is assumed to have the form

$$w^h(\boldsymbol{x}) = \sum_{A \in \eta - \eta_g} N_A(\boldsymbol{x}) c_A \tag{2.4.3}$$

where N_A is the shape function associated with node number A and c_A is a constant. We assume throughout that $w^h = 0$ if and only if $c_A = 0$ for each $A \in \eta - \eta_g$. Likewise

$$v^h(\boldsymbol{x}) = \sum_{A \in \eta - \eta_g} N_A(\boldsymbol{x}) d_A \tag{2.4.4}$$

where d_A is the unknown at node A (i.e., temperature) and

$$g^h(\boldsymbol{x}) = \sum_{A \in \eta_g} N_A(\boldsymbol{x}) g_A, \qquad g_A = g(\boldsymbol{x}_A) \tag{2.4.5}$$

From (2.4.5), we see that g^h has been *defined* to be the nodal interpolate of g by way of the shape functions.[5] Consequently, g^h will be, generally, only an approximation of g. See Fig. 2.4.2. Additional sources of error are (1) the use of approximations ℓ^h and h^h in place of ℓ and h, respectively; and (2) domain approximations in which the element boundaries do not exactly coincide with Γ. Analyses of these approximations are presented in Strang and Fix [2].

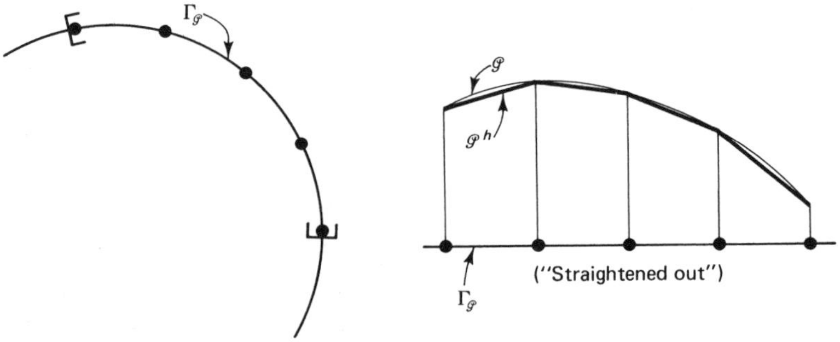

Figure 2.4.2 Piecewise linear approximation of boundary data (schematic).

Substituting (2.4.3)–(2.4.5) into (2.4.2) and arguing as in Sec. 1.6, results in

$$\sum_{B \in \eta - \eta_g} a(N_A, N_B) d_B = (N_A, \ell) + (N_A, h)_\Gamma - \sum_{B \in \eta_g} a(N_A, N_B) g_B, \qquad A \in \eta - \eta_g \tag{2.4.6}$$

To define the global stiffness matrix and force vector, we need to first specify the global ordering of equations. For this purpose we introduce the ID array, sometimes called the ***destination array,*** which assigns to node A the corresponding global equation number, viz.,

$$\text{ID}(A) = \begin{cases} P & \text{if } A \in \eta - \eta_g \\ 0 & \text{if } A \in \eta_g \end{cases} \qquad (P = \text{Global equation number}) \tag{2.4.7}$$

[5] This is not the only possibility, nor the best from the standpoint of accuracy. However, in practice it is generally the most convenient.

where $1 \leq P \leq n_{eq}$. The dimension of ID is n_{np}. As may be seen from (2.4.7), nodes at which g is prescribed are assigned "equation number" zero. An example of the setup of ID and other important data-processing arrays is presented in Sec. 2.6.

The matrix equivalent of (2.4.6) is given as follows:

$$\boldsymbol{K}\boldsymbol{d} = \boldsymbol{F} \tag{2.4.8}$$

$$\boldsymbol{K} = [K_{PQ}], \qquad \boldsymbol{d} = \{d_Q\}, \qquad \boldsymbol{F} = \{F_P\}, \qquad 1 \leq P, Q \leq n_{eq} \tag{2.4.9}$$

$$K_{PQ} = a(N_A, N_B), \qquad P = \mathrm{ID}(A), \qquad Q = \mathrm{ID}(B) \tag{2.4.10}$$

$$F_P = (N_A, \ell) + (N_A, h)_\Gamma - \sum_{B \in \eta_g} a(N_A, N_B) g_B \tag{2.4.11}$$

The main properties of $\boldsymbol{K}$ are established in the following theorem.

Theorem

1. $\boldsymbol{K}$ is symmetric.
2. $\boldsymbol{K}$ is positive definite.

Proof

1. The symmetry of $\boldsymbol{K}$ follows directly from the symmetry of $a(\cdot\,,\,\cdot)$, viz.,

$$\begin{aligned} K_{PQ} &= a(N_A, N_B) && \text{(by definition)} \\ &= a(N_B, N_A) && \text{(symmetry of } a(\cdot\,,\,\cdot)) \\ &= K_{QP} && \text{(by definition)} \end{aligned}$$

2. (Recall that we must show (i) $\boldsymbol{c}^T\boldsymbol{K}\boldsymbol{c} \geq 0$ and (ii) $\boldsymbol{c}^T\boldsymbol{K}\boldsymbol{c} = 0$ implies $\boldsymbol{c} = \boldsymbol{0}$.) To each n_{eq}-vector $\boldsymbol{c} = \{c_P\}$, we may associate a member $w^h \in \mathcal{V}^h$ by the expression $w^h = \sum_{A \in \eta - \eta_g} N_A \bar{c}_A$, where $\bar{c}_A = c_P$, $P = \mathrm{ID}(A)$.

i.

$$\begin{aligned} \boldsymbol{c}^T\boldsymbol{K}\boldsymbol{c} &= \sum_{P,Q=1}^{n_{eq}} c_P K_{PQ} c_Q \\ &= \sum_{A,B \in \eta - \eta_g} \bar{c}_A a(N_A, N_B) \bar{c}_B \\ &= a\Bigg(\sum_{A \in \eta - \eta_g} N_A \bar{c}_A, \sum_{B \in \eta - \eta_g} N_B \bar{c}_B \Bigg) && \text{(bilinearity of } a(\cdot\,,\,\cdot)) \\ &= a(w^h, w^h) && \text{(definition of } w^h) \\ &= \int_\Omega \underbrace{w^h_{,i} \kappa_{ij} w^h_{,j}}_{\geq 0} \, d\Omega && \text{(positive-definiteness of conductivities)} \\ &\geq 0 \end{aligned}$$

ii. Assume $\boldsymbol{c}^T\boldsymbol{K}\boldsymbol{c} = 0$. By the proof of part (i),

$$\int_\Omega \underbrace{w^h_{,i}\kappa_{ij}w^h_{,j}}_{\geq 0}\,d\Omega = 0$$

and thus it follows that

$$w^h_{,i}\kappa_{ij}w^h_{,j} = 0$$

By the positive-definiteness hypothesis on the conductivities, this requires $w^h_{,i} = 0$ and so w^h must be constant. However $w^h = 0$ on Γ_q (which is not empty) and so w^h must be zero throughout Ω. By the definition of w^h, it follows that each $c_P = 0$; that is $\boldsymbol{c} = \boldsymbol{0}$, which was to be proved. ∎

Remarks

1. Observe that it is the positive-definiteness hypothesis on the constitutive coefficients (i.e., κ_{ij}'s) *and* the boundary condition incorporated in the definition of $\mathcal{V}^h$ which together result in the positive-definiteness of $\boldsymbol{K}$ and thus ensure its invertibility.

2. The explicit structure of the shape functions, which will be delineated in Chapter 3, will also result in $\boldsymbol{K}$ being banded.

Exercise 1. (This exercise is a multidimensional analog of the one contained in Sec. 1.8.) Let

$$\Gamma_{\text{int}} = \left(\bigcup_{e=1}^{n_{el}} \Gamma^e\right) - \Gamma \qquad (\textit{\textbf{interior element boundaries}})$$

One side of Γ_{int} is (arbitrarily) designated to be the "+ side" and the other is the "− side." Let $\boldsymbol{n}^+$ and $\boldsymbol{n}^-$ be unit normals to Γ_{int} which point in the plus and minus directions, respectively. Clearly $\boldsymbol{n}^+ = -\boldsymbol{n}^-$. Let q_i^+ and q_i^- denote the values of q_i obtained by approaching $\boldsymbol{x} \in \Gamma_{\text{int}}$ from + and − sides, respectively. The "jump" in $q_n = q_i n_i$ at $\boldsymbol{x}$ is defined to be

$$\begin{aligned}[q_n] &= (q_i^+ - q_i^-)n_i^+ \\ &= q_i^+ n_i^+ + q_i^- n_i^-\end{aligned}$$

As may be easily verified, the jump is invariant with respect to reversing the + and − designations.

Consider the weak formulation (i.e., (2.3.6)) and assume w and u are smooth on the element interiors but may experience discontinuities in gradient across element boundaries. (Functions of this type contain the standard C^0 finite element interpolations; see Chapter 3.) Show that

$$\boxed{0 = \sum_{e+1}^{n_{el}} \int_{\Omega^e} w(q_{i,i} - \ell)\,d\Omega - \int_{\Gamma_h} w(q_n + h)\,d\Gamma + \int_{\Gamma_{\text{int}}} w[q_n]\,d\Gamma}$$

from which the Euler-Lagrange conditions may be readily deduced:

i. $q_{i,i} = \ell$ in $\bigcup_{e=1}^{n_{el}} \Omega^e$

ii. $-q_n = h$ on Γ_h

iii. $[q_n] = 0$ on Γ_{int}

As may be seen, (i) is the heat equation on the *element interiors* and (iii) is a continuity condition across element boundaries on the heat flux. Contrast the present results with those obtained assuming w and u are *globally* smooth.

The Galerkin finite element formulation obtains an *approximate* solution to (i) through (iii).

2.5 HEAT CONDUCTION: ELEMENT STIFFNESS MATRIX AND FORCE VECTOR

As before, we can break up the global arrays into sums of elemental contributions:

$$\boldsymbol{K} = \sum_{e=1}^{n_{el}} \boldsymbol{K}^e, \qquad \boldsymbol{K}^e = [K^e_{PQ}] \tag{2.5.1}$$

$$\boldsymbol{F} = \sum_{e=1}^{n_{el}} \boldsymbol{F}^e, \qquad \boldsymbol{F}^e = \{F^e_P\} \tag{2.5.2}$$

where

$$K^e_{PQ} = a(N_A, N_B)^e = \int_{\Omega^e} (\boldsymbol{\nabla} N_A)^T \boldsymbol{\kappa} (\boldsymbol{\nabla} N_B)\, d\Omega \tag{2.5.3}$$

$$\begin{aligned} F^e_P &= (N_A, \ell)^e + (N_A, h)^e_\Gamma - \sum_{B \in \eta_g} a(N_A, N_B)^e g_B \\ &= \int_{\Omega^e} N_A \ell \, d\Omega + \int_{\Gamma^e_h} N_A h \, d\Gamma - \sum_{B \in \eta_g} a(N_A, N_B)^e g_B \end{aligned} \tag{2.5.4}$$

$$\Gamma^e_h = \Gamma_h \cap \Gamma^e, \qquad P = \text{ID}(A), \qquad Q = \text{ID}(B) \tag{2.5.5}$$

See Fig. 2.5.1 for an illustration of Γ^e_h.

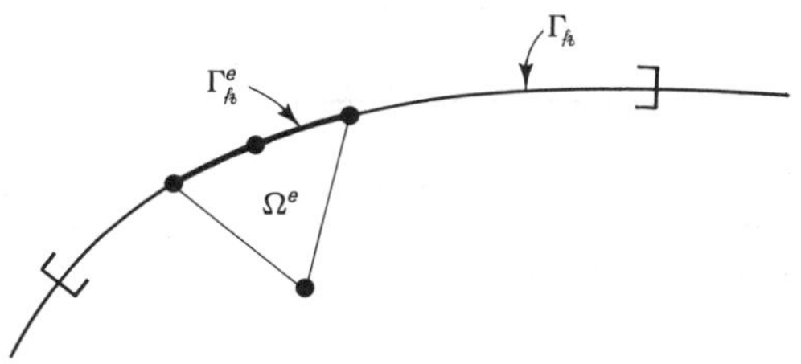

Figure 2.5.1

The element stiffness, $\boldsymbol{k}^e$, and element force vector, $\boldsymbol{f}^e$, may be deduced from these equations:

$$\boldsymbol{k}^e = [k^e_{ab}], \qquad \boldsymbol{f}^e = \{f^e_a\}, \qquad 1 \le a,\, b \le n_{en} \tag{2.5.6}$$

$$k^e_{ab} = a(N_a, N_b)^e = \int_{\Omega^e} (\boldsymbol{\nabla} N_a)^T \boldsymbol{\kappa} (\boldsymbol{\nabla} N_b)\, d\Omega \tag{2.5.7}$$

$$f^e_a = \int_{\Omega^e} N_a \ell\, d\Omega + \int_{\Gamma^e_h} N_a h\, d\Gamma - \sum_{b=1}^{n_{en}} k^e_{ab} q^e_b \tag{2.5.8}$$

where (recall) n_{en} is the number of element nodes, and $q^e_b = q(\boldsymbol{x}^e_b)$ if q is prescribed at node number b and *equals zero otherwise*.[6]

The global arrays, $\boldsymbol{K}$ and $\boldsymbol{F}$ may be formed from the element arrays $\boldsymbol{k}^e$ and $\boldsymbol{f}^e$, respectively, by way of an assembly algorithm as described in Sec. 1.14.

The element stiffness matrix can be written in a standard form convenient for programming:

$$\boldsymbol{k}^e = \int_{\Omega^e} \boldsymbol{B}^T \boldsymbol{D} \boldsymbol{B}\, d\Omega \tag{2.5.9}$$

where, in the present case,

$$\underbrace{\boldsymbol{D}}_{n_{sd} \times n_{sd}} = \boldsymbol{\kappa} \tag{2.5.10}$$

$$\underbrace{\boldsymbol{B}}_{n_{sd} \times n_{en}} = [\boldsymbol{B}_1, \boldsymbol{B}_2, \ldots, \boldsymbol{B}_{n_{en}}] \tag{2.5.11}$$

$$\underbrace{\boldsymbol{B}_a}_{n_{sd} \times 1} = \boldsymbol{\nabla} N_a \tag{2.5.12}$$

The component version of (2.5.9) is

$$k^e_{ab} = \int_{\Omega^e} \boldsymbol{B}_a^T \boldsymbol{D} \boldsymbol{B}_b\, d\Omega \tag{2.5.13}$$

[6] An implicit assumption in localizing the q-term is that if $\boldsymbol{x}_A$ is not a node attached to element e, then $N_A(\boldsymbol{x}) = 0$ for all $\boldsymbol{x} \in \overline{\Omega}^e$. Otherwise, the last term in (2.5.4) may involve q-data of nodes not attached to element e, which is not accounted for in (2.5.8).

Exercise 1. Let

$$\underbrace{\boldsymbol{d}^e}_{n_{en} \times 1} = \{d_a^e\} = \begin{Bmatrix} d_1^e \\ d_2^e \\ \vdots \\ d_{n_{en}}^e \end{Bmatrix} \tag{2.5.14}$$

where

$$d_a^e = u^h(\boldsymbol{x}_a^e) \tag{2.5.15}$$ [7]

$\boldsymbol{d}^e$ is called the ***element temperature vector.*** Show that the heat flux vector at point $\boldsymbol{x} \in \Omega^e$ can be calculated from the formula

$$\boldsymbol{q}(\boldsymbol{x}) = -\boldsymbol{D}(\boldsymbol{x})\boldsymbol{B}(\boldsymbol{x})\boldsymbol{d}^e = -\boldsymbol{D}(\boldsymbol{x}) \sum_{a=1}^{n_{en}} \boldsymbol{B}_a d_a^e \tag{2.5.16}$$

Exercise 2. Consider the strong statement of the boundary-value problem in classical linear heat conduction in which the h-type boundary condition (i.e., eq. (2.3.4)) is replaced by the following expression:

$$\lambda u - q_i n_i = h \qquad \text{on } \Gamma_h \tag{2.5.17}$$

where $\lambda \geq 0$ is a given function of $\boldsymbol{x} \in \Gamma_h$. Generalize the weak formulation to include (2.5.17) as a natural boundary condition. Obtain an expression for the additional contribution to k_{ab}^e (cf. (2.5.13)) arising from (2.5.17). Show that $\boldsymbol{K}$ is positive-definite.

The boundary condition (2.5.17) is equivalent to what is often called ***Newton's law of heat transfer;*** λ is called the ***coefficient of heat transfer.*** This boundary condition applies to the case in which the heat flux is proportional to the difference of the surface temperatures of the body and surrounding medium, the latter formally represented by h/λ in (2.5.17).

2.6 HEAT CONDUCTION: DATA PROCESSING ARRAYS; ID, IEN, AND LM

The element nodal data is stored in the array IEN, the ***element nodes array,*** which relates local node numbers to global node numbers, viz.,

$$\text{IEN}(\underset{\substack{\nearrow \\ \text{Local node number}}}{a}, \underset{\substack{\nwarrow \\ \text{Element number}}}{e}) = \underbrace{A}_{\text{Global node number}} \tag{2.6.1}$$

The relationship between global node numbers and global equation numbers as well as nodal boundary condition information is stored in the ID array (see 2.4.7). In

[7] The g-boundary conditions are accounted for in this definition.

practice, the IEN and ID arrays are set up from input data. The LM array, which was described in the context of the one-dimensional model problem in Sec. 1.14, may then be constructed from the relation

$$\boxed{\mathrm{LM}(a, e) = \mathrm{ID}(\mathrm{IEN}(a, e))} \tag{2.6.2}$$

Because of the previous relationship, we often think of LM as the element "localization" of ID. Strictly speaking the LM array is redundant. However, it is generally convenient in computing to set up LM once and for all, rather than make use of (2.6.2) repeatedly.

Example 1 illustrates the structure of the ID, IEN, and LM arrays.

Example 1

Consider the mesh of four-node, rectangular elements shown in Fig. 2.6.1. We assume that the local node numbering begins at the lower left-hand node of each element and proceeds in counterclockwise fashion. This is illustrated in Fig. 2.6.1 for element 2, which is typical. We also assume that essential boundary conditions (i.e., "g-type") are specified at nodes 1, 4, 7, and 10. Thus there will only be eight equations in the global system $\boldsymbol{Kd} = \boldsymbol{F}$. We adopt the usual convention that the global equation numbers run in ascending order with respect to the ascending order of global node numbers. The ID, IEN, and LM arrays are given in Fig. 2.6.2. The reader is urged to verify the details.

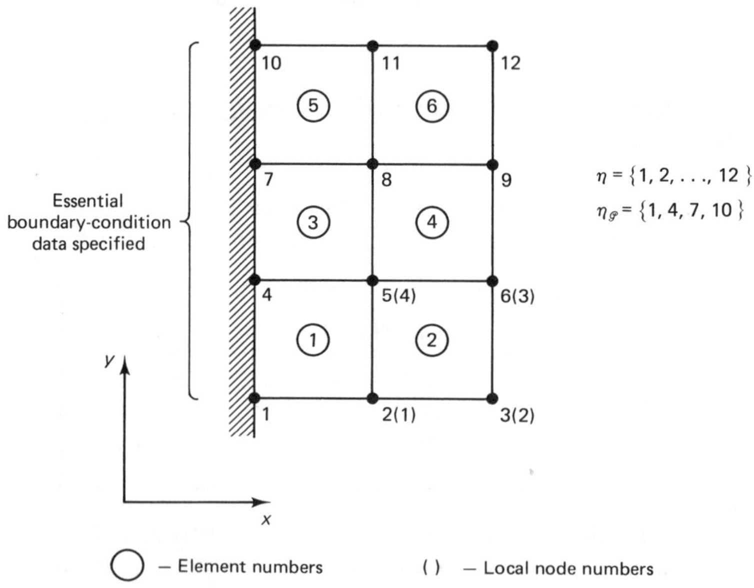

Figure 2.6.1 Mesh of four-node, rectangular elements; global and local node numbers, element numbers, and essential boundary condition nodes.

ID array:

Global node numbers (A)

1	2	3	4	5	6	7	8	9	10	11	12
0*	1	2	0	3	4	0	5	6	0	7	8

($n_{np} = 12$)

($n_{eq} = 8$)

$$P = \text{ID}(A)$$

IEN array:

Element numbers (e) ($n_{el} = 6$)

Local node numbers (a)	1	2	3	4	5	6
1	1	2	4	5	7	8
2	2	3	5	6	8	9
3	5	6	8	9	11	12
4	4	5	7	8	10	11

($n_{en} = 4$)

$$A = \text{IEN}(a, e)$$

LM array:

Element numbers (e) ($n_{el} = 6$)

Local node numbers (a)	1	2	3	4	5	6
1	0*	1	0	3	0	5
2	1	2	3	4	5	6
3	3	4	5	6	7	8
4	0	3	0	5	0	7

($n_{en} = 4$)

$$P = \text{LM}(a, e) = \text{ID}(\text{IEN}(a, e))$$

* Temperature boundary conditions ("g-type") denoted by zeros.

Figure 2.6.2. ID, IEN, and LM arrays for the mesh of Fig. 2.6.1.

In terms of the IEN and LM arrays, a precise definition of the g_a^e's may be given (see (2.5.8)):

$$g_a^e = \begin{cases} 0 & \text{if } \mathrm{LM}(a, e) \neq 0 \\ g_A & \text{if } \mathrm{LM}(a, e) = 0, \quad \text{where } A = \mathrm{IEN}(a, e) \end{cases} \tag{2.6.3}$$

This definition may be easily programmed.

In our final example of this section we shall illustrate the assembly procedure for a typical element subjected to essential boundary conditions.

Example 2

Consider a typical, four-noded element e. Assume values of the LM array, for this element, are given as follows:

$$\left.\begin{aligned} \mathrm{LM}(1, e) &= 5 \\ \mathrm{LM}(2, e) &= 0 \\ \mathrm{LM}(3, e) &= 0 \\ \mathrm{LM}(4, e) &= 9 \end{aligned}\right\} \tag{2.6.4}$$

We deduce from (2.6.4) that the contributions to the global arrays are given as follows:[8]

$$\left.\begin{aligned} K_{55} &\leftarrow K_{55} + k_{11}^e \\ K_{59} &\leftarrow K_{59} + k_{14}^e \\ K_{95} &\leftarrow K_{95} + k_{41}^e \\ K_{99} &\leftarrow K_{99} + k_{44}^e \end{aligned}\right\} \tag{2.6.5}$$

$$\left.\begin{aligned} F_5 &\leftarrow F_5 + f_1^e \\ F_9 &\leftarrow F_9 + f_4^e \end{aligned}\right\} \tag{2.6.6}$$

Note that all terms in the second and third rows and columns of $\boldsymbol{k}^e$ do *not* contribute to $\boldsymbol{K}$. However, they may contribute to $\boldsymbol{F}$ via f_1^e and f_4^e, since, by (2.5.8) we have

$$f_1^e = \cdots - k_{12}^e g_2^e - k_{13}^e g_3^e \tag{2.6.7}$$

$$f_4^e = \cdots - k_{42}^e g_2^e - k_{43}^e g_3^e \tag{2.6.8}$$

in which, for clarity, we have omitted the first two terms of the right-hand side of (2.5.8).

Special subroutines are easily programmed to carry out the operations indicated in (2.6.5)–(2.6.8).

It is instructive to visualize the contributions of the eth element to the global stiffness and force. These contributions are depicted in Fig. 2.6.3.

We note that all necessary element assembly information is provided by the LM array.

[8] Due to symmetry, k_{41}^e is not actually assembled in practice.

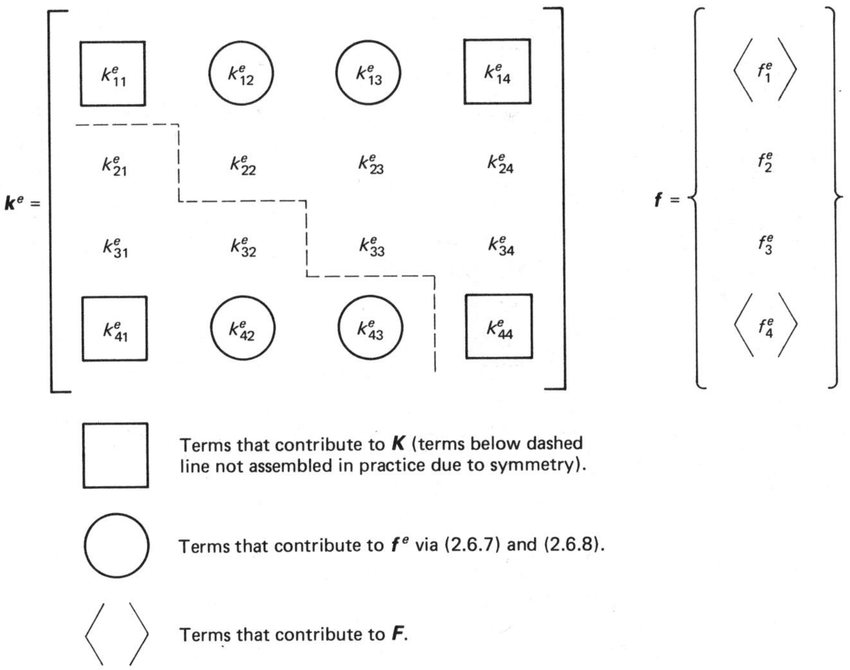

Figure 2.6.3 Contributions of heat conduction element in Example 2 to global arrays.

Exercise 1. Consider the accompanying mesh. Set up the ID, IEN, and LM arrays.

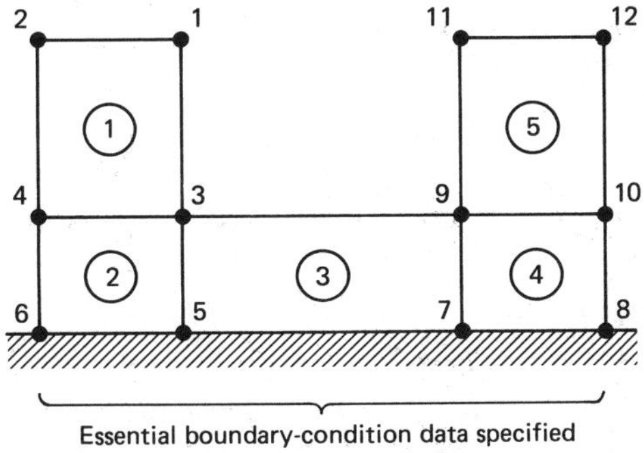

2.7 CLASSICAL LINEAR ELASTOSTATICS: STRONG AND WEAK FORMS; EQUIVALENCE

Classical elastostatics is a rich subject in its own right. See [3–7] for background and references to the literature. These references range from the very physical to the very mathematical. The most physical book is the one by Timoshenko and Goodier. In ascending order of mathematical content are Sokolnikoff, Gurtin, Duvaut-Lions, and Fichera.

The reader is reminded that indices i, j, k, and l take on values $1, \ldots, n_{sd}$, where n_{sd} is the number of spatial dimensions, and the summation convention applies to repeated indices i, j, k, and l *only*.

Let σ_{ij} denote (Cartesian components of) the (Cauchy) ***stress tensor,*** let u_i denote the ***displacement vector,*** and let ℓ_i be the ***prescribed body force per unit volume.*** The ***(infinitesimal) strain tensor,*** ϵ_{ij}, is defined to be the symmetric part of the displacement gradients, viz.,

$$\epsilon_{ij} = u_{(i,j)} \overset{\text{def.}}{=} \frac{u_{i,j} + u_{j,i}}{2} \qquad (\textbf{\textit{strain-displacement equations}}) \tag{2.7.1}$$

The stress tensor is defined in terms of the strain tensor by the ***generalized Hooke's law:***[9]

$$\sigma_{ij} = c_{ijkl}\epsilon_{kl} \tag{2.7.2}$$

where the c_{ijkl}'s, the ***elastic coefficients,*** are given functions of $\boldsymbol{x}$. (If the c_{ijkl}'s are constants throughout, the body is called ***homogeneous.***) The elastic coefficients are assumed to satisfy the following properties:

Symmetry

$$c_{ijkl} = c_{klij} \qquad \text{(major symmetry)} \tag{2.7.3}$$

$$\left.\begin{aligned} c_{ijkl} &= c_{jikl} \\ c_{ijkl} &= c_{ijlk} \end{aligned}\right\} \qquad \text{(minor symmetries)} \tag{2.7.4}$$

Positive-definiteness

$$c_{ijkl}(\boldsymbol{x})\psi_{ij}\psi_{kl} \geq 0 \tag{2.7.5}$$

$$c_{ijkl}(\boldsymbol{x})\psi_{ij}\psi_{kl} = 0 \quad \text{implies} \quad \psi_{ij} = 0 \tag{2.7.6}$$

for all $\boldsymbol{x} \in \overline{\Omega}$ and all $\psi_{ij} = \psi_{ji}$.

Note. The positive-definiteness condition is in terms of symmetric arrays, ψ_{ij}.

We shall see in Sec. 2.8 that a consequence of the major symmetry (2.7.3) is that $\boldsymbol{K}$ is symmetric. The first minor symmetry implies the symmetry of the stress tensor (i.e., $\sigma_{ij} = \sigma_{ji}$).[10] The positive-definiteness condition, when combined with appropriate boundary conditions on the displacement, leads to the positive-definiteness of $\boldsymbol{K}$.

[9] This is another ***constitutive equation,*** which reflects the elastic properties of the body under consideration.

[10] From a fundamental continuum mechanics standpoint, the symmetry of the Cauchy stress tensor emanates from the balance of angular momentum.

In the present theory, the unknown is a vector (i.e., the displacement vector). Consequently, a generalization of the boundary conditions considered previously is necessitated. We shall assume that Γ admits decompositions

$$\left.\begin{aligned} \Gamma &= \overline{\Gamma_{g_i} \cup \Gamma_{h_i}} \\ \varnothing &= \Gamma_{g_i} \cap \Gamma_{h_i} \end{aligned}\right\} \qquad i = 1, \ldots, n_{sd} \tag{2.7.7}$$

For example, in two dimensions the situation might appear as in Fig. 2.7.1. As can be seen there can be a different decomposition for each $i = 1, 2, \ldots, n_{sd}$.

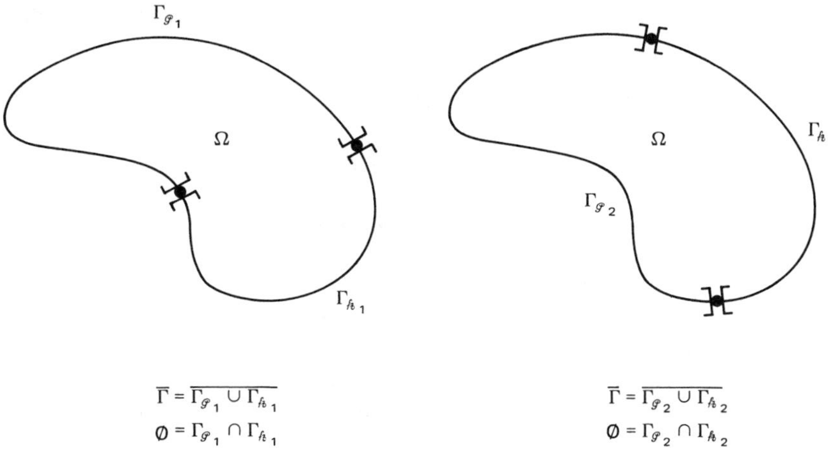

Figure 2.7.1

A formal statement of the strong form of the boundary-value problem goes as follows:

$$(S)\left\{\begin{array}{ll} \text{Given } f_i : \Omega \rightarrow \mathbb{R},\ g_i : \Gamma_{g_i} \rightarrow \mathbb{R}, \text{ and } h_i : \Gamma_{h_i} \rightarrow \mathbb{R}, \text{ find } u_i : \overline{\Omega} \rightarrow \mathbb{R} \text{ such that} & \\ \qquad \sigma_{ij,j} + f_i = 0 \quad \text{in } \Omega \quad (\textbf{\textit{equilibrium equations}}) & (2.7.8) \\ \qquad u_i = g_i \quad \text{on } \Gamma_{g_i} & (2.7.9) \\ \qquad \sigma_{ij} n_j = h_i \quad \text{on } \Gamma_{h_i} & (2.7.10) \\ \text{where } \sigma_{ij} \text{ is defined in terms of } u_i \text{ by (2.7.1) and (2.7.2).} & \end{array}\right.$$

Remarks

1. The functions g_i and h_i are called the ***prescribed boundary displacements*** and ***tractions,*** respectively.

2. (S) is sometimes referred to as the ***mixed boundary-value problem of linear elastostatics.*** Under appropriate hypotheses on the data, (S) possesses a unique solution (see Fichera [4]).

3. The additional complexity of the present theory, when compared with heat conduction, is that the unknown (i.e., $\boldsymbol{u} = \{u_i\}$) is vector-valued rather than scalar-valued.

4. In practice, it is important to be able to deal with somewhat more complicated boundary-condition specification. In order not to encumber the present exposition, this generalization will be considered later on in the form of an exercise (see Exercise 5 in Sec. 2.12).

Let $\mathcal{S}_i$ denote the trial solution space and $\mathcal{V}_i$ the variation space. Each member $u_i \in \mathcal{S}_i$ satisfies $u_i = g_i$ on Γ_{g_i}, whereas each $w_i \in \mathcal{V}_i$ satisfies $w_i = 0$ on Γ_{g_i}. Equation (2.7.10) will be a natural boundary condition.

The weak formulation goes as follows:

(W) Given $f_i : \Omega \to \mathbb{R}$, $g_i : \Gamma_{g_i} \to \mathbb{R}$ and $h_i : \Gamma_{h_i} \to \mathbb{R}$, find $u_i \in \mathcal{S}_i$ such that for all $w_i \in \mathcal{V}_i$,

$$\boxed{\int_\Omega w_{(i,j)}\sigma_{ij}\, d\Omega = \int_\Omega w_i f_i \, d\Omega + \sum_{i=1}^{n_{sd}} \left(\int_{\Gamma_{h_i}} w_i h_i \, d\Gamma \right)} \tag{2.7.11}$$

where σ_{ij} is defined in terms of u_i by (2.7.1) and (2.7.2).

Remarks

5. In the solid mechanics literature, (W) is sometimes referred to as the ***principle of virtual work,*** or ***principle of virtual displacements,*** w_i being the ***virtual displacements.***

6. The existence and uniqueness of weak solutions is discussed in Duvaut-Lions [3].

Note. The boundary integral in (2.7.11) takes on the explicit form:

$$\sum_{i=1}^{n_{sd}} \left(\int_{\Gamma_{h_i}} w_i h_i \, d\Gamma \right) = \int_{\Gamma_{h_1}} w_1 h_1 \, d\Gamma + \cdots + \int_{\Gamma_{h_{n_{sd}}}} w_{n_{sd}} h_{n_{sd}} \, d\Gamma \tag{2.7.12}$$

Theorem. Assume all functions are smooth enough to justify the manipulations. Then a solution of (S) is a solution of (W), and vice versa.

Remark

7. The proof of the equivalence theorem requires some preliminary results, which we shall establish in the following lemmas.

Lemma 1. *(Euclidean decomposition of a second-rank tensor.)* Let s_{ij} denote a general nonsymmetric, second-rank tensor. Then $s_{ij} = s_{(ij)} + s_{[ij]}$, where $s_{(ij)}$ is symmetric (i.e., $s_{(ij)} = s_{(ji)}$) and $s_{[ij]}$ is skew-symmetric (i.e., $s_{[ij]} = -s_{[ji]}$).

Proof. Define

$$s_{(ij)} = \frac{s_{ij} + s_{ji}}{2} \tag{2.7.13}$$

$$s_{[ij]} = \frac{s_{ij} - s_{ji}}{2} \tag{2.7.14}$$

It is easily verified that $s_{(ij)}$ is symmetric and $s_{[ij]}$ is skew-symmetric. ∎

Remark

8. $s_{(ij)}$ and $s_{[ij]}$ are called the ***symmetric*** and ***skew-symmetric parts*** of s_{ij}, respectively.

Lemma 2. Let s_{ij} be a nonsymmetric tensor and let t_{ij} be a symmetric tensor. Then

$$s_{ij}t_{ij} = s_{(ij)}t_{ij} \tag{2.7.15}$$

Proof. By Lemma 1, $s_{ij} = s_{(ij)} + s_{[ij]}$. Since

$$s_{ij}t_{ij} = s_{(ij)}t_{ij} + s_{[ij]}t_{ij}$$

the lemma will be established if we can show that $s_{[ij]}t_{ij} = 0$. We proceed as follows:

$$\begin{aligned} s_{[ij]}t_{ij} &= -s_{[ji]}t_{ij} && \text{(skew-symmetry of } s_{[ij]}\text{)} \\ &= -s_{[ji]}t_{ji} && \text{(symmetry of } t_{ij}\text{)} \\ &= -s_{[ij]}t_{ij} && \text{(redefinition of dummy indices)} \end{aligned}$$

from which the result follows. ∎

Proof of Theorem

1. Let u_i be a solution of (S). Thus $u_i \in \mathcal{S}_i$. Multiply (2.7.8) by $w_i \in \mathcal{V}_i$ and integrate over Ω, viz.,

$$0 = \int_\Omega w_i(\sigma_{ij,j} + f_i)\, d\Omega = -\int_\Omega w_{i,j}\sigma_{ij}\, d\Omega + \int_\Gamma w_i\sigma_{ij}n_j\, d\Gamma + \int_\Omega w_i f_i\, d\Omega$$

(integration by parts)

$$= -\int_\Omega w_{(i,j)}\sigma_{ij}\, d\Omega + \sum_{i=1}^{n_{sd}} \left(\int_{\Gamma_{h_i}} w_i h_i\, d\Gamma \right) + \int_\Omega w_i f_i\, d\Omega$$

(symmetry of σ_{ij}, Lemma 2, $w_i = 0$ on Γ_{g_i}, and (2.7.10))

Therefore, u_i is a solution of (W).

2. Assume u_i is a solution of (W). Since $u_i \in \mathcal{S}_i$, $u_i = g_i$ on Γ_{g_i}. From (2.7.11)

$$0 = -\int_\Omega \underbrace{w_{(i,j)}\sigma_{ij}}_{(= w_{i,j}\sigma_{ij} \text{ by Lemma 2})} d\Omega + \int_\Omega w_i f_i\, d\Omega + \sum_{i=1}^{n_{sd}} \int_{\Gamma_{h_i}} w_i h_i\, d\Gamma$$

$$= \int_\Omega w_i(\sigma_{ij,j} + f_i)\, d\Omega - \sum_{i=1}^{n_{sd}} \int_{\Gamma_{h_i}} w_i(\sigma_{ij}n_j - h_i)\, d\Gamma \qquad \begin{array}{l}\text{(integration by parts,} \\ w_i = 0 \text{ on } \Gamma_{g_i})\end{array} \tag{2.7.16}$$

Let

$$\alpha_i = \sigma_{ij,j} + f_i$$

$$\beta_i = \sigma_{ij}n_j - h_i$$

Thus to complete the proof we must show that

$$\alpha_i = 0 \qquad \text{on } \Omega$$

$$\beta_i = 0 \qquad \text{on } \Gamma_{h_i}$$

These conditions can be proved by the techniques used in Sec. 2.3. Let $w_i = \alpha_i \phi$, where

i. $\phi > 0$ on Ω;
ii. $\phi = 0$ on Γ; and
iii. ϕ is smooth.

(These conditions insure that $w_i \in \mathcal{V}_i$.) Substituting this w_i into (2.7.16) yields

$$0 = \int_\Omega \underbrace{\alpha_i \alpha_i}_{\geq 0} \; \underbrace{\phi}_{> 0} \; d\Omega$$

which implies $\alpha_i = 0$ on Ω.

Now take $w_i = \delta_{i1} \beta_1 \psi$, where

i′. $\psi > 0$ on Γ_{h_1};
ii′. $\psi = 0$ on Γ_{g_1}; and
iii′. ψ is smooth.

(Again, $w_i \in \mathcal{V}_i$.) Substituting this w_i into (2.7.16) and making use of $\alpha_i = 0$ results in

$$0 = \int_{\Gamma_{h_1}} \beta_1^2 \psi \, d\Gamma$$

from which it follows that $\beta_1 = 0$ on Γ_{h_1}.

We may proceed analogously to show $\beta_2 = 0$, and so on. Consequently, u_i is a solution of (S). ■

The abstract notation for the present case is

$$a(\boldsymbol{w}, \boldsymbol{u}) = \int_\Omega w_{(i,j)} c_{ijkl} u_{(k,l)} \, d\Omega \tag{2.7.17}$$

$$(\boldsymbol{w}, \boldsymbol{\ell}) = \int_\Omega w_i \ell_i \, d\Omega \tag{2.7.18}$$

$$(\boldsymbol{w}, \boldsymbol{h})_\Gamma = \sum_{i=1}^{n_{sd}} \left(\int_{\Gamma_{h_i}} w_i h_i \, d\Gamma \right) \tag{2.7.19}$$

Exercise 1. Verify that $a(\cdot\,, \cdot)$, $(\cdot\,, \cdot)$ and $(\cdot\,, \cdot)_\Gamma$, as just defined, are symmetric, bilinear forms (cf. Sec. 1.4).

Let $\mathcal{S} = \{\boldsymbol{u} \mid u_i \in \mathcal{S}_i\}$ and let $\mathcal{V} = \{\boldsymbol{w} \mid w_i \in \mathcal{V}_i\}$. Then the weak form can be concisely written in terms of (2.7.17–2.7.19) as follows:

(W) Given $\boldsymbol{\ell}$, $\boldsymbol{g}$, and $\boldsymbol{h}$ (in which the components are defined in (W)), find $\boldsymbol{u} \in \mathcal{S}$ such that for all $\boldsymbol{w} \in \mathcal{V}$

$$a(\boldsymbol{w}, \boldsymbol{u}) = (\boldsymbol{w}, \boldsymbol{\ell}) + (\boldsymbol{w}, \boldsymbol{h})_\Gamma \tag{2.7.20}$$

As discussed in Sec. 2.3, it is desirable to construct index-free counterparts of the expressions on the right-hand sides of (2.7.17)–(2.7.19). For concreteness we shall assume that $n_{sd} = 2$; thus $1 \le i, j, k, l \le 2$.

Let[11]

$$\boldsymbol{\epsilon}(\boldsymbol{u}) = \{\epsilon_I(\boldsymbol{u})\} = \begin{Bmatrix} u_{1,1} \\ u_{2,2} \\ u_{1,2} + u_{2,1} \end{Bmatrix} \tag{2.7.21}$$

$$\boldsymbol{\epsilon}(\boldsymbol{w}) = \{\epsilon_I(\boldsymbol{w})\} = \begin{Bmatrix} w_{1,1} \\ w_{2,2} \\ w_{1,2} + w_{2,1} \end{Bmatrix} \tag{2.7.22}$$

$$\boldsymbol{D} = [D_{IJ}] = \begin{bmatrix} D_{11} & D_{12} & D_{13} \\ & D_{22} & D_{23} \\ \text{symmetric} & & D_{33} \end{bmatrix} \tag{2.7.23}$$

[11] According to our previous notational conventions, $\boldsymbol{\epsilon} = [\epsilon_{ij}]$, the matrix of strain components. However, we will have no need for this matrix, and consequently we reserve $\boldsymbol{\epsilon}$ for the "strain vector" defined in (2.7.21). A similar notational conflict occurs with respect to the "stress vector," $\boldsymbol{\sigma}$, defined in Exercise 4. Note that factors of one-half have been eliminated from the *shearing components* (i.e., last components) of (2.7.21) and (2.7.22). (Compare (2.7.21) and (2.7.22) with (2.7.1).) This will considerably simplify subsequent writing.

where

$$D_{IJ} = c_{ijkl} \tag{2.7.24}$$

in which the indices are related by the following table:

TABLE 2.7.1

I \ J	i \ k	j \ l
1	1	1
3	1	2
3	2	1
2	2	2

As should be clear, we have "collapsed" pairs of indices (i, j, and k, l) into single indices (I and J, respectively) taking account of the symmetries of c_{ijkl}, $u_{(k,l)}$ and $w_{(i,j)}$. Observe that the indices I and J take on values 1, 2, and 3. In n_{sd} dimensions, the I and J indices will take on values $1, 2, \ldots, n_{sd}(n_{sd} + 1)/2$.

It can be shown that

$$w_{(i,j)}c_{ijkl}u_{(k,l)} = \boldsymbol{\epsilon}(\boldsymbol{w})^T \boldsymbol{D}\, \boldsymbol{\epsilon}(\boldsymbol{u}) \tag{2.7.25}$$

and so

$$\boxed{a(\boldsymbol{w}, \boldsymbol{u}) = \int_{\Omega} \boldsymbol{\epsilon}(\boldsymbol{w})^T \boldsymbol{D}\, \boldsymbol{\epsilon}(\boldsymbol{u})\, d\Omega} \tag{2.7.26}$$

Exercise 2. Verify (2.7.25) for $n_{sd} = 2$.

Exercise 3. Construct the analog of Table 2.7.1 for the case $n_{sd} = 3$. For definiteness of the ordering, take

$$\boldsymbol{\epsilon}(\boldsymbol{u}) = \begin{Bmatrix} u_{1,1} \\ u_{2,2} \\ u_{3,3} \\ u_{2,3} + u_{3,2} \\ u_{1,3} + u_{3,1} \\ u_{1,2} + u_{2,1} \end{Bmatrix} \tag{2.7.27}$$

Exercise 4. Show that

$$\boldsymbol{\sigma} = \boldsymbol{D}\, \boldsymbol{\epsilon}(\boldsymbol{u}) \tag{2.7.28}$$

where

$$\boldsymbol{\sigma} = \begin{Bmatrix} \sigma_{11} \\ \sigma_{22} \\ \sigma_{12} \end{Bmatrix}, \qquad n_{sd} = 2 \tag{2.7.29}$$

$$\boldsymbol{\sigma} = \begin{Bmatrix} \sigma_{11} \\ \sigma_{22} \\ \sigma_{33} \\ \sigma_{23} \\ \sigma_{13} \\ \sigma_{12} \end{Bmatrix}, \qquad n_{sd} = 3 \tag{2.7.30}$$

Exercise 5. If the body in question is ***isotropic***, then

$$c_{ijkl}(\boldsymbol{x}) = \mu(\boldsymbol{x})(\delta_{ik}\delta_{jl} + \delta_{il}\delta_{jk}) + \lambda(\boldsymbol{x})\ \delta_{ij}\delta_{kl} \tag{2.7.31}$$

where λ and μ are the ***Lamé parameters***; μ is often referred to as the shear modulus and denoted by G. The relationships of λ and μ to E, ***Young's modulus***, and ν, ***Poisson's ratio***, are given by

$$\lambda = \frac{\nu E}{(1 + \nu)(1 - 2\nu)} \tag{2.7.32}$$

$$\mu = \frac{E}{2(1 + \nu)} \tag{2.7.33}$$

(See Sokolnikoff [6], p. 71, for further relations with other equivalent moduli.) If (2.7.31) is not satisfied, the body is said to be ***anisotropic***. Using (2.7.31) set up $\boldsymbol{D}$ for $n_{sd} = 2$ and $n_{sd} = 3$. *Hint:* The answer for $n_{sd} = 2$ is

$$\boldsymbol{D} = \begin{bmatrix} \lambda + 2\mu & \lambda & 0 \\ & \lambda + 2\mu & 0 \\ \text{Symmetric} & & \mu \end{bmatrix} \tag{2.7.34}$$

This matrix manifests the ***plane strain*** hypothesis. (See Sokolnikoff [6] for elaboration.)

Remark

9. The case of isotropic ***plane stress*** may be determined from (2.7.34) by replacing λ by $\bar{\lambda}$, where

$$\bar{\lambda} = \frac{2\lambda\mu}{\lambda + 2\mu} \tag{2.7.35}$$

(See Sokolnikoff [6] or Timoshenko and Goodier [7] for elaboration on the physical ideas.)

Exercise 6. (See Exercise 1, Sec. 2.4 for background and an analogous result.)

Let the "jump" in $\sigma_{in} = \sigma_{ij}n_j$ be denoted by $[\sigma_{in}]$. Consider the weak formulation (i.e., (2.7.11) and assume w_i and u_i are smooth on element interiors, but experience gradient discontinuities across element boundaries. Show that

$$0 = \sum_{e=1}^{n_{el}} \int_{\Omega^e} w_i(\sigma_{ij,j} + \ell_i)\, d\Omega - \sum_{i=1}^{n_{sd}} \int_{\Gamma_{h_i}} w_i(\sigma_{in} - h_i)\, d\Gamma - \int_{\Gamma_{\text{int}}} w_i[\sigma_{in}]\, d\Gamma$$

from which the Euler-Lagrange conditions may be read:

i. $\sigma_{ij,j} + \ell_i = 0$ in $\bigcup_{e=1}^{n_{el}} \Omega^e$
ii. $\sigma_{in} = h_i$ on Γ_{h_i}
iii. $[\sigma_{in}] = 0$ on Γ_{int}

Here (i) is the equilibrium equation on *element interiors*, and (iii) is a traction continuity condition across element boundaries. Compare these results with those obtained assuming w_i and u_i are *globally* smooth.

2.8 ELASTOSTATICS: GALERKIN FORMULATION, SYMMETRY, AND POSITIVE-DEFINITENESS OF K

Let $\mathcal{S}^h$ and $\mathcal{V}^h$ be finite-dimensional approximations to $\mathcal{S}$ and $\mathcal{V}$, respectively. We assume members $\boldsymbol{w}^h \in \mathcal{V}^h$ result in satisfaction, or approximate satisfaction, of the boundary condition $w_i = 0$ on Γ_{g_i}, and members of $\mathcal{S}^h$ admit the decomposition

$$\boldsymbol{u}^h = \boldsymbol{v}^h + \boldsymbol{g}^h \tag{2.8.1}$$

where $\boldsymbol{v}^h \in \mathcal{V}^h$ and $\boldsymbol{g}^h$ results in satisfaction, or approximate satisfaction, of the boundary condition $u_i = g_i$ on Γ_{g_i}.

The Galerkin formulation of our problem is given as follows:

(G) Given $\boldsymbol{\ell}$, $\boldsymbol{g}$, and $\boldsymbol{h}$ (as in (W)), find $\boldsymbol{u}^h = \boldsymbol{v}^h + \boldsymbol{g}^h \in \mathcal{S}^h$ such that for all $\boldsymbol{w}^h \in \mathcal{V}^h$

$$a(\boldsymbol{w}^h, \boldsymbol{v}^h) = (\boldsymbol{w}^h, \boldsymbol{\ell}) + (\boldsymbol{w}^h, \boldsymbol{h})_\Gamma - a(\boldsymbol{w}^h, \boldsymbol{g}^h) \tag{2.8.2}$$

To define the global stiffness matrix and force vector for elasticity, it is necessary to introduce the ID array. This entails a generalization of the definition given in Sec. 2.6, since in the present case there will be more than 1 degree of freedom per node. For elasticity there are n_{sd} degrees of freedom per node, but in order to include in our definition cases such as heat conduction, we shall take the fully general situation in which it is assumed that there are n_{dof} ***degrees of freedom per node***.[12] In this case

Global equation number

$$\text{ID}(i, A) = \begin{cases} P & \text{if } A \in \eta - \eta_{g_i} \\ 0 & \text{if } A \in \eta_{g_i} \end{cases} \tag{2.8.3}$$

i: Degree of freedom number; A: Global node number

where $1 \leq i \leq n_{\text{dof}}$. Thus ID has dimensions $n_{\text{dof}} \times n_{np}$. If $n_{\text{dof}} = 1$, we reduce to the case considered previously in Sec. 2.6 (i.e., $\text{ID}(i, A) = \text{ID}(A)$).

Recall that $\eta = \{1, 2, \ldots, n_{np}\}$ denotes the set of global node numbers. Let $\eta_{g_i} \subset \eta$ be the set of nodes at which $u_i^h = g_i$ and let $\eta - \eta_{g_i}$ be the complement of η_{g_i}. For each node in $\eta - \eta_{g_i}$, the nodal value of u_i^h is to be determined.

The explicit representations of v_i^h and g_i^h, in terms of the shape functions and nodal values are

$$v_i^h = \sum_{A \in \eta - \eta_{g_i}} N_A d_{iA} \quad \text{(no sum on } i\text{)} \tag{2.8.4}$$

In d_{iA}: i = Degree of freedom number; A = Global node number

$$g_i^h = \sum_{A \in \eta_{g_i}} N_A g_{iA} \quad \text{(no sum on } i\text{)} \tag{2.8.5}$$

Let $\boldsymbol{e}_i$ denote the ith Euclidean basis vector for $\mathbb{R}^{n_{sd}}$; $\boldsymbol{e}_i$ has a 1 in slot i and zeros elsewhere. For example

$$(n_{sd} = 2) \qquad \boldsymbol{e}_1 = \begin{Bmatrix} 1 \\ 0 \end{Bmatrix}, \qquad \boldsymbol{e}_2 = \begin{Bmatrix} 0 \\ 1 \end{Bmatrix} \tag{2.8.6}$$

[12] In general, this is taken to mean the maximum number of degrees of freedom per node in the global model. It is possible in practice to have elements with fewer degrees of freedom per node contributing to the model.

$$(n_{sd} = 3) \qquad \boldsymbol{e}_1 = \begin{Bmatrix} 1 \\ 0 \\ 0 \end{Bmatrix}, \qquad \boldsymbol{e}_2 = \begin{Bmatrix} 0 \\ 1 \\ 0 \end{Bmatrix}, \qquad \boldsymbol{e}_3 = \begin{Bmatrix} 0 \\ 0 \\ 1 \end{Bmatrix} \tag{2.8.7}$$

The vector versions of (2.8.4) and (2.8.5) may be defined with the aid of $\boldsymbol{e}_i$, viz.,

$$\boldsymbol{v}^h = v_i^h \boldsymbol{e}_i \tag{2.8.8}$$

$$\boldsymbol{g}^h = g_i^h \boldsymbol{e}_i \tag{2.8.9}$$

Likewise, a typical member $\boldsymbol{w}^h \in \mathcal{V}^h$ has the representation

$$\boldsymbol{w}^h = w_i^h \boldsymbol{e}_i, \qquad w_i^h = \sum_{A \in \eta - \eta_{g_i}} N_A c_{iA} \qquad \text{(no sum on } i\text{)} \tag{2.8.10}$$

Substituting (2.8.4), (2.8.5), and (2.8.8)–(2.8.10) into (2.8.2) and arguing along the lines of Sec. 1.6 results in (verify!)

$$\begin{aligned} \sum_{j=1}^{n_{\text{dof}}} \left(\sum_{B \in \eta - \eta_{g_j}} a(N_A \boldsymbol{e}_i, N_B \boldsymbol{e}_j) d_{jB} \right) &= (N_A \boldsymbol{e}_i, \boldsymbol{\ell}) + (N_A \boldsymbol{e}_i, \boldsymbol{h})_\Gamma \\ &- \sum_{j=1}^{n_{\text{dof}}} \left(\sum_{B \in \eta_{g_j}} a(N_A \boldsymbol{e}_i, N_B \boldsymbol{e}_j) g_{jB} \right), \qquad A \in \eta - \eta_{g_i}, \qquad 1 \le i \le n_{sd} \end{aligned}$$

(2.8.11)[13]

This is equivalent to the matrix equation

[13]For correct interpretation of the meaning of these equations, the sum on j should be taken first. For example, in two dimensions

$$\sum_{j=1}^{2} \left(\sum_{B \in \eta - \eta_{g_j}} a(N_A \boldsymbol{e}_i, N_B \boldsymbol{e}_j) d_{jB} \right) = \sum_{B \in \eta - \eta_{g_1}} a(N_A \boldsymbol{e}_i, N_B \boldsymbol{e}_1) d_{1B} + \sum_{B \in \eta - \eta_{g_2}} a(N_A \boldsymbol{e}_i, N_B \boldsymbol{e}_2) d_{2B}$$

and

$$\sum_{j=1}^{2} \left(\sum_{B \in \eta_{g_j}} (N_A \boldsymbol{e}_i, N_B \boldsymbol{e}_j) g_{jB} \right) = \sum_{B \in \eta_{g_1}} a(N_A \boldsymbol{e}_i, N_B \boldsymbol{e}_1) g_{1B} + \sum_{B \in \eta_{g_2}} a(N_A \boldsymbol{e}_i, N_B \boldsymbol{e}_2) g_{2B}$$

$$\boldsymbol{Kd} = \boldsymbol{F} \tag{2.8.12}$$

where

$$\boldsymbol{K} = [K_{PQ}] \tag{2.8.13}$$

$$\boldsymbol{d} = \{d_Q\} \tag{2.8.14}$$

$$\boldsymbol{F} = \{F_P\} \tag{2.8.15}$$

$$K_{PQ} = a(N_A \boldsymbol{e}_i, N_B \boldsymbol{e}_j) \tag{2.8.16}$$

$$F_P = (N_A \boldsymbol{e}_i, \boldsymbol{f}) + (N_A \boldsymbol{e}_i, \boldsymbol{h})_\Gamma - \sum_{j=1}^{n_{\text{dof}}} \left(\sum_{B \in \eta_{g_j}} a(N_A \boldsymbol{e}_i, N_B \boldsymbol{e}_j) g_{jB} \right) \tag{2.8.17}$$

[13]

in which

$$P = \text{ID}(i, A), \qquad Q = \text{ID}(j, B) \tag{2.8.18}$$

Equation (2.8.16) may be written in more explicit form by using (2.7.26) and noting that (see (2.7.21) and (2.7.22)):

$$\boldsymbol{\epsilon}(N_A \boldsymbol{e}_i) = \boldsymbol{B}_A \boldsymbol{e}_i \tag{2.8.19}$$

where

$$(n_{sd} = 2) \qquad \boldsymbol{B}_A = \begin{bmatrix} N_{A,1} & 0 \\ 0 & N_{A,2} \\ N_{A,2} & N_{A,1} \end{bmatrix} \tag{2.8.20}$$

$$(n_{sd} = 3) \qquad \boldsymbol{B}_A = \begin{bmatrix} N_{A,1} & 0 & 0 \\ 0 & N_{A,2} & 0 \\ 0 & 0 & N_{A,3} \\ 0 & N_{A,3} & N_{A,2} \\ N_{A,3} & 0 & N_{A,1} \\ N_{A,2} & N_{A,1} & 0 \end{bmatrix} \tag{2.8.21}$$

Exercise 1. Verify (2.8.19)–(2.8.21).

With these, (2.8.16) becomes

$$\underbrace{K_{PQ}}_{\substack{\text{Global} \\ \text{equation} \\ \text{numbers}}} = \boldsymbol{e}_i^T \int_\Omega \boldsymbol{B}_A^T \boldsymbol{D} \boldsymbol{B}_B \, d\Omega \, \boldsymbol{e}_j \tag{2.8.22}$$

(A, B: Global node numbers; i, j: Degree of freedom numbers)

and the indices are related by (2.8.18).

Equation (2.8.17) is also amenable to explication. Note that, by (2.7.18)

$$(N_A \boldsymbol{e}_i, \boldsymbol{\ell}) = \int_\Omega N_A \ell_i \, d\Omega \tag{2.8.23}$$

and likewise by (2.7.19)

$$(N_A \boldsymbol{e}_i, \boldsymbol{h})_\Gamma = \int_{\Gamma_{h_i}} N_A h_i \, d\Gamma \qquad \text{(no sum)} \tag{2.8.24}$$

Thus (2.8.17) may be written as

$$F_P = \int_\Omega N_A \ell_i \, d\Omega + \int_{\Gamma_{h_i}} N_A h_i \, d\Gamma - \sum_{j=1}^{n_{\text{dof}}} \left(\sum_{B \in \eta_{g_j}} a(N_A \boldsymbol{e}_i, N_B \boldsymbol{e}_j) g_{jB} \right) \tag{2.8.25}$$ [14]

Now that we have defined $\boldsymbol{K}$, we can establish its fundamental properties. We shall need the following preliminary results.

Let $n_{sd} = 2$ or 3 and let $\boldsymbol{w} : \Omega \to \mathbb{R}^{n_{sd}}$. If $w_{(i,j)} = 0$ ("zero strains"), then $\boldsymbol{w}$ admits the representations:

$$(n_{sd} = 2) \qquad \boldsymbol{w}(\boldsymbol{x}) = \overbrace{\boldsymbol{c}}^{\text{Translation}} + \overbrace{c_3(x_1 \boldsymbol{e}_2 - x_2 \boldsymbol{e}_1)}^{\text{Rotation}} \tag{2.8.26}$$

$$(n_{sd} = 3) \qquad \boldsymbol{w}(\boldsymbol{x}) = \underbrace{\boldsymbol{c}_1}_{\text{Translation}} + \underbrace{\boldsymbol{c}_2 \times \boldsymbol{x}}_{\text{Rotation}} \tag{2.8.27}$$

where

$$\boldsymbol{c} = \begin{Bmatrix} c_1 \\ c_2 \end{Bmatrix} \qquad \boldsymbol{c}_1 = \begin{Bmatrix} c_{11} \\ c_{12} \\ c_{13} \end{Bmatrix} \qquad \boldsymbol{c}_2 = \begin{Bmatrix} c_{21} \\ c_{22} \\ c_{23} \end{Bmatrix} \tag{2.8.28}$$

and c_3 are constants; and $\times$ denotes the vector cross product. Equations (2.8.26) and (2.8.27) define ***infinitesimal rigid-body motions***.

Assumption R

We assume that the homogeneous boundary conditions incorporated into the definition of $\mathcal{V}^h$ *preclude* nontrivial infinitesimal rigid-body motions. In other words, we assume that if $\boldsymbol{w}^h \in \mathcal{V}^h$ is a rigid-body motion, then $\boldsymbol{w}^h$ is identically zero.

Theorem

1. $\boldsymbol{K}$ is symmetric.
2. If Assumption R holds, then $\boldsymbol{K}$ is also positive definite.

Proof of 1. **Symmetry** We may note that symmetry of $\boldsymbol{K}$ follows from (2.8.16) and the symmetry of $a(\cdot, \cdot)$. However, we shall provide an alternative proof in terms of (2.8.22).

[14] See footnote 13 of this chapter.

$$K_{PQ} = \boldsymbol{e}_i^T \int_\Omega \boldsymbol{B}_A^T \boldsymbol{D} \boldsymbol{B}_B \, d\Omega \, \boldsymbol{e}_j$$

$$= \boldsymbol{e}_j^T \int_\Omega \boldsymbol{B}_B^T \boldsymbol{D}^T \boldsymbol{B}_A \, d\Omega \, \boldsymbol{e}_i$$

$$= \boldsymbol{e}_j^T \int_\Omega \boldsymbol{B}_B^T \boldsymbol{D} \boldsymbol{B}_A \, d\Omega \, \boldsymbol{e}_i \qquad \text{(symmetry of } \boldsymbol{D}\text{)}$$

$$= K_{QP}$$ ■

Remark

Note that the symmetry of $\boldsymbol{K}$ followed from the symmetry of $\boldsymbol{D}$, which was a consequence of the major symmetry of the c_{ijkl}'s (see (2.7.3)).

Proof of 2. **Positive definite** (Recall from Sec. 1.9 that we must show (i) $\boldsymbol{c}^T\boldsymbol{K}\boldsymbol{c} \geq 0$ and (ii) $\boldsymbol{c}^T\boldsymbol{K}\boldsymbol{c} = 0$ implies $\boldsymbol{c} = \boldsymbol{0}$.)

Let $w_i^h = \sum_{A\in\eta-\eta_{g_i}} N_A c_{iA}$ be a member of $\mathcal{V}_i^h$. Then $c_P = c_{iA}$, where $P = \text{ID}(i, A)$, $1 \leq P \leq n_{eq}$, defines the components of an n_{eq}-vector $\boldsymbol{c}$.

i.

$$\boldsymbol{c}^T\boldsymbol{K}\boldsymbol{c} = \sum_{P,Q=1}^{n_{eq}} c_P K_{PQ} c_Q$$

$$= \sum_{i,j=1}^{n_{\text{dof}}} \left(\sum_{\substack{A\in\eta-\eta_{g_i} \\ B\in\eta-\eta_{g_j}}} c_{iA}\, a(N_A \boldsymbol{e}_i, N_B \boldsymbol{e}_j) c_{jB} \right) \qquad \text{(definition of } K_{PQ}\text{)}$$

$$= a\left(\sum_{i=1}^{n_{\text{dof}}} \left(\sum_{A\in\eta-\eta_{g_i}} c_{iA} N_A \boldsymbol{e}_i \right), \sum_{j=1}^{n_{\text{dof}}} \left(\sum_{B\in\eta-\eta_{g_j}} c_{jB} N_B \boldsymbol{e}_j \right) \right) \qquad \text{(bilinearity of } a(\cdot\,,\,\cdot)\text{)}$$

$$= a(\boldsymbol{w}^h, \boldsymbol{w}^h) \qquad \text{(definition of } \boldsymbol{w}^h\text{)}$$

$$= \int_\Omega \underbrace{w_{(i,j)}^h c_{ijkl} w_{(k,l)}^h}_{\geq 0} \, d\Omega \qquad \text{(by (2.7.5) and (2.7.17))}$$

$$\geq 0$$

ii. Assume $\boldsymbol{c}^T\boldsymbol{K}\boldsymbol{c} = 0$. By the proof of part (i), we deduce that

$$w_{(i,j)}^h c_{ijkl} w_{(k,l)}^h = 0$$

From (2.7.6), this means that $w_{(i,j)}^h = 0$, and so $\boldsymbol{w}^h$ is an infinitesimal rigid motion. By Assumption R, $\boldsymbol{w}^h = \boldsymbol{0}$, from which it follows that $c_p = 0$; hence $\boldsymbol{c} = \boldsymbol{0}$. ■

Remark

Positive definiteness of $\boldsymbol{K}$ is based upon two requirements: a positive-definiteness condition on the constitutive coefficients and suitable boundary conditions being incorporated into $\mathcal{V}^h$.

2.9 ELASTOSTATICS: ELEMENT STIFFNESS MATRIX AND FORCE VECTOR

As usual, $\boldsymbol{K}$ and $\boldsymbol{F}$ may be decomposed into sums of elemental contributions. These results will be omitted here as the reader should now be familiar with the ideas involved (cf. Sec. 2.5). We will proceed directly to the definitions of $\boldsymbol{k}^e$ and $\boldsymbol{f}^e$:

$$\boldsymbol{k}^e = [k^e_{pq}], \qquad \boldsymbol{f}^e = \{f^e_p\}, \qquad 1 \le p, q \le n_{ee} = n_{ed}n_{en} \tag{2.9.1}$$[15]

$$k^e_{pq} = \boldsymbol{e}_i^T \int_{\Omega^e} \boldsymbol{B}_a^T \boldsymbol{D} \boldsymbol{B}_b \, d\Omega \, \boldsymbol{e}_j, \qquad p = n_{ed}(a-1) + i, \quad q = n_{ed}(b-1) + j \tag{2.9.2}$$

$$(n_{sd} = 2) \qquad \boldsymbol{B}_a = \begin{bmatrix} N_{a,1} & 0 \\ 0 & N_{a,2} \\ N_{a,2} & N_{a,1} \end{bmatrix} \tag{2.9.3}$$

$$(n_{sd} = 3) \qquad \boldsymbol{B}_a = \begin{bmatrix} N_{a,1} & 0 & 0 \\ 0 & N_{a,2} & 0 \\ 0 & 0 & N_{a,3} \\ 0 & N_{a,3} & N_{a,2} \\ N_{a,3} & 0 & N_{a,1} \\ N_{a,2} & N_{a,1} & 0 \end{bmatrix} \tag{2.9.4}$$

and

$$f^e_p = \int_{\Omega^e} N_a f_i \, d\Omega + \int_{\Gamma^e_{h_i}} N_a h_i \, d\Gamma - \sum_{q=1}^{n_{ee}} k_{pq} g^e_q, \qquad \Gamma^e_{h_i} = \Gamma_{h_i} \cap \Gamma^e \quad \text{(no sum on } i\text{)} \tag{2.9.5}$$

where $g^e_q = g^e_{jb} = g_j(\boldsymbol{x}^e_b)$ if g_j is prescribed at node b, and equals zero otherwise.

It is useful for programming purposes to define the ***nodal submatrix***

$$\underbrace{\boldsymbol{k}^e_{ab}}_{n_{ed} \times n_{ed}} = \int_{\Omega^e} \boldsymbol{B}_a^T \boldsymbol{D} \boldsymbol{B}_b \, d\Omega \tag{2.9.6a}$$

From (2.9.2) we see that

[15] n_{ee} stands for the ***number of element equations*** and n_{ed} is the ***number of element degrees of freedom*** (per node). It is possible in practice to have $n_{ed} < n_{\text{dof}}$, although they are usually equal.

$$\boxed{k^e_{pq} = \boldsymbol{e}_i^T \boldsymbol{k}^e_{ab} \boldsymbol{e}_j} \tag{2.9.6b}$$

This means, "the pq-component of $\boldsymbol{k}^e$ is the ij-component of the submatrix $\boldsymbol{k}^e_{ab}$."

By (2.9.1) through (2.9.4), we see that $\boldsymbol{k}^e$ may be written as

$$\boxed{\boldsymbol{k}^e = \int_{\Omega^e} \boldsymbol{B}^T \boldsymbol{D} \boldsymbol{B} \, d\Omega} \tag{2.9.7}$$

where

$$\boldsymbol{B} = [\boldsymbol{B}_1, \boldsymbol{B}_2, \ldots, \boldsymbol{B}_{n_{en}}] \tag{2.9.8}$$

For example, in the case of a two-dimensional (i.e., $n_{sd} = n_{ed} = 2$), four-noded element, $\boldsymbol{k}^e$ takes the form

$$\underbrace{\boldsymbol{k}^e}_{8 \times 8} = \begin{bmatrix} \boldsymbol{k}^e_{11} & \boldsymbol{k}^e_{12} & \boldsymbol{k}^e_{13} & \boldsymbol{k}^e_{14} \\ \boldsymbol{k}^e_{21} & \boldsymbol{k}^e_{22} & \boldsymbol{k}^e_{23} & \boldsymbol{k}^e_{24} \\ \boldsymbol{k}^e_{31} & \boldsymbol{k}^e_{32} & \boldsymbol{k}^e_{33} & \boldsymbol{k}^e_{34} \\ \boldsymbol{k}^e_{41} & \boldsymbol{k}^e_{42} & \boldsymbol{k}^e_{43} & \boldsymbol{k}^e_{44} \end{bmatrix} \tag{2.9.9}$$

In practice, the submatrices above the dashed line are computed and those below, if required, are determined by symmetry.

The global arrays $\boldsymbol{K}$ and $\boldsymbol{F}$ may be formed from the element arrays $\boldsymbol{k}^e$ and $\boldsymbol{f}^e$, respectively, by way of an assembly algorithm as outlined in Sec. 1.14.

Exercise 1. Let

$$\underbrace{\boldsymbol{d}^e}_{n_{ee} \times 1} = \{\boldsymbol{d}^e_a\} = \begin{Bmatrix} \boldsymbol{d}^e_1 \\ \boldsymbol{d}^e_2 \\ \vdots \\ \boldsymbol{d}^e_{n_{en}} \end{Bmatrix} \tag{2.9.10}$$

$$(n_{ed} = 2) \qquad \boldsymbol{d}^e_a = \begin{Bmatrix} d^e_{1a} \\ d^e_{2a} \end{Bmatrix} \tag{2.9.11}$$

$$(n_{ed} = 3) \qquad \boldsymbol{d}^e_a = \begin{Bmatrix} d^e_{1a} \\ d^e_{2a} \\ d^e_{3a} \end{Bmatrix} \tag{2.9.12}$$

where

$$d_{ia}^e = u_i^h(\boldsymbol{x}_a^e) \tag{2.9.13}$$[16]

$\boldsymbol{d}^e$ is called the ***element displacement vector***. Show that the stress vector (see Exercise 4, Sec. 2.7.) at point $\boldsymbol{x} \in \Omega^e$ can be calculated from the formula

$$\boldsymbol{\sigma}(\boldsymbol{x}) = \boldsymbol{D}(\boldsymbol{x})\boldsymbol{B}(\boldsymbol{x})\boldsymbol{d}^e = \boldsymbol{D}(\boldsymbol{x}) \sum_{a=1}^{n_{en}} \boldsymbol{B}_a(\boldsymbol{x})\boldsymbol{d}_a^e \tag{2.9.14}$$

2.10 ELASTOSTATICS: DATA PROCESSING ARRAYS ID, IEN, AND LM

We have already noted that the definition of the ID array must be generalized for the present case as indicated in Sec. 2.8. We must also generalize our definition of the LM array. However, the IEN array remains the same as before.

In the present and fully general cases, the LM array is three-dimensional, with dimensions $n_{ed} \times n_{en} \times n_{el}$, and is defined by

$$\text{LM}(i, a, e) = \text{ID}(i, \text{IEN}(a, e)) \tag{2.10.1}$$

i: Degrees of freedom number; a: Local node number; e: Element number

Alternatively, it is sometimes convenient to think of LM as two-dimensional, with dimensions $n_{ee} \times n_{el}$, viz.,[17]

$$\text{LM}(p, e) = \text{LM}(i, a, e), \qquad p = n_{ed}(a - 1) + i \tag{2.10.2}$$

p: Local equation number; e: Element number

To see how everything works in practice, it is helpful to run through a simple example.

Example 1

Consider the mesh of four-node, rectangular elements illustrated in Fig. 2.10.1. We assume that the local node numbering begins in the lower left-hand corner for each element and proceeds in counterclockwise fashion.

[16] The g-boundary conditions are accounted for in this definition.

[17] The reader knowledgeable in FORTRAN will realize that the internal computer storage of (2.10.1) and (2.10.2) is identical.

This is shown for element 4, which is typical. Four displacement (i.e., "g-type") boundary conditions are specified; namely, the horizontal displacement is specified at nodes 1 and 10, and the vertical displacement is specified at nodes 1 and 3. Since $n_{np} = 12$, $n_{\text{dof}} = n_{ed} = 2$, and 4 displacement degrees of freedom are specified, we have $n_{eq} = 20$. As is usual, we adopt the convention that the global equation numbers run in ascending order with respect to the ascending order of global node numbers.[18] The ID, IEN, and LM arrays are given in Figure 2.10.2. The reader is urged to verify the results.

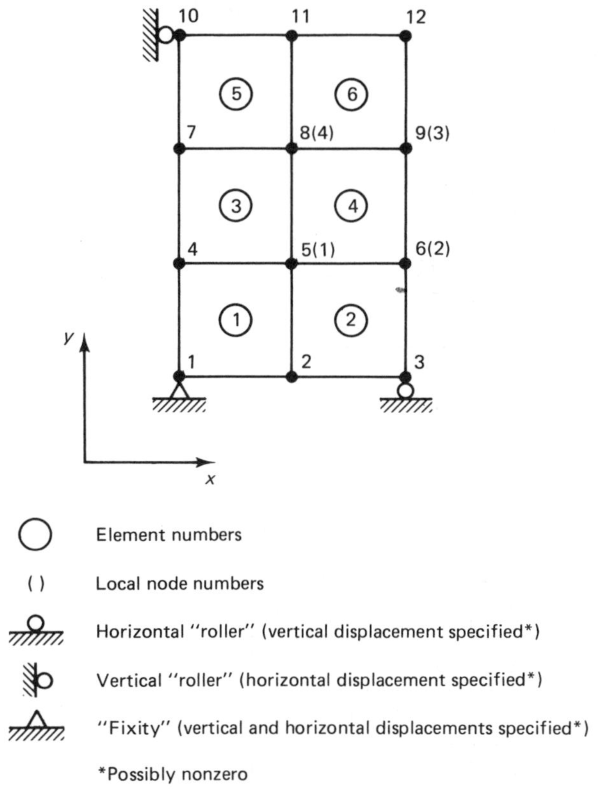

Figure 2.10.1 Mesh of four-node, rectangular, elasticity elements; global and local node numbers, element numbers, and displacement boundary conditions.

In terms of the IEN and LM arrays, a precise definition of the g_p^e's may be given (see (2.9.5)):

$$g_p^e = g_{ia}^e = \begin{cases} 0, & \text{if } \mathrm{LM}(i, a, e) \neq 0 \\ g_{iA}, & \text{where } A = \mathrm{IEN}(a, e), \text{ if } \mathrm{LM}(i, a, e) = 0 \end{cases} \tag{2.10.3}$$

This definition may be easily programmed.

[18] In practice, equation numbers are often renumbered internally to minimize the bandwidth of $\boldsymbol{K}$ and thus decrease storage and solution effort. This is especially important in analyzing large-scale systems involving tens of thousands of equations. An algorithm for reducing bandwidth is presented in [8].

ID array:

Global node numbers (A) ($n_{np} = 12$)

Global degrees of freedom (i)	1	2	3	4	5	6	7	8	9	10	11	12
1	0*	1	3	4	6	8	10	12	14	0	17	19
2	0	2	0	5	7	9	11	13	15	16	18	20

($n_{dof} = 2$) $P = \text{ID}(i, A)$ ($n_{eq} = 20$)

IEN array:

Element numbers (e) ($n_{el} = 6$)

Local node numbers (a)	1	2	3	4	5	6
1	1	2	4	5	7	8
2	2	3	5	6	8	9
3	5	6	8	9	11	12
4	4	5	7	8	10	11

($n_{en} = 4$)

$A = \text{IEN}(a, e)$

LM array:

Element numbers (e) ($n_{el} = 6$)

p	a	i	1	2	3	4	5	6
1	1	1	0*	1	4	6	10	12
2		2	0	2	5	7	11	13
3	2	1	1	3	6	8	12	14
4		2	2	0	7	9	13	15
5	3	1	6	8	12	14	17	19
6		2	7	9	13	15	18	20
7	4	1	4	6	10	12	0	17
8		2	5	7	11	13	16	18

($n_{ee} = 8$)

Element degree of freedom numbers (i, where $1 \le i \le n_{ed}$ and here $n_{ed} = 2$)

Local node numbers (a)

Local equation numbers ($p = n_{ed}(a - 1) + i$)

$$P = \text{LM}(p, e) = \text{LM}(i, a, e) = \text{ID}(i, \text{IEN}(a, e))$$

*Displacement boundary conditions are denoted by zeros.

Figure 2.10.2 ID, IEN, and LM arrays for the mesh of Figure 2.10.1.

Example 2

As a final example, we consider a typical four-node, elasticity element in some large mesh; see Fig. 2.10.3. We assume the pertinent entries of the ID array are given as follows:

$$\left.\begin{aligned} \mathrm{ID}(1,\ 32) &= 0 \\ \mathrm{ID}(2,\ 32) &= 0 \\ \mathrm{ID}(1,\ 59) &= 115 \\ \mathrm{ID}(2,\ 59) &= 116 \\ \mathrm{ID}(1,\ 164) &= 0 \\ \mathrm{ID}(2,\ 164) &= 325 \\ \mathrm{ID}(1,\ 168) &= 332 \\ \mathrm{ID}(2,\ 168) &= 333 \end{aligned}\right\} \tag{2.10.4}$$

The entries of IEN follow from Fig. 2.10.3:

$$\left.\begin{aligned} \mathrm{IEN}(1,\ e) &= 164 \\ \mathrm{IEN}(2,\ e) &= 32 \\ \mathrm{IEN}(3,\ e) &= 168 \\ \mathrm{IEN}(4,\ e) &= 59 \end{aligned}\right\} \tag{2.10.5}$$

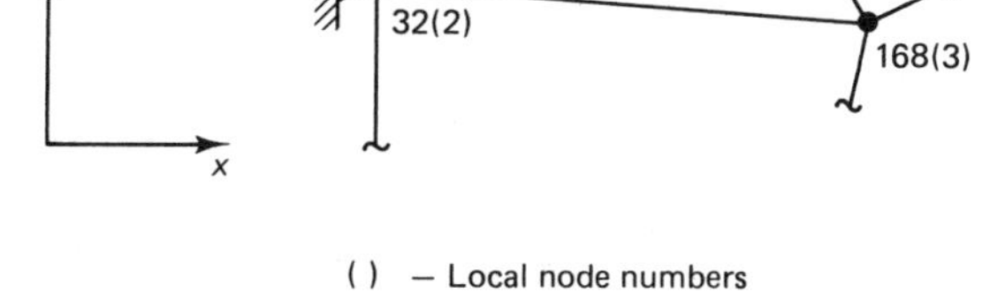

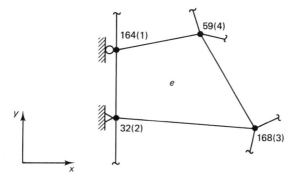

Figure 2.10.3 Typical four-node elasticity element; global and local node numbers.

Combining (2.10.4) and (2.10.5), by way of (2.10.1), yields entries of the LM array:

$$\left.\begin{aligned}
\mathrm{LM}(1, 1, e) &= 0\\
\mathrm{LM}(2, 1, e) &= 325\\
\mathrm{LM}(1, 2, e) &= 0\\
\mathrm{LM}(2, 2, e) &= 0\\
\mathrm{LM}(1, 3, e) &= 332\\
\mathrm{LM}(2, 3, e) &= 333\\
\mathrm{LM}(1, 4, e) &= 115\\
\mathrm{LM}(2, 4, e) &= 116
\end{aligned}\right\} \tag{2.10.6}$$

The contribution to the global arrays may be deduced from LM:

Stiffness (due to symmetry, only the upper triangular portion need be assembled.)

$$\left.\begin{aligned}
K_{115,\,115} &\leftarrow K_{115,\,115} + k^e_{77}\\
K_{115,\,116} &\leftarrow K_{115,\,116} + k^e_{78}\\
K_{115,\,325} &\leftarrow K_{115,\,325} + k^e_{72}\\
K_{115,\,332} &\leftarrow K_{115,\,332} + k^e_{75}\\
K_{115,\,333} &\leftarrow K_{115,\,333} + k^e_{76}\\
\\
K_{116,\,116} &\leftarrow K_{116,\,116} + k^e_{88}\\
K_{116,\,325} &\leftarrow K_{116,\,325} + k^e_{82}\\
K_{116,\,332} &\leftarrow K_{116,\,332} + k^e_{85}\\
K_{116,\,333} &\leftarrow K_{116,\,333} + k^e_{86}\\
\\
K_{325,\,325} &\leftarrow K_{325,\,325} + k^e_{22}\\
K_{325,\,332} &\leftarrow K_{325,\,332} + k^e_{25}\\
K_{325,\,333} &\leftarrow K_{325,\,333} + k^e_{26}\\
\\
K_{332,\,332} &\leftarrow K_{332,\,332} + k^e_{55}\\
K_{332,\,333} &\leftarrow K_{332,\,333} + k^e_{56}\\
\\
K_{333,\,333} &\leftarrow K_{333,\,333} + k^e_{66}
\end{aligned}\right\} \tag{2.10.7}$$

Force

$$\left.\begin{aligned}
F_{115} &\leftarrow F_{115} + f^e_7\\
F_{116} &\leftarrow F_{116} + f^e_8\\
F_{325} &\leftarrow F_{325} + f^e_2\\
F_{332} &\leftarrow F_{332} + f^e_5\\
F_{333} &\leftarrow F_{333} + f^e_6
\end{aligned}\right\} \tag{2.10.8}$$

where

$$f_p^e = \cdots - \sum_{q=1}^{n_{ee}} k_{pq}^e g_q^e \tag{2.10.9}$$

(We have omitted the first two terms in the right-hand side of (2.9.5) in writing (2.10.9).) In the present example, only g_1^e, g_3^e and g_4^e may be nonzero. Therefore (2.10.9) may be simplified to

$$f_p^e = \cdots -k_{p1}^e g_1^e - k_{p3}^e g_3^e - k_{p4}^e g_4^e \tag{2.10.10}$$

The multiplications indicated in (2.10.10) are only performed in practice if the g_p^e's are nonzero. A schematic representation of the contributions of $\boldsymbol{k}^e$ and $\boldsymbol{f}^e$ to $\boldsymbol{K}$ and $\boldsymbol{F}$ is shown in Figure 2.10.4.

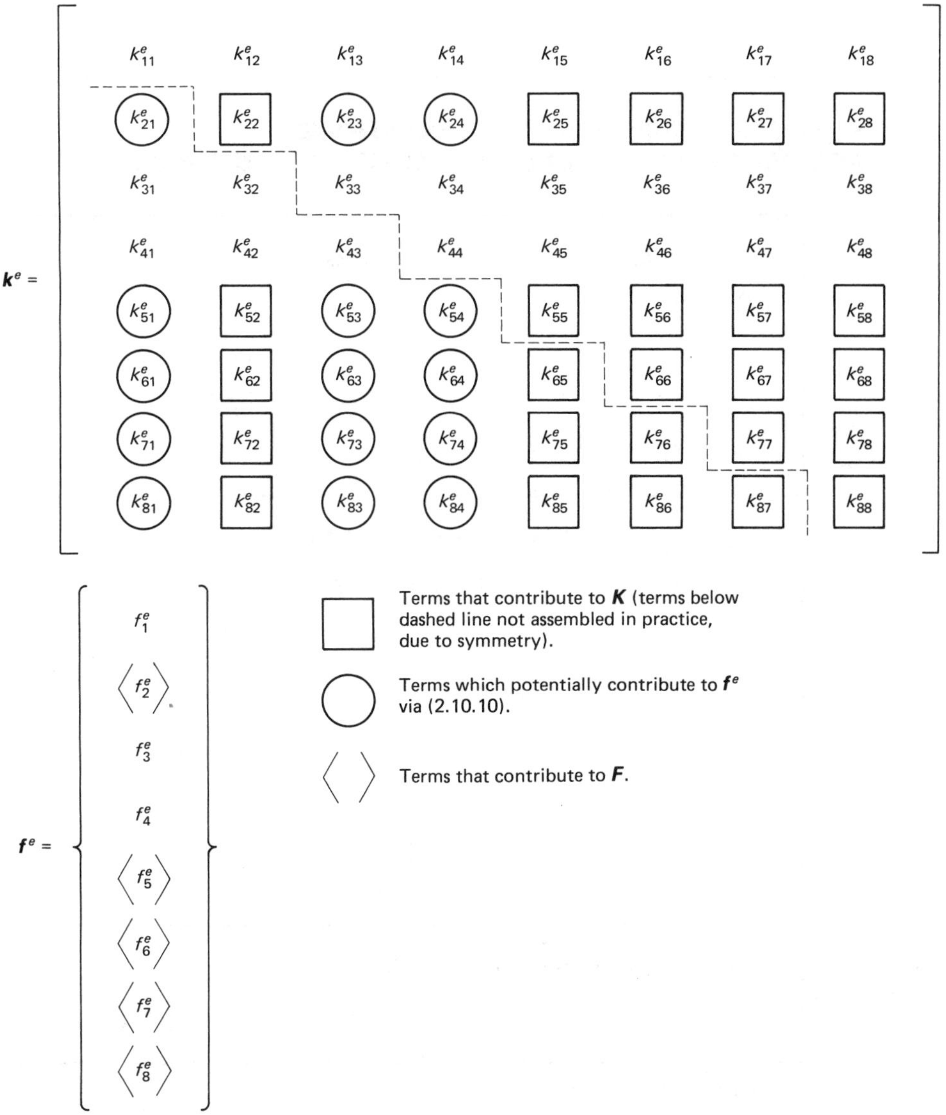

Figure 2.10.4 Contributions of elasticity element in Example 2 to global arrays.

Exercise 1. Consider a two-dimensional elastostatic boundary-value problem. Set up the ID, IEN, and LM arrays for the following mesh of four-node quadrilaterals:

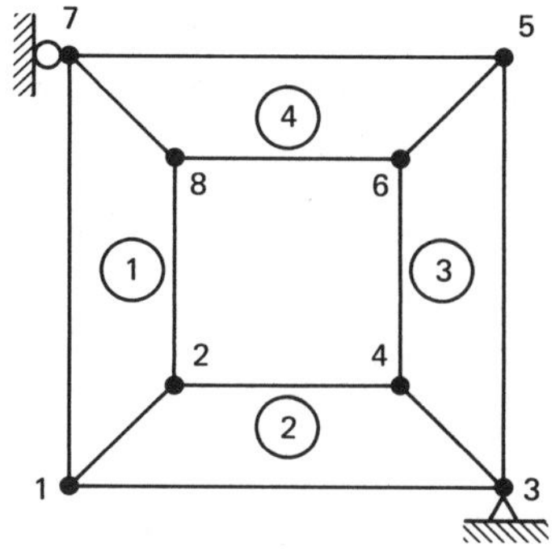

2.11 SUMMARY OF IMPORTANT EQUATIONS FOR PROBLEMS CONSIDERED IN CHAPTERS 1 AND 2

1. *Classical Linear Elastostatics*

(S)

$$\sigma_{ij,j} + f_i = 0 \qquad \text{on } \Omega$$

$$u_i = g_i \qquad \text{on } \Gamma_{g_i}$$

$$\sigma_{ij} n_j = h_i \qquad \text{on } \Gamma_{h_i}$$

where $\sigma_{ij} = c_{ijkl}\epsilon_{kl} = c_{ijkl}u_{(k,l)}$

(W)[19] Find $\boldsymbol{u} \in \mathcal{S}$, such that $\forall\ \boldsymbol{w} \in \mathcal{V}$

$$a(\boldsymbol{w}, \boldsymbol{u}) = (\boldsymbol{w}, \boldsymbol{f}) + (\boldsymbol{w}, \boldsymbol{h})_\Gamma$$

where

$$a(\boldsymbol{w}, \boldsymbol{u}) = \int_\Omega w_{(i,j)} c_{ijkl} u_{(k,l)}\, d\Omega$$

$$(\boldsymbol{w}, \boldsymbol{f}) = \int_\Omega w_i f_i\, d\Omega$$

$$(\boldsymbol{w}, \boldsymbol{h})_\Gamma = \sum_{i=1}^{n_{sd}} \left(\int_{\Gamma_{h_i}} w_i h_i\, d\Omega \right)$$

(G) Find $\boldsymbol{v}^h \in \mathcal{V}^h$, such that $\forall\ \boldsymbol{w}^h \in \mathcal{V}^h$

$$a(\boldsymbol{w}^h, \boldsymbol{v}^h) = (\boldsymbol{w}^h, \boldsymbol{f}) + (\boldsymbol{w}^h, \boldsymbol{h})_\Gamma - a(\boldsymbol{w}^h, \boldsymbol{g}^h)$$

[19] The notation $\forall$ means "for all."

$(M)\quad \boldsymbol{Kd} = \boldsymbol{F}$, where $\boldsymbol{K} = \underset{e=1}{\overset{n_{el}}{\mathbf{A}}}(\boldsymbol{k}^e)$, $\boldsymbol{F} = \boldsymbol{F}_{\text{nodal}} + \underset{e=1}{\overset{n_{el}}{\mathbf{A}}}(\boldsymbol{f}^e)$[20]

$$k^e_{pq} = \boldsymbol{e}_i^T \boldsymbol{k}^e_{ab} \boldsymbol{e}_j, \qquad \boldsymbol{k}^e_{ab} = \int_{\Omega^e} \boldsymbol{B}_a^T \boldsymbol{D} \boldsymbol{B}_b \, d\Omega$$

$$f^e_p = \int_{\Omega^e} N_a f_i \, d\Omega + \int_{\Gamma^e_{h_i}} N_a h_i \, d\Gamma - \sum_{q=1}^{n_{el}} k^e_{pq} g^e_q \qquad \text{(no sum on } i\text{)}$$

$$p = n_{ed}(a - 1) + i$$

$$q = n_{ed}(b - 1) + j$$

Stress at a point: $\boldsymbol{\sigma}(\boldsymbol{x}) = \boldsymbol{D}(\boldsymbol{x}) \sum_{a=1}^{n_{en}} \boldsymbol{B}_a(x) d^e_a$

2. *Classical Linear Heat Conduction*

(S)

$$q_{i,i} = f \quad \text{in } \Omega$$

$$u = g \quad \text{on } \Gamma_g$$

$$-q_i n_i = h \quad \text{on } \Gamma_h$$

where $q_i = \kappa_{ij} u_{,j}$

(W) Find $u \in \mathcal{S}$, such that $\forall\, w \in \mathcal{V}$

$$a(w, u) = (w, f) + (w, h)_\Gamma$$

where

$$a(w, u) = \int_\Omega w_{,i} \kappa_{ij} u_{,j} \, d\Omega$$

$$(w, f) = \int_\Omega w f \, d\Omega$$

$$(w, h)_\Gamma = \int_\Omega w h \, d\Gamma$$

(G) Find $v^h \in \mathcal{S}^h$, such that $\forall\, w^h \in \mathcal{V}^h$

$$a(w^h, v^h) = (w^h, f) + (w^h, h)_\Gamma - a(w^h, g^h)$$

[20] In defining $\boldsymbol{F}$ we have added to the element contributions the term $\boldsymbol{F}_{\text{nodal}}$, which is a vector of nodal applied forces. The reason for this is that it is often easier in practice to directly input concentrated forces at nodes rather than go through the element-by-element form and assemble procedure. The expression for $\boldsymbol{F}$ then emphasizes that both modes of constructing $\boldsymbol{F}$ are to be accommodated in the computer implementation of problems of this type.

(M) $\quad \boldsymbol{Kd} = \boldsymbol{F}$, where $\boldsymbol{K} = \mathop{\mathbf{A}}_{e=1}^{n_{el}}(\boldsymbol{k}^e), \qquad \boldsymbol{F} = \boldsymbol{F}_{\text{nodal}} + \mathop{\mathbf{A}}_{e=1}^{n_{el}}(\boldsymbol{f}^e)$[20]

$$k_{ab}^e = \int_{\Omega^e} \boldsymbol{B}_a^T \boldsymbol{D} \boldsymbol{B}_b \, d\Omega$$

$$f_a^e = \int_{\Omega^e} N_a \ell \, d\Omega + \int_{\Gamma_h^e} N_a h \, d\Gamma - \sum_{b=1}^{n_{el}} k_{ab}^e g_b^e$$

Heat flux vector at a point: $\quad \boldsymbol{q}(\boldsymbol{x}) = -\boldsymbol{D}(\boldsymbol{x}) \sum_{a=1}^{n_{en}} \boldsymbol{B}_a(\boldsymbol{x}) d_a^e$

3. *One-Dimensional Model Problem*

(S)

$$\begin{aligned} u_{,xx} + \ell &= 0 && \text{on } \Omega =]0, 1[\\ u(1) &= g && (\Gamma_g = \{1\}) \\ -u_{,x}(0) &= h && (\Gamma_h = \{0\}) \end{aligned}$$

(W) $\quad$ Find $u \in \mathcal{S}$, such that $\forall \; w \in \mathcal{V}$

$$a(w, u) = (w, \ell) + wh(0)$$

where

$$a(w, u) = \int_0^1 w_{,x} u_{,x} \, dx$$

$$(w, \ell) = \int_0^1 w \ell \, dx$$

(G) $\quad$ Find $v^h \in \mathcal{V}^h$, such that $\forall \; w^h \in \mathcal{V}^h$

$$a(w^h, v^h) = (w^h, \ell) + w^h(0)h - a(w^h, g^h)$$

(M) $\quad \boldsymbol{Kd} = \boldsymbol{F}$, where $\boldsymbol{K} = \mathop{\mathbf{A}}_{e=1}^{n_{el}}(\boldsymbol{k}^e), \qquad \boldsymbol{F} = \boldsymbol{F}_{\text{nodal}} + \mathop{\mathbf{A}}_{e=1}^{n_{el}}(\boldsymbol{f}^e)$[20]

$$k_{ab}^e = \int_{\Omega^e} N_{a,x} N_{b,x} \, dx$$

$$f_a^e = \int_{\Omega^e} N_a \ell \, dx + \begin{cases} h\delta_{a1} & e = 1 \\ 0 & e = 2, \ldots, n_{el} - 1 \\ -g k_{2a}^e & e = n_{el} \end{cases}$$

2.12 AXISYMMETRIC FORMULATIONS AND ADDITIONAL EXERCISES

Axisymmetric formulations are expressed in terms of cylindrical coordinates:

$$x_1 = r = \text{the radial coordinate}$$

$$x_2 = z = \text{the axial coordinate}$$

$$x_3 = \theta = \text{the circumferential coordinate}$$

The basic hypothesis of axisymmetry is that all functions under consideration are independent of θ. That is, they are functions of r and z only. Thus three-dimensional problem classes are reduced to two-dimensional ones.

Heat Conduction

The axisymmetric formulation for heat conduction is almost identical to the two-dimensional Cartesian case considered previously. The only difference is that a factor of $2\pi r$ needs to be included in each integrand of the variational equation to account for the correct volumetric weighting (e.g., $d\Omega = 2\pi r\, dr\, dz$ replaces $d\Omega = dx_1\, dx_2$). Since the constant 2π is common to all terms, it may be cancelled throughout if desired.

Elasticity

The displacement components in cylindrical coordinates are:

$$u_1 = u_r = \text{the radial displacement}$$

$$u_2 = u_z = \text{the axial displacement}$$

$$u_3 = u_\theta = \text{the circumferential displacement}$$

In addition to the basic hypothesis of axisymmetry, we further assume that $u_\theta = 0$, and thus

$$\epsilon_{r\theta} = \epsilon_{z\theta} = 0 \tag{2.12.1}$$

Note that $\epsilon_{\theta\theta} = u_r/r$ and therefore it is generally not zero. The constitutive moduli are assumed to be such that the preceding kinematical hypotheses result in

$$\sigma_{r\theta} = \sigma_{z\theta} = 0 \tag{2.12.2}$$

The preceding assumptions lead to what is called the ***torsionless axisymmetric case.*** This formulation is similar to but somewhat more complicated than the two-dimensional cases of plane strain and plane stress. Here there are four nonzero components of stress and strain:

$$\boldsymbol{\sigma} = \begin{Bmatrix} \sigma_{11} \\ \sigma_{22} \\ \sigma_{12} \\ \sigma_{33} \end{Bmatrix} = \begin{Bmatrix} \sigma_{rr} \\ \sigma_{zz} \\ \sigma_{rz} \\ \sigma_{\theta\theta} \end{Bmatrix} \tag{2.12.3}$$

$$\boldsymbol{\epsilon} = \begin{Bmatrix} \epsilon_{11} \\ \epsilon_{22} \\ 2\epsilon_{12} \\ \epsilon_{33} \end{Bmatrix} = \begin{Bmatrix} \epsilon_{rr} \\ \epsilon_{zz} \\ 2\epsilon_{rz} \\ \epsilon_{\theta\theta} \end{Bmatrix} \tag{2.12.4}$$

The ordering emanates from the following generalization of Table 2.7.1:

I / J	i / k	j / l
1	1	1
3	1	2
3	2	1
2	2	2
4	3	3

The $\boldsymbol{D}$ array takes on the following form:

$$\boldsymbol{D} = [D_{IJ}] = \underbrace{\begin{bmatrix} \boldsymbol{D}_{33} & \boldsymbol{D}_3 \\ \boldsymbol{D}_3^T & D_{44} \end{bmatrix}}_{4 \times 4} \tag{2.12.5}$$

$$\boldsymbol{D}_{33} = \begin{bmatrix} D_{11} & D_{12} & D_{13} \\ & D_{22} & D_{23} \\ \text{symmetric} & & D_{33} \end{bmatrix} \tag{2.12.6}$$

$$\boldsymbol{D}_3 = \begin{Bmatrix} D_{14} \\ D_{24} \\ D_{34} \end{Bmatrix} \tag{2.12.7}$$

where $D_{IJ} = c_{ijkl}$, in which the indices are related by the table. The $\boldsymbol{B}_a$-matrix takes on the form

$$\boldsymbol{B}_a = \begin{bmatrix} N_{a,1} & 0 \\ 0 & N_{a,2} \\ N_{a,2} & N_{a,1} \\ \hline \dfrac{N_a}{r} & 0 \end{bmatrix} \tag{2.12.8}$$

Again, a factor of $2\pi r$ needs to be included in all integrands.

Remark

The ***plane strain*** case may be obtained from the axisymmetric formulation by

i. Ignoring the $2\pi r$ factors; and
ii. Ignoring the fourth row of $\boldsymbol{B}_a$ and the fourth row and column of $\boldsymbol{D}$.

Furthermore, the ***plane stress*** case may be similarly obtained if, in addition, $\boldsymbol{D}_{33}$ is replaced by

$$\boxed{\boldsymbol{D}_{33} - \boldsymbol{D}_3 D_{44}^{-1} \boldsymbol{D}_3^T} \tag{2.12.9}$$

which directly follows from the plane stress condition, $\sigma_{44} = 0$. (References [6] and [7] may be consulted for further elaboration on the physical ideas.) Sometimes (2.12.9) is referred to as the ***statically condensed*** elastic coefficient matrix.

Consequently, in programming the axisymmetric case, for a small amount of additional effort both plane strain and plane stress may also be included.

Exercise 1. Under the assumption of isotropy, show that $\boldsymbol{D}_{33}$ is the same as the $\boldsymbol{D}$-matrix in (2.7.34). Furthermore, show that $D_{44} = \lambda + 2\mu$ and

$$\boldsymbol{D}_3 = \begin{Bmatrix} \lambda \\ \lambda \\ 0 \end{Bmatrix} \tag{2.12.10}$$

Exercise 2. Verify that for the isotropic case, (2.12.9) achieves a similar end to the procedure described in Remark 9 of Sec. 2.7.

Exercise 3. Consider the one-dimensional model problem discussed previously. Obtain exact expressions for $\boldsymbol{f} = \{f_a^e\}$, $a = 1, 2$, for the following cases (ignore g and h contributions):

i. $\ell = \text{constant}$.
ii. $\ell = \delta(x - \bar{x})$, the delta function, where $x_1^e \le \bar{x} \le x_2^e$. Specialize for the cases $\bar{x} = x_b^e$ and $\bar{x} = (x_1^e + x_2^e)/2$.

Solution

$$f_a^e = \int_{x_1^e}^{x_2^e} N_a(x)\ell(x)\, dx$$

i. $$f_a^e = \ell \int_{x_1^e}^{x_2^e} N_a(x)\, dx = \frac{\ell h^e}{2} \underbrace{\int_{-1}^{+1} N_a(\xi)\, d\xi}_{1}$$

$$\boldsymbol{f}^e = \frac{\ell h^e}{2} \begin{Bmatrix} 1 \\ 1 \end{Bmatrix}$$

ii. $f_a^e = \int_{x_1^e}^{x_2^e} N_a(x)\delta(x - \bar{x})\, dx = N_a(\bar{x})$

For $x = x_b^e$,

$$f_a^e = N_a(\bar{x}) = N_a(x_b^e) = \delta_{ab} \qquad \text{(Kronecker delta)}$$

$$f^e = \begin{Bmatrix} \delta_{1b} \\ \delta_{2b} \end{Bmatrix}$$

For $\bar{x} = (x_1^e + x_2^e)/2$,

$$f_a^e = N_a(\bar{x}) = N_a\left(\frac{x_1^e + x_2^e}{2}\right) = \frac{1}{2}$$

Therefore,

$$f^e = \frac{1}{2}\begin{Bmatrix} 1 \\ 1 \end{Bmatrix}$$

Exercise 4. Consider the boundary-value problem for classical linear elastostatics discussed previously. In the linearized theory of ***small displacements superposed upon large,*** the stiffness term in the variational equation,

$$\int_\Omega w_{(i,j)} c_{ijkl} u_{(k,l)}\, d\Omega$$

is replaced by

$$\int_\Omega w_{i,j} d_{ijkl} u_{k,l}\, d\Omega$$

where

$$d_{ijkl} = c_{ijkl} + \delta_{ik}\sigma_{jl}^0$$
$$\sigma_{jl}^0 = \sigma_{lj}^0$$

and the σ_{jl}^0's (i.e., ***initial stresses***) are given functions of $\boldsymbol{x} \in \Omega$. It follows from the symmetries of c_{ijkl} and σ_{jl}^0 that

$$d_{ijkl} = d_{klij}$$

Assume $n_{sd} = 2$. An index-free formulation of the stiffness term is given by

$$\int_\Omega \begin{Bmatrix} w_{1,1} \\ w_{2,2} \\ w_{1,2} + w_{2,1} \\ w_{1,2} - w_{2,1} \end{Bmatrix}^T \underbrace{\boldsymbol{D}}_{4 \times 4} \begin{Bmatrix} u_{1,1} \\ u_{2,2} \\ u_{1,2} + u_{2,1} \\ u_{1,2} - u_{2,1} \end{Bmatrix} d\Omega$$

which leads to the following definition of the element stiffness matrix:

$$k_{pq}^e = \boldsymbol{e}_i^T \int_{\Omega^e} \underbrace{\boldsymbol{B}_a^T}_{2 \times 4} \underbrace{\boldsymbol{D}}_{4 \times 4} \underbrace{\boldsymbol{B}_b}_{4 \times 2} d\Omega\, \boldsymbol{e}_j$$

Set up $\boldsymbol{B}_a$ in terms of the shape function N_a. Define the components of $\boldsymbol{D}$ in terms of the d_{ijkl}'s. (The σ_{jl}^0-contribution to the stiffness is sometimes called the ***initial-stress stiffness matrix.*** It is important to account for it in the solution of many nonlinear problems.)

Exercise 5. Let Ω be a region in $\mathbb{R}^2$ and let its boundary $\Gamma = \overline{\Gamma_1 \cup \Gamma_2 \cup \Gamma_3 \cup \Gamma_4}$ where $\Gamma_1, \ldots, \Gamma_4$ are nonoverlapping subregions of Γ. Let $\boldsymbol{n}$ be the unit outward normal vector to Γ such that $\boldsymbol{s}$ and $\boldsymbol{n}$ form a right-hand rule basis; see Fig. 2.12.1. Consider the following boundary-value problem in classical linear elastostatics: Given $f_i : \Omega \to \mathbb{R}$; $g_i : \Gamma_1 \to \mathbb{R}$; $h_i : \Gamma_2 \to \mathbb{R}$; g_n and $h_s : \Gamma_3 \to \mathbb{R}$; and g_s and $h_n : \Gamma_4 \to \mathbb{R}$; find $u_i : \overline{\Omega} \to \mathbb{R}$ such that

$$\sigma_{ij,j} + f_i = 0 \qquad \text{in } \Omega$$

$$u_i = g_i \qquad \text{on } \Gamma_1$$

$$\sigma_{ij} n_j = h_i \qquad \text{on } \Gamma_2$$

$$\left.\begin{aligned} u_i n_i &= g_n \\ \sigma_{ij} n_j s_i &= h_s \end{aligned}\right\} \quad \text{on } \Gamma_3 \quad \begin{pmatrix}\text{normal displacement} \\ \text{tangential traction}\end{pmatrix}$$

$$\left.\begin{aligned} u_i s_i &= g_s \\ \sigma_{ij} n_j n_i &= h_n \end{aligned}\right\} \quad \text{on } \Gamma_4 \quad \begin{pmatrix}\text{tangential displacement} \\ \text{normal traction}\end{pmatrix}$$

where $\sigma_{ij} = c_{ijkl} u_{(k,l)}$. Establish a weak formulation for this problem in which all "g-type" boundary conditions are essential and all "h-type" boundary conditions are natural. State all requirements on the spaces $\mathcal{S}$ and $\mathcal{V}$. *Hint:* $\boldsymbol{w} = w_n \boldsymbol{n} + w_s \boldsymbol{s}$; i.e., $w_i = w_n n_i + w_s s_i$.

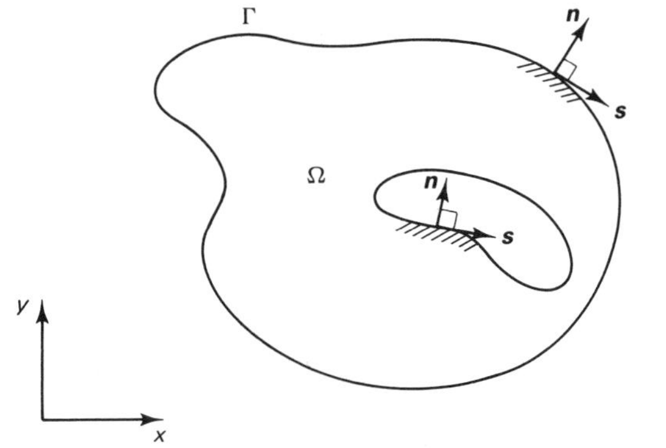

Figure 2.12.1

Exercise 6. In practice, it is often useful to generalize the constitutive equation of classical elasticity to the form

$$\sigma_{ij} = c_{ijkl}(\epsilon_{kl} - \epsilon_{kl}^0) + \sigma_{ij}^0 \tag{2.12.11}$$

where ϵ_{ij}^0 and σ_{ij}^0 are the ***initial strain*** and ***initial stress***, both given functions of $\boldsymbol{x}$. The initial strain term may be used to represent thermal expansion effects by way of

$$\epsilon_{kl}^0 = -\theta c_{kl} \tag{2.12.12}$$

where θ is the ***temperature*** and the c_{kl}'s are the ***thermal expansion coefficients*** (both given functions). Clearly, (2.12.11) will in no way change the stiffness matrix. However there will be additional contributions to f_p^e. Generalize the definition of f_p^e to account for these additional terms.

Solution We begin with the weak form

$$\int_\Omega w_{(i,j)} \sigma_{ij} \, d\Omega = \int_\Omega w_i f_i \, d\Omega + \sum_{i=1}^{n_{sd}} \left(\int_{\Gamma_{h_i}} w_i h_i \, d\Gamma \right)$$

Substituting (2.12.11) and (2.12.12) into the weak form leads to

$$\int_\Omega w_{(i,j)} c_{ijkl} \epsilon_{kl}\, d\Omega = \int_\Omega w_i \ell_i\, d\Omega + \sum_{i=1}^{n_{sd}} \left(\int_{\Gamma_{h_i}} w_i h_i\, d\Gamma \right)$$

↑ As before ↑

↓ New contributions to right-hand side: ↓

$$+ \int_\Omega w_{(i,j)} c_{ijkl} \epsilon^0_{kl}\, d\Omega$$

$$- \int_\Omega w_{(i,j)} \sigma^0_{ij}\, d\Omega$$

from which the additional terms in f^e_p may be deduced:

$$f^e_p = \cdots + e^T_i \int_{\Omega^e} B^T_a D \theta c\, d\Omega - e^T_i \int_{\Omega^e} B^T_a \sigma^0\, d\Omega$$

(↑ θ: Anywhere inside integral)

where

$$c = \begin{Bmatrix} c_{11} \\ c_{22} \\ 2c_{12} \end{Bmatrix}; \qquad \sigma^0 = \begin{Bmatrix} \sigma^0_{11} \\ \sigma^0_{22} \\ \sigma^0_{12} \end{Bmatrix}; \ \ldots$$

(↑ $2c_{12}$: Without loss of generality, we may assume symmetry, i.e., $c_{12} + c_{21} = 2c_{12}$)

Exercise 7. Consider the following boundary-value problem:

$$u_{,xx} - p(p-1)x^{p-2} = 0, \qquad 0 < x < 1$$

$$-u_{,x}(0) = 0$$

$$u(1) = 1$$

where p is a given constant.

i. Obtain the exact solution to this problem for $p = 5$. Sketch.
ii. State the weak formulation of the problem.
iii. State the Galerkin formulation.
iv. State the matrix formulation.
v. Solve the matrix problem assuming $p = 5$ and using the piecewise linear finite element space for the following cases:
   a. one element
   b. two equal-length elements
vi. Compare the exact value of $u_{,x}(1)$ with the approximate values computed in part v. Explain why it is impossible for these results to compare favorably.

Exercise 8. In heat conduction, it is often of interest to accurately calculate the boundary heat flux over a portion of the boundary where temperature is specified. Suppose we use the usual Galerkin finite element formulation to calculate the temperature. However, instead of calculating the heat flux in the usual way (i.e., by differentiating the temperature), we introduce a *post-processing* which derives from the following weak formulation:

Find $u \in \mathcal{S}$ and $h \in L_2(\Gamma_g)$ such that for all $w \in \mathcal{V}$,

$$-\int_\Omega w_{,i} q_i \, d\Omega = \int_\Omega w f \, d\Omega + \int_{\Gamma_h} w h \, d\Gamma + \int_{\Gamma_g} w h \, d\Gamma$$

where h is the *unknown* heat flux on Γ_g

(*Note:* In this formulation, it is *not* assumed that $w = 0$ on Γ_g!)

i. Show, in addition to the usual differential equations and boundary conditions, that

$$h = -q_i n_i \quad \text{on} \quad \Gamma_g$$

arises naturally from the new weak formulation.

ii. State the Galerkin and matrix formulations corresponding to the new weak formulation assuming h is approximated in the usual way, namely

$$h^h(\boldsymbol{x}) = \sum_{A \in \eta_g} N_A(\boldsymbol{x}) h_A$$

(*Hint:* The equations governing the termperature are unchanged.)

iii. Specialize this formulation to the one-dimensional problem described in Exercise 7 and calculate the boundary flux at $x = 1$ by the new procedure (cf. parts v and vi of Exercise 7).

(*Hint:* The new method should produce exact results for these cases.)

iv. Develop a counterpart of the new formulation for elasticity. That is, introduce the ith component of traction as an independent unknown on Γ_{g_i} and carefully state the weak formulation.

v. *Prove* that the new method is exact for the one-dimensional model problem of Chapter 1.

REFERENCES

Section 2.2

1. J. E. Marsden and A. J. Tromba, *Vector Calculus*. San Francisco: W. H. Freeman, 1976.

Section 2.4

2. G. Strang and G. J. Fix, *An Analysis of the Finite Element Method*. Englewood Cliffs, N.J.: Prentice-Hall, 1973.

Section 2.7

3. G. Duvaut and J. L. Lions, *Les Inéquations en Mécanique et en Physique*. Paris: Dunod, 1972.

4. G. Fichera, "Existence Theorems in Elasticity," in *Handbuch der Physik, Volume VIa/2, Mechanics of Solids II,* ed. C. Truesdell. New York: Springer-Verlag, 1972.

5. M. Gurtin, "The Linear Theory of Elasticity," in *Handbuch der Physik, Volume VIa/2, Mechanics of Solids II,* ed. C. Truesdell. New York: Springer-Verlag, 1972.

6. I. S. Sokolnikoff, *Mathematical Theory of Elasticity* (2nd ed.). New York: McGraw-Hill, 1956.

7. S. Timoshenko and J. N. Goodier, *Theory of Elasticity* (3rd ed.). New York: McGraw-Hill, 1969.

Section 2.10

8. N. E. Gibbs, W. G. Poole, Jr., and P. K. Stockmeyer, "An Algorithm for Reducing the Bandwidth and Profile of a Sparse Matrix," *SIAM Journal of Numerical Analysis,* 13 (1976), 236–250.

3

Isoparametric Elements and Elementary Programming Concepts

3.1 PRELIMINARY CONCEPTS

We wish to define the shape functions in such a way that, as the finite element mesh is refined, the approximate Galerkin solution converges to the exact solution. The following question arises: What conditions must the shape functions satisfy so that this property is guaranteed? We shall be content, for the time being, to state sufficient conditions for convergence. These conditions are possessed by the most prevalent and important finite element shape functions. However, we note that convergent elements can be constructed from shape functions which do not satisfy all these requirements. Nevertheless these conditions may be considered basic in that they provide the simplest criteria to ensure convergence for a wide class of problems.

The basic convergence requirements are that the shape functions be

C1. smooth (i.e., at least C^1) on each element interior, Ω^e;

C2. continuous across each element boundary Γ^e; and

C3. complete.

Remarks

1. Conditions C1 and C2 guarantee that first derivatives of the shape functions have, at worst, finite jumps across the element interfaces; see Fig. 3.1.1. This ensures that all integrals necessary for the computation of element arrays (see Sec. 2.11) are well defined, since at most first derivatives appear in the integrands. If we permit finite discontinuities in the shape functions on element boundaries, the derivatives possess delta functions (cf. Sec. 1.10) and we are unable to make sense out of the squares of these quantities that would appear in the stiffness integrands.

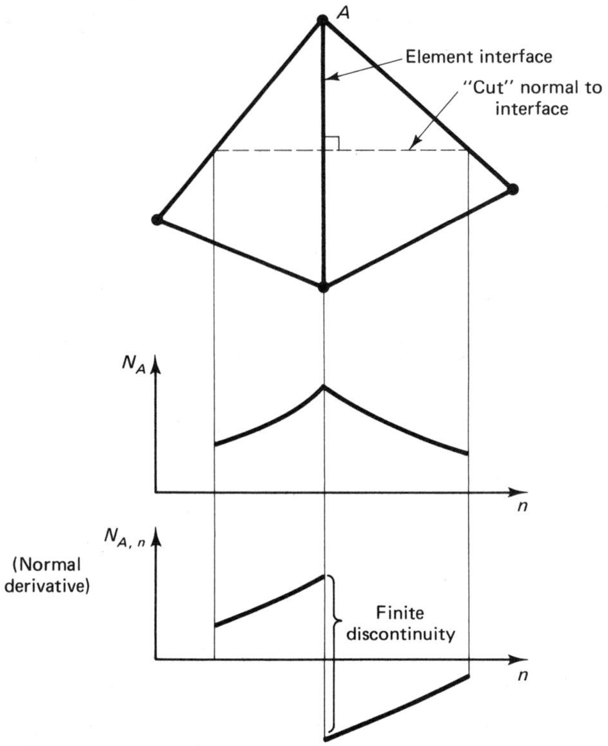

Figure 3.1.1 Example of a shape function that satisfies conditions C1 and C2.

2. Shape functions that satisfy C1 and C2 are of class $C^0(\overline{\Omega})$. Finite elements constructed from $C^0(\overline{\Omega})$ shape functions are often referred to as ***C^0-elements.***

3. Theories such as Bernoulli-Euler beam theory, which require higher-order derivatives in the stiffness integrands, necessitate strengthening condition C1 to C^2-continuity on Ω^e and condition C2 to C^1-continuity across Γ^e. Shape functions of this type are of class $C^1(\overline{\Omega})$ and lead to so-called ***C^1-elements.*** Examples of $C^1(\overline{\Omega})$ shape functions are the Hermite cubics discussed in Sec. 1.16.

4. If the stiffness integrands involve derivatives of order m, Condition C1 should be strengthened to C^m-continuity on Ω^e and Condition C2 should be strengthened to C^{m-1}-continuity across Γ^e. Finite elements that satisfy this property are called ***conforming***, or ***compatible.*** (The use of elements that violate this property, ***non-conforming*** or ***incompatible elements***, is, however, common. We say more about this later.)

5. It may be noted that C2 is equivalent to requiring that each function $u^h \in \mathcal{S}^h$ be continuous across Γ^e. In specific cases it is often easier to establish this fact directly rather than prove it for individual shape functions.

Completeness

To illustrate the completeness requirement, let us take the case of heat conduction in which the eth element interpolation function may be written

$$u^h = \sum_{a=1}^{n_{en}} N_a d_a^e \tag{3.1.1}$$

where $d_a^e = u^h(\boldsymbol{x}_a^e)$. Let $n_{sd} = 3$. In this case, the shape functions are said to be ***complete*** if

$$d_a^e = c_0 + c_1 x_a^e + c_2 y_a^e + c_3 z_a^e \tag{3.1.2}$$

implies that

$$u^h(\boldsymbol{x}) = c_0 + c_1 x + c_2 y + c_3 z \tag{3.1.3}$$

where $c_0, \ldots, c_3$ are arbitrary constants. In two dimensions, we omit the z-terms in (3.1.2) and (3.1.3). In words, *completeness requires that the element interpolation function is capable of exactly representing an arbitrary linear polynomial when the nodal degrees of freedom are assigned values in accordance with it.*

Completeness is a plausible requirement as the following argument indicates: As the finite element mesh is further and further refined, the exact solution and its derivatives approach constant values over each element domain. To ensure that these constant values are representable, the shape functions must contain all constant and linear monomials. This argument was originally given in [1] and has subsequently been proved to be the key mathematical idea for proving convergence theorems for finite element approximations (e.g., see [2]).

For elasticity , the requirement is essentially the same. In this case we may write

$$u_i^h = \sum_{a=1}^{n_{en}} N_a d_{ia}^e \tag{3.1.4}$$

where $d_{ia} = u_i^h(\boldsymbol{x}_a)$. The N_a are complete if

$$d_{ia}^e = c_0 + c_1 x_a^e + c_2 y_a^e + c_3 z_a^e \tag{3.1.5}$$

implies

$$u_i^h(\boldsymbol{x}) = c_0 + c_1 x + c_2 y + c_3 z \tag{3.1.6}$$

Remarks

6. In elasticity, the presence of all monomials through linear terms means that an element may exactly represent all rigid motions (i.e., rigid translations and rotations) and constant strain states. We thus often speak of completeness in terms of the ability to represent rigid motions and constant strains (e.g., see [3], p. 33).

7. For theories involving mth derivatives in the stiffness integrands, completeness must be strengthened to mth-order polynomials.

Example 1

The piecewise linear finite element space introduced in Chapter 1 satisfies the convergence conditions C1–C3. C1 and C2 follow from the definition (see Sec. 1.8). For C3 we proceed as follows: Assume $d_a^e = c_0 + c_1 x_a^e$ and substitute in $u^h = \Sigma_{a=1}^2 N_a d_a^e$, viz.,

$$\begin{aligned} u^h &= \sum_{a=1}^{2} N_a d_a^e \\ &= \sum_{a=1}^{2} N_a (c_0 + c_1 x_a^e) \\ &= \left(\sum_{a=1}^{2} N_a\right) c_0 + \left(\sum_{a=1}^{2} N_a x_a^e\right) c_1 \end{aligned} \tag{3.1.7}$$

Thus to establish completeness we must show that

$$\sum_{a=1}^{2} N_a = 1 \tag{3.1.8}$$

and

$$\sum_{a=1}^{2} N_a x_a^e = x \tag{3.1.9}$$

By (1.12.5), $N_a(\xi) = (1 + (-1)^a \xi)/2$, from which (3.1.8) follows. Equation (3.1.9) was established previously (see (1.12.6)).

3.2 BILINEAR QUADRILATERAL ELEMENT

This element is attributed to Taig [4]. The rectangular version was proposed earlier by Argyris in [5].

The domain of a straight-edged quadrilateral element is defined by the locations of its four nodal points $\boldsymbol{x}_a^e$, $a = 1, \ldots, 4$ in the $\mathbb{R}^2$-plane. We assume the nodal points are labeled in ascending order corresponding to the counterclockwise direction (see Fig. 3.2.1). We seek a change of coordinates which maps the given quadrilateral into the biunit square, as depicted in Fig. 3.2.1. The biunit square is sometimes called the ***parent domain.*** The coordinates of a point

$$\boldsymbol{\xi} = \begin{Bmatrix} \xi \\ \eta \end{Bmatrix} \tag{3.2.1}$$

in the biunit square are to be related to the coordinates of a point

$$\boldsymbol{x} = \begin{Bmatrix} x \\ y \end{Bmatrix} \tag{3.2.2}$$

in Ω^e by mappings of the form

$$x(\xi, \eta) = \sum_{a=1}^{4} N_a(\xi, \eta) x_a^e \tag{3.2.3}$$

$$y(\xi, \eta) = \sum_{a=1}^{4} N_a(\xi, \eta) y_a^e \tag{3.2.4}$$

ξ and η are sometimes called the ***natural coordinates.*** A more succinct representation of the above formulas is

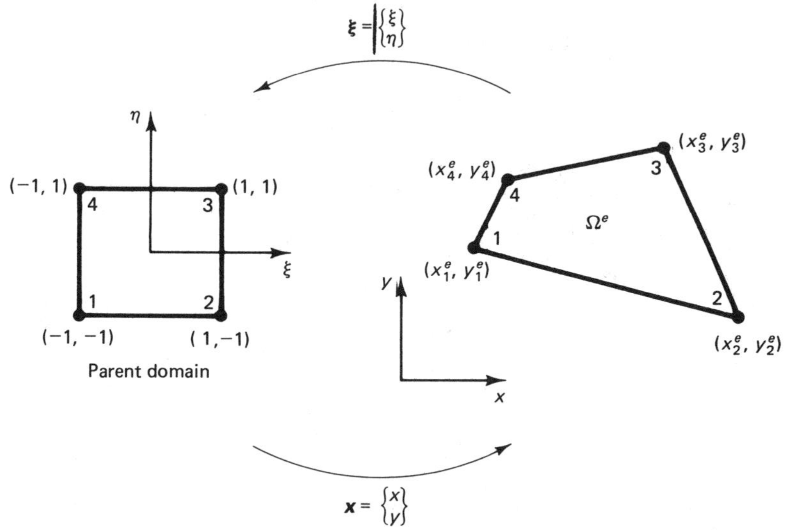

Figure 3.2.1 Bilinear quadrilateral element domain and local node ordering.

$$x(\boldsymbol{\xi}) = \sum_{a=1}^{4} N_a(\boldsymbol{\xi})\boldsymbol{x}_a^e \tag{3.2.5}$$

We obtain the functions N_a by first *assuming* the "bilinear" expansions

$$x(\xi, \eta) = \alpha_0 + \alpha_1\xi + \alpha_2\eta + \alpha_3\xi\eta \tag{3.2.6}$$

$$y(\xi, \eta) = \beta_0 + \beta_1\xi + \beta_2\eta + \beta_3\xi\eta \tag{3.2.7}$$

where the α's and β's are parameters to be determined; and second, by stipulating that (3.2.6) and (3.2.7) satisfy the (respective) conditions

$$x(\xi_a, \eta_a) = x_a^e \tag{3.2.8}$$

$$y(\xi_a, \eta_a) = y_a^e \tag{3.2.9}$$

where ξ_a and η_a are defined in Table 3.2.1.

TABLE 3.2.1 Nodal coordinates in $\boldsymbol{\xi}$-space

a	ξ_a	η_a
1	−1	−1
2	1	−1
3	1	1
4	−1	1

Conditions (3.2.8) and (3.2.9) impose restrictions on the functions N_a; namely,

$$N_a(\boldsymbol{\xi}_b) = \delta_{ab} \tag{3.2.10}$$[1]

To see this we can combine (3.2.3) and (3.2.8), viz.,

$$x_b^e = x(\xi_b, \eta_b) = \sum_{a=1}^{4} N_a(\xi_b, \eta_b) x_a^e \tag{3.2.11}$$

which holds only if $N_a(\xi_b, \eta_b) = \delta_{ab}$. This same restriction may be deduced by combining (3.2.4) and (3.2.9). Equations (3.2.6) through (3.2.9) lead to the following matrix equations:

$$\begin{Bmatrix} x_1^e \\ x_2^e \\ x_3^e \\ x_4^e \end{Bmatrix} = \begin{bmatrix} 1 & -1 & -1 & 1 \\ 1 & 1 & -1 & -1 \\ 1 & 1 & 1 & 1 \\ 1 & -1 & 1 & -1 \end{bmatrix} \begin{Bmatrix} \alpha_0 \\ \alpha_1 \\ \alpha_2 \\ \alpha_3 \end{Bmatrix} \tag{3.2.12}$$

$$\begin{Bmatrix} y_1^e \\ y_2^e \\ y_3^e \\ y_4^e \end{Bmatrix} = \begin{bmatrix} 1 & -1 & -1 & 1 \\ 1 & 1 & -1 & -1 \\ 1 & 1 & 1 & 1 \\ 1 & -1 & 1 & -1 \end{bmatrix} \begin{Bmatrix} \beta_0 \\ \beta_1 \\ \beta_2 \\ \beta_3 \end{Bmatrix} \tag{3.2.13}$$

In each case the coefficient matrix is the same. Solving these for the α's and β's, substituting the results into (3.2.6) and (3.2.7), and comparing with (3.2.3) and (3.2.4) reveals that

$$\boxed{N_a(\boldsymbol{\xi}) = N_a(\xi, \eta) = \tfrac{1}{4}(1 + \xi_a \xi)(1 + \eta_a \eta)} \tag{3.2.14}$$

Note that this is precisely the product of one-dimensional shape functions derived previously (see (1.12.5)).

Exercise 1. Verify (3.2.14).

Observe that coordinate lines in $\boldsymbol{\xi}$-space (i.e., lines such that ξ = constant or η = constant) are mapped into straight lines in $\boldsymbol{x}$-space. In particular, boundary lines are mapped into boundary lines (see Fig. 3.2.2).

Now we shall *assume* that the element functions are given by similar expansions in terms of the shape functions.

Heat conduction

$$u^h(\boldsymbol{\xi}) = \sum_{a=1}^{4} N_a(\boldsymbol{\xi}) d_a^e \tag{3.2.15}$$

[1]Equation (3.2.10) is sometimes referred to as the ***interpolation property*** since it implies that the assumed expansions (e.g., (3.2.6) and (3.2.7)) interpolate the nodal data (e.g., (3.2.8) and (3.2.9)).

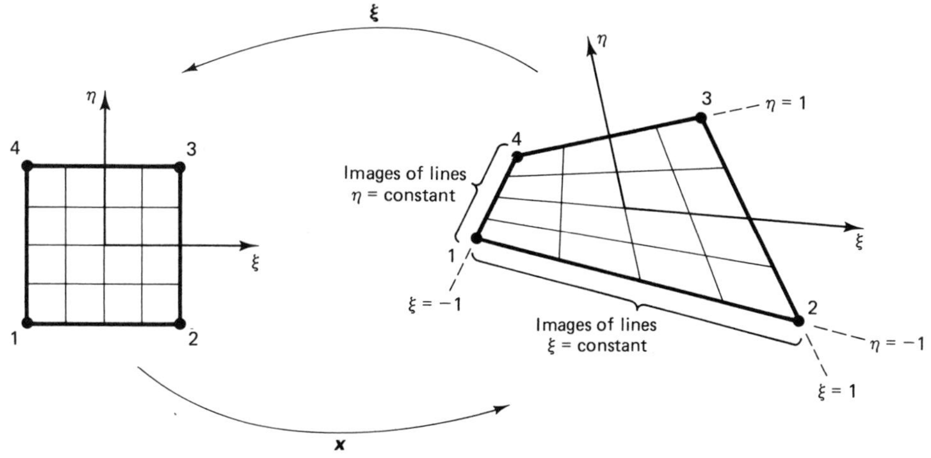

Figure 3.2.2

Elastostatics

$$u_i^h(\boldsymbol{\xi}) = \sum_{a=1}^{4} N_a(\boldsymbol{\xi}) d_{ia}^e \tag{3.2.16}$$

Henceforce, it will be sufficient to consider only (3.2.15). The case of elastostatics may be deduced by viewing (3.2.15) as any one of the n_{sd} component functions of (3.2.16).

(**C1**) **Smoothness on Ω^e.** It can be shown that N_a is a smooth function of x and y, if all interior angles formed by two adjacent edges are less than 180°; see Fig. 3.2.3. (Note that N_a is always a smooth function of ξ and η; see (3.2.14). When N_a is viewed as a function of x and y (i.e., $N_a(\xi(x, y), \eta(x, y))$), to ascertain smoothness we must consider the inverse functions $\xi(x, y)$ and $\eta(x, y)$, which may not be well defined if the element is too distorted. Figure 3.2.3 illustrates when this occurs for the four-node element. This issue is discussed from a more general perspective in the following section.)

(**C2**) **Continuity across Γ^e.** Let us examine a typical shape function N_a. For the moment, assume $a = 1$ or 2. The behavior of N_a along the edge connecting nodes 1 and 2 may be determined by substituting $\eta = -1$ into (3.2.14). The result is

$$N_a(\xi, -1) = \frac{1 + \xi_a \xi}{2}, \qquad a = 1, 2 \tag{3.2.17}$$

From Table 3.2.1, $\xi_1 = -1$ and $\xi_2 = +1$; thus the right-hand side of (3.2.17) is the familiar linear shape function of the one-dimensional theory (see Fig. 3.2.4). By virtue of the fact that (3.2.17) is typical of all four edges and by (3.2.10), we conclude

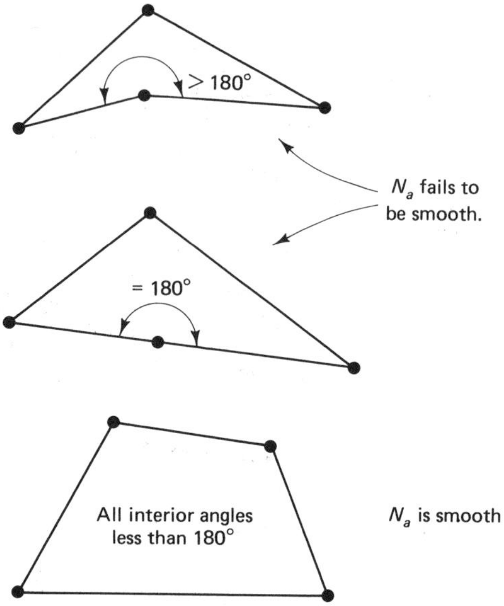

Figure 3.2.3 Smoothness criterion for bilinear shape functions.

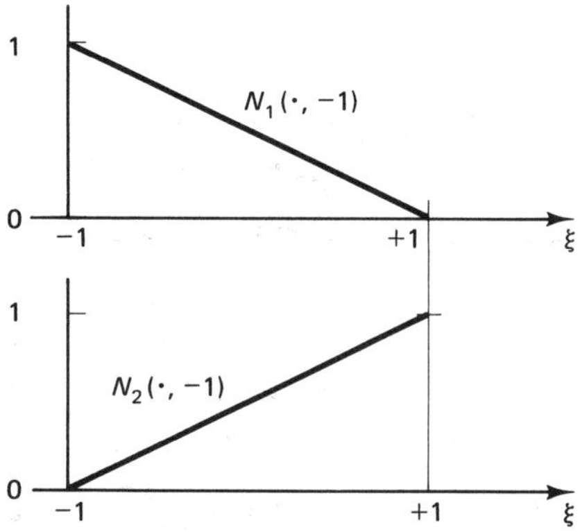

Figure 3.2.4 Bilinear shape functions along edge connecting nodes 1 and 2.

that N_a appears as in Fig. 3.2.5. N_a is said to be a *hyperbolic paraboloid*. Similar considerations for adjacent elements indicate that N_A is continuous and appears as a "tent" whose "pole" emanates from node A; see Fig. 3.2.6 for a typical situation at an internal node. Consequently, $\mathcal{S}^h$ consists of continuous functions as each member is a linear combination of the N_A's (cf. (2.4.3)–(2.4.5)). It is instructive to argue this fact from a slightly different point of view.

Consider the behavior of (3.2.15) along the boundary segment connecting $\boldsymbol{x}_1^e$ to $\boldsymbol{x}_2^e$. By (3.2.17), this may be written

$$u^h(\xi, -1) = \sum_{a=1}^{2} \tfrac{1}{2}(1 + \xi_a \xi) d_a^e \tag{3.2.18}$$

There are two important observations to be made about this result:

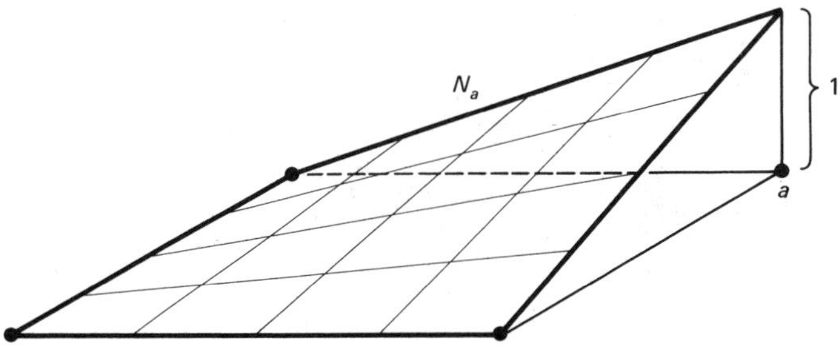

Figure 3.2.5 Bilinear shape function.

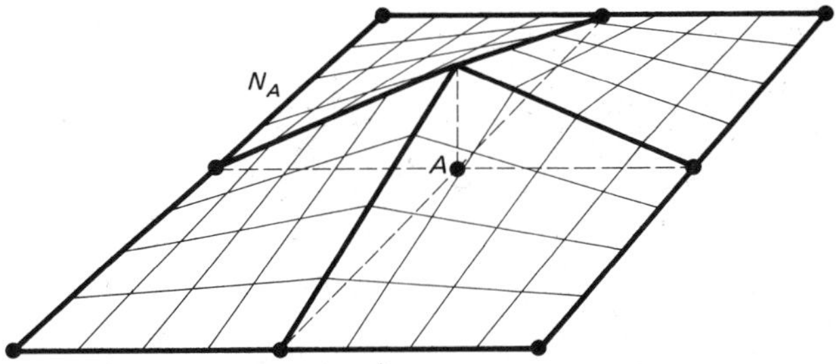

Figure 3.2.6 Global shape function constructed from element bilinear shape functions.

1. The behavior of u^h along $[\boldsymbol{x}_1^e, \boldsymbol{x}_2^e]$ is determined solely by the nodal values of u^h at the nodes $\boldsymbol{x}_1^e$ and $\boldsymbol{x}_2^e$.
2. The variation of u^h is linear in the natural coordinate ξ (see Fig. 3.2.7).

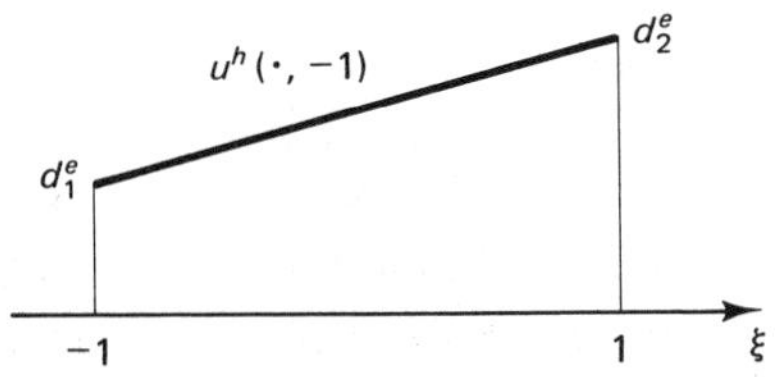

Figure 3.2.7 Graph of u^h along boundary segment joining nodes $\boldsymbol{x}_1^e$ and $\boldsymbol{x}_2^e$.

The same conclusion can be drawn for any other element of this type whose boundary contains $[\boldsymbol{x}_1^e, \boldsymbol{x}_2^e]$. Thus we see that continuity of u^h is automatic across element interfaces.

(C3) Completeness

$$
\begin{aligned}
u^h &= \sum_{a=1}^{n_{en}} N_a d_a^e \\
&= \sum_{a=1}^{n_{en}} N_a (c_0 + c_1 x_a^e + c_2 y_a^e) \\
&= \left(\sum_{a=1}^{n_{en}} N_a\right) c_0 + \underbrace{\left(\sum_{a=1}^{n_{en}} N_a x_a^e\right)}_{= x \text{ by } (3.2.3)} c_1 + \underbrace{\left(\sum_{a=1}^{n_{en}} N_a y_a^e\right)}_{= y \text{ by } (3.2.4)} c_2
\end{aligned}
$$

Thus it remains only to prove that the shape functions sum to value 1 at each point. This may be done by explicit calculation, viz.,

$$\begin{aligned}\sum_{a=1}^{n_{en}} N_a(\xi, \eta) &= \tfrac{1}{4}(1 - \xi)(1 - \eta) + \tfrac{1}{4}(1 + \xi)(1 - \eta) \\ &\quad + \tfrac{1}{4}(1 + \xi)(1 + \eta) + \tfrac{1}{4}(1 - \xi)(1 + \eta), \qquad \text{(by (3.2.14))} \\ &= 1 \qquad (3.2.20)\end{aligned}$$

3.3 ISOPARAMETRIC ELEMENTS

The isoparametric concept is generally attributed to Taig [6] and Irons [7].

Let $\square$ denote a parent domain in $\boldsymbol{\xi}$-space (e.g., the closed, biunit square in $\mathbb{R}^2$).

Definition 1. Let $\boldsymbol{x}: \square \to \overline{\Omega}^e$ be of the form

$$\boldsymbol{x}(\boldsymbol{\xi}) = \sum_{a=1}^{n_{en}} N_a(\boldsymbol{\xi})\boldsymbol{x}_a^e \qquad (3.3.1)$$

If the element interpolation function u^h can be written as

$$u^h(\boldsymbol{\xi}) = \sum_{a=1}^{n_{en}} N_a(\boldsymbol{\xi})d_a^e \qquad (3.3.2)$$

the element is said to be ***isoparametric.***

The key point to observe in the definition is that the shape functions which define (3.3.1) *also serve to define* (3.3.2).

The bilinear quadilateral, derived in the previous section, is an example of an isoparametric element.

We shall now discuss some fundamental mathematical properties of mappings. Based on this, we shall argue that convergence condition C1 is generally achieved for isoparametric elements. Furthermore, a criterion, which may be used in a computer program to determine whether or not individual elements satisfy C1, emanates from the discussion.

Convergence Condition C1

Definition 2. A mapping $\boldsymbol{x}: \square \to \overline{\Omega}^e \subset \mathbb{R}^{n_{sd}}$ is said to be ***one-to-one*** if for each pair of points $\boldsymbol{\xi}^{(1)}, \boldsymbol{\xi}^{(2)} \in \square$ such that $\boldsymbol{\xi}^{(1)} \neq \boldsymbol{\xi}^{(2)}$, then $\boldsymbol{x}(\boldsymbol{\xi}^{(1)}) \neq \boldsymbol{x}(\boldsymbol{\xi}^{(2)})$.

In words, this statement means that two different points of $\square$ do not get mapped into the same point in $\overline{\Omega}^e$.

Definition 3. $\boldsymbol{x} : \square \to \overline{\Omega}^e$ is said to be ***onto*** if $\overline{\Omega}^e = \boldsymbol{x}(\square)$ (i.e., each point in $\overline{\Omega}^e$ is the image of a point in $\square$ under the mapping $\boldsymbol{x}$).

Definition 4. Let $\boldsymbol{x} : \square \rightarrow \overline{\Omega}^e$ be a differentiable mapping. The determinant of the derivative, denoted by $j = \det(\partial \boldsymbol{x}/\partial \boldsymbol{\xi})$, is called the ***Jacobian determinant.***

The Jacobian determinants in two and three dimensions, respectively, are given explicitly by

$$j = \det\begin{bmatrix} x_{,\xi} & x_{,\eta} \\ y_{,\xi} & y_{,\eta} \end{bmatrix} \tag{3.3.3}$$

$$j = \det\begin{bmatrix} x_{,\xi} & x_{,\eta} & x_{,\zeta} \\ y_{,\xi} & y_{,\eta} & y_{,\zeta} \\ z_{,\xi} & z_{,\eta} & z_{,\zeta} \end{bmatrix} \tag{3.3.4}$$

Remark

1. It is a consequence of the inverse function theorem (e.g., see [8]) that if $\boldsymbol{x}$ is

i. one-to-one;
ii. onto;
iii. C^k, $k \geq 1$; and if
iv. $j(\boldsymbol{\xi}) > 0$ for all $\boldsymbol{\xi} \in \square$;

then the inverse mapping $\boldsymbol{\xi} = \boldsymbol{x}^{-1} : \overline{\Omega}^e \rightarrow \square$ exists and is C^k.

Proposition 1. Let the mapping defined by (3.3.1) satisfy (i) through (iv). Then the smoothness requirement (C1) is satisfied.

Proof. By virtue of the hypotheses, $N_a = N_a(\boldsymbol{\xi})$ is also a C^1 function. By Remark 1, $\boldsymbol{\xi} = \boldsymbol{\xi}(\boldsymbol{x})$ is also C^1. Thus $N_a(\boldsymbol{x}) = N_a(\boldsymbol{\xi}(\boldsymbol{x}))$ is a C^1 function of $\boldsymbol{x}$. (This last fact may be proved with the aid of the chain rule.) ■

Remarks

2. In practice, the mappings $\boldsymbol{x} : \square \rightarrow \overline{\Omega}^e$ usually satisfy (i) through (iv). However, there is one exception of practical importance. It is concerned with the technique of element "degeneration," in which nodes are coalesced. The simplest example of this procedure, in which two nodes of the bilinear quadilateral are coalesced to form a triangle, is presented in the next section. Further examples are presented in subsequent sections. When degeneration is performed, the Jacobian determinant vanishes at certain nodal points within the element. Away from these points it is positive, and the mapping $\boldsymbol{\xi} = \boldsymbol{\xi}(\boldsymbol{x})$ remains smooth (i.e., C1 is satisfied). For reasons that will be apparent later on, it is not usually required to calculate derivatives at these points.

3. In actual computations the sign of the Jacobian determinant is monitored at special points to ensure condition (iv) is satisfied. If a zero or negative value is encountered, computations are terminated. Generally this is an indication of an input-data error or an overly distorted element.

Convergence Condition (C3)

Proposition 2. If $\Sigma_{a=1}^{n_{en}} N_a = 1$, then completeness condition C3 is satisfied for isoparametric elements.

Proof. (We shall prove the assertion for the three-dimensional case.)

$$
\begin{aligned}
u^h &= \sum_{a=1}^{n_{en}} N_a d_a^e && \text{(by (3.3.2))} \\
&= \sum_{a=1}^{n_{en}} N_a(c_0 + c_1 x_a^e + c_2 y_a^e + c_3 z_a^e) \\
&= c_0\left(\sum_{a=1}^{n_{en}} N_a\right) + c_1\left(\sum_{a=1}^{n_{en}} N_a x_a^e\right) + c_2\left(\sum_{a=1}^{n_{en}} N_a y_a^e\right) + c_3\left(\sum_{a=1}^{n_{en}} N_a z_a^e\right) \\
&= c_0\left(\sum_{a=1}^{n_{en}} N_a\right) + c_1 x + c_2 y + c_3 z && \text{(by (3.3.1))}
\end{aligned}
$$

The condition that the shape functions sum to one, assumed in Proposition 2, is easily checked on a case-by-case basis. ■

The only remaining convergence condition is C2, the continuity requirement on Γ^e. This condition can be verified once the construction of the global shape functions from the element shape functions is explicated. It happens that if this procedure is done in the "obvious" way, continuity is achieved. In the sequel we shall consider this issue on a case-by-case basis.

Summary. The importance of the isoparametric concept is that the three basic convergence conditions are virtually automatic. In addition, isoparametric elements may be designed to take on convenient shapes for practical analysis, including curved boundaries, and lend themselves to concise computer implementation.

3.4 LINEAR TRIANGULAR ELEMENT; AN EXAMPLE OF "DEGENERATION"

The linear triangle was one of the first finite elements conceived (see Courant [9] and Turner et al. [10]) and is still widely used. In the structural mechanics literature it is often referred to as the constant stress/strain triangle, or simply the CST. We shall derive it from the bilinear quadrilateral by coalescing nodes 3 and 4; see Fig. 3.4.1. Specifically, we set $\boldsymbol{x}_4^e = \boldsymbol{x}_3^e$ in (3.2.5) and define new shape functions for the triangle N_a', $a = 1, 2, 3$, as follows:

$$
\begin{aligned}
\boldsymbol{x} &= \sum_{a=1}^{4} N_a \boldsymbol{x}_a^e \\
&= N_1 \boldsymbol{x}_a^e + N_2 \boldsymbol{x}_2^e + (N_3 + N_4)\boldsymbol{x}_3^e = \sum_{a=1}^{3} N_a' \boldsymbol{x}_a^e
\end{aligned} \tag{3.4.1}
$$

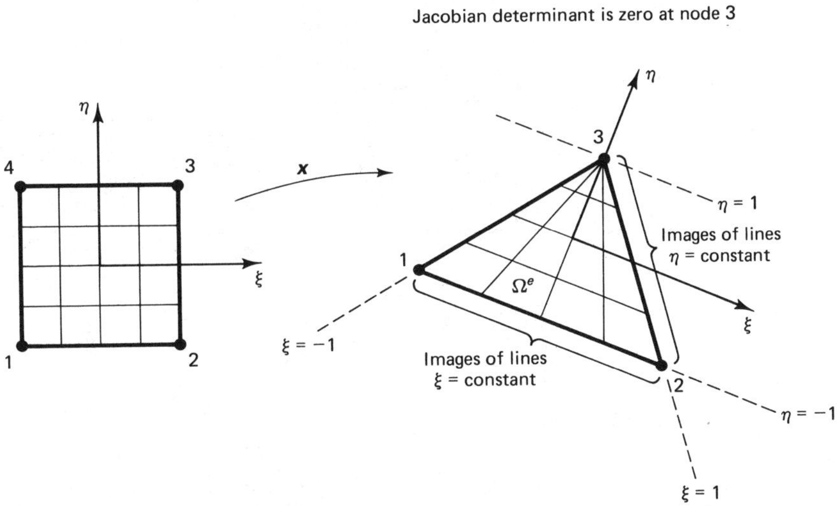

Figure 3.4.1

where

$$N'_a = \begin{cases} N_a = \frac{1}{4}[1 + (-1)^a \xi](1 - \eta) & a = 1, 2 \\ N_3 + N_4 = \frac{1}{2}(1 + \eta) & a = 3 \end{cases} \tag{3.4.2}$$

The shape functions are illustrated in Fig. 3.4.2; they are "planes" above Ω^e, and so the derivatives with respect to coordinates x and y are constants. In this case note that (3.4.1) is not one-to-one, since the boundary segment $\xi \in [-1, 1]$, $\eta = 1$ in $\boldsymbol{\xi}$-space is mapped into the point $\boldsymbol{x}_3^e$ in $\boldsymbol{x}$-space. In fact, the Jacobian determinant is zero at $\boldsymbol{x}_3^e$. However, as we have already seen, the derivatives with respect to x and y are well behaved. Thus condition C1 is satisfied.

The continuity condition C2 may be inferred form Fig. 3.4.2. A typical global shape function is illustrated in Fig. 3.4.3. Sometimes these functions are called *pyramids,* for obvious reasons. An alternative verification of continuity may be attained by evaluating the element interpolation function

$$u^h = \sum_{a=1}^{3} N'_a d_a^e \tag{3.4.3}$$

along the boundaries, viz.,

(*Side* 1, 2)

$$u^h(\xi, -1) = \tfrac{1}{2}(1 - \xi)d_1^e + \tfrac{1}{2}(1 + \xi)d_2^e \tag{3.4.4}$$

(*Side* 2, 3)

$$u^h(1, \eta) = \tfrac{1}{2}(1 - \eta)d_2^e + \tfrac{1}{2}(1 + \eta)d_3^e \tag{3.4.5}$$

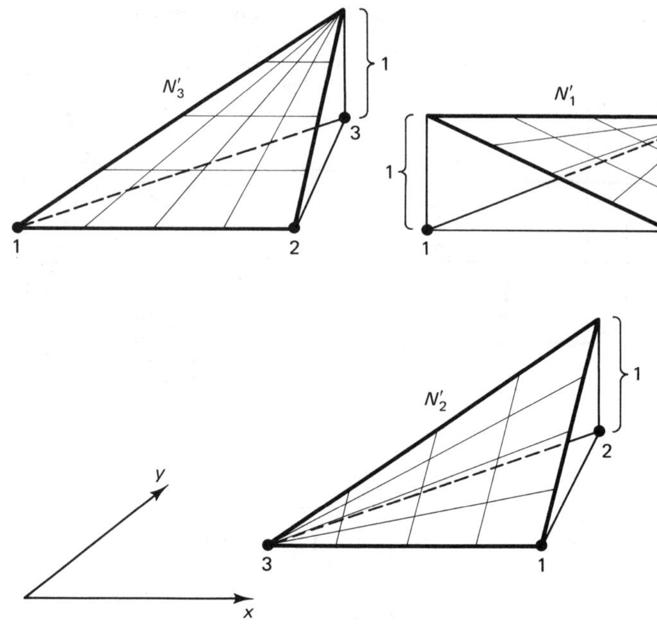

Figure 3.4.2 Element shape functions for the linear triangle.

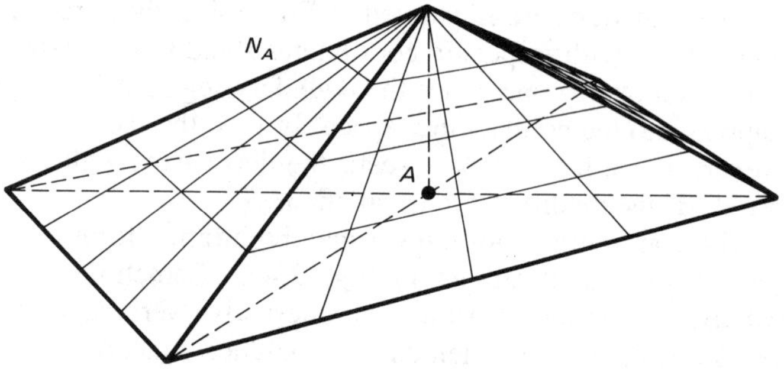

Figure 3.4.3 Global shape function constructed from linear triangular element shape functions.

(*Side* 3, 1)

$$u^h(-1, \eta) = \tfrac{1}{2}(1 + \eta)d_3^e + \tfrac{1}{2}(1 - \eta)d_1^e \qquad (3.4.6)$$

As is clear from (3.4.4)–(3.4.6), the behavior is linear along each edge. As this is the same for all contiguous elements, continuity is assured. We may also freely mix linear triangles and bilinear quadrilaterals in the same mesh and achieve continuity, since their edge behavior is identical.

The completeness condition C3 is automatic since we have employed the isoparametric concept in defining (3.4.3).

Exercise 1. Compute the Jacobian determinant at $\xi = \eta = 0$ for the element shown in Fig. 3.4.4. Plot the result as a function of $x_1^e \in [-2, 2]$. Comment.

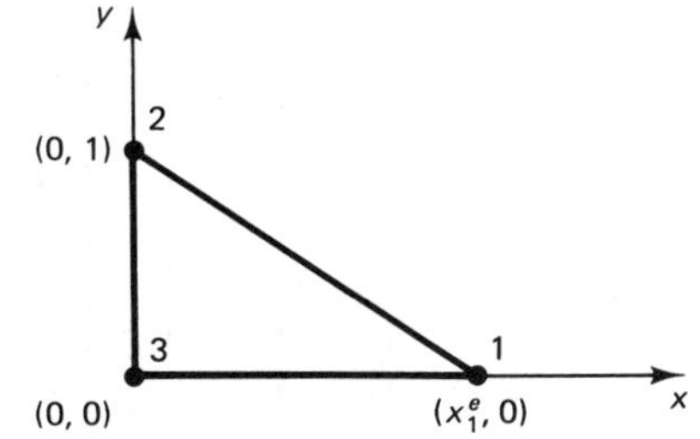

Figure 3.4.4

Summary. The linear triangular element may be obtained from the bilinear quadrilateral by *degeneration*. Computer implementation may be performed along similar lines once the bilinear quadrilateral is programmed.

3.5 TRILINEAR HEXAHEDRAL ELEMENT

The trilinear hexahedral element is the basic element for three-dimensional analysis. It is a straightforward generalization of the bilinear quadrilateral presented in Sec. 3.2. The domain Ω^e (see Fig. 3.5.1) is the image of the biunit cube in $\boldsymbol{\xi}$-space under the trilinear mapping

$$x(\boldsymbol{\xi}) = \alpha_0 + \alpha_1\xi + \alpha_2\eta + \alpha_3\zeta + \alpha_4\xi\eta + \alpha_5\eta\zeta + \alpha_6\xi\zeta + \alpha_7\xi\eta\zeta \tag{3.5.1}$$

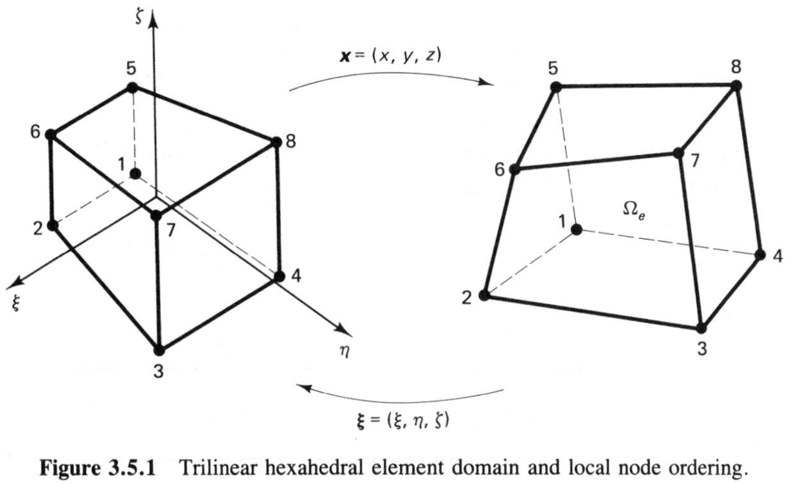

Figure 3.5.1 Trilinear hexahedral element domain and local node ordering.

with corresponding expressions for $y(\boldsymbol{\xi})$ and $z(\boldsymbol{\xi})$. The coefficients $\alpha_0, \ldots, \alpha_7$ are determined by the conditions

$$x(\boldsymbol{\xi}_a) = x_a^e, \qquad a = 1, \ldots, 8 \tag{3.5.2}$$

(The nodal coordinates in $\boldsymbol{\xi}$-space are defined in Table 3.5.1.) This gives rise to a system of linear algebraic equations, viz.,

$$\begin{bmatrix} 1 & -1 & -1 & -1 & 1 & 1 & 1 & -1 \\ 1 & 1 & -1 & -1 & -1 & 1 & -1 & 1 \\ 1 & 1 & 1 & -1 & 1 & -1 & -1 & -1 \\ 1 & -1 & 1 & -1 & -1 & -1 & 1 & 1 \\ 1 & -1 & -1 & 1 & 1 & -1 & -1 & 1 \\ 1 & 1 & -1 & 1 & -1 & -1 & 1 & -1 \\ 1 & 1 & 1 & 1 & 1 & 1 & 1 & 1 \\ 1 & -1 & 1 & 1 & -1 & 1 & -1 & -1 \end{bmatrix} \begin{Bmatrix} \alpha_0 \\ \alpha_1 \\ \alpha_2 \\ \alpha_3 \\ \alpha_4 \\ \alpha_5 \\ \alpha_6 \\ \alpha_7 \end{Bmatrix} = \begin{Bmatrix} x_1^e \\ x_2^e \\ x_3^e \\ x_4^e \\ x_5^e \\ x_6^e \\ x_7^e \\ x_8^e \end{Bmatrix} \tag{3.5.3}$$

TABLE 3.5.1 Nodal coordinates in $\boldsymbol{\xi}$-space

a	ξ_a	η_a	ζ_a
1	−1	−1	−1
2	1	−1	−1
3	1	1	−1
4	−1	1	−1
5	−1	−1	1
6	1	−1	1
7	1	1	1
8	−1	1	1

The matrix may be inverted with the aid of the computer. Solving for the α's and substituting in (3.5.3) yields

$$x(\boldsymbol{\xi}) = \sum_{a=1}^{8} N_a(\boldsymbol{\xi}) x_a^e \tag{3.5.4}$$

where

$$\boxed{N_a(\xi, \eta, \zeta) = \tfrac{1}{8}(1 + \xi_a \xi)(1 + \eta_a \eta)(1 + \zeta_a \zeta)} \tag{3.5.5}$$

with similar expressions for $y(\boldsymbol{\xi})$ and $z(\boldsymbol{\xi})$. Observe that the right-hand side of (3.5.5) consists of the product of the one-dimensional shape functions derived in Sec. 1.12.

We define u^h by invoking the isoparametric concept, i.e.,

$$u^h(\boldsymbol{\xi}) = \sum_{a=1}^{8} N_a(\boldsymbol{\xi}) d_a^e \tag{3.5.6}$$

The continuity of the interpolation functions may be seen by observing the behavior of (3.5.6) along a face of Ω^e. For example, if $\zeta = -1$, (3.5.6) becomes

$$u^h(\xi, \eta, -1) = \sum_{a=1}^{4} \tfrac{1}{4}(1 + \xi_a \xi)(1 + \eta_a \eta) d_a^e \tag{3.5.7}$$

As can be seen from (3.5.7), u^h has hyperbolic paraboloidal variation on the face $\zeta = -1$ and is uniquely defined by the four nodal values associated with that face. The same conclusion can be drawn for any other element of this type which shares this face, and so the continuity of u^h is assured. If the element domain, Ω^e, is not too distorted, the shape functions will be smooth functions of $\boldsymbol{x}$. The completeness condition is guaranteed by virtue of the isoparametric hypothesis.

The hexahedron, or "brick" as it is often referred to, is the most generally useful shape in modeling three-dimensional domains. However, it is often convenient to have a wedge-shaped element at our disposal. The technique of degeneration may be employed to derive a wedge-shaped element from the hexahedron. Specifically, we coalesce nodes 3 and 4, and 7 and 8 (see Fig. 3.5.2). In this case (3.5.4) degenerates to

$$x = \sum_{a=1}^{3} N'_a x_a^e + \sum_{a=5}^{7} N'_a x_a^e \tag{3.5.8}$$

where

$$N'_a = \begin{cases} N_a & a = 1, 2, 5, 6 \\ N_a + N_{a+1} & a = 3, 7 \end{cases} \tag{3.5.9}$$

The Jacobian determinant becomes zero along the edge connecting nodes 3 and 7, but the $\boldsymbol{x}$-derivatives of the shape functions are well behaved (cf. Sec. 3.4).

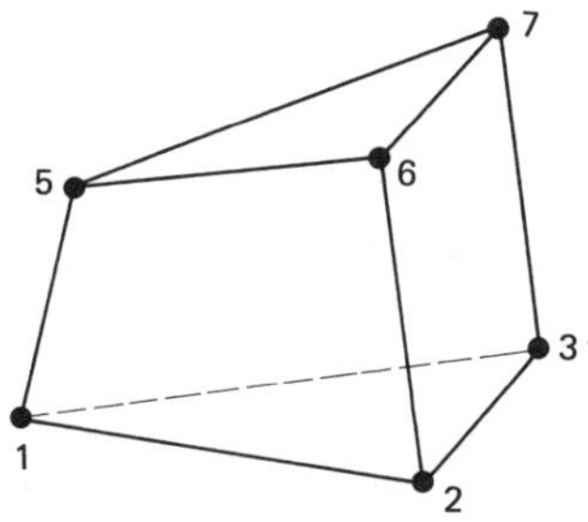

Figure 3.5.2 Wedge-shaped element formed by degenerating the eight-node hexahedral element.

A further coalescing of nodes 5, 6 and 7 yields the well-known linear tetrahedral element (see Fig. 3.5.3). Equation (3.5.8) becomes

$$x = \sum_{a=1}^{3} N'_a x^e_a + N''_5 x^e_5 \tag{3.5.10}$$

where

$$N''_5 = N'_5 + N'_6 + N'_7 \tag{3.5.11}$$

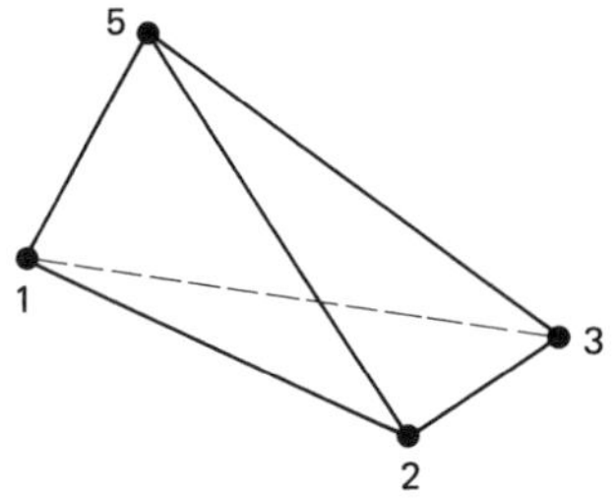

Figure 3.5.3 The linear tetrahedral element formed by degenerating the wedge-shaped element.

It can be shown that the $\boldsymbol{x}$-derivatives of the shape functions are constants in this case (cf. Sec. 3.4). The linear tetrahedron was first proposed by Gallagher et al. [11].

3.6 HIGHER-ORDER ELEMENTS; LAGRANGE POLYNOMIALS

Thus far we have considered elements for which the shape functions are constructed from products of the linear one-dimensional shape functions. In this section we begin consideration of *higher-order elements*. These elements are often capable of more accurate representations than the *linear* elements considered previously and also allow more faithful representations of element domains, as the boundary edges and surfaces may be curved. At the same time they are generally more expensive than the basic linear elements. Thus the cost-effectiveness of the various elements is often in dispute. It appears that the optimal choice is very problem-dependent, so no single element is to be preferred exclusively.

Systematic derivation of certain higher-order isoparametric elements may be performed with the aid of the one-dimensional Lagrange polynomials. We first present the definition of the Lagrange polynomials and then, by way of examples, indicate the construction for multidimensional elements.

Lagrange Polynomials

The Lagrange polynomials are defined by

Order of the polynomial

$$l_a^{n_{en}-1}(\xi) = \frac{\prod_{\substack{b=1 \\ b\neq a}}^{n_{en}} (\xi - \xi_b)}{\prod_{\substack{b=1 \\ b\neq a}}^{n_{en}} (\xi_a - \xi_b)}$$

node

a-term omitted

$$= \frac{(\xi - \xi_1)(\xi - \xi_2)\cdots(\xi - \xi_{a-1})(\xi - \xi_{a+1})\cdots(\xi - \xi_{n_{en}})}{(\xi_a - \xi_1)(\xi_a - \xi_2)\cdots(\xi_a - \xi_{a-1})(\xi_a - \xi_{a+1})\cdots(\xi_a - \xi_{n_{en}})}$$

a-term omitted

(3.6.1)

If the order of the polynomial is not important we omit the superscript and simply write l_a. It is simple to see from (3.6.1) that $l_a(\xi_a) = 1$ and, if $b \neq a$, $l_a(\xi_b) = 0$. In other words

$$l_a(\xi_b) = \delta_{ab} \tag{3.6.2}$$

We shall define the shape functions of an n_{en}-noded element in one dimension by the relation

$$\boxed{N_a = l_a^{n_{en}-1}} \tag{3.6.3}$$

The ξ_a's, as usual, denote the locations of the nodes in ξ-space. Equation (3.6.2) guarantees satisfaction of the interpolation property.

Example 1

Consider the two-noded, one-dimensional element presented in Sec. 1.12. We shall derive the shape functions by way of (3.6.3). From (3.6.3), we have that

$$l_1^1(\xi) = \frac{(\xi - \xi_2)}{(\xi_1 - \xi_2)} = \frac{(\xi - 1)}{-1 - (+1)} = \frac{1}{2}(1 - \xi) \tag{3.6.4}$$

$$l_2^1(\xi) = \frac{(\xi - \xi_1)}{(\xi_2 - \xi_1)} = \frac{(\xi + 1)}{1 - (-1)} = \frac{1}{2}(1 + \xi) \tag{3.6.5}$$

As the reader may see, (3.6.4) and (3.6.5) are the familiar shape functions of the one-dimensional, linear element.

Example 2

We shall now use (3.6.3) to derive the shape functions for a *quadratic* three-node element. We assume that the nodes are located at -1, 0, and $+1$ in ξ-space (see Fig.

3.6.1). The shape functions are given by

$$N_1(\xi) = l_1^2(\xi) = \frac{(\xi - \xi_2)(\xi - \xi_3)}{(\xi_1 - \xi_2)(\xi_1 - \xi_3)} = \frac{\xi(\xi - 1)}{(-1)(-2)} = \frac{1}{2}\xi(\xi - 1) \tag{3.6.6}$$

$$N_2(\xi) = l_2^2(\xi) = \frac{(\xi - \xi_1)(\xi - \xi_3)}{(\xi_2 - \xi_1)(\xi_2 - \xi_3)} = \frac{(\xi + 1)(\xi - 1)}{(1)(-1)} = 1 - \xi^2 \tag{3.6.7}$$

$$N_3(\xi) = l_3^2(\xi) = \frac{(\xi - \xi_1)(\xi - \xi_2)}{(\xi_3 - \xi_1)(\xi_3 - \xi_2)} = \frac{(\xi + 1)\xi}{(2)(1)} = \frac{1}{2}\xi(\xi + 1) \tag{3.6.8}$$

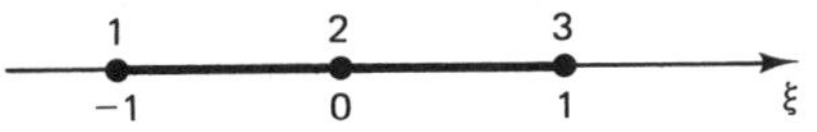

Figure 3.6.1 Node locations in ξ-space for the quadratic, three-node element.

These functions are depicted in Fig. 3.6.2.

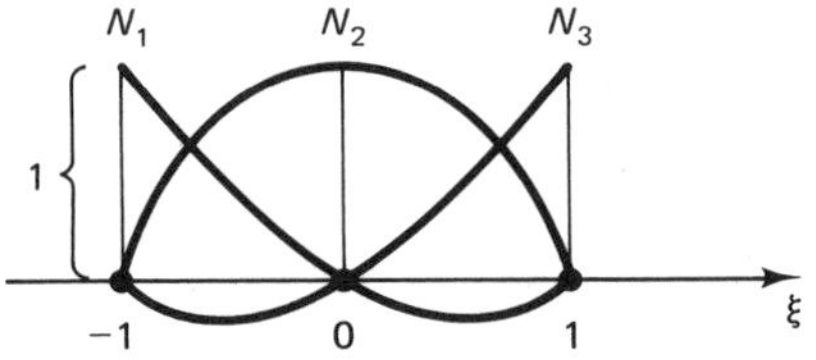

Figure 3.6.2 Shape functions for the quadratic, three-node element.

Exercise 1. Employ (3.6.1) to obtain the shape functions for the *cubic* four-noded (i.e., $n_{en} = 4$) element shown in Fig. 3.6.3. Sketch the results.

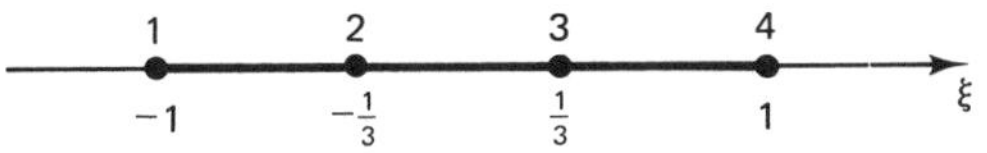

Figure 3.6.3 Node locations in ξ-space for the cubic, four-node element.

Example 3

The shape functions for the bilinear quadrilateral may be set up by taking products of the first order, Lagrange polynomials. The formula is

$$N_a(\xi, \eta) = l_b^1(\xi)l_c^1(\eta) \tag{3.6.9}$$

where the relationship among the indices is given in Table 3.6.1.

TABLE 3.6.1

a	b	c
1	1	1
2	2	1
3	2	2
4	1	2

The procedure is schematically illustrated in Fig. 3.6.4.

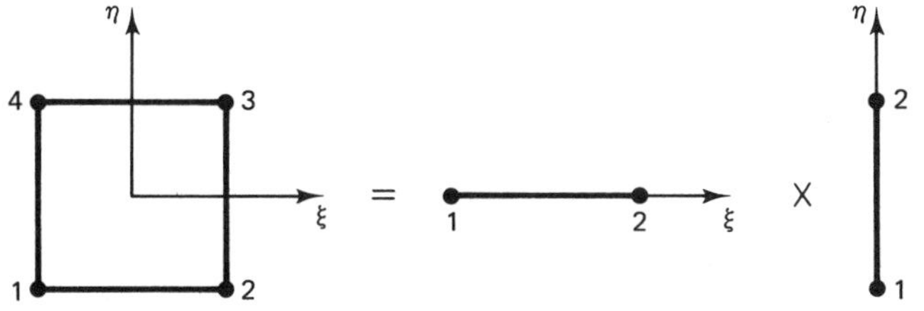

Figure 3.6.4 Schematic relationship between the linear, one-dimensional shape functions and the bilinear, two-dimensional shape functions.

Explicating (3.6.9), we have that

$$N_1(\xi, \eta) = l_1^1(\xi)l_1^1(\eta) = \tfrac{1}{4}(1 - \xi)(1 - \eta) \tag{3.6.10}$$

$$N_2(\xi, \eta) = l_2^1(\xi)l_1^1(\eta) = \tfrac{1}{4}(1 + \xi)(1 - \eta) \tag{3.6.11}$$

$$N_3(\xi, \eta) = l_2^1(\xi)l_2^1(\eta) = \tfrac{1}{4}(1 + \xi)(1 + \eta) \tag{3.6.12}$$

$$N_4(\xi, \eta) = l_1^1(\xi)l_2^1(\eta) = \tfrac{1}{4}(1 - \xi)(1 + \eta) \tag{3.6.13}$$

An analogous procedure may be employed to obtain the trilinear, three-dimensional shape functions of the previous section.

Example 4

Higher-order, two- and three-dimensional elements may be derived by taking products of Lagrange polynomials. We shall illustrate this idea by setting up the nine-node, two-dimensional Lagrange element. The element shape functions are the products of quadratic Lagrange polynomials. The setup is illustrated in Fig. 3.6.5. Note that the element domain Ω^e may have curved edges. The formula for the shape functions is

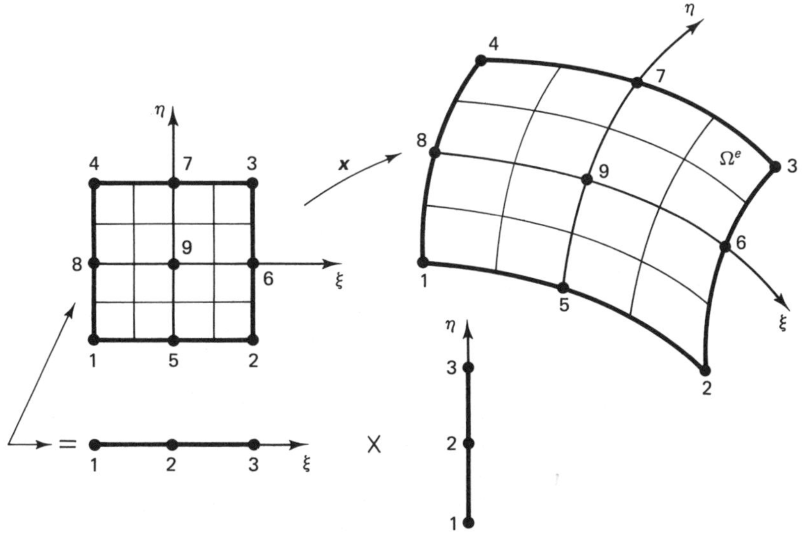

Figure 3.6.5 Nine-node, two-dimensional Lagrange element.

$$N_a(\xi, \eta) = l_b^2(\xi) l_c^2(\eta) \tag{3.6.14}$$

where the relationship among the indices is given in Table 3.6.2.

TABLE 3.6.2

a	b	c
1	1	1
2	3	1
3	3	3
4	1	3
5	2	1
6	3	2
7	2	3
8	1	2
9	2	2

Typical amongst (3.6.14) are

$$\begin{pmatrix}\text{Corner}\\ \text{node}\end{pmatrix} \quad N_1(\xi, \eta) = l_1^2(\xi) l_1^2(\eta) = \tfrac{1}{4}\xi\eta(\xi - 1)(\eta - 1) \tag{3.6.15}$$

$$\begin{pmatrix}\text{Midside}\\ \text{node}\end{pmatrix} \quad N_5(\xi, \eta) = l_2^2(\xi) l_1^2(\eta) = \tfrac{1}{2}\eta(1 - \xi^2)(\eta - 1) \tag{3.6.16}$$

$$\begin{pmatrix}\text{Middle}\\ \text{node}\end{pmatrix} \quad N_9(\xi, \eta) = l_2^2(\xi) l_2^2(\eta) = (1 - \xi^2)(1 - \eta^2) \tag{3.6.17}$$

These functions are shown in Fig. 3.6.6; N_9 is frequently referred to as the ***bubble function,*** for obvious reasons.

Remark

Elements for which the shape functions are formed from products of one-dimensional Lagrange polynomials are called ***Lagrange elements.***

Exercise 2. Use the results of Exercise 1 to derive the shape functions for the 16-node, two-dimensional Lagrange element.

Exercise 3. Use the quadratic one-dimensional shape functions to derive the shape functions for the triquadratic 27-node three-dimensional element.

Exercise 4. Use linear and quadratic shape functions to derive shape functions for the six-node quadrilateral depicted in Fig. 3.6.7.

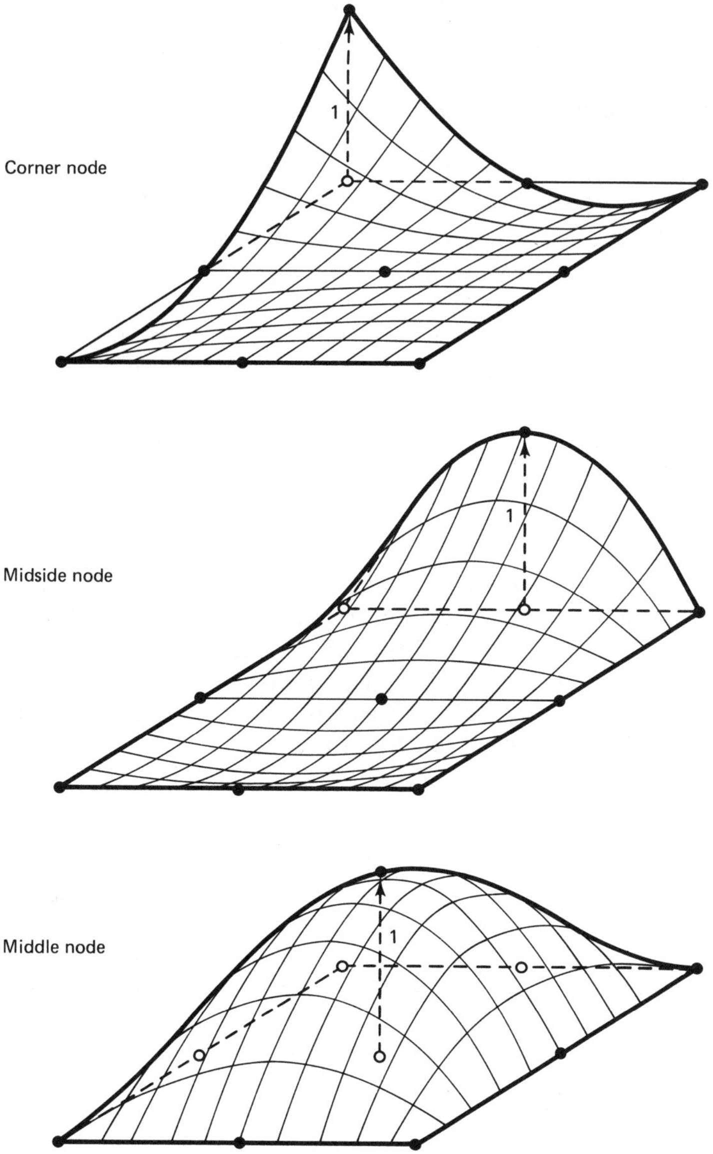

Figure 3.6.6 Typical Lagrange shape functions for the nine-node element.

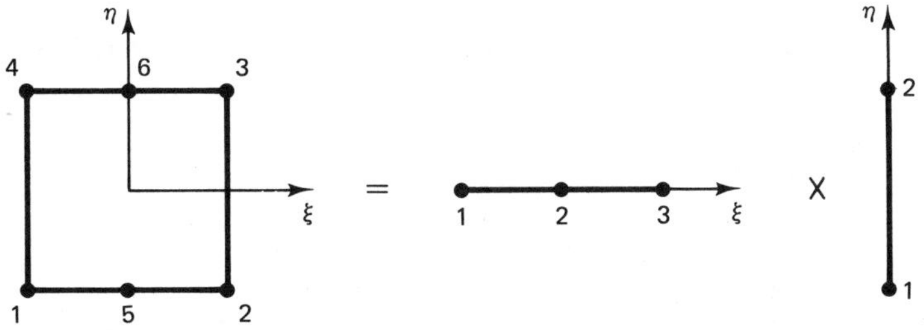

Figure 3.6.7

3.7 ELEMENTS WITH VARIABLE NUMBERS OF NODES

In this section we shall derive the shape functions of a two-dimensional "quadrilateral" element, which possesses four through nine nodes. When all nine nodes are present, the element is the Lagrange element considered in the previous section. When only four nodes are present, the element degenerates to the basic bilinear quadrilateral of Sec. 3.2. Any of nodes 5 through 9 (see Fig. 3.6.5) may be added or omitted. Thus either linear or quadratic behavior may be accommodated along any edge, which permits the "mixing" of basic linear and higher-order elements in one mesh.

We begin with the bilinear shape functions of the four-node quadrilateral element:

$$N_a(\xi, \eta) = \tfrac{1}{4}(1 + \xi_a \xi)(1 + \eta_a \eta), \qquad a = 1, 2, 3, 4 \tag{3.7.1}$$

Our first objective is to add the fifth node, permitting a curved edge and quadratic behavior along the edge connecting nodes 1 and 2 (see Fig. 3.7.1). Note that N_1 and N_2 do not vanish at node 5 (i.e., $N_a(\xi_5, \eta_5) = \tfrac{1}{2}$, $a = 1, 2$), an essential requirement of the shape functions of the five-node element. Shape functions 3 and 4 are satisfactory in this regard. Thus shape functions 1 and 2 will need to be modified. Let us first introduce a shape function for node 5. Let

$$\begin{aligned} N_5(\xi, \eta) &= \underbrace{l_2^2(\xi)}_{\substack{\text{Middle-node} \\ \text{quadratic} \\ \text{Lagrange} \\ \text{polynomial}}} \quad \underbrace{l_1^1(\eta)}_{\substack{\text{Left-node} \\ \text{linear} \\ \text{Lagrange} \\ \text{polynomial}}} \\ &= \frac{1}{2}(1 - \xi^2)(1 - \eta) \end{aligned} \tag{3.7.2}$$

Note that

$$N_5(\xi_a, \eta_a) = \delta_{a5}, \qquad a = 1, 2, \ldots, 5 \tag{3.7.3}$$

This shape function is illustrated in Fig. 3.7.2 and satisfies all requisite properties. In

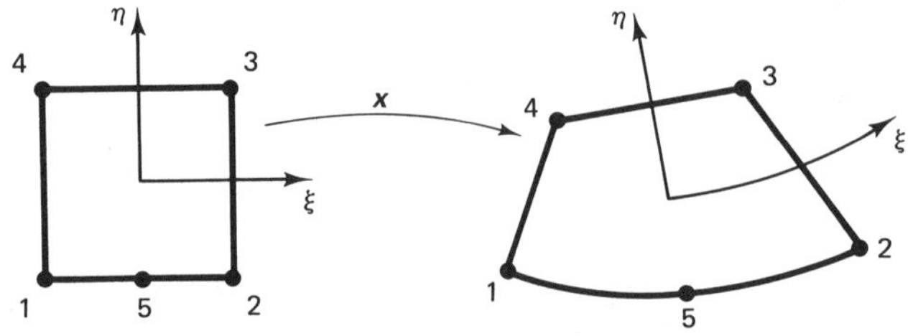

Figure 3.7.1 Five-node element.

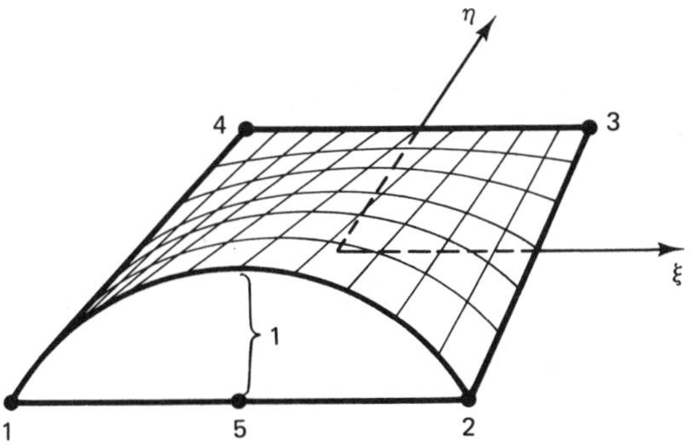

Figure 3.7.2

addition, we may employ it to correct the behavior of bilinear shape functions at node 5. To do this we subtract a multiple of N_5 from N_1 and N_2 so that the resulting functions vanish identically at node 5. Specifically, we define the new shape functions

$$N_a - \frac{1}{2}N_5, \qquad a = 1, 2 \tag{3.7.4}$$

It is easily seen that

$$N_a(\xi_b, \eta_b) - \frac{1}{2}N_5(\xi_b, \eta_b) = \delta_{ab}, \qquad a = 1, 2, \ . \ b = 1, 2, \ldots, 5 \tag{3.7.5}$$

and so (3.7.4) are appropriate shape functions for nodes 1 and 2 of the five-node element. These functions are illustrated in Fig. 3.7.3. In summary, the shape functions of the five-node element are given by (3.7.2), (3.7.4), and the original bilinear shape functions for nodes 3 and 4. Observe that in addition to introducing N_5, only the bilinear shape functions associated with the nodes along the edge containing node 5 needed to be modified. A similar situation occurs if we wish to add any of the other midside nodes. We may construct shape functions for midside nodes 6 through 8 in analogous fashion to the construction of N_5 (see (3.7.2)). These are

$$N_6(\xi, \eta) = \tfrac{1}{2}(1 - \eta^2)(1 + \xi) \tag{3.7.6}$$

$$N_7(\xi, \eta) = \tfrac{1}{2}(1 - \xi^2)(1 + \eta) \tag{3.7.7}$$

$$N_8(\xi, \eta) = \tfrac{1}{2}(1 - \eta^2)(1 - \xi) \tag{3.7.8}$$

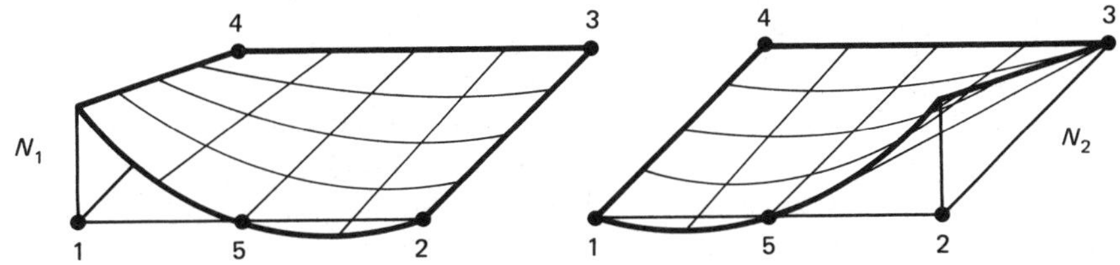

Figure 3.7.3

If any of nodes 5, 6, 7 and 8 are absent, we may formally define N_5, N_6, N_7, and N_8 to be identically zero, respectively. Then the modified shape functions for nodes 1 through 4 are

$$N_1 \leftarrow N_1 - \tfrac{1}{2}(N_5 + N_8) \tag{3.7.9}$$

$$N_2 \leftarrow N_2 - \tfrac{1}{2}(N_5 + N_6) \tag{3.7.10}$$

$$N_3 \leftarrow N_3 - \tfrac{1}{2}(N_6 + N_7) \tag{3.7.11}$$

$$N_4 \leftarrow N_4 - \tfrac{1}{2}(N_7 + N_8) \tag{3.7.12}$$

The reader may wish to verify that the modified shape functions, (3.7.9) through (3.7.12), satisfy

$$N_a(\xi_b, \eta_b) = \delta_{ab} \tag{3.7.13}$$

for all b corresponding to nodes present.

To include node 9 we introduce the ***bubble function:***

$$N_9(\xi, \eta) = (1 - \xi^2)(1 - \eta^2) \tag{3.7.14}$$

N_9 vanishes at nodes 1 through 8. However, $N_1, N_2, \ldots, N_8$ do not in general vanish at node 9, and therefore they must be modified. The procedure is carried out most expeditiously if the original bilinear shape functions and N_5, N_6, N_7, N_8 are first modified to account for the presence of node 9. This is accomplished as follows:

$a = 1, 2, 3, 4$:

$$N_a \leftarrow \underbrace{N_a}_{\substack{\text{Bilinear} \\ \text{shape functions; (3.7.1)}}} - \tfrac{1}{4}N_9 \tag{3.7.15}$$

$a = 5, 6, 7, 8$:

$$N_a = \begin{cases} \overbrace{N_a}^{\text{Defined by (3.7.2), (3.7.6)–(3.7.8)}} - \tfrac{1}{2}N_9 & \text{(if node } a \text{ is present)} \\ 0 & \text{(if node } a \text{ is absent)} \end{cases} \tag{3.7.16}$$

Following this modification, the corrections indicated in (3.7.9) through (3.7.12) need be performed. The procedure is most succinctly summarized in a flowchart; see Fig. 3.7.4.

Care must also be taken that high-order elements are not too distorted. For example, in the case of the eight-node serendipity quadrilateral (see Fig. 3.7.5), all interior angles must be less than 180° (as for the four-node quadrilateral) and, in addition, the midside nodes should not migrate too far off the center of the sides. It is recommended in [12], p. 186, that the safe zone is confined to the middle third of the side. Although rules of thumb are useful for initiatory considerations, in general, we need rely on numerical checks of the sign of the Jacobian determinant to determine if the element geometry is "too distorted."

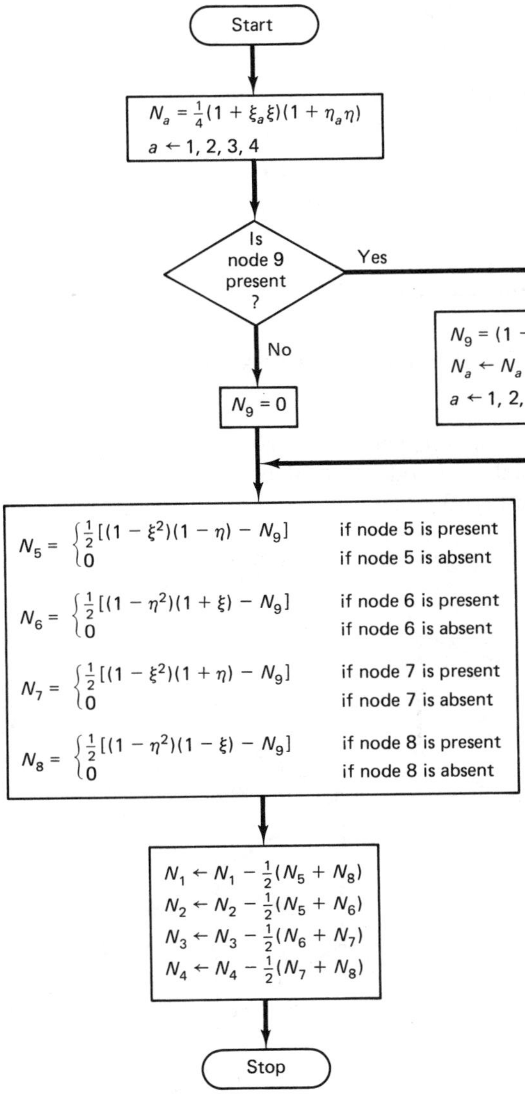

Figure 3.7.4 Calculation of a shape function for a four- through nine-node element.

Exercise 1. Use the flowchart in Fig. 3.7.4 to derive explicitly the shape functions of the eight-node ***serendipity quadrilateral*** for which the internal node (node 9) is omitted.

Exercise 2. Use the results of Exercise 1 and the degeneration technique to derive the shape functions of the six-node quadratic triangle; see Fig. 3.7.5.

Exercise 3. Generalize the flowchart of Fig. 3.7.4 so that the four- through nine-node element may be degenerated to a triangle.

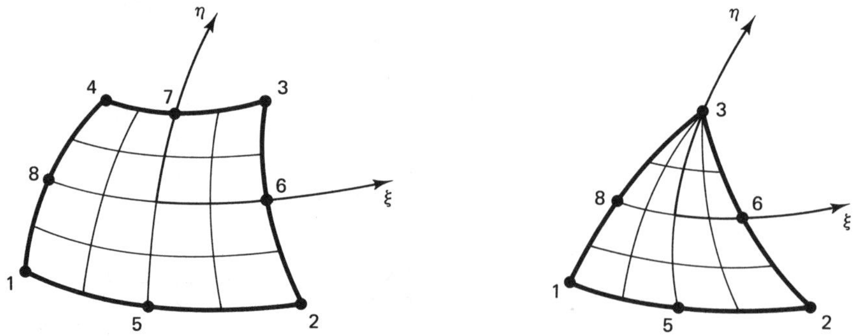

Figure 3.7.5 Degeneration of the eight-node serendipity element to the six-node triangle.

Exercise 4. Develop a flowchart analogous to the one in Fig. 3.7.4 for an 8- through 27-variable-number-of-nodes three-dimensional "brick" element. The node numbering is indicated in Fig. 3.7.6.

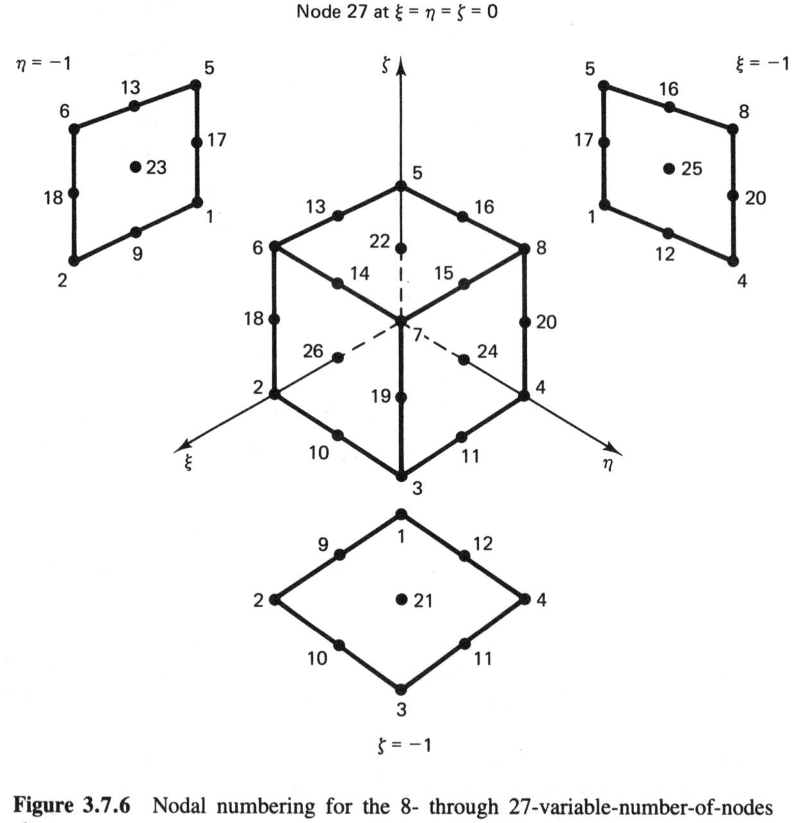

Figure 3.7.6 Nodal numbering for the 8- through 27-variable-number-of-nodes element.

Exercise 5. Generalize the flowchart of Exercise 4 so that the 8- through 27-node element may be degenerated to a wedge-shaped element.

Exercise 6. Develop a flowchart for a 4- through 16-variable-number-of-nodes quadrilateral element. (The 16-node element is the bicubic Lagrange element of Exercise 2, Sec. 3.6.)

Exercise 7. Develop shape functions for a six-node quadrilateral, which exhibits cubic behavior along one edge and linear behavior along the other three (see the accompanying figure).

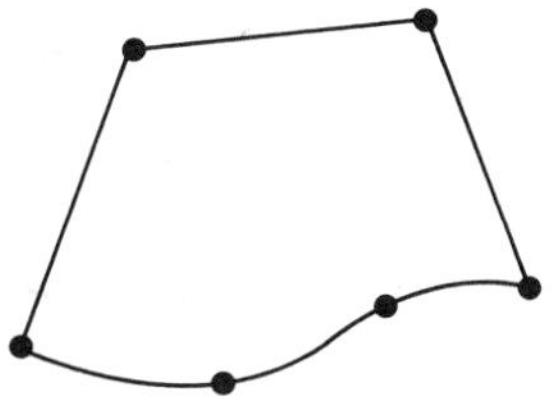

Standard Element Families

In Fig. 3.7.7, some members of standard two-dimensional element families are shown. (Brick, wedge, and tetrahedral three-dimension analogs may easily be envisioned.) Isoparametric shape functions for all these elements may easily be derived by the techniques described in this section.

We may note from Fig. 3.7.7 that the serendipity elements have nodeless interiors. Beyond cubic level, the serendipity elements may be enhanced by the inclusion of some internal nodes to achieve higher degrees of polynomial completeness. This may be understood by considering the ***Pascal triangles*** shown in Fig. 3.7.8. As may be seen, for quartic and higher-order serendipity elements, the degree of polynomial completeness is fixed at cubic. The degree of completeness is generally considered to be the most important measure of accuracy of an element. Thus it is seen to be necessary to include functions associated with internal nodes in order to achieve higher-order accurate elements.

3.8 NUMERICAL INTEGRATION; GAUSSIAN QUADRATURE

Let $f : \Omega^e \subset \mathbb{R}^{n_{sd}} \to \mathbb{R}$ be a given function. We are interested in computing

$$\int_{\Omega^e} f(\boldsymbol{x})\, d\Omega \tag{3.8.1}$$

for purposes of constructing element arrays. (f may be thought of as any term in an integrand that we have encountered so far. For example, $f = \boldsymbol{B}_a^T \boldsymbol{D} \boldsymbol{B}_b$, the integrand of the stiffness matrix in heat conduction; see Sec. 2.5.) Throughout we shall assume f is smooth and integrable, so there is no ambiguity as to the meaning of (3.8.1). We

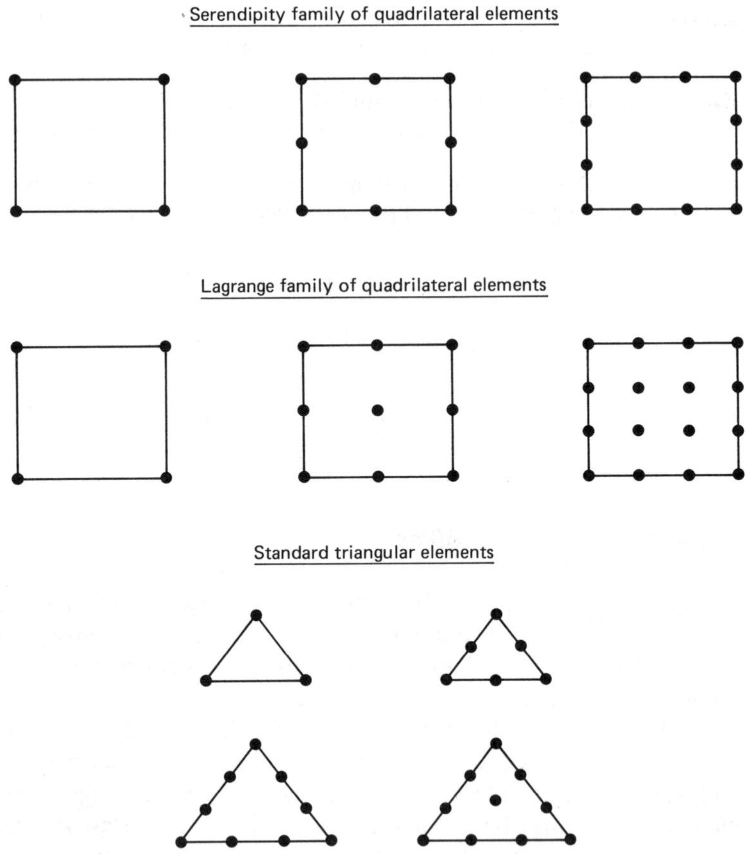

Figure 3.7.7 Standard two-dimensional element families. Note that the first members of the Lagrange and serendipity families are identical.

will need to make use of the change of variables formula. In each case we will pull back the integral over the element domain in $\boldsymbol{x}$-space to one over the parent domain (i.e., the biunit n_{sd}-cube). For completeness, we begin by recalling the one-dimensional case (see Sec. 1.15):

$n_{sd} = 1$:

$$\int_{\Omega^e} f(x)\, dx = \int_{-1}^{1} f(x(\xi))x_{,\xi}(\xi)\, d\xi \tag{3.8.2}$$

The two- and three-dimensional cases are given as follows:

$n_{sd} = 2$:

$$\boxed{\int_{\Omega^e} f(x, y)\, d\Omega = \int_{-1}^{1}\int_{-1}^{1} f(x(\xi, \eta), y(\xi, \eta))j(\xi, \eta)\, d\xi\, d\eta} \tag{3.8.3}$$

Lagrange quadrilaterals

$$1$$
$$\xi \quad \eta$$
$$\xi^2 \quad \xi\eta \quad \eta^2$$
$$\xi^3 \quad \xi^2\eta \quad \xi\eta^2 \quad \eta^3$$
$$\xi^4 \quad \xi^3\eta \quad \xi^2\eta^2 \quad \xi\eta^3 \quad \eta^4$$
$$\cdots \quad \xi^4\eta \quad \xi^3\eta^2 \quad \xi^2\eta^3 \quad \xi\eta^4 \quad \cdots$$
$$\xi^m \quad \xi^4\eta^2 \quad \xi^3\eta^3 \quad \xi^2\eta^4 \quad \eta^m$$
$$\xi^m\eta \quad \xi^4\eta^3 \quad \xi^3\eta^4 \quad \xi\eta^m$$
$$\xi^m\eta^2 \quad \xi^4\eta^4 \quad \xi^2\eta^m$$
$$\xi^m\eta^3 \quad \xi^3\eta^m$$
$$\xi^m\eta^4 \quad \xi^4\eta^m$$
$$\cdots \quad \cdots$$
$$\xi^m\eta^m$$

Serendipity quadrilaterals

$$1$$
$$\xi \quad \eta$$
$$\xi^2 \quad \xi\eta \quad \eta^2$$
$$\xi^3 \quad \xi^2\eta \quad \xi\eta^2 \quad \eta^3$$
$$\xi^4 \quad \xi^3\eta \quad \xi\eta^3 \quad \eta^4$$
$$\cdots \quad \xi^4\eta \quad \xi\eta^4 \quad \cdots$$
$$\xi^m \quad \cdots \quad \cdots \quad \eta^m$$
$$\xi^m\eta \quad \xi\eta^m$$

Triangles (internal nodal functions included)

$$1$$
$$\xi \quad \eta$$
$$\xi^2 \quad \xi\eta \quad \eta^2$$
$$\xi^3 \quad \xi^2\eta \quad \xi\eta^2 \quad \eta^3$$
$$\xi^4 \quad \xi^3\eta \quad \xi^2\eta^2 \quad \xi\eta^3 \quad \eta^4$$
$$\cdots \quad \cdots \quad \cdots \quad \cdots \quad \cdots \quad \cdots$$
$$\xi^m \quad \xi^{m-1}\eta \quad \xi^{m-2}\eta^2 \quad \cdots \quad \xi^2\eta^{m-2} \quad \xi\eta^{m-1} \quad \eta^m$$

Figure 3.7.8. Pascal triangles for standard two-dimensional element families.

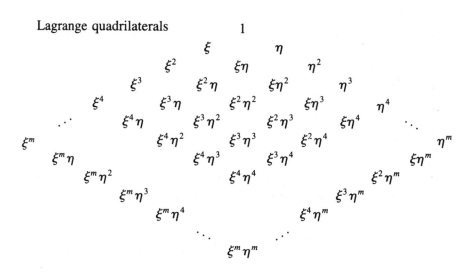

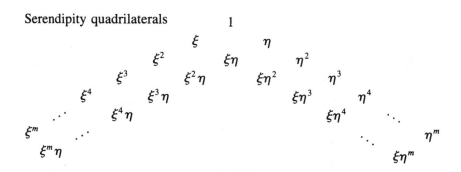

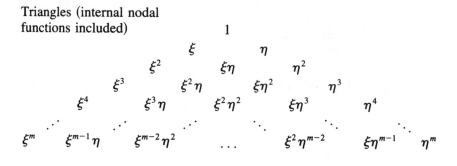

$n_{sd} = 3$:

$$\int_{\Omega^e} f(x, y, z)\, d\Omega = \int_{-1}^{1}\int_{-1}^{1}\int_{-1}^{1} f(x(\xi, \eta, \zeta), y(\xi, \eta, \zeta), z(\xi, \eta, \zeta))j(\xi, \eta, \zeta)\, d\xi\, d\eta\, d\zeta \tag{3.8.4}$$

Recall that $j = \det(\partial \boldsymbol{x}/\partial \boldsymbol{\xi})$, the Jacobian determinant.

In each case we have to evaluate an integral of the form:

$$\underbrace{\int_{-1}^{1} \cdots \int_{-1}^{1}}_{n_{sd} \text{ times}} \underbrace{g(\xi, \cdots)}_{n_{sd} \text{ arguments}} \quad \underbrace{d\xi \cdots}_{n_{sd} \text{ differentials}} \tag{3.8.5}$$

Consider the one-dimensional case, in which we wish to evaluate

$$\int_{-1}^{1} g(\xi)\, d\xi \tag{3.8.6}$$

This can be approximately computed by way of a numerical integration (quadrature) formula as follows:

$$\int_{-1}^{1} g(\xi)\, d\xi = \sum_{l=1}^{n_{\text{int}}} g(\tilde{\xi}_l)W_l + R \cong \sum_{l=1}^{n_{\text{int}}} g(\tilde{\xi}_l)W_l \tag{3.8.7}$$

where n_{int} is the number of integration (quadrature) points, $\tilde{\xi}_l$ is the coordinate of the lth integration point, W_l is the "weight" of the lth integration point, and R is the remainder. The reader is probably familiar with some of the classical numerical integration formulas, such as the trapezoidal and Simpson's rules. We review these now as examples.

Example 1 (Trapezoidal rule)

$$n_{\text{int}} = 2$$
$$\tilde{\xi}_1 = -1$$
$$\tilde{\xi}_2 = 1$$
$$W_l = 1, \qquad l = 1, 2$$
$$R = -\tfrac{2}{3} g_{,\xi\xi}(\bar{\xi})$$

$\bar{\xi}$ denotes some point in the interval $[-1, 1]$. The trapezoidal rule is exact for constants and linear polynomials but only approximate for quadratic and higher-order polynomials. It is said to be "second-order accurate."

Example 2 (Simpson's rule)

$$n_{\text{int}} = 3$$
$$\tilde{\xi}_1 = -1$$
$$\tilde{\xi}_2 = 0$$
$$\tilde{\xi}_3 = 1$$
$$W_1 = W_3 = \frac{1}{3}$$
$$W_2 = \frac{4}{3}$$
$$R = \frac{-g^{(4)}(\bar{\xi})}{90}$$

where $g^{(4)} = g_{,\xi\xi\xi\xi}$. Simpson's rule integrates general cubic polynomials exactly and is thus said to be "fourth-order accurate."

Although the above rules are widely used they are inefficient in the sense that there exist integration rules that are just as accurate but involve fewer integration points. *This issue is of great importance in practice since the fewer the integration points, the less the cost.* Considerable savings can be engendered by choosing an appropriate numerical integration rule.

Gaussian Quadrature

In one dimension the Gauss quadrature formulas are optimal. Accuracy of order $2n_{\text{int}}$ is achieved by n_{int} integration points. The locations of the quadrature points and values of associated weights are determined to attain maximum accuracy. The theory is thoroughly discussed in [13]. We give the first three Gaussian rules in the following:

Gaussian quadrature rules.

1. $n_{\text{int}} = 1$

$$\tilde{\xi}_1 = 0$$
$$W_1 = 2$$
$$R = \frac{g_{,\xi\xi}(\bar{\xi})}{3}$$

2. $n_{\text{int}} = 2$

$$\tilde{\xi}_1 = \frac{-1}{\sqrt{3}}$$

$$\tilde{\xi}_2 = \frac{1}{\sqrt{3}}$$

$$W_1 = W_2 = 1$$

$$R = \frac{g^{(4)}(\bar{\xi})}{135}$$

3. $n_{\text{int}} = 3$

$$\tilde{\xi}_1 = -\sqrt{\frac{3}{5}}$$

$$\tilde{\xi}_2 = 0$$

$$\tilde{\xi}_3 = \sqrt{\frac{3}{5}}$$

$$W_1 = W_3 = \frac{5}{9}$$

$$W_2 = \frac{8}{9}$$

$$R = \frac{g^{(6)}(\bar{\xi})}{15{,}750}$$

Exercise 1 Verify the accuracy of the preceding Gaussian rules by integrating the monomials $1, \xi, \xi^2, \ldots$ in turn.

The derivation of Gaussian quadrature formulas is illustrated by the following example.

Example 3 (Derivation of the Gaussian quadrature formula for $n_{\text{int}} = 2$)

We seek a rule involving two integration points that is exact for a general cubic polynomial. Let $g(\xi) = \alpha_0 + \alpha_1\xi + \alpha_2\xi^2 + \alpha_3\xi^3$, where the α's are arbitrary constants, and assume the integration points are equally weighted and symmetrically spaced (i.e., $W_1 = W_2$ and $\tilde{\xi}_1 = -\tilde{\xi}_2$, respectively). The exact integral of

$$\int_{-1}^{1} g(\xi)\, d\xi = 2\alpha_0 + \tfrac{2}{3}\alpha_2 \tag{3.8.8}$$

and this is to be equal to

$$\sum_{l=1}^{2} g(\tilde{\xi}_l)W_l = 2W_2(\alpha_0 + \alpha_2\tilde{\xi}_2^2) \tag{3.8.9}$$

This is to hold for arbitrary values of α_0 and α_2, and it thus follows that $W_2 = 1$ and $\tilde{\xi}_2 = 1/\sqrt{3}$.

Exercise 2. Derive the Gauss quadrature rule for $n_{\text{int}} = 3$. *Hint:* From considerations of symmetry, assume $\tilde{\xi}_1 = -\tilde{\xi}_3$, $\tilde{\xi}_2 = 0$ and $W_1 = W_3$.

General Gaussian Quadrature Rule

The general Gaussian quadrature rule is defined by the following:

$$W_l = \frac{2}{[(1 - \tilde{\xi}_l^2)(P'_{n_{\text{int}}}(\tilde{\xi}_l)^2]}, \qquad 1 \le l \le n_{\text{int}} \tag{3.8.10}$$

$$R = \frac{2^{2n_{\text{int}}+1}(n_{\text{int}}\,!)^4}{(2n_{\text{int}} + 1)[(2n_{\text{int}})!]^3} g_{,\underbrace{\xi\ldots\xi}_{2n_{\text{int}}\text{ times}}}(\bar{\xi}) \tag{3.8.11}$$

where $\tilde{\xi}_l$ is the lth zero of the Legendre polynomial $P_{n_{\text{int}}}(\xi)$, and $P'_{n_{\text{int}}}$ denotes the derivative of $P_{n_{\text{int}}}$. The Legendre polynomials are defined by

$$P_{n_{\text{int}}}(\xi) = \frac{1}{2^{n_{\text{int}}} n_{\text{int}}\,!} \frac{d^{n_{\text{int}}}}{d\xi^{n_{\text{int}}}} (\xi^2 - 1)^{n_{\text{int}}} \tag{3.8.12}$$

See [13] for tabulated values of $\tilde{\xi}_l$ and W_l.

Gaussian Quadrature in Several Dimensions

Gaussian rules for integrals in several dimensions are constructed by employing one-dimensional Gaussian rules on each coordinate separately. For example, in two dimensions

$$\int_{-1}^{1}\int_{-1}^{1} g(\xi, \eta)\, d\xi\, d\eta \cong \int_{-1}^{1} \left\{ \sum_{l^{(1)}=1}^{n_{\text{int}}^{(1)}} g(\tilde{\xi}_{l^{(1)}}^{(1)}, \eta) W_{l^{(1)}}^{(1)} \right\} d\eta \qquad \text{(Gaussian quadrature rule 1 applied to } \xi\text{-coordinate)}$$

$$\cong \sum_{l^{(1)}=1}^{n_{\text{int}}^{(1)}} \sum_{l^{(2)}=1}^{n_{\text{int}}^{(2)}} g(\tilde{\xi}_{l^{(1)}}^{(1)}, \tilde{\eta}_{l^{(2)}}^{(2)}) W_{l^{(1)}}^{(1)} W_{l^{(2)}}^{(2)} \qquad \text{(Gaussian quadrature rule 2 applied to } \eta\text{-coordinate)} \tag{3.8.13}$$

Example 4

If the one-point rule is used in each direction (3.8.13) specializes to

$$\int_{-1}^{1}\int_{-1}^{1} g(\xi, \eta)\, d\xi\, d\eta = 4g(0, 0) \tag{3.8.14}$$

Example 5

If the two-point rule is used in each direction, (3.8.13) specializes to

$$\int_{-1}^{1}\int_{-1}^{1} g(\xi, \eta)\, d\xi\, d\eta = g\left(\frac{-1}{\sqrt{3}}, \frac{-1}{\sqrt{3}}\right) + g\left(\frac{1}{\sqrt{3}}, \frac{-1}{\sqrt{3}}\right) + g\left(\frac{1}{\sqrt{3}}, \frac{1}{\sqrt{3}}\right) + g\left(\frac{-1}{\sqrt{3}}, \frac{1}{\sqrt{3}}\right) \tag{3.8.15}$$

It is frequently preferable to tabulate multidimensional rules in terms of a single index so that in place of (3.8.13) we would have

$$\int_{-1}^{1}\int_{-1}^{1} g(\xi, \eta)\, d\xi\, d\eta \cong \sum_{l=1}^{n_{\text{int}}} g(\tilde{\xi}_l, \tilde{\eta}_l) W_l \tag{3.8.16}$$

Comparison with (3.8.13) indicates that

$$n_{\text{int}} = n_{\text{int}}^{(1)} n_{\text{int}}^{(2)}, \qquad 1 \leq l \leq n_{\text{int}} \tag{3.8.17}$$

$$\tilde{\xi}_l = \tilde{\xi}_{l^{(1)}}^{(1)} \tag{3.8.18}$$

$$\tilde{\eta}_l = \tilde{\eta}_{l^{(2)}}^{(2)} \tag{3.8.19}$$

$$W_l = W_{l^{(1)}}^{(1)} W_{l^{(2)}}^{(2)} \tag{3.8.20}$$

where a tabular relationship among the indices must be established (e.g., see Table 3.8.1)

Example 6

If the one-point rule is used in each direction, corresponding to (3.8.17) through (3.8.20), we have

$$n_{\text{int}} = 1, \qquad \tilde{\xi}_1 = \tilde{\eta}_1 = 0, \qquad W_1 = 2 \cdot 2 = 4 \tag{3.8.21}$$

Thus (3.8.14), (3.8.16), and (3.8.21) are consistent.

Example 7

Consider use of the two-point rule in each direction. We assume the relationship among the indices is as given in Table 3.8.1.

TABLE 3.8.1

l	$l^{(1)}$	$l^{(2)}$
1	1	1
2	2	1
3	2	2
4	1	2

In this case (3.8.17) through (3.8.20) yield

$$n_{\text{int}} = 2 \cdot 2 = 4 \tag{3.8.22}$$

$$\tilde{\xi}_1 = \frac{-1}{\sqrt{3}}, \quad \tilde{\xi}_2 = \frac{1}{\sqrt{3}}, \quad \tilde{\xi}_3 = \frac{1}{\sqrt{3}}, \quad \tilde{\xi}_4 = \frac{-1}{\sqrt{3}} \tag{3.8.23}$$

$$\tilde{\eta}_1 = \frac{-1}{\sqrt{3}}, \quad \tilde{\eta}_2 = \frac{-1}{\sqrt{3}}, \quad \tilde{\eta}_3 = \frac{1}{\sqrt{3}}, \quad \tilde{\eta}_4 = \frac{1}{\sqrt{3}} \tag{3.8.24}$$

$$W_1 = W_2 = W_3 = W_4 = 1 \tag{3.8.25}$$

Clearly (3.8.15), (3.8.16), and (3.8.22) through (3.8.25) are consistent.

Exercise 3. Consider use of the three-point Gauss rule in each direction. Define a relationship between the indices l, $l^{(1)}$, and $l^{(2)}$ by constructing a table similar to Table 3.8.1. Determine $\tilde{\xi}_l$, $\tilde{\eta}_l$, and W_l, $1 \leq l \leq 9$.

Locations of the quadrature points for the one-point, 2 × 2, and 3 × 3 Gaussian rules are illustrated in Fig. 3.8.1.

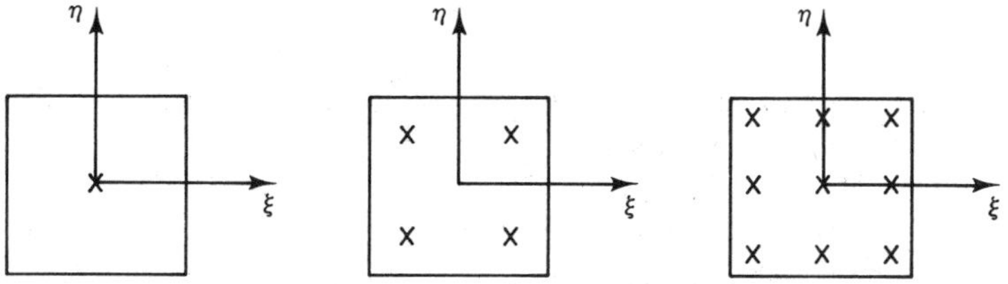

Figure 3.8.1 Locations of Gauss quadrature points for one-point, 2 × 2, and 3 × 3 rules.

Analogous arguments lead to the form of rules in three dimensions, viz.,

$$\begin{aligned}
\int_{-1}^{1}\int_{-1}^{1}\int_{-1}^{1} g(\xi, \eta, \zeta)\, d\xi\, d\eta\, d\zeta
&\cong \sum_{l^{(1)}=1}^{n_{\text{int}}^{(1)}} \sum_{l^{(2)}=1}^{n_{\text{int}}^{(2)}} \sum_{l^{(3)}=1}^{n_{\text{int}}^{(3)}} g(\tilde{\xi}^{(1)}_{l^{(1)}}, \tilde{\eta}^{(2)}_{l^{(2)}}, \tilde{\zeta}^{(3)}_{l^{(3)}}) W^{(1)}_{l^{(1)}} W^{(2)}_{l^{(2)}} W^{(3)}_{l^{(3)}} \\
&= \sum_{l=1}^{n_{\text{int}}} g(\tilde{\xi}_l, \tilde{\eta}_l, \tilde{\zeta}_l) W_l
\end{aligned} \tag{3.8.26}$$

Exercise 4. Determine $\tilde{\xi}_l$, $\tilde{\eta}_l$, $\tilde{\zeta}_l$, and W_l for the one-point and 2 × 2 × 2 Gaussian rules in three dimensions.

In multidimensional cases the Gaussian quadrature rules are no longer necessarily optimal. For example, in three dimensions the six-point (non-Gaussian) rule given in Table 3.8.2 is fourth-order accurate [14]. The quadrature points are located at the centers of the six faces of the cube. (This means that all monomials of the form

$\xi^i \eta^j \zeta^k$, $0 \leq i, j, k \leq 3$, $i + j + k \leq 3$, are exactly integrated.) The fourth-order Gaussian rule requires eight points ($2 \times 2 \times 2$ rule) and is thus more expensive to use.

TABLE 3.8.2 Fourth-Order Accurate, Six-Point Quadrature Rule in Three Dimensions

l	$\tilde{\xi}_l$	$\tilde{\eta}_l$	$\tilde{\zeta}_l$	W_l
1	1	0	0	4/3
2	−1	0	0	4/3
3	0	1	0	4/3
4	0	−1	0	4/3
5	0	0	1	4/3
6	0	0	−1	4/3

A 14-point rule for three dimensions that is sixth-order accurate is described in [15, 16]. This rule represents a considerable savings when compared with the 27-point ($3 \times 3 \times 3$) Gaussian rule [17].

Unfortunately, a general theory of optimal integration formulas for even such simple shapes as squares and cubes does not yet seem to be known. Despite some measure of inefficiency in multidimensional applications, Gaussian quadrature formulas are still widely used and will be sufficient for our present needs. Recent progress toward the development of improved rules is reported in [18].

Exercise 5. Consider the following four-point rule in two dimensions:

l	$\tilde{\xi}_l$	$\tilde{\eta}_l$	W_l
1	0	$+a$	1
2	0	$-a$	1
3	$+a$	0	1
4	$-a$	0	1

Determine a such that fourth-order accuracy is attained.

3.9 DERIVATIVES OF SHAPE FUNCTIONS AND SHAPE FUNCTION SUBROUTINES

We need to calculate explicitly the derivatives of the shape functions to construct the element stiffness matrices. For simplicity let us take the case $n_{sd} = 2$. The derivatives of N_a with respect to x and y may be evaluated with the aid of the chain rule:

$$N_{a,x} = N_{a,\xi}\xi_{,x} + N_{a,\eta}\eta_{,x} \tag{3.9.1}$$

$$N_{a,y} = N_{a,\xi}\xi_{,y} + N_{a,\eta}\eta_{,y} \tag{3.9.2}$$

It is worthwhile to recast these relations in the following matrix form:

$$\langle N_{a,x} N_{a,y}\rangle = \langle N_{a,\xi} N_{a,\eta}\rangle \begin{bmatrix} \xi_{,x} & \xi_{,y} \\ \eta_{,x} & \eta_{,y} \end{bmatrix} \tag{3.9.3}$$

The derivatives $N_{a,\xi}$ and $N_{a,\eta}$ may be explicitly computed. However, the terms in the matrix cannot be directly computed since we do not have explicit expressions $\xi = \xi(x, y)$ and $\eta = \eta(x, y)$. On the other hand, we do have the inverse relations

$$x(\xi, \eta) = \sum_{a=1}^{n_{en}} N_a(\xi, \eta) x_a^e \tag{3.9.4}$$

$$y(\xi, \eta) = \sum_{a=1}^{n_{en}} N_a(\xi, \eta) y_a^e \tag{3.9.5}$$

which enables us to calculate the matrix

$$\boldsymbol{x}_{,\boldsymbol{\xi}} = \begin{bmatrix} x_{,\xi} & x_{,\eta} \\ y_{,\xi} & y_{,\eta} \end{bmatrix} \tag{3.9.6}$$

The components are:

$$x_{,\xi} = \sum_{a=1}^{n_{en}} N_{a,\xi} x_a^e, \qquad x_{,\eta} = \sum_{a=1}^{n_{en}} N_{a,\eta} x_a^e \tag{3.9.7}$$

$$y_{,\xi} = \sum_{a=1}^{n_{en}} N_{a,\xi} y_a^e, \qquad y_{,\eta} = \sum_{a=1}^{n_{en}} N_{a,\eta} y_a^e \tag{3.9.8}$$

The matrix (3.9.6) is the inverse of the matrix in (3.9.3), i.e.,

$$\begin{bmatrix} \xi_{,x} & \xi_{,y} \\ \eta_{,x} & \eta_{,y} \end{bmatrix} = (\boldsymbol{x}_{,\boldsymbol{\xi}})^{-1} = \frac{1}{j}\begin{bmatrix} y_{,\eta} & -x_{,\eta} \\ -y_{,\xi} & x_{,\xi} \end{bmatrix} \tag{3.9.9}$$

where

$$j = \det(\boldsymbol{x}_{,\boldsymbol{\xi}}) = x_{,\xi} y_{,\eta} - x_{,\eta} y_{,\xi} \tag{3.9.10}$$

All the preceding relations are typically evaluated in ***shape function subroutines.*** It is convenient to segregate the calculations into those which can be performed once for all elements of the same type (i.e., an ***element group***) and those which need to be repeated for each element.

For the element group, calculate integration-rule weights, shape functions, and local derivatives:

> For $l = 1, \ldots, n_{\text{int}}$
>
> *Determine*: $W_l, \tilde{\xi}_l, \tilde{\eta}_l$
>
> For $a = 1, \ldots, n_{en}$
>
> *Calculate*: $N_a(\tilde{\xi}_l, \tilde{\eta}_l), N_{a,\xi}(\tilde{\xi}_l, \tilde{\eta}_l), N_{a,\eta}(\tilde{\xi}_l, \tilde{\eta}_l)$

Subroutine QDCSHL performs these calculations for the four-node bilinear quadrilateral element in program DLEARN (see Chapter 11).

Given the eth element's coordinates (x_a^e, y_a^e where $1 \leq a \leq n_{en}$), calculate global derivatives of the shape functions and Jacobian determinants:

> For $l = 1, \ldots, n_{\text{int}}$
>
> *Calculate*:
>
> $$x_{,\xi}(\tilde{\xi}_l, \tilde{\eta}_l) = \sum_{a=1}^{n_{en}} N_{a,\xi}(\tilde{\xi}_l, \tilde{\eta}_l)x_a^e$$
>
> $$x_{,\eta}(\tilde{\xi}_l, \tilde{\eta}_l) = \sum_{a=1}^{n_{en}} N_{a,\eta}(\tilde{\xi}_l, \tilde{\eta}_l)x_a^e$$
>
> $$y_{,\xi}(\tilde{\xi}_l, \tilde{\eta}_l) = \sum_{a=1}^{n_{en}} N_{a,\xi}(\tilde{\xi}_l, \tilde{\eta}_l)y_a^e$$
>
> $$y_{,\eta}(\tilde{\xi}_l, \tilde{\eta}_l) = \sum_{a=1}^{n_{en}} N_{a,\eta}(\tilde{\xi}_l, \tilde{\eta}_l)y_a^e$$
>
> $$j(\tilde{\xi}_l, \tilde{\eta}_l) = x_{,\xi}(\tilde{\xi}_l, \tilde{\eta}_l)y_{,\eta}(\tilde{\xi}_l, \tilde{\eta}_l) - x_{,\eta}(\tilde{\xi}_l, \tilde{\eta}_l)y_{,\xi}(\tilde{\xi}_l, \tilde{\eta}_l)$$
>
> For $a = 1, \ldots, n_{en}$
>
> *Calculate*:
>
> $$N_{a,x}(\tilde{\xi}_l, \tilde{\eta}_l) = \frac{N_{a,\xi}(\tilde{\xi}_l, \tilde{\eta}_l)y_{,\eta}(\tilde{\xi}_l, \tilde{\eta}_l) - N_{a,\eta}(\tilde{\xi}_l, \tilde{\eta}_l)y_{,\xi}(\tilde{\xi}_l, \tilde{\eta}_l)}{j(\tilde{\xi}_l, \tilde{\eta}_l)}$$
>
> $$N_{a,y}(\tilde{\xi}_l, \tilde{\eta}_l) = \frac{-[N_{a,\xi}(\tilde{\xi}_l, \tilde{\eta}_l)x_{,\eta}(\tilde{\xi}_l, \tilde{\eta}_l) - N_{a,\eta}(\tilde{\xi}_l, \tilde{\eta}_l)x_{,\xi}(\tilde{\xi}_l, \tilde{\eta}_l)]}{j(\tilde{\xi}_l, \tilde{\eta}_l)}$$

In obtaining the formulas for $N_{a,x}$ and $N_{a,y}$, we have combined (3.9.3) and (3.9.9). Subroutine QDCSHG performs these calculations in DLEARN for both the four-node bilinear quadrilateral and three-node linear triangle (using the "degeneration" technique of Sec. 3.4). The logical variable QUAD is used to indicate whether the element under consideration is a quadrilateral or triangle:

QUAD = .TRUE. (quadrilateral)

QUAD = .FALSE. (triangle)

Table 3.9.1 translates the text notation into the FORTRAN names used in QDCSHL and QDCSHG. The reader should study these routines carefully.

TABLE 3.9.1 Notation for Shape Function Subroutines QDCSHL and QDCSHG

Notation	FORTRAN name	Array dimensions
n_{int}	NINT (= 4)	
n_{sd}	NSD (= 2)	
n_{en}	NEN (= 4)	
W_l	W(L)	NINT
$\tilde{\xi}_l$	R	
$\tilde{\eta}_l$	S	
$N_a(\tilde{\xi}_l, \tilde{\eta}_l)$	SHL(3,I,L)	NROWSH × NEN × NINT (= 3 × 4 × 4)
$N_{a,\xi}(\tilde{\xi}_l, \tilde{\eta}_l)$	SHL(1,I,L)	
$N_{a,\eta}(\tilde{\xi}_l, \tilde{\eta}_l)$	SHL(2,I,L)	
x_a^e	XL(1,I)	NSD × NEN
y_a^e	XL(2,I)	
$N_a(\tilde{\xi}_l, \tilde{\eta}_l)$	SHG(3,I,L)	NROWSH × NEN × NINT
$N_{a,x}(\tilde{\xi}_l, \tilde{\eta}_l)$	SHG(1,I,L)	
$N_{a,y}(\tilde{\xi}_l, \tilde{\eta}_l)$	SHG(2,I,L)	
$\frac{\xi_a}{2}$	RA(I)	NEN
$\frac{\eta_a}{2}$	SA(I)	NEN
$\begin{bmatrix} x_{,\xi} & x_{,\eta} \\ y_{,\xi} & y_{,\eta} \end{bmatrix}$	XS(I,J)	NSD × NSD
$j(\tilde{\xi}_l, \tilde{\eta}_l)$	DET(L)	NINT

The case $n_{sd} = 3$ proceeds analogously. The main results are summarized as follows:

$$\mathrm{cof}_{11} = y_{,\eta}z_{,\zeta} - y_{,\zeta}z_{,\eta} \tag{3.9.11}$$

$$\mathrm{cof}_{12} = y_{,\zeta}z_{,\xi} - y_{,\xi}z_{,\zeta} \tag{3.9.12}$$

$$\mathrm{cof}_{13} = y_{,\xi}z_{,\eta} - y_{,\eta}z_{,\xi} \tag{3.9.13}$$

$$\mathrm{cof}_{21} = z_{,\eta}x_{,\zeta} - z_{,\zeta}x_{,\eta} \tag{3.9.14}$$

$$\mathrm{cof}_{22} = z_{,\zeta}x_{,\xi} - z_{,\xi}x_{,\zeta} \tag{3.9.15}$$

$$\mathrm{cof}_{23} = z_{,\xi}x_{,\eta} - z_{,\eta}x_{,\xi} \tag{3.9.16}$$

$$\mathrm{cof}_{31} = x_{,\eta}y_{,\zeta} - x_{,\zeta}y_{,\eta} \tag{3.9.17}$$

$$\mathrm{cof}_{32} = x_{,\zeta}y_{,\xi} - x_{,\xi}y_{,\zeta} \tag{3.9.18}$$

$$\mathrm{cof}_{33} = x_{,\xi}y_{,\eta} - x_{,\eta}y_{,\xi} \tag{3.9.19}$$

$$j = x_{,\xi}\mathrm{cof}_{11} + x_{,\eta}\mathrm{cof}_{12} + x_{,\zeta}\mathrm{cof}_{13} \tag{3.9.20}$$

$$N_{a,x} = \frac{N_{a,\xi}\mathrm{cof}_{11} + N_{a,\eta}\mathrm{cof}_{12} + N_{a,\zeta}\mathrm{cof}_{13}}{j} \tag{3.9.21}$$

$$N_{a,y} = \frac{N_{a,\xi}\mathrm{cof}_{21} + N_{a,\eta}\mathrm{cof}_{22} + N_{a,\zeta}\mathrm{cof}_{23}}{j} \tag{3.9.22}$$

$$N_{a,z} = \frac{N_{a,\xi}\mathrm{cof}_{31} + N_{a,\eta}\mathrm{cof}_{32} + N_{a,\zeta}\mathrm{cof}_{33}}{j} \tag{3.9.23}$$

The cof_{ij}'s are the cofactors of the matrix $\boldsymbol{x}_{,\boldsymbol{\xi}}$. (Recall the definition of matrix inverse: $(\boldsymbol{x}_{,\boldsymbol{\xi}})^{-1} = (\mathbf{cof})^T/j$.)

Exercise 1. Program a shape function subroutine for the eight-node trilinear brick. (To do this it will be helpful first to generalize explicitly the notation table (Table 3.9.1) to the three-dimensional case. Mimic the style of routines QDCSHL and QDCSHG.)

Exercise 2. Program shape function subroutines for the four- through nine-node element described in Fig. 3.7.4. Assume the IEN array is made available to the routines through the argument list and that if IEN(a) = 0, then node a is absent for the element under consideration. The IEN array may be taken to have dimension 9 within the routines. (Note that all operations performed to calculate the shape functions need also be performed to calculate the derivatives.)

Exercise 3. Generalize the subroutine of Exercise 1 to allow for the possibility of degeneration to a wedge-shaped element. Assume that the logical variable BRICK is used where

BRICK = .TRUE. (brick)

BRICK = .FALSE. (wedge)

The reader may set many additional shape-function-subroutine exercises by considering the different possibilities discussed in previous sections of this chapter.

3.10 ELEMENT STIFFNESS FORMULATION

Programming an element stiffness matrix is an essential step in learning the finite element method. There are several ways to go about this. In this section we describe three of the most important implementational styles of stiffness coding. Each has attributes and it is recommended that the reader fully understand each of these procedures.

Implementation 1. The first implementation employs the following form of the stiffness:

$$\boldsymbol{k}^e = \int_{\Omega^e} \boldsymbol{B}^T \boldsymbol{D} \boldsymbol{B} \, d\Omega = \int_{\square} \boldsymbol{B}^T \boldsymbol{D} \boldsymbol{B} j \, d\square \tag{3.10.1}$$

Applying numerical quadrature to (3.10.1) enables us to write

$$\boldsymbol{k}^e \cong \sum_{l=1}^{n_{\text{int}}} (\boldsymbol{B}^T \boldsymbol{D} \boldsymbol{B} j)\Big|_{\tilde{\xi}_l} W_l \tag{3.10.2}$$

It simplifies the coding to combine j and W_l with $\boldsymbol{D}$. In this case we write

$$\tilde{\boldsymbol{D}} = j(\tilde{\xi}_l) W_l \boldsymbol{D} \tag{3.10.3}$$

and thus

$$\boxed{\boldsymbol{k}^e \cong \sum_{l=1}^{n_{\text{int}}} (\boldsymbol{B}^T \tilde{\boldsymbol{D}} \boldsymbol{B})_l} \tag{3.10.4}$$

In the actual FORTRAN programming $\tilde{\boldsymbol{D}}$ is written over $\boldsymbol{D}$. Thus no additional storage for $\tilde{\boldsymbol{D}}$ is required. This implementation is used in subroutines QDCK (quadrilateral continuum $\boldsymbol{k}^e$) and TRUSK (truss $\boldsymbol{k}^e$) in DLEARN (see Chapter 11). The main steps are summarized next.

> For $l = 1, \ldots, n_{\text{int}}$
>
> Set up the strain-displacement matrix $\boldsymbol{B}$.
> Set up the constitutive matrix $\tilde{\boldsymbol{D}}$.
> Multiply $\tilde{\boldsymbol{D}} * \boldsymbol{B}$.
> Multiply $\boldsymbol{B}^T * (\tilde{\boldsymbol{D}}\boldsymbol{B})$, taking account of symmetry, and accumulate in $\boldsymbol{k}^e$.

Table 3.10.1 translates the text notation into the FORTRAN names used in QDCK and TRUSK (see also Table 3.9.1 for names of shape function variables).

TABLE 3.10.1 Notation for Element Stiffness Routines QDCK and TRUSK

Notation	FORTRAN name	Array dimensions
n_{ee}	NEE	
$\boldsymbol{B}$	B	NROWB[a] × NEE
$\boldsymbol{D}$	C	NROWB × NROWB
$\tilde{\boldsymbol{D}}$	DMAT	NROWB × NROWB
$\widetilde{\boldsymbol{DB}}$	DB	NROWB × NEE
$\boldsymbol{k}^e$	ELSTIF	NEE × NEE

[a] NROWB is a variable used to dimension the arrays. In subroutine QDCK, NROWB = 4. This dimension is needed, for example, in the axisymmetric case. In plane stress, however, only the first three rows are needed. To save calculations, another variable is introduced, namely, NSTR (3 in plane stress; 4 in the axisymmetric case), which is used to define the limits of the matrix multiplication loops. To appreciate this point, the reader should examine the calling statement to subroutine MULTAB in QDCK and subroutine MULTAB itself.

Exercise 1. Determine formulas for the numbers of multiplications and additions required to form an element stiffness by the above procedure.

Implementation 2. In the opinion of the author the method of coding just described is the most important because it is completely general and does not assume any special structure of $\boldsymbol{B}$ or $\boldsymbol{D}$. However, if special structure is present, then greater efficiency can be gained. For example, if there are zero entries in $\boldsymbol{B}$ and/or $\boldsymbol{D}$, then operations can be saved in the multiplications $\tilde{\boldsymbol{D}} * \boldsymbol{B}$ and $\boldsymbol{B}^T * (\widetilde{\boldsymbol{DB}})$. This requires programming special subroutines for these calculations.

Example 1

Consider two-dimensional, isotropic, plane-stress elasticity theory. The matrices $\boldsymbol{B}$ and $\boldsymbol{D}$ take the form:

$$\boldsymbol{B} = [\boldsymbol{B}_1, \ldots, \boldsymbol{B}_{n_{en}}] \tag{3.10.5}$$

$$\boldsymbol{B}_a = \begin{bmatrix} N_{a,x} & 0 \\ 0 & N_{a,y} \\ N_{a,y} & N_{a,x} \end{bmatrix}, \qquad 1 \le a \le n_{en} \tag{3.10.6}$$

$$\boldsymbol{D} = \begin{bmatrix} D_{11} & D_{12} & 0 \\ & D_{22} & 0 \\ \text{symmetric} & & D_{33} \end{bmatrix} \tag{3.10.7}$$

Recall that $\boldsymbol{k}^e$ can be written in terms of $n_{en} \times n_{en}$ nodal submatrices, $\boldsymbol{k}^e_{ab}$, where (see (2.9.6) through (2.9.9))

$$\underbrace{\boldsymbol{k}^e_{ab}}_{n_{ed} \times n_{ed}} = \int_{\Omega^e} \boldsymbol{B}_a^T \boldsymbol{D} \boldsymbol{B}_b \, d\Omega \tag{3.10.8}$$

We can write (3.10.8) in numerical-integration form as follows:

$$\boxed{\boldsymbol{k}^e_{ab} \cong \sum_{l=1}^{n_{\text{int}}} (\boldsymbol{B}_a^T \tilde{\boldsymbol{D}} \boldsymbol{B}_b)_l} \tag{3.10.9}$$

The nodal submatrix $\boldsymbol{k}^e_{ab}$ has dimension 2×2 in two dimensions, viz.,

$$\boldsymbol{k}^e_{ab} = \begin{bmatrix} k^e_{2a-1,\,2b-1} & k^e_{2a-1,\,2b} \\ k^e_{2a,\,2b-1} & k^e_{2a,\,2b} \end{bmatrix} \tag{3.10.10}$$

In the computer implementation, $\boldsymbol{k}^e$ may be computed by calculating the nodal submatrices, namely, (3.10.10), for each a and b such that $1 \leq a, b \leq n_{en}$. Only the nodal submatrices on or above the diagonal need be computed because the lower part is determined by symmetry. The procedure is given by the algorithm in the box.

For $l = 1, \ldots, n_{\text{int}}$

  Set up $\tilde{\boldsymbol{D}}$.

  For $b = 1, \ldots, n_{en}$

    Let $B_1 = N_{b,x}$ and $B_2 = N_{b,y}$.

    Multiply $\tilde{\boldsymbol{D}} * \boldsymbol{B}_b$, taking account of zeros in $\tilde{\boldsymbol{D}}$ and $\boldsymbol{B}_b$:

$$\widetilde{\boldsymbol{DB}}_b = \begin{bmatrix} \widetilde{DB}_{11} & \widetilde{DB}_{12} \\ \widetilde{DB}_{21} & \widetilde{DB}_{22} \\ \widetilde{DB}_{31} & \widetilde{DB}_{32} \end{bmatrix} = \begin{bmatrix} \tilde{D}_{11} B_1 & \tilde{D}_{12} B_2 \\ \tilde{D}_{12} B_1 & \tilde{D}_{22} B_2 \\ \tilde{D}_{33} B_2 & \tilde{D}_{33} B_1 \end{bmatrix}$$

    For $a = 1, \ldots, b$

      Let $B_1 = N_{a,x}$ and $B_2 = N_{a,y}$.

      Multiply $\boldsymbol{B}_a^T * (\widetilde{\boldsymbol{DB}}_b)$, taking account of zeros in $\boldsymbol{B}_a$, and accumulate in $\boldsymbol{k}^e$:

$$\begin{bmatrix} k^e_{2a-1,\,2b-1} & k^e_{2a-1,\,2b} \\ k^e_{2a,\,2b-1} & k^e_{2a,\,2b} \end{bmatrix} \leftarrow \begin{bmatrix} k^e_{2a-1,\,2b-1} & k^e_{2a-1,\,2b} \\ k^e_{2a,\,2b-1} & k^e_{2a,\,2b} \end{bmatrix} + \begin{bmatrix} (B_1 \widetilde{DB}_{11} + B_2 \widetilde{DB}_{31}) & (B_1 \widetilde{DB}_{12} + B_2 \widetilde{DB}_{32}) \\ (B_2 \widetilde{DB}_{21} + B_1 \widetilde{DB}_{31}) & (B_2 \widetilde{DB}_{22} + B_1 \widetilde{DB}_{32}) \end{bmatrix}$$

Exercise 2. Determine the numbers of multiplications and additions required to form a four-node element stiffness by the boxed procedure. Specialize the results of Exercise 1 to the present case and compare. (The savings in three dimensions are even more impressive!)

Implementation 3. This implementation was first proposed by Gupta and Mohraz [19]. It emanates from the following development. Consider two- and three-dimensional elasticity theory (see Secs. 2.7 through 2.11). Note that for these cases $n_{ed} = n_{sd}$ (i.e., the number of element degrees of freedom equals the number of space dimensions). Let

$$w_i^h = \sum_{a=1}^{n_{en}} c_{ia} N_a, \qquad u_i^h = \sum_{a=1}^{n_{en}} d_{ia} N_a \tag{3.10.11}$$

Recall the equivalent single-index representations of the coefficients,

$$c_p = c_{ia}, \qquad d_q = d_{jb} \tag{3.10.12}$$

where

$$p = n_{ed}(a-1) + i, \qquad q = n_{ed}(b-1) + j \tag{3.10.13}$$

With these we may write[2]

$$\begin{aligned}\int_{\Omega^e} w^h_{(i,k)} c_{ikjl} u^h_{(j,l)}\, d\Omega &= \int_{\Omega^e} w^h_{i,k} c_{ikjl} u^h_{j,l}\, d\Omega \quad \text{(by the minor symmetries of the } c_{ijkl}\text{'s and Lemma 2 of Sec. 2.7)}\\ &= \sum_{a,b=1}^{n_{en}} c_{ia}\left(\int_{\Omega^e} N_{a,k} c_{ikjl} N_{b,l}\, d\Omega\right) d_{jb}\\ &= \sum_{a,b=1}^{n_{en}} c_{ia} k^e_{iajb} d_{jb} = \sum_{p,q=1}^{n_{ee}} c_p k^e_{pq} d_q\end{aligned} \tag{3.10.14}$$

from which it follows that

$$k^e_{pq} = k^e_{iajb} = \int_{\Omega^e} N_{a,k} c_{ikjl} N_{b,l}\, d\Omega \tag{3.10.15}$$

It is convenient here to work with the four-indexed version of the element stiffness, namely, k^e_{iajb}. In FORTRAN, the two-indexed version and four-indexed version are stored in identical fashion in consecutive addresses in memory if the indices are arranged in the order given. This enables us to deal with the array ELSTIF (i.e., $\boldsymbol{k}^e$) as a two-dimensional array in one routine and as a four-dimensional array in another.

Two-dimensional version	Four-dimensional version
DIMENSION ELSTIF(NEE, NEE) ELSTIF(IP, JQ) where IP = NED*(IA − 1) + I JQ = NED*(JB − 1) + J	DIMENSION ELSTIF(NED, NEN, NED, NEN) ELSTIF(I, IA, J, JB) (3.10.16)

[2]Recall that the summation over repeated spatial indices (i.e., i, j, k, l) is in force.

Assume that the material occupying Ω^e is ***isotropic,*** that is,

$$c_{ikjl} = \mu(\delta_{ij}\delta_{kl} + \delta_{il}\delta_{kj}) + \lambda\delta_{ik}\delta_{jl} \tag{3.10.17}$$

and ***homogeneous,*** that is, λ and μ are constants. With these assumptions, (3.10.15) can be written as:

$$\begin{aligned} k^e_{iajb} &= (\mu(\delta_{ij}\delta_{kl} + \delta_{il}\delta_{kj}) + \lambda\delta_{ik}\delta_{jl}) \int_{\Omega^e} N_{a,k}N_{b,l}\, d\Omega \\ &= \mu\left(\delta_{ij}\int_{\Omega^e} N_{a,k}N_{b,k}\, d\Omega + \int_{\Omega^e} N_{a,j}N_{b,i}\, d\Omega\right) \\ &\quad + \lambda \int_{\Omega^e} N_{a,i}N_{b,j}\, d\Omega \end{aligned} \tag{3.10.18}$$

Thus to calculate the element stiffness, the expression

$$\int_{\Omega^e} N_{a,i}N_{b,j}\, d\Omega \cong \sum_{l=1}^{n_{\text{int}}} (N_{a,i}N_{b,j}j)\bigg|_{\tilde{\xi}_l} W_l \tag{3.10.19}$$

must be evaluated first. Observe that (3.10.19) is symmetric under the interchange of indices $(i, a) \leftrightarrow (j, b)$. This fact reduces the number of calculations involved. The algorithm in the following box calculates the element stiffness based upon (3.10.18) and (3.10.19). Note that the integrals, (3.10.19), are first stored in k^e_{iajb}, which is then overwritten by the stiffness. This procedure leads to additional savings compared with Implementation 2.

```
c1: = λ + μ
c2: = μ
c3: = λ
For l = 1, . . . , n_int
    const = j(ξ̃_l)W_l
    For b = 1, . . . , n_en
        For j = 1, . . . , n_ed
            temp = const · N_b,j(ξ̃_l)
            For a = 1, . . . , b
                For i = 1, . . . , j
                k^e_iajb ← k^e_iajb + temp · N_a,i(ξ̃_l)
For b = 1, . . . , n_en
    For a = 1, . . . , b
        temp = Σ_{k=1}^{n_ed} k^e_kakb
```

For $j = 1, \ldots, n_{ed}$
  For $i = 1, \ldots, j$
    if $i = j$, then
      $k^e_{iaib} \leftarrow c_1 k^e_{iaib} + c_2 \text{temp}$
    else
      if $a = b$, then
        $k^e_{iaja} \leftarrow c_1 k^e_{iaja}$
      else
        $k^e_{iajb} \leftarrow c_3 k^e_{iajb} + c_2 k^e_{jaib}$
      endif
    endif

Exercise 3. Determine the number of multiplications and additions for a four-node element stiffness and compare with results of Exercise 2.

Exercise 4. Determine the numbers of multiplications and additions for an eight-node brick element for each of the three implementations. Compare.

Discussion

The first implementation is the most general and easiest to program. The second is somewhat more specialized but registers a significant gain in efficiency. The third is the most specialized and most efficient. In anticipation of extending the present ideas to nonlinear situations, we may note that the matrix $\boldsymbol{D}$ is typically full. Furthermore, a popular method for improving the behavior of elements in both linear and nonlinear analysis engenders a full $\boldsymbol{B}$. (This method is described in Sec. 4.5.) Under these circumstances the first implementation is still applicable, whereas the others are not. However, the latter two implementations are preferable for the cases to which they pertain. The "best" implementation clearly depends upon what our goals are. For additional information on finite element programming see [20–24].

3.11 ADDITIONAL EXERCISES

Exercise 1. Consider a boundary-value problem in which

$$\Omega = \{(x, y) \mid x \in \,]a, b[\,, y \in \,]a, b[\}$$

A mesh of isoparametric elements is proposed for obtaining an approximate solution to the boundary-value problem (see Figure 3.11.1), in which elements 1 and 2 are eight-node quadrilaterals and elements 3 and 4 are four-node quadrilaterals. What fundamental requirement of the finite element spaces is not met by this mesh? Without changing the number of nodes, how would you fix the mesh?

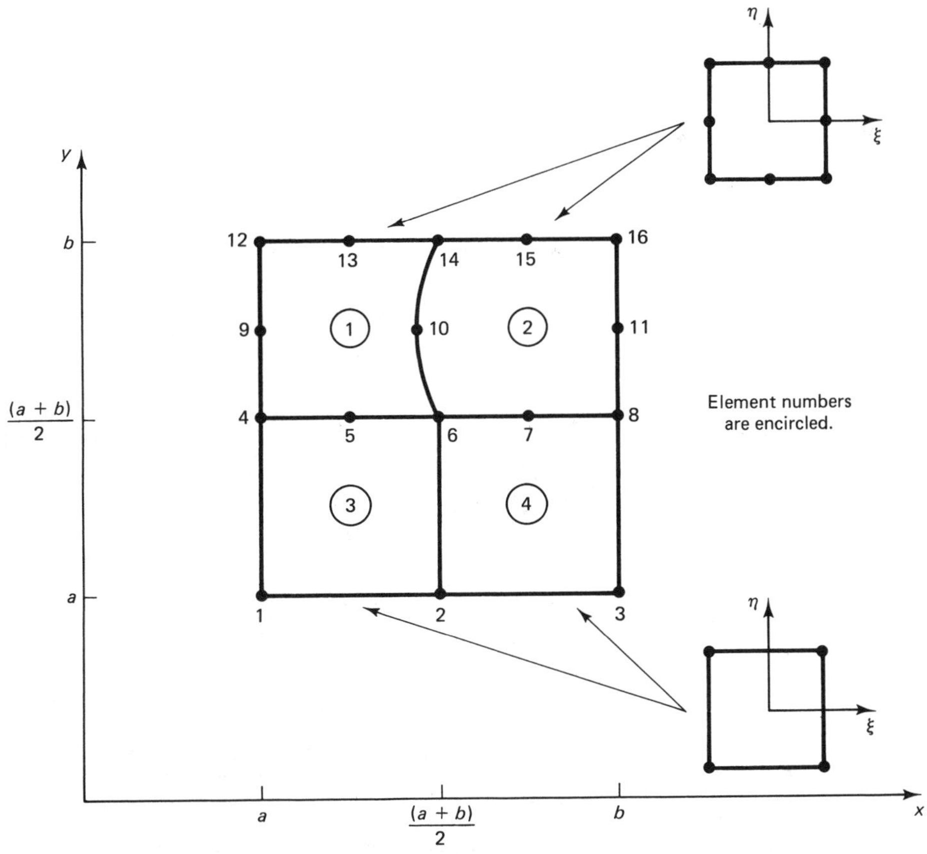

Figure 3.11.1

Exercise 2. Consider a one-dimensional, quadratic, three-node element shown in Fig. 3.11.2. The following relations hold:

$$x(\xi) = \sum_{a=1}^{3} N_a(\xi)x_a^e \qquad h^e = x_3^e - x_1^e$$

$$u^h(\xi) = \sum_{a=1}^{3} N_a(\xi)d_a^e \qquad \boldsymbol{f}^e = \{f_a^e\}$$

$$N_1(\xi) = \tfrac{1}{2}\xi(\xi - 1) \qquad f_a^e = \int_{x_1^e}^{x_3^e} N_a \ell \, dx$$

$$N_2(\xi) = 1 - \xi^2 \qquad \boldsymbol{k}^e = [k_{ab}^e]$$

$$N_3(\xi) = \tfrac{1}{2}\xi(\xi + 1) \qquad k_{ab}^e = \int_{x_1^e}^{x_3^e} N_{a,x}N_{b,x}\, dx, \quad 1 \leq a, b \leq 3$$

Given a "loading" ℓ, the elements of the consistently derived "force" vector $\boldsymbol{f}^e$ are sometimes surprising, especially for higher-order elements. This problem establishes

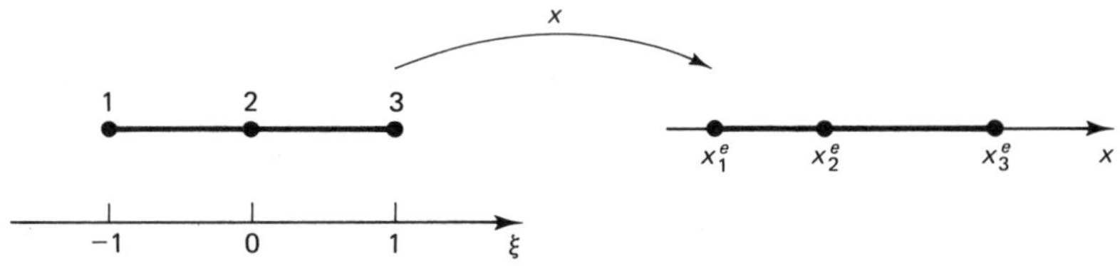

Figure 3.11.2

some basic results concerning the calculation of $\boldsymbol{f}^e$, appropriate quadrature formulas, and Barlow stress points for the three-node element.

a. Assume ℓ = constant and let $x_2^e = (x_1^e + x_3^e)/2$. Determine exact expressions for f_a^e, $a = 1, 2, 3$.

b.[3] Assume $\ell = \delta(x - \bar{x})$, $x_1^e \le \bar{x} \le x_3^e$, (the delta function) and let $x_2^e = (x_1^e + x_3^e)/2$. Obtain an exact expression for f_a^e, $a = 1, 2, 3$, and specialize for the cases $\bar{x} = x_1^e$ and $\bar{x} = x_2^e$.

c. Assume ℓ = constant, but make no assumption on the location of x_2^e other than $x_1^e < x_2^e < x_3^e$. Determine the lowest-order Gaussian quadrature formula ($n_{\text{int}} = ?$) which exactly integrates $\boldsymbol{f}^e$. Justify your answer.

d. Assume $x_2^e = (x_1^e + x_3^e)/2$. Determine the lowest-order Gaussian quadrature formula ($n_{\text{int}} = ?$) which exactly integrates $\boldsymbol{k}^e$. Justify your answer.

e. An analysis of the quadratic element is being performed to locate the Barlow stress points (i.e., the points at which $e_{,x} = u^h_{,x} - u_{,x}$ optimally converges). It is assumed in the analysis that $x_2^e = (x_1^e + x_3^e)/2$ and the nodal values are exact, that is, $d_a^e = u(x_a)$. Determine the Barlow stress points.

Solution of part (e) We present a simple solution to illustrate that we do not have to perform a lot of algebraic manipulations to solve problems of this type.

Note that the position of node 2 implies that

[3]We need to be careful in changing variables when delta functions are present. Consider the multidimensional case in which

$$d\Omega = j\, d\square$$

The delta function transforms by multiplication with the *inverse* Jacobian determinant:

$$\delta_{\bar{\boldsymbol{x}}} = \delta(\boldsymbol{x} - \bar{\boldsymbol{x}}) = j^{-1}\delta(\boldsymbol{\xi} - \bar{\boldsymbol{\xi}}) = j^{-1}\delta_{\bar{\boldsymbol{\xi}}}$$

where

$$\bar{\boldsymbol{x}} = \boldsymbol{x}(\bar{\boldsymbol{\xi}})$$

In one dimension, this specializes as follows:

$$dx = x_{,\xi}(\xi)\, d\xi$$

$$\delta_{\bar{x}} = \delta(x - \bar{x}) = (x_{,\xi})^{-1}\delta(\xi - \bar{\xi}) = (x_{,\xi})^{-1}\delta_{\bar{\xi}}$$

$$\bar{x} = x(\bar{\xi})$$

Further details concerning transformation rules for generalized functions may be found in Chapter 5 of I. Stakgold, *Boundary Value Problems of Mathematical Physics,* vol. II. New York: Macmillan, 1968.

$$e_{,x} = e_{,\xi}\xi_{,x} = \frac{2e_{,\xi}}{h^e}$$

Observe that if $u(\xi)$ is a quadratic polynomial, then $u^h(\xi) = u(\xi)$ by the assumption that the nodal values are exact. Therefore, take $u(\xi) = c\xi^3$, where c is an arbitrary constant. Compute as follows:

$$\begin{aligned} e_{,\xi} &= u^h_{,\xi} - u_{,\xi} \\ &= c(-\xi + \tfrac{1}{2} + \xi + \tfrac{1}{2} - 3\xi^2) \end{aligned}$$

Thus $e_{,\xi} = 0$ at $\xi = \pm 1\sqrt{3}$, the Gauss points of the two-point rule. If a formal Taylor-expansion approach is adopted, the coefficient of the $u_{,\xi\xi\xi}$ term leads to the same result after considerably more work!

f. Let

$$x_2^e = x_1^e + \frac{h^e}{4} \qquad \text{(quarter point)}$$

$$r = \frac{x - x_1^e}{h^e}$$

$$h^e = x_3^e - x_1^e$$

Determine an expression for $u^h_{,r}(r)$ and indicate the order of the singularity at $r = 0$ [i.e., determine α, where $u^h_{,r} = O(r^{-\alpha})$]. (This result is useful for the construction of ***crack elements***.)

Exercise 3. It is desired to develop a ***transition element,*** which will facilitate abrupt changes in mesh refinement between domains consisting of bilinear quadrilateral elements. The idea is illustrated in Fig. 3.11.3. The transition element is to be designed to maintain continuity across all interfaces. A shape function associated with node 5 which achieves this end is illustrated in Fig. 3.11.4. Starting with the usual bilinear shape functions, determine the appropriate modifications. Sketch the modified shape function associated with node 1.

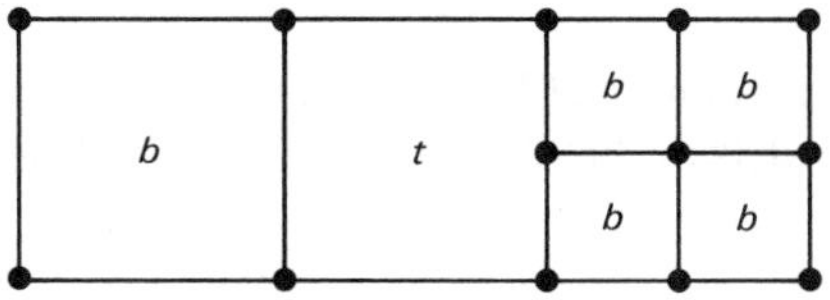

Key: b = bilinear quadrilateral
t = transition element

Figure 3.11.3

Exercise 4. Let

$$\begin{Bmatrix} r(\xi, \eta) \\ \theta(\xi, \eta) \end{Bmatrix} = \sum_{a=1}^{4} N_a(\xi, \eta) \begin{Bmatrix} r_a^e \\ \theta_a^e \end{Bmatrix}$$

Determine the N_a's for a ***sector element*** (see Fig. 3.11.5), so that the following conditions are met:

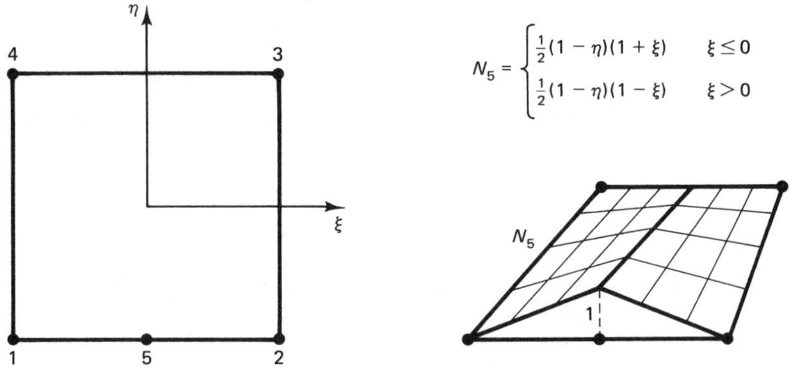
η
4
3
ξ
1
5
2
$N_5 = \begin{cases} \frac{1}{2}(1-\eta)(1+\xi) & \xi \le 0 \\ \frac{1}{2}(1-\eta)(1-\xi) & \xi > 0 \end{cases}$
N_5
1

Figure 3.11.4

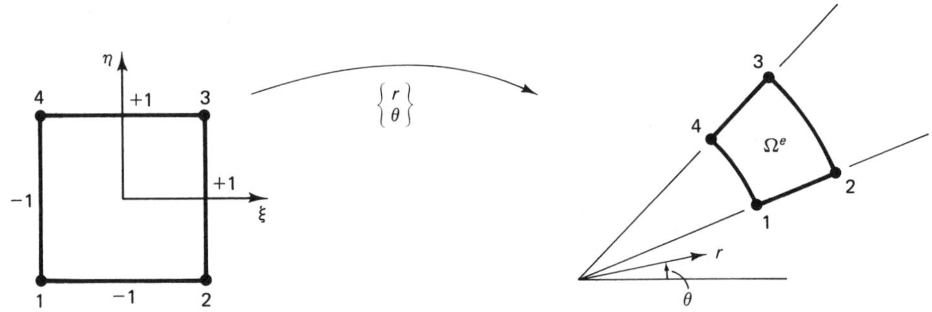

Figure 3.11.5

i. If $r_4 = r_1$, then $r(-1, \eta) = r_1$.
ii. If $r_3 = r_2$, then $r(+1, \eta) = r_2$.
iii. If $\theta_2 = \theta_1$, then $\theta(\xi, -1) = \theta_1$.
iv. If $\theta_4 = \theta_3$, then $\theta(\xi, +1) = \theta_3$.

Exercise 5. (See the solution of Exercise 2, part (e).) Determine the Barlow stress points for the four-node element in one dimension. Assume that

$$x_2^e = x_1^e + \frac{h^e}{3}$$

$$x_3^e = x_1^e + \frac{2h^e}{3}$$

$$h^e = x_4^e - x_1^e$$

$$d_a^e = u(x_a^e)$$

The regular positioning of the internal nodes implies $x_{,\xi} = h^e/2$. (*Ans.:* 0, $\pm\ (\frac{5}{9})^{1/2}$. The three Barlow points in this example do *not* coincide with the Gauss points of the three-point rule!)

Exercise 6. A boundary-value problem for the ***convection-diffusion equation*** is as follows: Given $f : \Omega \to \mathbb{R}$, $g : \Gamma_g \to \mathbb{R}$, and $h : \Gamma_h \to \mathbb{R}$, find $u : \overline{\Omega} \to \mathbb{R}$ such that

$$
\begin{aligned}
b_i u_{,i} &= (\kappa_{ij} u_{,j})_{,i} + f && \text{in } \Omega \\
u &= g && \text{on } \Gamma_g \\
\kappa_{ij} u_{,j} n_i &= h && \text{on } \Gamma_h
\end{aligned}
$$

a. Establish a weak formulation for this problem in which the boundary condition on Γ_g is essential and the boundary condition on Γ_h is natural.
b. Establish the Galerkin formulation and obtain expressions for K_{PQ} and F_P.
c. Show that if $b_{j,j} = 0$ on Ω and $b_j n_j = 0$ on Γ_h, then the b-term contribution to K_{PQ} is skew-symmetric (i.e., $K^b_{PQ} = -K^b_{QP}$).

Exercise 7.

a. Consider the four-node bilinear element. Prove that the one-point Gauss quadrature formula in two dimensions is sufficient to exactly integrate the area

$$\int_{\Omega^e} d\Omega$$

b. Let

$$u_i^h(\xi, \eta) = \sum_{a=1}^{4} N_a(\xi, \eta) d_{ia}^e, \qquad i = 1, 2$$

where the N_a's are the bilinear shape functions. Show that if

$$u_{i,i}^h(0, 0) = 0$$

then

$$\int_{\Omega^e} u_{i,i}^h \, d\Omega = 0 \qquad (\text{“mean incompressibility”})$$

(Part (b) requires some work.)

Exercise 8. Consider a bilinear quadrilateral element. Assume that the prescribed surface traction (i.e., h_i) along one edge is constant. What is the lowest-order, one-dimensional Gauss quadrature formula which will exactly integrate the prescribed traction contribution to $\boldsymbol{f}^e$? Justify your answer.

Solution The setup is shown in Fig. 3.11.6. Note that

Piecewise linear shape functions

$$s = N_a(\xi) s_a + N_b(\xi) s_b$$

Calculate as follows:

$$\int_{\Gamma_{h_i}} N_a h_i \, d\Gamma = \int_0^L N_a h_i \, ds \qquad \text{(continued)}$$

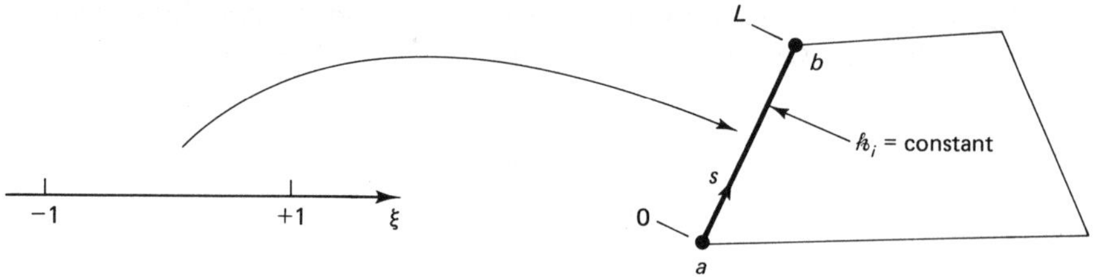

Figure 3.11.6

$$= h_i \int_{-1}^{+1} N_a(\xi)\frac{L}{2}\, d\xi$$

$$= \frac{L}{2} h_i \int_{-1}^{+1} \underbrace{N_a(\xi)}_{\text{Linear}}\, d\xi$$

$$= \frac{L}{2} h_i$$

Because the integral involves a linear polynomial in ξ, one-point Gauss quadrature suffices.

Exercise 9. Consider the eight-node isoparametric element shown:

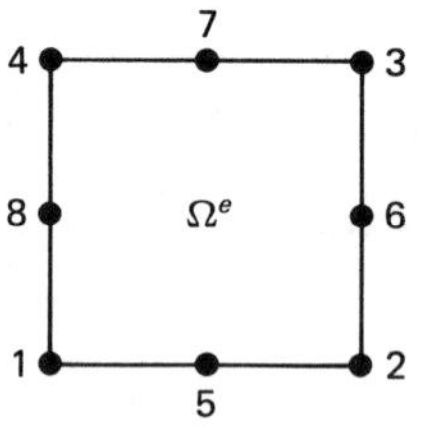

Assume the nodal coordinates are as follows:

a	x_a	y_a
1	$-\frac{h}{2}$	$-\frac{h}{2}$
2	$\frac{h}{2}$	$-\frac{h}{2}$
3	$\frac{h}{2}$	$\frac{h}{2}$
4	$-\frac{h}{2}$	$\frac{h}{2}$
5	0	$-\frac{h}{2}$
6	$\frac{h}{2}$	0
7	0	$\frac{h}{2}$
8	$-\frac{h}{2}$	0

Consider the following body force:

$$\ell = \begin{Bmatrix} 0 \\ -g \end{Bmatrix}; \qquad g \text{ constant}$$

Compute (by hand) the nodal forces for nodes 1 and 5, i.e.,

$$f_{21}^e = \int_{\Omega^e} N_1 \ell_2 \, dx$$

$$f_{25}^e = \int_{\Omega^e} N_5 \, \ell_2 \, dx$$

(Note that $f_{1a} = 0$, $1 \leq a \leq 8$, and, by symmetry, $f_{21}^e = f_{22}^e = f_{23}^e = f_{24}^e$ and $f_{25}^e = f_{26}^e = f_{27}^e = f_{28}^e$.)

The results of this simple computation should be somewhat surprising. Comment.

Exercise 10. Generalize Exercise 8 to the case in which the quadrilateral is of biquadratic type. Assume that the three nodes along an edge define a curved edge (i.e., they do not lie along a straight line). [*Ans.:* Two-point Gauss.]

Exercise 11. Generalize Exercise 10 to the case in which h_i represents normal pressure. That is,

$$h_i = -pn_i$$

where p, the applied pressure, is assumed constant and n_i is the unit outward normal to the boundary. [*Ans.:* Three-point Gauss.]

Exercise 12. Consider a trilinear brick element subjected to normal pressure along one surface. What is the lowest-order, two-dimensional Gauss quadrature formula which will exactly integrate the prescribed traction contribution to $\boldsymbol{f}^e$? Justify your answer. *Hint:* The following change-of-variables formula is useful in situations like these:

$$\int_{\Gamma_{h_i}} N_a h_i \, d\Gamma = \int_{-1}^{+1} \int_{-1}^{+1} N_a h_i \, \| \boldsymbol{x}_{,\xi} \times \boldsymbol{x}_{,\eta} \| \, d\xi \, d\eta$$

where $\times$ denotes cross product and $\| \cdot \|$ is the Euclidean length.

Exercise 13. Generalize Exercise 12 to the case of the triquadratic brick element.

Appendix 3.I

Triangular and Tetrahedral Elements

It is often stated that triangular and tetrahedral elements are responsible for the geometric flexibility of the finite element method. This is perhaps somewhat of an exaggeration as triangular and tetrahedral shapes are often not needed in practice. Most regions are conveniently discretized by arbitrary quadrilateral and brick-shaped elements, as described earlier.

It is also often stated that triangles and tetrahedra enable modeling of particularly intricate geometries and that these shapes facilitate transition from coarsely meshed zones of a grid to finely meshed zones. This is, of course, true, but quadrilaterals and bricks are capable of doing the same thing at least to some degree. To illustrate this point, consider Fig. 3.I.1 in which a triangular zone is discretized into three quadrilaterals. Thus we see that any triangular element could be replaced by quadrilaterals. (Mesh generation for triangular regions via quadrilaterals may be handled similarly; see Fig. 3.I.2.) Quadrilaterals may also be used to perform mesh transition as illustrated in Fig. 3.I.3.

Nevertheless, if a few triangles are desired (or tetrahedra, or wedges) in any particular situation, they may be easily obtained from quadrilaterals (respectively, bricks) by way of the element degeneration technique described previously. Consequently, no special finite element subroutines are required.

However, because many individuals are fond of particular triangular and tetrahedral elements and do not perceive them as being degenerated from quadrilaterals and bricks and because there are a number of useful triangular and tetrahedral elements for which there are no quadrilateral and brick counterparts, we shall describe the machinery necessary for the direct development of triangular and tetrahedral shape functions. The following techniques should also be employed if extensive use of triangles or tetrahedra is envisioned, because, in this case, the degeneration procedure proves inefficient.

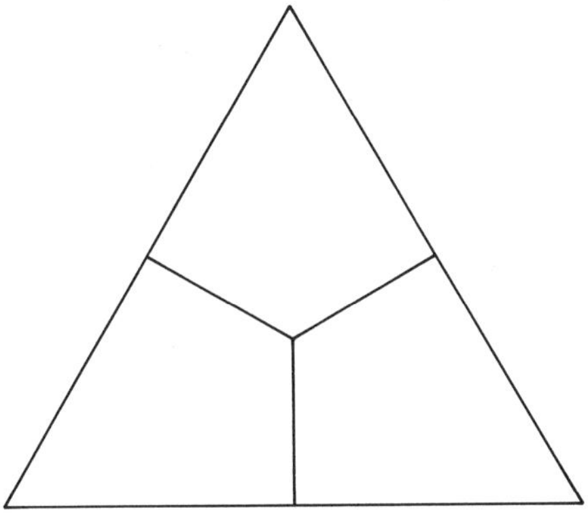

Figure 3.I.1

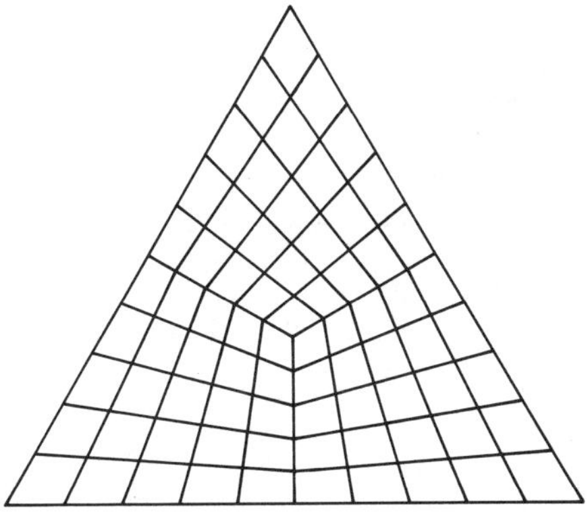

Figure 3.I.2

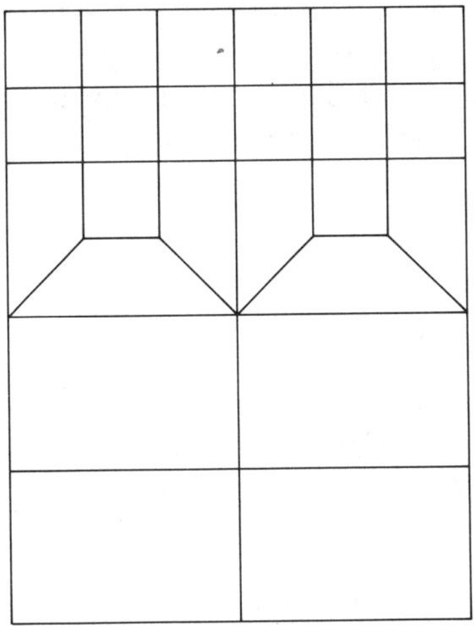

Figure 3.I.3

Shape Functions for Triangles

Consider the ***parent triangle*** illustrated in Fig. 3.I.4. This serves as the counterpart of the biunit square in ξ, η-space which was described in Sec. 3.2. We take r and s as the independent natural coordinates. Note that the equation of the inclined edge is

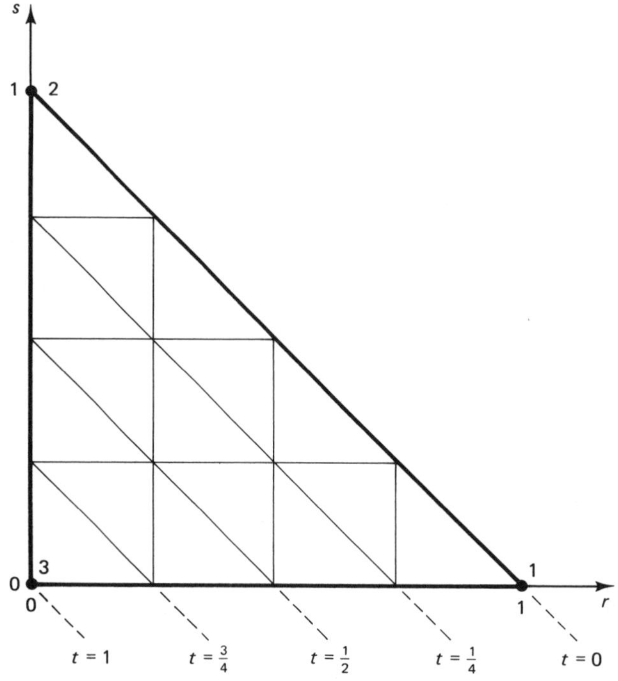

Figure 3.I.4

$1 - r - s = 0$. If we define

$$t = t(r, s) = 1 - r - s \tag{3.I.1}$$

then the family of t = constant lines are parallel to the inclined edge. The triple r, s, t defines ***triangular coordinates.*** Each is zero along one edge and takes on the value 1 at the opposite vertex. The shape functions of the piecewise linear triangle may be concisely expressed with the aid of the triangular coordinates, viz.,

$$N_1(r, s) = r \tag{3.I.2}$$

$$N_2(r, s) = s \tag{3.I.3}$$

$$N_3(r, s) = t(r, s) = 1 - r - s \tag{3.I.4}$$

These functions are equivalent to those computed by the degeneration technique in Sec. 3.4. Higher-order shape functions for triangles may be systematically defined through the use of triangular coordinates. The general formula for ***Lagrange-type interpolation*** over triangles is given by

$$N_a(r, s, t) = T_I(r)T_J(s)T_K(t) \tag{3.I.5}$$

where

$$T_I(r) = \begin{cases} l_I^{I-1}\left(\dfrac{2r}{r_I} - 1\right), & I \neq 1 \\ 1, & I = 1 \end{cases} \tag{3.I.6}$$

and $a = a(I, J, K)$ is the formula defining the single nodal index and l in (3.I.6) stands for the Lagrange interpolation formula (see Sec. 3.6). Illustrations of the use of (3.I.5) are given in the following examples.

Example 1

Consider the three-node triangle depicted in Fig. 3.I.5. The shape functions are:

$$N_1(r, s, t) = T_2(r)T_1(s)T_1(t) = l_2^1\left(\frac{2r}{r_2 - 1}\right) = r \tag{3.I.7}$$

$$N_2(r, s, t) = T_1(r)T_2(s)T_1(t) = l_2^1\left(\frac{2s}{s_2 - 1}\right) = s \tag{3.I.8}$$

$$N_3(r, s, t) = T_1(r)T_1(s)T_2(t) = l_2^1\left(\frac{2t}{t_2 - 1}\right) = t \tag{3.I.9}$$

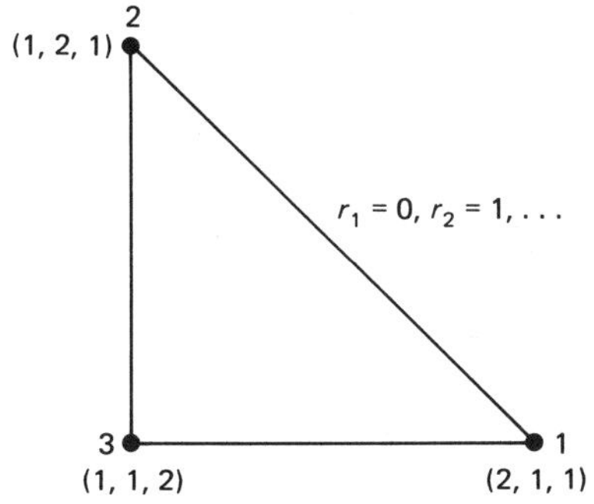

Figure 3.I.5

Example 2

Consider the six-node triangle shown in Fig. 3.I.6.

$$N_1(r, s, t) = T_3(r)T_1(s)T_1(t) = l_3^2\left(\frac{2r}{r_3 - 1}\right) = r(2r - 1) \tag{3.I.10}$$

$$N_2(r, s, t) = T_1(r)T_3(s)T_1(t) = l_3^2\left(\frac{2s}{s_3 - 1}\right)$$

(continued)

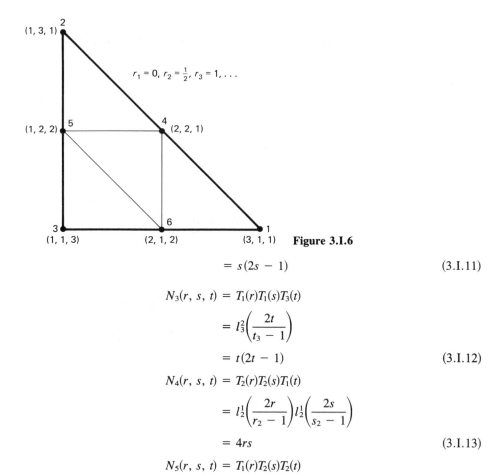

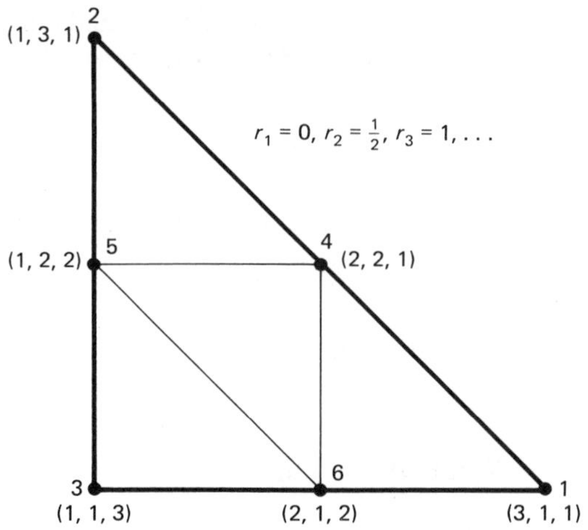

Figure 3.I.6

$$= s(2s - 1) \tag{3.I.11}$$

$$N_3(r, s, t) = T_1(r)T_1(s)T_3(t)$$

$$= l_3^2\left(\frac{2t}{t_3 - 1}\right)$$

$$= t(2t - 1) \tag{3.I.12}$$

$$N_4(r, s, t) = T_2(r)T_2(s)T_1(t)$$

$$= l_2^1\left(\frac{2r}{r_2 - 1}\right)l_2^1\left(\frac{2s}{s_2 - 1}\right)$$

$$= 4rs \tag{3.I.13}$$

$$N_5(r, s, t) = T_1(r)T_2(s)T_2(t)$$

$$= l_2^1\left(\frac{2s}{s_2 - 1}\right)l_2^1\left(\frac{2t}{t_2 - 1}\right)$$

$$= 4st \tag{3.I.14}$$

$$N_6(r, s, t) = T_2(r)T_1(s)T_2(t)$$

$$= l_2^1\left(\frac{2r}{r_2 - 1}\right)l_2^1\left(\frac{2t}{t_2 - 1}\right)$$

$$= 4rt \tag{3.I.15}$$

Example 3

Consider the 10-node triangle depicted in Fig. 3.I.7.

$$N_1(r, s, t) = T_4(r)T_1(s)T_1(t)$$

$$= l_4^3\left(\frac{2r}{r_4 - 1}\right) \qquad \text{(continued)}$$

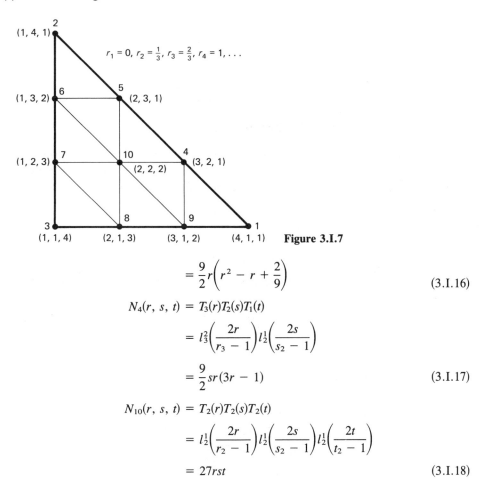

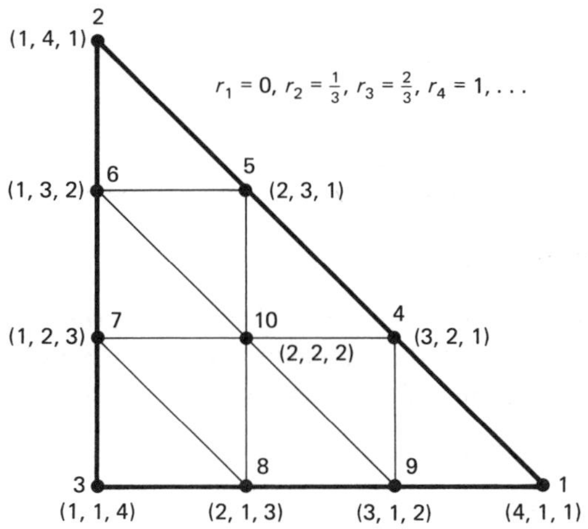

Figure 3.I.7

$$= \frac{9}{2}r\left(r^2 - r + \frac{2}{9}\right) \tag{3.I.16}$$

$$\begin{aligned} N_4(r, s, t) &= T_3(r)T_2(s)T_1(t) \\ &= l_3^2\left(\frac{2r}{r_3 - 1}\right)l_2^1\left(\frac{2s}{s_2 - 1}\right) \\ &= \frac{9}{2}sr(3r - 1) \end{aligned} \tag{3.I.17}$$

$$\begin{aligned} N_{10}(r, s, t) &= T_2(r)T_2(s)T_2(t) \\ &= l_2^1\left(\frac{2r}{r_2 - 1}\right)l_2^1\left(\frac{2s}{s_2 - 1}\right)l_2^1\left(\frac{2t}{t_2 - 1}\right) \\ &= 27rst \end{aligned} \tag{3.I.18}$$

The reader may wish to calculate the remaining shape functions.

Exercise 1. Determine the shape functions for the 15-node quartic Lagrange-type triangle.

Remark

Note that the Lagrange-type shape functions for triangles result in complete polynomials without any extraneous monomials (see the Pascal triangle, Fig. 3.7.8).

Serendipity-type triangular elements may also be derived with the aid of triangular coordinates. However, there does not seem to be much practical interest in elements of this type.

Exercise 2. Use triangular coordinates to develop shape functions for a four-node triangle which exhibits quadratic behavior along one edge and linear behavior along the other two (see the accompanying figure on page 170).

Shape Functions for Tetrahedra

The ***parent tetrahedron*** is illustrated in Fig. 3.I.8. The ***tetrahedral coordinates*** are r, s, t, u. Each is zero on one surface of the tetrahedron and takes on the value 1 at the opposite vertex. The shape functions of the linear tetrahedron are simply

$$N_1(r, s, t) = r \tag{3.I.19}$$

$$N_2(r, s, t) = s \tag{3.I.20}$$

$$N_3(r, s, t) = t \tag{3.I.21}$$

$$N_4(r, s. t) = u(r, s, t) = 1 - r - s - t \tag{3.I.22}$$

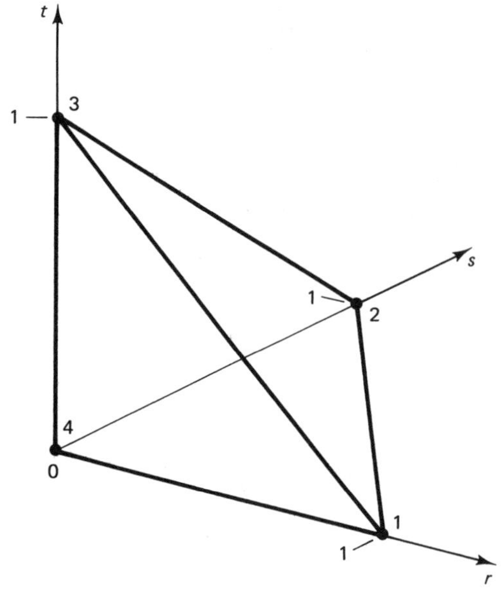

Figure 3.I.8

As in the case of triangles, shape functions for Lagrange-type tetrahedra may be derived with the aid of a simple formula:

$$N_a(r, s, t, u) = T_I(r)T_J(s)T_K(t)T_L(u) \tag{3.I.23}$$

where the T's are again defined by (3.I.6) and $a = a(I, J, K, L)$.

Exercise 3. Consider the 10-node quadratic tetrahedron shown in Fig. 3.I.9. Using (3.I.23) derive the shape functions and verify that they satisfy the interpolation property (i.e., $N_a(r_b, s_b, t_b, u_b) = \delta_{ab}$).

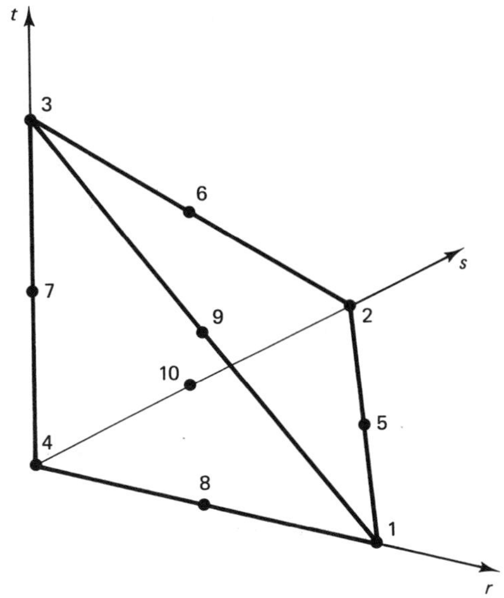

Figure 3.I.9

Shape Functions for Wedge-Shaped Elements

Shape functions of Lagrange-type for wedge-shaped elements may be derived by taking products of one-dimensional and triangular shape functions.

Example 4

Consider the basic six-node wedge element illustrated in Fig. 3.I.10. Shape functions for this element derive from the linear one-dimensional element and linear triangle, viz.,

$$N_1(\xi, r, s, t) = \tfrac{1}{2}(1 - \xi)r \tag{3.I.24}$$

$$N_2(\xi, r, s, t) = \tfrac{1}{2}(1 - \xi)s \tag{3.I.25}$$

$$N_3(\xi, r, s, t) = \tfrac{1}{2}(1 - \xi)t \tag{3.I.26}$$

$$N_4(\xi, r, s, t) = \tfrac{1}{2}(1 + \xi)r \tag{3.I.27}$$

$$N_5(\xi, r, s, t) = \tfrac{1}{2}(1 + \xi)s \tag{3.I.28}$$

$$N_6(\xi, r, s, t) = \tfrac{1}{2}(1 + \xi)t \tag{3.I.29}$$

Higher-order wedges may be similarly derived.

Integration

The integration of monomials over straight-edged triangles and flat-surfaced tetrahedra may be performed with the aid of the following exact expressions:

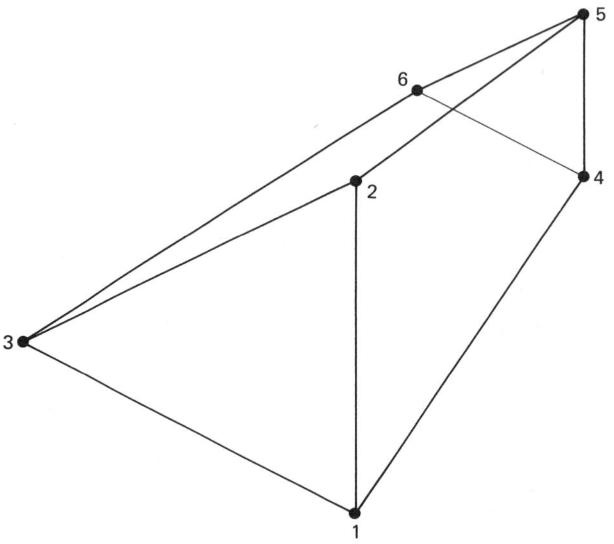

Figure 3.I.10

Triangles

$$\int_{\Omega} r^{\alpha} s^{\beta} t^{\gamma} \, d\Omega = \frac{\alpha! \; \beta! \; \gamma!}{(\alpha + \beta + \gamma + 2)!} 2A \tag{3.I.30}$$

where A, the area of Ω, is given by

$$2A = \det \begin{bmatrix} 1 & x_1 & y_1 \\ 1 & x_2 & y_2 \\ 1 & x_3 & y_3 \end{bmatrix} \tag{3.I.31}$$

in which x_a and y_a are the coordinates of vertex a.

Tetrahedra

$$\int_{\Omega} r^{\alpha} s^{\beta} t^{\gamma} u^{\delta} \, d\Omega = \frac{\alpha! \; \beta! \; \gamma! \; \delta!}{(\alpha + \beta + \gamma + \delta + 3)!} 6V \tag{3.I.32}$$

where V, the volume of Ω, is given by

$$6V = \det \begin{bmatrix} 1 & x_1 & y_1 & z_1 \\ 1 & x_2 & y_2 & z_2 \\ 1 & x_3 & y_3 & z_3 \\ 1 & x_4 & y_4 & z_4 \end{bmatrix} \tag{3.I.33}$$

in which x_a, y_a, and z_a are the coordinates of vertex a.

In the case of distorted elements, integrands will not consist simply of monomials. Consequently, numerical integration needs to be used. Formulas that are symmetrical with respect to the tetrahedral coordinates have been derived. References [25–27] may be consulted for results of this kind. A sampling of results is presented in Tables 3.I.1 and 3.I.2. Higher-order rules for tetrahedra are presented in [28].

TABLE 3.I.1 Integration Rules for Triangles [26, 27]

W_i	$\tilde{r}$	$\tilde{s}$	$\tilde{t}$	Multiplicity[(1)]
	3-point formula	*degree of precision 2*[(2)]		
0.33333 33333 33333	0.66666 66666 66667	0.16666 66666 66667	0.16666 66666 66667	3
	3-point formula	*degree of precision 2*		
0.33333 33333 33333	0.50000 00000 00000	0.50000 00000 00000	0.00000 00000 00000	3
	4-point formula	*degree of precision 3*		
−0.56250 00000 00000	0.33333 33333 33333	0.33333 33333 33333	0.33333 33333 33333	1
0.52083 33333 33333	0.60000 00000 00000	0.20000 00000 00000	0.20000 00000 00000	3
	6-point formula	*degree of precision 3*		
0.16666 66666 66667	0.65902 76223 74092	0.23193 33685 53031	0.10903 90090 72877	6
	6-point formula	*degree of precision 4*		
0.10995 17436 55322	0.81684 75729 80459	0.09157 62135 09771	0.09157 62135 09771	3
0.22338 15896 78011	0.10810 30181 68070	0.44594 84909 15965	0.44594 84909 15965	3
	7-point formula	*degree of precision 4*		
0.37500 00000 00000	0.33333 33333 33333	0.33333 33333 33333	0.33333 33333 33333	1
0.10416 66666 66667	0.73671 24989 68435	0.23793 23664 72434	0.02535 51345 51932	6
	7-point formula	*degree of precision 5*		
0.22503 00003 00000	0.33333 33333 33333	0.33333 33333 33333	0.33333 33333 33333	1
0.12593 91805 44827	0.79742 69853 53087	0.10128 65073 23456	0.10128 65073 23456	3
0.13239 41527 88506	0.47014 20641 05115	0.47014 20641 05115	0.05971 58717 89770	3
	9-point formula	*degree of precision 5*		
0.20595 05047 60887	0.12494 95032 33232	0.43752 52483 83384	0.43752 52483 83384	3
0.06369 14142 86223	0.79711 26518 60071	0.16540 99273 89841	0.03747 74207 50088	6
	12-point formula	*degree of precision 6*		
0.05084 49063 70207	0.87382 19710 16996	0.06308 90144 91502	0.06308 90144 91502	3
0.11678 62757 26379	0.50142 65096 58179	0.24928 67451 70910	0.24928 67451 70911	3
0.08285 10756 18374	0.63650 24991 21399	0.31035 24510 33785	0.05314 50498 44816	6

(1) The formulas are symmetric in the triangular coordinates. All permutations of the quadrature points need to be accounted for.

(2) The degree of precision refers to the order of the remainder. (One-point, centroidal quadrature achieves a degree of precision equal to two.)

TABLE 3.I.1 Integration Rules for Triangles [26, 27] (*cont.*)

W_l	$\tilde{r}$	$\tilde{s}$	$\tilde{t}$	Multi-plicity[1]
	13-point formula	*degree of precision 7*		
−0.14957 00444 67670	0.33333 33333 33333	0.33333 33333 33333	0.33333 33333 33333	1
0.17561 52574 33204	0.47930 80678 41923	0.26034 59660 79038	0.26034 59660 79038	3
0.05334 72356 08839	0.86973 97941 95568	0.06513 01029 02216	0.06513 01029 02216	3
0.07711 37608 90257	0.63844 41885 69809	0.31286 54960 04875	0.48690 31542 53160	6

TABLE 3.I.2 Integration Rules for Tetrahedra [25]

W_l	$\tilde{r}$	$\tilde{s}$	$\tilde{t}$	$\tilde{u}$	Multiplicity
	1-point formula		*degree of precision 2*		
1	$\frac{1}{4}$	$\frac{1}{4}$	$\frac{1}{4}$	$\frac{1}{4}$	1
	4-point formula		*degree of precision 3*		
$\frac{1}{4}$	0.58541020	0.13819660	0.13819660	0.13819660	4
	5-point formula		*degree of precision 4*		
$-\frac{4}{5}$	$\frac{1}{4}$	$\frac{1}{4}$	$\frac{1}{4}$	$\frac{1}{4}$	1
$\frac{9}{20}$	$\frac{1}{3}$	$\frac{1}{6}$	$\frac{1}{6}$	$\frac{1}{6}$	4

It is traditional to tabulate quadrature rules for triangles and tetrahedra such that the weights sum to 1. Because the area of the parent triangle equals $\frac{1}{2}$ and the volume of the parent tetrahedron equals $\frac{1}{6}$, the proper form for a quadrature rule in these cases is

$$\int_{\Omega} f \, d\Omega \cong c \sum_{l=1}^{n_{\text{int}}} f_l \, j_l W_l \tag{3.I.34}$$

where $c = \frac{1}{2}$ for triangles and $c = \frac{1}{6}$ for tetrahedra.

Derivatives of Shape Functions

To calculate derivatives of shape functions with respect to global coordinates we may employ the techniques described in Sec. 3.9. All formulas of Sec. 3.9 may be employed with the following substitutions.

Triangles

Replace ξ and η by r and s, respectively; set $t = 1 - r - s$ throughout.

Tetrahedra

Replace ξ, η, and ζ by r, s, and t, respectively; set $u = 1 - r - s - t$ throughout.

Appendix 3.II

Methodology for Developing Special Shape Functions with Application to Singularities

In many areas of finite element analysis, elements with special properties are required to achieve maximal accuracy. As examples, we may mention infinite elements for the representation of spatial domains that extend to infinity [29, 30] and singular elements for modeling point and line singularities engendered by geometric features such as reentrant corners and cracks [31–33]. In this appendix we are concerned with techniques for developing shape functions for special elements in general but with particular emphasis on singularities.

Geometric features of the spatial domain may give rise to highly singular response in the solution of a field problem [34]. Eigenfunction analyses are frequently available for purposes of characterizing the asymptotic strength of the singularity [35]. However, the actual solution depends upon the nature of the loading and there are situations when the singularity does not contribute significantly [36]. For this reason it has been suggested that, ideally, singular finite elements should strike some balance between being able to represent smooth behavior and potential singular behavior [36]. In this way the element will remain effective under all loading conditions. Translated into analytical terms, the finite element shape functions should be designed to represent the standard low-order polynomials as well as the singular functions emanating from asymptotic analysis. To construct shape functions of this type gives rise to an interpolation problem, which involves a system of linear simultaneous equations. In special circumstances explicit representations are available for the solution of the interpolation problem. The most well known example is the Lagrange interpolation formula, which is the solution of the interpolation problem for polynomials passing through distinct points. It appears that no such representation exists when different classes of functions are merged, as seems appropriate for singular elements. Although, in principle, the interpolation problem could be solved numerically for each singular element during formation, this would be highly inefficient due to the increased number of computations required. Furthermore, numerical sensitivity may manifest itself in certain configurations. Consequently, it is of practical interest to develop techniques for systematically defining shape functions for singularity modeling (and for developing special elements in general), which circumvent the interpolation problem. In what follows, a highly concise algorithm for generating shape functions from an arbitrary starting set of independent functions is presented [37]. A novel feature of the procedure is that *no* coefficient matrix is involved (i.e., no system of simultaneous equations need be solved). Some useful one-dimensional interpolatory schemes for modeling singularities are developed by employing the interpolation algorithm. These schemes are the basis for constructiong two- and three-dimensional elements for modeling point and line singularities.

Let $r = r(\xi) = (1 + \xi)/2$ and $r_a = r(\xi_a)$. Note that $r(\xi) \in [0, 1]$ for all $\xi \in [-1, 1]$. The r-coordinate description is convenient for modeling singularities. In multidimensional applications, we use boldfaced notations analogous to the preceding ones. For example, in two dimensions $\boldsymbol{r} = (r, s)$, where $s = (1 + \eta)/2$, $\eta \in [-1, 1]$.

The ***Lagrange polynomials*** are defined by

$$l_a(f) = \frac{\prod\limits_{\substack{b=1\\b\neq a}}^{m}(f - f_b)}{\prod\limits_{\substack{b=1\\b\neq a}}^{m}(f_a - f_b)}, \qquad a = 1, 2, \ldots, m \tag{3.II.1}$$

where the f_a's are distinct points in $\mathbb{R}$. It may be seen from (3.II.1) that $l_a(f)$ is an $(m - 1)$st order polynomial in f that satisfies the "interpolation property"(i.e., $l_a(f_b) = \delta_{ab}$, the Kronecker delta). *Note that f may be taken to be ξ, r, r^α, or any other convenient function of ξ.*

Algorithm for Constructing Interpolation Functions

Consider an n-node finite element. We assume we are given a set of "preliminary" shape functions, N_a, $a = 1, 2, \ldots, n$, which satisfy the interpolation property on the first m nodes only; viz., $N_a(\boldsymbol{r}_b) = \delta_{ab}$, where $a, b = 1, 2, \ldots, m < n$. The idea is to modify systematically the N_a's such that the interpolation property is satisfied on all n nodes. The procedure is given as follows:

Step 1
$$N_{m+1}(\boldsymbol{r}) \leftarrow \frac{N_{m+1}(\boldsymbol{r}) - \sum\limits_{a=1}^{m} N_{m+1}(\boldsymbol{r}_a)N_a(\boldsymbol{r})}{N_{m+1}(\boldsymbol{r}_{m+1}) - \sum\limits_{a=1}^{m} N_{m+1}(\boldsymbol{r}_a)N_a(\boldsymbol{r}_{m+1})} \tag{3.II.2}$$

Step 2
$$N_a(\boldsymbol{r}) \leftarrow N_a(\boldsymbol{r}) - N_a(\boldsymbol{r}_{m+1})N_{m+1}(\boldsymbol{r}), \qquad a = 1, 2, \ldots, m \tag{3.II.3}$$

Step 3 If $m+1 < n$, replace m by $m + 1$ and repeat Steps 1 to 3.
If $m+1 = n$, stop.

Remarks

1. After completing Steps 1 and 2, the shape functions satisfy the interpolation property on the first $m + 1$ nodes. Clearly, after completing the entire procedure, the desired properties are achieved.

2. In many cases, the algorithm enables the explicit hand calculation of shape functions. *However, there is no need to do this in practice, as the algorithm itself may be programmed as part of a shape function subroutine.*

3. The derivatives of shape functions may be constructed in similar fashion. The analogs of Steps 1 and 2 are respectively

$$\partial N_{m+1}(\boldsymbol{r}) \leftarrow \frac{\partial N_{m+1}(\boldsymbol{r}) - \sum_{a=1}^{m} N_{m+1}(\boldsymbol{r}_a)\partial N_a(\boldsymbol{r})}{N_{m+1}(\boldsymbol{r}_{m+1}) - \sum_{a=1}^{m} N_{m+1}(\boldsymbol{r}_a)N_a(\boldsymbol{r}_{m+1})} \tag{3.II.4}$$

and

$$\partial N_a(\boldsymbol{r}) \leftarrow \partial N_a(\boldsymbol{r}) - N_a(\boldsymbol{r}_{m+1})\partial N_{m+1}(\boldsymbol{r}) \tag{3.II.5}$$

In (3.II.4) and (3.II.5), ∂ represents the partial differentiation operator with respect to any of the coordinates (e.g., in two dimensions, ∂ may represent $\partial/\partial r$ or $\partial/\partial s$).

4. Interpolation problems are generally formulated in terms of a system of simultaneous linear equations. It may be noted that, if $m = 1$, the algorithm encompassed by Steps 1–3 solves the interpolation problem *without* engendering any coefficient matrix. Thus no "storage" is required (beyond that for the N_a's) in programming the procedure. Another nice feature of the algorithm is that the interpolation property possessed by the preliminary shape functions is fully exploited in that only $n - m$ passes through Steps 1–3 need be performed.

In practice, the Lagrange polynomial formula may be employed to generate a partial set of preliminary shape functions, which satisfies the interpolation property on the first m nodes. This is illustrated in the examples to follow.

Example 1

Consider a one-dimensional three-node element with nodes given by $r_1 = 0$, $r_2 = \frac{1}{2}$, and $r_3 = 1$. We wish to construct shape functions capable of exactly representing 1, r, and r^α. (If $\alpha = 2$, we have the usual three-node quadratic element. Other values of α enable the modeling of singularities.) We begin with the following preliminary shape functions:

$$N_1(r) = 1 - 2r, \qquad N_2(r) = 2r, \qquad N_3(r) = r^\alpha \tag{3.II.6}$$

Observe that $N_a(r_b) = \delta_{ab}$, $1 \le a, b \le 2$. In terms of our algorithm, $m = 2$ and $n = 3$, so only one pass through Steps 1 and 2 is required:

$$N_3(r) \leftarrow \frac{N_3(r) - N_3(\frac{1}{2})N_2(r)}{N_3(1) - N_3(\frac{1}{2})N_2(1)} = \frac{r^\alpha - 2(\frac{1}{2})^\alpha r}{1 - 2(\frac{1}{2})^\alpha} \tag{3.II.7}$$

$$N_1(r) \leftarrow N_1(r) - N_1(1)N_3(r) = 1 - 2r + \left[\frac{r^\alpha - 2(\frac{1}{2})^\alpha r}{1 - 2(\frac{1}{2})^\alpha}\right] \tag{3.II.8}$$

$$N_2(r) \leftarrow N_2(r) - N_2(1)N_3(r) = 2r - 2\left[\frac{r^\alpha - 2(\frac{1}{2})^\alpha r}{1 - 2(\frac{1}{2})^\alpha}\right] \tag{3.II.9}$$

Likewise, the derivatives are given by:

$$\partial N_3(r) \leftarrow \frac{\partial N_3(r) - N_3(\frac{1}{2})\partial N_2(r)}{N_3(1) - N_3(\frac{1}{2})N_2(1)} = \frac{\alpha r^{\alpha-1} - 2(\frac{1}{2})^\alpha}{1 - 2(\frac{1}{2})^\alpha} \tag{3.II.10}$$

$$\partial N_1(r) \leftarrow \partial N_1(r) - N_1(1)\partial N_3(r) = -2 + \left[\frac{\alpha r^{\alpha-1} - 2(\frac{1}{2})^\alpha}{1 - 2(\frac{1}{2})^\alpha}\right] \tag{3.II.11}$$

$$\partial N_2(r) \leftarrow \partial N_2(r) - N_2(1)\partial N_3(r) = 2 - 2\left[\frac{\alpha r^{\alpha-1} - 2(\frac{1}{2})^\alpha}{1 - 2(\frac{1}{2})^\alpha}\right] \tag{3.II.12}$$

(In this case ∂N is taken to mean $\partial N/\partial r$)

Example 2

Consider a one-dimensional four-node element for which the nodes are given by $r_1 = 0$, $r_2 = \frac{1}{3}$, $r_3 = \frac{2}{3}$, and $r_4 = 1$. We wish to construct shape functions capable of representing 1, r^α, $r^{2\alpha}$, and r^β. The following special cases are of practical importance:

$\alpha = 1$, $\beta = 3$:	$1, r, r^2, r^3$	cubic polynomial in r
$\alpha = 1$, β arbitrary:	$1, r, r^2, r^\beta$	quadratic polynomial in r plus linear polynomial in r^β
α arbitrary, $\beta = 1$:	$1, r, r^\alpha, r^{2\alpha}$	linear polynomial in r plus quadratic polynomial in r^α
α arbitrary, $\beta = 3\alpha$:	$1, r^\alpha, r^{2\alpha}, r^{3\alpha}$	cubic polynomial in r^α

To obtain a starting set of shape functions, we may employ the Lagrange interpolation formula with $f = r^\alpha$ and $m = 3$, viz.,

$$N_1(r) = l_1(r^\alpha) = \frac{(r^\alpha - r_2^\alpha)(r^\alpha - r_3^\alpha)}{(r_1^\alpha - r_2^\alpha)(r_1^\alpha - r_3^\alpha)} = \frac{(3^\alpha r^\alpha - 1)(3^\alpha r^\alpha - 2^\alpha)}{2^\alpha} \tag{3.II.13}$$

$$N_2(r) = l_2(r^\alpha) = \frac{(r^\alpha - r_1^\alpha)(r^\alpha - r_3^\alpha)}{(r_2^\alpha - r_1^\alpha)(r_2^\alpha - r_3^\alpha)} = \frac{3^\alpha r^\alpha(3^\alpha r^\alpha - 2^\alpha)}{(1 - 2^\alpha)} \tag{3.II.14}$$

$$N_3(r) = l_3(r^\alpha) = \frac{(r^\alpha - r_1^\alpha)(r^\alpha - r_2^\alpha)}{(r_3^\alpha - r_1^\alpha)(r_3^\alpha - r_2^\alpha)} = \frac{3^\alpha r^\alpha(3^\alpha r^\alpha - 1)}{2^\alpha(2^\alpha - 1)} \tag{3.II.15}$$

It may be verified that $N_a(r_b) = \delta_{ab}$ for all $a, b = 1, 2, 3$. The last preliminary shape function is taken to be

$$N_4(r) = r^\beta \tag{3.II.16}$$

With regard to the algorithm, $m = 3$ and $n = 4$, so only one pass through Steps 1 and 2 is needed, viz.,

$$N_4(r) \leftarrow \frac{N_4(r) - \sum_{a=2}^{3} N_4(r_a)N_a(r)}{1 - \sum_{a=2}^{3} N_4(r_a)N_a(1)} \tag{3.II.17}$$

$$N_a(r) \leftarrow N_a(r) - N_a(1)N_4(r), \qquad a = 1, 2, 3 \tag{3.II.18}$$

Equations (3.II.17) and (3.II.18) may be programmed in a shape function subroutine along with corresponding expressions for derivatives.

Remarks

5. The preceding one-dimensional examples cover most cases of practical interest. If the need arises, higher-order shape functions may be constructed along similar lines. The basic philosophy is to maximize m in order to minimize calculations in the shape function routines.

6. For completeness, we wish to mention the simplest one-dimensional shape functions that allow the modeling of singularities; namely, the two-node element with $r_1 = 0$, $r_2 = 1$, $N_1(r) = 1 - r^\alpha$, and $N_2(r) = r^\alpha$.

7. Let $u(r) = \Sigma_{a=1}^{n} N_a(r)d_a$, where (by construction) $d_a = u(r_a)$. The function u is capable of exactly representing the powers of r used in deriving the N_a's. (The d_a's need only be set accordingly.) In practice, it is usually desired that the same powers of the physical coordinate, $x = x(r)$, be representable. This will be attained if $x = x(r)$ is affine (i.e., of the form $x(r) = c_1 + c_2 r$). There are two ways of achieving this in practice.

The *first* depends upon 1 and r being present among the powers of r used in deriving the shape functions. In this case the isoparametric concept may be invoked; that is, we take $x(r) = \Sigma_{a=1}^{n} N_a(r)x_a$ and equally space the nodal coordinates (i.e., x_a's in x-space).

The *second* procedure does not require that 1 and r be present, but it involves an additional set of shape function calculations, so it is less efficient. In this case, we take $x(r) = \Sigma_{a=1}^{n} P_a(r)x_a$, where the P_a's are the standard polynomial shape functions, and the x_a's are, again, equally spaced.

To see the necessity of adopting the second procedure when 1 and r are absent from the N_a's, we need only consider the two-node element of Remark 6. If the isoparametric concept is adopted, then u varies linearly with x (rather than x^α). To see this we note that if $x_2 \neq x_1$, then we can write $d_a = c_1 + c_2 x_a$, where the c's are constants. Consequently

$$\begin{aligned} u &= \sum_{a=1}^{2} N_a d_a \\ &= \sum_{a=1}^{2} N_a(c_1 + c_2 x_a) \\ &= c_1\left(\sum_{a=1}^{2} N_a\right) + c_2\left(\sum_{a=1}^{2} N_a x_a\right) \\ &= c_1 + c_2 x \end{aligned} \tag{3.II.19}$$

We shall now employ the above results in deriving two-dimensional finite element shape functions for modeling line and point singularities.

Example 3

Consider the four-node quadrilateral illustrated in Fig. 3.II.1(a). The shape functions

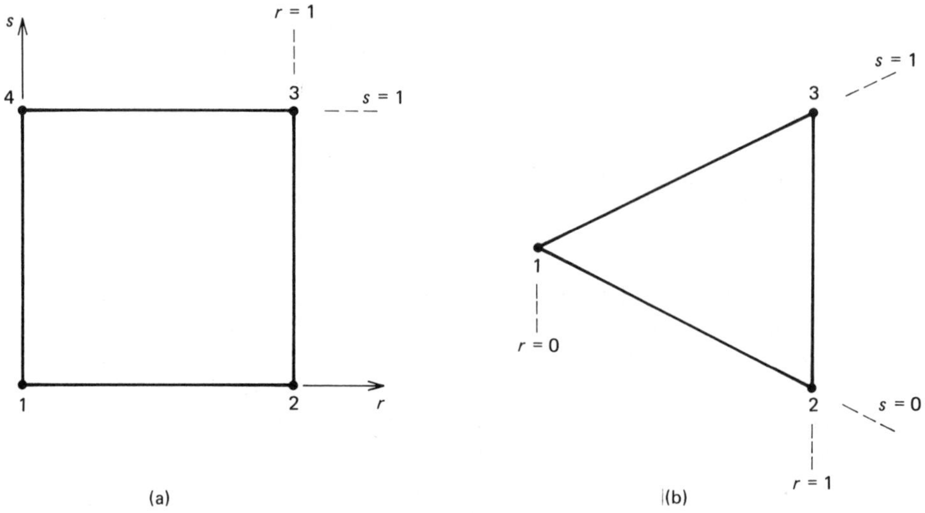

Figure 3.II.1 (a) Four-node quadrilateral element. (b) Three-node triangular element formed by degenerating the four-node quadrilateral.

may be constructed from products of linear polynomial shape functions in the s-direction and the shape functions of Remark 6 in the r-direction, viz.,

$$N_1(r, s) = (1 - r^\alpha)(1 - s) \tag{3.II.20}$$

$$N_2(r, s) = r^\alpha(1 - s) \tag{3.II.21}$$

$$N_3(r, s) = r^\alpha s \tag{3.II.22}$$

$$N_4(r, s) = (1 - r^\alpha)s \tag{3.II.23}$$

Thus $u(r, s) = \Sigma_{a=1}^4 N_a(r, s)d_a$ is capable of exactly representing a line singularity of order r^α along edge $\widehat{14}$. By virtue of Remark 7, the standard polynomial shape functions must be employed to define the geometry. That is $\boldsymbol{x}(r, s) = \Sigma_{a=1}^4 P_a(r, s)\boldsymbol{x}_a$, where

$$P_1(r, s) = (1 - r)(1 - s) \tag{3.II.24}$$

$$P_2(r, s) = r(1 - s) \tag{3.II.25}$$

$$P_3(r, s) = rs \tag{3.II.26}$$

$$P_4(r, s) = (1 - r)s \tag{3.II.27}$$

This element was first proposed by Akin [38].

Example 4

A three-node triangular element (see Fig. 3.II.1(b)) may be developed by combining the shape functions associated with nodes 1 and 4 in the previous example, viz.,

$$N_1(r, s) \leftarrow N_1(r, s) + N_4(r, s) = 1 - r^\alpha \tag{3.II.28}$$

$$P_1(r, s) \leftarrow P_1(r, s) + P_4(r, s) = 1 - r \tag{3.II.29}$$

The shape functions of nodes 2 and 3 remain the same. This element is capable of exactly representing a point singularity of order r^α. Similar interpolations were used in [33] and [39].

Example 5

Consider the nine-node Lagrange-type quadrilateral illustrated in Fig. 3.II.2(a). The shape functions shall be constructed from products of the three-node one-dimensional shape functions of Example 1. That is, we define $N_a(r)$, $a = 1, 2, 3$, by (3.II.7)—(3.II.9) and $P_a(s)$ by the same expressions, with r replaced by s and $\alpha = 2$. The two-dimensional shape functions may then be defined as follows:

$$N_1(r, s) = N_1(r)P_1(s) \tag{3.II.30}$$

$$N_2(r, s) = N_3(r)P_1(s) \tag{3.II.31}$$

$$N_3(r, s) = N_3(r)P_3(s) \tag{3.II.32}$$

$$N_4(r, s) = N_1(r)P_3(s) \tag{3.II.33}$$

$$N_5(r, s) = N_2(r)P_1(s) \tag{3.II.34}$$

$$N_6(r, s) = N_3(r)P_2(s) \tag{3.II.35}$$

$$N_7(r, s) = N_2(r)P_3(s) \tag{3.II.36}$$

$$N_8(r, s) = N_1(r)P_2(s) \tag{3.II.37}$$

$$N_9(r, s) = N_2(r)P_2(s) \tag{3.II.38}$$

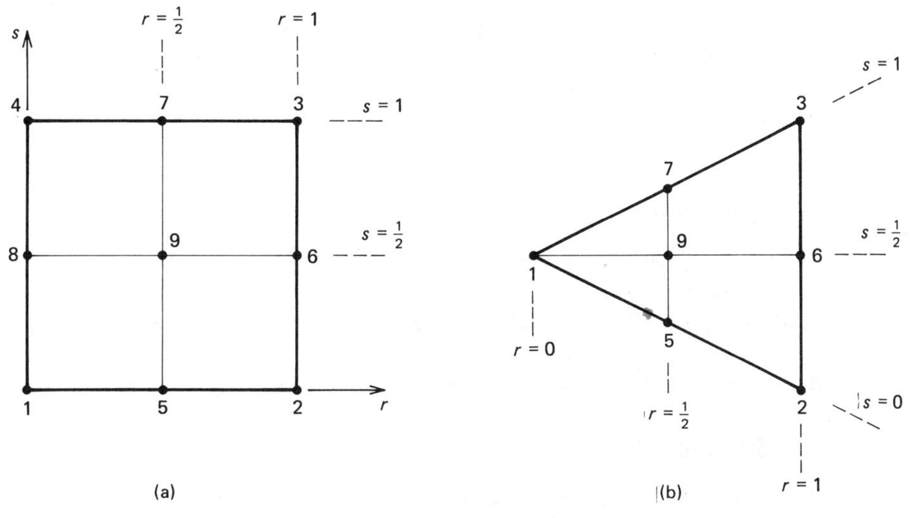

Figure 3.II.2 (a) Nine-node Lagrange quadrilateral element. (b) Seven-node triangular element formed by degenerating the nine-node Lagrange quadrilateral.

In this case $u(r, s) = \Sigma_{a=1}^{9} N_a(r, s)d_a$ is capable of exactly representing a line singularity of order r^α along edge $\widehat{14}$. Furthermore, the following monomials may be exactly represented: $1, r, s, r^\alpha, rs, s^2, r^\alpha s, s^2 r$, and $r^\alpha s^2$. By virtue of the presence of 1, r and s, either the isoparametric concept may be employed to define the geometry (i.e., $\boldsymbol{x}(r, s) = \Sigma_{a=1}^{9} N_a(r, s)\boldsymbol{x}_a$), or recourse may be made to the standard Lagrange polynomials (i.e., $\boldsymbol{x}(r, s) = \Sigma_{a=1}^{9} P_a(r, s)\boldsymbol{x}_a$, where the $P_a(r, s)$'s may be obtained from (3.II.30) through (3.II.38) by replacing $N_a(r)$ by $P_a(r)$, $a = 1, 2, 3$).

Remark

More complicated two- and three-dimensional elements developed along the lines sketched here are presented in [37].

REFERENCES

Section 3.1

1. G.P. Bazeley, Y.K. Cheung, B.M. Irons, and O.C. Zienkiewicz, "Triangular Elements in Bending—Conforming and Non-conforming Solutions," *Proceedings of the Conference on Matrix Methods in Structural Mechanics*, Wright-Patterson Air Force Base, Ohio, 1965.
2. G. Strang and G.J. Fix, *An Analysis of the Finite Element Method*. Englewood Cliffs, N.J.: Prentice-Hall, 1973.
3. O. C. Zienkiewicz, *The Finite Element Method*. London: McGraw-Hill, 1977.

Section 3.2

4. I.C. Taig, *Structural Analysis by the Matrix Displacement Method*, English Electric Aviation Report No. S017, 1961.
5. J.H. Argyris and S. Kelsey, *Energy Theorems and Structural Analysis*. London: Butterworths, 1960 (originally published in a series of articles in *Aircraft Engineering*, 1954–55).

Section 3.3

6. I.C. Taig, *Structural Analysis by the Matrix Displacement Method*, English Electric Aviation Report No. S017, 1961.
7. B. M. Irons, "Engineering Application of Numerical Integration in Stiffness Method," *Journal of the American Institute of Aeronautics and Astronautics*, 14 (1966), 2035–2037.
8. S. Lang, *Real Analysis*. Reading Mass.: Addison-Wesley, 1969.

Section 3.4

9. R. Courant, "Variational Methods for the Solution of Problems of Equilibrium and Vibration," *Bulletin of the American Mathematical Society*, 49 (1943), 1–23.
10. M. J. Turner, R. W. Clough, H. C. Martin, and L. P. Topp, "Stiffness and Deflection Analysis of Complex Structures," *Journal of Aeronautical Sciences*, 23, no. 9 (1956), 805–823.

Section 3.5

11. R. H. Gallagher, J. Padlog, and P. P. Bijlaard, "Stress Analysis of Heated Complex Shapes," *American Rocket Society Journal*, 32, no. 5 (1962), 700–707.

Section 3.7

12. O. C. Zienkiewicz, *The Finite Element Method.* London: McGraw-Hill, 1977.

Section 3.8

13. A. H. Stroud and D. Secrest, *Gaussian Quadrature Formulas.* Englewood Cliffs, N.J.: Prentice-Hall, 1966.
14. M. Abramowitz and I. A. Stegun, *Handbook of Mathematical Functions.* Washington, D.C.: National Bureau of Standards, 1964.
15. P. C. Hammer and A. H. Stroud, "Numerical Evaluation of Multiple Integrals II," *Mathematical Tables and Aids of Computation*, 12, no. 64 (1958), 272–280.
16. B. M. Irons, "Quadrature Rules for Brick Based Finite Elements," *International Journal of Numerical Methods in Engineering*, 3, no. 2 (1971), 293–294.
17. T. K. Hellen, "Effective Quadrature Rules for Quadratic Solid Isoparametric Finite Elements," *International Journal for Numerical Methods in Engineering*, 4, no. 4 (1972), 597–599.
18. C. Maforano, S. Odorizzi, and R. Vitaliani, "Shortened Quadrature Rules for Finite Elements," *Advances in Engineering Software*, 4, no. 2 (1982).

Section 3.10

19. A. K. Gupta and B. Mohraz, "A Method of Computing Numerically Integrated Stiffness Matrices," *International Journal for Numerical Methods in Engineering*, 5 (1972), 83–89.
20. R. L. Taylor, "Computer Procedures for Finite Element Analysis," Chapter 24 in O. C. Zienkiewicz, *The Finite Element Method.* London: McGraw-Hill, 1977.
21. E. Hinton and D. R. J. Owen, *Finite Element Programming.* London: Academic Press, 1977.
22. K. J. Bathe and E. L. Wilson, *Numerical Methods in Finite Element Analysis.* Englewood Cliffs, N. J.: Prentice-Hall, 1976.
23. D. R. J. Owen and E. Hinton, *Finite Elements in Plasticity: Theory and Practice.* Swansea, U.K.: Pineridge Press, 1980.
24. J. E. Akin, *Application and Implementation of Finite Element Methods.* London: Academic Press, 1982.

Appendix 3.I

25. P. C. Hammer, O. P. Marlowe, and A. H. Stroud, "Numerical Integration over Simplexes and Cones," *Mathematical Tables and Aids to Computation*, 10 (1956), 130–137.
26. G. R. Cowper, "Gaussian Quadrature Formulas for Triangles," *International Journal for Numerical Methods in Engineering*, 7 (1973), 405–408.
27. G. Strang and G. J. Fix, *An Analysis of the Finite Element Method.* Englewood Cliffs, N. J.: Prentice-Hall, 1973.
28. Y. Jinyun, "Symmetric Gaussian Quadrature Formulae for Tetrahedral Regions," *Computer Methods in Applied Mechanics and Engineering*, 43, no. 3 (1984), 349–353.

Appendix 3.II

29. D. Gartling and E. B. Becker, "Computationally Efficient Finite Element Analysis of

Viscous Flow Problems," *Computational Methods in Nonlinear Mechanics*, (ed. J. T. Oden). Austin: Texas Institute for Computational Mechanics, 1974, 603–614.

30. P. Bettess, "Infinite Elements," *International Journal for Numerical Methods in Engineering*, 11 (1977), 53–64.

31. W. S. Blackburn, "Calculation of Stress Intensity Factors at Crack Tips Using Special Finite Elements," *Mathematics of Finite Elements and Applications*, ed. J. R. Whiteman. London: Academic Press, (1973), 327–336.

32. R. D. Henshell and K. G. Shaw, "Crack Tip Elements are Unnecessary," *International Journal for Numerical Methods in Engineering*, 9 (1975), 495–507.

33. J. E. Akin, "Generation of Elements with Singularities," *International Journal for Numerical Methods in Engineering*, 10 (1976), 1249–1259.

34. R. E. Barnhill and J. R. Whiteman, "Error Analysis of Finite Element Methods with Triangles for Elliptic Boundary Value Problems," *Mathematics of Finite Element Methods and Applications*, ed. J. R. Whiteman. London: Academic Press, (1973), 83–112.

35. A. K. Rao, "Stress Concentrations and Singularities at Interface Corners," *ZAMM*, 51 (1971), 395–406.

36. R. S. Barsoum, "On the Use of Isoparametric Finite Elements in Linear Fracture Mechanics," *International Journal for Numerical Methods in Engineering*, 10 (1976), 25–37.

37. T. J. R. Hughes and J. E. Akin, "Techniques for Developing 'Special' Finite Element Shape Functions with Particular Reference to Singularities," *International Journal for Numerical Methods in Engineering*, 15 (1980), 733–751.

38. J. E. Akin, "Elements for Problems with Line Singularities," *Mathematics of Finite Elements and Applications III*, ed. J. R. Whiteman. London: Academic Press, 1978.

39. D. M. Tracey and T. S. Cook, "Analysis of Power Type Singularities Using Finite Elements," *International Journal for Numerical Methods in Engineering*, 10 (1976), 1249–1259.

4

Mixed and Penalty Methods, Reduced and Selective Integration, and Sundry Variational Crimes

4.1 "BEST APPROXIMATION" AND ERROR ESTIMATES: WHY THE STANDARD FEM USUALLY WORKS AND WHY SOMETIMES IT DOES NOT

In this section we shall establish the convergence of the finite element methodology considered so far and show, in addition, that in a sense it is an optimal technique. The main results require some mathematical preliminaries that are presented in Appendix 4.I for the uninitiated reader. We conclude the section with a caution regarding the application of these ideas to certain classes of problems.

We assume the ***exact problem*** can be written in the following abstract variational, or weak, form:[1] Find $\boldsymbol{u} \in \mathcal{S}$ such that for all $\boldsymbol{w} \in \mathcal{V}$,

$$a(\boldsymbol{w}, \boldsymbol{u}) = (\boldsymbol{w}, \boldsymbol{\ell}) + (\boldsymbol{w}, \boldsymbol{h})_\Gamma \tag{4.1.1}$$

Recall from Chapter 2 that functions in $\mathcal{S}$ satisfy the given essential boundary conditions, and functions in $\mathcal{V}$ satisfy corresponding ***homogeneous*** essential boundary conditions.

The ***approximate finite element problem*** can be stated analogously: Find $\boldsymbol{u}^h \in \mathcal{S}^h$ such that for all $\boldsymbol{w}^h \in \mathcal{V}^h$,

$$a(\boldsymbol{w}^h, \boldsymbol{u}^h) = (\boldsymbol{w}^h, \boldsymbol{\ell}) + (\boldsymbol{w}^h, \boldsymbol{h})_\Gamma \tag{4.1.2}$$

Throughout we shall assume that

i. $\mathcal{S}^h \subset \mathcal{S}$ and $\mathcal{V}^h \subset \mathcal{V}$.
ii. $a(\cdot\,,\cdot)$, $(\cdot\,,\cdot)$ and $(\cdot\,,\cdot)_\Gamma$ are symmetric and bilinear.

[1] With appropriate specialization, this includes all problems considered so far.

iii. $a(\cdot\,,\,\cdot)$ and $\|\cdot\|_m$ define ***equivalent norms*** on $\mathcal{V}$, that is,

$$c_1\|w\|_m \leq a(w,\, w)^{1/2} \leq c_2\|w\|_m \tag{4.1.3}$$ [2]

where

$$\|w\|_m = \left[\int_\Omega (w_i w_i + w_{i,j}w_{i,j} + \cdots + \underbrace{w_{i,jk\ldots l}}_{m \text{ indices}}\underbrace{w_{i,jk\ldots l}}_{m \text{ indices}})\, d\Omega\right]^{1/2} \tag{4.1.4}$$

and c_1 and c_2 are constants independent of w. In (4.1.3), m is the order of derivatives appearing in $a(\cdot\,,\,\cdot)$. For example, in heat conduction and elasticity, $m = 1$, whereas in Bernoulli-Euler beam theory $m = 2$. Equation (4.1.4) defines the ***mth Sobolev norm of w*** (i.e., H^m norm). Also, $a(w,\, w)^{1/2}$ is called the (***strain***) ***energy norm*** and $a(\cdot\,,\,\cdot)$ the (***strain***) ***energy inner product.***

Theorem. Let $e = u^h - u$ denote the *error* in the finite element approximation.

a. $a(w^h,\, e) = 0 \qquad$ for all $w^h \in \mathcal{V}^h$.

b. $a(e,\, e) \leq a(U^h - u,\, U^h - u) \qquad$ for all $U^h \in \mathcal{S}^h$.

Part (a) means that the error is orthogonal to the subspace $\mathcal{V}^h \subset \mathcal{V}$. Another way of putting this is to say "u^h is the projection of u onto $\mathcal{S}^h$ with respect to $a(\cdot\,,\,\cdot)$."

Part (b) means that there is no member of $\mathcal{S}^h$ that is a better approximation to u (with respect to the energy norm) than u^h, the solution of the Galerkin finite element problem. This is referred to as the ***best approximation property***. It may be interpreted as meaning that the approximate solution is a least-squares best fit of the exact solution in terms of $a(\cdot\,,\,\cdot)$. For the problem classes emphasized in previous chapters, this means that the mth derivatives of u^h best fit the mth derivates of u in a weighted sense. (In elasticity, this means the accuracy of strains, or stresses, is optimized.)

Proof

a. Because $\mathcal{V}^h \subset \mathcal{V}$, we may write the exact equation as

$$a(w^h,\, u) = (w^h,\, \ell) + (w^h,\, h)_\Gamma$$

for all $w^h \in \mathcal{V}^h$. Substracting this equation from (4.1.2), and using the bilinearity of $a(\cdot\,,\,\cdot)$, results in

$$0 = a(w^h,\, u^h) - a(w^h,\, u) = a(w^h,\, e)$$

b. Let $w^h \in \mathcal{V}^h$. Then, by the symmetry and bilinearity of $a(\cdot\,,\,\cdot)$,

$$a(e + w^h,\, e + w^h) = a(e, e) + \underbrace{2a(w^h,\, e)}_{0 \text{ by part (a)}} + a(w^h,\, w^h)$$

[2] This condition is verified for the cases considered so far in Appendix 4.I.

which, by virtue of $a(\boldsymbol{w}^h, \boldsymbol{w}^h) \geq 0$, implies

$$a(\boldsymbol{e}, \boldsymbol{e}) \leq a(\boldsymbol{e} + \boldsymbol{w}^h, \boldsymbol{e} + \boldsymbol{w}^h)$$

Any $\boldsymbol{U}^h \in \mathcal{S}^h$ can be written as

$$\boldsymbol{U}^h = \boldsymbol{u}^h + \boldsymbol{w}^h$$

for some $\boldsymbol{w}^h \in \mathcal{V}^h$. Because

$$\begin{aligned} \boldsymbol{e} + \boldsymbol{w}^h &= \boldsymbol{u}^h - \boldsymbol{u} + \boldsymbol{w}^h \\ &= \boldsymbol{U}^h - \boldsymbol{u} \end{aligned}$$

we immediately obtain the desired result:

$$\begin{aligned} a(\boldsymbol{e}, \boldsymbol{e}) &\leq a(\boldsymbol{e} + \boldsymbol{w}^h, \boldsymbol{e} + \boldsymbol{w}^h) \\ &= a(\boldsymbol{U}^h - \boldsymbol{u}, \boldsymbol{U}^h - \boldsymbol{u}) \end{aligned}$$ ■

Corollary. ("Pythagorean Theorem")
Assume $\mathcal{S}^h = \mathcal{V}^h$ (i.e., the essential boundary conditions are homogeneous). Then

$$a(\boldsymbol{u}, \boldsymbol{u}) = a(\boldsymbol{u}^h, \boldsymbol{u}^h) + a(\boldsymbol{e}, \boldsymbol{e})$$

Proof. The hypothesis and part (a) imply

$$a(\boldsymbol{u}^h, \boldsymbol{e}) = 0$$

Thus

$$\begin{aligned} a(\boldsymbol{u}, \boldsymbol{u}) &= a(\boldsymbol{u}^h - \boldsymbol{e}, \boldsymbol{u}^h - \boldsymbol{e}) \\ &= a(\boldsymbol{u}^h, \boldsymbol{u}^h) - \underbrace{2a(\boldsymbol{u}^h, \boldsymbol{e})}_{0} + a(\boldsymbol{e}, \boldsymbol{e}) \end{aligned}$$ ■

Remarks

1. Rearranging this result in the form $a(\boldsymbol{e}, \boldsymbol{e}) = a(\boldsymbol{u}, \boldsymbol{u}) - a(\boldsymbol{u}^h, \boldsymbol{u}^h)$ reveals that ***the energy of the error equals (minus) the error of the energy.***

2. It is a direct consequence of the corollary that

$$\boxed{a(\boldsymbol{u}^h, \boldsymbol{u}^h) \leq a(\boldsymbol{u}, \boldsymbol{u})}$$

that is, ***the approximate solution underestimates the strain energy.***

Exercise 1. Assume $\boldsymbol{g} = \boldsymbol{h} = \boldsymbol{0}$ and $\boldsymbol{f}(\boldsymbol{x}) = \delta(\boldsymbol{x} - \overline{\boldsymbol{x}})\boldsymbol{e}_i$, i.e., a Dirac delta function in the direction $\boldsymbol{e}_i$ located at $\overline{\boldsymbol{x}}_i$. Use Remark 2, above, to show that

$$u_i^h(\overline{\boldsymbol{x}}) \leq u_i(\overline{\boldsymbol{x}})$$

(In the parlance of structural mechanics, the "displacement under the load" is underestimated by the Galerkin finite element solution.)

Exercise 2. (***The Principle of Minimum Potential Energy***)

Let

$$\boldsymbol{U}_\epsilon = \boldsymbol{u} + \epsilon \boldsymbol{w} \tag{4.1.5}$$

where $\epsilon \in \mathbb{R}$. Note every member of $\mathcal{S}$ can be represented in the form (4.1.5) for some $\boldsymbol{w} \in \mathcal{V}$ and $\epsilon \in \mathbb{R}$. Define the ***potential energy*** by

$$I(\boldsymbol{U}_\epsilon) = \frac{1}{2}a(\boldsymbol{U}_\epsilon, \boldsymbol{U}_\epsilon) - (\boldsymbol{U}_\epsilon, \boldsymbol{f}) - (\boldsymbol{U}_\epsilon, \boldsymbol{h})_\Gamma$$

Establish the following results:

a. The potential energy is ***stationary*** (i.e., $(dI(\boldsymbol{U}_\epsilon)/d\epsilon)|_{\epsilon=0} = 0$) if and only if the variational equation (4.1.1) is satisfied.
b. The potential energy is ***minimized*** at $\boldsymbol{u}$, that is, $I(\boldsymbol{u}) \leq I(\boldsymbol{u} + \epsilon\boldsymbol{w})$ for all $\boldsymbol{w} \in \mathcal{V}$ and $\epsilon \in \mathbb{R}$, (*Hint*: Use part (a) and show that $(d^2I(\boldsymbol{U}_\epsilon)/d\epsilon^2)|_{\epsilon=0} = a(\boldsymbol{w}, \boldsymbol{w}) \geq 0$.)

   (Note that counterparts of (a) and (b) can be established for the Galerkin formulation (4.1.2) if $\boldsymbol{u}$, $\boldsymbol{w}$, $\boldsymbol{U}_\epsilon$, $\mathcal{S}$, and $\mathcal{V}$ are replaced by $\boldsymbol{u}^h$, $\boldsymbol{w}^h$, $\boldsymbol{U}_\epsilon^h$, $\mathcal{S}^h$, and $\mathcal{V}^h$, respectively.)
c. ***The approximate solution overestimates the potential energy***, i.e.,

$$I(\mathbf{u}^h) \geq I(\mathbf{u})$$

Hint: This follows immediately from $\mathcal{S}^h \subset \mathcal{S}$.

Part (a) of the problem indicates that the variational equations are derivable from the derivative of a potential (i.e., the potential energy). When this is the case, one says that a ***variational principle*** exists for the equations. Vainberg's theorem [1] asserts that ***the existence of a variational principle is equivalent to the symmetry of the bilinear form*** $a(\cdot, \cdot)$. The subjects of variational principles and methods are fascinating ones, which permeate the fields of solid and structural mechanics (e.g., see [2–4]). It must be emphasized, however, that the *finite element methodology presented is in no way dependent upon the existence of a variational principle*. On the other hand, the best approximation property is intimately linked with the fact that the weak forms considered so far emanate from the minimization of a potential.

Standard Error Estimates in Sobolev Norms

The finite element function spaces discussed in Chapter 3 are endowed with an approximation property that may be stated as follows: Given a function $\boldsymbol{u}$ possessing r square integrable generalized derivatives[3], i.e.,

[3] One says that $\boldsymbol{u}$ is of class H^r, or $\boldsymbol{u} \in H^r$.

$$\int_\Omega (u_i u_i + u_{i,j} u_{i,j} + \cdots + \underbrace{u_{i,jk\ldots l}}_{r \text{ indices}} \underbrace{u_{i,jk\ldots l}}_{r \text{ indices}})\, d\Omega < \infty$$

then there exists a function $\boldsymbol{U}^h \in \mathcal{S}^h$ (sometimes called the ***interpolate***) such that

$$\boxed{\|\boldsymbol{u} - \boldsymbol{U}^h\|_m \leq c\, h^\alpha \|\boldsymbol{u}\|_r} \tag{4.1.6}$$

where c is a constant *independent of* $\boldsymbol{u}$ *and* h, $\alpha = \min(k + 1 - m, r - m)$, k is the degree of complete polynomial appearing in the element shape functions, and h is the ***mesh parameter***, a scalar characterizing the refinement of the finite element mesh. The mesh parameter may be taken to be the diameter of the largest element in the mesh (see Fig. 4.1.1). A collection of finite element spaces $\{\mathcal{S}^h\}$ (i.e., meshes parameterized by h) possessing the approximation property (4.1.6) is called ***k, m-regular***. See Appendix 4.I for elaboration.

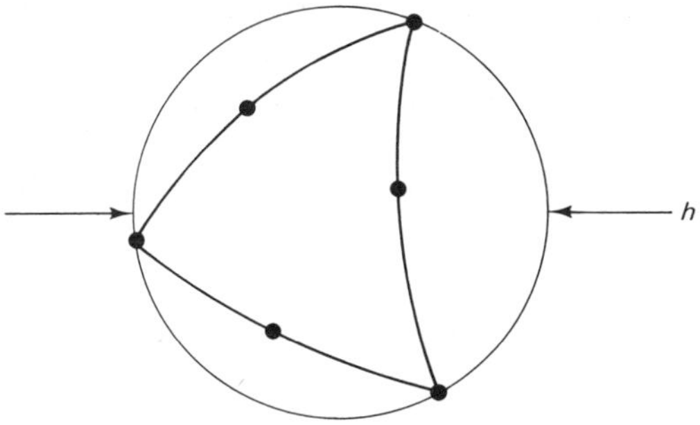

Figure 4.1.1 Schematic of the mesh parameter; h is the diameter of the smallest circle that contains the largest element of the mesh.

Theorem. The approximation result (4.1.6) and the best approximation property (part (b) of the last theorem) enable us to establish ***the fundamental error estimate for the elliptic boundary-value problem:***

$$\boxed{\|\boldsymbol{e}\|_m \leq \bar{c}\, h^\alpha \|\boldsymbol{u}\|_r}$$

where $\bar{c}$ is a constant independent of $\boldsymbol{u}$ and h.

Proof

$$\begin{aligned}\|\boldsymbol{e}\|_m &\leq \frac{1}{c_1} a(\boldsymbol{e}, \boldsymbol{e})^{1/2} \\ &\leq \frac{1}{c_1} a(\boldsymbol{u} - \boldsymbol{U}^h, \boldsymbol{u} - \boldsymbol{U}^h)^{1/2} \\ &\leq \frac{c_2}{c_1} \|\boldsymbol{u} - \boldsymbol{U}^h\|_m\end{aligned}$$

(continued)

$$\leq \bar{c} h^{\alpha} \|u\|_r$$

where $\bar{c} = cc_2/c_1$.

Lines 1 and 3 follow from the assumption that $a(\cdot, \cdot)^{1/2}$ and $\|\cdot\|_m$ define equivalent norms, i.e., (4.1.3). Line 2 follows from part (b) of the previous theorem. Line 4 follows from (4.1.6) by choosing U^h to be the interpolate. (The choice of U^h is arbitrary by virtue of part (b) of the previous theorem.) ■

Remarks

1. As long as $k + 1$ and r are greater than m, we have optimal convergence in the H^m norm.

2. Assume u is smooth in the sense that $u \in H^{k+1}$. Then the error satisfies

$$\boxed{\|e\|_m \leq c h^{k+1-m} \|u\|_{k+1}}$$

This is sometimes referred to as the ***standard error estimate***. (Remember that k is the degree of polynomial completeness of the element shape functions and m is the order of the highest derivatives appearing in the energy expression.)

3. Error estimates in lower H^s-norms, $0 \leq s \leq m$, can be developed by the ***Aubin-Nitsche method*** (see [5–7] for descriptions). Assuming $u \in H^{k+1}$, the main result is

$$\boxed{\|e\|_s \leq c\, h^{\beta} \|u\|_{k+1}}$$

where c is a constant, independent of u and h, and $\beta = \min\ (k + 1 - s, 2(k + 1 - m))$. Thus the finite element approximation is optimal in the H^s-norm for all s such that $0 \leq s \leq m$. For example, let $k = m = 1$. Then

$$\boxed{\begin{aligned} \|e\|_0 &\leq c\, h^2 \|u\|_2 \\ \|e\|_1 &\leq c\, h \|u\|_2 \end{aligned}}$$

Using mechanics terminology, the former relation gives the convergence rate in the L_2-norm for "displacements," and the latter gives the convergence rate in the L_2-norm for "displacement gradients" (i.e., strains or stresses). Increasing k increases the order of convergence, whereas increasing m decreases the order of convergence.

Exercise 3. Use the result of Remark 3 to determine the rates of convergence of e in L_2 and H^1 when $m = 1$ and $k = 2$ (i.e., quadratic elements).

Exercise 4. Determine the convergence rates of e in L_2, H^1, and H^2 for Bernoulli-Euler beam theory ($m = 2$) and Hermite cubic shape functions ($k = 3$).

Effect of Numerical Quadrature

The previously developed theory assumes that we rigorously adhere to the Galerkin recipe. In particular, this means that all integrals need to be calculated exactly. As indicated in Chapter 3, this would be virtually impossible for isoparametric elements in all but the simplest configurations. Because numerically integrated stiffnesses represent the rule rather than the exception, it is important to understand the accuracy implications of approximate numerical integration. A theory has been presented in [5] that provides *sufficient* conditions for quadrature rules to maintain the full rate of convergence of the exactly integrated formulation. A summary of the main results follows: Let $\bar{k}$ denote the order of the highest-order monomial present in the element shape functions—for example, for the bilinear quadrilateral $\bar{k} = 2$, due to the presence of $\xi\eta$; for the biquadratic quadrilateral $\bar{k} = 4$, due to $\xi^2\eta^2$; for the trilinear brick $\bar{k} = 3$, due to $\xi\eta\zeta$, and so on. The Pascal triangle can be consulted to determine $\bar{k}$ for standard elements. The full rate of convergence in the energy norm of the exactly integrated procedure is attained if the quadrature rule is capable of exactly integrating all monomials through degree $\bar{k} + k - 2m$. First-order convergence is maintained if the rule is exact for monomials through degree $\bar{k} - m$.

Caution. Sufficiently accurate quadrature rules turn out to be of lower order than we might expect. Low-order rules can have a deleterious side effect: ***rank deficiency***. This issue is discussed further toward the end of this chapter. For now it is sufficient to note that rank deficiency is an independent issue, which must also be considered in selecting a suitable quadrature rule.

Example 1

For the bilinear quadrilateral in heat conduction, or elasticity, full rate of convergence is maintained if the rule exactly integrates polynomials through degree $\bar{k} + k - 2m = 2 + 1 - 2 = 1$. So, for example, the one-point Gaussian rule is adequate from the standpoint of accuracy (but see Sec. 4.6 concerning rank deficiency!). Likewise, for the eight-node serendipity quadrilateral, $\bar{k} + k - 2m = 3 + 2 - 2 = 3$; thus 2×2 Gauss is required. Because $\bar{k} - m = 3 - 1 = 2$, the 2×2 Gauss rule is needed to maintain first-order convergence.

Example 2

One-dimensional elements, triangles, and tetrahedra typically employ complete polynomials. Thus $\bar{k} = k$, and so the rule must be capable of integrating polynomials of degree $2(k - m)$ to maintain full rate of convergence.

Remark

The quadrature rule determined by the above considerations can be applied to other element integrals (e.g., the body force integral) without altering the conclusions. This becomes important later on when we introduce mass matrices in dynamics.

Discussion

The preceding mathematical results should provide a degree of confidence in the theoretical soundness of the finite element method. In fact, in practice, for typical problems, the results obtained for the Galerkin finite element method generally are very good. The best approximation property is a significant guarantee of reasonable behavior even when very crude meshes are employed.

What could possibly go wrong? Well, suppose the best approximation is not very good. That is, there is no function in $\mathcal{S}^h$ that is a good approximation to $\boldsymbol{u}$. This can happen and, for typical finite elements which we have considered so far, does for ***constrained media problems*** such as ones involving incompressible, or nearly incompressible, behavior. In these cases the approximation may be so poor that it is practically useless. Fortunately, an alternative finite element formulation accommodates successful approximations. This is dealt with in the following sections.

4.2 INCOMPRESSIBLE ELASTICITY AND STOKES FLOW

Many problems of physical importance involve motions that essentially preserve volumes locally. That is, after deformation each small portion of the medium has the same volume as before deformation. Media that behave in this fashion are termed ***incompressible***. Rubber is often modeled as an incompressible elastic material, and many fluid flows are assumed incompressible.

In isotropic linear elasticity the condition of incompressibility may be expressed in terms of Poisson's ratio ν. As ν approaches $\frac{1}{2}$, resistance to volume change is greatly increased assuming resistance to shearing remains constant. This may be seen by calculating the ratio of bulk modulus, B, to shearing modulus, μ ([8], p. 71):

$$\frac{B}{\mu} = \frac{2(1+\nu)}{3(1-2\nu)} \tag{4.2.1}$$

Clearly, as $\nu \rightarrow \frac{1}{2}$, the ratio approaches infinity. The limiting value $\nu = \frac{1}{2}$ thus represents incompressibility.

This limit creates problems in the equations of (compressible) elasticity. Recall that the constitutive equation in the isotropic case may be written as (see eqs. (2.7.1), (2.7.2), and (2.7.31))

$$\sigma_{ij} = \lambda u_{k,k}\,\delta_{ij} + 2\mu\, u_{(i,j)} \tag{4.2.2}$$

where

$$\lambda = \frac{2\nu\mu}{1-2\nu} \tag{4.2.3}$$

Thus the Lamé parameter, λ, also becomes unbounded in the incompressible limit and an alternative formulation of the theory is therefore necessary. In this formulation the constitutive equation is written as

$$\sigma_{ij} = -p\delta_{ij} + 2\mu u_{(i,j)} \tag{4.2.4}$$

where $p = p(\boldsymbol{x})$ is the ***hydrostatic pressure.*** The pressure must be determined as part of the solution to the boundary-value problem and thus represents an additional unknown. The additional equation that needs to be introduced is the ***kinematic condition of incompressibility***, namely,

$$\text{div}\, \boldsymbol{u} = u_{i,i} = 0 \tag{4.2.5}$$

The boundary-value problem may then be stated as follows:

Incompressible Isotropic Elasticity

Given ℓ_i, g_i, and h_i (as in Sec. 2.7), find the displacement, u_i, and pressure, p, such that

$$\left.\begin{aligned} \sigma_{ij,j} + \ell_i &= 0 \\ u_{i,i} &= 0 \end{aligned}\right\} \quad \text{in } \Omega \tag{4.2.6, 4.2.7}$$

$$u_i = g_i \quad \text{on } \Gamma_{g_i} \tag{4.2.8}$$

$$\sigma_{ij} n_j = h_i \quad \text{on } \Gamma_{h_i} \tag{4.2.9}$$

where σ_{ij} is given by (4.2.4).

Remark

In the case of the displacement boundary-value problem, in which $\Gamma = \Gamma_{g_i}$ and $\Gamma_{h_i} = \varnothing$, a consistency condition on g_i follows from incompressibility:

$$\begin{aligned} 0 &= \int_\Omega u_{i,i}\, d\Omega && \text{(by (4.2.7))} \\ &= \int_\Gamma u_i n_i\, d\Gamma && \text{(divergence theorem, (2.2.11))} \\ &= \int_\Gamma g_i n_i\, d\Gamma && \text{(by (4.2.8))} \end{aligned} \tag{4.2.10}$$

The given g_i must satisfy (4.2.10) or else no solution to the boundary-value problem can exist. In the displacement boundary-value problem, the pressure is determinable only up to an arbitrary constant.

Stokes Flow

The equations of Stokes flow are identical to the equations of isotropic incompressible elasticity. Only the physical interpretation of the variables is different. In Stokes flow $\boldsymbol{u}$ is the velocity of the fluid and μ is the dynamic viscosity. Stokes flow governs highly viscous phenomena, often referred to as ***creeping flow.***

Exercise 1. Use (4.2.3) and (2.7.35) to show that the coefficients of the isotropic plane stress constitutive equation remain bounded as $\nu \rightarrow \frac{1}{2}$. Thus a special formulation is *not* required for the incompressible plane stress case.

4.2.1 Prelude To Mixed And Penalty Methods

The subject of finite element approximations to incompressible elasticity problems is replete with so-called mixed and penalty formulations. In order to introduce the essential aspects of these methods in a context that is as simple as possible, we shall consider the example of appending a constraint to a linear algebraic system. Suppose we wish to solve our standard matrix problem

$$\boldsymbol{Kd} = \boldsymbol{F} \tag{4.2.11}$$

where, as usual, $\boldsymbol{K}$ is symmetric and positive-definite, subject to a constraint on one of the degrees of freedom, namely,

$$d_Q = \mathcal{q} \tag{4.2.12}$$

where the subscript Q represents the equation number in the global ordering and $\mathcal{q}$ is a given constant. Physically, we can think of (4.2.12) as perhaps a modification to an original design. For example, the position of a boundary node may be altered to stiffen a structure for some purpose. The modified problem consisting of (4.2.11) and (4.2.12) may be formulated as a ***constrained variational problem.*** The essential character of mixed methods is exhibited in this framework. To develop this idea, it is helpful to think of $\boldsymbol{Kd} = \boldsymbol{F}$ as arising from the minimization of a function:

$$\mathcal{F}(\boldsymbol{d}) = \frac{\boldsymbol{d}^T\boldsymbol{Kd}}{2} - \boldsymbol{d}^T\boldsymbol{F} \tag{4.2.13}$$

which is called the ***total potential energy function.*** The vector that minimizes $\mathcal{F}$ satisfies (4.2.11). This can be seen as follows: Let ϵ be a real parameter. Form the one-parameter family of displacement vectors

$$\boldsymbol{d} + \epsilon\boldsymbol{c} \tag{4.2.14}$$

where $\boldsymbol{c}$ is arbitrary. $\mathcal{F}$ is minimized by $\boldsymbol{d}$ if

$$0 = \left(\frac{d}{d\epsilon}\mathcal{F}(\boldsymbol{d} + \epsilon\boldsymbol{c})\right)_{\epsilon=0} \tag{4.2.15}$$

for all vectors $\boldsymbol{c}$. This calculation is carried out as follows:

$$\begin{aligned}\left(\frac{d}{d\epsilon}\mathcal{F}(\boldsymbol{d} + \epsilon\boldsymbol{c})\right)_{\epsilon=0} &= \left(\frac{d}{d\epsilon}((\boldsymbol{d} + \epsilon\boldsymbol{c})^T\boldsymbol{K}(\boldsymbol{d} + \epsilon\boldsymbol{c})/2 - (\boldsymbol{d} + \epsilon\boldsymbol{c})^T\boldsymbol{F})\right)_{\epsilon=0} \\ &= \boldsymbol{c}^T\boldsymbol{Kd}/2 + \boldsymbol{d}^T\boldsymbol{Kc}/2 - \boldsymbol{c}^T\boldsymbol{F} \\ &= \boldsymbol{c}^T(\boldsymbol{Kd} - \boldsymbol{F}) \qquad \text{(symmetry of } \boldsymbol{K}\text{)}\end{aligned} \tag{4.2.16}$$

Thus we see that (4.2.16) must be zero for arbitrary $\boldsymbol{c}$, which implies $\boldsymbol{Kd} = \boldsymbol{F}$. Thus we have equivalent alternative characterizations of $\boldsymbol{d}$: It is the solution of problem (4.2.11) and also the minimizer of function (4.2.13). The fact that $\boldsymbol{d}$ minimizes $\mathcal{F}$ (i.e., solves a "variational problem") allows us to introduce standard calculus-of-variations methodology in order to formulate the modified problem consisting of (4.2.11) and (4.2.12). For this purpose it is convenient to rephrase (4.2.12) as

$$0 = \mathcal{G}(\boldsymbol{d}) = \mathbf{1}_Q^T \boldsymbol{d} - g \tag{4.2.17}$$

where

$$\mathbf{1}_Q^T = \langle 0 \ldots 0 \underset{\uparrow\ Q\text{th term}}{1} 0 \ldots 0 \rangle \tag{4.2.18}$$

The angle brackets signify a row vector. Clearly, (4.2.17) is equivalent to (4.2.12).

Lagrange-Multiplier Method

Consider the following function:

$$\mathcal{H}(\boldsymbol{d}, m) = \mathcal{F}(\boldsymbol{d}) + m\mathcal{G}(\boldsymbol{d}) \tag{4.2.19}$$

where m is a scalar parameter called the ***Lagrange multiplier.*** Rendering (4.2.19) "stationary" is equivalent to satisfaction of the constrained problem (i.e., eqs. (4.2.11) and (4.2.12)). The condition of stationarity is that

$$0 = \left(\frac{d}{d\epsilon}\mathcal{H}(\boldsymbol{d} + \epsilon\,\boldsymbol{c},\, m + \epsilon l)\right)_{\epsilon=0} \tag{4.2.20}$$

for all values of the vector $\boldsymbol{c}$ and scalar l. The multiplier m plays the role of the force that maintains the constraint (4.2.12). It is an additional unknown corresponding to the additional equation that must be satisfied, namely (4.2.12). Let us carry out the calculation indicated in (4.2.20):

$$\begin{aligned} 0 &= \left(\frac{d}{d\epsilon}(\mathcal{F}(\boldsymbol{d} + \epsilon\boldsymbol{c}) + (m + \epsilon l)\mathcal{G}(\boldsymbol{d} + \epsilon\boldsymbol{c}))\right)_{\epsilon=0} \\ &= \boldsymbol{c}^T(\boldsymbol{Kd} - \boldsymbol{F}) + l\mathcal{G}(\boldsymbol{d}) + m\mathbf{1}_Q^T\boldsymbol{c} \\ &= \boldsymbol{c}^T(\boldsymbol{Kd} + m\mathbf{1}_Q - \boldsymbol{F}) + l(\mathbf{1}_Q^T\boldsymbol{d} - g) \end{aligned} \tag{4.2.21}$$

Due to the arbitrariness of $\boldsymbol{c}$ and l, (4.2.21) implies

$$\boldsymbol{Kd} + m\mathbf{1}_Q = \boldsymbol{F} \tag{4.2.22}$$

and

$$\mathbf{1}_Q^T\boldsymbol{d} = g \tag{4.2.23}$$

or, equivalently,

$$\begin{bmatrix} \boldsymbol{K} & \mathbf{1}_Q \\ \mathbf{1}_Q^T & 0 \end{bmatrix} \begin{Bmatrix} \boldsymbol{d} \\ m \end{Bmatrix} = \begin{Bmatrix} \boldsymbol{F} \\ g \end{Bmatrix} \tag{4.2.24}$$

This is the equation system which determines $\boldsymbol{d}$ and m, the solution of the modified problem. To account for the constraint, the original system, (4.2.11), needed to be modified (see (4.2.22)). This is physically reasonable.

The equation system (4.2.24) is a prototype of a ***mixed method***, i.e., one in which there are both displacements and forces as unknowns. Its importance in regard to problems of incompressibility—a constraint—is that we have from the outset both displacements and pressure—the forcelike variable—as unknowns. Consequently, we may ultimately anticipate algebraic systems having features in common with (4.2.24). Note that the coefficient matrix in (4.2.24) is symmetric but not positive-definite. Symmetry follows from the definition of a transposed partitioned matrix:

$$\begin{bmatrix} A & B \\ C & D \end{bmatrix}^T = \begin{bmatrix} A^T & C^T \\ B^T & D^T \end{bmatrix} \tag{4.2.25}$$

Failure of the positive-definiteness condition (see (ii) of the definition in Sec. 1.9) follows from:

$$\begin{Bmatrix} \mathbf{0} \\ 1 \end{Bmatrix}^T \begin{bmatrix} \boldsymbol{K} & \mathbf{1}_Q \\ \mathbf{1}_Q^T & 0 \end{bmatrix} \begin{Bmatrix} \mathbf{0} \\ 1 \end{Bmatrix} = \begin{Bmatrix} \mathbf{0} \\ 1 \end{Bmatrix}^T \begin{Bmatrix} \mathbf{1}_Q \\ 0 \end{Bmatrix} = 0 \tag{4.2.26}$$

Penalty Method

The penalty method of formulating the constrained problem may be viewed as an approximation to the Lagrange-multiplier method. In the penalty formulation, the Lagrange multiplier is approximated as follows:

$$m \cong k\mathcal{G}(\boldsymbol{d}) \tag{4.2.27}$$

where k is a large positive number having the physical interpretation of a stiff spring constant. Note that k is *not* an unknown. With this approximation we may define a new function,

$$\mathcal{J}(\boldsymbol{d}) = \mathcal{F}(\boldsymbol{d}) + \frac{k}{2}\mathcal{G}(\boldsymbol{d})^2 \tag{4.2.28}$$

whose minimum defines an approximate solution to the constrained problem. Calculating, as before,

$$\begin{aligned} 0 &= \left(\frac{d}{d\epsilon}\mathcal{J}(\boldsymbol{d} + \epsilon\boldsymbol{c})\right)_{\epsilon=0} \\ &= \left(\frac{d}{d\epsilon}\left(\mathcal{F}(\boldsymbol{d} + \epsilon\boldsymbol{c}) + \frac{k}{2}\mathcal{G}(\boldsymbol{d} + \epsilon\boldsymbol{c})^2\right)\right)_{\epsilon=0} \\ &= \boldsymbol{c}^T(\boldsymbol{Kd} - \boldsymbol{F}) + k\mathcal{G}(\boldsymbol{d})\mathbf{1}_Q^T\boldsymbol{c} \\ &= \boldsymbol{c}^T\left((\boldsymbol{K} + k\mathbf{1}_Q\mathbf{1}_Q^T)\boldsymbol{d} - (\boldsymbol{F} + kq\mathbf{1}_Q)\right) \end{aligned} \tag{4.2.29}$$

which implies

$$(\boldsymbol{K} + k\mathbf{1}_Q\mathbf{1}_Q^T)\boldsymbol{d} = \boldsymbol{F} + kq\mathbf{1}_Q \tag{4.2.30}$$

Explicating (4.2.30) yields

$$\left(\boldsymbol{K} + \underbrace{\begin{bmatrix} 0 & \cdots & 0 \\ \vdots & k & \vdots \\ 0 & \cdots & 0 \end{bmatrix}}_{\substack{k \text{ appears in the} \\ Q\text{th diagonal entry}}}\right)\boldsymbol{d} = \boldsymbol{F} + \underbrace{\begin{Bmatrix} 0 \\ \vdots \\ 0 \\ kq \\ 0 \\ \vdots \\ 0 \end{Bmatrix}}_{\substack{kq \text{ appears in} \\ \text{the } Q\text{th row}}} \tag{4.2.31}$$

Thus it is clear that as $k \to \infty$, $d_Q \to q$ (i.e., the constraint is satisfied) and thus the right-hand side of (4.2.27) is an approximation to the constraining force (Lagrange multiplier). The larger the k, the better the approximation. This formulation is suggestive of general ways of approximating constrained problems. In the context of elasticity, one would interpret ideas like this as approximating the incompressible case by a slightly compressible formulation. That is, one for which the ratio B/μ, or equivalently λ/μ, is very large (see (4.2.1) and (4.2.3)).

4.3 A MIXED FORMULATION OF COMPRESSIBLE ELASTICITY CAPABLE OF REPRESENTING THE INCOMPRESSIBLE LIMIT

It is desirable to have a formulation of isotropic elasticity that is valid for both compressible and incompressible behavior. To this end we may introduce a pair of constitutive equations

$$\sigma_{ij} = -p\delta_{ij} + 2\mu u_{(i,j)} \tag{4.3.1}$$

$$0 = u_{i,i} + \frac{p}{\lambda} \tag{4.3.2}$$

where the pressure parameter, p, is viewed as an independent unknown. If $\nu = \frac{1}{2}$, (4.3.2) becomes the incompressibility condition and p is the hydrostatic pressure as in the previous section. If $\nu < \frac{1}{2}$, p may be eliminated from (4.3.1) by way of (4.3.2) to obtain the constitutive equation of the compressible case, namely,

$$\sigma_{ij} = \lambda u_{k,k}\delta_{ij} + 2\mu u_{(i,j)} \tag{4.3.3}$$

Thus we see that (4.3.1) and (4.3.2) are valid in both the compressible and incompressible cases.

Remark

Note that p may be interpreted as the hydrostatic pressure *only* in the incompressible case. In general, the hydrostatic pressure is $-\sigma_{ii}/3$. Thus in the compressible case, (4.3.3) yields

$$-\sigma_{ii}/3 = \underbrace{-(\lambda + 2\mu/3)}_{B} u_{i,i} \tag{4.3.4}$$

whereas (4.3.2) gives

$$p = -\lambda u_{i,i} \tag{4.3.5}$$

If $\mu \ll \lambda$ (nearly incompressible case), (4.3.5) is a good approximation to (4.3.4). On the other hand, in the incompressible limit

$$p = \frac{-\sigma_{ii}}{3} \tag{4.3.6}$$

follows directly from (4.3.1).

4.3.1 Strong Form

(S) With ℓ_i, g_i, and h_i given as before, we wish to find u_i and p such that

$$\left.\begin{aligned} \sigma_{ij,j} + \ell_i &= 0 \\ u_{i,i} + p/\lambda &= 0 \end{aligned}\right\} \quad \text{in } \Omega \tag{4.3.7, 4.3.8}$$

$$u_i = g_i \qquad \text{on } \Gamma_{g_i} \tag{4.3.9}$$

$$\sigma_{ij}n_j = h_i \qquad \text{on } \Gamma_{h_i} \tag{4.3.10}$$

where σ_{ij} is defined by (4.3.1).

Remark

The formulation presented in this section was first proposed by Herrmann [9]. Generalizations to anisotropic cases were proposed by Taylor, Pister, and Herrmann [10] and Key [11]. The subject of anisotropy and incompressibility is taken up in Sec. 4.6.

4.3.2 Weak Form

The weak formulation of the problem is similar to the one for compressible elasticity (see (2.7.11)) except we need to introduce a term that implies satisfaction of (4.3.8). In addition to the displacement weighting and trial solution spaces ($\mathcal{V}_i$ and $\mathcal{S}_i$, respectively) a ***space of pressures***, $\mathcal{P}$, is required. The functions in $\mathcal{P}$ are required to be square-integrable (i.e., "L_2-functions"; see Appendix 4.I). Because there are no explicit boundary conditions on the pressures, $\mathcal{P}$ suffices as both a trial solution space and a weighting function space.

The weak formulation may then be stated as follows.

(W) Given f_i, g_i, and h_i, as before, find $u_i \in \mathcal{S}_i$ and $p \in \mathcal{P}$, such that for all $w_i \in \mathcal{V}_i$ and $q \in \mathcal{P}$ (pressure weighting function)

$$\boxed{\begin{aligned}\int_\Omega w_{(i,j)}\sigma_{ij}\,d\Omega &- \int_\Omega q(u_{i,i} + p/\lambda)\,d\Omega \\ &= \int_\Omega w_i f_i\,d\Omega + \sum_{i=1}^{n_{sd}} \int_{\Gamma_{h_i}} w_i h_i\,d\Gamma\end{aligned}} \tag{4.3.11}$$

where σ_{ij} is given by (4.3.1).

To see what equations are implied by satisfaction of (4.3.11), we need to integrate (4.3.11) by parts (we assume all functions are smooth):

$$\begin{aligned}0 = &\int_\Omega w_i \underbrace{(\sigma_{ij,j} + f_i)}_{\text{equilibrium}}\,d\Omega \\ &+ \int_\Omega q\underbrace{(u_{i,i} + p/\lambda)}_{\text{eq. (4.3.8)}}\,d\Omega \\ &+ \sum_{i=1}^{n_{sd}} \int_{\Gamma_{h_i}} w_i \underbrace{(h_i - \sigma_{ij}n_j)}_{\text{traction boundary condition}}\,d\Gamma\end{aligned} \tag{4.3.12}$$

The usual arguments enable us to establish that the terms in parentheses in (4.3.12) vanish identically on their respective domains of definition (see Sec. 2.7 for detailed arguments of this type).

The variational equation (4.3.11) may be written in the abstract form:

$$\bar{a}(\boldsymbol{w}, \boldsymbol{u}) - (\operatorname{div} \boldsymbol{w}, p) - \left(q, \operatorname{div} \boldsymbol{u} + \frac{p}{\lambda}\right) = (\boldsymbol{w}, \boldsymbol{f}) + (\boldsymbol{w}, \boldsymbol{h})_\Gamma \tag{4.3.13}$$

where $\bar{a}(\cdot, \cdot)$ is the symmetric bilinear form defined by

$$\bar{a}(\boldsymbol{w}, \boldsymbol{u}) = \int_\Omega w_{(i,j)}\bar{c}_{ijkl}u_{(k,l)}\,d\Omega \tag{4.3.14}$$

in which

$$\bar{c}_{ijkl} = \mu(\delta_{ik}\delta_{jl} + \delta_{il}\delta_{jk}) \tag{4.3.15}$$

Exercise 1. Verify that (4.3.11) and (4.3.13) are equivalent.

Exercise 2. Show that $\bar{a}(\cdot, \cdot)$ satisfies the definition of a symmetric bilinear form (see (1.4.11) and (1.4.13)).

Exercise 3. Show that if $\mu > 0$, then

$$\bar{c}_{ijkl}\,\psi_{ij}\,\psi_{kl} \geq 0 \tag{4.3.16}$$

for all symmetric ψ_{ij} and furthermore

$$\bar{c}_{ijkl}\,\psi_{ij}\,\psi_{kl} = 0 \tag{4.3.17}$$

if and only if $\psi_{ij} = 0$ (positive definiteness, see (2.7.5) and (2.7.6)).

4.3.3 Galerkin Formulation

Recall that $\mathcal{S}^h$ and $\mathcal{V}^h$ are the finite-dimensional approximations to $\mathcal{S}$ and $\mathcal{V}$, respectively. Likewise, let $\mathcal{P}^h$ be the finite-dimensional space that approximates $\mathcal{P}$. The Galerkin formulation may then be stated as follows.

(G) Given $\boldsymbol{f}$, $\boldsymbol{g}$, and $\boldsymbol{h}$, as in (W), find $\boldsymbol{u}^h = \boldsymbol{v}^h + \boldsymbol{g}^h \in \mathcal{S}^h$ and $p^h \in \mathcal{P}^h$ such that for all $\boldsymbol{w}^h \in \mathcal{V}^h$ and $q^h \in \mathcal{P}^h$,

$$\begin{aligned}\bar{a}(\boldsymbol{w}^h, \boldsymbol{v}^h) - (\operatorname{div} \boldsymbol{w}^h, p^h) - (q^h, \operatorname{div} \boldsymbol{v}^h + p^h/\lambda) \\ = (\boldsymbol{w}^h, \boldsymbol{f}) + (\boldsymbol{w}^h, \boldsymbol{h})_\Gamma - \bar{a}(\boldsymbol{w}^h, \boldsymbol{g}^h) + (q^h, \operatorname{div} \boldsymbol{g}^h)\end{aligned} \tag{4.3.18}$$

4.3.4 Matrix Problem

In order to develop the matrix form of the problem we need to introduce interpolatory expansions for p^h. Since p^h just needs to be square-integrable, it may be discontinuous across element boundaries. (Note that no derivatives of p^h, or q^h, appear in the variational equation; see (4.3.18).) Thus a wider range of interpolations is permissible for pressure than for displacements. Possible combinations are illustrated in Fig. 4.3.1. *It must, however, be emphasized that arbitrary combinations of interpolations may lead to poor numerical performance and even nonconvergence.* It is difficult to give intuitive guidelines because seemingly "natural" combinations may fail in practice. The subject of appropriate combinations is dealt with in Sec. 4.5.

Let us denote the pressure interpolation by

$$p^h(\boldsymbol{x}) = \sum_{\tilde{A} \in \tilde{\eta}} \tilde{N}_{\tilde{A}}(\boldsymbol{x}) p_{\tilde{A}} \tag{4.3.19}$$

where $\tilde{\eta}$ is the set of pressure node numbers, $\tilde{N}_{\tilde{A}}$ is the pressure shape function associated with pressure node number $\tilde{A}$, and $p_{\tilde{A}}$ is the value of pressure at node number $\tilde{A}$.

Similarly, the pressure weighting function may be expressed as

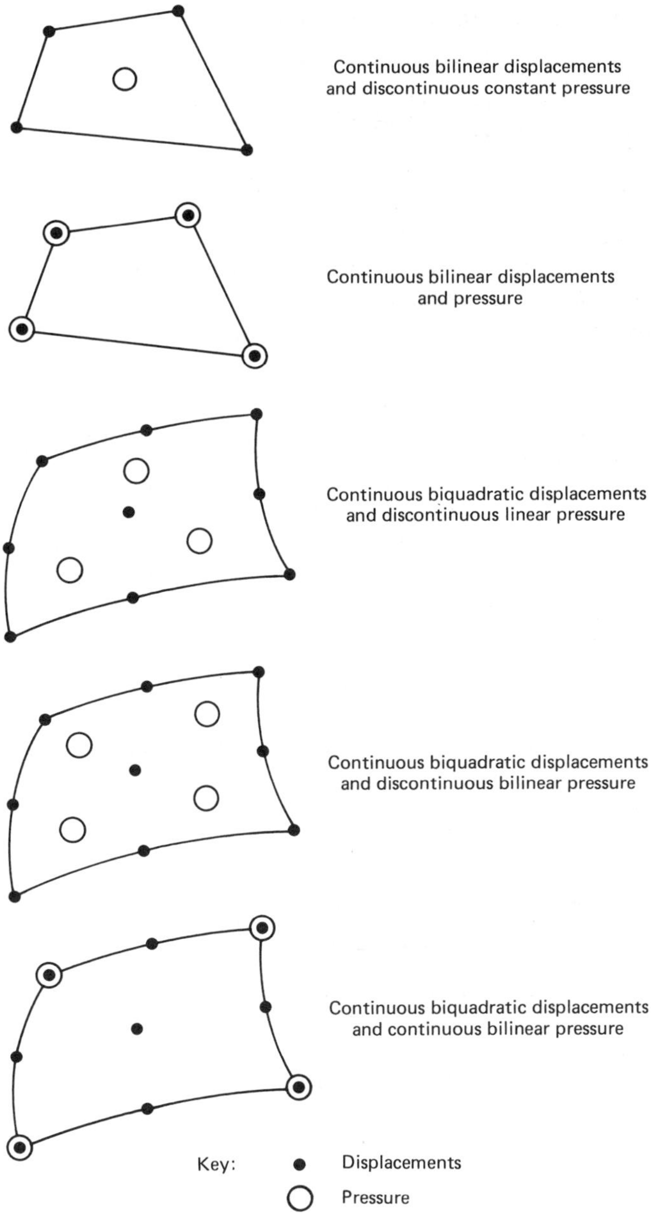

Figure 4.3.1 Examples of possible displacement and pressure interpolations in two dimensions. *Warning:* Not all are effective in practice.

$$q^h(\boldsymbol{x}) = \sum_{\widetilde{A} \in \widetilde{\eta}} \widetilde{N}_{\widetilde{A}}(\boldsymbol{x}) q_{\widetilde{A}} \tag{4.3.20}$$

Substitution of (4.3.19) and (4.3.20), along with the expressions for $\boldsymbol{v}^h$, $\boldsymbol{w}^h$, and q^h (see eqs. (2.8.4) through (2.8.10)) into (4.3.18) leads to the global matrix equation, which can be written in the following partitioned form.

Segregated *d, p*-form of the matrix equation

$$\begin{bmatrix} \bar{K} & G \\ G^T & M \end{bmatrix} \begin{Bmatrix} d \\ p \end{Bmatrix} = \begin{Bmatrix} \bar{F} \\ H \end{Bmatrix} \tag{4.3.21}$$

The arrays in (4.3.21) and corresponding terms in the variational equations are identified in Table 4.3.1. The matrix G is the discrete gradient operator, G^T is the discrete divergence operator, $\bar{K}$ and M are both symmetric, $\bar{K}$ is positive-definite, and M is negative-definite, except in the case $\nu = \frac{1}{2}$ in which $M = 0$.

TABLE 4.3.1

Global array	Term in Galerkin equation from which global array emanates
$\bar{K}$	$\bar{a}(w^h, v^h)$
G	$-(\text{div } w^h, p^h)$
G^T	$-(q^h, \text{div } v^h)$
M	$-(q^h, p^h/\lambda)$
$\bar{F}$	$(w^h, \mathcal{f}) + (w^h, \mathcal{h})_\Gamma - \bar{a}(w^h, \mathcal{g}^h)$
H	$(q^h, \text{div } \mathcal{g}^h)$

There are several possible ways of solving Eq. (4.3.21). Some of these procedures are described below:

Procedure 1

If $M \neq 0$ (compressible case), then p can be eliminated by expanding (4.3.21),

$$\bar{K}d + Gp = \bar{F} \tag{4.3.22}$$

$$G^T d + Mp = H \tag{4.3.23}$$

solving (4.3.23) for p,

$$p = M^{-1}(H - G^T d) \tag{4.3.24}$$

and substituting into (4.3.22):

$$\underbrace{(\bar{K} - GM^{-1}G^T)}_{K}d = \underbrace{\bar{F} - GM^{-1}H}_{F} \tag{4.3.25}$$

Equation (4.3.25) has the usual format. Observe that K is symmetric and positive definite. After solving (4.3.25) for d, p may be calculated from (4.3.24).

Procedure 2

If $\boldsymbol{M} = \boldsymbol{0}$ (incompressible case) the preceding elimination cannot be performed. However, the following procedure may be employed. Solve (4.3.22) for $\boldsymbol{d}$, premultiply by $\boldsymbol{G}^T$, and employ (4.3.23) to obtain the ***discrete Poisson equation for pressure:***

$$\boxed{\underbrace{(\boldsymbol{G}^T\overline{\boldsymbol{K}}^{-1}\boldsymbol{G})}_{\hat{\boldsymbol{K}}}\boldsymbol{p} = \underbrace{\boldsymbol{G}^T\overline{\boldsymbol{K}}^{-1}\overline{\boldsymbol{F}} - \boldsymbol{H}}_{\hat{\boldsymbol{F}}}} \tag{4.3.26}$$

After $\boldsymbol{p}$ is obtained by solving (4.3.26), (4.3.22) may be used to obtain $\boldsymbol{d}$.

Exercise 4. Generalize (4.3.26) to the case in which $\boldsymbol{M} \neq \boldsymbol{0}$.

Procedure 3

Both Procedures 1 and 2 are global in nature and are valid whether p^h is continuous or discontinuous. If pressure is *discontinuous* between elements then, in the compressible case, the pressure degrees of freedom may be eliminated on the element level. The element arrays that correspond to the global arrays are notationally defined in Table 4.3.2. In this case we may write the global system in the usual way as an assembly of element arrays, i.e.,

$$\boldsymbol{K}\boldsymbol{d} = \boldsymbol{F} \tag{4.3.27}$$

where

$$\boldsymbol{K} = \mathop{\mathbf{A}}_{e=1}^{n_{el}} (\boldsymbol{k}^e) \tag{4.3.28}$$

$$\boldsymbol{F} = \mathop{\mathbf{A}}_{e=1}^{n_{el}} (\boldsymbol{f}^e) \tag{4.3.29}$$

and

$$\boldsymbol{k}^e = \overline{\boldsymbol{k}}^e - \boldsymbol{g}^e(\boldsymbol{m}^e)^{-1}(\boldsymbol{g}^e)^T \tag{4.3.30}$$

$$\boldsymbol{f}^e = \overline{\boldsymbol{f}}^e - \boldsymbol{g}^e(\boldsymbol{m}^e)^{-1}\,\boldsymbol{h}^e \tag{4.3.31}$$

Note that (4.3.30) and (4.3.31) are the element analogs of the arrays that appear in (4.3.25). The above, although equivalent to Procedure 1, is much more convenient from a practical standpoint in that only operations on the element level are involved. Likewise, the element vector of nodal pressures, $\boldsymbol{p}^e$, can be calculated from the element nodal displacements once the latter are determined. This can be written as

$$\boldsymbol{p}^e = -(\boldsymbol{m}^e)^{-1}(\boldsymbol{g}^e)^T\boldsymbol{d}^e \tag{4.3.32}$$

where $\boldsymbol{d}^e$ is the element displacement vector. Recall that $\boldsymbol{d}^e$ includes specified displacement degrees of freedom (see eqs. (2.9.10) through (2.9.13)). This is the reason why no $\boldsymbol{h}^e$-term appears (compare the global counterpart of (4.3.32), i.e., (4.3.24)). By consulting Table 4.3.1, the rationale behind (4.3.32) may be verified.

TABLE 4.3.2

Global array	Corresponding element array
$\bar{\boldsymbol{K}}$	$\bar{\boldsymbol{k}}^e$
$\boldsymbol{G}$	$\boldsymbol{g}^e$
$\boldsymbol{M}$	$\boldsymbol{m}^e$
$\bar{\boldsymbol{F}}$	$\bar{\boldsymbol{f}}^e$
$\boldsymbol{H}$	$\boldsymbol{h}^e$

4.3.5 Definition of Element Arrays

The components of the element arrays introduced in the previous section are given in this section.

$$\bar{\boldsymbol{k}}^e = [\bar{k}^e_{pq}], \qquad 1 \le p, q \le n_{ee} \tag{4.3.33}$$

$$\bar{k}^e_{pq} = \boldsymbol{e}_i^T \bar{\boldsymbol{k}}^e_{ab} \boldsymbol{e}_j, \qquad 1 \le a, b \le n_{en} \tag{4.3.34}$$

$$p = n_{ed}(a-1) + i, \qquad q = n_{ed}(b-1) + j \tag{4.3.35}$$

$$\bar{\boldsymbol{k}}^e_{ab} = \int_{\Omega^e} \boldsymbol{B}_a^T \bar{\boldsymbol{D}} \boldsymbol{B}_b \, d\Omega \tag{4.3.36}$$

$$\bar{\boldsymbol{f}}^e = \{\bar{f}^e_p\} \tag{4.3.37}$$

$$\bar{f}^e_p = \int_{\Omega^e} N_a \ell_i \, d\Omega + \int_{\Gamma_{h_i}} N_a h_i \, d\Gamma - \sum_{q=1}^{n_{ee}} \bar{k}^e_{pq} g^e_q \tag{4.3.38}$$

The preceding formulas are close analogs of those given for elasticity in Chapter 2. The only difference involves the appearance of the matrix $\bar{\boldsymbol{D}}$ instead of $\boldsymbol{D}$. From (4.3.15), it may be concluded that $\bar{\boldsymbol{D}}$ is the part of $\boldsymbol{D}$ in which λ-terms are omitted. Therefore, we have the following (e.g., see eq. (2.7.34)).

Three dimensions

$$\bar{\boldsymbol{D}} = \mu \begin{bmatrix} 2 & & & & & \\ & 2 & & & & \\ & & 2 & & & \\ & & & 1 & & \\ & & & & 1 & \\ & & & & & 1 \end{bmatrix} \tag{4.3.39}$$

Plane strain

$$\overline{D} = \mu \begin{bmatrix} 2 & & \\ & 2 & \\ & & 1 \end{bmatrix} \tag{4.3.40}$$

Axisymmetry

$$\overline{D} = \mu \begin{bmatrix} 2 & & & \\ & 2 & & \\ & & 1 & \\ & & & 2 \end{bmatrix} \tag{4.3.41}$$

All omitted terms in (4.3.39) through (4.3.41) are zero.

In (4.3.33) through (4.3.38) the indexing pertains to displacement degrees of freedom only. We assume that the element in question possesses $\widetilde{n}_{en}$ pressure nodes and that $1 \leq \widetilde{a}, \widetilde{b} \leq \widetilde{n}_{en}$, where $\widetilde{a}$ and $\widetilde{b}$ are element pressure node numbers. With these we may write

$$\boldsymbol{m}^e = [m^e_{\widetilde{a}\widetilde{b}}] \tag{4.3.42}$$

$$m^e_{\widetilde{a}\widetilde{b}} = \int_{\Omega^e} \frac{1}{\lambda} \widetilde{N}_{\widetilde{a}} \widetilde{N}_{\widetilde{b}} \, d\Omega \tag{4.3.43}$$

$$\boldsymbol{g}^e = [g^e_{p\widetilde{a}}] \tag{4.3.44}$$

$$g^e_{p\widetilde{a}} = -\int_{\Omega^e} \operatorname{div}(N_a \boldsymbol{e}_i) \widetilde{N}_{\widetilde{a}} \, d\Omega \tag{4.3.45}$$

$$\boldsymbol{h}^e = \{h^e_{\widetilde{a}}\} \tag{4.3.46}$$

$$h^e_{\widetilde{a}} = -\sum_{p=1}^{n_{ee}} g^e_{p\widetilde{a}} \, g^e_p \tag{4.3.47}$$

In (4.3.45), $\operatorname{div}(N_a \boldsymbol{e}_i)$ is given by the following.

Three dimensions and plane strain

$$\operatorname{div}(N_a \boldsymbol{e}_i) = N_{a,i} \tag{4.3.48}$$

Axisymmetry (see Sec. 2.12)

$$\operatorname{div}(N_a \boldsymbol{e}_i) = \begin{cases} N_{a,1} + \dfrac{N_a}{r} & i = 1 \\ N_{a,2} & i = 2 \end{cases} \tag{4.3.49}$$

The stress "vector" in an element may be computed from the formula (compare eq. (2.9.14)):

$$\boldsymbol{\sigma}(\boldsymbol{x}) = -\left(\sum_{\tilde{a}=1}^{\tilde{n}_{en}} \tilde{N}_{\tilde{a}}(\boldsymbol{x}) p_{\tilde{a}}^{e}\right) \boldsymbol{V} + \bar{\boldsymbol{D}}(\boldsymbol{x}) \sum_{a=1}^{n_{en}} \boldsymbol{B}_a(\boldsymbol{x}) \boldsymbol{d}_a^{e} \tag{4.3.50}$$

where $\boldsymbol{V}$ is defined by

$$\boldsymbol{V} = \begin{Bmatrix} 1 \\ 1 \\ 1 \\ 0 \\ 0 \\ 0 \end{Bmatrix} \quad \text{(three dimensions)} \tag{4.3.51}$$

$$\boldsymbol{V} = \begin{Bmatrix} 1 \\ 1 \\ 0 \end{Bmatrix} \quad \text{(plane strain)} \tag{4.3.52}$$

$$\boldsymbol{V} = \begin{Bmatrix} 1 \\ 1 \\ 0 \\ 1 \end{Bmatrix} \quad \text{(axisymmetry)} \tag{4.3.53}$$

Remark

In practical computing, the segregated $\boldsymbol{d}, \boldsymbol{p}$-form of the global matrix equation, (4.3.21), is rarely employed. The reason for this is that the band-profile structure of the coefficient matrix is lost unless the displacement and pressure degrees of freedom associated with an element are grouped together in the overall equation-number ordering. The following exercises are useful in understanding the data processing aspects of the preceding formulation.

Exercise 5. Consider the mesh shown in Fig. 2.10.1. Assume that the present mixed formulation of elasticity is being employed and that both displacement and pressure are interpolated in continuous bilinear fashion over each element. Thus there are three degrees of freedom per node, less displacement boundary conditions. Assume the pressure degree of freedom at a node directly follows the displacement degrees of freedom. Set up the ID, IEN, and LM arrays.

Sketch the band-profile structure of the global coefficient matrix. Calculate the half-bandwidth (see Fig. 1.9.2). Reorder the rows and columns of the coefficient matrix so that the pressure degrees of freedom come last. This ordering puts the coefficient matrix into the segregated $\boldsymbol{d}, \boldsymbol{p}$-form, eq. (4.3.21). Calculate the new half-bandwidth and compare the result with the previous ordering.

Exercise 6. Repeat Exercise 5, but assume that pressure is piecewise constant on each element. Again assume three degrees of freedom per node and associate the single element pressure degree of freedom with the last node of the element in the local ordering. The third degree of freedom at each of the first three element nodes is a dummy degree of freedom and may be eliminated as if it were "prescribed" (i.e., set a zero in the appropriate position of the ID array).

Remark

It is important to realize that the matrix equation of the present formulation is somewhat different than the form considered heretofore. The present coefficient matrix is symmetric but not positive-definite. It possesses both positive and negative eigenvalues. In fact, in the incompressible case, improper interpolatory combinations may also lead to spurious zero eigenvalues, in which case the coefficient matrix is rendered singular and solution is impossible. These are frequently referred to as ***pressure modes*** in the literature; e.g., see Sani et al. [12]. (Equal-order interpolations, such as in Exercise 5, generally create this pathology. The interpolations of Exercise 6 do too under certain circumstances! See Sec. 4.5 for elaboration.)

In well-set cases, in which there are no spurious pressure modes, typical symmetric band-profile equation solvers are also capable of solving systems of the present type. However, some precautions must be taken. For example, an equation with a zero diagonal element must not appear first in the global ordering. Because the global ordering is arbitrary, this can always be accomplished. Due to the fact that some eigenvalues will be negative, equation-solving techniques that take square roots are inapplicable (e.g., the Cholesky decomposition). Alternatives that do not take square roots, such as the Crout algorithm, are acceptable however (see R.L. Taylor's chapter on computing in [13] and Sec. 11.2.2).

4.3.6 Illustration of a Fundamental Difficulty

We have already given some forewarning that arbitrary combinations of displacement and pressure interpolations may prove ineffective in incompressible cases. An example of one of the difficulties is illustrated by the mesh in Fig. 4.3.2. Suppose linear displacement–constant pressure triangular elements are being employed. Furthermore, assume the left-hand side and bottom edges of the mesh are fixed (i.e., the displacements are identically zero).

It may be concluded from the Galerkin equation, (4.3.18), that the condition of incompressibility is

$$(q^h, \operatorname{div} \boldsymbol{u}^h) = 0 \tag{4.3.54}$$

Because

$$(q^h, \operatorname{div} \boldsymbol{u}^h) = \sum_{e=1}^{n_{el}} \int_{\Omega^e} q^h \operatorname{div} \boldsymbol{u}^h \, d\Omega \tag{4.3.55}$$

and q^h is an arbitrary constant on each triangle, we infer from (4.3.54) and (4.3.55) that incompressibility is satisfied in the mean, that is

$$\int_{\Omega^e} \operatorname{div} \boldsymbol{u}^h \, d\Omega = 0, \qquad 1 \leq e \leq n_{el} \tag{4.3.56}$$

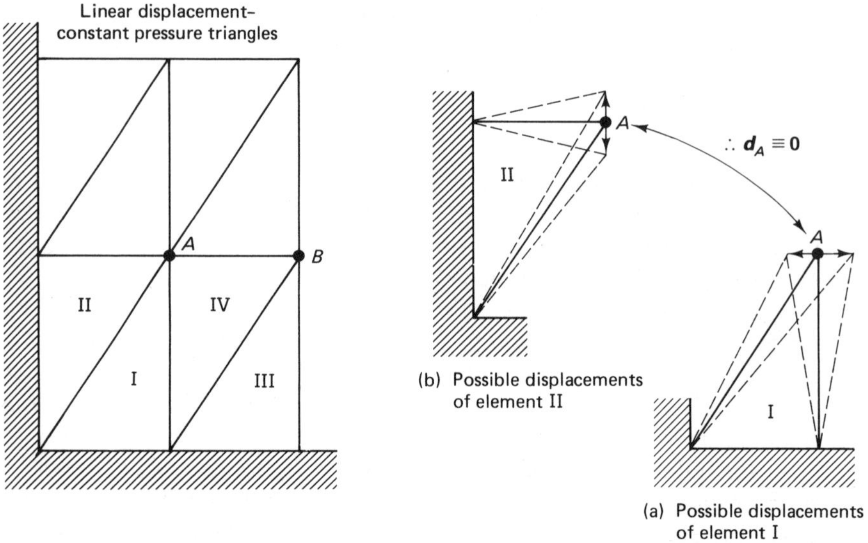

Figure 4.3.2 Mesh for which incompressibility dictates zero displacements.

Thus the area of each triangle must necessarily remain constant. (Due to the fact that $\boldsymbol{u}^h$ is a linear polynomial over each triangle and thus div $\boldsymbol{u}^h$ is constant, (4.3.56) actually implies the stronger pointwise condition

$$\text{div } \boldsymbol{u}^h = 0 \tag{4.3.57}$$

on each Ω^e. However, this result is not needed to illustrate the present difficulty.)

Let us examine what conditions (4.3.56) enforces on the kinematics of the mesh in Fig. 4.3.2. Consider element I. We see from Fig. 4.3.2(a) that constant volume prevents the displacement at node A, $\boldsymbol{d}_A$, from having a nonzero vertical component. Now consider element II. For this element the constant volume condition precludes horizontal motion of node A (see Fig. 4.3.2(b)). Taken together, $\boldsymbol{d}_A$ must be identically zero. We can now repeat the argument for elements III and IV to conclude $\boldsymbol{d}_B$ must also be zero. In fact, identical reasoning may be used to conclude that every node in the entire mesh must have zero displacement. Thus the only possible incompressible displacement is $\boldsymbol{u}^h \equiv \boldsymbol{0}$. This result holds no matter how many elements are present in each direction. Clearly, this type of mesh offers no approximation power whatsoever. This phenomenon is often referred to as ***mesh locking***. It is but one of the difficulties afflicting problems of incompressibility.

In the nearly incompressible case, the same phenomenon occurs, only this time $\boldsymbol{u}^h \cong \boldsymbol{0}$. Thus introducing slight compressibility does not make the problem go away. To varying degrees, a tendency to lock afflicts many standard elements.

A mathematical convergence theory for mixed finite element methods of the type under consideration has been established. The key technical ingredient is the celebrated ***Babuška-Brezzi, or LBB, stability condition.*** To establish whether or not this condition is satisfied for elements of interest is *not* a trivial task. The interested reader is urged to consult Oden and Carey [30] for a detailed presentation of the mathematics. For elements that satisfy the Babuška-Brezzi condition, error estimates of the following form may be established:

$$\|\boldsymbol{u}^h - \boldsymbol{u}\|_1 + \|p^h - p\|_0 = O(h^{\min\{k,\, l+1\}})$$

where k and l are the orders of the displacement and pressure interpolations, respectively. If $k = \min\{k,\ l + 1\}$, the rate of convergence is said to be "optimal." Clearly, elements that satisfy the Babuška-Brezzi condition will not lock.

The mathematics of mixed methods and, in particular, the Babuška-Brezzi condition are beyond the scope of this book. Thus it is desirable to have a simple procedure for assessing whether or not an element will lock. For this purpose the method of constraint counting proves quite effective [14–16].

4.3.7 Constraint Counts

This method is a heuristic approach for determining the ability of an element to perform well in incompressible and nearly incompressible applications. It should be emphasized that this is not a precise mathematical method for assessing elements but rather a quick and simple tool for obtaining an indication of element potential. However, it does seem to be able to predict a propensity for locking. There are, of course, other issues that need to be considered in an overall evaluation of element performance.

Let us introduce a standard mesh, which is illustrated in Fig. 4.3.3 for two-dimensional problems. Let n_{eq} represent the total number of displacement equations after boundary conditions have been imposed (i.e., the length of the vector $\boldsymbol{d}$ in (4.3.21)) and let n_c represent the total number of incompressibility constraints. As long as the pressure equations are linearly independent, n_c will equal $\tilde{n}_{eq}$, the number of pressure equations (i.e., length of the vector $\boldsymbol{p}$ in eq. (4.3.21)). We shall define the ***constraint ratio***, r, by

$$r = \frac{n_{eq}}{n_c} \tag{4.3.58}$$

We are interested in values of r as the number of elements per side, n_{es}, approaches infinity. The conjecture is that r should mimic the behavior of the number of equilibrium equations divided by the number of incompressibility conditions for the governing system of partial differential equations. These are n_{sd}, the number of space dimensions, and 1, respectively. So in two dimensions, the ideal value of r would be 2. A value of r less than 2 would indicate a tendency to lock. If $r \leq 1$, there are more constraints on $\boldsymbol{d}$ than there are displacement degrees of freedom available, and thus severe locking would be anticipated, such as was seen for the linear displacements–constant pressure triangle. A value of r much greater than 2 indicates that not enough incompressibility conditions are present, so the incompressibility condition may be poorly approximated in some problems. A summary of these ideas follows:

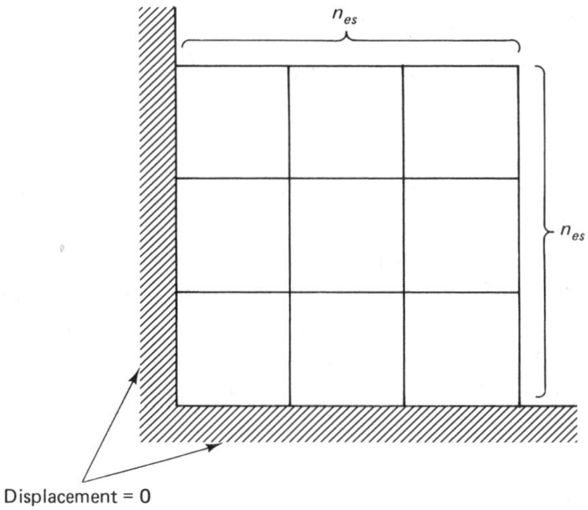

Figure 4.3.3 Standard mesh.

$r > 2$	too few incompressibility constraints
$r = 2$	optimal
$r < 2$	too many incompressibility constraints
$r \leq 1$	locking

We shall begin by calculating r for some two-dimensional elements in which ***pressure is discontinuous.*** In this case r is constant as a function of n_{es}, so r may be determined by considering a single element ($n_{es} = 1$). (The reader may wish to verify this statement for some of the cases considered.)

4.3.8 Discontinuous Pressure Elements

Example 1

Consider the case of the linear displacements–constant pressure triangle. See Fig. 4.3.4(a). In this case

$$r = \frac{n_{eq}}{n_c} = \frac{n_{eq}}{\widetilde{n}_{eq}} = \frac{2}{2} = 1$$

which is consistent with the behavior previously deduced. This element was first studied by Hughes and Allik [17].

Example 2

Consider the bilinear displacements–constant pressure element (see Fig. 4.3.4(b)). For this element, incompressibility is achieved in the mean, i.e.,

$$\int_{\Omega^e} \text{div}\, \boldsymbol{u}^h \, d\Omega = 0 \tag{4.3.59}$$

This may be shown by way of the same reasoning which led to (4.3.56). In this case we have

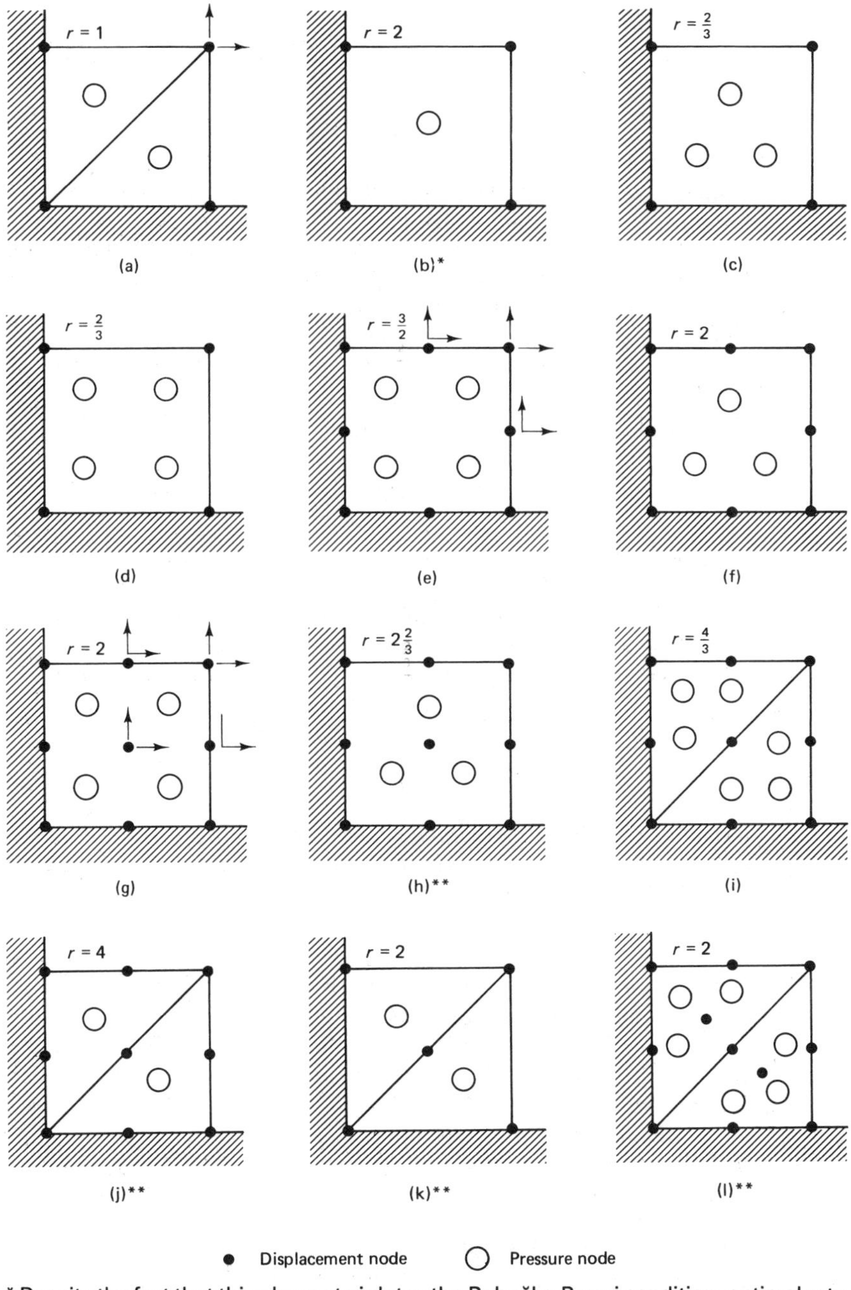

* Despite the fact that this element violates the Babuška-Brezzi condition, optimal rate of convergence (i.e., 1) can be proven under suitable assumptions, namely, the mesh must be composed of straight-edged quadrilateral macroelements consisting of four bilinear displacements–constant pressure elements; the pressure degrees of freedom need to be eliminated by way of regularization (i.e., the "nearly incompressible" approximation); and the pressure needs to be smoothed in an appropriate way (see [33] for further details).

** These elements satisfy the Babuška-Brezzi condition. The rate of convergence is 1 for the elements in (j) and (k) and 2 for the elements in (h) and (l). The remaining elements do not satisfy the Babuška-Brezzi condition.

Figure 4.3.4 Discontinuous pressure-field elements.

$$r = \frac{n_{eq}}{n_c} = \frac{n_{eq}}{\widetilde{n}_{eq}} = \frac{2}{1} = 2$$

Optimal behavior is indicated and this element is widely used. This element was first proposed by Hughes and Allik [17].

From Exercise 7 of Sec. 3.11 we recall that the mean value of div $\boldsymbol{u}^h$ occurs at the origin of the isoparametric coordinate system (i.e., $\xi = \eta = 0$). This is the only point in the element at which incompressibility is identically satisfied. The mean-value point shifts somewhat for the case of the axisymmetric version of this element.

Example 3

Consider the bilinear displacements–linear pressure element shown in Fig. 4.3.4(c). In this case,

$$r = \frac{n_{eq}}{n_c} = \frac{n_{eq}}{\widetilde{n}_{eq}} = \frac{2}{3}$$

which indicates severe locking.

Although no improvement could be expected by increasing the pressure interpolation to bilinear, we wish to consider this element (see Fig. 4.3.4(d)) since it illustrates a point. In this case[4]

$$\int_{\Omega^e} \underbrace{q^h}_{\text{bilinear}} \; \underbrace{\text{div}\, \boldsymbol{u}^h}_{\text{linear}} \, d\Omega = 0 \tag{4.3.60}$$

Thus there are more pressure weighting functions, and—consequently—incompressibility conditions, than there are independent monomials in div $\boldsymbol{u}^h$. As a result the four incompressibility conditions are linearly dependent and so

$$r = \frac{n_{eq}}{n_c} = \frac{n_{eq}}{\widetilde{n}_{eq} - 1} = \frac{2}{4 - 1} = \frac{2}{3}$$

as for the preceding case. Increasing the order of pressure interpolation further does not change r but increases the order of singularity of the matrix system. Clearly an approximation of this kind is useless, since the global matrix could not be inverted in the incompressible case.

Singularities of this type affect many elements in incompressible applications. it is not always obvious that this can occur for an element. For example, bilinear displacements–constant pressure elements exhibit a singularity in the global pressure equations for the mesh shown in Fig. 4.3.5(a) as long as n_{es} is even. This pathology is referred to as the ***checkerboard mode*** since the pressure degrees of freedom of the eigenvector of the global equations corresponding to the zero eigenvalue take the form $+1$ on the "red" squares and -1 on the "black" squares (Fig. 4.3.5(b)). When we come to the penalty method, we will show that this mode can be removed resulting in an

[4]For the standard mesh we can write $\xi = \xi(x)$ and $\eta = \eta(y)$, where each of these functions is linear. Thus a bilinear expansion in ξ and η can also be written as a bilinear expansion in x and y. For example, $u_i^h = \alpha_{0i} + \alpha_{1i}\xi + \alpha_{2i}\eta + \alpha_{3i}\xi\eta = \beta_{0i} + \beta_{1i}x + \beta_{2i}y + \beta_{3i}xy$ where the α's and β's are parameters which depend on the nodal values of u_i^h. The β's also depend on the lengths of the element edges. Clearly, div $\boldsymbol{u}^h = (\beta_{11} + \beta_{22}) + \beta_{32}x + \beta_{31}y$, which is identically zero pointwise if and only if $0 = \beta_{11} + \beta_{22} = \beta_{32} = \beta_{31}$, i.e., there are three independent constraints on $\boldsymbol{u}^h$.

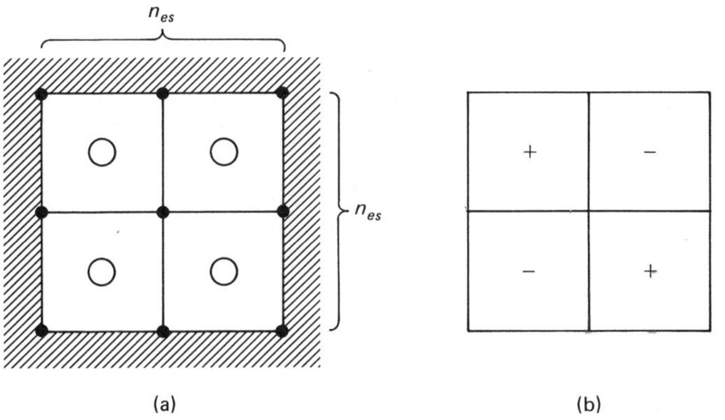

Figure 4.3.5 Checkerboard pressure mode.

effective formulation for this element. See [31–33] for mathematical results employing this idea.

In a sense all elements are subject to the problem of singularity in the incompressible limit. This can easily be seen from (4.3.21). If enough displacement degrees of freedom have been specified such that $n_{eq} < \tilde{n}_{eq}$, then the matrix *must* be singular at least to degree $\tilde{n}_{eq} - n_{eq}$.

Example 4

Consider the eight-node serendipity displacement–bilinear pressure quadrilateral shown in Fig. 4.3.4(e). A calculation of div $\boldsymbol{u}^h$ reveals that it can be expressed as a full quadratic polynomial in x and y and thus involves six independent coefficients, which must vanish for incompressibility to be satisfied pointwise. For a bilinear pressure expansion, the incompressibility condition takes the form

$$\int_{\Omega^e} \underbrace{q^h}_{\text{bilinear}} \underbrace{\text{div } \boldsymbol{u}^h}_{\text{quadratic}} \, d\Omega = 0 \tag{4.3.61}$$

and thus four conditions emanate from (4.3.61). The constraint ratio is, therefore,

$$r = \frac{n_{eq}}{n_c} = \frac{n_{eq}}{\tilde{n}_{eq}} = \frac{6}{4} = \frac{3}{2}$$

which indicates that there are too many incompressibility constraints.

Some ostensible improvement can be made by reducing the pressure interpolation to linear (see Fig. 4.3.4(f)). In this case

$$r = \frac{n_{eq}}{\tilde{n}_{eq}} = \frac{6}{3} = 2$$

the optimal value. However, neither of these elements (i.e., Fig. 4.3.4(e) and (f)) is currently favored.

Example 5

Consider the biquadratic displacements–bilinear pressure quadrilateral element shown in Fig. 4.3.4(g) for which

$$r = \frac{n_{eq}}{n_c} = \frac{n_{eq}}{\widetilde{n}_{eq}} = \frac{8}{4} = 2$$

Thus from a constraint ratio point of view this element appears ideal. However, it also gives rise to a pressure mode such as that for the bilinear displacements–constant pressure element.

Reducing the pressure interpolation to linear (Fig. 4.3.4(h)) removes this pressure mode and the constraint ratio increases to $r = \frac{8}{3} = 2\frac{2}{3}$. This element is currently felt to be one of the most effective quadrilateral elements for incompressible analysis.

Example 6

Consider the quadratic displacements–linear pressure triangle shown in Fig. 4.3.4(i). The constraint ratio is

$$r = \frac{n_{eq}}{n_c} = \frac{n_{eq}}{\widetilde{n}_{eq}} = \frac{8}{6} = \frac{4}{3}$$

Thus this element possesses too many incompressibility constraints. (Observe that pointwise satisfaction of the incompressibility constraint is attained.)

To reduce the number of incompressibility constraints, constant pressure may be employed (see Fig. 4.3.4(j)), resulting in incompressibility in the mean and a relatively high constraint ratio of $r = \frac{8}{2} = 4$. This element is, nevertheless, favored by some analysts [18] because it does not possess pressure modes of the kind described previously. The drawback, however, is the crude approximation of incompressibility (i.e., piecewise constant) relative to displacements (i.e., piecewise quadratic), which results in suboptimal convergence. (For optimal rate of convergence, the complete polynomial in the pressure field should be one order lower than the complete polynomial in the displacements.) The convergence of this element has been established in [19].

A more balanced approximation employing constant pressure is shown in Fig. 4.3.4(k) [20]. Each quadrilateral macroelement is composed of linear displacement–constant pressure triangles in which quadratic displacement modes are added along the diagonal edge. The constraint ratio is $r = \frac{4}{2} = 2$, which is optimal. Additionally, this element exhibits no spurious pressure modes.

Another way to "fix" the quadratic triangle of Fig. 4.3.4(i) is to add an internal displacement node, as in Fig. 4.3.4(l). (The displacement interpolations may be constructed using techniques described in Chapter 3.) This element possesses an optimal constraint ratio:

$$r = \frac{n_{eq}}{n_c} = \frac{n_{eq}}{\widetilde{n}_{eq}} = \frac{12}{6} = 2$$

Convergence and error estimates have been established by Crouzeix and Raviart [19].

The remaining examples involve ***continuous pressure fields.*** The first to study continuous pressure-field elements were Hughes and Allik [17]. In these cases r varies with n_{es}. However, it may be argued that

$$\lim_{n_{es}\to\infty} r$$

may be obtained by again considering the single corner element of the standard mesh and ignoring all degrees of freedom—pressure in addition to displacement—on the left and bottom boundaries. (The reader may wish to verify this statement as an exercise for some of the following elements.)

4.3.9 Continuous Pressure Elements

Example 7

Consider the bilinear displacements–(continuous!) bilinear pressure quadrilateral shown in Fig. 4.3.6(a). The constraint ratio is

$$\lim_{n_{es}\to\infty} r = 2$$

which is optimal. However, the convergence is from below and this element may exhibit spurious pressure modes. This appears to be a general fact for typical elements possessing identical displacement and pressure interpolations (e.g., see Fig. 4.3.6(b)).

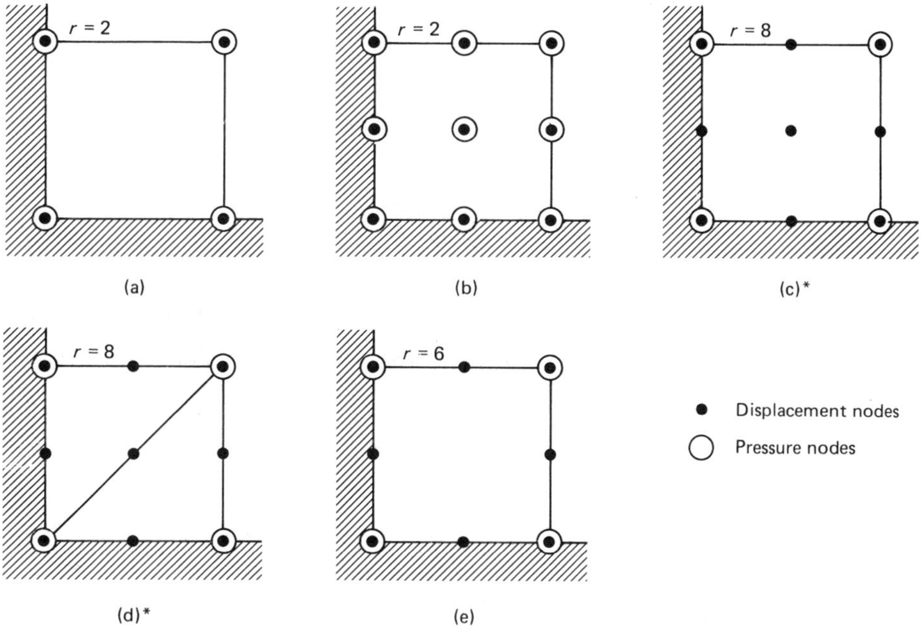

*These elements satisfy the Babuška-Brezzi condition. The convergence rate is 2. The elements in (a) and (b) violate the Babuška-Brezzi condition, whereas the issue is still open regarding the element in (e).

Figure 4.3.6 Continuous pressure-field elements.

Example 8

The deficiencies noted in the previous example can be corrected by lowering the pressure interpolation. Fig. 4.3.6(c) and (d) depicts elements of this type. For both these elements

$$\lim_{n_{es}\to\infty} r = 8$$

which is very high (i.e., there are too few incompressibility constraints). Although no pressure modes are exhibited by these elements and the convergence rate is theoretically optimal [21][5], very poor approximations of the incompressibility condition are frequently noted. Nevertheless, these elements are widely used in incompressible analysis.

The constraint ratio can be reduced somewhat by removing internal displacement degrees of freedom as, for example, in Fig. 4.3.6(e). For this element,

$$\lim_{n_{es}\to\infty} r = 6$$

which is still rather high. This element has also been widely used in incompressible analysis although, at the time of writing, the question of convergence is still open.

Remarks

1. It is an empirical observation that all elements which are effective in incompressible analysis possess constraint ratios greater than or equal to n_{sd}. We wish to reiterate, however, that some of these elements may give rise to so-called pressure modes (i.e., singularities in the global equations). Thus the analysis of incompressible media is a delicate matter and care must be exercised. We consider the matter further in subsequent sections.

2. A number of other interesting elements for incompressible media are described in Ruas [20], Thomasset [22], Fortin [23], Griffiths [24–27], Oden and Jacquotte [28], and Boland and Nicolaides [29].

3. The continuous-pressure macroelements illustrated in Fig. 4.3.7 have also been shown to converge. Note that the nodal patterns are identical to Fig. 4.3.6(c) and (d) and thus so are the constraint ratios (i.e., $\lim_{n_{es}\to\infty} r = 8$). These elements represent a trade-off compared with those of Fig. 4.3.6(c) and (d): The rate of convergence is one order higher for the elements of Fig. 4.3.6(c) and (d), but the elements of Fig. 4.3.7 are amenable to more efficient implementation. See Thomasset [22] and references therein for further details.

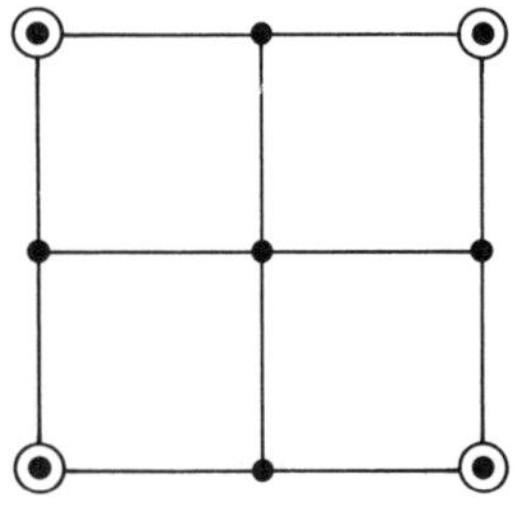

Four bilinear displacement quadrilaterals covered with a bilinear pressure field

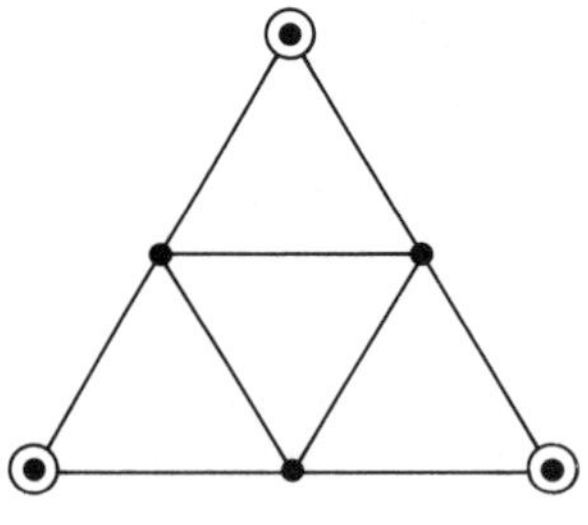

Four linear displacement triangles covered with a linear pressure field

Figure 4.3.7 Some convergent, continuous pressure-field macroelements.

[5] Certain cautions must be respected in specifying displacement boundary conditions. For the quadrilateral in Fig. 4.3.6(c), displacement should be specified on at most two edges, whereas for the triangle of Fig. 4.3.6(d), displacement should be specified on at most one edge.

4. In some recent works, convergence of equal-order interpolations is established by adopting modified formulations of the incompressible problem. Brezzi and Pitkäranta [34] propose a formulation for linear triangles and Hughes et al. [35] develop a formulation for general classes of elements.

4.4 PENALTY FORMULATION: REDUCED AND SELECTIVE INTEGRATION TECHNIQUES; EQUIVALENCE WITH MIXED METHODS

Let us recall the formulations we introduced previously for modeling a single constraint (Sec. 4.2). These were the ***Lagrange multiplier method*** and the ***penalty function formulation.*** In the context of the incompressibility constraint, the mixed formulation of Sec. 4.3 is a Lagrange-multiplier-type of formulation in which the pressure field plays the role of the Lagrange multiplier.

We also recall from Sec. 4.2 that in the penalty formulation, we simply approximated the single constraint by introducing a stiff elastic spring. In the context of incompressibility, this amounts to allowing for slight compressibility. That is, λ is taken finite, but large with respect to μ.

Nearly incompressible case

$$\frac{\lambda}{\mu} \gg 1 \tag{4.4.1}$$

This case can be easily done within the framework of Sec. 4.3 and has some advantages. For example, in the case of discontinuous pressure fields, the pressure degrees of freedom can be eliminated on the element level. The idea is to select λ sufficiently large so that compressibility errors are negligible, but not so large that numerical problems arise. The ratio λ/μ thus depends on the floating-point word length of the computer being utilized. In our experience with words of length 60–64 bits, we have found that the range

$$10^7 \leq \frac{\lambda}{\mu} \leq 10^9 \tag{4.4.2}$$

is effective. As noted in Sec. 4.3, allowing for slight compressibility does not change the situation with regard to elements. One must employ only those elements that are also effective in the incompressible limit, as slight compressibility does not eliminate the fundamental difficulties.

Once we are willing to introduce some compressibility, the theory of Chapter 2 is also applicable. The question naturally arises then as to the performance of the standard "displacement-only" elements of Chapter 3 in these circumstances.[6] The

[6] The finite element methodology of Chapters 2 and 3 is generally referred to as the ***displacement formulation.***

answer, as might be anticipated, is that these elements tend to perform poorly in nearly incompressible applications, frequently exhibiting a tendency to lock. However, a slight modification of the usual formulation enables the construction of elements that are identical to many of the discontinuous pressure field elements generated by the mixed formulation. The basic tools in this process are ***reduced and selective integration*** procedures, which are described as follows.

Reduced and Selective Integration

The expression for element stiffness in the displacement formulation is given by (we omit the element number superscript, e, for simplicity):

$$\boldsymbol{k} = [k_{pq}], \qquad 1 \leq p, q \leq n_{ee} \tag{4.4.3}$$

$$k_{pq} = \boldsymbol{e}_i^T \boldsymbol{k}_{ab} \boldsymbol{e}_j \tag{4.4.4}$$

$$p = n_{ed}(a - 1) + i, \qquad q = n_{ed}(b - 1) + j \tag{4.4.5}$$

$$\boldsymbol{k}_{ab} = \int_{\Omega^e} \boldsymbol{B}_a^T \boldsymbol{D} \boldsymbol{B}_b \, d\Omega \tag{4.4.6}$$

The material properties matrix $\boldsymbol{D}$ may be written as

$$\boldsymbol{D} = \overline{\boldsymbol{D}} + \overline{\overline{\boldsymbol{D}}} \tag{4.4.7}$$

where $\overline{\boldsymbol{D}}$ is the μ-part of $\boldsymbol{D}$, defined in Sec. 4.3, and $\overline{\overline{\boldsymbol{D}}}$, the remainder, is the λ-part. The reader may easily verify the following formulas.

Plane strain

$$\overline{\overline{\boldsymbol{D}}} = \lambda \begin{bmatrix} 1 & 1 & 0 \\ & 1 & 0 \\ \text{symm.} & & 0 \end{bmatrix} \tag{4.4.8}$$

Axisymmetry

$$\overline{\overline{\boldsymbol{D}}} = \lambda \begin{bmatrix} 1 & 1 & 0 & 1 \\ & 1 & 0 & 1 \\ \text{symm.} & & 0 & 0 \\ & & & 1 \end{bmatrix} \tag{4.4.9}$$

Three dimensions

$$\overline{\overline{\boldsymbol{D}}} = \lambda \begin{bmatrix} 1 & 1 & 1 & 0 & 0 & 0 \\ & 1 & 1 & 0 & 0 & 0 \\ & & 1 & 0 & 0 & 0 \\ & & & 0 & 0 & 0 \\ \text{symm.} & & & & 0 & 0 \\ & & & & & 0 \end{bmatrix} \tag{4.4.10}$$

Employing (4.4.7) in (4.4.6) enables us to write

$$\boxed{\boldsymbol{k}_{ab} = \bar{\boldsymbol{k}}_{ab} + \bar{\bar{\boldsymbol{k}}}_{ab}} \tag{4.4.11}$$

where

$$\bar{\boldsymbol{k}}_{ab} = \int_{\Omega^e} \boldsymbol{B}_a^T \bar{\boldsymbol{D}} \boldsymbol{B}_b \, d\Omega \tag{4.4.12}$$

$$\bar{\bar{\boldsymbol{k}}}_{ab} = \int_{\Omega^e} \boldsymbol{B}_a^T \bar{\bar{\boldsymbol{D}}} \boldsymbol{B}_b \, d\Omega \tag{4.4.13}$$

Note that $\bar{\boldsymbol{k}}_{ab}$ is the part of the stiffness that also appears in the mixed formulation (Sec. 4.3). Due to the fact that $\lambda/\mu \gg 1$ and $\bar{\bar{\boldsymbol{k}}}$ is proportional to λ, whereas $\bar{\boldsymbol{k}}$ is proportional to μ, the numerical values of terms in $\bar{\bar{\boldsymbol{k}}}$ tend to be very large compared with those of $\bar{\boldsymbol{k}}$. The $\bar{\bar{\boldsymbol{k}}}$-term is the part of the stiffness that attempts to maintain the volumetrically stiff behavior. Because typical finite elements tend to lock (i.e., there are proportionally too many incompressibility-type conditions), special treatment of $\bar{\bar{\boldsymbol{k}}}$ is required to alleviate this tendency. One simple and practically important way of going about this is to reduce the order of numerical quadrature employed to evaluate $\bar{\bar{\boldsymbol{k}}}$ below that "normally" used. The basic idea is illustrated in the following example.

Example 1

Consider the four-node bilinear displacement element in plane strain. "Normal" quadrature for this element is considered to be the 2×2 Gauss rule. The stiffness matrix turns out to be *identical* to that obtained in the mixed formulation for bilinear displacements and (discontinuous) linear pressures in which the pressure degrees of freedom are eliminated on the element level, as indicated in (4.3.30). This fact is known from an equivalence theorem due to Malkus and Hughes [36]. The constraint ratio of this element was calculated to be $\frac{2}{3}$ in Sec. 4.3, and thus locking-type behavior would be anticipated.

Recall also from Sec. 4.3 that by employing constant pressure over each element, the bilinear element attains an optimal constraint ratio of 2. The equivalent displacement model may be obtained by reducing the quadrature of the $\bar{\bar{\boldsymbol{k}}}$-term to one-point Gauss [36, 37]. Thus, as $\lambda/\mu \to \infty$, incompressibility in the mean is attained. The performance of this element is illustrated by the following numerical example.

Consider the equations of two-dimensional linear isotropic elasticity theory on the domain illustrated in Fig. 4.4.1. The boundary conditions are given as follows:

Displacement

$$u_1(0, 0) = u_2(0, 0) = 0$$

$$u_1(0, \pm c) = 0$$

Traction

$$h_1(x_1, \pm c) = h_2(x_1, \pm c) = 0, \qquad x_1 \in \,]0, L[$$

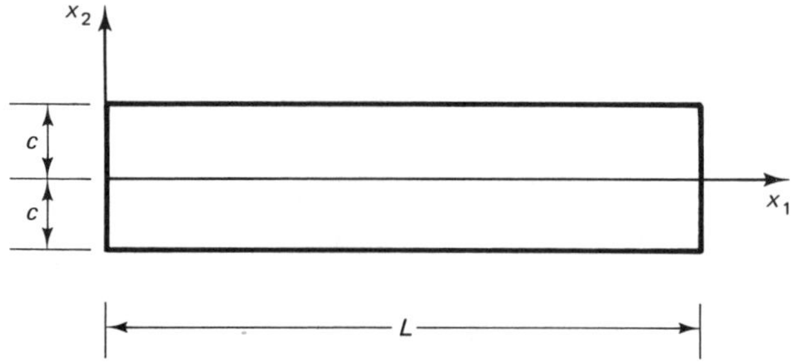

Figure 4.4.1 Domain for plane strain elasticity problem

$$\left.\begin{aligned} h_1(L, x_2) &= 0 \\ h_2(L, x_2) &= \frac{P}{2I}(c^2 - x_2^2) \end{aligned}\right\} \qquad x_2 \in \,]-c, c[$$

$$\left.\begin{aligned} h_1(0, x_2) &= \frac{PLx_2}{I} \\ h_2(0, x_2) &= -\frac{P}{2I}(c^2 - x_2^2) \end{aligned}\right\} \qquad x_2 \in \,]-c, 0[\,\cup\,]0, c[$$

where P is a given constant and $I = 2c^3/3$.

The traction boundary conditions are those encountered in simple bending theory for a cantilever beam with root section at $x = 0$—i.e., parabolically varying end shear and linearly varying bending stress at the root. The displacement boundary conditions allow the root section to warp. The following data were employed in the numerical calculations:

$$L = 16, \qquad c = 2$$

The mesh used is depicted in Fig. 4.4.2. (Only half the domain need be modeled since the x_1-axis is a line of antisymmetry.)

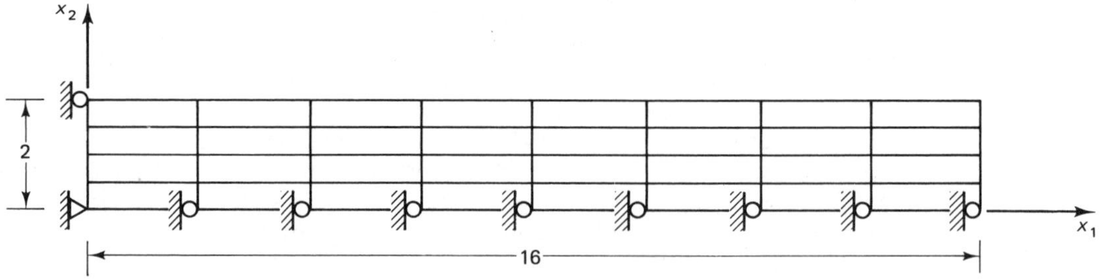

Figure 4.4.2 Element mesh and boundary conditions for plane strain elasticity problem.

Plane strain conditions were assumed and four-node quadrilateral elements were employed.

Vertical tip displacements [i.e., $u_2(16, 0)$] are compared in Table 4.4.1. Both the "standard" 2×2 Gauss quadrature element and the selective reduced element provide adequate results for $\nu = 0.3$. However, for the nearly incompressible case, the standard quadrilateral degenerates, whereas the selective integration element retains accuracy.

TABLE 4.4.1 Normalized Vertical Tip Displacements of Plane Strain Beam

ν	$U2$	$S1$
0.3	0.904	0.912
0.499	0.334	0.937

Key:
$U2$ 2 × 2 uniform integration (= exact in present case)
$S1$ selective integration (2 × 2 on μ-term, one-point on λ-term)

Definitions. ***Selective reduced integration*** refers to the case in which reduced integration is used only on the λ-term, whereas normal integration is used on the μ-term. ***Uniform reduced integration*** refers to the case when both terms are integrated with a reduced rule.

The ostensible advantage of uniform reduced integration is the economy of element formulation. The disadvantage is that the rank of the element stiffness may be reduced, resulting in singularity of the global matrix. Selective integration retains the correct rank of the element stiffness, and therefore the global stiffness also possesses correct rank. This follows from the fact that $\bar{a}(\cdot, \cdot)$ is positive definite. (The $\bar{\bar{a}}(\cdot, \cdot)$ term may be entirely ignored without affecting rank.)

Equivalence Theorem. The general theorem presented in Malkus and Hughes [36] established the equivalence of many mixed and reduced and selective integration elements. Some examples are presented in Fig. 4.4.3. It may thus be concluded that the reduced and selective integration procedures are very simple ways of attaining the performance of the mixed formulation without engendering the additional complications.

The equivalence theorem states that the element stiffness of the mixed method, namely (4.3.30), is identical to the element stiffness of the reduced or selective integration approach and consequently the displacements are also identical. At the Gauss points of the *reduced integration rule* in the reduced and selective integration approaches,

$$p^h = -\lambda \operatorname{div} \boldsymbol{u}^h \tag{4.4.14}$$

agrees with the pressure field of the mixed method. Elsewhere in the element, the pressure may be determined from the displacements by interpolating the values of (4.4.14) with the pressure shape functions. The equivalence theorem thus provides the interpretation rule by which the pressure should be defined within the reduced and selective integration approaches.

	Mixed element		Equivalent selective integration element	
	Displacement interpolation	Pressure interpolation	Normal Gaussian quadrature* $(\bar{\boldsymbol{k}})$	Reduced Gaussian quadrature $(\bar{\bar{\boldsymbol{k}}})$
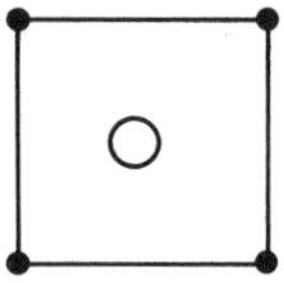	Bilinear	Constant	2 × 2	One-point
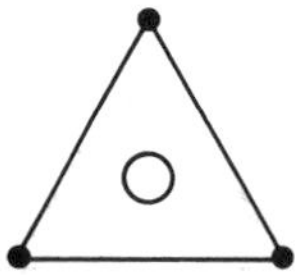	Linear	Constant	One-point	One-point
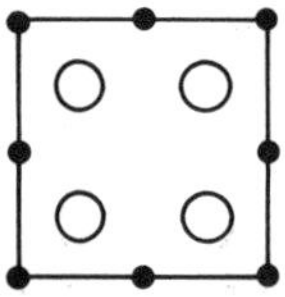	Serendipity	Bilinear	3 × 3	2 × 2
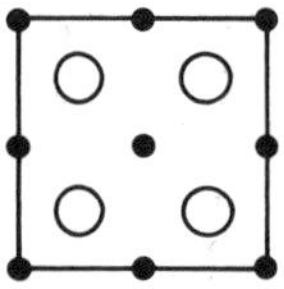	Biquadratic	Bilinear	3 × 3	2 × 2
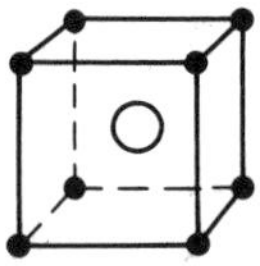	Trilinear	Constant	2 × 2 × 2	One-point

* Also used on all terms of mixed formulation.

Figure 4.4.3 Some equivalent elements.

For example, in the case of the selectively integrated four-node bilinear element, the constant element pressure of the mixed method agrees with the value computed from (4.4.14) at the origin of isoparametric coordinates (i.e., $\xi = \eta = 0$). As a second example, consider the selectively integrated nine-node element. In this case, the values of (4.4.14) computed at the 2×2 Gauss points need to be interpolated via bilinear shape functions. The resulting formula is

$$p^h(\xi, \eta) = \sum_{\tilde{a}=1}^{4} \tilde{N}_{\tilde{a}}(\xi, \eta) p_{\tilde{a}} \tag{4.4.15}$$

where

$$p_{\tilde{a}} = -\lambda(\operatorname{div} \boldsymbol{u}^h)(\tilde{\xi}_{\tilde{a}}, \tilde{\eta}_{\tilde{a}}) \tag{4.4.16}$$

and $\tilde{\xi}_{\tilde{a}}$, $\tilde{\eta}_{\tilde{a}}$ are the coordinates of the $\tilde{a}$th Gauss point. The shape functions in (4.4.15) are defined by

$$\tilde{N}_{\tilde{a}}(\xi, \eta) = \tfrac{1}{4}(1 + 3\tilde{\xi}_{\tilde{a}}\xi)(1 + 3\tilde{\eta}_{\tilde{a}}\eta) \tag{4.4.17}$$

Exercise 1. Verify that (4.4.17) satisfies the interpolation property at the reduced Gauss points, i.e.,

$$\tilde{N}_{\tilde{a}}(\tilde{\xi}_{\tilde{b}}, \tilde{\eta}_{\tilde{b}}) = \delta_{\tilde{a}\tilde{b}} \tag{4.4.18}$$

Exercise 2. Describe how to program the selective integration procedure.

Exercise 3. Show that for the isotropic case, the symmetric bilinear form $a(\cdot, \cdot)$ can be written as

$$a(\boldsymbol{w}, \boldsymbol{u}) = \bar{a}(\boldsymbol{w}, \boldsymbol{u}) + \bar{\bar{a}}(\boldsymbol{w}, \boldsymbol{u}) \tag{4.4.19}$$

where

$$\bar{\bar{a}}(\boldsymbol{w}, \boldsymbol{u}) = (\operatorname{div} \boldsymbol{w}, \lambda \operatorname{div} \boldsymbol{u}) \tag{4.4.20}$$

This result should reinforce the assertion that it is the $\bar{\bar{\boldsymbol{k}}}$-stiffness which is responsible for enforcing the volumetrically stiff behavior.

Remarks

1. Constraint ratios may be determined for reduced and selective integration elements as follows: Consider the expression that leads to $\bar{\bar{\boldsymbol{k}}}$, namely, (4.4.20). The maximum number of constraints possible is given by the number of independent monomials in div $\boldsymbol{u}^h$. Likewise, the maximum can be no greater than the number of quadrature points used to evaluate (4.4.20). Thus

$$n_c = \min\{\text{number of independent monomials present in div } \boldsymbol{u}^h\text{; number of quadrature points used to evaluate (4.4.20)}\}$$

As an example, consider the four-node bilinear element. Normal quadrature results in $n_c = 3$ (i.e., the number of independent monomials in div $\boldsymbol{u}^h$) and so $r = \frac{2}{3}$. On the other hand, if reduced one-point quadrature is used on (4.4.20), $n_c = 1$ (i.e., number of quadrature points) and so $r = 2$. These values of the constraint ratio agree with those computed for the equivalent mixed elements.

2. Fried [38] argued in a somewhat different way in favor of reducing the integration rule of the λ-term. Consider the case in which λ and μ are constants and let

$$\overline{\boldsymbol{K}} = \mu \boldsymbol{K}_1 \tag{4.4.21}$$

$$\overline{\overline{\boldsymbol{K}}} = \lambda \boldsymbol{K}_2 \tag{4.4.22}$$

Then

$$\begin{aligned} \boldsymbol{F} &= \boldsymbol{K}\boldsymbol{d} \\ &= (\overline{\boldsymbol{K}} + \overline{\overline{\boldsymbol{K}}})\boldsymbol{d} \\ &= (\mu \boldsymbol{K}_1 + \lambda \boldsymbol{K}_2)\boldsymbol{d} \end{aligned} \tag{4.4.23}$$

Fried noted that for typical elements $\boldsymbol{K}_2$ tended to have too great a rank (i.e., too many incompressibility constraints). In fact, in some situations $\boldsymbol{K}_2$ is nonsingular and therefore for $\lambda/\mu \gg 1$,

$$\begin{aligned} \boldsymbol{d} &= (\mu \boldsymbol{K}_1 + \lambda \boldsymbol{K}_2)^{-1}\boldsymbol{F} \\ &\cong \frac{1}{\lambda}\boldsymbol{K}_2^{-1}\boldsymbol{F} \end{aligned} \tag{4.4.24}$$

From (4.4.24), it can be seen that as $\lambda \to \infty$, $\boldsymbol{d} \to \boldsymbol{0}$. This situation may be seen to be equivalent to the locking phenomenon noted in the example of Fig. 4.3.2. Thus Fried argued that for a formulation to be successful in nearly incompressible applications, $\boldsymbol{K}_2$ ***must be singular*** so that (4.4.24) does not hold. Fried further argued that one way of achieving this end was to reduce the order of quadrature on the element contributions to $\overline{\overline{\boldsymbol{K}}}$.

3. The linear triangular displacement element, which is equivalent to the linear displacement–constant pressure triangular mixed element (Hughes and Allik [39]), exhibits pathological locking on the standard mesh (Fig. 4.3.3). However, Nagtegaal et al. [40] found that in the cross-diagonal pattern (see Fig. 4.4.4(a)) this element improves. Ostensibly, the constraint ratio for this pattern is still 1. However, what occurs is that the incompressibility conditions exhibit a linear dependency and thus there are only three incompressibility constraints per quadrilateral macroelement. Thus $r = \frac{4}{3}$.

Mercier [41] provided an elegant argument, which established similar improvement for quadratic triangles in the cross-diagonal pattern (see Fig. 4.4.4(b)).

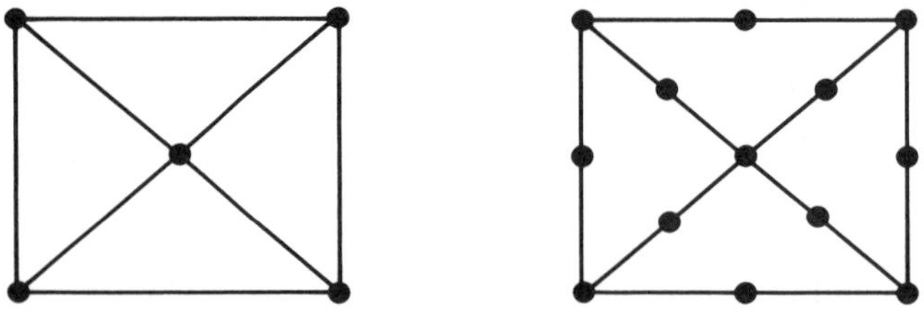

(a) (b)

Figure 4.4.4 Quadrilateral macroelements formed from triangles in the cross-diagonal pattern.

At this point, however, there appear to be several disadvantages in adopting the macroelement approach. First of all, since there is a necessity of assembling into quadrilateral macroelements, no additional flexibility is gained by the use of the triangular elements over standard quadrilateral elements. Second, when compared to the selectively integrated Lagrange quadrilaterals of equal interpolatory order, less accuracy is attained by the triangular macroelements despite the fact that approximately two times as many unknowns are involved.

On the other hand, a potential advantage of the triangular macroelements is that in the plane strain case, if the edges are straight, pointwise incompressibility may be obtained. Incompressibility, in general, occurs only at the quadrature points of the reduced integration rule for the selectively integrated Lagrange elements.

Exercise 4. Consider the Lagrange family of quadrilaterals and bricks in which normal integration is used for the entire stiffness. Determine expressions for the constraint ratio as functions of the number of nodes along an element side, say n. Deduce that the constraint ratio is poorer for lower-order members of the family than for higher-order members. Show that as $n \to \infty$, $r \to n_{sd}$ (optimal). (This result is consistent with the observation that fully integrated higher-order elements are more successful in nearly incompressible applications than fully integrated lower-order elements.) Contrast the results with those obtained using selective reduced integration.

Partial solution We shall do the two-dimensional case and leave the three-dimensional case for the reader.

$$n_{eq} = 2(n-1)^2$$

$$n_c = n^2 - 1 \qquad \text{(normal integration; exact)}$$

$$r = \frac{2(n-1)^2}{n^2-1} = 2\frac{(n-1)}{(n+1)}$$

For normal integration, r improves as the order of the element is increased. For example,

n	r
2	$\frac{2}{3}$
3	1
4	$\frac{6}{5}$
⋮	⋮
∞	$2\ (= n_{sd})$

Thus the necessity of reduced integration decreases as the order of the element increases.

In the case of reduced integration of the λ-term,

$$n_c = (n-1)^2$$

$$r = 2\ (= n_{sd}) \qquad \text{independent of } n$$

Some Historical Remarks on Mixed and Reduced and Selective Integration Methods

Mixed finite element formulations were first discussed by Fraeijs de Veubeke [42] and Herrmann [43]. Herrmann developed a reduced form of Reissner's variational principle particularly suited to problems of incompressible and nearly incompressible elasticity and, based upon this principle, established the first effective finite elements for such cases. This is the formulation given in Sec. 4.3. Prior to this development many displacement models were applied to these problems, and poor behavior was typically observed. The reasons for this were not understood at the time. Certain elements derived from Herrmann's formulation also failed. Hughes and Allik [39] traced this failure to a correspondence between mixed and displacement models, contained within Fraeijs de Veubeke's ***limitation principle*** [42].

The first example of a uniform reduced integration element was apparently the plate-shell element presented by Zienkiewicz et al. [44]. This element, among others, is discussed in Chapter 5. The same concept was employed in other areas by Zienkiewicz and colleagues. In particular, Naylor [45] and Zienkiewicz and Godbole [46] advocated the use of the eight-node serendipity element in problems involving incompressibility. The procedure, however, was viewed by many as more a "trick" than a method and some bad experiences were subsequently noted for the serendipity element.

The concept of selective integration was first employed by Doherty et al. [47] to obtain improved bending behavior in simple four-node elasticity elements. One-point Gauss quadrature was used on the shear-strain term, and 2×2 Gauss quadrature was used to integrate the remaining terms. Although improved behavior was noted in some configurations, lack of invariance opened the approach to criticism.

Studies performed by Fried [38], Nagtegaal et al. [40], and Argyris et al. [48] provided fresh insights into why the displacement approach failed in constrained problems. Malkus [49, 50] proved the equivalence of a class of mixed models with reduced selective integration single-field elements in linear elasticity theory. The equivalence results of Malkus and Hughes [36] elevated the reduced and selective integration approaches from the realm of tricks to a legitimate methodology. Considerable research on the behavior of mixed and reduced and selective integration elements has taken place in recent years. A summary of more recent developments is contained in the following sections.

4.4.1 Pressure Smoothing

The pressure field in the reduced and selective integration penalty function formulation is to be viewed as discontinuous from element to element. In fact, all displacement derivatives for C^0 isoparametric elements are, in general, discontinuous across element boundaries. Thus, for plotting purposes, it is desirable to employ a smoothing procedure, which redefines the field under consideration in terms of the displacement shape functions N_A.

With specific reference to the pressure, there is at least one other reason for employing a smoothing procedure. It was mentioned earlier that, in certain situations, discontinuous-pressure, mixed-method finite elements exhibit a rank-deficiency in the assembled pressure equations. By the equivalence theorem, "problems" are also to be expected with the pressure field of the penalty function formulation. These problems typically manifest themselves as pressure oscillations. For example, if four-node, quadrilateral elements are employed in a square mesh, with an even number of square elements in each direction, subjected to all velocity boundary conditions, then a checkerboard pressure oscillation is produced. Despite the pressure oscillations, the velocity field remains good.

Fortunately, smoothing procedures of a ***least squares*** type [51] seem to perform the necessary filtering as a byproduct. A comprehensive study of such techniques has been performed by Lee et al. [52]. The methods we prefer for constant-pressure elements [53], which involve slight modifications of schemes proposed in [52], are described next.

Let the discontinuous pressure field be written as

$$p^h = \sum_{e=1}^{n_{el}} \psi^e p^e \tag{4.4.25}$$

where p^e is the element mean pressure and ψ^e is the eth element "characteristic function," i.e.,

$$\psi^e(\boldsymbol{x}) = \begin{cases} 1 & \text{if } \boldsymbol{x} \in \Omega^e \\ 0 & \text{if } \boldsymbol{x} \not\in \Omega^e \end{cases} \tag{4.4.26}$$

The smoothed pressure is written

$$\widetilde{p} = \sum_{A=1}^{n_{np}} N_A \widetilde{p}_A \tag{4.4.27}$$

The standard least squares procedure gives rise to the following matrix problem:[7]

$$\boldsymbol{Y}\widetilde{\boldsymbol{p}} = \boldsymbol{P} \tag{4.4.28}$$

where

$$\boldsymbol{Y} = [Y_{AB}] \tag{4.4.29}$$

$$\widetilde{\boldsymbol{p}} = \{\widetilde{p}_B\} \tag{4.4.30}$$

and

$$\boldsymbol{P} = \{P_A\} \tag{4.4.31}$$

[7] The least squares procedure defines $\widetilde{p}$ by minimizing

$$\int_\Omega (\widetilde{p} - p^h)^2 \, d\Omega$$

with respect to the $\widetilde{p}_A$'s. The resulting equations emanate from

$$\frac{\partial}{\partial \widetilde{p}_A} \int_\Omega (\widetilde{p} - p^h)^2 \, d\Omega = 0$$

for $A = 1, 2, \ldots, n_{np}$.

The indices A, B take on the values 1, 2, . . . , n_{np}. The construction of $\boldsymbol{Y}$ and $\boldsymbol{P}$ is performed in the usual element-by-element fashion, viz.[8]

$$\boldsymbol{Y} = \mathop{\mathbf{A}}_{e=1}^{n_{el}} (\boldsymbol{y}^e), \qquad \boldsymbol{P} = \mathop{\mathbf{A}}_{e=1}^{n_{el}} (\boldsymbol{p}^e) \tag{4.4.32}$$

in which

$$\boldsymbol{y}^e = [y^e_{ab}], \qquad \boldsymbol{p}^e = \{p^e_a\}, \qquad 1 \le a,\ b \le n_{en} \tag{4.4.33}$$

$$y^e_{ab} = \int_{\Omega^e} N^e_a N^e_b \, d\Omega, \qquad p^e_a = p^e \int_{\Omega^e} N^e_a \, d\Omega \tag{4.4.34}$$

As it stands, the matrix $\boldsymbol{Y}$ is symmetric and positive-definite and possesses a band-profile structure. Additional simplification may be engendered by replacing $\boldsymbol{Y}$ by an associated diagonal matrix.[9] This is done by approximating the first of (4.4.34); effective procedures are summarized as follows:

$n = 2$; rectilinear case. The 2 × 2 product, trapezoidal integration rule may be used to diagonalize $\boldsymbol{y}^e$, i.e.,

$$y^e_{ab} = \delta_{ab} j^e(\xi_a, \eta_a) \qquad \text{(no sum on } a) \tag{4.4.35}$$

where

$$j^e = \det \begin{bmatrix} x^e_{1,\xi} & x^e_{1,\eta} \\ x^e_{2,\xi} & x^e_{2,\eta} \end{bmatrix} \qquad \text{(Jacobian determinant)} \tag{4.4.36}$$

$$\boldsymbol{x}^e = \sum_{a=1}^{n_{en}} N^e_a \boldsymbol{x}^e_a \tag{4.4.37}$$

and ξ_a and η_a are the coordinates of node a in the element "natural" coordinate system. Applying the same integration scheme to the second of (4.4.34) yields

$$p^e_a = p^e j^e(\xi_a, \eta_a) \tag{4.4.38}$$

Further simplification may be achieved by approximating $j^e(\xi_a, \eta_a)$ in (4.4.35) and (4.4.38) by $j^e(0, 0)$. (When Ω^e is a parallelogram, j^e is constant and no loss of accuracy is incurred by this procedure.)

The three-dimensional case is the straightforward generalization of the above, so we omit the details.

$n = 2$; axisymmetric case. If we attempt to apply the above procedure in the axisymmetric case, we encounter a difficulty due to the factor x_1 (i.e., r) in the integrands. Along the x_2-axis, $x_1 = 0$; hence the trapezoidal integration technique will

[8] The "assembly operators" in (4.4.32) are not the same as those used previously. Here, no boundary conditions are taken account of and there is only one degree of freedom per node.

[9] Lee et al. [52] have also found that higher accuracy is attained when $\boldsymbol{Y}$ is diagonal!

produce a zero diagonal entry in Y. In this case we employ a "row-sum" diagonalization technique in which

$$y^e_{ab} = \delta_{ab} \int_{\Omega^e} N^e_a \, d\Omega \qquad \text{(no sum on } a\text{)} \tag{4.4.39}$$

The above integration, which also suffices for the second of (4.4.34), may be performed by either one-point or 2 × 2 Gauss-Legendre integration—the latter scheme being exact.

The procedures just described render the formation, storage, and solution of the matrix equation (4.4.28) very efficient. The results produced tend to be very good at interior nodes but leave something to be desired at boundary nodes. To improve upon the results, a "correction" at each boundary node is performed. The procedure used for four-node elements may be described with the aid of an example.

Consider the mesh illustrated in Fig. 4.4.5(a). The nodes are segregated into four groups. The boundary node corrections are carried out in the following steps in order:

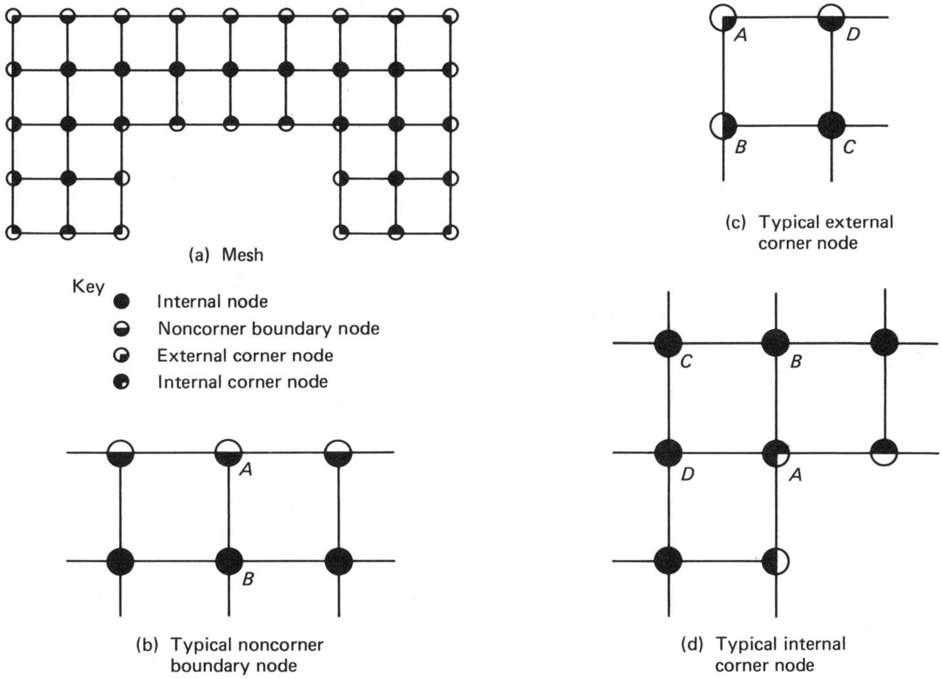

Figure 4.4.5 Example mesh for four-node element, pressure-smoothing algorithm.

Step 1: Noncorner, Boundary Nodes

A typical case of a noncorner, boundary node is depicted in Fig. 4.4.5(b). It may be observed that the unaltered value of $\tilde{p}_A$ is actually a higher-order approximation to the pressure at the midpoint of the line joining nodes A and B; see Barlow [54]. Thus we redefine the $\tilde{p}_A$ by way of linear extrapolation, i.e.,

$$\tilde{p}_A \longleftarrow 2\tilde{p}_A - \tilde{p}_B \tag{4.4.40}$$

Step 2: External Corner Nodes

A typical situation is depicted in Fig. 4.4.5(c). The unaltered value of $\tilde{p}_A$ is precisely the constant pressure p^e, because the above procedures reduce to "do-nothing" calculations at external corners. (If checkerboarding was occuring in the p^e's, the value of $\tilde{p}_A$ would be grossly in error.) In this case linear extrapolation is employed through nodes B, C, and D, i.e.,

$$\tilde{p}_A \longleftarrow \frac{\tilde{L}_B\tilde{p}_B + \tilde{L}_C\tilde{p}_C + \tilde{L}_D\tilde{p}_D}{L} \tag{4.4.41}$$

where

$$\tilde{L}_B = L_B + (x_{2C} - x_{2D})x_{1A} + (x_{1D} - x_{1C})x_{2A} \tag{4.4.42}$$

$$\tilde{L}_C = L_C + (x_{2D} - x_{2B})x_{1A} + (x_{1B} - x_{1D})x_{2A} \tag{4.4.43}$$

$$\tilde{L}_D = L_D + (x_{2B} - x_{2C})x_{1A} + (x_{1C} - x_{1B})x_{2A} \tag{4.4.44}$$

$$L_B = x_{1C}x_{2D} - x_{1D}x_{2C} \tag{4.4.45}$$

$$L_C = x_{1D}x_{2B} - x_{1B}x_{2D} \tag{4.4.46}$$

$$L_D = x_{1B}x_{2C} - x_{1C}x_{2B} \tag{4.4.47}$$

$$L = L_B + L_C + L_D \tag{4.4.48}$$

Step 3: Internal Corner Nodes

A typical configuration is shown in Fig. 4.4.5(d). In this case the unaltered $\tilde{p}_A$ is essentially a weighted average of the p^e's associated with the three elements which have node A in common. As in Step 2, if checkerboarding has occurred, the unaltered $\tilde{p}_A$ would be significantly in error. Again linear extrapolation is used; namely (4.4.41)–(4.4.48).

Generalizations of the above procedure may be used for smoothing pressures in some higher-order elements.

Example (Driven Cavity Flow)

A problem description is shown in Fig. 4.4.6. This problem is a much studied example of Stokes flow. Note that the boundary conditions are discontinuous at the upper corner. In the example problem the corner node velocity is set as illustrated in Fig. 4.4.7. For further discussion of the significance of the manner in which the corner discontinuity is modeled, see [53]. The calculation was performed in double precision (64 bits/floating-point word). The penalty parameter was defined by $\lambda/\mu = 10^7$. A 10×10 mesh of bilinear elements was employed with the $S1$ integration scheme. The unsmoothed pressures exhibit significant oscillations, which are removed by the method described above; see Fig. 4.4.8.

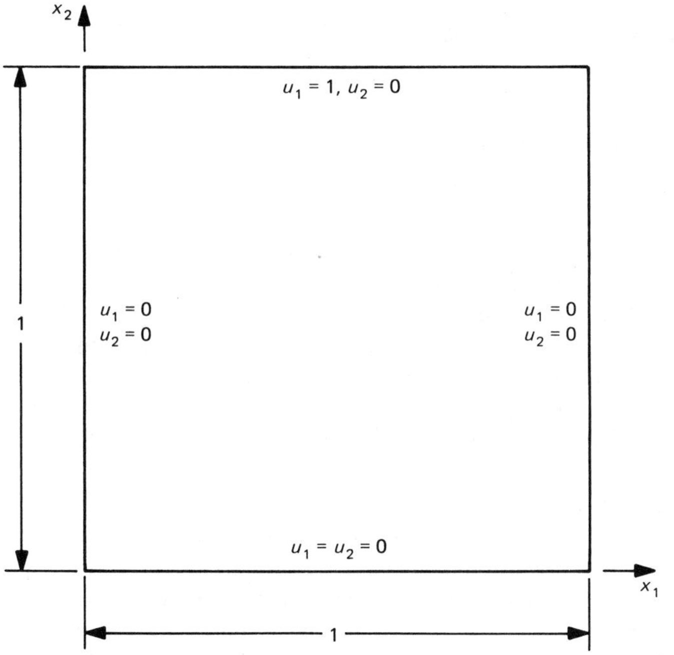

Figure 4.4.6 Driven cavity flow: problem description.

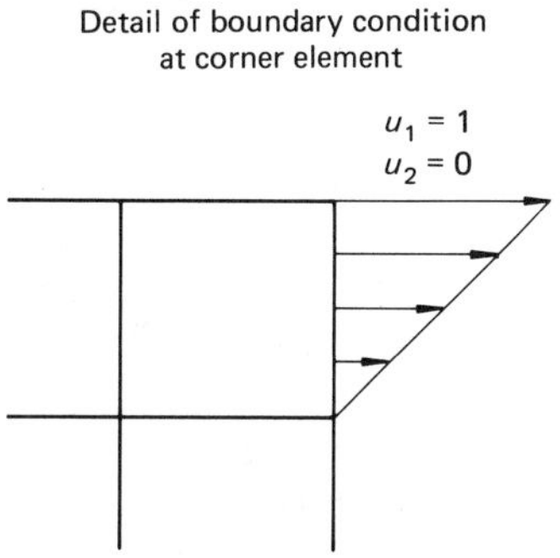

Figure 4.4.7

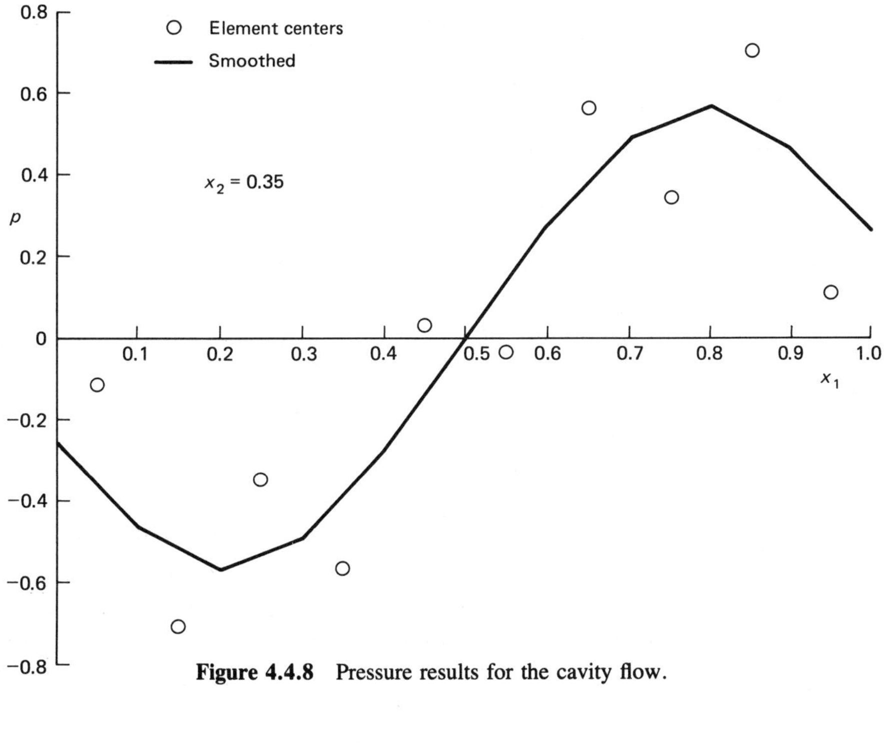

Figure 4.4.8 Pressure results for the cavity flow.

4.5. AN EXTENSION OF REDUCED AND SELECTIVE INTEGRATION TECHNIQUES

4.5.1 Axisymmetry and Anisotropy: Prelude to Nonlinear Analysis

As we pointed out in Sec. 4.4, reduced and selective integration procedures are in certain cases equivalent to the mixed formulations of Sec. 4.3. The equivalence typically holds in plane strain and three-dimensional analysis; however, it breaks down in the axisymmetric case. It is known in this case that the mixed formulation is superior, and thus it would be desirable to develop a pure displacement method, which achieves commensurate results. An approach achieving this end appeared initially in the paper of Nagtegaal et al. [55] (it is sometimes referred to as the ***mean-dilatation approach***).

Another deficiency of the selective integration procedure presented in Sec. 4.4 is that it is limited to the *isotropic case*. Extension to orthotropic and anisotropic cases is ambiguous and computationally inconvenient. Generalization of the mixed formulation to these cases [56, 57] also tends to be somewhat complicated. It thus would also be desirable to develop a simple, pure displacement approach, which simultaneously extends the selective integration procedure to anisotropic cases while attaining the theoretical coherency of the mixed formulation. This is also important for eventual application to nonlinear problems because in these cases the tangent moduli (i.e., the analogs of the c_{ijkl}'s) always exhibit anisotropic character.

A procedure of the kind desired has been presented by Hughes [58] and is described next. The approach may be simply implemented by a small change of the standard technique and is shown to specialize to the selective integration and mean-dilatation formulations under appropriate hypotheses.

4.5.2 Strain Projection: The $\bar{B}$-approach.

We begin by recalling the displacement-method definitions of the element arrays in which the strain-displacement matrix (i.e., $\boldsymbol{B}$) appears:

$$\boldsymbol{k}^e = \int_{\Omega^e} \boldsymbol{B}^T \boldsymbol{D} \boldsymbol{B} \, d\Omega \qquad \text{(element stiffness)} \tag{4.5.1}$$

$$\boldsymbol{f}^e = \int_{\Omega^e} \boldsymbol{B}^T \boldsymbol{\sigma} \, d\Omega \qquad \text{(element internal force)} \tag{4.5.2}$$

So far we have not had much occasion to use the element internal force vector; however, its importance increases in dynamic and nonlinear problems. We recall further that the strain-displacement matrix may be expanded in terms of nodal submatrices as follows:

$$\boldsymbol{B} = [\boldsymbol{B}_1, \boldsymbol{B}_2, \ldots, \boldsymbol{B}_{n_{en}}] \tag{4.5.3}$$

where n_{en} is the number of element nodes. In three-dimensional analysis, a typical submatrix, $\boldsymbol{B}_a$, $1 \leq a \leq n_{en}$, may be written as

$$\boldsymbol{B}_a = \begin{bmatrix} B_1 & 0 & 0 \\ 0 & B_2 & 0 \\ 0 & 0 & B_3 \\ \hline 0 & B_3 & B_2 \\ B_3 & 0 & B_1 \\ B_2 & B_1 & 0 \end{bmatrix} \tag{4.5.4}$$

in which

$$B_i = \partial N_a / \partial x_i, \qquad 1 \leq i \leq 3 \tag{4.5.5}$$

where N_a is the shape function associated with node a, and x_i is the ith Cartesian coordinate.

These expressions are standard but must be modified to be successful in application to nearly incompressible cases. Let $\boldsymbol{B}_a^{\text{dil}}$ denote the dilitational part of $\boldsymbol{B}_a$, *i.e.*,

$$\boldsymbol{B}_a^{\text{dil}} = \frac{1}{3} \begin{bmatrix} B_1 & B_2 & B_3 \\ B_1 & B_2 & B_3 \\ B_1 & B_2 & B_3 \\ \hline 0 & 0 & 0 \\ 0 & 0 & 0 \\ 0 & 0 & 0 \end{bmatrix} \tag{4.5.6}$$

Exercise 1. Derive (4.5.6). *Hint:* Employ the definition of ***dilatational components,*** $\frac{1}{3}\delta_{ij} u_{k,k}^h = \frac{1}{3}\delta_{ij} \sum_{a=1}^{n_{en}} N_{a,k} d_{ka}^e$, and consult Chapter 2 for necessary background.

The deviatoric part of $\boldsymbol{B}_a$ is then defined by

$$\boldsymbol{B}^{\text{dev}} = \boldsymbol{B}_a - \boldsymbol{B}_a^{\text{dil}} \tag{4.5.7}$$

(This follows from the definition of ***deviatoric components,*** $u_{(i,j)}^h - \frac{1}{3}\delta_{ij} u_{k,k}^h$.) To achieve an effective formulation for nearly incompressible applications, $\boldsymbol{B}_a^{\text{dil}}$ needs to be replaced by an "improved" dilatational contribution, which we shall denote by $\overline{\boldsymbol{B}}_a^{\text{dil}}$:

$$\overline{\boldsymbol{B}}_a^{\text{dil}} = \frac{1}{3} \begin{bmatrix} \bar{B}_1 & \bar{B}_2 & \bar{B}_3 \\ \bar{B}_1 & \bar{B}_2 & \bar{B}_3 \\ \bar{B}_1 & \bar{B}_2 & \bar{B}_3 \\ \hline 0 & 0 & 0 \\ 0 & 0 & 0 \\ 0 & 0 & 0 \end{bmatrix} \tag{4.5.8}$$

In place of $\boldsymbol{B}_a$ we now employ

$$\boxed{\overline{\boldsymbol{B}}_a = \boldsymbol{B}_a^{\text{dev}} + \overline{\boldsymbol{B}}_a^{\text{dil}}} \tag{4.5.9}$$

which is given explicitly by

$$\boxed{\overline{\boldsymbol{B}}_a = \begin{bmatrix} B_5 & B_6 & B_8 \\ B_4 & B_7 & B_8 \\ B_4 & B_6 & B_9 \\ \hdashline 0 & B_3 & B_2 \\ B_3 & 0 & B_1 \\ B_2 & B_1 & 0 \end{bmatrix}} \tag{4.5.10}$$

where

$$B_4 = \frac{\overline{B}_1 - B_1}{3} \tag{4.5.11}$$

$$B_5 = B_1 + B_4 \tag{4.5.12}$$

$$B_6 = \frac{\overline{B}_2 - B_2}{3} \tag{4.5.13}$$

$$B_7 = B_2 + B_6 \tag{4.5.14}$$

$$B_8 = \frac{\overline{B}_3 - B_3}{3} \tag{4.5.15}$$

$$B_9 = B_3 + B_8 \tag{4.5.16}$$

Clearly, the whole approach reduces to appropriate definitions of the $\overline{B}_i$'s. We shall present some examples after establishing some preliminary results.

We assume a quadrature rule is specified to integrate the element stiffness matrix and internal force vector. We wish to think of this rule as the "normal" one for the element.

In our first two examples, another quadrature rule is introduced, which may be thought of as a "reduced" rule. For this rule, $\tilde{n}_{\text{int}}$ and $\tilde{\xi}_{\tilde{a}}$ denote the number of points and locations, respectively. A special set of shape functions $\tilde{\boldsymbol{N}}_{\tilde{a}}$'s, are defined with nodal points at the $\tilde{\xi}_{\tilde{a}}$'s (i.e., $\tilde{N}_{\tilde{a}}(\tilde{\xi}_{\tilde{b}}) = \delta_{\tilde{a}\tilde{b}}$, $1 \leq \tilde{a}, \tilde{b} \leq \tilde{n}_{\text{int}}$).

For example, if the element under consideration was a quadrilateral and the reduced rule was the 2×2 Gauss rule, then the $\tilde{N}_{\tilde{a}}$'s would be bilinear functions interpolating the 2×2 Gauss points. (See (4.4.17) for an explicit representation.)

The general form of the $\overline{B}_i$'s is given by

$$\overline{B}_i(\boldsymbol{\xi}) = \sum_{\tilde{a}=1}^{\tilde{n}_{\text{int}}} \tilde{N}_{\tilde{a}}(\boldsymbol{\xi}) B_{i\tilde{a}} \tag{4.5.17}$$

where $\boldsymbol{\xi}$ represents the element natural coordinates and the $B_{i\tilde{a}}$'s are defined in the following examples.

Example 1 (A Generalization of Selective Integration)

The equivalent of selective integration may be attained by taking

$$B_{i\tilde{a}} = B_i(\tilde{\boldsymbol{\xi}}_{\tilde{a}}) \tag{4.5.18}$$

As an example, consider the four-node bilinear quadrilateral. The normal rule is the 2×2 Gauss rule. Take as the reduced rule the one-point Gauss rule, so that $\tilde{n}_{\text{int}} = 1$, $\tilde{N}_1 = 1$, and $\tilde{\boldsymbol{\xi}}_1 = \mathbf{0}$ (i.e., the element "center"). Then (4.5.17) reduces to

$$\overline{B}_i(\boldsymbol{\xi}) = B_i(\mathbf{0}) \tag{4.5.19}$$

That is, the value at the center of the element is used to compute the dilatational contribution.

Example 2 (A Generalization of the Mean-dilatation Formulation)

An approach that generalizes the mean-dilatation formulation of Nagtegaal et al. [55] is defined by

$$\boxed{B_{i\tilde{a}} = \sum_{\tilde{b}=1}^{\tilde{n}_{\text{int}}} (\boldsymbol{m}^{-1})_{\tilde{a}\tilde{b}} \int_{\Omega^e} \tilde{N}_{\tilde{b}}\, B_i \, d\Omega} \tag{4.5.20}$$

where

$$\boldsymbol{m} = [m_{\tilde{a}\tilde{b}}] \tag{4.5.21}$$

$$m_{\tilde{a}\tilde{b}} = \int_{\Omega^e} \tilde{N}_{\tilde{a}} \tilde{N}_{\tilde{b}} \, d\Omega, \qquad 1 \leq \tilde{a}, \tilde{b} \leq \tilde{n}_{\text{int}} \tag{4.5.22}$$

The $\tilde{\boldsymbol{\xi}}_{\tilde{a}}$'s play no role in this formulation. Note that (4.5.20) through (4.5.22) arise from the Galerkin equation $(\tilde{N}_{\tilde{b}}, \overline{B}_i - B_i) = 0$ and (4.5.17).

To see that (4.5.20) through (4.5.22) specialize to the mean-dilatation element of Nagtegaal, consider the same setup as in the previous example. In this case (4.5.17) becomes

$$\boxed{\overline{B}_i(\boldsymbol{\xi}) = \frac{\displaystyle\int_{\Omega^e} B_i \, d\Omega}{\displaystyle\int_{\Omega^e} d\Omega}} \tag{4.5.23}$$

Thus the mean value of B_i is used to compute the dilatational contribution.

In passing, we may note that this generalization of the mean-dilatation formulation [i.e., (4.5.20) through (4.5.22)] is somewhat more involved than the generalization of the selective integration formulation defined by (4.5.18).

Example 3 (Another Generalization of the Mean-dilatation Formulation)

In this example we show how the present formulation includes the higher-order generalization of the mean-dilatation formulation originally suggested in [55].

For this case there is no need to introduce a reduced quadrature rule. However, we may still employ (4.5.20) through (4.5.22), where the $\widetilde{N}_{\tilde{a}}$'s are here interpreted as an arbitrary set of $\widetilde{n}_{\text{int}}$ functions.

As an example of how this procedure might be used, consider the nine-node Lagrange quadrilateral. Assume that the 3×3 rule is to play the role of the normal quadrature rule. Take $\widetilde{n}_{\text{int}} = 3$ and assume the $\widetilde{N}_{\tilde{a}}$'s are 1, $x_1 - x_1(\mathbf{0})$, and $x_2 - x_2(\mathbf{0})$. This assumption was proposed in [55]. This element is an analog of the nine-node, linear pressure quadrilateral of Fig. 4.3.4(h).

The main significance of the preceding ideas is that the number of functions is not restricted to be the number of points of a quadrature rule. This restriction has limited the success of the selective integration procedure to specific elements.

Remark

Formulations of the type described in Examples 2 and 3 may be described as ***strain projections*** in that the dilatational strain is projected onto a simple set of functions by way of the Galerkin method. Example 1 is not generally a projection but is rather an interpolation procedure. In specific cases, however, it may be equivalent to a projection. This issue is discussed further in Hughes and Malkus [59].

Torsionless Axisymmetric Analysis

The preceding ideas may be extended to torsionless axisymmetry in a straightforward manner (compare Sec. 2.12). It is convenient in this case to introduce a cylindrical coordinate system in which

$$x_1 = \text{the radial coordinate}$$

$$x_2 = \text{the axial coordinate}$$

$$x_3 = \text{the circumferential coordinate}$$

Proceeding as in the development of (4.5.10), we are led to

$$\boxed{\overline{\boldsymbol{B}}_a = \begin{bmatrix} B_{12} & B_6 \\ B_{10} & B_7 \\ B_2 & B_1 \\ \hdashline B_{11} & B_6 \end{bmatrix}} \tag{4.5.24}$$

where

$$B_0 = \frac{N_a}{x_1} \tag{4.5.25}$$

$$B_{10} = B_4 + \frac{\overline{B}_0 - B_0}{3} \tag{4.5.26}$$

$$B_{11} = B_0 + B_{10} \tag{4.5.27}$$

$$B_{12} = B_1 + B_{10} \tag{4.5.28}$$

In (4.5.26) $\overline{B}_0$ is defined by (4.5.17).

Plane Strain Analysis

The plane strain case may be obtained from (4.5.24) through (4.5.28) by setting $\overline{B}_0 = B_0 = 0$ and interpreting the x_i's as Cartesian coordinates. It is important to note that for plane strain $\overline{\boldsymbol{B}}_a$ has dimension 4×3, whereas the standard $\boldsymbol{B}_a$ has dimension 3×3 (compare (2.9.3)).

Remarks

1. The preceding ideas, first proposed in Hughes [58, 60], apply to the nonlinear case if we employ an "updated" formulation for element array calculation.

2. Note that in all cases the $\overline{\boldsymbol{B}}$-formulation involves only one numerical integration "do loop" for element array calculations, and thus implementation is not significantly different from the standard uniform integration case (see Sec. 3.10).

3. The relationship of the present procedures with mixed formulations has been explored in Hughes and Malkus [59]. Simo and Hughes [61] and Hughes et al. [62] have established the equivalence of the $\overline{\boldsymbol{B}}$-approach with finite element methods derived from the Hu-Washizu variational formulation.

4. The $\overline{\boldsymbol{B}}$-formulation is now widely used in large-scale linear and nonlinear finite element systems. Due to the importance of the method, it was chosen for inclusion in program DLEARN (see Chapter 11). Specifically, the mean-dilatational four-node element is available for plane strain and axisymmetric analysis. The element mean values of the shape functions and their global derivatives are calculated in subroutine MEANSH and stored in an array SHGBAR. The setup of (4.5.24) for each node and integration point is performed in subroutine QDCB. Note that (4.5.24) is a full matrix, in contrast to the usual $\boldsymbol{B}_a$. The reader may wish to study QDCB and MEANSH in order to understand fully the implementation of the mean-dilatational element.

4.6 THE PATCH TEST; RANK DEFICIENCY

In the previous two sections we have presented methods of displacement-type which involve "violations" of the classical Galerkin finite element formulation described in Chapters 1 to 3 and analyzed in Sec. 4.1. The motivation for changing the standard formulation is clear: It simply does not always work for problems of interest. In subsequent sections we will consider even more drastic modifications to the standard formulations, in particular, so-called incompatible, or nonconforming, elements. Strang has aptly termed all these violations of the Galerkin code ***variational crimes.*** The theory of Sec. 4.1 simply does not apply to these cases. A more general theory is thus required, but this would take us beyond the scope of this book. It turns out that an essential ingredient in the mathematical analysis of nonstandard elements is based on an idea of Irons called the "patch test". Irons' original presentation [63] was very practically motivated and nonmathematical in nature. In fact, the test was described in terms of computational experiments, which would simultaneously assess the correctness of the formulation and its computer implementation. We shall refer to this form of the patch test as the "engineering version." A mathematical version was described and popularized in Strang and Fix [64]. Subsequently, the patch test has generated some mathematical controversy (see Stummel [65]) and undergone rumination (see Irons and Loikkanen [66] and Taylor et al. [67]). In addition, in the context of complicated theories, it is not always even clear how to pose patch tests. For these reasons faith in the patch test has eroded in some quarters. *This is unfortunate, for we firmly believe that, within the realm of problems dealt with so far in this book, the patch test is the most practically useful technique for assessing element behavior.* Thus we wish to avoid altogether the mathematically controversial facets of this subject and return to the spirit of Irons' original conception.

The Patch Test ("Engineering Version")

Basically, the patch test enables us to determine whether or not an element satisfies the completeness condition. The patch test may be applied to elements employing nonstandard features (e.g., selective integration, incompatible interpolations, etc.).

We shall say that the patch test is passed for a two-dimensional element if the states 1, x, and y are exactly representable by an arbitrary "patch" of elements whenever the exact solution behaves accordingly. What does this mean? We shall use an example to clarify the idea.

Example

Consider a two-dimensional elasticity problem in which the elastic moduli are constant and there are no body forces present. Consider the arbitrary patch of elements sketched in Fig. 4.6.1. We assume the *boundary* nodal displacements are set in accord with linear monomials as follows:

	Test number	u_{1a}^h	u_{2a}^h
1	1	1	0
	2	0	1
x	3	x_a	0
	4	0	x_a
y	5	y_a	0
	6	0	y_a

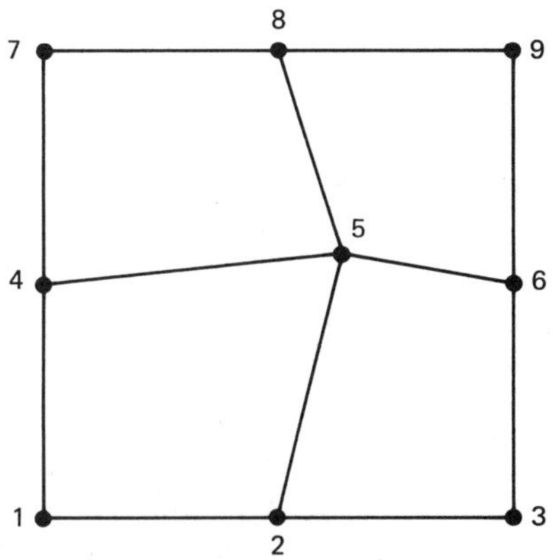

Figure 4.6.1 Arbitrary patch of two-dimensional elements.

Thus we wish to solve six displacement boundary-value problems. The patch test will be satisfied if in all cases: (1) the solution at node $a = 5$ is *exact*, i.e., takes on the value given in the above table, and (2) the displacement gradients, and consequently strains and stresses, are *exact* within each element.

In constructing an arbitrary patch, it is important to use an irregular geometry, because some elements pass the patch test in certain special configurations but not others. An example of this is described in Sec. 4.7.

The reader should have no trouble generalizing the above idea to three dimensions and specializing it to the scalar heat-conduction case.

Rank Deficiency

Consider a two-dimensional bilinear elasticity element. The element stiffness may be expressed as

$$\underbrace{\boldsymbol{k}^e}_{8 \times 8} = \int_{\Omega^e} \underbrace{\boldsymbol{B}^T}_{8 \times 3} \underbrace{\boldsymbol{D}}_{3 \times 3} \underbrace{\boldsymbol{B}}_{3 \times 8} \, d\Omega \tag{4.6.1}$$

If exact integration, or 2×2 Gaussian integration, is performed on the integral in (4.6.1), the resulting stiffness will have rank 5. This may be seen to be correct as follows: The rank will equal the number of degrees of freedom, namely, eight, minus the number of rigid-body modes, which in this case is three (i.e., two translations and one rotation). However, if instead of 2×2 Gaussian integration (or higher), we use the one-point Gauss rule, the rank will be reduced to three. This can be seen algebraically by examining the one-point stiffness matrix:

$$\boldsymbol{k}^e_{\text{1-pt.}} = 4(j\boldsymbol{B}^T\boldsymbol{D}\boldsymbol{B})\Big|_{\tilde{\xi}=\tilde{\eta}=0} \tag{4.6.2}$$

(Recall that j denotes the Jacobian determinant and 4 is the weight of the one-point rule.) By virtue of the fact that $\boldsymbol{D}$ is a 3×3 matrix, the rank of $\boldsymbol{k}^e_{\text{1-pt.}}$ can be no greater than three. It can be verified that it is, in fact, exactly three. This means that there are two additional (spurious) zero-energy modes. These are the ***hourglass modes*** depicted in Fig. 4.6.2. In a mesh formed from such elements, the hourglass modes can communicate to form a singular, or nearly singular, global stiffness $\boldsymbol{K}$. In the singular case, no solution is possible; in the nearly singular case, hourglass "instabilities" may

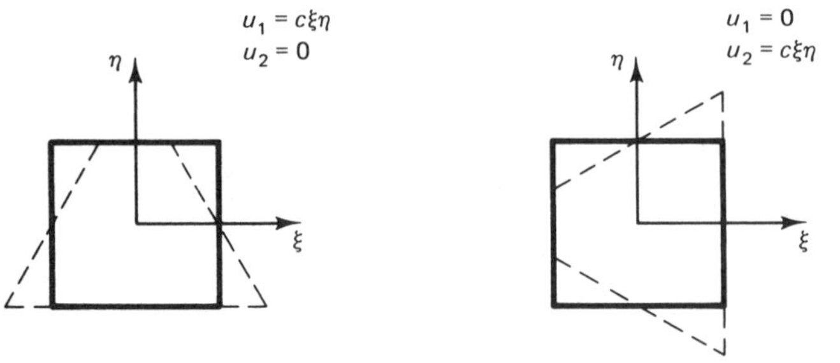

Figure 4.6.2 Hourglass modes; c is an arbitrary constant.

pollute the displacement field. Our initial reaction is, of course, to prudently avoid elements of this type. However, with appropriate modifications, one-point integration elements have proven to be very effective. We shall return to this theme in Sec. 4.8.

When reduced integration is used, we must be aware of the possibility of generating spurious zero-energy modes. (These are sometimes referred to in the literature as ***mechanisms***.) A way of numerically assessing the presence of zero-energy modes is to calculate the eigenvalues of the element stiffness, which are defined by

$$\det(\boldsymbol{k}^e - \lambda \boldsymbol{I}) = 0 \tag{4.6.3}$$

The number of zero eigenvalues equals the number of zero-energy modes. Another way is to perform a Crout factorization of $\boldsymbol{k}^e$ and count the number of zero "pivots." This procedure is employed in the DLEARN program and is described in Exercise 3 of Sec. 4.9.

Examples of Elements with Spurious Zero-energy Modes

Example 1

The correct rank for the eight-node brick element in three dimensions is 18 (i.e., 24 degrees of freedom minus 6 rigid-body modes). The $2 \times 2 \times 2$ Gaussian rule is sufficient to assure correct rank. One-point Gaussian integration produces a rank of 6. That is, there are 12 spurious zero-energy modes.

Example 2

The eight-node serendipity quadrilateral with reduced 2×2 quadrature possesses one spurious zero-energy mode; see Fig. 4.6.3. This mode is often described as "non-

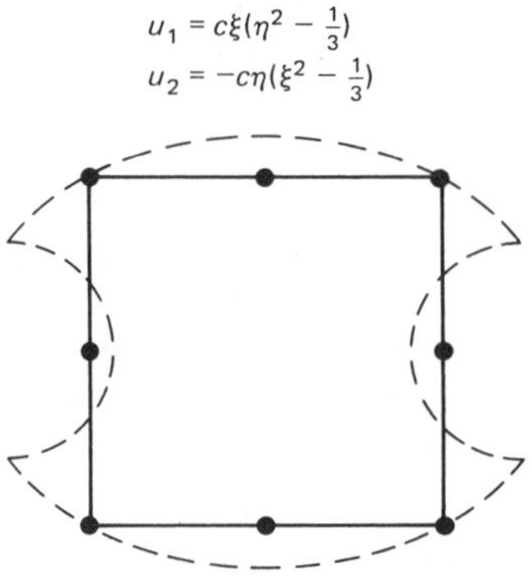

Figure 4.6.3 The spurious zero-energy mode of the reduced 2×2 Gaussian integration eight-node serendipity quadrilateral; c is an arbitrary constant.

communicable" because in an assembly of two or more elements no zero-energy modes are present. However, in certain situations problems can still occur. For example, consider the configuration illustrated in Fig. 4.6.4. The type of loading applied to the steel base can arouse the spurious zero-energy mode. The contiguous soil elements may be too soft to adequately resist it.

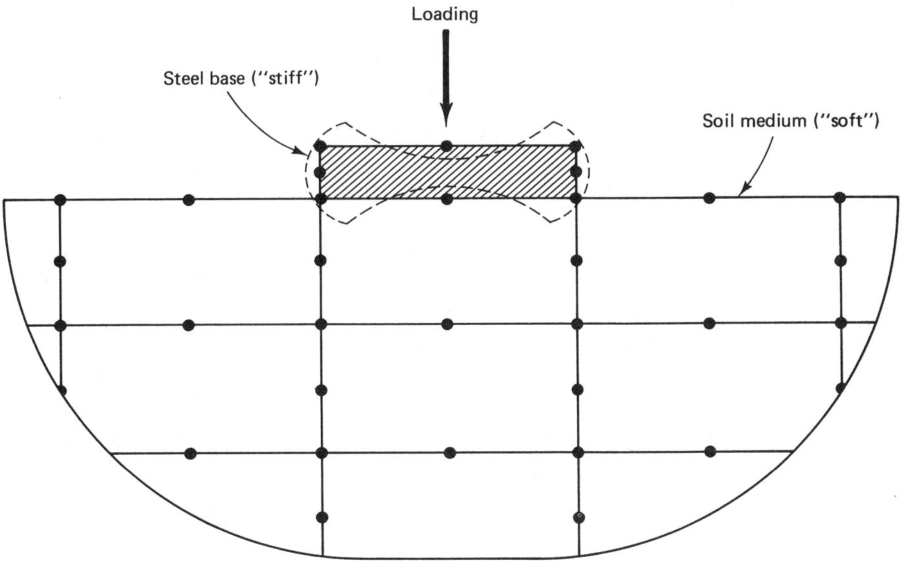

Figure 4.6.4 Stiff steel base on a soft soil medium. Illustration of potential modeling difficulties with the 2 × 2 reduced integration serendipity element.

Exercise. Consider a square bilinear two-dimensional element. Assume that the global coordinates are coincident with the local coordinates and differ only in length-scale. Thus we may write $x = h\xi/2$ and $y = h\eta/2$. All derivatives of the hourglass patterns (see Fig. 4.6.2) vanish identically at $\xi = \eta = 0$. This fact and the expression for strain energy,

$$\frac{1}{2}(\boldsymbol{d}^e)^T \boldsymbol{k}^e_{1\text{-pt.}} \boldsymbol{d}^e = 2(j\ u^h_{(i,j)} c_{ijkl} u^h_{(k,l)})\Big|_{\xi=\eta=0}$$

illustrate that hourglass modes possess zero strain energy when one-point quadrature is used to obtain the stiffness matrix. The following will enable the reader to verify this and related facts. Consider eight linearly independent deformation patterns for the square element:

1. $\begin{cases} u_1 = c \\ u_2 = 0 \end{cases}$ (x-translation)

2. $\begin{cases} u_1 = 0 \\ u_2 = c \end{cases}$ (y-translation)

3. $\begin{cases} u_1 = c\eta \\ u_2 = -c\xi \end{cases}$ (infinitesimal rotation)

4. $\begin{cases} u_1 = c\xi\eta \\ u_2 = 0 \end{cases}$ (x-hourglass)

5. $\begin{cases} u_1 = 0 \\ u_2 = c\xi\eta \end{cases}$ (y-hourglass)

6. $\begin{cases} u_1 = c\xi \\ u_2 = 0 \end{cases}$ (uniform x-extension)

7. $\begin{cases} u_1 = 0 \\ u_2 = c\eta \end{cases}$ (uniform y-extension)

8. $\begin{cases} u_1 = c\eta \\ u_2 = c\xi \end{cases}$ (uniform shear)

Calculate the strain energy in each pattern for an exactly integrated stiffness and a one-point quadrature stiffness. Deduce the following facts:

a. For both stiffnesses the rigid-body modes (i.e., 1 to 3) produce zero strain energy.
b. For both stiffnesses the homogeneous strain states (i.e., 6 to 8) produce nonzero strain energy.
c. One-point quadrature produces zero strain energy in the hourglass modes, whereas exact integration produces nonzero strain energy.

4.7 NONCONFORMING ELEMENTS

Thus far we have always constructed our finite element spaces such that they are proper subsets of the corresponding spaces employed in the exact weak statement of the problem. For example, this is written as follows:

$$\mathcal{S}^h \subset \mathcal{S} \tag{4.7.1}$$

$$\mathcal{V}^h \subset \mathcal{V} \tag{4.7.2}$$

Recall the meaning of (4.7.1) and (4.7.2): If $\boldsymbol{u}^h \in \mathcal{S}^h$, then $\boldsymbol{u}^h \in \mathcal{S}$, and so $\boldsymbol{u}^h$ inherits the properties defining $\mathcal{S}$. That is, if members of $\mathcal{S}$ satisfy certain boundary conditions and possess some degree of "regularity" (e.g., have square-integrable first derivatives), then members of $\mathcal{S}^h$ also possess these properties, and likewise for $\mathcal{V}$ and $\mathcal{V}^h$.

In Chapters 1 and 3 we constructed finite element shape functions, which were ***continuous*** across element boundaries. Thus, although derivatives are typically ***discontinuous*** across element boundaries, these functions still possess square-integrable first derivatives. This is the class of C^0 finite element interpolations and is appropriate for problems in which the expression for $a(\cdot, \cdot)$ contains products of first, but no higher, derivatives (such as heat conduction, or classical elasticity).

It is important to note that (4.7.1) and (4.7.2) are explicitly used in establishing convergence in Sec. 4.1. Thus there is no guarantee that convergence will be achieved if (4.7.1) and (4.7.2) are not respected. Nevertheless, finite elements have been proposed, and in fact are commonly used, in which C^0 continuity is violated, resulting in functions that do not possess square-integrable first derivatives. (Dirac delta distributions exist in the derivatives of such functions, and the product of these is not well defined.) Elements of this type are referred to as ***incompatible,*** or ***nonconforming.*** We shall illustrate the ideas involved by considering the ***incompatible modes element*** originally proposed by Wilson et al. [68].

Incompatible Modes Element

Version 1, Wilson et al. It may be recalled that the standard four-node quadrilaterals employed in the beam-bending elasticity example of Sec. 4.4 (see Figs. 4.4.1 and 4.4.2 and Table 4.4.1) attained a level of accuracy of approximately 90% in the case of Poisson's ratio = 0.3. In this example, 32 elements were used to model the upper half of the beam (see Fig. 4.4.2). For "bending" cases like this, the accuracy level attained by simple elements tends to be somewhat disappointing. If only one element is used through the thickness, the accuracy degrades further, leading to worthless results. (The reason for this will be made clear in a moment.) It thus would appear worthwhile to attempt to modify the basic four-node element so that improved behavior is attained in bending situations. This is the motivation that prompted Wilson et al. to propose the incompatible modes approach. They noted that if a bending moment is applied to a single element, the element responds in shear rather than bending, and this spurious shearing is responsible for the overly stiff behavior (see Fig. 4.7.1). They further observed that one way of producing correct behavior in this situation was to add quadratic modes of deformation corresponding to the exact pure bending solution.[10] This is the essence of the incompatible modes approach. Starting with the standard expansion in terms of bilinear shape functions for a typical element, quadratic modes are added as follows:

$$\boldsymbol{u}^h(\xi, \eta) = \sum_{a=1}^{4} N_a(\xi, \eta)\boldsymbol{d}_a^e + \sum_{a=5}^{6} N_a(\xi, \eta)\boldsymbol{\alpha}_a^e \tag{4.7.3}$$

where the first four N_a's are the standard bilinear shape functions (see Sec. 3.2),

$$\left.\begin{aligned} N_5(\xi, \eta) &= 1 - \xi^2 \\ N_6(\xi, \eta) &= 1 - \eta^2 \end{aligned}\right\} \quad \textbf{\textit{(incompatible modes)}} \tag{4.7.4}$$

and the $\boldsymbol{\alpha}_a$'s are ***generalized displacements*** (as distinguished from nodal displacements). The incompatible modes are sketched in Fig. 4.7.2. There are no nodal points associated with the additional modes and the $\boldsymbol{\alpha}_a$'s may be thought of as "internal" element degrees of freedom. The incompatible modes result in the element displacements being discontinuous between nodes (see Fig. 4.7.2).

Equation (4.7.3) is used in the definition of the element stiffness, or wherever the strain-displacement matrix, $\boldsymbol{B}$, *appears, but not elsewhere, i.e., not in the*

[10] Clearly, another possibility would be simply to employ higher-order elements, such as the eight-node serendipity, or nine-node Lagrange, quadrilateral. As might be anticipated, excellent results may be obtained in this fashion. Here, however, the objective is to achieve good behavior *without* introducing additional nodes to the basic four-node pattern.

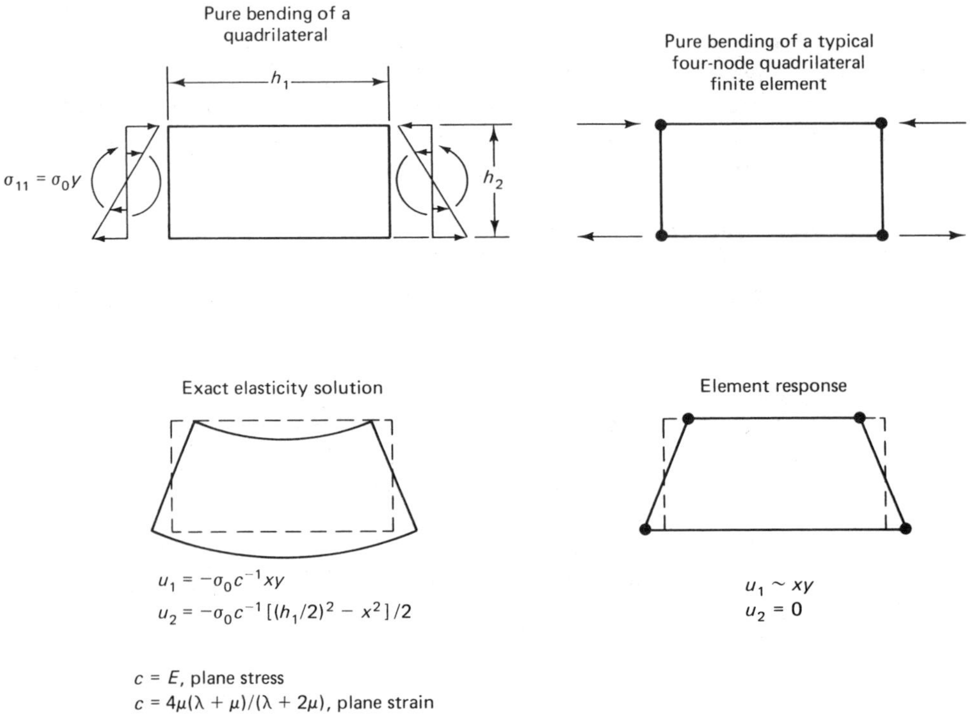

Pure bending of a
quadrilateral

Pure bending of a typical
four-node quadrilateral
finite element

$\sigma_{11} = \sigma_0 y$

Exact elasticity solution

Element response

$u_1 = -\sigma_0 c^{-1} xy$

$u_2 = -\sigma_0 c^{-1} [(h_1/2)^2 - x^2]/2$

$u_1 \sim xy$

$u_2 = 0$

$c = E$, plane stress

$c = 4\mu(\lambda + \mu)/(\lambda + 2\mu)$, plane strain

Figure 4.7.1

$u_1 = (1 - \xi^2)\alpha_{15}$

$u_1 = (1 - \eta^2)\alpha_{16}$

$u_2 = (1 - \xi^2)\alpha_{25}$

$u_2 = (1 - \eta^2)\alpha_{26}$

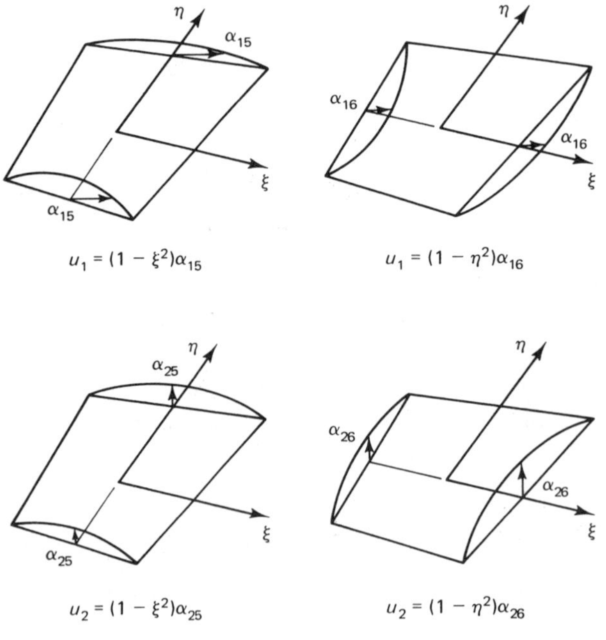

Figure 4.7.2   Incompatible displacement modes.

definition of element body and surface forces. To see why this is the case, consider a *constant* body force, $\boldsymbol{\ell}$. It is reasonable to insist that the sum of the element forces produced by $\boldsymbol{\ell}$ equals vol$(\Omega^e)\,\boldsymbol{\ell}$, where vol$(\Omega^e) = \int_{\Omega^e} d\Omega$. This is easily seen to be the case for any set of shape functions satisfying $\sum_{a=1}^{n_{en}} N_a = 1$. For example, in the case of the standard bilinear quadrilateral, we have

$$\begin{aligned}\sum_{a=1}^{4} \boldsymbol{f}_a^e &= \sum_{a=1}^{4} \int_{\Omega^e} N_a \boldsymbol{\ell}\, d\Omega \\ &= \int_{\Omega^e} \underbrace{\sum_{a=1}^{4} N_a}_{1}\, d\Omega\, \boldsymbol{\ell} \\ &= \int_{\Omega^e} d\Omega\, \boldsymbol{\ell} \\ &= \text{vol}\,(\Omega^e)\,\boldsymbol{\ell}\end{aligned} \tag{4.7.5}$$

If we insist on including forces associated with the additional modes, the sum total is too great, as may be seen by calculating as above:

$$\sum_{a=1}^{6} \int_{\Omega^e} N_a \boldsymbol{\ell}\, d\Omega = \left[\text{vol}\,(\Omega^e) + \underbrace{\sum_{a=5}^{6} \int_{\Omega^e} N_a\, d\Omega}_{>0}\right] \boldsymbol{\ell} \tag{4.7.6}$$

A similar conclusion may be drawn for surface forces.

Because the generalized displacements associated with the incompatible modes are unique to each element, they may be eliminated on the element level. (This process is analogous to the elimination of pressure degrees of freedom in the case of discontinuous pressure fields; see Sec. 4.3, Procedure 3.) This leads to a "condensed" element stiffness of normal size for a four-node quadrilateral, i.e., 8×8. The process of deriving the stiffness is given as follows:

We begin with the standard definition of $\boldsymbol{k}$, namely,

$$\boldsymbol{k} = \int_{\Omega^e} \boldsymbol{B}^T \boldsymbol{D} \boldsymbol{B}\, d\Omega \tag{4.7.7}$$

where

$$\boldsymbol{B} = [\boldsymbol{B}_1 \quad \boldsymbol{B}_2 \quad \boldsymbol{B}_3 \quad \boldsymbol{B}_4 \;\vdots\; \boldsymbol{B}_5 \quad \boldsymbol{B}_6] \tag{4.7.8}$$

The $\boldsymbol{B}_a$'s are defined in the usual way. The appearance of $\boldsymbol{B}_5$ and $\boldsymbol{B}_6$ is due to the presence of the incompatible modes. Each $\boldsymbol{B}_a$ has dimension 3×2; consequently, $\boldsymbol{B}$ has dimension 3×12 and $\boldsymbol{k}$ has dimension 12×12. It can be written in the following partitioned form:

$$\boldsymbol{k} = \left[\begin{array}{c|c} \boldsymbol{k}_{dd} & \boldsymbol{k}_{d\alpha} \\ \hline \boldsymbol{k}_{\alpha d} & \boldsymbol{k}_{\alpha\alpha} \end{array}\right] \tag{4.7.9}$$

Note that $\boldsymbol{k}_{dd}$ is the usual 8×8 stiffness of the four-node quadrilateral, $\boldsymbol{k}_{d\alpha} = \boldsymbol{k}_{\alpha d}^T$ has dimension 8×4, and $\boldsymbol{k}_{\alpha\alpha}$ has dimension 4×4. By virtue of the fact that there are no element forces corresponding to the incompatible modes, we can write

$$\boldsymbol{k}_{\alpha d}\boldsymbol{d}^e + \boldsymbol{k}_{\alpha\alpha}\boldsymbol{\alpha}^e = \boldsymbol{0} \qquad \text{where } \boldsymbol{\alpha}^e = \begin{Bmatrix} \boldsymbol{\alpha}_5^e \\ \boldsymbol{\alpha}_6^e \end{Bmatrix} \tag{4.7.10}$$

and solve for $\boldsymbol{\alpha}^e$ in terms of the nodal displacements[11]:

$$\boldsymbol{\alpha}^e = -\boldsymbol{k}_{\alpha\alpha}^{-1}\boldsymbol{k}_{\alpha d}\boldsymbol{d}^e \tag{4.7.11}$$

Equation (4.7.11) may then be used to eliminate the $\boldsymbol{\alpha}^e$'s from the equations. This process leads to the following 8×8 element stiffness:

$$\boxed{\tilde{\boldsymbol{k}} = \boldsymbol{k}_{dd} - \boldsymbol{k}_{d\alpha}\boldsymbol{k}_{\alpha\alpha}^{-1}\boldsymbol{k}_{\alpha d}} \tag{4.7.12}$$

Equation (4.7.12) is the stiffness of the incompatible modes element. The process used to obtain (4.7.12) is frequently referred to as ***static condensation*** in the structural mechanics literature. At this point, element stiffness assembly may proceed in the usual way.

Remark

The preceding developments define the original incompatible modes element of Wilson et al. A theoretical investigation was given by Lesaint [69]. Numerical studies indicated improved behavior for the incompatible modes element in bending applications. However, a flaw was soon discovered: When the element took the shape of an arbitrary quadrilateral (not a rectangle or a parallelogram), erratic behavior was noted. In this case the element was found to fail the patch test. Taylor et al. [70] further modified the incompatible modes element so that this deficiency was corrected. This is described next.

Version 2, Taylor et al. Taylor argued as follows: If the nodal displacements are set in accord with a given linear polynomial in the spatial coordinates, then the incompatible modes should not be activated (i.e., $\boldsymbol{\alpha}^e \equiv \boldsymbol{0}$). This is seen to be an alternative statement of the completeness condition, discussed in Sec. 3.1. Recall that completeness means that the displacement field throughout the element exactly coincides with the given linear polynomial. It may be concluded from (4.7.10) that $\boldsymbol{\alpha}^e = \boldsymbol{0}$ if $\boldsymbol{k}_{\alpha d}^e\boldsymbol{d}^e = \boldsymbol{0}$. By (4.7.7) and (4.7.8), (4.7.10) can be written in the form

$$\int_{\Omega^e} \boldsymbol{B}_a^T\boldsymbol{\sigma}^{(4)}\, d\Omega = -\sum_{b=5}^{6}\left(\int_{\Omega^e} \boldsymbol{B}_a^T\boldsymbol{D}\boldsymbol{B}_b\, d\Omega\right)\boldsymbol{\alpha}_b^e, \qquad a = 5, 6 \tag{4.7.13}$$

[11] It may be argued that $\boldsymbol{k}_{\alpha\alpha}$ is nonsingular.

where

$$\begin{aligned}\boldsymbol{\sigma}^{(4)}(\boldsymbol{x}) &= \boldsymbol{D}\boldsymbol{\epsilon}^{(4)}(\boldsymbol{x}) \\ &= \boldsymbol{D}\left(\sum_{b=1}^{4} \boldsymbol{B}_b(\boldsymbol{x})\boldsymbol{d}_b^e\right)\end{aligned} \tag{4.7.14}$$

is the stress attributable only to the nodal displacements.[12] Because the nodal values of displacement are set in accord with a linear polynomial and the *bilinear shape functions are complete* (see Chapter 3), the strains, $\boldsymbol{\epsilon}^{(4)}$, will be constant, and thus, by (4.7.14), so will $\boldsymbol{\sigma}^{(4)}$. The following conditions are thus equivalent:

$$\boldsymbol{0} = \int_{\Omega^e} \boldsymbol{B}_a^T \, d\Omega \, \boldsymbol{\sigma}^{(4)} \qquad \text{and} \qquad \boldsymbol{0} = \boldsymbol{k}_{\alpha d}\boldsymbol{d}^e \tag{4.7.15}$$

where $a = 5, 6$ and $\boldsymbol{\sigma}^{(4)}$ is arbitrary. Consequently, it is required that

$$\boldsymbol{0} = \int_{\Omega^e} \boldsymbol{B}_a^T \, d\Omega, \qquad a = 5, 6, \tag{4.7.16}$$

for *all* configurations of the element. Employing the results of Sec. 3.9, we obtain the following:

$$\begin{aligned}\int_{\Omega^e} \boldsymbol{B}_5 \, d\Omega &= \int_{\Omega^e} \begin{bmatrix} N_{5,x} & 0 \\ 0 & N_{5,y} \\ N_{5,y} & N_{5,x} \end{bmatrix} d\Omega \\ &= 2\int_{-1}^{+1}\int_{-1}^{+1} \begin{bmatrix} -\xi y_{,\eta} & 0 \\ 0 & \xi x_{,\eta} \\ \xi x_{,\eta} & -\xi y_{,\eta} \end{bmatrix} d\xi \, d\eta\end{aligned} \tag{4.7.17}$$

$$\begin{aligned}\int_{\Omega^e} \boldsymbol{B}_6 \, d\Omega &= \int_{\Omega^e} \begin{bmatrix} N_{6,x} & 0 \\ 0 & N_{6,y} \\ N_{6,y} & N_{6,x} \end{bmatrix} d\Omega \\ &= 2\int_{-1}^{+1}\int_{-1}^{+1} \begin{bmatrix} \eta y_{,\xi} & 0 \\ 0 & -\eta x_{,\xi} \\ -\eta x_{,\xi} & \eta y_{,\xi} \end{bmatrix} d\xi \, d\eta\end{aligned} \tag{4.7.18}$$

From (4.7.17) and (4.7.18) we see that $\boldsymbol{B}_5$ and $\boldsymbol{B}_6$ will integrate to zero if the derivatives of x and y with respect to ξ and η are *constants*. It may be verified that if the element is a rectangle or parallelogram, then the derivatives are constants. How-

[12] In general, the stress for the incompatible modes element should be computed in the usual way, i.e.,

$$\boldsymbol{\sigma}(\boldsymbol{x}) = \boldsymbol{D}(\boldsymbol{x})\boldsymbol{\epsilon}(\boldsymbol{x}) = \boldsymbol{D}(\boldsymbol{x})\left[\sum_{a=1}^{4} \boldsymbol{B}_a(\boldsymbol{x})\boldsymbol{d}_a^e + \sum_{a=5}^{6} \boldsymbol{B}_a(\boldsymbol{x})\boldsymbol{\alpha}_a^e\right]$$

Expression (4.7.14) is just an intermediate quantity introduced for purposes of derivation. However, if $\boldsymbol{\sigma}$ is computed at $\boldsymbol{x}(\xi, \eta) = \boldsymbol{x}(0, 0)$, then $\boldsymbol{B}_5$ and $\boldsymbol{B}_6$ may be omitted because they vanish identically at this location (see (4.7.17) and (4.7.18)) and, consequently, $\boldsymbol{\sigma}(\boldsymbol{x}(0, 0)) = \boldsymbol{\sigma}^{(4)}(\boldsymbol{x}(0, 0))$.

ever, in the general quadrilateral configuration, linear terms appear, and this is the reason why the element fails in these cases. Taylor proposed modifying the element by replacing the derivatives in (4.7.17) and (4.7.18) by their values at $\xi = \eta = 0$. The definitions for Taylor's $\boldsymbol{B}_5$ and $\boldsymbol{B}_6$ are then:

$$\boldsymbol{B}_5 = \frac{2}{j}\begin{bmatrix} -\xi y_{,\eta}(0, 0) & 0 \\ 0 & \xi x_{,\eta}(0, 0) \\ \xi x_{,\eta}(0, 0) & -\xi y_{,\eta}(0, 0) \end{bmatrix} \tag{4.7.19}$$

$$\boldsymbol{B}_6 = \frac{2}{j}\begin{bmatrix} \eta y_{,\xi}(0, 0) & 0 \\ 0 & -\eta x_{,\xi}(0, 0) \\ -\eta x_{,\xi}(0, 0) & \eta y_{,\xi}(0, 0) \end{bmatrix} \tag{4.7.20}$$

With the above modifications, the element passes the patch test for arbitrary configurations. Calculations in support of this assertion are presented in [70]. Clearly, version 2 is suitable for general analysis, whereas version 1 should be used only in rectangular or parallelogrammic configurations.

Example

The bending behavior of the incompatible modes element is assessed on the elasticity solution previously considered (see Figs. 4.4.1 and 4.4.2 and Table 4.4.1). Results are presented in Table 4.7.1 . The superiority of the incompatible modes element in the present situation is self-evident.

TABLE 4.7.1 Comparison of Bilinear Quadrilateral Elements on Beam-bending Problem. Tip Displacement Normalized by Exact Solution

	$U2$	$S1$	$U2$ with incompatible modes
$\nu = 0.3$	0.904	0.912	0.995
$\nu = 0.499$	0.334	0.937	0.992

Key:
$U2$ 2×2 uniform integration (exact in the present cases)
$S1$ selective integration (2×2 on μ-term, one-point on λ-term)

The behavior of the elements as Poisson's ratio approaches $\frac{1}{2}$ is shown in Fig. 4.7.3.[13] As may be seen, the standard four-node element and linear triangle deteriorate rapidly, as is to be expected. The selective integration element and incompatible modes element maintain their accuracy even for very large ratios of λ/μ. Deterioration commences at about $\lambda/\mu = 10^9$ due to the fact that rounding errors in equation solving

[13] Recall that $\lambda/\mu = 2\nu/(1 - 2\nu)$.

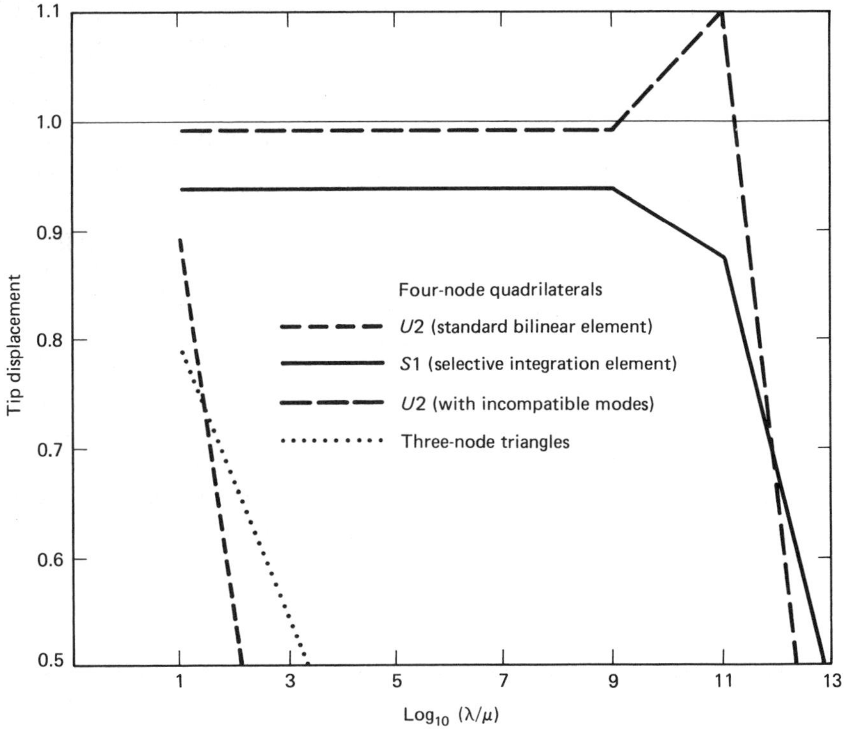

Figure 4.7.3

overwhelm the λ contributions, which are nine orders of magnitude smaller than the μ contributions. (A 64-bit floating-point word was used in the present calculations. This corresponds to approximately 16 digits.) The point at which deterioration begins is beyond the region of practical importance.[14]

To explain the good behavior of the incompatible modes element in nearly incompressible applications, we may attempt to calculate its constraint ratio. In this case the number of equations is six due to the presence of the four generalized displacements associated with the incompatible modes. Because the divergence of the displacement field is a linear polynomial, there are three incompressibility constraints. Consequently, the constraint ratio is $\frac{6}{3} = 2$, the optimal value.

Remarks

1. The incompatible modes element behaves well in bending, even with only one element through the thickness.

2. Due to the additional shape functions present and static condensation, the element is somewhat expensive to form. An alternative, but more economical element, which attains virtually identical results, is considered in the next section.

[14] The floating-point word length determines when deterioration commences in situations like this. If a shorter word is used, only smaller ratios of λ/μ can be effectively handled.

3. The following steps are required to generalize the implementation of the standard four-node element to the incompatible modes element:

i. Add shape functions N_5 and N_6 and their derivatives to the shape function subroutine. The derivatives of x and y with respect to ξ and η need to be evaluated at $\xi = \eta = 0$ in these calculations as per the Taylor modification in version 2.

ii. The $\boldsymbol{B}$ matrix needs to be enlarged to account for the incompatible modes. With particular reference to implementations 2 and 3 of Sec. 3.10, increase the row/column index of the element nodal do loops (i.e., NEN) from four to six to accommodate the incompatible modes.

iii. Statically condense the stiffness.

4. The generalization to three dimensions is straightforward.

5. The Wilson-Taylor element is used in several large linear structural analysis codes. Its use in nonlinear analysis proves somewhat awkward and thus it does not appear to be widely used in the nonlinear regime.

Exercise 1 (Linear Nonconforming Triangle). Consider a straight-edged triangle with three nodes located at the midsides (see Fig. 4.7.4). Assume the following linear shape functions:

$$N_1(r, s, t) = 1 - 2r \tag{4.7.21}$$

$$N_2(r, s, t) = 1 - 2s \tag{4.7.22}$$

$$N_3(r, s, t) = 1 - 2t \tag{4.7.23}$$

where r, s, t are triangular coordinates (see Appendix 3.I). Note that the shape functions are nonconforming. (Sketch them!)

a. Show that $\int_{\Omega^e} N_a N_b \, d\Omega = 0$ if $a \neq b$. (This orthogonality property is very useful in dynamics.)

b. Along with (4.7.21) through (4.7.23), assume a constant pressure field. Show that the incompressibility constraint ratio is 3. (Thus the element does not lock. It has been used successfully in incompressible problems.)

c. Derive shape functions for the analogous tetrahedron with nodes on the centers of the faces.

These elements have a number of other interesting properties. See Thomasset [71] for additional details. Unfortunately, the practical usefulness of elements of this type in elasticity is limited by the fact that they can give rise to spurious mechanisms as illustrated in Fig. 4.7.5. Element *II* is only attached to element *I* at node *A*. Therefore element *II* can rigidly rotate about node *A*.

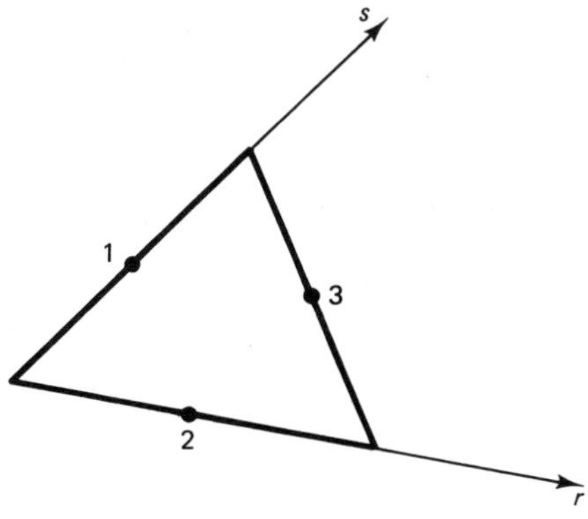

Figure 4.7.4

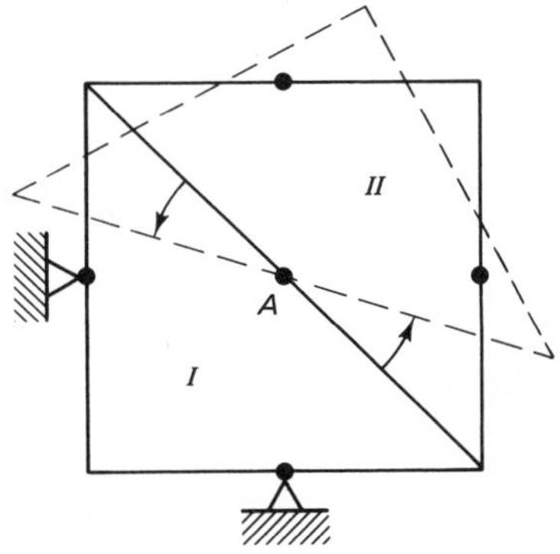

Figure 4.7.5

4.8 HOURGLASS STIFFNESS

Use of one-point quadrature to formulate the four-node bilinear element stiffness is quite economical (it requires approximately one-fourth the calculational effort of the standard 2 × 2 rule), but it results in rank deficiency as we observed in Sec. 4.6. The two modes that are rendered singular are the so-called hourglass patterns illustrated in Fig. 4.6.2. Kosloff and Frazier [72] have presented a very simple way to reintroduce the stiffness of the hourglass modes, which has two important attributes: (1) The superior bending behavior of the (modified) incompatible modes element is obtained;[15] (2) the element formulation costs are only a fraction greater than the one-point quadrature element and therefore are much less expensive than the incompatible modes and standard bilinear elements. The stiffness for the Kosloff-Frazier element consists of the one-point quadrature stiffness matrix plus an hourglass stiffness matrix which is constructed from the exact pure bending modes (see Fig. 4.7.1). The derivation proceeds as follows:

First, consider the exact stress and displacement field illustrated in Fig. 4.7.1. The nodal displacements may be written as

$$d_{1a} = \frac{-\sigma_0}{4c} \text{ vol } (\Omega^e)\phi_a \tag{4.8.1}$$

$$d_{2a} = 0 \tag{4.8.2}$$

$$\phi_a = (-1)^a \qquad \text{(hourglass mode)} \tag{4.8.3}$$

where $a = 1, 2, 3, 4$, and vol $(\Omega^e) = h_1 h_2$. Recall that $c = E$ in plane stress, and $c = 4\mu(\lambda + \mu)/(\lambda + 2\mu)$ in plane strain. By taking the dot product of (4.8.1) with

[15] In the isotropic case, in which the material properties are also constant, the Kosloff-Frazier element is identical to the incompatible modes element.

ϕ_a we can obtain the following explicit expression for σ_0:

$$\sigma_0 = -\frac{c}{\text{vol }(\Omega^e)}\left(\sum_{a=1}^{4} \phi_a d_{1a}\right) \tag{4.8.4}$$

The quantity $\Sigma_{a=1}^{4} \phi_a d_{1a}$ in (4.8.4) is a measure of the amplitude of the hourglass mode.

Next we wish to calculate the internal force vector at node a due to the given stress field:

$$\begin{aligned} \boldsymbol{f}_a^{\text{int}} &= \int_{\Omega^e} \boldsymbol{B}_a^T \boldsymbol{\sigma}\, d\Omega = \int_{\Omega^e} \begin{bmatrix} N_{a,x} & 0 & N_{a,y} \\ 0 & N_{a,y} & N_{a,x} \end{bmatrix} \begin{Bmatrix} \sigma_0 y \\ 0 \\ 0 \end{Bmatrix} d\Omega \\ &= \sigma_0 \int_{\Omega^e} \begin{Bmatrix} yN_{a,x} \\ 0 \end{Bmatrix} d\Omega \\ &= \begin{Bmatrix} \dfrac{c \text{ vol } (\Omega^e)}{48} (\xi_{,x})^2 \phi_a \displaystyle\sum_{b=1}^{4} \phi_b d_{1b} \\ 0 \end{Bmatrix} \end{aligned} \tag{4.8.5}$$

Exercise 1. Fill in the details to obtain (4.8.5).

An analogous calculation for the bending stress defined by

$$\sigma_{11} = \sigma_{12} = 0, \qquad \sigma_{22} = \sigma_0 x \tag{4.8.6}$$

leads to the following expression for internal force:

$$\boldsymbol{f}_a^{\text{int}} = \int_{\Omega^e} \boldsymbol{B}_a^T \boldsymbol{\sigma}\, d\Omega = \begin{Bmatrix} 0 \\ \dfrac{c \text{ vol } (\Omega^e)}{48} (\eta_{,y})^2 \phi_a \displaystyle\sum_{b=1}^{4} \phi_b d_{2b} \end{Bmatrix} \tag{4.8.7}$$

The next step is to combine (4.8.5) and (4.8.7), and generalize to the case in which the rectangular element is not aligned with the global x- and y-axes. After some algebraic manipulations the following expressions are obtained:

$$\boldsymbol{f}_a^{\text{int}} = \{f_{ia}^{\text{int}}\} \tag{4.8.8}$$

$$f_{ia}^{\text{int}} = s_{ij} \phi_a \sum_{b=1}^{4} \phi_b d_{jb} \qquad (\text{sum on } j = 1, 2) \tag{4.8.9}$$

$$s_{ij} = \frac{c \text{ vol } (\Omega^e)}{48} (\xi_{,x_i} \xi_{,x_j} + \eta_{,x_i} \eta_{,x_j}) \tag{4.8.10}$$

where $x_1 = x$ and $x_2 = y$. Clearly, (4.8.9) reduces to the previous formulas under the appropriate restrictions. In general, (4.8.10) is evaluated at $\xi = \eta = 0$.

Exercise 2. Derive (4.8.9) and (4.8.10).

From (4.8.9) we may immediately deduce the formula for the 8 × 8 hourglass stiffness:

$$\boldsymbol{k}^{\text{hourglass}} = [\boldsymbol{k}_{ab}^{\text{hourglass}}], \qquad 1 \leq a, b \leq 4 \tag{4.8.11}$$

$$\boldsymbol{k}_{ab}^{\text{hourglass}} = \phi_a \phi_b \boldsymbol{s} \tag{4.8.12}$$

where

$$\boldsymbol{s} = [s_{ij}], \qquad 1 \leq i, j \leq 2 \tag{4.8.13}$$

The final step in the derivation is to extend to the case of an arbitrary quadrilateral geometry. In order that the patch test be passed, the definition of the hourglass mode, (4.8.3), needs to be generalized. The requirement set forth by Kosloff is that the generalized hourglass mode be *orthogonal* to an arbitrary linear polynomial.[16] This may be expressed in the following way in terms of a "trial" hourglass mode $\widetilde{\phi}_a$:

$$\sum_{a=1}^{4} \widetilde{\phi}_a = 0 \tag{4.8.14}$$

$$\sum_{a=1}^{4} \widetilde{\phi}_a x_{1a} = 0 \tag{4.8.15}$$

$$\sum_{a=1}^{4} \widetilde{\phi}_a x_{2a} = 0 \tag{4.8.16}$$

Because the system (4.8.14) through (4.8.16) is underdetermined, we need to add an additional condition. For this purpose let us specify $\widetilde{\phi}_4 = 1$. Then (4.8.14) through (4.8.16) may be solved for $\widetilde{\phi}_1$, $\widetilde{\phi}_2$, and $\widetilde{\phi}_3$. In addition to the previously given orthogonality condition, ϕ_a needs to satisfy

$$\sum_{a=1}^{4} \phi_a^2 = 4 \tag{4.8.17}$$

which was used in the derivation of (4.8.4). Thus the desired ϕ_a may be obtained by normalizing $\widetilde{\phi}_a$:

$$\phi_a = 2\widetilde{\phi}_a \bigg/ \left(\sum_{a=1}^{4} \widetilde{\phi}_a^2 \right)^{1/2} \tag{4.8.18}$$

Equation (4.8.18) is then used in (4.8.9) and (4.8.12) to define the hourglass mode.

Remarks

1. The generalization to three dimensions is described in Kosloff and Frazier [72]. Several examples which illustrate the effectiveness of the hourglass stiffness elements are presented in [72].

[16] This guarantees that the generalized hourglass mode will not be aroused by rigid body modes and constant strain states, i.e., the patch test will be satisfied.

2. Because of the inherent economy of the hourglass stiffness approach, considerable attention has been drawn to it. Among the recent developments given by Flanagan and Belytschko [73] and Belytschko et al. [74] are techniques for developing ϕ_a for two- and three-dimensional problems without solving any equations; the fundamental idea is presented in the following exercise.

Exercise 3. (T. Belytschko)

a. Let

$$b_{1a} = \frac{\partial N_a}{\partial x_1}$$

$$b_{2a} = \frac{\partial N_a}{\partial x_2}$$

Show that

$$\sum_{a=1}^{4} b_{ia} x_{ja} = \delta_{ij}$$

Hint: Start with $x_{i,j} = \delta_{ij}$ and use the isoparametric expansion for x_i.

b. Let

$$s_a = 1, \qquad h_a = (-1)^a$$

Consider the following expression for $\widetilde{\phi}_a$:

$$\widetilde{\phi}_a = a_1 b_{1a} + a_2 b_{2a} + a_3 s_a + h_a$$

where the a_i's are arbitrary constants. Using equations (4.8.14) through (4.8.16) and part (a), show that

$$\widetilde{\phi}_a = h_a - \sum_{b=1}^{4} [(h_b x_{1b}) b_{1a} + (h_b x_{2b}) b_{2a}]$$

Note that this provides a direct solution for the "trial" hourglass mode; see [74] for application to three-dimensional problems.

Notes: In the papers of Belytschko and his colleagues, the terminology ***hourglass mode*** is reserved for $h_a = (-1)^a$ regardless of the shape of the element, and $\widetilde{\phi}_a$ is referred to as the ***hourglass stabilization*** (***or suppression***) ***operator***. The notation γ_a is used in place of $\widetilde{\phi}_a$ by Belytschko.

4.9 ADDITIONAL EXERCISES AND PROJECTS

Exercise 1. Consider the two-dimensional plane strain beam, Example 1 of Sec. 4.4. The exact solution of this problem is easily derived and is given as follows:

$$\sigma_{11} = -\frac{P\widetilde{x}y}{I}$$

$$\sigma_{22} = 0$$

$$\sigma_{12} = \frac{P}{2I}(c^2 - y^2)$$

$$\frac{6E^1 I}{P} u_1(x, y) = -y\{3(L^2 - \tilde{x}^2) + (2 + \nu^1)(y^2 - c^2)\}$$

$$\frac{6E^1 I}{P} u_2(x, y) = (\tilde{x}^3 - L^3) - \{(4 + 5\nu^1)c^2 + 3L^2\}(\tilde{x} - L) + 3\nu^1 \tilde{x} y^2$$

where $\tilde{x} = L - x$ and

	E^1	ν^1
Plane stress	E	ν
Plane strain	$\frac{E}{1 - \nu^2}$	$\frac{\nu}{1 - \nu}$

Obtain approximate solutions of the above elasticity problem by using the program DLEARN.

- Employ the following data in your calculations:

  $P = -1, \quad L = 16, \quad c = 2, \quad E = 1, \quad \nu = 0.3$ and 0.499

  (The latter value of Poisson's ratio corresponds to the "nearly incompressible case.")
- The mesh to be used is depicted in Fig. 4.4.2. You will need to compute nodal forces corresponding to the traction boundary conditions along $x = 0$ and $x = L$.
- Assume plane strain conditions.
- Use four-node quadrilaterals and employ each of the following quadrature rules:
  a. 2×2
  b. $\bar{\boldsymbol{B}}$ with 2×2
  c. one-point
- Use three-node triangles in the
  d. bisection pattern:

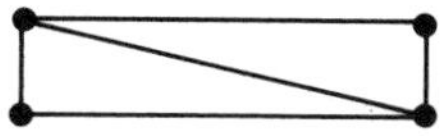

  e. cross-diagonal pattern:

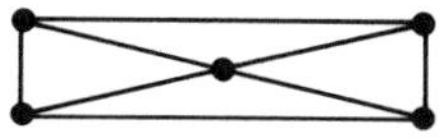

  (There are 10 cases in all.)
- In each case, compare the computed vertical tip displacement (i.e., $u_2(16, 0)$), the horizontal displacement of the root section (i.e., $u_1(0, y)$, $y \in [-c, c]$) and the bending and shear stress (i.e., σ_{11} and σ_{12}, respectively) for $x = 7$, $y \in [-c, c]$, with the exact solution.
- Write a brief report summarizing your findings and include an evaluation of the performance of each element.

Instructions for Using DLEARN

Consult the DLEARN User's Manual in Chapter 11 regarding set up. The data file for this problem, case (a), is presented in Table 11.5.2. Make sure you completely understand it. In particular, verify the consistent nodal forces in the data file by performing the calculations by hand. Your first run should be a data check run (i.e., IEXEC.EQ.0). Carefully check the output to ensure that you have input your data correctly. Keep using the data check mode until you are satisfied that your input data is correct. Then set IEXEC.EQ.1, which will cause the program to execute.

Run case (a) first and check the vertical tip displacement normalized by the exact value against the results presented in Table 4.4.1. If the comparison is satisfactory, proceed to the other cases. Otherwise, back to the drawing board—you've done something wrong!

Exercise 2. Consider the patch of four-node quadrilateral elements shown.

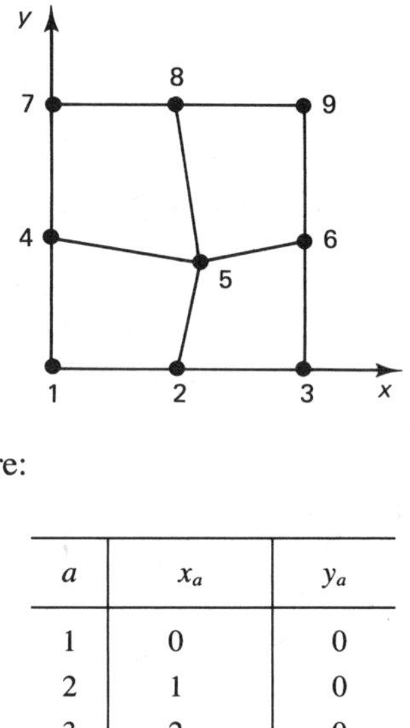

The nodal coordinates are:

a	x_a	y_a
1	0	0
2	1	0
3	2	0
4	0	1
5	1.1	0.8
6	2	1
7	0	2
8	1	2
9	2	2

Use DLEARN to execute the following displacement boundary-value problems ($1 \leq a \leq 4$, $6 \leq a \leq 9$):

Test number	d_{1a}	d_{2a}
1	1	0
2	0	1
3	x_a	0
4	0	x_a
5	y_a	0
6	0	y_a

Assume the plane strain option and use the following data: $E = 1.0$, $\nu = 0.3$. Test the following options:

a. 2×2 quadrature
b. $\bar{\boldsymbol{B}}$ with 2×2 quadrature
c. one-point quadrature

Based upon the results you obtain, determine whether or not each element passes the patch test. (Consider results "exact" if they are correct to one-half or more of the number of significant digits employed. Due to rounding errors, we cannot expect truly exact results on digital computers.)

Exercise 3. The rank check option in DLEARN can be used to determine the number of zero eigenvalues of the global stiffness matrix $\boldsymbol{K}$. That is, if IRANK.EQ.2, the diagonal elements of the diagonal matrix $\boldsymbol{D}$ in the factorization $\boldsymbol{K} = \boldsymbol{U}^T\boldsymbol{D}\boldsymbol{U}$, where $\boldsymbol{U}$ is an upper triangular matrix, are printed. The number of zeros in $\boldsymbol{D}$ equals the number of zero eigenvalues of $\boldsymbol{K}$, which corresponds to the number of zero-energy modes of the structure in question. Normally, there are, at most, three zero-energy modes—the rigid body modes. Consider the single element shown:

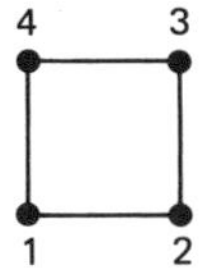

a	x_a	y_a
1	0	0
2	1	0
3	1	1
4	0	1

Assume the plane strain option and use the following data: $E = 1.0$ and $\nu = 0.3$. Note there are *no* kinematic boundary conditions. Perform rank check analyses for the following options:

a. 2×2 quadrature
b. $\bar{\boldsymbol{B}}$ with 2×2 quadrature
c. One-point quadrature

Comment on your results.

Exercise 4. It has been proposed to use C^0 triangular elements with complete quintic (i.e., fifth-order) polynomials for problems of two-dimensional elasticity. Determine the order of accuracy (i.e., a and b) in the following:

$$\|\boldsymbol{e}\|_0 \leq ch^a \|\boldsymbol{u}\|_{k+1}$$

$$\|\boldsymbol{e}\|_1 \leq ch^b \|\boldsymbol{u}\|_{k+1}$$

Justify your answers.

Exercise 5. A series of convergence rate studies on linear elastic boundary-value problems was performed. All results indicate that $\boldsymbol{e} = \boldsymbol{u}^h - \boldsymbol{u}$ behaves as follows:

$$\|e\|_1 = O(h^2)$$

which of the following elements was used in the studies?

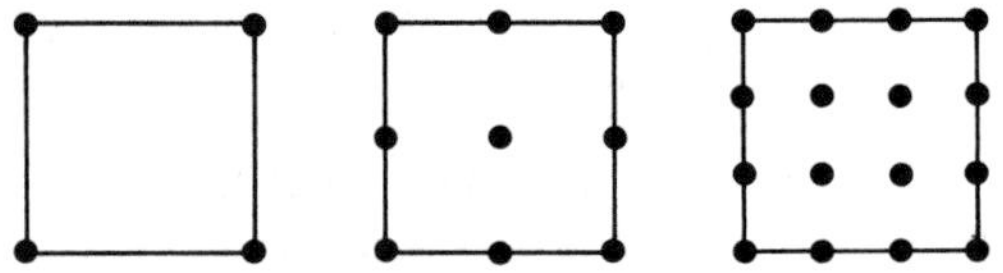

Justify your answer.

Exercise 6. Consider the following problem:

$$K\begin{bmatrix} 1 & \epsilon \\ \epsilon & 1 \end{bmatrix} \begin{Bmatrix} d_1 \\ d_2 \end{Bmatrix} = \begin{Bmatrix} F_1 \\ F_2 \end{Bmatrix}$$

subject to the constraint

$$d_1 + d_2 = 0$$

Assume $K = 1$ and $\epsilon = 0$.

a. Formulate the matrix problem by way of the Lagrange-multiplier method. Determine d_1, d_2, and the Lagrange multiplier.
b. Formulate the matrix problem by way of the penalty function approximation. Determine the approximations of d_1, d_2, and the Lagrange multiplier as functions of the penalty parameter.

Exercise 7. Consider the linear system $\boldsymbol{Kd} = \boldsymbol{F}$. It is desired to modify the system so that the following constraint is maintained:

$$\alpha d_P + \beta d_Q = \gamma$$

where α, β, and γ are given constants, and $1 \leq P < Q \leq n_{eq}$. Develop modified equation systems by

a. The Lagrange multiplier method
b. The penalty method

Exercise 8. Determine the constraint ratios for the following elements.

a. The "enriched" bilinear displacements–constant pressure quadrilateral shown, in which normal-displacement degrees of freedom are added to each edge:

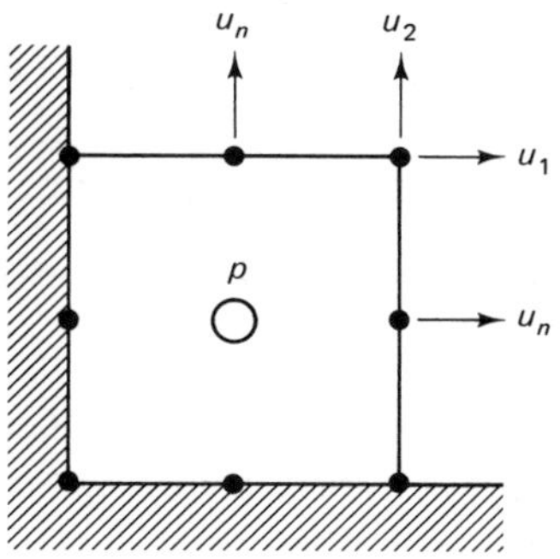

b. The "enriched" trilinear displacements–constant pressure brick in which normal-displacement degrees of freedom have been added to each face. This is the three-dimensional analog of the element in part (a).

Exercise 9. Calculate the constraint ratio for the accompanying macroelement. (This macroelement was developed by Patrick Le Tallec and was shown to converge.)

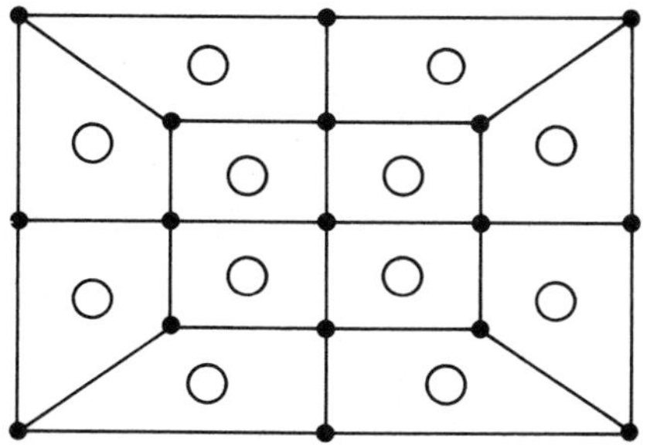

Exercise 10. To determine if a new finite element for two-dimensional heat conduction is convergent, a patch test is performed. The results are as follows:

a	x_a	y_a	d_a Problem 1	d_a Problem 2	d_a Problem 3
1	1.0	0.0	0.0	1.0	1
2	2.0	0.0	0.0	2.0	1
3	3.0	0.0	0.0	3.0	1
4	1.0	1.0	1.0	1.0	1
Interior node → 5	2.1	1.2	1.2	2.0	1
6	3.0	1.0	1.0	3.0	1
7	1.0	3.0	3.0	1.0	1
8	2.0	3.0	3.0	2.0	1
9	3.0	3.0	3.0	3.0	1

Nodes 1 through 4 and 6 through 9 are boundary nodes.
Does the element pass the patch test? Justify your answer.

Exercise 11. Consider the following patch of two-dimensional elasticity elements. Define six prescribed-displacement, boundary-value problems by setting the boundary nodes (i.e., 1 to 4) to values in accord with linear monomials. Determine the solution at nodes 5 to 8 that must be attained in order that the patch test is passed. The coordinates of the nodes are:

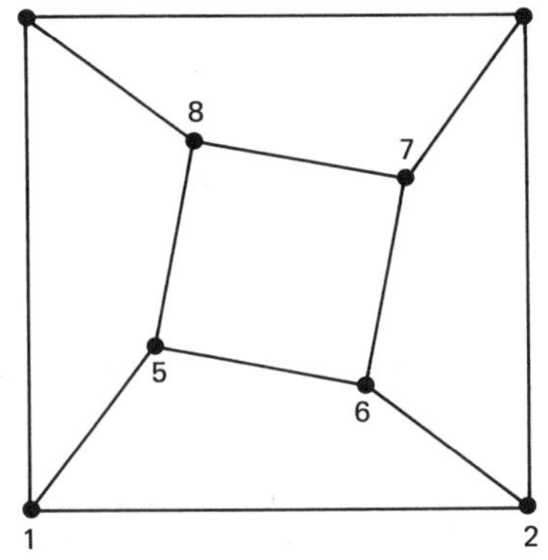

a	x_a	y_a
1	0	0
2	3	0
3	3	3
4	0	3
5	0.75	1
6	2	0.75
7	2.25	2
8	1	2.25

Projects

Various efforts have been made to design four-node quadrilateral elements with improved bending behavior. The Wilson, Taylor, and Kosloff-Frazier elements, discussed in the previous two sections, are prime examples. One of the first attempts employed selective integration [75]. The element was successful in the rectangular configuration if the rectangle was aligned with the global coordinate system but behaved poorly otherwise. Kavanagh and Key [76] traced the erratic behavior to a lack of "invariance" of the formulation. This defect can be easily corrected and is described next. The corrected element is now used in several structural analysis computer programs. A local rectangular Cartesian coordinate system (i.e., x', y') may be uniquely defined by constructing it so that its axes make equal angles with the natural ξ, η-system. Coordinates are rotated into the x', y'-system and the element stiffness and force vector are calculated with respect to x', y'. Before assembly, the stiffness and force are rotated into global coordinates. The improved bending behavior is facilitated by the use of a one-point quadrature on the shear term in the x', y'-system, namely,

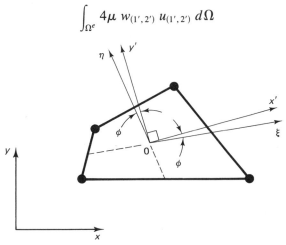

$$\int_{\Omega^e} 4\mu \, w_{(1',2')} \, u_{(1',2')} \, d\Omega$$

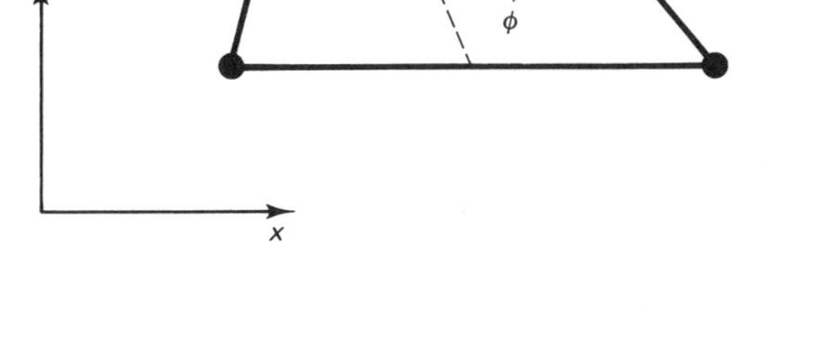

where the $1'$, $2'$-subscripts indicate differentiation with respect to x', y'-coordinates, respectively. The remaining stiffness terms may be integrated with the normal 2×2 rule. These terms may be written as

$$\int_{\Omega^e} \left(\lambda \, w_{(i',j')} \, u_{(i',j')} + 2\mu (w_{(1',1)} \, u_{(1',1')} + w_{(2',2')} \, u_{(2',2')}) \right) d\Omega$$

The reason for using the x', y'-system should be apparent: The shear term is not invariant with respect to change of reference frame. Thus use of a local preferred system, independent of the global x, y-system, is necessitated.

Project 1. Develop a finite element subroutine in DLEARN for the element just described. Check out your element by performing a rank check (correct rank of 5 should be attained), patch tests, and several sample problems. For example, perform a comparison of the element's behavior with other DLEARN elements on the beam bending problem considered previously. Make sure that you have correctly programmed the transformation between the x, y- and x', y'-systems. Print out the coordinates of the element in both systems for an irregular element configuration and check by hand. Perform a pure-bending single-element test (see Fig. 4.7.1) with a rectangular element aligned with the x, y-system and with one skew to it. The same results should be attained if transformed to a common frame. These tests will check the invariance of the formulation.

Project 2. A criticism that may be lodged at the previous element is its obvious inability to handle incompressible phenomena. This is due to the 2×2 Gauss quadrature treatment of the λ-term. A modified version designed to alleviate the incompressible locking problem would involve one-point quadrature on

$$\int_{\Omega^e} (\lambda \, w_{(i,j)} \, u_{(i,j)} + 4\mu \, w_{(1',2')} \, u_{(1',2')}) \, d\Omega$$

and 2×2 quadrature on the remainder, namely,

$$\int_{\Omega^e} 2\mu (w_{(1',1')} \, u_{(1',1')} + w_{(2',2')} \, u_{(2',2')}) \, d\Omega$$

Program this element and perform the check-out tests described under Project 1. In addition, assess the behavior in nearly incompressible situations by performing calculations in which the ratio λ/μ is varied. Use the plane strain two-dimensional beam problem as an example and compare your results with those presented in Fig. 4.7.3.

Project 3. The previous two projects involve elements that are applicable only to the isotropic case. For anisotropic applications, we may use the $\bar{\boldsymbol{B}}$-formulation described in Sec. 4.5. Set up the $\bar{\boldsymbol{B}}$ matrices corresponding to the selective integration schemes of Projects 1 and 2. In place of one-point quadrature treatment, use the element mean value in the definition of $\bar{\boldsymbol{B}}$. This will cause a difference only in the axisymmetric configuration. (The difference is beneficial; see discussion in Sec. 4.5.1.) Program the various $\bar{\boldsymbol{B}}$-elements and evaluate as described under Projects 1 and 2.

Appendix 4.1

Mathematical Preliminaries

In posing the variational form of the boundary-value problem, we found it useful to define collections of functions that satisfied certain conditions. This leads us to the study of ***linear spaces***. We begin with a description of the important properties of these spaces essential for finite element error analysis.

4.1.1 BASIC PROPERTIES OF LINEAR SPACES

We do not wish to discuss such dry and uninteresting material as the "axioms of a linear space." Basically, a ***real*** (as in real numbers) ***linear space*** is a collection of objects for which the operations of addition and scalar multiplication are defined and behave as follows: If x and y are members of the linear space and α and β are scalars, then $\alpha x + \beta y$ is also a member of the linear space. Because of this property we sometimes say a linear space is ***closed*** under addition and scalar multiplication.

Example 1 ($\mathbb{R}^n$ **Is a Linear Space**)

$\mathbb{R}^n$ consists of n-tuples (i.e., *points*) $x = (x_1, x_2, \ldots, x_n)$, where each $x_i \in \mathbb{R}$, $1 \leq i \leq n$. Addition means component by component addition:

$$x + y = (x_1, x_2, \ldots, x_n) + (y_1, y_2, \ldots, y_n)$$
$$= (x_1 + y_1, x_2 + y_2, \ldots, x_n + y_n)$$

Scalar multiplication is defined in the usual way: Let $\alpha \in \mathbb{R}$; then

$$\alpha x = (\alpha x_1, \alpha x_2, \ldots, \alpha x_n)$$

To show $\mathbb{R}^n$ is a linear space we must show $\alpha x + \beta y$ is in $\mathbb{R}^n$, where $x, y \in \mathbb{R}^n$, and α, $\beta \in \mathbb{R}$.

$$\alpha x + \beta y = (\alpha x_1, \alpha x_2, \ldots, \alpha x_n) + (\beta y_1, \beta y_2, \ldots, \beta y_n)$$
$$= (\alpha x_1 + \beta y_1, \alpha x_2 + \beta y_2, \ldots, \alpha x_n + \beta y_n)$$

The last result is an n-tuple of real numbers. Hence

$$\alpha x + \beta y \in \mathbb{R}^n$$

Linear spaces often come with important additional structure; namely, *inner products* and *norms*.

Definition. An ***inner product*** $\langle \cdot, \cdot \rangle$ on a real linear space A is a map $\langle \cdot, \cdot \rangle$: $A \times A \to \mathbb{R}$ (i.e., $\langle \cdot, \cdot \rangle$ assigns to an ordered pair $x, y \in A$ a real number denoted $\langle x, y \rangle$) with the following properties.

Let $x, y, z \in A$ and $\alpha \in \mathbb{R}$; then

i. $\langle x, y \rangle = \langle y, x \rangle$ (symmetry)

ii. $\langle \alpha x, y \rangle = \alpha \langle x, y \rangle$
iii. $\langle x + y, z \rangle = \langle x, z \rangle + \langle y, z \rangle$ (linearity)

iv. $\langle x, x \rangle \geq 0$, and
$\langle x, x \rangle = 0$ if and only if $x = 0$. (positive-definiteness)

Example 2

Without getting technical about the spaces involved, it can be easily shown that, under suitable hypotheses, all the $a(\cdot, \cdot)$, $(\cdot, \cdot)$, and $(\cdot, \cdot)_\Gamma$ encountered in previous chapters satisfy (i)–(iv) and therefore are inner products. Recall that the positive definiteness of $a(\cdot, \cdot)$ follows from positivity requirements on the constitutive tensor of the field theory in question *and* the homogeneous, essential (i.e., "kinematic") boundary conditions of the collection of functions under consideration. The latter condition is seen to be necessary. For example, suppose v, w: $[0, 1] \to \mathbb{R}$ and $a(w, v) = \int_0^1 w_{,x} v_{,x} dx$. Then $a(\cdot, \cdot)$ is positive-definite on the collection $H_0^1 = \{w \mid w \in H^1, w(1) = 0\}$ but not on the collection H^1. The function $u(x) = \text{constant} \neq 0$ is in H^1 and $a(u, u) = 0$.

Definition. Let $\{A, \langle \cdot, \cdot \rangle\}$ be an inner product space (i.e., a linear space A with an inner product $\langle \cdot, \cdot \rangle$ defined on A). Then $x, y \in A$ are said to be ***orthogonal*** or ***perpendicular*** (with respect to $\langle \cdot, \cdot \rangle$) if $\langle x, y \rangle = 0$.

Schwarz inequality. A key property of inner products is the ***Schwarz inequality***, which states that for all x and y

$$\langle x, y \rangle^2 \leq \langle x, x \rangle \langle y, y \rangle$$

The proof of this fact follows directly from the definition of an inner product, and is given as follows.

Proof. Let $\alpha \in \mathbb{R}$. Then by definition

$$\begin{aligned} 0 &\le \langle x + \alpha y, x + \alpha y\rangle \\ &= \langle x, x\rangle + 2\alpha\langle x, y\rangle + \alpha^2\langle y, y\rangle \end{aligned}$$

The last expression is a nonnegative quadratic in α. Hence, its discriminant is nonpositive, i.e.,

$$\langle x, y\rangle^2 - \langle x, x\rangle\langle y, y\rangle \le 0$$ ■

Definition. A ***norm*** $\|\cdot\|$ on a linear space A is a map $\|\cdot\|: A \to \mathbb{R}$, with the following properties.

Let $x, y \in A$ and $\alpha \in \mathbb{R}$; then

i. $\|x\| \ge 0$, and $\|x\| = 0$ if and only if $x = 0$ (positive-definiteness)

ii. $\|\alpha x\| = |\alpha|\,\|x\|$

iii. $\|x + y\| \le \|x\| + \|y\|$ (triangle inequality)

Note that an inner product space $\{A, \langle\cdot, \cdot\rangle\}$ possesses a ***natural norm*** defined by

$$\|x\| = \langle x, x\rangle^{1/2}$$

Exercise 1. Verify that the definition of a norm is satisfied by the above expression. *Hint:* Use the Schwarz inequality to prove the triangle inequality holds.

Example 3

$\mathbb{R}^n$ equipped with the dot product, defined by $\langle x, y\rangle = x \cdot y$, is an inner product space. The natural norm $\|x\| = (x \cdot x)^{1/2}$ is the length of the vector with tail at the origin and tip at x.

Example 4

$L_2(]0, 1[)$, equipped with the inner product $\langle u, v\rangle = \int_0^1 uv\, dx$, is an inner product space. The natural norm, denoted by $\|\cdot\|_0$, is defined by $\|u\|_0 = (\int_0^1 u^2\, dx)^{1/2}$.

Example 5

$H^1(]0, 1[)$, equipped with the inner product $\langle u, v\rangle = \int_0^1 (uv + u_{,x}v_{,x})\, dx$, is an inner product space. The natural norm, denoted by $\|\cdot\|_1$, is defined by

$$\|u\|_1 = \left(\int_0^1 (u^2 + u_{,x}^2)\, dx\right)^{1/2}$$

Remarks

1. Each of the spaces in the preceding examples satisfies a technical condition called ***completeness***. Roughly speaking, completeness means that the limit of a convergent sequence of elements in the space is also in the space. An inner product space which is complete, in its natural norm, is called a ***Hilbert space***. Hilbert spaces have a rich mathematical theory and are particularly important to the study of linear partial differential equations. Examples 3 through 5 are Hilbert spaces. For the purposes we have in mind, we shall not need to delve further into this subject.

2. In terms of a natural norm, the Schwarz inequality takes the form

$$|\langle x, y\rangle| \leq \|x\| \, \|y\|$$

This is the form frequently used in applications.

Definition. A ***seminorm*** $|\cdot|$ on a linear space A is a map $|\cdot| : A \to \mathbb{R}$ with the following properties.

Let $x, y \in A$ and $\alpha \in \mathbb{R}$; then

i. $|x| \geq 0$ (positive-semidefiniteness)

ii. $|\alpha x| = |\alpha| \, |x|$

iii. $|x + y| \leq |x| + |y|$ (triangle inequality)

Remark

The only difference between a norm and seminorm is that a norm is positive-definite, whereas a seminorm is positive-semidefinite.

Example 6

$|u| = (\int_0^1 u_{,x} u_{,x} dx)^{1/2}$ defines a seminorm on $H^1(]0, 1[)$ but not a norm.

4.1.2 SOBOLEV NORMS

Consider a domain $\Omega \subset \mathbb{R}^n$, $n \geq 1$, and let $u, v : \Omega \to \mathbb{R}$.

The ***$L_2(\Omega)$ inner product and norm*** are defined by

$$(u, v) = \int_\Omega uv \, d\Omega$$

and

$$\|u\|_0 = (u, u)^{1/2}$$

respectively.

The ***$H^1(\Omega)$ inner product and norm*** are defined by

$$(u, v)_1 = \int_\Omega (uv + u_{,i} v_{,i}) \, d\Omega \qquad (\text{sum, } 1 \leq i \leq n)$$

and

$$\|u\|_1 = (u, u)_1^{1/2}$$

respectively.

Analogously, we may define the ***$H^s(\Omega)$ inner product and norm*** by

$$(u, v)_s = \int_\Omega (uv + u_{,i}v_{,i} + u_{,ij}v_{,ij} + \cdots + \underbrace{u_{,ij\cdots k}}_{s\text{ indices}}\underbrace{v_{,ij\cdots k}}_{s\text{ indices}})\, d\Omega$$

$$(\text{sum, } 1 \leq i, j, \ldots, k \leq n)$$

and

$$\|u\|_s = (u, u)_s^{1/2}$$

respectively.

The case in which $\boldsymbol{u}, \boldsymbol{v} : \Omega \to \mathbb{R}^m$ (i.e., are vector-valued) is essentially the same.

Let $H^s(\Omega, \mathbb{R}^m)$ be the space of $\mathbb{R}^m$-valued functions $\boldsymbol{u} = (u_1, u_2, \ldots, u_m)$, for which each $u_i \in H^s(\Omega)$, $1 \leq i \leq m$.

The ***$H^s(\Omega, \mathbb{R}^m)$ inner product and norm*** are defined by

$$(\boldsymbol{u}, \boldsymbol{v})_s = \int_\Omega (u_i v_i + u_{i,j}v_{i,j} + u_{i,jk}v_{i,jk} + \cdots + \underbrace{u_{i,jk\ldots l}}_{s\text{ indices}}\underbrace{v_{i,jk\ldots l}}_{s\text{ indices}})\, d\Omega$$

$$(\text{sum, } 1 \leq i \leq m,\ 1 \leq j, k, \ldots, l \leq n)$$

and

$$\|u\|_s = (\boldsymbol{u}, \boldsymbol{u})_s^{1/2}$$

respectively.

We will also find it useful to define the ***$H^r(\Omega, \mathbb{R}^m)$ seminorm***:

$$|\boldsymbol{u}|_r = \left(\int_\Omega \underbrace{u_{i,jk\ldots l}}_{r\text{ indices}}\underbrace{u_{i,jk\ldots l}}_{r\text{ indices}}\, d\Omega\right)^{1/2} \qquad (\text{sum, } 1 \leq i \leq m,\ 1 \leq j, k, \ldots, l \leq n)$$

Remarks

1. It simplifies the writing considerably if we agree to drop the arguments in $H^s(\Omega, \mathbb{R}^m)$, the particular Ω and $\mathbb{R}^m$ being understood by the context.

2. Note that $H^0 = L_2$.

Definition. The ***sth Sobolev space***, denoted by H^s, is the collection of functions $u : \Omega \to \mathbb{R}^m$ with finite H^s norm.

Remarks

1. Sobolev spaces are, in particular, Hilbert spaces.

2. The inclusion property

$$H^r \subset H^s, \qquad r \geq s$$

follows directly from the definitions.

Sobolev Imbedding Theorems

The most-celebrated theorems concerning Sobolev spaces go under the name of ***Sobolev imbedding theorems***. A particularly useful result of this kind tells us when functions in H^s are smooth in the classical sense.

Definition. Let C_b^k, $k \geq 0$, be the space of functions $\boldsymbol{u}:\Omega \to \mathbb{R}^m$ which are

i. Bounded
ii. Continuous and have continuous and bounded derivatives of order j, $1 \leq j \leq k$.

Theorem. If Ω is an open set in $\mathbb{R}^n$ (not a hypersurface!) and $s > n/2 + k$,

$$H^s \subset C_b^k$$ ■

Example 7

Suppose $\Omega =]0, 1[\subset \mathbb{R}$. Then functions in H^1 are continuous and bounded (i.e., are C_b^0 functions). Functions in H^2 are C_b^1 functions, etc.

Example 8

Suppose Ω is an open set in $\mathbb{R}^2$. Then $H^2 \subset C_b^0$.

Remarks

1. The imbedding result is "sharp" in the sense that if $s \leq n/2 + k$, then there exists a function in H^s which is *not* in C_b^k.

2. Note that the range space $\mathbb{R}^m$ plays *no* role in the imbedding theorem.

Exercise 2. Let $\Omega = \{x, y \in \mathbb{R}^2 \mid r = (x^2 + y^2)^{1/2} \leq \frac{1}{2}\}$. Show that $\log(\log r^{-1})$ is in $H^1(\Omega)$ but not in $C_b^0(\Omega)$.

4.1.3 APPROXIMATION PROPERTIES OF FINITE ELEMENT SPACES IN SOBOLEV NORMS

The fundamental error estimates for finite element approximations to solutions of linear elliptic boundary-value problems are stated in terms of Sobolev norms of the error. The intent of this section is to answer the following question of approximation.

Given a Sobolev space H^r and a finite element space $\mathcal{S}^h$, how well can we approximate a given $u \in H^r$ if we are allowed to pick any member of $\mathcal{S}^h$? The answer to this question is given in the work of Ciarlet and Raviart [77, 78] (and others; see references therein). The book by Oden and Reddy [79] has a nice account of these "approximability" theorems. Rather than get into the details of this work, which can be rather technical, we shall just state the end results in the form of a set of hypotheses about the finite element space under consideration. We shall try then to communicate the practical meaning of the hypotheses and point out which finite element spaces that we are apt to employ actually satisfy these hypotheses.

Consider a collection of finite element spaces $\{\mathcal{S}^h | h \in]0, D], D = \text{diameter } \Omega\}$. Each $\mathcal{S}^h$ is finite dimensional. If $h_1 < h_2$, we are to think of $\mathcal{S}^{h_1}$ as a "refinement" of $\mathcal{S}^{h_2}$, with the dimension of $\mathcal{S}^{h_1}$, being greater than that of $\mathcal{S}^{h_2}$.

Remark

$\mathcal{S}^h$ need not be defined for every $h \in]0, D]$. However, for each $\epsilon \in]0, D]$, $\mathcal{S}^h$ must be defined for an $h < \epsilon$. That is, we want to be able to take arbitrarily small values of h.

Definition

A collection of finite element spaces will be called ***k, m-regular*** (or simply ***regular***) if for each fixed h

i. $\mathcal{S}^h \subset H^m$.

ii. For every $u \in H^r$, $r \geq 0$, and for all s such that $0 \leq s \leq \min \{r, m\}$, there exist a $U^h \in \mathcal{S}^h$ and constant c, independent of u and h, such that

$$\|u - U^h\|_s \leq ch^\alpha \|u\|_r$$

where $\alpha = \min (k + 1 - s, r - s)$, the rate of convergence.

Remarks

1. The essence of (ii) is that the more refined the mesh, the better U^h is able to approximate u. It is also clear that higher derivatives are approximated less accurately than lower derivatives.

2. To establish that a collection of finite element spaces is k, m-regular requires a "uniformity" condition on the mesh refinements as $h \to 0$. For example, consider $\Omega \subset \mathbb{R}^n$ and $\overline{\Omega} = \cup_{e=1}^{n_{el}} \overline{\Omega}^e$.
Let

$$h^e = \text{diameter } \Omega^e$$

$$\rho^e = \text{diameter of largest sphere contained in } \Omega^e$$

$$h = \max_{1 \leq e \leq n_{el}} (h^e)$$

$$\rho = \min_{1 \leq e \leq n_{el}} (\rho^e)$$

$$\sigma = \frac{h}{\rho} \qquad \text{(aspect ratio)}$$

A sufficient uniformity condition to obtain all the above results is that $\sigma \leq \sigma_0$ for all $\mathcal{S}^h$. When this is the case, the collection of spaces is called ***quasi-uniform.*** As may be seen, quasi-uniformity prevents the aspect ratio from "blowing up" as the mesh is refined. For example, this precludes the mesh refinement from being one-dimensional, as illustrated in the accompanying figure.

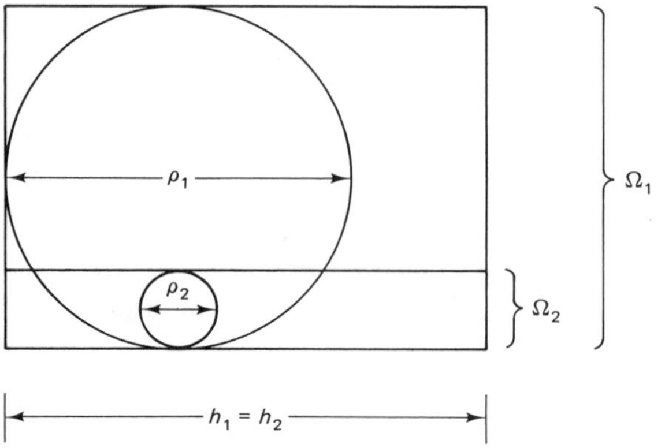

Without a condition such as quasi-uniformity, the best result typically obtainable takes the form (assuming $u \in H^{k+1}$)

$$\|u - U^h\|_s \leq C\frac{h^{k+1}}{\rho^s}\|u\|_{k+1}$$

The deleterious effect of "slivers" is apparent from the above relation. Quasi-uniformity allows us to replace the ratio h^{k+1}/ρ^s by h^{k+1-s} (with appropriate adjustment to the constant) and hence the definition of k, m-regular collections.

In the following examples, we give without proof some examples of k, m-regular collections of finite element spaces. (For further reading, see the papers of Ciarlet and Raviart [77, 78], and the book by Oden and Reddy [79].)

Example 9

Let $\mathcal{S}^h$ denote the piecewise linear finite element space on $]0, 1[$ consisting of N elements. The collection $\{\mathcal{S}^h \mid h = 1/N, N \text{ a positive integer}\}$ is 1, 1-regular. That is, for each $h = 1/N$

i. $\mathcal{S}^h \subset H^1$.

ii. For every $u \in H^2$ there exists a $U^h \in \mathcal{S}^h$ such that

$$\|u - U^h\|_1 \leq ch|u|_2$$

$$\|u - U^h\|_0 \leq ch^2|u|_2$$

where c is a constant independent of u and h. (If the given u is in H^1, but not in H^2, then instead of the above, we have only that $\|u - U^h\|_0 \leq ch|u|_1$.)

Example 10

The collection of piecewise quadratic finite element spaces on $]0, 1[$ defined in identical fashion to Example 9 is 2, 1 regular. Assuming $u \in H^3$, condition (ii) for this case becomes

$$\|u - U^h\|_1 \le ch^2\|u\|_3$$
$$\|u - U^h\|_0 \le ch^3\|u\|_3$$

If each "middle node" is located exactly at the midpoint of the element, then the norms on the right-hand side of the above may be replaced by seminorms.

Example 11

The collection of Hermite finite element spaces on $]0, 1[$, defined as in Example 9, is 3, 2-regular. Assuming $u \in H^4$, condition (ii) becomes

$$\|u - U^h\|_2 \le ch^2|u|_4$$
$$\|u - U^h\|_1 \le ch^3|u|_4$$
$$\|u - U^h\|_0 \le ch^4|u|_4$$

Example 12 (Isoparametric Elements)

a. Consider a collection of finite element spaces consisting of isoparametric elements on a domain $\Omega \subset \mathbb{R}^n$, $n \ge 1$. We assume there is no geometric error in modeling the boundary of the domain, i.e., $\overline{\Omega} \equiv \cup_{e=1}^{n_{el}} \overline{\Omega}^e$. Such a collection is *at least* 1, 1-regular. In particular

$$\|u - U^h\|_1 \le ch|u|_2$$
$$\|u - U^h\|_0 \le ch^2|u|_2$$

for every $u \in H^2$.

b. Assume the same conditions as in part (a). In addition, assume each isoparametric element mesh is such that its isoparametric transformation $\boldsymbol{x} = \boldsymbol{x}(\boldsymbol{\xi})$ is a linear polynomial in $\boldsymbol{\xi}$ and that the element shape functions are capable of exactly representing a complete kth-degree polynomial in $\boldsymbol{\xi}$. Then the collection is k, 1-regular. That is,

$$\|u - U^h\|_1 \le ch^k|u|_{k+1}$$
$$\|u - U^h\|_0 \le ch^{k+1}|u|_{k+1}$$

for every $u \in H^{k+1}$.

For example, a collection of spaces consisting of the nine-node quadrilaterals in rectangular, or parallelogram (see the accompanying figure), configurations is 2, 1-regular.

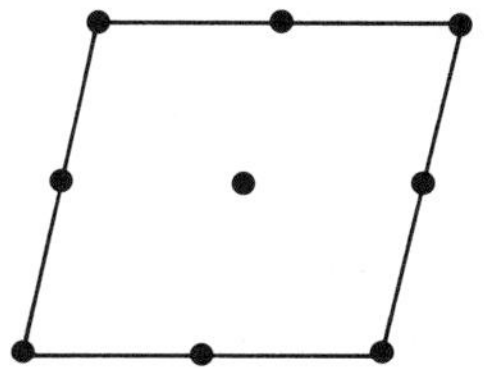

c. Assume the same conditions as in part (a). Consider the case in which the isoparametric functions are not necessarily linear polynomials in $\boldsymbol{\xi}$ but for which the deviation from linearity is "small."

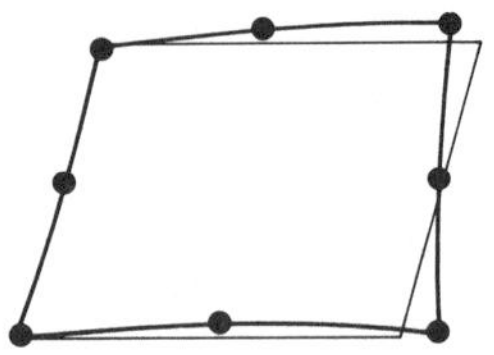

For example, the accompanying curved quadrilateral element illustrates what we mean by a small deviation. As in the previous case, if the element shape functions are capable of exactly representing a complete kth degree polynomial in $\boldsymbol{\xi}$ then the family of spaces in question is k, 1-regular, but here

$$\|u - U^h\|_1 \leq ch^k\|u\|_{k+1}$$

$$\|u - U^h\|_0 \leq ch^{k+1}\|u\|_{k+1}$$

That is, in the curved case the seminorms on the right-hand side of the estimates are replaced by norms. Thus we have the optimal interpolation estimates for curved elements, if they are not too distorted. However, as the magnitude of distortion is increased, the approximation capability of the elements deteriorates (see [79], p. 283). What happens is the "constant" c becomes progressively larger as the element is more and more distorted.

Remark

As is clear from the previous discussion, the main condition to ascertain in order to determine the approximation capability of a finite element space is the highest-degree complete polynomial that the element is capable of exactly representing. Values of k for some of the standard C^0-elements are given in the accompanying figures. As is clear, there are a multitude of k, 1-regular families of finite element spaces. Constructing k, m-regular families, $m \geq 2$, is possible but not as straightforward.

$k = 1$

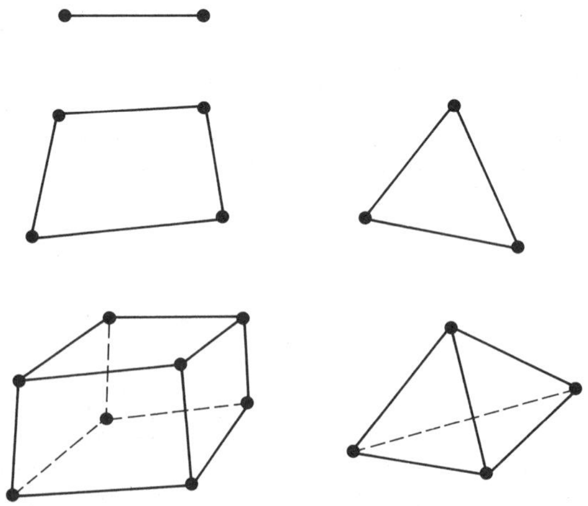

$k = 2$

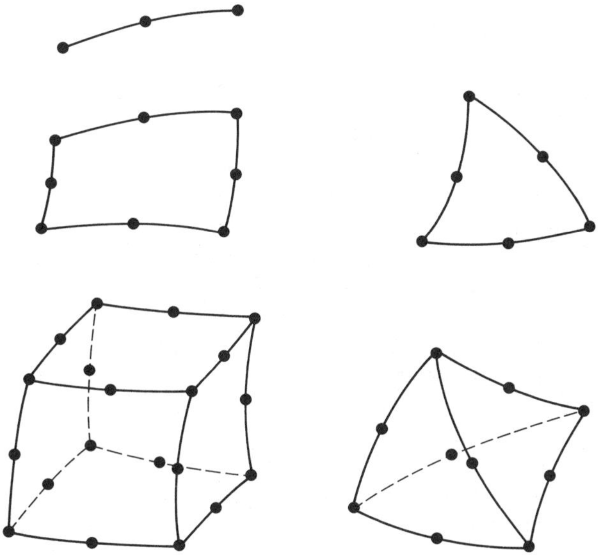

4.I.4 HYPOTHESES ON $a(\cdot, \cdot)$

All the problem classes considered thus far can be written in the form

$$a(\boldsymbol{w}, \boldsymbol{u}) = (\boldsymbol{w}, \boldsymbol{\ell}) + (\boldsymbol{w}, \boldsymbol{h})_\Gamma$$

This equation is to hold for all $\boldsymbol{w} \in \mathcal{V} \subset H_0^m$, for some $m \geq 1$, where the zero subscript on H_0^m indicates that certain homogeneous essential boundary conditions are built into the definition of H_0^m.

Definition. Two norms, $\|\cdot\|^{(1)}$ and $\|\cdot\|^{(2)}$, on a linear space A, are said to be ***equivalent*** if there exist constants $c_1, c_2 > 0$, such that for all $x \in A$,

$$c_1\|x\|^{(1)} \leq \|x\|^{(2)} \leq c_2\|x\|^{(1)}$$

Remarks

1. Our main hypothesis is that we assume $\|\cdot\|_m$ and $a(\cdot, \cdot)^{1/2}$ define equivalent norms on H_0^m, i.e., for all $\boldsymbol{w} \in H_0^m$

$$c_1\|\boldsymbol{w}\|_m \leq a(\boldsymbol{w}, \boldsymbol{w})^{1/2} \leq c_2\|\boldsymbol{w}\|_m$$

2. $a(\cdot, \cdot)$ is sometimes called the ***energy inner product,*** and the induced natural norm, as defined, is called the ***energy norm.***

The following examples are stated without proof.

Example 13

Consider $a(w, v) = \int_0^1 w_{,x}v_{,x}\, dx$. Then $\|\cdot\|_1$ and $a(\cdot, \cdot)^{1/2}$ define equivalent norms on

$H_0^1(]0, 1[)$. (Recall that the subscript zero is taken to mean that members of this space are zero at $x = 1$.)

Example 14

Consider the generalized heat conduction problem. Let

$$a(w, v) = \int_\Omega w_{,i} \kappa_{ij} v_{,j} \, d\Omega$$

where $\Omega \subset \mathbb{R}^n$, $1 \le i, j \le n$, and repeated indices are to be summed. The conductivities $\kappa_{ij} = \kappa_{ji}$ are assumed positive definite in the sense that

$$\kappa_{ij}(\boldsymbol{x}) y_i y_j \ge \delta y_i y_i$$

for a positive constant δ and all $y_i \in \mathbb{R}$, $1 \le i \le n$, and all $\boldsymbol{x} \in \Omega$. In addition, we require

$$\kappa_{ij}(\boldsymbol{x}) \le M, \qquad 1 \le i, j \le n$$

where M is some positive constant. Under these circumstances, $\|\cdot\|_1$ and $a(\cdot\,,\,\cdot)^{1/2}$ define equivalent norms on $H_0^1(\Omega)$.

Example 15

Consider classical elasticity theory. Let

$$a(\boldsymbol{w}, \boldsymbol{v}) = \int_\Omega w_{(i,j)} c_{ijkl} v_{(k,l)} \, dx$$

where $\Omega \subset \mathbb{R}^n$, $1 \le i, j, k, l \le n$, and repeated indices are to be summed. The hypotheses on the elastic coefficients read as follows: For all $\boldsymbol{x} \in \Omega$ and $\psi_{ij} = \psi_{ji}$, there exist positive constants δ and M such that

$$c_{ijkl}(\boldsymbol{x}) = c_{jikl}(\boldsymbol{x}) = c_{ijlk}(\boldsymbol{x}) = c_{klij}(\boldsymbol{x})$$

$$c_{ijkl}(\boldsymbol{x}) \psi_{ij} \psi_{kl} \ge \delta \psi_{ij} \psi_{ij}$$

$$c_{ijkl}(\boldsymbol{x}) \le M$$

Under these circumstances $\|\cdot\|$ and $a(\cdot\,,\,\cdot)^{1/2}$ are equivalent norms on $H_0^1(\Omega, \mathbb{R}^n)$.

Example 16

Consider the case of Bernoulli-Euler beam theory. Assume the domain is simply $]0, 1[$. Define

$$a(w, v) = \int_0^1 w_{,xx} v_{,xx} \, dx$$

Then $\|\cdot\|_2$ and $a(\cdot\,,\,\cdot)^{1/2}$ define equivalent norms on $H_0^2(]0, 1[)$. The subscript zero may denote any physically reasonable homogeneous, essential boundary conditions, e.g.,

$$w(0) = w(1) = 0 \qquad \text{(simply supported)}$$

$$w(0) = w(1) = w_{,x}(0) = w_{,x}(1) = 0 \qquad \text{(built in)}$$

$$\left.\begin{aligned} w(0) = w_{,x}(0) = 0 \\ w(1) = w_{,x}(1) = 0 \end{aligned}\right\} \qquad \text{(cantilever)}$$

etc.

We have asserted that all classes of boundary-value problems studied thus far, under suitable hypotheses on constitutive coefficients and boundary conditions, define energy norms equivalent to an appropriate Sobolev norm. To prove some of the above results is *not* so easy! The interested reader may profitably consult Mikhlin [80] and Fichera [81].

Appendix 4.II

Advanced Topics in the Theory of Mixed and Penalty Methods: Pressure Modes and Error Estimates

David S. Malkus

4.II.1 PRESSURE MODES, SPURIOUS AND OTHERWISE

There seems to be a perplexing variety of finite elements that are candidates for employment in problems with the incompressibility constraint. As has been shown earlier, each element seems to have drawbacks of some kind, and none seems to be ideally suited to such problems. Consider the case of the bilinear displacement–constant pressure isoparametric element: It has an "optimal" constraint ratio; it is a simple element, well suited to nonlinear problems because of relatively low (re-)assembly cost. However, until recently, no error estimates could be derived for this element in the incompressible case, even for the simplest meshes. Fortin [82] discovered the spurious pressure modes associated with this element and observed that state-of-the-art methods for obtaining error estimates could not be applied to the element. Even more disturbing was the development of simple, physically reasonable test problems in which the element gave obviously meaningless results [83]. Even though the bilinear displacement–constant pressure element was widely used in practice and seemed to be effective in many important physical applications, the seeds of doubt were sown.

Many investigators tended to shy away from using any element with spurious modes and opted for elements that were "safe" but had obvious drawbacks. A common feature of the "safe" elements is that many seem to have a limiting constraint ratio r, larger than two, whereas many of the elements with $r = 2$ exhibit spurious modes—at least on some types of meshes (usually very simple ones), with some types of boundary conditions. Many of the safe elements have continuous pressure fields as well, which appears to rule out employment in penalty methods. We will shortly investigate the reasons for safe elements being classified as such, but it might be useful to recall some safe elements:

1. Quadratic displacement triangle–constant pressure (Fig. 4.3.4(j))
2. Quadratic displacement triangle–continuous linear pressure (Fig. 4.3.6(d))
3. Biquadratic quadrilateral–constant pressure
4. Biquadratic quadrilateral–linear discontinuous pressure (Fig. 4.3.4(h))
5. Biquadratic quadrilateral–continuous bilinear pressure (Fig. 4.3.6(c))

This list is not exhaustive, but it is representative. Numbers 2, 4, and 5 have optimal error estimates but have drawbacks that have either been mentioned or will be explored. The reason that the standard error estimates for incompressible elements [84] steer the practitioner toward the underconstrained and inconvenient elements is that the standard estimates demand too much from a Lagrange multiplier (or related penalty pressure). The role of the Lagrange multiplier has been seen to be a twofold role of enforcer of the constraint and of pressure solution. The choice made in all five of the safe elements is to choose elements in which the role of enforcer has to some extent been sacrificed to avoid pressure modes. Before we can make a convincing argument along these lines, however, we must say something more definitive about the observations of Sani et al. [83]. At first sight, their results seem to say, simply, that reasonable problems can be concocted for which bilinear displacement–constant pressure elements will not work at all. To see that this is not really the case, we need to investigate the much maligned "pressure mode" in more detail. We will show conclusively that pressure modes are natural and even useful phenomena, more like sharks or buzzards than the finite element demons they have been made out to be. Our investigation will be largely confined to two-dimensional elements, because these are ones most easily visualized, and these are the ones for which current theory is most completely worked out. Much of what we say either carries over to three dimensions or can be easily generalized to such cases. The reader should proceed to three-dimensional problems with caution, however, as there *are* indeed important questions unresolved about, for example, integration accuracy in isoparametric three-dimensional reduced and selective formulations, and the possible generalization of macro-elements which work in two dimensions to three dimensions. In this investigation, we will try to get as much insight as possible from an "algebraic" approach to the problem at hand rather than a "functional analytic" approach. Inevitably, some functional analysis will creep in toward the end, but we hope not more than the reader has already been exposed to in this text.

A ***pressure mode*** is defined to be any pressure trial function, $q^h \not\equiv 0$, which satisfies

$$\int_\Omega q^h \operatorname{div} \boldsymbol{u}^h \, d\Omega = 0 \qquad \text{for all } \boldsymbol{u}^h \in \mathcal{S}^h \tag{4.II.1}$$

Although this might seem to be a very restrictive condition on q^h, the existence of such nonzero q^h can be crucially dependent on boundary conditions: Let q^h = constant in Ω, and apply the divergence theorem to (4.II.1):

$$\int_{\Gamma} \boldsymbol{u}^h \cdot \boldsymbol{n}\, d\Gamma = 0 \qquad \text{for all } \boldsymbol{u}^h \in \mathcal{S}^h \tag{4.II.2}$$

Thus q^h = constant in Ω satisfies (4.II.1) whenever the essential boundary conditions imposed on $\mathcal{S}^h$ imply $\boldsymbol{u}^h \cdot \boldsymbol{n} = 0$ on Γ, that is, whenever volume is *globally* preserved by the boundary conditions. It should be noted that (4.II.1) and (4.II.2) are the same as (4.2.10). Whenever $\boldsymbol{u}^h \cdot \boldsymbol{n}$ is specified on all Γ, preserving volume leads to a pressure mode; failing to preserve volume leads to no solution. Many elements used in practice also allow other nonzero solutions to (4.II.1). Any such pressure mode, other than the constant, has been dubbed a ***spurious mode.***

If we consider (4.3.21) with $\boldsymbol{M} = \boldsymbol{0}$, we can see that a pressure mode satisfies

$$\begin{bmatrix} \overline{\boldsymbol{K}} & \boldsymbol{G} \\ \boldsymbol{G}^T & \boldsymbol{0} \end{bmatrix} \begin{Bmatrix} \boldsymbol{0} \\ \boldsymbol{q} \end{Bmatrix} = \begin{Bmatrix} \boldsymbol{0} \\ \boldsymbol{0} \end{Bmatrix} \tag{4.II.3}$$

where $\boldsymbol{q}$ is the vector of nodal values of the pressure mode. To see why this is the case, consider the matrix form of (4.II.1):

$$0 = (q^h, \operatorname{div} \boldsymbol{u}^h) = \boldsymbol{q}^T \boldsymbol{G}^T \boldsymbol{d} \qquad \text{for all } \boldsymbol{d}$$

where $\boldsymbol{d}$ is the vector of nodal values of a $\boldsymbol{u}^h \in \mathcal{S}^h$, and the product $(\cdot\,,\,\cdot)$ denotes the L_2 inner product of (4.II.1) and will be used wherever possible hereafter. Thus $\boldsymbol{Gq} = \boldsymbol{0}$ as required. Note that by (4.II.3) a pressure mode is a pressure distribution that can arise as a nonzero response to zero applied loads. The constant pressure mode in the case when $\boldsymbol{u}^h \cdot \boldsymbol{n} = 0$ on Γ has a physical interpretation. Spurious modes do not, hence their name.

4.II.2 EXISTENCE AND UNIQUENESS OF SOLUTIONS IN THE PRESENCE OF MODES

As we indicated earlier, (4.II.3) shows that the matrix equation (4.3.21) is singular when the element has pressure modes. Before we try to obtain a solution, it behooves us to consider whether there *are* any solutions. Ominously, the answer is: *not always*. Much of the controversy about spurious modes has resulted from this fact, which can easily be overlooked—particularly when the related penalty method is employed in place of (4.3.21). There is a subtlety about (4.3.21) as it relates to what follows. Numerical or exact quadrature may have been used to evaluate the matrices involved. In particular, the former may make $\boldsymbol{Q} \equiv -\boldsymbol{M}$ (slightly) different from the L_2-inner product. In all cases, we assume that $\boldsymbol{Q}$ is positive-definite. We also use two additional notions of orthogonality: that implied by the usual vector dot product, and $\boldsymbol{Q}$-orthogonality, when quadrature is consistently applied to compute $\boldsymbol{M}$ and the vectors involved in the dot product. We will be careful to say "$\boldsymbol{Q}$-orthogonal" when that is what is intended. The key algebraic fact is contained in the following lemma.

Lemma. Using the notation of (4.3.21), given any vector of nodal values, $\boldsymbol{H}$, there is a vector of displacement nodal values, $\boldsymbol{d}$, such that

$$\boldsymbol{H} = \boldsymbol{G}^T \boldsymbol{d}$$

if and only if for any q such that

$$Gq = 0 \tag{4.II.4}$$

(i.e., q is a pressure mode),

$$H^T q = 0 \tag{4.II.5}$$

Proof. We seek a solution to

$$G^T d = H \tag{4.II.6}$$

That is, we wish to know when H is in the range of G^T. But the range of G^T is precisely the orthogonal complement of the solutions of (4.II.4), as we will argue: If H is in the range of G^T, then $H = G^T d$ for some d; thus $q^T H = q^T G^T d = 0$ for all solutions to (4.II.4). Conversely, let $v_1, v_2, \ldots, v_k$ denote the columns of G^T and suppose (4.II.5) is satisfied for all solutions to (4.II.4). The statement that $q^T G^T = 0^T$ is the statement that q is in the orthogonal complement of $v_1, v_2, \ldots, v_k$, which clearly span the range of G^T (though the v_i's need not be independent). Thus (4.II.5) says that H is in the orthogonal complement of the orthogonal complement of the range of G^T, or the range of G^T itself. This concludes the proof. ■

The lemma provides the following existence condition.

Theorem 1. The incompressible case of (4.3.21), i.e, the $M = 0$ case, has solutions if and only if (4.II.5) is satisfied for all solutions to (4.II.4).

Proof. The second equation of (4.3.21) is just (4.II.6), which, in view of the lemma, is satisfied only if (4.II.5) is satisfied for all solutions to (4.II.4). Therefore, we know that there is a d_1 with $G^T d_1 = H$. We will attempt to find a solution of the first equation of (4.3.21), in the form $d_1 + d_0$, where $G^T d_0 = 0$. This can be done by solving

$$\overline{K}(d_1 + d_0) + Gp = \overline{F}$$

using the "discrete Poisson equation," (4.3.26), which we rederive here so we can examine the intermediate steps:

$$\begin{aligned} \overline{K}d_0 + Gp &= \overline{F} - \overline{K}d_1 \\ d_0 + \overline{K}^{-1}Gp &= \overline{K}^{-1}\overline{F} - d_1 \end{aligned} \tag{4.II.7}$$

Obviously,more than one p can satisfy (4.II.7), since d_0 is as yet undetermined (e.g., $p = 0$ will work). But we also have the requirement that $G^T d_0 = 0$; if we can find p such that this is satisfied, then (4.II.7) defines an incompressible d_0 such that $d_1 + d_0$ satisfies the requirements of the theorem. This implies

$$G^T \overline{K}^{-1} Gp = G^T \overline{K}^{-1} \overline{F} - H \tag{4.II.8}$$

The argument of the lemma shows that this has a solution p if and only if the right-hand side is orthogonal to the kernel of $G^T \overline{K}^{-1} G$. (The form or dimension of G^T was in no way taken into account in the proof of the lemma. Here $G^T \overline{K}^{-1} G$ is symmetric, so it plays the role of both G and G^T in the lemma.) The kernel of $G^T \overline{K}^{-1} G$ is precisely the subspace of solutions to (4.II.4). (This can be seen by taking $LL^T = \overline{K}$ and writing $G^T\overline{K}^{-1}G = (L^{-1}G)^T L^{-1}G$.) By the definition of q and assumption of the theorem, for any solution, q, to (4.II.4),

$$q^T G^T \overline{K}^{-1} \overline{F} - q^T H = 0$$

Thus there are solutions to (4.II.8) of the form $p + q$, where q is any solution to (4.II.4) and p is orthogonal to all such q. Given any such $p + q$, d_0 is well determined by (4.II.7) (and depends only on p). This completes the proof. ■

Corollary 1.1. When the solutions of Theorem 1 exist, $\{d, p\}$ is always unique if p is taken to be Q-orthogonal to all solutions to (4.II.4). When the only solution to (4.II.4) is $q \equiv 0$, $\{d, p\}$ is the only solution; otherwise any $\{d, p + q\}$ is a solution, where q solves (4.II.4).

Proof. First we show that d is unique (if it exists) in all cases: Suppose that there are two solutions, $\{d_a, p_a\}$ and $\{d_b, p_b\}$. Subtraction of equations shows

$$\overline{K}(d_a - d_b) + G(p_a - p_b) = 0$$
$$G^T(d_a - d_b) = 0$$

Thus

$$d_a - d_b + \overline{K}^{-1}G(p_a - p_b) = 0 \tag{4.II.9}$$

and, going again to a discrete Poisson equation:

$$G^T\overline{K}^{-1}G(p_a - p_b) = 0 \tag{4.II.10}$$

We deduce that $p_a - p_b$ is in the fact a solution to (4.II.4) (by taking the Cholesky decomposition of $G^T\overline{K}^{-1}G$, as before). This has two important ramifications. First, it shows that $G(p_a - p_b) = 0$; thus from (4.II.9), $d_a = d_b$, which proves the desired uniqueness.

To discuss the second ramification of (4.II.10), we use the uniqueness result we have just proved to write any solution of Theorem 1 as $\{d, \overline{p}\}$ for the fixed, unique d and some pressure solution. The Hilbert projection theorem (with respect to Q-orthogonality) enables us to write $\overline{p}$ as

$$\overline{p} = p_0 + q_0$$

where

$$p_0^T Q q_0 = 0$$
$$G q_0 = 0$$

by taking q_0 to be the projection of $\overline{p}$ onto the null-space of G. Observe that

$$\overline{K}d + G(p_0 + q_0) = F$$
$$G^T d = H$$

but also

$$\begin{aligned} \bar{\boldsymbol{K}}\boldsymbol{d} + \boldsymbol{G}\boldsymbol{p}_0 &= \bar{\boldsymbol{F}} \\ \boldsymbol{G}^T\boldsymbol{d} &= \boldsymbol{H} \end{aligned}$$

since $\boldsymbol{G}\boldsymbol{q}_0 = \boldsymbol{0}$. Thus $\{\boldsymbol{d}, \boldsymbol{p}_0\}$ is also a solution. Let $\{\boldsymbol{d}, \bar{\bar{\boldsymbol{p}}}\}$ be some other solution. Similar arguments show

$$\begin{aligned} \bar{\bar{\boldsymbol{p}}} &= \boldsymbol{p}_1 + \boldsymbol{q}_1 \\ \boldsymbol{p}_1^T\boldsymbol{Q}\boldsymbol{q}_1 &= 0 \\ \boldsymbol{G}\boldsymbol{q}_1 &= \boldsymbol{0} \end{aligned}$$

and the results of (4.II.10) shows that

$$\boldsymbol{G}(\bar{\bar{\boldsymbol{p}}} - \bar{\boldsymbol{p}}) = \boldsymbol{0}$$

but

$$\boldsymbol{G}(\bar{\bar{\boldsymbol{p}}} - \bar{\boldsymbol{p}}) = \boldsymbol{G}(\boldsymbol{p}_1 - \boldsymbol{p}_0) = \boldsymbol{0}$$

A simple linear algebra argument (left to the reader) shows that if both $\boldsymbol{p}_1$ and $\boldsymbol{p}_0$ are $\boldsymbol{Q}$-orthogonal to the null space of $\boldsymbol{G}$, then $\boldsymbol{G}(\boldsymbol{p}_1 - \boldsymbol{p}_0) = \boldsymbol{0}$ only if $\boldsymbol{p}_1 = \boldsymbol{p}_0$. Thus the representative pressure solution, $\boldsymbol{Q}$-orthogonal to the solutions of (4.II.4), call it $\boldsymbol{p}$, is unique as required, and all other solutions are of the form $\boldsymbol{p} + \boldsymbol{q}$ with $\boldsymbol{G}\boldsymbol{q} = \boldsymbol{0}$. Thus the corollary is proved in its entirety. ■

We summarize developments to this point:

1. For solutions to (4.3.21) to exist, the right-hand side $\boldsymbol{H}$-vector must be orthogonal to the "pressure modes" in the sense of (4.II.5).
2. Orthogonality to the constant pressure mode, when it exists, is assured by global volume preservation of the boundary conditions.
3. When the solvability condition of (4.II.5) is satisfied, the "$\boldsymbol{d}$-part" of the solution to (4.3.21) is unique, and there is a unique pressure solution $\boldsymbol{Q}$-orthogonal to the pressure modes, though any particular pressure solution may consist of this $\boldsymbol{p}$ plus an arbitrary pressure mode.

4.II.3 TWO SIDES OF PRESSURE MODES

We should remind ourselves at this point just what the $\boldsymbol{H}$-vector is and why it will automatically satisfy the solvability condition with respect to the constant pressure mode but may have trouble with other pressure modes: Following Table 4.3.1, (4.II.5) requires

$$\boxed{(q^h, \operatorname{div} g^h) = 0} \qquad (4.\text{II}.11)$$

for all modes q^h with nodal-values $\boldsymbol{q}$ satisfying (4.II.4). Suppose q^h = constant in Ω is a solution to (4.II.4); then (4.II.11) becomes (4.II.2). The solvability condition with respect to the constant mode implies that the (small-strain) volume change is zero for an incompressible elastic body. This must be arranged for by the analyst and (4.II.5) says that (4.II.2) must be satisfied exactly. We must take into account the exact volume change induced by $\boldsymbol{g}^h$. This may be tricky if the boundary is complicated and may require a priori evaluation of boundary integrals of $\boldsymbol{g}^h$, because the interpolated volume change implied by $\boldsymbol{g}^h$ may be slightly different from the volume change in the exact boundary conditions we are interpolating. But the point is that preserving volume exactly makes physical sense and is something most analysts would try to do—even if it did pose some routine technical difficulties.

When q^h is a spurious pressure mode, there is no obvious physical interpretation of (4.II.11). Spurious pressure modes place additional, apparently unphysical, constraints on $\boldsymbol{g}^h$, which we would be unlikely to satisfy as a matter of course. The problem in dealing with such constraints is twofold: First the pressure modes for the element, mesh, and boundary conditions under consideration must be characterized. Second, the essential boundary conditions may have to be modified in such a way that (4.II.11) is satisfied, *exactly*. This seems to pose formidable obstacles for the analyst, and at this point we might wonder whether opting for safe elements without spurious modes might indeed be the wisest course. The fact is that we have given all the bad news about spurious modes first, and while we cannot sweep the technical difficulties of using elements which have them under the rug, there now is good news: First, elements which have spurious modes seem to have some distinct advantages not available with safe elements; second, employment of elements with spurious modes in a penalty formulation seems to alleviate many of the potential difficulties; and third, new analysis and error estimates have shown the way systematically to obtain good results for some elements with spurious modes. Before we pursue these ideas further, it seems appropriate to look at a now-classical example of the interaction between pressure modes and boundary conditions. The problem considered is that of driven cavity flow described in Sec. 4.4.1 and illustrated in Fig. 4.4.6. The reader more familiar with elasticity is reminded of the formal equivalence of Stokes flow and linear elasticity discussed in Sec. 4.2.

Example 1 (Driven Cavity Problem—A Stokes-flow Problem with Inhomogeneous Boundary Conditions)

Let $N_1, N_2, \ldots, N_M$ be the global shape functions associated with nodes $1, 2, \ldots, M$ of Fig. 4.II.1. Consider three ways of specifying the unit velocity of the driven lid, obtained from three choices for $\boldsymbol{g}^h$ (see Table 4.3.1):

Case 1: Leaky lid

$$\boldsymbol{g}^h = \begin{Bmatrix} \sum_{i=1}^{M} N_i \\ 0 \end{Bmatrix} = \begin{Bmatrix} g_1^h \\ g_2^h \end{Bmatrix}$$

Note that

$$g_1^h(x_1, x_2) = \begin{cases} \dfrac{x_2}{h} & x_2 \geq 0 \\ 0 & x_2 < 0 \end{cases}$$

Thus div $g^h \equiv 0$, and (4.II.11) is automatically satisfied for any possible modes. This was the g^h used in Sec. 4.4.1.

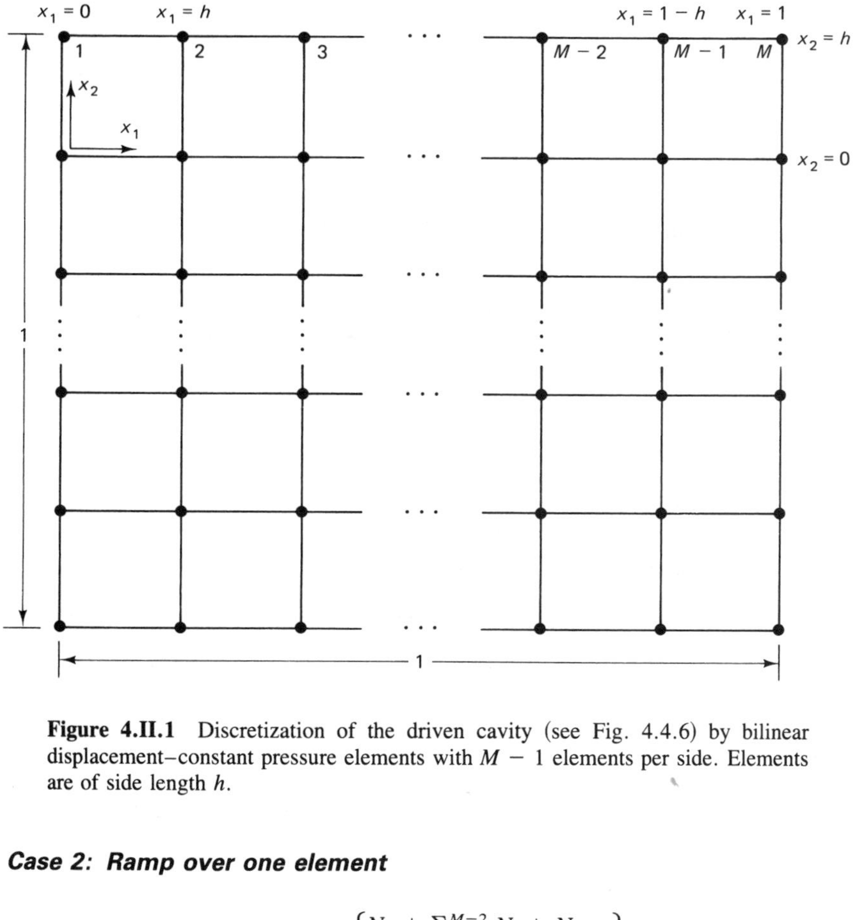

Figure 4.II.1 Discretization of the driven cavity (see Fig. 4.4.6) by bilinear displacement–constant pressure elements with $M - 1$ elements per side. Elements are of side length h.

Case 2: Ramp over one element

$$g^h = \begin{Bmatrix} N_2 + \Sigma_{i=3}^{M-2} N_i + N_{M-1} \\ 0 \end{Bmatrix}$$

In this case, we assume that $M \geq 4$ and that the summation term is zero if $M = 4$. Observe that instead of allowing flux through the domain as in case 1, the velocity component, u_1, is zero at nodes 1 and M and linearly rises to $u_1 = 1$ in the first element and linearly falls from $u_1 = 1$ to zero in the $(M - 1)$st element. In this case

$$g_1^h = \begin{cases} \dfrac{x_1 x_2}{h^2} & 0 \le x_1 \le h,\ x_2 \ge 0 \\ x_2/h & h < x_1 < 1 - h,\ x_2 \ge 0 \\ \dfrac{(1 - x_1)x_2}{h^2} & 1 - h \le x_1 \le 1,\ x_2 \ge 0 \\ 0 & x_2 < 0 \end{cases}$$

and thus

$$\text{div } \boldsymbol{g}^h = \begin{cases} \dfrac{x_2}{h^2} & 0 \le x_1 \le h,\ x_2 \ge 0 \\ 0 & h < x_1 < 1 - h,\ x_2 \ge 0 \\ \dfrac{-x_2}{h^2} & 1 - h \le x_1 \le 1,\ x_2 \ge 0 \\ 0 & x_2 < 0 \end{cases}$$

Note that

$$\int_\Omega \text{div } \boldsymbol{g}^h \, d\Omega = 0$$

But the bilinear velocity–constant pressure element has an additional mode, the ***checkerboard mode*** (see [86], for example):

$$c_B^h \equiv (-1)^{i+j-M+1} \qquad i, j = 1, 2, \ldots, M - 1$$

in the element which is the ith element in the x_1-direction and the jth in the x_2-direction. In particular, for the top row of elements, the checkerboard mode may be expressed as

$$c_B^h = (-1)^i \qquad i = 1, 2, \ldots, M - 1$$

Thus a simple calculation shows that

$$\boxed{\int_\Omega c_B^h \text{ div } \boldsymbol{g}^h \, d\Omega = -\frac{h}{2}[1 + (-1)^{M-1}]}$$

So for an even number of elements in the x_1-direction (i.e., M odd), (4.II.11) is *not* satisfied. Evidently, *no* solution to the discrete problem exists in this case.

Case 3: Ramp over two elements

In this case we take $g_2^h \equiv 0$ *and*

$$g_1^h = \begin{cases} \dfrac{x_1 x_2}{2h^2} & 0 \le x_1 \le 2h,\ x_2 \ge 0 \\ x_2/h & 2h < x_1 < 1 - 2h,\ x_2 \ge 0 \\ \dfrac{(1 - x_1)x_2}{2h^2} & 1 - 2h \le x_1 \le 1,\ x_2 \ge 0 \\ 0 & x_2 < 0 \end{cases}$$

We assume here that $M \geq 5$, so that at the very least we have nodes at $x_1 = 0, h, 2h, 1 - h$, and 1. Equivalently, we could write g_1^h in terms of the N_i's as before, using nodal values of 0, 0.5, and 1 at the first three nodes on the lid, to "ramp" u_1 linearly from zero to one over the first two elements; the same nodal values in reverse order are used at the last three nodes to bring u_1 down from 1 to 0. We find here that

$$\operatorname{div} \boldsymbol{g}^h = \begin{cases} \dfrac{x_2}{2h^2} & 0 \leq x_1 \leq 2h,\ x_2 \geq 0 \\ 0 & 2h < x_1 < 1 - 2h,\ x_2 \geq 0 \\ \dfrac{-x_2}{2h^2} & 1 - 2h \leq x_1 \leq 1,\ x_2 \geq 0 \\ 0 & x_2 < 0 \end{cases}$$

Consequently

$$\int_\Omega \operatorname{div} \boldsymbol{g}^h \, d\Omega = 0$$

and

$$\begin{aligned} \int_\Omega c_B^h \operatorname{div} \boldsymbol{g}^h \, d\Omega &= \int_0^h \int_0^{2h} c_B^h \operatorname{div} \boldsymbol{g}^h \, dx_1 dx_2 + \int_0^h \int_{1-2h}^1 c_B^h \operatorname{div} \boldsymbol{g}^h \, dx_1 dx_2 \\ &= -\int_0^h \left[\int_0^h \frac{x_2}{2h^2} \, dx_1 - \int_h^{2h} \frac{x_2}{2h^2} \, dx_1 \right] dx_2 \\ &\quad - (-1)^M \int_0^h \left[\int_{1-2h}^{1-h} \frac{x_2}{2h^2} \, dx_1 - \int_{1-h}^1 \frac{x_2}{2h^2} \, dx_1 \right] dx_2 = 0 \end{aligned}$$

We get zero because each bracketed term is zero independently. Unlike case 2, the value of $(-1)^M$ is not crucial, and (4.II.11) is always satisfied for any $M \geq 5$.

We summarize the important observations of Example 1:

1. In all three cases, the boundary condition is the same in the limit $h \rightarrow 0$.
2. In the leaky-lid case, $\boldsymbol{u}^h \cdot \boldsymbol{n}$ is nonzero in the elements at the top corners of Ω, but (4.II.2) is satisfied because $\boldsymbol{u}^h \cdot \boldsymbol{n}$ is of equal magnitude and opposite in sign at corresponding points of the first and last elements of the top row, i.e., *mass is globally conserved*. In other cases, mass is globally conserved because there is no flow through the domain boundary. These cases have obvious analogies to preservation of volume in elasticity.

There is strong evidence to suggest that essential boundary conditions can always be modified to satisfy (4.II.11) without loss of accuracy, because it can be shown—at least for interior approximation—that the checkerboard modes cannot contribute to approximation accuracy [85]. We suspect that this result can be generalized to approximation of boundary conditions.

What we have seen so far is only part of the story about spurious modes. If the reader wonders why we dwell at such length on spurious modes and why we will continue to do so for much of this appendix, we should say at this point that not only is learning to understand and to live with spurious modes a practical matter—it allows the analyst a much wider choice of finite elements for problems with the incompressibility constraint—but understanding spurious modes is also the key to understanding the whole subject of constraints in the finite element method. We now are in a position to appreciate this fact. What we have observed about spurious modes up to now suggests that they are invariably tied to boundary conditions, and are some sort of unphysical artifact of elements which would best be avoided. The only spurious modes we have actually *seen* so far *are* indeed associated with boundary conditions. However, there is another kind of spurious mode which is illustrated by the following example.

Example 2 (The Crossed-Triangle Element and Redundant Constraints)

This example concerns the linear crossed-triangle macroelement discussed in Remark 3 of Sec. 4.4, which is illustrated in Fig. 4.4.4(a). Given $\boldsymbol{u}^h \in \mathcal{S}^h$, consider a typical macroelement and let e_1, e_2, e_3, e_4 denote the four subtriangles, ordered in counterclockwise fashion. It is important to note that the typical macroelement under consideration can be an arbitrary quadrilateral, but the central node *must* fall on the intersection of the diagonals of the quadrilateral. Mercier proved that

$$\boxed{\operatorname{div} \boldsymbol{u}^h|_{e1} - \operatorname{div} \boldsymbol{u}^h|_{e2} + \operatorname{div} \boldsymbol{u}^h|_{e3} - \operatorname{div} \boldsymbol{u}^h|_{e4} = 0}$$

Mercier then argued that for $\mathcal{P}^h$ based on piecewise constants on the triangles, only three of the four incompressibility constraints are independent, since setting $\operatorname{div} \boldsymbol{u}^h = 0$ on three triangles forces it to be zero on the fourth. As explained in Sec. 4.4, this has the effect of modifying the apparent constraint ratio of $r = 1$ to a more acceptable $r = \frac{4}{3}$.

This argument, which shows that the macroelement is not overconstrained, can be turned around to show that the macroelement also has many modes. Consider again our typical macroelement, labeled M, and define

$$c_M^h \equiv \begin{cases} (-1)^i/a^{e_i} & \text{on triangle } e_i \\ 0 & \text{outside of macroelement } M \end{cases}$$

where a^{e_i} is the area of traingle e_i. Now

$$\int_\Omega c_M^h \operatorname{div} \boldsymbol{u}^h \, d\Omega = \int_{\Omega^M} c_M^h \operatorname{div} \boldsymbol{u}^h \, d\Omega = \sum_{i=1}^{4} a^{e_i}(c_M^h \operatorname{div} \boldsymbol{u}^h)_{e_i}$$

since c_M^h and $\operatorname{div} \boldsymbol{u}^h$ are constant on triangles. But the definition of c_M^h and Mercier's equation immediately lead to the conclusion that

$$\int_\Omega c_M^h \operatorname{div} \boldsymbol{u}^h \, d\Omega = 0$$

for any $\boldsymbol{u}^h \in \mathcal{S}^h$, and thus c_M^h is a pressure mode by (4.II.1). Such a c_M^h was constructed without the imposition of any boundary conditions, and it could have been constructed for a mesh consisting of a single macroelement. Such a mode, then, must be a mode of the macroelement's "element" matrix. Note that there is one such mode for each macroelement.

To summarize the salient points of Example 2:

1. There are evidently two kinds of spurious modes: There are global ones. introduced by the imposition of certain types of boundary conditions when certain elements are used. These are the kind discovered by Fortin and they do not appear to be very useful. Then there is the possibility of ***local spurious modes,*** which are spurious modes of the macroelement matrix without any imposed boundary conditions; these are responsible for the favorable constraint ratio of crossed triangles.
2. The crossed triangle element has the incompressibility constraint enforced in the strongest possible way by the Lagrange multiplier—*pointwise exactly*.
3. A price that must be paid, it seems, for having an element that satisfies the incompressibility constraint exactly is the presence of a global checkerboard mode. This occurs for the crossed triangle on meshes which would have such a mode for bilinear rectangles. The mode is identical to the mode associated with the bilinear–constant pressure element. (The proof of this may be found in [86].)

We will refer to spurious modes of the kind observed by Fortin, induced by boundary conditions as "(spurious) modes of the first kind," or "global (spurious) modes." Those of the type that we just encountered in Example 2, which are due to redundant constraints on the element level, we refer to as "(spurious) modes of the second kind," or "local (spurious) modes." The discovery of the redundancy of constraints is due to Nagtegaal et al. [87], and Mercier [88, 89]. Mercier showed that crossing triangles also leads to a dramatic improvement in the constraint ratio for quadratic triangles as well as linear ones (more about this later). These authors did not phrase their discovery in what now can easily be seen to be a nearly paradoxical statement: Some elements seemed to work *only* because they had spurious modes, but standard error analysis predicted their failure precisely *because* they had spurious modes. Something had to give, and it turns out that it was standard error analysis which gave way. Still, at this point, it may seem a dubious proposition to employ elements with local spurious modes: To increase the constraint ratio in any meaningful way with local modes seems to require a very substantial number of such modes. To arrange satisfaction of (4.II.11) for all such modes at first sight seems a tedious task at best and hopeless at worst. Then, even if we could resolve that problem, according to (4.II.3), each independent pressure mode leads to an additional rank deficiency of the system of equations to be solved. This could pose formidable computational problems. Evidently, however, it was intended by some grand design that we should be able to use elements with spurious modes, and each of these problems can be dealt with easily. In what follows we use the word "element," but with Example 2 in mind, the element could just as well be a macroelement.

Theorem 2. Consider the incompressible form of (4.3.21), with a pressure trial space, $\mathcal{P}^h$, based on the same nodal pattern as the displacement space, $\mathcal{S}^h$, but with *discontinuous* pressures (no sharing of nodal values between elements at all). If the element gradient matrix, $\boldsymbol{g}^e$, has a spurious mode, $\boldsymbol{q}^e$, i.e., if

$$\boldsymbol{g}^e \boldsymbol{q}^e = \boldsymbol{0} \tag{4.II.12}$$

there is a corresponding spurious mode $q^h \in \mathcal{P}^h$ for each such element in the mesh such that

$$\boldsymbol{Gq} = \boldsymbol{0}$$

However, such a spurious mode satisfies (4.II.11) for any $\mathcal{q}^h$, i.e., a spurious mode arising from a local, or element, spurious mode is always consistent with any inhomogeneous boundary conditions imposed via "$\mathcal{q}$-node" specification.

Proof: To construct a pressure mode from an element mode, just extend the element mode outside the element with a zero value on all the rest of Ω; this can be done because the pressures admit a discontinuity at the element boundary. One such mode is thus constructed for each element in the mesh, and such modes are $\boldsymbol{Q}$-orthogonal since they have disjoint supports. Thus we have constructed the desired number of independent modes.

To see the second part of the theorem, consider the constraint matrix, $\overline{\boldsymbol{G}}^T$, associated with the space $\mathcal{S}^h \times \mathcal{P}^h$, i.e., without any essential boundary conditions. We may deduce that $\overline{\boldsymbol{G}}\boldsymbol{q} = \boldsymbol{0}$ for any local $\boldsymbol{q}$ constructed as in the first part of this proof. This follows since

$$\overline{\boldsymbol{G}}\boldsymbol{q} = \boldsymbol{0} \Leftrightarrow \boldsymbol{g}^e \boldsymbol{q}^e = \boldsymbol{0} \tag{4.II.13}$$

for the $\boldsymbol{g}^e$ and $\boldsymbol{q}^e$ from which $\boldsymbol{q}$ was constructed, which may in turn be attributed to the fact that there is no nodal-value sharing. But then, $\mathcal{q}^h \in \mathcal{S}^h$, and thus (4.II.13) implies (4.II.11) when q^h in (4.II.11) is taken to be the mode whose nodal values appear in (4.II.13). This concludes the proof of the theorem. ■

Theorem 2 is very important in practice: It means that *no modification of essential boundary conditions is required to accommodate the spurious modes of the second kind,* which must be numerous to be useful. These spurious modes—indeed any spurious modes—always have a dual displacement field which is weakly incompressible. It is not possible to identify the spurious mode with a specific weakly incompressible field, but rather we infer that there is such a field because each spurious mode increases the effective constraint ratio referred to in Example 2. To see this, we need to sharpen the definition of constraint ratio given in Sec. 4.3. The new definition explicitly counts the redundancy of constraints and incorporates this number in the computation of n_c of the earlier definition. We note also that the new definition is based on the constraint matrix, not the mesh; therefore it is not tied to the special test mesh of Fig. 4.3.3. We do lose the nice property of that mesh which assured that for discontinuous pressures, r could be found by considering only one element; it will differ slightly because of the possibility of more or fewer "$\mathcal{q}$-nodes", and the possibility of spurious modes of the first kind introduced by some boundary conditions. It will agree with the previous r in the limit of mesh refinement if the spurious modes are properly taken into account in computing n_c.

Definition. (Redefinition of the constraint ratio.) The ***constraint ratio,*** $\bar{r}$, is defined as the dimension of $\mathcal{S}^h$ divided by the number of effective constraints implied by $\boldsymbol{G}^T$, which is the number of independent rows (equations) of $\boldsymbol{G}^T$ or the number of independent columns of $\boldsymbol{G}$.

Theorem 3. Let "dim" denote "dimension of" (as a vector space), and "ker" denote "kernel (null space) of". The *constraint ratio* (as redefined above) is given by

$$\bar{r} = \frac{\dim \mathcal{S}^h}{n_z}$$

where

$$n_z \equiv \dim \mathcal{P}^h - \dim \ker \boldsymbol{G} \tag{4.II.14}$$

Proof: To establish the result requires only that n_z be the number of independent rows of $\boldsymbol{G}^T$. The law of nullity of linear algebra states that the rank of $\boldsymbol{G}$ is equal to the dimension of the domain space, dim $\mathcal{P}^h$, minus the nullity of $\boldsymbol{G}$, dim ker $\boldsymbol{G}$. But then the number of independent equations in $\boldsymbol{G}^T$ is equal to its rank, which is also the rank of $\boldsymbol{G}$, by an equally basic result of linear algebra. This concludes the proof. ■

So actually the theorem is a linear-algebraic triviality, but by writing the triviality in the way we did in (4.II.14) we learn something nontrivial: *Each spurious mode reduces the number of constraints by one—in general, not just for the crossed triangle*. We next deal with the second objection to spurious modes voiced earlier; the rank deficiency of the matrix equations, which is evidently increased by one for each mode.

4.II.4 PRESSURE MODES IN THE PENALTY FORMULATION

While there may be ways to solve the resultant equations in Lagrange-multiplier form, despite the singularity, we believe that the most useful elements possessing spurious modes of the second kind are ideally suited for penalty methods. These are the crossed triangular elements of arbitrary degree which are useful because, as we observed in Example 2 with crossed linear triangles, exact incompressibility can be obtained by the Lagrange-multiplier solution. For these elements the exact incompressibility of the Lagrange-multiplier solution translates into an $O(\frac{\mu}{\lambda})$ compressibility error of the penalty solution. This is a much better compressibility error than can be expected from most elements, which will usually have $O(\frac{\mu}{\lambda}) + O(h^m)$ compressibility error, for some m dependent on both $\mathcal{S}^h$ and $\mathcal{P}^h$. This means that in most cases the h-dependent term will dominate, and a higher degree of satisfaction of the constraint can only be obtained by mesh refinement, which is unlikely to lead to a dominance of the penalty term, even for the finest meshes. We will amplify on this theme as we go along, but first we will pursue a question which may seem puzzling at the moment: What happens to pressure modes in a penalty formulation?

We have seen in an earlier section that the penalty formulation produces positive-definite matrices of standard finite element form; the only obstacle to a solution procedure is penalty-induced ill-conditioning, which can be minimized by judicious choice of the penalty. This applies to reduced and selective integration techniques and $\bar{\boldsymbol{B}}$ methods. It should be noted that in plane problems using Cartesian coordinates, the crossed triangle family of elements can be integrated *exactly*, as was implied in Example 2. (All integrands consist of polynomials as long as the global triangle domain is an affine mapping of the parent domain.) The spurious modes of the second kind in the limiting Lagrange-multiplier method give a favorable constraint ratio for the penalty method. One can easily deduce that the integration *must* be exact for the discontinuous pressure trial space based on the integration points to contain the divergences of the displacements. Actually, in such cases, the reduced and selective, $\bar{\boldsymbol{B}}$, and exactly integrated slightly compressible formulations are identical. This brings us squarely up against the question of axisymmetry: Current wisdom seems to suggest that the $\bar{\boldsymbol{B}}$ approach is the best approach here; use the pressure shape functions which give the best constraint ratio versus pressure accuracy (for the crossed triangle, for example, this would be the pressures based on the nodes of the quadrature formula which would have been exact in Cartesian coordinates). Use a quadrature formula to evaluate the integrals of (4.5.1) and (4.5.2) which is "accurate enough" near the axis of symmetry (i.e., the singularity) to give confidence in the accuracy of the evaluation of the integrals. Use as many points—at least for the elements near the symmetry line—as experience leads you to believe you need. You may have to experiment to balance cost of using many points with the quality of solutions you desire. One thing you need not worry about is the constraint ratio, since the $\bar{\boldsymbol{B}}$ formulation has freed the choice of quadrature formula from assumed pressure interpolation. If reduced integration is used, or in the case of the crossed triangles, the formula which would have been exact in Cartesian coordinates is used, the equivalence theorem of Sec. 4.4 between penalty method and *numerically integrated* mixed method still holds, but the quadrature formula might lead to excessive integration error in the region of the singularity at the axis of symmetry in the numerically integrated mixed method. Thus, although the penalty method is equivalent to a mixed method, it might not be a particularly good mixed method. It was this difficulty that the original mean-dilatational methods were intended to address.

The astute reader may have already guessed why we are up against the problem of axisymmetry: Whether we use accurately integrated $\bar{\boldsymbol{B}}$ or try to get away with reduced or selective integration, the advantages of the crossed triangle elements seem to disappear in axisymmetry. Because the space "div $\mathcal{S}^h$" of divergences of members of $\mathcal{S}^h$ contains u^h/r terms, div $\mathcal{S}^h$ is no longer contained in $\mathcal{P}^h$ and thus the exact incompressibility of the Lagrange-multiplier solution is lost. It does appear that this problem can be circumvented by using trial functions that are not polynomials and are specially suited to the coordinate system. This would also lead to quadrature formulas that would be exact for the crossed triangles in axisymmetry. What we say subsequently will be true for axisymmetry, but the advantage of using the elements with spurious modes of the second kind is yet to be realized in axisymmetry.

The next theorem, about modes in the penalty formulation, gives a partial answer to the question of what happens to modes in penalty methods: They are invisible, at least as far as the penalty pressure is concerned. Part (b) of the following theorem requires some techniques beyond the scope of the present discussion. A proof can be constructed based on arguments like those used in [89] to prove a similar result for the continuous rather than discrete problem. The reader who has digested the material in Appendix 4.I and this appendix might be able to see how this is done, and the bravest of readers are invited to try a proof as an exercise.

Theorem 4. Consider either a reduced or selective penalty method or a $\overline{\boldsymbol{B}}$ method and the corresponding mixed method of (4.3.21), with all matrices involved in the mixed and penalty formulations evaluated with the appropriate numerical quadrature to make an exact equivalence between the *appropriately integrated* mixed method and penalty method (exact integration allowed), in the sense of Malkus-Hughes [36]. Use the same quadrature on the relevant matrices of the limiting Lagrange-multiplier method. Assume the algebraic consistency condition, (4.II.5), is satisfied with respect to all modes of the Lagrange-multiplier method, for the penalty method as well as the limiting Lagrange-multiplier method. Let $\epsilon = \mu/\lambda$ and p_ϵ^h be the penalty pressure solution corresponding to that value of ϵ. Let p^h be the Lagrange multiplier, or $\epsilon = 0$, solution of Corollary 1.1, which is $\boldsymbol{Q}$-orthogonal to all pressure modes. Then

a. p_ϵ^h is $\boldsymbol{Q}$-orthogonal to all the pressure modes of the corresponding Lagrange-multiplier method.
b. $\|p_\epsilon^h - p^h\|_0 < C\epsilon$, for a constant, C, independent of ϵ.

Proof of Part (a). In any of the penalty methods under consideration—reduced or selective or $\overline{\boldsymbol{B}}$—the pressure is computed by a *weak form* of (4.4.14).

$$\langle q^h, p_\epsilon^h \rangle = -\lambda \langle q^h, \text{div } \boldsymbol{u}^h \rangle \qquad \text{for all } q^h \in \mathcal{P}^h \tag{4.II.15}$$

where $\langle \cdot, \cdot \rangle$ denotes the ***appropriately integrated*** $\boldsymbol{L_2}$***-inner product,*** which implies satisfaction of Eq. (4.4.14) at the integration points (i.e., $\mathcal{P}^h$-nodes) in the reduced and selective cases. In general, (4.II.15) is equivalent to

$$\boldsymbol{p}_\epsilon = \boldsymbol{M}^{-1}\boldsymbol{G}^T\boldsymbol{d} \tag{4.II.16}$$

where the use of numerical integration on the matrices corresponds to the use of the appropriate $\langle \cdot, \cdot \rangle$; $\boldsymbol{p}_\epsilon$ is the vector of nodal values of p_ϵ^h and $\boldsymbol{d}$ the nodal values of the penalty displacement solution. Taking the $\boldsymbol{Q}$-inner product of a pressure mode with nodal values, $\boldsymbol{q}$, involves premultiplying (4.II.16) by $\boldsymbol{q}^T\boldsymbol{Q}^T$, from which the desired result follows immediately. This concludes the proof of part (a). ■

There are a few important points related to the theorem that should be emphasized:

1. In almost all cases, the quadrature applied to $\boldsymbol{M}$ is accurate enough to imply that the penalty p_ϵ^h is L_2-orthogonal to a constant pressure mode, should one exist. This is particularly true when the quadrature is exact on $\boldsymbol{M}$, which is true in many practical cases.
2. While the algebraic consistency condition is assumed to hold and the penalty pressures are free of modes, these facts alone will be seen to be not enough to guarantee the convergence of the penalty pressures to the exact solution. We will see that a postprocessing of the penalty pressures is often wise [85, 90].
3. It is possibie to show that the same $O(\epsilon)$ convergence rate holds for penalty displacement solutions to corresponding Lagrange-multiplier solutions. The readers who try to prove this and part (b) of the theorem should be warned that if the element fails to satisfy the "LBB, or Babuška-Brezzi condition," (discussed further in the next section) the constant C may appear to be unfavorably dependent on h. However, this does not appear to be the case in practice.

Finally, there is a neat heuristic way to describe what happens to pressure modes in a penalty formulation. Recall that the coefficient matrix of the linear system to be solved in such cases is

$$\overline{\boldsymbol{K}} - \boldsymbol{G}\boldsymbol{M}^{-1}\boldsymbol{G}^T$$

So the pressure modes are "trapped inside" the penalty matrix and visible on the outside only as extra weakly incompressible modes. According to Theorem 4, the pressure modes remain invisible even when the penalty pressure is recovered from the displacements. The only effect of the modes is observed if we fail to modify the essential boundary conditions to make the $\boldsymbol{H}$-vector orthogonal to the modes. Then the effects can produce dramatically nonsensical results [83], which caused much confusion and consternation among researchers until all this was sorted out (to the extent it is now).

4.II.5 THE BIG PICTURE

In what follows, we want to put together what we have observed about pressure modes in an overall heuristic picture of the incompressibility constraint in the finite element method. We want our picture to be painted with broad strokes, overlooking but not denying the kind of detailed structure we have observed to this point. We can make things easy for ourselves by thinking of exactly integrated Lagrange-multiplier methods in Cartesian coordinates; this is the underlying prototypical method from whose success or failure the success or failure of all subtle variants and closely related methods will follow.

We argue that there are two types of finite elements, *both* of which are useful in the constrained finite element method, *both* of which can lead to optimal convergence rates in specific cases: "locking elements" and "nonlocking elements." To make this distinction, we need to define two subspaces of the trial solution space. The first is composed of solutions satisfying the ***weak constraint,***

$$W^h \equiv \{u^h \in \mathcal{S}^h \,|\, \text{div } u^h \perp \mathcal{P}^h\}$$

where "$\perp$" refers to L_2-orthogonality. Since we are assuming exact integration, we can see that W^h is the subspace of all weakly incompressible trial solutions with respect to the constraint as enforced by the Lagrange-multiplier method. The next space is one to which the reader may not have given much thought: the space of all trial solutions that are exactly pointwise-incompressible,

$$V^h \equiv \{u^h \in \mathcal{S}^h \,|\, \text{div } u^h = 0\}$$

Clearly $V^h \subset W^h$. Note that the definition of V^h is intrinsic to the displacement space alone and makes no reference to a pressure trial space.

Definition. Consider the test mesh of Fig. 4.3.3; an element is a ***locking element*** if there is a constant, C_1, independent of n_{es} such that

$$\dim V^h \leq C_1 n_{es}$$

An element is a ***nonlocking element*** if there exists a constant, C_2, independent of n_{es} such that

$$\dim V^h \geq C_2 {n_{es}}^2$$

We note that the determination of whether an element is of the locking or nonlocking type is independent of any choice of pressure trial space. The prototypical examples of the two kinds of element are as follows:

Locking elements

1. Linear triangles in nonredundant arrangements (such as Fig. 4.3.4(a))
2. All Lagrange elements
3. All serendipity elements.

Nonlocking elements

1. All triangles of degree greater than or equal to two
2. Linear crossed triangles

We are now ready to shop for a space of Lagrange multipliers and pressures, knowing that if we try to enforce the incompressibility constraint too vigorously with locking elements, the results can reflect what the name implies. With the nonlocking elements we evidently can be more vigorous in enforcement of the constraint; how vigorous we can or should be will be tempered by what follows. The first caution we have is contained in the next theorem, which covers only the special cases in which there is a containment relation one way or another between div $\mathcal{S}^h$ and $\mathcal{P}^h$:

Theorem 5. For an exactly integrated Lagrange-multiplier method,

a. If $\mathcal{P}^h \subset \text{div } \mathcal{S}^h$, then there are no modes and incompressibility is imposed only approximately by the weak constraint when the containment is proper.

b. If $\mathcal{P}^h \equiv \text{div } \mathcal{S}^h$, then there are no modes and incompressibility is imposed exactly.

c. If $\mathcal{P}^h \supset \text{div } \mathcal{S}^h$, then there are modes if the containment is proper and incompressibility is imposed exactly.

Proof: The result follows easily from reconsidering the equation

$$(q^h, \text{div } \boldsymbol{u}^h) = 0 \tag{4.II.17}$$

In case (a), if (4.II.17) is to hold for fixed q^h and all $\boldsymbol{u}^h \in \boldsymbol{S}^h$, there is a $\boldsymbol{u}^h$ with div $\boldsymbol{u}^h = q^h$ by assumption; thus any q^h satisfying (4.II.17) is zero. If $\mathcal{P}^h$ is not all of div $\mathcal{S}^h$ then there is a nonzero member of div $\mathcal{S}^h$ orthogonal to all of $\mathcal{P}^h$. Such a field is not exactly incompressible but is weakly incompressible. This takes care of (a). In case (b), we easily deduce that there are no modes for the same reason there are none in case (a). This time there is nothing in div $\mathcal{S}^h$ orthogonal to all of $\mathcal{P}^h$ except the zero field. Thus all weakly incompressible fields are exactly incompressible fields. This takes care of case (b). In case (c), weakly incompressible implies exactly incompressible for the same reason as in case (b). When the containment is proper, there is something nonzero in $\mathcal{P}^h$ orthogonal to all of div $\mathcal{S}^h$, because the latter is a subspace of the former, but any such pressure is a mode by definition. This takes care of case (c) and proves the theorem. ■

The proof of the following corollary is left to the reader:

Corollary 5.1. When the global constant is a pressure mode, Theorem 5 holds if $\mathcal{P}^h$ is replaced by the subspace of $\mathcal{P}^h$ consisting of members of $\mathcal{P}^h$ orthogonal to constants, and the word "mode" is replaced by the words "mode other than the global constant."

What the theorem and corollary say is that in the special case of containment holding one way or another as described, modes result from having more pressures than needed to enforce exact incompressibility. We are at the heart of the matter; now is the time to bear down and hold fast to the unraveling threads.

One of the first questions which the reader is likely to ask is: Are there any elements for which the containment relations hold? We have to hedge the answer somewhat by saying that it is *easy* to find elements for which case (c) of Theorem 5 and Corollary 5.1 hold, but it is very hard to verify cases (a) and (b). The reason we have to hedge is a *crucial* point: It is easy to identify the possible polynomial terms in div $\mathcal{S}^h$, and in most cases, there *seem* to be standard finite element spaces that could be taken for $\mathcal{P}^h$ and contain *precisely* those terms. For example, div $\mathcal{S}^h$ for quadratic triangles contains pure linear terms, for quadratic serendipity elements div $\mathcal{S}^h$ contains pure quadratic terms, for biquadratic elements div $\mathcal{S}^h$ contains just those terms associated with quadratic serendipity elements, and so on. But div $\mathcal{S}^h$ always contains fewer independent members than the total number of possible polynomial terms. The reason is that interelement continuity enforced in $\mathcal{S}^h$ constrains the polynomials in the divergences to be related between elements in some complicated and hard-to-characterize way, so that fewer than all possible combinations of piecewise discontinuous polynomial terms available are actually allowed. If one attempts to strike the balance of case (b)—which seems to be the golden mean for nonlocking elements—by assuring that the correct polynomial terms are independently present in $\mathcal{P}^h$ on each element, the results will almost surely be modes. Since it is an extremely difficult task to assure that $\mathcal{P}^h$ is entirely composed of members which are divergences of continuous displacements, it is no more easy to arrange case (a) either. The only practical course, if exact incompressibility is desired, is to match all possible polynomial terms independently in $\mathcal{P}^h$ and engender modes. Recent work of L. R. Scott and M. Vogelius appears to show that with higher-degree triangles and a particular choice of boundary condition, there are—almost miraculously—examples of the occurrence of case (b) [91].

Theorem 5 and its corollary are of practical importance for nonlocking elements and triangles, in particular. It forces us to subtly reinterpret spurious modes of the second kind: The picture we got in the wake of Theorem 3 was a picture in which $\mathcal{P}^h$ was given as if from on high. The members of $\mathcal{P}^h$ were assumed to be independent constraints until proven otherwise—if there were too many constraints, we could buy back an incompressible field only at the price of a mode. Now the point of view adopted starts by looking at $\boldsymbol{V}^h$ and tries to get a $\mathcal{P}^h$ that is accurate enough and as close as possible to div $\mathcal{S}^h$; this can be done easily only by throwing in all possible polynomial terms, and in so doing we usually get too many pressures, at least as far as enforcing incompressibility without modes is concerned. These excess pressures just get dumped into the mode pool; as long as we have assured at least case (b) of Theorem 5, we could continue (uselessly) to pour on the pressures, and they would continue to run off into the mode pool. Those readers who are familiar with Fraeijs de Veubeke's limitation principle [42] will now recognize our new interpretation of modes of the second kind. We are saying that div $\mathcal{S}^h$ usually falls inconveniently "in between" possible choices for $\mathcal{P}^h$, and modes result from crossing the limitation line in choosing a convenient and accurate $\mathcal{P}^h$.

Many of the locking elements can be successfully employed by backing off on the incompressibility constraint to obtain some kind of analogy to case (a) of Theorem 5. The most successful of these elements appear to be the Lagrange elements. Taking these examples , we see that an element of the "bi-k degree" is really an element of degree k in terms of attainable accuracy, but it contains terms of degree $2k$. Apparently optimal accuracy can be obtained by choosing pressure trial spaces that contain complete polynomials of degree $k - 1$. The best example of this is the biquadratic displacement–pure linear pressure element of Fig. 4.3.4(h): an optimal, safe element, and probably the best of all safe elements. The key to the success of Lagrange elements is that optimal pressure accuracy can be obtained without including all the polynomial terms from the divergences of the "partial polynomials" of degree higher than $k - 1$. Inclusion of such terms would, of course, lock the element. Note that the same cannot be said of triangles: A triangle of degree k has only terms of degree $k - 1$ in div $\mathcal{S}^h$, and it seems that they all need to be included to get optimal accuracy. Backing off on the constraints for triangles seems doomed to suboptimality—*unless continuous pressure fields are used.* Nodal-value sharing can reduce the constraint ratio to avoid modes while retaining full accuracy of the pressures. Continuous pressure fields are not very suitable for pushing to the ultimately low level of compressibility, because there is no recourse to penalty methods using globally connected pressure fields if modes are incurred. And as has been observed for equal-order interpolations (Fig. 4.3.6(a) and (b)) modes can result without enforcing a high degree of incompressibility. Continuous pressure fields can be employed successfully with quadrilaterals (see Fig. 4.3.4(c)), but they seem most appropriately used to retain optimal accuracy with triangles and avoid spurious modes. (Of course, the price of sharpness of enforcement of the constraint is extra unknowns and lack of positive-definiteness.)

Back to biquadratics with discontinuous pressure fields: Biquadratic displacement–linear pressure elements back off on the constraint as far as possible, and there are no modes of either the first or second kind [92]. The biquadratic displacement–bilinear pressure element (Fig. 4.3.4(g)) moves closer to the edge and can give a spurious mode of the first kind [92]. The advantages of the first of these two elements is that it is safe and optimal; the advantage of the second is that it enforces the constraint more strongly and can be used in a reduced selective formulation. The biquadratic displacement–linear pressure element must be used in $\bar{\boldsymbol{B}}$-form to avoid the integration error possible using the three-point formula. This is no real drawback and can be done very efficiently, but it does mean that the analyst cannot implement the element just by changing the integration formula in a pre-existing code. We do not know if the biquadratic displacement–linear pressure element precisely satisfies case (a) of Theorem 5 or its corollary; that is difficult to check. The spirit of the element is very much the same as case (a): By taking a subspace of the polynomial terms in the divergences, the effect is like taking a $\mathcal{P}^h$ contained in div $\mathcal{S}^h$, even if precise containment is not achieved. For the biquadratic displacement–bilinear pressure element, in many circumstances where there is a constant pressure mode there is a partner spurious mode [92]; when this happens it means that containment cannot hold either way between div $\mathcal{S}^h$ and $\mathcal{P}^h$. (The reader should try to prove this as an exercise.) The biquadratic displacement–bilinear pressure element and the bilinear displacement–constant pressure elements are sitting on the edge of over-constraining. In fact the spurious mode of the first kind, which often accompanies the global constant mode, can be seen to be symptomatic of this. If no boundary conditions are enforced, $\bar{\boldsymbol{G}}$ has no spurious mode. When, say homogeneous boundary conditions are enforced all around Ω, $\mathcal{S}^h$—and thus div $\mathcal{S}^h$—are correspondingly reduced in dimension, but $\mathcal{P}^h$ is not reduced in size. For the two elements under consideration, exactly two modes result: the checkerboard and the constant, which are precisely the modes that should have been removed from $\mathcal{P}^h$ in a corresponding reduction in size. In general, modes of the first kind result because enforcing some essential boundary conditions engenders one or more incompressibility constraints, rendering redundant one or more of those imposed by $\mathcal{P}^h$. Modes of the second kind can arise because added degrees of freedom in excess of those required to force exact incompressibility are, of necessity, redundant. Modes are the inevitable partner of redundant constraints.

4.II.6 ERROR ESTIMATES AND PRESSURE SMOOTHING

We would like to be able to say that now that we have come to grips with modes, the way to error estimates is clear. The way to error estimates for safe elements is clear and has been for some time, but to say the same for some of the other elements discussed so far would amount to journalistic excess. The fact is that recent theoretical advances [85, 86, 90, 91, 93–95] have led to the etablishment of error estimates for elements such as the bilinear displacement–constant pressure element, which at one time appeared to be a lost cause—in theory, at least [93, 95]. These results apply in the most generality to the bilinear displacement–constant pressure element, but require quite restrictive assumptions in other cases. Also these results have so far been established for homogeneous boundary conditions on all of Γ, which of course, are automatically consistent with all modes in the sense of Theorem 1. It appears to be an easy matter to extend these results to consistent inhomogeneous boundary conditions, but we are not aware that anyone has done so to date. Still, these results tend to inspire the authors to believe that more comprehensive estimates are possible. Even though the theoretical advances are limited in scope, when they are supported by a reinterpretation of computational experience in the light of what is now known about pressure modes, a coherent picture seems to emerge, which suggests that many more elements than just the safe ones can lead to convergent schemes. Now that we understand the role of the algebraic consistency condition imposed by modes, we know that Example 1 and other related puzzles uncovered in [83] do *not* mean that the elements are useless but mean that some essential boundary conditions need to be modified. In fact, spurious modes of the first kind are observed for some elements only on geometrically simple meshes, and these are precisely the sorts of meshes on which error estimates can be obtained by the new techniques. A detailed exposition of the new analysis is beyond the scope of the present discussion. That is probably just as well; it is often the case in mathematics that new discoveries are presented rather awkwardly at first and neat exposition results only after repeated reworking and refining. Here we would like to give a brief summary of an approach for obtaining estimates in the spirit of the new analysis, because there are some important practical consequences. The most important is that convergence of the pressures in the L_2-norm cannot be guaranteed—even though it is often observed in practice. Because the pressures may not converge, we will propose that "smoothing" or "filtering" of the pressures be undertaken as a routine procedure when using all but the safe elements.

To make things simple and to stick to conditions for which estimates have been worked out rigorously, let us consider a simple elasticity problem—the problem of Sec. 4.2 with $\Gamma = \Gamma_g$ and $\boldsymbol{g} \equiv \mathbf{0}$. Some of the results require Ω to be a rectangle, but the general prescription does not put any restriction on the domain. Restrictions arise from trying to apply the prescription to specific elements, so we will mention them only when the need arises. The analysis we summarize here was developed by Malkus and Olsen [85, 86]. Much of it was inspired by a desire to synthesize the ideas found in [88, 89, 93, 94] in a way that relates those works to the more heuristic criteria for successful constrained finite elements that we have studied here. We wanted an error analysis that somehow recovered the notions of "overconstraining" and "underconstraining" and related those notions to actual convergence rates. A fuller explanation of this synthesis may be found, as first presented, in [85] and, as later refined, in [86]. We will deal with the estimates for the Lagrange-multiplier method of Sec. 4.3, which leads to (4.3.21) with $\boldsymbol{M} = \boldsymbol{0}$ in the discrete form. We assume that the integrations are exact for the matrices involved. Again, to apply our general prescription to specific elements, severe restrictions on the mesh are often required but will be mentioned only when necessary. In most, if not all, situations of practical import, the deviatoric part of the strain energy, $\bar{a}(\boldsymbol{u}, \boldsymbol{u})$ of (4.3.14) is positive-definite and bounded; that is, it satisfies (4.1.3). Thus the energy is equivalent to the $\|\cdot\|_1$-norm and will be used in its place from here on. Whenever we use the symbol $\|\boldsymbol{u}\|_E$, it will be understood to mean $\bar{a}(\boldsymbol{u}, \boldsymbol{u})^{1/2}$ in what follows.

The central idea in relating our more heuristic concept of constraint balance to the more-rigorous notion of convergence rate estimate is projection onto the subspace of weakly divergenceless trial functions, $\boldsymbol{W}^h$, given by

$$Z_h : H^1 \rightarrow \boldsymbol{W}^h \tag{4.II.18}$$

which assigns to each $\boldsymbol{u} \in H^1$ its (deviatoric) energy best approximation from $\boldsymbol{W}^h$. (H^s denotes the Sobolev spaces defined in Appendix 4.I; they consist of vector-valued functions when associated with displacements and scalar-valued functions when associated with pressures.) In the present case we mean projection in the sense of the Hilbert projection theorem—not projection onto $\mathcal{S}^h$ as was discussed in Sec. 4.1, but projection onto the special subspace of $\mathcal{S}^h$ whose members satisfy the weak constraint imposed by the Lagrange-multiplier method. We will argue that if we could estimate the accuracy, in the energy norm, of $\boldsymbol{W}^h$ as an approximating space of functions, then we could estimate the accuracy of the finite element displacement solution. The functions that $\boldsymbol{W}^h$ needs to approximate are possible exact solutions; as such, they are exactly incompressible and smoother than arbitrary functions in H^1. A key point is that, under such circumstances, the estimates of the displacements can be uncoupled from the convergence rate of the pressure solution, which cannot be guaranteed. In [85], it is proved that if $\boldsymbol{u}$ is the exact displacement solution, $\boldsymbol{u}^h$ is the finite element solution, and p is the exact pressure solution (L_2-orthogonal to constants), then

$$\boxed{\bar{a}(Z_h\boldsymbol{u} - \boldsymbol{u}^h, Z_h\boldsymbol{u} - \boldsymbol{u}^h)^{1/2} \leq C\|q^h - p\|_0 \qquad \text{for all } q^h \in \mathcal{P}^h} \tag{4.II.19}$$

for a constant, C, which is equal to zero if $\mathcal{P}^h \supset \operatorname{div} \mathcal{S}^h$. Some points here bear strong emphasis:

1. The finite element pressure solution appears *nowhere* in (4.II.19) *nor* is its accuracy required to derive the result.
2. The arbitrariness of the q^h in the right-hand side of (4.II.19) means that it could be taken to be the nodal interpolate to p (should that exist) or the L_2-best approximation to p, still retaining the inequality. Thus the right-hand side can be made as small as the pressures are accurate.

3. Equation (4.II.19) says that the finite element displacement solution is as close to being the energy best approximation from $\boldsymbol{W}^h$ to the exact displacement solution as is constrained by the *accuracy* of the pressure trial space—except in case (b) or (c) of Corollary 5.1, when the finite element solution *is* that best approximation.
4. Equation (4.II.19) should be compared and contrasted to the results of Sec. 4.1, where it was found that the unconstrained finite element solution was the best energy approximation from all of $\mathcal{S}^h$ to the exact solution: here we have an "almost best approximation" (best when $C = 0$) from a constrained subspace, $\boldsymbol{W}^h$.

By (4.II.19) and the triangle inequality,

$$\|\boldsymbol{u} - \boldsymbol{u}^h\|_E \leq \|\boldsymbol{u} - Z_h\boldsymbol{u}\|_E + C\|q^h - p\|_0 \qquad \text{for all } q^h \in \mathcal{P}^h \tag{4.II.20}$$

So we have a displacement estimate if we could just estimate $\|\boldsymbol{u} - Z_h\boldsymbol{u}\|_E$. This is precisely what the constraint ratio was designed to give us an idea of, in its own humble way. The constraint ratio is a measure of whether $\boldsymbol{W}^h$ has a big enough dimension to be an accurate constrained trial space. The ratio compares the dimension of $\boldsymbol{W}^h$ to the dimension of a trial space assumed to be accurate—$\mathcal{S}^h$—and asks the question: Is the dimension of $\boldsymbol{W}^h$ sufficiently large to be at least half the dimension of $\mathcal{S}^h$? For heuristic reasons, it is assumed that the ratio one-half is optimal for two-dimensional problems because the exact incompressibility constraint seems to "use up" half the degrees of freedom of the exact solution, in that it makes one vector component linearly dependent on the other. The three-dimensional generalization is obvious. A delicate balance is implied by (4.II.19) and (4.II.20), as it was with our earlier more heuristic discussion in Sec. 4.3. The displacement solution can be only as accurate as the pressure solution allows, owing to a pressure accuracy term in the displacement estimate. *Pressure accuracy is what does the constraining*. But we cannot take arbitrarily accurate pressures for reasons we have already observed—too many pressures may overconstrain the problem for locking elements. Too many pressures cause many modes, which cannot sharpen the constraint after a certain point. Most importantly, the best that can be done is to force $\boldsymbol{V}^h \equiv \boldsymbol{W}^h$, which can only work for a nonlocking element; this does not improve the accuracy of the displacement solution beyond the corresponding estimate for $\|\boldsymbol{u} - Z_h\boldsymbol{u}\|_E$, no matter how many pressures we pour on. We will not be able to discuss it in detail here, but error estimates for the smoothed pressures contain a displacement *solution* accuracy term, so pouring on the pressures cannot produce a more-accurate pressure solution, no matter how accurate the pressure trial space has the potential to be. By the way, this last remark points out the beauty of this kind of analysis: It gives an estimate for the displacements first, then the pressures can be considered, with a displacement estimate already in hand. All of this presumes that we can indeed estimate $\|\boldsymbol{u} - Z_h\boldsymbol{u}\|_E$.

Before we go on to discuss ways of estimating the accuracy of $\boldsymbol{W}^h$ as an approximating space, we pause to sharpen our notion of over- and underconstrained elements. Our base accuracy level will be the accuracy in energy of $\mathcal{S}^h$ as an approximating space for approximation of all—not just incompressible—fields, $\boldsymbol{u}$ such that $\|\boldsymbol{u}\|_{k+1}$ is finite, with k as in Sec. 4.1. This level of accuracy is thus $O(h^k)$ and is what will be referred to as "optimal".

Definition. (Constrained Optimality) A finite element for incompressible media—displacement and pressure taken into consideration—will be called ***underconstrained*** if the pressure trial space is less than $O(h^k)$ accurate in approximating pressures for which $\|p\|_k$ is finite. The element will be called ***overconstrained*** if $\|\boldsymbol{u} - Z_h\boldsymbol{u}\|_E$ is less than $O(h^k)$ accurate, no matter how smooth $\boldsymbol{u}$ is in terms of the Sobolev norm. An element will be called ***optimally constrained*** if it is neither over- nor underconstrained.

Equations (4.II.19) and (4.II.20) imply that optimally constrained elements have optimal error estimates, at least as far as convergence rate is concerned. Some specific error estimates require a higher than optimal degree of smoothness of the exact displacement solution to get an optimal convergence rate. The most recent estimates for the bilinear displacement–constant pressure element suggests that this extra smoothness requirement is an artifact of proof technique, as opposed to a genuine restriction. Compare the refined results of [95], which require no such extra smoothness, to the earlier proofs for that element which did require it [93]. However, the extra smoothness is still formally required to get estimates for other elements; this is the reason for the terminology "no matter how smooth" in the definition of overconstrained. We want to allow elements that require extra smoothness to show that $\|\boldsymbol{u} - Z_h\boldsymbol{u}\|_E$ is optimal to be counted as optimally constrained.

Examples of optimally constrained elements are bilinear displacement–constant pressure elements and linear crossed triangles. For the former, the techniques of [93] are refined in [86] and applied to rectangular elements "rectangulating" a domain, which is essentially composed of unions of rectangles, in a grid-type called "rectangularly regular." For the second element, it is shown that for meshes of rectangular elements, whenever the bilinear displacement–constant pressure element is optimal, so is the crossed triangle element. Since the bilinear displacement–constant pressure element estimate of [93] requires the exact displacement solution to be in H^3, so does the crossed triangle estimate of [86]. The newer estimates of [95] for the bilinear element require that the exact solution be only in H^2, and this translates directly into a similar estimate for the crossed triangles on rectangularly regular meshes, and most probably on more complicated ones. The delicate matter in showing this optimality is that neither of the two elements satisfy the LBB-condition [85, 86, 90]. This was proposed in its most cogent form by Brezzi [84], and Fortin [82] was the first to recognize that the bilinear rectangle did not satisfy it. The condition in most practical situations requires that there be no pressure modes—other than the global constant, when appropriate—and that the operator represented by $\boldsymbol{G}\boldsymbol{Q}^{-1}\boldsymbol{G}^T$ be a uniformly invertible operator when restricted to the energy orthocomplement of the incompressible trial functions. Such a restriction *always* makes the operator represented by $GQ^{-1}G^T$ algebraically invertible, but the notion of uniformity means that the energy norm of its restricted inverse is bounded as the mesh is refined. This can be determined from a relatively simple eigenvalue problem [85, 96], which can actually be solved numerically to find out whether an element satisfies the LBB condition on a given sequence of meshes. One of the observations that was at first most puzzling to those trying to untangle this whole business was that it is possible to have boundary conditions for which, say, the bilinear displacement–constant pressure element would have no modes, yet the uniformity requirement would not be met. On other sequences of meshes with the same boundary conditions, numerical evidence could be interpreted to say only that there were no modes *and* the uniformity condition was met. A convergence theory that could hold with or without the LBB condition seemed to be required. If an element satisfies the LBB condition, then it is automatically not overconstrained, because an immediate consequence of the LBB condition is that $\|\boldsymbol{u} - Z_h\boldsymbol{u}\|_E$ is optimal, with no requirement of extra smoothness on $\boldsymbol{u}$ [82, 84–86]. All the elements that we have enumerated as "safe" satisfy the LBB condition (or a subtle variant of it, which can take its place for continuous pressure fields). Indeed all the elements flagged as satisfying the LBB condition in Figs. 4.3.4 and 4.3.6 satisfy the condition or its variant and have no spurious modes. So indeed, elements that satisfy the LBB condition can be, and often are, underconstrained. We have already discussed that matter thoroughly.

We close this brief discussion of displacement estimates for the simple elasticity problem by observing that nonlocking elements can be either overconstrained, underconstrained, or optimal, depending on the choice of $\mathcal{P}^h$. One class of displacement elements can serve to illustrate all the possibilities: quadratic triangles. With piecewise constant pressures, they satisfy the LBB condition and are underconstrained (Fig. 4.3.4(j)); with pure linear, discontinuous pressures, they do not satisfy LBB, are exactly incompressible, and in arrangements without many redundant constraints (like Fig. 4.3.4(i)) are overconstrained and have six modes (five spurious) in the simple elasticity problem on a square domain with square elements [85]. Crossing the quadratic triangles, as with linear crossed triangles, gives an element that is optimally constrained but has one mode of the second kind for each macroelement and an unknown number of the first kind [88]. Finally, as observed earlier, using a continuous linear pressure field with quadratic triangles (Fig. 4.3.6(d)) yields an optimal element with no spurious modes and convergent pressures.

At last, we close this section with the topic that is least clear to us at the time of this writing—pressure smoothing. There is no big picture we know of that gives a heuristic guide to what should work and what will not. In many problems with exact displacement solutions in $H^{k+1}(\Omega)$, optimal convergence to the "raw" pressure representative, $\boldsymbol{p}$, of Corollary 1.1 is observed, in spite of the fact that the uniformity requirement of the LBB condition is not met. It is not easy to see how this can be, since the best the estimates of [85, 86, 90, 93–95] can say is that the blowup of the norm of the inverse of the relevant operator is multiplying the error term for the finite element pressure solution. We do have examples of problems in which the exact displacement solution is less smooth than $H^{k+1}(\Omega)$, so that a loss of convergence rate due to a singularity in the exact solution would be predicted by standard analysis, and where an apparent nonconvergence of the pressures is observed [85]. We also know that the only way a pressure error estimate can be obtained for elements that do not satisfy the LBB condition is by using some kind of pressure "filtering" or "smoothing." This amounts to approximating the raw pressure computed by solving the Lagrange-multiplier or penalty problem by projecting it into an auxiliary pressure trial space, call it $\mathcal{X}^h$. The idea is to choose $\mathcal{X}^h$ so that the L_2-projection operator smooths out the spurious oscillations in the raw pressures. These oscillations may be components of spurious modes, and in many smoothing schemes, $\mathcal{X}^h$ is chosen so that spurious modes in $\mathcal{P}^h$ are projected onto the zero member of $\mathcal{X}^h$. This would have the effect of wiping out the $\boldsymbol{q}$-component of Corollary 1.1. But theoretically and in practice, at least for the problems with rougher exact solutions, there is more to project out than modes. This can be fairly certainly deduced, because oscillation of penalty pressures can be observed in these rougher problems [85, 86], and penalty pressures cannot be infected with a mode component, according to Theorem 4. Presumably, the oscillations of penalty pressures can only be due to the failure of the uniformity portion of the LBB condition. If the LBB condition is satisfied, the raw pressures converge, and there is no need for any smoothing. We know that it is possible in theory to choose $\mathcal{X}^h$ so that it fulfills all the remarkable requirements outlined here, because for some elements and meshes proofs to that effect have been devised [85, 86, 93–95]. The proofs we know of work only with even numbers of elements per side, and $\mathcal{X}^h$ must be composed of 2×2 composite quadrilaterals.This is hardly a desirable choice of pressure element from the computational point of view, since it mixes global and element level computations. There is an abstract condition, a kind of generalized LBB condition [85], which can be applied to $\mathcal{X}^h$ and can predict when a smoothing scheme should work, but it is so hard to prove that it has been successfully applied only in very restrictive circumstances [85].

We and others have found, through numerical experimentation, that many smoothing schemes will work fairly well, in spite of our inability to prove anything about them. The one we have come to favor is the one proposed in [97] and described in Sec. 4.4.1. It was proposed long before we had any idea of all the subtleties that have recently emerged. This is the choice of $\mathcal{X}^h$ as a continuous pressure field based on the displacement element nodes. In the case of linear crossed triangles, the continuous pressure field $\mathcal{X}^h$ is based on the macroelement corner nodes [85, 90]. This scheme has the advantage of referring pressures to displacement nodes, so that both quantities are referred to the same spatial locations. The pressure approximation is continuous, so that some of the good aspects of incompressible elements with continuous pressure fields are recovered by the smoothing scheme. It should be pointed out that since the smoothed pressure can, at best, be expected to converge to the exact pressure in L_2, there is no requirement for convergence at given points. These schemes leave remnants of the oscillations at completely unshared nodes (domain corners), and the pressures in the strip of elements nearest to boundaries are less accurate than could be hoped [85, 90, 97]. There are ways to correct the boundary glitches, which are described in Sec. 4.4.1.

So we have come to the end of this not-so-brief discussion of the subtleties involved in the choice of finite elements for problems with the incompressibility constraint. We have presented an overview of the subject, but we cannot hope to be definitive at this point because this area is one in which new ideas are rapidly developing, and old ideas are being explained in ever simpler and more unified ways. The reader with a strong background and interest in functional analysis might prepare himself or herself to keep abreast of these developments by further reading. We particularly recommend study of [82, 84–86, 91, 93, 95], which provide a deeper analysis of the problems discussed here.

REFERENCES

Section 4.1

1. M. M. Vainberg, *Variational Methods for the Study of Nonlinear Operators*. San Francisco: Holden-Day, 1964.
2. K. Washizu, *Variational Methods in Elasticity and Plasticity*. Oxford: Pergamon Press, 1968.
3. J. T. Oden and J. N. Reddy, *Variational Methods in Theoretical Mechanics*. Heidelberg: Springer-Verlag, 1976.
4. G. Duvaut and J. L. Lions, *Les Inéquations en Mécanique et en Physique*. Paris: Dunod, 1972.
5. G. Strang and G. Fix, *An Analysis of the Finite Element Method*. Englewood Cliffs, N. J.: Prentice-Hall, 1973.
6. P. G. Ciarlet, *The Finite Element Method for Elliptic Problems*. Amsterdam: North-Holland, 1978.
7. J. T. Oden and G. F. Carey, *Finite Elements: Mathematical Aspects*, Vol. IV. Englewood Cliffs, N. J.: Prentice-Hall, 1983.

Section 4.2

8. I. S. Sokolnikoff, *Mathematical Theory of Elasticity* (2nd ed.). New York: McGraw-Hill, 1956.

Section 4.3

9. L. R. Herrmann, "Elasticity Equations for Nearly Incompressible Materials by a Variational Theorem," *AIAA J.*, 3 (1965), 1896–1900.
10. R. L. Taylor, K. S. Pister, and L. R. Herrmann, "On a Variational Theorem for Incompressible and Nearly Incompressible Orthotropic Elasticity," *International Journal of Solids and Structures*, 4 (1968), 875–883.
11. S. W. Key, "A Variational Principle for Incompressible and Nearly Incompressible Anisotropic Elasticity," *International Journal of Solids and Structures*, 5 (1969), 951–964.
12. R. L. Sani, P. M. Gresho, R. L. Lee, D. F. Griffiths, and M. Engleman, "The Cause and Cure (?) of the Spurious Pressures Generated by Certain FEM Solutions of the Navier-Stokes Equations, Parts I and II," *International Journal for Numerical Methods in Fluids*, 1 (1981), 17–43 and 171–204.

13. O. C. Zienkiewicz, *The Finite Element Method.* London: McGraw-Hill, 1977.

14. J. C. Nagtegaal, D. M. Parks, and J. R. Rice, "On Numerically Accurate Finite Element Solutions in the Fully Plastic Range," *Computer Methods in Applied Mechanics and Engineering,* 4 (1974), 153–178.

15. J. H. Argyris, P. C. Dunne, T. Angelopoulos, and B. Bichat, "Large Natural Strains and Some Special Difficulties Due to Nonlinearity and Incompressibility in Finite Elements," *Computer Methods in Applied Mechanics and Engineering,* 4 (1974), 219–278.

16. D. S. Malkus and T. J. R. Hughes, "Mixed Finite Element Methods—Reduced and Selective Integration Techniques: A Unification of Concepts," *Computer Methods in Applied Mechanics and Engineering,* 15, no. 1 (1978), 68–81.

17. T. J. R. Hughes and H. Allik, "Finite Elements for Compressible and Incompressible Continua," *Proceedings of the Symposium on Civil Engineering,* Vanderbilt University, Nashville, Tenn. (1969), 27–62.

18. E. G. Thompson, "Average and Complete Incompressibility in the Finite Element Method," *International Journal for Numerical Methods in Engineering,* 9 (1975), 925–932.

19. M. Crouzeix and P. A. Raviart, "Conforming and Nonconforming Finite Element Methods for Solving the Stationary Stokes Equation," *Revue Française d'Automatique Informatique et Recherche Opérationelle,* R-3 (1973), 33–76.

20. V. Ruas, "A Class of Asymmetric Simplicial Finite Element Methods for Solving Finite Incompressible Elasticity Problems," *Computer Methods in Applied Mechanics and Engineering,* 27 (1981), 319–343.

21. M. Bercovier and O. Pironneau, "Error Estimates for Finite Element Solutions of the Stokes Problem in Primitive Variables," *Numerische Mathematik,* 33 (1979), 211–224.

22. F. Thomasset, *Implementation of Finite Element Methods for Navier-Stokes Equations.* New York: Springer-Verlag, 1981.

23. M. Fortin, "Old and New Finite Elements for Incompressible Flows," *International Journal for Numerical Methods in Fluids,* 1 (1979) 347–364.

24. D. F. Griffiths, "Finite Elements for Incompressible Flows," *Mathematical Methods in Applied Science,* 1 (1979), 16–31.

25. D. F. Griffiths, "The Construction of Approximately Divergence-free Finite Elements," *Mathematics of Finite Elements and Applications III,* ed. J. Whiteman. London: Academic Press, 1979, 239–245.

26. D. F. Griffiths, "An Approximately Divergence-free 9-node Velocity Element (with Variations) for Incompressible Flows," *International Journal for Numerical Methods in Fluids,* 1 (1981), 232–346.

27. D. F. Griffiths, "The Effect of Pressure Approximations on Finite Element Calculations of Incompressible Flows," in *Numerical Methods for Fluid Dynamics,* eds. K. W. Morton and M. J. Baines. London: Academic Press, 1982.

28. J. T. Oden and O. Jacquotte, "A Stable Second-Order Accurate Finite Element Scheme for the Analysis of Two-Dimensional Incompressible Viscous Flows," in *Proceedings of the Fourth International Symposium on Finite Element Methods in Flow Problems,* ed. T. Kawai, Chuo University, Tokyo, July 26–29, 1982.

29. J. Boland and R. Nicolaides, "Stability of Finite Elements Under Divergence Constraints," *SIAM Journal of Numerical Analysis,* 20, no. 4 (1983), 722–731.

30. J. T. Oden and G. F. Carey, *Finite Elements: Mathematical Aspects,* Vol. IV. Englewood Cliffs, N. J.: Prentice-Hall, 1984.

31. C. Johnson and J. Pitkäranta, "Analysis of Some Mixed Finite Element Methods Related to Reduced Intergration," *Mathematics of Computation,* 38 (1982), 375–400.

32. J. Boland and R. Nicolaides, "Stable and Semistable Low Order Finite Elements for Viscous Flow," *SIAM Journal of Numerical Analysis,* 22 (1985), 474–492.

33. J. Pitkäranta and R. Stenberg, "Error Bounds for the Approximation of the Stokes Problem Using Bilinear/Constant Elements on Irregular Quadrilateral Meshes," Report-MAT-A222, Helsinki University of Technology, Institute of Mathematics, Finland, 1984.

34. F. Brezzi and J. Pitkäranta, "On the Stabilization of Finite Element Approximations of the Stokes Equations," Report-MAT-A219, Helsinki University of Technology, Institute of Mathematics, Finland, 1984.

35. T. J. R. Hughes, L. P. Franca, and M. Balestra, "Circumventing the Babuška-Brezzi Condition: A Stable Petrov-Galerkin Formulation of the Stokes Problem Accommodating Equal-Order Interpolation," *Computer Methods in Applied Mechanics and Engineering,* 59(1986), 85–99.

Section 4.4

36. D. S. Malkus and T. J. R. Hughes, "Mixed Finite Element Methods—Reduced and Selective Integration Techniques: A Unification of Concepts," *Computer Methods in Applied Mechanics and Engineering,* 15, no. 1 (1978), 63–81.

37. T. J. R. Hughes, "Equivalence of Finite Elements for Nearly Incompressible Elasticity," *Journal of Applied Mechanics,* 44 (1977), 181–183.

38. I. Fried, "Finite Element Analysis of Incompressible Material by Residual Energy Balancing," *International Journal of Solids and Structures,* 10 (1974), 993–1002.

39. T. J. R. Hughes and H. Allik, "Finite Elements for Compressible and Incompressible Continua," *Proceedings of the Symposium on Civil Engineering,* Vanderbilt University, Nashville, Tenn. (1969), 27–62.

40. J. C. Nagtegaal, D. M. Parks, and J. R. Rice, "On Numerically Accurate Finite Element Solutions in the Fully Plastic Range," *Computer Methods in Applied Mechanics and Engineering,* 4 (1974), 153–178.

41. B. Mercier, "A Conforming Finite Element Method for Two-Dimensional Incompressible Elasticity," *International Journal for Numerical Methods in Engineering,* 14, no. 6 (1979), 942–945.

42. B. Fraeijs de Veubeke, "Displacement and Equilibrium Models in the Finite Element Method," in *Stress Analysis,* eds. O. C. Zienkiewicz and G. S. Holister. London: John Wiley, 1965.

43. L. R. Herrmann, "Elasticity Equations for Incompressible and Nearly Incompressible Materials by a Variational Theorem," *AIAA J.,* 3 (1965), 1896–1900.

44. O. C. Zienkiewicz, R. L. Taylor, and J. M. Too, "Reduced Integration Technique in General Analysis of Plates and Shells," *International Journal for Numerical Methods in Engineering,* 3 (1971), 275–290.

45. D. J. Naylor, "Stresses in Nearly Incompressible Materials by Finite Elements with Application to the Calculation of Excess Pore Pressures," *International Journal for Numerical Methods in Engineering,* 8 (1974), 443–460.

46. O. C. Zienkiewicz and P. N. Godbole, "Viscous Incompressible Flow with Special Reference to Non-Newtonian (Plastic) Fluids," in *Finite Element Methods in Fluids,* Vol. 1. London: John Wiley, 1975.

47. W. P. Doherty, E. L. Wilson, and R. L. Taylor, "Stress Analysis of Axisymmetric Solids Utilizing Higher Order Quadrilateral Finite Elements," SESM Report No. 69-3, Department of Civil Engineering, University of California, Berkeley, 1969.

48. J. H. Argyris, P. C. Dunne, T. Angelopoulus, and B. Bichat, "Large Natural Strains and Some Special Difficulties Due to Nonlinearity and Incompressibility in Finite Elements," *Computer Methods in Applied Mechanics and Engineering,* 4 (1974), 219–278.

49. D. S. Malkus, "Finite Element Analysis of Incompressible Solids," Ph.D. Thesis, Boston University, Boston (1975).

50. D. S. Malkus, "A Finite Element Displacement Model Valid for Any Value of the Compressibility," *International Journal of Solids and Structures,* 12 (1976), 731–738.

51. E. Hinton, "Least Squares Analysis Using Finite Elements," M.Sc.Thesis, Civil Engineering Department, University of Wales, Swansea, 1968.

52. R. L. Lee, P. M. Gresho, and R. L. Sani, "Numerical Smoothing Techniques Applied to Some Finite Element Solutions of the Navier-Stokes Equations," Second International Conference on Finite Elements in Water Resources, London, England, July 10–14, 1978. See also: "Smoothing Techniques for Certain Primitive Variable Solutions of the Navier-Stokes Equations," *International Journal for Numerical Methods in Engineering,* 14 (1979), 1785–1804.

53. T. J. R. Hughes, W. K. Liu, and A. Brooks, "Review of Finite Element Analysis of Incompressible Viscous Flows by the Penalty Function Formulation," *Journal of Computational Physics,* 30, no. 1 (1979), 1–60.

54. J. Barlow, "Optimal Stress Locations in Finite Element Models," *International Journal for Numerical Methods in Engineering,* 10 (1976), 243–251.

Section 4.5

55. J. C. Nagtegaal, D. M. Parks, and J. R. Rice, "On Numerically Accurate Finite Element Solutions in the Fully Plastic Range," *Computer Methods in Applied Mechanics and Engineering,* 4 (1974), 153–178.

56. R. L. Taylor, K. S. Pister, and L. R. Herrmann, "On a Variational Theorem for Incompressible and Nearly-Incompressible Orthotropic Elasticity," *International Journal of Solids and Structures,* 4 (1968), 875–883.

57. S. W. Key, "A Variational Principle for Incompressible and Nearly Incompressible Anisotropic Elasticity," *International Journal of Solids and Structures,* 5 (1969), 951–964.

58. T. J. R. Hughes, "Generalization of Selective Integration Procedures to Anisotropic and Nonlinear Media," *International Journal for Numerical Methods in Engineering,* 15 (1980), 1413–1418.

59. T. J. R. Hughes and D. S. Malkus, "A General Penalty/Mixed Equivalence Theorem for Anisotropic, Incompressible Finite Elements," in *Hybrid and Mixed Finite Element Methods,* eds. S. N. Atluri, R. H. Gallagher, and O. C. Zienkiewicz. London: John Wiley, 1983, 487–496.

60. T. J. R. Hughes, "Recent Developments in Computer Methods for Structural Analysis" *Nuclear Engineering and Design,* 57, no. 2 (1980), 427–439.

61. J. C. Simo and T. J. R. Hughes, "On the Variational Foundations of Assumed Strain Methods," *Journal of Applied Mechanics,* 53, no. 1 (1986), 51–54.

62. T. J. R. Hughes, J. C. Simo, T. Belytschko, and H. Stolarski, "Foundations of Assumed Strain Methods in Finite Element Analysis," *Computer Methods in Applied Mechanics and Engineering,* to appear.

Section 4.6

63. G. P. Bazeley, Y. K. Cheung, B. M. Irons, and O.C. Zienkiewicz, "Triangular Elements in Plate Bending—Conforming and Nonconforming Solutions," *Proceedings of the First Conference on Matrix Methods in Structural Mechanics,* Wright-Patterson AFB, Ohio, 1965.
64. G. Strang and G. J. Fix, *An Analysis of the Finite Element Method.* Englewood Cliffs, N. J.: Prentice-Hall, 1973.
65. F. Stummel, "The Limitations of the Patch Test," *International Journal for Numerical Methods in Engineering,* 15 (1980), 177–188.
66. B. Irons and M. Loikkanen, "An Engineer's Defense of the Patch Test," *International Journal for Numerical Methods in Engineering,* 19, no. 9 (1983), 1391–1401.
67. R. L. Taylor, J. C. Simo, O. C. Zienkiewicz, and A. C. Chan, "The Patch Test: A Condition for Assessing Finite Element Convergence," *International Journal for Numerical Methods in Engineering,* 22, no. 1 (1986), 39–62.

Section 4.7

68. E. L. Wilson, R. L. Taylor, W. P. Doherty, and J. Ghaboussi, "Incompatible Displacement Models," in *Numerical and Computer Models in Structural Mechanics,* eds. S. J. Fenves, N. Perrone, A. R. Robinson, and W. C. Schnobrich. New York: Academic Press, 1973, 43–57.
69. P. Lesaint, "On the Convergence of Wilson's Non-conforming Element for Solving the Elastic Problem," *Computer Methods in Applied Mechanics and Engineering,* 7 (1976), 1–16.
70. R. L. Taylor, P. J. Beresford, and E. L. Wilson, "A Nonconforming Element for Stress Analysis," *International Journal for Numerical Methods in Engineering,* 10, no. 6 (1976), 1211–1219.
71. F. Thomasset, *Implementation of Finite Element Methods for the Navier-Stokes Equations.* New York: Springer-Verlag, 1981.

Section 4.8

72. D. Kosloff and G. A. Frazier, "Treatment of Hourglass Patterns in Low Order Finite Element Codes," *Numerical and Analytical Methods in Geomechanics* 2 (1978), 57–72.
73. D. P. Flanagan and T. Belytschko, "A Uniform Strain Hexahedron and Quadrilateral with Orthogonal Hourglass Control," *International Journal for Numerical Methods in Engineering,* 17 (1981), 679–706.
74. T. Belytschko, J. S-J. Ong, W. K. Liu, and J. M. Kennedy, "Hourglass Control in Linear and Nonlinear Problems," *Computer Methods in Applied Mechanics and Engineering,* 43 (1984), 251–276.

Section 4.9

75. W. P. Doherty, E. L. Wilson, and R. L. Taylor, "Stress Analysis of Axisymmetric Solids Utilizing Higher Order Quadrilateral Finite Elements," SESM Report 69-3, Department of Civil Engineering, University of California, Berkeley, 1969.

76. K. Kavanagh and S. W. Key, "A Note on Selective and Reduced Integration Techniques in the Finite Element Method," *International Journal for Numerical Methods in Engineering,* 4, no. 1 (1972) 148–150.

Appendix 4.I

77. P. G. Ciarlet and P. A. Raviart, "General Lagrange and Hermite Interpolation in $\mathbb{R}^n$ with Application to the Finite Element Method," *Archive for Rational Mechanics and Analysis,* 46 (1972), 177–199.

78. P. G. Ciarlet and P. A. Raviart, "Interpolation Theory over Curved Elements," *Computer Methods in Applied Mechanics and Engineering,* 1, (1972), 217–249.

79. J. T. Oden and J. N. Reddy, *The Mathematical Theory of Finite Elements.* New York: Wiley-Interscience, 1976.

80. S. G. Mikhlin, *Variational Methods in Mathematical Physics.* Oxford; Pergamon Press, 1964.

81. G. Fichera, "Existence Theorems in Elasticity," in *Mechanics of Solids* II, Vol. VIa/2, *Encyclopedia of Physics.* Berlin: Springer-Verlag, 1972.

Appendix 4.II

82. M. Fortin, "An Analysis of the Convergence of Mixed Finite Element Methods," *Revue Française d'Automatique Informatique et Recherche Opérationelle,* 11 (1977), 341–354.

83. R. L. Sani, P. M. Gresho, R. L. Lee, D. F. Griffiths, and M. Engelman, "The Cause and Cure (?) of the Spurious Pressures Generated by Certain FEM Solutions to the Navier-Stokes Equations, Parts I and II," *International Journal for Numerical Methods in Fluids,* 1 (1981), 17–43 and 171–204.

84. F. Brezzi, "On the Existence, Uniqueness and Approximation of Saddle Point Problems Arising From Lagrangian Multipliers," *Revue Française d'Automatique Informatique et Recherche Opérationelle,* 8 (1974), 129–151.

85. D. S. Malkus and E. T. Olsen, "Obtaining Error Estimates for Optimally Constrained Finite Elements," *Computer Methods in Applied Mechanics and Engineering,* 42 (1984), 331–353.

86. E. T. Olsen, "Stable Finite Elements for Non-Newtonian Flows; First-Order Elements which Fail the LBB Condition," Ph.D. Thesis, Illinois Institute of Technology, Chicago, August, 1983.

87. J. C. Nagtegaal, D. M. Parks, and J. R. Rice, "On Numerically Accurate Finite Element Solutions in the Fully Plastic Range," *Computer Methods in Applied Mechanics and Engineering,* 4 (1974), 153–178.

88. B. Mercier, "A Conforming Finite Element for Two-Dimensional Incompressible Elasticity," *International Journal for Numerical Methods in Engineering,* 14, no. 6 (1979), 942–945.

89. B. Mercier, *Topics in the Finite Element Solution of Elliptic Problems,* Tata Institute of Fundamental Research Lecture Series. Berlin: Springer-Verlag, 1979.

90. B. Bernstein, D. S. Malkus, and E. T. Olsen, "A Finite Element for Incompressible Plane Flows of Fluids with Memory," *International Journal for Numerical Methods in Fluids,* 5 (1985), 43–70.

91. L. R. Scott and M. Vogelius, "Norm Estimates for a Maximal Right Inverse of the Divergence Operator in Spaces of Piecewise Polynomials," Technical Note BN-1013, Institute for Physical Science and Technology, University of Maryland, November, 1983.

92. M. Engelman, R. L. Sani, P. M. Gresho, and M. Bercovier, "Consistent vs. Reduced Integration Penalty Methods for Incompressible Media Using Several Old and New Elements," *International Journal for Numerical Methods in Fluids,* 2 (1981), 25–42.

93. C. Johnson and J. Pitkäranta, "Analysis of Some Mixed Methods Related to Reduced Integration," *Mathematics of Computation,* 38 (1982), 375–400.

94. J. T. Oden, N. Kikuchi, and Y. J. Song, "Penalty Finite Element Methods for Stokesian Flows," *Computational Methods in Applied Mechanics and Engineering,* 31 (1982), 297–329.

95. J. Pitkäranta and R. Stenberg, "Error Bounds for the Approximation of the Stokes Problem Using Bilinear/Constant Elements on Irregular Quadrilateral Meshes," Report MAT-A222, Helsinki University of Technology, Institute of Mathematics, Finland, 1984.

96. D. S. Malkus, "Eigenproblems Associated With the Discrete LBB Condition for Incompressible Finite Elements," *International Journal of Engineering Science,* 19 (1981), 1299–1310.

97. T. J. R. Hughes, W. K. Liu, and A. Brooks, "Review of Finite Element Analysis of Incompressible Viscous Flows by the Penalty Function Formulation," *Journal of Computational Physics,* 30, no. 1 (1979), 1–60.

5

The C^0-Approach to Plates and Beams

5.1 INTRODUCTION

The classical Poisson-Kirchhoff theory of plates requires C^1-continuity, just as does the classical Bernoulli-Euler beam theory (see Sec. 1.16). Continuous (i.e., C^0) finite element interpolations are easily constructed. The same cannot be said for multi-dimensional C^1-interpolations. It has taken considerable ingenuity to develop compatible C^1-interpolation schemes for two-dimensional plate elements based on classical theory, and the resulting schemes have always been extremely complicated in one way or another.

More and more, there is a turning away from Poisson-Kirchhoff type elements to elements based upon theories which accommodate transverse shear strains (Reissner and Mindlin theories) and require only C^0-continuity. This approach opens the way to a greater variety of interpolatory schemes but is not without its own inherent difficulties. Recently, displacement-type elements have been derived based upon Reissner–Mindlin theory, which seem to be superior to plate elements derived heretofore. This chapter discusses the basic techniques and considerations involved and summarizes recent developments in this area.

Following this, a similar approach is discussed in the context of beams and frames in which transverse shearing strains are accounted for. This also proves to be extremely simple and effective.

5.2 REISSNER–MINDLIN PLATE THEORY

5.2.1 Main Assumptions

All quantities are referred to a fixed system of rectangular, Cartesian coordinates. A general point in this system is denoted by (x_1, x_2, x_3) or (x, y, z), whichever is more convenient. Throughout, Latin and Greek indices take on the values 1, 2, 3 and 1, 2, respectively.

The main assumptions of the plate theory are

1. The domain Ω is of the following special form:

$$\Omega = \{(x, y, z) \in \mathbb{R}^3 \,|\, z \in \left[\frac{-t}{2}, \frac{t}{2}\right], (x, y) \in A \subset \mathbb{R}^2\}$$

   where t is the plate thickness and A is its area. The boundary of A is denoted by s.
2. $\sigma_{33} = 0$.
3. $u_\alpha(x, y, z) = -z\theta_\alpha(x, y)$.
4. $u_3(x, y, z) = w(x, y)$.

Remarks

1. In Assumption 1, we may take the plate thickness t to be a function of x and y, if desired.

2. Assumption 2 is the plane stress hypothesis. It contradicts Assumption 4 but ultimately causes no problem. The justification of the present theory is its usefulness in practical structural engineering applications. No plate theory is completely consistent with the three-dimensional theory and, at the same time, both simple and useful. Assumption 2 is to be substituted into the constitutive equation; ϵ_{33} is to be solved for and subsequently eliminated.

3. Assumption 3 implies that plane sections remain plane. θ_α is interpreted as the rotation of a fiber initially normal to the plate midsurface (i.e., $z = 0$).

4. By Assumption 4, the transverse displacement, w, does not vary through the thickness.

The sign convention is illustrated in Fig. 5.2.1. "Right-hand-rule" rotations $\hat{\theta}_\alpha$ are defined by $\theta_\alpha = -e_{\alpha\beta}\hat{\theta}_\beta$, where $e_{\alpha\beta}$ is the alternator tensor, viz.,

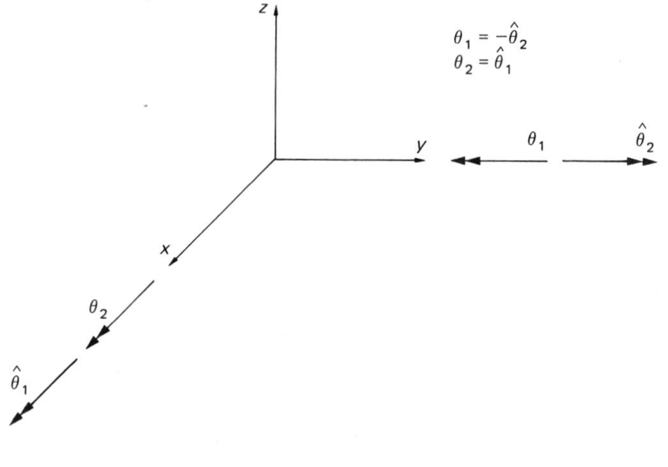

Figure 5.2.1 Sign conventions for rotations. $\hat{\theta}_1$, $\hat{\theta}_2$ are right-hand-rule rotations; θ_1, θ_2 are rotations that simplify the development of the plate theory.

$$[e_{\alpha\beta}] = \begin{bmatrix} 0 & 1 \\ -1 & 0 \end{bmatrix} \tag{5.2.1}$$

We prefer to develop the theory in terms of θ_α rather than $\hat{\theta}_\alpha$ because the algebra is greatly simplified, due to the absence of alternator tensors. In typical structural analysis computer programs, the right-hand-rule convention is usually, but not always, adopted. Consequently, it is the responsibility of the analyst to determine which convention is being employed. It is common when analysts use a new program that errors are made because of lack of careful attention to this point. *Whenever a plate or shell analysis is being undertaken the analyst should check the rotation–bending moment sign convention of the computer program being used before embarking upon that analysis.*

Plate kinematics are summarized in Fig. 5.2.2.

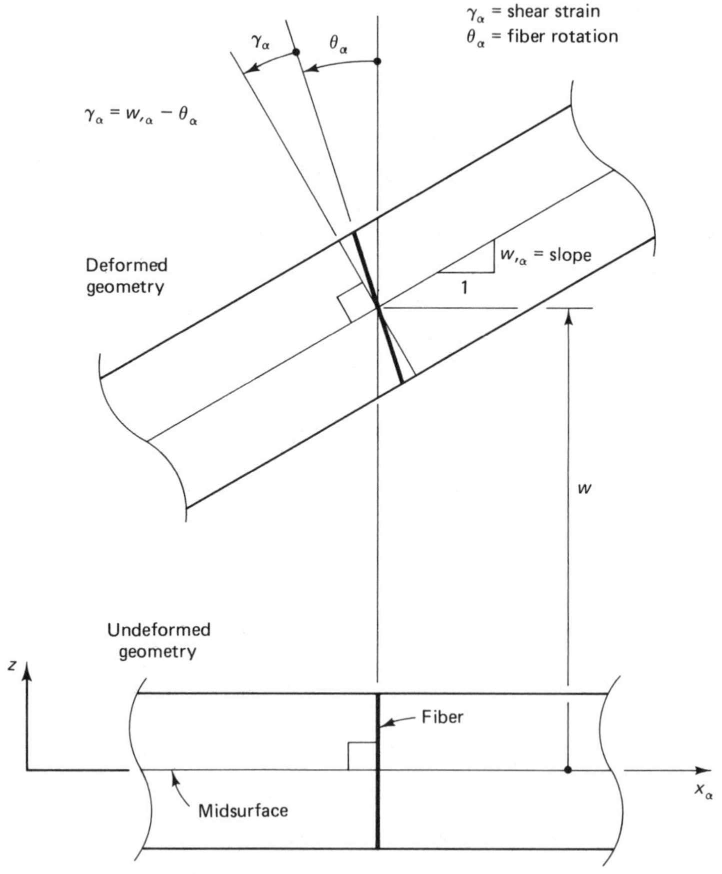

Figure 5.2.2 Plate kinematics. Transverse shear strains need not vanish in the present theory.

5.2.2 Constitutive Equation

The reduced form of the constitutive equation used in the plate theory is determined by substituting Assumption 2 into the three-dimensional constitutive equation and eliminating ϵ_{33}. For simplicity, we shall consider the isotropic case in which

$$\sigma_{ij} = \lambda \delta_{ij} \epsilon_{kk} + 2\mu \epsilon_{ij} \tag{5.2.2}$$

where λ and μ are the Lamé coefficients and δ_{ij} is the Kronecker delta. Assumption 2 implies

$$\epsilon_{33} = \frac{-\lambda}{\lambda + 2\mu} \epsilon_{\alpha\alpha} \tag{5.2.3}$$

$$\sigma_{\alpha\beta} = \bar{\lambda} \delta_{\alpha\beta} \epsilon_{\gamma\gamma} + 2\mu \epsilon_{\alpha\beta} \tag{5.2.4}$$

$$\sigma_{\alpha 3} = 2\mu \epsilon_{\alpha 3} \tag{5.2.5}$$

where

$$\bar{\lambda} = \frac{2\lambda\mu}{\lambda + 2\mu} \tag{5.2.6}$$

$\bar{\lambda}$ and μ may be eliminated in favor of E and ν (Young's modulus and Poisson's ratio, respectively):

$$\bar{\lambda} = \frac{\nu E}{1 - \nu^2} \tag{5.2.7}$$

$$\mu = \frac{E}{2(1 + \nu)} \tag{5.2.8}$$

5.2.3 Strain-Displacement Equations

Assumptions 3 and 4 lead to the following form of the strain-displacement equations:

$$\epsilon_{\alpha\beta} = u_{(\alpha,\beta)} = -z\theta_{(\alpha,\beta)} \tag{5.2.9}$$

$$\epsilon_{\alpha 3} = u_{(\alpha,3)} = \frac{-\theta_\alpha + w_{,\alpha}}{2} \tag{5.2.10}$$

Note that the normal-fiber rotation (i.e., θ_α) and slope (i.e., $w_{,\alpha}$) are not necessarily the same and thus transverse shear strains are accommodated. This is to be contrasted with *classical Poisson-Kirchhoff (i.e., "thin plate") theory* in which $\theta_\alpha = w_{,\alpha}$ and, consequently, $\epsilon_{\alpha 3} = 0$. In the thin plate limit, we usually expect very small transverse shear strains.

5.2.4 Summary of Plate Theory Notations

w	(transverse displacement)
θ_α	(rotation vector)
$\kappa_{\alpha\beta} = \theta_{(\alpha,\beta)}$	(curvature tensor)
$\gamma_\alpha = -\theta_\alpha + w_{,\alpha}$	(shear strain vector)
$m_{\alpha\beta} = \int_{-t/2}^{t/2} \sigma_{\alpha\beta} z \, dz$	(moment tensor)
$q_\alpha = \int_{-t/2}^{t/2} \sigma_{\alpha 3} \, dz$	(shear force vector)
W	(prescribed boundary displacement)
Θ_α	(prescribed boundary rotations)
$F = \int_{-t/2}^{t/2} \ell_3 dz + \langle h_3 \rangle$[1]	(total applied transverse force per unit area)
$C_\alpha = \int_{-t/2}^{t/2} \ell_\alpha \, z \, dz + \langle h_\alpha z \rangle$	(total applied couple per unit area)
$M_\alpha = \int_{-t/2}^{t/2} h_\alpha z \, dz$	(prescribed boundary moments)
$Q = \int_{-t/2}^{t/2} h_3 \, dz$	(prescribed boundary shear force)

5.2.5 Variational Equation

The variational equation of the plate theory is derived from the variational equation of the three-dimensional theory by making use of the preceding relations. The main steps are as follows.

i. Let s_q and s_h be subregions of s which satisfy $\overline{s_q \cup s_h} = s$ and $s_q \cap s_h = \varnothing$. The integrals appearing in the three-dimensional variational equations are replaced by the following iterated integrals:

$$\int_\Omega \ldots d\Omega = \int_A \int_{-t/2}^{t/2} \ldots dz \, dA \tag{5.2.11}$$

$$\int_{\Gamma_h} \ldots d\Gamma = \int_A \langle \ldots \rangle \, dA + \int_{s_h} \int_{-t/2}^{t/2} \ldots dz \, ds \tag{5.2.12}$$

[1] The operator $\langle \ \rangle$ is defined as follows: Let f be an arbitrary function of x, y, and z. Then $\langle f(x,y,z) \rangle = f(x,y,-t/2) + f(x,y,t/2)$.

The kinematic relations are also employed, yielding

$$
\begin{aligned}
0 = & \int_A \int_{-t/2}^{t/2} [\overline{u}_{(\alpha,\beta)}\sigma_{\alpha\beta} + 2\overline{u}_{(\alpha,3)}\sigma_{\alpha 3}]\, dz\, dA \\
& - \int_A \int_{-t/2}^{t/2} (\overline{u}_\alpha \ell_\alpha + \overline{u}_3 \ell_3)\, dz\, dA \\
& - \int_A (\langle \overline{u}_\alpha h_\alpha \rangle + \langle \overline{u}_3 h_3 \rangle)\, dA \\
& - \int_{s_h} \int_{-t/2}^{t/2} (\overline{u}_\alpha h_\alpha + \overline{u}_3 h_3)\, dz\, ds \\
= & \int_A \int_{-t/2}^{t/2} (-\overline{\kappa}_{\alpha\beta}\sigma_{\alpha\beta} z + \overline{\gamma}_\alpha \sigma_{\alpha 3})\, dz\, dA \\
& - \int_A \int_{-t/2}^{t/2} (-\overline{\theta}_\alpha \ell_\alpha z + \overline{w}\ell_3)\, dz\, dA \\
& - \int_A (-\overline{\theta}_\alpha \langle h_\alpha z \rangle + \overline{w}\langle h_3 \rangle)\, dA \\
& - \int_{s_h} \int_{-t/2}^{t/2} (-\overline{\theta}_\alpha h_\alpha z + \overline{w} h_3)\, dz\, ds
\end{aligned} \tag{5.2.13}
$$

where

$$\overline{\kappa}_{\alpha\beta} = \overline{\theta}_{(\alpha,\beta)} \tag{5.2.14}$$

$$\overline{\gamma}_\alpha = -\overline{\theta}_\alpha + \overline{w}_{,\alpha} \tag{5.2.15}$$

Note. In the preceding relations we have used quantities with superposed bars to denote weighting functions in order to avoid notational conflicts and a proliferation of new notations.

ii. The definitions of force resultants are used, yielding

$$
\boxed{
\begin{aligned}
0 = & \int_A (-\overline{\kappa}_{\alpha\beta} m_{\alpha\beta} + \overline{\gamma}_\alpha q_\alpha)\, dA \\
& - \int_A (-\overline{\theta}_\alpha C_\alpha + \overline{w} F)\, dA \\
& - \int_{s_h} (-\overline{\theta}_\alpha M_\alpha + \overline{w} Q)\, ds
\end{aligned}
} \tag{5.2.16}
$$

iii. Integration by parts indicates, under the usual hypotheses, the differential equations and boundary conditions that are satisfied:

$$0 = \int_A \bar{\theta}_\alpha \underbrace{(m_{\alpha\beta,\beta} - q_\alpha + C_\alpha)}_{\text{moment equilibrium}} dA$$

$$- \int_A \bar{w} \underbrace{(q_{\alpha,\alpha} + F)}_{\substack{\text{transverse}\\ \text{equilibrium}}} dA$$

$$+ \int_{s_h} \{\bar{\theta}_\alpha \underbrace{(-m_{\alpha n} + M_\alpha)}_{\substack{\text{moment}\\ \text{boundary}\\ \text{conditions}}} + \bar{w} \underbrace{(q_n - Q)}_{\substack{\text{shear}\\ \text{boundary}\\ \text{condition}}}\} \, ds \tag{5.2.17}$$

where

$$m_{\alpha n} = m_{\alpha\beta} n_\beta \tag{5.2.18}$$

$$q_n = q_\alpha n_\alpha \tag{5.2.19}$$

iv. Explicit forms of the constitutive equations in terms of the plate-theory variables are computed as follows:

$$\begin{aligned} m_{\alpha\beta} &= \int_{-t/2}^{t/2} \sigma_{\alpha\beta} z \, dz \\ &= \int_{-t/2}^{t/2} (\bar{\lambda} \delta_{\alpha\beta} \epsilon_{\gamma\gamma} + 2\mu \epsilon_{\alpha\beta}) z \, dz \\ &= -\frac{t^3}{12} [\bar{\lambda} \delta_{\alpha\beta} \theta_{\gamma,\gamma} + 2\mu \theta_{(\alpha,\beta)}] \\ &= -c_{\alpha\beta\gamma\delta} \kappa_{\gamma\delta} \end{aligned} \tag{5.2.20}$$

where

$$c_{\alpha\beta\gamma\delta} = \frac{t^3}{12} [\mu(\delta_{\alpha\gamma}\delta_{\beta\delta} + \delta_{\alpha\delta}\delta_{\beta\gamma}) + \bar{\lambda}\delta_{\alpha\beta}\delta_{\gamma\delta}] \tag{5.2.21}$$

$$\begin{aligned} q_\alpha &= \int_{-t/2}^{t/2} \sigma_{\alpha 3} \, dz \\ &= \int_{-t/2}^{t/2} 2\mu \epsilon_{\alpha 3} \, dz \\ &= t\mu(-\theta_\alpha + w_{,\alpha}) \\ &= c_{\alpha\beta} \gamma_\beta \end{aligned} \tag{5.2.22}$$

$$c_{\alpha\beta} = t\mu \delta_{\alpha\beta} \tag{5.2.23}$$

Remarks

1. Symmetry of the stiffness matrix will follow from the symmetries

$$c_{\alpha\beta\gamma\delta} = c_{\gamma\delta\alpha\beta} \tag{5.2.24}$$

$$c_{\alpha\beta} = c_{\beta\alpha} \tag{5.2.25}$$

The additional symmetries

$$c_{\alpha\beta\gamma\delta} = c_{\beta\alpha\gamma\delta} = c_{\alpha\beta\delta\gamma} \tag{5.2.26}$$

also hold.

2. To achieve results consistent with classical bending theory it is necessary to introduce a shear correction factor, κ, in the shear force–shear strain constitutive equation. This can be done by replacing $c_{\alpha\beta}$ by $\kappa c_{\alpha\beta}$. Throughout it is assumed that $\kappa = \frac{5}{6}$.

3. More-general material behavior (e.g., orthotropy) can be considered by appropriately redefining the elastic coefficients $c_{\alpha\beta\gamma\delta}$ and $c_{\alpha\beta}$.

5.2.6 Strong Form

The formal statement of the strong form of the plate theory boundary-value problem is as follows.

Given F, C_α, M_α, Q, W, and Θ_α, find w and θ_α such that

$$\left.\begin{aligned} m_{\alpha\beta,\beta} - q_\alpha + C_\alpha &= 0 && (5.2.27)\\ q_{\alpha,\alpha} + F &= 0 && (5.2.28)\\ m_{\alpha\beta} &= -c_{\alpha\beta\gamma\delta}\kappa_{\gamma\delta} && (5.2.29)\\ q_\alpha &= c_{\alpha\beta}\gamma_\beta && (5.2.30)\\ \kappa_{\alpha\beta} &= \theta_{(\alpha,\beta)} && (5.2.31)\\ \gamma_\alpha &= -\theta_\alpha + w_{,\alpha} && (5.2.32) \end{aligned}\right\} \text{ in } A$$

$$\left.\begin{aligned} \theta_\alpha &= \Theta_\alpha && (5.2.33)\\ w &= W && (5.2.34) \end{aligned}\right\} \text{ on } s_g$$

$$\left.\begin{aligned} m_{\alpha n} &= m_{\alpha\beta}n_\beta = M_\alpha && (5.2.35)\\ q_n &= q_\alpha n_\alpha = Q && (5.2.36) \end{aligned}\right\} \text{ on } s_h$$

Sign conventions for stress resultants are depicted in Fig. 5.2.3.

5.2.7 Weak Form

The statement of the variational, or weak, form of the boundary-value problem is as follows.

Given F, C_α, M_α, Q, W, and Θ_α, find $\{\theta_1, \theta_2, w\} \in \mathcal{S}$ such that, for all $\{\overline{\theta}_1, \overline{\theta}_2, \overline{w}\} \in \mathcal{V}$

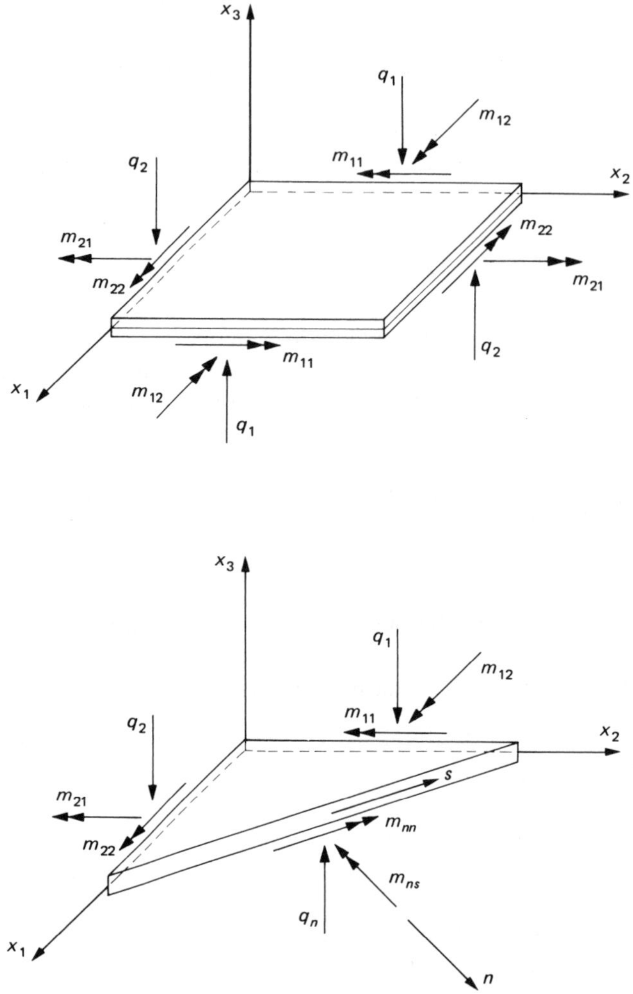

Figure 5.2.3 Sign convention for stress resultants.

$$0 = \int_A [\overline{\theta}_{(\alpha,\beta)} c_{\alpha\beta\gamma\delta}\, \theta_{(\gamma,\delta)} + \overline{\gamma}_\alpha c_{\alpha\beta} \gamma_\beta]\, dA + \int_A (\overline{\theta}_\alpha C_\alpha - \overline{w} F)\, dA + \int_{s_h} (\overline{\theta}_\alpha M_\alpha - \overline{w} Q)\, ds \tag{5.2.37}$$

We assume that if

$$\boldsymbol{u} = \begin{Bmatrix} w \\ \theta_1 \\ \theta_2 \end{Bmatrix} \in \mathcal{S} \qquad \text{(the trial solution space)}$$

then

$$\begin{Bmatrix} w \\ \theta_1 \\ \theta_2 \end{Bmatrix} = \begin{Bmatrix} W \\ \Theta_1 \\ \Theta_2 \end{Bmatrix} \qquad \text{on } s_g$$

and that if

$$\boldsymbol{u} = \begin{Bmatrix} \overline{w} \\ \overline{\theta}_1 \\ \overline{\theta}_2 \end{Bmatrix} \in \mathcal{V} \qquad \text{(the weighting function space)}$$

then

$$\begin{Bmatrix} \overline{w} \\ \overline{\theta}_1 \\ \overline{\theta}_2 \end{Bmatrix} = \begin{Bmatrix} 0 \\ 0 \\ 0 \end{Bmatrix} \qquad \text{on } s_g$$

Exercise 1. Show that (5.2.27), (5.2.28), (5.2.35), and (5.2.36) are implied by (5.2.37).

Exercise 2. Put (5.2.37) into abstract notation: $a(\overline{\boldsymbol{u}}, \boldsymbol{u}) = (\overline{\boldsymbol{u}}, \boldsymbol{\ell}) + (\overline{\boldsymbol{u}}, \boldsymbol{h})_\Gamma$. Define $\boldsymbol{\ell}$ and $\boldsymbol{h}$ and show that $a(\cdot, \cdot)$; $(\cdot, \cdot)$; and $(\cdot, \cdot)_\Gamma$ are symmetric, bilinear forms.

5.2.8 Matrix Formulation

The matrix formulation of the variational equation is given as follows.

$$\boxed{\begin{aligned} 0 = &\int_A (\overline{\boldsymbol{\kappa}}^T \boldsymbol{D}^b \boldsymbol{\kappa} + \overline{\boldsymbol{\gamma}}^T \boldsymbol{D}^s \boldsymbol{\gamma})\, dA \\ &+ \int_A (\overline{\boldsymbol{\theta}}^T \boldsymbol{C} - \overline{w} F)\, dA \\ &+ \int_{s_h} (\overline{\boldsymbol{\theta}}^T \boldsymbol{M} - \overline{w} Q)\, ds \end{aligned}} \tag{5.2.38}$$

where

$$\boldsymbol{\theta} = \begin{Bmatrix} \theta_1 \\ \theta_2 \end{Bmatrix} \qquad \overline{\boldsymbol{\theta}} = \begin{Bmatrix} \overline{\theta}_1 \\ \overline{\theta}_2 \end{Bmatrix} \tag{5.2.39}$$

$$\boldsymbol{\gamma} = \begin{Bmatrix} \gamma_1 \\ \gamma_2 \end{Bmatrix} \qquad \overline{\boldsymbol{\gamma}} = \begin{Bmatrix} \overline{\gamma}_1 \\ \overline{\gamma}_2 \end{Bmatrix} \tag{5.2.40}$$

$$\boldsymbol{\kappa} = \begin{Bmatrix} \kappa_{11} \\ \kappa_{22} \\ 2\kappa_{12} \end{Bmatrix} \qquad \overline{\boldsymbol{\kappa}} = \begin{Bmatrix} \overline{\kappa}_{11} \\ \overline{\kappa}_{22} \\ 2\overline{\kappa}_{12} \end{Bmatrix} \tag{5.2.41}$$

$$\boldsymbol{D}^b = \begin{bmatrix} D_{11}^b & D_{12}^b & D_{13}^b \\ & D_{22}^b & D_{23}^b \\ \text{symmetric} & & D_{33}^b \end{bmatrix} \tag{5.2.42}$$

$$D_{IJ}^b = c_{\alpha\beta\gamma\delta}, \tag{5.2.43}$$

I/J	α/γ	β/δ
1	1	1
2	2	2
3	1	2

$$\boldsymbol{D}^s = \begin{bmatrix} D_{11}^s & D_{12}^s \\ \text{symm.} & D_{22}^s \end{bmatrix} \tag{5.2.44}$$

$$D_{\alpha\beta}^s = c_{\alpha\beta} \tag{5.2.45}$$

5.2.9 Finite Element Stiffness Matrix and Load Vector

The finite element stiffness matrix and load vector may be obtained directly from the matrix form of the variational equation. The finite element approximations of w, $\overline{w}$, θ_α, and $\overline{\theta}_\alpha$ are denoted by w^h, $\overline{w}^h$, θ_α^h, and $\overline{\theta}_\alpha^h$, respectively. In a typical element, possessing n_{en} nodes,

$$w^h = \sum_{a=1}^{n_{en}} N_a w_a^h \tag{5.2.46}$$

$$\overline{w}^h = \sum_{a=1}^{n_{en}} N_a \overline{w}_a^h \tag{5.2.47}$$

$$\theta_\alpha^h = \sum_{a=1}^{n_{en}} N_a \theta_{\alpha a}^h \tag{5.2.48}$$

$$\overline{\theta}_\alpha^h = \sum_{a=1}^{n_{en}} N_a \overline{\theta}_{\alpha a}^h \tag{5.2.49}$$

where N_a is the shape function associated with node a, and w_a^h, $\overline{w}_a^h$, $\theta_{\alpha a}^h$, and $\overline{\theta}_{\alpha a}^h$ are the ath nodal values of w^h, $\overline{w}^h$, θ_α^h, and $\overline{\theta}_\alpha^h$, respectively.

Remark

It is not necessary to assume θ_α^h and w^h are defined in terms of the same shape functions and nodal patterns. However, in the applications we have in mind, this will be the case.

Define

$$\boldsymbol{d}^e = \{d_p^e\} \tag{5.2.50}$$

$$\bar{\boldsymbol{d}}^e = \{\bar{d}_p^e\} \tag{5.2.51}$$

$$d_p^e = \begin{cases} w_a^h & p = 3a - 2 \\ \theta_{1a}^h & p = 3a - 1 \\ \theta_{2a}^h & p = 3a \end{cases} \tag{5.2.52}$$

$$\bar{d}_p^e = \begin{cases} \bar{w}_a^h & p = 3a - 2 \\ \bar{\theta}_{1a}^h & p = 3a - 1 \\ \bar{\theta}_{2a}^h & p = 3a \end{cases} \tag{5.2.53}$$

$$\boldsymbol{\kappa} = \boldsymbol{B}^b \boldsymbol{d}^e \qquad \bar{\boldsymbol{\kappa}} = \boldsymbol{B}^b \bar{\boldsymbol{d}}^e \tag{5.2.54}$$

$$\boldsymbol{\gamma} = \boldsymbol{B}^s \boldsymbol{d}^e \qquad \bar{\boldsymbol{\gamma}} = \boldsymbol{B}^s \bar{\boldsymbol{d}}^e \tag{5.2.55}$$

$$\boldsymbol{B}^b = [\boldsymbol{B}_1^b, \boldsymbol{B}_2^b, \ldots, \boldsymbol{B}_{n_{en}}^b] \tag{5.2.56}$$

$$\boldsymbol{B}^s = [\boldsymbol{B}_1^s, \boldsymbol{B}_2^s, \ldots, \boldsymbol{B}_{n_{en}}^s] \tag{5.2.57}$$

$$\boldsymbol{B}_a^b = \begin{bmatrix} 0 & N_{a,x} & 0 \\ 0 & 0 & N_{a,y} \\ 0 & N_{a,y} & N_{a,x} \end{bmatrix} \tag{5.2.58}$$

$$\boldsymbol{B}_a^s = \begin{bmatrix} N_{a,x} & -N_a & 0 \\ N_{a,y} & 0 & -N_a \end{bmatrix} \tag{5.2.59}$$

With these definitions, the following expressions for the element stiffness and load may be obtained:

$$\boldsymbol{k}^e = \boldsymbol{k}_b^e + \boldsymbol{k}_s^e \tag{5.2.60}$$

$$\boldsymbol{k}_b^e = \int_{A^e} \boldsymbol{B}^{b^T} \boldsymbol{D}^b \boldsymbol{B}^b \, dA \qquad \text{(bending stiffness)} \tag{5.2.61}$$

$$\boldsymbol{k}_s^e = \int_{A^e} \boldsymbol{B}^{s^T} \boldsymbol{D}^s \boldsymbol{B}^s \, dA \qquad \text{(shear stiffness)} \tag{5.2.62}$$

$$\boldsymbol{f}^e = \{f_p^e\} \tag{5.2.63}$$

$$f_p^e = \begin{cases} \displaystyle\int_{A^e} N_a F \, dA + \int_{s^e \cap s_h} N_a Q \, ds & p = 3a - 2 \\ \displaystyle -\int_{A^e} N_a C_1 \, dA - \int_{s^e \cap s_h} N_a M_1 \, ds & p = 3a - 1 \\ \displaystyle -\int_{A^e} N_a C_2 \, dA - \int_{s^e \cap s_h} N_a M_2 \, ds & p = 3a \end{cases} \tag{5.2.64}$$

A^e and s^e are the area and the boundary, respectively, of the eth element. The adjustment to f_p^e for prescribed displacements is given by

$$f_p^e \leftarrow f_p^e - \sum_{q=1}^{n_{ee}} k_{pq}^e g_q, \qquad n_{ee} = 3n_{en} \tag{5.2.65}$$

where

$$g_p = \begin{cases} W(x_a, y_a) & p = 3a - 2 \\ \Theta_1(x_a, y_a) & p = 3a - 1 \\ \Theta_2(x_a, y_a) & p = 3a \end{cases} \tag{5.2.66}$$

The element stresses may be obtained from the following relations:

$$\begin{Bmatrix} m_{xx} \\ m_{yy} \\ m_{xy} \end{Bmatrix} = -\boldsymbol{D}^b \boldsymbol{B}^b \boldsymbol{d}^e \qquad \text{(bending moments)} \tag{5.2.67}$$

$$\begin{Bmatrix} q_x \\ q_y \end{Bmatrix} = \boldsymbol{D}^s \boldsymbol{B}^s \boldsymbol{d}^e \qquad \text{(shear resultants)} \tag{5.2.68}$$

Exercise 3. (Arrays with Respect to Right-hand-rule Rotations.) Show that if right-hand-rule rotations are being employed, i.e., if $\theta_{\alpha a}^h \leftarrow \hat{\theta}_{\alpha a}^h$ in (5.2.48), and, likewise, if $\overline{\theta}_{\alpha a}^h \leftarrow \hat{\overline{\theta}}{}_{\alpha a}^h$ in (5.2.49), then in place of (5.2.58), (5.2.59), and (5.2.64), respectively, we need to use

$$\boldsymbol{B}_a^b = \begin{bmatrix} 0 & 0 & -N_{a,x} \\ 0 & N_{a,y} & 0 \\ 0 & N_{a,x} & -N_{a,y} \end{bmatrix}$$

$$\boldsymbol{B}_a^s = \begin{bmatrix} N_{a,x} & 0 & N_a \\ N_{a,y} & -N_a & 0 \end{bmatrix}$$

$$f_p^e = \begin{cases} \text{same as in (5.2.64),} & p = 3a - 2 \\ -\int_{A^e} N_a C_2 \, dA - \int_{s^e \cap s_h} N_a M_2 \, ds & p = 3a - 1 \\ \int_{A^e} N_a C_1 \, dA + \int_{s^e \cap s_h} N_a M_1 \, ds & p = 3a \end{cases}$$

5.3 PLATE-BENDING ELEMENTS

5.3.1 Some Convergence Criteria

It is important to realize that convergence criteria for elements derived from the present theory are quite different than those for elements derived from thin plate theory. *Necessary* conditions in the present case are:

1. All three rigid body modes must be exactly representable
2. The following five constant strain states must be exactly representable:

$$\left.\begin{aligned}&\theta_{1,1}\\&\theta_{2,2}\\&\tfrac{1}{2}(\theta_{1,2}+\theta_{2,1})\end{aligned}\right\}\quad\text{(curvatures)}$$

$$\left.\begin{aligned}&-\theta_1+w_{,1}\\&-\theta_2+w_{,2}\end{aligned}\right\}\quad\text{(transverse shear strains)}$$

These conditions are satisfied for standard isoparametric elements and for the non-standard isoparametric elements described in the following sections.

5.3.2 Shear Constraints and Locking

An important consideration in the development of plate-bending elements, based upon the present theory, is the number of shear strain constraints engendered in the thin plate limit (i.e., as $t \to 0$). To see this, we consider a heuristic example.

Example 1

Assume a four-node isoparametric quadrilateral element and, for simplicity, assume the element is of rectangular plan and the sides are aligned with the global x- and y-axes. In this case, the element expansions may be written as

$$w^h = \beta_0 + \beta_1 x + \beta_2 y + \beta_3 xy \tag{5.3.1}$$

$$\theta_\alpha^h = \gamma_{\alpha 0} + \gamma_{\alpha 1} x + \gamma_{\alpha 2} y + \gamma_{\alpha 3} xy \tag{5.3.2}$$

where β_i and $\gamma_{\alpha i}$, $0 \le i \le 3$, are constants that depend upon the nodal parameters w_a^h and $\theta_{\alpha a}^h$, $1 \le a \le 4$, respectively. The conditions

$$\begin{aligned}0 &= \gamma_1\\&= -\theta_1^h + w_{,1}^h\\&= (-\gamma_{10} + \beta_1) - \gamma_{11}x + (-\gamma_{12} + \beta_3)y - \gamma_{13}xy\end{aligned} \tag{5.3.3}$$

$$\begin{aligned}0 &= \gamma_2\\&= -\theta_2^h + w_{,2}^h\\&= (-\gamma_{20} + \beta_2) + (-\gamma_{21} + \beta_3)x - \gamma_{22}y - \gamma_{23}xy\end{aligned} \tag{5.3.4}$$

impose eight constraints per element and are approximately in force as $t \to 0$ if exact integration of $\boldsymbol{k}_s^e$ is performed. (Two-by-two Gauss integration is exact in this case.) In a large rectangular mesh, there are approximately three degrees-of-freedom per element, and thus the element tends to be overly constrained. In practice, worthless numerical results are obtained [1]. To alleviate the "locking" effect, one might consider using one-point Gauss quadrature for $\boldsymbol{k}_s^e$. Clearly, this results in only two constraints per element, and now there are more degrees of freedom than there are constraints. This element, with one-point shear integration and 2 × 2 bending integration, was proposed and shown to be effective by Hughes et al. [2].

Arguments similar to those in Example 1 have been used to evaluate other possibilities (see Pugh et al. [1, 3] and Malkus and Hughes [4]).

The situation is seen to be similar to that for the incompressible problem discussed in Chapter 4. Again we shall define the ***constraint ratio,*** r, for the standard mesh of Fig. 4.3.3 by

$$r = \frac{n_{eq}}{n_c}$$

where, in the present case, n_{eq} is the total number of displacement and rotation equations after boundary conditions have been imposed and n_c is the total number of shear strain constraints. Again, the idea is that as the number of equations in the standard mesh approaches infinity, r should approximate the ratio of equilibrium equations to constraints for the governing system of partial differential equations (in the present case, 3 and 2, respectively). Consequently, here the ideal value of r would be $\frac{3}{2}$. Smaller values would indicate the presence of too many shear strain constraints and a potential for locking. A larger value would indicate too few shear strain constraints and suggest that the Kirchhoff limit might be poorly approximated. Note that for the fully integrated four-node element discussed above, $r = \frac{3}{8}$, indicative of locking, whereas if one-point Gaussian quadrature is used for $\boldsymbol{k}_s^e$, then $r = \frac{3}{2}$, the optimal value.

We wish to emphasize again that the constraint ratio is only a quick device for estimating an element's propensity to lock. (See the discussion in Sec. 4.3.7.) In fact, the constraint ratio is not as successful for plates as for incompressible continuum elements. (There are excellent plate elements with constraint ratios less than $\frac{3}{2}$. See, for example, Sec. 5.3.7.) A superior, yet still simple, methodology for assessing the tendency of plates to lock is based upon the ***Kirchhoff mode concept*** [17, 18]. Much of the recent work on plate element design explicitly or implicitly employs this concept. The interested reader should consult [17] for a complete description.

5.3.3 Boundary Conditions

It is important to realize that boundary conditions in the present theory are not always the same as those for the classical thin plate theory. The differences occur in the specification of the "simply supported" case. In the present theory, there are two ways of going about this, depending on the actual physical constraint. Rather than being an additional complication, this freedom turns out to be a considerable benefit, for it enables the solution of problems in which thin plate finite elements have heretofore failed (see Rossow [5] and Scott [6]).

Consider a smooth portion of the plate boundary and a local s, n-coordinate system to it (s denotes the tangential direction and n the outward normal direction; see Fig. 5.3.1). The most common boundary conditions encountered in practice are given as follows.

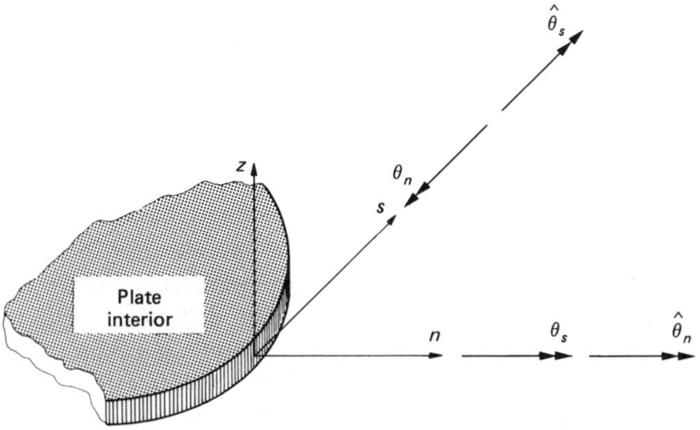

Figure 5.3.1 Local tangential-normal coordinate system at plate boundary.

Clamped

$$w = 0$$
$$\theta_s = 0$$
$$\theta_n = 0$$

Free

$$Q = 0$$
$$M_s = 0$$
$$M_n = 0$$

Simply supported

$$SS_1 \qquad w = 0$$
$$M_s = 0$$
$$M_n = 0$$
$$SS_2 \qquad w = 0$$
$$\theta_s = 0$$
$$M_n = 0$$

Symmetric

$$Q = 0$$
$$M_s = 0$$
$$\theta_n = 0$$

Skew symmetric[2]

$$w = 0$$
$$\theta_s = 0$$
$$M_n = 0$$

In thin-plate theory, SS_2 is the appropriate simply supported boundary condition, since $w = 0$ along the boundary necessitates $\partial w/\partial s = 0$, and the absence of shear strains requires $\theta_s = \partial w/\partial s$. When curved-boundary, simply supported plates are approximated by straight-edged, thin plate elements, SS_2 leads to difficulties. For in this case, specifying $\partial w/\partial s = 0$ at interelement boundaries, for which s is not collinear (see Fig. 5.3.2), implies $\partial w/\partial n = 0$ as well. Thus, as the mesh is refined, the clamped boundary condition is achieved. In other words, the correct solution to the *wrong* problem is attained! Strategies for circumventing this "paradox" tend to be inconvenient from an implementational standpoint (for further details, see Scott [6]).

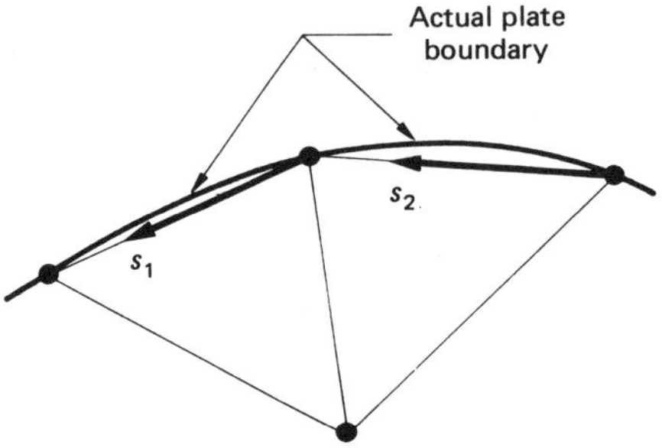

Figure 5.3.2 Approximation to curved plate boundary by straight-edged finite elements.

Another "paradox", discussed by Rossow [5], concerns the solution of a uniformly loaded, simply supported, rhombic thin plate. The analytic solution is singular in the vicinity of the obtuse vertices, with moments of *opposite* sign. However, Sander [7] has reported solutions in which thin-plate elements yield moments of the *same* sign. This phenomenon may also be traced to an overconstraining of thin plate elements caused by SS_2. (Some numerical results for the rhombic plate are presented in Sec. 5.3.8.)

A pleasant feature of the present theory is that the alternative, simply supported boundary condition, SS_1, completely obviates these "paradoxes." In cases in which there is no danger of overconstraining, such as simply supported rectangular plates, SS_2 may be employed to further eliminate degrees of freedom, for greater economy of solution. Computations in support of these assertions are presented in the following sections.

Rhee [8] has discussed the polygonal approximation to simply supported, curved-edged, ***thin plates.*** Rhee has demonstrated convergence of thin plate elements for this case by setting boundary-node values of w, but not $\partial w/\partial x_\alpha$, equal to zero. This simple technique appears to be the most computationally attractive strategy for overcoming the convergence difficulties described above. We note, however, that this

[2] Observe that SS_2 and the skew-symmetric case are the same.

approach does not achieve $w = 0$ along the edge of the plate, except in the fine-mesh limit.

5.3.4 Reduced and Selective Integration Lagrange Plate Elements

To avoid shear "locking" in thin plates, some form of reduced integration is suggested. The two most obvious possibilities are ***uniform reduced integration*** and ***selective reduced integration.***

In uniform reduced integration, both the bending and shear terms are integrated with the same rule, which is of lower order than the "normal" one. Apparently, the first example of a uniform reduced integration plate element was the eight-node serendipity element of Zienkiewicz et al. [9], in which 2×2 Gauss quadrature was employed. Although this element has received wide use, it has now been shown to behave poorly in the thin plate limit [1, 3]. Nevertheless, reduced integration still represents a considerable improvement over normal integration.

In selective reduced integration, the bending term is integrated with the normal rule, whereas the shear term is integrated with a lower-order rule.

Integration rules for Lagrange plate-bending elements are indicated in Fig. 5.3.3. These elements have been investigated by Hughes et al. [2, 10] and Pugh et al. [1, 3], among others.

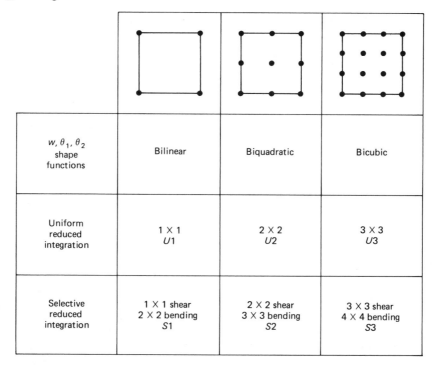

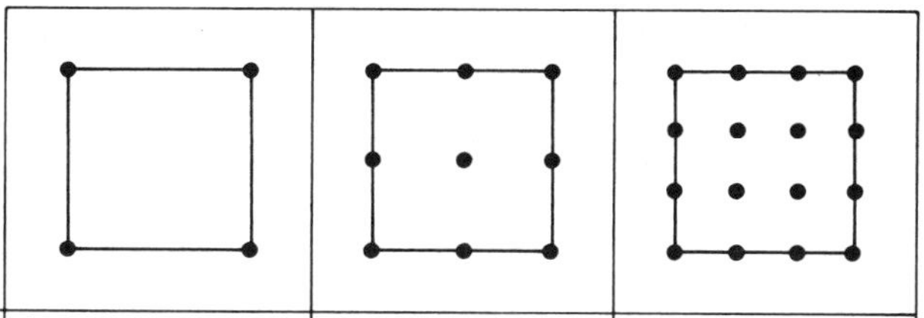

w, θ_1, θ_2 shape functions	Bilinear	Biquadratic	Bicubic
Uniform reduced integration	1×1 $U1$	2×2 $U2$	3×3 $U3$
Selective reduced integration	1×1 shear 2×2 bending $S1$	2×2 shear 3×3 bending $S2$	3×3 shear 4×4 bending $S3$

Figure 5.3.3 Lagrange plate elements. Three degrees of freedom per node: w, θ_1, θ_2.

Numerical Examples

Several numerical examples are presented, which indicate the accuracy attainable by the reduced integration Lagrange elements. Bending moments are reported at "optimal points" [11] (i.e., the Gauss points of the shear integration rule). A Poisson's ratio of 0.3 and Young's modulus of 10.92×10^5 were used throughout. Generally, the meshes are constructed so that there are four $U1$, or $S1$, elements for every $U2$, or $S2$, element, resulting in the same number of equations. The plates analyzed herein may be considered thin, and so comparison is made with classical thin plate theory. However, since shear deformations are included in the present theory, convergence to slightly greater displacements is to be expected.

Square plate

Convergence studies were carried out for both simply supported and clamped, thin, square plates, which were subjected to uniform and concentrated loads. Mesh types are depicted in Fig. 5.3.4 and results are presented in Fig. 5.3.5. The geometric data employed were $L = 10$ and $t = 0.1$. Refinements were constructed by bisection. In this problem, the SS_2 simply supported boundary condition produces convergent solutions and results in fewer equations.

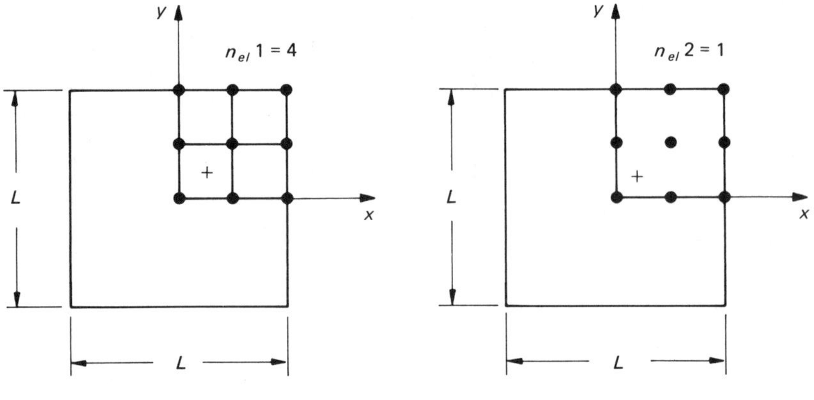

$n_{el}1$ = number of four-node elements $n_{el}2$ = number of nine-node elements

+ denotes location of Gauss point nearest center
(1 X 1 rule for four-node elements and 2 X 2 rule for nine-node elements)

Figure 5.3.4 Square plate. Due to symmetry, only one quadrant is discretized.

Circular plate

A convergence study, similar to the preceding one, was carried out for thin circular plates. Meshes are depicted in Fig. 5.3.6 and results are presented in Fig. 5.3.7. The geometric data employed were $R = 5$ and $t = 0.1$. In this case, it is necessary to use

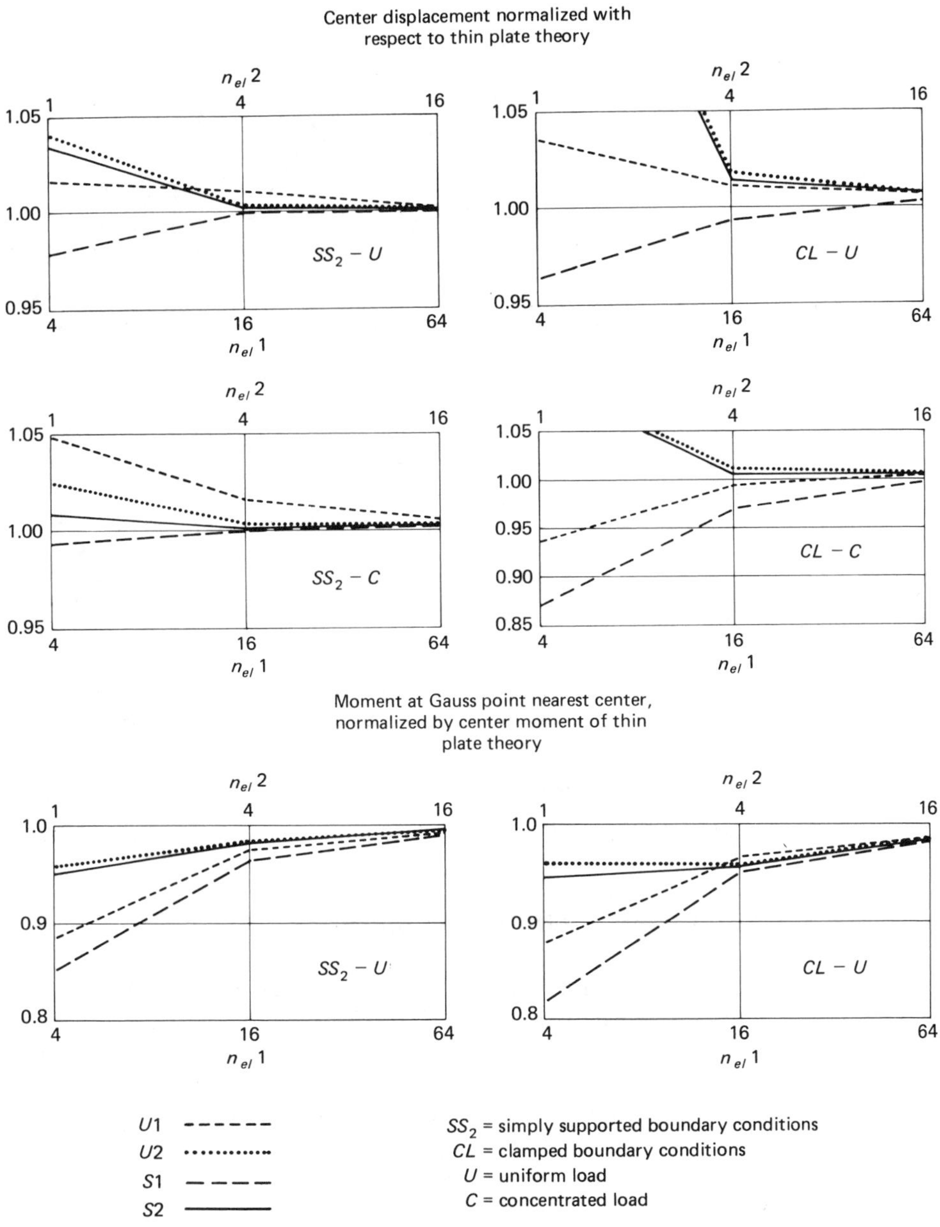

Figure 5.3.5 Convergence study for thin, square plate.

the SS_1 boundary condition to achieve convergence in the simply supported case (see Sec. 5.3.3).

Bending moments for elements $U1$ and $S1$ are plotted in Fig. 5.3.8.

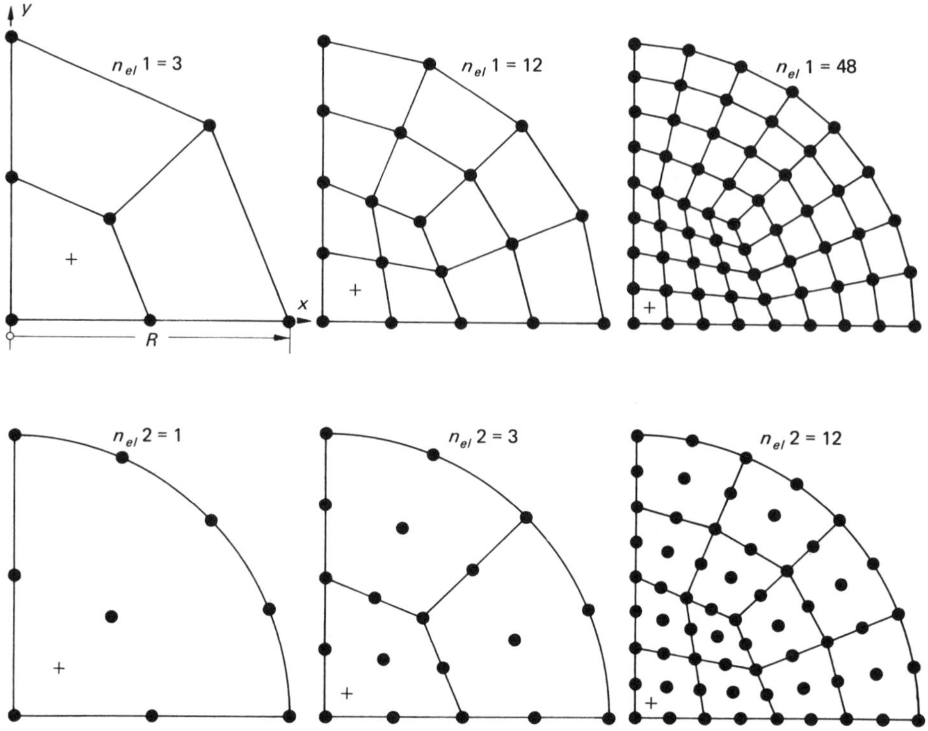

+ denotes location of Gauss point nearest center
(1 × 1 rule for four-node elements and 2 × 2 rule for nine-node elements)

Figure 5.3.6 Circular plate. Due to symmetry, only one quadrant is discretized.

5.3.5 Equivalence with Mixed Methods

The reduced and selective integration Lagrange plate elements have been shown to be identical to elements derived from a mixed formulation, in which the shear forces, q_α, are variables in addition to w and θ_α; see Malkus and Hughes [4]. The shear variables, with nodes at the Gauss points of the shear integration rules, are discontinuous between elements. Some examples of equivalent elements are given below.

Example 2

Let w and θ_α be approximated by bilinear shape functions and let q_α take on constant values within each element. If one-point Gauss quadrature is used to construct the stiffness, it is equivalent to $U1$. If 2×2 Gauss quadrature is used on the bending term and one-point Gauss quadrature is used on the remaining shear term, then the element is equivalent to $S1$. In each case, the constant value of q_α in the mixed formulation is the same as that computed from w and θ_α at the centroid of $U1$ or $S1$, respectively.

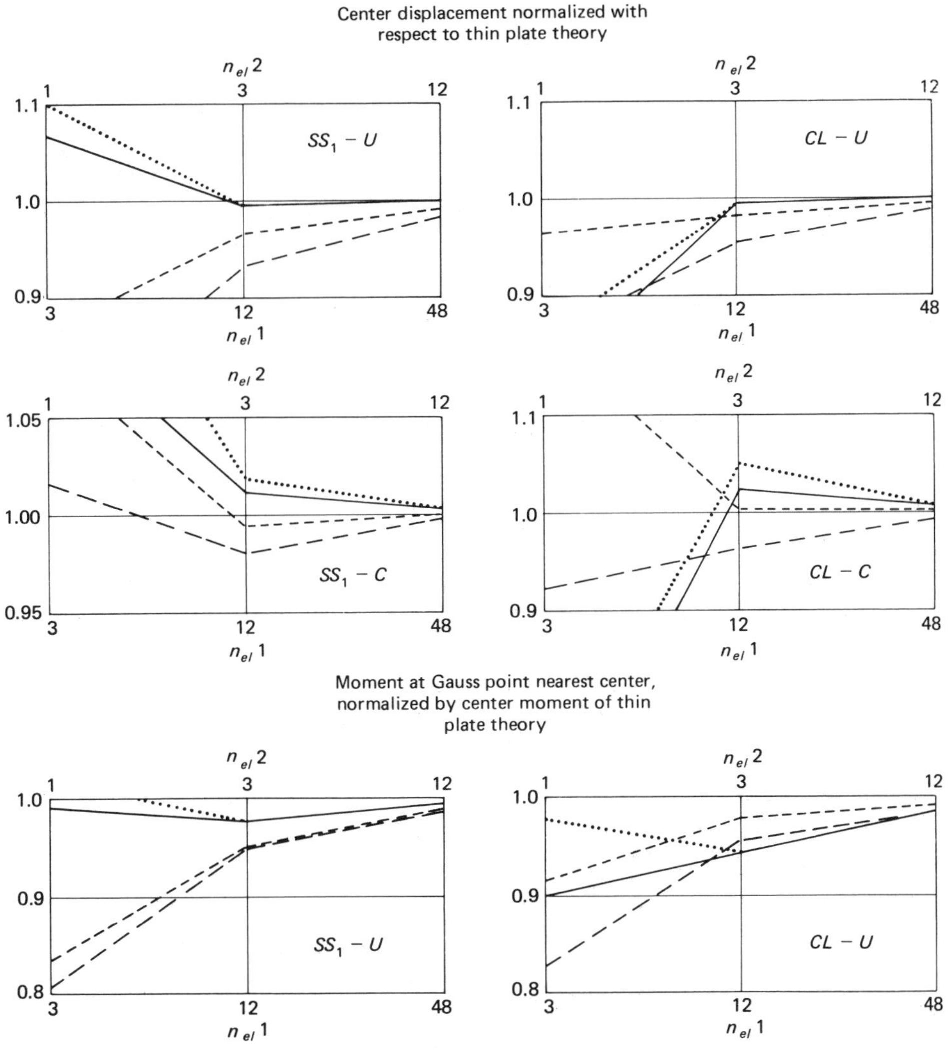

Figure 5.3.7 Convergence study for thin circular plate.

Example 3

Let w and θ_α be approximated by biquadratic shape functions and let q_α be approximated by bilinear shape functions. The nodes for q_α are the 2×2 Gauss points. If the 2×2 Gauss rule is used for all terms, the element is equivalent to $U2$. If the 3×3 rule is used on the bending term and the 2×2 rule is used on the shear terms, the element is equivalent to $S2$. In each case, the bilinear variation of q_α within each element in the mixed formulation may be recaptured in $U2$ and $S2$, respectively, by passing a bilinear function through the 2×2 Gauss point values of q_α computed from w and θ_α.

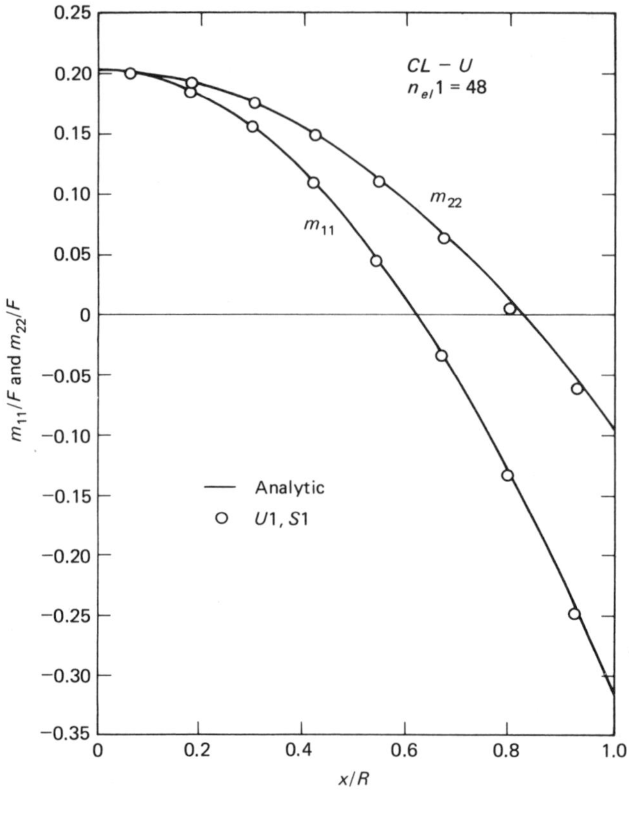

Figure 5.3.8 Circular plate. Bending moments at Gauss points nearest x-axis.

Additional details on these and related matters may be found in Malkus and Hughes [4].

5.3.6 Rank Deficiency

Although reduced integration of Lagrange elements alleviates the shear-locking problem, it has a deleterious side effect: rank deficiency (i.e., there are zero-energy modes in excess of the three rigid-body modes; see Table 5.3.1). This can result in oscillatory errors in certain cases (see Sec. 5.3.7) and, occasionally, even in a singular global stiffness matrix. As can be clearly seen, the number is lower for each selective reduced integration element than for the corresponding uniform reduced integration element, an advantage of the former.

The four zero-energy modes of a square $U1$ element are depicted in Fig. 5.3.9. Element $S1$ possesses only the w hourglass and the in-plane twist modes.

The zero-energy modes for a square $U2$ element are:

$$w = x_1^2x_2^2 - \tfrac{1}{3}(x_1^2 + x_2^2), \qquad \theta_1 = \theta_2 = 0 \tag{5.3.5}$$

$$w = 0, \qquad \theta_1 = 0, \qquad \theta_2 = (x_1^2 - \tfrac{1}{3})(x_2^2 - \tfrac{1}{3}) \tag{5.3.6}$$

$$w = 0, \qquad \theta_1 = (x_1^2 - \tfrac{1}{3})(x_2^2 - \tfrac{1}{3}), \qquad \theta_2 = 0 \tag{5.3.7}$$

$$w = 0, \qquad \theta_1 = x_1(x_2^2 - \tfrac{1}{3}), \qquad \theta_2 = -x_2(x_1^2 - \tfrac{1}{3}) \qquad (5.3.8)$$

Only (5.3.5) is present in $S2$.

TABLE 5.3.1 Zero-energy Modes, in Excess of Rigid-body Modes, for Reduced Integration, Lagrange, Plate Elements

Element	$U1$	$U2$	$U3$	$S1$	$S2$	$S3$
Number of zero-energy modes	4	4	4*	2	1	1**

* Pugh et al. [3].
** Wong [28].

$w = xy$
$\theta_1 = 0$
$\theta_2 = 0$

w-hourglass mode

$w = 0$
$\theta_1 = -y$
$\theta_2 = x$

In-plane twist mode

$w = 0$
$\theta_1 = xy$
$\theta_2 = 0$

θ_1-hourglass mode

$w = 0$
$\theta_1 = 0$
$\theta_2 = xy$

θ_2-hourglass mode

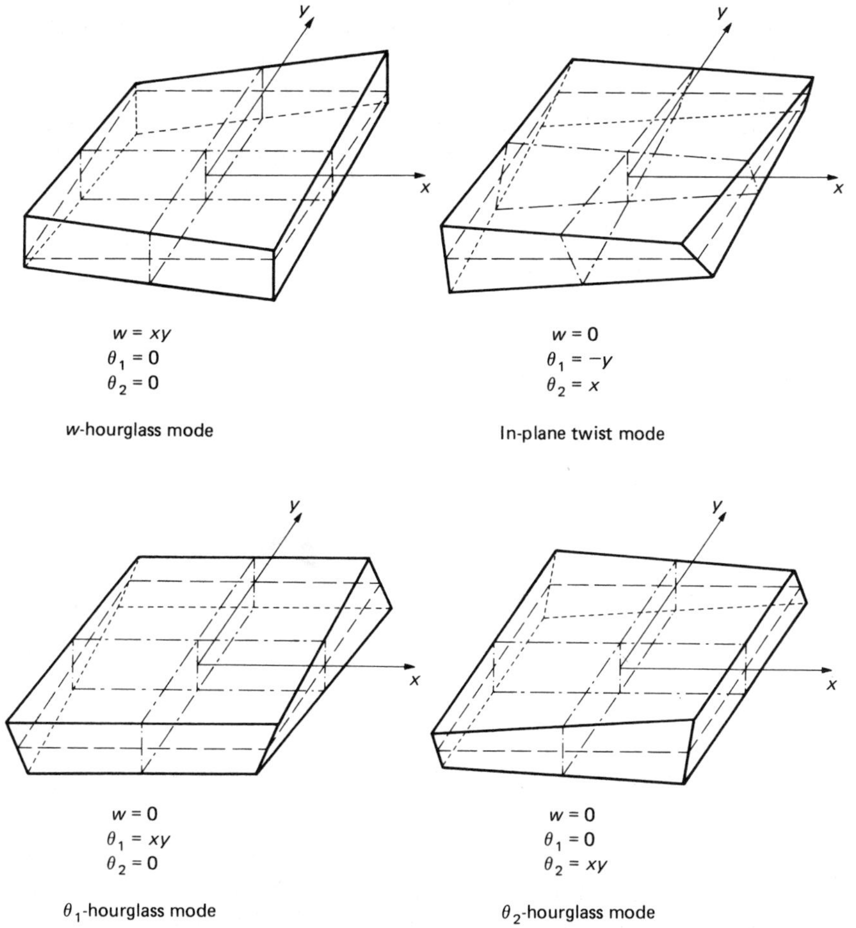

Figure 5.3.9 Zero-energy modes of element $U1$.

In general, boundary conditions render the assembled stiffness matrix positive-definite, and so the zero-energy modes are not globally present. A guideline that we have employed with success for selective integration elements is: If the boundary conditions preclude the rigid mode from forming in one element, it is also precluded in the remainder of the mesh.

It turns out that a sufficient condition for satisfying the guideline for elements $S1$ and $S2$ is that two adjacent nodal values of w be specified in at least one element.

A critical test is provided by a point-supported, pergola roof. (We thank P. C. Jennings for suggesting this problem to us.) The roof is assumed to be a square plate, subjected to uniform transverse load, and "fixed" at the center. A mesh of square $S1$ elements was constructed with the center node clamped (i.e., $w = \theta_1 = \theta_2 = 0$). Rank-check analyses have revealed that the in-plane twist mode disappears in a mesh of two or more $S1$ elements, so it is not a problem. However, the w hourglass is present unless eliminated via boundary conditions. As is clear, a global pattern, in which each element hourglasses, is possible in the present situation, since only one w degree of freedom is specified. (A rank-check analysis confirmed this to be the case, indicating one zero mode.) An alternative mesh was employed, in which the point support was distributed in a diamond pattern; see Fig. 5.3.10. (In fact, this is more in keeping with the physical situation.) Since $w = 0$ along an edge of each of the four central elements, these elements cannot hourglass. A rank-check analysis revealed a positive-definite stiffness, in confirmation of this. Note that it is crucial to have $w = 0$ at the center; otherwise the w hourglass reappears.

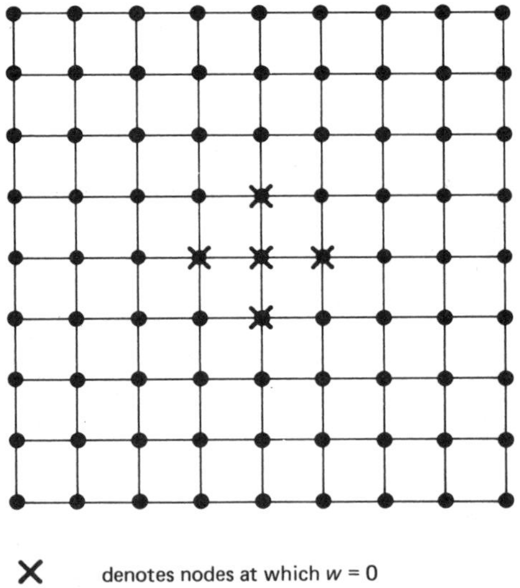

Figure 5.3.10 Pergola roof. Mesh of $S1$ elements exhibiting no zero-energy modes.

Due to the potential problems created by rank deficiency, a number of research efforts have been undertaken to develop Reissner-Mindlin-type elements which have correct rank and maintain accuracy in both thin and thick plate applications. Several elements of this type are discussed in the following sections.

Before going further, we note that there has been considerable interest in the one-point quadrature element $U1$ because of its obvious economy and simplicity.

Some efforts have been made to stabilize its spurious rigid body modes by way of the hourglass-stiffness technique. The interested reader may consult [12] and [13] for details.

5.3.7 The Heterosis Element

The first higher-order Mindlin-type element to simultaneously achieve correct rank and high accuracy was the heterosis element of Hughes and Cohen [14]. The heterosis element represents a synthesis of the selectively integrated nine-node Lagrange element and eight-node serendipity element. The heterosis element combines the attributes of its progenitors but avoids the shortcominings. It may be recalled that the selectively integrated nine-node Lagrange element is rank deficient; and the eight-node serendipity element tends to be overconstrained—the shear-locking phenomenon—occasionally resulting in poor convergence characteristics [1, 10].

The heterosis element has consistently performed well on numerical tests, including cases in which the serendipity and Lagrange elements are poor. Pertinent information for the element is presented in Fig. 5.3.11, along with descriptions of the serendipity and Lagrange elements. The name heterosis was chosen because, in genetics, the improvement in characteristics exhibited by hybrids over those of the parents is called heterosis.

Implementation

The use of two interpolatory schemes, as ostensibly necessitated by the heterosis elements, would be a drawback. However, implementation can be made without such considerations [15]. Specifically, we start with routines that generate the stiffness and internal force for the Lagrange element (denoted by $\boldsymbol{k}^e_{\text{Lag}}$, and $\boldsymbol{f}^e_{\text{Lag}}$, respectively). Then a transformation matrix, $\boldsymbol{H}$, is constructed, which restrains the transverse displacements of the internal nodes to interpolate the serendipity shape functions. The heterosis stiffness, $\boldsymbol{k}^e_{\text{het}}$, and internal force, $\boldsymbol{f}^e_{\text{het}}$, are then defined by

$$\boldsymbol{k}^e_{\text{het}} = \boldsymbol{H}^T \boldsymbol{k}^e_{\text{Lag}} \boldsymbol{H} \tag{5.3.9}$$

$$\boldsymbol{f}^e_{\text{het}} = \boldsymbol{H}^T \boldsymbol{f}^e_{\text{Lag}} \tag{5.3.10}$$

The matrix $\boldsymbol{H}$ is defined as follows:

$$\boldsymbol{H} = \begin{bmatrix} \boldsymbol{I}_{24} & \boldsymbol{0}_{25,2} \\ \boldsymbol{h} & \boldsymbol{I}_2 \end{bmatrix} \tag{5.3.11}$$

$$\boldsymbol{h} = [\boldsymbol{h}_1, \boldsymbol{h}_2, \ldots, \boldsymbol{h}_8] \tag{5.3.12}$$

$$\boldsymbol{h}_a = \begin{bmatrix} N_a^{\text{ser}}(0, 0) & 0 & 0 \\ 0 & 0 & 0 \\ 0 & 0 & 0 \end{bmatrix} \qquad 1 \leq a \leq 8 \tag{5.3.13}$$

where $\boldsymbol{I}_m$ denotes the $m \times m$ identity matrix; $\boldsymbol{0}_{m,n}$ denotes the $m \times n$ zero matrix; and $N_a^{\text{ser}}(0, 0)$ is the serendipity shape function associated with node a evaluated at the origin of isoparametric coordinates (i.e., location of node 9).

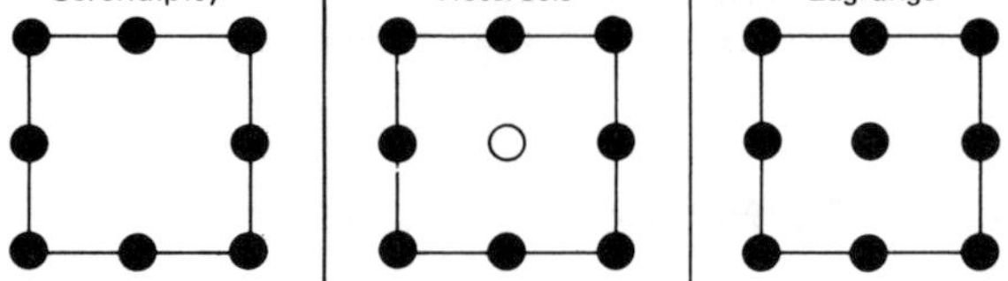

	Serendipity	Heterosis	Lagrange
w-shape functions	Serendipity	Serendipity	Lagrange
θ_1, θ_2-shape functions	Serendipity	Lagrange	Lagrange
Integration scheme	$U2$	$S2$	$S2$
Number of spurious zero-energy modes	1*	0	1
Constraint ratio	1.125	1.375	1.5

Key:

● w, θ_1, θ_2 degrees of freedom

○ θ_1, θ_2 degrees of freedom

$U2$ = 2 × 2 Gauss

$S2 = \begin{cases} \text{bending } 3 \times 3 \text{ Gauss} \\ \text{shear } \quad 2 \times 2 \text{ Gauss} \end{cases}$

*Not communicable in a mesh of two or more elements.

Figure 5.3.11 Quadrilateral plate elements.

The matrix multiplications in (5.3.9) and (5.3.10) are formal, and in actual practice sparse coding may be employed to obtain $\boldsymbol{k}^e_{\text{het}}$ and $\boldsymbol{f}^e_{\text{het}}$ efficiently. Stresses for the heterosis element may again be obtained via (5.2.67) and (5.2.68), where the center transverse displacement is defined by

$$w_9 = \sum_{a=1}^{8} N_a^{\text{ser}}(0, 0) w_a \tag{5.3.14}$$

Note that the derivation of the heterosis element arrays depends upon the serendipity shape functions only through their values at node 9. These eight numbers are independent of the particular element and may be stored once and for all in an element routine through the use of a FORTRAN DATA statement.

An alternative implementation of the heterosis element, based upon the "hierarchical concept," has been described in [16].

Higher-order versions of the heterosis element have been proposed in [15]; see Fig. 5.3.12.

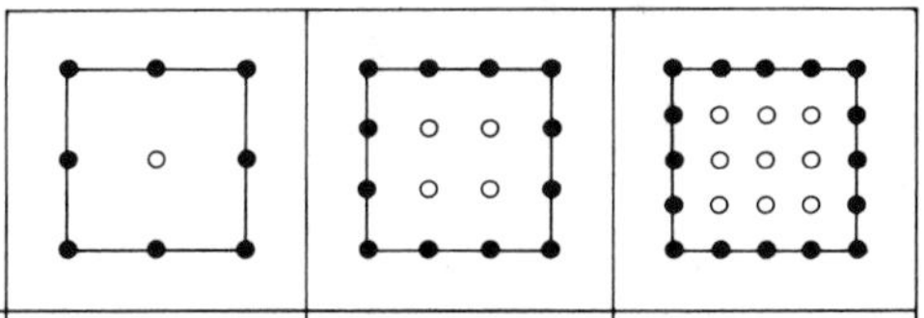

Integration scheme	*S*2	*S*3	*S*4
Constraint ratio	1.375	1.278	1.219

Key: ● w, θ_1, θ_2 degrees of freedom

○ θ_1, θ_2 degrees of freedom

Sn $\begin{cases} \text{bending } (n+1) \times (n+1) \text{ Gauss} \\ \text{shear } \quad n \times n \text{ Gauss} \end{cases}$

Figure 5.3.12 Heterosis plate elements.

Numerical Examples

A Poisson's ratio of 0.3 and a Young's modulus of 10.92×10^5 were used throughout.

Square plate

Mesh patterns are depicted in Fig. 5.3.13. The geometric data employed were $L = 10$ and t (thickness) $= 0.1$.

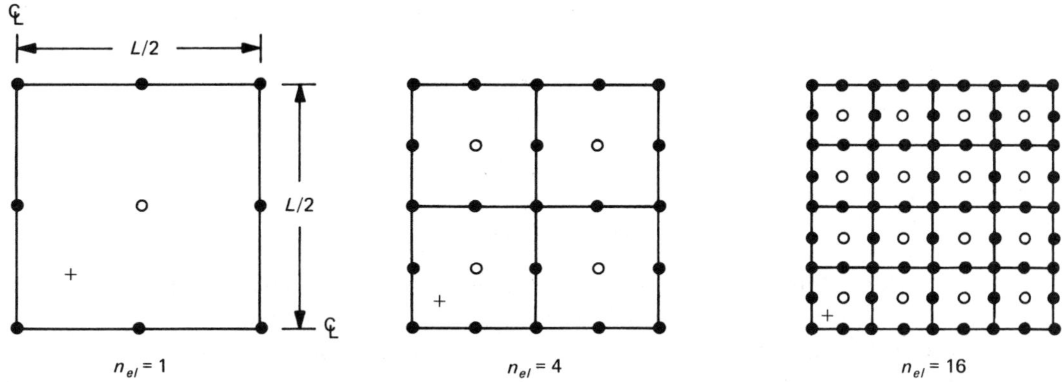

+ denotes location of Gauss point (2 × 2 rule) nearest center

Figure 5.3.13 Square plate. Due to symmetry, only one quadrant is discretized.

The results of a convergence study are presented in Fig. 5.3.14. The serendipity element exhibits particularly poor convergence characteristics for the clamped cases. The Lagrange and heterosis elements demonstrate good convergence properties, with the former being somewhat more accurate.

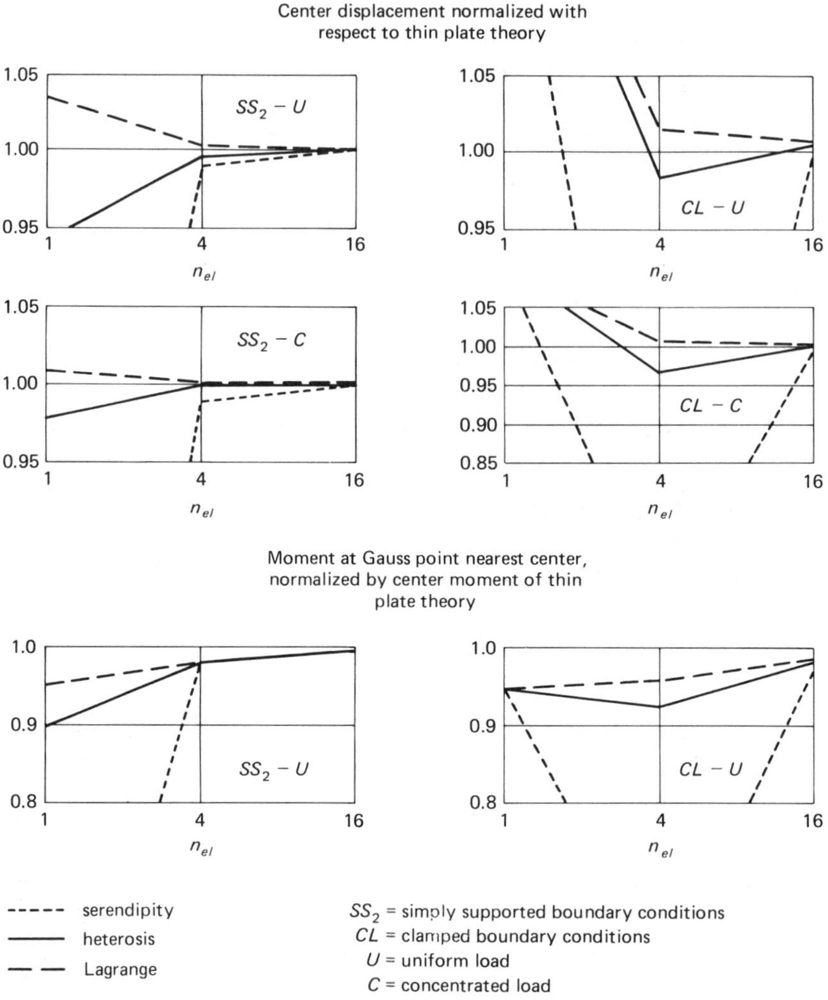

Figure 5.3.14 Convergence study for thin square plate.

The ability of the elements to handle extremely thin plates is illustrated in Fig. 5.3.15. Both the Lagrange and heterosis elements exhibit "stable" convergence characteristics for large length-to-thickness ratios. Numerical deterioration commences at $L/t = 10^6$ because the shear stiffness is $O(L^2/t^2)$ times the bending stiffness, resulting in the bending effects being overwhelmed by rounding errors. The "plateaus" exhibited by the Lagrange and heterosis elements, up to $L/t = 10^6$, may be extended *ad infinitum* by the technique proposed in [2]. On the other hand, the serendipity element diverges rapidly for relatively small values of L/t (i.e., $O(10^2)$).

Rectangular plate

Convergence studies for rectangular plates are presented in [15]. Sample results for the meshes shown in Fig. 5.3.16 are presented in Fig. 5.3.17.

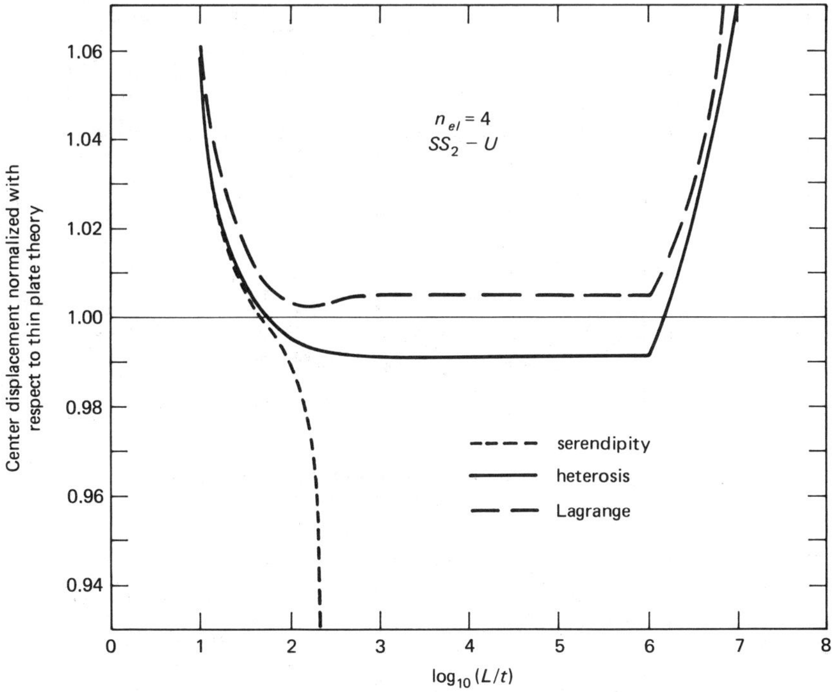

Figure 5.3.15 Aspect ratio study for square plate.

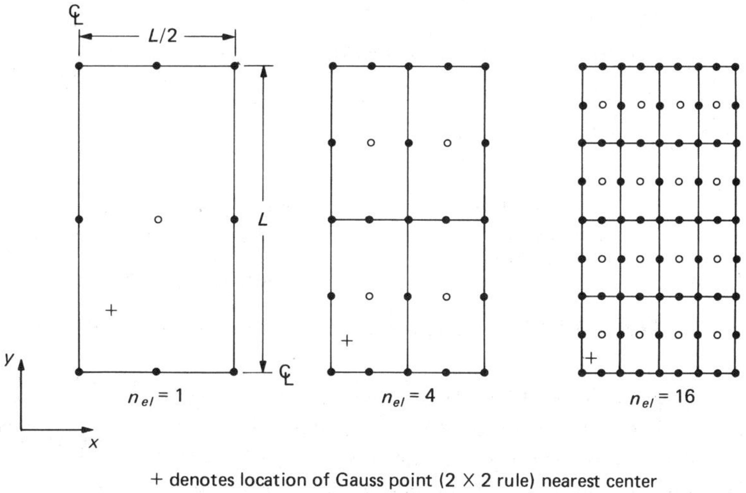

Figure 5.3.16 Rectangular plate (aspect ratio = 2). Due to symmetry, only one quadrant is discretized.

Circular plate

The meshes for this problem are depicted in Fig. 5.3.18. A value of $R = 5$ is used

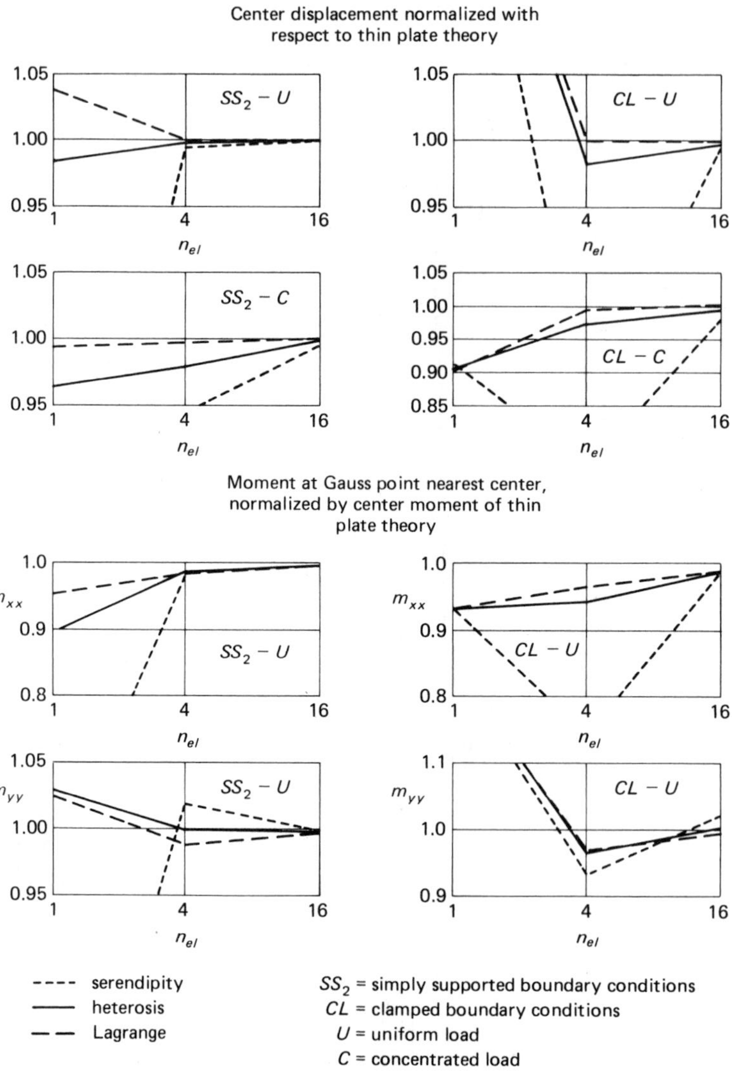

Figure 5.3.17 Convergence study for thin rectangular plate (aspect ratio = 2).

for all cases. Thin plate (i.e., $t = 0.1$) convergence results are presented in Fig. 5.3.19. The heterosis element is consistently superior to the serendipity element.

As we have seen so far, the presence of the spurious zero-energy mode generally does not adversely affect the accuracy of the Lagrange element. The reason for this is that when two adjacent nodal values of transverse displacement are set to zero, the mode is no longer present (i.e., the global stiffness matrix is not rank deficient). However, when the element thickness-to-length ratio increases beyond a certain value, the mode is only weakly coupled to the boundary and may become manifest in certain singular situations. This is illustrated in Fig. 5.3.20, where significant oscillations in transverse displacement may be noted. (In Fig. 5.3.20, r denotes the radial

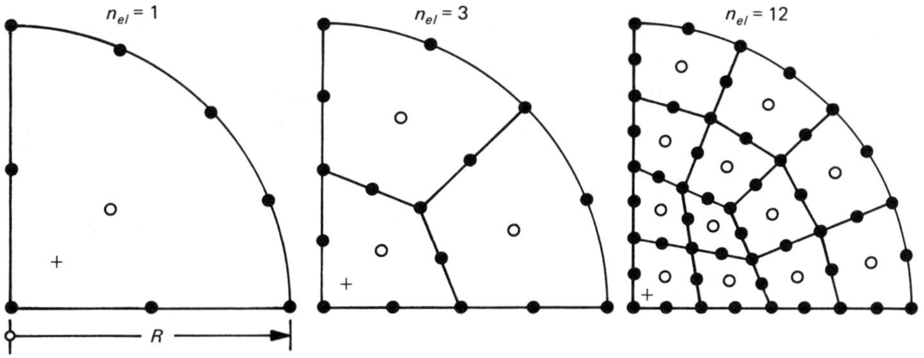

+ denotes location of Gauss point (2 × 2 rule) nearest center

Figure 5.3.18 Circular plate. Due to symmetry, only one quadrant is discretized.

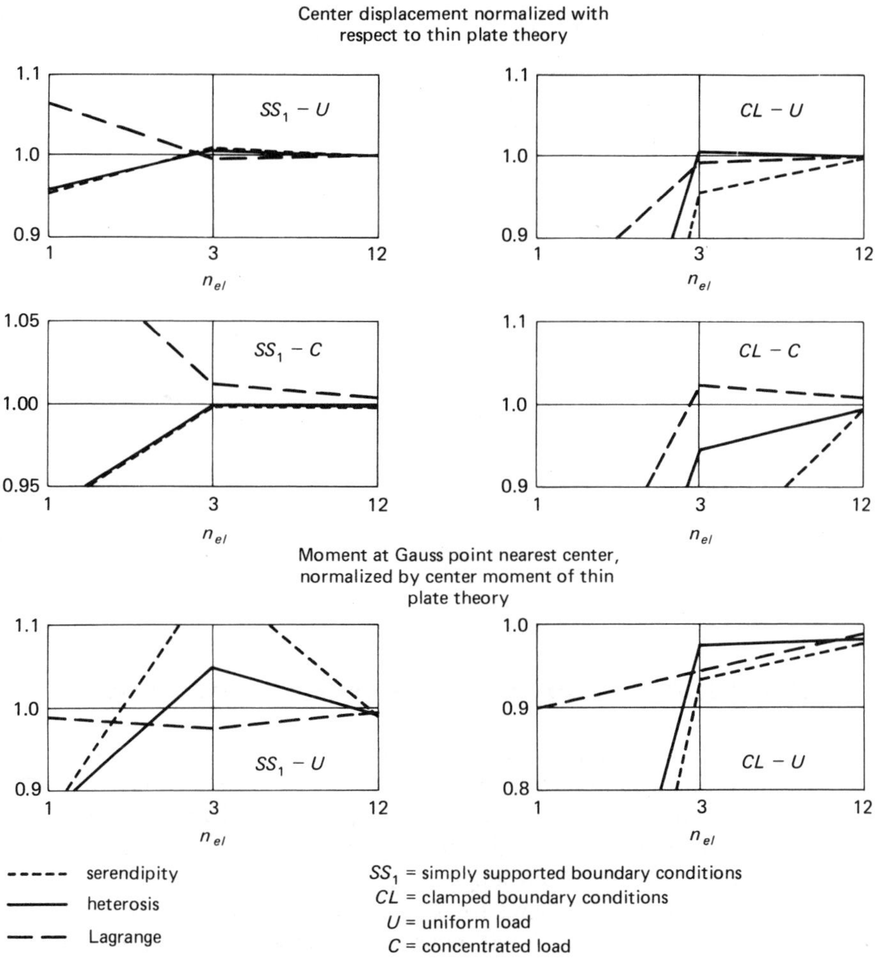

Figure 5.3.19 Convergence study for thin circular plate.

coordinate, D is the flexural rigidity, and P is the amplitude of the uniform load.) The heterosis element, having correct rank, does not exhibit this deficiency. (For this problem, the serendipity results are almost identical to the heterosis results. At points at which the plot symbols tended to overlap, we included only the heterosis symbol.)

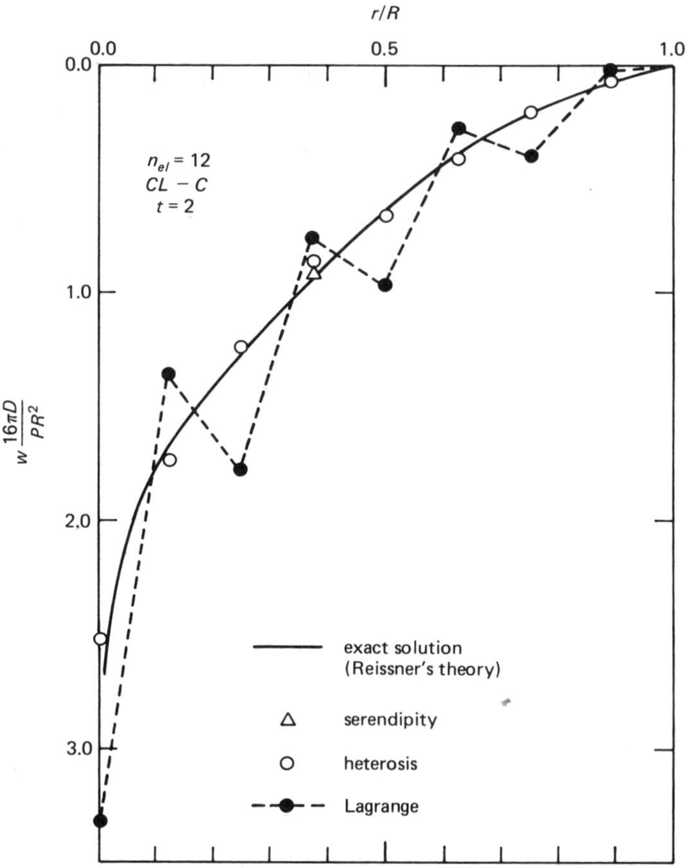

Figure 5.3.20 Comparison of transverse displacements for thick circular plate.

5.3.8 T1: A Correct-rank, Four-node Bilinear Element

A special procedure for interpolating transverse shear strains is used in element $T1$ (Hughes-Tezduyar [17]) to achieve correct rank. The technique falls within the framework of the $\overline{\boldsymbol{B}}$-approach, described in the preceding chapter, as will be shown shortly.

Pertinent geometric and kinematic data are shown in Fig. 5.3.21. Note that the direction vectors have unit length (e.g., $\|\boldsymbol{e}_{11}\| = 1$, etc.). Let w_a and $\boldsymbol{\theta}_a$ denote the transverse displacement and rotation vector, respectively, associated with node a. Throughout, a subscript b will equal $a + 1$ modulo 4. That is

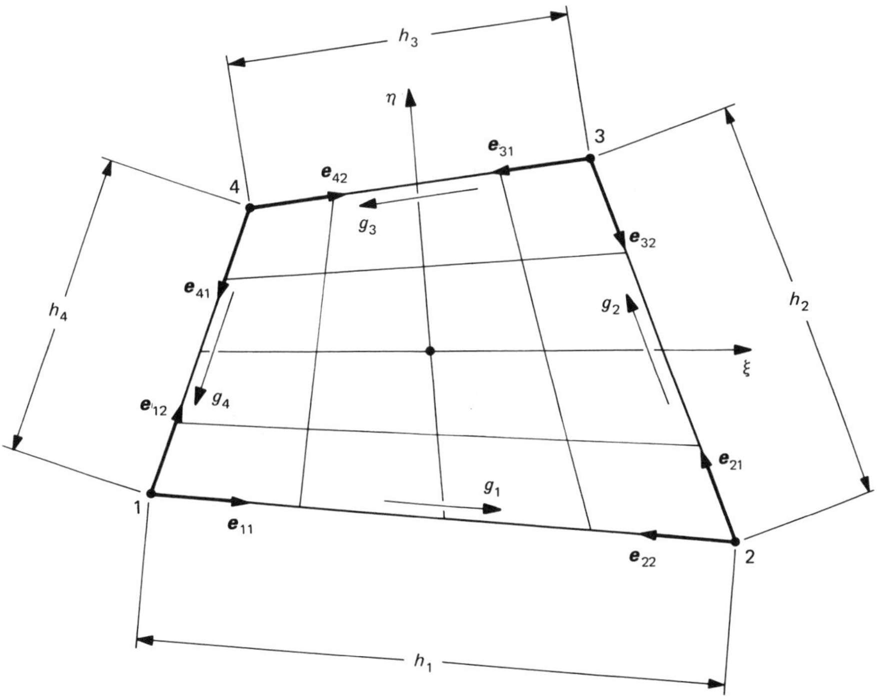

Figure 5.3.21 Geometric and kinematic data for the four-node, bilinear, quadrilateral element, $T1$.

a	b
1	2
2	3
3	4
4	1

(5.3.15)

The definition of the element shear strains is facilitated by the following steps:

1. For each element side, define a shear strain component, located at the midpoint, in a direction parallel to the side, viz.,

$$g_a = \frac{w_b - w_a}{h_a} - \boldsymbol{e}_{a1} \cdot \frac{\boldsymbol{\theta}_b + \boldsymbol{\theta}_a}{2} \tag{5.3.16}$$

The g_a's are schematically illustrated in Fig. 5.3.21.

2. For each node, define a shear strain vector (see Fig. 5.3.22 for a geometric interpretation of this process).

$$\boldsymbol{\gamma}_b = \gamma_{b1}\boldsymbol{e}_{b1} + \gamma_{b2}\boldsymbol{e}_{b2} \tag{5.3.17}$$

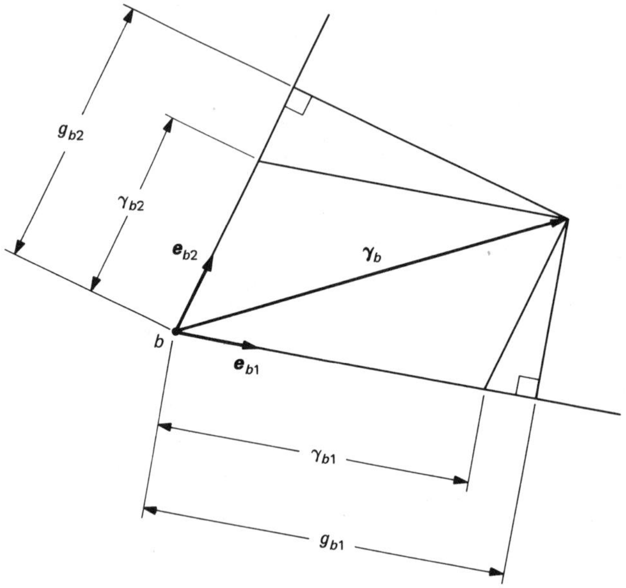

Figure 5.3.22 Definition of nodal transverse shear strain vector.

$$\gamma_{b2} = (1 - \alpha_b^2)^{-1}(g_{b2} - g_{b1}\alpha_b) \tag{5.3.18}$$

$$\gamma_{b1} = (1 - \alpha_b^2)^{-1}(g_{b1} - g_{b2}\alpha_b) \tag{5.3.19}$$

$$\alpha_b = \boldsymbol{e}_{b1} \cdot \boldsymbol{e}_{b2} \tag{5.3.20}$$

$$g_{b1} = g_b \tag{5.3.21}$$

$$g_{b2} = -g_a \tag{5.3.22}$$

3. Interpolate the nodal values by way of the bilinear shape functions (N_a's):

$$\boldsymbol{\gamma} = \sum_{a=1}^{4} N_a \boldsymbol{\gamma}_a \tag{5.3.23}$$

Remarks

1. If the nodal transverse displacements and rotations are specified to consistently interpolate a constant transverse shear strain field—for example, $\bar{\boldsymbol{\gamma}}$—then the preceding steps will result in $\boldsymbol{\gamma} = \bar{\boldsymbol{\gamma}}$. ***That is, constant transverse shear deformation modes are exactly representable in the general quadrilateral geometry.***

2. In the rectangular configuration, the strains take on the following form (we assume the origin of coordinates coincides with the element center):

$$\gamma_1(x_1, x_2) = w_{,1}(0, 0) - \theta_1(0, 0) + x_2[w_{,12} - \theta_{1,2}(0, 0)] \tag{5.3.24}$$

$$\gamma_2(x_1, x_2) = w_{,2}(0, 0) - \theta_2(0, 0) + x_1[w_{,21} - \theta_{2,1}(0, 0)] \tag{5.3.25}$$

where $w_{,12} = w_{,21} =$ constant. In this case the linear variations of γ_1 with x_2 and γ_2 with x_1 may be clearly seen. Note that there are four scalar transverse shear strain modes. (This may be concluded in general from Steps 1 to 3, which amount to an interpolation of the four scalar parameters g_1, g_2, g_3, and g_4.) These modes include the two constant transverse shear modes and the hourglass and in-plane twist modes, thus enabling the element to achieve correct rank. In the rectangular configuration, the transverse shear strain variation is equivalent to the selective integration scheme of MacNeal's QUAD4 [18]. The generalizations to quadrilateral configurations differ somewhat.

3. The constraint ratio (as defined in Sec. 5.3.2) for the present element is $\frac{3}{2}$, the optimal value. This figure is arrived at by observing that the tangential component of transverse shear strain is *continuous* across element boundaries, effectively halving the number of ostensible constraints. (Spilker and Munir [19–21] have proposed an alternative constraint-counting measure for plates called a *rotational constraint index*.)

Implementation

All aspects of the element formulation are identical to the standard four-node, bilinear quadrilateral except that in place of the usual transverse shear strain–nodal displacement matrix, $\boldsymbol{B}^s$ (see (5.2.55), (5.2.57), (5.2.59), (5.2.62), and (5.2.68)), we need to use $\overline{\boldsymbol{B}}^s$, which is defined as follows (recall the relation between subscripts a and b; see (5.3.15)):

$$\overline{\boldsymbol{B}}^s = [\overline{\boldsymbol{B}}_1^s\ \overline{\boldsymbol{B}}_2^s\ \overline{\boldsymbol{B}}_3^s\ \overline{\boldsymbol{B}}_4^s] \tag{5.3.26}$$

$$\overline{\boldsymbol{B}}_b^s = [\overline{\boldsymbol{B}}_{b1}^s\ \overline{\boldsymbol{B}}_{b2}^s\ \overline{\boldsymbol{B}}_{b3}^s], \qquad 1 \le b \le 4 \tag{5.3.27}$$

$$\overline{\boldsymbol{B}}_{b1}^s = h_a^{-1}\boldsymbol{G}_a - h_b^{-1}\boldsymbol{G}_b \tag{5.3.28}$$

$$\overline{\boldsymbol{B}}_{b2}^s = \frac{e_{b2}^1\boldsymbol{G}_a - e_{b1}^1\boldsymbol{G}_b}{2} \tag{5.3.29}$$

$$\overline{\boldsymbol{B}}_{b3}^s = \frac{e_{b2}^2\boldsymbol{G}_a - e_{b1}^2\boldsymbol{G}_b}{2} \tag{5.3.30}$$

$$\boldsymbol{G}_a = (1 - \alpha_a^2)^{-1}N_a(\boldsymbol{e}_{a1} - \alpha_a\boldsymbol{e}_{a2}) - (1 - \alpha_b^2)^{-1}N_b(\boldsymbol{e}_{b2} - \alpha_b\boldsymbol{e}_{b1}) \tag{5.3.31}$$

$$\boldsymbol{e}_{b1} = \begin{Bmatrix} e_{b1}^1 \\ e_{b1}^2 \end{Bmatrix} \tag{5.3.32}$$

$$\vdots$$

Two-by-two Gaussian quadrature is used to integrate all element contributions.

Numerical Examples

$T1$ is compared with the four-node reduced integration elements $S1$ and $U1$, described previously and, in one case, the "twisted ribbon," comparison is made with an element proposed by Robinson [22], dubbed LORA, and MacNeal's QUAD4 [18].

Thin Square and Rectangular Plates

In the cases studied, the differences between $S1$ and $T1$ are indiscernible on the scale of the plots. Square-plate results for $T1$ are the same as the $S1$ results of Fig. 5.3.5. The meshes for rectangular plates are shown in Figs. 5.3.23 and 5.3.24, and results are presented in Figs. 5.3.25 and 5.3.26. These problems test the response of the elements to changes in planar aspect ratio. As can be seen, there is some deterioration of accuracy with planar aspect ratio—a common but not well-understood phenomenon for virtually all finite elements.

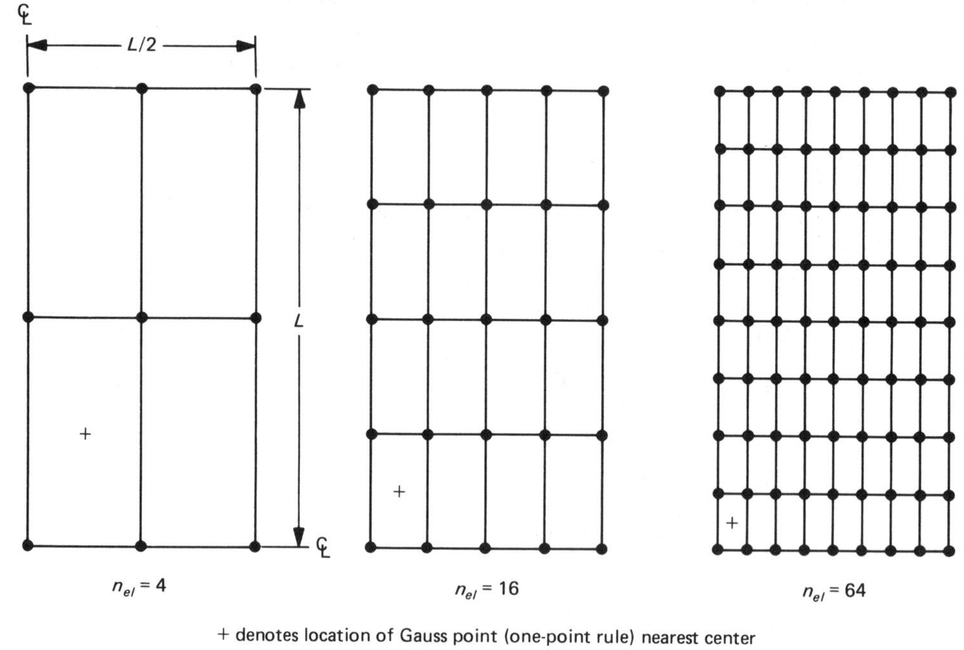

Figure 5.3.23 Rectangular plate meshes (aspect ratio = 2). Due to symmetry, only one quadrant is discretized.

Thin Circular Plate

These problems test the behavior of the elements in nonrectangular configurations. The meshes are the same as shown in Fig. 5.3.6 and convergence results are presented in Fig. 5.3.27. In this case, $T1$ is generally the best performer, although all elements perform well.

Thin Rhombic Plate

The configuration and mesh are shown in Fig. 5.3.28. The length parameter $a = 100$. The plate is uniformly loaded and simply supported boundary conditions (SS_1) are employed. This problem is a difficult one, since there is a singularity at the obtuse vertex. The analytical solution reveals that the x_1 and x_2 bending moments have

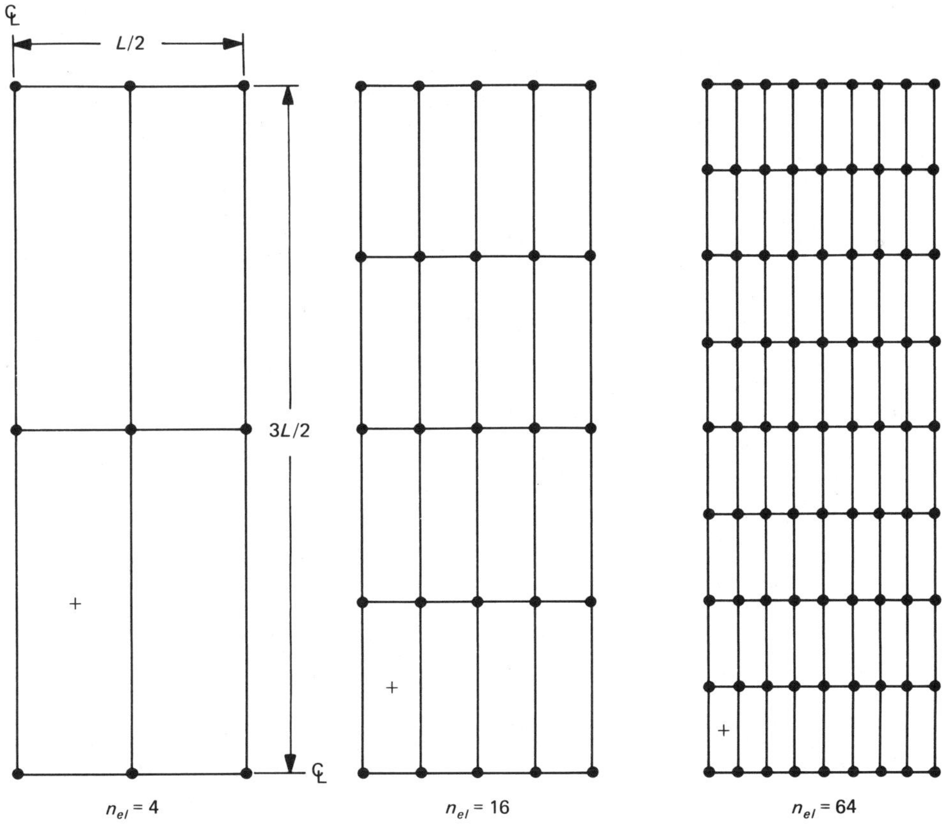

Figure 5.3.24 Rectangular plate meshes (aspect ratio = 3). Due to symmetry, only one quadrant is discretized.

opposite signs in the vicinity of the obtuse vertex. Many thin plate elements yield pathological results for this problem in that moments with the *same* sign are obtained (see Sec. 5.3.3 for a discussion and references). Moment results are presented in Fig. 5.3.29. The general trend for each element is correct. However, the elements have a tendency to oscillate somewhat, as may be seen. The worst oscillations are produced by $U1$. Considering that the mesh is not biased to favor the singularity and that the problem is a numerically difficult one, the accuracy of the results obtained for $S1$ and $T1$ is considered to be fairly good.

Thick Circular Plate

This problem was considered previously for the heterosis, Lagrange, and serendipity elements (see Fig. 5.3.20). In the present case, the 48-element mesh shown in Fig. 5.3.6 was employed. The singularity gives rise to almost identical oscillatory patterns for the rank-deficient elements $S1$ and $U1$, as may be seen in Fig. 5.3.30. On the other hand, element $T1$ produces very accurate results.

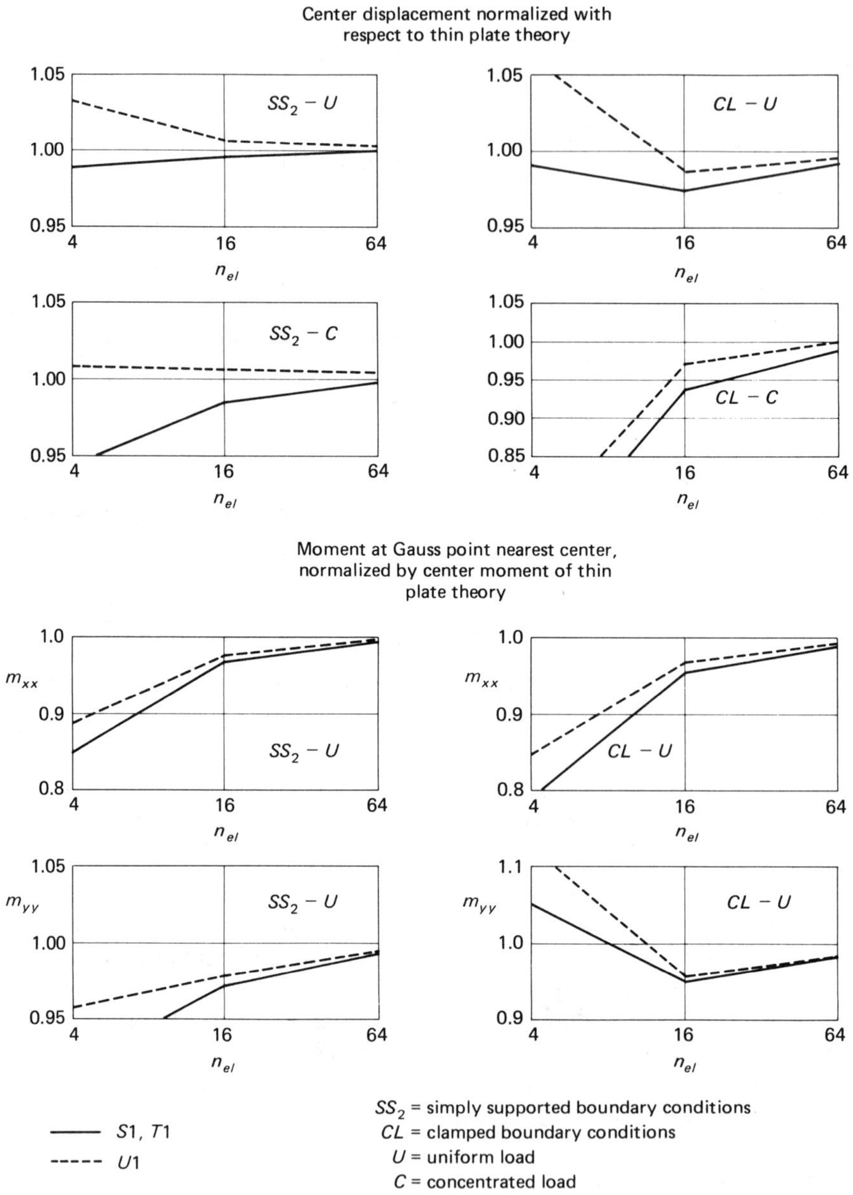

Figure 5.3.25 Convergence study for thin rectangular plate (aspect ratio = 2).

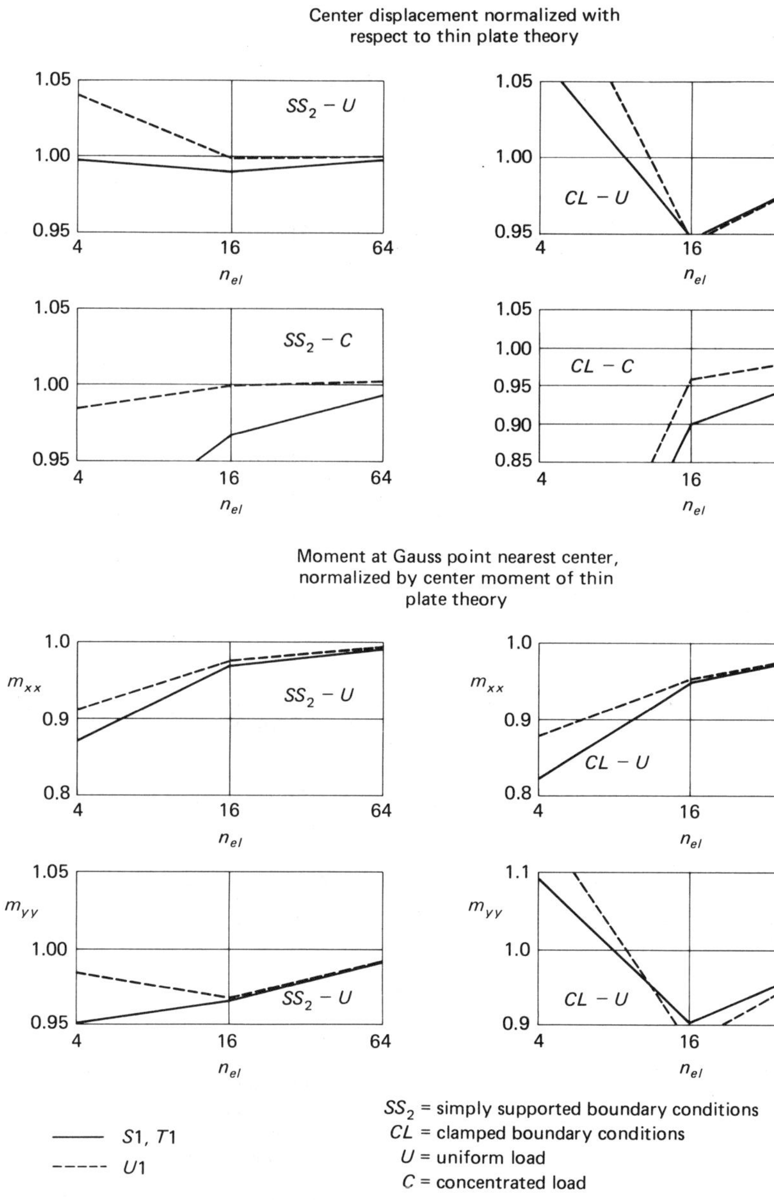

Figure 5.3.26 Convergence study for thin rectangular plate (aspect ratio = 3).

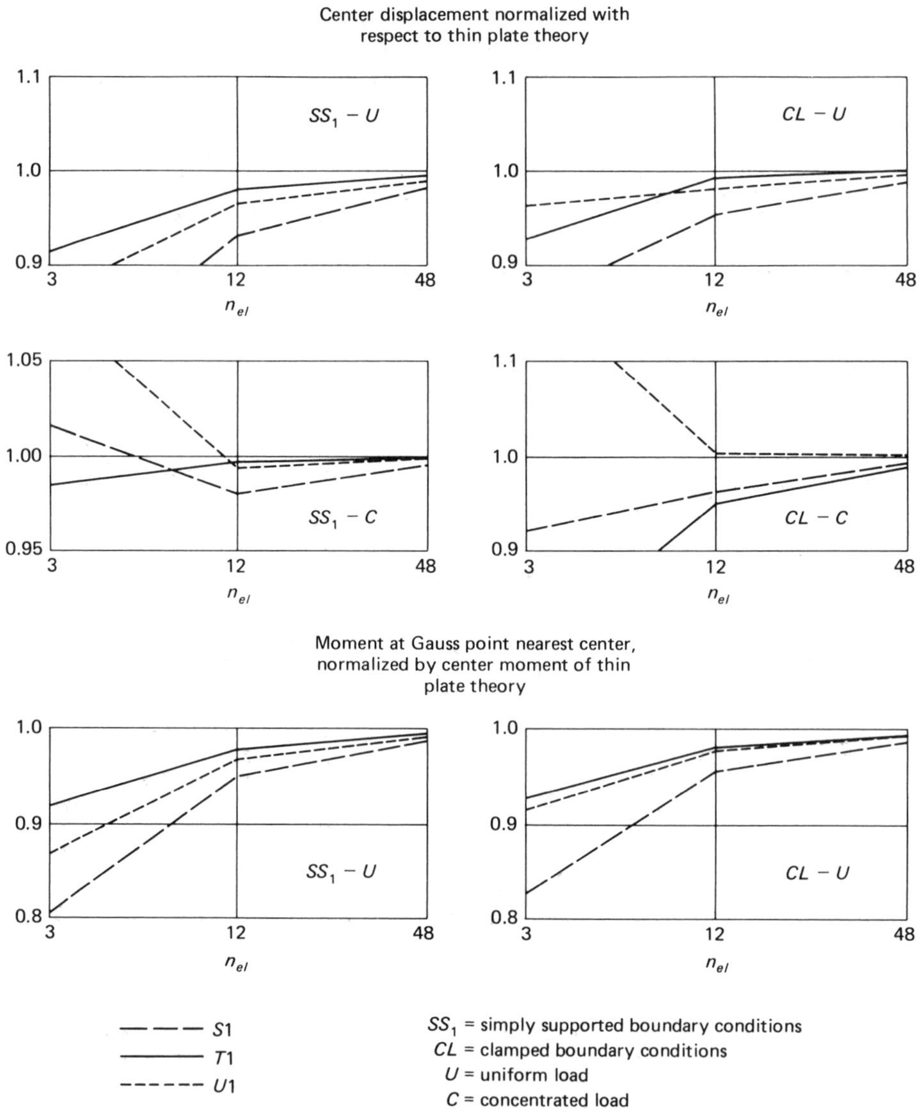

Figure 5.3.27 Convergence study for thin circular plate.

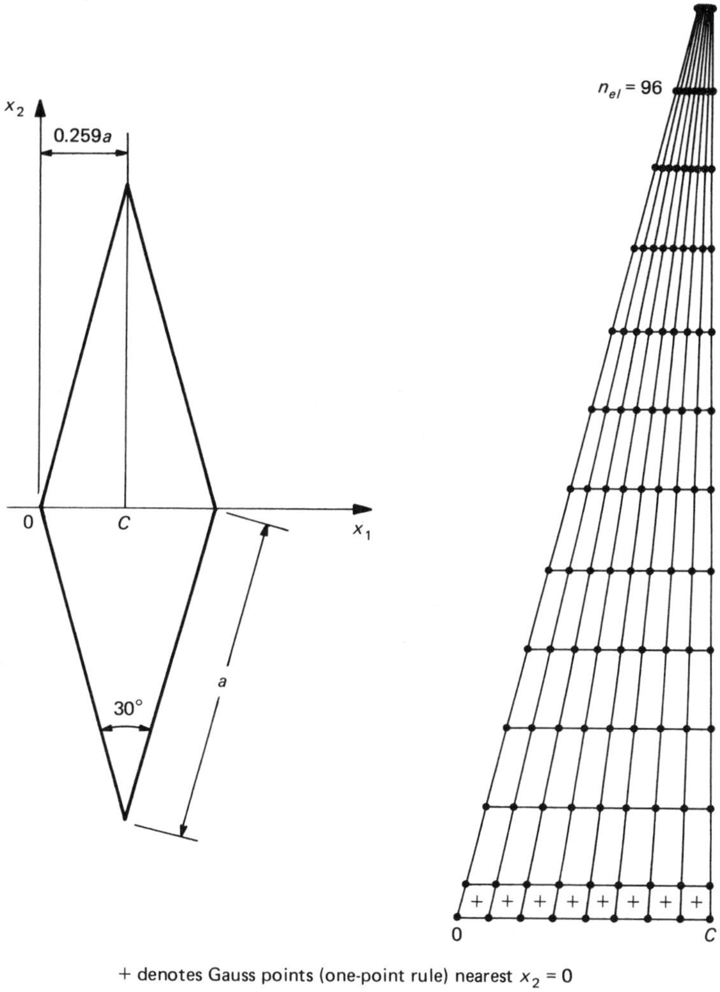

Figure 5.3.28 Rhombic plate mesh. Due to symmetry, only one quadrant is discretized.

Twisted Ribbon

Configurations, data and results for this problem are shown in Fig. 5.3.31. In each analysis, only one element is employed. Robinson [22] has proposed this as a critical single element test for plate-bending elements. Comparisons are made with data presented in [22] for Robinson's element, LORA, and MacNeal's QUAD4 [18].

For cases A and B (fully fixed boundary), comparison is made with respect to a benchmark analysis, reported upon in [22], involving 16 high-precision elements. As may be seen, the results for $T1$ are superior to the results for both LORA and QUAD4. Furthermore, no deterioration with increasing aspect ratio is detected. For this case, elements $S1$ and $U1$ exhibit pathological behavior due to rank deficiency (not shown).

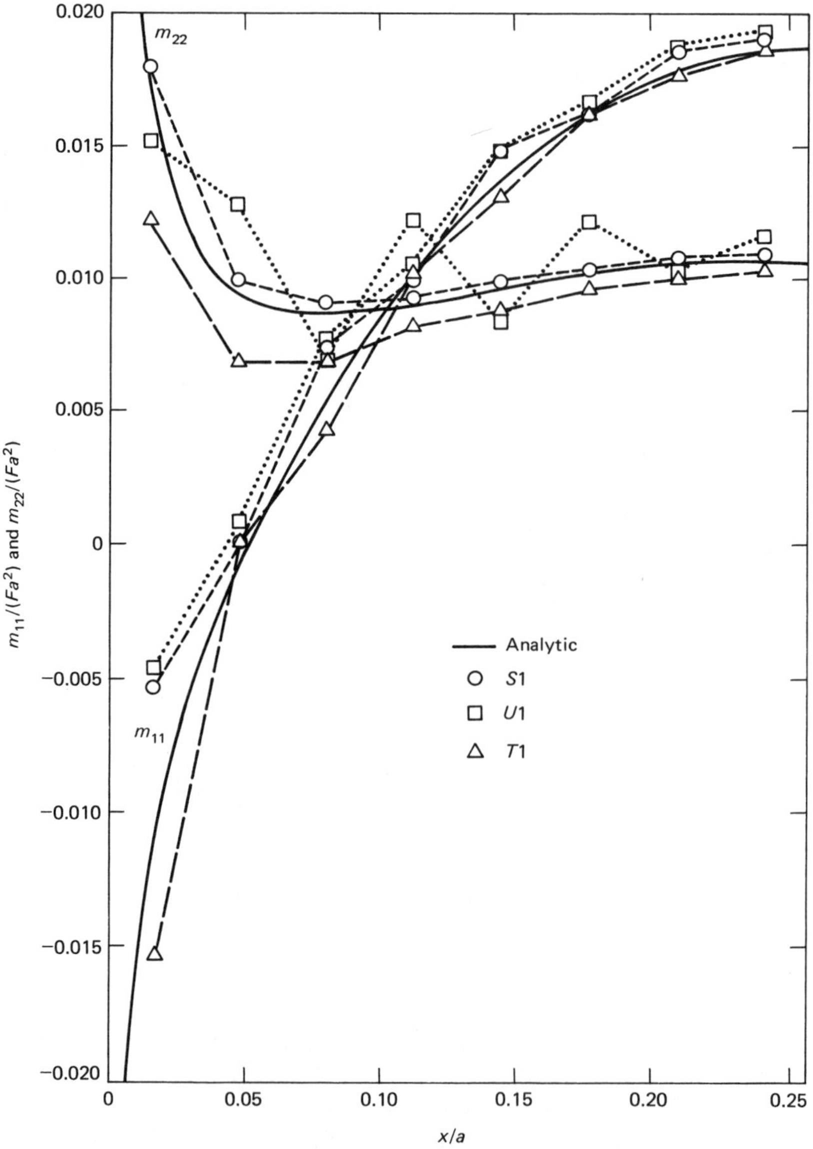

Figure 5.3.29 Bending moments for rhombic plate.

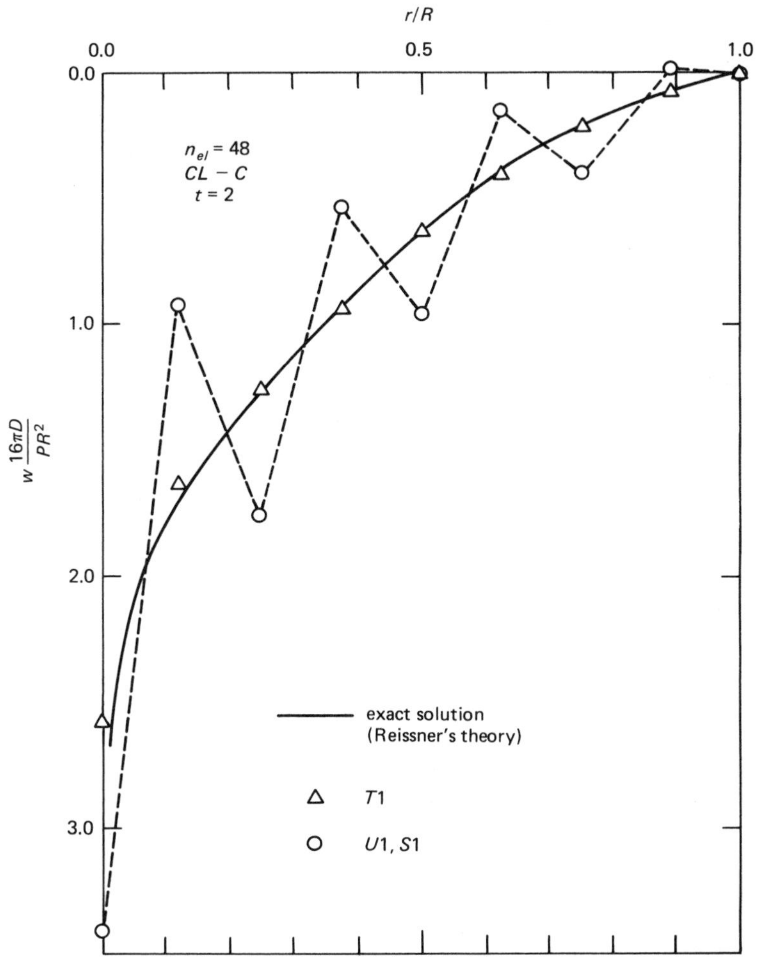

Figure 5.3.30 Displacement results for thick circular plate.

If only half the domain is modeled and antisymmetrical boundary conditions are enforced (cases C and D), the exact solution is one of pure twist. For these cases, $S1$ and $T1$ yield exact solutions, whereas $U1$ still behaves pathologically (not shown).

Remarks

1. A rationale for the development of $T1$, based upon the ***Kirchhoff mode concept,*** is presented in [17]. The basic idea emanates from [18].

2. MacNeal [23] has more recently presented a new version of QUAD4 in which the definition of the transverse shear strain field has essential features in common with $T1$.

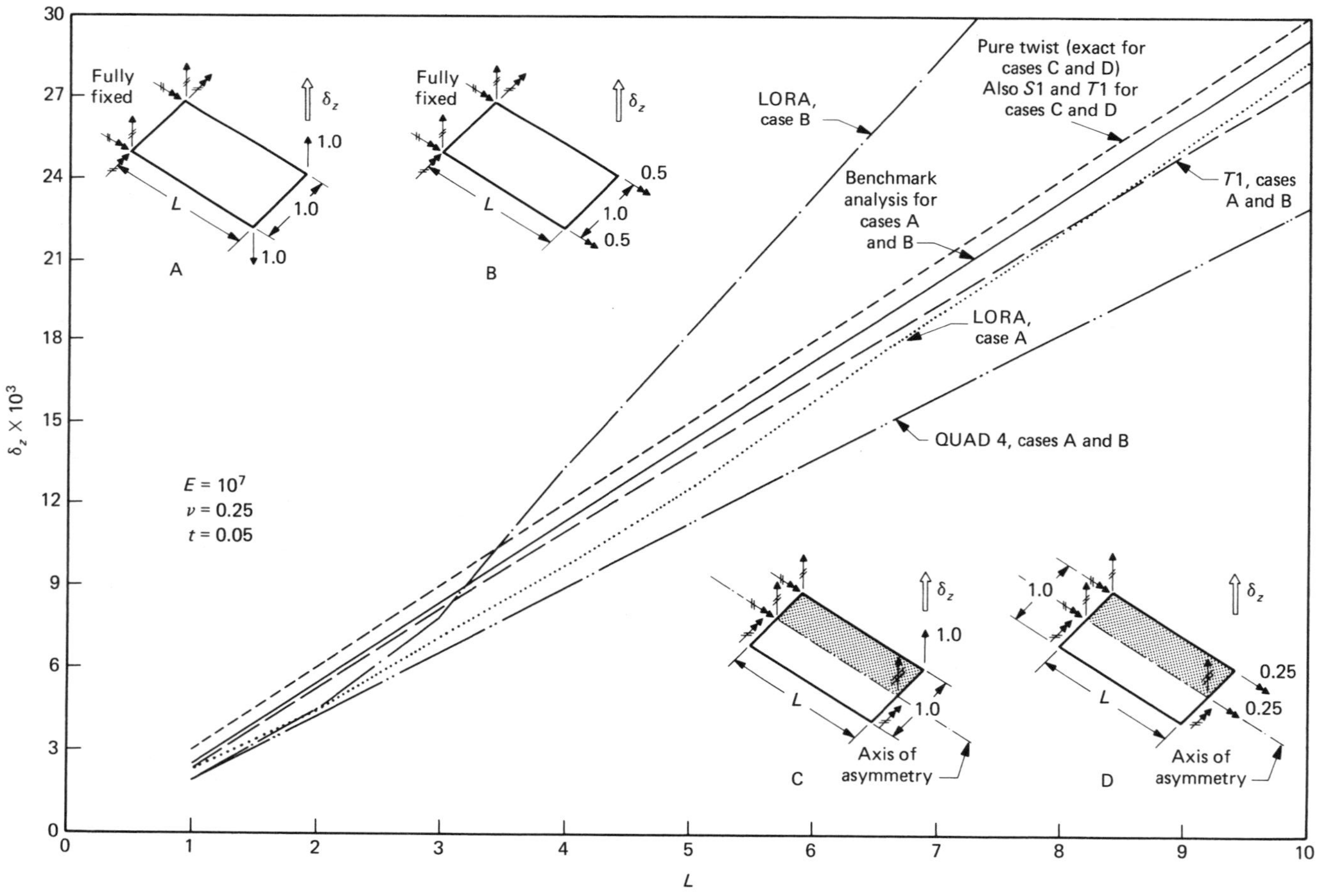

Figure 5.3.31 Displacement results for the twisted ribbon. One element used in all cases.

5.3.9 The Linear Triangle

A three-node triangle employing linear shape functions for w and θ_α exhibits severe shear locking under normal circumstances. The technique used in the previous section to define the transverse shear strain field for $T1$ may also be used in this case to alleviate locking while maintaining correct rank [24]. The development follows (5.3.15) through (5.3.23) closely. The only changes involve the nodal indices, which this time are defined by the relation

a	b
1	2
2	3
3	1

(5.3.33)

which replaces (5.3.15), and the summation in (5.3.23) is taken over $a = 1, 2, 3$, wherein the N_a's represent linear (instead of bilinear) shape functions. Figure 5.3.32 depicts element geometric and kinematic quantities. Implementation follows along the same lines as for $T1$ (see (5.3.26) through (5.3.32)). The only changes involve replacing (5.3.26) by

$$\overline{\boldsymbol{B}}^s = [\overline{\boldsymbol{B}}_1^s \; \overline{\boldsymbol{B}}_2^s \; \overline{\boldsymbol{B}}_3^s] \tag{5.3.34}$$

and limiting $b \le 3$ in (5.3.27).

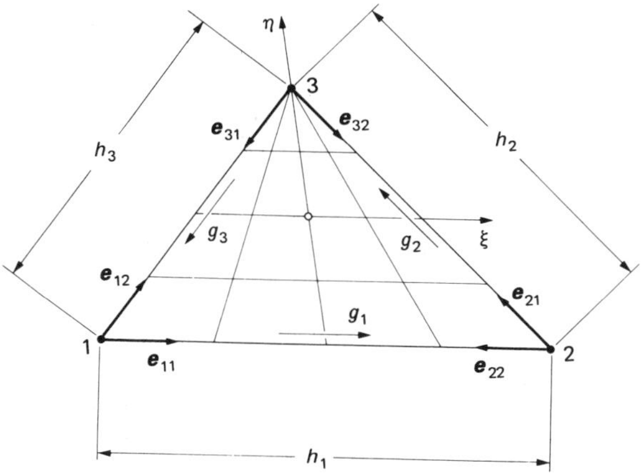

Figure 5.3.32 Geometric and kinematic data for the three-node linear triangular element.

Numerical Examples

Two different quadrature treatments of the linear triangle have been used: one-point centroidal quadrature and 2 × 2 Gauss quadrature in the ξ,η-system (see Fig. 5.3.32), which is exact in the present circumstances. (Note that centroidal quadrature is not the same as one-point Gauss quadrature.) A single triangle possesses correct rank (i.e., 6) when integrated exactly, but possesses one spurious mechanism when underintegrated by one-point centroidal quadrature. However, in any assemblage of two or more elements, the spurious mechanism disappears, and thus it does not appear to be of serious consequence in practical computing.

Calculations were performed for simply supported, thin, square and circular plates subjected to uniform loads. The edge length of the square plate and radius of the circular plate were taken to be 10.0 and 5.0, respectively. In each case, due to symmetry, only one quadrant of the plate was discretized. Meshes are depicted in Figs. 5.3.33 and 5.3.34. Note that the cross-diagonal meshes involve approximately twice the number of unknowns as the other mesh types. The SS_1 boundary condition was used in each case. The quadrature treatment of the consistent load was the same as used for the stiffness.

Numerical results for the cases studied are presented in Tables 5.3.2 and 5.3.3. Moment results were obtained at the centroids of the triangular elements nearest the plate center. When the vertices of two triangles coincided with the plate center, moments were averaged over the two elements. It is immediately apparent from Table 5.3.2 that exact quadrature behaves very poorly for mesh types A and B.

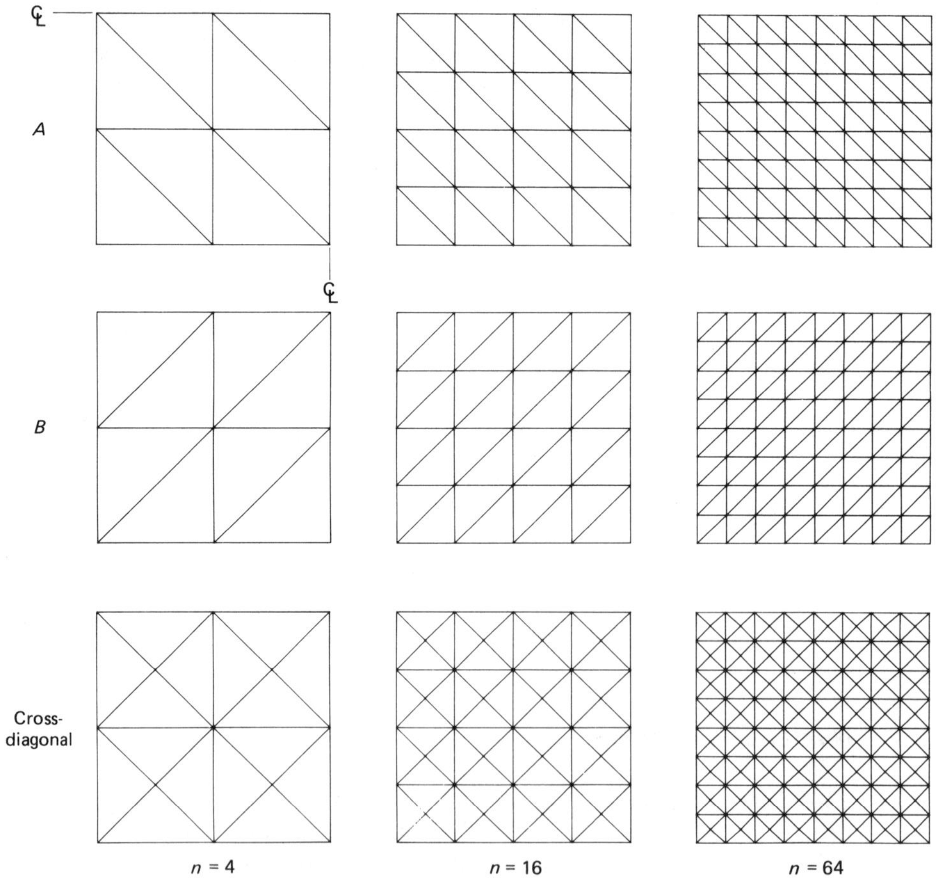

Figure 5.3.33 Square plate meshes. Due to symmetry, only one quadrant is discretized.

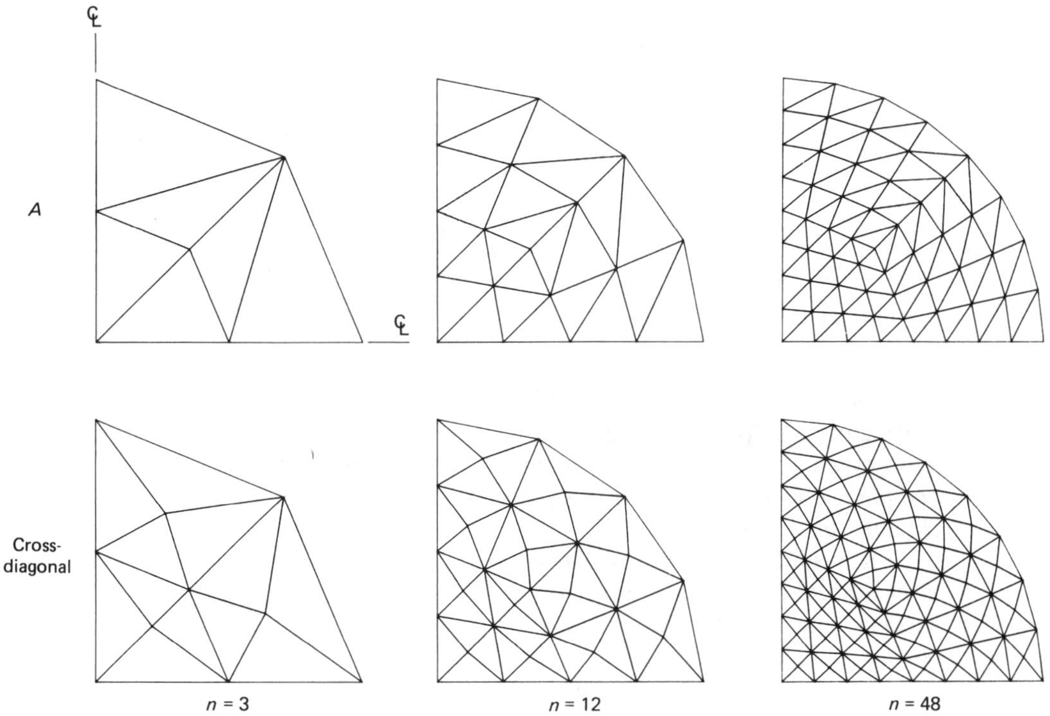

Figure 5.3.34 Circular plate meshes. Due to symmetry, only one quadrant is discretized.

TABLE 5.3.2 Center Displacement and Moment for Simply Supported, Thin, Square Plate Subjected to Uniform Load

a. Displacement

	One-point centroidal quadrature			Exact quadrature		
n	A	B	cross-diagonal	A	B	cross-diagonal
4	0.681	0.883	0.912	0.681	0.002	0.912
16	0.784	0.989	0.978	0.773	0.036	0.978
64	0.947	0.999	0.994	0.827	0.363	0.994

b. Moments

	One-point centroidal quadrature			Exact quadrature		
n	A	B	cross-diagonal	A	B	cross-diagonal
4	0.555	0.904	0.919	0.555	0.002	0.919
16	0.564	1.127	0.979	0.539	0.036	0.979
64	0.835	1.098	0.996	0.580	0.386	0.996

TABLE 5.3.3 Center Displacement and Moment for Simply Supported, Thin, Circular Plate Subjected to Uniform Load

	A		Cross-diagonal	
n	Displacement	Moment	Displacement	Moment
3	0.703	0.576	0.927	0.885
12	0.912	0.878	0.981	0.976
48	0.948	0.975	0.976	0.986

Results presented are for one-point centroidal quadrature.

On the other hand, for the cross-diagonal meshes, exact quadrature yields quite satisfactory results. One-point centroidal quadrature represents some improvement for mesh types *A* and *B*, but still yields moments that are not satisfactory. Note that one-point quadrature for the cross-diagonal pattern yields identical results to those obtained by exact quadrature.

It is quite clear from the results of Table 5.3.2 that only the cross-diagonal meshes yield consistently satisfactory results. Furthermore, there is no advantage to exact quadrature. The economic superiority of one-point quadrature makes it the obvious choice for practical use.

The results for the circular plate, presented in Table 5.3.3, are satisfactory for both mesh patterns studied.

Results for cross-diagonal meshes are compared with quadrilateral *T*1 elements in Fig. 5.3.35. The moments for *T*1 are computed at the Gauss point (2×2 rule) nearest the center of the plate. As may be seen, the displacement results in this case favor *T*1, whereas the moment results for the linear triangle are superior. However, the sampling points for moments is closer to the center for the triangles than for *T*1, and this favors the triangle results, especially on coarse meshes.

Remarks

1. MacNeal [25] has developed a linear triangular bending element, TRIA3, for the MSC (MacNeal-Schwendler Corporation) version of the NASTRAN program. This element is very similar to the linear triangle presented herein, except a "residual bending flexibility" is introduced to further improve behavior. The concept of residual bending flexibility, introduced by MacNeal in [18], will be illustrated in Section 5.5 in the context of the linear beam element.

2. A similar idea involving a "*c*-factor" modification [26], has been employed by Garnet et al. [27] to alleviate the tendency to lock for the standard three-node linear triangle. (This element does not employ special interpolation of the transverse shear strain.)

Discussion

Numerical results for the exactly integrated linear triangle described in this section are somewhat erratic for certain mesh configurations. However, the cross-diagonal pattern

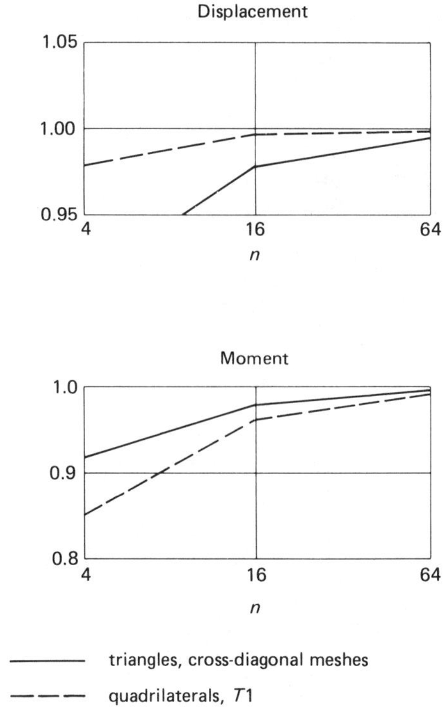

Figure 5.3.35 Comparison of center displacement and moment for triangular and quadrilateral elements; simply supported, thin, square plate subjected to uniform load.

was found to be successful for both exact and reduced integration (i.e., one-point centroidal quadrature). Although one-point quadrature results in rank deficiency for a single element, this is not a practical detriment, since the rank deficiency disappears in an assemblage of two or more elements. Due to the greater economy of one-point quadrature, it appears the obvious choice for practical use.

Comparison of results with those for the bilinear quadrilateral $T1$ suggests that linear triangles in the cross-diagonal pattern may be competitive. In the comparison, cross-diagonal triangulated meshes required approximately twice the number of unknowns as for bilinear quadrilaterals. On the other hand, the center-node degrees of freedom are easily eliminated via elementwise static condensation, and the one-point evaluation over each triangle is amenable to very efficient element programming.

Techniques described in Remarks 1 and 2 have been introduced to improve the performance of linear triangular bending elements. A firm theoretical foundation is, however, not yet available, but it is to be hoped that this will be forthcoming so that a coherent exposition of these ideas may be presented in the near future.

5.3.10 The Discrete Kirchhoff Approach

In the discrete Kirchhoff approach, thin plate elements are developed by insisting that the Kirchhoff-theory constraint of zero transverse shear strains be satisfied at a discrete number of points within the element. To illustrate the approach, we will consider the development of the 12-degree-of-freedom, four-node quadrilateral DKQ (discrete Kirchhoff quadrilateral [29]):

i. The domain of the element is a straight-edged quadrilateral. The rotations are initially defined by

$$\boldsymbol{\theta} = \sum_{a=1}^{8} N_a \boldsymbol{\theta}_a \tag{5.3.35}$$

where the N_a's are the eight-node serendipity shape functions (see Sec. 3.7). Thus we begin with 16 rotational degrees of freedom. The midside degrees of freedom are eventually eliminated.

ii. Let $\boldsymbol{n}$ and $\boldsymbol{s}$ denote normal and tangential unit vectors along a side, with orientation as defined in Fig. 5.3.1. The normal component of $\boldsymbol{\theta}$ is required to vary linearly along each side, i.e.,

$$\boldsymbol{n} \cdot \boldsymbol{\theta}_c = \frac{\boldsymbol{n} \cdot (\boldsymbol{\theta}_a + \boldsymbol{\theta}_b)}{2} \tag{5.3.36}$$

where $c = a + 4$, $a = 1, 2, 3, 4$, and $b = 2, 3, 4, 1$, respectively. This amounts to 4 constraints.

iii. The transverse displacement is defined *only* along the element boundary. It is assumed to vary cubically along each edge, according to Hermite interpolation (see Sec. 1.16). Thus nodal values of w and $w_{,\alpha}$ are required to define the edge displacement. This introduces 12 more degrees of freedom, for a total of 28.

iv. Kirchhoff constraints are imposed as follows:

1. Corner nodes:

$$w_{,\alpha} = \theta_\alpha \qquad (\text{i.e., } \gamma_\alpha = 0) \tag{5.3.37}$$

2. Midside nodes[3]

$$w_{,s} = s_\alpha \theta_\alpha \qquad (\text{i.e., } s_\alpha \gamma_\alpha = 0) \tag{5.3.38}$$

where $w_{,s} = s_\alpha w_{,\alpha}$ (sum). Equation (5.3.37) represents 8 constraints and (5.3.38) gives 4 more. These 12 Kirchhoff constraints, and the 4 normal rotation conditions under (ii), enable reduction to 12 degrees of freedom, namely, the corner node displacements and rotations. Furthermore, explicit formulas may then be obtained for the rotations:

$$\theta_\alpha = \sum_{p=1}^{12} N_{\alpha p} d_p^e \tag{5.3.39}$$

where

$$d_p^e = \begin{cases} w_a & p = 3a - 2 \\ \theta_{1a} & p = 3a - 1 \\ \theta_{2a} & p = 3a \end{cases} \tag{5.3.40}$$

[3] A handy formula for $w_{,s}$ at the midside nodes is given by:

$$w_{,s_c} = \frac{3}{2h_a}(w_b - w_a) + \frac{1}{4}(w_{,s_b} - w_{,s_a})$$

where $w_{,s_c} = w_{,s}(\boldsymbol{x}_c)$; the nodal indexing is as defined under (ii); and h_a is the edge length (see Fig. 5.3.21).

(Expressions for the shape functions, $N_{\alpha p}$, are somewhat lengthy. The interested reader is referred to the original sources for further details.)

v. In the development of the element stiffness, the transverse shear terms are simply ignored. Thus the bending stiffness is completely defined by the derivatives of the rotation interpolations derived under (iv). (See (5.2.56) through (5.2.62)).

Remarks

1. By virtue of the fact that the tangential shear strain component vanishes at three distinct points along each edge, it vanishes identically along the element boundary. (This follows from the observation that it varies quadratically without imposition of the Kirchhoff constraints.)

2. The exact integration of the element stiffness (in the rectangular configuration) requires 3×3 Gaussian quadrature. However, it is observed that the 2×2 rule maintains correct rank and so, due to its greater economy, it is recommended in practice [29].

3. No internal interpolation for transverse displacement is defined. This creates ambiguities in the correct definition of consistent transverse applied forces. (The same is true for the definition of element inertial properties in dynamics.) In [29], two ad hoc possibilities are studied: bilinear variation, and an incomplete quartic scheme. Both seem to perform adequately in practice.

4. It is asserted that convergence to the thin plate solution is to be expected because the transverse shear stiffness is neglected and because the Kirchhoff hypothesis is satisfied (tangentially) along the boundary.

5. A triangular element, "DKT," employing virtually identical concepts, was the historical predecessor of DKQ (see [30–32]). References [31, 32] should be consulted for additional discussion about discrete Kirchhoff elements and pertinent references to the literature.

In addition to discrete Kirchhoff constraints, elements have been derived that also employ integral Kirchhoff constraints (e.g., the area, or boundary, mean shear strain may be required to vanish). Most prominent among these are perhaps Irons' SEMILOOF [33] and Lyons' ISOFLEX [34] elements. SEMILOOF is rather complicated; however, one of Lyons' elements is a four-node 12-degree-of-freedom quadrilateral. Crisfield [35] has developed a modified version of Lyons' four-node element in which the Kirchhoff constraints are given in explicit algebraic form, thus avoiding cumbersome element-level calculations.

A somewhat related approach has been employed by Tessler and Hughes in the development of a four-node quadrilateral [36] and three-node triangle [37]. For the quadrilateral, bilinear rotations are employed, and, initially, an eight-node serendipity interpolation is used for transverse displacement. The midside transverse-displacement degrees of freedom are removed by constraining the tangential component of transverse shear strain to be constant along each edge. The triangle is developed along similar lines, starting with linear rotations and quadratic transverse displacements. By virtue of the fact that transverse-shear deformation modes are retained, these elements differ from discrete-Kirchhoff elements in that they are applicable to moderately thick as well as thin plates. The good behavior of the elements requires a "c-factor," or "residual bending flexibility," type of modification. See [36, 37] for further details.

5.3.11 Discussion of Some Quadrilateral Bending Elements

$T1$ and the latest version of QUAD4 [23] have much in common. The treatment of transverse shear is equivalent in the rectangular configuration and differs only slightly in the general quadrilateral configuration. Three additional devices are employed in an effort to further improve the behavior of QUAD4:

i. The shearing properties of the plate are modified to account for a "residual bending flexibility" correction. The concept of residual bending flexibility is an intriguing one. In simple cases, such as beams, a clear exposition of the basic idea is possible and its beneficial effect is obvious (see Sec. 5.5 and [18]). However, for arbitrary geometries and in envisioning extension to nonlinear cases, it becomes less clear how to proceed systematically. This modification tends to make QUAD4 more flexible than $T1$, which is a step in the right direction because $T1$ tends to err on the stiff side.

ii. The component of twisting curvature, κ_{12}, at the 2×2 Gauss points is replaced by $2\kappa_{12}(\tilde{\boldsymbol{\xi}}_l) - \kappa_{12}(\mathbf{0})$, where $\tilde{\boldsymbol{\xi}}_l$, $l = 1, 2, 3, 4$, represent the 2×2 Gauss point coordinates. (κ_{12} is defined with respect to a locally defined Cartesian coordinate system, thus making sense invariantly.)

iii. A "tuning parameter" is introduced into the material properties to improve planar aspect ratio behavior.

The interested reader may consult [18, 23] for rationale and additional details. In some instances, these techniques lead to improved behavior compared with $T1$ (see Fig. 5.3.36), whereas, occasionally, the opposite is true (see Fig. 5.3.31). Whether or not the improvements engendered justify the additional complications is a somewhat subjective matter.

Recently, Bathe and Dvorkin [38] proposed a four-node plate element. It is immediately apparent that this element is identical to $T1$ for rectangular configurations. Results presented in Bathe and Dvorkin [39] indicated superiority of their element over $T1$ in rhombic configurations, but these results proved to be in error. It has been subsequently established that the Bathe-Dvorkin element is also identical to $T1$ in rhombic configurations.

In comparing DKQ with $T1$ and QUAD4, one must first note that DKQ is a thin plate element and thus does not apply to cases in which shear deformations must be accounted for. In most test cases the convergence of DKQ seems somewhat slower than $T1$ and QUAD4. However, in some cases it is superior. DKQ seems to be particularly insensitive to distortion, so overall the accuracy level appears commensurate. The shape functions are somewhat more complicated than the bilinear ones used for $T1$ and QUAD4. Nevertheless, 2×2 quadrature suffices, so the cost of element calculations is probably not substantially different.

There is considerable interest in the development of methodology for controlling the spurious zero-energy modes of the one-point quadrature element (i.e., $U1$). Efficient methods have been proposed by Belytschko and Tsay [40] and Park and

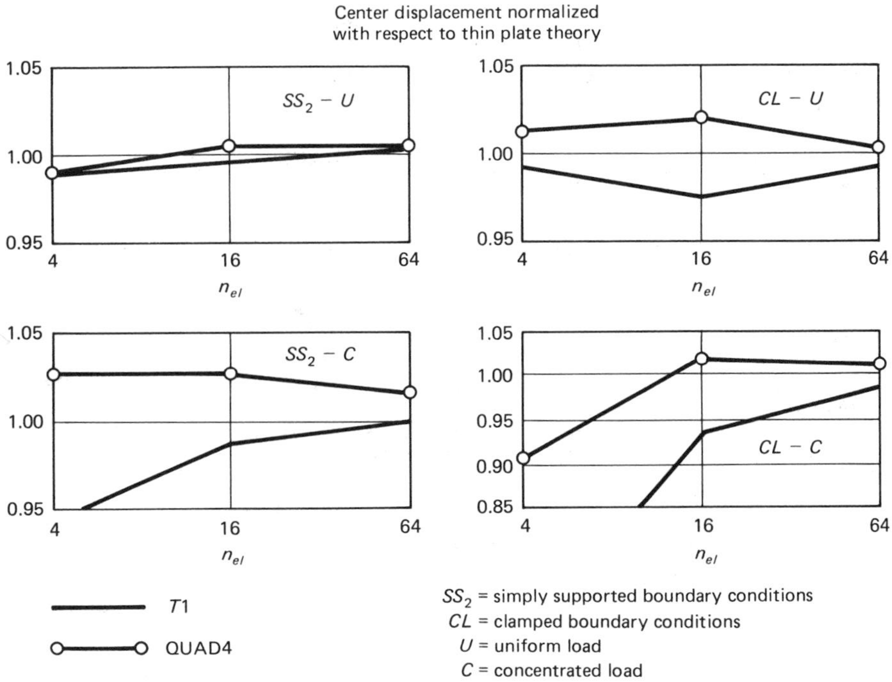

Figure 5.3.36 Comparison of $T1$ and QUAD4 for thin rectangular plate (aspect ratio = 2).

Flaggs [41]. Several new improved, nine-node elements have also been proposed recently; see Belytschko, Ong, and Liu [42], Crisfield [43], and Huang and Hinton [44].

Plate element design is still an active area of research and we anticipate many additional developments in the future.

5.4 BEAMS AND FRAMES

In this section we present a simple theory for finite element approximations of beam and frame structures. Throughout, Latin indices take on the values 1, 2, 3, and Greek indices take on the values 1, 2.

5.4.1 Main Assumptions

1. Domain. We assume from the start that the domain is divided into segments, or elements, interconnected at nodal points. An example of such a structure is depicted in Fig. 5.4.1.

$$\Omega = \bigcup_{e=1}^{n_{el}} \Omega^e$$

$$\Omega^e = \{(x_1^e, x_2^e, x_3^e) \in \mathbb{R}^3 \,|\, x_3^e \in [0, h^e],\ (x_1^e, x_2^e) \in A^e \subset \mathbb{R}^2\}$$

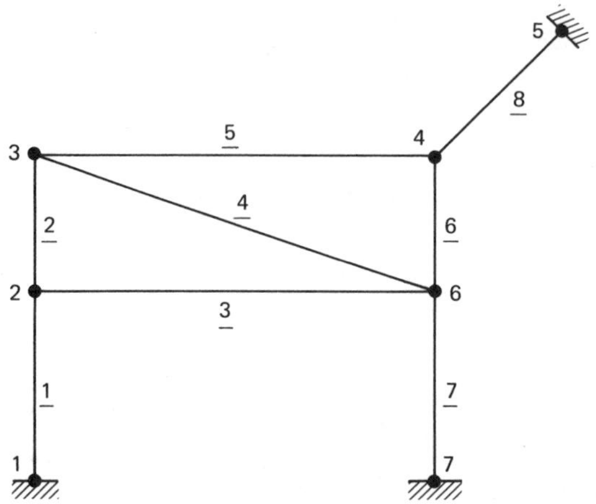

Figure 5.4.1 A structural model consisting of beam elements; $n_{el} = 8$, $n_{np} = 7$.

$$h^e = \text{length of the } e\text{th beam element}$$

$$A^e = \text{cross-section area}$$

We assume the (x_1^e, x_2^e, x_3^e)-axes are *locally* defined with respect to the beam segment and are principal axes, i.e.,

$$0 = \int_A x_1^e \, dA = \int_A x_2^e \, dA = \int_A x_1^e x_2^e \, dA$$

Note. To save some writing, we shall frequently omit the superscript e in the sequel.

2. $\sigma_{\alpha\beta} = 0$

3. $u_1(x_1, x_2, x_3) = w_1(x_3) - x_2\theta_3(x_3)$

$$u_2(x_1, x_2, x_3) = w_2(x_3) + x_1\theta_3(x_3)$$

$$u_3(x_1, x_2, x_3) = w_3(x_3) - x_1\theta_2(x_3) + x_2\theta_1(x_3)$$

Remarks

1. In Assumption 1, the beam is taken to be prismatic. However, no essential difficulty is encountered if we take A^e to be a function of x_3^e.

2. Assumption 2 is used in the constitutive equation to eliminate $\epsilon_{\alpha\beta}$.

3. The kinematic conditions of Assumption 3 do not include warping (i.e., plane sections remain plane).

The sign convention on θ_i follows the right-hand rule and is depicted in Fig. 5.4.2.

For convenience in the sequel, we recall the equations of classical linear elasticity theory:

$$\left.\begin{aligned} \sigma_{ij,j} + \ell_i &= 0 \qquad (5.4.1)\\ \sigma_{ij} &= c_{ijkl}\epsilon_{kl} \qquad (5.4.2)\\ \epsilon_{ij} &= u_{(i,j)} \qquad (5.4.3)\end{aligned}\right\} \text{ in } \Omega$$

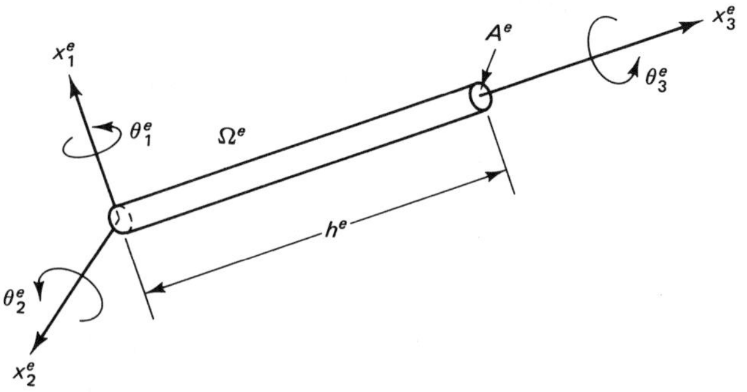

Figure 5.4.2

$$u_i = g_i \qquad \text{on } \Gamma_g \tag{5.4.4}$$

$$\sigma_{ij}n_j = h_i \qquad \text{on } \Gamma_h \tag{5.4.5}$$

5.4.2 Constitutive Equation

For simplicity, we assume the homogeneous isotropic case:

$$\sigma_{ij} = \lambda \epsilon_{kk}\delta_{ij} + 2\mu\epsilon_{ij} \tag{5.4.6}$$

For this case, Assumption 2 becomes

$$0 = \sigma_{\alpha\beta} = \lambda\epsilon_{kk}\,\delta_{\alpha\beta} + 2\mu\epsilon_{\alpha\beta} \tag{5.4.7}$$

Contracting this relation leads to

$$\epsilon_{\alpha\alpha} = \frac{-\lambda}{\lambda + \mu}\epsilon_{33} \tag{5.4.8}$$

Substituting this expression into the preceding one enables us to solve for $\epsilon_{\alpha\beta}$, viz.,

$$\epsilon_{\alpha\beta} = \frac{-\lambda\epsilon_{33}}{2(\lambda + \mu)}\delta_{\alpha\beta} \tag{5.4.9}$$

The expression for $\epsilon_{\alpha\alpha}$ can be used to obtain the classical expression for σ_{33},

$$\begin{aligned}\sigma_{33} &= \lambda\epsilon_{kk} + 2\mu\epsilon_{33} \\ &= \lambda\epsilon_{\alpha\alpha} + (\lambda + 2\mu)\epsilon_{33}\end{aligned}$$

$$\boxed{\sigma_{33} = E\epsilon_{33}} \tag{5.4.10}$$

In addition, we will need expressions for $\sigma_{\alpha 3}$, which by definition are

$$\boxed{\sigma_{\alpha 3} = 2\mu\epsilon_{\alpha 3}} \tag{5.4.11}$$

5.4.3 Strain-displacement Equations

Employing the kinematic conditions stated in Assumption 3, we obtain the following results:

$$\epsilon_{\alpha\beta} = u_{(\alpha,\beta)} = 0 \tag{5.4.12}$$

$$\epsilon_{13} = \frac{w_1' - x_2\theta_3' - \theta_2}{2} \tag{5.4.13}$$

$$\epsilon_{23} = \frac{w_2' + x_1\theta_3' + \theta_1}{2} \tag{5.4.14}$$

$$\epsilon_{33} = w_3' - x_1\theta_2' + x_2\theta_1' \tag{5.4.15}$$

where the primes denote differentiation with respect to x_3.

Remark

As is often the case in beam and plate theories, the stress and kinematic assumptions lead to "microscopic" inconsistencies. For example, from Assumption 2 we have deduced $\epsilon_{\alpha\beta} = -\lambda_{33}\delta_{\alpha\beta}/[2(\lambda + \mu)]$, whereas from Assumption 3 we have $\epsilon_{\alpha\beta} = 0$. For purposes of calculating $\epsilon_{\alpha\beta}$, the former expression is preferred.

These inconsistencies ultimately cause no harm, and, as we have remarked previously, the ultimate justification of a "macroscopic" theory such as this one is its usefulness in practical structural engineering applications.

5.4.4 Definitions of Quantities Appearing in the Theory

Let $\mathcal{F}$ be the set of node numbers at which forces and moments are applied, and let $\mathcal{D}$ be the set of node numbers at which displacements and rotations are prescribed. (For simplicity, we shall assume $\mathcal{F} \cap \mathcal{D} = \emptyset$. In practice, as is always the case, we may specify forces and moments and displacements and rotations to mutually exclusive degrees of freedom at a node.) Let $\boldsymbol{Q}_A$ and $\boldsymbol{M}_A$ denote the 3×1 vectors of applied point loads and moments, respectively, at node $A \in \mathcal{F}$. (We assume the components of $\boldsymbol{M}_A$ are defined in terms of the right-hand rule.) Let $\boldsymbol{W}_A$ and $\boldsymbol{\Theta}_A$ denote the vectors of prescribed displacements and rotations, respectively, at node $A \in \mathcal{D}$.

Quantity	Description
w_i	Displacement
θ_i	Rotation
$\kappa_\alpha = \theta_\alpha'$	Curvature

Quantity	Description
$\left.\begin{aligned}\gamma_1 &= w_1' - \theta_2 \\ \gamma_2 &= w_2' + \theta_1\end{aligned}\right\}$	Shear strains
$\epsilon = w_3'$	Axial strain
$\psi = \theta_3'$	Twist
$\left.\begin{aligned} m_1 &= \int_{A^e} \sigma_{33} x_2 \, dA \\ m_2 &= \int_{A^e} \sigma_{33} x_1 \, dA \end{aligned}\right\}$	Bending moments
$m_3 = \int_{A^e} (\sigma_{23} x_1 - \sigma_{13} x_2) \, dA$	Twisting moment
$q_\alpha = \int_{A^e} \sigma_{\alpha 3} \, dA$	Shear force
$q_3 = \int_{A^e} \sigma_{33} \, dA$	Axial force
$W_i = \{W_{iA}\}, \quad A \in \mathcal{D}$	Prescribed nodal displacements
$\Theta_i = \{\Theta_{iA}\}, \quad A \in \mathcal{D}$	Prescribed nodal rotations
$Q_i = \{Q_{iA}\}, \quad A \in \mathcal{F}$	Prescribed nodal forces
$M_i = \{M_{iA}\}, \quad A \in \mathcal{F}$	Prescribed nodal moments
$\left.\begin{aligned} F_i &= \{F_i^e\}, \quad 1 \le e \le n_{el} \\ F_i^e &= \int_{A^e} \ell_i \, dA \end{aligned}\right\}$	Element applied forces per unit length
$\left.\begin{aligned} C_i &= \{C_i^e\}, \quad 1 \le e \le n_{el} \\ C_1^e &= \int_{A^e} \ell_3 x_2 \, dA \\ C_2^e &= \int_{A^e} \ell_3 x_1 \, dA \\ C_3^e &= \int_{A^e} (\ell_2 x_1 - \ell_1 x_2) \, dA \end{aligned}\right\}$	Element applied couples per unit length
$\left.\begin{aligned} I_1^e &= \int_{A^e} x_2^2 \, dA \\ I_2^e &= \int_{A^e} x_1^2 \, dA \\ J^e &= I_1^e + I_2^e \end{aligned}\right\}$	Element cross-section properties

Local-Global Transformations

Let $\hat{x}_1$, $\hat{x}_2$, $\hat{x}_3$ denote the coordinates of the global system. Similarly, a "hat" above a quantity will indicate components with respect to the global coordinate system. For example,

$$\hat{\mathbf{w}} = \begin{Bmatrix} \hat{w}_1 \\ \hat{w}_2 \\ \hat{w}_3 \end{Bmatrix} \quad \text{and} \quad \hat{\boldsymbol{\theta}} = \begin{Bmatrix} \hat{\theta}_1 \\ \hat{\theta}_2 \\ \hat{\theta}_3 \end{Bmatrix} \tag{5.4.16}$$

are the global representations of $\boldsymbol{w}$ and $\boldsymbol{\theta}$, respectively. Let $\boldsymbol{t}^e$ denote the transformation matrix which rotates vector quantities in the eth local system into the global system. For example,

$$\hat{\boldsymbol{w}}^e = \boldsymbol{t}^e \boldsymbol{w}^e \qquad \hat{\boldsymbol{\theta}}^e = \boldsymbol{t}^e \boldsymbol{\theta}^e \tag{5.4.17}$$

$\boldsymbol{t}^e$ is given by

$$\boldsymbol{t}^e = [t_{ij}^e] \qquad 1 \leq i, j \leq 3 \tag{5.4.18}$$

$$t_{ij}^e = \boldsymbol{e}_i \cdot \boldsymbol{e}_j^e \tag{5.4.19}$$

where $\boldsymbol{e}_i$ is a global unit basis vector and $\boldsymbol{e}_j^e$ is a local unit basis vector. Since $\boldsymbol{t}^e$ is orthogonal (i.e., $\boldsymbol{t}^{e^{-1}} = \boldsymbol{t}^{e^T}$), it follows that

$$\boldsymbol{w}^e = \boldsymbol{t}^{e^T} \hat{\boldsymbol{w}}^e \qquad \boldsymbol{\theta}^e = \boldsymbol{t}^{e^T} \hat{\boldsymbol{\theta}}^e \tag{5.4.20}$$

and so on.

5.4.5 Variational Equation

We begin with the variational equation for the three-dimensional theory and incorporate the preceding relations:

$$0 = \int_\Omega \overline{u}_{(i,j)} \sigma_{ij} \, d\Omega - \int_\Omega \overline{u}_i \ell_i \, d\Omega - \int_{\Gamma_h} \overline{u}_i h_i \, d\Gamma \tag{5.4.21}$$

i. Iterate integrals and employ kinematic relations. Let $dA = dx_1 dx_2$; then

$$\int_\Omega \ldots d\Omega = \sum_{e=1}^{n_{el}} \int_{\Omega^e} \ldots d\Omega = \sum_{e=1}^{n_{el}} \int_0^{h^e} \int_{A^e} \ldots dA \, dx_3 \tag{5.4.22}$$

Without loss of generality, we shall assume $\Gamma_h = \varnothing$.

$$\begin{aligned} 0 = \sum_{e=1}^{n_{el}} \Bigg\{ & \int_0^{h^e} \int_{A^e} (2\overline{u}_{(\alpha,3)} \sigma_{\alpha 3} + \overline{u}_{3,3} \sigma_{33} \, dA \, dx_3 \\ & - \int_0^{h^e} \int_{A^e} \overline{u}_i f_i \, dA \, dx_3 \Bigg\} \\ = \sum_{e=1}^{n_{el}} \Bigg\{ & \int_0^{h^e} \int_{A^e} [(\overline{w}_1' - x_2 \overline{\theta}_3' - \overline{\theta}_2) \sigma_{13} + (\overline{w}_2' + x_1 \overline{\theta}_3' + \overline{\theta}_1) \sigma_{23} \\ & + (\overline{w}_3' - x_1 \overline{\theta}_2' + x_2 \overline{\theta}_1') \sigma_{33}] \, dA \, dx_3 - \int_0^{h^e} \int_{A^e} [(\overline{w}_1 - x_2 \overline{\theta}_3) \ell_1 \\ & + (\overline{w}_2 + x_1 \overline{\theta}_3) \ell_2 + (\overline{w}_3 - x_1 \overline{\theta}_2 + x_2 \overline{\theta}_1) \ell_3] \, dA \, dx_3 \Bigg\} \end{aligned} \tag{5.4.23}$$

ii. Use definitions of force resultants.

$$0 = \sum_{e=1}^{n_{el}} \left\{ \int_0^{h^e} [(\overline{w}_1' - \overline{\theta}_2)q_1 + (\overline{w}_2' + \overline{\theta}_1)q_2 + \overline{w}_3'q_3 + \overline{\theta}_1'm_1 - \overline{\theta}_2'm_2 + \overline{\theta}_3'm_3]\, dx_3 \right.$$
$$\left. - \int_0^{h^e} [\overline{w}_1F_1 + \overline{w}_2F_2 + \overline{w}_3F_3 + \overline{\theta}_1C_1 - \overline{\theta}_2C_2 + \overline{\theta}_3C_3]\, dx_3 \right\} \tag{5.4.24}$$

iii. Integration by parts indicates the differential equations which are satisfied, viz.,

$$\begin{aligned} 0 = \sum_{e=1}^{n_{el}} \Bigg\{ \int_0^{h^e} - [&\overline{w}_1 \underbrace{(q_1' + F_1)}_{\substack{\text{transverse}\\ \text{equilibrium}}} \\ &+ \overline{w}_2 \underbrace{(q_2' + F_2)}_{\substack{\text{transverse}\\ \text{equilibrium}}} \\ &+ \overline{w}_3 \underbrace{(q_3' + F_3)}_{\substack{\text{axial}\\ \text{equilibrium}}} \\ &+ \overline{\theta}_1 \underbrace{(m_1' - q_2 + C_1)}_{\text{moment equilibrium}} \\ &- \overline{\theta}_2 \underbrace{(m_2' - q_1 + C_2)}_{\text{moment equilibrium}} \\ &+ \overline{\theta}_3 \underbrace{(m_3' + C_3)}_{\substack{\text{torsional}\\ \text{equilibrium}}}]\, dx_3 \Bigg\} \\ &+ \text{contributions to nodal force and moment} \\ &\quad \text{equilibrium conditions} \end{aligned} \tag{5.4.25}$$

Remark

Each element contributes end forces and moments to the nodes. Since the "barred" kinematic quantities are assumed continuous at the nodes, the sum of the element contributions must vanish, implying the six nodal equilibrium conditions. We shall add nodal applied forces and moments to the theory later on.

iv. Obtain explicit forms of the constitutive equations in terms of beam theory variables.

$$\begin{aligned} q_1 &= \int_A \sigma_{13}\, dA \\ &= \mu \int_A (w_1' - x_2\theta_3' - \theta_2)\, dA \\ &= \mu\, A\, \gamma_1 \end{aligned} \tag{5.4.26}$$

$$\begin{aligned} q_2 &= \int_A \sigma_{23}\, dA \\ &= \mu \int_A (w_2' + x_1\theta_3' + \theta_1)\, dA \\ &= \mu\, A\, \gamma_2 \end{aligned} \tag{5.4.27}$$

$$\begin{aligned} q_3 &= \int_A \sigma_{33}\, dA \\ &= \int_A (w_3' - x_1\,\theta_2' + x_2\,\theta_1')\, dA \\ &= E\, A\, \epsilon \end{aligned} \tag{5.4.28}$$

$$\begin{aligned} m_1 &= \int_A \sigma_{33}x_2\, dA \\ &= E \int_A (w_3' - x_1\,\theta_2' + x_2\,\theta_1')x_2\, dA \\ &= EI_1\kappa_1 \end{aligned} \tag{5.4.29}$$

$$\begin{aligned} m_2 &= \int_A \sigma_{33}x_1\, dA \\ &= E \int_A (w_3' - x_1\,\theta_2' + x_2\,\theta_1')x_1\, dA \\ &= -EI_2\kappa_2 \end{aligned} \tag{5.4.30}$$

$$\begin{aligned} m_3 &= \int_A (\sigma_{23}x_1 - \sigma_{13}x_2)\, dA \\ &= \mu \int_A [(w_2' + x_1\,\theta_3' + \theta_1)x_1 - (w_1' - x_2\,\theta_3' - \theta_2)x_2]\, dA \\ &= \mu\, J\, \psi \end{aligned} \tag{5.4.31}$$

Remark

"Shear correction factors" may be introduced by suitably modifying the appropriate terms. That is, replace

$$q_\alpha = \mu A \gamma_\alpha \tag{5.4.32}$$

by

$$q_\alpha = \mu A_\alpha^s \gamma_\alpha \qquad \text{(no sum)} \tag{5.4.33}$$

where A_α^s is the "effective shear area" for the α direction.

5.4.6 Strong Form

The domain of the beam theory boundary-value problem consists of n_{el} line segments in $\mathbb{R}^3$, interconnected at nodes. Each line segment corresponds to the $[0, h^e]$ portion of the x_3^e axis in the local, element coordinate system. By a slight abuse of notation, we shall write the element domain $[0, h^e]$. For each node A, let $\mathcal{E}_A$ be the set of element numbers of elements attached to node A.

Let

$$\boldsymbol{q}_A^e = \begin{cases} \boldsymbol{q}^e(0) & \text{(node } A \text{ corresponds to } x_3^e = 0) \\ -\boldsymbol{q}^e(h^e) & \text{(node } A \text{ corresponds to } x_3^e = h^e) \end{cases} \tag{5.4.34}$$

$$\boldsymbol{q}^e = \begin{Bmatrix} q_1^e \\ q_2^e \\ q_3^e \end{Bmatrix} \tag{5.4.35}$$

$$\boldsymbol{m}_A^e = \begin{cases} \boldsymbol{m}^e(0) & \text{(node } A \text{ corresponds to } x_3^e = 0) \\ -\boldsymbol{m}^e(h^e) & \text{(node } A \text{ corresponds to } x_3^e = h^e) \end{cases} \tag{5.4.36}$$

$$\boldsymbol{m}^e = \begin{Bmatrix} m_1^e \\ -m_2^e \\ m_3^e \end{Bmatrix} \tag{5.4.37}$$

Given $F_i\colon \bigcup_{e=1}^{n_{el}} [0, h^e] \to \mathbb{R}$ and $C_i\colon \bigcup_{e=1}^{n_{el}} [0, h^e] \to \mathbb{R}$, $\boldsymbol{Q}_A$ and $\boldsymbol{M}_A$, $A \in \mathcal{F}$, and $\boldsymbol{W}_A$ and $\boldsymbol{\Theta}_A$, $A \in \mathcal{D}$, find $w_i\colon \bigcup_{e=1}^{n_{el}} [0, h^e] \to \mathbb{R}$ and $\theta_i\colon \bigcup_{e=1}^{n_{el}} [0, h^e] \to \mathbb{R}$ such that

$$q_1' + F_1 = 0 \tag{5.4.38}$$

$$q_2' + F_2 = 0 \tag{5.4.39}$$

$$q_3' + F_3 = 0 \tag{5.4.40}$$

$$m_1' - q_2 + C_1 = 0 \tag{5.4.41}$$

$$m_2' - q_1 + C_2 = 0 \tag{5.4.42}$$

$$m_3' + C_3 = 0 \tag{5.4.43}$$

$$q_1 = \mu A_1^s \gamma_1 \tag{5.4.44}$$

$$q_2 = \mu A_2^s \gamma_2 \qquad \text{in }]0, h^e[, \; 1 \le e \le n_{el} \tag{5.4.45}$$

$$q_3 = EA\epsilon \tag{5.4.46}$$

$$m_1 = EI_1 \kappa_1 \tag{5.4.47}$$

(cont. top of page 372)

$$m_2 = -EI_2\kappa_2 \tag{5.4.48}$$

$$m_3 = \mu J\psi \tag{5.4.49}$$

$$\gamma_1 = w_1' - \theta_2 \tag{5.4.50}$$

$$\gamma_2 = w_2' + \theta_1 \tag{5.4.51}$$

$$\epsilon = w_3' \tag{5.4.52}$$

$$\kappa_1 = \theta_1' \tag{5.4.53}$$

$$\kappa_2 = \theta_2' \tag{5.4.54}$$

$$\psi = \theta_3' \tag{5.4.55}$$

$$\left.\begin{aligned} \boldsymbol{t}^e\boldsymbol{w}_A^e &= \boldsymbol{W}_A && (5.4.56)\\ \boldsymbol{t}^e\boldsymbol{\theta}_A^e &= \boldsymbol{\Theta}_A && (5.4.57)\end{aligned}\right\} \quad A \in \mathcal{D},\ e \in \mathcal{E}_A$$

$$\left.\begin{aligned} \sum_{e\in\mathcal{E}_A} \boldsymbol{t}^e\boldsymbol{q}_A^e + \boldsymbol{Q}_A &= \boldsymbol{0} && (5.4.58)\\ \sum_{e\in\mathcal{E}_A} \boldsymbol{t}^e\boldsymbol{m}^e + \boldsymbol{M}_A &= \boldsymbol{0} && (5.4.59)\end{aligned}\right\} \quad A \in \mathcal{F}$$

5.4.7 Weak Form

The spaces we need are as follows:

$$\begin{aligned} \mathcal{S} &= \mathcal{S}\left(\bigcup_{e=1}^{n_{el}} [0, h^e]\right)\\ &= \{(\boldsymbol{w}, \boldsymbol{\theta}) \mid \boldsymbol{w}_A = \boldsymbol{W}_A,\ \boldsymbol{\theta}_A = \boldsymbol{\Theta}_A,\ A \in \mathcal{D}\} \end{aligned} \tag{5.4.60}$$

$$\begin{aligned} \mathcal{V} &= \mathcal{V}\left(\bigcup_{e=1}^{n_{el}} [0, h^e]\right)\\ &= \{(\overline{\boldsymbol{w}}, \overline{\boldsymbol{\theta}}) \mid \boldsymbol{w}_A = \boldsymbol{0},\ \boldsymbol{\theta}_A = \boldsymbol{0},\ A \in \mathcal{D}\} \end{aligned} \tag{5.4.61}$$

Given $\boldsymbol{F}$, $\boldsymbol{C}$, $\boldsymbol{Q}_A$, $\boldsymbol{M}_A$, $\boldsymbol{W}_A$, and $\boldsymbol{\Theta}_A$, as above, find $(\boldsymbol{w}, \boldsymbol{\theta}) \in \mathcal{S}$ such that for all $(\overline{\boldsymbol{w}}, \overline{\boldsymbol{\theta}}) \in \mathcal{V}$.

$$\begin{aligned} 0 = \sum_{e=1}^{n_{el}} \Bigg\{ &\int_0^{h^e} (\overline{\gamma}_1 \mu A_1^s \gamma_1 + \overline{\gamma}_2 \mu A_2^s \gamma_2 + \overline{\kappa}_1 EI_1 \kappa_1 + \overline{\kappa}_2 EI_2 \kappa_2 + \overline{\epsilon} EA \epsilon \\ &+ \overline{\psi} \mu J \psi)\, dx_3^e \\ &- \int_0^{h^e} (\overline{w}_1 F_1 + \overline{w}_2 F_2 + \overline{w}_3 F_3 + \overline{\theta}_1 C_1 - \overline{\theta}_2 C_2 + \overline{\theta}_3 C_3)\, dx_3^e \Bigg\} \\ &- \sum_{A\in\mathcal{F}} (\overline{\boldsymbol{w}}_A^T \boldsymbol{Q}_A + \overline{\boldsymbol{\theta}}_A^T \boldsymbol{M}_A) \end{aligned} \tag{5.4.62}$$

This can be written in the usual way:

$$a(\bar{\boldsymbol{u}}, \boldsymbol{u}) = (\bar{\boldsymbol{u}}, \boldsymbol{\ell}) + (\bar{\boldsymbol{u}}, \boldsymbol{h})_\Gamma \tag{5.4.63}$$

where

$$\boldsymbol{u} = \begin{Bmatrix} w_1 \\ w_2 \\ w_3 \\ \theta_1 \\ \theta_2 \\ \theta_3 \end{Bmatrix} \quad \bar{\boldsymbol{u}} = \begin{Bmatrix} \bar{w}_1 \\ \bar{w}_2 \\ \bar{w}_3 \\ \bar{\theta}_1 \\ \bar{\theta}_2 \\ \bar{\theta}_3 \end{Bmatrix} \quad \boldsymbol{\ell} = \begin{Bmatrix} F_1 \\ F_2 \\ F_3 \\ C_1 \\ -C_2 \\ C_3 \end{Bmatrix} \quad \boldsymbol{h} = \begin{Bmatrix} Q_1 \\ Q_2 \\ Q_3 \\ M_1 \\ M_2 \\ M_3 \end{Bmatrix} \tag{5.4.64}$$

$$a(\bar{\boldsymbol{u}}, \boldsymbol{u}) = \sum_{e=1}^{n_{el}} \left\{ \int_0^{h^e} (\bar{\gamma}_1 \mu A_1^s \gamma_1 + \bar{\gamma}_2 \mu A_2^s \gamma_2 + \bar{\kappa}_1 EI_1 \kappa_1 + \bar{\kappa}_2 EI_2 \kappa_2 + \bar{\epsilon} EA \epsilon + \bar{\psi} EJ \psi)\, dx_3^e \right\} \tag{5.4.65}$$

$$(\bar{\boldsymbol{u}}, \boldsymbol{\ell}) = \sum_{e=1}^{n_{el}} \int_0^{h^e} (\bar{w}_1 F_1 + \bar{w}_2 F_2 + \bar{w}_3 F_3 + \bar{\theta}_1 C_1 - \bar{\theta}_2 C_2 + \bar{\theta}_3 C_3)\, dx_3^e \tag{5.4.66}$$

$$(\bar{\boldsymbol{u}}, \boldsymbol{h})_\Gamma = \sum_{A \in \mathcal{F}} (\bar{\boldsymbol{w}}_A^T \boldsymbol{Q}_A + \bar{\boldsymbol{\theta}}_A^T \boldsymbol{M}_A) \tag{5.4.67}$$

As usual, $a(\cdot, \cdot)$, $(\cdot, \cdot)$, and $(\cdot, \cdot)_\Gamma$ are symmetric bilinear forms.

5.4.8 Matrix Formulation of the Variational Equation

$$0 = \sum_{e=1}^{n_{el}} \left\{ \int_0^{h^e} [\underbrace{\bar{\boldsymbol{\gamma}}^T \boldsymbol{D}^s \boldsymbol{\gamma}}_{\text{shear}} + \underbrace{\bar{\boldsymbol{\kappa}}^T D^b \boldsymbol{\kappa}}_{\text{bending}} + \underbrace{\bar{\epsilon}(EA)\epsilon}_{\text{axial}} + \underbrace{\bar{\psi}(\mu J)\psi}_{\text{torsional}}]\, dx_3^e \right\} - \int_0^{h^e} [\bar{\boldsymbol{w}}^T \boldsymbol{F} + \bar{\boldsymbol{\theta}}^T \boldsymbol{C}]\, dx_3^e - \sum_{A \in \mathcal{F}} (\bar{\boldsymbol{w}}_A^T \boldsymbol{Q}_A + \bar{\boldsymbol{\theta}}_A^T \boldsymbol{M}_A) \tag{5.4.68}$$

where

$$\boldsymbol{\theta} = \begin{Bmatrix} \theta_1 \\ \theta_2 \\ \theta_3 \end{Bmatrix} \quad \bar{\boldsymbol{\theta}} = \begin{Bmatrix} \bar{\theta}_1 \\ \bar{\theta}_2 \\ \bar{\theta}_3 \end{Bmatrix} \tag{5.4.69}$$

$$\boldsymbol{w} = \begin{Bmatrix} w_1 \\ w_2 \\ w_3 \end{Bmatrix} \qquad \overline{\boldsymbol{w}} = \begin{Bmatrix} \overline{w}_1 \\ \overline{w}_2 \\ \overline{w}_3 \end{Bmatrix} \tag{5.4.70}$$

$$\boldsymbol{\gamma} = \begin{Bmatrix} \gamma_1 \\ \gamma_2 \end{Bmatrix} \qquad \overline{\boldsymbol{\gamma}} = \begin{Bmatrix} \overline{\gamma}_1 \\ \overline{\gamma}_2 \end{Bmatrix} \tag{5.4.71}$$

$$\boldsymbol{\kappa} = \begin{Bmatrix} \kappa_1 \\ \kappa_2 \end{Bmatrix} \qquad \overline{\boldsymbol{\kappa}} = \begin{Bmatrix} \overline{\kappa}_1 \\ \overline{\kappa}_2 \end{Bmatrix} \tag{5.4.72}$$

$$\boldsymbol{D}^s = \begin{bmatrix} \mu A_1^s & 0 \\ 0 & \mu A_2^s \end{bmatrix} \tag{5.4.73}$$

$$\boldsymbol{D}^b = \begin{bmatrix} EI_1 & 0 \\ 0 & EI_2 \end{bmatrix} \tag{5.4.74}$$

5.4.9 Finite Element Stiffness Matrix and Load Vector

We shall assume that

$$w_i^h = \sum_{a=1}^{n_{en}} N_a w_{ia}^h \tag{5.4.75}$$

$$\theta_i^h = \sum_{a=1}^{n_{en}} N_a \theta_{ia}^h \tag{5.4.76}$$

Remark

It is not necessary to assume the same shape functions for transverse and extensional displacements or for bending and torsional rotations. Arguments can be made, in fact, that there are some conceptual and practical advantages to employing different interpolations in the present context (see Tessler [45] and Hughes and Tezduyar [46]). However, for the elements we wish to emphasize in Sec. 5.5, this generality is unnecessary.

Define

$$\boldsymbol{d}^e = \{d_p^e\} \tag{5.4.77}$$

$$\overline{\boldsymbol{d}}^e = \{\overline{d}_p^e\} \tag{5.4.78}$$

$$d_p^e = \begin{cases} w_{ia}^h & p = 6a - 6 + i \\ \theta_{ia}^h & p = 6a - 3 + i \end{cases} \tag{5.4.79}$$

$$\overline{d}_p^e = \begin{cases} \overline{w}_{ia}^h & p = 6a - 6 + i \\ \overline{\theta}_{ia}^h & p = 6a - 3 + i \end{cases} \tag{5.4.80}$$

Thus there are six degrees of freedom per node and we can write

$$\boldsymbol{\kappa} = \boldsymbol{B}^b \boldsymbol{d}^e \qquad \overline{\boldsymbol{\kappa}} = \boldsymbol{B}^b \overline{\boldsymbol{d}}^e \tag{5.4.81}$$

$$\boldsymbol{\gamma} = \boldsymbol{B}^s \boldsymbol{d}^e \qquad \overline{\boldsymbol{\gamma}} = \boldsymbol{B}^s \overline{\boldsymbol{d}}^e \tag{5.4.82}$$

$$\boldsymbol{\epsilon} = \boldsymbol{B}^a \boldsymbol{d}^e \qquad \overline{\boldsymbol{\epsilon}} = \boldsymbol{B}^a \overline{\boldsymbol{d}}^e \tag{5.4.83}$$

$$\boldsymbol{\psi} = \boldsymbol{B}^t \boldsymbol{d}^e \qquad \overline{\boldsymbol{\psi}} = \boldsymbol{B}^t \overline{\boldsymbol{d}}^e \tag{5.4.84}$$

$$\boldsymbol{B}^b = [\boldsymbol{B}_1^b, \ldots, \boldsymbol{B}_{n_{en}}^b] \tag{5.4.85}$$

$$\boldsymbol{B}^s = [\boldsymbol{B}_1^s, \ldots, \boldsymbol{B}_{n_{en}}^s] \tag{5.4.86}$$

$$\boldsymbol{B}^a = [\boldsymbol{B}_1^a, \ldots, \boldsymbol{B}_{n_{en}}^a] \tag{5.4.87}$$

$$\boldsymbol{B}^t = [\boldsymbol{B}_1^t, \ldots, \boldsymbol{B}_{n_{en}}^t] \tag{5.4.88}$$

$$\boldsymbol{B}_c^b = \begin{bmatrix} 0 & 0 & 0 & N_c' & 0 & 0 \\ 0 & 0 & 0 & 0 & N_c' & 0 \end{bmatrix} \tag{5.4.89}$$

$$\boldsymbol{B}_c^s = \begin{bmatrix} N_c' & 0 & 0 & 0 & -N_c & 0 \\ 0 & N_c' & 0 & N_c & 0 & 0 \end{bmatrix} \tag{5.4.90}$$

$$\boldsymbol{B}_c^a = [0 \quad 0 \quad N_c' \quad 0 \quad 0 \quad 0] \tag{5.4.91}$$

$$\boldsymbol{B}_c^t = [0 \quad 0 \quad 0 \quad 0 \quad 0 \quad N_c'] \tag{5.4.92}$$

where $1 \le c \le n_{en}$.

With these definitions, we obtain the following expressions for the stiffness and load (with respect to the local coordinate system):

$$\boldsymbol{k}^e = \boldsymbol{k}_b^e + \boldsymbol{k}_s^e + \boldsymbol{k}_a^e + \boldsymbol{k}_t^e \tag{5.4.93}$$

$$\boldsymbol{k}_b^e = \int_0^{he} \boldsymbol{B}^{b^T} \boldsymbol{D}^b \boldsymbol{B}^b \, dx_3^e \qquad \text{(bending stiffness)} \tag{5.4.94}$$

$$\boldsymbol{k}_s^e = \int_0^{he} \boldsymbol{B}^{s^T} \boldsymbol{D}^s \boldsymbol{B}^s \, dx_3^e \qquad \text{(shear stiffness)} \tag{5.4.95}$$

$$\boldsymbol{k}_a^e = \int_0^{he} \boldsymbol{B}^{a^T} (EA) \boldsymbol{B}^a \, dx_3^e \qquad \text{(axial stiffness)} \tag{5.4.96}$$

$$\boldsymbol{k}_t^e = \int_0^{he} \boldsymbol{B}^{t^T} (\mu J) \boldsymbol{B}^t \, dx_3^e \qquad \text{(torsional stiffness)} \tag{5.4.97}$$

$$\boldsymbol{f}^e = \{f_p^e\} \tag{5.4.98}$$

$$f_p^e = \begin{cases} \displaystyle\int_0^{he} N_a F_i \, dx_3^e & p = 6a - 6 + i \\ \displaystyle(-1)^{i+1} \int_0^{he} N_a C_i \, dx_3^e & p = 6a - 3 + i \end{cases} \tag{5.4.99}$$

5.4.10 Representation of Stiffness and Load in Global Coordinates

Before assembly it is necessary to transform $\boldsymbol{k}^e$ and $\boldsymbol{f}^e$ into global coordinates. Assume all "internal" degrees of freedom (if any) have been statically eliminated and only the 12 degrees of freedom corresponding to the two end nodes remain (i.e., $\boldsymbol{k}^e$ and $\boldsymbol{f}^e$ have dimension 12 × 12 and 12 × 1, respectively, with the usual ordering). Let

$$\underset{12\times 12}{\boldsymbol{T}^e} = \begin{bmatrix} \boldsymbol{t}^e & \boldsymbol{0} & \boldsymbol{0} & \boldsymbol{0} \\ \boldsymbol{0} & \boldsymbol{t}^e & \boldsymbol{0} & \boldsymbol{0} \\ \boldsymbol{0} & \boldsymbol{0} & \boldsymbol{t}^e & \boldsymbol{0} \\ \boldsymbol{0} & \boldsymbol{0} & \boldsymbol{0} & \boldsymbol{t}^e \end{bmatrix} \tag{5.4.100}$$

Then

$$\hat{\boldsymbol{k}}^e = \boldsymbol{T}^e \boldsymbol{k}^e \boldsymbol{T}^{e^T} \tag{5.4.101}$$

and

$$\hat{\boldsymbol{f}}^e = \boldsymbol{T}^e \boldsymbol{f}^e \tag{5.4.102}$$

are the globally oriented counterparts of $\boldsymbol{k}^e$ and $\boldsymbol{f}^e$, respectively. (These formulas can be derived as follows. First, note that $\hat{\boldsymbol{d}}^e = \boldsymbol{T}^e \boldsymbol{d}^e$, $\hat{\boldsymbol{c}}^e = \boldsymbol{T}^e \boldsymbol{c}^e$ and $\boldsymbol{T}^e$ is orthogonal. Compute:

$$\begin{aligned} \hat{\boldsymbol{c}}^{e^T} \hat{\boldsymbol{k}}^e \hat{\boldsymbol{d}}^e &= \boldsymbol{c}^{e^T} \boldsymbol{k}^e \boldsymbol{d}^e \\ &= \hat{\boldsymbol{c}}^{e^T} (\boldsymbol{T}^e \boldsymbol{k}^e \boldsymbol{T}^{e^T}) \hat{\boldsymbol{d}}^e \end{aligned} \tag{5.4.103}$$

$$\begin{aligned} \hat{\boldsymbol{c}}^{e^T} \hat{\boldsymbol{f}}^e &= \boldsymbol{c}^{e^T} \boldsymbol{f}^e \\ &= \hat{\boldsymbol{c}}^{e^T} (\boldsymbol{T}^e \boldsymbol{f}^e) \end{aligned} \tag{5.4.104}$$

The above are to hold for all $\hat{\boldsymbol{c}}^e$ and $\hat{\boldsymbol{d}}^e$. Hence, the results follow.)

Remark

It may be noted that "trusses" are special cases of the preceding theory. For a truss, only axial extension effects are accounted for (i.e., bending, transverse shear, and torsion are neglected). Thus the rotational degrees of freedom may be ignored.

5.5 REDUCED INTEGRATION BEAM ELEMENTS

Given the assumption made in Sec. 5.4 that the transverse displacements and transverse rotations are interpolated with the *same* shape functions, there is a tendency to "lock" if the shear term is not handled appropriately. An effective strategy is to employ uniform reduced integration in which the quadrature rule is taken to be one order lower than the "normal" one. The situation is summarized Fig. 5.5.1. Correct rank is attained in all cases. Note that the reduced rule exactly integrates all terms of the stiffness except the transverse shear contributions.

The basic one-point quadrature, two-noded element, in which linear inter-

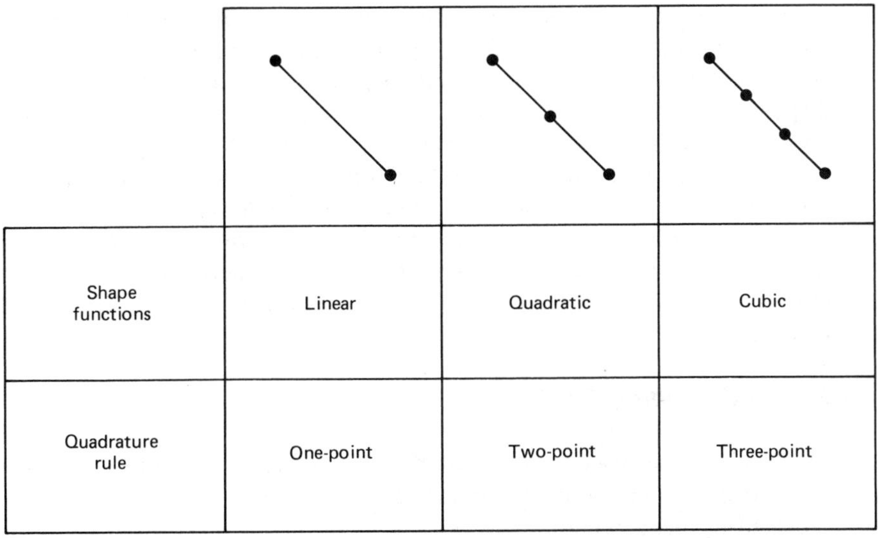

Figure 5.5.1 Uniform reduced integration beam elements.

polation functions are used for displacements and rotations, was first proposed in [47]. Although it is apparent that such low-order interpolations are incapable of exactly representing typical bending behavior, the improvement engendered by reduced integration is dramatic, as may be seen from the example presented in Fig. 5.5.2. The

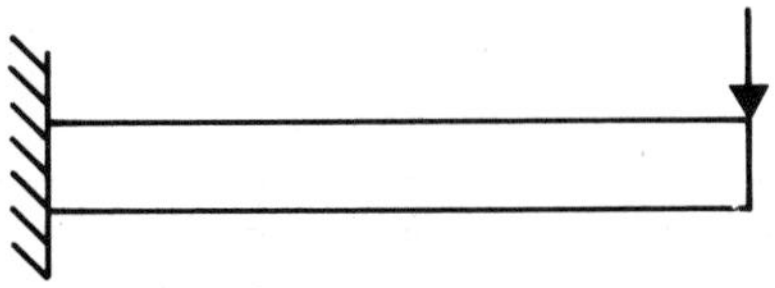

Normalized Tip Displacement

Number of elements	One-point	Two-point
	Thick beam	
1	0.762	0.416×10^{-1}
2	0.940	0.445
4	0.985	0.762
8	0.996	0.927
16	0.999	0.981
	Thin beam	
1	0.750	0.200×10^{-4}
2	0.938	0.800×10^{-4}
4	0.984	0.320×10^{-4}
8	0.996	0.128×10^{-3}
16	0.999	0.512×10^{-3}

Figure 5.5.2 Comparison of reduced integration (one-point) and full integration (two-point) for linear beam element.

reduced-integration element may be shown to be equivalent to an exactly integrated one in which quadratic shape functions are used for transverse displacement and linear shape functions are used for axial displacement and rotations. The middle-node displacement needs to be statically condensed to achieve identical stiffnesses. This observation was apparently first made by H. Allik [48].

The three-node, quadratic reduced-integration element has been in use for some time. It is interesting to note that, upon static condensation of the internal node, the corresponding beam element yields the "exact" stiffness matrix of structural theory, which is based upon cubic transverse displacement interpolation (e.g., see [49]). (This observation is also due to H. Allik [48].) The salubrious effect of reduced quadrature is again apparent.

Residual Bending Flexibility

MacNeal [50] has used the concept of "residual bending flexibility" to further improve the two-node, one-point quadrature beam element. He has shown that if μA_α is replaced by $\overline{\mu A^s_\alpha}$, where

$$\overline{\mu A^s_\alpha} \overset{\text{def}}{=} \left(\frac{1}{\mu A^s_\alpha} + \frac{h^2}{12EI_\alpha} \right)^{-1}$$

then the exact stiffness of structural analysis theory is also obtained. The term $h^2/(12EI_\alpha)$ is called the ***residual bending flexibility.*** Thus by combining the reduced integration and residual bending flexibility concepts, linear functions can be made to achieve the accuracy of cubics in bending. This simple and efficient element seems ideally suited for most practical applications.

Exercise 1. Consider the case of two-dimensional beam bending. Neglect axial and torsional effects as well as out-of-plane bending and shear. Assume further that linear shape functions are employed; and the nodal degrees of freedom are ordered as follows: w_1, θ_1, w_2, θ_2 (see Fig. 5.5.3).

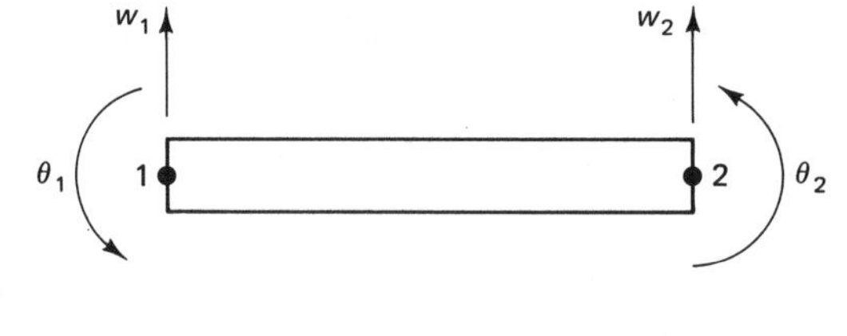

Figure 5.5.3

i. Derive the following expression for bending stiffness:

$$k_b = \frac{EI}{h} \begin{bmatrix} 0 & 0 & 0 & 0 \\ 0 & 1 & 0 & -1 \\ 0 & 0 & 0 & 0 \\ 0 & -1 & 0 & 1 \end{bmatrix}$$

ii. Use one-point Gaussian quadrature to derive the following shear stiffness:

$$k_s^{(1)} = \frac{\mu A^s}{h}\begin{bmatrix} 1 & \frac{h}{2} & -1 & \frac{h}{2} \\ \frac{h}{2} & \frac{h^2}{4} & -\frac{h}{2} & \frac{h^2}{4} \\ -1 & -\frac{h}{2} & 1 & -\frac{h}{2} \\ \frac{h}{2} & \frac{h^2}{4} & -\frac{h}{2} & \frac{h^2}{4} \end{bmatrix}$$

iii. Use two-point Gaussian quadrature to obtain the following shear stiffness:

$$k_s^{(2)} = \frac{\mu A^s}{h}\begin{bmatrix} 1 & \frac{h}{2} & -1 & \frac{h}{2} \\ \frac{h}{2} & \frac{h^2}{3} & -\frac{h}{2} & \frac{h^2}{6} \\ -1 & -\frac{h}{2} & 1 & -\frac{h}{2} \\ \frac{h}{2} & \frac{h^2}{6} & -\frac{h}{2} & \frac{h^2}{3} \end{bmatrix}$$

iv. Show that the rank of $\boldsymbol{k}_s^{(1)}$ is one and the rank of $\boldsymbol{k}_s^{(2)}$ is two.

v. Use the residual bending flexibility technique to modify the one-point shear stiffness of part (ii). Combine this result with the bending stiffness derived in part (i) to form the total stiffness. (This is the "exact" stiffness to which we alluded previously.) Take the limit $\mu \rightarrow \infty$ ("infinite shear stiffness") and show that the resulting stiffness is identical to the one obtained from Bernoulli-Euler theory with Hermite cubics (see Sec. 1.16).

Exercise 2. Use the results of Problem 1 to calculate the three-dimensional 12×12 beam stiffness in the local coordinate system accounting for axial and torsional effects as well as bending and transverse shear about two planes.

REFERENCES

Section 5.3

1. E. D. L. Pugh, "The Static and Dynamic Analysis of Mindlin Plates by Isoparametric Finite Elements," M.Sc. Thesis, C/M/125/76, Department of Civil Engineering, University College of Swansea, U.K., 1976.

2. T. J. R. Hughes, R. L. Taylor, and W. Kanoknukulchai, "A Simple and Efficient Element for Plate Bending," *International Journal for Numerical Methods in Engineering*, 11, no. 10 (1977), 1529–1543.

3. E. D. L. Pugh, E. Hinton, and O. C. Zienkiewicz, "A Study of Quadrilateral Plate Bending Elements with 'Reduced' Integration," *International Journal for Numerical Methods in Engineering*, 12, no. 7 (1978), 1059–1079.

4. D. S. Malkus and T. J. R. Hughes, "Mixed Finite Element Methods—Reduced and Selective Integration Techniques: A Unification of Concepts," *Computer Methods in Applied Mechanics and Engineering*, 15, no. 1 (1978), 63–81.

5. M. Rossow, "Efficient C^0 Finite Element Solution of Simply Supported Plates of Polygonal Shape," *Journal of Applied Mechanics*, 44 (1977), 347–349.

6. L. R. Scott, "A Survey of Displacement Methods for the Plate Bending Problem," U.S.–Germany Symposium on Formulations and Computational Algorithms in Finite Element Analysis, Massachusetts Institute of Technology, Cambridge, Massachusetts, August 9–13, 1976.

7. G. Sander, "Application of the Dual Analysis Principle," in *High Speed Computing of Elastic Structures, Proceedings of IUTAM Symposium*, Liege, Belgium (1971), 167–207.

8. H. C. Rhee, Ph.D. Thesis, Georgia Institute of Technology, Atlanta (1976).

9. O. C. Zienkiewicz, R. L. Taylor, and J. M. Too, "Reduced Integration Technique in General Analysis of Plates and Shells," *International Journal for Numerical Methods in Engineering*, 3 (1971), 275–290.

10. T. J. R. Hughes, M. Cohen, and M. Haroun, "Reduced and Selective Integration Techniques in the Finite Element Analysis of Plates," *Nuclear Engineering and Design*, 46 (1978), 203–222.

11. J. Barlow, "Optimal Stress Locations in Finite Element Models," *International Journal for Numerical Methods in Engineering*, 10 (1976), 243–251.

12. T. Belytschko, C. S. Tsay, and W. K. Liu, "A Stabilization Matrix for the Bilinear Mindlin Plate Element," *Computer Methods in Applied Mechanics and Engineering*, 29 (1981), 313–327.

13. R. L. Taylor, "Finite Elements for General Shell Analysis," *Preprints of the 5th International Seminar on Computational Aspects of the Finite Element Method*, West Berlin, Germany, August 20–21, 1979.

14. T. J. R. Hughes and M. Cohen, "The 'Heterosis' Family of Plate Finite Elements," *Proceedings of the ASCE Electronic Computations Conference*, St. Louis, Missouri, August 6–8, 1979.

15. T. J. R. Hughes and M. Cohen, "The 'Heterosis' Finite Element for Plate Bending," *Computers and Structures*, 9 (1978), 445–450.

16. D. R. J. Owen and E. Hinton, *Finite Elements in Plasticity: Theory and Practice*, Swansea, U.K.: Pineridge Press, 1980.

17. T. J. R. Hughes and T. E. Tezduyar, "Finite Elements Based Upon Mindlin Plate Theory With Particular Reference to the Four-node Bilinear Isoparametric Element," *Journal of Applied Mechanics*, (September 1981), 587–596.

18. R. H. MacNeal, "A Simple Quadrilateral Shell Element," *Computers and Structures*, 8 (1978), 175–183.

19. R. L. Spilker and N. I. Munir, "The Hybrid-Stress Model for Thin Plates," *International Journal for Numerical Methods in Engineering*, 15, no. 8 (1980), 1239–1260.

20. R. L. Spilker and N. I. Munir, "A Serendipity Cubic-Displacement Hybrid-Stress Element for Thin and Moderately Thick Plates," *International Journal for Numerical Methods in Engineering*, 15, no. 8 (1980), 1261–1278.

21. R. L. Spilker and N. I. Munir, "A Hybrid-Stress Quadratic Serendipity Displacement Mindlin Plate Bending Element," *Computers and Structures*, 12 (1980), 11–21.

22. J. Robinson, "LORA—An Accurate Four Node Stress Plate Bending Element, " *International Journal for Numerical Methods in Engineering*, 14, no. 2 (1979), 296–306.

23. R. H. MacNeal, "Derivation of Element Stiffness Matrices by Assumed Strain Distributions," *Nuclear Engineering and Design*, 70 (1982), 3–12.

24. T. J. R. Hughes and R. L. Taylor, "The Linear Triangular Bending Element," in *The Mathematics of Finite Elements and Applications IV, MAFELAP 1981* London: Academic Press, 1982, pp. 127–142.

25. R. H. MacNeal, "The TRIA3 Plate Element," Memo RHM–37, MacNeal Schwendler Corporation, October 1976.

26. I. Fried and S. K. Yang, "Triangular, Nine–degree–of–freedom, C^0 Plate Bending Element of Quadratic Accuracy," *Quarterly Journal of Applied Mathematics*, 31 (1973), 303–312.

27. H. Garnet, J. Crouzet-Pascal, and A. B. Pifko, "Aspects of a Simple Triangular Plate Bending Finite Element," *Computers and Structures*, 12 (1980), 783–789.

28. T. K. Wong, "Nonlinear Analysis of Reinforced Concrete Slab Systems Using Cubic Mindlin Plate Elements," M.Sc. Thesis, Department of Civil Engineering, University College of Swansea, U.K., January 1981.

29. J. L. Batoz and M. Ben Tahar, "Formulation et Evaluation d'un Nouvel Elément Quadrilatéral à 12 D.L. pour la Flexion des Plaques Minces," Département de Génie Mécanique, Université de Technologie, Compiègne, France.

30. J. A. Stricklin, W. Haisler, P. Tisdale, and R. Gunderson, "A Rapidly Converging Triangular Plate Element," *AIAA J.*, 7, no. 1 (1969), 180–181.

31. J. L. Batoz, K. J. Bathe, and L. W. Ho, "A Study of Three-node Triangular Plate Bending Elements," *International Journal for Numerical Methods in Engineering*, 15 (1980), 1771–1812.

32. J. L. Batoz, "An Explicit Formulation for an Efficient Triangular Plate-Bending Element," *International Journal for Numerical Methods in Engineering*, 18 (1982), 1077–1089.

33. B. M. Irons, "The Semiloof Shell Element," *Finite Elements for Thin Shells and Curved Members*, (eds. D. G. Ashwell and R. H. Gallagher) London: John Wiley, 1976, Chapter 11, pp. 197–222.

34. L. P. R. Lyons, "A General Finite Element System with Special Reference to the Analysis of Cellular Structures," Ph.D. Thesis, Imperial College, London, 1977.

35. M. A. Crisfield, "A Four-noded Plate Bending Element Using Shear Constraints; A Modified Version of Lyons' Element," *Computer Methods in Applied Mechanics and Engineering*, 38 (1983), 93–120.

36. A. Tessler and T. J. R. Hughes, "An Improved Treatment of Transverse Shear in the Mindlin-type Four-node Quadrilateral Element," *Computer Methods in Applied Mechanics and Engineering*, 39 (1983), 311–335.

37. A. Tessler and T. J. R. Hughes, "Three-node Mindlin Plate Element with Improved Transverse Shear," *Computer Methods in Applied Mechanics and Engineering*, 50 (1985), 71–101.

38. K. J. Bathe and E. N. Dvorkin, "A Four-node Plate Bending Element Based on Mindlin/Reissner Plate Theory and a Mixed Interpolation," *International Journal for Numerical Methods in Engineering*, 21 (1985), 367–383.

39. K. J. Bathe and E. N. Dvorkin, "A Formulation of General Shell Elements —The Use of Mixed Interpolation of Tensorial Components," *Proceedings of the Conference on Numerical Methods in Engineering: Theory and Applications*, University College of Swansea, U.K., January 1985.
40. T. Belytschko and C. S. Tsay, "A Stabilization Procedure for the Quadrilateral Plate Bending Element with One-point Quadrature," *International Journal for Numerical Methods in Engineering*, 19 (1983), 405–420.
41. K. C. Park and D. L. Flaggs, "A Symbolic Fourier Synthesis of a One-point Integrated Quadrilateral Plate Element," *Computer Methods in Applied Mechanics and Engineering*, 48, no. 2 (1985), 203–236.
42. T. Belytschko, J. S.-J. Ong, and W. K. Liu, "A Consistent Control of Spurious Singular Modes in the 9-node Lagrange Element for the Laplace and Mindlin Plate Equations," *Computer Methods in Applied Mechanics and Engineering*, 44 (1984), 269–295.
43. M. A. Crisfield, "A Quadratic Mindlin Element Using Shear Constraints," *Computers and Structures*, 18 (1984), 833–852.
44. H. C. Huang and E. Hinton, "A Nine-node Lagrangian Plate Element with Enhanced Shear Interpolation," *Engineering Computations*, 1 (1984), 369–379.

Section 5.4

45. A. Tessler and S. B. Dong, "On a Hierarchy of Conforming Timoshenko Beam Elements," *Computers and Structures*, 14 (1981), 335–344.
46. T. J. R. Hughes and T. E. Tezduyar, "Finite Elements Based Upon Mindlin Plate Theory With Particular Reference to the Four-node Bilinear Isoparametric Element," *Journal of Applied Mechanics*, (September 1981), 587–596.

Section 5.5

47. T. J. R. Hughes, R. L. Taylor, and W. Kanoknukulchai, "A Simple and Efficient Element for Plate Bending," *International Journal for Numerical Methods in Engineering*, 11, no. 10 (1977), 1529–1543.
48. H. Allik, Private communications, 1976.
49. R. D. Cook, *Concepts and Applications of Finite Element Analysis*, New York: John Wiley, 1974.
50. R. H. MacNeal, "A Simple Quadrilateral Shell Element," *Computers and Structures*, 8, (1978), 175–183.

6

The C^0-Approach to Curved Structural Elements

6.1 INTRODUCTION

In this chapter we present a general formulation for curved structural elements. Throughout, transverse shear deformations are accounted for. This enables the use of C^0 interpolations as in the plate and beam theories of Chapter 5. Despite the fact that for over 20 years, intense interest has been focused on the development of shell finite elements, there is still some dissatisfaction with available methodology. Currently, a number of research efforts are directed at deepening the understanding of and improving shell finite element capabilities. Many different approaches have been developed in finite element shell analysis, and an enormous literature now exists. No attempt will be made to review the literature in this brief chapter. (A literature review in finite element shell analysis would entail in itself a major work!) The approach presented herein is felt to be quite general and the one currently gaining favor. The extension of the present ideas to nonlinear analysis is relatively straightforward (see, e.g., [1–3] and references therein). This is viewed as an important attribute of this type of approach because practical shell analysis often involves consideration of nonlinear effects such as buckling.

By now the reader should be familiar with the scheme of developing finite element formulations advocated herein—namely, a classical statement of the boundary-value problem is posed first. Then a corresponding weak formulation is developed and subsequently discretized by introducing finite element interpolation functions. The "shell theory" presented in this chapter is simply three-dimensional elasticity with certain kinematic and mechanical assumptions built in. The situation is essentially the same as in Chapter 5, where Reissner-Mindlin plate theory was developed from three-dimensional theory.[1] To avoid a cumbersome presentation of the basic theory, we will take advantage of the reader's familiarity with the material in Chapters 1–5 and move ahead to the development of the finite element formulation.

In Sec. 6.2 we present the three-dimensional theory. The reduction to practically important two-dimensional cases is treated in Sec. 6.3. The two-dimensional formulation applies to shells of revolution, tubes, rings, and curved beams.

6.2 DOUBLY CURVED SHELLS IN THREE DIMENSIONS

6.2.1 Geometry

The geometry of a typical quadrilateral shell element is defined by the following relations:

$$\boldsymbol{x}(\xi, \eta, \zeta) = \bar{\boldsymbol{x}}(\xi, \eta) + \boldsymbol{X}(\xi, \eta, \zeta) \tag{6.2.1}$$

$$\bar{\boldsymbol{x}}(\xi, \eta) = \sum_{a=1}^{n_{en}} N_a(\xi, \eta)\bar{\boldsymbol{x}}_a \tag{6.2.2}$$

$$\boldsymbol{X}(\xi, \eta, \zeta) = \sum_{a=1}^{n_{en}} N_a(\xi, \eta)\boldsymbol{X}_a(\zeta) \tag{6.2.3}$$

$$\boldsymbol{X}_a(\zeta) = z_a(\zeta)\hat{\boldsymbol{X}}_a \qquad \text{(no sum)} \tag{6.2.4}$$

$$z_a(\zeta) = N_+(\zeta)z_a^+ + N_-(\zeta)z_a^- \tag{6.2.5}$$

$$N_+(\zeta) = \tfrac{1}{2}(1 + \zeta), \qquad N_-(\zeta) = \tfrac{1}{2}(1 - \zeta) \tag{6.2.6}$$

In (6.2.1) through (6.2.6), $\boldsymbol{x}$ denotes the position vector of a generic point of the shell; $\bar{\boldsymbol{x}}$ is the position vector of a point in the reference surface; $\boldsymbol{X}$ is a position vector based at a point in the reference surface which defines the "fiber direction" through the point; $\bar{\boldsymbol{x}}_a$ is the position vector of nodal point a; N_a denotes a two-dimensional shape function associated with node a; n_{en} is the number of element nodes; $\hat{\boldsymbol{X}}_a$ is a unit vector emanating from node a in the fiber direction; and z_a is a "thickness function," associated with node a, which is defined by the location of the reference surface.

Equations (6.2.1) through (6.2.6) represent a smooth mapping of the biunit cube into the physical shell domain; see Fig. 6.2.1. For ζ fixed, the surface defined by (6.2.1) is called a ***lamina***; for ξ, η fixed, the line defined by (6.2.1) is called the ***fiber***. The fibers are generally *not* perpendicular to the laminae. Sometimes the fiber is referred to as the "pseudonormal".

For a particular choice of two-dimensional shape functions, (6.2.1) through (6.2.6) are precisely defined upon specification of $\bar{\boldsymbol{x}}_a$, $\hat{\boldsymbol{X}}_a$, z_a^+, and z_a^- ($a = 1, 2, \ldots, n_{en}$). It is convenient in practice to take as input the coordinates of the top and bottom surfaces of the shell along each nodal fiber ($\boldsymbol{x}_a^+$ and $\boldsymbol{x}_a^-$, respectively) and a parameter $\bar{\zeta} \in [-1, +1]$, which defines the location of the reference surface. For example, if $\bar{\zeta} = -1, 0, +1$ (respectively), then the reference surface is taken to be the bottom,

[1] In the present context the approach is sometimes described as the "degenerated shell element" procedure [4].

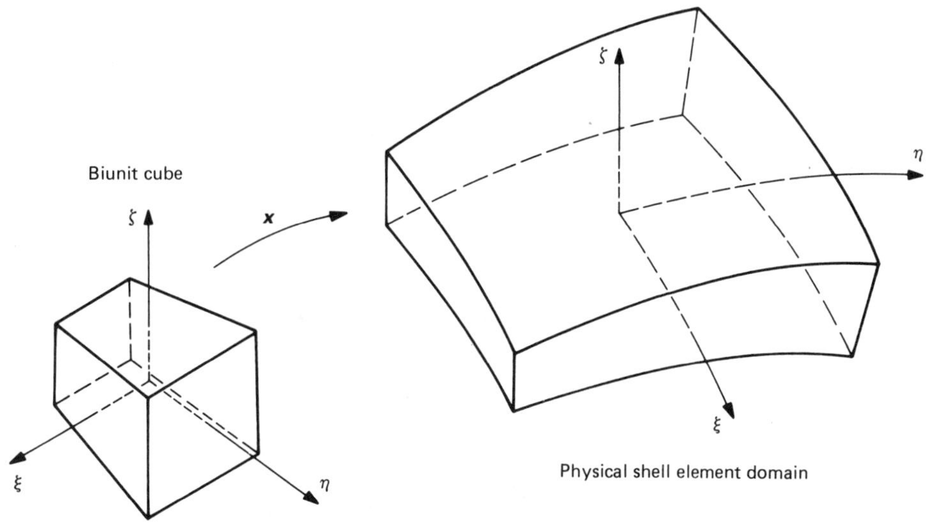

Figure 6.2.1

middle, top (respectively) of the shell. This option enables compatible modeling of shell-continuum interfaces. From these data we may calculate

$$\bar{\boldsymbol{x}}_a = \frac{1}{2}(1 - \bar{\zeta})\boldsymbol{x}_a^- + \frac{1}{2}(1 + \bar{\zeta})\boldsymbol{x}_a^+ \tag{6.2.7}$$

$$\hat{\boldsymbol{X}}_a = \frac{\boldsymbol{x}_a^+ - \boldsymbol{x}_a^-}{\|\boldsymbol{x}_a^+ - \boldsymbol{x}_a^-\|} \tag{6.2.8}$$

$$z_a^+ = \frac{1}{2}(1 - \bar{\zeta})\|\boldsymbol{x}_a^+ - \boldsymbol{x}_a^-\| \tag{6.2.9}$$

$$z_a^- = -\frac{1}{2}(1 + \bar{\zeta})\|\boldsymbol{x}_a^+ - \boldsymbol{x}_a^-\| \tag{6.2.10}$$

where $\|\cdot\|$ denotes the Euclidean norm (i.e., $\|\boldsymbol{x}\| = (x_1^2 + x_2^2 + x_3^2)^{1/2}$). An illustration of these ideas is presented in Fig. 6.2.2.

The top- and bottom-surface coordinates are uniquely defined at element interfaces. Consequently, there are no gaps or overlaps along element boundaries.

6.2.2 Lamina Coordinate Systems

At each integration point in the element, a Cartesian reference frame is erected so that two axes are tangent to the lamina through the point. The frame is defined by its orthonormal basis vectors $\boldsymbol{e}_1^l$, $\boldsymbol{e}_2^l$, $\boldsymbol{e}_3^l$ in which $\boldsymbol{e}_3^l$ is perpendicular to the lamina (see Fig. 6.2.3). The basis vectors are calculated as follows.

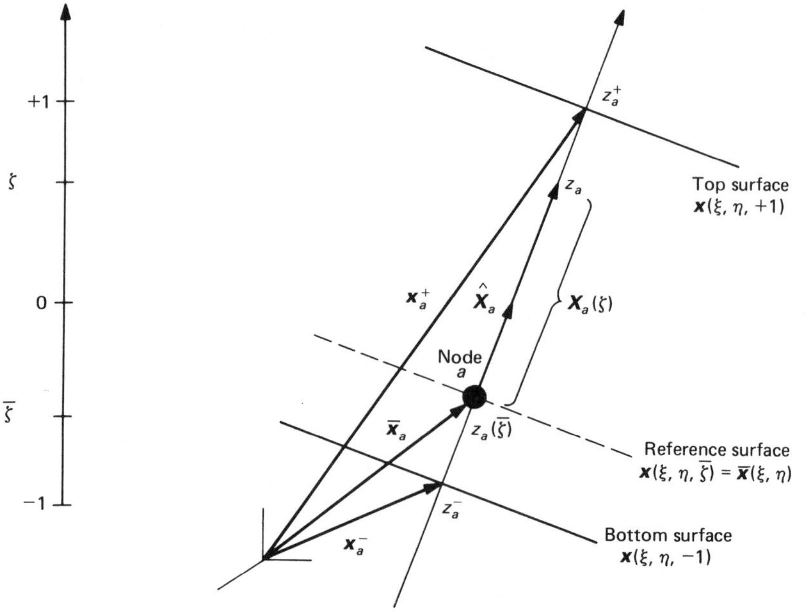

Figure 6.2.2

Construct unit tangent vectors to the ξ- and η-coordinate directions:

$$e_\xi = \frac{x_{,\xi}}{\|x_{,\xi}\|} \tag{6.2.11}$$

$$e_\eta = \frac{x_{,\eta}}{\|x_{,\eta}\|} \tag{6.2.12}$$

(6.2.11) and (6.2.12) suffice to define e_3^l:

$$e_3^l = \frac{e_\xi \times e_\eta}{\|e_\xi \times e_\eta\|} \tag{6.2.13}$$

The vectors tangent to the lamina are selected so that the angle between e_1^l and e_ξ is

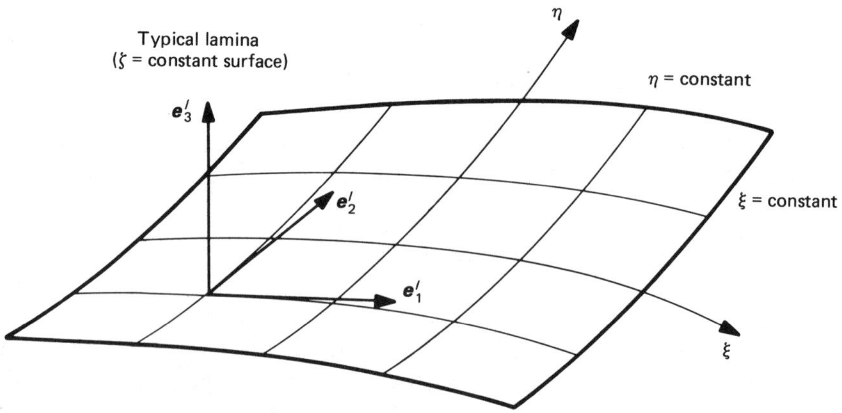

Figure 6.2.3 Typical lamina coordinate system.

the same as the angle between $\boldsymbol{e}_\eta$ and $\boldsymbol{e}_2^l$ and so that the $\boldsymbol{e}_1^l$, $\boldsymbol{e}_2^l$-basis is as "close" as possible to the $\boldsymbol{e}_\xi$, $\boldsymbol{e}_\eta$-basis. Thus

$$\boldsymbol{e}_1^l = \frac{\sqrt{2}}{2}(\boldsymbol{e}_\alpha - \boldsymbol{e}_\beta) \tag{6.2.14}$$

$$\boldsymbol{e}_2^l = \frac{\sqrt{2}}{2}(\boldsymbol{e}_\alpha + \boldsymbol{e}_\beta) \tag{6.2.15}$$

where

$$\boldsymbol{e}_\alpha = \frac{\frac{1}{2}(\boldsymbol{e}_\xi + \boldsymbol{e}_\eta)}{\|\frac{1}{2}(\boldsymbol{e}_\xi + \boldsymbol{e}_\eta)\|} \tag{6.2.16}$$

$$\boldsymbol{e}_\beta = \frac{\boldsymbol{e}_3^l \times \boldsymbol{e}_\alpha}{\|\boldsymbol{e}_3^l \times \boldsymbol{e}_\alpha\|} \tag{6.2.17}$$

Note that $\boldsymbol{e}_3^l$ is not generally tangent to the fiber direction; see Fig. 6.2.4. The $\boldsymbol{e}_3^l$ direction is used for purposes of invoking the plane stress hypothesis.

In the sequel, it will be necessary to transform quantities from the global coordinate system to the lamina system. This is facilitated by the following matrix:

$$\boldsymbol{q} = [q_{ij}] = [\boldsymbol{e}_1^l \quad \boldsymbol{e}_2^l \quad \boldsymbol{e}_3^l]^T: \text{ global} \rightarrow \text{lamina} \tag{6.2.18}$$

in which the superscript T denotes *transpose*. Clearly, $\boldsymbol{q}$ is orthogonal.

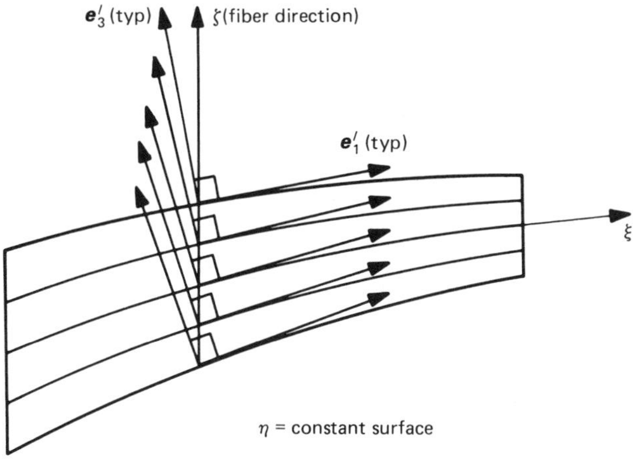

Figure 6.2.4 Lamina coordinate systems along a fiber.

6.2.3 Fiber Coordinate Systems

At each node a unique local Cartesian coordinate system is constructed, which is used as a reference frame for rotations. The only requirement that the frame must satisfy is that one direction coincide with the fiber direction. This in itself is not sufficient to define the frame. In many cases, the physical situation may be such that an "obvious" choice presents itself. For example, in the case of a cylindrical shell, it is natural to choose the cylindrical basis vectors: $\boldsymbol{e}_\theta$, $\boldsymbol{e}_z$, $\boldsymbol{e}_r$. When there is no obvious choice, an algorithm may be employed to select the basis. For example, let $\hat{\boldsymbol{X}}$ denote the unit basis vector in the fiber direction. (We omit the nodal subscript a throughout this discussion for notational clarity.) Let $\boldsymbol{e}_1$, $\boldsymbol{e}_2$, $\boldsymbol{e}_3$ denote the global Cartesian basis, i.e.,

$$\boldsymbol{e}_1 = \begin{Bmatrix} 1 \\ 0 \\ 0 \end{Bmatrix} \qquad \boldsymbol{e}_2 = \begin{Bmatrix} 0 \\ 1 \\ 0 \end{Bmatrix} \qquad \boldsymbol{e}_3 = \begin{Bmatrix} 0 \\ 0 \\ 1 \end{Bmatrix} \tag{6.2.19}$$

The global Cartesian components of $\hat{\boldsymbol{X}}$ are denoted by $\hat{X}_i$, $i = 1, 2, 3$.

Algorithm

1. Let $a_i = |\hat{X}_i|$, $i = 1, 2, 3$.
2. $j = 1$.
3. If $a_1 > a_3$, then $a_3 = a_1$, and $j = 2$.
4. If $a_2 > a_3$, $j = 3$.
5. $\boldsymbol{e}_3^f = \hat{\boldsymbol{X}}$.
6. $\boldsymbol{e}_2^f = (\hat{\boldsymbol{X}} \times \boldsymbol{e}_j)/\|\hat{\boldsymbol{X}} \times \boldsymbol{e}_j\|$.
7. $\boldsymbol{e}_1^f = \boldsymbol{e}_2^f \times \hat{\boldsymbol{X}}$.

The orthonormal fiber basis obtained (i.e., $\boldsymbol{e}_1^f$, $\boldsymbol{e}_2^f$, $\boldsymbol{e}_3^f$) satisfies the condition that if $\hat{\boldsymbol{X}}$ is "close" to $\boldsymbol{e}_3$, then $\boldsymbol{e}_1^f$, $\boldsymbol{e}_2^f$, $\boldsymbol{e}_3^f$ will be "close" to $\boldsymbol{e}_1$, $\boldsymbol{e}_2$, $\boldsymbol{e}_3$, respectively (see Fig. 6.2.5).

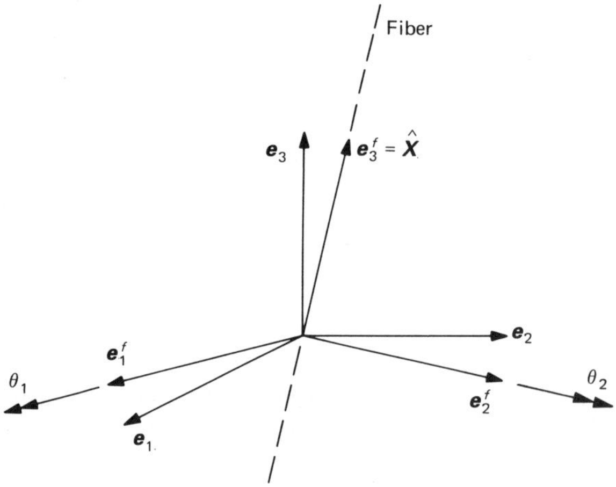

Figure 6.2.5 Nodal fiber basis and right-hand-rule sign convention for rotations.

6.2.4 Kinematics

The displacement of the shell is assumed to take the following form:

$$\boldsymbol{u}(\xi, \eta, \zeta) = \bar{\boldsymbol{u}}(\xi, \eta) + \boldsymbol{U}(\xi, \eta, \zeta) \tag{6.2.20}$$

$$\bar{\boldsymbol{u}}(\xi, \eta) = \sum_{a=1}^{n_{en}} N_a(\xi, \eta)\bar{\boldsymbol{u}}_a \tag{6.2.21}$$

$$U(\xi, \eta, \zeta) = \sum_{a=1}^{n_{en}} N_a(\xi, \eta) U_a(\zeta) \tag{6.2.22}$$

$$U_a(\zeta) = z_a(\zeta) \hat{U}_a \qquad \text{(no sum)} \tag{6.2.23}$$

where $\boldsymbol{u}$ is the displacement of a generic point; $\bar{\boldsymbol{u}}$ is the displacement of a point on the reference surface; and $\boldsymbol{U}$ is the "fiber displacement". The vector $\hat{\boldsymbol{U}}_a$ is constructed such that the fiber may rotate, but not stretch, viz.[2],

$$\hat{U}_a = \theta_{a2} \boldsymbol{e}^f_{a1} - \theta_{a1} \boldsymbol{e}^f_{a2} \tag{6.2.24}$$

The quantities θ_{a1} and θ_{a2} represent the rotations of the fiber about the basis vectors $\boldsymbol{e}^f_{a1}$ and $\boldsymbol{e}^f_{a2}$, respectively. The right-hand-rule sign convention is illustrated in Fig. 6.2.5. The "fiber inextensibility hypothesis," manifested by (6.2.24), is analogous to plate Assumption 4, Sec. 5.2.1.

6.2.5 Reduced Constitutive Equation

We begin with the constitutive equation of three-dimensional elasticity *written with respect to the lamina coordinate system at the point under consideration*:

$$\boldsymbol{\sigma}^l = \boldsymbol{D}^l \boldsymbol{\epsilon}^l \tag{6.2.25}$$

The superscript l is used to emphasize that the components are in the lamina system. The ordering of components is given as follows:

$$\boldsymbol{\sigma}^l = \{\sigma^l_I\} = \begin{Bmatrix} \sigma^l_{11} \\ \sigma^l_{22} \\ \sigma^l_{12} \\ \sigma^l_{23} \\ \sigma^l_{31} \\ \sigma^l_{33} \end{Bmatrix} \tag{6.2.26}$$

$$\boldsymbol{\epsilon}^l = \{\epsilon^l_I\} = \begin{Bmatrix} \dfrac{\partial u^l_1}{\partial x^l_1} \\ \dfrac{\partial u^l_2}{\partial x^l_2} \\ \dfrac{\partial u^l_1}{\partial x^l_2} + \dfrac{\partial u^l_2}{\partial x^l_1} \\ \dfrac{\partial u^l_2}{\partial x^l_3} + \dfrac{\partial u^l_3}{\partial x^l_2} \\ \dfrac{\partial u^l_3}{\partial x^l_1} + \dfrac{\partial u^l_1}{\partial x^l_3} \\ \dfrac{\partial u^l_3}{\partial x^l_3} \end{Bmatrix} \tag{6.2.27}$$

[2] Equation (6.2.24) is the linearized version of fiber inextensibility.

$$D^l_{IJ} = c^l_{ijkm} \tag{6.2.28}$$

I/J	i/k	j/m
1	1	1
2	2	2
3	1	2
4	2	3
5	3	1
6	3	3

(6.2.29)

The constitutive equation needs to be modified in order to enforce the ***zero normal-stress condition***[3] in the 3-direction of the lamina system. This amounts to solving for the 33-component of strain in terms of the remaining components:

$$\sigma^l_6 = 0 \Rightarrow \epsilon^l_6 = \frac{-\left(\sum_{I=1}^{5} D^l_{6I}\epsilon^l_I\right)}{D^l_{66}} \tag{6.2.30}$$

Equivalently, this may be expressed in tensor form:

$$\sigma^l_{33} = 0 \Rightarrow \epsilon^l_{33} = \frac{-\left(\sum_{ij\neq 33} c^l_{33ij}\epsilon^l_{ij}\right)}{c^l_{3333}} \tag{6.2.31}$$

Substituting (6.2.30) in (6.2.25) results in

$$\begin{aligned}\sigma^l_I &= \sum_{J=1}^{5} D^l_{IJ}\epsilon^l_J + D^l_{I6}\epsilon^l_6 \\ &= \sum_{J=1}^{5}\left(\frac{D^l_{IJ} - D^l_{I6}D^l_{6J}}{D^l_{66}}\right)\epsilon^l_J\end{aligned} \tag{6.2.32}$$

and thus

$$\tilde{\boldsymbol{\sigma}}^l = \tilde{\boldsymbol{D}}^l\tilde{\boldsymbol{\epsilon}}^l \tag{6.2.33}$$

where

[3] This is the analog of plate Assumption 2, Sec. 5.2.1.

$$\tilde{\boldsymbol{\sigma}}^l = \begin{Bmatrix} \sigma_1^l \\ \sigma_2^l \\ \sigma_3^l \\ \sigma_4^l \\ \sigma_5^l \end{Bmatrix} \tag{6.2.34}$$

$$\tilde{\boldsymbol{\epsilon}}^l = \begin{Bmatrix} \epsilon_1^l \\ \epsilon_2^l \\ \epsilon_3^l \\ \epsilon_4^l \\ \epsilon_5^l \end{Bmatrix} \tag{6.2.35}$$

$$\tilde{\boldsymbol{D}}_{IJ}^l = [\tilde{D}_{IJ}^l], \qquad 1 \le I, J \le 5 \tag{6.2.36}$$

$$\tilde{D}_{IJ}^l = D_{IJ}^l - \frac{D_{I6}^l D_{6J}^l}{D_{66}^l} \tag{6.2.37}$$

(6.2.33) is called the ***reduced constitutive equation***. The reason for placing the 33-components in the last entries of $\boldsymbol{\sigma}$ and $\boldsymbol{\epsilon}$ should now be apparent.

Exercise 1. Show that for the isotropic case

$$\tilde{\boldsymbol{D}}^l = \frac{E}{(1-\nu^2)} \begin{bmatrix} 1 & \nu & 0 & 0 & 0 \\ & 1 & 0 & 0 & 0 \\ & & \dfrac{1-\nu}{2} & 0 & 0 \\ & \text{symmetric} & & \dfrac{1-\nu}{2} & 0 \\ & & & & \dfrac{1-\nu}{2} \end{bmatrix} \tag{6.2.38}$$

where E is Young's modulus and ν is Poisson's ratio.

Shear Correction Factors

To attain results consistent with classical bending theory, ***shear correction factors*** need to be introduced. In (6.2.38), this amounts to multiplying the transverse shearing moduli by $\kappa = \frac{5}{6}$, viz.,

$$\widetilde{D}^l = \frac{E}{(1-\nu^2)} \begin{bmatrix} 1 & \nu & 0 & 0 & 0 \\ & 1 & 0 & 0 & 0 \\ & & \dfrac{1-\nu}{2} & 0 & 0 \\ & \text{symmetric} & & \dfrac{\kappa(1-\nu)}{2} & 0 \\ & & & & \dfrac{\kappa(1-\nu)}{2} \end{bmatrix} \tag{6.2.39}$$

6.2.6 Strain-displacement Matrix

In application to shells, special treatment needs to be given to transverse shear and membrane terms to prevent "mesh-locking" phenomena [5–13]. A particularly effective treatment may be performed by employing the reduced-selective integration concept. In the present formulation, we make use of the $\bar{\boldsymbol{B}}$-method introduced in [14, 15] and presented in Chapter 4, which enables implementation of selective integration type procedures by a simple modification of the strain-displacement matrix. This technique has advantages in anisotropic situations because it engenders only a minor change to the standard element-array implementation.

The definition of the strain-displacement matrix adopted is given as follows:

$$\boldsymbol{B} = [\boldsymbol{B}_1, \boldsymbol{B}_2, \ldots, \boldsymbol{B}_{n_{en}}] \tag{6.2.40}$$

The strain-displacement transformation may be written as

$$\underbrace{\widetilde{\boldsymbol{\epsilon}}}_{5\times 1} = \sum_{a=1}^{n_{en}} \underbrace{\boldsymbol{B}_a}_{5\times 5} \underbrace{\begin{Bmatrix} \bar{\boldsymbol{u}}_a \\ \theta_{a1} \\ \theta_{a2} \end{Bmatrix}}_{5\times 1} \tag{6.2.41}$$

where each $\boldsymbol{B}_a$ has the form:

$$\boldsymbol{B}_a = [\boldsymbol{b}_I^u \quad \boldsymbol{b}_I^\theta] = \begin{bmatrix} \boldsymbol{b}_1^u & \boldsymbol{b}_1^\theta \\ \boldsymbol{b}_2^u & \boldsymbol{b}_2^\theta \\ \boldsymbol{b}_3^u & \boldsymbol{b}_3^\theta \\ \hdashline \boldsymbol{b}_4^u & \boldsymbol{b}_4^\theta \\ \boldsymbol{b}_5^u & \boldsymbol{b}_5^\theta \end{bmatrix} \tag{6.2.42}$$

The $\boldsymbol{b}_I^u$'s and $\boldsymbol{b}_I^\theta$'s are row vectors:

$$\boldsymbol{b}_I^u = \langle b_{Im}^u \rangle = \langle b_{I1}^u \quad b_{I2}^u \quad b_{I3}^u \rangle \tag{6.2.43}$$

$$\boldsymbol{b}_I^\theta = \langle b_{I\alpha}^\theta \rangle = \langle b_{I1}^\theta \quad b_{I2}^\theta \rangle \tag{6.2.44}$$

Explicit formulas may be obtained by calculating the displacement gradients:

$$\frac{\partial u_i^l}{\partial x_j^l} = \sum_{m=1}^{3} q_{im} \frac{\partial u_m}{\partial x_j^l}$$

$$= \sum_{m=1}^{3} q_{im} \sum_{a=1}^{n_{en}} \left(\frac{\partial N_a}{\partial x_j^l} \bar{u}_{am} + \frac{\partial (N_a z_a)}{\partial x_j^l} (\theta_{a2} e_{am1}^f - \theta_{a1} e_{am2}^f) \right) \tag{6.2.45}$$

The notation

$$\boldsymbol{e}_{a\alpha}^f = \{e_{am\alpha}^f\} = \begin{Bmatrix} e_{a1\alpha}^f \\ e_{a2\alpha}^f \\ e_{a3\alpha}^f \end{Bmatrix} \tag{6.2.46}$$

has been used in (6.2.45). From (6.2.45), we may read the following definitions:

$$b_{1m}^u = q_{1m} \frac{\partial N_a}{\partial x_1^l} \tag{6.2.47}$$

$$b_{2m}^u = q_{2m} \frac{\partial N_a}{\partial x_2^l} \tag{6.2.48}$$

$$b_{3m}^u = q_{1m} \frac{\partial N_a}{\partial x_2^l} + q_{2m} \frac{\partial N_a}{\partial x_1^l} \tag{6.2.49}$$

$$b_{4m}^u = q_{2m} \frac{\partial N_a}{\partial x_3^l} + q_{3m} \frac{\partial N_a}{\partial x_2^l} \tag{6.2.50}$$

$$b_{5m}^u = q_{3m} \frac{\partial N_a}{\partial x_1^l} + q_{1m} \frac{\partial N_a}{\partial x_3^l} \tag{6.2.51}$$

$$b_{1\alpha}^\theta = \omega_{a1\alpha} \frac{\partial (N_a z_a)}{\partial x_1^l} \tag{6.2.52}$$

$$b_{2\alpha}^\theta = \omega_{a2\alpha} \frac{\partial (N_a z_a)}{\partial x_2^l} \tag{6.2.53}$$

$$b_{3\alpha}^\theta = \omega_{a1\alpha} \frac{\partial (N_a z_a)}{\partial x_2^l} + \omega_{a2\alpha} \frac{\partial (N_a z_a)}{\partial x_1^l} \tag{6.2.54}$$

$$b_{4\alpha}^\theta = \omega_{a2\alpha} \frac{\partial (N_a z_a)}{\partial x_3^l} + \omega_{a3\alpha} \frac{\partial (N_a z_a)}{\partial x_2^l} \tag{6.2.55}$$

$$b_{5\alpha}^\theta = \omega_{a3\alpha} \frac{\partial (N_a z_a)}{\partial x_1^l} + \omega_{a1\alpha} \frac{\partial (N_a z_a)}{\partial x_3^l} \tag{6.2.56}$$

where

$$\omega_{ai1} = -\sum_{m=1}^{3} q_{im} e_{am2}^f \tag{6.2.57}$$

$$\omega_{ai2} = \sum_{m=1}^{3} q_{im} e_{am1}^f \tag{6.2.58}$$

The differentiations indicated in (6.2.47) through (6.2.56) are calculated by transforming corresponding global quantities:

$$\frac{\partial}{\partial x_j^l} = \sum_{m=1}^{3} q_{jm} \frac{\partial}{\partial x_m} \tag{6.2.59}$$

Let b denote any term in the matrix $\boldsymbol{B}_a$. Suppose we wish to treat b with a reduced lamina quadrature rule. In this case b will be replaced by its reduced quadrature counterpart $\bar{b}$, which is defined as follows: Assume a lamina quadrature rule is specified to integrate the element stiffness. Think of this rule as the "normal" one for the element. Another quadrature rule is introduced which may be thought of as the "reduced" rule. For this rule, $\bar{n}_{\text{int}}$ and $\bar{\xi}_l$, $\bar{\eta}_l$ denote the number of points and locations, respectively. A special set of shape functions, $\bar{N}_l$'s, is defined with nodes at the quadrature points (i.e., $N_k(\bar{\xi}_l, \bar{\eta}_l) = \delta_{kl}$, $1 \leq k, l \leq \bar{n}_{\text{int}}$).

For example, if the element under consideration was a quadrilateral and the reduced rule was the 2×2 Gauss-Legendre rule, then the $\bar{N}_l$'s would be bilinear functions interpolating the 2×2 Gauss points.

The general form of $\bar{b}$ is given by

$$\bar{b}(\xi, \eta) = \sum_{l=1}^{\bar{n}_{\text{int}}} \bar{N}_l(\xi, \eta) b(\bar{\xi}_l, \bar{\eta}_l) \tag{6.2.60}$$

As an example, consider the four-node bilinear quadrilateral. The normal rule is the 2×2 Gauss-Legendre rule. Take as the reduced rule the one-point Gauss-Legendre rule, so that $\bar{n}_{\text{int}} = 1$, $\bar{N}_1 = 1$, and $\bar{\xi}_1 = \bar{\eta}_1 = 0$ (i.e., the element center). Then (6.2.60) reduces to

$$\bar{b}(\xi, \eta) = b(0, 0) \tag{6.2.61}$$

That is, the value at the center of the element is used to compute $\bar{b}$.

In shell elements, selective reduced integration has been used on membrane and transverse-shear terms to avoid locking. The terms in the matrix $\boldsymbol{B}_a$ that are affected are indicated next.

Reduced integration of transverse shear

$$\boldsymbol{B}_a = \left[\begin{array}{c|c} \boldsymbol{b}_1^u & \boldsymbol{b}_1^\theta \\ \boldsymbol{b}_2^u & \boldsymbol{b}_2^\theta \\ \boldsymbol{b}_3^u & \boldsymbol{b}_3^\theta \\ \hline \bar{\boldsymbol{b}}_4^u & \bar{\boldsymbol{b}}_4^\theta \\ \bar{\boldsymbol{b}}_5^u & \bar{\boldsymbol{b}}_5^\theta \end{array}\right] = \left[\begin{array}{c|c} \cdot & \cdot \\ \cdot & \cdot \\ \cdot & \cdot \\ \hline \times & \times \\ \times & \times \end{array}\right] \quad \begin{array}{l}(\times \text{ signifies} \\ \text{location of} \\ \text{affected terms)}\end{array} \tag{6.2.62}$$

Uniform reduced integration of membrane effects

$$\boldsymbol{B}_a = \left[\begin{array}{c|c} \bar{\boldsymbol{b}}_1^u & \boldsymbol{b}_1^\theta \\ \bar{\boldsymbol{b}}_2^u & \boldsymbol{b}_2^\theta \\ \bar{\boldsymbol{b}}_3^u & \boldsymbol{b}_3^\theta \\ \hline \boldsymbol{b}_4^u & \boldsymbol{b}_4^\theta \\ \boldsymbol{b}_5^u & \boldsymbol{b}_5^\theta \end{array}\right] = \left[\begin{array}{c|c} \times & \cdot \\ \times & \cdot \\ \times & \cdot \\ \hline \cdot & \cdot \\ \cdot & \cdot \end{array}\right] \tag{6.2.63}$$

Selective reduced integration of membrane effects

In this case, only the membrane shear-strain term is underintegrated:

$$\boldsymbol{B}_a = \left[\begin{array}{c:c} \boldsymbol{b}_1^u & \boldsymbol{b}_1^\theta \\ \boldsymbol{b}_2^u & \boldsymbol{b}_2^\theta \\ \bar{\boldsymbol{b}}_3^u & \boldsymbol{b}_3^\theta \\ \hdashline \boldsymbol{b}_4^u & \boldsymbol{b}_4^\theta \\ \boldsymbol{b}_5^u & \boldsymbol{b}_5^\theta \end{array}\right] = \left[\begin{array}{c:c} \cdot & \cdot \\ \cdot & \cdot \\ \times & \cdot \\ \hdashline \cdot & \cdot \\ \cdot & \cdot \end{array}\right] \tag{6.2.64}$$

Note that there is no "invariance" problem in this case, since the entries are defined with respect to lamina coordinates, which in turn are defined intrinsically by the element geometry.

Cases (6.2.62) and (6.2.63), or (6.2.62) and (6.2.64), may be combined. For example,

Reduced integration of transverse shear and membrane effects:

$$\boldsymbol{B}_a = \left[\begin{array}{c:c} \bar{\boldsymbol{b}}_1^u & \boldsymbol{b}_1^\theta \\ \bar{\boldsymbol{b}}_2^u & \boldsymbol{b}_2^\theta \\ \bar{\boldsymbol{b}}_3^u & \boldsymbol{b}_3^\theta \\ \hdashline \bar{\boldsymbol{b}}_4^u & \bar{\boldsymbol{b}}_4^\theta \\ \bar{\boldsymbol{b}}_5^u & \bar{\boldsymbol{b}}_5^\theta \end{array}\right] = \left[\begin{array}{c:c} \times & \cdot \\ \times & \cdot \\ \times & \cdot \\ \hdashline \times & \times \\ \times & \times \end{array}\right] \tag{6.2.65}$$

Remark

In certain situations, different reduced rules may be employed on different terms. For example, suppose that two different reduced rules are used, one for membrane effects ("$\bar{b}$-treatment") and one for transverse shear ("$\bar{\bar{b}}$-treatment"). Then

$$\boldsymbol{B}_a = \left[\begin{array}{c:c} \bar{\boldsymbol{b}}_1^u & \boldsymbol{b}_1^\theta \\ \bar{\boldsymbol{b}}_2^u & \boldsymbol{b}_2^\theta \\ \bar{\boldsymbol{b}}_3^u & \boldsymbol{b}_3^\theta \\ \hdashline \bar{\bar{\boldsymbol{b}}}_4^u & \bar{\bar{\boldsymbol{b}}}_4^\theta \\ \bar{\bar{\boldsymbol{b}}}_5^u & \bar{\bar{\boldsymbol{b}}}_5^\theta \end{array}\right] = \left[\begin{array}{c:c} \times & \cdot \\ \times & \cdot \\ \times & \cdot \\ \hdashline \times & \times \\ \times & \times \end{array}\right] \tag{6.2.66}$$

Quite elaborate schemes, involving several different rules, have been developed to produce effective shell elements (e.g., see MacNeal's description of the QUAD4 shell element used in NASTRAN [16]).

Exercise 2. The displacement interpolation is not form-identical to the geometric mapping. Thus the elements may not be classified as isoparametric. However, it may be shown that the displacement interpolation entails all rigid body modes and constant strain states. Present an argument which corroborates this assertion.

6.2.7 Stiffness Matrix

The element stiffness is defined as follows:

$$\underbrace{\boldsymbol{k}}_{5n_{en} \times 5n_{en}} = [\boldsymbol{k}_{ab}] \tag{6.2.67}$$

$$\underbrace{\boldsymbol{k}_{ab}}_{5 \times 5} = \int_{\square} \underbrace{\int_{-1}^{+1} \boldsymbol{B}_a^T \widetilde{\boldsymbol{D}}^l \boldsymbol{B}_b j \, d\zeta}_{\text{fiber integral}} \, d\square \tag{6.2.68}$$

where

$$\int_{\square} \ldots d\square = \int_{-1}^{+1}\int_{-1}^{+1} \ldots d\xi \, d\eta \qquad \text{(lamina integral)} \tag{6.2.69}$$

$$j = \det \begin{bmatrix} x_{1,\xi} & x_{1,\eta} & x_{1,\zeta} \\ x_{2,\xi} & x_{2,\eta} & x_{2,\zeta} \\ x_{3,\xi} & x_{3,\eta} & x_{3,\zeta} \end{bmatrix} \tag{6.2.70}$$

6.2.8 External Force Vector

We allow for body, surface, and edge force vectors.

Body force

The element body force vector is given by

$$\underbrace{\boldsymbol{f}^{\text{body}}}_{5n_{en} \times 1} = \{\boldsymbol{f}_a^{\text{body}}\} \tag{6.2.71}$$

$$\underbrace{\boldsymbol{f}_a^{\text{body}}}_{5 \times 1} = \int_{\square}\int_{-1}^{+1} \boldsymbol{N}_a^T \boldsymbol{\ell} j \, d\zeta \, d\square \tag{6.2.72}$$

where

$$\boldsymbol{N}_a = \left[\begin{array}{ccc:cc} N_a & 0 & 0 & N_a z_a e^f_{a12} & -N_a z_a e^f_{a11} \\ 0 & N_a & 0 & N_a z_a e^f_{a22} & -N_a z_a e^f_{a21} \\ 0 & 0 & N_a & N_a z_a e^f_{a32} & -N_a z_a e^f_{a31} \end{array}\right] \tag{6.2.73}$$

The definition of $\boldsymbol{N}_a$ follows directly from the kinematic assumptions.

Surface force

The element surface force vector is defined by

$$\boldsymbol{f}^{\text{surf}} = \{\boldsymbol{f}_a^{\text{surf}}\} \tag{6.2.74}$$

$$\boldsymbol{f}_a^{\text{surf}} = \int_{\square} \boldsymbol{N}_a^T \boldsymbol{h}\, j_s \, d\square, \qquad \zeta = \begin{cases} +1 \text{ top} \\ -1 \text{ bottom} \end{cases} \tag{6.2.75}$$

where

$$j_s = \|\boldsymbol{x}_{,\xi} \times \boldsymbol{x}_{,\eta}\| \qquad \text{(surface Jacobian)} \tag{6.2.76}$$

and $\boldsymbol{h}$ is the surface force vector (per unit surface area).

It is convenient to allow for pressure and shear surface force vectors as separate cases.

Pressure

In this case

$$\boldsymbol{h} = -\zeta p \boldsymbol{n}, \qquad (\zeta = +1 \text{ or } -1) \tag{6.2.77}$$

$$\boldsymbol{n} = \frac{\boldsymbol{e}_\xi \times \boldsymbol{e}_\eta}{\|\boldsymbol{e}_\xi \times \boldsymbol{e}_\eta\|} \tag{6.2.78}$$

where p is the pressure and $\boldsymbol{n}$ is the unit normal vector to the surface.

Shear

We assume that the shear is specified in the ξ and/or η directions on the surface in question. In this case the surface force vector is given by

$$\boldsymbol{h} = h_\xi \boldsymbol{e}_\xi + h_\eta \boldsymbol{e}_\eta \tag{6.2.79}$$

where h_ξ and h_η are the shears in the ξ and η directions, respectively.

Edge force

Suppose we wish to apply a distributed loading along an $\eta = +1$ or -1 edge. Let $\boldsymbol{h}$ denote the distributed surface force. The nodal forces are

$$\boldsymbol{f}_a^{\text{edge}} = \int_{-1}^{+1}\int_{-1}^{+1} (\boldsymbol{N}_a^T \boldsymbol{h}\, j_e)\Big|_{\eta=+1 \text{ or } -1} d\zeta\, d\xi \tag{6.2.80}$$

where

$$j_e = \|\boldsymbol{x}_{,\xi} \times \boldsymbol{x}_{,\zeta}\| \qquad \text{(edge surface Jacobian)} \tag{6.2.81}$$

The case of loading along an $\xi = +1$ or -1 edge is handled by interchanging ξ and

η in (6.2.80) and (6.2.81). Note that when the reference surface is not taken to be the midsurface, nodal moments are produced even when h is constant (i.e., in general, $f_{a4} \neq 0, f_{a5} \neq 0$).

If edge forces or moments are specified per unit edge length, then nodal forces are computed as follows: Consider an $\eta = +1$ or -1 edge. Let $\ell_i^{\text{line}} = \ell_i^{\text{line}}(\xi)$ denote the edge force and let $m_i^{\text{line}} = m_i^{line}(\xi)$ denote the edge moment. The nodal forces are then given by

$$f_a^{\text{edge}} = \int_{-1}^{+1} N_a \bigg|_{\eta=+1 \text{ or } -1} \begin{Bmatrix} \ell_1^{\text{line}} \\ \ell_2^{\text{line}} \\ \ell_3^{\text{line}} \\ \hline m_1^{\text{line}} \\ m_2^{\text{line}} \end{Bmatrix} \|x_{,\xi}\| \, d\xi \qquad (6.2.82)$$

Note that m_1^{line} and m_2^{line} must have the same sense as θ_1 and θ_2. The result is made applicable to an $\xi = +1$ or -1 edge if ξ and η are interchanged in (6.2.82).

The element external force vector is defined by

$$f^{\text{ext}} = f^{\text{body}} + f^{\text{surf}} + f^{\text{edge}} \qquad (6.2.83)$$

6.2.9 Fiber Numerical Integration

In the general case, fiber integrals need be evaluated by a numerical integration technique. Several ways of going about this present themselves, each having advantages in certain circumstances.

If the integrand is a smooth function of ζ (e.g., when the shell consists of one homogeneous layer), then Gaussian quadrature is most efficient. If the reference surface is taken to be the midsurface, then the one-point Gauss rule (i.e., midpoint rule) senses membrane and transverse-shear effects, whereas at least two points are required to manifest the bending behavior. In the one-layer case, an efficient alternative is to perform the fiber integration analytically as in [12]. See also [17] for recent developments.

If the shell is built up from a series of layers of different materials such that the material properties and stresses are discontinuous functions of ζ, then Gaussian rules may be effectively used over each layer. If there are a large number of approximately equal-sized layers, then the midpoint rule on each layer should suffice. If, on the other hand, there are a small number of layers or if the layers vary considerably in thickness, then different Gaussian rules should be assigned to individual layers. This may be facilitated by allowing the user to input the location and weights of the fiber quadrature rule. Thus any special set of circumstances may be accommodated.

6.2.10 Stress Resultants

Bending moments, membrane forces and transverse shear resultants may be computed at any lamina point, say, $\boldsymbol{\xi} = (\xi, \eta)$.

Moments

The moments may be calculated from the following expressions:

$$m_{\alpha\beta}(\boldsymbol{\xi}) = \int_{-1}^{+1} \sigma^{l}_{\alpha\beta}(\boldsymbol{\xi}, \zeta)\, z(\boldsymbol{\xi}, \zeta)\, d\zeta\, z_{,\zeta}(\boldsymbol{\xi}), \qquad 1 \leq \alpha, \beta \leq 2 \tag{6.2.84}$$

$$z(\boldsymbol{\xi}, \zeta) = N_{+}(\zeta)\, z^{+}(\boldsymbol{\xi}) + N_{-}(\zeta)\, z^{-}(\boldsymbol{\xi}) \tag{6.2.85}$$

$$z^{\pm}(\boldsymbol{\xi}) = \sum_{a=1}^{n_{en}} N_a(\boldsymbol{\xi}) z_a^{\pm} \tag{6.2.86}$$

$$z_{,\zeta}(\boldsymbol{\xi}) = \frac{z^{+}(\boldsymbol{\xi}) - z^{-}(\boldsymbol{\xi})}{2} \tag{6.2.87}$$

Membrane forces

The membrane forces may be computed as follows:

$$n_{\alpha\beta}(\boldsymbol{\xi}) = \int_{-1}^{+1} \sigma^{l}_{\alpha\beta}(\boldsymbol{\xi}, \zeta)\, d\zeta\, z_{,\zeta}(\boldsymbol{\xi}), \qquad 1 \leq \alpha, \beta \leq 2 \tag{6.2.88}$$

Transverse shears

The shears may be computed from the following formula:

$$q_{\alpha}(\boldsymbol{\xi}) = \int_{-1}^{+1} \sigma^{l}_{\alpha 3}(\boldsymbol{\xi}, \zeta)\, d\zeta\, z_{,\zeta}(\boldsymbol{\xi}), \qquad 1 \leq \alpha \leq 2 \tag{6.2.89}$$

The integrations of (6.2.84), (6.2.88), and (6.2.89) are performed using the quadrature points of the fiber rule.

The sign conventions for the stress resultants are illustrated in Fig. 6.2.6.

6.2.11 Shell Elements

The development of shell elements is still an active area of research. We shall describe only some of the simpler elements currently used.

Lagrange Elements

The lamina shape functions and quadrature rules for the Lagrange elements are shown in Fig. 6.2.7. In each case the appropriate reduced rule is one order lower than the

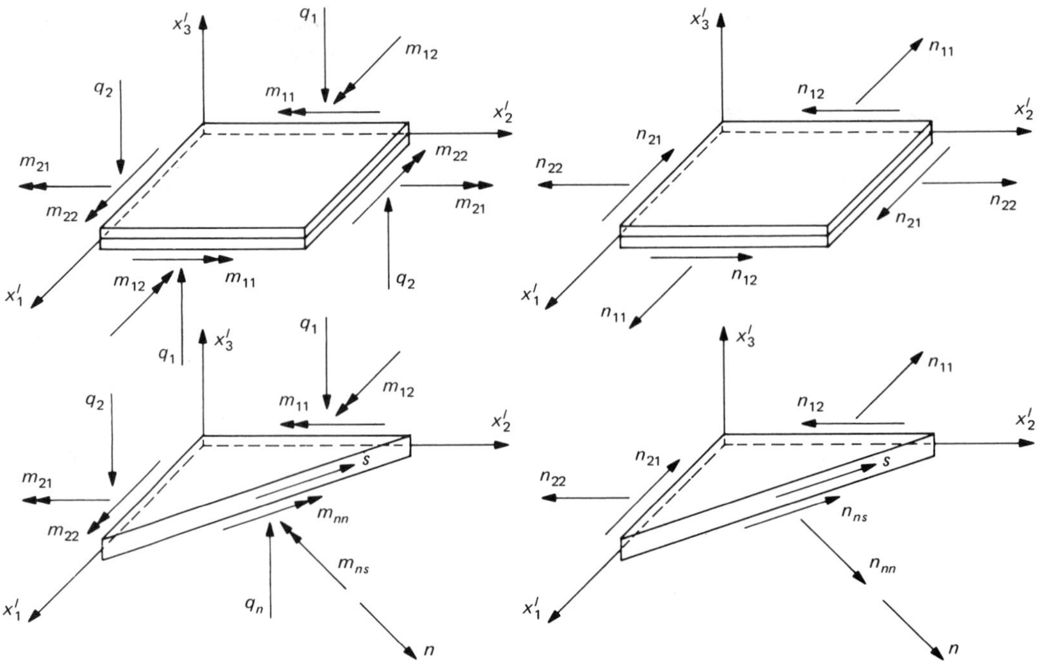

Figure 6.2.6 Sign conventions for stress resultants.

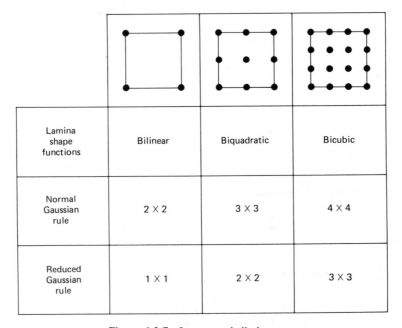

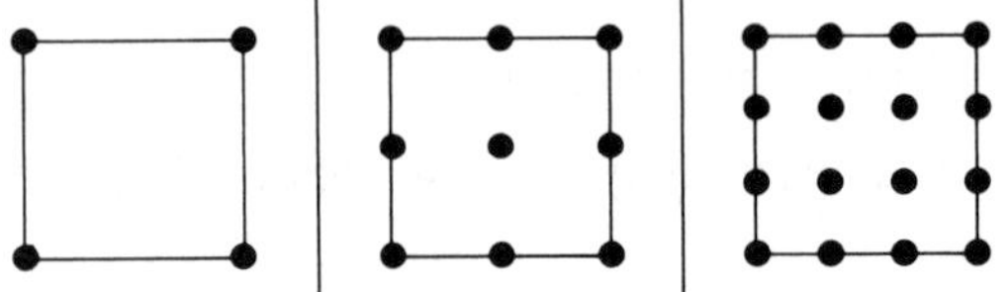

Lamina shape functions	Bilinear	Biquadratic	Bicubic
Normal Gaussian rule	2 X 2	3 X 3	4 X 4
Reduced Gaussian rule	1 X 1	2 X 2	3 X 3

Figure 6.2.7 Lagrange shell elements.

normal rule. If the normal rule and reduced rule are combined, as described in Sec. 6.2.6, the element is called a ***selective integration element.*** If either the normal or reduced rule is used exclusively, then the element is called a ***uniform integration element.*** Uniform normal integration tends to cause elements to "lock" in thin shell applications. This phenomenon is especially pronounced for low-order elements but lessens as the order of interpolation is increased (e.g., the normally integrated bicubic element is felt by some to be a fairly good performer). On the other hand, selective and uniform reduced integration elements behave well in thin shell applications but may occasionally engender rank deficiency (i.e., spurious "mechanisms"). The problem is less acute for the selective integration elements than for the uniform reduced integration elements. In some situations, the mechanisms are precluded from globally forming by boundary conditions, but nevertheless they represent a potentially dangerous deficiency.

Research has been undertaken to efficiently remove mechanisms. One successful procedure is described next.

Heterosis Element

The heterosis concept was originally developed in [5] to eliminate the spurious zero-energy mode in the nine-node, selectively integrated Lagrange plate element (see also Chapter 5). The resulting element possesses correct rank and behaves well in the thin plate limit.

Implementation of the heterosis shell element begins by constructing the arrays for the selectively integrated Lagrange element (e.g., say $\boldsymbol{k}_{\text{Lag}}$, $\boldsymbol{f}_{\text{Lag}}$, etc.) Then a projection matrix, $\boldsymbol{H}$, is constructed from the serendipity shape functions associated with the element-boundary nodes. The role $\boldsymbol{H}$ plays is to eliminate all translational degrees of freedom from the internal (i.e., nonboundary) nodes by restraining them to interpolate the serendipity shape functions (see Sec. 5.3.7 for analogous details for the plate). The heterosis arrays are then defined by

$$\boldsymbol{k}_{\text{het}} = \boldsymbol{H}^T \boldsymbol{k}_{\text{Lag}} \boldsymbol{H} \tag{6.2.90}$$

$$\boldsymbol{f}_{\text{het}} = \boldsymbol{H}^T \boldsymbol{f}_{\text{Lag}} \tag{6.2.91}$$

At this point, if desired, the internal rotational degrees of freedom may be removed by static condensation. Both transverse shear and membrane effects need to be treated with reduced quadrature to avoid locking. The heterosis shell element has been around for a while but still remains one of the better elements for general shell analysis [17].

Example. (Pinched Cylinder [17])

A thin circular cylinder of length L, radius R, and thickness t was subjected to equal and opposite point loads P as shown in Fig. 6.2.8. Due to symmetry, only one octant of the cylinder needed to be modeled. Two end conditions were considered: free ends and ends restrained by a rigid diaphragm, Figs. 6.2.8(a) and 6.2.8(b), respectively. Uniform and selective integration elements were employed in a convergence study. For the selective integration elements, bending terms were treated with the "normal rule" and membrane and shear terms were treated with the "reduced rule"; see (6.2.65). Results for the following elements are presented:

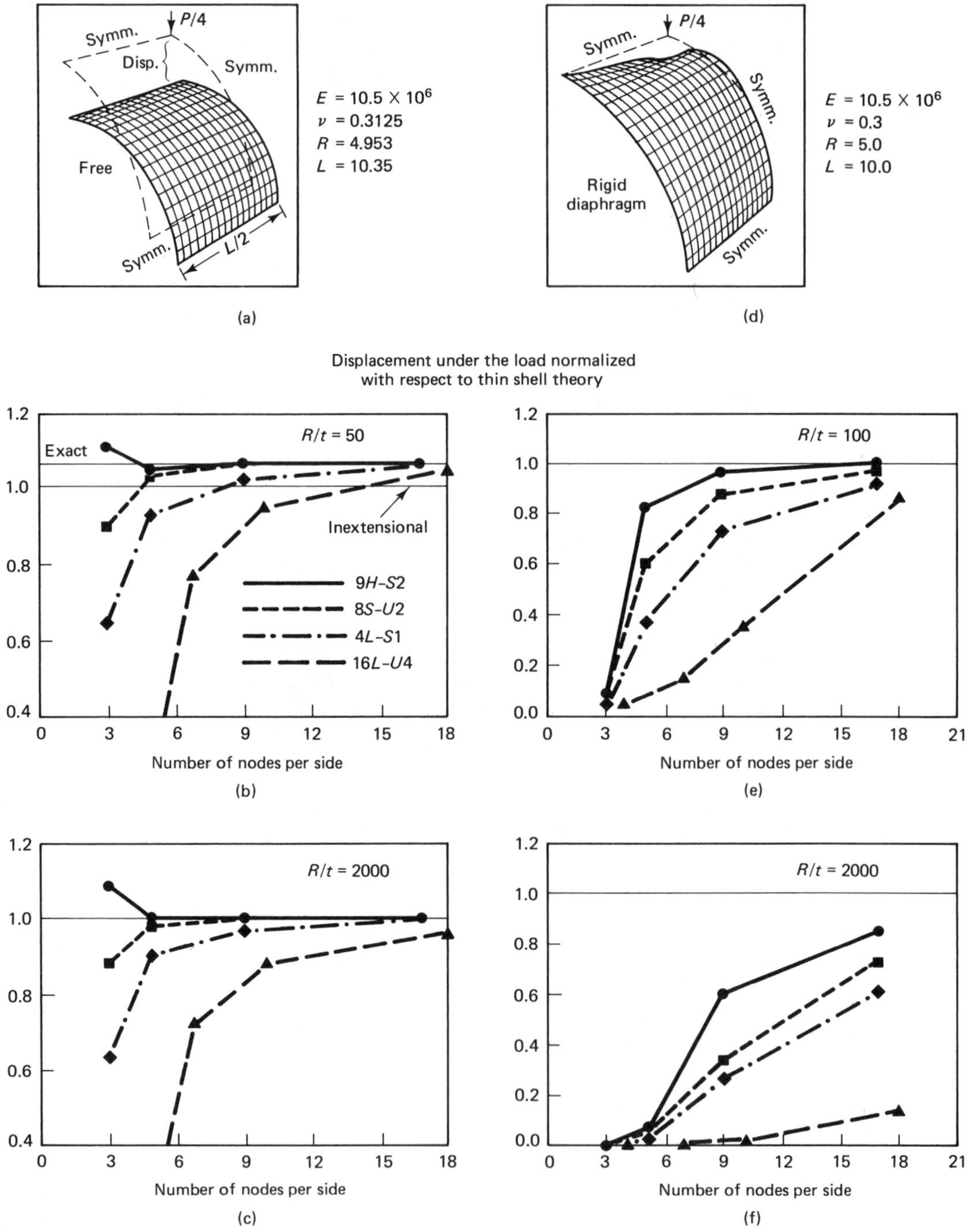

Figure 6.2.8 Convergence study for pinched cylinder.

$9H$-$S2$	Nine-node heterosis element with selective integration.
$8S$-$U2$	Eight-node serendipity element with uniform integration.
$4L$-$S1$	Four-node Lagrange element with selective integration.
$16L$-$U4$	Sixteen-node Lagrange element with uniform integration.

The integration rules are designated by:

Un $\quad n \times n$ Gaussian quadrature.

$$Sn \quad \begin{cases} (n + 1) \times (n + 1) \text{ Gaussian quadrature normal rule.} \\ n \times n \text{ Gaussian quadrature reduced rule.} \end{cases}$$

It may be seen from the results that convergence for the open-ended case is more rapid than for the rigid-diaphragm case. In the latter case, when $R/t = 2000$, convergence is particularly slow; see Fig. 6.2.8(f). The reason for this is the exact solution involves highly localized deformations in the vicinity of the load, which are poorly resolved by the uniform meshes employed. Shells often exhibit fine-scale behavior, and the need for locally refined meshes is apparent.

In all cases presented, the heterosis element converges fastest while the fully integrated Lagrange element ($16L$-$U4$) converges slowest. The latter element possesses correct rank. Due to the underintegration of membrane terms, the heterosis element possesses one mechanism (see Fig. 4.6.3). However, this mechanism is noncommunicable in a mesh of two or more elements so, for all practical purposes, this element may be viewed as having correct rank. The four-node Lagrange element ($4L$-$S1$) possesses four mechanisms: two membrane hourglass modes (see Fig. 4.6.2); and the in-plane twist mode and transverse-displacement hourglass mode (see Fig. 5.3.9). The three hourglass modes are communicable and, under certain circumstances, can be aroused globally. Consequently, extreme caution is advised in the use of this element. For further evaluations of these and other elements see [17]. Likewise, the serendipity element ($8S$-$U2$) possesses two noncommunicable mechanisms.

6.2.12 Some References to the Recent Literature

The thesis of Stanley [17] compares many existing shell elements and includes evaluation of some recently developed elements, such as the nine-node element of Park and Stanley [19] and the four-node elements of Dvorkin and Bathe [20] and Park, Stanley, and Flaggs [21]. See also Stanley, Park, and Hughes [22]. Belytschko, Liu, and their associates have developed several new improved elements in recent years. See, for example, Belytschko et al. [23, 24]. Most of the new elements use some form of strain projection or interpolation and thus may be viewed as "$\bar{\boldsymbol{B}}$-elements" (see Sec. 4.5.2). Another successful element within this category is the nine-node element of Huang and Hinton [25]. Hallquist and his colleagues at the Lawrence Livermore National Laboratory have developed an unprecedented capability for fully nonlinear transient analysis based upon one-point quadrature elements with hourglass control (e.g., see Hallquist et al. [26]).

There is great research interest in the development of shell finite element analysis procedures. The state of the art is summarized in Hughes and Hinton [27].

6.2.13 Simplifications: Shells as an Assembly of Flat Elements

A simple but crude approximation to curved shells may be constructed by way of flat elements. For example, consider the shell geometry discretized into flat triangles. A local Cartesian coordinate system is constructed in the plane of each triangle. A plate bending element stiffness and "membrane" element stiffness (i.e., plane-stress two-dimensional element stiffness) are generated and combined in this coordinate system. Note that there is no coupling in the local system because the plate and membrane have different degrees of freedom. The stiffness matrix and force vector are then transformed from local coordinates to global coordinates before assembly. The procedure is the same as for the straight beam-frame element described in Chapter 5. Further details follow:

Consider an n_{en}-noded flat element. Assume an element Cartesian coordinate system defined by a basis $\{\boldsymbol{e}_1^e, \boldsymbol{e}_2^e, \boldsymbol{e}_3^e\}$ in which $\boldsymbol{e}_3^e$ is normal to the element and $\boldsymbol{e}_1^e$, and $\boldsymbol{e}_2^e$ are in the plane of the element. The global-local transformation is defined by

$$\boldsymbol{t}^e = [t_{ij}^e] \qquad 1 \le i, j \le 3 \tag{6.2.92}$$

$$t_{ij}^e = \boldsymbol{e}_i \cdot \boldsymbol{e}_j^e \tag{6.2.93}$$

where $\boldsymbol{e}_1$, $\boldsymbol{e}_2$, $\boldsymbol{e}_3$ is the global Cartesian basis. Let

$$\boldsymbol{k}_m^e = 2n_{en} \times 2n_{en} \text{ membrane stiffness}$$

$$\boldsymbol{k}_b^e = 3n_{en} \times 3n_{en} \text{ bending stiffness}$$

These are placed in an element matrix of dimension $6n_{en} \times 6n_{en}$:

$$\boldsymbol{k}^e = \begin{bmatrix} \boldsymbol{k}_m^e & \boldsymbol{0} & \boldsymbol{0} \\ \boldsymbol{0} & \boldsymbol{k}_b^e & \boldsymbol{0} \\ \boldsymbol{0} & \boldsymbol{0} & \boldsymbol{0} \end{bmatrix} \tag{6.2.94}$$

If adjacent elements are in the same plane, rank deficiency can clearly occur by virtue of the zero stiffness associated with rotations about the normal (see (6.2.94)). Ill-conditioning may also occur if the planes defined by adjacent elements almost coincide. For these reasons, a "fictitious" $n_{en} \times n_{en}$ stiffness, $\boldsymbol{k}_f^e$, is often added to stabilize the assemblage:

$$\boldsymbol{k}^e = \begin{bmatrix} \boldsymbol{k}_m^e & \boldsymbol{0} & \boldsymbol{0} \\ \boldsymbol{0} & \boldsymbol{k}_b & \boldsymbol{0} \\ \boldsymbol{0} & \boldsymbol{0} & \boldsymbol{k}_f^e \end{bmatrix} \tag{6.2.95}$$

The fictitious stiffness resists rotations about the normal to the plane of the element. These are sometimes referred to as the ***drilling degrees of freedom.*** See [28] for a study of this approach. The rows and columns of $\boldsymbol{k}^e$ are now reordered so that the three displacement degrees of freedom at each node precede the three rotational degrees of freedom. The situation is very much like that for the straight beam element in three-dimensional space presented in Chapter 5. The transformation from local to global is performed as follows:

$$\underbrace{\hat{k}^e}_{6n_{en} \times 6n_{en}} = T^e \widetilde{k}^e T^{eT} \tag{6.2.96}$$

where $\widetilde{k}^e$ is the reordered version of k^e and

$$\underbrace{T^e}_{6n_{en} \times 6n_{en}} = \begin{bmatrix} t^e & 0 & \cdots & 0 \\ 0 & t^e & & \vdots \\ \vdots & & \ddots & \\ 0 & \cdots & & t^e \end{bmatrix} \tag{6.2.97}$$

The (reordered) element force vector is similarly transformed:

$$\underbrace{\hat{f}^e}_{6n_{en} \times 1} = T^e \widetilde{f}^e \tag{6.2.98}$$

Assembly may now be performed in the usual fashion.

An intriguing alternative to the use of a fictitious stiffness is to employ a membrane element that incorporates drilling degrees of freedom. Until recently, a successful formulation of this type had never been developed. In an excellent paper, Bergan and Felippa [29] have developed an accurate membrane element with drilling degrees of freedom. Taylor and Simo [30] have also developed a triangular element of this kind and linked it with the DKT bending element to create a C^0-compatible shell element.

6.3 SHELLS OF REVOLUTION; RINGS AND TUBES IN TWO DIMENSIONS

6.3.1 Geometric and Kinematic Descriptions

The geometric and kinematic assumptions for the two-dimensional cases follow directly from the three-dimensional case treated in Sec. 6.2 by omitting the ξ-dependence and ignoring the third (out-of-plane) component of all vectors. In this case, the geometry of the shell is represented by a two-dimensional quadrilateral section as shown in Fig. 6.3.1. As in Sec. 6.2, the ζ-coordinate defines the fiber direction, and the η-coordinate lines coincide with shell laminae. This may be expressed by

$$\begin{aligned} x(\eta, \zeta) &= \bar{x}(\eta) + X(\eta, \zeta) \\ &= \sum_{a=1}^{n_{en}} N_a(\eta)\bar{x}_a + \sum_{a=1}^{n_{en}} N_a(\eta) z_a(\zeta)\hat{X}_a \\ &= \sum_{a=1}^{n_{en}} N_a(\eta)(\bar{x}_a + z_a(\zeta)\hat{X}_a) \end{aligned} \tag{6.3.1}$$

where $N_a(\eta)$ represents a one-dimensional lamina shape function and $\bar{x}_a$, $z_a(\zeta)$, and $\hat{X}_a$ are defined by (6.2.5) through (6.2.10).

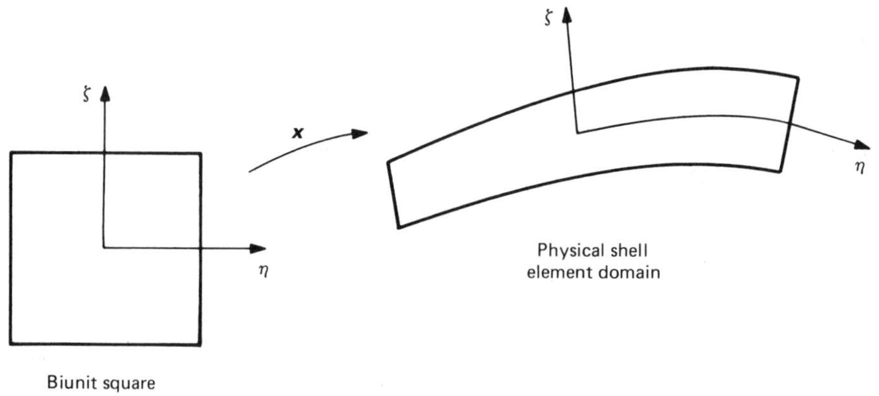

Figure 6.3.1

Lamina coordinate systems

Lamina coordinate systems are defined by (see Fig. 6.3.2)

$$\boldsymbol{e}_1^l = \begin{Bmatrix} e_{11}^l \\ e_{12}^l \end{Bmatrix} = \frac{\boldsymbol{x}_{,\eta}}{\|\boldsymbol{x}_{,\eta}\|} \tag{6.3.2}$$

$$\boldsymbol{e}_2^l = e_{11}^l \boldsymbol{e}_2 - e_{12}^l \boldsymbol{e}_1 \tag{6.3.3}$$

where

$$\boldsymbol{e}_1 = \begin{Bmatrix} 1 \\ 0 \end{Bmatrix}, \qquad \boldsymbol{e}_2 = \begin{Bmatrix} 0 \\ 1 \end{Bmatrix} \tag{6.3.4}$$

The global-lamina transformation is defined by

$$\boldsymbol{q} = [q_{ij}] = [\boldsymbol{e}_1^l \boldsymbol{e}_2^l]^T : \text{global} \rightarrow \text{lamina} \tag{6.3.5}$$

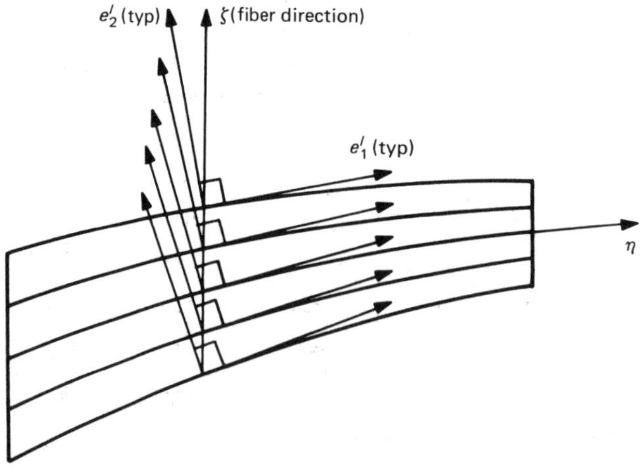

Figure 6.3.2 Lamina coordinate systems along a fiber.

Fiber coordinate systems

The fiber coordinate basis is defined as follows:

$$e_2^f = \hat{X} \tag{6.3.6}$$

$$e_1^f = \hat{X}_2 e_1 - \hat{X}_1 e_2 \tag{6.3.7}$$

The kinematics are defined by

$$u(\eta, \zeta) = \sum_{a=1}^{n_{en}} N_a(\eta)(\bar{u}_a - z_a(\zeta)\theta_a e_{a1}^f) \tag{6.3.8}$$

See Fig. 6.3.3 for an interpretation of the nodal quantities.

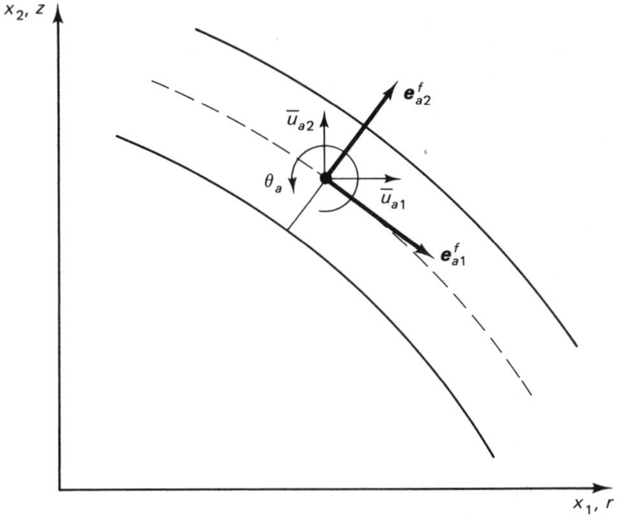

Figure 6.3.3 Nodal kinematic quantities for the two-dimensional cases.

6.3.2 Reduced Constitutive Equations

We begin with the three-dimensional constitutive equation written in lamina coordinates

$$\boldsymbol{\sigma}^l = \boldsymbol{D}^l \boldsymbol{\epsilon}^l \tag{6.3.9}$$

The ordering is indicated by the following

$$\boldsymbol{\sigma}^l = \{\sigma_I^l\} = \begin{Bmatrix} \sigma_{11}^l \\ \sigma_{12}^l \\ \sigma_{33}^l \\ \sigma_{22}^l \\ \sigma_{23}^l \\ \sigma_{31}^l \end{Bmatrix} \tag{6.3.10}$$

In what follows, we will assume that

$$\sigma_{22}^{l} = 0 \qquad \text{(zero normal stress in lamina coordinates)} \tag{6.3.11}$$

$$\sigma_{23}^{l} = \sigma_{31}^{l} = 0 \qquad \text{(zero out-of-plane shear stresses)} \tag{6.3.12}$$

As in the three-dimensional case, the corresponding strain components can be eliminated,

$$\begin{Bmatrix} \epsilon_4^l \\ \epsilon_5^l \\ \epsilon_6^l \end{Bmatrix} = -\begin{bmatrix} D_{44}^l & D_{45}^l & D_{46}^l \\ D_{54}^l & D_{55}^l & D_{56}^l \\ D_{64}^l & D_{65}^l & D_{66}^l \end{bmatrix}^{-1} \begin{bmatrix} D_{41}^l & D_{42}^l & D_{43}^l \\ D_{51}^l & D_{52}^l & D_{53}^l \\ D_{61}^l & D_{62}^l & D_{63}^l \end{bmatrix} \begin{Bmatrix} \epsilon_1^l \\ \epsilon_2^l \\ \epsilon_3^l \end{Bmatrix} \tag{6.3.13}$$

resulting in a statically condensed matrix of elastic coefficients:

$$\widetilde{\boldsymbol{D}}^l = \begin{bmatrix} D_{11}^l & D_{12}^l & D_{13}^l \\ D_{21}^l & D_{22}^l & D_{23}^l \\ D_{31}^l & D_{32}^l & D_{33}^l \end{bmatrix} - \begin{bmatrix} D_{14}^l & D_{15}^l & D_{16}^l \\ D_{24}^l & D_{25}^l & D_{26}^l \\ D_{34}^l & D_{35}^l & D_{36}^l \end{bmatrix} \begin{bmatrix} D_{44}^l & D_{45}^l & D_{46}^l \\ D_{54}^l & D_{55}^l & D_{56}^l \\ D_{64}^l & D_{65}^l & D_{66}^l \end{bmatrix}^{-1} \begin{bmatrix} D_{41}^l & D_{42}^l & D_{43}^l \\ D_{51}^l & D_{52}^l & D_{53}^l \\ D_{61}^l & D_{62}^l & D_{63}^l \end{bmatrix} \tag{6.3.14}$$

The reduced constitutive equation is

$$\widetilde{\boldsymbol{\sigma}}^l = \widetilde{\boldsymbol{D}}^l \widetilde{\boldsymbol{\epsilon}}^l \tag{6.3.15}$$

where

$$\widetilde{\boldsymbol{\sigma}}^l = \begin{Bmatrix} \sigma_1^l \\ \sigma_2^l \\ \sigma_3^l \end{Bmatrix} = \begin{Bmatrix} \sigma_{11}^l \\ \sigma_{12}^l \\ \sigma_{33}^l \end{Bmatrix}, \text{ etc.} \tag{6.3.16}$$

Exercise 1. Consider the isotropic case. Show that

$$\widetilde{\boldsymbol{D}}^l = \frac{E}{(1-\nu^2)} \left[\begin{array}{cc|c} 1 & 0 & \nu \\ 0 & \dfrac{1-\nu}{2} & 0 \\ \hline \nu & 0 & 1 \end{array}\right] \tag{6.3.17}$$

As always, a shear correction factor, $\kappa = \frac{5}{6}$, needs to be introduced to match classical bending results:

$$\widetilde{\boldsymbol{D}}^l = \frac{E}{(1-\nu^2)} \left[\begin{array}{cc|c} 1 & 0 & \nu \\ 0 & \dfrac{\kappa(1-\nu)}{2} & 0 \\ \hline \nu & 0 & 1 \end{array}\right] \tag{6.3.18}$$

The above result may be used in the ***axisymmetric*** and ***plane strain*** cases. In the case of ***plane stress***, further simplifications ensue. In this case the out-of-plane normal stress, σ^l_{33}, is also assumed zero. Proceeding along now familiar lines:

$$\epsilon^l_3 = -\left(\sum_{J=1}^{2} \frac{\widetilde{D}^l_{3J}\epsilon^l_J}{\widetilde{D}^l_{33}}\right) \tag{6.3.19}$$

$$\widetilde{D}^l_{IJ} \leftarrow \widetilde{D}^l_{IJ} - \frac{\widetilde{D}^l_{I3}\widetilde{D}^l_{3J}}{\widetilde{D}^l_{33}} \tag{6.3.20}$$

Exercise 2. Show that in the isotropic case, the plane stress assumption causes (6.3.18) to be replaced with:

$$\boxed{\widetilde{\boldsymbol{D}}^l = \left[\begin{array}{cc|c} E & 0 & 0 \\ 0 & \dfrac{\kappa E}{2(1+\nu)} & 0 \\ \hline 0 & 0 & 0 \end{array}\right]} \tag{6.3.21}$$

Remark

The various two-dimensional cases have the following physical applications (see Fig. 6.3.4):

Axisymmetric: shells of revolution subjected to axisymmetric loading.

Plane strain: long tubes loaded in the plane perpendicular to the generating axis with the load not dependent upon the axial coordinate.

Plane stress: rings and curved beams with rectangular cross sections loaded in plane.

6.3.3 Strain-displacement Matrix

The strains are defined in terms of the displacement by

$$\widetilde{\boldsymbol{\epsilon}}^l = \{\widetilde{\epsilon}^l\} = \left\{\begin{array}{c} \dfrac{\partial u^l_1}{\partial x^l_1} \\ \dfrac{\partial u^l_1}{\partial x^l_2} + \dfrac{\partial u^l_2}{\partial x^l_1} \\ \hline \dfrac{u_1}{x_1} \end{array}\right\} \quad \text{(axisymmetric)} \tag{6.3.22}$$

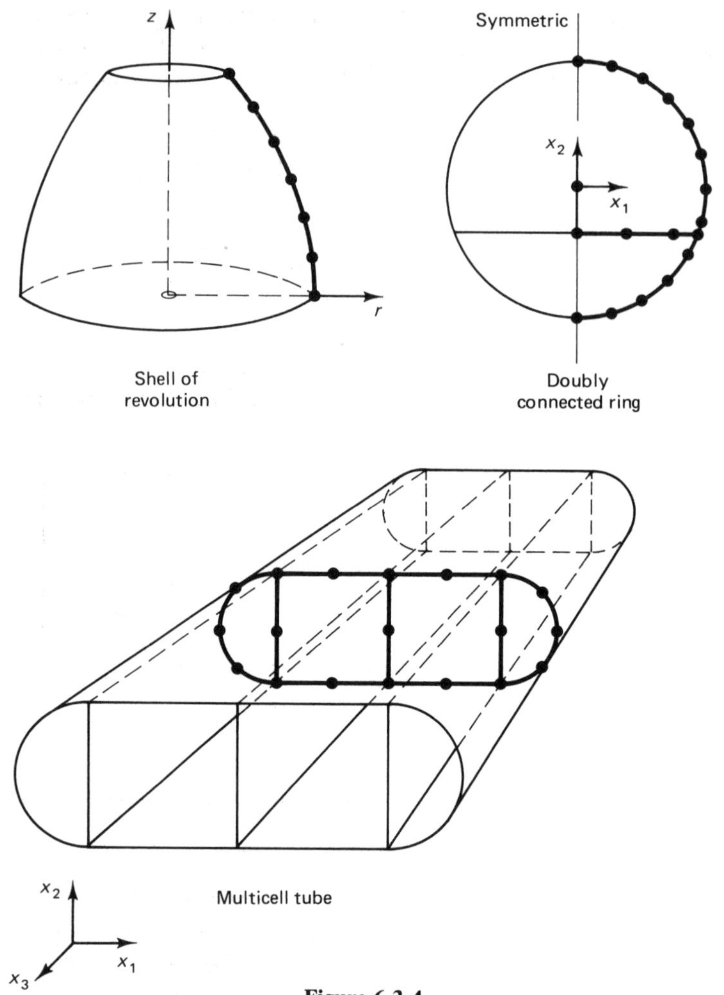

Figure 6.3.4

$$\tilde{\boldsymbol{\epsilon}}^l = \{\tilde{\epsilon}^l\} = \left\{ \begin{array}{c} \dfrac{\partial u_1^l}{\partial x_1^l} \\ \dfrac{\partial u_1^l}{\partial x_2^l} + \dfrac{\partial u_2^l}{\partial x_1^l} \\ \text{- - - - - - - - - -} \\ 0 \end{array} \right\} \quad \begin{array}{l}\text{(plane strain and} \\ \text{plane stress)}\end{array} \tag{6.3.23}$$

where $x_1 = r$ and $x_2 = z$ in axisymmetric analysis. It is our aim to derive an explicit representation of the matrix $\boldsymbol{B}_a$, where

$$\tilde{\boldsymbol{\epsilon}}^l = \sum_{a=1}^{n_{en}} \boldsymbol{B}_a \left\{ \begin{array}{c} \overline{u}_{a1} \\ \overline{u}_{a2} \\ \theta_a \end{array} \right\} \tag{6.3.24}$$

$$\boldsymbol{B}_a = \left[\begin{array}{cc|c} b^u_{11} & b^u_{12} & b^\theta_1 \\ b^u_{21} & b^u_{22} & b^\theta_2 \\ \hline b^u_{31} & 0 & b^\theta_3 \end{array}\right] \quad \text{(axisymmetric)} \tag{6.3.25}$$

$$= \left[\begin{array}{cc|c} b^u_{11} & b^u_{12} & b^\theta_1 \\ b^u_{21} & b^u_{22} & b^\theta_2 \\ \hline 0 & 0 & 0 \end{array}\right] \quad \text{(plane strain and plane stress)} \tag{6.3.26}$$

To derive formulas for the b's we need first to form the displacement gradients:

$$\begin{aligned} \frac{\partial u^l_i}{\partial x^l_j} &= \sum_{m=1}^{2} q_{im} \frac{\partial u_m}{\partial x^l_j} \\ &= \sum_{m=1}^{2} q_{im} \sum_{a=1}^{n_{en}} \left(\frac{\partial N_a}{\partial x^l_j} \bar{u}_{am} - \frac{\partial (N_a z_a)}{\partial x^l_j} \right) \theta_a e^f_{am1} \end{aligned} \tag{6.3.27}$$

In (6.3.27) we have used the notation:

$$\boldsymbol{e}^f_{a1} = \{e^f_{am1}\} = \begin{Bmatrix} e^f_{a11} \\ e^f_{a21} \end{Bmatrix} \tag{6.3.28}$$

From (6.3.8) and (6.3.27) we may read the desired expressions:

$$b^u_{1m} = q_{1m} \frac{\partial N_a}{\partial x^l_1} \tag{6.3.29}$$

$$b^u_{2m} = q_{1m} \frac{\partial N_a}{\partial x^l_2} + q_{2m} \frac{\partial N_a}{\partial x^l_1} \tag{6.3.30}$$

$$b^u_{31} = \frac{N_a}{x_1} \tag{6.3.31}$$

$$b^\theta_1 = -\left(\sum_{m=1}^{2} q_{1m} e^f_{am1} \right) \frac{\partial (N_a z_a)}{\partial x^l_1} \tag{6.3.32}$$

$$b^\theta_2 = -\left(\sum_{m=1}^{2} q_{1m} e^f_{am1} \right) \frac{\partial (N_a z_a)}{\partial x^l_2} - \left(\sum_{m=1}^{2} q_{2m} e^f_{am1} \right) \frac{\partial (N_a z_a)}{\partial x^l_1} \tag{6.3.33}$$

$$b^\theta_3 = -\frac{N_a z_a e^f_{a11}}{x_1} \tag{6.3.34}$$

In (6.3.31) and (6.3.34), x_1 is given by (6.3.1). The differentiations indicated in (6.3.29), (6.3.30), (6.3.32), and (6.3.33) are performed in global coordinates and then transformed:

$$\frac{\partial}{\partial x^l_j} = \sum_{n=1}^{2} q_{jn} \frac{\partial}{\partial x_n} \tag{6.3.35}$$

6.3.4 Stiffness Matrix

The element stiffness matrix is defined by

$$\underbrace{\boldsymbol{k}}_{3n_{en} \times 3n_{en}} = [\boldsymbol{k}_{ab}] \tag{6.3.36}$$

$$\underbrace{\boldsymbol{k}_{ab}}_{3 \times 3} = \underbrace{\int_{-1}^{+1} \underbrace{\int_{-1}^{+1} \boldsymbol{B}_a^T \widetilde{\boldsymbol{D}} \boldsymbol{B}_b j \, d\zeta}_{\text{fiber integral}} \, d\eta}_{\text{lamina integral}} \tag{6.3.37}$$

where

$$j = \begin{cases} \bar{j} & \text{(plane strain and plane stress)} \\ x_1 \bar{j} & \text{(axisymmetric)} \end{cases} \tag{6.3.38}$$

$$\bar{j} = \det \begin{bmatrix} x_{1,\eta} & x_{1,\zeta} \\ x_{2,\eta} & x_{2,\zeta} \end{bmatrix} \tag{6.3.39}$$

6.3.5 External Force Vector

Body force

The element body force vector is given by

$$\underbrace{\boldsymbol{f}^{\text{body}}}_{3n_{en} \times 1} = \{\boldsymbol{f}_a^{\text{body}}\} \tag{6.3.40}$$

$$\underbrace{\boldsymbol{f}_a^{\text{body}}}_{3 \times 1} = \int_{-1}^{+1} \int_{-1}^{+1} \boldsymbol{N}^T \boldsymbol{\ell} j \, d\zeta \, d\eta \tag{6.3.41}$$

where j is defined by (6.3.38) and (6.3.39), and

$$\boldsymbol{N}_a = \left[\begin{array}{cc:c} N_a & 0 & -N_a z_a e_{a11}^f \\ 0 & N_a & -N_a z_a e_{a21}^f \end{array} \right] \tag{6.3.42}$$

Surface force

The element surface force vector is given by

$$\boldsymbol{f}^{\text{surf}} = \{\boldsymbol{f}_a^{\text{surf}}\} \tag{6.3.43}$$

$$\boldsymbol{f}_a^{\text{surf}} = \int_{-1}^{+1} \boldsymbol{N}_a^T \boldsymbol{h} \, j_s \, d\eta, \qquad \zeta = \begin{cases} +1 & \text{top} \\ -1 & \text{bottom} \end{cases} \tag{6.3.44}$$

where

$$j_s = \begin{cases} \bar{j}_s & \text{(plane stress and plane strain)} \\ x_1 \bar{j}_s & \text{(axisymmetric)} \end{cases} \tag{6.3.45}$$

$$\bar{j}_s = \|\boldsymbol{x}_{,\eta}\| \tag{6.3.46}$$

and $\boldsymbol{h}$ is the surface force vector (per unit surface area), which includes both normal pressure and tangential shear as special cases, viz.,

Pressure

$$\boldsymbol{h} = -\zeta p \boldsymbol{e}_2^l, \qquad (\zeta = +1 \text{ or } -1) \tag{6.3.47}$$

In (6.3.47), p is the pressure.

Shear

$$\boldsymbol{h} = h_\eta \boldsymbol{e}_1^l \tag{6.3.48}$$

In (6.3.48), h_η is the tangential shear in the η direction.

6.3.6 Stress Resultants

The stress resultants are given as follows.

Moment

$$m_{11}(\eta) = \int_{-1}^{+1} \sigma_{11}^l(\eta, \zeta) z(\eta, \zeta)\, d\zeta\, z_{,\zeta}(\eta) \tag{6.3.49}$$

$$z(\eta, \zeta) = N_+(\zeta)\, z^+(\eta) + N_-(\zeta) z^-(\eta) \tag{6.3.50}$$

$$z^\pm(\eta) = \sum_{a=1}^{n_{en}} N_a(\eta)\, z_a^\pm \tag{6.3.51}$$

$$z_{,\zeta}(\eta) = \frac{z^+(\eta) - z^-(\eta)}{2} \tag{6.3.52}$$

Membrane forces

$$n_{ii}(\eta) = \int_{-1}^{+1} \sigma_{ii}^l(\eta, \zeta)\, d\zeta\, z_{,\zeta}(\eta), \qquad ii = 11 \text{ or } 33 \quad \text{(no sum)} \tag{6.3.53}$$

Shear

$$q(\eta) = \int_{-1}^{+1} \sigma_{12}^l(\eta, \zeta)\, d\zeta\, z_{,\zeta}(\eta) \tag{6.3.54}$$

The fiber integrations may be calculated by a variety of quadrature rules. A discussion of various possibilities is presented in Sec. 6.2.9.

The sign conventions for the stress resultants are illustrated in Fig. 6.3.5.

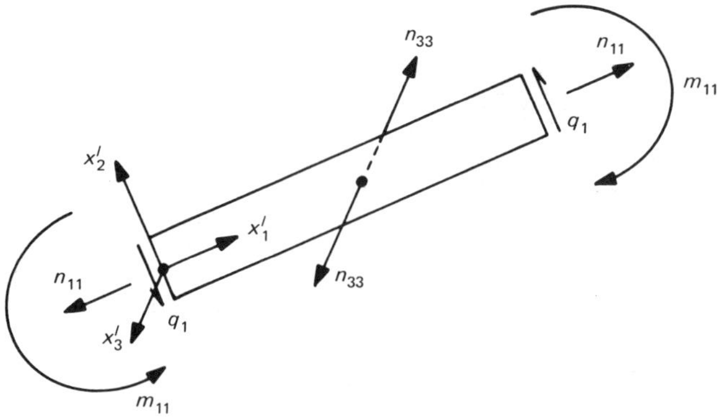

Figure 6.3.5 Sign conventions for stress resultants.

6.3.7 Boundary Conditions

The boundary conditions are essentially the same as for the three-dimensional case. That is, the kinematic conditions are the displacements and rotation of the node, and the mechanical conditions are the edge forces and moment. These latter quantities are specified per unit out-of-plane thickness in the plane strain and plane stress cases, and per unit radian in the axisymmetric case.

6.3.8 Shell Elements

The shell elements advocated herein employ one-dimensional Lagrangian interpolation in the η direction and ***uniform reduced quadrature*** (see Fig. 6.3.6). It is essential to use reduced integration to avoid shear and membrane-bending "locking" phenomena. *Rank deficiency problems, which afflict many three-dimensional shell*

Shape functions	Linear	Quadratic	Cubic
Quadrature rule	One-point	Two-point	Three-point

Figure 6.3.6 Uniform reduced integration two-dimensional shell elements.

elements derived along similar lines, are not present at all in the two-dimensional uniform reduced integration Lagrange elements.

The two-node element was first suggested in [31] and developed subsequently for linear applications in [32] and [33]. In [33] a number of problems were solved, and the accuracy and economy of the element were demonstrated. The three-node element seems to have been employed by several investigators and its origin is uncertain.

A possible improvement in the two-node element in axisymmetric applications might be made by using a "mean value" of the matrix $\boldsymbol{B}_a$ (with respect to integration) rather than the value at $\eta = 0$. This is in keeping with the approach originally presented for incompressibility by Nagtegaal et al. [34] and discussed more recently in [35] (see also Chapter 4).

REFERENCES

Section 6.1

1. T. J. R. Hughes and W. K. Liu, "Nonlinear Finite Element Analysis of Shells: Part I. Three-dimensional Shells," *Computer Methods in Applied Mechanics and Engineering*, 26 (1981), 331–362.
2. T. J. R. Hughes and W. K. Liu, "Nonlinear Finite Element Analysis of Shells: Part II. Two-dimensional Shells," *Computer Methods in Applied Mechanics and Engineering*, 27 (1981), 167–181.
3. T. J. R. Hughes and E. Carnoy, "Nonlinear Finite Element Shell Formulation Accounting for Large Membrane Strains," *Computer Methods in Applied Mechanics and Engineering*, 39 (1983), 69–82.
4. S. Ahmad, B. M. Irons, and O. C. Zienkiewicz, "Analysis of Thick and Thin Shell Structures by Curved Finite Elements," *International Journal for Numerical Methods in Engineering*, 2 (1970), 419–451.

Section 6.2

5. T. J. R. Hughes and M. Cohen, "The 'Heterosis' Finite Element for Plate Bending," *Computers and Structures*, 9 (1978), 445–450.
6. T. J. R. Hughes and M. Cohen, "The 'Heterosis' Family of Plate Finite Elements," *Proceedings of the ASCE Electronic Computations Conference*, St. Louis, Missouri, August 6–8, 1979.
7. T. J. R. Hughes, M. Cohen, and M. Haroun, "Reduced and Selective Integration Techniques in the Finite Element Analysis of Plates," *Nuclear Engineering and Design*, 46, no. 1 (1978), 203–222.
8. T. J. R. Hughes, R. L. Taylor, and W. Kanoknukulchai, "A Simple and Efficient Element for Plate Bending," *International Journal for Numerical Methods in Engineering*, 11, no. 10 (1977), 1529–1543.
9. D. S. Malkus and T. J. R. Hughes, "Mixed Finite Element Methods—Reduced and Selective Integration Techniques: A Unification of Concepts," *Computer Methods in Applied Mechanics and Engineering*, 15, no. 1 (1978), 63–81.
10. E. Oñate, E. Hinton, and N. Glover, *Techniques for Improving the Performance of Ahmad*

Shell Elements, C/R/313/78, Department of Civil Engineering, University of Wales, Swansea, U.K., 1978.

11. E. D. L. Pugh, E. Hinton, and O. C. Zienkiewicz, "A Study of Quadrilateral Plate Bending Elements with 'Reduced' Integration," *International Journal for Numerical Methods in Engineering*, 12, no. 7 (1978), 1059–1079.

12. O. C. Zienkiewicz, R. L. Taylor, and J. M. Too, "Reduced Integration Technique in General Analysis of Plates and Shells," *International Journal for Numerical Methods in Engineering*, 3 (1971), 275–290.

13. H. Stolarski and T. Belytschko, "Shear and Membrane Locking in Curved C^0 Elements," *Computer Methods in Applied Mechanics and Engineering*, 41 (1983), 279–296.

14. T. J. R. Hughes, "Recent Developments in Computer Methods for Structural Analysis," *Nuclear Engineering and Design*, 57, no. 2 (1980), 427–439.

15. T. J. R. Hughes, "Generalization of Selective Integration Procedures to Anistropic and Nonlinear Media," *International Journal for Numerical Methods in Engineering*, 15, no. 9 (1980), 1413–1418.

16. R. H. MacNeal, "A Simple Quadrilateral Shell Element," *Computers and Structures*, 8 (1978), 175–183.

17. G. M. Stanley, "Continuum-based Shell Elements," Ph.D. Thesis, Division of Applied Mechanics, Stanford University, August, 1985.

18. T. J. R. Hughes, W. K. Liu, and I. Levit, "Nonlinear Dynamic Finite Element Analysis of Shells," in *Nonlinear Finite Element Analysis in Structural Mechanics*, eds. W. Wunderlich et al., Berlin: Springer-Verlag, 1981, 151–168.

19. K. C. Park and G. M. Stanley, "A Curved C^0 Shell Element based on Assumed Natural-Coordinate Strains", *Journal of Applied Mechanics*, 53 (1986), 278–290.

20. E. N. Dvorkin and K. J. Bathe, "A Continuum Mechanics based Four-node Shell Element for General Nonlinear Analysis," *Engineering Computations*, 1 (1984), 77–88.

21. K. C. Park, G. M. Stanley, and D. L. Flaggs, "A Uniformly Reduced Four-noded C^0 Shell Element with Consistent Rank Corrections," *Computers and Structures*, 20 (1985), 129–139.

22. G. M. Stanley, K. C. Park, and T. J. R. Hughes, "Continuum-Based Resultant Shell Elements," in *Finite Element Methods for Plate and Shell Structures 1: Element Technology*, eds. T. J. R. Hughes and E. Hinton. Swansea, U.K.: Pineridge Press, 1986.

23. T. Belytschko, W. K. Liu, J. S-J. Ong, and D. Lam, "Implementation and Application of a 9-node Lagrange Shell Element with Spurious Mode Control," *Computers and Structures*, 20 (1985), 121–128.

24. T. Belytschko, H. Stolarski. W. K. Liu, N. Carpenter, and J. S-J. Ong, "Stress Projection for Membrane and Shear Locking in Shell Finite Elements," *Computer Methods in Applied Mechanics and Engineering*, 51 (1985), 221–258.

25. E. C. Huang and E. Hinton, "Elastic-Plastic and Geometrically Nonlinear Analysis of Plates and Shells Using a New Nine-Node Element," *Finite Elements for Nonlinear Problems*, eds. P. Bergan et al., Berlin: Springer-Verlag, 1986, 283–297.

26. J. O. Hallquist, D. J. Benson, and G. L. Goudreau, "Implementation of a Modified Hughes-Liu Shell into a Fully Vectorized Explicit Finite Element Code," *Finite Elements for Nonlinear Problems*, eds. P. Bergan et al., Berlin: Springer-Verlag, 1986, 465–479.

27. T. J. R. Hughes and E. Hinton, eds., *Finite Element Methods for Plate and Shell Structures 1: Element Technology* and *Finite Element Methods for Plate and Shell Structures 2: Formulations and Algorithms*. Swansea, U.K.: Pineridge Press, 1986.

28. W. Kanoknukulchai, "A Simple and Efficient Finite Element for General Shell Analysis," *International Journal for Numerical Methods in Engineering*, 14 (1979), 179–200.

29. P. G. Bergan and C. A. Felippa, "A Triangular Membrane Element with Rotational Degrees of Freedom," *Computer Methods in Applied Mechanics and Engineering*, 50 (1985), 25–69.

30. R. L. Taylor and J. C. Simo, "Bending and Membrane Elements for Analysis of Thick and Thin Shells," pp. 587–591, *Proceedings of the NUMETA '85 Conference* (eds. J. Middleton and G. N. Pande) Swansea, 7–11 January 1985. Published by A. A. Balkema, Rotterdam.

Section 6.3

31. T. J. R. Hughes, R. L. Taylor, and W. Kanoknukulchai, "A Simple and Efficient Element for Plate Bending," *International Journal for Numerical Methods in Engineering*, 11, no. 10 (1977), 1529–1543.

32. G. L. Goudreau, *A Computer Module for One-Step Dynamic Response of an Axisymmetric Plane Linear Elastic Thin Shell*, Lawrence Livermore Laboratory Report No. UCID-17730, February 1978.

33. O. C. Zienkiewicz, J. Bauer, K. Morgan, and E. Oñate, "A Simple Element for Axisymmetric Shells with Shear Deformation," *International Journal for Numerical Methods in Engineering*, 11, no. 10 (1977), 1545–1558.

34. J. C. Nagtegaal, D. M. Parks, and J. R. Rice, "On Numerically Accurate Finite Element Solutions in the Fully Plastic Range," *Computer Methods in Applied Mechanics and Engineering*, 4 (1974) 153–178.

35. J. C. Nagtegaal and J. E. de Jong, "Some Computational Aspects of Elastic-Plastic Large Strain Analysis," *International Journal for Numerical Methods in Engineering*, 17 (1981), 15–41.

7

Formulation of Parabolic, Hyperbolic, and Elliptic-Eigenvalue Problems

In order to understand this chapter, a detailed understanding of the material and notational conventions of Chapter 2 is required.

7.1. PARABOLIC CASE: HEAT EQUATION

The heat equation is the prototypical "parabolic" partial differential equation of mathematical physics (e.g., see Stakgold [1]). The formulation given here generalizes the formulation of Chapter 2, which was restricted to steady heat conduction, to time-dependent heat conduction processes. We view all the data to be functions of time, denoted by t, as well as space. For example, the volumetric heat source ℓ is written as

$$\ell : \Omega \times]0, T[\rightarrow \mathbb{R} \tag{7.1.1}$$

which means ℓ is a function of $\boldsymbol{x} \in \Omega$ and $t \in]0, T[$, the open time interval of length $T > 0$. Likewise, we write $\ell = \ell(\boldsymbol{x}, t)$. Similarly, the boundary data is also time-dependent:

$$q : \Gamma_q \times]0, T[\rightarrow \mathbb{R} \tag{7.1.2}$$

$$h : \Gamma_h \times]0, T[\rightarrow \mathbb{R} \tag{7.1.3}$$

Note that Γ_q and Γ_h are assumed *not* to vary with time. To render the initial/boundary-value problem well posed, an ***initial condition*** on temperature must also be specified. We denote the initial temperature by

$$u_0 : \Omega \to \mathbb{R} \tag{7.1.4}$$

Two other material properties come into play: The ***density***, ρ, and ***capacity***, c; both are assumed positive functions of $\boldsymbol{x} \in \Omega$.

The ***strong form*** of the initial/boundary-value problem may now be stated:

(S) Given f, g, h, and u_0, as in (7.1.1) through (7.1.4), find $u : \overline{\Omega} \times [0, T] \to \mathbb{R}$ such that

$$\rho c u_{,t} + q_{i,i} = f \quad \text{on} \quad \Omega \times]0, T[\qquad (\textbf{\textit{heat equation}}) \tag{7.1.5}$$

$$u = g \quad \text{on} \quad \Gamma_g \times]0, T[\tag{7.1.6}$$

$$-q_i n_i = h \quad \text{on} \quad \Gamma_h \times]0, T[\tag{7.1.7}$$

$$u(\boldsymbol{x}, 0) = u_0(\boldsymbol{x}) \qquad \boldsymbol{x} \in \Omega \tag{7.1.8}$$

Recall from Chapter 2 that

$$q_i = -\kappa_{ij} u_{,j} \tag{7.1.9}$$

where $\kappa_{ij} = \kappa_{ij}(\boldsymbol{x})$ denotes the conductivity tensor. The notation $u_{,t}$ denotes the time derivative of u (i.e., $u_{,t} = \partial u / \partial t$).

Let $\mathcal{V}$ denote the usual space of weighting functions satisfying zero-temperature boundary conditions. Note that functions in $\mathcal{V}$ do *not* depend on time in any way. Let $\mathcal{S} = \mathcal{S}_t$ denote the space of trial solutions. Note that $\mathcal{S}$ varies as a function of t due to the temperature boundary condition, (7.1.2),

$$\mathcal{S} = \mathcal{S}_t = \left\{ u(\cdot\,, t) \,|\, u(\boldsymbol{x}, t) = g(\boldsymbol{x}, t),\ \boldsymbol{x} \in \Gamma_g,\ u(\cdot\,, t) \in H^1(\Omega) \right\} \tag{7.1.10}$$

The ***weak formulation*** consists of the following:

(W) Given f, g, h, and u_0, find $u(t) \in \mathcal{S}_t$, $t \in [0,T]$ such that for all $w \in \mathcal{V}$,

$$\boxed{(w, \rho c \dot{u}) + a(w, u) = (w, f) + (w, h)_\Gamma} \tag{7.1.11}$$

$$(w, \rho c u(0)) = (w, \rho c u_0) \tag{7.1.12}$$

Remark

In the weak formulation, $\boldsymbol{x}$ is suppressed as an argument of u. Thus $u(0)$ represents the function u of $\boldsymbol{x}$ at time 0 [i.e., $u(0) = u(\cdot\,, 0)$]. The superposed dot is used to denote time differentiation. These notations, among other things, enable us to remove commas from the weak formulation which might be confused with those normally appearing in the bilinear forms. Equation (7.1.11) is the weak formulation of the heat equation (7.1.5) and heat-conduction boundary condition (7.1.7), and (7.1.12) is the weak form of the initial condition (7.1.8).

Exercise 1. Verify the formal equivalence of (S) and (W) by proceeding along the lines of Sec. 2.3. The procedure is virtually identical to Sec. 2.3 if (7.1.11) is written as

$$a(w, u) = (w, \tilde{\ell}) + (w, h)_\Gamma$$

where

$$\tilde{\ell} = \ell - \rho c \dot{u}$$

To develop the analogous Galerkin formulation, we proceed as in Sec. 2.4. The approximate weighting function space, $\mathcal{V}^h$, is identical to before. Functions $u^h \in \mathcal{S}^h$ are built up in the usual way:

$$u^h = v^h + g^h \tag{7.1.13}$$

where $v^h(t) \in \mathcal{V}^h$ (i.e, for each fixed t, $v^h(\cdot\ , t)$ is a function in $\mathcal{V}^h$) and g^h enables satisfaction of the temperature boundary condition. Note that $g^h(t) \in \mathcal{S}_t^h$. All the functions in (7.1.13) depend upon both space and time, i.e.,

$$u^h(\boldsymbol{x}, t) = v^h(\boldsymbol{x}, t) + g^h(\boldsymbol{x}, t) \tag{7.1.14}$$

The ***Galerkin formulation*** is stated as follows:

(G) Given ℓ, g, h, and u_0, find $u^h = v^h + g^h$, $u^h(t) \in \mathcal{S}_t^h$, such that for all $w^h \in \mathcal{V}^h$,

$$\boxed{(w^h, \rho c \dot{v}^h) + a(w^h, v^h) = (w^h, \ell) + (w^h, h)_\Gamma - (w^h, \rho c \dot{g}^h) - a(w^h, g^h)} \tag{7.1.15}$$

$$(w^h, \rho c v^h(0)) = (w^h, \rho c u_0) - (w^h, \rho c g^h(0)) \tag{7.1.16}$$

Note that all given quantities appear on the right-hand sides of (7.1.15) and (7.1.16). The Galerkin equation given by (7.1.15) is called a ***semidiscrete*** equation because time is left continuous.

The matrix equations require representations of v^h and g^h in terms of basis functions. These are given by

$$v^h(\boldsymbol{x}, t) = \sum_{A \in \eta - \eta_g} N_A(\boldsymbol{x}) d_A(t) \tag{7.1.17}$$

$$g^h(\boldsymbol{x}, t) = \sum_{A \in \eta_g} N_A(\boldsymbol{x}) g_A(t) \tag{7.1.18}$$

Note that the entire time-dependence is carried by the nodal values (i.e., the d_A's and g_A's). The shape functions are identical to those used previously (see Chapters 2 and 3). In particular, they are *not* time-dependent.

Remark

Upon encountering representations such as (7.1.17) and (7.1.18) for the first time, we may wonder if the approximation is valid only when the exact solution is "separable" in $\boldsymbol{x}$ and t (e.g., see [1] for a discussion of the separation-of-variables technique). This is not the case. Even nonseparable exact solutions may be approximated arbitrarily closely by representations like (7.1.17) and (7.1.18).

To develop the matrix equations, we substitute (7.1.17) and (7.1.18) into (7.1.15) and (7.1.16) and proceed along the same lines as in Secs. 2.4 and 2.5. The end result is the following ***matrix problem***:

(M)

Given $\boldsymbol{F}:]0, T[\rightarrow \mathbb{R}^{n_{eq}}$, find $\boldsymbol{d}: [0, T] \rightarrow \mathbb{R}^{n_{eq}}$ such that

$$\boldsymbol{M}\dot{\boldsymbol{d}} + \boldsymbol{K}\boldsymbol{d} = \boldsymbol{F} \qquad t \in]0, T[\tag{7.1.19}$$

$$\boldsymbol{d}(0) = \boldsymbol{d}_0 \tag{7.1.20}$$

where

$$\boldsymbol{M} = \mathop{\mathbf{A}}_{e=1}^{n_{el}} (\boldsymbol{m}^e) \tag{7.1.21}$$

$$\boldsymbol{m} = [m_{ab}^e] \tag{7.1.22}$$

$$m_{ab}^e = \int_{\Omega^e} N_a \rho c N_b \, d\Omega \tag{7.1.23}$$

$$\boldsymbol{K} = \mathop{\mathbf{A}}_{e=1}^{n_{el}} (\boldsymbol{k}^e) \tag{7.1.24}$$

$$\boldsymbol{k}^e = [k_{ab}^e] \tag{7.1.25}$$

$$k_{ab}^e = \int_{\Omega^e} \boldsymbol{B}_a^T \boldsymbol{D} \boldsymbol{B}_b \, d\Omega \tag{7.1.26}$$

$$\boldsymbol{F}(t) = \boldsymbol{F}_{\text{nodal}}(t) + \mathop{\mathbf{A}}_{e=1}^{n_{el}} (\boldsymbol{f}^e(t)) \tag{7.1.27}$$

$$\boldsymbol{f}^e = \{f_a^e\} \tag{7.1.28}$$

$$f_a^e = \int_{\Omega^e} N_a \ell \, d\Omega + \int_{\Gamma_h^e} N_a h \, d\Gamma - \sum_{b=1}^{n_{en}} (k_{ab}^e g_b^e + m_{ab}^e \dot{g}_b^e) \tag{7.1.29}$$

$$\boldsymbol{d}_0 = \boldsymbol{M}^{-1} \mathop{\mathbf{A}}_{e=1}^{n_{el}} (\hat{\boldsymbol{d}}^e) \tag{7.1.30}$$

$$\hat{\boldsymbol{d}}^e = \{\hat{d}_a^e\} \tag{7.1.31}$$

$$\hat{d}_a^e = \int_{\Omega^e} N_a \rho c u_0 \, d\Omega - \sum_{b=1}^{n_{en}} m_{ab}^e g_b^e(0) \tag{7.1.32}$$

Remarks

1. If we compare with the steady formulation of Chapter 2, we see that the only new matrix which appears is $\boldsymbol{M}$, the ***capacity matrix.*** That $\boldsymbol{M}$ is symmetric and positive-definite follows directly from its definition.

2. Equation (7.1.19) is a coupled system of ordinary differential equations. Thus to solve the initial-value problem, i.e., (7.1.19) and (7.1.20), we need to introduce algorithms for solving systems of ordinary differential equations. This subject is taken up in subsequent chapters.

3. Observe in (7.1.29) that the element capacity matrices come into play in accounting for the effect of specified boundary temperatures.

4. It is common in practice to simplify the specification of initial conditions. That is, rather than employ (7.1.30) through (7.1.32), which emanate from (7.1.16), we directly specify $\boldsymbol{d}_0$ such that the given nodal values are interpolated [i.e., $d_{0A} = u_0(\boldsymbol{x}_A)$, $A \in \eta - \eta_g$].

5. The element capacity matrix $\boldsymbol{m}^e$ turns out to be virtually identical to the element "mass" matrix, which will be introduced in the next section. Consequently, we shall postpone more detailed consideration of the structure of $\boldsymbol{m}^e$ until later.

6. Recall that $g_b^e(t) = 0$ if degree of freedom b of element e is *not* specified. The same rule applies to $\dot{g}_b^e$.

Exercise 2. The details of arriving at (7.1.19) through (7.1.32) are straightforward given familiarity with the analogous developments in Chapter 2. If the results are not "obvious," the reader should fill in all details in step-by-step fashion.

Example

The one-dimensional heat equation is, of course, just a special case of the general formulation above. A strong form of the initial/boundary-value problem can be set which is a generalization of the one-dimensional model problem considered in Chapter 1. The equations are

$$(S)\quad \begin{cases} \rho c u_{,t} - (\kappa u_{,x})_{,x} = \ell & \text{on }]0, L[\times]0, T[\quad (7.1.33) \\ u(L, t) = g(t) & t \in]0, T[\quad (7.1.34) \\ -\kappa u_{,x}(0, t) = h(t) & t \in]0, T[\quad (7.1.35) \\ u(x, 0) = u_0(x) & x \in]0, L[\quad (7.1.36) \end{cases}$$

The element arrays are

$$m_{ab}^e = \int_{\Omega^e} N_a \rho c N_b \, d\Omega \tag{7.1.37}$$

$$k_{ab}^e = \int_{\Omega^e} N_{a,x} \kappa N_{b,x} \, d\Omega \tag{7.1.38}$$

$$f_a^e = \int_{\Omega^e} N_a \ell \, d\Omega + N_a(0)\delta_{e1} h - \sum_{b=1}^{n_{en}} (k_{ab}^e g_b^e + m_{ab}^e \dot{g}_b^e) \tag{7.1.39}$$

7.2 HYPERBOLIC CASE: ELASTODYNAMICS AND STRUCTURAL DYNAMICS

The developments of this section generalize those of Secs. 2.7 through 2.10. The nature of the generalization is similar in format to that of the previous section and the reader should first become familiar with Sec. 7.1 before considering this section even if he or she is uninterested in heat conduction.

The initial conditions this time involve specification of both displacements and velocities. Thus

$$u_{0i} : \Omega \rightarrow \mathbb{R} \tag{7.2.1}$$

and

$$\dot{u}_{0i} : \Omega \rightarrow \mathbb{R} \tag{7.2.2}$$

are given functions for each i, $1 \leq i \leq n_{sd}$. The superposed-dot notation in (7.2.2) is symbolic rather than operational.

The remaining prescribed data are

$$\ell_i : \Omega \times]0, T[\rightarrow \mathbb{R} \tag{7.2.3}$$

$$g_i : \Gamma_{g_i} \times]0, T[\rightarrow \mathbb{R} \tag{7.2.4}$$

$$h_i : \Gamma_{h_i} \times]0, T[\rightarrow \mathbb{R} \tag{7.2.5}$$

The ***density***, $\rho : \Omega \rightarrow \mathbb{R}$, assumed to be positive, needs also to be specified in the present case.

The ***strong form*** of the initial/boundary-value problem is

(S) Given ℓ_i, g_i, h_i, u_{0i} and $\dot{u}_{0i}$, as in (7.2.1) through (7.2.5), find $u_i : \overline{\Omega} \times [0, T] \rightarrow \mathbb{R}$ such that

$$\rho u_{i,tt} = \sigma_{ij,j} + \ell_i \quad \text{on } \Omega \times]0, T[\quad (\textit{equation of motion}) \tag{7.2.6}$$

$$u_i = g_i \quad \text{on } \Gamma_{g_i} \times]0, T[\tag{7.2.7}$$

$$\sigma_{ij} n_j = h_i \quad \text{on } \Gamma_{h_i} \times]0, T[\tag{7.2.8}$$

$$u_i(\boldsymbol{x}, 0) = u_{0i}(\boldsymbol{x}) \quad \boldsymbol{x} \in \Omega \tag{7.2.9}$$

$$u_{i,t}(\boldsymbol{x}, 0) = \dot{u}_{0i}(\boldsymbol{x}) \quad \boldsymbol{x} \in \Omega \tag{7.2.10}$$

Recall that $\sigma_{ij} = c_{ijkl}\, u_{(k,l)}$ and $c_{ijkl} = c_{ijkl}(\boldsymbol{x})$. Note that the second time derivative (i.e., acceleration) appears in (7.2.6). This is the reason that two initial conditions are required.

The corresponding ***weak formulation*** is: [1]

[1] We now use "direct notation"; see Chapter 2 for elaboration.

(W) Given $\boldsymbol{\ell}$, $\boldsymbol{g}$, $\boldsymbol{h}$, $\boldsymbol{u}_0$, and $\dot{\boldsymbol{u}}_0$, find $\boldsymbol{u}(t) \in \mathcal{S}_t$, $t \in [0, T]$, such that for all $\boldsymbol{w} \in \mathcal{V}$,

$$(\boldsymbol{w}, \rho\ddot{\boldsymbol{u}}) + a(\boldsymbol{w}, \boldsymbol{u}) = (\boldsymbol{w}, \boldsymbol{\ell}) + (\boldsymbol{w}, \boldsymbol{h})_\Gamma \tag{7.2.11}$$

$$(\boldsymbol{w}, \rho\boldsymbol{u}(0)) = (\boldsymbol{w}, \rho\boldsymbol{u}_0) \tag{7.2.12}$$

$$(\boldsymbol{w}, \rho\dot{\boldsymbol{u}}(0)) = (\boldsymbol{w}, \rho\dot{\boldsymbol{u}}_0) \tag{7.2.13}$$

Exercise 1. Verify the formal equivalence of (S) and (W). The arguments of Chapter 2 may be used virtually unaltered if (7.2.11) is written as

$$a(\boldsymbol{w}, \boldsymbol{u}) = (\boldsymbol{w}, \tilde{\boldsymbol{\ell}}) + (\boldsymbol{w}, \boldsymbol{h})_\Gamma$$

where

$$\tilde{\boldsymbol{\ell}} = \boldsymbol{\ell} - \rho\ddot{\boldsymbol{u}}$$

The ***semidiscrete Galerkin formulation of elastodynamics*** is:

(G) Given $\boldsymbol{\ell}$, $\boldsymbol{g}$, $\boldsymbol{h}$, $\boldsymbol{u}_0$, and $\dot{\boldsymbol{u}}_0$, find $\boldsymbol{u}^h = \boldsymbol{v}^h + \boldsymbol{g}^h$, $\boldsymbol{u}^h(t) \in \mathcal{S}_t^h$, such that for all $\boldsymbol{w}^h \in \mathcal{V}^h$,

$$(\boldsymbol{w}^h, \rho\ddot{\boldsymbol{v}}^h) + a(\boldsymbol{w}^h, \boldsymbol{v}^h) = (\boldsymbol{w}^h, \boldsymbol{\ell}) + (\boldsymbol{w}^h, \boldsymbol{h})_\Gamma - (\boldsymbol{w}^h, \rho\ddot{\boldsymbol{g}}^h) - a(\boldsymbol{w}^h, \boldsymbol{g}^h) \tag{7.2.14}$$

$$(\boldsymbol{w}^h, \rho\boldsymbol{v}^h(0)) = (\boldsymbol{w}^h, \rho\boldsymbol{u}_0) - (\boldsymbol{w}^h, \rho\boldsymbol{g}^h(0)) \tag{7.2.15}$$

$$(\boldsymbol{w}^h, \rho\dot{\boldsymbol{v}}^h(0)) = (\boldsymbol{w}^h, \rho\dot{\boldsymbol{u}}_0) - (\boldsymbol{w}^h, \rho\dot{\boldsymbol{g}}^h(0)) \tag{7.2.16}$$

The representations of $\boldsymbol{v}^h$ and $\boldsymbol{g}^h$ are given by

$$v_i^h(\boldsymbol{x}, t) = \sum_{A \in \eta - \eta_{g_i}} N_A(\boldsymbol{x}) d_{iA}(t) \tag{7.2.17}$$

$$g_i^h(\boldsymbol{x}, t) = \sum_{A \in \eta_{g_i}} N_A(\boldsymbol{x}) g_{iA}(t) \tag{7.2.18}$$

These and the usual arguments lead to the ***matrix problem:***

Given $\boldsymbol{F}$: $]0, T[\rightarrow \mathbb{R}^{n_{eq}}$, find $\boldsymbol{d}$: $]0, T[\rightarrow \mathbb{R}^{n_{eq}}$ such that

$$\boldsymbol{M}\ddot{\boldsymbol{d}} + \boldsymbol{K}\boldsymbol{d} = \boldsymbol{F} \qquad t \in\,]0, T[\tag{7.2.19}$$

$$\boldsymbol{d}(0) = \boldsymbol{d}_0 \tag{7.2.20}$$

$$\dot{\boldsymbol{d}}(0) = \dot{\boldsymbol{d}}_0 \tag{7.2.21}$$

where

$$\boldsymbol{M} = \mathop{\mathbf{A}}_{e=1}^{n_{el}} (\boldsymbol{m}^e)$$

$$\boldsymbol{m}^e = [m_{pq}^e] \tag{7.2.23}$$

$$m_{pq}^e = \delta_{ij} \int_{\Omega^e} N_a \rho N_b \, d\Omega \tag{7.2.24}$$

(Recall $p = n_{ed}(a-1) + i$ and $q = n_{ed}(b-1) + j$)

$$\boldsymbol{K} = \mathop{\mathbf{A}}_{e=1}^{n_{el}} (\boldsymbol{k}^e) \tag{7.2.25}$$

$$\boldsymbol{k}^e = [k_{pq}^e] \tag{7.2.26}$$

$$k_{pq}^e = \boldsymbol{e}_i^T \int_{\Omega^e} \boldsymbol{B}_a^T \boldsymbol{D} \boldsymbol{B}_b \, d\Omega \, \boldsymbol{e}_j \tag{7.2.27}$$

(M)

$$\boldsymbol{F}(t) = \boldsymbol{F}_{\text{nodal}}(t) + \mathop{\mathbf{A}}_{e=1}^{n_{el}} (\boldsymbol{f}^e(t)) \tag{7.2.28}$$

$$\boldsymbol{f}^e = \{f_p^e\} \tag{7.2.29}$$

$$f_p^e = \int_{\Omega} N_a \ell_i \, d\Omega + \int_{\Gamma_{h_i}^e} N_a h_i \, d\Gamma - \sum_{q=1}^{n_{ee}} (k_{pq}^e g_q^e + m_{pq}^e \ddot{g}_q^e) \quad \text{(no sum on } i\text{)} \tag{7.2.30}$$

$$\boldsymbol{d}_0 = \boldsymbol{M}^{-1} \mathop{\mathbf{A}}_{q=1}^{n_{el}} (\hat{\boldsymbol{d}}^e) \tag{7.2.31}$$

$$\hat{\boldsymbol{d}}^e = \{\hat{d}_p^e\} \tag{7.2.32}$$

$$\hat{d}_p^e = \int_{\Omega^e} N_a \rho u_{0i} \, d\Omega - \sum_{q=1}^{n_{ee}} m_{pq}^e g_q^e(0) \tag{7.2.33}$$

$$\dot{\boldsymbol{d}}_0 = \boldsymbol{M}^{-1} \mathop{\mathbf{A}}_{e=1}^{n_{el}} (\hat{\dot{d}}^e) \tag{7.2.34}$$

$$\hat{\dot{d}}^e = \{\hat{\dot{d}}_p^e\} \tag{7.2.35}$$

$$\hat{\dot{d}}_p^e = \int_{\Omega^e} N_a \rho \dot{u}_{0i} \, d\Omega - \sum_{q=1}^{n_{ee}} m_{pq}^e \dot{g}_q^e(0) \tag{7.2.36}$$

Remarks

1. The main addition to what we have encountered previously in the elastostatics formulation of Chapter 2 is the ***mass matrix***, $\boldsymbol{M}$. The reader should verify that $\boldsymbol{M}$ is symmetric and positive-definite. Except for the Kronecker delta and different material parameter inside the integrand, the mass and capacity matrix [(7.1.21) through (7.1.23)] are identical. To appreciate the origin of the Kronecker delta, we will sketch part of the calculation that leads to the matrix formulation. If we restrict attention to the eth element, then

$$\begin{aligned}(\boldsymbol{w}^h, \rho\ddot{\boldsymbol{u}}^h)^e &= \int_{\Omega^e} w_i^h\ \rho\ \ddot{u}_i^h\ d\Omega \\ &= \delta_{ij} \int_{\Omega^e} w_i^h\ \rho\ \ddot{u}_j^h\ d\Omega \qquad (7.2.37) \\ &= \sum_{A,B} c_{iA}\ \delta_{ij} \int_{\Omega^e} N_A\ \rho\ N_B\ d\Omega\ \ddot{d}_{jB} \quad \text{(summation of } i \text{ and } j \text{ implied)}\end{aligned}$$

2. Equation (7.2.19) is a coupled system of second-order ordinary differential equations. Algorithms for solving equations of this type are described in Chapter 9.

3. Note that the element mass is involved in the adjustment of forces due to nonzero boundary accelerations (see (7.2.30)).

4. As mentioned in the previous section, we usually simplify the specification of initial conditions in practice. Nodal interpolates in this case take the form

$$d_{0P} = u_{0i}(x_A) \qquad (7.2.38)$$

$$\dot{d}_{0P} = \dot{u}_{0i}(x_A) \qquad (7.2.39)$$

where $P = \text{LM}(i, A)$. Recall LM is the array described in Chapter 2.

Exercise 2. Fill in the omitted details leading to the matrix formulation of elastodynamics.

Viscous Damping

In structural dynamics we often work with systems of the form

$$\boxed{\boldsymbol{M}\ddot{\boldsymbol{d}} + \boldsymbol{C}\dot{\boldsymbol{d}} + \boldsymbol{K}\boldsymbol{d} = \boldsymbol{F}} \qquad (7.2.40)$$

where $\boldsymbol{C}$ is the ***viscous damping matrix***. A particularly convenient form of $\boldsymbol{C}$ is the ***Rayleigh damping matrix***

$$\boxed{\boldsymbol{C} = a\boldsymbol{M} + b\boldsymbol{K}} \qquad (7.2.41)$$

where a and b are parameters. The two constituents of Rayleigh damping are seen to be mass and stiffness proportional. We would like to enlarge our theoretical framework to include Rayleigh damping. The necessary modifications are as follows: Replace the equation of motion, (7.2.6), by

$$\rho u_{i,tt} + a\rho u_{i,t} = \sigma_{ij,j} + f_i \tag{7.2.42}$$

where the generalized Hooke's law is modified to account for the stiffness proportional effect, namely,

$$\sigma_{ij} = c_{ijkl}(u_{(k,l)} + b\ \dot{u}_{(k,l)}) \tag{7.2.43}$$

In addition to the appearance of the $\boldsymbol{C}\dot{\boldsymbol{d}}$-term in (7.2.40), the effect of the viscous damping matrix is also felt in modifying the forces due to prescribed displacement boundary conditions. Specifically,

$$f_p^e = \text{right-hand side of (7.2.30)} - \sum_{q=1}^{n_{ee}} c_{pq}^e \dot{g}_q^e \tag{7.2.44}$$

where

$$\boldsymbol{c}^e = a\boldsymbol{m}^e + b\boldsymbol{k}^e \tag{7.2.45}$$

Everything else remains the same. The parameters a and b may be selected to produce desired damping characteristics.

Example

In one dimension, the above formulation leads to a ***wave equation***. This also may be viewed as a generalization of the one-dimensional model problem of Chapter 1. Various interpretations are possible. For example, the axial motion of an elastic rod of length L is governed by the equation

$$\rho u_{,tt} = \sigma_{,x} + f \qquad \text{on }]0, L[\times]0, T[\tag{7.2.46}$$

where

$$\sigma = E\ u_{,x} \tag{7.2.47}$$

and $E = E(x)$ is Young's modulus. Boundary and initial conditions may be specified in analogous fashion to Chapter 1, namely,

$$u(L, t) = g(t) \qquad t \in]0, T[\tag{7.2.48}$$

$$-Eu_{,x}(0, t) = h(t) \qquad t \in]0, T[\tag{7.2.49}$$

$$u(x, 0) = u_0(x) \qquad x \in]0, L[\tag{7.2.50}$$

$$\dot{u}(x, 0) = \dot{u}_0(x) \qquad x \in]0, L[\tag{7.2.51}$$

The resulting element arrays are virtually identical to the ones encountered in the one-dimensional heat equation example at the end of Sec. 7.1, viz.,

$$m_{ab}^e = \int_{\Omega^e} N_a \rho N_b \, d\Omega \tag{7.2.52}$$

$$k^e_{ab} = \int_{\Omega^e} N_{a,x} E\, N_{b,x}\, d\Omega \tag{7.2.53}$$

$$f^e_a = \int_{\Omega^e} N_a \ell\, d\Omega - N_a(0)\delta_{e1} h - \sum_{b=1}^{n_{en}} (k^e_{ab} g^e_b + m^e_{ab} \ddot{g}^e_b) \tag{7.2.54}$$

Note that in the present case, element equation and node numbers coincide (i.e., $p = a$ and $q = b$). Compare (7.1.37) through (7.1.39) with (7.2.52) through (7.2.54).

Exercise 3. In previous chapters, the element stiffness matrix, (7.2.53), was evaluated (without E) for typical C^0 shape functions. Assuming ρ is constant, evaluate the element mass matrix, $\boldsymbol{m}^e$, for the following shape functions:

i. Piecewise linears.

$$\left[\textit{Answer}: \boldsymbol{m}^e = \frac{\rho h}{6} \begin{bmatrix} 2 & 1 \\ 1 & 2 \end{bmatrix}. \right]$$

ii. Piecewise quadratics.

$$\left[\textit{Answer}: \boldsymbol{m}^e = \frac{\rho h}{30} \begin{bmatrix} 4 & 2 & -1 \\ 2 & 16 & 2 \\ -1 & 2 & 4 \end{bmatrix}. \right]$$

Exercise 4. The initial/boundary-value problem for the deflection of an ***elastic membrane on a Winkler foundation*** is stated as follows:

Given $\ell : \Omega \times]0, T[\to \mathbb{R}$, $g : \Gamma_g \times]0, T[\to \mathbb{R}$, $h : \Gamma_h \times]0, T[\to \mathbb{R}$, $u_0 : \Omega \to \mathbb{R}$, and $\dot{u}_0 : \Omega \to \mathbb{R}$, find $u : \overline{\Omega} \times [0, T] \to \mathbb{R}$ such that

$$u_{,ii} - \alpha u + \ell = u_{,tt} \qquad \text{on } \Omega \times]0, T[$$

$$u = g \qquad x \in \Gamma_g,\ t \in]0, T[$$

$$u_{,n} = h \qquad x \in \Gamma_h,\ t \in]0, T[$$

$$u = u_0 \qquad x \in \Omega,\ t = 0$$

$$u_{,t} = \dot{u}_0 \qquad x \in \Omega,\ t = 0$$

where $u_{,n} = \partial u / \partial n$ is the derivative in the direction of the outward unit normal vector and $\alpha > 0$.

Set up the following finite element paraphernalia:

a. A semidiscrete weak form of the problem in which the h-boundary condition is "natural."
b. The corresponding Galerkin form of the problem.
c. The matrix ordinary-differential-equation problem.

Solution

a. Find $u \in \mathcal{S}_t$ such that for all $w \in \mathcal{V}$

$$(w, \ddot{u}) + a(w, u) = (w, \ell) + (w, h)_\Gamma$$

$$(w, u(0) - u_0) = 0$$
$$(w, \dot{u}(0) - \dot{u}_0) = 0$$

where

$$(w, \ddot{u}) = \int_\Omega w\ddot{u}\, d\Omega$$

$$a(w, u) = \int_\Omega (w_{,i}u_{,i} + \alpha wu)\, d\Omega$$

$$(w, \ell) = \int_\Omega w\ell\, d\Omega$$

$$(w, h)_\Gamma = \int_{\Gamma_h} wh\, d\Gamma$$

$$\vdots$$

b. Find $u^h = v^h + g^h \in \mathcal{S}^h$, $v^h \in \mathcal{V}^h$ such that for all $w^h \in \mathcal{V}^h$ (as usual)

$$(w^h, \ddot{u}^h) + a(w^h, u^h) = (w^h, \ell) + (w, h)_\Gamma$$

$$(w^h, \ddot{v}^h) + a(w^h, v^h) = (w^h, \ell) + (w^h, h)_\Gamma - (w^h, \ddot{g}^h) - a(w^h, g^h)$$

(Likewise for initial conditions.)

c. $\boldsymbol{M}\ddot{\boldsymbol{d}} + \boldsymbol{K}\boldsymbol{d} = \boldsymbol{F}$; $\boldsymbol{d}(0) = \boldsymbol{d}_0$; $\dot{\boldsymbol{d}}(0) = \dot{\boldsymbol{d}}_0$

where

$$m^e_{ab} = \int_{\Omega^e} N_a N_b\, d\Omega$$

$$k^e_{ab} = \int_{\Omega^e} (N_{a,i} N_{b,i} + \alpha N_a N_b) d\Omega$$

$$f^e_a = \int_{\Omega^e} N_a \ell\, d\Omega + \int_{\Gamma^e_h} N_a h\, d\Gamma - \sum_{b=1}^{n_{en}} (m^e_{ab}\, \ddot{g}^e_b + k^e_{ab}\, g^e_b)$$

(Usual assembly algorithm plus initial conditions.)

7.3 EIGENVALUE PROBLEMS: FREQUENCY ANALYSIS AND BUCKLING

To each of the classes of time-dependent problems considered in Sec. 7.1 and 7.2 there are corresponding eigenvalue problems. Eigenvalue problems play particularly important roles in the analysis of elastic solids and structures. The natural frequencies and mode shapes of free vibration are determined by eigenvalue problems as are buckling loads and modes. Some examples illustrating potential applications are as follows.

Free Vibration of an Elastic Rod

The frequencies and mode shapes are governed by the following equation:

$$\lambda \rho u + (Eu_{,x})_{,x} = 0 \qquad x \in \,]0, L[\tag{7.3.1}$$

Various homogeneous boundary conditions may be employed. For example, corresponding to our "standard" one-dimensional case we have

$$u(L) = 0 \qquad \text{(fixed)} \tag{7.3.2}$$

$$-Eu_{,x}(0) = 0 \qquad \text{(free)} \tag{7.3.3}$$

The nontrivial solutions of (7.3.1) through (7.3.3) are countably infinite. That is, for each $k = 1, 2, \ldots, \infty$, there is an eigenvalue λ_k and corresponding eigenfunction $u_{(k)}:]0, L[\rightarrow \mathbb{R}$ which satisfy (7.3.1) through (7.3.3). It is a simple exercise to show that each λ_k is positive, $0 < \lambda_1 \le \lambda_2 \le \lambda_3 \le \ldots$, and that the eigenfunctions are orthogonal in the sense that

$$\int_0^L u_{(k)}\, \rho u_{(l)}\, dx = \delta_{kl} \tag{7.3.4}$$

The eigenvalue $\lambda_k = \omega_k^2$, where ω_k is the kth natural frequency and $u_{(k)}$ is the corresponding normal mode shape. The eigenfunctions are determined only up to a multiplicative constant. The arbitrariness may be removed by the normalization

$$\int_0^L \rho u^2\, dx = 1 \qquad \text{or} \qquad \max_{x \in [0, L]} |u(x)| = 1.$$

A ***weak formulation*** of this problem is given by the following:

(W) Find all eigenpairs $\{\lambda, u\}$, $u \in \mathcal{S}(\equiv \mathcal{V})$, such that for all $w \in \mathcal{V}$

$$\boxed{a(w, u) - \lambda (w, \rho u) = 0} \tag{7.3.5}$$

where

$$a(w, u) = \int_0^L w_{,x}\, Eu_{,x}\, dx \tag{7.3.6}$$

$$(w, \rho u) = \int_0^L w \rho u\, dx \tag{7.3.7}$$

The essential boundary condition, $u(L) = 0$, needs to be built into the definition of $\mathcal{V}$.

The equivalence of strong and weak forms follows from by now standard arguments. Note

$$a(w, u) - \lambda (w, \rho u) = -\int_0^L w(\lambda \rho u + (Eu_{,x})_{,x})\, dx - (wEu_{,x})\Big|_{x=0} \tag{7.3.8}$$

The remaining details are left as an exercise for the reader.

The corresponding ***Galerkin formulation*** is as follows:
Find all $\lambda^h \in \mathbb{R}$ and $u^h \in \mathcal{S}^h (\equiv \mathcal{V}^h)$ such that for all $w^h \in \mathcal{V}^h$

$$a(w^h, u^h) - \lambda^h(w^h, \rho u^h) = 0 \tag{7.3.9}$$

Substitution of shape-function expansions for w^h and u^h gives rise to the ***matrix eigenvalue problem***.

Find eigenvalues λ_k^h and eigenvectors $\boldsymbol{\psi}_k$, $k = 1, 2, \ldots, n_{eq}$, that satisfy

$$(\boldsymbol{K} - \lambda_k^h \boldsymbol{M})\boldsymbol{\psi}_k = \boldsymbol{0} \qquad \text{(no sum on } k) \tag{7.3.10}$$

The eigenvectors $\boldsymbol{\psi}_k$ contain the nodal values of $u_{(k)}^h$. That is, the Ath component of $\boldsymbol{\psi}_k$ is $u_{(k)}^h(\boldsymbol{x}_A)$. In terms of the matrix problem, the orthogonality condition may be expressed as [see (7.3.4)]

$$\boldsymbol{\psi}_k^T \boldsymbol{M} \boldsymbol{\psi}_l = \delta_{kl} \tag{7.3.11}$$

The element matrices that contribute to $\boldsymbol{M}$ and $\boldsymbol{K}$ are defined by (7.2.52) and (7.2.53), respectively.

Buckling of a Thin Beam ("Elastic Instability")

Consider a thin (i.e., Bernoulli-Euler) beam subjected to simply supported boundary conditions (see Fig. 7.3.1). We assume the beam is subjected to a uniform axial compression, λ, and wish to calculate the critical value of λ (i.e., "buckling load"). The equations are

$$\lambda u_{,xx} + (EIu_{,xx})_{,xx} = 0 \tag{7.3.12}$$

$$u(0) = u(L) = 0 \tag{7.3.13}$$

$$u_{,xx}(0) = u_{,xx}(L) = 0 \tag{7.3.14}$$

There are an infinite number of eigenpairs that satisfy (7.3.12) through (7.3.14). The smallest eigenvalue, $\lambda_1 > 0$, is the critical buckling load and the corresponding eigenfunction, $u_{(1)}$, is the buckling mode. An exact analysis reveals that $\lambda_k = k^2\pi^2 EI/L^2$ and $u_{(k)} = \sin k\pi x/L$.

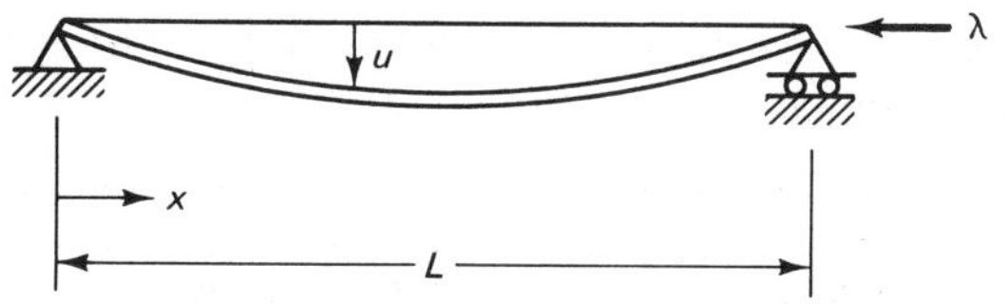

Figure 7.3.1 Simply supported beam under axial compression.

A ***weak formulation*** of the problem is:

(W) Find $u \in \mathcal{S}(\equiv \mathcal{V})$ such that for all $w \in \mathcal{V}$

$$\boxed{a(w, u) - \lambda b(w, u) = 0} \tag{7.3.15}$$

where

$$a(w, u) = \int_0^L w_{,xx} EIu_{,xx}\, dx \tag{7.3.16}$$

$$b(w, u) = \int_0^L w_{,x}\, u_{,x}\, dx \tag{7.3.17}$$

An appropriate space of functions consists of H^2-functions (see Sec. 1.16) with (7.3.13) as essential boundary conditions. Integration by parts and $w(0) = w(L) = 0$ reveals

$$a(w, u) - \lambda b(w, u) = \int_0^L w(\lambda u_{,xx} + (EIu_{,xx})_{,xx})\, dx + (w_{,x} EIu_{,xx})\Big|_0^L \tag{7.3.18}$$

The equivalence of strong and weak forms may be argued in the usual way with the aid of this expression.

The corresponding ***Galerkin formulation*** is:
Find $u^h \in \mathcal{S}^h(\equiv \mathcal{V}^h)$ such that for all $w^h \in \mathcal{V}^h$,

$$\boxed{a(w^h, u^h) - \lambda^h b(w^h, u^h) = 0} \tag{7.3.19}$$

The C^1-continuous Hermite cubics (see Sec. 1.16) serve to define $\mathcal{V}^h$.

The matrix problem has the same form as (7.3.10). The element contributions are defined by

$$k_{pq}^e = \int_{\Omega^e} N_{p,xx} EI N_{q,xx}\, dx \tag{7.3.20}$$

$$m_{pq}^e = \int_{\Omega^e} N_{p,x} N_{q,x}\, dx \tag{7.3.21}$$

Strictly speaking, (7.3.21) is not a mass matrix. It is usually refered to as an ***initial-stress***, or ***geometric***, ***stiffness*** in this context.

Exercise 1. Evaluate $\boldsymbol{m}^e$ [i.e., (7.3.21)] using Hermite cubics. (Observe that three-point Gauss quadrature enables exact integration.) Recall that the exact expression for $\boldsymbol{k}^e$ was obtained in Sec. 1.16.

Free Vibration of a Thin Beam

The vibration of a thin beam is governed by

$$-\lambda \rho A u + (EI u_{,xx})_{,xx} = 0 \tag{7.3.22}$$

where $A = A(x)$ is the cross-sectional area. Comparison with (7.3.12) reveals that in place of $\lambda u_{,xx}$, the term $-\lambda \rho A u$ appears. In the simply supported case, the weak formulation is

$$\boxed{a(w, u) - \lambda(w, \rho A u) = 0} \tag{7.3.23}$$

where $a(w, u)$ is again defined by (7.3.16) and

$$(w, \rho A u) = \int_0^L w \rho A u \, dx \tag{7.3.24}$$

The matrix problem takes the form (7.3.10) and the element arrays are defined by (7.3.20) and

$$\boxed{m_{pq}^e = \int_{\Omega^e} N_p \rho A N_q \, dx} \tag{7.3.25}$$

Exercise 2. Evaluate (7.3.25) using Hermite cubics.

7.3.1 Standard Error Estimates

For the types of problems considered in Chapters 1 through 3, standard error estimates are available for eigenvalues and eigenfunctions. These estimates are the counterparts of those given in Sec. 4.1. The more-sophisticated methods presented after Sec. 4.1 and in Chapters 5 and 6 lie outside the realm of these estimates and therefore must be considered on a separate basis. The present estimates indicate the optimality of the Galerkin finite element method for the standard classes of elliptic eigenvalue problems. The main results are [2]:

$$\lambda_l \le \lambda_l^h \le \lambda_l + c h^{2(k+1-m)} \lambda_l^{(k+1)/m} \tag{7.3.26}$$

$$\| \boldsymbol{u}_{(l)}^h - \boldsymbol{u}_{(l)} \|_m \le c h^{(k+1-m)} \lambda_l^{(k+1)/2m} \tag{7.3.27}$$

where c is a constant independent of h and λ_l.

Remarks

1. Note that the first inequality in (7.3.26) says that the lth approximate eigenvalue is bounded from below by the lth exact eigenvalue. This "upper-bound" property is *not* preserved once the Galerkin rules are violated (e.g., when reduced integration or incompatible modes are employed).

2. The rate of convergence (i.e., power of h) of eigenvalues is *twice* that of eigenfunctions in the H^m-norm [compare (7.3.26) with (7.3.27)].

3. The appearance of powers of the eigenvalue on the right-hand sides of (7.3.26) and (7.3.27) suggests that the quality of approximation *deteriorates* for higher modes (recall $0 < \lambda_1 \leq \lambda_2 \leq \lambda_3 \leq \ldots$). This will be shown to be the case in specific examples.

4. In frequency analysis we are interested in the error in $\omega_l^h = (\lambda_l^h)^{1/2}$. It turns out that the order of error in the frequency is the same as that in the eigenvalue. This can be seen as follows: Let

$$\epsilon = \frac{\lambda_l^h}{\lambda_l} - 1 \tag{7.3.28}$$

Then,

$$\frac{\omega_l^h}{\omega_l} = \left(\frac{\lambda_l^h}{\lambda_l}\right)^{1/2} = (1 + \epsilon)^{1/2} = 1 + \frac{\epsilon}{2} + O(\epsilon^2) \tag{7.3.29}$$

and so

$$\frac{\omega_l^h}{\omega_l} - 1 = O(\epsilon) = O(h^{2(k+1-m)}) \tag{7.3.30}$$

Thus we can expect the same rate of convergence for frequencies and eigenvalues.

5. An L_2-estimate for eigenfunctions is also available [2]:

$$\boxed{\|\boldsymbol{u}_{(l)}^h - \boldsymbol{u}_{(l)}\|_0 \leq c h^{\sigma} \lambda_l^{(k+1)/(2m)}} \tag{7.3.31}$$

where

$$\sigma = \min\{k + 1, 2(k + 1 - m)\} \tag{7.3.32}$$

6. The accuracy of the eigenvalues and eigenfunctions are measures of the quality of both $\boldsymbol{M}$ and $\boldsymbol{K}$, in other words, the entire spatial discretization.

Example 1

Consider an elastic boundary value problem ($m = 1$) and assume linear elements are employed ($k = 1$). The error estimates take the form:

$$\frac{\omega_l^h}{\omega_l} - 1 = O(h^2)$$

$$\|\boldsymbol{u}_{(l)}^h - \boldsymbol{u}_{(l)}\|_1 = O(h)$$

For quadratic-level elements ($k = 2$) we have

$$\frac{\omega_l^h}{\omega_l} - 1 = O(h^4)$$

$$\| \boldsymbol{u}_{(l)}^h - \boldsymbol{u}_{(l)} \|_1 = O(h^2)$$

Example 2

Consider the Hermite cubic beam element ($m = 2$, $k = 3$):

$$\frac{\omega_l^h}{\omega_l} - 1 = O(h^4)$$

$$\| \boldsymbol{u}_{(l)}^h - \boldsymbol{u}_{(l)} \|_2 = O(h^2)$$

Remark

If numerical quadrature is employed to integrate the element mass and/or stiffness, the same conditions apply as described in Sec. 4.1. That is, a rule accurate enough to exactly integrate all monomials through degree $\bar{k} + k - 2m$ is sufficient to maintain full rate of convergence. (Recall $\bar{k}$ is the order of the highest-order monomial appearing in the element shape functions.) A sufficient condition for convergence is that the rule exactly integrate monomials through degree $\bar{k} - m$.

Exercise 3. Use the "minimax" characterization of eigenvalues, i.e.,

$$\lambda_l = \min_{E_l} \{ \max_{\boldsymbol{u} \in E_l} \mathcal{R}(\boldsymbol{u}) \} \tag{7.3.33}$$

$$\lambda_l^h = \min_{E_l^h} \{ \max_{\boldsymbol{u}^h \in E_l^h} \mathcal{R}(\boldsymbol{u}^h) \} \tag{7.3.34}$$

where $\mathcal{R}$ denotes the ***Rayleigh quotient***, defined by

$$\mathcal{R}(\boldsymbol{u}) = \frac{a(\boldsymbol{u}, \boldsymbol{u})}{(\boldsymbol{u}, \rho \boldsymbol{u})} \tag{7.3.35}$$

and E_l and E_l^h are l-dimensional subspaces of H^m and $\mathcal{S}^h$, respectively, to establish $\lambda_l \le \lambda_l^h$, $l = 1, 2, \ldots, n_{eq}$. (The bilinear form in the denominator of (7.3.35) may be $b(u, u)$, $(u, \rho c u)$, etc., as in previously discussed cases.) This result is obtained in [2].

Exercise 4. An isoparametric finite element is used in a series of convergence rate studies on linear elastic eigenvalue problems. All tests indicate that the fundamental frequency behaves as follows:

$$\log\left(\frac{\omega^h}{\omega} - 1\right) \sim 2 \log h$$

where ω is the exact frequency, ω^h is the finite element frequency, and h is the mesh parameter.

a. Which of the following elements was used in the study?

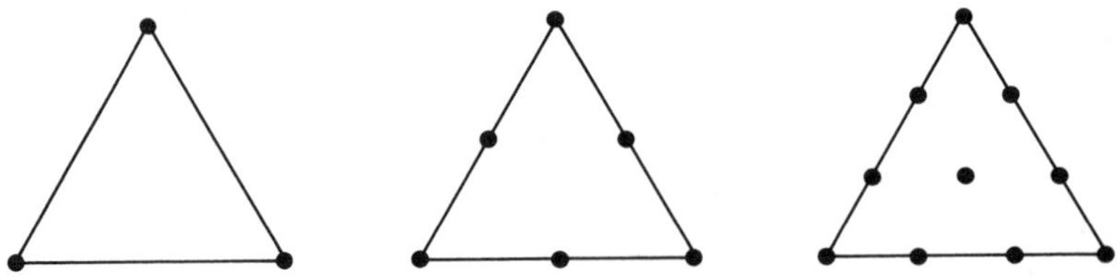

Justify your answer.

b. What is the rate of convergence of eigenvectors in the L_2 and H^1 norms?

Exercise 5. Consider the application of bilinear quadrilateral elements to frequency calculations in elasticity. The mass matrix is to be constructed using a Gaussian quadrature formula. The choice is between the one-point, 2 × 2, and 3 × 3 rules. Pick the rule involving the minimum number of points that guarantees that the full rate of convergence of consistent mass is maintained. Explain how you arrived at your answer.

Solution for Exercise 5 The rule must exactly integrate polynomials of degree $\bar{k} + k - 2m = 2 + 1 - 2 = 1$ (i.e., constants and linear terms must be integrated exactly). The ***one-point Gauss rule*** achieves this accuracy.

7.3.2 Alternative Definitions of the Mass Matrix; Lumped and Higher-order Mass

The recipe given by the Galerkin formulation for the mass matrix is sometimes referred to as variationally "consistent." Prior to establishment of the variational foundations of the finite element method, mass matrices were concocted in ad hoc fashion. Archer [3] is generally acknowledged as having first pointed out the correctness of ***consistent mass***. The consistent-mass matrix leads to the optimal error estimates described in the preceding section. However, as is often the case in finite element analysis, there is strong motivation for breaking the basic rules. This is particularly true in the development of mass matrices. In practice, diagonal or "lumped" mass matrices are often employed due to their general economy and because they lead to some especially attractive time-integration schemes (so-called explicit methods; see Chapters 8 and 9). There are several ways of going about the construction of lumped-mass matrices. One way is to employ nodal quadrature rules.

Mass Lumping by Nodal Quadrature

An integration formula with integration points at the nodes takes the form (see Chapter 3):

$$\int_{\square} g(\boldsymbol{\xi})\, d\square \cong \sum_{a=1}^{n_{en}} g(\boldsymbol{\xi}_a) W_a \tag{7.3.36}$$

The effect of such a rule on the mass matrix is to diagonalize it, viz.,

$$\begin{aligned} m^e_{pq} &= \delta_{ij} \int_{\Omega^e} N_a \rho N_b \, d\Omega \\ &= \delta_{ij} \int_{\square} N_a \rho N_b j \, d\square \\ &\cong \delta_{ij} \sum_{c=1}^{n_{en}} \underbrace{N_a(\xi_c)}_{\delta_{ac}} \rho(\xi_c) \underbrace{N_b(\xi_c)}_{\delta_{bc}} j(\xi_c) W_c \\ &= \begin{cases} \delta_{ij} \rho(\xi_a) j(\xi_a) W_a & a = b \quad \text{(no sum on } a\text{)} \\ 0 & a \neq b \end{cases} \end{aligned} \tag{7.3.37}$$

A theory of numerical quadrature presented in [2] and described earlier defines under what conditions the nodal quadrature mass matrix is convergent and retains full order of accuracy. In the one-dimensional case, full rate of convergence is maintained if the rule is capable of exactly integrating polynomials of degree $2(k - m)$; the minimum condition for convergence is a rule that can exactly integrate polynomials of degree $k - m$. (It should be pointed out that these are *sufficient* conditions. They may not be *necessary*.)

Example 1

Consider piecewise linear shape functions in one dimension and assume $\rho =$ constant. The trapezoidal rule, namely,

$$\xi_1 = -1, \qquad \xi_2 = 1, \qquad W_1 = W_2 = 1$$

is sufficiently accurate to maintain full rate of convergence (i.e., $2(k - m) = 2(1 - 1) = 0$; thus only constants need to be exactly integrated). It may be verified that in this case

$$\boldsymbol{m}^e_{\text{lumped}} = \frac{\rho h^e}{2} \begin{bmatrix} 1 & 0 \\ 0 & 1 \end{bmatrix} \tag{7.3.38}$$

Example 2

Consider piecewise quadratics. Simpson's rule,

$$\xi_1 = -1, \qquad \xi_2 = 0, \qquad \xi_3 = 1, \qquad W_1 = W_3 = \tfrac{1}{3}, \qquad W_2 = \tfrac{4}{3}$$

is capable of exactly integrating cubic polynomials and so attains full rate of convergence. (The requirement is $2(k - m) = 2(2 - 1) = 2$; i.e., quadratic polynomials.) The lumped-mass matrix is (verify!):

$$\boldsymbol{m}^e_{\text{lumped}} = \frac{\rho h^e}{6} \begin{bmatrix} 1 & 0 & 0 \\ 0 & 4 & 0 \\ 0 & 0 & 1 \end{bmatrix} \tag{7.3.39}$$

Numerical confirmation of the convergence rate of the fundamental frequency is presented in Fig. 7.3.2. Frequency spectra are presented in Fig. 7.3.3. The upper-bound property of consistent mass (i.e., $\omega^h/\omega \geq 1$) is clearly in evidence in Fig. 7.3.3.

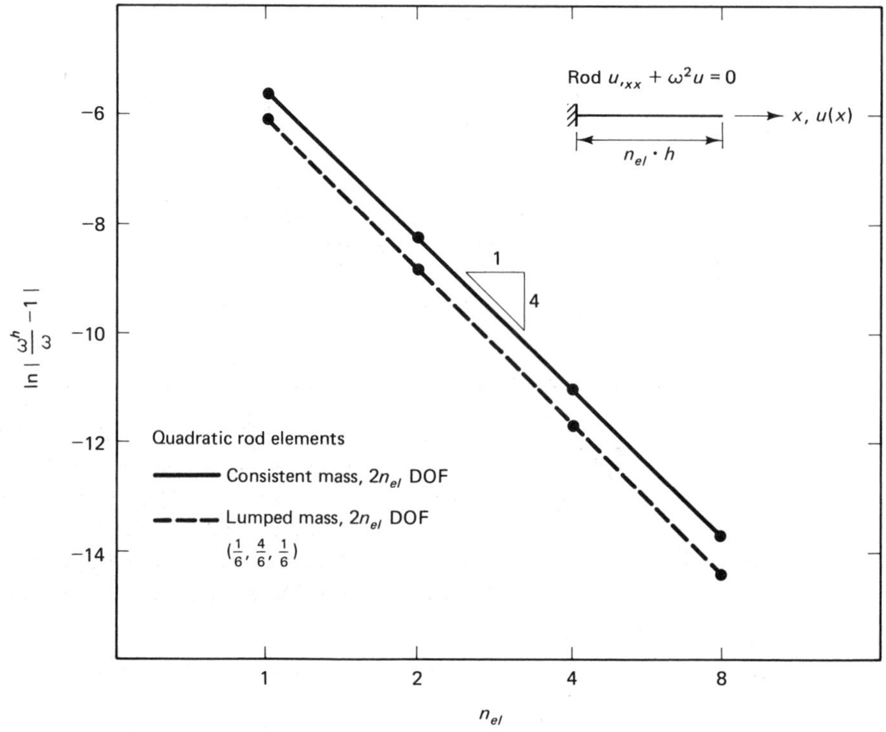

Figure 7.3.2 Convergence of the fundamental frequency for quadratic rod elements (Hughes et al. [4]).

Example 3

For the standard four-node cubic element we need a rule capable of exactly integrating quartic polynomials (i.e., $2(k - m) = 2(3 - 1) = 4$). Unfortunately, for the usual setup in ξ-space, this is impossible. The restriction of interior nodal placement at $\xi = \pm\frac{1}{3}$ limits the order of convergence to cubic:

g	$\int_{-1}^{+1} g\, d\xi$	$\sum_{a=1}^{4} g(\xi_a) W_a$
1	2	$= W_1 + W_2 + W_3 + W_4$
ξ	0	$= -W_1 - \frac{1}{3}W_2 + \frac{1}{3}W_3 + W_4$
ξ^2	$\frac{2}{3}$	$= W_1 + \frac{1}{9}W_2 + \frac{1}{9}W_3 + W_4$
ξ^3	0	$= -W_1 - \frac{1}{27}W_2 + \frac{1}{27}W_3 + W_4$
ξ^4	$\frac{2}{5}$	$\neq W_1 + \frac{1}{81}W_2 + \frac{1}{81}W_3 + W_4$

$$\Rightarrow \begin{cases} W_1 = \frac{1}{4} \\ W_2 = \frac{3}{4} \\ W_3 = W_2 \\ W_4 = W_1 \end{cases}$$

Example 4

Let us reconsider the four-node cubic element. We need one additional parameter in order to create a nodal rule that exactly integrates quartic polynomials. If we relax the condition that the interior nodes are located at $\xi = \pm\frac{1}{3}$, a sufficiently accurate rule can

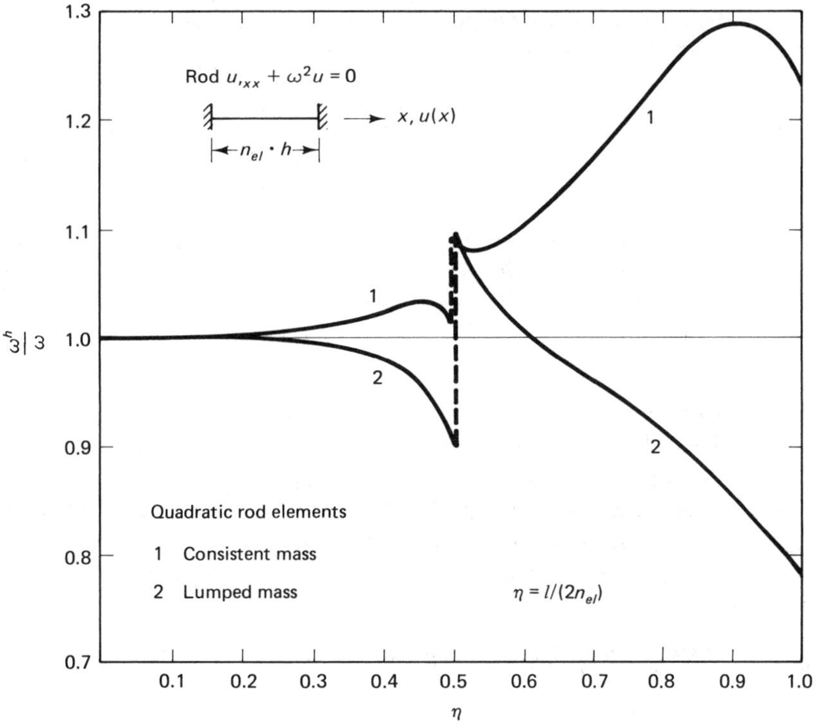

Figure 7.3.3 Frequency spectra for quadratic rod elements (Hughes et al. [4]). η is the mode number.

be developed. Consider the ξ-space setup shown in Fig. 7.3.4. The object is to pick c along with the weights so that an arbitrary quartic polynomial is exactly integrated. The idea is similar to that which led to the Gauss rules. Assume $W_4 = W_1$ and $W_3 = W_2$. Then

g	$\int_{-1}^{+1} g \, d\xi$	$W_1(g(-1) + g(+1)) + W_2(g(-c) + g(+c))$
1	2	$= 2(W_1 + W_2)$
ξ	0	$= 0$
ξ^2	$\frac{2}{3}$	$= 2(W_1 + c^2 W_2)$
ξ^3	0	$= 0$
ξ^4	$\frac{2}{5}$	$= 2(W_1 + c^4 W_2)$
ξ^5	0	$= 0$

There are three equations for the three unknowns. The first equation is used to eliminate

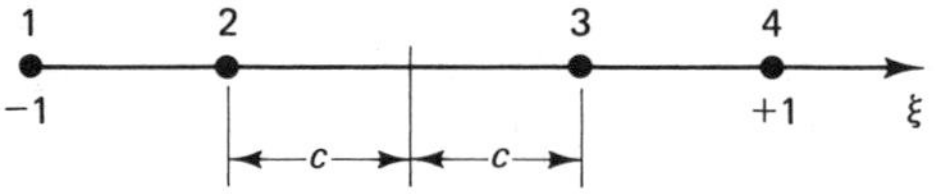

Figure 7.3.4

W_1 from the second and third. Dividing the third equation by the second yields $c^2 = \frac{1}{5}$. Thus $W_1 = \frac{1}{6}$ and $W_2 = \frac{5}{6}$. The mass then is

$$m^e_{\text{lumped}} = \frac{\rho h^e}{12}\begin{bmatrix} 1 & 0 & 0 & 0 \\ 0 & 5 & 0 & 0 \\ 0 & 0 & 5 & 0 \\ 0 & 0 & 0 & 1 \end{bmatrix} \tag{7.3.40}$$

This element mass was originally developed by Fried and Malkus [5]. Their numerical tests confirmed the theoretically predicted convergence rate (i.e., $\lambda^h - \lambda = O(h^6)$).

Note that the relocation of the interior nodes necessitates the derivation of new shape functions.

Remark

The rule derived above is an example of a ***Lobatto quadrature rule***. The Lobatto rules are similar to the Gauss rules, except in Lobatto rules the end points of the interval are always included. The locations of the interior points are determined to maximize the accuracy of the rule. The trapezoidal rule and Simpson's rule are the first two Lobatto rules. Lobatto rules for $n_{en} = 3$ through $n_{en} = 10$ are listed in Table 7.3.1.

TABLE 7.3.1 Lobatto quadrature rules

	$\pm\xi_a$	W_a		$\pm\xi_a$	W_a
			7	1.00000 000	0.04761 904
3	1.00000 000	0.33333 333		0.83022 390	0.27682 604
	0.00000 000	1.33333 333		0.46884 879	0.43174 538
				0.00000 000	0.48761 904
			8	1.00000 000	0.03571 428
4	1.00000 000	0.16666 667		0.87174 015	0.21070 422
	0.44721 360	0.83333 333		0.59170 018	0.34112 270
				0.20929 922	0.41245 880
			9	1.00000 00000	0.02777 77778
5	1.00000 000	0.10000 000		0.89975 79954	0.16549 53616
	0.65465 367	0.54444 444		0.67718 62795	0.27453 87126
	0.00000 000	0.71111 111		0.36311 74638	0.34642 85110
				0.00000 00000	0.37151 92744
			10	1.00000 00000	0.02222 22222
6	1.00000 000	0.06666 667		0.91953 39082	0.13330 59908
	0.76505 532	0.37847 495		0.73877 38651	0.22488 93420
	0.28523 152	0.55485 838		0.47792 49498	0.29204 26836
				0.16527 89577	0.32753 97612

Exercise 6. Derive the cubic shape functions of the four-node ***Lobatto element***.

Example 5

Consider the Hermite cubic beam element. Because the nodes are located at the end points, the trapezoidal rule will produce a lumped mass:

$$m^e_{\text{lumped}} = \frac{\rho h^e}{2}\begin{bmatrix} 1 & 0 & 0 & 0 \\ 0 & 0 & 0 & 0 \\ 0 & 0 & 1 & 0 \\ 0 & 0 & 0 & 0 \end{bmatrix} \tag{7.3.41}$$

Note that half the mass is lumped at each translational degree of freedom and that there are zero masses at the rotational (i.e., slope) degrees of freedom. In order for the full rate of convergence of consistent mass to be guaranteed, the rule should be able to integrate exactly polynomials of degree $2(k - m) = 2(3 - 2) = 2$. The trapezoidal rule is capable of integrating only linear polynomials. Nevertheless, computational results suggest full accuracy is achieved, i.e., (see Fig. 7.3.5),

$$\frac{\omega_l^h}{\omega_l} - 1 = O(h^4) \tag{7.3.42}$$

(This does not contradict the numerical integration theory, which provides sufficient, rather than necessary, conditions.)

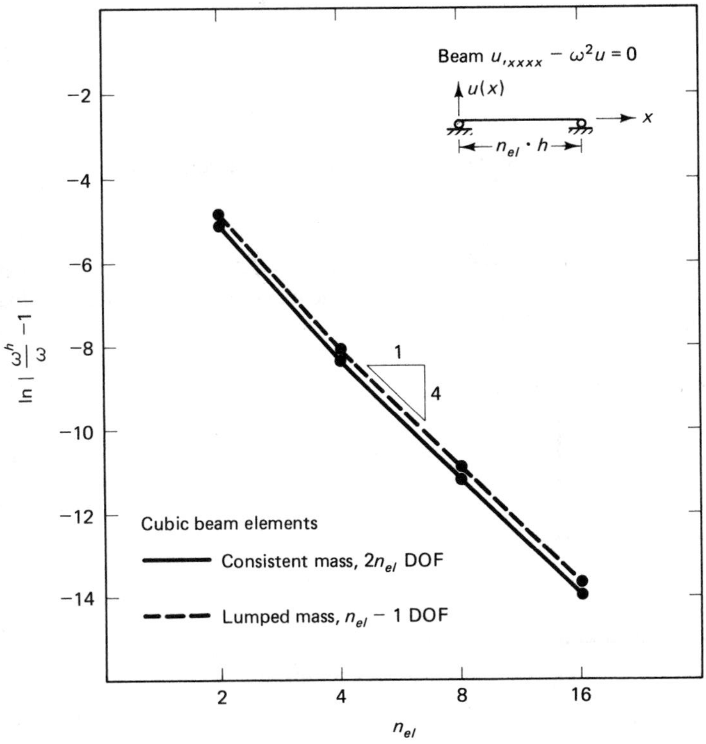

Figure 7.3.5 Convergence of the fundamental frequency for a simply supported beam (Hughes et al. [4]).

Example 6

Optimal quadrature rules have been developed for triangles by Fried and Malkus [5]. The rules are summarized in Fig. 7.3.6.

The theoretical eigenvalue convergence rate for each of the three triangles has been verified numerically in [5]. Note that for the quadratic triangle there are zero masses at the vertices. For the cubic triangle the edge nodes need to be relocated as shown. The vertex masses in this case are negative. Zero and negative masses pose significant impediments in practical applications. (See Remark 1, following the examples.)

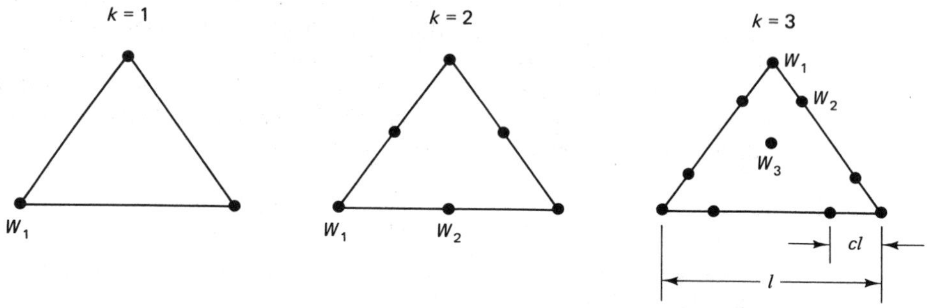

$W_1 = \frac{1}{3}$

$W_1 = 0$
$W_2 = \frac{1}{3}$

$W_1 = -\frac{1}{60}$
$W_2 = \frac{1}{10}$
$W_3 = \frac{9}{20}$
$c = \frac{\sqrt{3}(\sqrt{3} - 1)}{6}$

Figure 7.3.6 Optimally accurate nodal quadrature rules for triangles.

Example 7

For nodal patterns that are products of one-dimensional Lobatto patterns, products of Lobatto rules may be employed. Some examples are illustrated in Fig. 7.3.7 below and on the top of p. 443.

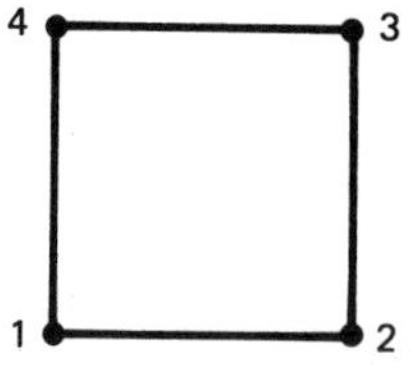

Product trapezoidal $\quad W_a = 1, \ a = 1, 2, 3, 4$

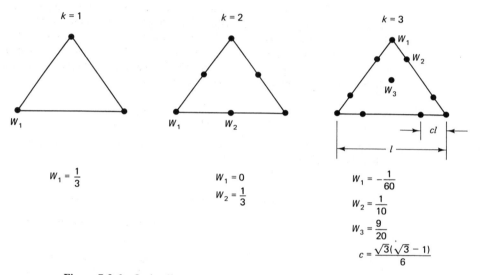

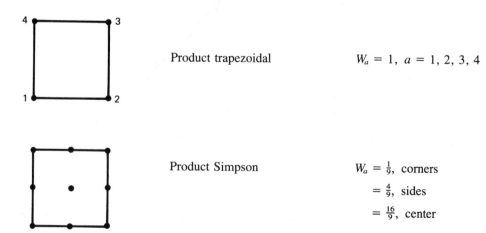

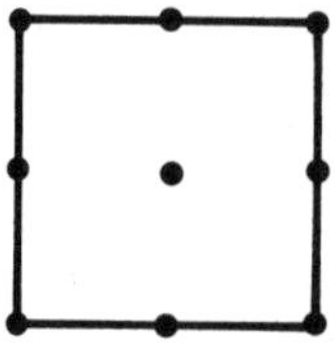

Product Simpson $\quad W_a = \frac{1}{9}$, corners

$= \frac{4}{9}$, sides

$= \frac{16}{9}$, center

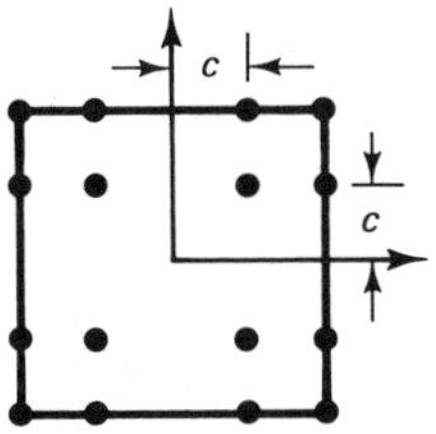

Product Lobatto $W_a = \frac{1}{36}$, corners

$= \frac{5}{36}$, sides

$= \frac{25}{36}$, interior

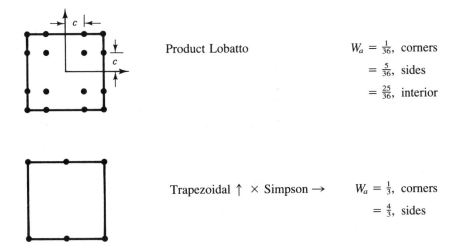

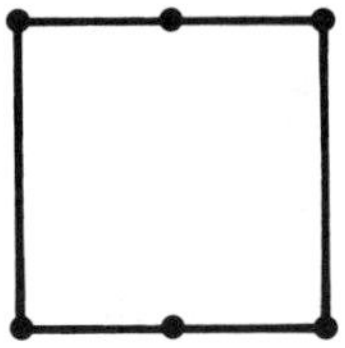

Trapezoidal ↑ × Simpson → $W_a = \frac{1}{3}$, corners

$= \frac{4}{3}$, sides

Figure 7.3.7 Optimal nodal quadrature rules for product Lobatto patterns.

Example 8

For nodal patterns that are not products of one-dimensional patterns, optimal nodal quadrature rules must be worked out on a case-by-case basis. As an example, consider the eight-node serendipity element (see Fig. 7.3.8). Let the weights assigned to corner and side nodes be denoted W_1 and W_2, respectively. A rule can be developed that is accurate for all cubic polynomials. This turns out to be of high enough precision to guarantee full rate of convergence (i.e., $\bar{k} + k - 2m = 3 + 2 - 2 = 3$). However, the corner masses are negative and thus this mass matrix is unsuitable for practical use. The calculations leading to the rule are summarized below:

g	$\int_{\square} g \, d\square$	$W_1 \sum_{a=1}^{4} g_a + W_2 \sum_{a=5}^{8} g_a$
1	4	$= 4(W_1 + W_2)$
ξ	0	$= 0$
η	0	$= 0$
$\xi\eta$	0	$= 0$
ξ^2	$\frac{4}{3}$	$= 4W_1 + 2W_2$
η^2	$\frac{4}{3}$	$= 4W_1 + 2W_2$
ξ^3	0	$= 0$
$\xi^2\eta$	0	$= 0$
$\xi\eta^2$	0	$= 0$
η^3	0	$= 0$

$\therefore W_1 = -\frac{1}{3}, \; W_2 = \frac{4}{3}$

Remarks

1. We have asserted that negative masses are unacceptable in practice. The reason for this can be seen by considering single degree-of-freedom model equations of parabolic and hyperbolic type:

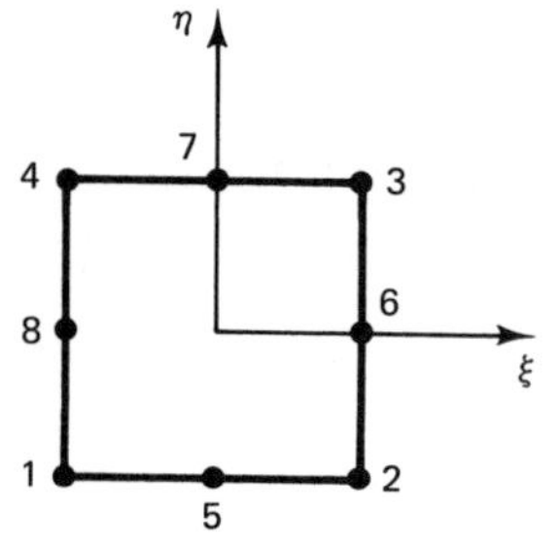

Figure 7.3.8 Eight-node serendipity element.

$$\begin{cases} \dot{d} + \lambda\, d = 0, & d = \exp(-\lambda t) d_0 \\ -\dot{d} + \lambda\, d = 0, & d = \exp(\lambda t) d_0 \end{cases} \quad \begin{matrix}\text{(exponential decay)} \\ \text{(unbounded)}\end{matrix}$$

$$\begin{cases} \ddot{d} + \omega^2 d = 0, & d = A \sin \omega t + B \cos \omega t, \\ -\ddot{d} + \omega^2 d = 0, & d = A \sinh \omega t + B \cosh \omega t \end{cases} \quad \begin{matrix}\text{(oscillatory and bounded)} \\ \text{(unbounded)}\end{matrix}$$

In each case negative mass produces nonphysical behavior. Modes in a multidegree of freedom system will behave in this fashion, vitiating the true solution, when negative masses are present.

2. If nodal quadrature is used in axisymmetric analysis, zero masses result along the axis of symmetry for all elements. This can be seen from the following:

$$m_{pq}^e = 2\pi\, \delta_{ij} \int_{\Omega^e} \rho\, N_a N_b r\; dr\; dz \tag{7.3.43}$$

$$\cong \begin{cases} 2\pi\, \delta_{ij}\, \rho(\boldsymbol{\xi}_a) r(\boldsymbol{\xi}_a) j(\boldsymbol{\xi}_a) W_a, & a = b \\ 0 & a \neq b \end{cases} \tag{7.3.44}$$

Along the axis of symmetry, $r = 0$, and consequently so are the masses of nodes located there.

As we have seen, due to a tendency to generate zero and negative masses, nodal quadrature is not always an effective tool for producing useful lumped-mass matrices. Thus, other methods have been considered.

Row-sum technique

In the row-sum technique the lumped mass is defined by

$$\boxed{m_{pq}^e = \begin{cases} \delta_{ij} \int_{\Omega^e} \rho N_a \, d\Omega & a = b \\ 0 & a \neq b \end{cases}} \tag{7.3.45}$$

The name derives from the fact that

$$\sum_{b=1}^{n_{en}} \int_{\Omega^e} N_a \rho N_b \, d\Omega = \int_{\Omega^e} N_a\, \rho \underbrace{\left(\sum_{b=1}^{n_{en}} N_b \right)}_{1} d\Omega = \int_{\Omega^e} N_a \rho \, d\Omega \tag{7.3.46}$$

That is, the elements in each row are summed and lumped on the diagonal.

The row-sum technique eliminates the problem of zero masses at nodes along the z-axis in problems of axisymmetry. However, like nodal quadrature, it sometimes produces negative masses. This is the case for corner nodes of the eight-node serendipity element.

"Special Lumping Technique"

The special lumping technique was developed by Hinton et al. [6]. The name is not too revealing, but the method has an important attribute: *It always produces positive lumped masses*. The idea is to set the entries of the lumped-mass matrix proportional to the diagonal entries of the consistent mass. (By virtue of positive-definiteness, the diagonal entries of consistent mass are necessarily positive.) The constant of proportionality is selected to conserve the total element mass. The formula is

$$\boxed{m_{pq}^{e} = \begin{cases} \alpha\delta_{ij} \displaystyle\int_{\Omega^e} \rho N_a^2 \, d\Omega & a = b \\ 0 & a \neq b \end{cases}} \tag{7.3.47}$$

where

$$\alpha = \underbrace{\int_{\Omega^e} \rho \, d\Omega}_{\text{total element mass}} \Bigg/ \underbrace{\left(\sum_{a=1}^{n_{en}} \int_{\Omega^e} \rho N_a^2 \, d\Omega \right)}_{\text{sum of diagonal entries of consistent mass}} \tag{7.3.48}$$

The special lumping procedure has been shown numerically to work well on structural and solid mechanics problems [6]. Optimal rates of convergence are typically achieved. Presently, it is the only lumping method that can be recommended for arbitrary elements. Unfortunately, no mathematical theory in support of it has been forthcoming.

Exercise 7. For simple elements, the lumping procedures described above tend to produce similar if not identical masses. Verify this assertion by calculating element masses for the linear triangle and bilinear quadrilateral. Assume the quadrilateral is in the rectangular configuration and in all cases take $\rho =$ constant. Use nodal quadrature, row-sum, and the special lumping techniques.

Exercise 8. Calculate diagonal mass matrices for the one-dimensional three-node element by the row-sum and special lumping techniques: Compare the results with the diagonal mass matrix obtained by Simpson's rule. Comment.

Remark

Although lumped masses have often been used successfully in solid and structural mechanics and heat conduction, some disappointing results have been obtained in fluid mechanics (see Gresho et al. [7]).

Higher-Order Mass

The accuracy of consistent mass can often be achieved by a much simpler lumped mass. A question naturally arises: Is consistent mass the best nondiagonal element mass? The answer appears to be: not always. There are examples of element mass matrices that exhibit superior accuracy to consistent mass. However, no general theory of obtaining higher-order accurate mass matrices exists yet.

Example 1

The first and simplest example of a higher-order accurate mass matrix was presented by Goudreau [8]. The matrix is to be viewed as an alternative for the two-node rod element. Recall the consistent and lumped mass matrices for this element are (respectively)

$$\boldsymbol{m}^e_{\text{consistent}} = \frac{\rho h^e}{6}\begin{bmatrix} 2 & 1 \\ 1 & 2 \end{bmatrix} \tag{7.3.49}$$

$$\boldsymbol{m}^e_{\text{lumped}} = \frac{\rho h^e}{2}\begin{bmatrix} 1 & 0 \\ 0 & 1 \end{bmatrix} \tag{7.3.50}$$

Both of these mass matrices result in second-order accurate eigenvalues. The higher-order mass matrix is simply the average of (7.3.49) and (7.3.50):

$$\boldsymbol{m}^e_{\text{higher-order}} = \frac{\rho h^e}{12}\begin{bmatrix} 5 & 1 \\ 1 & 5 \end{bmatrix} \tag{7.3.51}$$

For a uniform mesh, it can be analytically shown that (7.3.51) leads to fourth-order accurate frequencies [8]. Numerical results supporting this assertion are presented for a fixed-fixed rod in Fig. 7.3.9. The same order of convergence is exhibited for a non-uniform mesh in Fig. 7.3.10 and for fixed-free boundary conditions in Fig. 7.3.11. (Frequency spectra are presented in Fig. 9.1.4.)

For purposes of generalization to other elements it may be observed that (7.3.51) may be produced by employing the quadrature rule $\xi_1 = -\xi_2 = \sqrt{\tfrac{2}{3}}$, $W_1 = W_2 = 1$, on the element mass integral [10]. Clearly this rule can be iterated for multidimensional applications.

Another way of deriving (7.3.51) is to take a linear combination of the consistent mass and stiffness:

$$\begin{aligned} \boldsymbol{m}^e &= \boldsymbol{m}^e_{\text{consistent}} + \left(\frac{1}{6} - r\right)\frac{\rho (h^e)^2}{E}\boldsymbol{k}^e \\ &= \frac{\rho h^e}{6}\begin{bmatrix} 2 & 1 \\ 1 & 2 \end{bmatrix} + \rho h^e\left(\frac{1}{6} - r\right)\begin{bmatrix} 1 & -1 \\ -1 & 1 \end{bmatrix} \end{aligned} \tag{7.3.52}$$

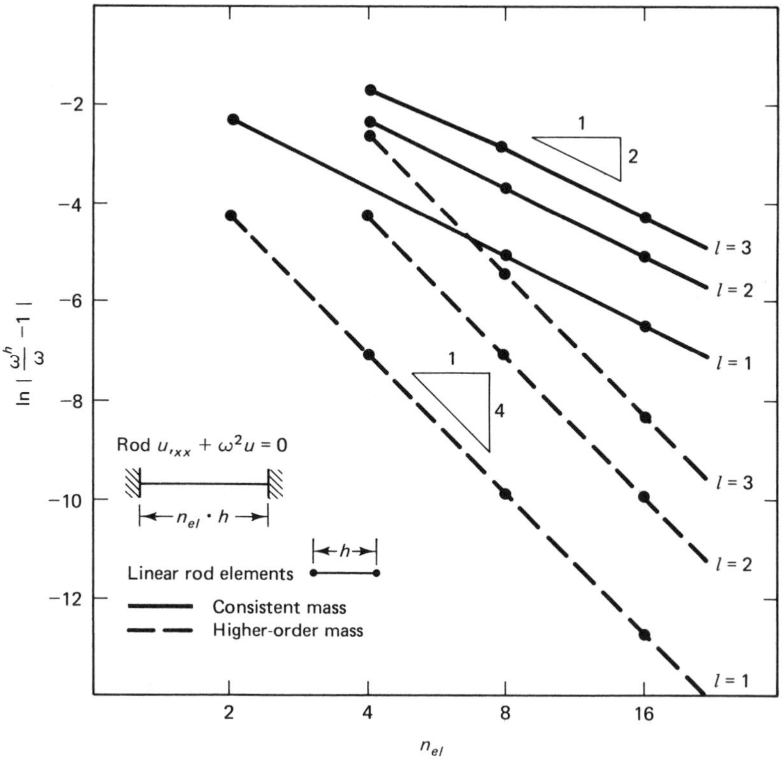

Figure 7.3.9 Convergence of the first three frequencies for linear rod elements [9].

Clearly, for $r = \frac{1}{12}$ we arrive at (7.3.51). The consistent and lumped mass matrices correspond to $r = \frac{1}{6}$ and $r = 0$, respectively.

Example 2

A higher-order mass matrix for the Hermite cubic beam element may be derived by taking a linear combination of the consistent mass and stiffness:

$$\boldsymbol{m}^e_{\text{higher-order}} = \boldsymbol{m}^e_{\text{consistent}} + \alpha \frac{\rho A (h^e)^4}{EI} \boldsymbol{k}^e \tag{7.3.53}$$

Taylor and Iding [11] first experimented with matrices of this form, but did not obtain an optimal value of α. The value $\alpha = \frac{1}{720}$ was derived in [9] and shown to lead to sixth-order accurate frequencies for a uniform mesh. Numerical confirmation is presented in Fig. 7.3.12 for a simply supported beam. Frequency spectra are presented in Fig. 7.3.13.

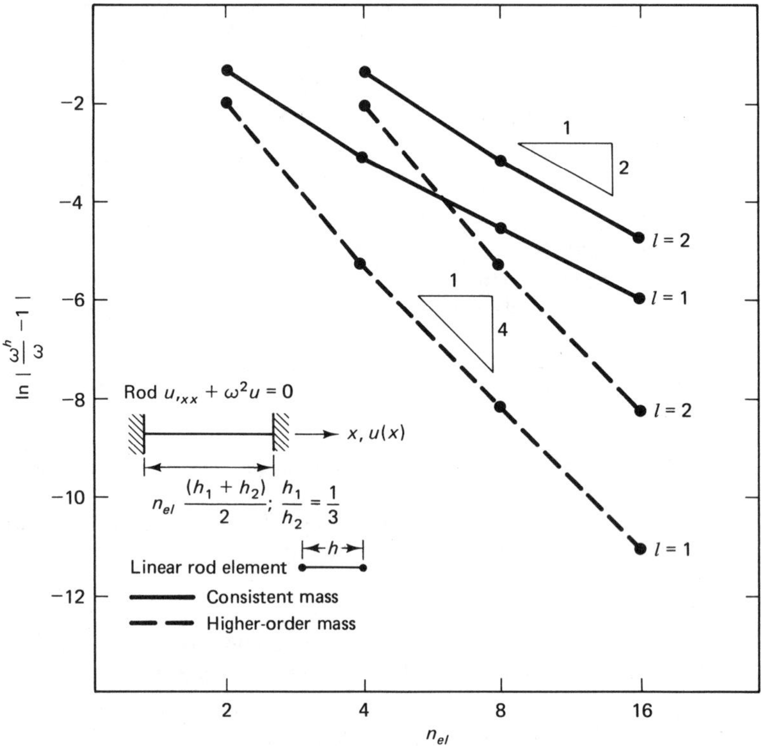

Figure 7.3.10 Convergence of the first two frequencies for linear rod elements with nonuniform mesh [9].

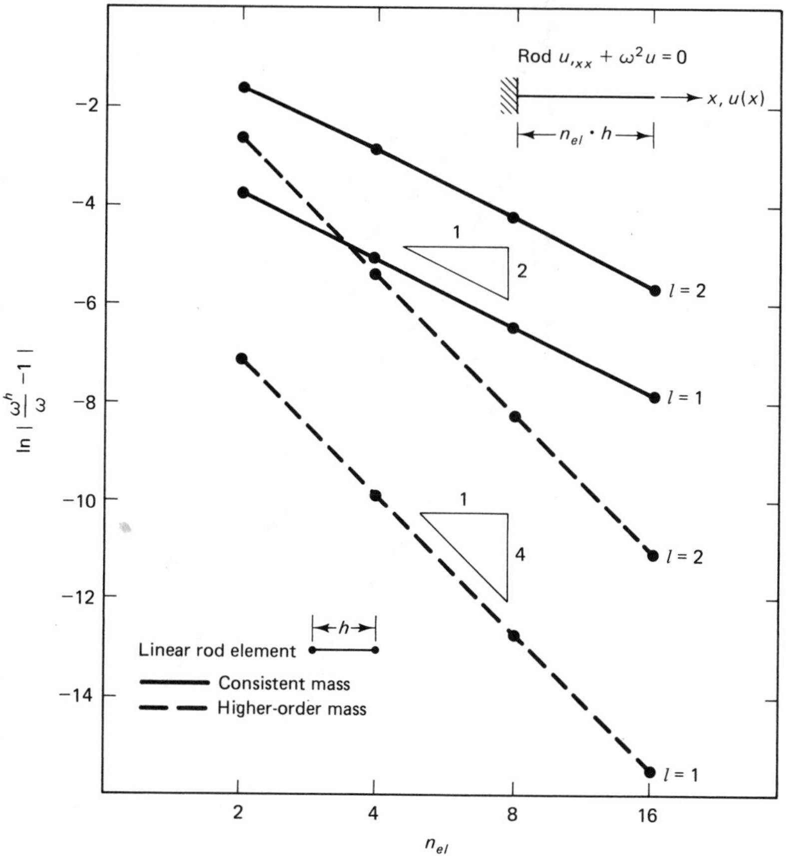

Figure 7.3.11 Convergence of the first two frequencies for linear rod elements with one end fixed and one end free [9].

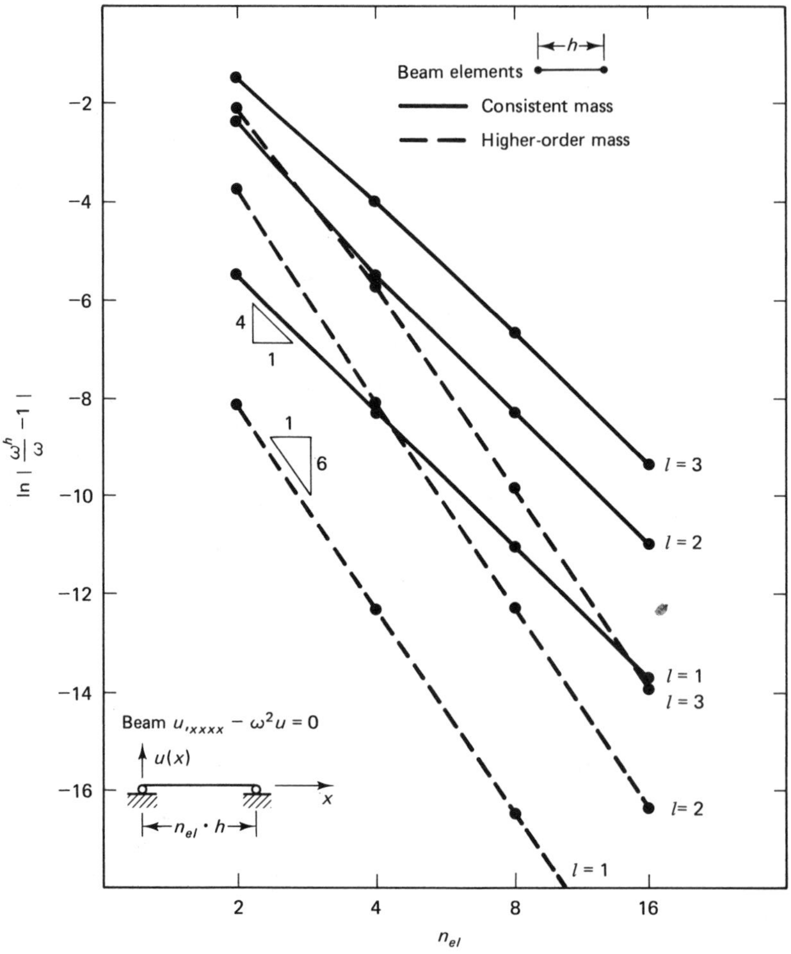

Figure 7.3.12 Convergence of the first three frequencies of a simply supported beam [9].

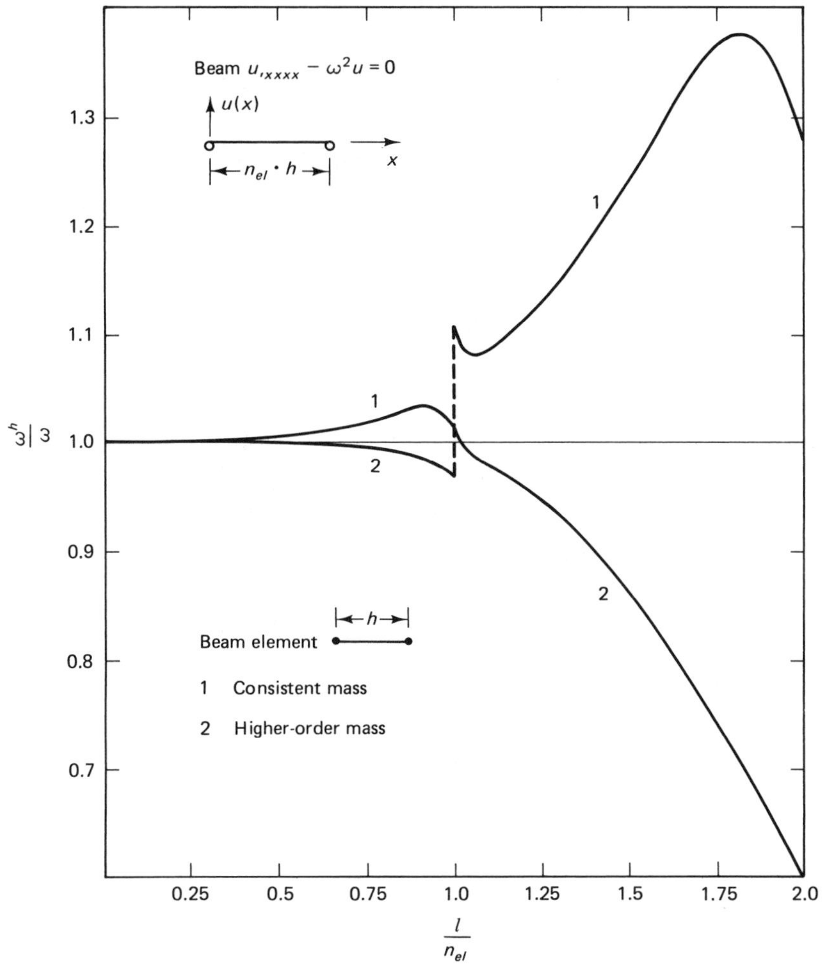

Figure 7.3.13 Frequency spectra for the cubic beam element [9].

7.3.3 Estimation of Eigenvalues

In time-dependent calculations it is often necessary to have a conservative estimate of the maximum eigenvalue (i.e., $\lambda^h_{n_{eq}}$). (We will return to this point in subsequent chapters.) It can be shown that [12,13]

$$\lambda^h_{n_{eq}} \leq \max_e \, (\lambda^e_{\max}) \tag{7.3.54}$$

where $\lambda^e_{\max}$ is the maximum eigenvalue of element e. The eigenvalues of element e satisfy[2]

$$(\boldsymbol{k}^e - \lambda^e \boldsymbol{m}^e)\boldsymbol{\psi}^e = \boldsymbol{0} \tag{7.3.55}$$

In the remainder of this subsection we will present an argument in support of (7.3.54).

Lemma

$$\lambda^h_{n_{eq}} = \max_{\substack{\boldsymbol{\phi} \in \mathbb{R}^{n_{eq}} \\ \boldsymbol{\phi} \neq 0}} \mathcal{R}(\boldsymbol{\phi}) \tag{7.3.56}$$

where $\mathcal{R}(\boldsymbol{\phi}) = \boldsymbol{\phi}^T\boldsymbol{K}\boldsymbol{\phi}/\boldsymbol{\phi}^T\boldsymbol{M}\boldsymbol{\phi}$ is the Rayleigh quotient, and the maximum occurs for $\boldsymbol{\phi} = c\boldsymbol{\psi}_{n_{eq}}$, where c is an arbitrary nonzero constant.

Proof. The eigenvectors $\boldsymbol{\psi}_1, \boldsymbol{\psi}_2, \ldots, \boldsymbol{\psi}_{n_{eq}}$ form a basis for $\mathbb{R}^{n_{eq}}$. Thus we can expand any vector, say $\boldsymbol{\phi}$, as

$$\boldsymbol{\phi} = \sum_{l=1}^{n_{eq}} c_l \, \boldsymbol{\psi}_l \tag{7.3.57}$$

and therefore

$$\boldsymbol{\phi}^T\boldsymbol{M}\boldsymbol{\phi} = \sum_{l=1}^{n_{eq}} |c_l|^2 \tag{7.3.58}$$

$$\boldsymbol{\phi}^T\boldsymbol{K}\boldsymbol{\phi} = \sum_{l=1}^{n_{eq}} \lambda^h_l |c_l|^2 \tag{7.3.59}$$

For $\boldsymbol{\phi} \neq \boldsymbol{0}$,

$$\mathcal{R}(\boldsymbol{\phi}) = \frac{\sum_{l=1}^{n_{eq}} \lambda^h_l |c_l|^2}{\sum_{l=1}^{n_{eq}} |c_l|^2} \leq \lambda^h_{n_{eq}} \tag{7.3.60}$$

by virtue of $\lambda^h_1 \leq \lambda^h_2 \leq \ldots \leq \lambda^h_{n_{eq}}$. Because $\mathcal{R}(c\boldsymbol{\psi}_{n_{eq}}) = \lambda^h_{n_{eq}}$, the proof of the lemma is complete. ■

[2]Solution of the element eigenvalue problem is actually rarely done. Whenever possible an exact formula is developed for $\lambda^e_{\max}$. Otherwise a conservative estimate is developed. This subject will be discussed further in Chapter 9.

Let $\boldsymbol{K}$ and $\boldsymbol{M}$ be global matrices formed by assembling the collections of matrices $\{\boldsymbol{k}^e\}_1^{n_{el}}$ and $\{\boldsymbol{m}^e\}_1^{n_{el}}$ according to some connectivity array (i.e., an "LM-array"; see Chapter 2). Consider the symmetric, block-diagonal matrices

$$\widetilde{\boldsymbol{K}} = \begin{bmatrix} \boldsymbol{k}^1 & \boldsymbol{0} & \cdots & \boldsymbol{0} \\ \boldsymbol{0} & \boldsymbol{k}^2 & \ddots & \vdots \\ \vdots & \ddots & \ddots & \boldsymbol{0} \\ \boldsymbol{0} & \cdots & \boldsymbol{0} & \boldsymbol{k}^{n_{el}} \end{bmatrix} \tag{7.3.61}$$

$$\widetilde{\boldsymbol{M}} = \begin{bmatrix} \boldsymbol{m}^1 & \boldsymbol{0} & \cdots & \boldsymbol{0} \\ \boldsymbol{0} & \boldsymbol{m}^2 & \ddots & \vdots \\ \vdots & \ddots & \ddots & \boldsymbol{0} \\ \boldsymbol{0} & \cdots & \boldsymbol{0} & \boldsymbol{m}^{n_{el}} \end{bmatrix} \tag{7.3.62}$$

The dimension of $\widetilde{\boldsymbol{K}}$ and $\widetilde{\boldsymbol{M}}$, say $\widetilde{n}_{eq}$, clearly satisfies $\widetilde{n}_{eq} \geq n_{eq}$.

The eigenvalue problems for the $\boldsymbol{K}$, $\boldsymbol{M}$, and $\widetilde{\boldsymbol{K}}$, $\widetilde{\boldsymbol{M}}$ systems are (respectively)

$$(\boldsymbol{K} - \lambda^h \boldsymbol{M})\boldsymbol{\psi} = \boldsymbol{0} \tag{7.3.63}$$

and

$$(\widetilde{\boldsymbol{K}} - \widetilde{\lambda}^h \widetilde{\boldsymbol{M}})\widetilde{\boldsymbol{\psi}} = \boldsymbol{0} \tag{7.3.64}$$

This lemma may also be applied to the eigenvalue problem (7.3.64). Thus

$$\widetilde{\lambda}^h_{\widetilde{n}_{eq}} = \max_{\substack{\widetilde{\boldsymbol{\phi}} \in \mathbb{R}^{\widetilde{n}_{eq}} \\ \widetilde{\boldsymbol{\phi}} \neq \boldsymbol{0}}} \frac{\widetilde{\boldsymbol{\phi}}^T \widetilde{\boldsymbol{K}} \widetilde{\boldsymbol{\phi}}}{\widetilde{\boldsymbol{\phi}}^T \widetilde{\boldsymbol{M}} \widetilde{\boldsymbol{\phi}}} \tag{7.3.65}$$

Let $\mathbb{R}_c^{\widetilde{n}_{eq}}$ denote the set of vectors $\widetilde{\boldsymbol{\phi}} \in \mathbb{R}^{\widetilde{n}_{eq}}$ such that components of $\widetilde{\boldsymbol{\phi}}$ that correspond to the same degree of freedom in $\mathbb{R}^{n_{eq}}$ under the identification of the connectivity array used in assembling $\boldsymbol{K}$ and $\boldsymbol{M}$ are to be equal. With this definition, it is clear that

$$\lambda^h_{n_{eq}} = \max_{\substack{\widetilde{\boldsymbol{\phi}} \in \mathbb{R}_c^{\widetilde{n}_{eq}} \\ \widetilde{\boldsymbol{\phi}} \neq \boldsymbol{0}}} \frac{\widetilde{\boldsymbol{\phi}}^T \widetilde{\boldsymbol{K}} \widetilde{\boldsymbol{\phi}}}{\widetilde{\boldsymbol{\phi}}^T \widetilde{\boldsymbol{M}} \widetilde{\boldsymbol{\phi}}} \leq \widetilde{\lambda}^h_{\widetilde{n}_{eq}} \tag{7.3.66}$$

Remarks

1. The equality in (7.3.66) may be obvious to some readers. If it is not, the following simple example should be helpful:

Let

$$\boldsymbol{k}^e = \begin{bmatrix} k_{11}^e & k_{12}^e \\ k_{21}^e & k_{22}^e \end{bmatrix}$$

$$\boldsymbol{m}^e = \begin{bmatrix} m_{11}^e & m_{12}^e \\ m_{21}^e & m_{22}^e \end{bmatrix}$$

$$e = 1, 2$$

$$\boldsymbol{K} = \begin{bmatrix} k_{11}^1 & k_{12}^1 & 0 \\ k_{21}^1 & (k_{22}^1 + k_{11}^2) & k_{12}^2 \\ 0 & k_{21}^2 & k_{22}^2 \end{bmatrix}$$

$$\boldsymbol{M} = \begin{bmatrix} m_{11}^1 & m_{12}^1 & 0 \\ m_{21}^1 & (m_{22}^1 + m_{11}^2) & m_{12}^2 \\ 0 & m_{21}^2 & m_{22}^2 \end{bmatrix}$$

$$\underbrace{\widetilde{\boldsymbol{K}}}_{4 \times 4} = \begin{bmatrix} \boldsymbol{k}^1 & \boldsymbol{0} \\ \boldsymbol{0} & \boldsymbol{k}^2 \end{bmatrix}$$

$$\underbrace{\widetilde{\boldsymbol{M}}}_{4 \times 4} = \begin{bmatrix} \boldsymbol{m}^1 & \boldsymbol{0} \\ \boldsymbol{0} & \boldsymbol{m}^2 \end{bmatrix}$$

$$\boldsymbol{\phi} = \begin{Bmatrix} \phi_1 \\ \phi_2 \\ \phi_3 \end{Bmatrix}$$

$$\widetilde{\boldsymbol{\phi}} = \begin{Bmatrix} \phi_1 \\ \phi_2 \\ \phi_2 \\ \phi_3 \end{Bmatrix} \in \mathbb{R}_c^4$$

The equality in (7.3.66) is established for the example under consideration by the following calculations:

$$\begin{aligned} \widetilde{\boldsymbol{\phi}}^T \widetilde{\boldsymbol{K}}\, \widetilde{\boldsymbol{\phi}} &= \begin{Bmatrix} \phi_1 \\ \phi_2 \end{Bmatrix}^T \boldsymbol{k}^1 \begin{Bmatrix} \phi_1 \\ \phi_2 \end{Bmatrix} + \begin{Bmatrix} \phi_2 \\ \phi_3 \end{Bmatrix}^T \boldsymbol{k}^2 \begin{Bmatrix} \phi_2 \\ \phi_3 \end{Bmatrix} \\ &= \boldsymbol{\phi}^T \boldsymbol{K} \boldsymbol{\phi} \end{aligned}$$

$$\begin{aligned} \widetilde{\boldsymbol{\phi}}^T \widetilde{\boldsymbol{M}}\, \widetilde{\boldsymbol{\phi}} &= \begin{Bmatrix} \phi_1 \\ \phi_2 \end{Bmatrix}^T \boldsymbol{m}^1 \begin{Bmatrix} \phi_1 \\ \phi_2 \end{Bmatrix} + \begin{Bmatrix} \phi_2 \\ \phi_3 \end{Bmatrix}^T \boldsymbol{m}^2 \begin{Bmatrix} \phi_2 \\ \phi_3 \end{Bmatrix} \\ &= \boldsymbol{\phi}^T \boldsymbol{M} \boldsymbol{\phi} \end{aligned}$$

2. The inequality in (7.3.66) follows directly from (7.3.65).

In terms of the element matrices, the eigenvalue problem (7.3.64) can be written as

$$\begin{bmatrix} \boldsymbol{k}^1 - \widetilde{\lambda}^h \boldsymbol{m}^1 & \boldsymbol{0} & \cdots & \boldsymbol{0} \\ \boldsymbol{0} & \boldsymbol{k}^2 - \widetilde{\lambda}^h \boldsymbol{m}^2 & & \vdots \\ \vdots & & \ddots & \boldsymbol{0} \\ \boldsymbol{0} & \cdots & \boldsymbol{0} & \boldsymbol{k}^{n_{el}} - \widetilde{\lambda}^h \boldsymbol{m}^{n_{el}} \end{bmatrix} \begin{Bmatrix} \widetilde{\boldsymbol{\psi}}^1 \\ \widetilde{\boldsymbol{\psi}}^2 \\ \vdots \\ \widetilde{\boldsymbol{\psi}}^{n_{el}} \end{Bmatrix} = \begin{Bmatrix} \boldsymbol{0} \\ \boldsymbol{0} \\ \vdots \\ \boldsymbol{0} \end{Bmatrix} \tag{7.3.67}$$

From this expression it can be seen that the eigenvalues of (7.3.64) are the same as the totality of eigenvalues of the uncoupled element problems

$$(\boldsymbol{k}^e - \lambda^e \boldsymbol{m}^e)\boldsymbol{\psi}^e = \boldsymbol{0}, \qquad e = 1, \ldots, n_{el} \tag{7.3.68}$$

Therefore

$$\widetilde{\lambda}^h_{\widetilde{n}_{eq}} = \max_e (\lambda^e_{\max}) \tag{7.3.69}$$

where

$$\lambda^e_{\max} = \max_{\substack{\boldsymbol{\psi}^e \in \mathbb{R}^{n_{ee}} \\ \boldsymbol{\psi}^e \neq 0}} \frac{\boldsymbol{\psi}^{e^T} \boldsymbol{k}^e \boldsymbol{\psi}^e}{\boldsymbol{\psi}^{e^T} \boldsymbol{m}^e \boldsymbol{\psi}^e}$$

Combining (7.3.66) with (7.3.69), we have that

$$\lambda^h_{n_{eq}} \leq \max_e (\lambda^e_{\max})$$

which completes the proof. ■

Appendix 7.I

Error Estimates for Semidiscrete Galerkin Approximations

In [14] the following error estimates are obtained (c denotes a constant).

Parabolic case

$$\| e(t) \|_0 \leq ch^{\mu}\Big[\| u(t) \|_{k+1} + \exp(-\lambda_1^h t) \| u(0) \|_{k+1} + \int_0^t \exp(\lambda_1^h(\tau - t)) \| \dot{u}(\tau) \|_{k+1} \, d\tau \Big] \tag{7.I.1}$$

where $e = u^h - u$ and $\mu = \min\{k + 1, 2(k + 1 - m)\}$

For each fixed t, the term in (7.I.1) in square brackets will be bounded: Thus this result establishes the convergence of $u^h(t)$ to $u(t)$ in the L_2-norm as $h \rightarrow 0$.

Hyperbolic case

Let $E(\boldsymbol{u}, \dot{\boldsymbol{u}}) = \frac{1}{2}[(\dot{\boldsymbol{u}}, \rho\dot{\boldsymbol{u}}) + a(\boldsymbol{u}, \boldsymbol{u})]$ denote the ***total energy***. The square root of E defines a norm on $\mathcal{V} \times L_2$ equivalent to the $H^m \times L_2$ norm. The main result is

$$E(\boldsymbol{e}(t), \dot{\boldsymbol{e}}(t))^{1/2} \leq c\Big\{ h^{\nu}\Big[\| \boldsymbol{u}(0) \|_{k+1} + \| \boldsymbol{u}(t) \|_{k+1} \Big]$$

$$+ h^{\mu}\left[\|\dot{\boldsymbol{u}}(0)\|_{k+1} + \|\dot{\boldsymbol{u}}(t)\|_{k+1} + \int_0^t \|\ddot{\boldsymbol{u}}(\tau)\|_{k+1}\,d\tau\right]\Bigg\} \tag{7.I.2}$$

where $\nu = k + 1 - m$ and $\mu = \min\{k + 1, 2(k + 1 - m)\}$. For each fixed t, the terms in square brackets are bounded. Thus $E(\boldsymbol{e}(t), \dot{\boldsymbol{e}}(t))^{1/2} \to 0$ as $h \to 0$, which in turn implies the convergence of $\boldsymbol{u}^h(t)$ to $\boldsymbol{u}(t)$ in H^m and $\dot{\boldsymbol{u}}^h(t)$ to $\dot{\boldsymbol{u}}(t)$ in L_2. Because $\nu \leq \mu$, the rate of convergence is ν. However, note that the integral in (7.I.2) is $O(t)$. Therefore, this estimate is only good for times no smaller than $O(h^{-m})$[3]. Otherwise, the error is $O(h^{\mu}t)$.

If you are interested in learning how to obtain results of this kind, consult [14].

Exercise 1. In the parabolic case, assume ℓ, q, and h are zero. Obtain the following ***growth/decay estimates***:

$$\|u(t)\|_0 \leq \exp(-\lambda_1 t)\|u(0)\|_0$$

$$\|u^h(t)\|_0 \leq \exp(-\lambda_1^h t)\|u^h(0)\|_0$$

Exercise 2. In the hyperbolic case, assume $\boldsymbol{\ell}$, $\boldsymbol{q}$, and $\boldsymbol{h}$ are zero. Establish the ***conservation of total energy***:

$$E(\boldsymbol{u}(t), \dot{\boldsymbol{u}}(t)) = E(\boldsymbol{u}(0), \dot{\boldsymbol{u}}(0))$$

$$E(\boldsymbol{u}^h(t), \dot{\boldsymbol{u}}^h(t)) = E(\boldsymbol{u}^h(0), \dot{\boldsymbol{u}}^h(0))$$

REFERENCES

Section 7.1

1. I. Stakgold, *Boundary-Value Problems of Mathematical Physics*, Vol. II. New York: Macmillan, 1968.

Section 7.3

2. G. Strang and G. J. Fix, *An Analysis of the Finite Element Method.* Englewood Cliffs, N.J.: Prentice-Hall, 1973.
3. J. S. Archer, "Consistent Mass Matrix for Distributed Systems," *Proceedings of ASCE*, 89, ST4 (1963) 161–178.
4. T. J. R. Hughes, H. M. Hilber, and R. L. Taylor, "A Reduction Scheme for Problems of Structural Dynamics," *International Journal of Solids and Structures*, 12 (1976), 749–767.

[3]This statement needs to be suitably nondimensionalized to be made precise. For example, consider the axial motion of an elastic rod (see (7.2.46) and (7.2.47)). In this case the ***bar wave velocity*** is $c = \sqrt{E/\rho}$, assumed constant. Note h/c has dimensions of time. Thus the rate of convergence ν is good up to times $\sim (h/c)^{-m}$.

5. I. Fried and D. S. Malkus, "Finite Element Mass Matrix Lumping by Numerical Integration Without Convergence Rate Loss," *International Journal of Solids and Structures*, 11 (1976) 461–466.

6. E. Hinton, T. Rock, and O. C. Zienkiewicz, "A Note on Mass Lumping and Related Processes in the Finite Element Method," *Earthquake Engineering and Structural Dynamics*, 4 (1976), 245–249.

7. P. Gresho, R. Lee, and R. Sani, "Advection-dominated Flows, with Emphasis on the Consequences of Mass Lumping," in *Finite Elements in Fluids*, Vol. 3, London: John Wiley, 1978, 335–350.

8. G. L. Goudreau, "Evaluation of Discrete Methods for the Linear Dynamic Response of Elastic and Viscoelastic Solids," UC SESM Report 69–15, University of California, Berkeley, June 1970.

9. H. M. Hilber, T. J. R. Hughes, and R. L. Taylor, "Mass Matrices for Improved Finite Element Eigenvalue Approximations," unpublished report, December 1975.

10. T. J. R. Hughes and T. E. Tezduyar, "Stability and Accuracy Analysis of Some Fully Discrete Algorithms for the One-Dimensional Second-Order Wave Equation," *Computers and Structures*, 19 (1984), 665–668.

11. R. L. Taylor and R. Iding, "Application of Extended Variational Principles to Finite Element Analysis," in *Variational Methods in Engineering*, Vol. I, Southampton University Press, 1973, 2.54–2.67.

12. B. M. Irons, "Applications of a Theorem on Eigenvalues to Finite Element Problems," (CR/132/70), University of Wales, Department of Civil Engineering, Swansea, U.K., 1970.

13. T. J. R. Hughes, K. S. Pister, and R. L. Taylor, "Implicit-Explicit Finite Elements in Nonlinear Transient Analysis," *Computer Methods in Applied Mechanics and Engineering*, 17/18 (1979), 159–182.

Appendix 7.I

14. G. Strang and G. J. Fix, *An Analysis of the Finite Element Method*, Englewood Cliffs, N.J.: Prentice-Hall, 1973.

8

Algorithms for Parabolic Problems

8.1 ONE-STEP ALGORITHMS FOR THE SEMIDISCRETE HEAT EQUATION: GENERALIZED TRAPEZOIDAL METHOD

Recall that the semidiscrete heat equation can be written as

$$\boldsymbol{M}\dot{\boldsymbol{d}} + \boldsymbol{K}\boldsymbol{d} = \boldsymbol{F} \tag{8.1.1}$$

where $\boldsymbol{M}$ is the capacity matrix, $\boldsymbol{K}$ is the conductivity matrix, $\boldsymbol{F}$ is the heat supply vector, $\boldsymbol{d}$ is the temperature vector, and $\dot{\boldsymbol{d}}$ is the time (t) derivative of $\boldsymbol{d}$. The matrices $\boldsymbol{M}$ and $\boldsymbol{K}$ are assumed symmetric, $\boldsymbol{M}$ is positive-definite, and $\boldsymbol{K}$ is positive-semidefinite (usually $\boldsymbol{K}$ is positive-definite too). The heat supply is a prescribed function of t. We write $\boldsymbol{F} = \boldsymbol{F}(t)$ for $t \in [0, T]$. Equation (8.1.1) is to be thought of as a coupled system of n_{eq} ordinary differential equations.

The initial-value problem consists of finding a function $\boldsymbol{d} = \boldsymbol{d}(t)$ satisfying (8.1.1), and the initial condition

$$\boldsymbol{d}(0) = \boldsymbol{d}_0 \tag{8.1.2}$$

where $\boldsymbol{d}_0$ is given.

Perhaps the most well known and commonly used algorithms for solving (8.1.1) are members of the ***generalized trapezoidal family of methods,*** which consists of the following equations:

$$\boldsymbol{M}\boldsymbol{v}_{n+1} + \boldsymbol{K}\boldsymbol{d}_{n+1} = \boldsymbol{F}_{n+1} \tag{8.1.3}$$

$$\boldsymbol{d}_{n+1} = \boldsymbol{d}_n + \Delta t\ \boldsymbol{v}_{n+\alpha} \tag{8.1.4}$$

$$\boldsymbol{v}_{n+\alpha} = (1 - \alpha)\boldsymbol{v}_n + \alpha\boldsymbol{v}_{n+1} \tag{8.1.5}$$

where $\boldsymbol{d}_n$ and $\boldsymbol{v}_n$ are the approximations to $\boldsymbol{d}(t_n)$ and $\dot{\boldsymbol{d}}(t_n)$, respectively; $\boldsymbol{F}_{n+1} = \boldsymbol{F}(t_{n+1})$; Δt is the time step, assumed constant for the time being; and α is a parameter, taken to be in the interval $[0, 1]$.

Some well-known members of the generalized trapezoidal family are identified below.

α	Method
0	Forward differences; forward Euler
$\frac{1}{2}$	Trapezoidal rule; midpoint rule; Crank-Nicolson
1	Backward differences; backward Euler

So that the reader has a basic appreciation of the computations entailed by the algorithm, we will present next a brief overview of implementational considerations.

Implementation 1: v-form

The computational problem is to determine $\boldsymbol{d}_{n+1}$ and $\boldsymbol{v}_{n+1}$ given $\boldsymbol{d}_n$ and $\boldsymbol{v}_n$. The procedure begins at $n = 0$ with $\boldsymbol{d}_0$ known. The initial value, $\boldsymbol{v}_0$, may be determined from the time-discrete heat equation evaluated at $t = 0$:

$$\boldsymbol{M}\boldsymbol{v}_0 = \boldsymbol{F}_0 - \boldsymbol{K}\boldsymbol{d}_0 \tag{8.1.6}$$

There are several ways of implementing the recursion relationship that takes us from step n to step $n + 1$. Let the ***predictor value*** of $\boldsymbol{d}_{n+1}$ be defined by

$$\tilde{\boldsymbol{d}}_{n+1} = \boldsymbol{d}_n + (1 - \alpha)\Delta t\ \boldsymbol{v}_n \tag{8.1.7}$$

Note that (8.1.4) and (8.1.5) can be combined and expressed in terms of (8.1.7):

$$\boldsymbol{d}_{n+1} = \tilde{\boldsymbol{d}}_{n+1} + \alpha\Delta t\ \boldsymbol{v}_{n+1} \tag{8.1.8}$$

Substituting this expression into (8.1.3) results in an equation that may be solved for $\boldsymbol{v}_{n+1}$:

$$(\boldsymbol{M} + \alpha\Delta t\ \boldsymbol{K})\boldsymbol{v}_{n+1} = \boldsymbol{F}_{n+1} - \boldsymbol{K}\tilde{\boldsymbol{d}}_{n+1} \tag{8.1.9}$$

Observe that the terms on the right-hand side of (8.1.9) are known. Once $\boldsymbol{v}_{n+1}$ is determined, (8.1.8) serves to define $\boldsymbol{d}_{n+1}$, and so on.

Remarks

1. In the case of $\alpha = 0$, the method is said to be ***explicit***. The attribute of explicit methods may be seen from (8.1.9) if $\boldsymbol{M}$ is assumed "lumped" (i.e., diagonal). In this case the solution may be advanced without the necessity of equation solving.

2. If $\alpha \neq 0$, the method is said to be ***implicit***. In these cases a system of equations, with coefficient matrix $(\boldsymbol{M} + \alpha\Delta t\, \boldsymbol{K})$, needs to be solved at each step to advance the solution. As long as Δt is constant, only one factorization is required.

3. The right-hand-side vector is formed one element at a time. Recall that the definition of $\boldsymbol{F}$ is given in this way (see (7.1.27) through (7.1.29)). The product

$$\boldsymbol{K}\tilde{\boldsymbol{d}}_{n+1} = \mathop{\mathbf{A}}_{e=1}^{n_{el}} (\boldsymbol{k}^e\, \tilde{\boldsymbol{d}}^e_{n+1}) \tag{8.1.10}$$

where $\tilde{\boldsymbol{d}}^e_{n+1}$ contains zeros in degrees of freedom that correspond to prescribed boundary conditions. The effect of the boundary-condition terms is already accounted for in the definition of $\boldsymbol{F}$ [see (7.1.29)]. This aspect of the implementation warrants further discussion. We will postpone the details until we consider a general implementation of a class of hyperbolic and hyperbolic-parabolic algorithms in Chapter 9. The general case can be specialized to the present circumstances.

4. It is useful to note that the implementation manifested by (8.1.9) can easily be generalized to so-called implicit-explicit methods, where part of $\boldsymbol{K}$ is treated implicitly and part explicitly. The *only* required change is to replace $\boldsymbol{K}$ on the *left-hand side* of (8.1.9) with the part to be treated implicitly, say $\boldsymbol{K}^I$. This could be the assembly of a subset of elements, which leads to ***implicit-explicit element mesh partitions*** (see Hughes et al. [1–3]). It may be observed that the implementation is trivial. We shall put off a more detailed discussion until Chapter 9.

Implementation 2: d-form

Another possibility is to eliminate $\boldsymbol{v}_{n+1}$ from (8.1.3) via (8.1.8). Thus in place of (8.1.9) we have

$$\frac{1}{\alpha\Delta t}(\boldsymbol{M} + \alpha\Delta t\, \boldsymbol{K})\boldsymbol{d}_{n+1} = \boldsymbol{F}_{n+1} + \frac{1}{\alpha\Delta t}\boldsymbol{M}\tilde{\boldsymbol{d}}_{n+1} \tag{8.1.11}$$

In this implementation, (8.1.11) is used to determine $\boldsymbol{d}_{n+1}$ and then $\boldsymbol{v}_{n+1}$ may be determined from (8.1.8), i.e.,

$$\boldsymbol{v}_{n+1} = \frac{\boldsymbol{d}_{n+1} - \tilde{\boldsymbol{d}}_{n+1}}{\alpha\Delta t} \tag{8.1.12}$$

The advantage of this implementation occurs when $\boldsymbol{M}$ is diagonal. In this case the calculation of the right-hand side of (8.1.11) may be performed much more economically than the right-hand side of (8.1.9). The equation-solving burden is of course the same for (8.1.9) and (8.1.11).

Remark

Note that if $\alpha \Delta t \to \infty$ in (8.1.11), the equilibrium, or steady-state solution (i.e., one for which $\dot{\boldsymbol{d}} = \boldsymbol{0}$) is approached, viz.,

$$\boldsymbol{K}\boldsymbol{d}_{n+1} \to \boldsymbol{F}_{n+1} \tag{8.1.13}$$

This fact can be exploited in a transient analysis computer program if the equilibrium solution is desired. Just select a value of $\alpha \Delta t$ large enough so that (8.1.13) holds to desired precision.

Exercise 1. Derive an implementation in which the $\boldsymbol{v}_n$'s are unnecessary. This results in a saving of computer storage. [*Answer*: $(\boldsymbol{M} + \alpha \Delta t\, \boldsymbol{K})\boldsymbol{d}_{n+1} = (\boldsymbol{M} - (1 - \alpha)\Delta t\, \boldsymbol{K})\boldsymbol{d}_n + \Delta t(\alpha \boldsymbol{F}_{n+1} + (1 - \alpha)\boldsymbol{F}_n)$.]

8.2 ANALYSIS OF THE GENERALIZED TRAPEZOIDAL METHOD

The primary requirement of the algorithms given in the last section is that they converge. We shall call an algorithm ***convergent*** if for t_n fixed and $\Delta t = t_n/n$, $\boldsymbol{d}_n \to \boldsymbol{d}(t_n)$ as $\Delta t \to 0$. To establish the convergence of an algorithm, two additional notions must be considered: ***stability*** and ***consistency***. We shall show later on that once stability and consistency are verified, convergence is automatic. In addition, we shall be concerned with the ***accuracy*** of an algorithm, i.e., the rate of convergence as $\Delta t \to 0$, and allied topics such as the behavior of the (spurious) higher modes of the semidiscrete system. There are several techniques that can be employed to study the characteristics of an algorithm. In the present context the most revealing approach appears to be the "modal approach" (sometimes called spectral, or Fourier, analysis) in which the problem is decomposed into n_{eq} uncoupled scalar equations. It can be rigorously established that the behavior of the entire coupled system reduces to consideration of the individual modal equations that comprise it. Our first step in analyzing the family of algorithms introduced in Sec. 8.1 will be to perform the reduction to single-degree-of-freedom (SDOF) form.

8.2.1 Modal Reduction to SDOF Form

The essential property used in reducing to SDOF form is the orthogonality of the eigenvectors of the associated eigenvalue problem. Recall that

$$(\boldsymbol{K} - \lambda_l^h \boldsymbol{M})\boldsymbol{\psi}_l = \boldsymbol{0}, \qquad l \in \{1, 2, \ldots, n_{eq}\} \tag{8.2.1}$$

where

$$0 \leq \lambda_1^h \leq \lambda_2^h \leq \cdots \leq \lambda_{n_{eq}}^h \tag{8.2.2}$$

and

$$\boldsymbol{\psi}_l^T \boldsymbol{M}\, \boldsymbol{\psi}_m = \delta_{lm} \qquad \text{(orthonormality)} \tag{8.2.3}$$

where δ_{lm} is the Kronecker delta. Furthermore, the eigenvectors $\{\boldsymbol{\psi}_l\}_1^{n_{eq}}$ constitute a ***basis*** for $\mathbb{R}^{n_{eq}}$, meaning that any element in $\mathbb{R}^{n_{eq}}$ can be written as a linear combination of the $\boldsymbol{\psi}_l$'s. From the orthonormality property, it immediately follows that

$$\boldsymbol{\psi}_l^T \boldsymbol{K} \boldsymbol{\psi}_m = \lambda_l^h \delta_{lm} \qquad \text{(no sum)} \tag{8.2.4}$$

These properties will be used to decompose *both* the semidiscrete heat equation and the generalized trapezoidal algorithm.

Semidiscrete heat equation. Let

$$\boldsymbol{d}(t) = \sum_{m=1}^{n_{eq}} d_{(m)}(t)\boldsymbol{\psi}_m \tag{8.2.5}$$

from which it follows that

$$\dot{\boldsymbol{d}}(t) = \sum_{m=1}^{n_{eq}} \dot{d}_{(m)}(t)\boldsymbol{\psi}_m \tag{8.2.6}$$

The scalar-valued functions $d_{(m)}(t)$ (***Fourier coefficients***) are obtained by premultiplying (8.2.5) by $\boldsymbol{\psi}_l^T\boldsymbol{M}$ and invoking the orthonormality property. The result is

$$d_{(l)}(t) = \boldsymbol{\psi}_l^T \boldsymbol{M}\boldsymbol{d}(t) \tag{8.2.7}$$

(We have included the subscript in parentheses to avoid notational confusion.) The coefficients in (8.2.6) are obtained by differentiating (8.2.7):

$$\dot{d}_{(l)}(t) = \boldsymbol{\psi}_l^T \boldsymbol{M}\dot{\boldsymbol{d}}(t) \tag{8.2.8}$$

Employing (8.2.5) and (8.2.6) in (8.1.1) and premultiplying by $\boldsymbol{\psi}_l^T$ yields

$$\sum_{m=1}^{n_{eq}} (\dot{d}_{(m)}\boldsymbol{\psi}_l^T\boldsymbol{M}\boldsymbol{\psi}_m + d_{(m)}\boldsymbol{\psi}_l^T\boldsymbol{K}\boldsymbol{\psi}_m) = \boldsymbol{\psi}_l^T\boldsymbol{F} \tag{8.2.9}$$

from which it follows that

$$\dot{d}_{(l)} + \lambda_l^h d_{(l)} = F_{(l)} \tag{8.2.10}$$

the lth-modal equation, where $F_{(l)}(t) = \boldsymbol{\psi}_l^T\boldsymbol{F}(t)$.

The initial condition for (8.2.10) is obtained by premultiplying (8.1.2) by $\boldsymbol{\psi}_l^T\boldsymbol{M}$:

$$\begin{aligned} d_{(l)}(0) &= \boldsymbol{\psi}_l^T\boldsymbol{M}\,\boldsymbol{d}(0) \\ &= \boldsymbol{\psi}_l^T\boldsymbol{M}\,\boldsymbol{d}_0 \\ &\overset{\text{def}}{=} d_{0(l)} \end{aligned} \tag{8.2.11}$$

Solution of (8.2.10) for each mode allows us to construct the solution $\boldsymbol{d}$ of the original problem via (8.2.5).

To simplify the subsequent writing further we shall omit the lth -modal subscript in (8.2.10) and (8.2.11). Thus our typical modal initial-value problem consists of the following equations:

$$\left.\begin{aligned} \dot{d} + \lambda^h d &= F, \qquad t \in [0, T] \\ d(0) &= d_0 \end{aligned}\right\} \quad \textbf{\textit{(SDOF model problem)}} \qquad \begin{aligned} (8.2.12) \\ (8.2.13) \end{aligned}$$

Generalized trapezoidal algorithm. The procedure for decomposing the generalized trapezoidal algorithm is similar to that for the semidiscrete equation. The main results are collected here:

$$\boldsymbol{d}_n = \sum_{m=1}^{n_{eq}} d_{n(m)}\boldsymbol{\psi}_m \tag{8.2.14}$$

$$\boldsymbol{d}_{n+1} = \sum_{m=1}^{n_{eq}} d_{n+1(m)}\boldsymbol{\psi}_m \tag{8.2.15}$$

$$d_{n(l)} = \boldsymbol{\psi}_l^T \boldsymbol{M}\boldsymbol{d}_n \tag{8.2.16}$$

$$d_{n+1(l)} = \boldsymbol{\psi}_l^T \boldsymbol{M}\boldsymbol{d}_{n+1} \tag{8.2.17}$$

$$\sum_{m=1}^{n_{eq}} [d_{n+1(m)}\boldsymbol{\psi}_l^T(\boldsymbol{M} + \alpha\Delta t\,\boldsymbol{K})\boldsymbol{\psi}_m - d_{n(m)}\boldsymbol{\psi}_l^T(\boldsymbol{M} - (1-\alpha)\Delta t\boldsymbol{K})\boldsymbol{\psi}_m] = \Delta t\,\boldsymbol{\psi}_l^T\boldsymbol{F}_{n+\alpha} \tag{8.2.18}$$

$$\boldsymbol{F}_{n+\alpha} = (1-\alpha)\boldsymbol{F}_n + \alpha\boldsymbol{F}_{n+1} \tag{8.2.19}$$

$$(1 + \alpha\Delta t\,\lambda_l^h)d_{n+1(l)} = (1 - (1-\alpha)\Delta t\,\lambda_l^h)d_{n(l)} + \Delta t\,F_{n+\alpha(l)} \tag{8.2.20}$$

The initial value, $d_{0(l)}$, is defined by (8.2.11). As before, we will omit the lth-modal subscript in (8.2.20). Thus we may write

$$\left.\begin{aligned} (1 + \alpha\Delta t\,\lambda^h)d_{n+1} &= (1 - (1-\alpha)\Delta t\,\lambda^h)d_n + \Delta t\,F_{n+\alpha} \\ d_0 &\text{ given} \end{aligned}\right\} \quad \textit{(temporally discretized SDOF model problem)} \tag{8.2.21}$$

Remarks

1. The convergence of $\boldsymbol{d}_n$ to $\boldsymbol{d}(t_n)$ can be established by showing the Fourier coefficients converge [i.e., $d_{n(l)}$ converges to $d_{(l)}(t_n)$ for each $l \in \{1, 2, \ldots, n_{eq}\}$]. The proof of this goes as follows. Let $\boldsymbol{e}(t_n) = \boldsymbol{d}_n - \boldsymbol{d}(t_n)$ denote the error in $\boldsymbol{d}_n$, and let $e_{(l)}(t_n) = d_{n(l)} - d_{(l)}(t_n)$ denote the lth-Fourier component of $\boldsymbol{e}(t_n)$. Then

$$\begin{aligned} \boldsymbol{e}(t_n)^T\boldsymbol{M}\boldsymbol{e}(t_n) &= \sum_{l,m=1}^{n_{eq}} (e_{(l)}(t_n)\boldsymbol{\psi}_l)^T\boldsymbol{M}(e_{(m)}(t_n)\boldsymbol{\psi}_m) \\ &= \sum_{l,m=2}^{n_{eq}} e_{(l)}(t_n)e_{(m)}(t_n)\boldsymbol{\psi}_l^T\boldsymbol{M}\boldsymbol{\psi}_m \\ &= \sum_{l,m=1}^{n_{eq}} e_{(l)}(t_n)e_{(m)}(t_n)\delta_{lm} \qquad \text{(orthonormality)} \\ &= \sum_{l=1}^{n_{eq}} (e_{(l)}(t_n))^2 \end{aligned} \tag{8.2.22}$$

and so $\boldsymbol{e}(t_n)^T \boldsymbol{M} \boldsymbol{e}(t_n) \rightarrow 0$ if and only if $e_{(l)}(t_n) \rightarrow 0$ for each $l \in \{1, 2, \ldots . n_{eq}\}$. Because $\boldsymbol{M}$ is assumed positive-definite, $\boldsymbol{e}(t_n)^T \boldsymbol{M} \boldsymbol{e}(t_n) \rightarrow 0$ if and only if $\boldsymbol{e}(t_n) \rightarrow \boldsymbol{0}$. Consequently, we need consider only the SDOF problems in subsequent discussion.

2. Note that directly applying the generalized trapezoidal method to (8.2.12) also results in (8.2.20). This fact is depicted in the following commutative diagram:

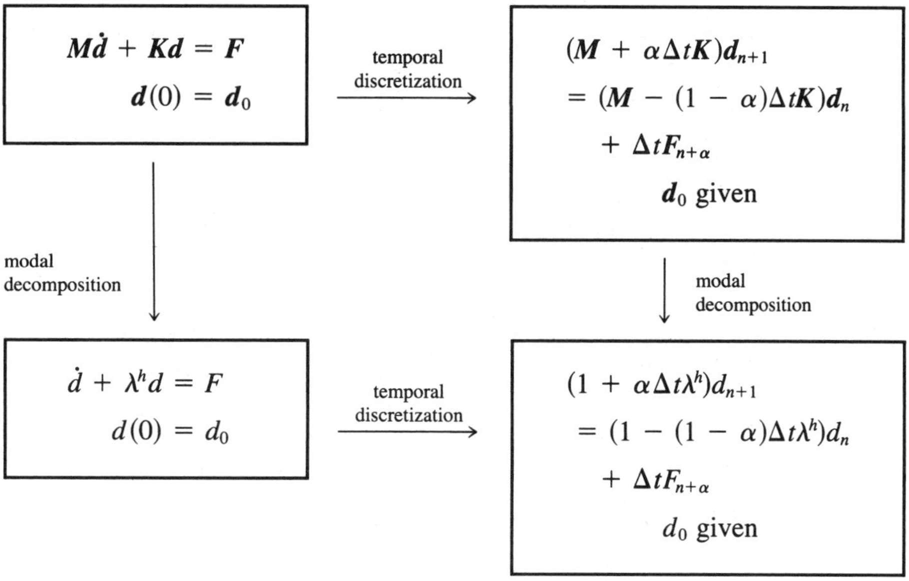

8.2.2 Stability

To motivate the appropriate notion of stability for the case under consideration, we shall investigate the behavior of the homogeneous model equation

$$\dot{d} + \lambda^h d = 0 \tag{8.2.23}$$

This is a first-order ordinary differential equation, which can be easily solved. The solution at time t_{n+1} for initial value $d(t_n)$, $t_{n+1} > t_n$, is

$$d(t_{n+1}) = \exp\left(-\lambda^h (t_{n+1} - t_n)\right) d(t_n) \tag{8.2.24}$$

from which it follows that

$$\left.\begin{aligned} |d(t_{n+1})| &< |d(t_n)|, & \lambda^h &> 0 \\ d(t_{n+1}) &= d(t_n), & \lambda^h &= 0 \end{aligned}\right\} \tag{8.2.25}$$

The conditions in (8.2.25) are what we wish to mimic in the temporally discrete case.

The homogeneous temporally discrete model equation is

$$(1 + \alpha\Delta t\ \lambda^h)d_{n+1} = (1 - (1 - \alpha)\Delta t\ \lambda^h)d_n \tag{8.2.26}$$

Noting that $(1 + \alpha\Delta t\ \lambda^h) > 0$ for all allowable values of the parameters, we can write (8.2.26) as

$$d_{n+1} = Ad_n \tag{8.2.27}$$

where $A = (1 - (1 - \alpha)\Delta t\ \lambda^h)/(1 + \alpha\Delta t\ \lambda^h)$ is the ***amplification factor.***

Our stability requirements will be that

$$\left.\begin{aligned} |d_{n+1}| < |d_n|, &\qquad \lambda^h > 0 \\ d_{n+1} = d_n, &\qquad \lambda^h = 0 \end{aligned}\right\} \tag{8.2.28}$$

From the definition of A, the second of (8.2.28) is automatic. The first condition is equivalent to insisting that

$$|A| < 1 \tag{8.2.29}$$

for $\lambda^h > 0$. To determine the restrictions imposed upon α, Δt, and λ^h, it is convenient to rewrite (8.2.29) as $-1 < A < 1$, viz.,

$$\boxed{-1 < \frac{(1 - (1 - \alpha)\Delta t\ \lambda^h)}{(1 + \alpha\Delta t\ \lambda^h)} < 1} \tag{8.2.30}$$

The right-hand inequality is satisfied for all allowable values of the parameters, and the left-hand inequality is satisfied whenever $\alpha \geq \frac{1}{2}$. However, if $\alpha < \frac{1}{2}$, the left-hand inequality requires $\lambda^h\Delta t < 2/(1 - 2\alpha)$. For a given λ^h this imposes an upper bound on the size of the allowable time step. The greater the λ^h, the smaller the time step required.

Remark

An algorithm for which stability imposes a time step restriction is called ***conditionally stable.*** An algorithm for which there is no time step restriction imposed by stability is called ***unconditionally stable.***

The significance of the stability concept introduced may be seen from the following example:

Example

Note that the solution of (8.2.27) may be written

$$d_n = A^n d_0 \tag{8.2.31}$$

If d_n is to behave like the solution of (8.2.23) (i.e., decay), A^n must converge (i.e., as $n \rightarrow \infty$, $A^n \rightarrow 0$). If $|A| < 1$, clearly this will be the case. On the other hand, if

$|A| > 1$, growth will occur. Even if the value of $|A|$ is only slightly larger than 1, disastrous growth can occur. To see what can happen, consider the numerical data in Table 8.2.1. The necessity of keeping $|A| < 1$ should be clearly apparent, or else virtually unbounded errors will enter a computation.

TABLE 8.2.1 A^n for Various Values of A and n

A \ n	100	1000
.99	.37	4.32×10^{-5}
1.01	2.70	2.09×10^{4}
.9	2.66×10^{-5}	1.75×10^{-46}
1.1	1.39×10^{4}	2.47×10^{41}

Remarks

1. In the conditionally stable case, the stability condition $\Delta t < 2/[(1 - 2\alpha)\lambda^h]$ must hold for all modes (i.e., all λ_l^h, $l \in \{1, 2, \ldots, n_{eq}\}$) in the system. The greatest λ_l^h, namely, $\lambda_{n_{eq}}^h$, imposes the most stringent restriction upon the time step (i.e., $\Delta t < 2/[(1 - 2\alpha)\lambda_{n_{eq}}^h]$). In fact, for the heat conduction problem, it can be shown that $\lambda_{n_{eq}}^h = O(h^{-2})$, where h is the mesh parameter, and thus the critical time step must satisfy $\Delta t <$ constant $\cdot\, h^2$. In a large system of equations (i.e., $n_{eq} \gg 1$), this condition is a severe constraint; thus unconditionally stable algorithms are generally preferred.

2. In Fig. 8.2.1 the behavior of A as a function of $\lambda^h \Delta t$ is depicted for several values of α. The value of $\lambda^h \Delta t$ for which $A = 0$ is called the ***oscillation limit*** because for greater values, the sign of A^n changes from step to step. For the unconditionally stable algorithms (i.e., ones for which $\alpha \geq \frac{1}{2}$), the asymptotic value of the amplification factor satisfies $|A_\infty| \leq 1$. Thus for all fixed $\lambda^h \Delta t$, $|A| < 1$ and all spurious high modal components decay. However, for $\alpha = \frac{1}{2}$ (or very near $\frac{1}{2}$) and $\lambda^h \Delta t \gg 1$, $A \cong -1$, and thus high modal components will behave like $(-1)^n$. This "sawtooth" pattern in time manifests itself frequently in computations. In reporting data for a calculation in which this is the case, these spurious higher modes may be filtered out by reporting the step-to-step averages, $(\boldsymbol{d}_{n+1} + \boldsymbol{d}_n)/2$, because $(-1)^{n+1} + (-1)^n = 0$.

Summary: Stability for the generalized trapezoidal methods

Amplification factor: $$A = \frac{1 - (1 - \alpha)\Delta t \lambda^h}{1 + \alpha \Delta t \lambda^h}$$

Stability requirement: $|A| < 1$ for $\lambda^h = \lambda_{n_{eq}}^h$ (= maximum eigenvalue)

Unconditional stability: $\alpha \geq \frac{1}{2}$

Conditional stability: $\alpha < \frac{1}{2}$, $$\Delta t < \frac{2}{(1 - 2\alpha)\lambda_{n_{eq}}^h}$$

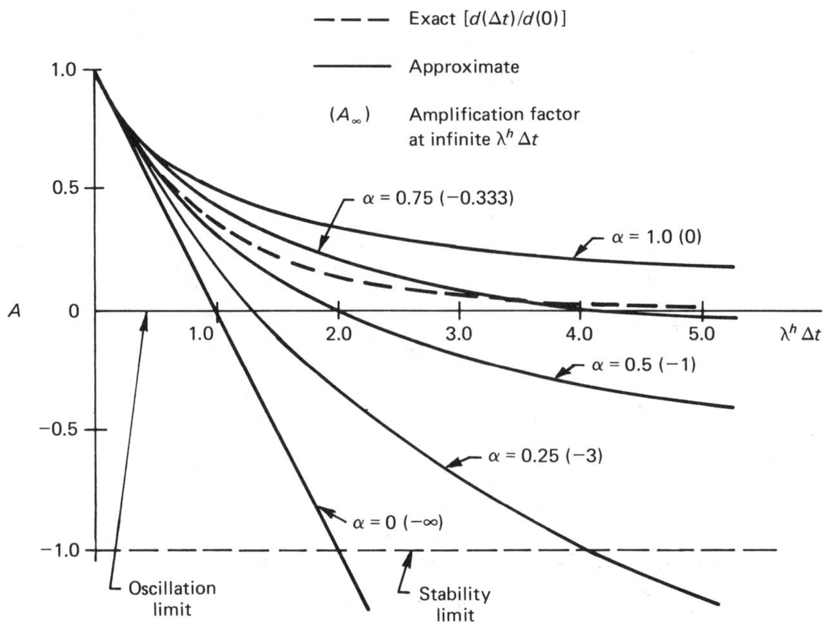

Figure 8.2.1 Amplification factor for typical one-step methods.

8.2.3 Convergence

The temporally discrete model problem may be written in the form

$$d_{n+1} - Ad_n - L_n = 0 \tag{8.2.32}$$

where the ***load*** $L_n = \Delta t F_{n+\alpha}/(1 + \alpha \Delta t \lambda^h)$. If we replace d_n and d_{n+1} in the left-hand side of (8.2.32) by the corresponding exact values, we obtain an expression of the form

$$d(t_{n+1}) - Ad(t_n) - L_n = \Delta t \cdot \tau(t_n) \tag{8.2.33}$$

where $\tau(t_n)$ is called the ***local truncation error***. If $|\tau(t)| \leq c\Delta t^k$, for all $t \in [0, T]$, where c is a constant independent of Δt, and $k > 0$, the algorithm defined by (8.2.32) is called ***consistent***; k is called the ***order of accuracy*** or ***rate of convergence***.

Proposition. The generalized trapezoidal methods are consistent, and furthermore $k = 1$ for all $\alpha \in [0, 1]$, except $\alpha = \frac{1}{2}$, in which case $k = 2$.

Proof. Expand $d(t_{n+1})$ and $d(t_n)$ about $t_{n+\alpha}$ in finite Taylor expansions and use the model equation to eliminate t-derivatives of $d(t_{n+\alpha})$:

$$d(t_{n+1}) = d(t_{n+\alpha}) + (1 - \alpha)\Delta t \dot{d}(t_{n+\alpha}) + \frac{((1 - \alpha)\Delta t)^2}{2}\ddot{d}(t_{n+\alpha}) + \frac{((1 - \alpha)\Delta t)^3}{3!}\dddot{d}(t_{n+\alpha}) + O(\Delta t^4)$$

$$d(t_n) = d(t_{n+\alpha}) + (-\alpha\Delta t)\,\dot{d}(t_{n+\alpha}) + \frac{(-\alpha\Delta t)^2}{2}\ddot{d}(t_{n+\alpha}) + \frac{(-\alpha\Delta t)^3}{3!}\dddot{d}(t_{n+\alpha}) + O(\Delta t^4)$$

$$\begin{aligned}
\Delta t(1 + \alpha\Delta t\,\lambda^h)\,\tau(t_n)
&= (1 + \alpha\Delta t\,\lambda^h)\,d(t_{n+1}) - (1 - (1 - \alpha)\Delta t\,\lambda^h)\,d(t_n) - \Delta t F_{n+\alpha} \\
&= \{(1 + \alpha\Delta t\,\lambda^h) - (1 - (1 - \alpha)\Delta t\,\lambda^h)\}d(t_{n+\alpha}) \\
&\quad + \{(1 + \alpha\Delta t\,\lambda^h)(1 - \alpha)\Delta t - (1 - (1 - \alpha)\Delta t\,\lambda^h)(-\alpha\Delta t)\}\dot{d}(t_{n+\alpha}) \\
&\quad + \left\{(1 + \alpha\Delta t\,\lambda^h)\frac{((1 - \alpha)\Delta t)^2}{2} - (1 - (1 - \alpha)\Delta t\,\lambda^h)\frac{(-\alpha\Delta t)^2}{2}\right\}\ddot{d}(t_{n+\alpha}) \\
&\quad + \left\{(1 + \alpha\Delta t\,\lambda^h)\frac{((1 - \alpha)\Delta t)^3}{3!} - (1 - (1 - \alpha)\Delta t\,\lambda^h)\frac{(-\alpha\Delta t)^3}{3!}\right\}\dddot{d}(t_{n+\alpha}) \\
&\quad - \Delta t\Bigg([\alpha + (1 - \alpha)]F(t_{n+\alpha}) + [\alpha(1 - \alpha)\Delta t + (1 - \alpha)(-\alpha\Delta t)]\dot{F}(t_{n+\alpha}) \\
&\quad + \left\{\alpha\frac{((1 - \alpha)\Delta t)^2}{2} + (1 - \alpha)\frac{(-\alpha\Delta t)^2}{2}\right\}\ddot{F}(t_{n+\alpha}) \\
&\quad + \left\{\alpha\frac{((1 - \alpha)\Delta t)}{3!} + (1 - \alpha)\frac{(-\alpha\Delta t)^3}{3!}\right\}\dddot{F}(t_{n+\alpha})\Bigg) + O(\Delta t^4)
\end{aligned}$$

Using $\dot{d} + \lambda^h d = F$ and time derivatives of same, it can be shown that

$$\tau = (1 - 2\alpha)O(\Delta t^1) + O(\Delta t^2)$$

(Fill in the remaining details as an exercise.) ■

Remark

Thus the trapezoidal rule ($\alpha = \frac{1}{2}$) is the only member of the family of methods that is second-order accurate.

Theorem. Consider equations (8.2.32) and (8.2.33). Let t_n be fixed (n, and consequently Δt, are allowed to vary), and assume the following conditions hold:

i. $|A| \leq 1$ (stability)

ii. $|\tau(t)| \leq c\Delta t^k$, $t \in [0, T]$, $k > 0$ (consistency)

Then $e(t_n) \to 0$ as $\Delta t \to 0$.

Proof. Subtract (8.2.33) from (8.2.32):

$$\boxed{e(t_{n+1}) = Ae(t_n) - \Delta t \cdot \tau(t_n) \qquad \textbf{\textit{error equation}}} \tag{8.2.34}$$

Replace $e(t_n)$ on the right-hand side by the error equation for the previous step, i.e.,

$$e(t_n) = Ae(t_{n-1}) - \Delta t \cdot \tau(t_{n-1})$$

to obtain

$$e(t_{n+1}) = A^2 e(t_{n-1}) - \Delta t A \tau(t_{n-1}) - \Delta t \cdot \tau(t_n)$$

Now repeat this procedure to eliminate $e(t_{n-1})$ from the right-hand side, viz.,

$$e(t_{n+1}) = A^3 e(t_{n-2}) - \Delta t A^2 \tau(t_{n-2}) - \Delta t A \tau(t_{n-1}) - \Delta t \cdot \tau(t_n)$$

and so on. The final result is

$$\boxed{e(t_{n+1}) = A^{n+1} e(0) - \Delta t \sum_{i=0}^{n} A^i \tau(t_{n-i})}$$

The first term on the right vanishes since $e(0) = d_0 - d(0) = 0$. Taking absolute values evaluated at t_n instead of t_{n+1} and using some elementary facts:

$$\begin{aligned}
|e(t_n)| &= \Delta t \left| \sum_{i=0}^{n-1} A^i \tau(t_{n-1-i}) \right| \\
&\leq \Delta t \sum_{i=0}^{n-1} |A|^i \, |\tau(t_{n-1-i})| \\
&\leq \Delta t \sum_{i=0}^{n-1} |\tau(t_{n-1-i})| && \text{(stability)} \\
&\leq t_n \max |\tau(t)| && t \in [0, T] \\
&\leq t_n c \, \Delta t^k && \text{(consistency)}
\end{aligned}$$

Therefore $e(t_n) \to 0$ as $\Delta t \to 0$, and furthermore the rate of convergence is k (i.e., $e(t_n) = O(\Delta t^k)$). ■

Remarks

1. This result establishes the convergence of the generalized trapezoidal methods. The maximal rate of convergence is 2 and is attained by the trapezoidal rule.

2. This theorem is a particular example of perhaps the most celebrated theorem in numerical analysis, the Lax equivalence theorem, which may be stated as "consistency plus stability is necessary and sufficient for convergence."

Exercise 1. Suppose that in (8.2.32) $F_{n+\alpha}$ is replaced by an approximation, $F_{n+\alpha}^{\Delta t}$, which satisfies $|F_{n+\alpha} - F^{\Delta t}_{n+\alpha}| \leq \text{constant} \cdot \Delta t^q$. What condition must q satisfy to retain kth-order accuracy?

8.2.4 An Alternative Approach to Stability: The Energy Method

An alternative approach to stability is the energy method [4]. In the present circumstances the results are identical to those obtained by the modal approach. However, the energy method is often applicable to situations in which the modal approach is unsuccessful. We shall illustrate the energy method on the generalized trapezoidal family of algorithms.

It simplifies the analysis if the following discrete operators are introduced:

$$[\boldsymbol{d}_n] = \boldsymbol{d}_{n+1} - \boldsymbol{d}_n \qquad \text{(undivided forward-difference operator)} \qquad (8.2.35)$$

$$\langle \boldsymbol{d}_n \rangle = \frac{\boldsymbol{d}_{n+1} + \boldsymbol{d}_n}{2} \qquad \text{(mean-value operator)} \qquad (8.2.36)$$

Note that

$$\langle \boldsymbol{d}_n \rangle^T \boldsymbol{M} [\boldsymbol{d}_n] = \frac{1}{2}[\boldsymbol{d}_n^T \boldsymbol{M} \boldsymbol{d}_n] \qquad (8.2.37)$$

Equation (8.1.4) may be written as

$$\begin{aligned} [\boldsymbol{d}_n] &= \Delta t \, \boldsymbol{v}_{n+\alpha} \\ &= \Delta t(\langle \boldsymbol{v}_n \rangle + (\alpha - \tfrac{1}{2})[\boldsymbol{v}_n]) \end{aligned} \qquad (8.2.38)$$

By combining the temporally discrete heat equations at steps n and $n + 1$, we arrive at

$$\boldsymbol{M}\boldsymbol{v}_{n+\alpha} + \boldsymbol{K}\boldsymbol{d}_{n+\alpha} = \boldsymbol{F}_{n+\alpha} \qquad (8.2.39)$$

This may also be written as

$$\frac{1}{\Delta t}\boldsymbol{M}[\boldsymbol{d}_n] + \boldsymbol{K}\boldsymbol{d}_{n+\alpha} = \boldsymbol{F}_{n+\alpha} \qquad (8.2.40)$$

Premultiplying (8.2.40) by $\langle \boldsymbol{d}_n \rangle^T$ leads to

$$\begin{aligned} \frac{1}{2\Delta t}[\boldsymbol{d}_n^T \boldsymbol{M} \boldsymbol{d}_n] &= -\langle \boldsymbol{d}_n \rangle^T \boldsymbol{K} \boldsymbol{d}_{n+\alpha} + \langle \boldsymbol{d}_n \rangle^T \boldsymbol{F}_{n+\alpha} \\ &= -\langle \boldsymbol{d}_n \rangle^T \boldsymbol{K}\left(\langle \boldsymbol{d}_n \rangle + \left(\alpha - \frac{1}{2}\right)[\boldsymbol{d}_n]\right) + \langle \boldsymbol{d}_n \rangle^T \boldsymbol{F}_{n+\alpha} \\ &= -\langle \boldsymbol{d}_n \rangle^T \boldsymbol{K} \langle \boldsymbol{d}_n \rangle - \frac{(\alpha - \frac{1}{2})}{2}[\boldsymbol{d}_n^T \boldsymbol{K} \boldsymbol{d}_n] + \langle \boldsymbol{d}_n \rangle^T \boldsymbol{F}_{n+\alpha} \end{aligned} \qquad (8.2.41)$$

Rearranging (8.2.41) results in

$$\boxed{d_{n+1}^T A d_{n+1} = d_n^T A d_n - 2\Delta t \langle d_n \rangle^T K \langle d_n \rangle + 2\Delta t \langle d_n \rangle^T F_{n+\alpha}} \tag{8.2.42}$$

where

$$A = M + \left(\alpha - \frac{1}{2}\right) \Delta t K \tag{8.2.43}$$

By positive-semidefiniteness,

$$d_{n+1}^T A d_{n+1} \leq d_n^T A d_n + 2\Delta t \langle d_n \rangle^T F_{n+\alpha} \tag{8.2.44}$$

For purposes of stability, we may consider the ***homogeneous case***:

$$d_{n+1}^T A d_{n+1} \leq d_n^T A d_n \tag{8.2.45}$$

From which we may conclude

$$d_n^T A d_n \leq d_0^T A d_0, \qquad \text{for all } n \tag{8.2.46}$$

As long as A is positive-definite d_n will not amplify. We take this as the definition of ***energy stability***. Let c be an arbitrary vector. We may expand c in terms of the eigenvectors:

$$c = \sum_{l=1}^{n_{eq}} c_{(l)} \psi_l \tag{8.2.47}$$

Therefore,

$$c^T A c = \sum_{l=1}^{n_{eq}} \left(1 + \left(\alpha - \frac{1}{2}\right) \Delta t \lambda_l^h\right) c_{(l)}^2 \tag{8.2.48}$$

and so

$$1 + \left(\alpha - \frac{1}{2}\right) \Delta t \lambda_l^h > 0, \qquad \text{for all } l = 1, 2, \ldots, n_{eq} \tag{8.2.49}$$

is the stability condition. This condition is satisfied if $\alpha \geq \frac{1}{2}$, or if

$$\lambda_{n_{eq}}^h \Delta t < \frac{2}{1 - 2\alpha} \tag{8.2.50}$$

These are the same stability conditions arrived at previously.

Equation (8.2.42) is a discrete analog of the growth/decay identity satisfied by the exact solution $d(t)$ of the semidiscrete heat equation. This can be seen as follows. Multiply (8.1.1) by d^T and integrate over the time interval $[t_n, t_{n+1}]$:

$$\begin{aligned} 0 &= \int_{t_n}^{t_{n+1}} d^T (M\dot{d} + Kd - F)\, dt \\ &= \int_{t_n}^{t_{n+1}} \left(\frac{1}{2} \frac{d}{dt}(d^T M d) + d^T K d - d^T F\right) dt \end{aligned} \tag{8.2.51}$$

from which it follows that (cf. (8.2.42)):

$$\boxed{\boldsymbol{d}(t_{n+1})^T \boldsymbol{M}\boldsymbol{d}(t_{n+1}) = \boldsymbol{d}(t_n)^T \boldsymbol{M}\boldsymbol{d}(t_n) - 2\int_{t_n}^{t_{n+1}} (\boldsymbol{d}^T \boldsymbol{K}\boldsymbol{d} - \boldsymbol{d}^T \boldsymbol{F})\, dt} \tag{8.2.52}$$

8.2.5 Additional Exercises

Exercise 2. Consider the following one-parameter (α) family of ***predictor-corrector algorithms:***

$$\boxed{\begin{aligned} \boldsymbol{M}\boldsymbol{v}_{n+1} + \boldsymbol{K}\tilde{\boldsymbol{d}}_{n+1} &= \boldsymbol{F}_{n+1} \\ \tilde{\boldsymbol{d}}_{n+1} &= \boldsymbol{d}_n + (1-\alpha)\Delta t \boldsymbol{v}_n \qquad \text{(predictor)} \\ \boldsymbol{d}_{n+1} &= \tilde{\boldsymbol{d}}_{n+1} + \alpha \Delta t\, \boldsymbol{v}_{n+1} \qquad \text{(corrector)} \end{aligned}}$$

Assume $\alpha \in [0, 1]$.

i. Determine an expression for the amplification factor A and plot A versus $\lambda^h \Delta t$.
ii. Determine under what circumstances the algorithms are stable.
iii. Obtain an expression for the local truncation error.
iv. Determine the rate of convergence.
v. If $\boldsymbol{M}$ is diagonal, is the algorithm implicit or explicit?

Answers: The modal equations are:

$$v_{n+1} + \lambda^h \tilde{d}_{n+1} = F_{n+1} \tag{a}$$

$$\tilde{d}_{n+1} = d_n + (1-\alpha)\Delta t v_n \tag{b}$$

$$d_{n+1} = \tilde{d}_{n+1} + \alpha \Delta t v_{n+1} \tag{c}$$

(a) and (c) imply $(1 - \alpha\Delta t\lambda^h)v_{n+1} + \lambda^h d_{n+1} = F_{n+1}$ (d)

Likewise at n: $(1 - \alpha\Delta t\lambda^h)v_n + \lambda^h d_n = F_n$ (e)

(b) and (c) imply $d_{n+1} = d_n + (1-\alpha)\Delta t v_n + \alpha\Delta t v_{n+1}$ (f)

Multiply (f) by $(1 - \alpha\Delta t\lambda^h)$:

$$(1 - \alpha\Delta t\lambda^h)d_{n+1} = (1 - \alpha\Delta t\lambda^h)d_n + \Delta t(1-\alpha)\overbrace{(1 - \alpha\Delta t\lambda^h)v_n}^{(e)} + \alpha\Delta t\underbrace{(1 - \alpha\Delta t\lambda^h)v_{n+1}}_{(d)}$$

$$(1 - \cancel{\alpha\Delta t\lambda^h})d_{n+1} = (1 - \cancel{\alpha\Delta t\lambda^h})d_n + \Delta t(1-\alpha)[F_n - \lambda^h d_n] + \alpha\Delta t[F_{n+1} - \lambda^h \cancel{d_{n+1}}]$$

$$d_{n+1} = (1 - \Delta t\lambda^h)d_n + \Delta t\underbrace{[(1-\alpha)F_n + \alpha F_{n+1}]}_{F_{n+\alpha}}$$

i. $A = 1 - \Delta t \lambda^h$

ii. $|A| < 1$ implies $\Delta t \lambda^h < 2$

iii. $\Delta t \tau(t_n) = d(t_{n+1}) - (1 - \Delta t \lambda^h)\, d(t_n) - \Delta t F_{n+\alpha}$

$$= \Delta t \underbrace{(\dot{d}(t_n) + \lambda^h d(t_n) - F(t_n))}_{0} + \Delta t^2 \left(\frac{\ddot{d}(t_n)}{2} - \alpha \dot{F}(t_n) \right) + O(\Delta t^3)$$

Therefore $|\tau(t)| \leq c\, \Delta t$

iv. $k = 1$

v. Explicit

Exercise 3. Consider the following ***fractional-step*** algorithm for the homogeneous model heat equation:

$$\frac{d_{n+1/2} - d_n}{(\Delta t/2)} + \lambda^h d_n = 0 \tag{a}$$

$$\frac{d_{n+1} - d_{n+1/2}}{(\Delta t/2)} + \lambda^h d_{n+1} = 0 \tag{b}$$

i. Obtain an expression for the amplification factor.
ii. Determine the conditions under which the algorithm is stable.
iii. What is the rate of convergence of this algorithm with respect to solutions of $\dot{d} + \lambda^h d = 0$?

Answers:

i. Solve (a) for $d_{n+1/2}$ and eliminate from (b) to arrive at

$$d_{n+1} = A d_n; \qquad A = \frac{1 - \Delta t \lambda^h/2}{1 + \Delta t \lambda^h/2}$$

Observe that A is identical to the one obtained for $\alpha = \frac{1}{2}$ (i.e., trapezoidal rule) in the generalized trapezoidal family.

ii. Unconditionally stable

iii. $k = 2$

Exercise 4. Consider the homogeneous semidiscrete heat equation,

$$\boldsymbol{M\dot{d}} + \boldsymbol{Kd} = \boldsymbol{0}$$

and the following algorithm:

$$\boldsymbol{d}_{n+1} = \boldsymbol{d}_n + \Delta t \boldsymbol{v}_n + \frac{\Delta t^2}{2} \boldsymbol{a}_n$$

$$\boldsymbol{M v}_{n+1} + \boldsymbol{K d}_{n+1} = \boldsymbol{0}$$

$$\boldsymbol{M a}_{n+1} + \boldsymbol{K v}_{n+1} = \boldsymbol{0}$$

i. Determine the amplification factor for this algorithm and plot it versus $\lambda^h \Delta t$. *Hint:*

$$\boldsymbol{\psi}_m^T \boldsymbol{M}(\boldsymbol{M}^{-1}\boldsymbol{K})^2 \boldsymbol{\psi}_m = \lambda_m^2.$$

ii. Determine the stability condition.

iii. Assuming $\boldsymbol{M}$ is diagonal, is this algorithm implicit or explicit? Justify your answer.

iv. Determine an expression for the local truncation error. *Hint:* Expand about t_n.

v. Based on the result of part (iv), determine the order of accuracy of this algorithm.

Answers:

i. $A = 1 - \lambda^h + (\Delta t \lambda^h)^2/2$

ii. $\Delta t < 2/\lambda_{n_{eq}}^h$

iii. Explicit

iv. $\tau(t_n) = \Delta t^2 \dddot{d}(\bar{t})/6, \quad \text{where} \quad \bar{t} \in [t_n, t_{n+1}]$

v. $k = 2$

Exercise 5. Modal decomposition of the algorithm

$$\boldsymbol{d}_{n+1} = \boldsymbol{d}_n + \frac{\Delta t}{2}(\boldsymbol{v}_n + \boldsymbol{v}_{n+1}) + \frac{\Delta t^2}{12}(\boldsymbol{a}_n - \boldsymbol{a}_{n+1})$$

$$\boldsymbol{M}\boldsymbol{v}_{n+1} = -\boldsymbol{K}\boldsymbol{d}_{n+1} + \boldsymbol{F}_{n+1}$$

$$\boldsymbol{M}\boldsymbol{a}_{n+1} = -\boldsymbol{K}\boldsymbol{v}_{n+1} + \dot{\boldsymbol{F}}_{n+1}$$

leads to the difference equation

$$\left(1 + \frac{\Delta t \lambda^h}{2} + \frac{(\Delta t \lambda^h)^2}{12}\right) d_{n+1} = \left(1 - \frac{\Delta t \lambda^h}{2} + \frac{(\Delta t \lambda^h)^2}{12}\right) d_n + \frac{\Delta t}{2}(F_n + F_{n+1})$$
$$- \frac{\Delta t^2 \lambda^h}{12}(F_n - F_{n+1}) + \frac{\Delta t^2}{12}(\dot{F}_n - \dot{F}_{n+1}) \qquad \text{(a)}$$

i. Plot the amplification factor A versus $\Delta t \lambda^h$. Comment on the stability of the algorithm. [*Answer*: Unconditionally stable.]

ii. Determine the order of accuracy. [*Answer*: $k = 4$. This one is a lot of work!]

Remark

The matrix version of (a), which would be implemented on the computer to obtain $\boldsymbol{d}_{n+1}$, involves the product $\boldsymbol{K}\boldsymbol{M}^{-1}\boldsymbol{K}$ in the coefficient matrix. This product has a profile structure involving approximately twice the number of terms as the original $\boldsymbol{K}$ if $\boldsymbol{M}$ is diagonal and is full if $\boldsymbol{M}$ is consistent. Thus the algorithm involves considerably more storage and computations than does the generalized trapezoidal method, for example. It is also inconvenient to obtain the time derivatives of $\boldsymbol{F}$ required at each step.

Exercise 6. The SDOF model problem under consideration consists of

$$\dot{d} + (\lambda^h + \widetilde{\lambda}^h)d = F$$

$$d(0) = d_0$$

Assume

$$\lambda^h > 0, \qquad \widetilde{\lambda}^h > 0$$

Consider the following one-parameter family of algorithms:

$$\begin{aligned} v_{n+1} + \lambda^h d_{n+1} + \widetilde{\lambda}^h \widetilde{d}_{n+1} &= F_{n+1} \\ \widetilde{d}_{n+1} &= d_n + (1-\alpha)\Delta t v_n \\ d_{n+1} &= \widetilde{d}_{n+1} + \alpha \Delta t v_{n+1} \end{aligned}$$

Assume $\alpha \in [0, 1]$.

i. Determine an expression for the amplification factor.
ii. Determine under what circumstances the algorithm is stable.
iii. Obtain an expression for the local truncation error.
iv. Determine the rate of convergence.

Answers:

i. $A = [1 - \alpha \Delta t\, \widetilde{\lambda}^h - (1-\alpha)\Delta t(\lambda^h + \widetilde{\lambda}^h)]/(1 + \alpha \Delta t\, \lambda^h)$
ii. $\Delta t[\widetilde{\lambda}^h + \lambda^h(1 - 2\alpha)] < 2$
iii. $\tau(t_n) = \Delta t\left[\left(\alpha - \frac{1}{2}\right)\ddot{d}(t_{n+\alpha}) + \alpha\, \widetilde{\lambda}^h \dot{d}(t_{n+\alpha})\right]$
iv. $k = 1$

Exercise 7. Consider the following one-parameter (α) family of iterative, predictor-corrector algorithms:

$$\begin{aligned} \boldsymbol{M}\boldsymbol{v}_{n+1}^{i+1} + \boldsymbol{K}\boldsymbol{d}_{n+1}^{i} &= \boldsymbol{F}_{n+1} \\ \boldsymbol{d}_{n+1}^{0} &= \boldsymbol{d}_n + (1-\alpha)\Delta t \boldsymbol{v}_n \\ \boldsymbol{d}_{n+1}^{i+1} &= \boldsymbol{d}_{n+1}^{0} + \alpha \Delta t \boldsymbol{v}_{n+1}^{i+1} \end{aligned}$$

Assume $\alpha \in [0, 1]$ and $i = 0, 1, \ldots, I$, where $I + 1$ is the total number of iterations. Define

$$\begin{aligned} \boldsymbol{d}_{n+1} &= \boldsymbol{d}_{n+1}^{I+1} \\ \boldsymbol{v}_{n+1} &= \boldsymbol{v}_{n+1}^{I+1} \end{aligned}$$

(With $I = 0$, this algorithm is the same as the one considered in Exercise 2.) Take the case $I = 1$.

i. Determine an expression for the amplification factor.
ii. Determine under what circumstances the algorithms are stable.
iii. Obtain an expression for the local truncation error.
iv. Determine the rate of convergence as a function of α.

Solution:

$$M v_{n+1}^1 + K d_{n+1}^0 = F_{n+1} \tag{a}$$

$$M v_{n+1}^2 + K d_{n+1}^1 = F_{n+1} \tag{b}$$

$$d_{n+1}^0 = d_n + (1 - \alpha)\Delta t\, v_n \tag{c}$$

$$d_{n+1}^1 = d_{n+1}^0 + \alpha \Delta t\, v_{n+1}^1 \tag{d}$$

$$d_{n+1}^2 = d_{n+1}^0 + \alpha \Delta t\, v_{n+1}^2 \tag{e}$$

$$\text{(a)} \Rightarrow v_{n+1}^1 = M^{-1}(F_{n+1} - K d_{n+1}^0) \tag{f}$$

$$\text{(d) and (f)} \Rightarrow d_{n+1}^1 = d_{n+1}^0 + \alpha \Delta t M^{-1}(F_{n+1} - K d_{n+1}^0) \tag{g}$$

$$\text{(b) and (g)} \Rightarrow M v_{n+1}^2 + K d_{n+1}^0 + \alpha \Delta t K M^{-1} F_{n+1} - \alpha \Delta t K M^{-1} K d_{n+1}^0 = F_{n+1}$$

That is,

$$M v_{n+1}^2 + K(I - \alpha \Delta t M^{-1} K) d_{n+1}^0 = (I - \alpha \Delta t K M^{-1}) F_{n+1}$$

Let

$$B = I - \alpha \Delta t M^{-1} K$$

$$\widetilde{B} = I - \alpha \Delta t K M^{-1}.$$

Therefore,

$$M v_{n+1}^2 + K B d_{n+1}^0 = \widetilde{B} F_{n+1} \tag{h}$$

$$\text{(e) and (h)} \Rightarrow M v_{n+1}^2 + K B (d_{n+1}^2 - \alpha \Delta t\, v_{n+1}^2) = \widetilde{B} F_{n+1}$$

Therefore,

$$(M - \alpha \Delta t\, K B) v_{n+1}^2 + K B d_{n+1}^2 = \widetilde{B} F_{n+1}$$

$$v_{n+1}^2 = v_{n+1}, \qquad d_{n+1}^2 = d_{n+1}$$

Thus

$$\widetilde{M} v_{n+1} + \widetilde{K} d_{n+1} = \widetilde{F}_{n+1}$$

$$d_{n+1} = d_n + (1 - \alpha)\Delta t\, v_n + \alpha \Delta t\, v_{n+1}$$

where

$$\widetilde{M} = M - \alpha \Delta t K B, \qquad \widetilde{K} = K B, \quad \text{and} \quad \widetilde{F}_{n+1} = \widetilde{B} F_{n+1}$$

Hence, the modal equation is

$$(1 - \alpha \Delta t \lambda^h z) v_{n+1} + \lambda^h z\, d_{n+1} = z F_{n+1}$$

where

$$d_{n+1} = d_n + (1 - \alpha)\Delta t\, v_n + \alpha \Delta t v_{n+1}$$

and

$$z = 1 - \alpha \Delta t \lambda^h$$

Therefore

$$(1 - \alpha\Delta t\,\lambda^h z)\cdot\frac{1}{\alpha\Delta t}(d_{n+1} - d_n - (1-\alpha)\Delta t\, v_n) + \lambda^h z\, d_{n+1} = zF_{n+1}$$

$$d_{n+1} - (1 - \alpha\Delta t\,\lambda^h z)d_n - (1 - \alpha\Delta t\,\lambda^h z)(1-\alpha)\Delta t\, v_n = \alpha\Delta t\, zF_{n+1}$$

but

$$(1 - \alpha\,\Delta t\,\lambda^h z)v_n + \lambda^h z\, d_n = zF_n$$

Thus

$$d_{n+1} - (1 - \alpha\Delta t\,\lambda^h z)d_n + \Delta t(1-\alpha)(\lambda^h z\, d_n - zF_n) = \alpha\,\Delta t\, zF_{n+1}$$

Therefore,

$$\boxed{d_{n+1} - (1 - \Delta t\,\lambda^h z)\, d_n = \Delta t\, zF_{n+\alpha}}$$

i. *Amplification factor*: $A = 1 - \Delta t\,\lambda^h(1 - \alpha\Delta t\,\lambda^h) = 1 - \Delta t\,\lambda^h + \alpha(\Delta t\,\lambda^h)^2$

ii. *Stability conditions:* $|A| < 1$, i.e., $-1 < A < 1$.

$$A < 1 \Rightarrow -\Delta t\,\lambda^h + \alpha(\Delta t\,\lambda^h)^2 < 0$$

Because $\lambda^h > 0$ and $\Delta t > 0$, it follows that $\alpha\Delta t\,\lambda^h < 1$ and $\Delta t < 1/\alpha\lambda^h$.

$$A > -1 \Rightarrow \alpha(\Delta t\,\lambda^h)^2 - \Delta t\,\lambda^h + 2 > 0$$

i.e.,

$$(\Delta t\,\lambda^h)^2 - \frac{\Delta t\,\lambda^h}{\alpha} + \frac{2}{\alpha} > 0 \qquad (\alpha > 0)$$

$$f(\Delta t\,\lambda^h) = \left[\Delta t\,\lambda^h - \frac{1}{2\alpha}(1 - \sqrt{1-8\alpha})\right]\left[\Delta t\,\lambda^h - \frac{1}{2\alpha}(1 + \sqrt{1-8\alpha})\right] > 0$$

If $\alpha \le \frac{1}{8}$, $\Delta t\,\lambda^h > \dfrac{1}{2\alpha}\left(1 + \sqrt{1-8\alpha}\right)$ or $\Delta t\,\lambda^h < \dfrac{1}{2\alpha}\left(1 - \sqrt{1-8\alpha}\right)$.

If $\alpha > \frac{1}{8}$, $f(\Delta t\,\lambda^h) > 0$ for all $\Delta t\,\lambda^h$.

> *Stability conditions*
>
> If $\alpha \le \frac{1}{8}$, $\Delta t < \dfrac{1}{2\alpha\lambda^h}\left(1 - \sqrt{1-8\alpha}\right)$.
>
> If $\alpha > \frac{1}{8}$, $\Delta t < \dfrac{1}{\alpha\lambda^h}$.

iii. Let $d(t_n) = d$, $\dot{d}(t_n) = \dot{d}$, ...

$$\begin{aligned}\Delta t\,\tau(t_n) &= d(t_{n+1}) - (1 - \Delta t\,\lambda^h + \alpha(\Delta t\,\lambda^h)^2)d(t_n) \\ &\quad -\alpha\Delta t(1 - \alpha\Delta t\,\lambda^h)F_{n+1} - \Delta t(1-\alpha)(1 - \alpha\Delta t\lambda^h)F_n \\ &= \not{d} + \Delta t\,\dot{d} + \frac{\Delta t^2}{2}\ddot{d} + O(\Delta t^3) - \not{d} + \Delta t\,\lambda^h d - \alpha(\Delta t\,\lambda^h)^2 d\end{aligned}$$

$$- (1 - \alpha\Delta t\lambda^h)\Delta t\{\alpha \dot{F}_n + \alpha\Delta t\dot{F}_n + \alpha\frac{\Delta t^2}{2}\ddot{F}_n + (1 - \alpha)F_n\}$$

$$= \Delta t(\dot{d} + \lambda^h d - F_n) + \frac{\Delta t^2}{2}(\ddot{d} - 2\alpha(\lambda^h)^2 d - 2\alpha\dot{F}_n + 2\alpha\lambda^h F_n)$$

$$+ O(\Delta t^3)$$

But $\dot{d} + \lambda^h d = F_n$, so

$$\Delta t\tau(t_n) = \frac{\Delta t^2}{2}(\ddot{d} - 2\alpha\dot{F}_n + 2\alpha\lambda^h(F_n - \lambda^h d)) + O(\Delta t^3)$$

$$= \frac{\Delta t^2}{2}(\ddot{d} - 2\alpha(\dot{F}_n - \lambda^h\dot{d})) + O(\Delta t^3)$$

$$= \frac{\Delta t^2}{2}(1 - 2\alpha)\ddot{d} + O(\Delta t^3)$$

iv. If $\alpha \neq \frac{1}{2}$, $|\tau(t_n)| \leq c\Delta t$

$k = 1$, first order

If $\alpha = \frac{1}{2}$, $|\tau(t_n)| \leq c\Delta t^2$

$k = 2$, second order

8.3 ELEMENTARY FINITE DIFFERENCE EQUATIONS FOR THE ONE-DIMENSIONAL HEAT EQUATION; THE VON NEUMANN METHOD OF STABILITY ANALYSIS

In order to appreciate some of the strengths and weaknesses of recently proposed finite element algorithms, a basic understanding of finite difference equations and their methods of analysis is required. All of the results obtained in this section are in the form of examples and exercises.

The von Neumann Method

A clear description of this technique, and others used for stability analysis of difference equations, is presented in Mitchell and Griffiths [5]. To illustrate its use, we shall consider only a simple example. (The reader interested in further details is urged to consult [5] and references therein.)

Example

Consider the one-dimensional heat equation

$$u_{,t} = ku_{,xx} \tag{8.3.1}$$

where $k = \kappa/\rho c$, and the algorithm (***FTCS: forward in time, centered in space***)

$$u_{n+1}^h(m) = (1 - 2r)u_n^h(m) + r(u_n^h(m + 1) + u_n^h(m - 1)) \tag{8.3.2}$$

where $u_n^h(m) = u_n^h(x_m)$, and so on, and $r = k\Delta t/h^2$, where h is the node spacing. Equation (8.3.2) is explicit, and so we would like to know for which values of r the algorithm is stable. The error induced by small perturbations in initial data, rounding, and so on, is denoted by $\delta_n(m)$ and satisfies the algorithmic equation, namely,

$$\delta_{n+1}(m) = (1 - 2r)\delta_n(m) + r(\delta_n(m+1) + \delta_n(m-1)) \tag{8.3.3}$$

Furthermore, $\delta_n(m)$ is assumed to take on the following form:

$$\delta_n(m) = \zeta^n \exp(im\xi), \qquad i = \sqrt{-1} \tag{8.3.4}$$

Substitution of (8.3.4) into (8.3.3) leads to

$$\zeta = 1 - 4r\sin^2\left(\frac{\xi}{2}\right) \tag{8.3.5}$$

The condition of stability is $|\zeta| \leq 1$ for all ξ, which requires

$$r \leq \frac{1}{2\sin^2(\xi/2)} \tag{8.3.6}$$

for all ξ, and so $r \leq \frac{1}{2}$ is the stability result (see Remark 1 under Table 8.2.1).

Exercise 1. Consider the following algorithms for the one-dimensional heat equation:

$$u_{n+1}^h(m) = u_n^h(m) + r(u_{n+1}^h(m+1) - 2u_{n+1}^h(m) + u_{n+1}^h(m-1))$$
$$\textbf{(\textit{BTCS: backward in time, centered in space})} \tag{8.3.7}$$

$$u_{n+1}^h(m) = u_n^h(m) + \frac{r}{2}\Big(u_{n+1}^h(m+1) - 2u_{n+1}^h(m) + u_{n+1}^h(m-1)$$
$$+ u_n^h(m+1) - 2u_n^h(m) + u_n^h(m-1)\Big) \qquad \textbf{(\textit{Crank-Nicolson})} \tag{8.3.8}$$

$$u_{n+1}^h(m) = u_{n-1}^h(m) + 2r(u_n^h(m+1) - 2u_n^h(m) + u_n^h(m-1)) \qquad \textbf{(\textit{leap frog})} \tag{8.3.9}$$

Determine the stability condition for each algorithm based upon the von Neumann approach. [*Answer:* BTCS and Crank-Nicolson are unconditionally stable; leap frog is unconditionally unstable.]

Local Truncation Error

The local truncation error measures how well the exact solution satisfies the algorithmic equation. For example, consider the FTCS scheme (8.3.2). In this case,

$$u(x_m, t_{n+1}) = (1 - 2r)u(x_m, t_n) + r(u(x_{m+1}, t_n) + u(x_{m-1}, t_n)) + \Delta t \cdot \tau \tag{8.3.10}$$

defines the local truncation error τ. To determine the form of τ, Taylor expansions of u about (x_m, t_n) may be employed. It can be shown that τ in (8.3.10) has the form $O(\Delta t^k, h^l)$ and $k = 1, l = 2$ (verify!). If the algorithm in question is stable, then the exponents k and l govern the rate of convergence of $u_n^h(m)$ to $u(x_m, t_n)$. As long as k and l are greater than zero, the algorithm is called consistent.

Exercise 2. Determine k and l for the algorithms considered in the previous exercise. [*Answers:* BTCS, $k = 1$, $l = 2$; Crank-Nicolson and leap frog, $k = l = 2$.]

Exercise 3. The ***DuFort-Frankel scheme*** [6] is defined by

$$u^h_{n+1}(m) = u^h_{n-1}(m) + 2r(u^h_n(m+1) - u^h_{n-1}(m) - u^h_{n+1}(m) + u^h_n(m-1)) \tag{8.3.11}$$

Show that this scheme is unconditionally stable, but the local truncation error has the form

$$\tau = O\left(\Delta t^2, h^2, \frac{\Delta t^2}{h^2}\right) \tag{8.3.12}$$

Thus convergence will only be achieved if $\Delta t \to 0$ faster than $h \to 0$. Such a scheme is called ***conditionally consistent.***

Remark

The DuFort-Frankel scheme is frequently described as an unconditionally stable, explicit method. Rearranging (8.3.11) exhibits the fact that the solution may be obtained explicitly:

$$(1 + 2r)u^h_{n+1}(m) = (1 - 2r)u^h_{n-1}(m) + 2r(u^h_n(m+1) + u^h_n(m-1)) \tag{8.3.13}$$

Conditional consistency seems the price one has to pay to achieve an explicit, unconditionally stable method. The time-step restriction imposed is often much smaller than that needed for accuracy in typical implicit methods such as Crank-Nicolson.

Exercise 4. ***Saul'yev's method*** [7] is given by

$$u^h_{n+1}(m) = u^h_n(m) + r(u^h_n(m+1) - u^h_n(m) - u^h_{n+1}(m) + u^h_{n+1}(m-1)) \tag{8.3.14}$$

Assume that u^h is specified at node x_0 (i.e., $m = 0$). Further, assume that the equations are solved in the order $m = 1, 2, 3, \ldots$, and so on. In this case (8.3.14) can be rearranged so that the solution may be obtained explicitly:

$$(1 + r)u^h_{n+1}(m) = u^h_n(m) + r(u^h_n(m+1) - u^h_n(m) + \underbrace{u^h_{n+1}(m-1)}_{\text{known}}) \tag{8.3.15}$$

Show that Saul'yev's method is unconditionally stable but conditionally consistent.

Exercise 5. Assume piecewise linear finite element functions in one dimension. Assuming $\ell = 0$ and equal node spacing, show that the semidiscrete equations at an internal node take the form

$$\frac{\rho c h}{6}\left(\dot{d}_{m-1}(t) + 4\dot{d}_m(t) + \dot{d}_{m+1}(t)\right) - \frac{\kappa}{h}\left(d_{m-1}(t) - 2d_m(t) + d_{m+1}(t)\right) = 0 \tag{8.3.16}$$

where here the subscripts refer to the node number.

Exercise 6. Apply the generalized trapezoidal family of ordinary differential equation algorithms to (8.3.16). Use the von Neumann method to determine the stability conditions. Obtain local truncation error expressions. Contrast this family of methods with the other difference schemes presented in this section. (Note that $d_m(t_n) = u_n^h(x_m)$ for comparison with the methods of this section.)

Exercise 7. Show that if the consistent mass is replaced by (7.3.52), then (8.3.16) becomes

$$\rho ch\left(\nu\dot{d}_{m-1}(t) + (1 - 2\nu)\dot{d}_m(t) + \nu\dot{d}_{m+1}(t)\right) - \frac{\kappa}{h}\left(d_{m-1}(t) - 2d_m(t) + d_{m+1}(t)\right) = 0 \tag{8.3.17}$$

Combine (8.3.17) with the trapezoidal algorithm (i.e., $\alpha = \frac{1}{2}$) in time. Assuming $\nu = \frac{1}{12}$ (i.e., "higher-order mass"; see Sec. 7.3.2), show that the local truncation error is $O(\Delta t^2, h^4)$.

Exercise 8. Generalize (8.3.17) to include a source term:

$$\rho ch\left(\nu\dot{d}_{m-1}(t) + (1 - 2\nu)\dot{d}_m(t) + \nu\dot{d}_{m+1}(t)\right) - \frac{\kappa}{h}\left(d_{m-1}(t) - 2d_m(t) + d_{m+1}(t)\right)$$
$$= h[\nu\ell_{m-1}(t) + (1 - 2\nu)\ell_m(t) + \nu\ell_{m+1}(t)] \tag{8.3.18}$$

Show that trapezoidal rule and $\nu = \frac{1}{12}$ result in $\tau = O(\Delta t^2, h^4)$.

Remark

Exercise 8 illustrates that the idea of higher-order mass treatment applies to the source term as well. The right-hand side of (8.3.18), with $\nu = \frac{1}{12}$, can be constructed by assuming a piecewise-linear continuous distribution of ℓ, and by applying the quadrature rule, $\xi_2 = -\xi_1 = \sqrt{\frac{2}{3}}$, $W_1 = W_2 = 1$. This is the rule that generates the higher-order element mass matrix.

Project. Program and compare the methods described in this section. The explicit methods are easy to program. For the implicit methods, you will need to solve a system of equations at each time step. The equation solver in the DLEARN program can be used for this purpose. Define, set up, and solve a simple problem of one-dimensional heat conduction. Vary Δt and h. Compare the performances of the methods on the basis of stability and accuracy.

Remarks

1. Analyses of some other fully discrete schemes for the one-dimensional heat equation may be found in [8].

2. A multidimensional finite element generalization of the Saul'yev scheme has been developed by Trujillo [9]. The capacity matrix is lumped and the conductivity is split into upper and lower triangular parts, $\boldsymbol{K}_U$ and $\boldsymbol{K}_L$, respectively, such that $\boldsymbol{K}_U = \boldsymbol{K}_L^T$. Solution proceeds in a two-step format in which upper and lower triangular matrices are used in alternating fashion. No factorizations are required and the process has features in common with the Gauss-Seidel iterative procedure for solving linear equations.

8.4. ELEMENT-BY-ELEMENT (EBE) IMPLICIT METHODS

Unconditionally stable, second-order accurate, implicit methods, such as the trapezoidal rule, perform very well in heat conduction analysis. The drawbacks are the storage and equation-solving burden engendered by the coefficient matrix. In recent years attempts have been made to achieve the desirable properties of implicit methods in a simpler computational setting. Methods described in this section employ the element-by-element concept. A product approximation of the element assembly is made so that the inversion of the coefficient matrix is replaced by sequential inversions of element matrices.

Method 1 (Hughes et al. [10]):

Recall that the generalized trapezoidal algorithm can be written in the form

$$(\boldsymbol{M} + \alpha\Delta t\, \boldsymbol{K})\, \boldsymbol{d}_{n+1} = (\boldsymbol{M} - (1 - \alpha)\, \Delta t\, \boldsymbol{K})\boldsymbol{d}_n + \Delta t\, \boldsymbol{F}_{n+\alpha} \qquad (8.4.1)$$

For future reference, it is convenient to rewrite (8.4.1) as

$$\boldsymbol{M}^{1/2}\boldsymbol{d}_{n+1} = \boldsymbol{A}\boldsymbol{M}^{1/2}\, \boldsymbol{d}_n + \Delta t\, \boldsymbol{B}\boldsymbol{M}^{-1/2}\boldsymbol{F}_{n+\alpha} \qquad (8.4.2)$$

where

$$\boldsymbol{A} = \boldsymbol{B}(\boldsymbol{I} - (1 - \alpha)\, \Delta t\, \boldsymbol{C}) \qquad (8.4.3)$$

$$\boldsymbol{B} = (\boldsymbol{I} + \alpha\Delta t\, \boldsymbol{C})^{-1} \qquad (8.4.4)$$

$$\boldsymbol{C} = \boldsymbol{M}^{-1/2}\, \boldsymbol{K}\boldsymbol{M}^{-1/2} \qquad (8.4.5)$$

and $\boldsymbol{M}^{1/2}$ is the square root of $\boldsymbol{M}$ and $\boldsymbol{M}^{-1/2} = (\boldsymbol{M}^{1/2})^{-1}$. The square root of $\boldsymbol{M}$ can be defined for the general case. However, we will restrict attention to the case in which $\boldsymbol{M}$ is diagonal and $\boldsymbol{M}^{1/2}$ has the obvious definition. For the method being described, this is necessary in order to achieve computational efficiency. We wish to view $\boldsymbol{A}$ and $\boldsymbol{B}$ as functions of α and Δt,

$$\boldsymbol{A} = \boldsymbol{A}(\alpha, \Delta t) \qquad (8.4.6)$$

$$\boldsymbol{B} = \boldsymbol{B}(\alpha, \Delta t) \qquad (8.4.7)$$

One-pass EBE algorithm. Element counterparts of (8.4.3) through (8.4.5) are, respectively,

$$\boldsymbol{A}^e = \boldsymbol{B}^e(\boldsymbol{I} - (1 - \alpha)\, \Delta t\, \boldsymbol{C}^e) \qquad (8.4.8)$$

$$\boldsymbol{B}^e = (\boldsymbol{I} + \alpha\Delta t\, \boldsymbol{C}^e)^{-1} \qquad (8.4.9)$$

$$\boldsymbol{C}^e = \boldsymbol{M}^{-1/2}\boldsymbol{K}^e\boldsymbol{M}^{-1/2} \qquad (8.4.10)$$

Consider the following analog of (8.4.2)

$$\boldsymbol{M}^{1/2}\boldsymbol{d}_{n+1} = \left(\prod_{e=1}^{n_{el}} \boldsymbol{A}^e\right)\boldsymbol{M}^{1/2}\, \boldsymbol{d}_n + \Delta t\left(\prod_{e=1}^{n_{el}} \boldsymbol{B}^e\right)\boldsymbol{M}^{-1/2}\boldsymbol{F}_{n+\alpha} \qquad (8.4.11)$$

where

$$\prod_{e=1}^{n_{el}} \boldsymbol{A}^e = \boldsymbol{A}^1 \boldsymbol{A}^2 \cdots \boldsymbol{A}^{n_{el}} \tag{8.4.12}$$

$$\prod_{e=1}^{n_{el}} \boldsymbol{B}^e = \boldsymbol{B}^1 \boldsymbol{B}^2 \cdots \boldsymbol{B}^{n_{el}} \tag{8.4.13}$$

The matrices (8.4.12) and (8.4.13) may be thought of as approximations to $\boldsymbol{A}$ and $\boldsymbol{B}$, respectively. Equation (8.4.11) defines an algorithm in which the calculations may be performed on an element-by-element basis. Specifically, $\boldsymbol{K}$ need never be formed, and equation solving involves arrays of element size only. Thus the storage requirements of the algorithm are greatly reduced compared with typical implicit methods. With $\alpha \geq \frac{1}{2}$, unconditional stability and first-order accuracy in Δt are attained by (8.4.11). Note that the ordering of the elements is completely arbitrary and thus no limitation is placed on the topology of the mesh.

Two-pass EBE algorithm. Consider the following algorithm:

$$\boldsymbol{M}^{1/2}\boldsymbol{d}_{n+1} = \left(\prod_{e=1}^{n_{el}} \boldsymbol{A}^e\left(\alpha, \frac{\Delta t}{2}\right)\right)\left(\prod_{e=n_{el}}^{1} \boldsymbol{A}^e\left(\alpha, \frac{\Delta t}{2}\right)\right)\boldsymbol{M}^{1/2}\boldsymbol{d}_n$$

$$+ \Delta t\left(\prod_{e=1}^{n_{el}} \boldsymbol{B}^e\left(\alpha, \frac{\Delta t}{2}\right)\right)\left(\prod_{e=n_{el}}^{1} \boldsymbol{B}^e\left(\alpha, \frac{\Delta t}{2}\right)\right)\boldsymbol{M}^{-1/2}\boldsymbol{F}_{n+\gamma} \tag{8.4.14}$$

If $\alpha = \frac{1}{2}$, unconditional stability and second-order accuracy in Δt are attained [10].

Remark

The computational properties of this method seem very attractive. Unfortunately, conditional consistency afflicts this procedure just as it does the DuFort-Frankel and Saul'yev methods described in Sec. 8.3. In an effort to retain the present computational architecture but achieve the accuracy behavior of standard implicit methods, the following method has been introduced.

Method 2 (Hughes et al. [11, 12], Winget and Hughes [13])

In this method we employ the element-by-element concept in an iterative equation-solving format. The equations of the generalized trapezoidal algorithm are iteratively solved using ***preconditioned conjugate gradients*** (***PCG***) with an element-by-element approximate factorization preconditioner. Thus the properties of the trapezoidal algorithm are inherited. In order to describe the details, let the algorithmic equation, for example,

$$(\boldsymbol{M} + \alpha\Delta t\, \boldsymbol{K})\boldsymbol{v}_{n+1} = \boldsymbol{F}_{n+1} - \boldsymbol{K}\tilde{\boldsymbol{d}}_{n+1} \tag{8.4.15}$$

be written as

$$\boldsymbol{A}\boldsymbol{x} = \boldsymbol{b} \tag{8.4.16}$$

where

$$A = M + \alpha \Delta t K \tag{8.4.17}$$

$$x = v_{n+1} \tag{8.4.18}$$

$$b = F_{n+1} - K\tilde{d}_{n+1} \tag{8.4.19}$$

Note that M may be nondiagonal in this case. Equation (8.4.16) is solved by the PCG method (see Table 8.4.1).

TABLE 8.4.1 Flowchart of Preconditioned Conjugate Gradients

Step 1.	Initialization: $m = 0, \quad x_0 = 0$ $r = b$ $p_0 = z_0 = B^{-1} r_0$
Step 2.	$\alpha_m = r_m^T z_m / p_m^T A p_m$
Step 3.	$x_{m+1} = x_m + \alpha_m p_m$
Step 4.	$r_{m+1} = r_m - \alpha_m A p_m$
Step 5.	Convergence check: $\Vert r_{m+1} \Vert < \delta \Vert r_0 \Vert$? Yes: Return No: Continue
Step 6.	$z_{m+1} = B^{-1} r_{m+1}$
Step 7.	$\beta_m = r_{m+1}^T z_{m+1} / r_m^T z_m$
Step 8.	$p_{m+1} = z_{m+1} + \beta_m p_m$
Step 9.	$m \leftarrow m + 1$, go to Step 2.

The preconditioning matrix is denoted by B in Table 8.4.1. If $B = I$ (the identity matrix), then we have the classical conjugate gradients procedure. The more closely B approximates A, the faster the convergence will be. An element-by-element procedure will be used to define B. We have developed a number of these procedures in [12, 13] but will just present here what we view to be the most effective one. To do so it is helpful to introduce the following notations: Let C be a symmetric, positive-definite matrix. Then we have

$$C = L_p(C) D_p(C) U_p(C) \qquad \text{(product decomposition)} \tag{8.4.20}$$

$$C = L_s(C) + D_s(C) + U_s(C) \qquad \text{(sum decomposition)} \tag{8.4.21}$$

where the subscripts p and s indicate product and sum, respectively.

Equation (8.4.20) represents the ***Crout factorization*** of C. Thus L_p and U_p are lower and upper triangular matrices, respectively, with diagonal entries equal to 1, and D_p is a diagonal matrix.

In equation (8.4.21), L_s and U_s, are lower and upper triangular arrays, respectively, with diagonal entries equal to 0, and D_s is diagonal.

By the symmetry of $\boldsymbol{C}$,

$$\boldsymbol{L}_p(\boldsymbol{C}) = \boldsymbol{U}_p(\boldsymbol{C})^T \tag{8.4.22}$$

$$\boldsymbol{L}_s(\boldsymbol{C}) = \boldsymbol{U}_s(\boldsymbol{C})^T \tag{8.4.23}$$

Corresponding to $\boldsymbol{A}^e$, we introduce a ***scaled, regularized element array*** as follows [13]:

$$\widetilde{\boldsymbol{A}}^e = \boldsymbol{I} + \boldsymbol{D}_s(\boldsymbol{A})^{-1/2}(\boldsymbol{A}^e - \boldsymbol{D}_s(\boldsymbol{A}^e))\boldsymbol{D}_s(\boldsymbol{A})^{-1/2} \tag{8.4.24}$$

With these notations we can define the ***reordered Crout EBE preconditioner:***

$$\boldsymbol{B} = \boldsymbol{D}_s(\boldsymbol{A})^{1/2}\left(\prod_{e=1}^{n_{el}} \boldsymbol{L}_p(\widetilde{\boldsymbol{A}}^e)\right)\left(\prod_{e=1}^{n_{el}} \boldsymbol{D}_p(\widetilde{\boldsymbol{A}}^e)\right)\left(\prod_{e=n_{el}}^{1} \boldsymbol{U}_p(\widetilde{\boldsymbol{A}}^e)\right)\boldsymbol{D}_s(\boldsymbol{A})^{1/2} \tag{8.4.25}$$

Remarks

1. $\boldsymbol{B}$ is symmetric and positive-definite. Aside from the trivial scaling manifested by the $\boldsymbol{D}_s(\boldsymbol{A})^{1/2}$ terms in (8.4.25), all computations may be performed on the element level. This involves processing full, small matrices. The major cost constituents are element-level forward reductions–back substitutions and the function evaluation required in Step 2 of Table 8.4.1, i.e., the element-level calculation

$$\boldsymbol{A}\boldsymbol{p}_m = \sum_{e=1}^{n_{el}} (\boldsymbol{A}^e \boldsymbol{p}_m) \tag{8.4.26}$$

The element factorization cost is not significant because it is amortized over the number of time steps and iterations per step required for convergence.

2. If elements are segregated into noncontiguous subgroups, then calculations are parallelizable. For example, bricklike domains can be decomposed into eight noncontiguous element groups (see Fig. 8.4.1). Because the elements in each subgroup have no common degrees of freedom, they can be processed in parallel. The eight groups, however, need to be processed sequentially. For analogous two-dimensional domains, four element groups need to be employed. In our experiences so far, parallel orderings have converged about as fast as sequential orderings [13]. This means that for bricklike domains the theoretical speed-up factor on a parallel computer is $n_{el}/8$ in three dimensions and $n_{el}/4$ in two dimensions.

3. The potential of this method is greatest in three-dimensional applications where the bandwidth of the coefficient matrix is large. The element-by-element procedure is independent of bandwidth and thus achieves significant operation count advantages [13]. The advantage increases in nonlinear applications where frequent refactorizations are typically necessary [13].

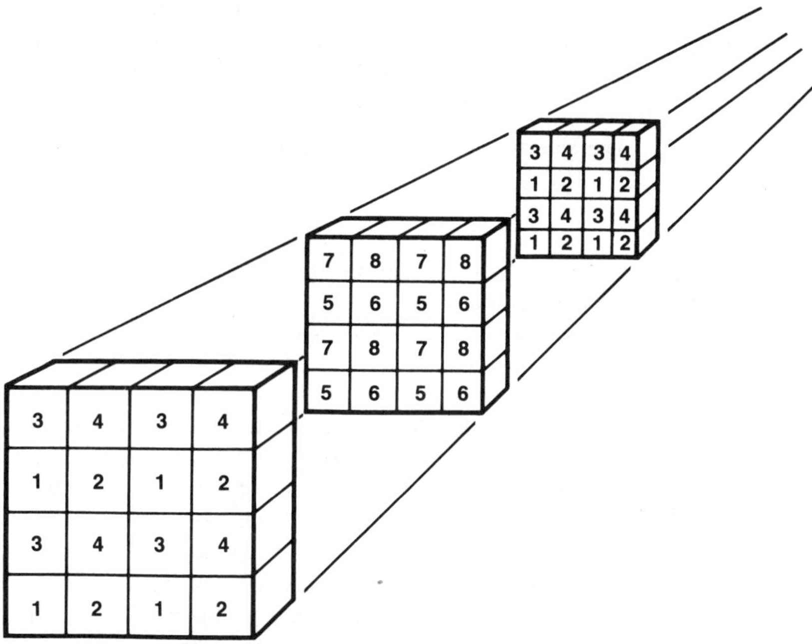

Figure 8.4.1 Decomposition of three-dimensional domain into eight groups of brick elements for parallel processing.

8.5 MODAL ANALYSIS

The technique of modal analysis is often used in place of step-by-step integration procedures such as those described in the previous sections. For the problem under consideration, modal analysis involves the following steps:

Step 1. Solve the eigenvalue problem for the pairs $\{\lambda_l^h, \boldsymbol{\psi}_l\}$, $1 \leq l \leq n_{\text{modes}}$, where $n_{\text{modes}} \leq n_{eq}$ is the desired number of modes to participate in the subsequent calculations. The determination of n_{modes} is a matter of engineering judgment, which involves the following considerations:

a. For each mode l selected, λ_l^h and $\boldsymbol{\psi}_l$ should be good approximations of the exact eigenvalue and eigenvector. This generally requires that $n_{\text{modes}} \ll n_{eq}$.

b. The spatial variation of $\boldsymbol{d}$ and $\boldsymbol{F}$ must be adequately represented by an expansion in the first n_{modes}.

Step 2. Solve the scalar ordinary differential equations

$$\left.\begin{aligned} \dot{d}_{(l)} &= -\lambda_l^h d_{(l)} + F_{(l)} \\ d_{(l)}(0) &= d_{0(l)} \end{aligned}\right\} \quad 1 \leq l \leq n_{\text{modes}} \tag{8.5.1}$$

The solutions may be obtained exactly if each $F_{(l)}$ is a simple function of t. However, it is generally more convenient to use a step-by-step method. To ensure essentially exact results, a very small time step may be employed. By virtue of the fact that the problems are of a scalar nature, the computational expenditure is generally negligible compared with Step 1.

Step 3. The approximation to $\boldsymbol{d}(t)$ is synthesized from the modes, viz.,

$$\boldsymbol{d}(t) \cong \sum_{l=1}^{n_{\text{modes}}} d_{(l)}(t)\boldsymbol{\psi}_l \tag{8.5.2}$$

The calculations on the right-hand side of (8.5.2) must be performed for each time t at which the solution is required.

Remarks

Step-by-step methods of integration and modal analysis each have attributes that may make them advantageous in particular circumstances. Step-by-step methods are (1) easily coded, (2) felt to be more efficient for short-time calculations, and (3) are generalizable to nonlinear situations. Modal analysis is felt to be more efficient (1) if many analyses of the same configuration are necessary, (2) for long-time calculations, and (3) if only a small number of modes are participating in the solution. The appropriate technique depends heavily on the problem under consideration and thus both approaches should be within the repertoire of the analyst.

REFERENCES

Section 8.1

1. T. J. R. Hughes and W. K. Liu, "Implicit-Explicit Finite Elements in Transient Analysis: Stability Theory," *Journal of Applied Mechanics*, 45 (1978), 371–374.
2. T. J. R. Hughes and W. K. Liu, "Implicit-Explicit Finite Elements in Transient Analysis: Implementation and Numerical Examples," *Journal of Applied Mechanics*, 45 (1978), 375–378.
3. T. J. R. Hughes, K. S. Pister, and R. L. Taylor, "Implicit-Explicit Finite Elements in Nonlinear Transient Analysis," *Computer Methods in Applied Mechanics and Engineering*, 17/18 (1979), 159–182.

Section 8.2

4. R. D. Richtmyer and K. W. Morton, *Difference Methods for Initial-Value Problems* (2nd ed). New York: Interscience, 1967.

Section 8.3

5. A. R. Mitchell and D. F. Griffiths, *The Finite Difference Method in Partial Differential Equations*. London: John Wiley, 1980.
6. W. F. Ames, *Numerical Methods for Partial Differential Equations*, (2nd ed.). New York: Academic Press, 1977.

7. V. K. Saul'yev, *Integration of Equations of the Parabolic Type by the Method of Nets*. New York: Pergamon Press, 1964.

8. T. J. R. Hughes and T. E. Tezduyar, "Analysis of Some Fully Discrete Algorithms for the One-Dimensional Heat Equation," *International Journal for Numerical Methods in Engineering*, 21 (1985), 163–168.

9. D. M. Trujillo, "An Unconditionally Stable Explicit Algorithm for Finite Element Heat Conduction Analysis," *Nuclear Engineering and Design*, 32 (1975), 110–120.

Section 8.4

10. T. J. R. Hughes, I. Levit, and J. Winget, "Element-by-Element Implicit Algorithms for Heat Conduction," *Journal of Engineering Mechanics*, 109, no. 2 (1983), 576–585.

11. T. J. R. Hughes, I. Levit, and J. Winget, "An Element-by-Element Solution Algorithm for Problems of Structural and Solid Mechanics," *Computer Methods in Applied Mechanics and Engineering*, 36, no. 2 (1983), 241–254.

12. T. J. R. Hughes, J. Winget, I. Levit, and T. E. Tezduyar, "New Alternating Direction Procedures in Finite Element Analysis based upon EBE Approximate Factorizations", in *Recent Developments in Computer Methods for Nonlinear Solid and Structural Mechanics*, eds. S. Atluri and N. Perrone, ASME Applied Mechanics Symposia Series, New York, 1983, 75–109.

13. J. M. Winget and T. J. R. Hughes, "Solution Algorithms for Nonlinear Transient Heat Conduction Analysis Employing Element-by-Element Iterative Strategies," *Computer Methods in Applied Mechanics and Engineering*, 52 (1985), 711–815.

9

Algorithms for Hyperbolic and Parabolic-Hyperbolic Problems

9.1 ONE-STEP ALGORITHMS FOR THE SEMIDISCRETE EQUATION OF MOTION

9.1.1 The Newmark Method

Recall from Chapter 7 that the semidiscrete equation of motion is written as

$$\boldsymbol{M}\ddot{\boldsymbol{d}} + \boldsymbol{C}\dot{\boldsymbol{d}} + \boldsymbol{K}\boldsymbol{d} = \boldsymbol{F} \tag{9.1.1}$$

where $\boldsymbol{M}$ is the mass matrix, $\boldsymbol{C}$ is the viscous damping matrix, $\boldsymbol{K}$ is the stiffness matrix, $\boldsymbol{F}$ is the vector of applied forces, and $\boldsymbol{d}$, $\dot{\boldsymbol{d}}$, and $\ddot{\boldsymbol{d}}$ are the displacement, velocity and acceleration vectors, respectively. We take $\boldsymbol{M}$, $\boldsymbol{C}$, and $\boldsymbol{K}$ to be symmetric; $\boldsymbol{M}$ is positive-definite, and $\boldsymbol{C}$ and $\boldsymbol{K}$ are positive-semidefinite.

The initial-value problem for (9.1.1) consists of finding a displacement, $\boldsymbol{d} = \boldsymbol{d}(t)$, satisfying (9.1.1) and the given initial data:

$$\boldsymbol{d}(0) = \boldsymbol{d}_0 \tag{9.1.2}$$

$$\dot{\boldsymbol{d}}(0) = \boldsymbol{v}_0 \tag{9.1.3}$$

Perhaps the most widely used family of direct methods for solving (9.1.1) to (9.1.3) is the ***Newmark family*** [1], which consists of the following equations:

$$\boldsymbol{M}\boldsymbol{a}_{n+1} + \boldsymbol{C}\boldsymbol{v}_{n+1} + \boldsymbol{K}\boldsymbol{d}_{n+1} = \boldsymbol{F}_{n+1} \tag{9.1.4}$$

$$\boldsymbol{d}_{n+1} = \boldsymbol{d}_n + \Delta t\boldsymbol{v}_n + \frac{\Delta t^2}{2}[(1 - 2\beta)\boldsymbol{a}_n + 2\beta\boldsymbol{a}_{n+1}] \tag{9.1.5}$$

$$\boldsymbol{v}_{n+1} = \boldsymbol{v}_n + \Delta t[(1-\gamma)\boldsymbol{a}_n + \gamma \boldsymbol{a}_{n+1}] \tag{9.1.6}$$

where $\boldsymbol{d}_n$, $\boldsymbol{v}_n$, and $\boldsymbol{a}_n$ are the approximations of $\boldsymbol{d}(t_n)$, $\dot{\boldsymbol{d}}(t_n)$, and $\ddot{\boldsymbol{d}}(t_n)$, respectively. Equation (9.1.4) is simply the equation of motion in terms of the approximate solution, and (9.1.5) and (9.1.6) are finite difference formulas describing the evolution of the approximate solution. The parameters β and γ determine the stability and accuracy characteristics of the algorithm under consideration. Equations (9.1.4) to (9.1.6) may be thought of as three equations for determining the three unknowns $\boldsymbol{d}_{n+1}$, $\boldsymbol{v}_{n+1}$, and $\boldsymbol{a}_{n+1}$, it being assumed that $\boldsymbol{d}_n$, $\boldsymbol{v}_n$, and $\boldsymbol{a}_n$ are known from the previous step's calculations. The Newmark family contains as special cases many well-known and widely used methods.

Implementation: a-form

There are several possible implementations. We will sketch one, but we leave further details until Sec. 9.4, which deals with operator and mesh partitions. The results in Sec. 9.4 include the Newmark method as a special case. Define ***predictors***:

$$\tilde{\boldsymbol{d}}_{n+1} = \boldsymbol{d}_n + \Delta t \boldsymbol{v}_n + \frac{\Delta t^2}{2}(1-2\beta)\boldsymbol{a}_n \tag{9.1.7}$$

$$\tilde{\boldsymbol{v}}_{n+1} = \boldsymbol{v}_n + (1-\gamma)\Delta t \boldsymbol{a}_n \tag{9.1.8}$$

Equations (9.1.5) and (9.1.6) may then be written as

$$\boldsymbol{d}_{n+1} = \tilde{\boldsymbol{d}}_{n+1} + \beta \Delta t^2 \boldsymbol{a}_{n+1} \tag{9.1.9}$$

$$\boldsymbol{v}_{n+1} = \tilde{\boldsymbol{v}}_{n+1} + \gamma \Delta t \boldsymbol{a}_{n+1} \tag{9.1.10}$$

To start the process, $\boldsymbol{a}_0$ may be calculated from

$$\boldsymbol{M}\boldsymbol{a}_0 = \boldsymbol{F} - \boldsymbol{C}\boldsymbol{v}_0 - \boldsymbol{K}\boldsymbol{d}_0 \tag{9.1.11}$$

or specified directly. The recursion relation determines $\boldsymbol{a}_{n+1}$:

$$(\boldsymbol{M} + \gamma \Delta t \boldsymbol{C} + \beta \Delta t^2 \boldsymbol{K})\boldsymbol{a}_{n+1} = \boldsymbol{F}_{n+1} - \boldsymbol{C}\tilde{\boldsymbol{v}}_{n+1} - \boldsymbol{K}\tilde{\boldsymbol{d}}_{n+1} \tag{9.1.12}$$

Equations (9.1.9) and (9.1.10) may then be used to calculate $\boldsymbol{d}_{n+1}$ and $\boldsymbol{v}_{n+1}$, respectively.

This form of implementation is convenient for generalization to algorithms that employ "mesh partitions" (see Sec. 9.4) but is not the most efficient implementation.

Exercise 1. Develop an implementation in which $\boldsymbol{a}_{n+1}$ and $\boldsymbol{v}_{n+1}$ are eliminated from (9.1.4) by way of (9.1.9) and (9.1.10). In this implementation the right-hand side of the equation system does not entail stiffness calculations.

9.1.2 Analysis

The convergence analysis of the Newmark family of methods is similar to the convergence analysis given in the previous chapter. The main steps are (1) reduction to an SDOF model problem; (2) definition of a suitable notion of stability, which is shown to hold under certain circumstances; (3) determination of the local truncation error, from which the order of accuracy is obtained, and (4) use of the latter two conditions to prove convergence for the SDOF problem. Some assumption on the form of $\boldsymbol{C}$ is necessary to perform the modal reduction. For example, if Rayleigh damping is assumed, i.e.,

$$\boldsymbol{C} = a\boldsymbol{M} + b\boldsymbol{K} \tag{9.1.13}$$

then the symmetry and positive-definiteness of $\boldsymbol{M}$ and symmetry and positive-semidefiniteness of $\boldsymbol{K}$ enable the decomposition indicated in the commutative diagram on page 494, where ω is the undamped frequency of vibration and $\xi = (a/\omega + b\omega)/2$ is the damping ratio.

The SDOF model problem can be recast in a first-order form analogous to (8.2.32):

$$\boxed{\boldsymbol{y}_{n+1} = \boldsymbol{A}\boldsymbol{y}_n + \boldsymbol{L}_n} \tag{9.1.14}$$

where $\boldsymbol{A}$ is the ***amplification matrix*** and

$$\boldsymbol{y}_n = \begin{Bmatrix} d_n \\ v_n \end{Bmatrix} \tag{9.1.15}$$

Difference equations such as (9.1.14) are often referred to as ***one-step, multivalue methods***, the number of values being equal to the dimension of the vector $\boldsymbol{y}$, which in the present case is 2. The complete analysis of the algorithm reduces to consideration of (9.1.14). Further details will be described in this section and in the following sections.

A summary of ***stability conditions for the Newmark method*** follows [2, 3, 4].

Unconditional

$$2\beta \geq \gamma \geq \frac{1}{2} \tag{9.1.16}$$

Conditional

$$\gamma \geq \frac{1}{2} \tag{9.1.17}$$

$$\beta < \frac{\gamma}{2} \tag{9.1.18}$$

$$\omega^h \Delta t \leq \Omega_{\text{crit}} \tag{9.1.19}$$

where

$$\Omega_{\text{crit}} = \frac{\xi(\gamma - \frac{1}{2}) + [\gamma/2 - \beta + \xi^2(\gamma - \frac{1}{2})^2]^{1/2}}{(\gamma/2 - \beta)} \quad \textit{(critical sampling frequency)} \tag{9.1.20}$$

The stability conditions must be satisfied for each mode in the system. Consequently, the maximum natural frequency, $\omega^h_{n_{eq}}$, is critical and therefore must satisfy (9.1.19). As described in Chapter 7, $\omega^h_{n_{eq}}$ may be bounded by the maximum frequency of the individual elements. Note that if $\gamma = \frac{1}{2}$ viscous damping has *no* effect on stability. Furthermore, when $\gamma > \frac{1}{2}$ the effect of viscous damping is to *increase* the critical time step of conditionally stable Newmark methods. ***Thus the undamped critical sampling frequency, i.e.,*** $\Omega_{\text{crit}} = (\gamma/2 - \beta)^{-1/2}$, ***serves as a conservative value when an estimate of the modal damping coefficient is not available***. (Recall that, in the case of Rayleigh damping, ξ is determined by ω^h.) In practice it is often more convenient to express (9.1.19) in terms of the ***period of vibration***, $T = 2\pi/\omega$, in which case (9.1.19) becomes $\Delta t/T \leq \Omega_{\text{crit}}/(2\pi)$.

The Newmark family contains as special cases many well-known and widely used methods. Properties of some classical methods are summarized in Table 9.1.1. The average acceleration method is one of the most widely used methods for structural dynamics applications. Note that the linear acceleration and Fox-Goodwin methods are implicit and *conditionally* stable. They are not felt to be economically competitive for large-scale systems when compared to implicit and *unconditionally* stable techniques such as the average acceleration method. The central difference method is

TABLE 9.1.1 Properties of Well-known Members of the Newmark Family of Methods

Method	Type	β	γ	Stability condition(2)	Order of accuracy(3)
Average acceleration (trapezoidal rule)	Implicit	$\frac{1}{4}$	$\frac{1}{2}$	Unconditional	2
Linear acceleration	Implicit	$\frac{1}{6}$	$\frac{1}{2}$	$\Omega_{\text{crit}} = 2\sqrt{3} \cong 3.464$	2
Fox-Goodwin (royal road)	Implicit	$\frac{1}{12}$	$\frac{1}{2}$	$\Omega_{\text{crit}} = \sqrt{6} \cong 2.449$	2
Central difference	Explicit(1)	0	$\frac{1}{2}$	$\Omega_{\text{crit}} = 2$	2

Notes: 1. Strictly speaking, $\boldsymbol{M}$ and $\boldsymbol{C}$ need to be diagonal for the central-difference method to be explicit.
2. Stability is based upon the undamped case, in which $\xi = 0$.
3. Second-order accuracy is achieved if and only if $\gamma = \frac{1}{2}$.

$$\boldsymbol{M}\ddot{\boldsymbol{d}} + \boldsymbol{C}\dot{\boldsymbol{d}} + \boldsymbol{K}\boldsymbol{d} = \boldsymbol{F}$$
$$\boldsymbol{d}(0) = \boldsymbol{d}_0$$
$$\dot{\boldsymbol{d}}(0) = \boldsymbol{v}_0$$

Temporal discretization →

$$\boldsymbol{M}\boldsymbol{a}_{n+1} + \boldsymbol{C}\boldsymbol{v}_{n+1} + \boldsymbol{K}\boldsymbol{d}_{n+1} = \boldsymbol{F}_{n+1}$$
$$\boldsymbol{d}_{n+1} = \boldsymbol{d}_n + \Delta t\boldsymbol{v}_n + \frac{\Delta t^2}{2}\{(1 - 2\beta)\boldsymbol{a}_n + 2\beta\boldsymbol{a}_{n+1}\}$$
$$\boldsymbol{v}_{n+1} = \boldsymbol{v}_n + \Delta t\{(1 - \gamma)\boldsymbol{a}_n + \gamma\boldsymbol{a}_{n+1}\}$$
$\boldsymbol{d}_0$, $\boldsymbol{v}_0$ given

Modal decomposition ↓ (both columns)

$$\ddot{d} + 2\xi\omega^h\dot{d} + (\omega^h)^2 d = F$$
$$d(0) = d_0$$
$$\dot{d}(0) = v_0$$

Temporal discretization →

$$a_{n+1} + 2\xi\omega^h v_{n+1} + (\omega^h)^2 d_{n+1} = F_{n+1}$$
$$d_{n+1} = d_n + \Delta t v_n + \frac{\Delta t^2}{2}\{(1 - 2\beta)a_n + 2\beta a_{n+1}\}$$
$$v_{n+1} = v_n + \Delta t\{(1 - \gamma)a_n + \gamma a_{n+1}\}$$
d_0, v_0 given

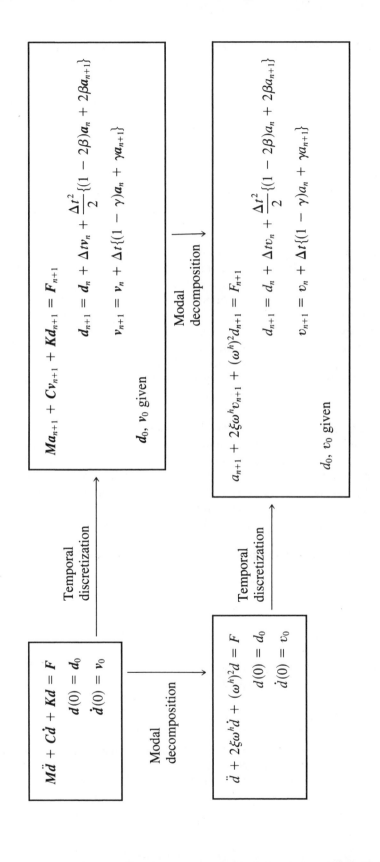

conditionally stable. However, when $\boldsymbol{M}$ and $\boldsymbol{C}$ are diagonal, it is explicit. When the time step restriction is not too severe, as is often the case in elastic wave-propagation problems, the central difference method is generally the most economical direct integration procedure and is thus widely used. The methods enumerated in Table 9.1.1 are numerically compared in Goudreau and Taylor [2].

Exercise 2. The name "central difference method" derives from the fact that (9.1.5) and (9.1.6) may be combined to yield

$$\boldsymbol{v}_n = \frac{\boldsymbol{d}_{n+1} - \boldsymbol{d}_{n-1}}{2\Delta t} \tag{9.1.21}$$

$$\boldsymbol{a}_n = \frac{\boldsymbol{d}_{n+1} - 2\boldsymbol{d}_n + \boldsymbol{d}_{n-1}}{\Delta t^2} \tag{9.1.22}$$

the classical first and second central-difference approximations, respectively. Verify this assertion.

Exercise 3. The average acceleration method may also be derived by applying the trapezoidal rule to the first-order form of the equations of motion. Let

$$\boldsymbol{y} = \begin{Bmatrix} \boldsymbol{d} \\ \dot{\boldsymbol{d}} \end{Bmatrix} \tag{9.1.23}$$

and

$$\boldsymbol{f}(\boldsymbol{y}, t) = \begin{Bmatrix} \dot{\boldsymbol{d}} \\ \hdashline \boldsymbol{M}^{-1}(\boldsymbol{F}(t) - \boldsymbol{C}\dot{\boldsymbol{d}} - \boldsymbol{K}\boldsymbol{d}) \end{Bmatrix} \tag{9.1.24}$$

Then

$$\dot{\boldsymbol{y}} = \boldsymbol{f}(\boldsymbol{y}, t) \tag{9.1.25}$$

defines a first-order system of $2n_{eq}$ ordinary differential equations equivalent to (9.1.1). The trapezoidal-rule approximation of (9.1.25) may be written in terms of the following two equations:

$$\boldsymbol{z}_{n+1} = \boldsymbol{f}(\boldsymbol{y}_{n+1}, t_{n+1}) \tag{9.1.26}$$

$$\boldsymbol{y}_{n+1} = \boldsymbol{y}_n + \frac{\Delta t}{2}(\boldsymbol{z}_n + \boldsymbol{z}_{n+1}) \tag{9.1.27}$$

Here $\boldsymbol{y}_n$ and $\boldsymbol{z}_n$ are the approximations of $\boldsymbol{y}(t_n)$ and $\dot{\boldsymbol{y}}(t_n)$. Show that (9.1.26) and (9.1.27) are equivalent to (9.1.4) through (9.1.6) with $\beta = \frac{1}{4}$ and $\gamma = \frac{1}{2}$.

To analyze the Newmark algorithm, we need to derive explicit formulas for $\boldsymbol{A}$ and $\boldsymbol{L}_n$ in (9.1.14). These may be obtained by eliminating a_n and a_{n+1} from

$$d_{n+1} = d_n + \Delta t v_n + \frac{\Delta t^2}{2}[(1 - 2\beta)a_n + 2\beta a_{n+1}] \tag{9.1.28}$$

$$v_{n+1} = v_n + \Delta t[(1 - \gamma)a_n + \gamma a_{n+1}] \tag{9.1.29}$$

by using

$$a_n + 2\xi\omega^h v_n + (\omega^h)^2 d_n = F_n \tag{9.1.30}$$

$$a_{n+1} + 2\xi\omega^h v_{n+1} + (\omega^h)^2 d_{n+1} = F_{n+1} \tag{9.1.31}$$

This leads to the following expressions:

$$\boldsymbol{A} = \boldsymbol{A}_1^{-1}\boldsymbol{A}_2 \tag{9.1.32}$$

$$\boldsymbol{L}_n = \boldsymbol{A}_1^{-1}\begin{Bmatrix} \dfrac{\Delta t^2}{2}[(1-2\beta)F_n + 2\beta F_{n+1}] \\ \Delta t[(1-\gamma)F_n + \gamma F_{n+1}] \end{Bmatrix} \tag{9.1.33}$$

$$\boldsymbol{A}_1 = \begin{bmatrix} 1 + \Delta t^2\beta(\omega^h)^2 & 2\Delta t^2\beta\xi\omega^h \\ \Delta t\gamma(\omega^h)^2 & 1 + 2\Delta t\gamma\xi\omega^h \end{bmatrix} \tag{9.1.34}$$

$$\boldsymbol{A}_2 = \begin{bmatrix} 1 - \dfrac{\Delta t^2}{2}(1-2\beta)(\omega^h)^2 & \Delta t(1 - \Delta t(1-2\beta)\xi\omega^h) \\ -\Delta t(1-\gamma)(\omega^h)^2 & 1 - 2\Delta t(1-\gamma)\xi\omega^h \end{bmatrix} \tag{9.1.35}$$

Convergence. The vector of ***local truncation errors***, $\boldsymbol{\tau}$ is defined by

$$\boxed{\boldsymbol{y}(t_{n+1}) = \boldsymbol{A}\boldsymbol{y}(t_n) + \boldsymbol{L}_n + \Delta t \cdot \boldsymbol{\tau}(t_n)} \tag{9.1.36}$$

where

$$\boldsymbol{y}(t_n) = \begin{Bmatrix} d(t_n) \\ \dot{d}(t_n) \end{Bmatrix} \tag{9.1.37}$$

Obtaining an explicit expression for $\boldsymbol{\tau}$ is tedious, due to the complicated nature of $\boldsymbol{A}$ and $\boldsymbol{L}_n$. However, we can use an analysis similar to the one described in Chapter 8 to infer the correct exponent of Δt in the entries of $\boldsymbol{\tau}$. It may be argued that $\boldsymbol{\tau}(t) = O(\Delta t^k)$ for all $t \in [0,T]$, where $k = 2$ if $\gamma = \frac{1}{2}$ and $k = 1$ otherwise.

The stability of the Newmark methods is determined by properties of the amplification matrix. The convergence proof follows along the lines of the theorem presented in Sec. 8.2.3. Here, the equation

$$\boxed{\boldsymbol{e}(t_n) = \boldsymbol{A}^n\boldsymbol{e}(0) - \Delta t \sum_{i=0}^{n-1} \boldsymbol{A}^i\,\boldsymbol{\tau}(t_{n-1-i})} \tag{9.1.38}$$

determined from (9.1.14) and (9.1.36), is used in place of the scalar analog in Sec.

8.2.3. Again, stability plus consistency implies convergence. The rate of convergence is k. This result holds for an arbitrary one-step, multivalue method. Multistep methods, which will be studied subsequently, may be recast as one-step multivalue methods although the dimension of the vector $\boldsymbol{y}$ will generally be greater than 2.

Stability. Let $\lambda_i(\boldsymbol{A})$ denote the eigenvalues of $\boldsymbol{A}$. The modulus of $\lambda_i(\boldsymbol{A})$ is written $|\lambda_i(\boldsymbol{A})| = (\lambda_i(\boldsymbol{A})\overline{\lambda_i(\boldsymbol{A})})^{1/2}$, where $\overline{\lambda_i(\boldsymbol{A})}$ is the complex conjugate of $\lambda_i(\boldsymbol{A})$. The ***spectral radius*** of $\boldsymbol{A}$, $\rho(\boldsymbol{A})$, is defined by

$$\boxed{\rho(\boldsymbol{A}) = \max_i |\lambda_i(\boldsymbol{A})|} \tag{9.1.39}$$

The stability condition is needed to prevent amplification of $\boldsymbol{A}^n$ as n becomes large. The following conditions will be required:

i. $\rho(\boldsymbol{A}) \leq 1$.
ii. Eigenvalues of $\boldsymbol{A}$ of multiplicity greater than one are strictly less than one in modulus.

Conditions (i) and (ii) define a ***spectrally stable*** $\boldsymbol{A}$.

That spectral stability bounds the growth of $\boldsymbol{A}^n$ can be seen as follows. Consider a 2×2 matrix $\boldsymbol{A}$ such as for the Newmark method. Then $\boldsymbol{A}$ can be written as either

$$\boldsymbol{A} = \boldsymbol{P}\boldsymbol{\Lambda}(\boldsymbol{A})\boldsymbol{P}^{-1}, \qquad \boldsymbol{\Lambda}(\boldsymbol{A}) = \begin{bmatrix} \lambda_1(\boldsymbol{A}) & 0 \\ 0 & \lambda_2(\boldsymbol{A}) \end{bmatrix} \tag{9.1.40}$$

or

$$\boldsymbol{A} = \boldsymbol{Q}\boldsymbol{J}(\boldsymbol{A})\boldsymbol{Q}^{-1}, \qquad \boldsymbol{J}(\boldsymbol{A}) = \begin{bmatrix} \lambda(\boldsymbol{A}) & 1 \\ 0 & \lambda(\boldsymbol{A}) \end{bmatrix} \tag{9.1.41}$$

Equations (9.1.40) and (9.1.41) correspond to the cases in which $\boldsymbol{A}$ has linearly independent eigenvectors and linearly dependent eigenvectors, respectively. Note that in (9.1.41) the eigenvalue $\lambda(\boldsymbol{A})$ has multiplicity two. The nth power of $\boldsymbol{A}$ may be computed with the aid of (9.1.40) or (9.1.41):

$$\boldsymbol{A}^n = \boldsymbol{P}\boldsymbol{\Lambda}(\boldsymbol{A})^n\boldsymbol{P}^{-1}, \qquad \boldsymbol{\Lambda}(\boldsymbol{A})^n = \begin{bmatrix} \lambda_1(\boldsymbol{A})^n & 0 \\ 0 & \lambda_2(\boldsymbol{A})^n \end{bmatrix} \tag{9.1.42}$$

$$\boldsymbol{A}^n = \boldsymbol{Q}\boldsymbol{J}(\boldsymbol{A})^n\boldsymbol{Q}^{-1}, \qquad \boldsymbol{J}(\boldsymbol{A})^n = \begin{bmatrix} \lambda(\boldsymbol{A})^n & n\lambda(\boldsymbol{A})^{n-1} \\ 0 & \lambda(\boldsymbol{A})^n \end{bmatrix} \tag{9.1.43}$$

In the former case $\boldsymbol{A}^n$ will be bounded as long as $\rho(\boldsymbol{A}) = \max\{|\lambda_1(\boldsymbol{A})|, |\lambda_2(\boldsymbol{A})|\} \leq 1$. In the latter case we need to invoke the stronger condition $|\lambda(\boldsymbol{A})| < 1$ so that the off-diagonal term does not exhibit growth. Observe that $|\lambda(\boldsymbol{A})| < 1$ implies $|n\lambda(\boldsymbol{A})^{n-1}| \to 0$ as $n \to \infty$, whereas if $|\lambda(\boldsymbol{A})|$ was allowed to be equal to one, then

$n|\lambda(\boldsymbol{A})|^{n-1} = n$. Thus the necessity of condition (ii) in the definition of spectral stability can be appreciated.

Violations of condition (i) produce "explosive" instabilities, which grow like $\rho(\boldsymbol{A})^n$ (cf. Table 8.2.1). If (i) is in force but (ii) is violated, the growth is much weaker (e.g., $O(n)$ as above). These are referred to as ***weak instabilities***. In the general case when the multiplicity of a modulus-one eigenvalue is m, the growth rate is $O(n^{m-1})$.

By virtue of the fact that the stability condition depends only upon the eigenvalues of $\boldsymbol{A}$, it can be expressed in terms of the ***principal invariants*** of $\boldsymbol{A}$. For a 2×2 amplification matrix, we have that

$$0 = \det(\boldsymbol{A} - \lambda(\boldsymbol{A})\boldsymbol{I}) = \lambda(\boldsymbol{A})^2 - 2A_1\,\lambda(\boldsymbol{A}) + A_2 \tag{9.1.44}$$

where

$$A_1 = \tfrac{1}{2}\text{trace}\,\boldsymbol{A} = \tfrac{1}{2}(A_{11} + A_{22}) \tag{9.1.45}$$

$$A_2 = \det\boldsymbol{A} = A_{11}A_{22} - A_{12}A_{21} \tag{9.1.46}$$

are the principal invariants. The eigenvalues are thus

$$\lambda_{1,2}(\boldsymbol{A}) = A_1 \pm (A_1^2 - A_2)^{1/2} \tag{9.1.47}$$

Exercise 4. For the Newmark method, use (9.1.32), (9.1.34), and (9.1.35) to show that

$$A_1 = 1 - \frac{[\xi\Omega + \Omega^2(\gamma + \frac{1}{2})/2]}{D} \tag{9.1.48}$$

$$A_2 = 1 - \frac{[2\xi\Omega + \Omega^2(\gamma - \frac{1}{2})]}{D} \tag{9.1.49}$$

where

$$D = 1 + 2\gamma\xi\Omega + \beta\Omega^2 \tag{9.1.50}$$

$$\Omega = \omega^h\Delta t \tag{9.1.51}$$

For a 2×2 amplification matrix, the most convenient characterization of spectral stability has been devised by Hilber [4] (see also [5] for the derivation): The stability region in A_1, A_2-space satisfies (see Fig. 9.1.1)

$$\frac{-(A_2 + 1)}{2} \le A_1 \le \frac{(A_2 + 1)}{2}, \qquad -1 \le A_2 < 1 \tag{9.1.52}$$

$$-1 < A_1 < 1, \qquad A_2 = 1 \tag{9.1.53}$$

Combining (9.1.48) through (9.1.53) leads to the stability conditions quoted previously, namely (9.1.16) through (9.1.20).

High-frequency behavior. Because the higher modes of semidiscrete structural equations are artifacts of the discretization process and not representative of the behavior of the governing partial differential equations, it is generally viewed as desirable and often is considered absolutely necessary to have some form of algorithmic damping present to remove the participation of the high-frequency modal

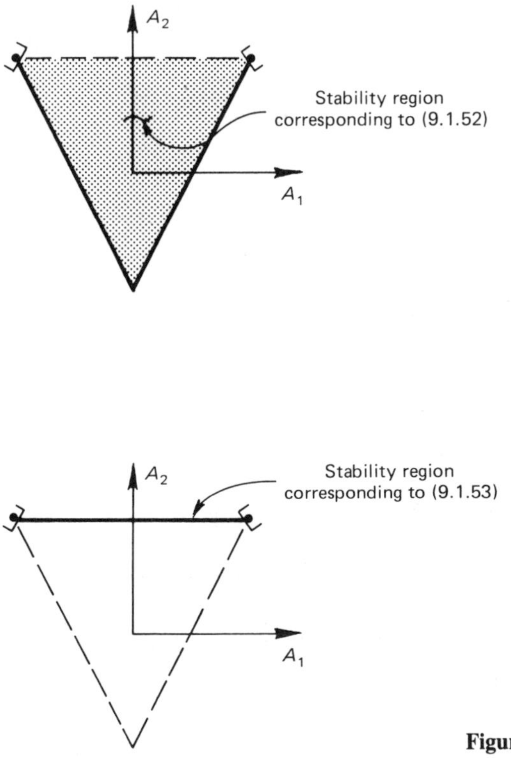

Figure 9.1.1

components. In terms of the Newmark method, $\gamma > \frac{1}{2}$ is necessary to introduce high-frequency dissipation. For a fixed $\gamma > \frac{1}{2}$, one can select β such that high-frequency dissipation is maximized. It turns out that this condition is created by restricting the eigenvalues of the amplification matrix to be complex conjugate. From (9.1.47) it can be seen that

$$A_1^2 < A_2 \tag{9.1.54}$$

is required. The subregion of the spectrally stable region of A_1, A_2-space is shown in Fig. 9.1.2. The Newmark methods fall within this region if:

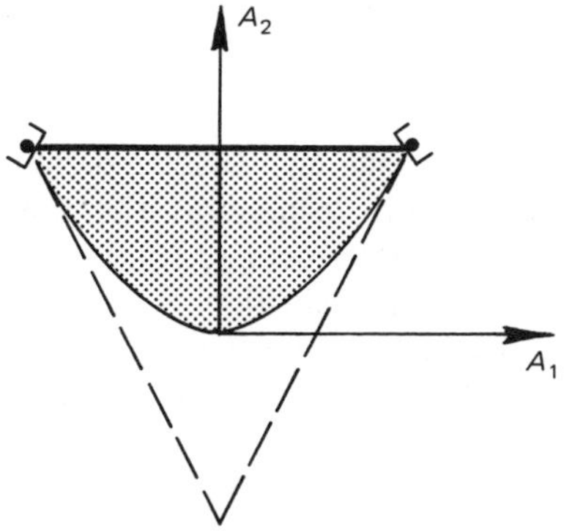

Figure 9.1.2 Stability region for which roots of the amplification matrix are complex conjugates.

Unconditional

$$0 \le \xi < 1, \qquad \gamma \ge \frac{1}{2}, \qquad \beta \ge \frac{(\gamma + 1/2)^2}{4} \tag{9.1.55}$$

Conditional

$$0 \le \xi < 1, \qquad \gamma \ge \frac{1}{2}, \qquad \Omega < \Omega_{\text{bif}} \tag{9.1.56}$$

where

$$\Omega_{\text{bif}} = \frac{\xi(\gamma - \frac{1}{2})/2 + [(\gamma + \frac{1}{2})^2/4 - \beta + \xi^2(\beta - \gamma/2)]^{1/2}}{(\gamma + \frac{1}{2})^2/4 - \beta} \tag{9.1.57}$$

In the undamped case, (9.1.57) becomes simply

$$\Omega_{\text{bif}} = \left[\frac{(\gamma + 1/2)^2}{4} - \beta\right]^{-1/2} \tag{9.1.58}$$

Ω_{bif} is the value of Ω at which complex conjugate eigenvalues bifurcate into real, distinct eigenvalues. Figure 9.1.3 illustrates the significance of this condition. The high modes decay like $\rho(\boldsymbol{A})^n$. Therefore the minimum value of $\rho_\infty = \lim_{\Delta t/T \to \infty} \rho(\boldsymbol{A})$ is the most effective in filtering out the spurious high frequencies. This is achieved by selecting $\beta = (\gamma + \frac{1}{2})^2/4$ (i.e., $0.49 = (0.9 + 0.5)^2/4$). Increasing or decreasing β reduces ρ_∞. The minimum value of β that retains unconditional stability is $\beta = \gamma/2 = 0.45$. However, as may be seen from Fig. 9.1.3, the high modes are not damped because $\rho_\infty = 1$. For values of β such that $\gamma/2 < \beta < (\gamma + \frac{1}{2})^2/4$, the eigenvalues of $\boldsymbol{A}$ bifurcate and become real beyond some $\Delta t/T$. This occurs at the minimum points of the $\beta = 0.47$ and 0.45 curves in Fig. 9.1.3. For $\beta = 0.55$, no bifurcation occurs. It may be deduced from this discussion that the ideal value of β is indeed $\beta = (\gamma + \frac{1}{2})^2/4$.

Viscous damping. It may seem that an obvious damping mechanism—viscous damping—has been neglected in this discussion. The reason for this is that viscous damping can be shown to damp an intermediate band of frequencies without significant effect in the all-important high modes [5].

In introducing high-frequency dissipation by selecting $\gamma > \frac{1}{2}$, it is, of course, desirable to maintain good accuracy in the low modes. Unfortunately $\gamma \ne \frac{1}{2}$ results in a drop to first-order accuracy, and for this reason there has been considerable research into the development of alternative methods for algorithmically damping the higher modes.

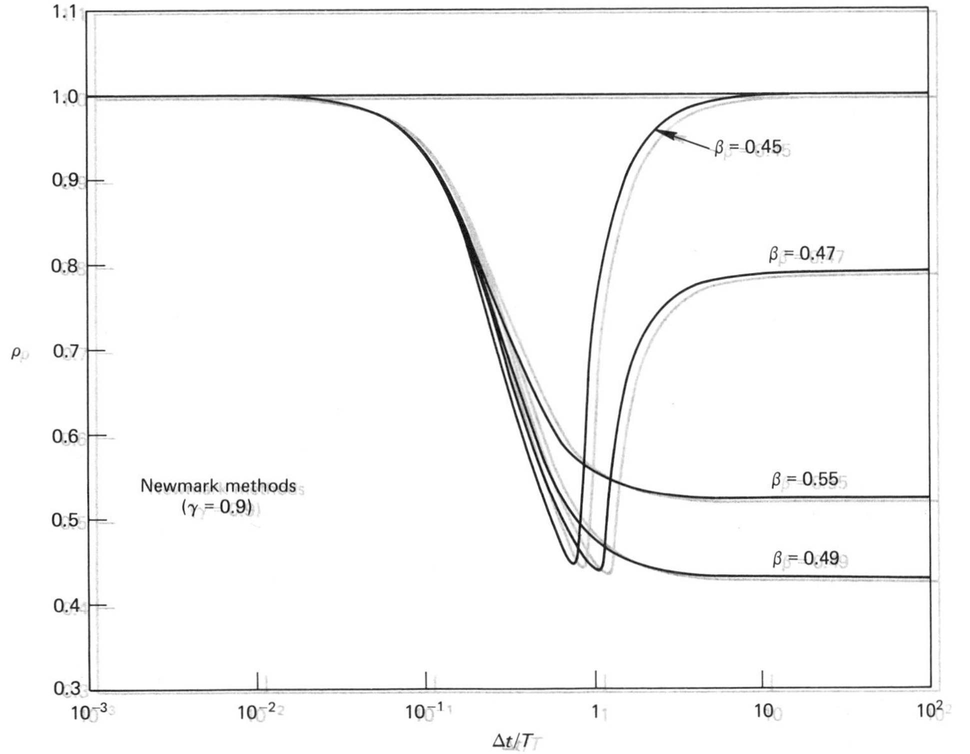

Figure 9.1.3 Spectral radii for Newmark methods for varying β [9].

Exercise 5. Consider the second-order ordinary differential equation

$$\ddot{d} + (\omega^h)^2 d = 0$$

The generalized trapezoidal method (see Sec. 8.1) for this equation consists of the following:

$$a_{n+1} + (\omega^h)^2 d_{n+1} = 0$$

$$d_{n+1} = d_n + \Delta t\{(1 - \alpha)v_n + \alpha v_{n+1}\}$$

$$v_{n+1} = v_n + \Delta t\{(1 - \alpha)a_n + \alpha a_{n+1}\}$$

Set up the 2×2 amplification matrix for this method and determine the stability behavior and the order of accuracy for all $\alpha \in [0, 1]$. In particular, comment on the stability of the forward Euler ($\alpha = 0$) and backward Euler ($\alpha = 1$) methods. [*Ans.*: $\alpha \geq \frac{1}{2}$, unconditionally stable; $\alpha < \frac{1}{2}$, unconditionally unstable; $\alpha = \frac{1}{2}$, second-order accurate, otherwise first-order accurate.]

Exercise 6. Consider the trapezoidal rule (i.e., $\gamma = \frac{1}{2}$, $\beta = \frac{1}{4}$) and assume $F = 0$. Show that as $\Omega \to \infty$,

$$d_{n+1} \to -d_n$$

$$v_{n+1} \to -v_n$$

$$a_{n+1} \to -a_n$$

and thus

$$d_n = (-1)^n d_0$$
$$v_n = (-1)^n v_0$$
$$a_n = (-1)^n a_0$$

Remark

The results of the preceding exercise indicate that spurious high-frequency components may be filtered for output purposes by reporting step-to-step averages, e.g., $d_{n+1/2} = (d_{n+1} + d_n)/2$; cf. Sec. 8.2.2.

Exercise 7. Consider unconditionally stable Newmark schemes (i.e., those schemes for which $2\beta \geq \gamma \geq \frac{1}{2}$). Let ρ_∞ denote the asymptotic value of the spectral radius of $\boldsymbol{A}$ as $\Omega \rightarrow \infty$.

i. Show that if $2\beta = \gamma$, $\rho_\infty = 1$.
ii. Let $\beta = (\gamma + \frac{1}{2})^2/4$ and determine an expression for ρ_∞. Call this value $\overline{\rho}_\infty$. [*Ans*.: $\overline{\rho}_\infty = |1 - 2/(\gamma + \frac{1}{2})|$.]
iii. Show that if $\gamma/2 \leq \beta < (\gamma + \frac{1}{2})^2/4$, $\rho_\infty > \overline{\rho}_\infty$.
iv. Show that if $\beta > (\gamma + \frac{1}{2})^2/4$, $\rho_\infty > \overline{\rho}_\infty$.
v. Determine values of γ and β such that $\rho_\infty = 0$. [*Ans*.: From part (ii), $\beta = 1$ and $\gamma = \frac{3}{2}$.]

Remark

The preceding exercise analytically establishes the trends shown in Figure 9.1.3 and shows in particular that maximal high-frequency numerical dissipation is provided by picking $\beta = (\gamma + \frac{1}{2})^2/4$ for a given $\gamma > \frac{1}{2}$. In addition, it shows that the algorithm $\beta = 1$ and $\gamma = \frac{3}{2}$ provides the strongest possible high-frequency dissipation. Unfortunately, this algorithm achieves very poor accuracy in the low modes and thus it is not useful for transient analysis. However, the strong damping characteristics may be exploited in quickly bringing a transient solution to equilibrium.

Exercise 8. Consider the following predictor-corrector algorithm:

$$\left.\begin{aligned} \widetilde{d}_{n+1} &= d_n + \Delta t v_n + \frac{\Delta t^2}{2} a_n \\ \widetilde{a}_{n+1} + (\omega^h)^2 \widetilde{d}_{n+1} &= 0 \end{aligned}\right\} \quad \text{predictors}$$

$$\left.\begin{aligned} d_{n+1} &= d_n + \Delta t v_n + \frac{\Delta t^2}{2}\{(1 - 2\beta)a_n + 2\beta \widetilde{a}_{n+1}\} \\ v_{n+1} &= v_n + \Delta t\{(1 - \gamma)a_n + \gamma \widetilde{a}_{n+1}\} \\ a_{n+1} + (\omega^h)^2 d_{n+1} &= 0 \end{aligned}\right\} \quad \text{correctors}$$

Assume $\gamma = \frac{1}{2}$.

(A) Determine the order of accuracy.
(B) Determine an expression for the critical time step in terms of β and ω^h.

Solution

(A)
$$\tilde{d}_{n+1} = d_n + \Delta t v_n + \frac{\Delta t^2}{2} a_n \tag{a}$$

$$\tilde{a}_{n+1} + (\omega^h)^2 \tilde{d}_{n+1} = 0 \tag{b}$$

$$d_{n+1} = d_n + \Delta t v_n + \frac{\Delta t^2}{2}\{(1-2\beta)a_n + 2\beta \tilde{a}_{n+1}\} \tag{c}$$

$$v_{n+1} = v_n + \frac{\Delta t}{2}\{a_n + \tilde{a}_{n+1}\} \tag{d}$$

$$a_{n+1} + (\omega^h)^2 d_{n+1} = 0 \tag{e}$$

The truncation errors contributed by (a) through (e) are

$$\tau_{(a)} = O(\Delta t^2), \qquad \tau_{(b)} = 0, \qquad \tau_{(c)} = O(\Delta t^2), \qquad \tau_{(d)} = O(\Delta t^2), \qquad \tau_{(e)} = 0$$

Therefore, on combining the equations as indicated above, it is clear that the algorithm is ***second-order accurate.***

(B)
$$\begin{Bmatrix} d_{n+1} \\ \Delta t v_{n+1} \end{Bmatrix} = \boldsymbol{A} \begin{Bmatrix} d_n \\ \Delta t v_n \end{Bmatrix}$$

Absorbing Δt into the velocity degrees of freedom has no effect on the stability calculation. Eliminating $\tilde{d}_{n+1}$, $\tilde{a}_{n+1}$, and a_n, we get

$$\boldsymbol{A} = \begin{bmatrix} 1 - \dfrac{\Omega^2}{2} + \beta\dfrac{\Omega^4}{2} & 1 - \beta\Omega^2 \\ -\Omega^2 + \gamma\dfrac{\Omega^4}{2} & 1 - \gamma\Omega^2 \end{bmatrix}$$

$$2A_1 = \operatorname{tr} \boldsymbol{A} = 2 - \left(\frac{1}{2} + \gamma\right)\Omega^2 + \frac{\beta}{2}\Omega^4$$

$$A_2 = \det \boldsymbol{A} = 1 + \left(\frac{1}{2} - \gamma\right)\Omega^2 - \frac{\beta}{2}\Omega^4$$

Stability conditions:

$$\text{I.} \quad \begin{cases} \overset{\text{(i)}}{-(A_2 + 1) \le} \, 2A_1 \overset{\text{(ii)}}{\le} A_2 + 1 \\ \overset{\text{(iii)}}{-1 \le} \, A_2 \overset{\text{(iv)}}{<} 1 \end{cases}$$

or

$$\text{II.} \quad \begin{cases} \overset{\text{(i)}}{-1 <} \, A_1 \overset{\text{(ii)}}{<} 1 \\ \text{if } A_2 = 1 \end{cases}$$

$$\text{Ii.}\quad -2 - \left(\frac{1}{2} - \gamma\right)\Omega^2 + \frac{\beta}{2}\Omega^4 \leq 2 - \left(\frac{1}{2} + \gamma\right)\Omega^2 + \frac{\beta}{2}\Omega^4$$

$$\gamma\Omega^2 \leq 2 \qquad \left(\Omega^2 \leq 4 \text{ for } \gamma = \frac{1}{2}\right)$$

$$\text{Iii.}\quad 2 - \left(\frac{1}{2} + \gamma\right)\Omega^2 + \frac{\beta}{2}\Omega^4 \leq 2 + \left(\frac{1}{2} - \gamma\right)\Omega^2 - \frac{\beta}{2}\Omega^4$$

$$\beta\Omega^4 \leq \Omega^2$$

$$\beta\Omega^2 \leq 1$$

$$\text{Iiii.}\quad -1 \leq 1 + \left(\frac{1}{2} - \gamma\right)\Omega^2 - \frac{\beta}{2}\Omega^4 \qquad \left(\beta\Omega^4 \leq 4 \text{ for } \gamma = \frac{1}{2}\right)$$

$$\text{Iiv.}\quad 1 + \left(\frac{1}{2} - \gamma\right)\Omega^2 - \frac{\beta}{2}\Omega^4 < 1 \qquad \left(\beta > 0 \text{ for } \gamma = \frac{1}{2}\right)$$

$$\text{IIi.}\quad \left(\text{Note } \gamma = \frac{1}{2} \text{ and } A_2 = 1 \Rightarrow \beta = 0\right)$$

$$-1 < \frac{1}{2}(2 - \Omega^2)$$

$$\Omega^2 < 4$$

$$\text{IIi.}\quad \frac{1}{2}(2 - \Omega^2) < 1 \qquad \text{(automatic)}$$

Summary ($\gamma = \frac{1}{2}$):

- $\beta \geq 0$
- If $\beta = 0$, $\Omega^2 < 4$
- If $\beta > 0$,

$$\Omega^2 \leq \min\left\{\left(\frac{4}{\beta}\right)^{1/2}, \frac{1}{\beta}, 4\right\}$$

Note. If $\beta = \frac{1}{4}$, $\Omega^2 \leq 4$, which is the same as for the central difference method.

9.1.3 Measures of Accuracy: Numerical Dissipation and Dispersion

In the present application it is useful to introduce accuracy measures, other than local truncation error, which measure numerical dissipation and dispersion. To motivate these measures observe that the continuous SDOF model problem can be exactly solved:

$$d(t) = e^{-\xi\omega^h t}(d_0 \cos \omega_d^h t + c \sin \omega_d^h t) \tag{9.1.59}$$

where

$$c = \frac{v_0 + \xi\omega^h d_0}{\omega_d^h} \tag{9.1.60}$$

$$\omega_d^h = (1 - \xi^2)^{1/2}\omega^h \qquad \text{(damped natural frequency)} \tag{9.1.61}$$

In arriving at (9.1.59) through (9.1.61), we have assumed the homogeneous and "underdamped" case (i.e., $F = 0$ and $0 \leq \xi < 1$, respectively). Likewise, the discrete solution can be put in the following similar form:

$$d_n = e^{-\bar{\xi}\bar{\omega}^h t_n}(d_0 \cos \bar{\omega}_d^h t_n + \bar{c} \sin \bar{\omega}_d^h t_n) \tag{9.1.62}$$

$$\bar{c} = \frac{d_1 - A_1 d_0}{(A_2 - A_1^2)^{1/2}} = \frac{\frac{1}{2}(A_{11} - A_{22})d_0 + A_{12}v_0}{(A_2 - A_1^2)^{1/2}} \tag{9.1.63}$$

$$\bar{\omega}_d^h = (1 - \xi^2)^{1/2}\bar{\omega}^h \tag{9.1.64}$$

Details of the derivation may be found in Hughes [5]. Comparing (9.1.59) and (9.1.62), we see that $\bar{\xi}$ and $\bar{\omega}^h$ are the algorithmic counterparts of ξ and ω^h, respectively. The measures of numerical dissipation and dispersion chosen are $\bar{\xi}$, the ***algorithmic damping ratio,*** and $(\bar{T} - T)/T$, the ***relative period error,*** where $\bar{T} = 2\pi/\bar{\omega}^h$ and $T = 2\pi/\omega^h$. These measures have some advantages over others that have been introduced in the literature (see [5] for elaboration). It is difficult analytically to obtain expressions for $\bar{\xi}$ and $(\bar{T} - T)/T$. Consequently, computer evaluation is usually necessary. Nevertheless, a few analytical results have been obtained for stable Newmark methods, which we collect here:

$$\bar{\xi} = \xi + \frac{1}{2}\left(\gamma - \frac{1}{2}\right)\underbrace{[\Omega + O(\Omega^2)]}_{\geq 0} \tag{9.1.65}$$

$$\frac{\bar{T} - T}{T} = O(\Omega^2) \tag{9.1.66}$$

Results (9.1.65) and (9.1.66) illustrate that first-order errors created by $\gamma \neq \frac{1}{2}$ enter only as dissipation errors and not period errors. Furthermore, $\gamma = \frac{1}{2}$ leads to no numerical dissipation, whereas values of $\gamma > \frac{1}{2}$ create numerical dissipation. Later on we shall graphically compare the Newmark method with other commonly used step-by-step methods on the basis of numerical dissipation and dispersion.

Let $t_n = \bar{T}$. Then

$$\begin{aligned} \exp(-\bar{\xi}\bar{\omega}^h t_n) &= \exp(-2\pi\bar{\xi}) \\ &\cong 1 - 2\pi\bar{\xi} + \cdots \end{aligned} \tag{9.1.67}$$

and so, after one period, for small $\bar{\xi}$, the ***amplitude decay*** is

$$AD \cong 2\pi\bar{\xi} \tag{9.1.68}$$

The accuracy measures are illustrated in the figure on page 506.

9.1.4 Matched Methods

Let us consider an elastodynamics calculation in which a member of the Newmark family of algorithms is to be employed. Assume $\xi = 0$ and $\gamma = \frac{1}{2}$, which implies $\bar{\xi} = 0$ (i.e., no amplitude decay). The periods of individual modes of the finite element model will, however, be distorted by the particular integrator. For example, if the trapezoidal rule is employed ($\beta = \frac{1}{4}$), periods (frequencies, respectively) will be

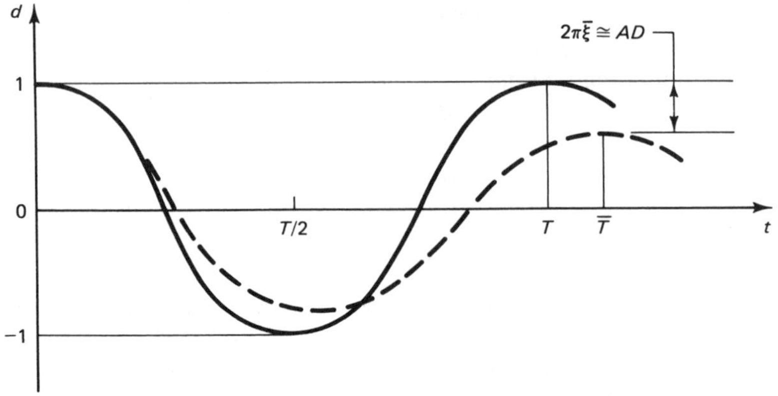

elongated (decreased, respectively). On the other hand, if the central difference method is employed ($\beta = 0$), periods (frequencies, respectively) will be shortened (increased, respectively).[1]

It is important to keep in mind that the periods of the finite element model (i.e., "spatially discrete system") are already in error when compared with the original, "exact" problem. We recall that if the consistent mass matrix is employed, an upper (lower, respectively) bound property may be established for the finite element frequencies (periods, respectively). Experience has indicated that lumped mass matrices tend to behave in the opposite fashion.

These observations suggest that transient integrators and mass matrices should be "matched" so that the induced period errors may tend to cancel [3]. For example, trapezoidal rule and consistent mass, and central differences and lumped mass, would be appropriate matches in that sense.

The following example, which is taken from [3], makes these ideas precise in the context of a model problem.

Example

Consider a finite element model for the wave equation[2]

$$u_{,tt} = u_{,xx} \tag{9.1.69}$$

consisting of n_{el} linear elements of equal length $h = 1/n_{el}$. Recall that in this case the element stiffness is given by (we drop the e superscript since all elements are the same):

$$\boldsymbol{k} = \frac{1}{h}\begin{bmatrix} 1 & -1 \\ -1 & 1 \end{bmatrix} \tag{9.1.70}$$

A one-parameter (r) family of element mass matrices discussed in Chapter 7 may be written as

[1] See Fig. 9.3.4.

[2] We may think of (9.1.69) as governing the transient response of a linear elastic rod in which the density, ρ, and Young's modulus, E, are chosen to attain unit speed of propagation. This simplifies the development somewhat. The general case is described in Remark 1 at the end of this section.

$$m_{(r)} = h\begin{bmatrix} \frac{1}{2} - r & r \\ r & \frac{1}{2} - r \end{bmatrix} \tag{9.1.71}$$

Recall that particular values of r yield the commonly used mass matrices, e.g.,

i. *consistent mass* ($r = \frac{1}{6}$)

$$m_{(1/6)} = \frac{h}{6}\begin{bmatrix} 2 & 1 \\ 1 & 2 \end{bmatrix} \tag{9.1.72}$$

ii. *lumped mass* ($r = 0$)

$$m_{(0)} = \frac{h}{2}\begin{bmatrix} 1 & 0 \\ 0 & 1 \end{bmatrix} \tag{9.1.73}$$

iii. *higher-order mass* ($r = \frac{1}{12}$)

$$m_{(1/12)} = \frac{h}{12}\begin{bmatrix} 5 & 1 \\ 1 & 5 \end{bmatrix} \tag{9.1.74}$$

In what follows we will have to distinguish carefully between three different frequency parameters, defined as follows:

ω = exact frequency

ω^h = frequency produced by finite element spatial discretization

$\overline{\omega}^h$ = frequency produced by time integration algorithm in conjunction with finite element spatial discretization

Both the finite element spatial discretization and time integration algorithm introduce errors. In practice we see $\overline{\omega}^h$, but we would like to see ω.

The finite element matrix equation corresponding to (9.1.69) is obtained by assembling (9.1.70) and (9.1.71) in the usual way. The result is

$$\boldsymbol{M}_{(r)}\ddot{\boldsymbol{d}} + \boldsymbol{K}\boldsymbol{d} = \boldsymbol{0} \tag{9.1.75}$$

Let $d_A(t)$ denote the displacement at node number A, and let

$$\mathscr{C}d_A = d_{A+1} - 2d_A + d_{A-1} \tag{9.1.76}$$

$\mathscr{C}$ is called the (undivided) second central difference operator. (We assume that the nodes are labeled in ascending order from left to right.) If A represents an interior node, the scalar equation in (9.1.75) associated with node A may be written as

$$h(1 + r\mathscr{C})\ddot{d}_A - \frac{1}{h}\mathscr{C}d_A = 0 \tag{9.1.77}$$

Our aim is to solve (9.1.77) *exactly*. To this end we assume the solution has the form

$$d_A(t) = S_A T(t) \tag{9.1.78}$$

Substituting (9.1.78) in (9.1.77) and adding and subtracting $(\omega^h)^2 T(1 + r\mathscr{C})S_A$ yields an equation convenient for subsequent developments, namely,

$$[\ddot{T} + (\omega^h)^2 T](1 + r\mathscr{C})S_A - \left[(\omega^h)^2(1 + r\mathscr{C})S_A + \frac{1}{h^2}\mathscr{C}S_A\right]T = 0 \tag{9.1.79}$$

Satisfaction of (9.1.79) is achieved by selecting S_A and T such that

$$[(\omega^h h)^2(1 + r\mathscr{C}) + \mathscr{C}]S_A = 0 \tag{9.1.80}$$

and

$$\ddot{T} + (\omega^h)^2 T = 0 \tag{9.1.81}$$

We assume the general solution of (9.1.80) takes the form

$$S_A = c_1 \sin \frac{A\lambda}{n_{el}} + c_2 \cos \frac{A\lambda}{n_{el}} \tag{9.1.82}$$

Permissible values of λ are determined by imposing homogeneous boundary conditions. The results are

$$\left.\begin{matrix}\text{Fixed-fixed}\\ \text{Free-free}\end{matrix}\right\} \quad \lambda = l\pi \tag{9.1.83}$$

$$\left.\begin{matrix}\text{Free-fixed}\\ \text{Fixed-free}\end{matrix}\right\} \quad \lambda = \frac{(2l-1)\pi}{2} \tag{9.1.84}$$

where l is an integer. It is easily shown that the above values of λ coincide with the exact frequencies, ω, obtained by solving the eigenvalue problem associated with (9.1.69).

We shall provide a detailed study of the fixed-fixed case in what follows. The other cases may be carried out analogously and the interested reader may wish to provide the details as an exercise.

A consequence of the boundary conditions in the fixed-fixed case is that $c_2 = 0$ in (9.1.82). Under these circumstances, (9.1.80) becomes

$$(\omega^h h)^2 \sin \frac{A\omega}{n_{el}} + [1 + r(\omega^h h)^2]\left(\sin \frac{(A-1)\omega}{n_{el}} - 2 \sin \frac{A\omega}{n_{el}} + \sin \frac{(A+1)\omega}{n_{el}}\right) = 0 \tag{9.1.85}$$

Making use of the trigonometric identity $\sin(a \pm b) = \sin a \cos b \pm \sin b \cos a$ in (9.1.85) results in

$$\left(\frac{\omega^h h}{2}\right)^2 \left[1 + 2r\left(\cos \frac{\omega}{n_{el}} - 1\right)\right] + \frac{1}{2}\left(\cos \frac{\omega}{n_{el}} - 1\right) = 0 \tag{9.1.86}$$

The half-angle formula, $\sin^2 a/2 = (1 - \cos a)/2$, can be used to simplify (9.1.86), viz.,

$$\left(\frac{\omega^h h}{2}\right)^2 \left[1 - 4r \sin^2 \frac{\omega}{2n_{el}}\right] - \sin^2 \frac{\omega}{2n_{el}} = 0 \tag{9.1.87}$$

which, upon replacing n_{el} by $1/h$, leads to

$$\frac{\omega^h}{\omega} = \frac{\sin \omega h/2}{(\omega h/2)[1 - 4r \sin^2(\omega h/2)]^{1/2}} \tag{9.1.88}$$

The other boundary conditions result in identical expressions. Equation (9.1.88) is plotted in Fig. 9.1.4 for the three cases cited previously.

We now turn our attention to a frequency expression for the Newmark method, derived in [5], in which we assume $\xi = 0$ and $\gamma = \frac{1}{2}$, viz.,

$$\overline{\omega}^h \Delta t = \arctan \left[\frac{(A_2 - A_1^2)^{1/2}}{A_1}\right] \tag{9.1.89}$$

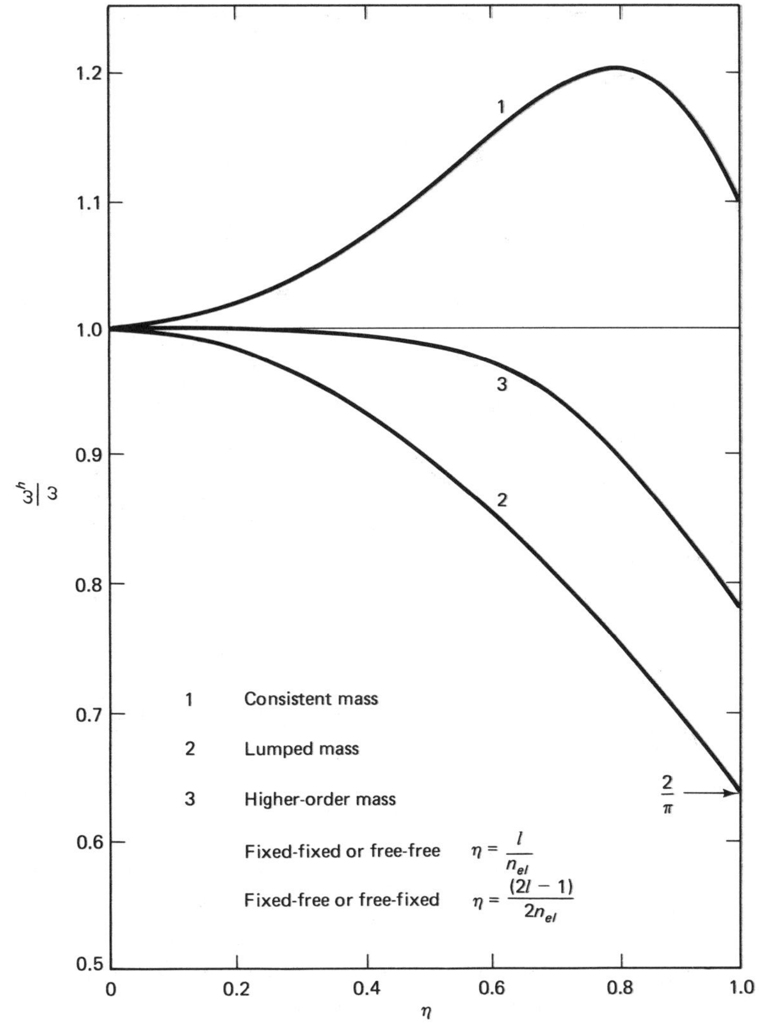

Figure 9.1.4 Error in discrete spectrum of simple bar [2].

where

$$\left.\begin{aligned} A_1 &= 1 - \frac{(\omega^h \Delta t)^2}{[2(1 + \beta(\omega^h \Delta t)^2)]} \\ A_2 &= 1 \end{aligned}\right\} \tag{9.1.90}$$

For reasons which will become obvious, we wish to obtain an expression for $\sin^2(\bar{\omega}^h \Delta t/2)$. To do this we make use of the trigonometric identities $\cot a = \cot 2a + \csc 2a$ and $\csc a = (\cot^2 a + 1)^{1/2}$. These, (9.1.89), and (9.1.90) enable us to write

$$\cot^2 \frac{\bar{\omega}^h \Delta t}{2} = \frac{1 + A_1}{1 - A_1} = -1 + 4\beta + \left(\frac{2}{\omega^h \Delta t}\right)^2 \tag{9.1.91}$$

Now we use the identity $\sin^2 a = 1/(1 + \cot^2 a)$ to achieve the desired result

$$\sin^2 \frac{\overline{\omega}^h \Delta t}{2} = \frac{1}{4\beta + (2/\omega^h \Delta t)^2} \tag{9.1.92}$$

Equation (9.1.88) is substituted into (9.1.92) to obtain the main result:

$$\boxed{\sin^2 \frac{\overline{\omega}^h \Delta t}{2} = \frac{\sin^2(\omega h/2)}{4[\beta - r(h/\Delta t)^2] \sin^2(\omega h/2) + (h/\Delta t)^2}} \tag{9.1.93}$$

Note that if $h = \Delta t$ and $\beta = r$, then (9.1.93) implies that $\overline{\omega}^h = \omega$, i.e., *the errors introduced by finite element spatial discretization, the particular mass matrix and temporal algorithm all cancel to yield exact results.* The time step $\Delta t = h$ is called the ***characteristic time step.*** It is equal to the transit time for a "wave" moving at unit speed to traverse one element.

Remarks

1. If equation (9.1.69) is replaced by

$$u_{,tt} = c^2 u_{,xx} \tag{9.1.94}$$

where $c^2 = E/\rho$, then h should be replaced by h/c in (9.1.93). c is called the ***characteristic velocity,*** or ***speed of sound;*** it represents the propagation speed of solutions of (9.1.94). (See [6] for basic results on wave propagation.) Thus if we compute at the characteristic time step, $\Delta t = h/c$, and select $\beta = r$, the numerical results will be exact at the nodes no matter how few elements are employed ("superconvergence"). The combination $\beta = r = 0$ (i.e., central difference method and lumped mass) is the simplest in practice. We note also that for $r = 0$, as $\beta \to 0^+$, we approach the exact solution. Numerical results along these lines are presented in [7] and Fig. 9.1.5.

2. It is interesting to note that reducing the size of the time step while holding the mesh length fixed can only *worsen* the results. In this case we converge to the exact solution of the spatially discrete, temporally continuous system (i.e., "mass points and springs") rather than the exact solution of (9.1.94). The characteristic time step also represents the stability limit (i.e., critical time step) of the lumped mass, central difference method, and, therefore, computations cannot be performed at larger time steps. For these reasons, it is generally considered advantageous to compute at a time step as close to critical as possible.

3. So that there is no chance of confusion, we wish to emphasize that in more general settings (e.g., unequal element lengths, variable material properties, multidimensional problems, etc.), results obtained by matched methods, such as central differences and lumped mass, will not be exact. However, it is felt that results obtained by matched methods will generally be superior to inappropriate combinations, such as consistent mass and central differences.

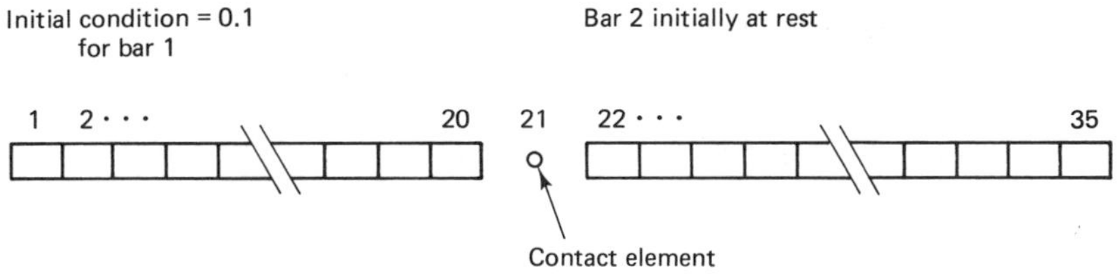

Data:

$L_1 = 10$	$L_2 = 10$	(length/bar)
$E_1 = 49$	$E_2 = 100$	(Young's modulus)
$\rho_1 = 0.01$	$\rho_2 = 0.01$	(density)
$A_1 = 1$	$A_2 = 1$	(area)
$\Delta x_1 = 0.5$	$\Delta x_2 = 0.714$	(element length)
	$\Delta t = 0.00714$	(time step)
	$\beta = 0.001001$	(Newmark parameters)
	$\gamma = 0.502$	

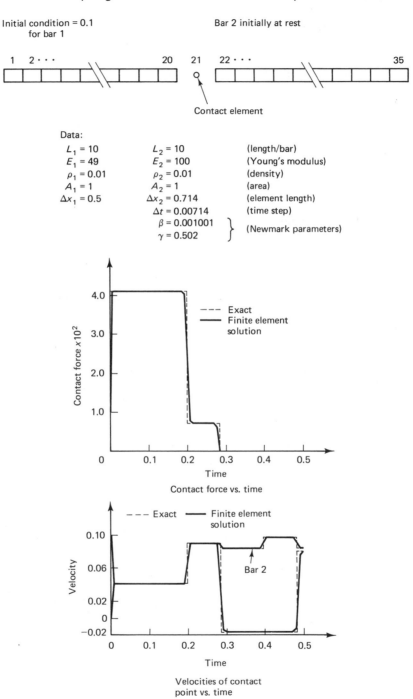

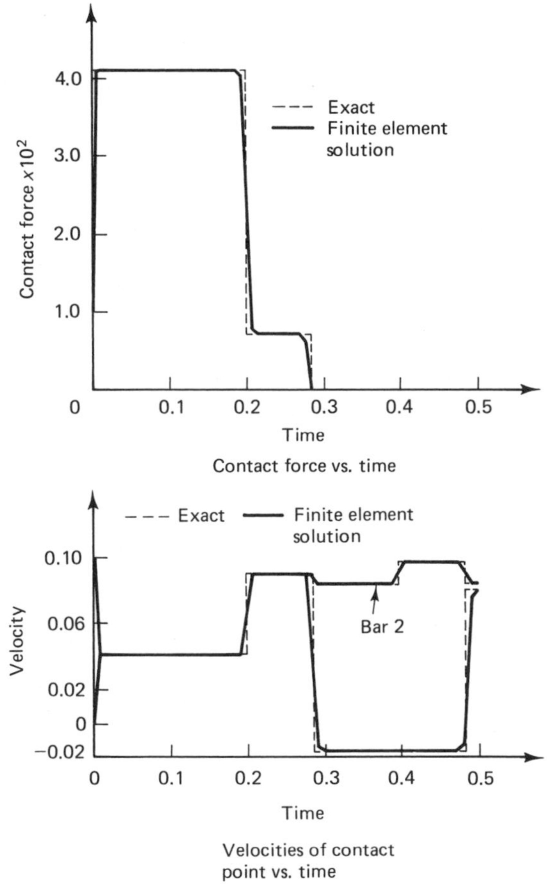

Figure 9.1.5 Impact of two dissimilar bars.

9.1.5 Additional Exercises

Exercise 9. Show that the trapezoidal rule (i.e., $\beta = \frac{1}{4}$, $\gamma = \frac{1}{2}$) exactly conserves total energy when $\boldsymbol{C} = \boldsymbol{0}$ and $\boldsymbol{F} = \boldsymbol{0}$.

Solution

In terms of undivided forward-difference and mean-value operators (i.e., [ ] and $\langle\ \rangle$, respectively, see Sec. 8.2.4), the Newmark formulas, namely, (9.1.5) and (9.1.6), may be written as

$$[\boldsymbol{v}_n] = \Delta t \boldsymbol{a}_{n+\gamma} \tag{9.1.95}$$

$$[\boldsymbol{d}_n] = \Delta t \langle \boldsymbol{v}_n \rangle + \Delta t^2(\beta - \gamma/2)[\boldsymbol{a}_n] \tag{9.1.96}$$

The ***total energy*** is defined by

$$E(\boldsymbol{d}_n, \boldsymbol{v}_n) = T(\boldsymbol{v}_n) + U(\boldsymbol{d}_n) \tag{9.1.97}$$

where

$$T(\boldsymbol{v}_n) = \frac{1}{2}\boldsymbol{v}_n^T \boldsymbol{M} \boldsymbol{v}_n \qquad (\textit{kinetic energy}) \tag{9.1.98}$$

$$U(\boldsymbol{d}_n) = \frac{1}{2}\boldsymbol{d}_n^T \boldsymbol{K} \boldsymbol{d}_n \qquad (\textit{strain energy}) \tag{9.1.99}$$

For future use, note that (9.1.4) and the hypotheses result in

$$\boldsymbol{M}[\boldsymbol{a}_n] + \boldsymbol{K}[\boldsymbol{d}_n] = \boldsymbol{0} \tag{9.1.100}$$

$$\boldsymbol{M}\langle \boldsymbol{a}_n \rangle + \boldsymbol{K}\langle \boldsymbol{d}_n \rangle = \boldsymbol{0} \tag{9.1.101}$$

Collecting results, we obtain the following:

$$\begin{aligned}
[T(\boldsymbol{v}_n)] &= [\boldsymbol{v}_n]^T \boldsymbol{M} \langle \boldsymbol{v}_n \rangle \qquad (\text{by } \boldsymbol{M} = \boldsymbol{M}^T) \\
&= \boldsymbol{a}_{n+\gamma}^T \boldsymbol{M}\left([\boldsymbol{d}_n] - \Delta t^2\left(\beta - \frac{\gamma}{2}\right)[\boldsymbol{a}_n]\right) \\
&= \left(\langle \boldsymbol{a}_n \rangle + \left(\gamma - \frac{1}{2}\right)[\boldsymbol{a}_n]\right)^T \boldsymbol{M}[\boldsymbol{d}_n] - \Delta t^2\left(\beta - \frac{\gamma}{2}\right)[\boldsymbol{a}_n] \\
&= -\langle \boldsymbol{d}_n \rangle^T \boldsymbol{K}[\boldsymbol{d}_n] - \left(\gamma - \frac{1}{2}\right)[\boldsymbol{d}_n]^T \boldsymbol{K}[\boldsymbol{d}_n] \\
&\quad - \Delta t^2\left(\beta - \frac{\gamma}{2}\right)\left([T(\boldsymbol{a}_n)] + 2\left(\gamma - \frac{1}{2}\right)T([\boldsymbol{a}_n])\right) \\
&= -[U(\boldsymbol{d}_n)] - 2\left(\gamma - \frac{1}{2}\right)U([\boldsymbol{d}_n]) \\
&\quad - \Delta t^2\left(\beta - \frac{\gamma}{2}\right)\left([T(\boldsymbol{a}_n)] + 2\left(\gamma - \frac{1}{2}\right)T([\boldsymbol{a}_n])\right)
\end{aligned} \tag{9.1.102}$$

Therefore

$$[T(\boldsymbol{v}_n)] + [U(\boldsymbol{d}_n)] + \Delta t^2\left(\beta - \frac{\gamma}{2}\right)[T(\boldsymbol{a}_n)]$$
$$= -2\left(\gamma - \frac{1}{2}\right)U([\boldsymbol{d}_n]) - \Delta t^2\left(\gamma - \frac{1}{2}\right)(2\beta - \gamma)T([\boldsymbol{a}_n]) \tag{9.1.103}$$

It follows from (9.1.103) that if $\beta = \frac{1}{4}$ and $\gamma = \frac{1}{2}$,

$$[E(\boldsymbol{d}_n, \boldsymbol{v}_n)] = [T(\boldsymbol{v}_n)] + [U(\boldsymbol{d}_n)] = 0 \tag{9.1.104}$$

That is, ***total energy is conserved.***

Remarks

1. From (9.1.103) it follows that if $\gamma = \frac{1}{2}$, $E(\boldsymbol{d}_n, \boldsymbol{v}_n) + \Delta t^2(\beta - \gamma/2)T(\boldsymbol{a}_n)$ is conserved.

2. If $\gamma > \frac{1}{2}$ and $\beta = \gamma/2$, then, from (9.1.103), total energy is nonincreasing. That is $[E(\boldsymbol{d}_n, \boldsymbol{v}_n)] \leq 0$.

3. If $\gamma > \frac{1}{2}$ and $\beta > \gamma/2$, then $E(\boldsymbol{d}_n, \boldsymbol{v}_n) + \Delta t^2(\beta - \gamma/2)T(\boldsymbol{a}_n)$ is nonincreasing.

Exercise 10. (T. K. Caughey). Consider application of the trapezoidal algorithm to $\ddot{d} + (\omega^h)^2 d = 0$. Define $x_n = \langle\langle d_{n-1}\rangle\rangle = (d_{n+1} + 2d_n + d_{n-1})/4$. Show that

$$x_n = \frac{d_n}{1 + (\omega^h \Delta t)^2/4}$$

It may be concluded that x_n is a second-order accurate approximation to d_n and acts as a high-frequency filter because $x_n \to 0$ as $\omega^h \Delta t \to \infty$.

9.2 SUMMARY OF TIME-STEP ESTIMATES FOR SOME SIMPLE FINITE ELEMENTS

As we remarked previously, sufficient conditions for stability in finite element analysis may be obtained from estimates of the maximum eigenvalues of individual elements. We give some examples.

Example 1 (Two-node linear rod element)

If we assume a ***lumped*** (i.e., diagonal) mass matrix, then

$$\omega^h_{\max} = \frac{2c}{h} \tag{9.2.1}$$

where h is the element length and $c = \sqrt{E/\rho}$ is the so-called bar-wave velocity, in which E is Young's modulus and ρ is density. The critical time step for the Newmark method with $\beta = 0,\ \gamma = \frac{1}{2}$ (central-difference method) is

$$\boxed{\Delta t \leq \frac{2}{\omega^h_{\max}} = \frac{h}{c}} \tag{9.2.2}$$

which is the time required for a bar wave to traverse one element.

If we assume a ***consistent*** mass matrix, then

$$\omega^h_{\max} = \frac{2\sqrt{3}\,c}{h} \tag{9.2.3}$$

resulting in a reduced critical time step, viz.,

$$\boxed{\Delta t \leq \frac{h}{\sqrt{3}\,c}} \tag{9.2.4}$$

This result is typical: ***Consistent-mass matrices tend to yield smaller critical time steps than lumped-mass matrices.***

Example 2 (Three-node quadratic rod element)

For this case, if we assume a lumped mass matrix based on a Simpson's rule weighting [8] (i.e., the ratio of the middle node mass to the end node masses is 4), we get

$$\omega^h_{\max} = \frac{2\sqrt{6}\,c}{h} \tag{9.2.5}$$

$$\boxed{\Delta t \leq \frac{h}{\sqrt{6}\,c}} \tag{9.2.6}$$

Comparison of (9.2.6) with (9.2.2) reveals that the allowable time step is about 0.4082 that for linear elements with lumped mass. This is based upon equal element lengths. Perhaps a more equitable comparison is one based upon equal nodal spacing. In this case the ratio doubles to 0.8164, but still the advantage is with linear elements.

Remark

Results for one-dimensional heat conduction may be deduced from the preceding cases by employing

$$\Delta t \leq \frac{2}{\lambda^h_{\max}} \tag{9.2.7}$$

where $\lambda^h_{\max} = (\omega^h_{\max})^2$ and E/ρ is replaced by $k = \kappa/(\rho c)$, the diffusivity coefficient. For example, corresponding to (9.2.2) we have the classical result

$$\boxed{\Delta t \le \frac{h^2}{2k}} \tag{9.2.8}$$

Example 3 (The linear beam element)

This element has been described in [9], [10], and Sec. 5.5. Transverse displacement, w, and face rotation, θ, are assumed to vary linearly over the element. One-point Gauss quadrature exactly integrates the bending stiffness and appropriately underintegrates the shear stiffness to avoid "locking" [9]. We assume the trapezoidal rule is used to develop the lumped mass matrix [10]. The matrices for a typical element describing bending in the plane are:

$$\boldsymbol{k}_b = \frac{EI}{h}\begin{bmatrix} 0 & 0 & 0 & 0 \\ 0 & 1 & 0 & -1 \\ 0 & 0 & 0 & 0 \\ 0 & -1 & 0 & 1 \end{bmatrix} \quad \text{(bending stiffness)} \tag{9.2.9}$$

$$\boldsymbol{k}_s = \frac{\mu \hat{A}^s}{h}\begin{bmatrix} 1 & \frac{h}{2} & -1 & \frac{h}{2} \\ \frac{h}{2} & \frac{h^2}{4} & \frac{-h}{2} & \frac{h^2}{4} \\ -1 & \frac{-h}{2} & 1 & \frac{-h}{2} \\ \frac{h}{2} & \frac{h^2}{4} & \frac{-h}{2} & \frac{h^2}{4} \end{bmatrix} \quad \text{(shear stiffness)} \tag{9.2.10}$$

$$\boldsymbol{m} = \rho\hat{A}\frac{h}{2}\begin{bmatrix} 1 & 0 & 0 & 0 \\ 0 & \frac{I}{\hat{A}} & 0 & 0 \\ 0 & 0 & 1 & 0 \\ 0 & 0 & 0 & \frac{I}{\hat{A}} \end{bmatrix} \quad \text{(mass)} \tag{9.2.11}$$

where here $\hat{A}$ is the cross-section area, $\hat{A}^s$ is the shear area, I is the moment of inertia, and μ is the shear modulus. The degree-of-freedom ordering is w_1, θ_1, w_2, θ_2, where the subscript is the node number.

Solution of the eigenvalue problem results in

$$\omega^h_{\max} = \max\left\{\frac{2c}{h}, \left(\frac{2c_s}{h}\right)\left[1 + \frac{\hat{A}}{I}\left(\frac{h}{2}\right)^2\right]^{1/2}\right\} \tag{9.2.12}$$

where c is the bar-wave velocity and $c_s^2 = \mu \hat{A}^s/(\rho \hat{A})$, the beam shear-wave velocity. Thus the critical time step for the central difference operator is given by

$$\boxed{\Delta t \le \min\left\{\frac{h}{c}, \left(\frac{h}{c_s}\right)\left[1 + \frac{\hat{A}}{I}\left(\frac{h}{2}\right)^2\right]^{-1/2}\right\}} \tag{9.2.13}$$

To get a feeling for these quantities, we shall take a typical situation. Assume the cross section is rectangular with depth t and width 1. This results in $\hat{A} = t$ and $I = t^3/12$. We assume the ratio of wave speeds $c/c_s = \sqrt{3}$. This corresponds to a Poisson's ratio of $\frac{1}{4}$ and shear correction factor $\kappa = \hat{A}^s/\hat{A} = \frac{5}{6}$, so it is a reasonable approximation for most metals. The time step incurred by the bending mode will be critical when

$$\frac{h}{t} \le \sqrt{\frac{2}{3}} \cong 0.8165 \tag{9.2.14}$$

This would be the case only for a very deep beam or an extremely fine mesh and is thus unlikely in practice. The more typical situation in structural analysis is when $t \ll h$ (i.e., very thin beams or coarse meshing). In this case the critical time step is slightly less than t/c, the time for a bar wave to traverse the thickness. As this is an extremely small time step, the cost of explicit integration becomes prohibitive.

A more favorable condition can be derived by adopting different values for the rotational lumped-mass coefficients. Specifically, take

$$\boldsymbol{m} = \rho \hat{A} \frac{h}{2} \begin{bmatrix} 1 & 0 & 0 & 0 \\ 0 & \alpha & 0 & 0 \\ 0 & 0 & 1 & 0 \\ 0 & 0 & 0 & \alpha \end{bmatrix} \tag{9.2.15}$$

and select α to achieve a more desirable critical time step, without upsetting convergence. Beam mass matrices of the above type were introduced by Key and Beisinger [11]. This time, solution of the eigenvalue problem yields

$$\omega_{\max}^h = \max\left\{\left(\frac{2c}{h}\right)\left(\frac{I}{\alpha \hat{A}}\right)^{1/2}, \left(\frac{2c_s}{h}\right)\left[1 + \frac{1}{\alpha}\left(\frac{h}{2}\right)^2\right]^{1/2}\right\} \tag{9.2.16}$$

and thus

$$\boxed{\Delta t \le \min\left\{\left(\frac{h}{c}\right)\left(\frac{\alpha \hat{A}}{I}\right)^{1/2}, \left(\frac{h}{c_s}\right)\left[1 + \frac{1}{\alpha}\left(\frac{h}{2}\right)^2\right]^{-1/2}\right\}} \tag{9.2.17}$$

Taking the value $\alpha = h^2/8$ in (9.2.17) and again assuming $\hat{A}/I = 12/t^2$ and $c/c_s = \sqrt{3}$ results in

$$\boxed{\Delta t \le \min\left\{\left(\frac{h}{c}\right)\left(\sqrt{\frac{3}{2}} \cdot \frac{h}{t}\right), \frac{h}{c}\right\}} \tag{9.2.18}$$

As long as

$$\frac{h}{t} > \sqrt{\frac{2}{3}} \cong 0.8165 \tag{9.2.19}$$

bar-wave transit time or better is achieved. Condition (9.2.19) will almost certainly be the case in practice. Analogous procedures may be applied to plate and shell elements (see Sec. 9.5 and Hughes, Liu, and Levit [12]). The original paper of Key and Beisinger [11] may be consulted for a treatment of the cubic, Bernoulli-Euler beam element, and analogous shell element considerations.

Example 4 (Quadrilateral and hexahedral elements)

Flanagan and Belytschko [13] have performed a valuable analysis of the one-point quadrature (four-node) quadrilateral and (eight-node) hexahedron, applicable to arbitrary geometric configurations of the elements. They obtain the following estimate of maximum element frequency:

$$\omega_{\max}^h \le c_d g^{1/2} \tag{9.2.20}$$

where $c_d^2 = (\lambda + 2\mu)/\rho$, dilatational wave velocity; λ and μ are the Lamé parameters; and g is a geometric parameter. For example, in the case of the quadrilateral, g is defined as follows:

$$g = \frac{4}{A^2} \sum_{i=1}^{2} \sum_{a=1}^{4} B_{ia} B_{ia} \tag{9.2.21}$$

$$[B_{ia}] = \frac{1}{2} \begin{bmatrix} (y_2 - y_4) & (y_3 - y_1) & (y_4 - y_2) & (y_1 - y_3) \\ (x_4 - x_2) & (x_1 - x_3) & (x_2 - x_4) & (x_3 - x_1) \end{bmatrix} \tag{9.2.22}$$

where x_a and y_a are the coordinates of node a and A is the area. (For the definition of g in the case of the hexahedron, see [13].) The estimate, (9.2.20), leads to a *sufficient* condition for stability. For the central difference method (9.2.20) results in

$$\boxed{\Delta t \le \frac{2}{c_d g^{1/2}}} \tag{9.2.23}$$

As an example of the restriction imposed by (9.2.23), consider a rectangular element with side lengths h_1 and h_2. In this case (9.2.23) becomes

$$\boxed{\Delta t \le \frac{1}{c_d(1/h_1^2 + 1/h_2^2)^{1/2}}} \tag{9.2.24}$$

For higher-order elements, there appears that little of a precise nature has been done. Most results are of the form (9.2.23) where the geometric factor is approximated by trial and error. One would hope that, with the aid of automatic symbolic manipulators, improved time-step estimates will become available in the ensuing years for more complex elements, material properties, etc.

Exercises

1. Verify the results obtained in Example 1.
2. Verify the results obtained in Example 2.

Solution of Exercise 2

The stiffness matrix and Simpson's rule lumped mass are:

$$\boldsymbol{k} = \frac{E}{h}\begin{bmatrix} 7 & -8 & 1 \\ -8 & 16 & -8 \\ 1 & -8 & 7 \end{bmatrix}$$

$$\boldsymbol{m} = \frac{\rho h}{6}\begin{bmatrix} 1 & 0 & 0 \\ 0 & 4 & 0 \\ 0 & 0 & 1 \end{bmatrix}$$

The eigenvalues and eigenvectors are:

Eigenvalue	Eigenvector	
0	$\{\ 1 \quad 1 \quad 1\}^T$	Rigid translation
$12\ c^2/h^2$	$\{-1 \quad 0 \quad 1\}^T$	Uniform extension
$24\ c^2/h^2$	$\left\{1 \quad \frac{-1}{2} \quad 1\right\}^T$	Quadratic extension

Discussion

It is instructive to know how the eigenvalues and eigenvectors may be deduced without going through the setup and calculation of the full eigenvalue problem. These ideas prove extremely useful in more complicated two- and three-dimensional situations. For example, consider the linear beam element described by (9.2.9) through (9.2.11). The eigenvalues and eigenvectors are:

Eigenvalue	Eigenvector	
0	$\{\ 1 \quad 0 \quad 1 \quad 0\}^T$	Rigid translation
0	$\left\{-1 \quad \frac{2}{h} \quad 1 \quad \frac{2}{h}\right\}^T$	Rigid rotation
$\frac{4c^2}{h^2}$	$\{\ 0 \quad 1 \quad 0 \quad -1\}^T$	Bending
$\left(\frac{4c_s^2}{h^2}\right)\left(1 + \frac{\hat{A}}{I}\left(\frac{h}{2}\right)^2\right)$	$\left\{1 \quad \frac{\hat{A}h}{2I} \quad -1 \quad \frac{\hat{A}h}{2I}\right\}^T$	Shear

They may be obtained as follows: First, note that from physical considerations two of the eigenvectors must correspond to rigid motions. The rigid translation mode is easily guessed, and it can be verified that it is indeed an eigenvector. The rigid rotation clearly must have the form $\{-1 \quad a \quad 1 \quad a\}^T$, where a is a constant determined by the requirement that the shear force engendered by this mode (i.e., the product of the shear stiffness and this vector) must be zero. This condition entails $a = 2/h$. The bending mode was guessed and it was easily verified to be an eigenvector, viz.,

$$(\boldsymbol{k}_b + \boldsymbol{k}_s - \lambda^h \boldsymbol{m}) \begin{Bmatrix} 0 \\ 1 \\ 0 \\ -1 \end{Bmatrix} = \left(\frac{2EI}{h} - \frac{\rho I h}{2}\lambda^h\right) \begin{Bmatrix} 0 \\ 1 \\ 0 \\ -1 \end{Bmatrix} = \begin{Bmatrix} 0 \\ 0 \\ 0 \\ 0 \end{Bmatrix} \Leftrightarrow \lambda^h = \frac{4c^2}{h^2}$$

The three eigenvectors computed so far are orthogonal with respect to $\boldsymbol{m}$. Since the fourth eigenvector must be orthogonal to these three (with respect to $\boldsymbol{m}$), it can be computed by assuming it has the form $\{a_1\ a_2\ a_3\ a_4\}^T$, where $a_1, \ldots, a_4$ are to be identified by the three orthogonality conditions

$$a_1 + a_3 = 0$$

$$-\hat{A}a_1 + \frac{2}{h}Ia_2 + \hat{A}a_3 + \frac{2}{h}Ia_4 = 0$$

$$a_2 - a_4 = 0$$

Taking $a_1 = 1$, these equations yield the mode $\left\{1 \quad \frac{\hat{A}h}{2I} \quad -1 \quad \frac{\hat{A}h}{2I}\right\}^T$. The product of this vector with the bending stiffness is zero, and therefore this is a pure shear deformation. The verification that this mode is an eigenvector, and the computation of the eigenvalue is:

$$(\boldsymbol{k}_b + \boldsymbol{k}_s - \lambda^h \boldsymbol{m}) \begin{Bmatrix} 1 \\ \dfrac{\hat{A}h}{2I} \\ -1 \\ \dfrac{\hat{A}h}{2I} \end{Bmatrix} = \left(\hat{A}^s G \frac{2}{h}\left[1 + \frac{\hat{A}}{I}\left(\frac{h}{2}\right)^2\right] - \rho\hat{A}\frac{h}{2}\lambda^h\right) \begin{Bmatrix} 1 \\ \dfrac{h}{2} \\ -1 \\ \dfrac{h}{2} \end{Bmatrix} = \begin{Bmatrix} 0 \\ 0 \\ 0 \\ 0 \end{Bmatrix}$$

$$\Leftrightarrow \lambda^h = \left(\frac{4c_s^2}{h^2}\right)\left(1 + \frac{\hat{A}}{I}\left(\frac{h}{2}\right)^2\right)$$

Exercise 3. Generalize the argument of the preceding discussion to pertain to the lumped-mass matrix given by (9.2.15). Show that the rigid modes are the same, whereas the eigenvalues and eigenvectors of the other modes become:

Eigenvalue	Eigenvector	
$\dfrac{4c^2}{h^2}\left(\dfrac{I}{\alpha\hat{A}}\right)$	$\{0 \quad 1 \quad 0 \quad -1\}^T$	Bending
$\left(\dfrac{4c_s^2}{h^2}\right)\left[1 + \dfrac{1}{\alpha}\left(\dfrac{h}{2}\right)^2\right]$	$\left\{1 \quad \alpha^{-1}\dfrac{h}{2} \quad -1 \quad \alpha^{-1}\dfrac{h}{2}\right\}^T$	Shear

Exercise 4. (Key and Beisinger [11]). Consider a thin beam element based upon the Bernoulli-Euler theory in which a Hermite cubic is assumed for the displacement function. The element matrices are given by

$$\boldsymbol{k} = \frac{2EI}{h^3}\begin{bmatrix} 6 & 3h & -6 & 3h \\ & 2h^2 & -3h & h^2 \\ & & 6 & -3h \\ \text{symmetric} & & & 2h^2 \end{bmatrix},$$

$$\boldsymbol{m}_1 = \frac{\rho\hat{A}h}{420}\begin{bmatrix} 156 & 22h & 54 & -13h \\ & 4h^2 & 13h & -3h^2 \\ & & 156 & -22h \\ \text{symmetric} & & & 4h^2 \end{bmatrix} \quad \text{(consistent mass)}$$

$$\boldsymbol{m}_2 = \frac{\rho\hat{A}h}{2}\begin{bmatrix} 1 & 0 & 0 & 0 \\ & ah^2 & 0 & 0 \\ & & 1 & 0 \\ \text{symmetric} & & & ah^2 \end{bmatrix} \quad \text{(lumped mass)}$$

Determine critical time-step expressions for the consistent and lumped-mass cases. Provide a discussion of the consistent mass time step comparing it with that obtained for lumped mass.

Hint: The answer for the lumped case is

$$\boxed{\Delta t_{crit} = \frac{h^2}{c}\sqrt{\frac{a\hat{A}}{(12a+3)I}}}$$

Discussion of Exercise 4

Let us again assume a rectangular cross section, as in Example 3. If we also assume $a = \frac{1}{12}$, then

$$\boxed{\Delta t_{\text{crit}} = \left[\frac{h}{2t}\right]\frac{h}{c}}$$

Since $\Delta t_{\text{crit}} = O(h^2)$, the critical time step will be extremely small for fine meshes. Comparison with the critical time step engendered by the axial beam mode (see Example 1) reveals that bending will be critical if

$$\frac{h}{t} < 2$$

This condition is in force only in extremely fine meshes or when the beam is "deep."

(And for this case, the present theory is wholly inappropriate!) As long as $h/t \geq 2$, bar-wave transit time or better is achieved. This is the practically important case.

Exercise 5. In a transient frame analysis program, the critical time step must be computed accounting for the following deformation modes:

Axial mode
Torsional mode
Bending modes
Shear modes (if present)

The smallest time step resulting from the above modes determines the stable step for the element.

Assuming linear elements and lumped mass, determine the critical time step for the torsional mode. (For background information, see Secs. 5.4 and 5.5.) Compare this result with that for the axial mode.

Answer: $\Delta t_{\text{crit}} = h/c_t$, $c_t = \sqrt{\mu/\rho}$. By virtue of $\mu = E/[2(1 + \nu)]$, $\mu < E$; therefore $(\Delta t_{\text{crit}})_{\text{torsional}} > (\Delta t_{\text{crit}})_{\text{axial}}$.

Remark

From the preceding examples, exercises and discussions, it may be concluded that use of special lumped-mass matrices for beam and frame elements will produce critical time steps no worse than that governed by the axial mode (i.e., bar-wave transit time).

Exercise 6. A finite element computer program is being written to solve the wave equation in two-space dimensions. Bilinear quadrilaterals are to be employed. The element stiffness matrices are to be evaluated with one-point Gaussian quadrature, viz.,

$$\boldsymbol{k}^e = [k^e_{ab}], \qquad 1 \leq a, b \leq 4$$

$$k^e_{ab} = 4c^2(jN_{a,i}\, N_{b,i})\big|_{(0,0)}$$

(c is a wave speed parameter.) Element lumped-mass matrices are to be evaluated by the product trapezoidal rule, viz.,

$$\boldsymbol{m}^e = [m^e_{ab}]$$

$$m^e_{ab} = j(\xi_a, \eta_a)\delta_{ab} \qquad \text{(no sum)}$$

The Newmark method with $\beta = 0$, $\gamma = \frac{1}{2}$ is to be employed for the temporal discretization (i.e., "central differences").

i. Perform an analysis to determine the critical time step, making use of the following simplifications: Assume that the elements are rectangular and aligned parallel to the global Cartesian coordinate system. Denote by h^e_1 and h^e_2 the lengths of the sides in the x_1 and x_2 directions, respectively. Under the present circumstances, the isoparametric paraphernalia simplifies as follows:

$$j(\xi, \eta) = \frac{h_1^e h_2^e}{4}$$

$$N_{a,1}(0, 0) = \frac{\xi_a}{2h_1^e}$$

$$N_{a,2}(0, 0) = \frac{\eta_a}{2h_2^e}$$

ii. Physically interpret the result you have obtained.

Solution

i. $$k_{ab}^e = 4c^2(jN_{a,i}\, N_{b,i})\big|_{(0,0)}$$

$$= 4c^2\left(\frac{h_1^e h_2^e}{4}\right)\left(\frac{\xi_a}{2h_1^e}\frac{\xi_b}{2h_1^e} + \frac{\eta_a}{2h_2^e}\frac{\eta_b}{2h_2^e}\right)$$

$$m_{ab}^e = j(\xi_a, \eta_a)\delta_{ab} \qquad \text{(no sum)}$$

$$= \left(\frac{h_1^e h_2^e}{4}\right)\delta_{ab}$$

Given an eigenvector $\underbrace{\boldsymbol{\psi}}_{4\times 1} = \{\psi_a\}$, the eigenvalue may be determined from the Rayleigh quotient:

$$\mathcal{R}(\boldsymbol{\psi}) = \frac{\sum_{a,b} k_{ab}^e\, \psi_a \psi_b}{\sum_{a,b} m_{ab}^e\, \psi_a \psi_b} = \frac{c^2 \sum_{a,b} (\xi_a \xi_b + \eta_a \eta_b)\psi_a \psi_b}{\sum_a \psi_a \psi_a}$$

It is a simple matter to deduce the eigenvectors. There must be a rigid translation and an hourglass mode (one-point quadrature!). These are (respectively)

$$\{1\} = \begin{Bmatrix} 1 \\ 1 \\ 1 \\ 1 \end{Bmatrix}, \qquad \{\xi_a \eta_a\} = \begin{Bmatrix} 1 \\ -1 \\ 1 \\ -1 \end{Bmatrix}$$

The eigenvalue corresponding to each is zero. The remaining eigenvectors are orthogonal to these and therefore must have the form

$$\begin{Bmatrix} \alpha \\ \beta \\ -\alpha \\ -\beta \end{Bmatrix}, \qquad \text{where } \alpha \text{ and } \beta \text{ are constants}$$

By setting up the eigenproblem, it can be verified that the eigenvectors are:

$$\{\eta_a\} = \begin{Bmatrix} -1 \\ -1 \\ 1 \\ 1 \end{Bmatrix} \qquad \{\xi_a\} = \begin{Bmatrix} -1 \\ 1 \\ 1 \\ -1 \end{Bmatrix}$$

Substituting into the Rayleigh quotient yields the eigenvalues (respectively)

$$\lambda^h = \left(\frac{2c}{h_2}\right)^2, \left(\frac{2c}{h_1}\right)^2$$

The critical time step for $\beta = 0,\ \gamma = \frac{1}{2}$ is $2/\omega_{\max}^h$. Therefore

$$\boxed{\Delta t_{\text{crit}} = \frac{h}{c}, \qquad \text{where } h = \min\{h_1, h_2\}}$$

ii. The time step equals the transit time for a wave traveling at velocity c to traverse the shortest distance across the element.

Remark

Further time-step estimates may be found in Belytschko [14].

Exercise 7. Consider the one-dimensional second-order wave equation

$$u_{,tt} = c^2 u_{,xx}$$

Perform a von Neumann stability analysis of the central-difference finite-difference method:

$$u_{n+1}^h(m) - 2u_n^h(m) + u_{n-1}^h(m) = r^2(u_n^h(m+1) - 2u_n^h(m) + u_n^h(m-1))$$

where $r = c\Delta t/h$. Obtain an expression for the local truncation error. (See Sec. 8.3 for notational definitions and a description of the von Neumann method.) The central-difference finite-difference method is identical to the use of $\beta = 0,\ \gamma = \frac{1}{2}$ in Newmark's method in conjunction with piecewise linear finite elements and lumped mass.

9.3 LINEAR MULTISTEP (LMS) METHODS

9.3.1 LMS Methods for First-Order Equations

Consider a system of first-order, linear, ordinary differential equations

$$\boxed{\dot{\boldsymbol{y}} = \boldsymbol{f}(\boldsymbol{y}, t) = \boldsymbol{G}\boldsymbol{y} + \boldsymbol{H}(t)} \tag{9.3.1}$$

where $\boldsymbol{G}$ is a given constant matrix and $\boldsymbol{H}$ is a given vector-valued function of t. A ***k-step linear multistep method*** for (9.3.1) is defined by the following expression:

$$\boxed{\sum_{i=0}^{k} \{\alpha_i \boldsymbol{y}_{n+1-i} + \Delta t\, \beta_i \boldsymbol{f}(\boldsymbol{y}_{n+1-i}, t_{n+1-i})\} = 0} \tag{9.3.2}$$

The α_i's and β_i's are parameters which define the method. Note that the word "linear" in linear multistep method has nothing to do with the linearity of (9.3.1). Indeed, (9.3.2) is perfectly well defined for a nonlinear function $\boldsymbol{f}(\boldsymbol{y}, t)$. Rather, the linearity pertains to the form of (9.3.2). An excellent reference for LMS methods of this type is Gear [15]. An LMS method is called ***explicit*** if $\beta_0 = 0$; otherwise it is called ***implicit.*** It is called a ***backward-difference method*** if $\beta_i = 0$ for all $i \geq 1$. An example of a one-step LMS method of first-order type is the generalized trapezoidal family discussed in Chapter 8 for which $\alpha_0 = -\alpha_1 = 1$, $\beta_0 = -\alpha$, and $\beta_1 = \alpha - 1$. Recall tht $\alpha = 0$ defines an explicit method, and $\alpha = 1$ defines a backward difference method. The greater the number of steps involved in the definition of an LMS method, the greater the "historical data pool" that must be stored in computing—a practical disadvantage.

Exercise 1. Put the semidiscrete heat equation into the first-order form defined by (9.3.1). Conclude that the eigenvalues of $\boldsymbol{G}$ are nonpositive real numbers. *Thus the spectrum of* $\boldsymbol{G}$ *resides upon the negative real axis in the complex plane* $\mathbb{C}$ (i.e., $\lambda(\boldsymbol{G}) \leq 0$).

Exercise 2. Put the semidiscrete equation of motion in first-order form. Take $\boldsymbol{y} = \{\boldsymbol{d}, \dot{\boldsymbol{d}}\}^T$. Conclude that the $2n_{eq}$ eigenvalues of $\boldsymbol{G}$ take on the following forms:

i. $0 \leq \xi < 1$, ***underdamped***:

$$\lambda_{1,2}(\boldsymbol{G}) = -\xi\omega^h \pm i\omega^h\sqrt{1 - \xi^2}$$

ii. $\xi = 1$, ***critically damped:***

$$\lambda_{1,2}(\boldsymbol{G}) = -\omega^h \qquad \text{(double root)}$$

iii. $\xi > 1$, ***overdamped:***

$$\lambda_{1,2}(\boldsymbol{G}) = -\xi\omega^h \pm \omega^h\sqrt{\xi^2 - 1}$$

When $\xi = 0$, the eigenvalues are complex conjugate, $\pm i\omega^h$, and thus reside on the imaginary axis. *In all cases, the eigenvalues of* $\boldsymbol{G}$ *are confined to the negative half-plane of* $\mathbb{C}$ *including the negative real axis* (i.e., $\text{Re}(\lambda(\boldsymbol{G})) \leq 0$).

Exercise 3. Make a sketch of the complex plane and identify where the eigenvalues, $\lambda(\boldsymbol{G})$, as determined in Exercises 1 and 2, fall.

Many algorithms of practical interest take the form (9.3.2). The analysis of LMS methods may be performed using modal techniques, as described previously. Stability is generally phrased in terms of the roots of the polynomial associated to (9.3.2), namely,

$$\boxed{\sum_{i=0}^{k} (\alpha_i + \Delta t\, \lambda\, \beta_i)\, \zeta^{n+1-i} = 0} \qquad (9.3.3)$$

where λ is an eigenvalue of $\boldsymbol{G}$. (To derive (9.3.3), assume $\boldsymbol{G}$ admits the representation $\boldsymbol{G} = \boldsymbol{P}\boldsymbol{\Lambda}\boldsymbol{P}^{-1}$, where $\boldsymbol{\Lambda} = \text{diag}(\lambda_1, \lambda_2, \ldots, \lambda_N)$ and N is the dimension of $\boldsymbol{G}$. Show that (9.3.2) can be reduced to an uncoupled system of scalar equations having the form $\Sigma_{i=0}^{k} (\alpha_i + \Delta t\, \lambda\, \beta_i) z_{n+1-i} = 0$, which has solutions $z_n \sim c\,\zeta^n$.) An LMS method of the form (9.3.2) is said to be ***absolutely stable***, at a fixed $\lambda\Delta t$, if all ζ satisfying (9.3.3) are such that $|\zeta| \leq 1$. The ***region of absolute stability*** of an LMS method is the set of $\lambda\Delta t \in \mathbb{C}$ at which it is absolutely stable. An LMS method is ***A-stable*** if solutions of the SDOF analog of (9.3.2) $\to 0$ as $n \to \infty$ when $\text{Re}(\lambda) < 0$. Physically speaking, an A-stable algorithm is one which produces solutions which decay to zero whenever the corresponding exact solutions decay to zero. Clearly, if an algorithm is A-stable then the region of absolute stability contains the left half-plane of $\mathbb{C}$. The condition of A-stability places no limitation on the size of Δt, consequently it is closely related to what we termed "unconditional stability" previously. The only difference is that spectral stability allows eigenvalues of unit modulus which potentially could conserve a solution rather than contract it to zero. It should be apparent that A-stable algorithms are important for many physical problem classes. However, within the class of LMS methods, the subclass of A-stable methods is severely limited. This is the content of a celebrated theorem:

Dahlquist's Theorem [16]

1. An explicit A-stable LMS method does not exist.
2. A third-order accurate A-stable LMS method does not exist.
3. The second-order accurate A-stable LMS method with the smallest error constant[3] is the trapezoidal rule.

The upshot of this theorem is that if we seek an A-stable LMS method that possesses some special feature (such as high-frequency numerical dissipation), we necessarily entail some loss of accuracy with respect to the trapezoidal rule. Thus it would appear that the trapezoidal rule is the canonical A-stable LMS method for structural dynamics. The widespread use and popularity of the trapezoidal rule in this context appears to confirm this observation. However, there are other useful A-stable LMS methods for structural dynamics; for example, Park's method [17]. Many structural dynamics algorithms do not fall within the class of first-order LMS methods considered so far. The Newmark family of methods may be mentioned in this regard. Another class of LMS methods for second-order equations will be described subsequently; it contains the Newmark family, among others.

Since A-stable LMS methods are at most second-order accurate, the question arises, "Is there some weakened notion of A-stability which pertains to a physically relevant class of problems and allows for the development of higher-order-accurate methods?" An affirmative answer is provided by the concept of stiff stability proposed

[3] The error constant is the coefficient of Δt^k in the local truncation error.

by Gear [15]. A method is said to be ***stiffly stable*** if it is absolutely stable in the region of the $\lambda\Delta t$-plane defined by Re $(\lambda\Delta t) < -\delta$, where δ is a positive constant. Gear has developed a family of k-step, kth-order-accurate LMS methods which are stiffly stable for $k = 2, 3, \ldots, 6$. Each method is a backward-difference LMS method. The second-order method is, in addition, A-stable; however, it is significantly less accurate than the trapezoidal rule. For $k \geq 3$, the regions of instability include portions of the imaginary axis. Consequently, these stiffly stable algorithms also appear inappropriate for typical structural dynamics applications. Because the region of stability of each of the methods, $k = 2, 3, \ldots, 6$, includes the negative real axis, one would anticipate good behavior in application to problems of heat conduction.

Park [17] has developed a second-order accurate, A-stable algorithm, which retains good accuracy in the low frequencies and strong dissipative characteristics in the high frequencies, by combining Gear's two-step and three-step stiffly stable algorithms. The resultant scheme is defined by

Park's Method

$$k = 3, \quad \alpha_0 = -1, \quad \alpha_1 = \frac{15}{10}, \quad \alpha_2 = \frac{-6}{10}, \quad \alpha_3 = \frac{1}{10}, \quad \beta_0 = \frac{6}{10},$$

$$\beta_1 = \beta_2 = \beta_3 = 0$$

(9.3.4)

9.3.2 LMS Methods for Second-order Equations

Consider a system of second-order, linear, ordinary differential equations written in the form:

$$\ddot{\boldsymbol{y}} = \boldsymbol{f}(\boldsymbol{y}, \dot{\boldsymbol{y}}, t) = \boldsymbol{G}_0\boldsymbol{y} + \boldsymbol{G}_1\dot{\boldsymbol{y}} + \boldsymbol{H}(t) \tag{9.3.5}$$

where $\boldsymbol{G}_0$ and $\boldsymbol{G}_1$ are given constant matrices. A ***k-step LMS method*** for (9.3.5) is given by (Geradin [18]):

$$\sum_{i=0}^{k} \{\alpha_i \boldsymbol{y}_{n+1-i} + \Delta t\, \beta_i \boldsymbol{G}_1 \boldsymbol{y}_{n+1-i} + \Delta t^2\, \gamma_i [\boldsymbol{G}_0 \boldsymbol{y}_{n+1-i} + \boldsymbol{H}(t_{n+1-i})]\} = \boldsymbol{0}$$

(9.3.6)

The method is defined by the values of α_i, β_i and γ_i, $i = 0, 1, \ldots, k$. An LMS method of this type is called ***explicit*** if β_0 and $\gamma_0 = 0$. (If $\boldsymbol{G}_1 = \boldsymbol{0}$, then clearly only γ_0 need be zero.) The method is a ***backward-difference method*** if β_i and $\gamma_i = 0$,

$i \geq 1$. (Likewise, if $\boldsymbol{G}_1 = \boldsymbol{0}$, only the γ_i's, $i \geq 1$, need be zero.) The equation of motion may be put in the form of (9.3.5) simply by multiplying through by $\boldsymbol{M}^{-1}$. Observe that with $\boldsymbol{y} = \boldsymbol{d}$, (9.3.6) is a ***displacement difference-equation form of the algorithm.*** The commonly used algorithms of structural dynamics can all be put into this form, although very few are naturally cast this way.

In the sense of an LMS method for second-order systems, Newmark's method is a two-step method. LMS methods applied to the first-order form of the equation of motion may likewise be recast as second-order-type LMS methods (e.g., Park's method).

The stability properties of second-order-type LMS methods may be investigated in similar fashion to first-order LMS methods. Briefly, modal reduction may be employed which, for the equation of motion, leads to the following polynomial:

$$\sum_{i=0}^{k} (\alpha_i + 2\xi\omega^h \Delta t\, \beta_i + (\omega^h \Delta t)^2\, \gamma_i)\zeta^{n+1-i} = 0 \tag{9.3.7}$$

An analog of Dahlquist's theorem also holds for LMS methods for second-order equations (see Krieg [19]).

Exercise 4. Derive the following ***displacement difference-equation form of Newmark's algorithm:***

$$(\boldsymbol{M} + \gamma\Delta t\, \boldsymbol{C} + \beta\Delta t^2 \boldsymbol{K})\boldsymbol{d}_{n+1} + \left[-2\boldsymbol{M} + (1 - 2\gamma)\Delta t\, \boldsymbol{C} + \left(\frac{1}{2} - 2\beta + \gamma\right)\Delta t^2 \boldsymbol{K}\right]\boldsymbol{d}_n$$
$$+ \left[\boldsymbol{M} - (1 - \gamma)\Delta t\, \boldsymbol{C} + \left(\frac{1}{2} + \beta - \gamma\right)\Delta t^2 \boldsymbol{K}\right]\boldsymbol{d}_{n-1} + \Delta t^2 \overline{\boldsymbol{F}}_n = \boldsymbol{0} \tag{9.3.8}$$

Define $\overline{\boldsymbol{F}}_n$ in terms of $\boldsymbol{F}_{n-1}$, $\boldsymbol{F}_n$, and $\boldsymbol{F}_{n+1}$.

Exercise 5. Modally reduce the homogeneous version of (9.3.8) to the form (9.3.7). Define α_i and β_i, $i = 0, 1, 2$. Show that this results in an identical equation to the one derived from the 2×2 amplification matrix form, (9.1.44). Deduce that the roots of the stability polynomial (9.3.7) are identical to the eigenvalues of the amplification matrix (i.e., $\zeta = \lambda(\boldsymbol{A})$).

Exercise 6. Consider the scalar difference equation that emanates from (9.3.1), (9.3.2) after modal reduction:

$$\sum_{i=0}^{k} (\alpha_i + \Delta t \lambda \beta_i) z_{n+1-i} = 0 \tag{9.3.9}$$

Derive a one-step, multivalue method corresponding to (9.3.9), which has the form

$$\boldsymbol{Z}_{n+1} = \boldsymbol{A}\boldsymbol{Z}_n \tag{9.3.10}$$

where $\boldsymbol{Z}_n = \{z_n, z_{n-1}, \ldots, z_{n+1-k}\}^T$. That is, explicitly define the $k \times k$ amplification matrix $\boldsymbol{A}$. Establish that the eigenvalues of $\boldsymbol{A}$ are the same as the roots, ζ, of the stability polynomial, (9.3.3), emanating from (9.3.9).

Exercise 7. Following along the lines of the previous exercise, consider the stability polynomial:

$$\zeta^3 - 2A_1\zeta^2 + A_2\zeta - A_3 = 0 \tag{9.3.11}$$

Derive a 3×3 amplification matrix, as in (9.3.10), which gives rise to (9.3.11). Show that

$$A_1 = \frac{1}{2}\text{ trace }\boldsymbol{A} \tag{9.3.12}$$

$$A_2 = \text{sum of the principal minors of }\boldsymbol{A} \tag{9.3.13}$$

$$A_3 = \det \boldsymbol{A} \tag{9.3.14}$$

(Equations (9.3.12) through (9.3.14) define the ***principal invariants of A***. They arise naturally in calculating the stability polynomial by way of $\det(\boldsymbol{A} - \zeta\boldsymbol{I}) = 0$; cf. (9.3.11).)

Exercise 8. The region of absolute stability of a fourth-order explicit Runge-Kutta LMS method for the model equation $\dot{y} = \lambda y$, where λ is a complex number, is depicted in the accompanying figure. (Assume the eigenvalues of the amplification matrix are distinct.)

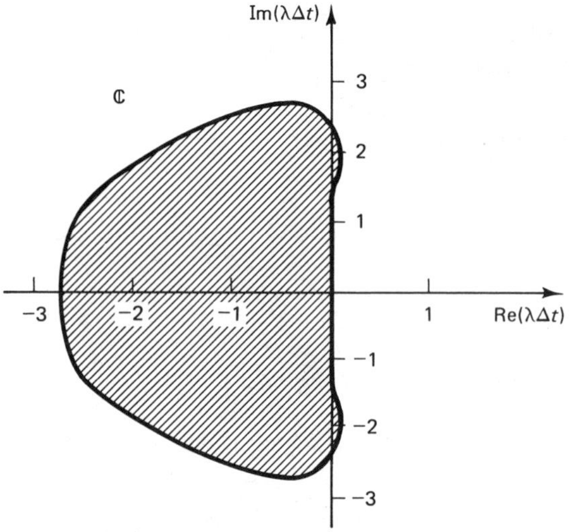

Determine the stability conditions for the following cases:

i. Parabolic case. [*Ans.*: $-2.7 \leq \lambda^h\Delta t \leq 0$]
ii. Undamped hyperbolic case. [*Ans.*: $\omega^h\Delta t \leq 2.5$]

Exercise 9. Consider the trapezoidal-rule algorithm. Show that it is A-stable. Determine its region of absolute stability. [*Ans.*: The left-hand complex plane including the imaginary axis; i.e., it is stable for all $\lambda^h\Delta t$ such that $\text{Re}(\lambda^h\Delta t) \leq 0$.]

Exercise 10. Show that the Euler backward-difference method is A-stable. Determine its region of absolute stability.

Exercise 11. Determine the region of absolute stability of the Euler forward-difference algorithm.

Exercise 12. Determine local truncation error expressions for LMS methods of type (9.3.2) and (9.3.6).

Answer:

$$\Delta t \tau = \sum_{i=0}^{k} \{\alpha_i \boldsymbol{y}(t_{n+1-i}) + \Delta t \beta_i \boldsymbol{f}(\boldsymbol{y}(t_{n+1-i}), t_{n+1-i})\}$$

$$\Delta t^2 \tau = \sum_{i=0}^{k} \{\alpha_i \boldsymbol{y}(t_{n+1-i}) + \Delta t \beta_i \boldsymbol{G}_1 \boldsymbol{y}(t_{n+1-i}) + \Delta t^2 \gamma_i [\boldsymbol{G}_0 \boldsymbol{y}(t_{n+1-i}) + \boldsymbol{H}(t_{n+1-i})]\}$$

9.3.3 Survey of Some Commonly Used Algorithms in Structural Dynamics

This section surveys and compares some algorithms which have been proposed for structural dynamics applications. The algorithms considered are all LMS methods of first-order or second-order type, and the comparison is based upon both stability behavior and accuracy. Accuracy characteristics of LMS methods may be studied in similar fashion to the study of accuracy for Newmark's method. The only complication is that one needs to ignore so-called *spurious roots*, focusing attention on the *principal roots*. For a method possessing an amplification matrix of rank k, assuming linearly independent eigenvectors, the analog of (9.1.62) can be written as

$$d_n = \sum_{i=1}^{k} c_j \lambda_j^n = e^{-\bar{\xi}\bar{\omega}^h t_n}(\bar{c}_1 \cos \bar{\omega}_d^h t_n + \bar{c}_2 \sin \bar{\omega}_d^h t_n) + \sum_{i=3}^{k} c_i \lambda_i^n \qquad (9.3.15)$$

The first two roots, λ_1 and λ_2, are the principal roots, and the remaining $k - 2$ are the spurious roots. The principal roots define $\bar{\xi}$ and $\bar{\omega}^h$ through

$$\lambda_{1,2} = e^{(-\bar{\xi}\bar{\omega}^h \pm i\bar{\omega}_d^h)\Delta t} \qquad (9.3.16)$$

The comparisons that follow are for the physically undamped case (i.e., $\xi = 0$).

Houbolt's Method

Houbolt's method [20] was one of the earliest employed in structural dynamics computations. It is defined by the following formulas:

$$\boldsymbol{M}\boldsymbol{a}_{n+1} + \boldsymbol{C}\boldsymbol{v}_{n+1} + \boldsymbol{K}\boldsymbol{d}_{n+1} = \boldsymbol{F}_{n+1} \qquad (9.3.17)$$

$$\boldsymbol{a}_{n+1} = \frac{2\boldsymbol{d}_{n+1} - 5\boldsymbol{d}_n + 4\boldsymbol{d}_{n-1} - \boldsymbol{d}_{n-2}}{\Delta t^2} \qquad (9.3.18)$$

$$\boldsymbol{v}_{n+1} = \frac{11\boldsymbol{d}_{n+1} - 18\boldsymbol{d}_n + 9\boldsymbol{d}_{n-1} - 2\boldsymbol{d}_{n-2}}{6\Delta t} \qquad (9.3.19)$$

Combining (9.3.17) through (9.3.19) leads to a three-step LMS method of second-order type which is unconditionally stable and second-order accurate. Furthermore, if $\boldsymbol{C} = \boldsymbol{0}$, it is a backward-difference, stiffly A-stable method and thus has many features in common with Park's method. The advantage of Park's method is that it is more accurate. However, it entails a significantly greater historical data pool. A disadvantage of both the Houbolt and Park methods is that they require special *starting procedures*.

Exercise 13. Apply Park's method to the first-order form of the equations of motion. Develop the displacement-difference equation form of the algorithm by eliminating v_{n+1-i}, $i = 0, 1, 2, 3$. What is k when the algorithm is viewed as a k-step LMS method of second-order type?

Exercise 14

i. Derive an amplification matrix for Houbolt's method. That is, define $\boldsymbol{A}$ such that

$$\begin{Bmatrix} d_{n+1} \\ d_n \\ d_{n-1} \end{Bmatrix} = \underbrace{\boldsymbol{A}}_{3 \times 3} \begin{Bmatrix} d_n \\ d_{n-1} \\ d_{n-2} \end{Bmatrix}$$

ii. Write a computer program that calculates the spectral radius of $\boldsymbol{A}$. Plot $\rho(\boldsymbol{A})$ versus $\Delta t/T$.

iii. Put Houbolt's method into displacement difference equation form and derive an expression for local truncation error. *Hint*: See Exercise 12.

iv. The ***Routh-Hurwitz criterion*** gives sufficient conditions for the roots of a stability polynomial, or associated amplification matrix, to be less than or equal to one in modulus. In the present case, the Routh-Hurwitz criterion takes the form:

$$1 - 2A_1 + A_2 - A_3 \geq 0$$

$$3 - 2A_1 - A_2 + 3A_3 \geq 0$$

$$3 + 2A_1 - A_2 - 3A_3 \geq 0$$

$$1 + 2A_1 + A_2 + A_3 \geq 0$$

$$1 - A_2 + A_3(2A_1 - A_3) \geq 0$$

Use these expressions to show that Houbolt's method is unconditionally stable.

Collocation Schemes

Collocation methods generalize and combine aspects of the Newmark method and Wilson-θ method [21]. A systematic analysis of collocation is contained in Hilber and Hughes [22]. The collocation schemes are defined by the following equations:

$$\boldsymbol{M}\boldsymbol{a}_{n+\theta} + \boldsymbol{C}\boldsymbol{v}_{n+\theta} + \boldsymbol{K}\boldsymbol{d}_{n+\theta} = \boldsymbol{F}_{n+\theta} \tag{9.3.20}$$

$$a_{n+\theta} = (1 - \theta)a_n + \theta a_{n+1} \tag{9.3.21}$$

$$F_{n+\theta} = (1 - \theta)F_n + \theta F_{n+1} \tag{9.3.22}$$

$$d_{n+\theta} = d_n + \theta\Delta t v_n + \frac{(\theta\Delta t)^2}{2}\{(1 - 2\beta)a_n + 2\beta a_{n+\theta}\} \tag{9.3.23}$$

$$v_{n+\theta} = v_n + \theta\Delta t\{(1 - \gamma)a_n + \gamma a_{n+\theta}\} \tag{9.3.24}$$

To (9.3.20) through (9.3.24) are appended the Newmark formulas (i.e., (9.1.5) and (9.1.6)) for purposes of defining d_{n+1} and v_{n+1}, respectively. θ is called the ***collocation parameter.*** The collocation schemes can be put into the form of a three-step LMS method of second-order type. Alternatively, the collocation schemes' model equation can be put in amplification matrix form; see [23]. In this case, the rank of the amplification matrix is 3.

If $\theta = 1$, the scheme reverts to Newmark's method. If $\beta = \frac{1}{6}$ and $\gamma = \frac{1}{2}$, the Wilson-θ methods are obtained. A necessary and sufficient condition for second-order accuracy is that $\gamma = \frac{1}{2}$. Unconditionally stable, second-order accurate schemes are defined by

$$\gamma = \frac{1}{2}, \qquad \theta \geq 1, \qquad \frac{\theta}{2(\theta + 1)} \geq \beta \geq \frac{2\theta^2 - 1}{4(2\theta^3 - 1)} \tag{9.3.25}$$

The best-behaved collocation schemes were determined in [22]. This amounts to a one-parameter subfamily of methods with $\gamma = \frac{1}{2}$, and $\theta = \theta^*(\beta)$ defined by Table 9.3.1. These methods are referred to as ***optimal collocation methods*** and are the only ones considered henceforth.

TABLE 9.3.1 Smallest collocation parameter, θ^*, which ensures complex conjugate principle roots as $\Delta t/T \to \infty$. Corresponding values of algorithmic damping ratio and relative period error for $\Delta t/T = 0.1$ [22]

β	θ^*	$\bar{\xi}$	$(\bar{T}-T)/T$
$\frac{1}{4}$	1	0	0.032
0.24	1.021712	0.60×10^{-4}	0.032
0.23	1.047364	0.27×10^{-3}	0.033
0.22	1.077933	0.70×10^{-3}	0.034
0.21	1.114764	0.14×10^{-2}	0.036
0.20	1.159772	0.27×10^{-2}	0.039
0.19	1.215798	0.46×10^{-2}	0.043
0.18	1.287301	0.77×10^{-2}	0.050
0.17	1.381914	0.13×10^{-1}	0.060
$\frac{1}{6}$	1.420815	0.15×10^{-1}	0.064
0.16	1.514951	0.21×10^{-1}	0.075

α-Method (Hilber-Hughes-Taylor Method)

Numerical damping cannot be introduced in the Newmark method without degrading the order of accuracy. To improve upon this situation, Hilber, Hughes, and Taylor [24] introduced the α-method. In the α-method, the finite difference formulas of the Newmark method are retained (i.e., (9.1.5) and (9.1.6)) whereas the time-discrete equation of motion is modified as follows:

$$\boldsymbol{M}\boldsymbol{a}_{n+1} + (1 + \alpha)\boldsymbol{C}\boldsymbol{v}_{n+1} - \alpha\boldsymbol{C}\boldsymbol{v}_n + (1 + \alpha)\boldsymbol{K}\boldsymbol{d}_{n+1} - \alpha\boldsymbol{K}\boldsymbol{d}_n = \boldsymbol{F}(t_{n+\alpha}) \tag{9.3.26}$$

where $t_{n+\alpha} = (1 + \alpha)t_{n+1} - \alpha t_n = t_{n+1} + \alpha\Delta t$. Clearly, if $\alpha = 0$ we reduce to the Newmark method. If the parameters are selected such that $\alpha \in [-\frac{1}{3}, 0]$, $\gamma = (1 - 2\alpha)/2$, and $\beta = (1 - \alpha)^2/4$, an unconditionally stable, second-order accurate scheme results [25]. At $\alpha = 0$, we have the trapezoidal rule. Decreasing α increases the amount of numerical dissipation. The α-method can be put in the form of a three-step LMS method for second-order equations, or equivalently, in a rank-3 amplification matrix form; see [23]. The α-method is incorporated in program DLEARN (see Chapter 11).

Comparison of Algorithms

The algorithms compared in Figs. 9.3.1 through 9.3.4 are

1. Houbolt's method
2. Park's method
3. Optimal collocation methods
4. α-methods

These methods have the following features in common: They are implicit, unconditionally stable, second-order accurate, and dissipative. All require approximately the same computational effort. However, Park's method requires a larger historical data pool than the others. The collocation and α-methods permit a parametric control of numerical dissipation, whereas the Houbolt and Park methods do not. (For comparison purposes we also include some results for the Newmark and Wilson-θ methods.)

In Fig. 9.3.1 the spectral radii of the various cases are presented. The spectral radii of the Houbolt and Park methods approach zero as $\Delta t/T \to \infty$, as is typical of backward-difference schemes. Collocation schemes and α-methods are seen also to possess strong damping in the high-frequency regime. Recall that the effect of $\rho < 1$ is accumulative, e.g., $\rho^n \leq e^{-n(1-\rho)}$, which rapidly approaches zero as n is increased.

In Fig. 9.3.2 algorithmic damping ratios are compared. The Houbolt and collocation methods are seen to affect the low modes (i.e., $\Delta t/T \leq 0.1$) too strongly. The inefficiency of the damping in the collocation scheme versus the α-method can

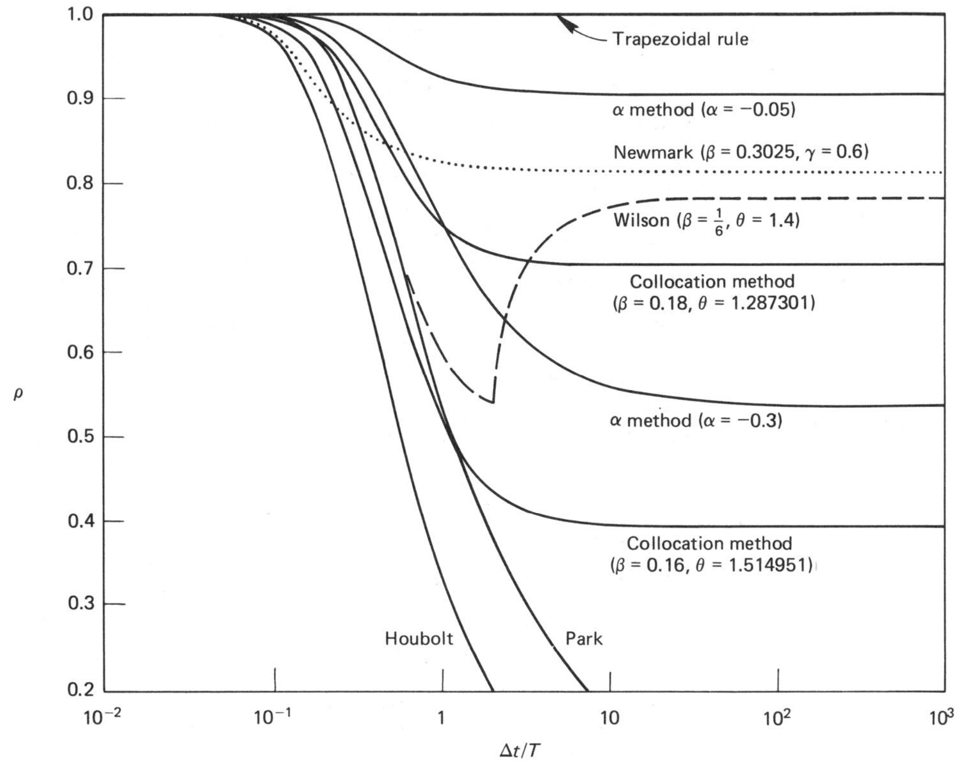

Figure 9.3.1 Spectral radii for α-methods, optimal collocation schemes, and Houbolt, Newmark, Park, and Wilson methods [22].

be seen by comparing the cases $\beta = 0.18$ and $\alpha = -0.3$, respectively. From Fig. 9.3.1, $\alpha = -0.3$ is seen to damp the high modes more strongly than $\beta = 0.18$. On the other hand, from Fig. 9.3.2 the low modes are affected less by $\alpha = -0.3$ than by $\beta = 0.18$.

Relative period errors are compared in Fig. 9.3.3. The collocation schemes and α-methods have smaller errors than the Houbolt and Park methods.

The following observations summarize the comparisons: All methods are capable of sufficiently damping out the high modes. The Houbolt method affects the low modes much too strongly, both from the point of view of damping ratio and of period error. Park's method possesses good low mode damping properties; however, its period error is higher than the collocation schemes and α-methods. The collocation schemes damp the low modes too strongly.

In Fig. 9.3.4, period error results are presented for undamped Newmark methods ($\gamma = \frac{1}{2}$), and comparison is made with results for the Houbolt and Wilson algorithms. Note that for $\beta < \frac{1}{4}$, the methods presented are conditionally stable, which is made evident by the abrupt increases in period error at finite $\Delta t/T$. Notice also that the central difference method ($\beta = 0$) tends to shorten periods, whereas the trapezoidal rule ($\beta = \frac{1}{4}$) increases periods.

As long as the method is stable, $\gamma = \frac{1}{2}$ implies $\bar{\xi} = 0$ for physically undamped

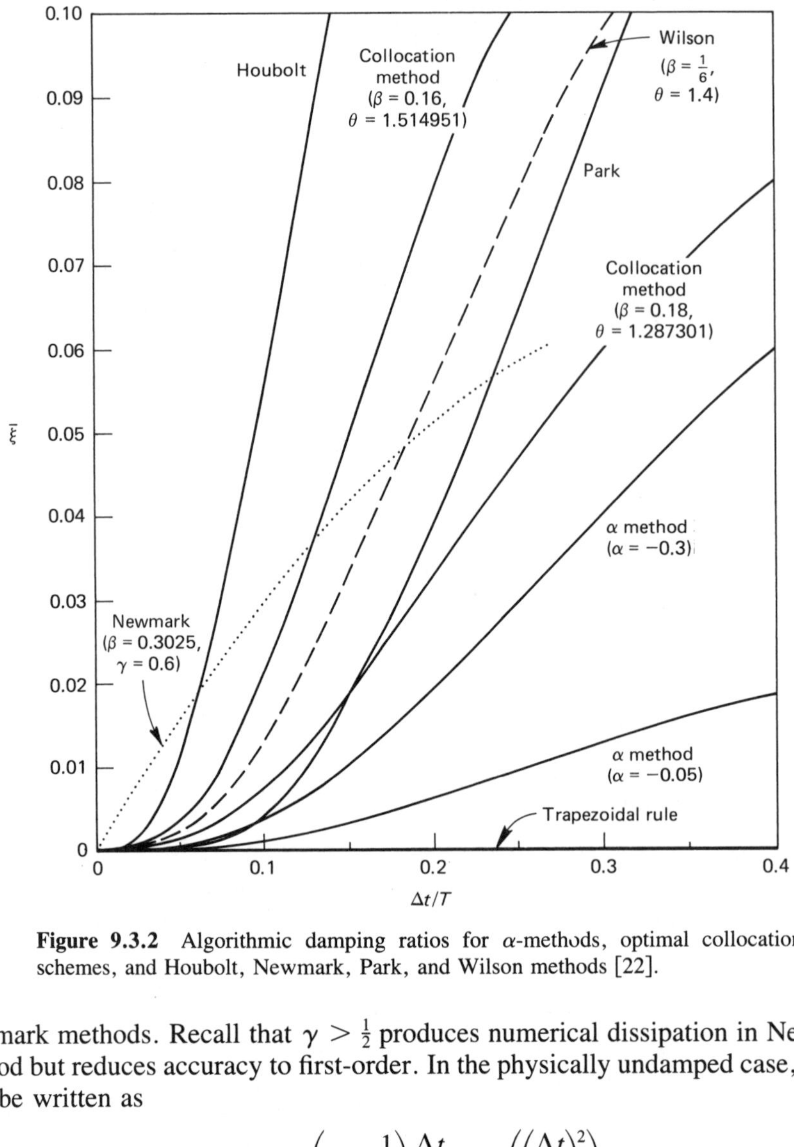

Figure 9.3.2 Algorithmic damping ratios for α-methods, optimal collocation schemes, and Houbolt, Newmark, Park, and Wilson methods [22].

Newmark methods. Recall that $\gamma > \frac{1}{2}$ produces numerical dissipation in Newmark's method but reduces accuracy to first-order. In the physically undamped case, (9.1.65) may be written as

$$\bar{\xi} = \pi\left(\gamma - \frac{1}{2}\right)\frac{\Delta t}{T} + O\left(\left(\frac{\Delta t}{T}\right)^2\right) \tag{9.3.27}$$

Thus the slope of the $\bar{\xi}$ versus $\Delta t/T$ curve is positive at $\Delta t/T = 0$, an indication of first-order accuracy. This may be contrasted with the zero slopes of the second-order methods compared in Fig. 9.3.2. To estimate (9.3.27), assume $\gamma = 0.6$ and $\Delta t/T = 0.1$, and neglect the quadratic term. This results in $\bar{\xi} \cong 0.01\pi$. Connecting the point $(\Delta t/T, \bar{\xi}) = (0.1, 0.01\pi)$ to the origin with a straight line gives a fairly accurate representation of the behavior of $\bar{\xi}$ for the $\gamma = 0.6$ Newmark method, $0 \leq \Delta t/T \leq 0.1$ (cf. Fig. 9.3.2). This excessive low-mode numerical dissipation is the main shortcoming of damped Newmark methods.

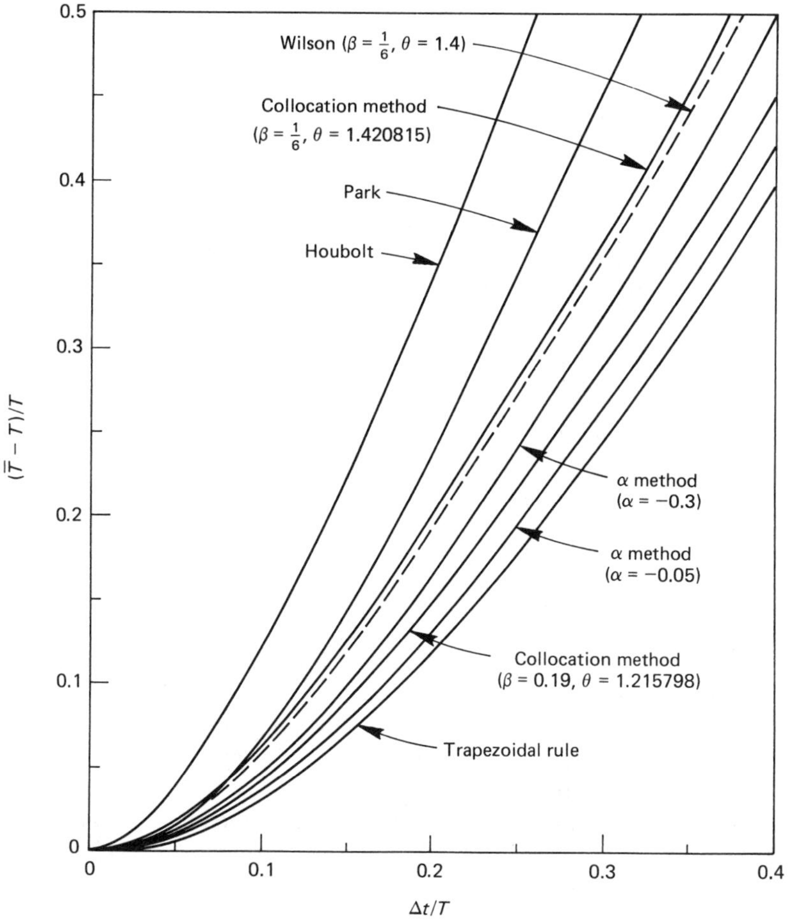

Figure 9.3.3 Relative period errors for α-methods, optimal collocation schemes, and Houbolt, Park, and Wilson methods [22].

Discussion Concerning Time-Stepping Algorithms in Structural Dynamics

Considerable effort has gone into the development of efficient computational methods for the step-by-step integration of the equations of structural dynamics. Although there is no universal consensus, it is generally agreed that for a method to be competitive it should possess the following attributes:

1. Unconditional stability when applied to linear problems.
2. No more than one set of implicit equations to be solved at each step.
3. Second-order accuracy.
4. Controllable algorithmic dissipation in the higher modes.
5. Self-starting.

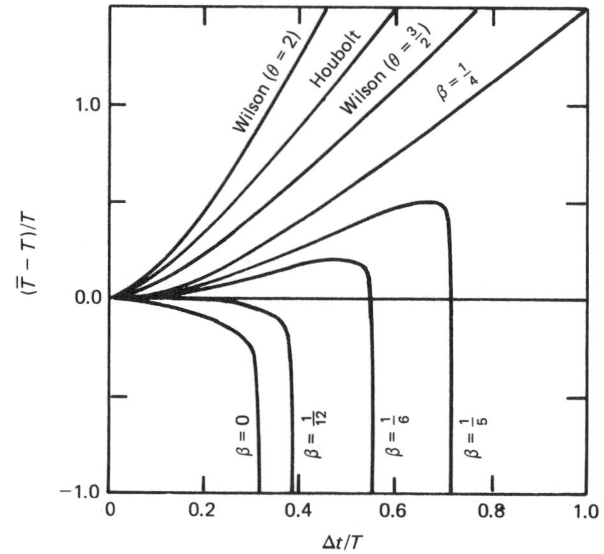

Figure 9.3.4 Period errors for undamped Newmark methods ($\gamma = \frac{1}{2}$) compared with Wilson and Houbolt schemes [26].

Conditionally stable algorithms require that a time step be taken that is less than a constant times the smallest period of the structure. In complicated structural models, containing slender members exhibiting bending effects, this restriction is a stringent one and often entails using time steps that are much smaller than those needed for accuracy, especially when only low-mode response is of interest. For these reasons unconditionally stable algorithms are generally preferrred.

Dahlquist's theorem [16] (see also Krieg [19]) asserts that there is no unconditionally stable explicit method amongst the class of linear multistep methods. Thus attribute 1 necessitates the use of implicit methods, and this engenders considerable computational effort since coefficient matrices must be stored and factored. It is important for purposes of efficiency that this work be kept to a minimum. This can be done by insisting that the solution of no more than one implicit system, of the size of the mass and stiffness matrices, be required at each time step. More elaborate schemes have been proposed that require larger implicit systems, or two or more implicit systems of the size of the stiffness and mass to be solved at each step, and improved properties have been obtained (see for example Argyris, Dunne and Angelopoulos [29], Geradin [18], Hilber [23], Krieg and Key [30], and Kurdi [31]). However, these techniques require at least twice the storage and computational effort of the simpler methods and thus have not been widely adopted.

Experience has indicated that in structural dynamics, second-order accurate methods are vastly superior to first-order accurate methods. Another consequence of Dahlquist's theorem is that there is no third-order accurate unconditionally stable linear multistep method. Thus we must be content with second-order accuracy. Furthermore, the second-order method with the smallest error constant is the trapezoidal rule. Hence any effort to obtain some specific property within the context of second-order accurate unconditionally stable methods must result in some degradation of accuracy when compared with the trapezoidal rule. It is important to be aware of this since numerical dissipation is another desirable characteristic and one not possessed by trapezoidal rule. (Recall, numerical dissipation is used to damp out any spurious participation of the higher modes.)

It is also desirable that an algorithm be self-starting, since our understanding of ones that are not is generally obscured. For example, if an algorithm requires a distinct starting procedure, we must analyze the starting procedure in addition to the algorithm and we must also analyze the interaction of the algorithm with all possible values generated by the starting procedure. A distinct starting procedure may also engender additional coding and computational effort.

Algorithms that are not self-starting involve data from more than two time steps to advance the solution. This requires additional storage, and since no matter how many steps we include we are restricted to at most second-order accuracy, a point of diminishing return is quickly reached. Furthermore, it is possible to achieve all the aforementioned attributes within a "one-step method" (i.e., a method for which the displacements, velocities, and accelerations at one step, along with the given forces, suffice to advance the solution to the next step). *Thus it seems that the second-order accurate, unconditionally stable, one-step method which achieves the optimal balance between effective numerical dissipation and loss of accuracy compared with the trapezoidal rule is the desired algorithm for most problems of structural dynamics.*

Overshoot

In [26] Goudreau and Taylor discovered a peculiar property of the Wilson-θ method. Despite the method's unconditional stability, numerical experiments revealed a tendency to significantly "overshoot" exact solutions in the first few steps. Argyris et al. [27] and Ghose [28] have also drawn attention to this behavior.

The overshoot phenomenon can occur despite the fact that the method's amplification matrix is spectrally stable. This suggests that a method's amplification matrix should be scrutinized more carefully. To see the genesis of the overshoot phenomenon in terms of a spectrally stable, hypothetical amplification matrix, consider

$$\boldsymbol{A} = \begin{bmatrix} \epsilon & k \\ 0 & \epsilon \end{bmatrix} \tag{9.3.28}$$

where $0 < \epsilon < 1$ and $k \gg 1$. The spectral radius of $\boldsymbol{A}$ is ϵ. The effect of the spectral radius as $n \rightarrow \infty$ is evident from

$$\boldsymbol{A}^n = \begin{bmatrix} \epsilon^n & n\epsilon^{n-1}k \\ 0 & \epsilon^n \end{bmatrix} \tag{9.3.29}$$

in which all terms go to zero as $n \rightarrow \infty$. However, due to the presence of k, the term $n\epsilon^{n-1}k$, is very large for n small enough. From this example we see that the long-term, or asymptotic, behavior is governed by the spectral properties of $\boldsymbol{A}$. However, the short-term potential for overshoot requires scrutiny of all the entries of $\boldsymbol{A}$. To assess an algorithm's tendency to overshoot, we may consider the SDOF model problem subjected to arbitrary initial conditions d_0 and v_0 and calculate d_1 and v_1 as a function of Ω. The numerical solution may then be compared with the exact solution. Because we naturally consider only convergent schemes, there can be no overshoot as $\Omega \to 0$. The other end of the spectrum, $\Omega \to \infty$, gives an indication of the behavior of the high frequencies in a system in which the time step is large compared with the shortest periods present. This is a typical situation when implicit, unconditionally stable methods are applied to large multidegree-of-freedom systems.

Example

Consider the undamped, homogeneous case (i.e., $\xi = 0$ and $F = 0$) and the limit $\Omega \to \infty$. Then, the collocation and α-methods behave as follows:

Collocation schemes

$$\left(\frac{\theta}{2(\theta + 1)} \geq \beta \geq \frac{2\theta^2 - 1}{4(2\theta^3 - 1)},\ \theta \geq 1\right)$$

$$d_1 \cong -\frac{1}{2}\left(1 - \frac{1}{\theta}\right)\Omega^2 d_0 + \left(1 - \frac{1}{\theta^2}\right)\Delta t v_0 \tag{9.3.30}$$

$$v_1 \cong \left(\frac{1}{4\beta\theta} - 1\right)\Omega \omega^h d_0 + \left(1 - \frac{1}{2\beta\theta^2}\right)v_0 \tag{9.3.31}$$

α-methods

$$(\beta = (1 - \alpha)^2/4,\ \gamma = \tfrac{1}{2} - \alpha,\ \alpha \in [-\tfrac{1}{3}, 0])$$

$$d_1 \cong \left(1 - \frac{1}{2\beta(1 + \alpha)}\right)d_0 \tag{9.3.32}$$

$$v_1 \cong \left(\frac{\gamma}{2\beta} - 1\right)\Omega\omega^h d_0 + (1 - \gamma/\beta)v_0 \tag{9.3.33}$$

Retaining only significant terms according to order of magnitude and assuming neither $\theta = 1$ nor $\alpha = 0$, we get

Collocation schemes

$$d_1 \cong O(\Omega^2)d_0 + O(\Delta t)v_0 \tag{9.3.34}$$

$$v_1 \cong O(\Omega)\omega^h d_0 + O(1)v_0 \tag{9.3.35}$$

α-methods

$$d_1 \cong O(1)d_0 \tag{9.3.36}$$

$$v_1 \cong O(\Omega)\omega^h d_0 + O(1)v_0 \tag{9.3.37}$$

Both methods exhibit a tendency to overshoot linearly in Ω in the velocity equation due to the initial displacement terms. The α-methods exhibit no overshoot in displacement. On the other hand, d_1 of the collocation schemes overshoots quadratically in Ω due to initial displacement and linearly in Δt due to initial velocity. The superiority of the α-methods over the collocation schemes with regard to overshoot is evident.

For both families the critical condition is caused by initial displacement. In Fig. 9.3.5 numerical results for some of the present schemes are exhibited for initial data $d_0 = 1$ and $v_0 = 0$, and $\Delta t/T = 10$. The trapezoidal rule and representatives of the Newmark and Wilson methods are included for comparison purposes. The trapezoidal rule exhibits no overshoot in either the displacements or velocities. Mild overshooting is exhibited by the

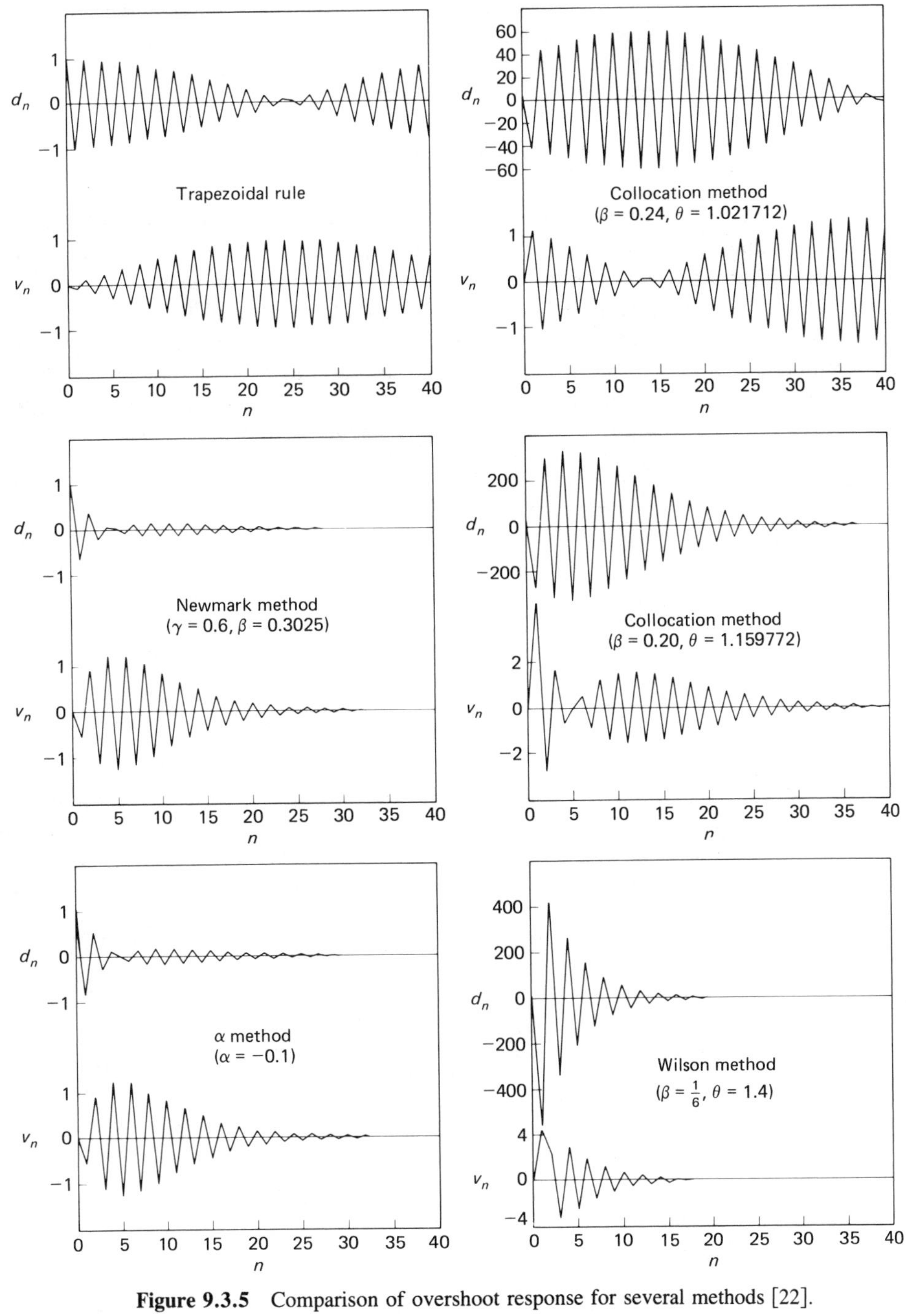

Figure 9.3.5 Comparison of overshoot response for several methods [22].

velocities in the Newmark and α-methods but none in displacements. The pathological displacement overshoot characteristics of two optimal collocation schemes and the Wilson method are manifest, whereas the velocities for these cases overshoot only mildly. Further discussion may be found in Hilber and Hughes [22].

Modal Analysis in Structural Dynamics

In comparing the use of step-by-step integration algorithms with the modal analysis technique, the points articulated in Sec. 8.5 may be reiterated here. In addition, there are other reasons that favor modal analysis for problems of structural dynamics. For example, the frequencies and mode shapes usually need to be computed for design purposes. Their availability effectively eliminates that most costly constituent in modal analysis—solution of the eigenproblem. Modal analysis is also the ideal method of "damping" the higher modes without adversely affecting the lower modes. The retained low modes may be exactly integrated. The contribution of the rest of the modes is eliminated completely.

Exercise 15. (This exercise illustrates how the analytical information determined for algorithms can be used to select stable and accurate time steps prior to performing a full-scale transient analysis.) A transient analysis is being performed of an undamped multidegree-of-freedom structure. The initial-value problem consists of the following:

$$\boldsymbol{M}\ddot{\boldsymbol{d}} + \boldsymbol{K}\boldsymbol{d} = \boldsymbol{0}$$

$$\boldsymbol{d}(0) = \boldsymbol{d}_0$$

$$\dot{\boldsymbol{d}}(0) = \boldsymbol{0}$$

Engineering insight reveals that the response will be primarily in the first six modes. Assume $0 < \omega_1^h \le \omega_2^h \le \ldots \le \omega_{n_{eq}}^h$, $\omega_6^h = 10\omega_1^h$, and $\omega_{n_{eq}}^h = 1000\omega_1^h$. (Recall the period of the lth mode is $T_l = 2\pi/\omega_l^h$. Engineering accuracy dictates that relative period error and amplitude decay (per cycle) be no more than 5 percent for any of the first six modes.)

Determine the largest time step permissible, as a fraction of T_1, for the following algorithms:

a. Central difference method ($\beta = 0$, $\gamma = \frac{1}{2}$).
b. Trapezoidal rule ($\beta = \frac{1}{4}$, $\gamma = \frac{1}{2}$).
c. Damped Newmark method ($\beta = 0.3025$, $\gamma = 0.6$). (It is safe to assume that dissipation errors will require a smaller time step than period errors.)
d. α-method ($\alpha = -0.3$).
e. Wilson-θ method ($\theta = 1.4$).
f. Houbolt method.
g. Park method.

Hint: You need to consider the stability condition, and dissipation and dispersion errors of each algorithm. Much of the necessary information is contained in Figs. 9.3.1–9.3.4.

Solution

a. Central difference method

$$\bar{\xi} = 0 \Rightarrow AD = 0$$

The time step for stability is so small that we do not worry about relative period error.

$$2 = \Omega_{\text{crit}} = \omega^h_{n_{eq}}\,\Delta t_{\text{crit}} = \frac{2\pi}{T_{n_{eq}}}\Delta t_{\text{crit}} = \frac{2\pi}{T_1} 1000\,\Delta t_{\text{crit}}$$

Therefore,

$$\frac{\Delta t}{T_1} \le \frac{1}{(1000\pi)} \cong \underline{\underline{0.000318}}$$

b. Trapezoidal rule

$$\bar{\xi} = 0 \Rightarrow AD = 0$$

Unconditionally stable.

$$\frac{\bar{T} - T}{T} \le 0.05 \Rightarrow \frac{\Delta t}{T} \le 0.125$$

Therefore

$$\frac{\Delta t}{T_1} \le 0.1 \times 0.125 = \underline{\underline{0.0125}}$$

c. Damped Newmark method ($\gamma = 0.6$, $\beta = 0.3025$)

$$\gamma > \frac{1}{2}, \qquad \beta > \frac{\gamma}{2}$$

Therefore, unconditionally stable.

$$AD \cong 2\pi\bar{\xi}, \qquad \bar{\xi} \le \frac{0.05}{2\pi} \cong 0.008 \Rightarrow \frac{\Delta t}{T_6} \le 0.03$$

Therefore,

$$\frac{\Delta t}{T_1} \le \underline{\underline{0.003}}$$

Cases (d), (e), (f), and (g) involve *unconditionally stable* algorithms, so accuracy determines Δt:

d. α-method ($\alpha = -0.3$)

$$\bar{\xi} \le 0.008 \Rightarrow \frac{\Delta t}{T_6} \le 0.13 \Rightarrow \frac{\Delta t}{T_1} \le 0.013$$

$$\frac{\bar{T} - T}{T} \le 0.05 \Rightarrow \frac{\Delta t}{T_6} \le 0.10 \Rightarrow \frac{\Delta t}{T_1} \le \underline{\underline{0.010}}$$

e. Wilson-θ method

$$\bar{\xi} \le 0.008 \Rightarrow \frac{\Delta t}{T_6} \le 0.08 \Rightarrow \frac{\Delta t}{T_1} \le 0.008$$

$$\frac{\bar{T} - T}{T} \leq 0.05 \Rightarrow \frac{\Delta t}{T_6} \leq 0.08 \Rightarrow \frac{\Delta t}{T_1} \leq \underline{\underline{0.008}}$$

f. Houbolt method

$$\bar{\xi} \Rightarrow \frac{\Delta t}{T_1} \leq \underline{\underline{0.004}}, \qquad \frac{\bar{T} - T}{T} \Rightarrow \frac{\Delta t}{T_1} \leq 0.006$$

g. Park method

$$\bar{\xi} = \frac{\Delta t}{T_1} \leq 0.012, \qquad \frac{\bar{T} - T}{T} \Rightarrow \frac{\Delta t}{T_1} \leq \underline{\underline{0.008}}$$

Remark

From the results we see that the largest time step is permitted by the trapezoidal rule (0.0125). The second largest is attained by the α-method (0.010). The Wilson-θ and Park methods are tied for third (0.008). Houbolt's method is fifth (0.004), the damped Newmark is sixth (0.003), and the central-difference method is seventh (0.000318). Keep in mind that there may be other considerations: For example, the trapezoidal rule may produce some undesirable high-frequency oscillations; potential "overshoot" phenomena must be investigated for the Wilson-θ method; the fact that the central-difference method is explicit may compensate for the small time-step restriction.

Exercise 16. Consider the following system:

$$\boldsymbol{M}\ddot{\boldsymbol{d}} + \boldsymbol{K}\boldsymbol{d} = \boldsymbol{0}$$

$$\boldsymbol{d} = \begin{Bmatrix} d_1 \\ d_2 \end{Bmatrix}, \qquad \boldsymbol{M} = \begin{bmatrix} m_1 & 0 \\ 0 & m_2 \end{bmatrix}, \qquad \boldsymbol{K} = \begin{bmatrix} (k_1 + k_2) & -k_2 \\ -k_2 & k_2 \end{bmatrix}$$

Assume $k_1 = 10^4$, $k_2 = 1$, $m_1 = 1$, and $m_2 = 1$.

The intent of this two-degree-of-freedom model is to represent the character of typical large systems. The first mode is intended to represent those modes that are physically important and must be accurately integrated. The second mode represents the spurious high frequencies. It is desirable that the step-by-step integrator filter these high modes from the response of the system.

a. Determine the natural frequencies ω_1 and ω_2 of this system.
b. Consider the initial-value problem for the system above with initial data given by

$$\boldsymbol{d}_0 = \begin{Bmatrix} 1 \\ 10 \end{Bmatrix}$$

$$\boldsymbol{v}_0 = \begin{Bmatrix} 0 \\ 0 \end{Bmatrix}$$

(The initial condition is designed to exacerbate overshoot phenomena.)

Write a computer program to solve this problem employing methods 1–7.

1. Central difference method ($\beta = 0$, $\gamma = \frac{1}{2}$)
2. Trapezoidal rule ($\beta = \frac{1}{4}$, $\gamma = \frac{1}{2}$)
3. Damped Newmark method ($\beta = 0.3025$, $\gamma = 0.6$)
4. α-method ($\alpha = -0.3$)
5. Wilson-θ method ($\theta = 1.4$)
6. Houbolt method
7. Park method

Run the program at $\Delta t = T_1/20$ over a time interval of $[0, 5T_1]$. For the Houbolt and Park methods, use the trapezoidal rule to establish starting values. Obtain time-history plots for the displacements and velocities in each case. To facilitate data reduction use some form of computer plotting. Comment on the relative effectiveness of the algorithms.

Solution Computer response is presented in Figure 9.3.6. The calculations were performed by Alec Brooks. The reader should realize that $\Delta t = T_1/20$ leads to an unstable calculation with central difference. The critical time step for central differences is $2 \div 100.005 \cong 0.019999$. (Verify!) The central-difference calculations were run at 0.01999, just below the critical value. Note that this leads to a "beating" phenomenon in the high-frequency components (see Fig. 9.3.6(a)), which has been analytically explained by Hilber [23].

Exercise 17. Repeat Exercise 16 with $k_1 = 1$ and $k_2 = 10^4$ and the initial conditions

$$\boldsymbol{d}_0 = \begin{Bmatrix} 10 \\ 11 \end{Bmatrix}$$

$$\boldsymbol{v}_0 = \begin{Bmatrix} 0 \\ 0 \end{Bmatrix}$$

9.3.4 Some Recently Developed Algorithms for Structural Dynamics

In this section we briefly mention some more recently developed algorithms for structural dynamics.

Bossak's Method

Bossak's method [32] is defined by the Newmark formulas (i.e., (9.1.5) and (9.1.6)) and the following version of the time-discrete equation of motion (see page 551):

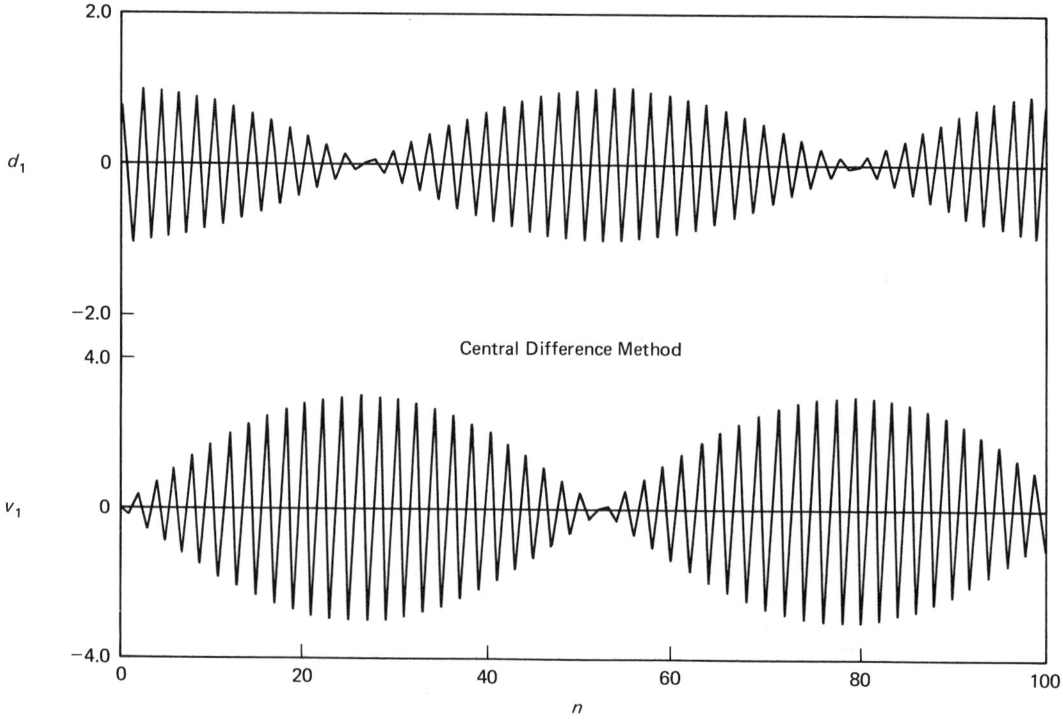

Figure 9.3.6(a)

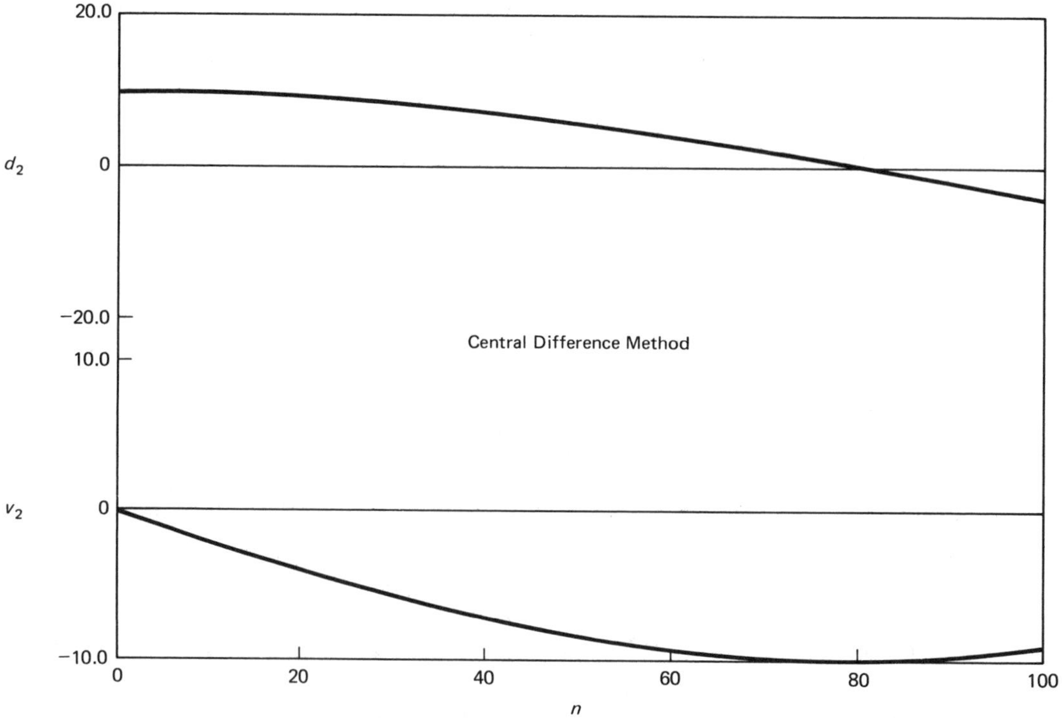

Figure 9.3.6(b)

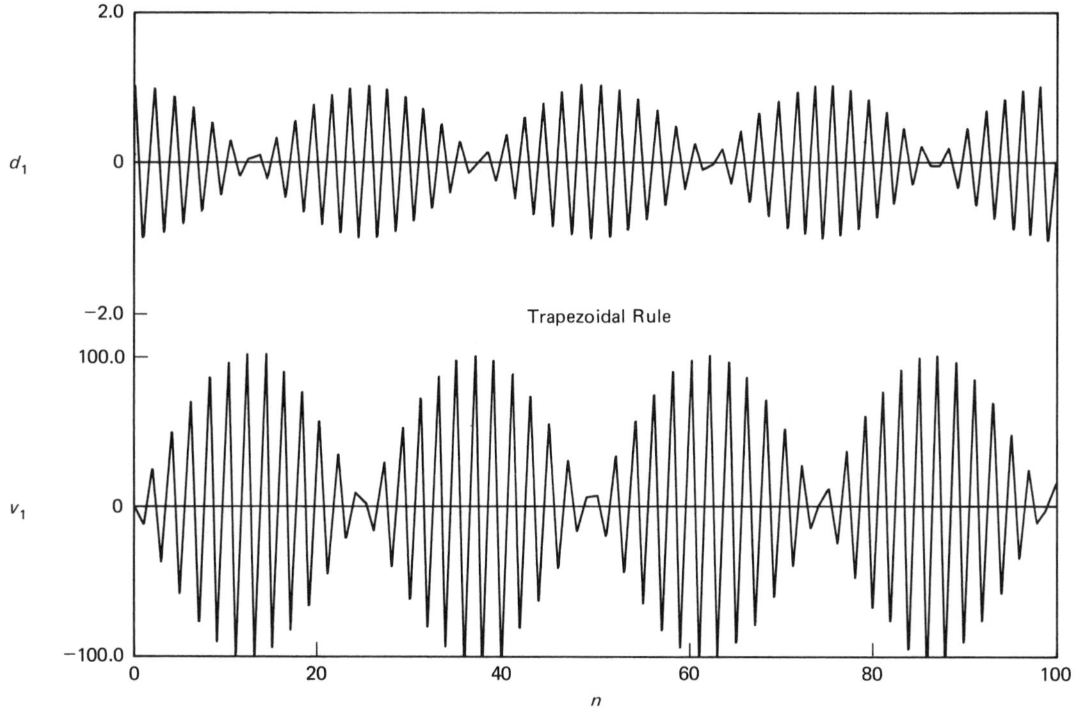

Figure 9.3.6(c)

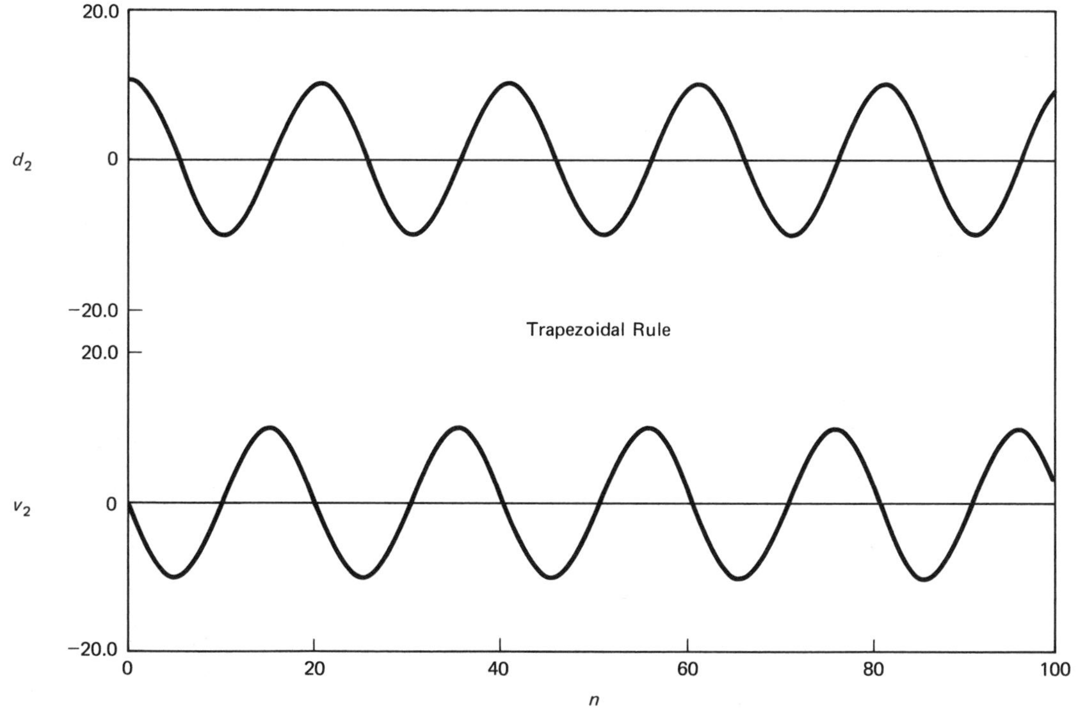

Figure 9.3.6(d)

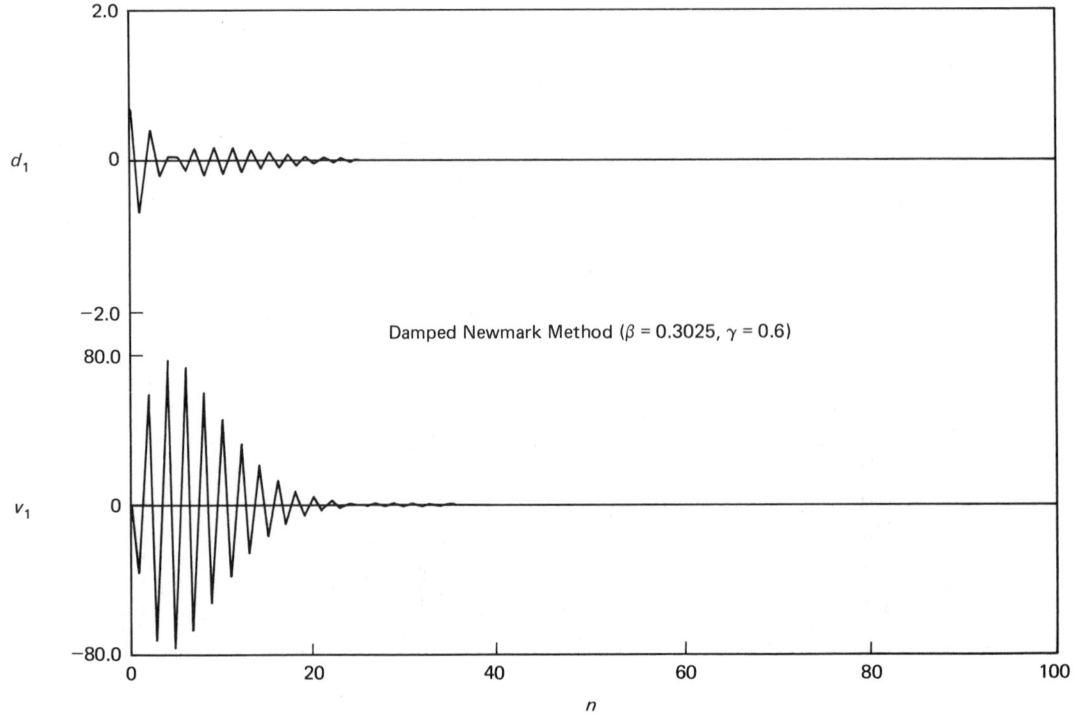

Figure 9.3.6(e)

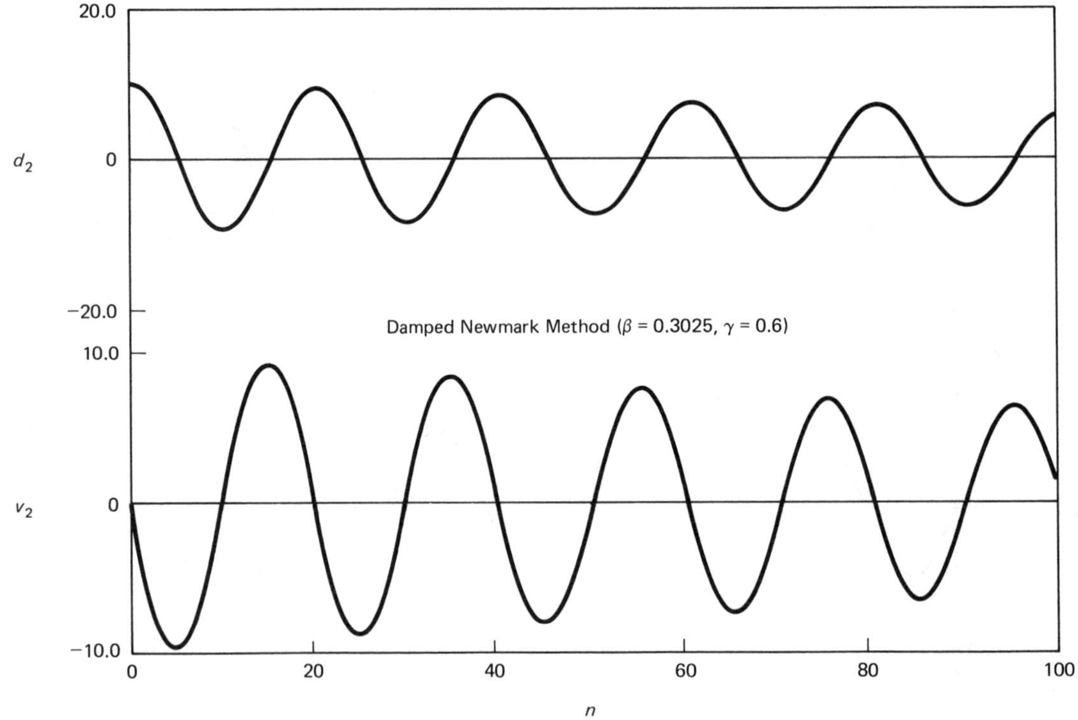

Figure 9.3.6(f)

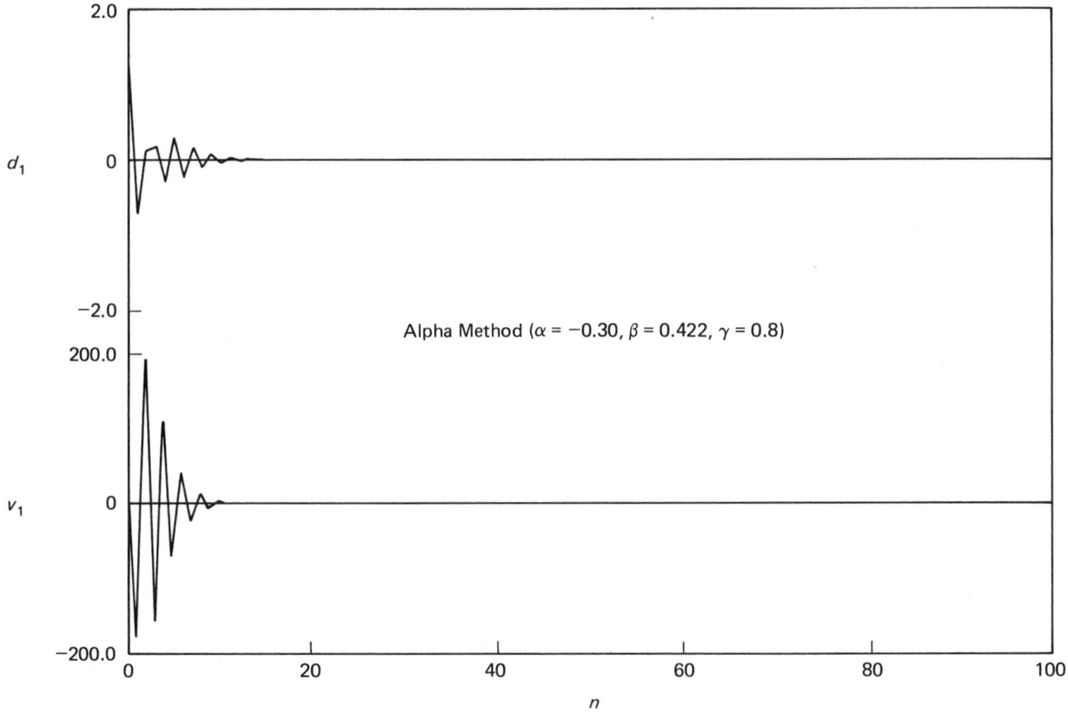

Figure 9.3.6(g)

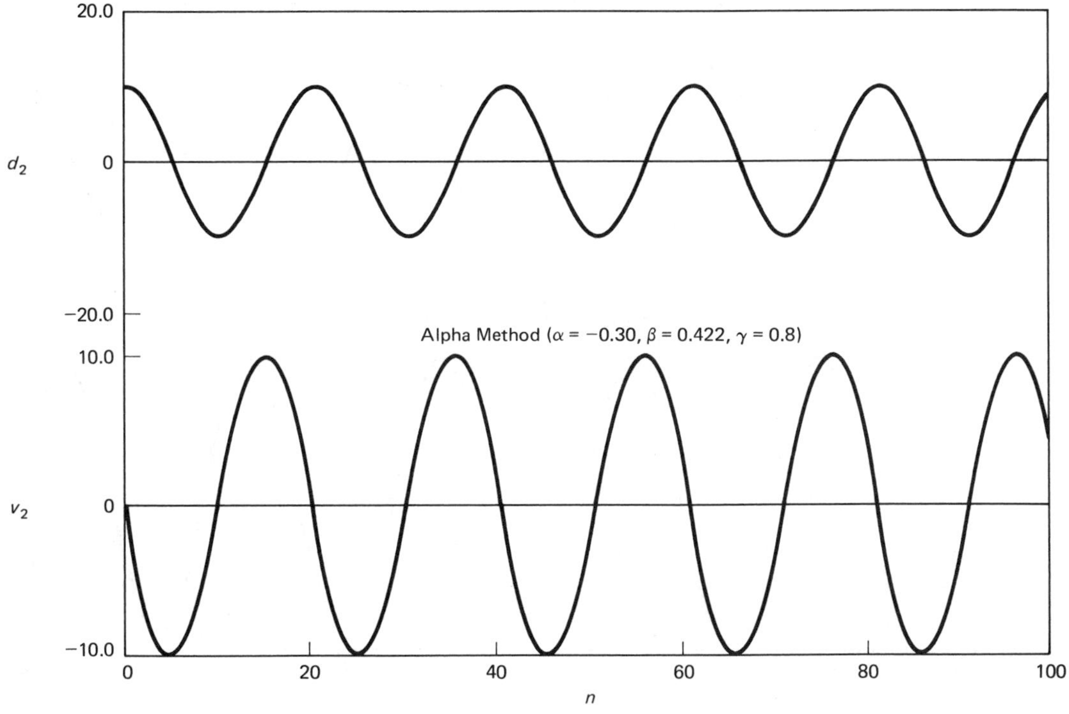

Figure 9.3.6(h)

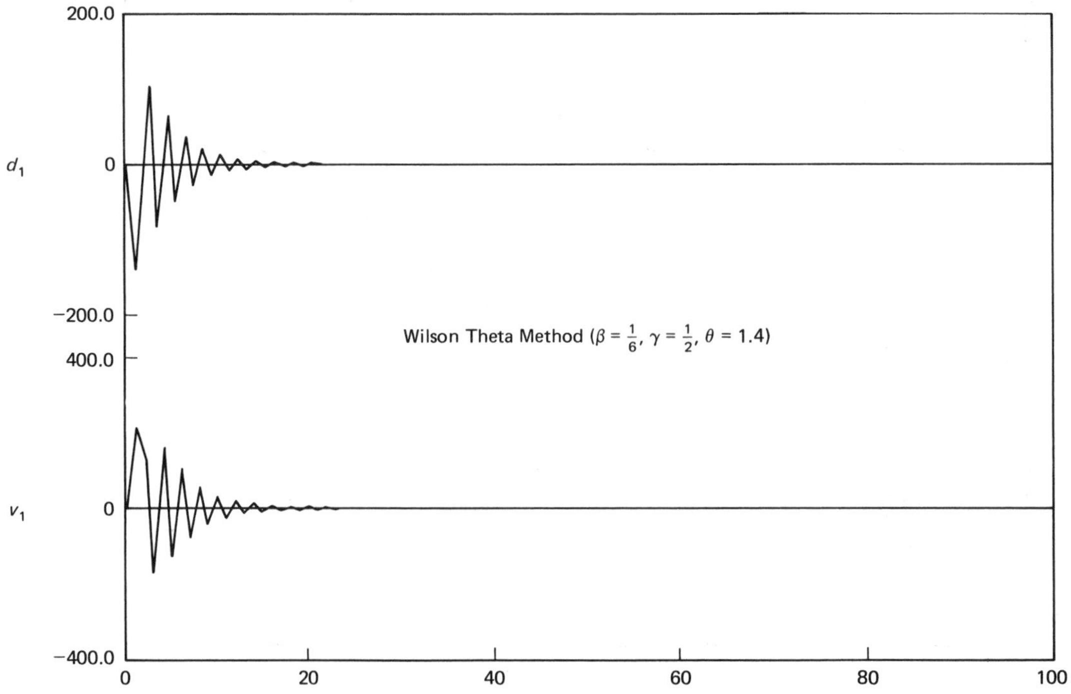

Figure 9.3.6(i)

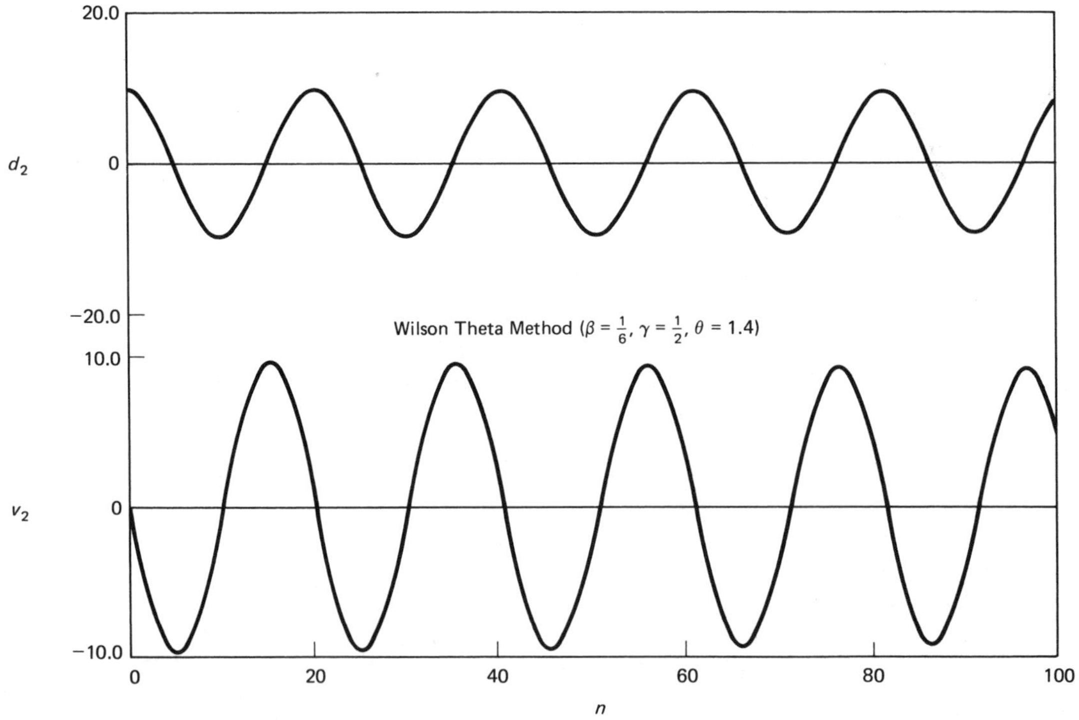

Figure 9.3.6(j)

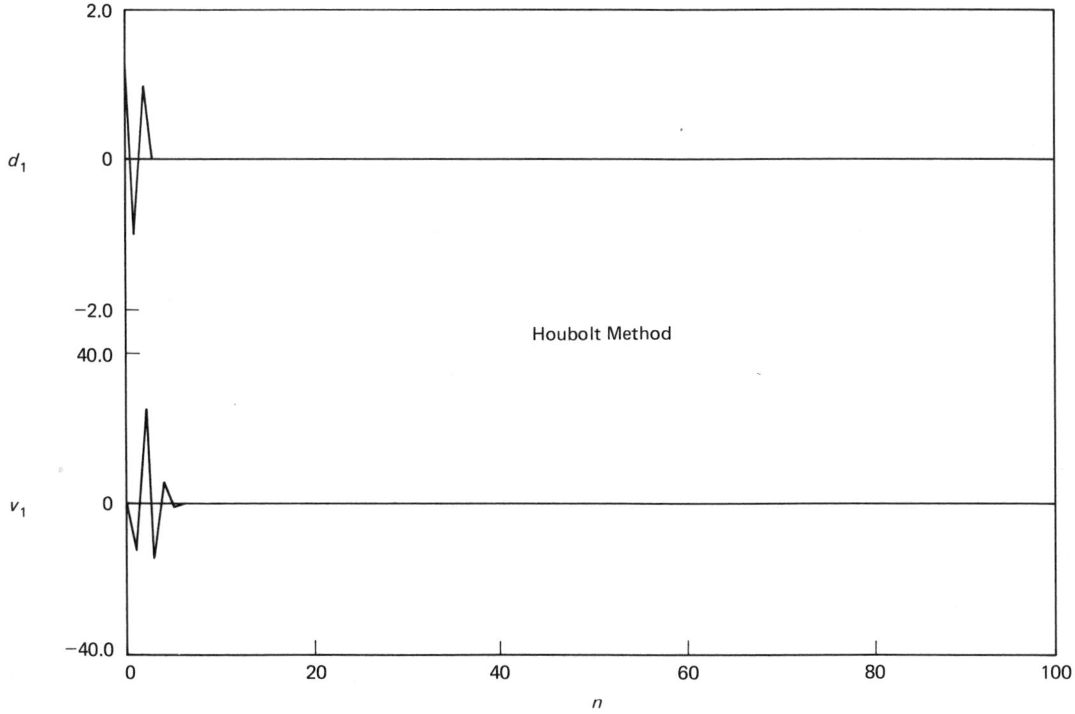

Figure 9.3.6(k)

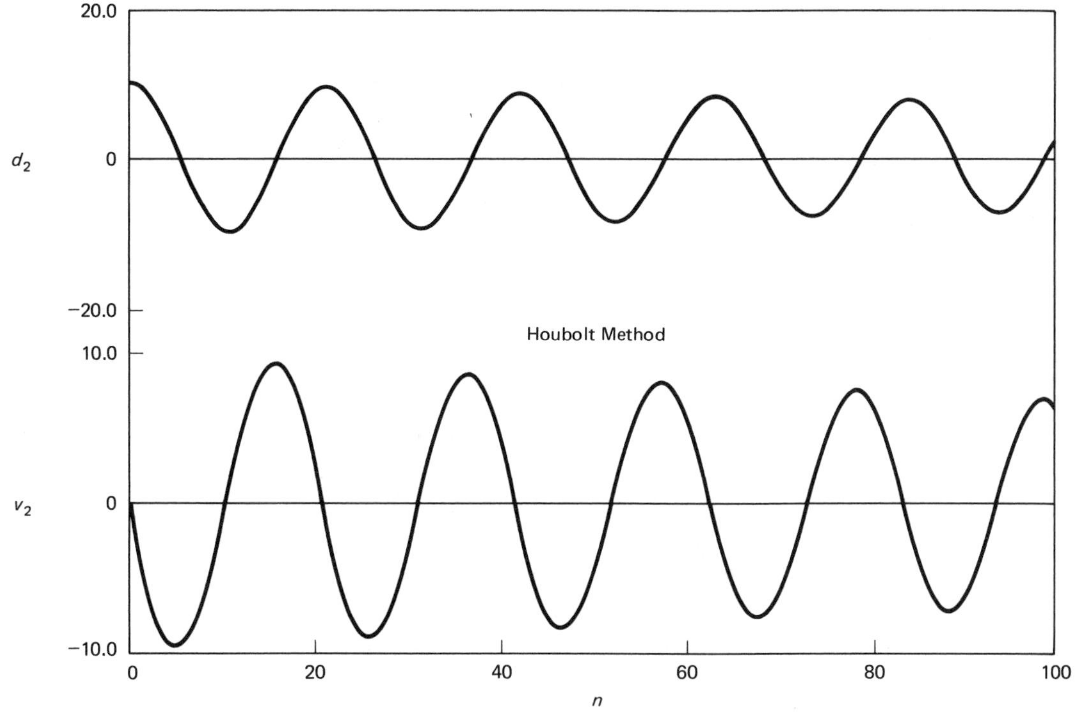

Figure 9.3.6(l)

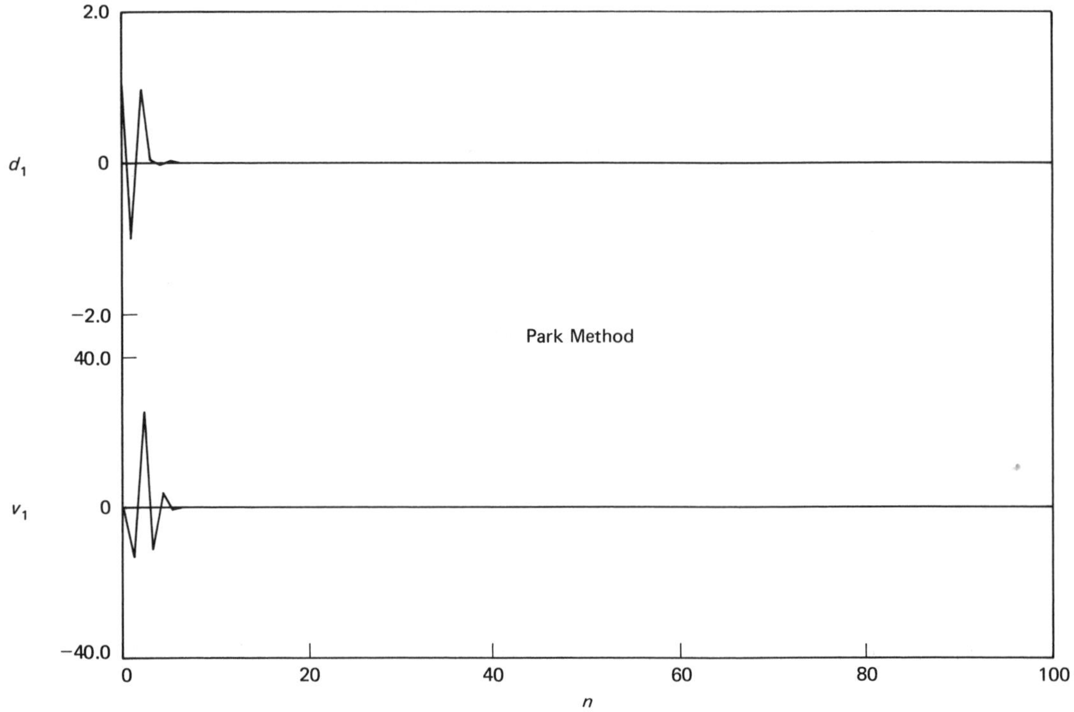

Figure 9.3.6(m)

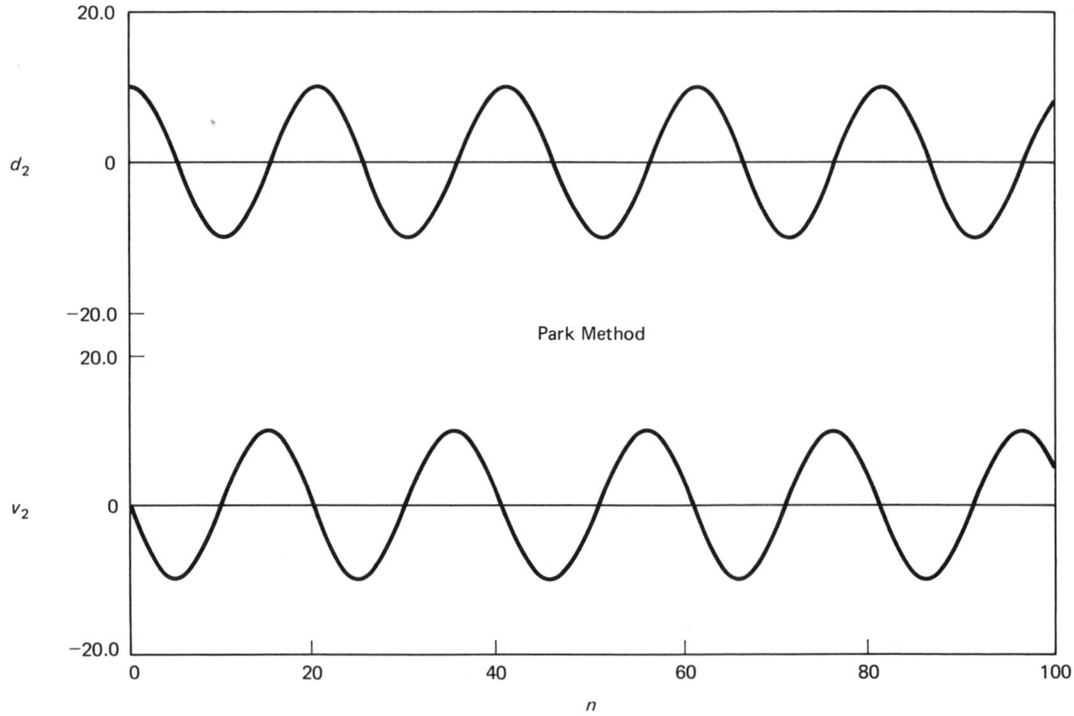

Figure 9.3.6(n)

$$\boxed{(1-\alpha_\beta)\boldsymbol{Ma}_{n+1}+\alpha_B\boldsymbol{Ma}_n+\boldsymbol{Cv}_{n+1}+\boldsymbol{Kd}_{n+1}=\boldsymbol{F}_{n+1}} \tag{9.3.38}$$

where α_B is an algorithmic parameter. If $\alpha_B = 0$, we reduce to Newmark's method. Unconditional stability, second-order accuracy and maximal high-frequency dissipation are achieved if

$$\alpha_B \leq 0 \tag{9.3.39}$$

$$\gamma = \tfrac{1}{2} - \alpha_B \tag{9.3.40}$$

and

$$\beta = \tfrac{1}{4}(1-\alpha_B)^2 \tag{9.3.41}$$

Bossak's method was inspired by and has features in common with the Hilber-Hughes-Taylor α-method. A comparison of the two methods is presented in Adams and Wood [33]. If $\alpha = \alpha_B$, both methods possess the same high-frequency dissipation properties, but the Hilber-Hughes-Taylor scheme registers an advantage in accuracy that becomes significant at values of $\alpha = \alpha_B$ near $-1/3$. For additional details the interested reader is urged to consult [32] and [33]. (The "B_1-curve" in Fig. 1, page 1565, of [32] is wrong. A corrected version is presented in [33].)

Bazzi-Anderheggen ρ-method

The Bazzi-Anderheggen method [34] can be written in the following form:[4]

$$\boldsymbol{d}_{n+1} = \boldsymbol{d}_n + \Delta t\boldsymbol{v}_n + (1-\alpha)\frac{\Delta t^2}{2}\{(1-2\beta)\boldsymbol{a}_n + 2\beta\boldsymbol{a}_{n+1}\} \tag{9.3.42}$$

$$\boldsymbol{v}_{n+1} = \boldsymbol{v}_n + \Delta t\{(1-\gamma)\boldsymbol{a}_n + \gamma\boldsymbol{a}_{n+1}\} \tag{9.3.43}$$

$$\boldsymbol{M}[\boldsymbol{v}_n] + \boldsymbol{C}[\boldsymbol{d}_n] + \Delta t\boldsymbol{Kd}_{n+\alpha} = \Delta t\boldsymbol{F}_{n+\alpha} \tag{9.3.44}$$

where

$$[\boldsymbol{d}_n] = \boldsymbol{d}_{n+1} - \boldsymbol{d}_n, \ldots \tag{9.3.45}$$

$$\boldsymbol{d}_{n+\alpha} = \alpha\boldsymbol{d}_{n+1} + (1-\alpha)\boldsymbol{d}_n, \ldots \tag{9.3.46}$$

$$\alpha = \frac{1}{1+\rho} \tag{9.3.47}$$

$$\beta = \frac{1}{\rho(\rho^2+1)} \tag{9.3.48}$$

$$\gamma = \frac{2}{(\rho^2+1)(\rho+1)} \tag{9.3.49}$$

[4]The style of presentation here is somewhat different than in [34].

where ρ is an algorithmic parameter that equals the value of the spectral radius of the amplification matrix as $\Delta t/T \to \infty$. Thus values of ρ that ensure unconditional stability are confined to the interval $[0, 1]$, with $\rho = 0$ reducing to the trapezoidal rule. If physical damping is absent, (i.e., $\boldsymbol{C} = \boldsymbol{0}$), the algorithm is second-order accurate. However, if $\boldsymbol{C} \neq \boldsymbol{0}$, the algorithm is limited to first-order accuracy. For the undamped case, the algorithm compares favorably with other commonly used methods. It is recommended that $\rho = 0.8$. See [34] for additional details.

Unified Set of Single-Step Methods

In [35], Zienkiewicz et al. present a unified formulation of single-step methods, which contains as special cases many of the popular structural dynamics schemes. The compact form of the algorithm enables a number of different methods to be easily implemented in one program. See [35] for further details.

9.4 ALGORITHMS BASED UPON OPERATOR SPLITTING AND MESH PARTITIONS

In recent years, methods outside the LMS realm have been introduced with considerable success. In this section we present the procedures developed in [36–38], which allow part of the mesh, or operator, to be treated implicitly and part to be treated explicitly. This has considerable practical advantages in that "stiff" subdomains of large finite element models can be treated economically with an implicit integrator. A stiff subdomain might arise from very fine local meshing, which reduces the Δt_{crit} considerably below what is allowable in the more coarsely meshed region. Penalty constraints, such as described in Sec. 4.2, also entail very small Δt_{crit} and thus it is advantageous to treat them with an unconditionally stable integrator even if the remainder of the model is to be treated explicitly. Earlier work on this subject endeavored to merge different integration methods, such as trapezoidal rule and central differences, but implementation proved somewhat cumbersome [39, 40]. In the approach presented in [36–38], only one integration procedure is used and the distinction between implicit and explicit zones is handled entirely by the finite element assembly algorithm. Implementation thus becomes trivially simple, and, at the same time, the procedure is amenable to rigorous stability [36] and convergence analyses [41]. Because the method does not fall within the LMS framework, the modal approach to stability seems inapplicable. However, an energy analysis provides necessary and sufficient conditions.

The basic ingredients in the implicit-explicit procedure advocated in [36–38] are (1) a given implicit integrator, (2) a predictor-corrector explicit scheme, which is constructed to be "compatible" with the given implicit integrator, and (3) a synthesis of the implicit and explicit schemes by way of a modified time-discrete equation of motion. We can start with *any* implicit integrator. For simplicity we shall employ the Newmark family with β assumed positive, which gives rise to implicit methods. We recall the integration formulas employed:

$$\left.\begin{aligned} \boldsymbol{d}_{n+1} &= \widetilde{\boldsymbol{d}}_{n+1} + \beta \Delta t^2 \boldsymbol{a}_{n+1} \quad (9.4.1) \\ \boldsymbol{v}_{n+1} &= \widetilde{\boldsymbol{v}}_{n+1} + \gamma \Delta t \boldsymbol{a}_{n+1} \quad (9.4.2) \end{aligned}\right\} \quad \textbf{\textit{(correctors)}}$$

$$\left.\begin{aligned} \widetilde{\boldsymbol{d}}_{n+1} &= \boldsymbol{d}_n + \Delta t \boldsymbol{v}_n + \frac{\Delta t^2}{2}(1 - 2\beta)\boldsymbol{a}_n \quad (9.4.3) \\ \widetilde{\boldsymbol{v}}_{n+1} &= \boldsymbol{v}_n + \Delta t(1 - \gamma)\boldsymbol{a}_n \quad (9.4.4) \end{aligned}\right\} \quad \textbf{\textit{(predictors)}}$$

Implicit Newmark methods. The Newmark family is defined by (9.4.1) through (9.4.4) and the temporally discretized equation of motion in the form (cf. (9.1.4) through (9.1.6)):

$$\boldsymbol{M}\boldsymbol{a}_{n+1} + \boldsymbol{C}\boldsymbol{v}_{n+1} + \boldsymbol{K}\boldsymbol{d}_{n+1} = \boldsymbol{F}_{n+1} \tag{9.4.5}$$

For $\beta > 0$, (9.4.1) through (9.4.5) defines an implicit method. This is ingredient (1).

Explicit predictor-corrector methods. The "compatible" explicit scheme is defined by (9.1.1) through (9.1.4) and a temporally discrete equation of motion in which the stiffness and damping terms are rendered explicit:

$$\boxed{\boldsymbol{M}\boldsymbol{a}_{n+1} + \boldsymbol{C}\widetilde{\boldsymbol{v}}_{n+1} + \boldsymbol{K}\widetilde{\boldsymbol{d}}_{n+1} = \boldsymbol{F}_{n+1}} \tag{9.4.6}$$

Note that, compared with (9.4.5), (9.4.6) amounts to replacing corrector values of $\boldsymbol{d}_{n+1}$ and $\boldsymbol{v}_{n+1}$ with corresponding predictor values. Clearly, as long as $\boldsymbol{M}$ is diagonal, $\boldsymbol{a}_{n+1}$ may be determined from (9.4.6) without solving equations; $\boldsymbol{d}_{n+1}$ and $\boldsymbol{v}_{n+1}$ are then defined by (9.4.1) and (9.4.2). This defines ingredient (2).

Exercise 1. Assume $\gamma = \frac{1}{2}$ and $\boldsymbol{F} = \boldsymbol{0}$. Show that the displacement difference equation for the explicit predictor-corrector method is independent of β and that if in addition $\boldsymbol{C} = \boldsymbol{0}$ then it is the same as that for the Newmark method with $\beta = 0$ and $\gamma = \frac{1}{2}$. (This means that under the stated assumptions the explicit-predictor corrector scheme and central-difference method become identical up to the starting procedure.)

Implicit-explicit methods. Consider a finite element model in which the elements are divided into two groups: the implicit elements and the explicit elements. Let I and E superscripts refer to the implicit and explicit groups, respectively. In particular let $\boldsymbol{M}^I$, $\boldsymbol{C}^I$, $\boldsymbol{K}^I$, and $\boldsymbol{F}^I$ (respectively, $\boldsymbol{M}^E$, $\boldsymbol{C}^E$, $\boldsymbol{K}^E$, and $\boldsymbol{F}^E$) be the assembled mass, damping, stiffness and load for the implicit (respectively, explicit) group. We assume $\boldsymbol{M}^E$ is diagonal. Each of the aforementioned matrices is assumed positive-semidefinite. The implicit-explicit methods, composites of the implicit Newmark methods, and explicit, predictor-corrector methods, are given by (9.4.1) through (9.4.4) and

$$\boldsymbol{M a}_{n+1} + \boldsymbol{C}^I \boldsymbol{v}_{n+1} + \boldsymbol{C}^E \tilde{\boldsymbol{v}}_{n+1} + \boldsymbol{K}^I \boldsymbol{d}_{n+1} + \boldsymbol{K}^E \tilde{\boldsymbol{d}}_{n+1} = \boldsymbol{F}_{n+1} \tag{9.4.7}$$

where

$$\boldsymbol{M} = \boldsymbol{M}^I + \boldsymbol{M}^E \tag{9.4.8}$$

$$\boldsymbol{C} = \boldsymbol{C}^I + \boldsymbol{C}^E \tag{9.4.9}$$

$$\boldsymbol{K} = \boldsymbol{K}^I + \boldsymbol{K}^E \tag{9.4.10}$$

$$\boldsymbol{F} = \boldsymbol{F}^I + \boldsymbol{F}^E \tag{9.4.11}$$

Note that in (9.4.7) the implicit arrays multiply corrector values, whereas the explicit arrays multiply predictor values. By using (9.4.1) and (9.4.2) in (9.4.7), we arrive at the equation that determines $\boldsymbol{a}_{n+1}$:

$$\boldsymbol{M}^* \boldsymbol{a}_{n+1} = \boldsymbol{F}_{n+1} - \boldsymbol{C} \tilde{\boldsymbol{v}}_{n+1} - \boldsymbol{K} \tilde{\boldsymbol{d}}_{n+1} \tag{9.4.12}$$

$$\boldsymbol{M}^* = \boldsymbol{M} + \gamma \Delta t \boldsymbol{C}^I + \beta \Delta t^2 \boldsymbol{K}^I \tag{9.4.13}$$

Note that the only manifestation of the implicit-explicit partition is in the left-hand side coefficient matrix where only the implicit parts of $\boldsymbol{C}$ and $\boldsymbol{K}$ appear. The Newmark methods correspond to the choices:

$$\left.\begin{aligned} \boldsymbol{C}^I &= \boldsymbol{C}, & \boldsymbol{K}^I &= \boldsymbol{K} \\ \boldsymbol{C}^E &= \boldsymbol{0}, & \boldsymbol{K}^E &= \boldsymbol{0} \end{aligned}\right\} \tag{9.4.14}$$

The explicit predictor-corrector methods correspond to:

$$\left.\begin{aligned} \boldsymbol{C}^I &= \boldsymbol{0}, & \boldsymbol{K}^I &= \boldsymbol{0} \\ \boldsymbol{C}^E &= \boldsymbol{C}, & \boldsymbol{K}^E &= \boldsymbol{K} \end{aligned}\right\} \tag{9.4.15}$$

The implicit stiffness and damping contributions engender a band-profile structure, which corresponds to the connectivity of the implicit group *only*. Consequently, $\boldsymbol{M}^*$, which is symmetric and positive-definite, will have diagonal subregions emanating from the explicit elements. A simple mesh that illustrates these properties is shown in Fig. 9.4.1.

The structure of $\boldsymbol{M}^*$ is fully exploited by active (compacted) column equation solvers, in which zeros outside the profile are neither stored nor operated upon; see Chapter 11.

It may be clear to persons familiar with the computer implementation of implicit transient schemes that the implicit-explicit element methods advocated herein represent no additional complications and may be implemented into many existing implicit codes with relative ease.

Many other possibilities fit within this general framework. For example, (9.4.8) through (9.4.11) may be viewed as general definitions of "splitting" an operator. In

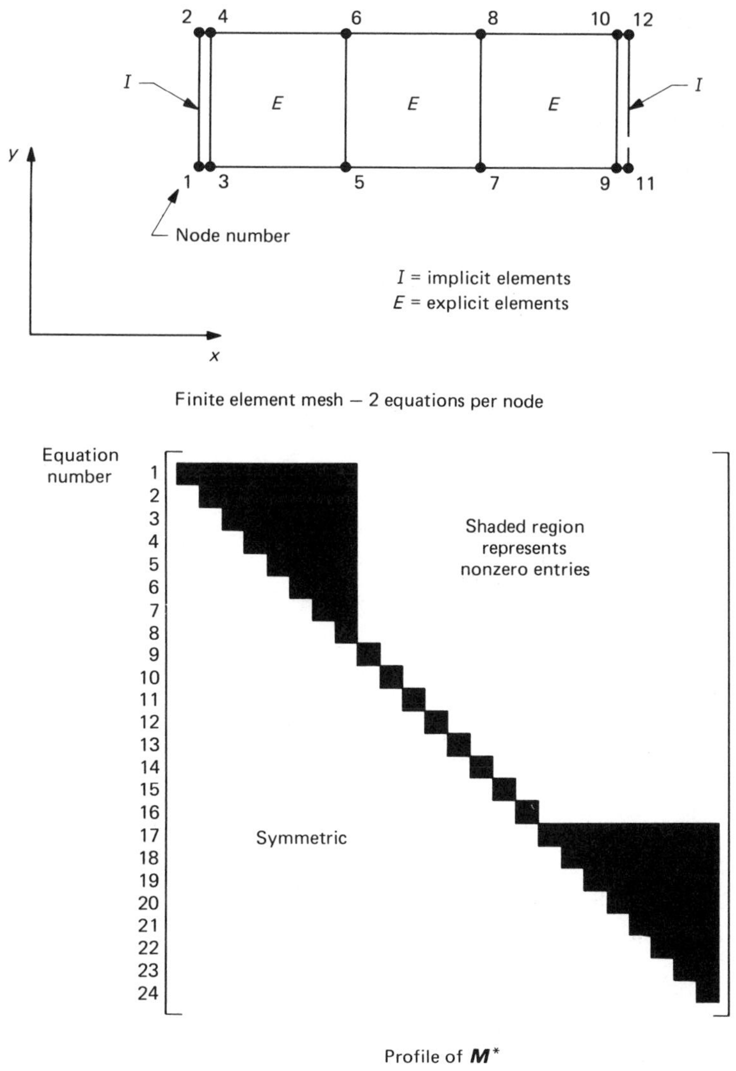

Figure 9.4.1 Profile of effective mass matrix for implicit-explicit finite element mesh partition.

this way, the procedures described in [42, 43] (sometimes referred to as "semi-implicit schemes," or "triangular splittings") may be subsumed. As observed by Park and Felippa [42], "nodal" partitions may also be encompassed by formulations of this type.

Exercise 2. Develop an implicit-explicit algorithm based upon the α-method described in Sec. 9.3.3. *Ans.*: The equations consist of (9.4.1) through (9.4.4) and

$$\begin{aligned} \boldsymbol{M}\boldsymbol{a}_{n+1} &+ (1+\alpha)\boldsymbol{C}^I\boldsymbol{v}_{n+1} + (1+\alpha)\boldsymbol{C}^E\widetilde{\boldsymbol{v}}_{n+1} - \alpha\boldsymbol{C}\boldsymbol{v}_n \\ &+ (1+\alpha)\boldsymbol{K}^I\boldsymbol{d}_{n+1} + (1+\alpha)\boldsymbol{K}^E\widetilde{\boldsymbol{d}}_{n+1} - \alpha\boldsymbol{K}\boldsymbol{d}_n = \boldsymbol{F}(t_{n+\alpha}) \end{aligned} \tag{9.4.16}$$

9.4.1 Stability via the Energy Method

The plan for obtaining stability conditions for the implicit-explicit algorithms is first to apply the energy method to the Newmark and predictor-corrector algorithms and then to show that the implicit-explicit algorithms may be reduced to a combination of the previous cases. It suffices to restrict attention to the case in which $\boldsymbol{F} = \boldsymbol{0}$ for purposes of stability.

Newmark Methods

Equations (9.4.1) through (9.4.5) may be combined to form the following identity:

$$\begin{aligned} \boldsymbol{a}_{n+1}^T \boldsymbol{A} \boldsymbol{a}_{n+1} + \boldsymbol{v}_{n+1}^T \boldsymbol{K} \boldsymbol{v}_{n+1} = \boldsymbol{a}_n^T \boldsymbol{A} \boldsymbol{a}_n + \boldsymbol{v}_n^T \boldsymbol{K} \boldsymbol{v}_n \\ - (2\gamma - 1)[\boldsymbol{a}_n]^T \boldsymbol{B} [\boldsymbol{a}_n] \\ - 2\Delta t \langle \boldsymbol{a}_n \rangle^T \boldsymbol{C} \langle \boldsymbol{a}_n \rangle \end{aligned} \tag{9.4.17}$$

where

$$\boldsymbol{B} = \boldsymbol{M} + \Delta t\left(\gamma - \frac{1}{2}\right)\boldsymbol{C} + \Delta t^2\left(\beta - \frac{\gamma}{2}\right)\boldsymbol{K} \tag{9.4.18}$$

$$\boldsymbol{A} = \boldsymbol{B} + \Delta t\left(\gamma - \frac{1}{2}\right)\boldsymbol{C} \tag{9.4.19}$$

(There should be no confusion over the present definition of the matrix $\boldsymbol{A}$ and earlier use for denoting an amplification matrix.)

Theorem. If $\gamma \geq \frac{1}{2}$ and $\boldsymbol{B}$ is positive definite, then $\boldsymbol{a}_n$ and $\boldsymbol{v}_n$ are bounded.

Proof: The hypotheses imply $\boldsymbol{A}$ is positive definite and

$$\boldsymbol{a}_{n+1}^T \boldsymbol{A} \boldsymbol{a}_{n+1} + \boldsymbol{v}_{n+1}^T \boldsymbol{K} \boldsymbol{v}_{n+1} \leq \boldsymbol{a}_n^T \boldsymbol{A} \boldsymbol{a}_n + \boldsymbol{v}_n^T \boldsymbol{K} \boldsymbol{v}_n \tag{9.4.20}$$

which in turn implies

$$\boldsymbol{a}_n^T \boldsymbol{A} \boldsymbol{a}_n + \boldsymbol{v}_n^T \boldsymbol{K} \boldsymbol{v}_n \leq \boldsymbol{a}_0^T \boldsymbol{A} \boldsymbol{a}_0 + \boldsymbol{v}_0^T \boldsymbol{K} \boldsymbol{v}_0, \qquad n = 1, 2, 3, \ldots \tag{9.4.21}$$

from which the conclusions follow. ■

Remarks

1. If $\boldsymbol{K}^{-1}$ exists, then $\boldsymbol{d}_n$ is also bounded. This follows from (9.4.5).

2. To establish the stability conditions we need only to determine when $\boldsymbol{B}$ is positive-definite. We may use the normal modes to diagonalize $\boldsymbol{B}$, from which it follows that

$$1 + 2\xi\left(\gamma - \frac{1}{2}\right)\omega^h \Delta t + \left(\beta - \frac{\gamma}{2}\right)(\omega^h \Delta t)^2 \tag{9.4.22}$$

must be positive for each mode in the system. Equation (9.4.22) leads directly to the conditions obtained earlier [cf. (9.1.16) through (9.1.20)].

Predictor-Corrector Methods

Equations (9.4.1) through (9.4.4) and (9.4.6) lead to the identity

$$\begin{aligned} \boldsymbol{a}_{n+1}^T \overline{\boldsymbol{A}} \boldsymbol{a}_{n+1} + \boldsymbol{v}_{n+1}^T \boldsymbol{K} \boldsymbol{v}_{n+1} = \boldsymbol{a}_n^T \overline{\boldsymbol{A}} \boldsymbol{a}_n &+ \boldsymbol{v}_n^T \boldsymbol{K} \boldsymbol{v}_n \\ &- (2\gamma - 1)[\boldsymbol{a}_n]^T \overline{\boldsymbol{B}} [\boldsymbol{a}_n] \\ &- 2\Delta t \langle \boldsymbol{a}_n \rangle^T \boldsymbol{C} \langle \boldsymbol{a}_n \rangle \end{aligned} \tag{9.4.23}$$

where

$$\overline{\boldsymbol{B}} = \boldsymbol{B} - \Delta t \gamma \boldsymbol{C} - \Delta t^2 \beta \boldsymbol{K} \tag{9.4.24}$$

$$\overline{\boldsymbol{A}} = \overline{\boldsymbol{B}} + \Delta t\left(\gamma - \frac{1}{2}\right)\boldsymbol{C} \tag{9.4.25}$$

Theorem: If $\gamma \geq \frac{1}{2}$ and $\overline{\boldsymbol{B}}$ is positive-definite, then $\boldsymbol{a}_n$ and $\boldsymbol{v}_n$ are bounded.

Proof: The proof is identical to that for the previous theorem, with $\overline{\boldsymbol{A}}$ in place of $\boldsymbol{A}$. ■

Remarks

1. If $\boldsymbol{K}^{-1}$ exists, then $\boldsymbol{d}_n$ is also bounded. This follows from (9.4.6), where $\tilde{\boldsymbol{d}}_{n+1}$ and $\tilde{\boldsymbol{v}}_{n+1}$ are eliminated by (9.4.1) and (9.4.2), respectively.

2. The stability characteristics of the predictor-corrector methods are determined by the conditions that render $\overline{\boldsymbol{B}}$ positive-definite. The same argument as before leads to the condition that

$$1 - \xi\omega^h \Delta t - \frac{\gamma}{2}(\omega^h \Delta t)^2 \tag{9.4.26}$$

must be positive for each mode of the system. Thus the following two conditions must be satisfied:

$$\gamma \geq \frac{1}{2} \tag{9.4.27}$$

$$\Omega \leq \Omega_{\text{crit}} \overset{\text{def}}{=} \frac{(\xi^2 + 2\gamma)^{1/2} - \xi}{\gamma} \tag{9.4.28}$$

Observe that (9.4.27) and (9.4.28) are independent of β.

3. It is to be expected that there are no unconditionally stable predictor-corrector schemes, since they are all explicit linear multistep methods.

4. Spectral radii are presented in Fig. 9.4.2 for the case $\xi = 0$. The Ω_{crit} obtained from the diagram is consistent with (9.4.28). The slope discontinuity in the spectral radius curves corresponds to the point at which the complex conjugate, principal roots of the amplification matrix become real and bifurcate. The value of Ω at which bifurcation occurs is

$$\boxed{\Omega_{bif} = \frac{2(1-\xi)}{\gamma + \frac{1}{2}}} \tag{9.4.29}$$

To ensure high-frequency numerical dissipation when $\zeta < 1$, Ω should be kept below Ω_{bif}, resulting in a somewhat smaller time step than that required by Ω_{crit}; cf. (9.4.28).

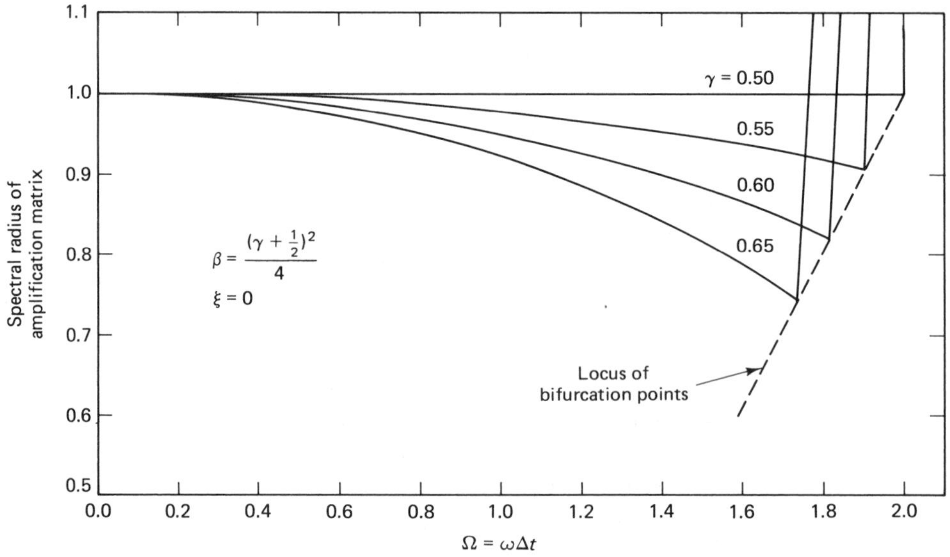

Figure 9.4.2 Spectral radius of amplification matrix for predictor-corrector algorithms [3].

5. Increasing ξ and γ *decreases* the critical time step for the predictor-corrector algorithms. Thus an estimate of ξ is essential in determining a stable time step.

6. When viscous damping is absent, $\Omega_{crit} = (2/\gamma)^{1/2}$. The maximum occurs for $\gamma = \frac{1}{2}$. That is $\Omega_{crit} = 2$, which is the same result as for the central difference method.

7. A local truncation-error analysis of the predictor-corrector schemes indicates that if $\boldsymbol{C} = \boldsymbol{0}$, $\gamma = \frac{1}{2}$ is a necessary and sufficient condition for second-order accuracy. If $\boldsymbol{C} \neq \boldsymbol{0}$, then first-order accuracy is attained for all $\gamma \geq \frac{1}{2}$. (A generalization of the basic scheme, which enables second-order accuracy to be regained under these circumstances, will be discussed subsequently.)

8. Stability conditions for the first-order case (i.e., when $\boldsymbol{K} = \boldsymbol{0}$) are easily deducible from the theorem. In this case, $\bar{\boldsymbol{B}}$ is positive-definite when $\lambda \Delta t < 2$, where λ is the maximum eigenvalue of $\boldsymbol{M}^{-1}\boldsymbol{C}$. (*Note:* $\gamma \geq \frac{1}{2}$ is still required.)

Implicit-Explicit Algorithms

An identity analogous to those for the previous cases may be derived from (9.4.1) through (9.4.4) and (9.4.7):

$$\begin{aligned}\boldsymbol{a}_{n+1}^T(\boldsymbol{A}^I + \bar{\boldsymbol{A}}^E)\boldsymbol{a}_{n+1} + \boldsymbol{v}_{n+1}^T\boldsymbol{K}\boldsymbol{v}_{n+1} = {} & \boldsymbol{a}_n^T(\boldsymbol{A}^I + \bar{\boldsymbol{A}}^E)\boldsymbol{a}_n + \boldsymbol{v}_n^T\boldsymbol{K}\boldsymbol{v}_n \\ & - (2\gamma - 1)[\boldsymbol{a}_n]^T(\boldsymbol{B}^I + \bar{\boldsymbol{B}}^E)[\boldsymbol{a}_n] \\ & - 2\Delta t\langle \boldsymbol{a}_n\rangle^T\boldsymbol{C}\langle \boldsymbol{a}_n\rangle \end{aligned} \tag{9.4.30}$$

where

$$\boldsymbol{B}^I = \boldsymbol{M}^I + \Delta t\left(\gamma - \frac{1}{2}\right)\boldsymbol{C}^I + \Delta t^2\left(\beta - \frac{\gamma}{2}\right)\boldsymbol{K}^I \tag{9.4.31}$$

$$\boldsymbol{A}^I = \boldsymbol{B}^I + \Delta t\left(\gamma - \frac{1}{2}\right)\boldsymbol{C}^I \tag{9.4.32}$$

$$\bar{\boldsymbol{B}}^E = \boldsymbol{B}^E - \Delta t\gamma\boldsymbol{C}^E - \Delta t^2\beta\boldsymbol{K}^E \tag{9.4.33}$$

$$\boldsymbol{B}^E = \boldsymbol{M}^E + \Delta t\left(\gamma - \frac{1}{2}\right)\boldsymbol{C}^E + \Delta t^2\left(\beta - \frac{\gamma}{2}\right)\boldsymbol{K}^E \tag{9.4.34}$$

$$\bar{\boldsymbol{A}}^E = \bar{\boldsymbol{B}}^E + \Delta t\left(\gamma - \frac{1}{2}\right)\boldsymbol{C}^E \tag{9.4.35}$$

Theorem: If $\gamma \geq \frac{1}{2}$ and $\boldsymbol{B}^I + \bar{\boldsymbol{B}}^E$ is positive definite, then $\boldsymbol{a}_n$ and $\boldsymbol{v}_n$ are bounded.

Proof: Again, the proof is identical to that for the first theorem with $\boldsymbol{A}^I + \bar{\boldsymbol{A}}^E$ in place of $\boldsymbol{A}$. ■

Remarks

1. If $\boldsymbol{K}^{-1}$ exists, then $\boldsymbol{d}_n$ is also bounded.

2. $\boldsymbol{B}^I + \bar{\boldsymbol{B}}^E$ is rendered positive definite, and stability is thereby achieved, when β, γ, and Δt satisfy (9.1.16) or (9.1.17)–(9.1.20) and (9.4.28). The time-step restrictions, (9.1.19), (9.1.20), and (9.4.28), pertain only to the implicit and explicit element groups, respectively.

3. In practice, one might as well achieve unconditional stability in the implicit element group. Thus for a given value of $\gamma \geq \frac{1}{2}$, β could be selected according to (9.1.55). The time-step restriction then emanates from satisfaction of (9.4.28), or (9.4.29), in the explicit element group. For example, if $\gamma = \frac{1}{2}$, (9.1.55) yields $\beta = \frac{1}{4}$, resulting in the trapezoidal rule algorithm for the implicit group. In the absence of viscous damping in the explicit element group, (9.4.28) or (9.4.29) results in

$\Omega_{\text{crit}} = 2$. A sufficient condition for stability then is that $\Omega < 2$ for the maximum natural frequency of the explicit elements.

Derivation of (9.4.17), (9.4.23), and (9.4.30)

We work with the undivided forward-difference and mean-value operators (i.e., [ ] and $\langle\ \rangle$) introduced earlier. Assume we have a symmetric array $\boldsymbol{A}$. Then it may be easily verified that

$$\boldsymbol{x}_{n+\alpha}^T \boldsymbol{A}[\boldsymbol{x}_n] = \langle \boldsymbol{x}_n \rangle^T \boldsymbol{A}[\boldsymbol{x}_n] + \left(\alpha - \frac{1}{2}\right)[\boldsymbol{x}_n]^T \boldsymbol{A}[\boldsymbol{x}_n] \tag{9.4.36}$$

$$\langle \boldsymbol{x}_n \rangle^T \boldsymbol{A}[\boldsymbol{x}_n] = \frac{1}{2}[\boldsymbol{x}_n^T \boldsymbol{A} \boldsymbol{x}_n] \tag{9.4.37}$$

Equations (9.4.1) through (9.4.4) may be written in the form:

$$[\boldsymbol{v}_n] = \Delta t \boldsymbol{a}_{n+\gamma} \tag{9.4.38}$$

$$[\boldsymbol{d}_n] = \Delta t \langle \boldsymbol{v}_n \rangle + \frac{\Delta t^2}{2}(2\beta - \gamma)[\boldsymbol{a}_n] \tag{9.4.39}$$

Let

$$T(\boldsymbol{x}) = \frac{1}{2}\boldsymbol{x}^T \boldsymbol{M} \boldsymbol{x} \tag{9.4.40}$$

$$U(\boldsymbol{x}) = \frac{1}{2}\boldsymbol{x}^T \boldsymbol{K} \boldsymbol{x} \tag{9.4.41}$$

$$D(\boldsymbol{x}) = -\boldsymbol{x}^T \boldsymbol{C} \boldsymbol{x} \tag{9.4.42}$$

Apply the [ ]-operator to the time-discrete equation of motion, assuming $\boldsymbol{F} = \boldsymbol{0}$, and premultiply by $[\boldsymbol{v}_n]^T$:

$$\begin{aligned}
0 &= [\boldsymbol{v}_n]^T (\boldsymbol{M}[\boldsymbol{a}_n] + \boldsymbol{C}[\boldsymbol{v}_n] + \boldsymbol{K}[\boldsymbol{d}_n]) \\
&= \Delta t\, \boldsymbol{a}_{n+\gamma}^T \boldsymbol{M}[\boldsymbol{a}_n] + [\boldsymbol{v}_n]^T \boldsymbol{C}[\boldsymbol{v}_n] + \langle \boldsymbol{v}_n^T \rangle \boldsymbol{K}\, \Delta t[\boldsymbol{v}_n] \\
&\quad + \Delta t\, \boldsymbol{a}_{n+\gamma}^T \boldsymbol{K} \frac{\Delta t^2}{2}(2\beta - \gamma)[\boldsymbol{a}_n] \qquad \text{(by (9.4.38) and (9.4.39))} \\
&= \Delta t\left\{[T(\boldsymbol{a}_n)] + 2\left(\gamma - \frac{1}{2}\right)T([\boldsymbol{a}_n])\right\} - D([\boldsymbol{v}_n]) + \Delta t[U(\boldsymbol{v}_n)] \\
&\quad + \frac{\Delta t^3}{2}(2\beta - \gamma)\left\{[U(\boldsymbol{a}_n)] + 2\left(\gamma - \frac{1}{2}\right)U([\boldsymbol{a}_n])\right\} \\
&\qquad \text{(by (9.4.36), (9.4.37), (9.4.40)–(9.4.42))}
\end{aligned} \tag{9.4.43}$$

Note that

$$\begin{aligned}
-D([\boldsymbol{v}_n]) &= \Delta t^2 \boldsymbol{a}_{n+\gamma}^T \boldsymbol{C} \boldsymbol{a}_{n+\gamma} \\
&= \Delta t^2\left\{\langle \boldsymbol{a}_n \rangle^T \boldsymbol{C} \langle \boldsymbol{a}_n \rangle + 2\left(\gamma - \frac{1}{2}\right)\langle \boldsymbol{a}_n \rangle^T \boldsymbol{C}[\boldsymbol{a}_n] + \left(\gamma - \frac{1}{2}\right)^2 [\boldsymbol{a}_n]^T \boldsymbol{C}[\boldsymbol{a}_n]\right\}
\end{aligned}$$

$$= -\Delta t^2\left\{D(\langle \boldsymbol{a}_n\rangle) + \left(\gamma - \frac{1}{2}\right)[D(\boldsymbol{a}_n)] + \left(\gamma - \frac{1}{2}\right)^2 D([\boldsymbol{a}_n])\right\} \tag{9.4.44}$$

Employing (9.4.44) in (9.4.43) enables us to write

$$[T(\boldsymbol{a}_n)] + \frac{\Delta t^2}{2}(2\beta - \gamma)[U(\boldsymbol{a}_n)] - \Delta t\left(\gamma - \frac{1}{2}\right)[D(\boldsymbol{a}_n)] + [U(\mathbf{v}_n)]$$

$$= -2\left(\gamma - \frac{1}{2}\right)T([\boldsymbol{a}_n]) - \frac{\Delta t^2}{2}(2\beta - \gamma)\, 2\left(\gamma - \frac{1}{2}\right)U([\boldsymbol{a}_n])$$

$$+ \Delta t\left(\gamma - \frac{1}{2}\right)^2 D([\boldsymbol{a}_n]) + \Delta t\, D(\langle \boldsymbol{a}_n\rangle) \tag{9.4.45}$$

Rearranging:

$$\langle \boldsymbol{a}_n\rangle^T\left\{\boldsymbol{M} + \frac{\Delta t^2}{2}(2\beta - \gamma)\boldsymbol{K} + 2\Delta t\left(\gamma - \frac{1}{2}\right)\boldsymbol{C}\right\}[\boldsymbol{a}_n] + [U(\boldsymbol{v}_n)]$$

$$= -\left(\gamma - \frac{1}{2}\right)[\boldsymbol{a}_n]^T\left\{\boldsymbol{M} + \frac{\Delta t^2}{2}(2\beta - \gamma)\boldsymbol{K} + \Delta t\left(\gamma - \frac{1}{2}\right)\boldsymbol{C}\right\}[\boldsymbol{a}_n]$$

$$-\Delta t\langle \boldsymbol{a}_n\rangle^T \boldsymbol{C}\langle \boldsymbol{a}_n\rangle \tag{9.4.46}$$

It follows directly from (9.4.46) that

$$\frac{1}{2}\boldsymbol{a}_{n+1}^T\boldsymbol{A}\boldsymbol{a}_{n+1} + \frac{1}{2}\boldsymbol{v}_{n+1}^T\boldsymbol{K}\boldsymbol{v}_{n+1}$$

$$= \frac{1}{2}\boldsymbol{a}_n^T\boldsymbol{A}\boldsymbol{a}_n + \frac{1}{2}\boldsymbol{v}_n^T\boldsymbol{K}\boldsymbol{v}_n - \left(\gamma - \frac{1}{2}\right)[\boldsymbol{a}_n]^T\boldsymbol{B}[\boldsymbol{a}_n] - \Delta t\langle \boldsymbol{a}_n\rangle^T\boldsymbol{C}\langle \boldsymbol{a}_n\rangle \tag{9.4.47}$$

which is identical to (9.4.17) where $\boldsymbol{A}$ and $\boldsymbol{B}$ are defined by (9.4.19) and (9.4.18), respectively.

To derive (9.4.23) we observe that $\widetilde{\boldsymbol{d}}_{n+1}$ and $\widetilde{\boldsymbol{v}}_{n+1}$ can be eliminated from (9.4.6) by way of (9.4.1) and (9.4.2) respectively. This results in

$$\overline{\boldsymbol{M}}\boldsymbol{a}_{n+1} + \boldsymbol{C}\boldsymbol{v}_{n+1} + \boldsymbol{K}\boldsymbol{d}_{n+1} = \boldsymbol{F}_{n+1} \tag{9.4.48}$$

where

$$\overline{\boldsymbol{M}} = \boldsymbol{M} - \gamma\Delta t\,\boldsymbol{C} - \beta\Delta t^2\,\boldsymbol{K} \tag{9.4.49}$$

Thus the predictor-corrector schemes consist of (9.4.1) through (9.4.4), (9.4.48) and (9.4.49). Note that (9.4.1) through (9.4.4) and (9.4.48) are Newmark's algorithm for an equation of motion in which the mass matrix is given by (9.4.49). Therefore the analysis leading to (9.4.47) applies to the predictor-corrector method if $\boldsymbol{M}$ is replaced by $\overline{\boldsymbol{M}}$. This immediately gives rise to (9.4.23) through (9.4.25).

A similar strategy is used to derive (9.4.30). Substitute (9.4.1) and (9.4.2) in (9.4.7) to arrive at

$$\overline{\overline{\boldsymbol{M}}}\boldsymbol{a}_{n+1} + \boldsymbol{C}\boldsymbol{v}_{n+1} + \boldsymbol{K}\boldsymbol{d}_{n+1} = \boldsymbol{F}_{n+1} \tag{9.4.50}$$

where

$$\bar{\bar{M}} = M - \gamma\Delta t\, C^E - \beta\Delta t^2\, K^E \tag{9.4.51}$$

Equation (9.4.47) with M replaced by $\bar{\bar{M}}$ results in (9.4.30). (The details are left as an exercise.)

9.4.2 Predictor/Multicorrector Algorithms [38]

Generalization of the previously described schemes to a predictor/multicorrector format leads to improved characteristics in some circumstances and is suggestive of further algorithmic generalizations. The implementational sequence is given as follows:

$$i = 0 \qquad (i \text{ is the iteration counter}) \tag{9.4.52}$$

$$\left.\begin{aligned} d_{n+1}^{(i)} &= \tilde{d}_{n+1} && (9.4.53)\\ v_{n+1}^{(i)} &= \tilde{v}_{n+1} && (9.4.54)\\ a_{n+1}^{(i)} &= 0 && (9.4.55)\end{aligned}\right\} \quad \text{(predictor phase)}$$

$$\Delta F_{n+1}^{(i)} = \mathop{\mathbf{A}}_{e=1}^{n_{el}} \left(f_{n+1}^{\text{ext}} - m^e a_{n+1}^{e(i)} - c^e v_{n+1}^{e(i)} - k^e d_{n+1}^{e(i)}\right) \tag{9.4.56}$$

$$M^*\Delta a = \Delta F_{n+1}^{(i)} \tag{9.4.57}$$

$$\left.\begin{aligned} a_{n+1}^{(i+1)} &= a_{n+1}^{(i)} + \Delta a && (9.4.58)\\ v_{n+1}^{(i+1)} &= \tilde{v}_{n+1} + \Delta t\, \gamma\, a_{n+1}^{(i+1)} && (9.4.59)\\ d_{n+1}^{(i+1)} &= \tilde{d}_{n+1} + \Delta t^2\, \beta\, a_{n+1}^{(i+1)} && (9.4.60)\end{aligned}\right\} \quad \text{(corrector phase)}$$

If additional iterations are to be performed, i is replaced by $i + 1$, and calculations resume with (9.4.56). When the iterative phase is completed, the solution is defined by the last iterates (i.e., (9.4.58) through (9.4.60)).

The definition of M^* emanates from the time-discrete equation of motion, which for the predictor/multicorrector algorithms (allowing for operator and/or mesh partitions), takes the form:

$$M a_{n+1}^{(i+1)} + C^I v_{n+1}^{(i+1)} + C^E v_{n+1}^{(i)} + K^I d_{n+1}^{(i+1)} + K^E d_{n+1}^{(i)} = F_{n+1} \tag{9.4.61}$$

In this case

$$M^* = M + \Delta t\, \gamma\, C^I + \Delta t^2 \beta\, K^I \tag{9.4.62}$$

Remarks

1. If only one pass is made through (9.4.52) through (9.4.60), the algorithm is equivalent to the implicit-explicit scheme given earlier.

2. If any part of $\boldsymbol{C}$ is treated explicitly, an additional pass through (9.4.56) through (9.4.60) can increase accuracy.

To see this in a simple setting, it suffices to consider the following special case:

$$\boldsymbol{M}\boldsymbol{a}_{n+1}^{(i+1)} + \boldsymbol{C}\boldsymbol{v}_{n+1}^{(i)} = \boldsymbol{0} \tag{9.4.63}$$

It may then be shown that one additional pass through (9.4.56) through (9.4.60) results in

$$\boldsymbol{v}_{n+1} = (\boldsymbol{I} - \Delta t\boldsymbol{M}^{-1}\boldsymbol{C} + \gamma(\Delta t\boldsymbol{M}^{-1}\boldsymbol{C})^2)\boldsymbol{v}_n \tag{9.4.64}$$

which is clearly second-order accurate if $\gamma = \dfrac{1}{2}$.

The stability condition is easily determined from (9.4.64): $\Omega < \Omega_{\text{crit}} = 1/\gamma = 2$, where $\Omega = \lambda\Delta t$ and λ is the maximum eigenvalue of $\boldsymbol{M}^{-1}\boldsymbol{C}$. (This may be seen by noting that the eigenvectors of $\boldsymbol{M}^{-1}\boldsymbol{C}$ diagonalize all matrices of the form $\boldsymbol{M}(\boldsymbol{M}^{-1}\boldsymbol{C})^j$, where j is a positive integer. That is, if $(\boldsymbol{C} - \lambda\boldsymbol{M})\boldsymbol{\psi} = \boldsymbol{0}$, then $\boldsymbol{\psi}^T\boldsymbol{M}(\boldsymbol{M}^{-1}\boldsymbol{C})^j\boldsymbol{\psi} = \lambda^j$.) Thus there is no adverse effect on stability compared with a one-pass formulation, which amounts to the forward-difference algorithm (cf. Sec. 8.2.2).

3. A completely general stability analysis of partitioned predictor/multi-corrector algorithms is not available yet. The unpartitioned case is simpler, however, since all pertinent equations are diagonalizable. Thus one may work in terms of an SDOF model equation. The greater the number of passes, the higher the order of the stability polynomial.

4. Note the way in which the residual, or "out-of-balance," force, (9.4.56), has been written. The term $\boldsymbol{f}_{n+1}^{\text{ext}}$ refers to the prescribed body and surface force contributions to $\boldsymbol{f}_{n+1}^e$, but *not* the prescribed displacement, velocity, and acceleration contributions. These effects are accounted for in the remaining terms. That is, for all kinematic degrees of freedom that are prescribed, the appropriate quantities are included in the element vectors $\boldsymbol{a}_{n+1}^e$, $\boldsymbol{v}_{n+1}^e$, and $\boldsymbol{d}_{n+1}^e$.

5. Kinematic boundary conditions can be specified in terms of displacement, velocity, or acceleration. Whichever quantity is given, the other two can be determined from the integration formulas which define the algorithm. For example, suppose acceleration is given; that is, assume we know g_n, $\dot{g}_n$, $\ddot{g}_n$, and $\ddot{g}_{n+1}$ for a particular degree of freedom. Then g_{n+1} and $\dot{g}_{n+1}$ are given by

$$g_{n+1} = g_n + \Delta t\dot{g}_n + \frac{\Delta t^2}{2}\{(1 - 2\beta)\ddot{g}_n + 2\beta\ddot{g}_{n+1}\} \tag{9.4.65}$$

$$\dot{g}_{n+1} = \dot{g}_n + \Delta t\{(1 - \gamma)\ddot{g}_n + \gamma\ddot{g}_{n+1}\} \tag{9.4.66}$$

Likewise, given any one of g_{n+1}, $\dot{g}_{n+1}$ and $\ddot{g}_{n+1}$, the remaining two may be determined from (9.4.65) and (9.4.66). Quantities computed in this manner are set once and for all in the element vectors $\boldsymbol{d}_{n+1}^e$, $\boldsymbol{v}_{n+1}^e$, and $\boldsymbol{a}_{n+1}^e$ throughout the iterative phase during a time step (see Chapter 11).

6. Note that by ignoring contributions to $\boldsymbol{K}$, the predictor/multicorrector schemes are applicable to the parabolic case considered in Chapter 8. Here $\boldsymbol{C}$ must be taken to the conductivity matrix.

7. An implicit-explicit predictor/multicorrector version of the Hilber-Hughes-Taylor α-method is implemented in program DLEARN (see Chapter 11). The reader may wish to study the details of this implementation in order fully to appreciate the data structure and treatment of initial and boundary conditions in transient analysis.

Exercise 3. Consider an algorithm governed by

$$\boldsymbol{X}_{n+1} = \boldsymbol{A}\boldsymbol{X}_n$$

where $\boldsymbol{A}$, the amplification matrix, is spectrally stable. Let us assume the eigenvalues of $\boldsymbol{A}$ are distinct so that $\boldsymbol{A}$ admits the representation

$$\boldsymbol{A} = \boldsymbol{P}\boldsymbol{\Lambda}\boldsymbol{P}^{-1}$$

where $\boldsymbol{P}$ is a matrix of eigenvectors and $\boldsymbol{\Lambda}$ is a diagonal matrix of eigenvalues.

Prove that if $\rho(\boldsymbol{A}) \leq 1$, then $\|\boldsymbol{X}_{n+1}\|_N \leq \|\boldsymbol{X}_n\|_N$, where $\|\boldsymbol{X}\|_N = (\boldsymbol{X}^T\boldsymbol{N}\boldsymbol{X})^{1/2}$, $\boldsymbol{N}$ is the symmetric real part of $(\boldsymbol{P}^{-1})^*\boldsymbol{P}^{-1}$, and $*$ denotes Hermitian transpose.

Solution

$$\boldsymbol{P}^{-1}\boldsymbol{X}_{n+1} = \boldsymbol{\Lambda}\boldsymbol{P}^{-1}\boldsymbol{X}_n$$

and

$$\boldsymbol{X}_{n+1}^T(\boldsymbol{P}^{-1})^*\boldsymbol{P}^{-1}\boldsymbol{X}_{n+1} = \boldsymbol{X}_n^T(\boldsymbol{P}^{-1})^*\,\boldsymbol{\Lambda}^*\,\boldsymbol{\Lambda}\boldsymbol{P}^{-1}\boldsymbol{X}_n$$

where

$$\boldsymbol{\Lambda}^*\,\boldsymbol{\Lambda} = |\boldsymbol{\Lambda}|^2$$

which implies

$$\|\boldsymbol{X}_{n+1}\|_N \leq \|\boldsymbol{X}_n\|_N$$

$$\|\boldsymbol{X}\|_N^2 = \boldsymbol{X}(\boldsymbol{P}^{-1})^*\boldsymbol{P}^{-1}\boldsymbol{X}$$

$(\boldsymbol{P}^{-1})^*\boldsymbol{P}^{-1}$ is a Hermitian matrix. That is, it consists of a symmetric, real, positive definite matrix, plus imaginary, skew-symmetric matrix (which can be ignored in the norm).

Remark

The above result indicates that any spectrally stable algorithm possesses a conservation law, or growth inequality, in terms of a norm defined by the eigenvectors $\boldsymbol{P}$, and thus a spectrally stable algorithm is stable with respect to the norm defined by $\boldsymbol{P}$.

9.5 MASS MATRICES FOR SHELL ELEMENTS

In Chapter 6 we discussed a class of C^0 shell elements. Consistent- and lumped-mass matrices for these elements are presented next. The lumped matrix has scaled rotational degress of freedom to achieve effective critical time steps in explicit analysis. These ideas generalize those for the linear beam presented in Example 3 of Sec. 9.2.

Consistent Mass

The element consistent-mass matrix is defined by

$$\boldsymbol{m} = [\boldsymbol{m}_{ab}]$$
$$\boldsymbol{m}_{ab} = \int_{\square}\int_{-1}^{+1} \boldsymbol{N}_a^T \boldsymbol{N}_b \rho j \, d\zeta \, d\square \tag{9.5.1}$$

where $1 \le a, b \le n_{en}$, the number of element nodes; ρ is the mass density; j is the Jacobian determinant; and

$$\int_{\square} \cdots d\square = \begin{cases} \displaystyle\int_{-1}^{+1}\int_{-1}^{+1} \cdots d\xi \, d\eta & \text{(three dimensions)} \\ \displaystyle\int_{-1}^{+1} \cdots d\eta & \text{(two dimensions)} \end{cases} \tag{9.5.2}$$

In (9.5.1), $\boldsymbol{N}_a$ is the shape function matrix (see (6.2.73) and Chapter 6 for further details).

Lumped Mass

Element lumped-mass matrices may be computed by performing the following steps:

1. Calculate

$$\bar{j} = \int_{-1}^{+1} j \, d\zeta \tag{9.5.3}$$

$$m_a^{\text{rot}} = \int_{\square} N_a^2 \rho \bar{j} \, d\square \tag{9.5.4}$$

(ρ is assumed to be independent of ζ.)

2. For elements other than *heterosis:*

$$m_a^{\text{disp}} = m_a^{\text{rot}}, \qquad a = 1, 2, \ldots, n_{en} \tag{9.5.5}$$

For *heterosis*:

$$m^{\text{disp}} = \int_{\square} (N_a + N_a^{\text{ser}}(0, 0)N_9)^2 \rho \bar{j} \, d\square \qquad a = 1, 2, \ldots, 8 \tag{9.5.6}$$

$$m_9^{\text{disp}} = 0$$

(N_a^{ser} is the eight-node serendipity shape function.) (9.5.7)

3.

$$V = \int_{\square} \bar{j}\, d\square, \qquad M = \int_{\square} \rho \bar{j}\, d\square \tag{9.5.8}$$

4. Normalization

$$\widetilde{M}^{\text{rot}} = \sum_{a=1}^{n_{en}} m_a^{\text{rot}} \tag{9.5.9}$$

$$\widetilde{M}^{\text{disp}} = \sum_{a=1}^{n_{en}} m_a^{\text{disp}} \tag{9.5.10}$$

$$m_a^{\text{rot}} \leftarrow (M/\widetilde{M}^{\text{rot}})\, m_a^{\text{rot}} \tag{9.5.11}$$

$$m_a^{\text{disp}} \leftarrow (M/\widetilde{M}^{\text{disp}})\, m_a^{\text{disp}} \tag{9.5.12}$$

5. Adjustment to rotational inertia:

$$\langle z_a \rangle = \frac{z_a^+ + z_a^-}{2} \tag{9.5.13}$$

$$[z_a] = z_a^+ - z_a^- \tag{9.5.14}$$

$$\alpha_a = \int_{-1}^{+1} z_a^2\, d\zeta/2 = \langle z_a \rangle^2 + \frac{1}{12}[z_a]^2 \tag{9.5.15}$$

$$h = \frac{1}{n_{en}} \sum_{a=1}^{n_{en}} [z_a], \qquad A = V/h \tag{9.5.16}$$

$$\alpha_a = \max\left\{\alpha_a, \frac{A}{8}\right\} \tag{9.5.17}$$

$$m_a^{\text{rot}} \leftarrow m_a^{\text{rot}} \alpha_a \qquad \text{(no sum)} \tag{9.5.18}$$

Remarks

1. In the axisymmetric case, j needs to be replaced throughout by rj, where r is the radial coordinate.

2. The mass matrices may be specialized for application to the plate elements of Chapter 5.

REFERENCES

Section 9.1

1. N. M. Newmark, "A Method of Computation for Structural Dynamics," *Journal of the Engineering Mechanics Division,* ASCE, (1959), 67–94.
2. G. L. Goudreau and R. L. Taylor, "Evaluation of Numerical Integration Methods in Elastodynamics," *Computer Methods in Applied Mechanics and Engineering,* 2 (1972), 69–97.
3. R. D. Krieg and S. W. Key, "Transient Shell Response by Numerical Time Integration," *International Journal for Numerical Methods in Engineering,* 7 (1973), 273–286.
4. H. M. Hilber, "Analysis and Design of Numerical Integration Methods in Structural Dynamics," EERC Report No. 77-29, Earthquake Engineering Research Center, University of California, Berkeley, California, November 1976.
5. T. J. R. Hughes, "Analysis of Transient Algorithms with Particular Reference to Stability Behavior," pp. 67–155 in *Computational Methods for Transient Analysis,* eds. T. Belytschko and T. J. R. Hughes. Amsterdam: North-Holland, 1983.
6. H. Kolsky, *Stress Waves in Solids.* New York: Dover, 1963.
7. T. J. R. Hughes, R. L. Taylor, J. L. Sackman, A. Curnier, and W. Kanoknukulchai, "A Finite Element Method for a Class of Contact-Impact Problems," *Computer Methods in Applied Mechanics and Engineering,* 8 (1976), 249–276.

Section 9.2

8. I. Fried and D. S. Malkus, "Finite Element Mass Matrix Lumping by Numerical Integration Without Convergence Rate Loss," *International Journal of Solids and Structures,* 11 (1976), 461–466.
9. T. J. R. Hughes, R. L. Taylor, and W. Kanoknukulchai, "A Simple and Efficient Element for Plate Bending," *International Journal for Numerical Methods in Engineering,* 11, no. 10 (1977), 1529–1543.
10. T. J. R. Hughes, M. Cohen, and M. Haroun, "Reduced and Selective Integration Techniques in the Finite Element Analysis of Plates," *Nuclear Engineering and Design,* 46, no. 1 (1978), 203–222.
11. S. W. Key and Z. E. Beisinger, "The Transient Dynamic Analysis of Thin Shells by the Finite Element Method," *Proceedings of the 3rd Conference on Matrix Methods in Structural Mechanics,* Wright-Patterson Air Force Base, Ohio, 1971.
12. T. J. R. Hughes, W. K. Liu, and I. Levit, "Nonlinear Dynamic Finite Element Analysis," pp. 151–168 in *Nonlinear Finite Element Analysis in Structural Mechanics,* eds. W. Wunderlich et al. Berlin: Springer-Verlag, 1981. (Proceedings of the Europe-U.S. Workshop, Ruhr-Universität, Bochum, Germany, July 28–31, 1980.)
13. D. P. Flanagan and T. Belytschko, "A Uniform Strain Hexahedron and Quadrilateral with Orthogonal Hourglass Control," *International Journal for Numerical Methods in Engineering,* 17 (1981), 679–706.
14. T. Belytschko, "An Overview of Semidiscretization and Time Integration Procedures," pp. 1–65 in *Computational Methods for Transient Analysis,* eds. T. Belytschko and T. J. R. Hughes. Amsterdam: North-Holland, 1983.

Section 9.3

15. C. W. Gear, *Numerical Initial Value Problems in Ordinary Differential Equations*. Englewood Cliffs, N.J.: Prentice-Hall, 1971.

16. G. Dahlquist, "A Special Stability Problem for Linear Multistep Methods," *BIT*, 3 (1963), 27–43.

17. K. C. Park, "Evaluating Time Integration Methods for Nonlinear Dynamic Analysis," pp. 35–58 in *Finite Element Analysis of Transient Nonlinear Behavior*, eds. T. Belytschko et al., AMD 14. New York: ASME, 1975.

18. M. Geradin, "A Classification and Discussion of Integration Operators for Transient Structural Response," AIAA Paper 74-105, presented at AIAA 12th Aerospace Sciences Meeting, Washington, D.C., Jan. 30 through Feb. 1, 1974.

19. R. D. Krieg, "Unconditional Stability in Numerical Time Integration Methods," *Journal of Applied Mechanics*, 40 (1973), 417–421.

20. J. C. Houbolt, "A Recurrence Matrix Solution for the Dynamic Response of Elastic Aircraft," *Journal of the Aeronautical Sciences*, 17 (1950), 540–550.

21. E. L. Wilson, "A Computer Program for the Dynamic Stress Analysis of Underground Structures," SESM Report No. 68-1, Division of Structural Engineering and Structural Mechanics, University of California, Berkeley, 1968.

22. H. M. Hilber and T. J. R. Hughes, "Collocation, Dissipation and 'Overshoot' for Time Integration Schemes in Structural Dynamics," *Earthquake Engineering and Structural Dynamics*, 6 (1978), 99–118.

23. H. M. Hilber, "Analysis and Design of Numerical Integration Methods in Structural Dynamics," EERC Report No. 76-29, Earthquake Engineering Research Center, University of California, Berkeley, November 1976.

24. H. M. Hilber, T. J. R. Hughes, and R. L. Taylor, "Improved Numerical Dissipation for Time Integration Algorithms in Structural Dynamics," *Earthquake Engineering and Structural Dynamics*, 5 (1977), 283–292.

25. G. McVerry, "Numerical Integration Schemes for Structural Dynamics," unpublished manuscript.

26. G. L. Goudreau and R. L. Taylor, "Evaluation of Numerical Methods in Elastodynamics," *Journal of Computer Methods in Applied Mechanics and Engineering*, 2 (1973), 69–97.

27. J. H. Argyris, P. C. Dunne, and T. Angelopoulos, "Nonlinear Oscillations Using the Finite Element Technique," *Computer Methods in Applied Mechanics and Engineering*, 2 (1973), 203–250.

28. A. Ghose, "Computational Procedures for Inelastic Dynamic Analysis," Ph.D. Dissertation, Division of Structural Engineering and Structural Mechanics, University of California, Berkeley (1974).

29. J. H. Argyris, P. C. Dunne, and T. Angelopoulos, "Dynamic Response by Large Step Integration," *Earthquake Engineering and Structural Dynamics*, 2 (1973), 185–203.

30. R. D. Krieg and S. W. Key, "Transient Shell Response by Numerical Time Integration," *International Journal for Numerical Methods in Engineering*, 7 (1973), 273–286.

31. M. A. Kurdi, "Stable High Order Methods for Time Discretizations of Stiff Differential Equations," Ph.D. Dissertation, Department of Mathematics, University of California, Berkeley (1974).

32. W. L. Wood, M. Bossak, and O. C. Zienkiewicz, "An Alpha Modification of Newmark's

Method," *International Journal for Numerical Methods in Engineering,* 15 (1980), 1562–1566.

33. D. D. Adams and W. L. Wood, "Comparison of Hilber-Hughes-Taylor and Bossak 'α-methods' for the Numerical Integration of Vibration Equations," *International Journal for Numerical Methods in Engineering*, 19, no. 5 (1983), 765–771.

34. G. Bazzi and E. Anderheggen, "The ρ-family of Algorithms for Time-step Integration with Improved Numerical Dissipation," *Earthquake Engineering and Structural Dynamics,* 10 (1982), 537–550.

35. O. C. Zienkiewicz, W. L. Wood, N. W. Hine, and R. L. Taylor, "A Unified Set of Single Step Algorithms. Part 1: General Formulation and Applications," *International Journal for Numerical Methods in Engineering,* 20 (1984), 1529–1552.

Section 9.4

36. T. J. R. Hughes and W. K. Liu, "Implicit-Explicit Finite Elements in Transient Analysis: Stability Theory," *Journal of Applied Mechanics,* 45 (1978), 371–374.

37. T. J. R. Hughes and W. K. Liu, "Implicit-Explicit Finite Elements in Transient Analysis: Implementation and Numerical Examples," *Journal of Applied Mechanics,* 45 (1978), 375–378.

38. T. J. R. Hughes, K. S. Pister, and R. L. Taylor, "Implicit-Explicit Finite Elements in Nonlinear Transient Analysis," *Computer Methods in Applied Mechanics and Engineering,* 17/18 (1979), 159–182.

39. T. Belytschko and R. Mullen, "Mesh Partitions of Explicit-Implicit Time Integration," *Proceedings, U.S.-Germany Symposium on Formulations and Computational Algorithms in Finite Element Analysis,* Massachussetts Institute of Technology, Cambridge, Mass., August 1976.

40. T. Belytschko and R. Mullen, "Stability of Explicit-Implicit Mesh Partitions in Time Integration," *International Journal for Numerical Methods in Engineering,* 12, no. 10 (1978), 1575–1586.

41. T. J. R. Hughes and R. S. Stephenson, "Convergence of Implicit-Explicit Finite Elements in Nonlinear Transient Analysis," *International Journal of Engineering Science,* 19 (1981), 295–302.

42. K. C. Park and C.A. Felippa, "Partitioned Analysis of Coupled Systems," pp. 157–219 in *Computational Methods for Transient Analysis,* eds. T. Belytschko and T. J. R. Hughes. Amsterdam: North-Holland, 1983.

43. D. M. Trujillo, "An Unconditionally Stable Explicit Algorithm for Structural Dynamics," *International Journal for Numerical Methods in Engineering,* 11 (1977), 1579–1592.

10

Solution Techniques for Eigenvalue Problems

10.1 THE GENERALIZED EIGENPROBLEM

In this chapter we are concerned with algorithms for solving the so-called generalized eigenproblem arising from finite element discretizations. Recall that this takes the form:[1]

$$(\boldsymbol{K} - \lambda \boldsymbol{M})\boldsymbol{\psi} = \boldsymbol{0} \tag{10.1.1}$$

where $\boldsymbol{K}$ is symmetric and positive-semidefinite, $\boldsymbol{M}$ is symmetric and positive-definite, and both matrices possess typical band/profile structure. For convenience we recall some basic properties of the solution of (10.1.1). We assume as always that the dimensions of $\boldsymbol{K}$ and $\boldsymbol{M}$ are $n_{eq} \times n_{eq}$. There exist n_{eq} eigenvalues and corresponding eigenvectors[2] that satisfy (10.1.1). That is,

$$\boxed{(\boldsymbol{K} - \lambda_l \boldsymbol{M})\boldsymbol{\psi}_l = 0, \qquad \text{(no sum)}} \tag{10.1.2}$$

where $l = 1, 2, \ldots, n_{eq}$ denotes the mode number. Furthermore,

$$\boxed{0 \leq \lambda_1 \leq \lambda_2 \leq \cdots \leq \lambda_{n_{eq}}} \tag{10.1.3}$$

[1] Problems of this type are motivated and formulated in Chapter 7. To simplify subsequent writing we omit the superscript h on λ.

[2] The eigenvectors may be nonunique.

and

$$\boldsymbol{\psi}_k^T \boldsymbol{M} \boldsymbol{\psi}_l = \delta_{kl} \qquad (\boldsymbol{M} \text{ orthonormality}) \tag{10.1.4}$$

$$\boldsymbol{\psi}_k^T \boldsymbol{K} \boldsymbol{\psi}_l = \lambda_l \, \delta_{kl} \quad (\text{no sum}) \quad (\boldsymbol{K} \text{ orthogonality}) \tag{10.1.5}$$

The eigenvectors $\{\boldsymbol{\psi}_l\}_{l=1}^{n_{eq}}$ constitute a ***basis*** for $\mathbb{R}^{n_{eq}}$. If, additionally, $\boldsymbol{K}$ is positive-definite, as is often the case, then (10.1.3) may be strengthened as follows:

$$0 < \lambda_1 \leq \lambda_2 \leq \cdots \leq \lambda_{n_{eq}} \tag{10.1.6}$$

In practice, we are typically interested only in the lower modes. These are the most important from a physical standpoint. For example, the lowest mode is the most important in a structural stability (i.e., "buckling") calculation and the lower frequencies and corresponding mode shapes are usually the most important in considerations of dynamic response. Additionally, the higher modes of finite element discretizations are not accurate renditions of physical behavior but rather spurious artifacts of the discretization process. Consequently, they are of no engineering interest. For these reasons, we are interested in economical computational algorithms for extracting $\{\lambda_l, \boldsymbol{\psi}_l\}$, $1 \leq l \leq n_{\text{modes}}$, where $n_{\text{modes}} \ll n_{eq}$ is the number of desired eigenpairs. In practice n_{eq} may be very large, and thus solution of the eigenvalue problem, even for only a few eigenpairs, may entail an extensive and costly calculation.

Most procedures used for solving the large-scale generalized eigenproblem, (10.1.1), involve a ***reduced system*** of the form

$$(\boldsymbol{K}^* - \lambda^* \boldsymbol{M}^*)\boldsymbol{\psi}^* = \boldsymbol{0} \tag{10.1.7}$$

where $\boldsymbol{K}^*$ and $\boldsymbol{M}^*$ are *small, full, symmetric matrices*. Algorithms, such as the ***generalized Jacobi method*** [1], are available for directly solving (10.1.7).

Systems such as (10.1.7) may also be solved by first transforming to ***standard form***:

$$(\overline{\boldsymbol{K}}^* - \lambda^* \boldsymbol{I})\overline{\boldsymbol{\psi}}^* = \boldsymbol{0} \tag{10.1.8}$$

where

$$\overline{\boldsymbol{K}}^* = \overline{\boldsymbol{U}}^{-T} \boldsymbol{K}^* \overline{\boldsymbol{U}}^{-1} \tag{10.1.9}$$

$$\overline{\boldsymbol{\psi}}^* = \overline{\boldsymbol{U}} \, \boldsymbol{\psi}^* \tag{10.1.10}$$

and $\overline{\boldsymbol{U}}$ is the upper-triangular Cholesky factor of $\boldsymbol{M}^*$, i.e.,

$$\boldsymbol{M}^* = \overline{\boldsymbol{U}}^T \overline{\boldsymbol{U}} \tag{10.1.11}$$

(The Cholesky factor is related to the Crout factor by $\overline{\boldsymbol{U}} = \boldsymbol{D}^{1/2}\boldsymbol{U}$, where $\boldsymbol{M}^* = \boldsymbol{U}^T\boldsymbol{D}\,\boldsymbol{U}$, $\boldsymbol{U}$ is upper-triangular with ones on the diagonal and $\boldsymbol{D}$ is a diagonal matrix of positive numbers; see Chapter 11 for further information on the Crout factorization.)

Observe that the eigenvalues of the standard form, (10.1.8), are identical to the generalized form, (10.1.7). However, the eigenvectors need to be transformed as indicated by (10.1.10)

Exercise 1. Derive (10.1.7) from (10.1.8) through (10.1.11).

There are many classical and widely available procedures for solving (10.1.8). For example, the ***Jacobi***, ***Givens***, and ***Householder-QR methods*** may be mentioned in this regard (e.g., see Bathe [1] and Noble [2]). For general matrices, it is currently felt that the most efficient strategy for solving the generalized eigenproblem, (10.1.7), is to first transform to standard form, (10.1.8), and then use the Householder-QR algorithm. However, under certain circumstances, such as when the subspace iteration procedure is employed (see Sec. 10.5), for example, direct use of the generalized Jacobi method proves very effective. It is very difficult to make sweeping statements about efficiency because the type of computer (e.g., sequential, vector, or parallel) strongly influences the performance of algorithms.

Exercise 2. Consider the undamped equation of motion,

$$\boldsymbol{M}\ddot{\boldsymbol{d}} + \boldsymbol{K}\boldsymbol{d} = \boldsymbol{F} \tag{10.1.12}$$

subject to zero initial displacement and velocity. Expand the solution in terms of the eigenvectors of the associated eigenproblem. Diagonalize the system and exactly solve the individual modal equations. Show that

$$\boldsymbol{d}(t) = \sum_{l=1}^{n_{eq}} \left\{ \frac{1}{\omega_l} \int_0^t F_{(l)}(\tau) \sin \omega_l (t - \tau)\, d\tau\, \boldsymbol{\psi}_l \right\} \tag{10.1.13}$$

The $1/\omega_l$ factor in the expansion illustrates the diminishing influence of the higher modes. This analysis reveals why low mode response is viewed as "most important" and therefore, why, in practical calculations the summation in (10.1.13) is truncated at $n_{\text{modes}} \ll n_{eq}$.

Exercise 3. Obtain an exact solution for the static problem,

$$\boldsymbol{K}\boldsymbol{d} = \boldsymbol{F} \tag{10.1.14}$$

by way of an eigenvector expansion. Discuss the relative importance of low and high modes for this case.

Exercise 4. Generalize Exercise 2 to account for Rayleigh damping and non-zero initial conditions. Discuss the influence of low and high modes.

10.2 STATIC CONDENSATION

In practice one often encounters eigenvalue problems which can be written in the following ***partitioned form***:

$$\left(\begin{bmatrix} \boldsymbol{K}_{11} & \boldsymbol{K}_{12} \\ \boldsymbol{K}_{21} & \boldsymbol{K}_{22} \end{bmatrix} - \lambda \begin{bmatrix} \boldsymbol{M}_{11} & \boldsymbol{0} \\ \boldsymbol{0} & \boldsymbol{0} \end{bmatrix}\right) \begin{Bmatrix} \boldsymbol{\psi}_1 \\ \boldsymbol{\psi}_2 \end{Bmatrix} = \boldsymbol{0} \tag{10.2.1}$$

where $\boldsymbol{M}_{11}$ is symmetric and positive-definite. That is, many degrees of freedom are "massless." Problems of this type arise naturally when a relatively light structure is used to support heavy nonstructural masses which can be "lumped" at a few degrees of freedom. In such situations the structural mass is often insignificant and may be neglected, resulting in a system like (10.2.1). As it stands, (10.2.1) is ill-posed in the sense that the zero-diagonal masses give rise to infinite eigenvalues. The mass matrix can be "regularized" by the addition of small positive nonzero diagonal masses, in which case we return to the format of the generalized eigenvalue problem originally considered, or, alternatively, the zero-mass degrees of freedom can be eliminated by ***static condensation.*** To arrive at the statically condensed form, we expand (10.2.1):

$$\boldsymbol{K}_{11}\boldsymbol{\psi}_1 + \boldsymbol{K}_{12}\boldsymbol{\psi}_2 - \lambda\, \boldsymbol{M}_{11}\boldsymbol{\psi}_1 = \boldsymbol{0} \tag{10.2.2}$$

$$\boldsymbol{K}_{21}\boldsymbol{\psi}_1 + \boldsymbol{K}_{22}\boldsymbol{\psi}_2 = \boldsymbol{0} \tag{10.2.3}$$

The next step is to solve (10.2.3) for $\boldsymbol{\psi}_2$ and then substitute in (10.2.2), which results in

$$(\boldsymbol{K}_{11}^{*} - \lambda \boldsymbol{M}_{11})\boldsymbol{\psi}_1 = \boldsymbol{0} \tag{10.2.4}$$

where

$$\boldsymbol{K}_{11}^{*} = \boldsymbol{K}_{11} - \boldsymbol{K}_{12}\boldsymbol{K}_{22}^{-1}\boldsymbol{K}_{21} \qquad (\textit{statically condensed stiffness}) \tag{10.2.5}$$

The advantage of transforming to statically condensed form is that the problem size is reduced. However, $\boldsymbol{K}_{11}^{*}$ tends to be full. Thus, unless the size of the nonzero-mass matrix is rather small, the reduction to statically condensed form may be uneconomical because the profile structure of $\boldsymbol{K}$ is lost.

Note that to calculate the statically condensed stiffness (10.2.5) efficiently, $\boldsymbol{K}_{22}$ is never actually inverted.

The following steps may be used in practice:

$$\boldsymbol{K}_{22} = \boldsymbol{U}^{T}\boldsymbol{D}\boldsymbol{U} \qquad \text{(Crout factorization)} \tag{10.2.6}$$

$$U\,D^{1/2}Z = K_{21} \qquad \text{(solve for } Z\text{)} \tag{10.2.7}$$

$$K_{12}K_{22}^{-1}K_{21} = Z^T Z \tag{10.2.8}$$

Equation (10.2.7) amounts to solution of an equation system with upper triangular coefficient array and multiple right-hand sides (i.e., the columns of K_{21}).

For further computational considerations regarding static condensation, see [3].

10.3 DISCRETE RAYLEIGH-RITZ REDUCTION

In the discrete Rayleigh-Ritz approach, ***static load patterns***, P, are selected and corresponding displacement vectors, R, are calculated from:

$$\underbrace{K}_{n_{eq}\times n_{eq}} \quad \underbrace{R}_{n_{eq}\times n_{lp}} = \underbrace{P}_{n_{eq}\times n_{lp}}, \qquad n_{lp} < n_{eq} \tag{10.3.1}$$

where n_{lp} refers to the number of load patterns. The displacements R, referred to as the ***trial vectors***, are used to form the reduced eigenproblem, i.e., (10.1.7) by defining

$$K^* = R^T K R \tag{10.3.2}$$

$$M^* = R^T M R \tag{10.3.3}$$

The load patterns, P, are usually subject to the following criteria:

i. The columns of P should be linearly independent.

ii. The columns of P should be selected to arouse the low modes by activating the heaviest masses and most flexible areas of the model.

Remarks

1. If n_{lp} is small, K^* and M^* will be full, but small.

2. Equation (10.3.1) amounts to solution of a multiple right-hand-side system with profile coefficient matrix.

3. The discrete Rayleigh-Ritz procedure is a general formalism for obtaining the reduced system. However, the guidelines for selecting P are somewhat vague and thus a more systematic strategy is required for practical use.

4. The discrete Rayleigh-Ritz reduction is often referred to as a "projection method."

5. The eigenvector approximations are defined by $\psi \cong R\,\psi^*$.

6. By virtue of the fact that the calculation of trial vectors from load patterns requires solution of (10.3.1), K must be nonsingular. This precludes application to cases in which there are zero eigenvalues (e.g., structures which possess rigid body modes). A simple reformulation of the original problem involving a positive *shifting* of the eigenvalues allows us to handle this case: Note that by adding and subtracting $\alpha M\psi$, where α is a positive number, we produce an eigenproblem

$$(\bar{\boldsymbol{K}} - \bar{\lambda}\boldsymbol{M})\boldsymbol{\psi} = \boldsymbol{0} \tag{10.3.4}$$

$$\bar{\boldsymbol{K}} = \boldsymbol{K} + \alpha\boldsymbol{M} \tag{10.3.5}$$

$$\bar{\lambda} = \lambda + \alpha \tag{10.3.6}$$

in which $\bar{\boldsymbol{K}}$ is positive-definite, and the eigenvectors are unchanged. The eigenvalues are related by (10.3.6).

This stratagem may also be employed if some of the λ's are negative. Suppose an estimate of the smallest eigenvalue, λ_1, is available. Then select $\alpha > 0$ such that $-\alpha < \lambda_1 \leq 0$. The $\bar{\lambda}$'s are then all positive and solution may proceed as usual.

7. Shifting is a particular example of a general class of ***spectral transformations***. Let

$$\boldsymbol{S} = [\boldsymbol{\psi}_1, \boldsymbol{\psi}_2, \ldots, \boldsymbol{\psi}_{n_{eq}}] \tag{10.3.7}$$

and

$$\boldsymbol{\Lambda} = \text{diag}(\lambda_1, \lambda_2, \ldots, \lambda_{n_{eq}}) \tag{10.3.8}$$

Then it is easily verified that

$$\boldsymbol{K} = \boldsymbol{S}\,\boldsymbol{\Lambda}\,\boldsymbol{S}^T \tag{10.3.9}$$

and

$$\boldsymbol{M} = \boldsymbol{S}\boldsymbol{S}^T \tag{10.3.10}$$

Let $f = f(\lambda)$ represent a scalar-valued function of λ. Define a matrix-valued function $\boldsymbol{f} = \boldsymbol{f}(\boldsymbol{K})$ by way of

$$\boldsymbol{f}(\boldsymbol{K}) = \boldsymbol{S}\ \text{diag}(f(\lambda_1), f(\lambda_2), \ldots, f(\lambda_{n_{eq}}))\ \boldsymbol{S}^T \tag{10.3.11}$$

Then it is readily established that solutions of the eigenproblem

$$(\boldsymbol{f}(\boldsymbol{K}) - \bar{\lambda}\boldsymbol{M})\boldsymbol{\phi} = \boldsymbol{0} \tag{10.3.12}$$

are related to solutions of

$$(\boldsymbol{K} - \lambda\boldsymbol{M})\boldsymbol{\psi} = \boldsymbol{0} \tag{10.3.13}$$

by

$$\bar{\lambda} = f(\lambda) \tag{10.3.14}$$

$$\boldsymbol{\phi} = \boldsymbol{\psi} \tag{10.3.15}$$

Exercise 1. Consider the partitioned eigenproblem defined by (10.2.1). Show that the statically condensed eigenproblem, (10.2.4) and (10.2.5), can be obtained from the discrete Rayleigh-Ritz approach by selecting the trial vectors to be

$$\boldsymbol{R} = \left[\begin{array}{c} \boldsymbol{I} \\ \hline -\boldsymbol{K}_{22}^{-1}\ \boldsymbol{K}_{21} \end{array}\right] \tag{10.3.16}$$

10.4 IRONS-GUYAN REDUCTION

The Irons-Guyan reduction [4–6] fits into the discrete Rayleigh-Ritz framework described in the previous section. To understand the method it is helpful to rewrite the generalized eigenproblem in partitioned form:

$$\left(\begin{bmatrix} \boldsymbol{K}_{11} & \boldsymbol{K}_{12} \\ \boldsymbol{K}_{21} & \boldsymbol{K}_{22} \end{bmatrix} - \lambda \begin{bmatrix} \boldsymbol{M}_{11} & \boldsymbol{M}_{12} \\ \boldsymbol{M}_{21} & \boldsymbol{M}_{22} \end{bmatrix}\right)\begin{Bmatrix} \boldsymbol{\psi}_1 \\ \boldsymbol{\psi}_2 \end{Bmatrix} = \boldsymbol{0} \tag{10.4.1}$$

The degrees of freedom in $\boldsymbol{\psi}_1$ are to be retained in the reduced eigenproblem, whereas those in $\boldsymbol{\psi}_2$ are to be eliminated. For these purposes $\boldsymbol{R}$ is defined by (10.3.16), which leads to the definitions of the reduced arrays (verify!):

$$\boldsymbol{K}^* = \boldsymbol{R}^T \boldsymbol{K} \boldsymbol{R} = \boldsymbol{K}_{11} - \boldsymbol{K}_{12}\boldsymbol{K}_{22}^{-1}\boldsymbol{K}_{21} \tag{10.4.2}$$

$$\boldsymbol{M}^* = \boldsymbol{R}^T \boldsymbol{M} \boldsymbol{R} = \boldsymbol{M}_{11} - \boldsymbol{M}_{12}\boldsymbol{K}_{22}^{-1}\boldsymbol{K}_{21} - \boldsymbol{K}_{12}\boldsymbol{K}_{22}^{-1}\,(\boldsymbol{M}_{21} - \boldsymbol{M}_{22}\boldsymbol{K}_{22}^{-1}\boldsymbol{K}_{21}) \tag{10.4.3}$$

Note that (10.4.2) is the statically condensed stiffness and, if $\boldsymbol{M}_{22} = \boldsymbol{0}$ and $\boldsymbol{M}_{12} = \boldsymbol{M}_{21}^T = \boldsymbol{0}$, then $\boldsymbol{M}^* = \boldsymbol{M}_{11}$ as in the statically condensed problem of Sec. 10.2.

The Irons-Guyan reduction may be thought of as invoking a "static relation," namely

$$\boldsymbol{\psi}_2 = -\boldsymbol{K}_{22}^{-1}\boldsymbol{K}_{21}\boldsymbol{\psi}_1 \tag{10.4.4}$$

to eliminate the unwanted variables. Sometimes $\boldsymbol{\psi}_1$ and $\boldsymbol{\psi}_2$ are referred to as the "master" and "slave" degrees of freedom, respectively. In practice, the following guidelines are suggested:

i. Determine the number of eigenvalues and eigenvectors required, say n_{modes}.
ii. Retain a multiple of n_{modes} degrees of freedom (e.g., $5 \times n_{\text{modes}}$ degrees of freedom).
iii. The degrees of freedom to be retained are identified by calculating the ratios of diagonal elements in $\boldsymbol{M}$ and $\boldsymbol{K}$, namely, K_{pp}/M_{pp} $1 \le p \le n_{eq}$. The degrees of freedom with the smallest ratios are retained (i.e., they are the degrees of freedom in $\boldsymbol{\psi}_1$; [7]).

Remark

The calculation of the reduced mass, (10.4.3), is costly!

10.5 SUBSPACE ITERATION

A disadvantage of reduction techniques such as the Irons-Guyan procedure is that there is no guarantee that the eigenvalues and eigenvectors of the reduced problem, namely, λ_l^* and $R\psi_l^*$, will be good approximations of those of the original problem, λ_l and ψ_l, respectively. Consequently, methods have been developed in which a reduced problem is used along with an iterative strategy to obtain exactly the lower modes of the generalized eigenproblem. This is the underlying idea of the ***subspace iteration,*** or ***block power, method,*** which is widely used for large-scale finite element calculations. Roughly speaking, the procedure is as follows: Load patterns are selected and trial vectors are calculated. The trial vectors are used to form a reduced problem, which is solved. New load patterns are calculated from the "inertial" forces engendered by the eigenvectors of the reduced problem, namely,

$$\underbrace{\boldsymbol{P}}_{n_{eq} \times n_{lp}} = \boldsymbol{M}\boldsymbol{R}[\boldsymbol{\psi}_1^*, \boldsymbol{\psi}_2^*, \ldots, \boldsymbol{\psi}_{n_{lp}}^*] \tag{10.5.1}$$

With these load patterns the process is repeated until convergence is achieved. The calculations are summarized in the flowchart contained in Table 10.5.1. Extensive discussion and further details are presented in [8].

TABLE 10.5.1 Subspace Iteration Procedure

I. Initialization

1. Form $\boldsymbol{K}$ and $\boldsymbol{M}$.
2. Factorize $\boldsymbol{K} = \boldsymbol{U}^T\boldsymbol{D}\boldsymbol{U}$.
3. Specify initial load patterns, $\boldsymbol{P}$.

II. Iteration

1. Solve for trial vectors:

$$(\boldsymbol{U}^T\, \boldsymbol{D}\, \boldsymbol{U})\boldsymbol{R} = \boldsymbol{P}$$

2. Compute reduced matrices:

$$\boldsymbol{K}^* = \boldsymbol{R}^T\boldsymbol{K}\boldsymbol{R} = \boldsymbol{R}^T\boldsymbol{P}$$

$$\boldsymbol{M}^* = \boldsymbol{R}^T\boldsymbol{M}\boldsymbol{R}$$

3. Solve the reduced eigenproblem:

$$(\boldsymbol{K}^* - \lambda_k^*\boldsymbol{M}^*)\boldsymbol{\psi}_k^* = \boldsymbol{0}, \qquad k = 1, 2, \ldots, n_{lp}$$

4. Calculate approximations to the eigenvectors of the original system:

$$\boldsymbol{R}[\boldsymbol{\psi}_1^*, \boldsymbol{\psi}_2^*, \ldots, \boldsymbol{\psi}_{n_{lp}}^*]$$

5. Perform convergence checks. If the desired eigenvalues have converged, stop. Otherwise continue.
6. Calculate improved load patterns:

$$\boldsymbol{P} = \boldsymbol{M}\boldsymbol{R}[\boldsymbol{\psi}_1^*, \boldsymbol{\psi}_2^*, \ldots, \boldsymbol{\psi}_{n_{lp}}^*]$$

7. Go to Step II-1.

Remarks

1. It is recommended from practical experience [8] that:

i. The number of load patterns employed should be calculated from

$n_{lp} = \min\{2\,n_{\text{modes}},\ n_{\text{modes}} + 8\}$, where n_{modes} is the number of eigenpairs desired.

ii. In the first load pattern (i.e., first column of $\boldsymbol{P}$) a 1 should be placed in the entry corresponding to the minimum value of K_{pp}/M_{pp}, and 0 in the remaining entries. Likewise, a 1 should be placed in the entry of the second column corresponding to the next smallest value of K_{pp}/M_{pp}, and so forth.

2. Wilson [9] recommends that one load pattern be generated from random numbers in the interval [0, 1].

3. The convergence condition may be specified in terms of consecutive iterates:

$$\frac{(\lambda_k^*)_{I+1} - (\lambda_k^*)_I}{(\lambda_k^*)_I} \leq \epsilon, \qquad k = 1, 2, \ldots, n_{\text{modes}}$$

where I is the iteration number and ϵ is a preassigned error tolerance.

4. Solution of the generalized eigenproblem in the subspace iteration algorithm (Step II-3 in Table 10.5.1) is efficiently carried out by way of the ***generalized Jacobi method*** [8]. The reason for this is that after a number of iteration steps, the generalized eigenproblem possesses diagonally dominant matrices for which the generalized Jacobi method proves very effective.

10.5.1 Spectrum Slicing

The convergence of a procedure such as subspace iteration does not necessarily guarantee that the first n_{modes} eigenpairs of the original system have been found. An eigenvalue can be missed if the original trial vectors are orthogonal to the corresponding eigenvector. To ascertain whether or not this has occurred, a ***spectrum slicing*** may be performed.[3] The steps involved are as follows: Perform a Crout factorization of the matrix $\boldsymbol{K} - \alpha\boldsymbol{M}$, where $\alpha = (1 + \delta)\lambda_{n_{\text{modes}}}$ and δ is a small positive number such that $(1 + \delta)\lambda_{n_{\text{modes}}} < \lambda_{n_{\text{modes}+1}}$. Let $\boldsymbol{D}_\alpha$ denote the diagonal matrix of pivots obtained. It follows from ***Sylvester's inertia theorem*** that the number of eigenvalues smaller than α will equal the number of negative entries in $\boldsymbol{D}_\alpha$. If this number, say n_α, is greater than n_{modes}, then $n_\alpha - n_{\text{modes}}$ eigenvalues smaller than α were missed in the calculation. If this is the case, a revised set of load patterns must be employed and the eigensolution repeated. If $n_\alpha = n_{\text{modes}}$, then a degree of confidence in the solution is attained.

Remark

Spectrum slicing can be used to determine the number of eigenvalues in an interval $]\alpha, \beta[$, where $\beta > \alpha$. Factorize $\boldsymbol{K} - \alpha\boldsymbol{M}$ and $\boldsymbol{K} - \beta\boldsymbol{M}$. If $\boldsymbol{D}_\alpha$ and $\boldsymbol{D}_\beta$ are the corresponding pivots and n_α and n_β are the number of negative entries of $\boldsymbol{D}_\alpha$ and $\boldsymbol{D}_\beta$, respectively, then $n_\beta - n_\alpha$ is the number of eigenvalues in $]\alpha, \beta[$.

[3] Spectrum slicing is sometimes referred to as a ***Sturm sequence check*** [8].

10.5.2 Inverse Iteration

When the number of load patterns used in the subspace iteration procedure is one, the process is called ***inverse iteration***. In this case convergence to the lowest eigenvalue occurs as long as the initial trial vector is not orthogonal to the corresponding eigenvector.

By reversing the roles of $\boldsymbol{K}$ and $\boldsymbol{M}$, convergence to the largest eigenvalue, $\lambda_{n_{eq}}$, can be achieved. This may be seen from:

$$\mathbf{0} = (\boldsymbol{K} - \lambda\, \boldsymbol{M})\boldsymbol{\psi} = (\boldsymbol{M} - \lambda^{-1}\boldsymbol{K})\boldsymbol{\psi}$$

Note that when working with the latter form, subspace iteration requires $\boldsymbol{M}$ to be nonsingular.

Exercise 1. Consider the following eigenvalue problem:

$$(\boldsymbol{K} - \lambda \boldsymbol{M})\boldsymbol{\psi} = \mathbf{0}, \qquad \lambda = \omega^2$$

where

$$\boldsymbol{M} = \begin{bmatrix} m_1 & 0 \\ 0 & m_2 \end{bmatrix}$$

$$\boldsymbol{K} = \begin{bmatrix} (k_1 + k_2) & -k_2 \\ -k_2 & k_2 \end{bmatrix}$$

$$\boldsymbol{\psi} = \begin{Bmatrix} \psi_1 \\ \psi_2 \end{Bmatrix}$$

Assume $k_1 = k_2 = 1$, $m_1 = 3$, $m_2 = 2$.

a. Calculate the frequencies (i.e., ω_1, ω_2) and mode shapes (i.e., $\boldsymbol{\psi}_1$, $\boldsymbol{\psi}_2$).

b. Assuming that the fundamental mode load pattern is given approximately by

$$\boldsymbol{P} = \begin{Bmatrix} \dfrac{m_1}{(k_1 + k_2)} \\ \dfrac{m_2}{k_2} \end{Bmatrix}$$

use the discrete Rayleigh-Ritz reduction procedure to obtain an estimate of the fundamental frequency and mode shape.

c. Use the Irons-Guyan procedure to reduce the problem to one degree of freedom. Pick the degree of freedom to be retained according to the criterion presented in Sec. 10.4. Determine the approximate fundamental frequency and mode shape.

d. Use the subspace iteration procedure to calculate the fundamental frequency and mode shape. Initialize the computations with the load pattern

$$\boldsymbol{P} = \begin{Bmatrix} 0 \\ 1 \end{Bmatrix}$$

Employ two iterations.

Solution

a.
$$\boldsymbol{M} = \begin{bmatrix} 3 & 0 \\ 0 & 2 \end{bmatrix} \qquad \boldsymbol{K} = \begin{bmatrix} 2 & -1 \\ -1 & 1 \end{bmatrix}$$

$$0 = \det \begin{bmatrix} 2 - 3\lambda & -1 \\ -1 & 1 - 2\lambda \end{bmatrix} = 6\lambda^2 - 7\lambda + 1$$

$$\lambda_{1,2} = \tfrac{1}{6},\ 1$$

$$\omega_{1,2} = \frac{1}{\sqrt{6}},\ 1 = 0.40825,\ 1$$

$$\boldsymbol{\psi}_1 = \begin{Bmatrix} \frac{2}{3} \\ 1 \end{Bmatrix} \qquad \boldsymbol{\psi}_2 = \begin{Bmatrix} 1 \\ -1 \end{Bmatrix}$$

b.
$$\boldsymbol{P} = \begin{Bmatrix} \frac{3}{2} \\ 2 \end{Bmatrix}$$

$$\boldsymbol{KR} = \boldsymbol{P}$$

$$\boldsymbol{R} = \begin{Bmatrix} \frac{7}{2} \\ \frac{11}{2} \end{Bmatrix} \qquad \left(\text{the } \frac{1}{2} \text{ factors may be neglected}\right)$$

$$\boldsymbol{K}^* = \boldsymbol{R}^T\boldsymbol{K}\boldsymbol{R} = \langle 7 \quad 11 \rangle \begin{bmatrix} 2 & -1 \\ -1 & 1 \end{bmatrix} \begin{Bmatrix} 7 \\ 11 \end{Bmatrix} = 65$$

$$\boldsymbol{M}^* = \boldsymbol{R}^T\boldsymbol{M}\boldsymbol{R} = \langle 7 \quad 11 \rangle \begin{bmatrix} 3 & 0 \\ 0 & 2 \end{bmatrix} \begin{Bmatrix} 7 \\ 11 \end{Bmatrix} = 389$$

$$(\boldsymbol{K}^* - \lambda^*\boldsymbol{M}^*)\boldsymbol{\psi}^* = \boldsymbol{0} \Rightarrow \lambda^* = \frac{65}{389}, \qquad \boldsymbol{\psi}^* = 1$$

$$\boldsymbol{\psi}_1 \cong \boldsymbol{R}\boldsymbol{\psi}_1^* = \frac{1}{11}\begin{Bmatrix} 7 \\ 11 \end{Bmatrix} = \begin{Bmatrix} 0.64 \\ 1 \end{Bmatrix}; \qquad \omega^* = \left(\frac{65}{389}\right)^{1/2} = 0.4087$$

c.
$$\left.\begin{aligned} \frac{K_{11}}{M_{11}} &= \frac{2}{3} \\ \frac{K_{22}}{M_{22}} &= \frac{1}{2} \end{aligned}\right\} \Rightarrow \text{retain degree of freedom number 2}$$

Reorder equations into the standard partitioned form.

$$\boldsymbol{K} = \begin{bmatrix} 1 & -1 \\ -1 & 2 \end{bmatrix} \qquad \boldsymbol{M} = \begin{bmatrix} 2 & 0 \\ 0 & 3 \end{bmatrix}$$

$$\boldsymbol{R} = \begin{Bmatrix} 1 \\ -K_{22}^{-1}K_{21} \end{Bmatrix} = \begin{Bmatrix} 1 \\ \frac{1}{2} \end{Bmatrix}$$

$$\boldsymbol{K}^* = \boldsymbol{R}^T\boldsymbol{K}\boldsymbol{R} = \frac{1}{2}$$

$$\boldsymbol{M}^* = \boldsymbol{R}^T\boldsymbol{M}\boldsymbol{R} = \frac{11}{4}$$

$$\lambda^* = \frac{2}{11}, \qquad \boxed{\omega^* = \left(\frac{2}{11}\right)^{1/2} = 0.4264}, \qquad \boldsymbol{\psi}_1^* = 1$$

$$\boldsymbol{\psi}_1 \cong \boldsymbol{R}\,\boldsymbol{\psi}_1^* = \begin{Bmatrix} 1 \\ \frac{1}{2} \end{Bmatrix}, \qquad \text{(reorder):} \quad \boxed{\boldsymbol{\psi}_1 = \begin{Bmatrix} \frac{1}{2} \\ 1 \end{Bmatrix}}$$

d. $\boldsymbol{P} = \begin{Bmatrix} 0 \\ 1 \end{Bmatrix}$; $\quad \boldsymbol{K}\,\boldsymbol{R} = \boldsymbol{P}$; $\quad \boldsymbol{R} = \begin{Bmatrix} 1 \\ 2 \end{Bmatrix}$

$\boldsymbol{K}^* = 2$; $\quad \boldsymbol{M}^* = 11$; $\quad \lambda^* = \frac{2}{11}$ (same as (c)) } iteration number 1

$\omega^* = 0.4264 \quad \boldsymbol{\psi}_1 \cong \begin{Bmatrix} \frac{1}{2} \\ 1 \end{Bmatrix}$

$\boldsymbol{P} = \boldsymbol{M}\,\boldsymbol{\psi}_1 = \begin{Bmatrix} \frac{3}{2} \\ 2 \end{Bmatrix}$ (same as (b))

Therefore } iteration number 2

$\omega^* = 0.4087 \quad \boldsymbol{\psi}_1 \cong \begin{Bmatrix} 0.64 \\ 1 \end{Bmatrix}$

10.6 THE LANCZOS ALGORITHM FOR SOLUTION OF LARGE GENERALIZED EIGENPROBLEMS

BAHRAM NOUR-OMID

10.6.1 Introduction

The Lanczos algorithm was first proposed in 1950. Lanczos intended his algorithm to be used for computing a few of the extreme eigenvalues and corresponding eigenvectors of a symmetric matrix. However, the algorithm was taken up as a method for reducing a symmetric matrix to tridiagonal form. Lanczos himself observed that round-off error has a significant effect on the algorithm's performance and expensive modifications seemed to be necessary to overcome this defect. By 1955 the Householder method replaced the Lanczos algorithm as a more efficient method for tridiagonalizing a matrix.

Most engineering applications require only a few eigenvalues at one end of the spectrum, but for small problems the Householder-QR algorithm can find all the eigenvalues almost as fast as a few. For large problems however it is a waste to tridiagonalize the whole matrix. The Lanczos iteration has the property that well-isolated eigenvalues are approximated accurately after a comparatively small number of steps. For example, the largest eigenvalues will often be captured after 20 steps, almost independently of the order of the problem, n. Moreover, the number of steps taken by the Lanczos method to compute these eigenvalues will always be less than that for the power method. This property makes the Lanczos algorithm very efficient. However, the eigenvalues most frequently wanted are the smaller ones. The Lanczos algorithm is so powerful that it can compute the smaller eigenvalues of a matrix without any factorization. However, they will not be well approximated until nearly n steps have been taken. Consequently it is necessary, in these cases, to apply the iteration to an inverted form of the matrix. In this respect the Lanczos algorithm is just like the subspace iteration method. See [15] for a comparison of Lanczos and subspace iteration algorithms.

In this section we are concerned with the application of the Lanczos algorithm to the generalized symmetric eigenproblem

$$(\mathbf{K} - \lambda \mathbf{M})\mathbf{z} = \mathbf{0} \tag{10.6.1}$$

where $\mathbf{K}$ and $\mathbf{M}$ are $n \times n$ real symmetric matrices. We assume for the moment that the eigenvalues of (10.6.1) are real. Later we state the conditions on $\mathbf{K}$ and $\mathbf{M}$ that must hold to ensure real eigenvalues. We will derive the generalized Lanczos algorithm for the solution of the eigenproblem (10.6.1). The relation between the Lanczos method and vector iteration methods will be established. We then look at the effect of round-off on the algorithm and the resulting loss of orthogonality. We consider two possible modifications to the algorithm that can maintain a desired level of orthogonality among the Lanczos vectors. Finally, we give a detailed description of the algorithm we prefer together with a listing of the computer subprograms.

10.6.2 Spectral Transformation

We turn now to an important misconception concerning the Lanczos algorithm. It is usually thought of as a way of computing eigenpairs, (λ, z), of the standard eigenproblem

$$(\boldsymbol{A} - \lambda \boldsymbol{I})z = \boldsymbol{0} \tag{10.6.2}$$

The Lanczos method is so powerful that one can work directly with $\boldsymbol{A}$ to evaluate eigenvalues at both ends of the spectrum of (10.6.2) without solving any system of equations. Only products of $\boldsymbol{A}$ with a sequence of vectors need be computed. This virtue blinded certain users of the Lanczos method to the great advantages to be gained by a shift-and-invert procedure. The matrix $(\boldsymbol{A} - \sigma \boldsymbol{I})^{-1}$ has the same eigenvectors as $\boldsymbol{A}$. Using $(\boldsymbol{A} - \sigma \boldsymbol{I})^{-1}$ instead of $\boldsymbol{A}$ will cause eigenvectors corresponding to eigenvalues close to σ to converge much more quickly. Of course there are cases when factoring of $\boldsymbol{A}$ is not possible (e.g., when the matrix is only known implicitly) or not desirable (e.g., when $\boldsymbol{A}$ has a given sparsity structure that can be destroyed when factored).

However, we are more interested in the generalized eigenproblem (10.6.1). For this problem some form of inversion or factoring of a matrix, either explicitly or implicitly, is required. It is interesting to note that when the structure of $\boldsymbol{M}$ is the same as that of $\boldsymbol{K}$, as in the case of a consistent mass matrix, then the cost of one step of the power iteration method is exactly the same as that for the inverse iteration method. Each requires one triangular factorization.

There are two different procedures for transforming the generalized eigenproblem to the standard form:

i. Factor $\boldsymbol{M}$ into $\boldsymbol{M} = \boldsymbol{C}\boldsymbol{C}^T$. This is the Cholesky factorization of $\boldsymbol{M}$. Then

$$(\boldsymbol{K} - \lambda \boldsymbol{M}) = \boldsymbol{C}(\boldsymbol{C}^{-1}\boldsymbol{K}\boldsymbol{C}^{-T} - \lambda \boldsymbol{I})\boldsymbol{C}^T \tag{10.6.3}$$

and the eigenproblem of (10.6.1) reduces to the standard form $(\boldsymbol{A}_1 - \lambda \boldsymbol{I})\hat{z} = \boldsymbol{0}$ where $\boldsymbol{A}_1 = \boldsymbol{C}^{-1}\boldsymbol{K}\boldsymbol{C}^{-T}$ and $\hat{z} = \boldsymbol{C}^T z$. The eigenvalues of the standard problem are the same as those of the generalized problem. Note that if $\boldsymbol{M}$ is positive-semidefinite, then its Cholesky factors are singular and this transformation can not be performed.

ii. Using a similar procedure, (10.6.1) can be transformed into

$$(\boldsymbol{A}_2 - \mu \boldsymbol{I})\hat{z} = \boldsymbol{0} \tag{10.6.4}$$

where $\boldsymbol{A}_2 = \boldsymbol{C}^T\boldsymbol{K}^{-1}\boldsymbol{C}$ and $\mu = 1/\lambda$. Note that in this case both the eigenvalues and the eigenvectors of $\boldsymbol{A}_2$ are different from those of (10.6.1).

A more general form of the second transformation is to first perform a shift of the origin, $(\boldsymbol{K}_\sigma - (\lambda - \sigma)\boldsymbol{M})z = \boldsymbol{0}$, where $\boldsymbol{K}_\sigma = \boldsymbol{K} - \sigma \boldsymbol{M}$, and then perform (ii). This results in a standard eigenproblem with $\boldsymbol{A}_\sigma = \boldsymbol{C}^T\boldsymbol{K}_\sigma^{-1}\boldsymbol{C}$. The spectrum of $\boldsymbol{A}_\sigma$ is related to the original spectrum through

$$\boxed{\nu = \frac{1}{\lambda - \sigma}} \tag{10.6.5}$$

where ν is the eigenvalue of $\boldsymbol{A}_\sigma$[11].

The second procedure requires two triangular factorizations, one for $\boldsymbol{M}$ and one for $\boldsymbol{K}$. It is possible to avoid the factorization of $\boldsymbol{M}$ by working with

$$(\boldsymbol{K}_\sigma^{-1}\boldsymbol{M} - \nu\boldsymbol{I})\boldsymbol{z} = \boldsymbol{0} \tag{10.6.6}$$

Although the matrix of this transformation is not symmetric, it is self-adjoint with respect to the ***inertial inner product*** defined by

$$(\boldsymbol{u},\boldsymbol{v})_M = \boldsymbol{v}^T\boldsymbol{M}\boldsymbol{u} \tag{10.6.7}$$

which is shown in the following steps:

$$\begin{aligned}(\boldsymbol{K}_\sigma^{-1}\boldsymbol{M}\boldsymbol{u},\ \boldsymbol{v})_M &= \boldsymbol{v}^T\boldsymbol{M}\ \boldsymbol{K}_\sigma^{-1}\boldsymbol{M}\boldsymbol{u} \\ &= (\boldsymbol{u},\ \boldsymbol{K}_\sigma^{-1}\boldsymbol{M}\boldsymbol{v})_M\end{aligned}$$

This unsymmetric form is particularly advantageous since it has the same eigenvectors as the original problem [16]. The algorithm that is derived later employs the transformation of equation (10.6.6).

10.6.3 Conditions for Real Eigenvalues

Contrary to common belief, the fact that $\boldsymbol{K}$ and $\boldsymbol{M}$ are symmetric is not a sufficient condition to ensure real eigenvalues for (10.6.1). This can be illustrated using the following simple example.

Example 1

Let $\boldsymbol{K} = \begin{bmatrix} 1 & 1 \\ 1 & 0 \end{bmatrix}$ and $\boldsymbol{M} = \begin{bmatrix} 1 & 0 \\ 0 & -1 \end{bmatrix}$. Then the eigenvalues of (10.6.1) with these matrices are $\frac{1}{2}(1 \pm i\sqrt{3})$, where $i^2 = -1$.

The eigenvalues of (10.6.1) are all real if some linear combination of $\boldsymbol{K}$ and $\boldsymbol{M}$ is positive-definite; that is $\rho\boldsymbol{K} + \tau\boldsymbol{M}$ is positive-definite for some choice of real ρ and τ. This is only a sufficient condition that guarantees real eigenvalues, but there is no easy way of computing ρ and τ.

Exercise 1. Prove that $\rho\boldsymbol{K} + \tau\boldsymbol{M}$ being positive definite for some choice of ρ and τ is only a sufficient condition for (10.6.1) to have real eigenvalues.

Solution Multiply (10.6.1) by ρ and shift the spectrum by τ:

$$\begin{aligned}\boldsymbol{0} &= (\rho\boldsymbol{K} - \rho\lambda\boldsymbol{M})\boldsymbol{z} \\ &= (\rho\boldsymbol{K} + \tau\boldsymbol{M} - (\rho\lambda + \tau)\boldsymbol{M})\boldsymbol{z}\end{aligned}$$

Because $\rho\boldsymbol{K} + \tau\boldsymbol{M}$ is positive definite, it possesses a Cholesky decomposition; i.e., $\rho\boldsymbol{K} + \tau\boldsymbol{M} = \boldsymbol{C}\boldsymbol{C}^T$. Thus, the above equation can be reduced to standard form:

$$(\boldsymbol{A} - \mu\boldsymbol{I})\hat{\boldsymbol{z}} = \boldsymbol{0}$$

where $\boldsymbol{A} = \boldsymbol{C}^{-1}\boldsymbol{M}\boldsymbol{C}^{-T}$, $\hat{\boldsymbol{z}} = \boldsymbol{C}^T\boldsymbol{z}$, and $\mu = 1/(\rho\lambda + \tau)$. By the symmetry of $\boldsymbol{A}$, the μ's are real. Consequently, so are the λ's. This establishes "sufficiency". Now consider the following case. Let

$$K = \begin{bmatrix} 1 & 0 & 0 \\ 0 & -2 & 0 \\ 0 & 0 & 3 \end{bmatrix}, \quad \text{and} \quad M = \begin{bmatrix} -3 & 0 & 0 \\ 0 & 2 & 0 \\ 0 & 0 & -1 \end{bmatrix}$$

Clearly, the eigenvalues of this problem are real. However, there is no linear combination of $\boldsymbol{K}$ and $\boldsymbol{M}$ that is positive definite. This establishes that the above is not a necessary condition.

Fortunately, in many problems encountered in structural mechanics, either $\boldsymbol{K}$ or $\boldsymbol{M}$ or both are positive-definite.

10.6.4 The Rayleigh-Ritz Approximation

Consider a given set of vectors, $\boldsymbol{X}_m = [\boldsymbol{x}_1, \boldsymbol{x}_2, \cdots, \boldsymbol{x}_m]$, with $m \ll n$. We refer to these as ***trial vectors***. We proceed to obtain an approximation to some of the eigenvectors of (10.6.1) by taking a linear combination of the trial vectors. Let $\boldsymbol{y} = \boldsymbol{X}_m\, \boldsymbol{s} = \sum_{i=1}^{m} \boldsymbol{x}_i s_i$ be the desired approximation. The ith component of $\boldsymbol{s}$ is the coefficient of $\boldsymbol{x}_i$ in this representation of $\boldsymbol{y}$. We denote the corresponding approximation to the eigenvalues of (10.6.1) by θ. The ***residual***, $\boldsymbol{r}$, associated with the approximating pair $\{\theta, \boldsymbol{y}\}$ is given by

$$\boxed{\boldsymbol{r} = \boldsymbol{K}\boldsymbol{y} - \theta\boldsymbol{M}\boldsymbol{y}} \tag{10.6.8}$$

The Rayleigh-Ritz method requires the residual vector be orthogonal to each of the trial vectors; i.e.,

$$\boldsymbol{X}_m^T\boldsymbol{r} = \boldsymbol{X}_m^T\boldsymbol{K}\boldsymbol{y} - \theta\boldsymbol{X}_m^T\boldsymbol{M}\boldsymbol{y} = \boldsymbol{0}$$

Substituting the representation of $\boldsymbol{y}$ in terms of the trial vectors, in the above orthogonality condition for $\boldsymbol{r}$, we obtain the reduced eigenproblem

$$(\boldsymbol{K}_m - \theta\boldsymbol{M}_m)\boldsymbol{s} = \boldsymbol{0}$$

where $\boldsymbol{K}_m = \boldsymbol{X}_m^T\boldsymbol{K}\boldsymbol{X}_m$ and $\boldsymbol{M}_m = \boldsymbol{X}_m^T\boldsymbol{M}\boldsymbol{X}_m$. The eigenvector of the reduced eigenproblem, $\boldsymbol{s}^{(k)}$, will determine the approximation, $\boldsymbol{y}^{(k)}$, to the eigenvector of (10.6.1). We refer to $\boldsymbol{y}^{(k)}$ as the ***Ritz vector*** and to its corresponding $\theta^{(k)}$ as the ***Ritz value***, for each $k = 1, \cdots, m$.[4]

[4] Some authors refer to the trial vectors as Ritz vectors. In keeping with more prevalent usage, we prefer to reserve the latter terminology for the $\boldsymbol{y}$'s.

10.6.5 Derivation of The Lanczos Algorithm

The Lanczos method can be thought of as a means of constructing an orthogonal set of vectors, known as Lanczos vectors, for use in the Rayleigh-Ritz procedure. The algorithm is closely related to the inverse iteration and power methods for calculating a single eigenpair. Given a pair of matrices $\boldsymbol{K}_\sigma = \boldsymbol{K} - \sigma\boldsymbol{M}$ and $\boldsymbol{M}$, and a starting vector $\boldsymbol{r}$, these basic methods generate a sequence of vectors, $\{\boldsymbol{r}, \boldsymbol{K}_\sigma^{-1}\boldsymbol{M}\boldsymbol{r}, (\boldsymbol{K}_\sigma^{-1}\boldsymbol{M})^2\boldsymbol{r}, \ldots, (\boldsymbol{K}_\sigma^{-1}\boldsymbol{M})^j\boldsymbol{r}\}$, during j iterations. These vectors are referred to as the ***Krylov sequence***; the sequence converges (as $j \rightarrow \infty$) to the eigenvector corresponding to the eigenvalue, λ, of (10.6.1) closest to the shift σ.

The basic difference between the Lanczos method and the other two is that the information contained in each successive vector of the Krylov sequence is used to obtain the best approximation to the wanted eigenvectors instead of using only the last vector in the sequence. To be more specific, the Lanczos algorithm is equivalent to obtaining the Rayleigh-Ritz approximation with the vectors in the Krylov sequence as the trial vectors. This involves supplementing the Krylov sequence with the Gram-Schmidt orthogonalization process at each step. The result is a set of $\boldsymbol{M}$-orthonormal vectors (the ***Lanczos vectors***) that is used in the Rayleigh-Ritz procedure to reduce the dimension of the eigenproblem. We show below that orthogonalization is required only with respect to two preceding vectors; a fact recognized by Lanczos. The Rayleigh-Ritz procedure with the $\boldsymbol{M}$-orthonormal basis of the Krylov subspace leads to a standard eigenproblem with a tridiagonal matrix.

To derive the Lanczos algorithm it will be assumed for the moment that the first j Lanczos vectors, $\{\boldsymbol{q}_1, \boldsymbol{q}_2, \cdots, \boldsymbol{q}_j\}$ have been found, and the construction of the $j + 1$ vector will be described. The resulting vectors all satisfy the condition $\boldsymbol{q}_i^T\boldsymbol{M}\boldsymbol{q}_j = \delta_{ij}$, where δ_{ij} is the Kronecker delta; that is, the vectors are orthonormal with respect to the mass matrix. To calculate $\boldsymbol{q}_{j+1}$, we must orthogonalize $\boldsymbol{v}_j = (\boldsymbol{K}_\sigma^{-1}\boldsymbol{M})^j\boldsymbol{r}$ against the j Lanczos vectors computed so far. From the definition of $\boldsymbol{v}_j$ we obtain $\boldsymbol{v}_j = \boldsymbol{K}_\sigma^{-1}\boldsymbol{M}\boldsymbol{v}_{j-1}$. Now, $\boldsymbol{v}_{j-1}$ is the vector that is $\boldsymbol{M}$-orthonormalized against the first $j - 1$ Lanczos vectors to obtain $\boldsymbol{q}_j$. Therefore

$$\boldsymbol{v}_{j-1} = \sum_{i=1}^{j} v_i \boldsymbol{q}_i$$

where v_i is the component of $\boldsymbol{v}_{j-1}$ along $\boldsymbol{q}_i$. This result is used to eliminate $\boldsymbol{v}_{j-1}$ in the above recursive relation for $\boldsymbol{v}_j$. We then get

$$\begin{aligned}\boldsymbol{v}_j &= \sum_{i=1}^{j} v_i \boldsymbol{K}_\sigma^{-1}\boldsymbol{M}\boldsymbol{q}_i \\ &= v_j\boldsymbol{K}_\sigma^{-1}\boldsymbol{M}\boldsymbol{q}_j + \sum_{i=1}^{j-1} v_i\boldsymbol{K}_\sigma^{-1}\boldsymbol{M}\boldsymbol{q}_i\end{aligned}$$

Observe that each vector, $\boldsymbol{K}_\sigma^{-1}\boldsymbol{M}\boldsymbol{q}_i$, in the above summation can be written as a linear combination of the first $i + 1$ Lanczos vectors. Therefore the sum can be written as a linear combination of the first j Lanczos vectors. Consequently

$$\boldsymbol{v}_j = v_j\boldsymbol{K}_\sigma^{-1}\boldsymbol{M}\boldsymbol{q}_j + \sum_{i=1}^{j} \overline{v}_i \boldsymbol{q}_i$$

The $\boldsymbol{M}$-orthogonalization of $\boldsymbol{v}_j$ against the preceding j Lanczos vectors will purge the component of $\boldsymbol{v}_j$ along each of the $\boldsymbol{q}_i$'s, and therefore the final result will be unaffected by the sum in the last equation. Therefore the next Lanczos vector, $\boldsymbol{q}_{j+1}$, will be obtained by first computing a preliminary vector $\bar{\boldsymbol{r}}_j$ from the previous vector, $\boldsymbol{q}_j$,

$$\bar{\boldsymbol{r}}_j = \boldsymbol{K}_\sigma^{-1}\boldsymbol{M}\boldsymbol{q}_j \tag{10.6.9}$$

and $\boldsymbol{M}$-orthonormalizing it against all the previous Lanczos vectors. Now, in general it may be assumed that this preliminary vector contains components from each of the preceding vectors. Thus

$$\bar{\boldsymbol{r}}_j = \boldsymbol{r}_j + \alpha_j \boldsymbol{q}_j + \beta_j \boldsymbol{q}_{j-1} + \gamma_j \boldsymbol{q}_{j-2} + \cdots \tag{10.6.10}$$

where $\boldsymbol{r}_j$ is the "pure" component of $\bar{\boldsymbol{r}}_j$ orthogonal to all previous Lanczos vectors, and $\alpha_j, \beta_j, \gamma_j, \cdots$ are the amplitudes of the previous vectors contained in $\bar{\boldsymbol{r}}_j$. These amplitude coefficients are evaluated from the orthonormality of the Lanczos vectors. Thus, if both sides of (10.6.10) are multiplied by $\boldsymbol{q}_j^T\boldsymbol{M}$, the result is

$$\boldsymbol{q}_j^T\boldsymbol{M}\bar{\boldsymbol{r}}_j = \boldsymbol{q}_j^T\boldsymbol{M}\boldsymbol{r}_j + \alpha_j \boldsymbol{q}_j^T\boldsymbol{M}\boldsymbol{q}_j + \beta_j \boldsymbol{q}_j^T\boldsymbol{M}\boldsymbol{q}_{j-1} + \gamma_j \boldsymbol{q}_j^T\boldsymbol{M}\boldsymbol{q}_{j-2} + \cdots \tag{10.6.11}$$

Here the first term on the right-hand side vanishes by *definition*, and all terms beyond the second vanish similarly due to $\boldsymbol{M}$-orthogonality. The normalizing definition applied to the second term then reduces (10.6.11) to an expression for the amplitude of $\boldsymbol{q}_j$ along $\bar{\boldsymbol{r}}_j$:

$$\boxed{\alpha_j = \boldsymbol{q}_j^T\boldsymbol{M}\bar{\boldsymbol{r}}_j} \tag{10.6.12}$$

The amplitude of $\boldsymbol{q}_{j-1}$ contained in $\bar{\boldsymbol{r}}_j$ may be found similarly by multiplying (10.6.10) by $\boldsymbol{q}_{j-1}^T\boldsymbol{M}$. In this case all terms except the third vanish by orthogonality, and the coefficient of β_j is unity, so $\beta_j = \boldsymbol{q}_{j-1}^T\boldsymbol{M}\bar{\boldsymbol{r}}_j$. But, using (10.6.9) to eliminate $\bar{\boldsymbol{r}}_j$ this gives $\beta_j = \boldsymbol{q}_{j-1}^T\boldsymbol{M}\boldsymbol{K}_\sigma^{-1}\boldsymbol{M}\boldsymbol{q}_j$ and applying the transpose of (10.6.9) to the $\boldsymbol{q}_{j-1}^T$ vector gives

$$\beta_j = \bar{\boldsymbol{r}}_{j-1}^T\boldsymbol{M}\boldsymbol{q}_j \tag{10.6.13}$$

Finally, expanding $\bar{\boldsymbol{r}}_{j-1}$ in terms of its pure component, $\boldsymbol{r}_{j-1}$, and the preceding Lanczos vectors, as in (10.6.10), the transpose of (10.6.13) becomes

$$\beta_j = \boldsymbol{q}_j^T\boldsymbol{M}\boldsymbol{r}_{j-1} + \alpha_{j-1}\boldsymbol{q}_j^T\boldsymbol{M}\boldsymbol{q}_{j-1} + \beta_{j-1}\boldsymbol{q}_j^T\boldsymbol{M}\boldsymbol{q}_{j-2} + \gamma_{j-1}\boldsymbol{q}_j^T\boldsymbol{M}\boldsymbol{q}_{j-3} + \cdots \tag{10.6.14}$$

It is evident that all terms except the first vanish on the right-hand side. Now $\boldsymbol{q}_j$ is the vector obtained by normalizing $\boldsymbol{r}_{j-1}$, i.e.,

$$\boldsymbol{q}_j = \frac{1}{\|\boldsymbol{r}_{j-1}\|_M}\boldsymbol{r}_{j-1} \tag{10.6.15}$$

where $\|\boldsymbol{r}_{j-1}\|_M = (\boldsymbol{r}_{j-1}^T\boldsymbol{M}\boldsymbol{r}_{j-1})^{1/2}$. Using this expression for $\boldsymbol{q}_j$ in (10.6.14), we obtain β_j as $\beta_j = (1/\|\boldsymbol{r}_{j-1}\|_M)\, \boldsymbol{r}_{j-1}^T\boldsymbol{M}\boldsymbol{r}_{j-1}$, or

$$\boxed{\beta_j^2 = \boldsymbol{r}_{j-1}^T \boldsymbol{M} \boldsymbol{r}_{j-1}} \tag{10.6.16}$$

This is an alternative to (10.6.13) for evaluating β_j. In [26] Scott established that computing β_j using (10.6.16) is preferable to (10.6.13) and numerical experiments also confirm his results. Using (10.6.16) ensures that the Lanczos vectors are properly normalized even if $\boldsymbol{q}_{j+1}^T \boldsymbol{M} \boldsymbol{q}_{j-1}$ is not exactly zero.

Continuing in the same way, the amplitude of $\boldsymbol{q}_{j-2}$ contained in $\bar{\boldsymbol{r}}_j$ is found to be

$$\gamma_j = \boldsymbol{q}_{j-2}^T \boldsymbol{M} \bar{\boldsymbol{r}}_j \tag{10.6.17}$$

Following the procedure used to derive (10.6.14), this leads to

$$\gamma_j = \boldsymbol{q}_j^T \boldsymbol{M} \boldsymbol{r}_{j-2} + \alpha_{j-2} \boldsymbol{q}_j^T \boldsymbol{M} \boldsymbol{q}_{j-2} + \beta_{j-2} \boldsymbol{q}_j^T \boldsymbol{M} \boldsymbol{q}_{j-3} + \gamma_{j-2} \boldsymbol{q}_j^T \boldsymbol{M} \boldsymbol{q}_{j-4} + \cdots \tag{10.6.18}$$

But, using the normalizing relationship equivalent to (10.6.15), $\boldsymbol{r}_{j-2} = \beta_{j-1} \boldsymbol{q}_{j-1}$. Hence, when this is substituted into (10.6.18) all terms on the right-hand side vanish, with the result that $\gamma_j = 0$. A corresponding procedure could be used to demonstrate that all further terms in the expansion for $\bar{\boldsymbol{r}}_j$, (10.6.10), vanish; in other words, ***the orthogonalization procedure used in generating each Lanczos vector need be applied only to the previous two vectors.***

A summary of the Lanczos algorithm is presented in Table 10.6.1.

TABLE 10.6.1. The Lanczos Algorithm

Given an arbitrary vector $\boldsymbol{r}_0$ then:

1. Set
   a. $\boldsymbol{q}_0 = \boldsymbol{0}$
   b. $\beta_1 = (\boldsymbol{r}_0^T \boldsymbol{M} \boldsymbol{r}_0)^{1/2}$
   c. $\boldsymbol{q}_1 = \dfrac{\boldsymbol{r}_0}{\beta_1}$
   d. $\boldsymbol{p}_1 = \boldsymbol{M} \boldsymbol{q}_1$
2. For $j = 1, 2, \ldots,$ repeat:
   a. $\bar{\boldsymbol{r}}_j = \boldsymbol{K}_\sigma^{-1} \boldsymbol{p}_j$
   b. $\hat{\boldsymbol{r}}_j = \bar{\boldsymbol{r}}_j - \boldsymbol{q}_{j-1} \beta_j$
   c. $\alpha_j = \boldsymbol{q}_j^T \boldsymbol{M} \hat{\boldsymbol{r}}_j = \boldsymbol{p}_j^T \hat{\boldsymbol{r}}_j$
   d. $\boldsymbol{r}_j = \hat{\boldsymbol{r}}_j - \boldsymbol{q}_j \alpha_j$
   e. $\bar{\boldsymbol{p}}_j = \boldsymbol{M} \boldsymbol{r}_j$
   f. $\beta_{j+1} = (\boldsymbol{r}_j^T \boldsymbol{M} \boldsymbol{r}_j)^{1/2} = (\bar{\boldsymbol{p}}_j^T \boldsymbol{r}_j)^{1/2}$
   g. if enough vectors, then terminate the loop
   h. $\boldsymbol{q}_{j+1} = \dfrac{1}{\beta_{j+1}} \boldsymbol{r}_j$
   i. $\boldsymbol{p}_{j+1} = \dfrac{1}{\beta_{j+1}} \bar{\boldsymbol{p}}_j$

The above process may be started from a random vector, $\boldsymbol{r}_0$, with $\boldsymbol{q}_0 = \boldsymbol{0}$ and $\beta_1 = (\boldsymbol{r}_0^T \boldsymbol{M} \boldsymbol{r}_0)^{1/2}$. At a typical step, j, the Lanczos algorithm computes α_j, β_{j+1}, and $\boldsymbol{q}_{j+1}$, in order. In addition to the storage needs of $\boldsymbol{K}_\sigma$ and $\boldsymbol{M}$, the algorithm requires storage for five vectors of length n; one for each of the vectors, $\boldsymbol{q}_{j-1}$, $\boldsymbol{q}_j$, $\boldsymbol{M}\boldsymbol{r}_j$, $\boldsymbol{p}_j$, and $\boldsymbol{r}_j$. The total cost for one step of the algorithm involves a multiply with $\boldsymbol{M}$, the solution of a system of equations with $\boldsymbol{K}_\sigma$ as the coefficient matrix, two inner products and four products of a scalar by a vector.

10.6.6 Reduction to Tridiagonal Form

Using the results of the previous section, (10.6.10) can be rewritten as the three-term relation

$$\boxed{\boldsymbol{r}_j = \beta_{j+1}\, \boldsymbol{q}_{j+1} = \boldsymbol{K}_\sigma^{-1} \boldsymbol{M} \boldsymbol{q}_j - \boldsymbol{q}_j\, \alpha_j - \boldsymbol{q}_{j-1}\, \beta_j} \qquad (10.6.19)$$

where $\alpha_j = \boldsymbol{q}_j^T \boldsymbol{M} \boldsymbol{K}_\sigma^{-1} \boldsymbol{M} \boldsymbol{q}_j$ and $\boldsymbol{r}_j$ is normalized with respect to the mass matrix to obtain $\boldsymbol{q}_{j+1}$ with normalizing factor $\beta_{j+1} = (\boldsymbol{r}_j^T \boldsymbol{M} \boldsymbol{r}_j)^{1/2}$. After m Lanczos steps all the quantities obtained from equation (10.6.19) can be arranged in a global matrix form

$$\left[\boldsymbol{K}_\sigma^{-1}\boldsymbol{M}\right]\left[\boldsymbol{Q}_m\right] - \left[\boldsymbol{Q}_m\right][\boldsymbol{T}_m] = \left[\boldsymbol{0} \,\vdots\, \boldsymbol{0} \,\vdots\, \cdots \,\vdots\, \boldsymbol{0} \,\vdots\, \boldsymbol{r}_m\right] = \boldsymbol{r}_m\, \boldsymbol{e}_m^T \qquad (10.6.20)$$

Here $\boldsymbol{e}_m^T = \langle 0,0, \cdots ,0,1 \rangle$, $\boldsymbol{Q}_m$ is an $n \times m$ matrix with columns $\boldsymbol{q}_i$, $i = 1,2, \cdots ,m$, and $\boldsymbol{T}_m$ is a ***tridiagonal matrix*** of the form

$$\boldsymbol{T}_m = \begin{bmatrix} \alpha_1 & \beta_2 & & & & \\ \beta_2 & \alpha_2 & \beta_3 & & & \\ & \beta_3 & \cdot & \cdot & & \\ & & \cdot & \cdot & \cdot & \\ & & & \cdot & \cdot & \cdot \\ & & & & \cdot & \cdot\ \beta_m \\ & & & & & \beta_m\ \alpha_m \end{bmatrix} \qquad (10.6.21)$$

The orthogonality property of the Lanczos vectors, $\boldsymbol{Q}_m^T \boldsymbol{M} \boldsymbol{Q}_m = \boldsymbol{I}_m$, where $\boldsymbol{I}_m$ is the $m \times m$ identity matrix, can be used in (10.6.20) to obtain

$$\boxed{\boldsymbol{Q}_m^T \boldsymbol{M} \boldsymbol{K}_\sigma^{-1} \boldsymbol{M} \boldsymbol{Q}_m = \boldsymbol{T}_m} \qquad (10.6.22)$$

Choosing the set of Lanczos vectors, $\boldsymbol{Q}_m$, for the trial vectors, the Rayleigh-Ritz procedure can be used to obtain the best approximation to the eigenvectors of (10.6.6). The approximating Ritz vectors will then be of the form

$$\boldsymbol{y}_i^{(m)} = \boldsymbol{Q}_m \boldsymbol{s}_i^{(m)}, \quad i = 1, \ldots, m \qquad (10.6.23)$$

When a residual vector, $\{K_\sigma^{-1} M y_i^{(m)} - \theta_i^{(m)} y_i^{(m)}\}$, associated with the pair $\{\theta_i^{(m)}, y_i^{(m)}\}$ is M-orthogonal to the set of Lanczos vectors, then $\{\theta_i^{(m)}, y_i^{(m)}\}$ is a Ritz pair. Accordingly

$$Q_m^T M \{K_\sigma^{-1} M y_i^{(m)} - \theta_i^{(m)} y_i^{(m)}\} = \mathbf{0}$$

Using the relation between the Ritz vector, $y_i^{(m)}$ and the Lanczos vectors, given in (10.6.23), together with the orthonormality condition of the q's, and the tridiagonal properties of the Lanczos vectors, (10.6.22), the above equation reduces to the tridiagonal eigenproblem

$$\boxed{T_m \, s_i^{(m)} - \theta_i^{(m)} s_i^{(m)} = \mathbf{0}} \qquad (10.6.24)$$

Thus $\{\theta_i^{(m)}, s_i^{(m)}\}$ is an eigenpair of the tridiagonal matrix, T_m. As the total number of Lanczos vectors increases, i.e., as we take more Lanczos steps, the size of the tridiagonal matrix increases and the eigenvalues of T_m converge to the eigenvalues of the transformed problem (10.6.6), $1/(\lambda_i - \sigma)$. When $m = n$, the order of K, then $\theta_i^{(n)} = 1/(\lambda_i - \sigma)$ for all i, but we hope to stop long before $m = n$. The steps enumerated in Step 2 of Table 10.6.1 are repeated until the Ritz pairs $\{\theta_i^{(m)}, y_i^{(m)}\}$ have sufficiently converged to the desired eigenpairs.

Example 2

$$K = \frac{1}{80}\begin{bmatrix} 39 & -9 & 21 & -11 \\ -9 & 39 & -11 & 21 \\ 21 & -11 & 39 & -9 \\ -11 & 21 & -9 & 39 \end{bmatrix}$$

For this example we assume the mass matrix is the identity. Then the eigenvalue matrix for this problem is $\Lambda = \text{diag}\,(\frac{1}{5}, \frac{1}{4}, \frac{1}{2}, 1)$. The Lanczos algorithm presented here requires the solution of a linear system of equations with K. We therefore give K^{-1} explicitly for convenience. Accordingly,

$$K^{-1} = \frac{1}{2}\begin{bmatrix} 6 & 0 & -3 & 1 \\ 0 & 6 & 1 & -3 \\ -3 & 1 & 6 & 0 \\ 1 & -3 & 0 & 6 \end{bmatrix}$$

Choosing a starting vector $r_0 = \langle 1, 0, 0, 0 \rangle$, then

Step 1:

$$\beta_1 = \|r_0\| = 1, \qquad q_1 = \frac{r_0}{\beta_1} = \begin{Bmatrix} 1 \\ 0 \\ 0 \\ 0 \end{Bmatrix}, \qquad K^{-1} q_1 = \frac{1}{2}\begin{Bmatrix} 6 \\ 0 \\ -3 \\ 1 \end{Bmatrix}$$

$$\alpha_1 = \boldsymbol{q}_1^T \boldsymbol{K}^{-1} \boldsymbol{q}_1 = 3, \qquad \boldsymbol{r}_1 = \boldsymbol{K}^{-1}\boldsymbol{q}_1 - \alpha_1 \boldsymbol{q}_1 = \frac{1}{2}\begin{Bmatrix} 0 \\ 0 \\ -3 \\ 1 \end{Bmatrix}, \qquad \boldsymbol{T}_1 = [3]$$

Step 2:

$$\beta_2 = \|\boldsymbol{r}_1\| = \frac{\sqrt{10}}{2}, \qquad \boldsymbol{q}_2 = \frac{\boldsymbol{r}_1}{\beta_2} = \frac{1}{\sqrt{10}}\begin{Bmatrix} 0 \\ 0 \\ -3 \\ 1 \end{Bmatrix}, \qquad \boldsymbol{K}^{-1}\boldsymbol{q}_2 = \frac{1}{\sqrt{10}}\begin{Bmatrix} 5 \\ -3 \\ -9 \\ 3 \end{Bmatrix}$$

$$\alpha_2 = \boldsymbol{q}_2^T \boldsymbol{K}^{-1} \boldsymbol{q}_2 = 3, \qquad \boldsymbol{r}_2 = \boldsymbol{K}^{-1}\boldsymbol{q}_2 - \alpha_2 \boldsymbol{q}_2 - \beta_2 \boldsymbol{q}_1 = \frac{1}{\sqrt{10}}\begin{Bmatrix} 0 \\ -3 \\ 0 \\ 0 \end{Bmatrix}$$

$$\boldsymbol{T}_2 = \begin{bmatrix} 3 & \dfrac{\sqrt{10}}{2} \\ \dfrac{\sqrt{10}}{2} & 3 \end{bmatrix}$$

Step 3:

$$\beta_3 = \|\boldsymbol{r}_2\| = \frac{3}{\sqrt{10}}, \qquad \boldsymbol{q}_3 = \frac{\boldsymbol{r}_2}{\beta_3} = \begin{Bmatrix} 0 \\ -1 \\ 0 \\ 0 \end{Bmatrix}, \qquad \boldsymbol{K}^{-1}\boldsymbol{q}_3 = \frac{1}{2}\begin{Bmatrix} 0 \\ -6 \\ -1 \\ 3 \end{Bmatrix}$$

$$\alpha_3 = \boldsymbol{q}_3^T \boldsymbol{K}^{-1} \boldsymbol{q}_3 = 3, \qquad \boldsymbol{r}_3 = \boldsymbol{K}^{-1}\boldsymbol{q}_3 - \alpha_3 \boldsymbol{q}_3 - \beta_3 \boldsymbol{q}_2 = \frac{1}{5}\begin{Bmatrix} 0 \\ 0 \\ 2 \\ 6 \end{Bmatrix}$$

$$\boldsymbol{T}_3 = \begin{bmatrix} 3 & \dfrac{\sqrt{10}}{2} & 0 \\ \dfrac{\sqrt{10}}{2} & 3 & \dfrac{3}{\sqrt{10}} \\ 0 & \dfrac{3}{\sqrt{10}} & 3 \end{bmatrix}$$

Step 4:

$$\beta_4 = \|\boldsymbol{r}_3\| = \frac{4}{\sqrt{10}}, \qquad \boldsymbol{q}_4 = \frac{\boldsymbol{r}_3}{\beta_4} = \frac{1}{\sqrt{10}}\begin{Bmatrix} 0 \\ 0 \\ 1 \\ 3 \end{Bmatrix}, \qquad \boldsymbol{K}^{-1}\boldsymbol{q}_4 = \frac{1}{\sqrt{10}}\begin{Bmatrix} 0 \\ -4 \\ 3 \\ 9 \end{Bmatrix}$$

$$\alpha_4 = \boldsymbol{q}_4^T \boldsymbol{K}^{-1} \boldsymbol{q}_4 = 3, \qquad \boldsymbol{r}_4 = \boldsymbol{K}^{-1}\boldsymbol{q}_4 - \alpha_4 \boldsymbol{q}_4 - \beta_4 \boldsymbol{q}_3 = \begin{Bmatrix} 0 \\ 0 \\ 0 \\ 0 \end{Bmatrix}$$

$$\boldsymbol{T}_4 = \begin{bmatrix} 3 & \frac{\sqrt{10}}{2} & 0 & 0 \\ \frac{\sqrt{10}}{2} & 3 & \frac{3}{\sqrt{10}} & 0 \\ 0 & \frac{3}{\sqrt{10}} & 3 & \frac{4}{\sqrt{10}} \\ 0 & 0 & \frac{4}{\sqrt{10}} & 3 \end{bmatrix}$$

The eigenvalues of the tridiagonal matrices converge to the *inverses* of the eigenvalues of $\boldsymbol{K}$ in this example. This can be demonstrated by computing the eigenvalues of the tridiagonal at each step as shown in Table 10.6.2

TABLE 10.6.2 Convergence of the Ritz Values for the Small 4 × 4 Example

j	Eigenvalues of $\boldsymbol{T}_j$
1	3.0000
2	1.4189, 4.5811
3	1.1561, 3.0000, 4.8439
4	1.0000, 2.0000, 4.0000, 5.0000

In Figure 10.6.1 we plot the Ritz values for a larger example. Notice the early convergence at both ends of the spectrum.

10.6.7 Convergence Criterion for Eigenvalues

The eigenvalues, $\theta_i^{(j)}$, $i = 1, \ldots, j$, of the tridiagonal $\boldsymbol{T}_j$ are the Rayleigh-Ritz approximations to eigenvalues of (10.6.6). As j increases these Ritz values get closer to the eigenvalues they are approximating. Often users wait until the change in the computed quantities is within a specified tolerance; that is

$$\frac{|\theta^{(j-1)} - \theta^{(j)}|}{|\theta^{(j)}|} \leq \text{tol} \tag{10.6.25}$$

Here we demonstrate the weakness of such convergence tests. Suppose one is performing inverse iteration with a specified shift, σ, in an interval of interest $[\sigma_1, \sigma_2]$; i.e.,

$$\sigma = \gamma\sigma_1 + (1 - \gamma)\sigma_2$$

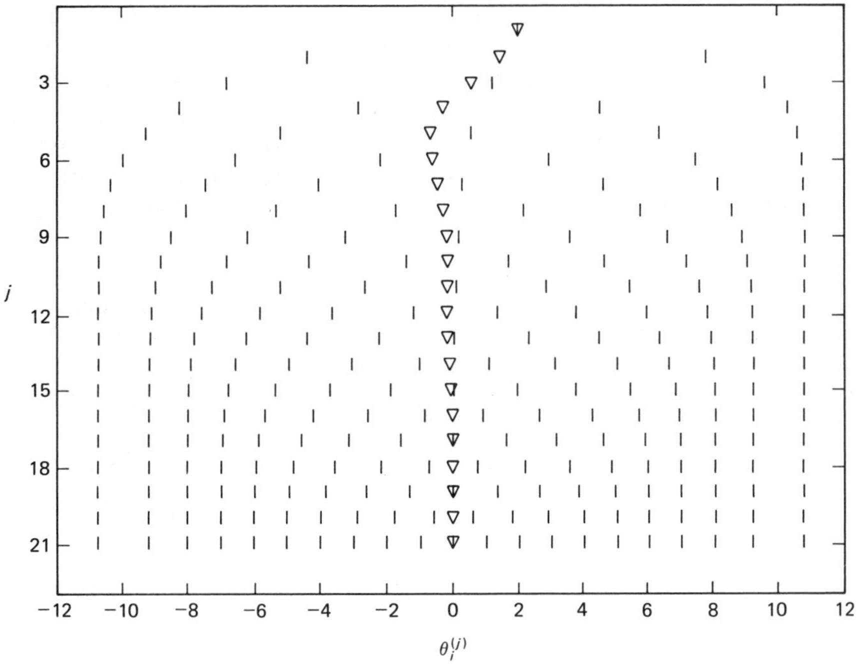

Figure 10.6.1 A typical pattern for the progress of the eigenvalues of the tridiagonal matrix, $\boldsymbol{T}_j$. The symbol ∇ indicates the last diagonal entry of $\boldsymbol{T}_j$.

where γ is a positive number less than 1. Further, assume that the problem has eigenvalues, λ_a and λ_b outside this interval with $\lambda_a < \sigma_1$ and $\lambda_b > \sigma_2$. Then the corresponding eigenvalues of the transformed problem, (10.6.6), are $\nu_a = 1/(\lambda_a - \sigma)$ and $\nu_b = 1/(\lambda_b - \sigma)$, with $\nu_a < 0$ and $\nu_b > 0$. Now, let us perform a step of the inverse iteration method with an unfortunate starting vector, $\boldsymbol{x}^{(1)} = \boldsymbol{z}_a \sin\psi + \boldsymbol{z}_b \cos\psi$. Here, $\boldsymbol{z}_a$ and $\boldsymbol{z}_b$ are eigenvectors corresponding to λ_a and λ_b, respectively. We assume that the eigenvectors are normalized with respect to the mass matrix. Then $\boldsymbol{x}^{(1)}$ will also be normalized. The Ritz value due to $\boldsymbol{x}^{(1)}$ is

$$\begin{aligned}\theta^{(1)} &= \boldsymbol{x}^{(1)T}\boldsymbol{K}\boldsymbol{x}^{(1)} \\ &= \lambda_a \sin^2\psi + \lambda_b \cos^2\psi\end{aligned}$$

After one step of inverse iteration the improved vector is

$$\boldsymbol{x}^{(2)} = \frac{1}{[\nu_a^2 \sin^2\psi + \nu_b^2 \cos^2\psi]^{1/2}}(\nu_a \boldsymbol{z}_a \sin\psi + \nu_b \boldsymbol{z}_b \cos\psi)$$

and the Ritz value associated with $\boldsymbol{x}^{(2)}$ is

$$\theta^{(2)} = \frac{1}{[\nu_a^2 \sin^2\psi + \nu_b^2 \cos^2\psi]^{1/2}}(\nu_a^2 \lambda_a \sin^2\psi + \nu_b^2 \lambda_b \cos^2\psi)$$

Now if $\lambda_a = \sigma - \tau$ and $\lambda_b = \sigma + \tau$, in which case $\nu_a = -1/\tau$ and $\nu_b = 1/\tau$, then

$$\theta^{(1)} = \theta^{(2)} = (\sigma - \tau)\sin^2\psi + (\sigma + \tau)\cos^2\psi = \sigma + \tau\cos 2\psi$$

These θ's, referred to as "ghost" eigenvalues in [25], satisfy (10.6.25) and therefore would be accepted as eigenvalues. Further, an appropriate choice for ψ can result in a Ritz value that is in the interval of interest. This behavior, although rare, has been observed in certain implementations of the subspace iteration method [25] and reflects a serious difficulty. We refer to this as ***misconvergence*** in [23] and, as we observed therein, it cannot be detected by a natural criterion such as (10.6.25). A rigorous inexpensive termination criterion can be obtained using the residual error bound given later.

Consider a pair $\{\theta, \boldsymbol{y}\}$, which approximates an eigenpair $\{\nu, \boldsymbol{z}\}$, of (10.6.6), where ν is the closest eigenvalue of (10.6.6) to θ. For simplicity we consider only normalized Ritz vectors, $\|\boldsymbol{y}\|_M = 1$ where $\|\boldsymbol{y}\|_M = (\boldsymbol{y}^T\boldsymbol{M}\boldsymbol{y})^{1/2}$. The residual vector associated with this approximation is given by

$$\boldsymbol{r} = \boldsymbol{K}_\sigma^{-1}\boldsymbol{M}\boldsymbol{y} - \theta\boldsymbol{y} \tag{10.6.26}$$

The norm of the residual vector can then be used to assess the accuracy of θ through the inequality

$$\boxed{|\nu - \theta| \le \|\boldsymbol{r}\|_M} \tag{10.6.27}$$

For the proof and an in-depth study see [21]. This result allows us to predict the accuracy of any candidate eigenpair of a matrix.

We use the residual vector associated with the Ritz pair $\{\theta_i^{(m)}, \boldsymbol{y}_i^{(m)}\}$ given by (10.6.23) and (10.6.24) to check for their convergence. The tridiagonal property of the Lanczos algorithm, (10.6.20), greatly simplifies the computation of the residual norm. Postmultiplying (10.6.20) by $\boldsymbol{s}_i^{(m)}$ leads to

$$\boldsymbol{K}_\sigma^{-1}\boldsymbol{M}\boldsymbol{Q}_m\boldsymbol{s}_i^{(m)} - \boldsymbol{Q}_m\boldsymbol{T}_m\boldsymbol{s}_i^{(m)} = \boldsymbol{r}_m\boldsymbol{e}_m^T\boldsymbol{s}_i^{(m)} \tag{10.6.28}$$

Since $\boldsymbol{s}_i^{(m)}$ is an eigenvector of $\boldsymbol{T}_m$ (see (10.6.24)), we can replace $\boldsymbol{T}_m\boldsymbol{s}_i^{(m)}$ by $\theta_i^{(m)}\boldsymbol{s}_i^{(m)}$. Further, by definition (10.6.23), $\boldsymbol{Q}_m\boldsymbol{s}_i^{(m)}$ is simply the Ritz vector $\boldsymbol{y}_i^m$ and so (10.6.28) will reduce to

$$\boldsymbol{K}_\sigma^{-1}\boldsymbol{M}\boldsymbol{y}_i^{(m)} - \theta_i^{(m)}\boldsymbol{y}_i^{(m)} = \boldsymbol{r}_m\boldsymbol{e}_m^T\boldsymbol{s}_i^{(m)} \tag{10.6.29}$$

Taking norms,

$$\boxed{\begin{aligned} \|\boldsymbol{K}_\sigma^{-1}\boldsymbol{M}\boldsymbol{y}_i^{(m)} - \theta_i^{(m)}\boldsymbol{y}_i^{(m)}\|_M &= \|\boldsymbol{r}_m\boldsymbol{e}_m^T\boldsymbol{s}_i^{(m)}\|_M \\ &= \|\boldsymbol{r}_m\|_M|\boldsymbol{e}_m^T\boldsymbol{s}_i^{(m)}| \\ &= \beta_{m+1}|\zeta_i| \end{aligned}} \tag{10.6.30}$$

where $\zeta_i = \boldsymbol{e}_m^T\boldsymbol{s}_i^{(m)}$ is the bottom element of $\boldsymbol{s}_i^{(m)}$, the normalized eigenvector of $\boldsymbol{T}_m$. β_{m+1} is a scalar quantity that is computed in the course of the Lanczos process. The bottom element of the eigenvector of $\boldsymbol{T}_m$ can be obtained at little cost once the associated eigenvalue is found. Therefore, the residual norm associated with a Ritz pair can be computed as $\rho_{mi} = \beta_{m+1} | \zeta_i |$ and the Ritz value is considered converged once $\rho_{mi} < \text{tol}$. There is no need to compute $\boldsymbol{y}_i^{(m)}$ until it has converged.

Multiple Eigenvalues

In exact arithmetic the simple Lanczos algorithm cannot compute a second copy of a multiple eigenvalue. Suppose λ is an eigenvalue with multiplicity two. Then there is no unique eigenvector associated with λ. In fact one can obtain two vectors, z_1 and z_2, such that any linear combination of these vectors is also an eigenvector of λ. Moreover, it is possible to obtain a linear combination of z_1 and z_2, $\bar{z} = \xi_1 z_1 + \xi_2 z_2$, that is orthogonal to $\boldsymbol{q}_1$. Then the eigenvector $\bar{z}$ will be orthogonal to all the subsequent Lanczos vectors. Therefore $\bar{z}$ will never be found.

Fortunately, this argument only holds in exact arithmetic. In finite precision, roundoff comes to the rescue. If $\boldsymbol{q}_1$ is perfectly orthogonal to an eigenvector, $\bar{z}$, then because of roundoff error, $\boldsymbol{q}_2$ will have a small component along $\bar{z}$. This component, although tiny, will eventually grow such that $\bar{z}$ can be represented by a linear combination of the computed Lanczos vectors. However, the second copy of λ will converge some steps after the first has converged.

10.6.8 Loss of Orthogonality

In a previous section we derived the equations governing the Lanczos algorithm. These relations are only satisfied by quantities obtained in exact arithmetic. In finite precision, however, each computation introduces a small error and therefore each computed quantity will differ from its exact counterpart. Our objective here is to describe the effect of roundoff on the Lanczos process. For this purpose we need to introduce an important quantity that measures the accuracy of the arithmetic. Let ϵ be the smallest number in the computer such that $1 + \epsilon > 1$. It is known as the ***unit roundoff error***.

Although the tridiagonal relation, (10.6.20), is preserved to within roundoff, the $\boldsymbol{M}$-orthogonality property of the Lanczos vectors completely breaks down after a certain number of steps, depending on ϵ and the distribution of the eigenvalues of (10.6.6). The Lanczos vectors, which are orthogonal in exact arithmetic, not only lose their orthogonality, but may even become linearly dependent. Initially it was believed that loss of orthogonality is due to cancellations that occur each time $\boldsymbol{r}_j$ is evaluated using (10.6.19). This step is simply $\boldsymbol{M}$-orthogonalization of $\boldsymbol{K}_\sigma^{-1}\boldsymbol{M}\boldsymbol{q}_j$ against $\boldsymbol{q}_j$ and $\boldsymbol{q}_{j-1}$. Comparing the final vector resulting from this computation to the starting vector, one can obtain a measure of the cancellation that occurs in this step; i.e., the ratio

$$\chi_j^2 = \frac{\|\boldsymbol{r}_j\|_M^2}{\|\boldsymbol{K}_\sigma^{-1}\boldsymbol{M}\boldsymbol{q}_j\|_M^2} = \frac{\beta_{j+1}^2}{\beta_j^2 + \alpha_j^2 + \beta_{j+1}^2} \tag{10.6.31}$$

indicates how much cancellation has occurred. χ_j is the sine of the angle between the vector $\boldsymbol{K}_\sigma^{-1}\boldsymbol{M}\boldsymbol{q}_j$ and the plane containing the vectors $\boldsymbol{q}_j$ and $\boldsymbol{q}_{j-1}$. When χ_j is zero it indicates that in the jth Lanczos step we are orthogonalizing a vector that is already in this plane against $\boldsymbol{q}_j$ and $\boldsymbol{q}_{j-1}$, and therefore complete cancellation occurs. In practice χ_j rarely drops below $\frac{1}{10}$. When χ_j drops to $\frac{1}{100}$, for instance, it indicates that $\boldsymbol{K}_\sigma^{-1}\boldsymbol{M}\boldsymbol{q}_j$ is nearly parallel to this plane, and therefore at the end of this Lanczos step the computed $\boldsymbol{r}_j$ may not be orthogonal to the plane containing $\boldsymbol{q}_j$ and $\boldsymbol{q}_{j-1}$ to working accuracy. In such cases $\boldsymbol{r}_j$ should be orthogonalized against $\boldsymbol{q}_j$ and $\boldsymbol{q}_{j-1}$ a second time. For a long time it was believed that loss of orthogonality was solely due to this cancellation. Indeed if this were the case, then we would have no option other than a complete reorthogonalization at each step. However, as we soon shall see, a completely different mechanism is at work.

Suppose for the moment the algorithm was executed in exact arithmetic for j steps, except that at some step $k < j$ a small error was introduced into the computation of $\boldsymbol{q}_k$. The first $k - 1$ Lanczos vectors will be perfectly $\boldsymbol{M}$-orthonormal, but they will not be orthogonal to all the vectors computed after the kth step. The error introduced at step k will be amplified in the subsequent steps to such an extent that linear independency may also be lost.

Hence, the loss of orthogonality can be viewed as the subsequent amplification of the error introduced after each computation. To analyze the way in which orthogonality deteriorates, we let $\boldsymbol{Q}_m$ denote the computed Lanczos vectors and define the following matrix:

$$\boldsymbol{H}_m = \boldsymbol{Q}_m^T\boldsymbol{M}\boldsymbol{Q}_m \tag{10.6.32}$$

where the i,j-component of $\boldsymbol{H}_m$ is $\eta_{i,j} = \boldsymbol{q}_i^T\boldsymbol{M}\boldsymbol{q}_j$. In exact arithmetic $\boldsymbol{H}_m$ is the identity matrix. The off-diagonals of $\boldsymbol{H}_m$ will depend on ϵ, the unit roundoff error. Further, (10.6.19) will be satisfied by the computed quantities only to within roundoff error. Now, multiply (10.6.19) by $\boldsymbol{q}_i^T\boldsymbol{M}$ and use the above definition to get the approximate relation

$$\boldsymbol{q}_i^T\boldsymbol{M}\boldsymbol{K}_\sigma^{-1}\boldsymbol{M}\boldsymbol{q}_j \cong \alpha_j\,\eta_{i,j} + \beta_j\,\eta_{i,j-1} + \beta_{j+1}\,\eta_{i,j+1} \tag{10.6.33}$$

A similar equation can be obtained when the above procedure is repeated for the ith Lanczos step

$$\boldsymbol{q}_j^T\boldsymbol{M}\boldsymbol{K}_\sigma^{-1}\boldsymbol{M}\boldsymbol{q}_i \cong \alpha_i\,\eta_{j,i} + \beta_i\,\eta_{j,i-1} + \beta_{i+1}\,\eta_{j,i+1} \tag{10.6.34}$$

By symmetry of $\boldsymbol{M}\boldsymbol{K}_\sigma^{-1}\boldsymbol{M}$, the terms on the left-hand side of (10.6.33) and (10.6.34) are equal and therefore can be eliminated by subtraction, resulting in the relation

$$\beta_{j+1}\,\eta_{j+1,i} \cong \beta_{i+1}\,\eta_{j,i+1} + (\alpha_i - \alpha_j)\eta_{j,i} + \beta_i\,\eta_{j,i-1} - \beta_j\,\eta_{j-1,i} \tag{10.6.35}$$

This recursion holds for $j \geq 2$ and $1 \leq i \leq j - 1$ and starts by assuming that the diagonal of $\boldsymbol{H}_m$ is the identity, $\eta_{j,j} = 1$ for all $j \geq 1$, and the first off-diagonal of $\boldsymbol{H}_m$ is at roundoff level, $\eta_{j,j-1} = \epsilon$ for $j \geq 2$. The above relation is due to Simon [27]. It provides a means of estimating the elements of a column of $\boldsymbol{H}_m$ from the elements of $\boldsymbol{T}_m$ and the elements in the previous columns of $\boldsymbol{H}_m$. This recursion can be restated in the vector form

$$\boxed{\beta_{j+1}\,\boldsymbol{h}_{j+1} \cong \boldsymbol{T}_{j-1}\,\boldsymbol{h}_j - \alpha_j\boldsymbol{h}_j - \beta_j\,\boldsymbol{h}_{j-1}} \tag{10.6.36}$$

where h_{j-1}, h_j and h_{j+1} are vectors of length $j - 1$ containing the top $j - 1$ elements of the $(j - 1)$st, jth, and $(j + 1)$st columns of $(H_m - I_m)$. Here, the bottom element of h_{j-1} is zero. Then, all the terms in h_j depend on ϵ. Taking norms we can show

$$\begin{aligned} \beta_{j+1} \| h_{j+1} \| &\leq (\| T_{j-1} \| + | \alpha_j |) \| h_j \| + \beta_j \| h_{j-1} \| \\ &\leq 2 \| T_j \| \max(\| h_j \|, \| h_{j-1} \|) \end{aligned} \tag{10.6.37}$$

This result shows that the level of nonorthogonality can grow by at most a factor of $2\| T_j \| / \beta_{j+1}$ after each step. A drop in the value of β_{j+1} can result in a sudden loss of orthogonality but this rarely occurs.

An alternative characterization of the pattern in which orthogonality is lost was presented by Paige [18–20]. Instead of examining the vector $h_{j+1} = Q_j^T M q_{j+1}$, he looked at a linear combination of the components in this vector. To be more specific, he examined the inner product between each Ritz vector $y_i^{(j)}$ given by (10.6.23) and q_{j+1}. That is

$$\begin{aligned} y_i^{(j)\,T} M q_{j+1} &= s_i^{(j)\,T} Q_j^T M q_{j+1} \\ &= s_i^{(j)\,T} h_{j+1} \end{aligned} \tag{10.6.38}$$

In exact arithmetic this value is zero. However, in his work Paige showed that

$$y_i^{(j)\,T} M q_{j+1} = \frac{\gamma_{ji}\, \epsilon \| T_j \|}{\beta_{j+1} | \zeta_i |} \tag{10.6.39}$$

Recall that ζ_i is the bottom element of $s_i^{(j)}$, the ith eigenvector of T_j. γ_{ji} is a scalar quantity usually close to unity. We omit the derivation of this result and refer the interested reader to [21]. Note that Paige's result also shows that a sudden drop in β_{j+1} can result in a severe loss of orthogonality. Moreover, recall the quantity in the denominator is also a measure of the convergence of the Ritz value $\theta_i^{(j)}$ (see (10.6.30)). The only way the left side of (10.6.39) can rise up to values like 0.1 is for $\rho_{ji} (= \beta_{j+1} | \zeta_i |)$ to drop down to $10\gamma_{ji}\epsilon \| T_j \|$, so

Loss of orthogonality ⇒ convergence of a Ritz value

When only a single Ritz value converges at step j, then q_{j+1} loses orthogonality by tilting toward the converged Ritz vector, which in turn is a linear combination of the previous Lanczos vectors. When more than one Ritz vector converges simultaneously, the picture is more complicated. In this case q_{j+1} tilts toward a linear combination of these vectors.

Return of Banished Ritz Vectors

In theory, if two successive Lanczos vectors, q_{j-1} and q_j, are orthogonal to an eigenvector, z, then all the subsequent Lanczos vectors will also be orthogonal to z. In practice, however, a converged Ritz vector, y_i, will not remain orthogonal to all the subsequent Lanczos vectors. This is a consequence of the same mechanism by which multiple eigenvalues are found. Roundoff errors will add to each Lanczos vector a small component of $\boldsymbol{y}_i$, which will grow eventually to such a magnitude that orthogonality to $\boldsymbol{y}_i$ is lost. This phenomenon can also be observed in the inverse iteration method; i.e., orthogonalizing the starting vector against the first eigenvector does not guarantee convergence of the iteration vectors to the next eigenvector.

The state of orthogonality between a converged Ritz vector, $\boldsymbol{y}_i$, and the current Lanczos vector, $\boldsymbol{q}_j$, can be measured by the component of $\boldsymbol{y}_i$ along $\boldsymbol{q}_j$. We define $\tau_j = \boldsymbol{y}_i^T\boldsymbol{M}\boldsymbol{q}_j$. Then multiplying (10.6.19) by $\boldsymbol{y}_i^T\boldsymbol{M}$ and considering the effect of round-off on (10.6.19) yields

$$\beta_{j+1}\boldsymbol{y}_i^T\boldsymbol{M}\boldsymbol{q}_{j+1} \cong \boldsymbol{y}_i^T\boldsymbol{M}\boldsymbol{K}_\sigma^{-1}\boldsymbol{M}\boldsymbol{q}_j - \alpha_j\boldsymbol{y}_i^T\boldsymbol{M}\boldsymbol{q}_j - \beta_j\boldsymbol{y}_i^T\boldsymbol{M}\boldsymbol{q}_{j-1}$$

Using the fact that $\boldsymbol{y}_i$ is a converged Ritz value, i.e., $\boldsymbol{K}_\sigma^{-1}\boldsymbol{M}\boldsymbol{y}_i = \theta_i\boldsymbol{y}_i$, together with the definition of τ_j we obtain

$$\tau_{j+1} \cong \frac{(\theta_i - \alpha_j)\tau_j - \beta_j\tau_{j-1}}{\beta_{j+1}} \tag{10.6.40}$$

This recurrence can be updated for each converged Ritz value, θ_i. The magnitude of τ_j can be used as an indicator for loss of orthogonality against a converged Ritz vector. The growth of the components of the Lanczos vectors along a converged Ritz vector is referred to as the ***return of a banished Ritz vector***.

10.6.9 Restoring Orthogonality

In this section we look at a number of preventive measures that we can adopt to maintain a certain level of orthogonality. Lanczos [14] was aware of the effects of round-off on the algorithm when he presented his work. He proposed that the newly computed Lanczos vector, $\boldsymbol{q}_{j+1}$, be explicitly orthogonalized against all the preceding vectors at the end of each step j. We will refer to this technique as the "full reorthogonalization" method. This scheme is adopted in [12, 29]. With this procedure the Lanczos vectors will meet the stringent requirement

$$|\boldsymbol{q}_i^T\boldsymbol{M}\boldsymbol{q}_j| < n\epsilon, \quad i \neq j \tag{10.6.41}$$

Although this scheme increases the overall cost of an eigenvalue computation, for short Lanczos runs (when the number of Lanczos steps is less than the half-bandwidth of $\boldsymbol{K}$) the increase in cost is small compared to the cost of solving a system of equations with $\boldsymbol{K}_\sigma$. For longer runs the cost of a full reorthogonalization step will begin to dominate the cost of a Lanczos step, although vector computers will delay this effect.

The orthogonality condition of (10.6.41) can be replaced by a more relaxed condition,

$$|\boldsymbol{q}_i^T\boldsymbol{M}\boldsymbol{q}_j| < \sqrt{n\epsilon}, \quad i \neq j \tag{10.6.42}$$

We refer to this as the ***semiorthogonality condition*** and we refer to procedures that adopt the weaker condition as ***selective orthogonalization methods***. Imposing the more relaxed condition can result in considerable reduction in the number of operations in the reorthogonalization step and semiorthogonality is sufficient to make T_j exact to working precision.

We consider two different reorthogonalization schemes that adopt the more relaxed condition of (10.6.42).

A. ***Orthogonalization against Ritz vectors:*** This procedure is a consequence of the result of Paige [18–20]. As soon as a Ritz value converges, its corresponding Ritz vector is computed and the component of this vector along q_{j+1} is purged. Further, for each converged Ritz value, the three-term recurrence for τ_j is updated and whenever τ_j becomes greater than $\sqrt{\epsilon}$ in absolute value, it signals that the component of the corresponding eigenvector has grown too much. So the new Lanczos vector is orthogonalized against this known eigenvector [22, 25].

B. ***Orthogonalization against previous Lanczos vectors***: This scheme can be based on either of the techniques given in [13,27]. In [27] the vector h_{j+1} is updated using (10.6.36) and the magnitudes of its elements are monitored. Whenever the ith element of h_{j+1} is greater than $\sqrt{\epsilon}$ then semiorthogonality is lost between q_{j+1} and q_i. At this step the appropriate Lanczos vectors are brought in from secondary storage and their components along q_{j+1} are removed. This scheme is also adopted in [17].

Both schemes indicate loss of orthogonality at about the same step. The first method performs orthogonalization against fewer vectors and therefore costs less. However, alone it has some shortcomings. The result in (10.6.39) gives an accurate picture at the early stages of a Lanczos run when one or two Ritz values converge. When for a number of steps, a few Ritz values converge at the same time, then a gradual loss of semiorthogonality may occur. This appears most often in a Lanczos run with spectral transformation. For this reason we do not use scheme A on its own. However, for short Lanczos runs where only a very few (less than 10) eigenpairs are wanted, scheme A is very effective.

We advocate a combination of the above two schemes. Whenever possible we orthogonalize against a Ritz vector (scheme A) because of lower costs. We update h_{j+1} using (10.6.36) and monitor its elements. If any of the elements of h_{j+1} is greater than $\sqrt{\epsilon}$, then semiorthogonality may have been lost. This is the only procedure we use to determine loss of semiorthogonality, which can occur in two possible ways:

i. Convergence of a Ritz value.
ii. Growth of components of computed eigenvectors along the Lanczos vector.

We take different actions for each of these. If the observed orthogonality loss is due to (i), then we perform a step of scheme B. The Lanczos vectors are brought in from secondary storage and the newly computed Lanczos vector, r_j, is orthogonalized against them. Further, to minimize data transfer from secondary storage, we also compute some of the converged Ritz vectors at the same time. Therefore the computation of newly converged Ritz vectors is delayed until the Lanczos vectors are brought back for reorthogonalization. We should note that when (i) occurs, both methods A and B would require recalling the Lanczos vectors from secondary storage but for different reasons: the first for computing Ritz vectors and the second for reorthogonalization. We perform both. Our method would clearly require more operations than scheme A but has the following two advantages:

a. An eigenvector is computed in a few steps after the convergence of its eigenvalue and therefore has more than half correct digits.
b. The gradual loss of orthogonality just mentioned will not occur with the combined procedure.

However, if loss of orthogonality is due to (ii), then we can use scheme A, i.e., orthogonalize against a computed Ritz vector. For this reason, the τ recurrence, (10.6.40), is also monitored to establish which Ritz vector, $\boldsymbol{y}_k$, has contaminated $\boldsymbol{q}_{j+1}$.

The orthogonalization of $\boldsymbol{q}_{j+1}$ against $\boldsymbol{y}_k$ alters the state of orthogonality among the Lanczos vectors. The modified $\overline{\boldsymbol{q}}_{j+1}$ is given by

$$\overline{\boldsymbol{q}}_{j+1} = \boldsymbol{q}_{j+1} - \xi_k \boldsymbol{y}_k$$

where $\xi_k = \boldsymbol{y}_k^T \boldsymbol{M} \boldsymbol{q}_{j+1}$. Let $\overline{\boldsymbol{h}}_{j+1} = \boldsymbol{Q}_j^T \boldsymbol{M} \overline{\boldsymbol{q}}_{j+1}$. Then

$$\begin{aligned} \overline{\boldsymbol{h}}_{j+1} &= \boldsymbol{Q}_j^T \boldsymbol{M} (\boldsymbol{q}_{j+1} - \xi_k \boldsymbol{y}_k) \\ &= \boldsymbol{h}_{j+1} - \xi_k \boldsymbol{s}_k \end{aligned} \tag{10.6.43}$$

An approximation to ξ_k can be obtained via

$$\xi_k = \boldsymbol{y}_k^T \boldsymbol{M} \boldsymbol{q}_{j+1} = \boldsymbol{s}_k^T \boldsymbol{Q}_j^T \boldsymbol{M} \boldsymbol{q}_{j+1} = \boldsymbol{s}_k^T \boldsymbol{h}_{j+1}$$

If the approximate result for ξ_k is used then (10.6.43) reduces to the orthogonalization of $\boldsymbol{h}_{j+1}$ against $\boldsymbol{s}_k$. This simple step is used to modify $\boldsymbol{h}_{j+1}$ to reflect the changes caused by orthogonalizing $\boldsymbol{q}_{j+1}$ against a Ritz vector. The flowchart in Fig. 10.6.5 (see p. 615) gives a global view of our implementation of the selective orthogonalization algorithm.

When it becomes necessary to restore semiorthogonality, the computed vector $\boldsymbol{q}_{j+1}$ must be orthogonalized against some linear combination of the previous Lanczos vectors. $\boldsymbol{q}_j$ and $\boldsymbol{q}_{j-1}$ remain unchanged at the end of this step. But, at the next step, $\boldsymbol{q}_j$ appears again in the computation of $\boldsymbol{q}_{j+2}$. If no action is taken, then $\boldsymbol{q}_{j+2}$ will be contaminated by $\boldsymbol{q}_j$ and the reorthogonalization efforts of the previous step would be wasted. Therefore a second reorthogonalization step must be performed. If the Lanczos vectors or the converged Ritz vectors reside in secondary storage, they must be retrieved in two successive steps.

Alternatively, at the same time the orthogonality of $\boldsymbol{q}_{j+1}$ is being restored, we can also perform similar modifications on $\boldsymbol{q}_j$. Then at the end of the next step, no reorthogonalization of $\boldsymbol{q}_{j+2}$ will be necessary. The number of operations for this scheme is the same as that of the scheme above, but vectors are retrieved only once and therefore the I/O overhead is halved.

10.6.10 LANSEL Package

In this section we describe all the ingredients that go into the LANSEL eigenpackage and provide sufficient detail for installing LANSEL into a finite element program. The steps in each subroutine are described in sequence followed by the listing. Higher level routines are described first.

User-Supplied Subroutines

The only interface between the algorithm and the eigenproblem is through two user-supplied subroutines. In general, the structures of the matrices $\boldsymbol{K}$ and $\boldsymbol{M}$ vary greatly from problem to problem. No single subroutine can be designed to take advantage of the special structure of $\boldsymbol{K}$ and $\boldsymbol{M}$ for all problems. Furthermore, the characteristics of a given equation solver depend strongly on the computing environment. Only the user is aware of the properties of his or her matrices and the computer system that is being used. Therefore, the job of solving the linear system of equations $\boldsymbol{K}_\sigma \bar{\boldsymbol{r}}_j = \boldsymbol{p}_j$ and multiplication of a vector by the mass matrix, $\bar{\boldsymbol{p}}_j = \boldsymbol{M}\boldsymbol{r}_j$ is relegated to the user. This has the added advantage that a shifting strategy may be implemented without modifying any part of the routines presented here. The two routines that connect $\boldsymbol{K}$ and $\boldsymbol{M}$ with our program must take the form

1. SUBROUTINE OPK (X, Y, N). This solves the linear system of equations $\boldsymbol{K}_\sigma \boldsymbol{y} = \boldsymbol{x}$. N is the length of the vectors X and Y.
2. SUBROUTINE OPM (X, Y, N). This forms the matrix-vector product $\boldsymbol{y} = \boldsymbol{M}\boldsymbol{x}$. N is the length of the vectors X and Y.

A third subroutine must also be provided for the management of the Lanczos vectors. From time to time the program calls the subroutine STORE. The subroutine statement for STORE is

SUBROUTINE STORE (V, N, J, ISW)

When ISW = 1 the routine must store the column vector V of length N and identifier J in secondary storage. When ISW = 2, a vector V with identifier J that is in secondary storage must be fetched. This way the user can take full advantage of any data management system that is available.

TABLE 10.6.3 List of Vector Operation Routines used by LANSEL

Name	Description
IDAMAX	Finds the index of the element of a vector with maximum absolute value; $i = \arg(\max_{i=1,n} \lvert y_i \rvert)$.
DAXPY	Computes the product of a scalar, a, by a vector, $\boldsymbol{x}$, adding the result to a vector, $\boldsymbol{y}$; $\boldsymbol{y} = a\boldsymbol{x} + \boldsymbol{y}$.
DCOPY	Copies a vector, $\boldsymbol{x}$, to a vector $\boldsymbol{y}$; $\boldsymbol{y} = \boldsymbol{x}$.
DDOT	Computes the Euclidean inner product of two vectors, $\boldsymbol{x}$ and $\boldsymbol{y}$; dot $= \boldsymbol{x}^T\boldsymbol{y}$.
DSCAL	Scales the elements of a vector, $\boldsymbol{y}$, by a scalar factor, a; $\boldsymbol{y} = a\boldsymbol{y}$.
DZERO	Resets the elements of a vector, $\boldsymbol{y}$, to zero; $\boldsymbol{y} = \boldsymbol{0}$.

LANSEL also makes use of subroutines for performing basic vector operations, such as addition of two vectors. Some of these routines may be found in [10]. Many computers have libraries of carefully written assembly language implementations of these subroutines, which are much faster. In Table 10.6.3 we give a list of vector operations used by LANSEL. Alternatively, the user may wish to write his or her own subroutines to perform these tasks.

Project 1. Write a program to set up and solve the 4 × 4 example problem of Sec. 10.6.6. The program should employ the LANSEL package described herein. You will have to provide "user-supplied subroutines" that are general enough to handle the data of the 4 × 4 problem. Print out all intermediate calculations and compare with the results presented in Sec. 10.6.6.

Project 2. Modify program DLEARN (see Chapter 11) to perform eigenvalue/eigenvector calculations by interfacing it with the LANSEL package. DLEARN contains several utility routines (e.g., equation-solving routines, a variety of matrix-vector manipulation routines, etc.) which can be helpful in developing the "user-supplied subroutines" for the LANSEL package. You will also need to write input and output routines for the computed eigenvalues and eigenvectors. (This project requires a fairly comprehensive knowledge of the DLEARN program.)

Subroutine LANDRV

LANDRV is the driver for the main subroutine LANSEL. Table 10.6.4 gives a brief description of the list of parameters for LANDRV.

TABLE 10.6.4 Description of the Parameters for LANDRV

Name	Type	Dimension	Description
N	I	scalar	The dimension of the eigenproblem. $\boldsymbol{K}$ and $\boldsymbol{M}$ are N×N matrices.
LANMAX	I	scalar	Upper limit to the number of Lanczos steps. LANMAX must not exceed N.
MAXPRS	I	scalar	Upper limit to the number of wanted eigenpairs. MAXPRS must not exceed LANMAX.
ENDL	F	scalar	ENDL $= \nu_L = 1/(\lambda_L - \sigma)$, where λ_L is the left end of the interval containing the wanted eigenvalues (see Fig. 10.6.2).
ENDR	F	scalar	ENDR $= \nu_R = 1/(\lambda_R - \sigma)$, where λ_R is the right end of the interval containing the wanted eigenvalues (see Fig. 10.6.2).
NW	I	scalar	Length of the work array W. NW must be at least 6×N + 2×MAXPRS + 10×LANMAX. A larger NW can result in faster computation time. NW need not be greater than 6×N + 2×MAXPRS + LANMAX×(6 + LANMAX).
W	F	NW	Work array of length NW. The first N words of W hold a user-supplied starting vector.

Name	Type	Dimension	Description
IW	I	MAXPRS	Work array of length MAXPRS.
EIG	F	MAXPRS	EIG will hold the converged Ritz values on return. The Ritz values are in the order they were computed.
Y	F	MAXPRS×N	Y will hold the converged Ritz vectors on return. Y is dimensioned as a two-dimensional array Y(N,MAXPRS) and the Ith column of Y holds the Ritz vector corresponding to EIG(I).
IERR	I	scalar	IERR is an error flag. A successful execution is indicated when IERR = 0. For a list of error flags see Table 10.6.5.
NEIG	I	scalar	The number of computed eigenpairs.

LANDRV acts as an interface between the user and LANSEL. It performs the following tasks before calling LANSEL:

i. Checks the control parameters for possible error.
ii. Allocates the working memory for LANSEL.
iii. If user does not supply a starting vector in W then LANDRV supplies a random vector.
iv. Performs the first step of the Lanczos algorithm by calling STPONE.

If an error is detected, IERR is reset and returned as soon as the checking of the control parameters is complete. Each bit of the integer IERR is used as an error flag. IERR is set to zero when entering LANDRV. The Ith bit of IERR can then be reset to 1 using the command IERR = IERR + 2**I. For example the third bit is set to 1 by adding 8 to IERR. This way all the errors in the control parameters can be detected at the same time. Table 10.6.5 gives a description of possible errors indicated by IERR.

TABLE 10.6.5 List of Errors Indicated by IERR

Bit	Indicates
1	N < 0
2	LANMAX < 0
3	ENDR < ENDL
4	MAXPRS < 0
5	MAXPRS > LANMAX
6	LANMAX > N
7	NW too small

In addition, IERR = −1 indicates that the maximum number of Lanczos steps in LANSEL has been reached.

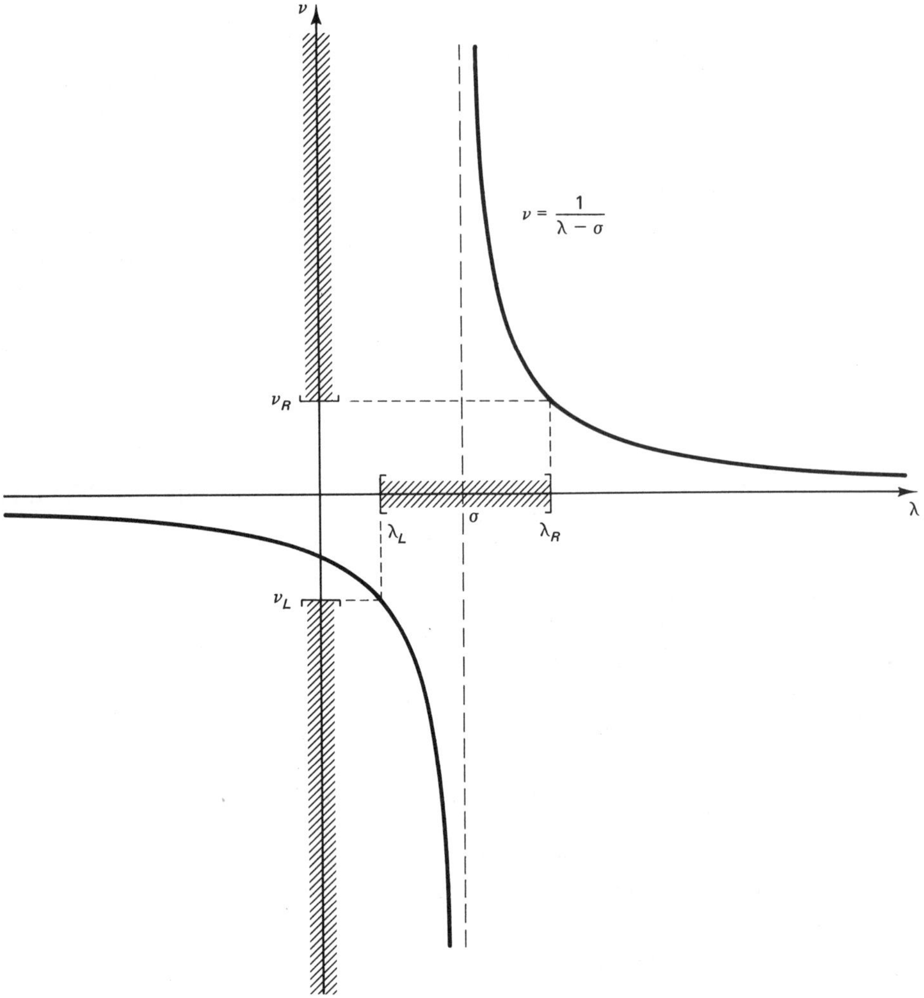

Figure 10.6.2 Spectral transformation of the interval containing the wanted eigenvalues; σ denotes the shift.

Remarks

1. LANSEL has no knowledge of the shift. The eigenvalues that it computes are those of the transformed problem (10.6.6). The eigenvalues of the original problem may be obtained by applying the inverse transformation $\lambda = \sigma + 1/\nu$. This is the responsibility of the user. The λ's closest to σ *usually* converge first. In particular, for a system with positive-definite $\boldsymbol{K}$ and $\boldsymbol{M}$, and $\sigma = 0$, the λ's *usually* converge in ascending order starting with the smallest.

2. LANSEL may return only a single copy of multiple eigenvalues. To obtain additional copies, restart LANSEL with a new starting vector orthogonal to the computed eigenvectors. (The number of missed eigenvalues may be determined by the technique of spectrum slicing; see Sec. 10.5.1.).

3. Certain rare execution errors, beyond the control of the Lanczos algorithm may occur in subsidiary routines such as GIVENS. These are indicated by a returned value of IERR greater than 127. The only recourse is to rerun LANSEL with a different shift and/or starting vector.

```
C
      SUBROUTINE LANDRV(N,LANMAX,MAXPRS,ENDL,ENDR,NW,W,IW,EIG,Y,IERR,
     1                  NEIG)
C
C            ****************************************
C            *                                      *
C            *          L  A  N  S  E  L            *
C            *                                      *
C            *       LANCZOS ALORITHM WITH          *
C            *     SELECTIVE ORTHOGONALIZATION      *
C            *                                      *
C            *        B. N O U R - O M I D          *
C            *                                      *
C            ****************************************
C
C.... INPUTS
C
C     N       DIMENSION OF THE EIGENPROBLEM
C     LANMAX  UPPER LIMIT TO THE NUMBER OF LANCZOS STEPS
C     MAXPRS  UPPER LIMIT TO THE NUMBER OF WANTED EIGENPAIRS
C     ENDL    LEFT END OF THE INTERVAL CONTAINING THE WANTED EIGENVALUES
C     ENDR    RIGHT END OF THE INTERVAL CONTAINING THE WANTED EIGENVALUES
C     NW      LENGTH OF THE WORK ARRAY W
C     W       WORK ARRAY OF LENGTH NW
C     IW      WORK ARRAY OF LENGTH MAXPRS
C
C.... OUTPUTS
C
C     EIG     ARRAY OF LENGTH MAXPRS TO HOLD THE CONVERGED RITZ VALUES
C     Y       ARRAY OF LENGTH MAXPRS*N TO HOLD THE CONVERGED RITZ VECTORS
C     IERR    ERROR FLAG
C     NEIG    TOTAL NUMBER OF CONVERGED EIGENPAIRS
C
C.... SUBROUTINES: DDOT,GETEPS,LANSEL,OPM,RAN,STPONE
C
      IMPLICIT DOUBLE PRECISION (A-H,O-Z)
      DIMENSION NQ(5),Y(1),EIG(1),W(1),IW(1)
      INTEGER*2 IRANDOM1,IRANDOM2
C
C.... COMMON IDATA
C
C     EIGL    INNER MOST EIGENVALUE CONVERGED FROM LEFT END OF
C             TRANSFORMED SPECTRUM
C     EIGR    INNER MOST EIGENVALUE CONVERGED FROM RIGHT END OF
C             TRANSFORMED SPECTRUM
C     NEIG1   TOTAL NUMBER OF CONVEGED EIGENVALUES
C
      COMMON /IDATA/ EIGL, EIGR, NEIG1
C
C.... COMMON RDATA
C
C     RNM     NORM OF THE RESIDUAL VECTOR IN R(NQ(1))
C     RNM2    SQUARE OF RNM
C     SPREAD  WIDTH OF THE INTERVAL CONTAINING THE UNCONVERGED
C             EIGENVALUES
C     TOL     TOLERANCE FOR CONVERGENCE OF THE EIGENVALUES
C     EPS     COMPUTER PRECISION
C     EPS1    EPS*SQRT(N)
C     REPS    SQUARE ROOT OF EPS
C
      COMMON /RDATA/ RNM, RNM2, SPREAD, TOL, EPS, EPS1, REPS
C
      COMMON /LANCON/ ZERO, TENTH, EIGHTH, FOURTH, HALF, ONE, TWO,
     1                FOUR, TEN, ONE28, TWO56, FIVE12, ORTFAC, OVRFLW
C
      DATA ZERO, TENTH, EIGHTH, FOURTH, HALF, ONE, TWO, FOUR, TEN,
     1     ONE28, TWO56, FIVE12, ORTFAC, OVRFLW /
     2     0.0D0, 0.1D0, 0.125D0, 0.25D0, 0.5D0, 1.0D0, 2.0D0, 4.0D0,
     3     10.0D0, 128.0D0, 256.0D0, 512.0D0, 4.0D0, 1.7014D+38/
C
C.... CHECK INPUT DATA
C
      IERR = 0
      MT = 6*N + 2*MAXPRS + 10*LANMAX
      IF ( N .LE. 0 )            IERR = IERR + 1
      IF ( LANMAX .LE. 0 )       IERR = IERR + 2
      IF ( ENDR .LE. ENDL )      IERR = IERR + 4
      IF ( MAXPRS .LE. 0 )       IERR = IERR + 8
      IF ( MAXPRS .GT. LANMAX )  IERR = IERR + 16
```

```
      IF ( LANMAX .GT. N )        IERR = IERR + 32
      IF ( MT  .GT. NW )          IERR = IERR + 64
      IF ( IERR .GT. 0 ) RETURN
C
C.... COMPUTE THE MACHINE PRECISION
C
      EPS   = GETEPS(IBETA,IT,IRND)
C
      REPS = SQRT(EPS)
      EPS1 = EPS*SQRT(FLOAT(N))
C
C.... SET POINTERS AND INITIALIZE
C
      M1  = 1       + 6*N
      M2  = MAXPRS  + M1
      M3  = MAXPRS  + M2
      M4  = LANMAX  + M3
      M5  = LANMAX  + M4
      M6  = LANMAX  + M5
      M7  = LANMAX  + M6
      M8  = LANMAX  + M7
      M9  = LANMAX  + M8
      M10 = LANMAX  + M9
      M11 = LANMAX  + M10
C
C.... IF MORE STORAGE IS TO BE ALLOCATED, MUST CHANGE "MT" ABOVE
C
      NS = NW - MT + 2*LANMAX
      NQ(1) = N + 1
      DO 20 I = 2,5
         NQ(I) = NQ(I-1) + N
20    CONTINUE
C
C.... CHECK FOR STARTING VECTOR
C
      RNM2 = ZERO
      DO 30 I = 1,N
         RNM2 = RNM2 + ABS(W(I))
30    CONTINUE
C
      IF (RNM2 .EQ. ZERO) THEN
C
C....    GET RANDOM STARTING VECTOR
C
         IRAND = N + LANMAX + MAXPRS + NW
       IRANDOM1=IRAND
         DO 40 I = 1,N
          IRANDOM2=I
            W(I) = RAN(IRANDOM1,IRANDOM2)
40       CONTINUE
      END IF
C
      CALL OPM(W,W(NQ(3)),N)
      RNM2 = DDOT(N,W,1,W(NQ(3)),1)
      CALL STPONE(N,W(M3),W(M4),W(M5),W(M6),W,NQ)
      CALL LANSEL(N,LANMAX,MAXPRS,NS,ENDL,ENDR,W,W(M3),W(M4),W(M5),
     1            W(M6),EIG,W(M1),W(M2),W(M10),W(M11),W(M7),W(M8),
     2            IW,Y,W(M9),NQ,IERR)
      IF (IERR.GT.0) THEN
         RETURN
      ENDIF
      NEIG = NEIG1
C
      RETURN
      END
```

Function GETEPS

GETEPS determines the precision of the computer being used. It evaluates the unit roundoff error (i.e., the smallest number, ϵ, in the computer such that $1 + \epsilon > 1$).

```
C
      FUNCTION GETEPS(IBETA, IT, IRND)
C
      IMPLICIT DOUBLE PRECISION (A-H,O-Z)
C
      COMMON /LANCON/ ZERO, TENTH, EIGHTH, FOURTH, HALF, ONE, TWO,
     1                FOUR, TEN, ONE28, TWO56, FIVE12, ORTFAC, OVRFLW
C
C.... DETERMINE IBETA, BETA
C
```

```
      A = ONE
10    A = A + A
      IF (((A + ONE) - A) - ONE .EQ. ZERO) GO TO 10
      B = ONE
20    B = B + B
      IF ((A + B) - A .EQ. ZERO) GO TO 20
      IBETA = INT(SNGL((A + B) - A))
      BETA = FLOAT(IBETA)
C
C.... DETERMINE IT, IRND
C
      IT = 0
      B = ONE
30    IT = IT + 1
      B = B * BETA
      IF (((B + ONE) - B) - ONE .EQ. ZERO) GO TO 30
      IRND = 0
      BETAM1 = BETA - ONE
      IF ((A + BETAM1) - A .NE. ZERO) IRND = 1
C
C.... DETERMINE EPS
C
      BETAIN = ONE / BETA
      A = ONE
      DO 40 I = 1, IT + 3
         A = A * BETAIN
40    CONTINUE
C
50    IF ((ONE + A) - ONE .NE. ZERO) GO TO 60
      A = A * BETA
      GO TO 50
60    EPS = A
      IF ((IBETA .EQ. 2) .OR. (IRND .EQ. 0)) GO TO 70
      A = (A*(ONE + A)) / (ONE + ONE)
      IF ((ONE + A) - ONE .NE. ZERO) EPS = A
70    GETEPS = EPS
C
      RETURN
      END
```

Subroutine STPONE

This routine performs the initialization of ALF, BET, ALPH and BET2 as well as the first step of the Lanczos algorithm. See Sec. 10.6.5 for notation. Since $\boldsymbol{q}_{j-1} = \boldsymbol{0}$ at the first step the algorithm reduces to an orthogonalization of $\boldsymbol{K}_{\sigma}^{-1}\boldsymbol{M}\boldsymbol{q}_1$ against $\boldsymbol{q}_1$. $\boldsymbol{q}_1$ is obtained by normalizing $\boldsymbol{r}$ and is stored in the array R starting at location NQ(1). $\boldsymbol{Mr}$, stored in starting location NQ(3) of R, is then normalized to get $\boldsymbol{M}\boldsymbol{q}_1$ which is put in R starting from NQ(4). NQ(2) and NQ(5) are the location of $\boldsymbol{q}_{j-1}$ and a temporary vector in R, respectively.

```
C
      SUBROUTINE STPONE(N,ALF,BET,ALPH,BET2,R,NQ)
C
C.... THIS ROUTINE PERFORMS THE FIRST STEP OF THE LANCZOS ALGORITHM.
C     IT PERFORMS A STEP OF LOCAL REORTHOGONALIZATION IF NEEDED.
C
C.... INPUT/OUTPUT
C
C     N       DIMENSION OF THE EIGENPROBLEM
C     ALF     THE NEW DIAGONAL OF T
C     BET     THE NEW OFF-DIAGONAL OF T
C     ALPH    THE NEW DIAGONAL OF THE DEFLATED T
C     BET2    THE NEW OFF-DIAGONAL SQUARED OF THE DEFLATED T
C     R       AN ARRAY CONTAINING [R(J),Q(J),Q(J-1),P(J),MR(J)]
C     NQ(5)   LOCATION POINTERS FOR THE ARRAY R
C
C.... SUBROUTINES: DAXPY,DCOPY,DDOT,DSCAL,OPK,OPM
C
      IMPLICIT DOUBLE PRECISION (A-H,O-Z)
      DIMENSION NQ(1),R(1)
C
      COMMON /RDATA/ RNM,RNM2,SPREAD,TOL,EPS,EPS1,REPS
      COMMON /LANCON/ ZERO, TENTH, EIGHTH, FOURTH, HALF, ONE, TWO,
     1                FOUR, TEN, ONE28, TWO56, FIVE12, ORTFAC, OVRFLW
C
C.... MODIFY R TO SATISFY THE CORRECT CONDITIONS FOR SINGULAR M
C
```

```
      T = ONE/SQRT(RNM2)
      CALL DCOPY(N,R(NQ(3)),1,R(NQ(4)),1)
      CALL DSCAL(N,T,R(NQ(4)),1)
C
      CALL OPK(R(NQ(4)),R,N)
      CALL OPM(R,R(NQ(3)),N)
      RNM2 = DDOT(N,R,1,R(NQ(3)),1)
      RNM  = SQRT(RNM2)
C
C.... BET2(1) OUGHT TO BE ZERO.
C     BET(1) STORES THE M-NORM OF THE STARTING VECTOR.
C
      BET2 = ZERO
      BET = RNM
      T = ONE/RNM
      CALL DCOPY(N,R,1,R(NQ(1)),1)
      CALL DSCAL(N,T,R(NQ(1)),1)
      CALL DCOPY(N,R(NQ(3)),1,R(NQ(4)),1)
      CALL DSCAL(N,T,R(NQ(4)),1)
      CALL OPK(R(NQ(4)),R,N)
C
      ALF = ZERO
      DALF = DDOT(N,R,1,R(NQ(4)),1)
      DO 10 I=1,2
         CALL DAXPY(N,-DALF,R(NQ(1)),1,R,1)
         ALF = ALF +DALF
         ALPH = ALF
         CALL OPM(R,R(NQ(3)),N)
         RNM2 = DDOT(N,R,1,R(NQ(3)),1)
         IF (I .EQ. 2) RETURN
C
         DALF = DDOT(N,R(NQ(1)),1,R(NQ(3)),1)
         DBET = DDOT(N,R(NQ(2)),1,R(NQ(3)),1)
         CALL DAXPY(N,-DBET,R(NQ(2)),1,R,1)
10    CONTINUE
C
      RETURN
      END
```

Subroutine LANSEL

LANSEL, LANczos algorithm with SELective orthogonalization, performs the main steps of the procedure described in the previous sections. The flowchart of Fig. 10.6.3 gives a global structure of LANSEL. After some initializations the algorithm checks if any orthogonalization is necessary by calling PURGE. Then it calls LANSIM to perform a step of simple Lanczos. The orthogonality estimates and the eigenvalues of the tridiagonal matrix are updated by calling ORTBND and ANALZT, respectively. When enough eigenvalues have been computed loop 10 is terminated and finally the wanted eigenvectors are computed using RITVEC.

LANSEL terminates the iteration if any of the following conditions are satisfied:

a. The norm of $\mathbf{r}_j$ becomes small (indicating an invariant subspace).
b. The number of Lanczos steps reaches LANMAX, the maximum allowed. In this case IERR is set to -1.
c. The desired eigenvalues are computed (see LOGICAL FUNCTION ENOUGH).

If the interval [ENDL, ENDR] contains no eigenvalue, then LANSEL will continue until an eigenvalue has converged outside each end of this interval. If the first MAXPRS eigenvalues are wanted, then set ENDL $= -\infty$, and ENDR $= +\infty$ (i.e., set to very large negative and positive numbers, respectively).

Structure of R: The first vector starting at location 1 is $\mathbf{r}_j$ when entering the subroutine. The remaining five vectors are stored at locations NQ(1) through NQ(5).

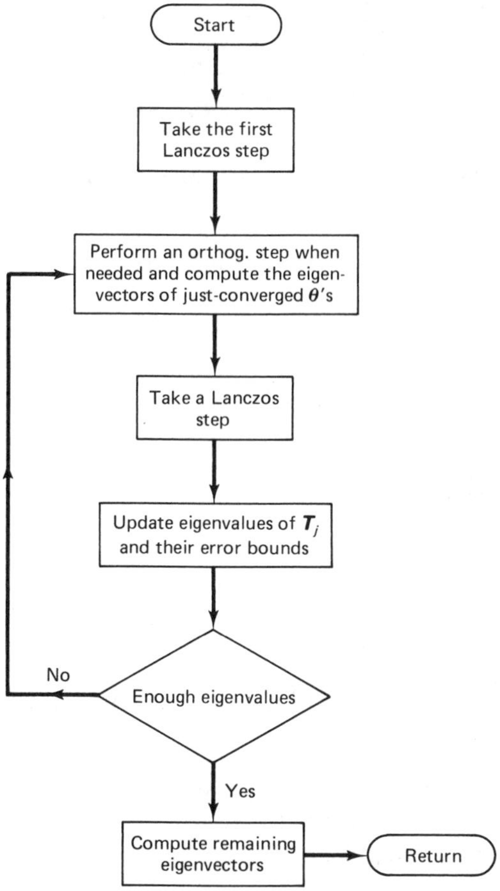

Figure 10.6.3 A global picture of the Lanczos method with selective orthogonalization indicating the way the individual modules are used.

This eliminates the need for moving the content of one array to another for swapping. Only the pointers of the arrays are changed. The vectors q_j, q_{j-1}, Mr_j and Mq_j are held in R starting from locations NQ(1) through NQ(4), respectively. The space in R starting from NQ(5) is used for a working vector in other parts of the program.

```
C
      SUBROUTINE LANSEL(N,LANMAX,MAXPRS,NS,ENDL,ENDR,R,ALF,BET,ALPH,
     1                  BET2,EIG,TAU,OLDTAU,RHO,WORK,ETA,OLDETA,INFO,
     2                  Y,S,NQ,IERR)
C
C.... INPUTS
C
C     N         DIMENSION OF THE EIGENPROBLEM
C     LANMAX    UPPER LIMIT TO THE NUMBER OF LANCZOS STEPS
C     MAXPRS    UPPER LIMIT TO THE NUMBER OF WANTED EIGENPAIRS
C     NS        LENGTH OF THE ARRAY S
C     ENDL      LEFT END OF THE INTERVAL CONTAINING THE WANTED EIGENVALUES
C     ENDR      RIGHT END OF THE INTERVAL CONTAINING THE WANTED EIGENVALUES
C
C.... WORK SPACE
C
C     R         HOLDS 6 VECTORS OF LENGTH N. SEE THE TEXT FOR DETAILS.
C     NQ(5)     CONTAINS THE POINTERS TO THE BEGINING OF EACH VECTOR IN R.
C     ALF       ARRAY OF LENGTH LANMAX TO HOLD DIAGONAL OF THE TRIDIAGONAL T
C     BET       ARRAY OF LENGTH LANMAX TO HOLD OFF-DIAGONAL OF T
C     ALPH      DIAGONAL OF THE DEFLATED TRIDIAGONAL
C     BET2      SQUARE OF THE OFF-DIAGONALS OF THE DEFLATED TRIDIAGONAL
C     TAU       ORTHOGONALITY ESTIMATE OF RITZ VECTORS AT STEP J
```

```
C      OLDTAU   ORTHOGONALITY ESTIMATE OF RITZ VECTORS AT STEP J-1
C      RHO      WORKING ARRAY USED IN DEFLAT()
C      ETA      ORTHOGONALITY ESTIMATE OF LANCZOS VECTORS AT STEP J
C      OLDETA   ORTHOGONALITY ESTIMATE OF LANCZOS VECTORS AT STEP J-1
C      INFO     INFORMATION ARRAY ABOUT EIGENVECTORS OF T
C      S        ARRAY FOR COMPUTING THE EIGENVECTORS OF THE TRIDIAGONAL
C      WORK     WORKING ARRAY TO HOLD SQUARES OF ARRAY BETA
C
C.... OUTPUTS
C
C      EIG      ARRAY OF LENGTH MAXPRS TO HOLD THE CONVERGED RITZ VALUES
C      Y        ARRAY OF LENGTH MAXPRS*N TO HOLD THE CONVERGED RITZ VECTORS
C      NEIG     NUMBER OF COMPUTED EIGENPAIRS
C      IERR     ERROR FLAG
C
C.... SUBROUTINES: ANALZT,ENOUGH,LANSIM,ORTBND,PURGE,RITVEC
C
      IMPLICIT DOUBLE PRECISION (A-H,O-Z)
      DIMENSION R(1),Y(N,1),EIG(1),TAU(1),OLDTAU(1),ALF(1),BET(1)
      DIMENSION S(1),NQ(1),ALPH(1),BET2(1),ETA(1),OLDETA(1),INFO(1)
      DIMENSION RHO(1),WORK(1)
      LOGICAL LASTEP,ENOUGH
C
      COMMON /RDATA/ RNM,RNM2,SPREAD,TOL,EPS,EPS1,REPS
      COMMON /IDATA/ EIGL,EIGR,NEIG
C
      JJ     = 1
      ETA(1) = EPS1
      EIGL   = ENDL - (ENDR - ENDL)
      EIGR   = ENDR + (ENDR - ENDL)
      NEIG   = 0
C
C.... LANCZOS LOOP
C
      DO 10 J = 2,LANMAX
         NBUF = NS/J
         RNM  = SQRT(RNM2)
C
C....    RESTORE THE ORTHOGONALITY STATE WHEN NEEDED
C
         CALL PURGE(R,R(NQ(1)),R(NQ(3)),R(NQ(4)),R(NQ(5)),Y,ALF,BET,S,
     1              EIG,ETA,OLDETA,TAU,OLDTAU,WORK,INFO,N,J-1,NBUF,IERR)
         IF (IERR .GT. 0) RETURN
C
C....    UPDATE THE RITZ VALUES
C
         CALL ANALZT(JJ,ALPH,BET2,EIG,TAU,OLDTAU,RHO,INFO)
C
         IF ( ENOUGH(ENDL,ENDR,MAXPRS) ) GO TO 30
C
         IF (RNM .LT. REPS*SPREAD) GO TO 20
         JJ   = JJ + 1
C
C....    TAKE A LANCZOS STEP
C
         CALL LANSIM(R,ALF(J),BET(J),ALPH(JJ),BET2(JJ),RNM,RNM2,NQ,
     1               N,J)
C
C....    UPDATE THE ORTHOGONALITY BOUNDS
C
         CALL ORTBND(ALF,BET,J,EPS1,ETA,OLDETA,TAU,OLDTAU,EIG,INFO,
     1                RNM,NEIG,N)
C
10    CONTINUE
20    J = J - 1
C
C.... COMPUTE THE REMAINING RITZ VECTORS
C
30    DO 50 I=1,NEIG
         M = 1
         DO 40 K = 1,NEIG
            IF ( INFO(K) .GT. 0 ) INFO(K) = -INFO(K)
            M =  MIN(ABS(INFO(K)),M)
40       CONTINUE
C
         IF ( M .EQ. 0 ) THEN
            CALL RITVEC(R,R(NQ(1)),R(NQ(3)),R(NQ(4)),R(NQ(5)),Y,ALF,
     1             BET,EIG,S,INFO,N,J,NEIG,NBUF,.TRUE.,WORK,IERR)
            IF (IERR .GT. 0) THEN
               RETURN
            ENDIF
         ELSE
            GO TO 60
         END IF
C
50    CONTINUE
```

```
C
60    CONTINUE
      IF( J .EQ. LANMAX ) THEN
      IERR = -1
      END IF
C
      RETURN
      END
```

Logical Function ENOUGH

ENOUGH determines if all the desired eigenvalues have converged. This is established if one of the following conditions is satisfied:

a. The number of computed eigenvalues exceeds the maximum number requested, MAXPRS.
b. An eigenvalue is found outside the interval [ENDL, ENDR] at each end.

```
C
      LOGICAL FUNCTION ENOUGH(ENDL,ENDR,MAXPRS)
C
C.... EXAMINE IF ENOUGH EIGENVALUES HAVE CONVERGED
C
C.... INPUT
C
C     ENDL    LEFT END OF THE INTERVAL CONTAINING THE WANTED EIGENVALUES
C     ENDR    RIGHT END OF THE INTERVAL CONTAINING THE WANTED EIGENVALUES
C     MAXPRS  UPPER LIMIT TO THE NUMBER OF WANTED EIGENPAIRS
C
      IMPLICIT DOUBLE PRECISION(A-H,O-Z)
      PARAMETER (NMAX = 128)
C
      COMMON/ATDATA/THET(NMAX),BJ(NMAX),WINDOW,NBD(2),NDST
      COMMON/IDATA/EIGL,EIGR,NEIG
C
      ENOUGH = .TRUE.
      IF ( NEIG .GT. MAXPRS ) RETURN
C
      ENOUGH = ( THET(1) .GT. ENDL .AND. THET(NDST) .LT. ENDR ) .AND.
     1         (( EIGL .GT. ENDL .AND. EIGL .LT. ENDR ) .OR.
     2          ( EIGR .GT. ENDL .AND. EIGR .LT. ENDR ))
C
      RETURN
      END
```

Subroutine LANSIM

This routine performs a single step of the simple Lanczos process. The flowchart in Fig. 10.6.4 gives a global view of the algorithm. All the vectors required in this routine are stored in array R.

```
C
      SUBROUTINE LANSIM(R,ALF,BET,ALPH,BET2,RNM,RNM2,NQ,N,J)
C
C.... THIS ROUTINE PERFORMS A SINGLE STEP OF THE LANCZOS ALGORITHM,
C     FOLLOWED BY A STEP OF LOCAL REORTHOGONALIZATION IF NEEDED.
C
C.... INPUT/OUTPUT
C
C     R      AN ARRAY CONTAINING [R(J),Q(J),Q(J-1),P(J),MR(J)]
C     ALF    THE NEW DIAGONAL OF T
C     BET    THE NEW OFF-DIAGONAL OF T
C     ALPH   THE NEW DIAGONAL OF THE DEFLATED T
C     BET2   THE NEW OFF-DIAGONAL SQUARED OF THE DEFLATED T
C     RNM    NORM OF R(J)
C     RNM2   RNM**2
C     NQ(5)  LOCATION POINTERS FOR THE ARRAY R
C     N      DIMENSION OF THE EIGENPROBLEM
C     J      CURRENT LANCZOS STEP
C
C.... SUBROUTINES: DAXPY,DCOPY,DDOT,DSCAL,OPK,OPM,STORE
C
```

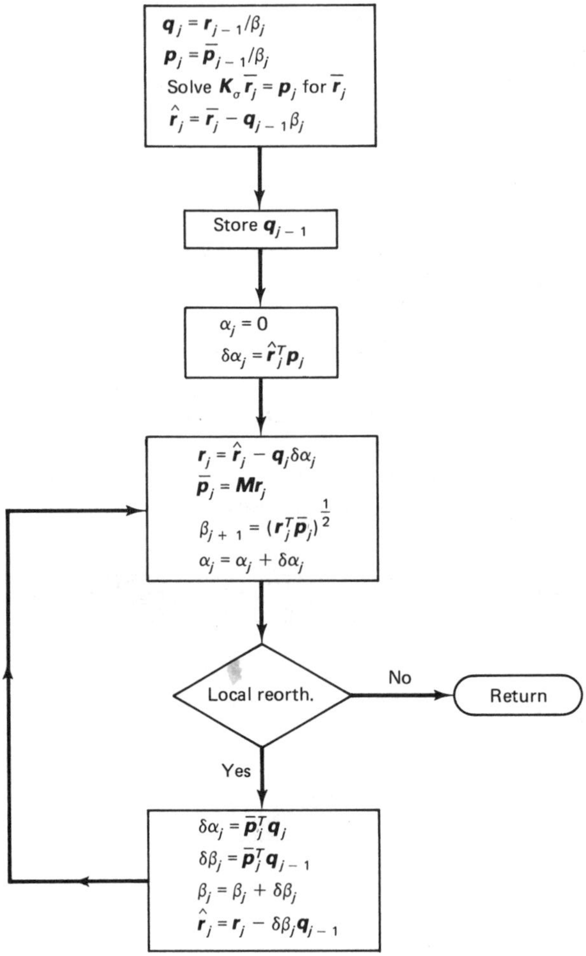

Figure 10.6.4 Flowchart description of the jth step of the simple Lanczos algorithm.

```
      IMPLICIT DOUBLE PRECISION (A-H,O-Z)
      DIMENSION NQ(1),R(1)
C
      COMMON /LANCON/ ZERO, TENTH, EIGHTH, FOURTH, HALF, ONE, TWO,
     1                FOUR, TEN, ONE28, TWO56, FIVE12, ORTFAC, OVRFLW
C
C.... SWAP Q(J) AND Q(J-1)
C
      NTMP = NQ(2)
      NQ(2) = NQ(1)
      NQ(1) = NTMP
C
C.... Q = R/BETA
C
      T = ONE/RNM
      CALL DCOPY(N,R,1,R(NQ(1)),1)
      CALL DSCAL(N,T,R(NQ(1)),1)
C
C.... P = PBAR/BETA
C
      CALL DCOPY(N,R(NQ(3)),1,R(NQ(4)),1)
      CALL DSCAL(N,T,R(NQ(4)),1)
C
C.... R = ( K INVERSE ) * Q
C
      CALL OPK(R(NQ(4)),R,N)
C
C.... R = R - Q(J-1)*BETA
```

```
C
      T = -RNM
      CALL DAXPY(N,T,R(NQ(2)),1,R,1)
C
C.... STORE Q(J-1)
C
      CALL STORE(R(NQ(2)),N,J-1,1)
C
C.... START LOCAL REORTHOGONALIZATION
C
      BET = RNM
      BET2 = RNM2
      ALF = ZERO
C
C.... ALF = ( R TRANSPOSE ) * P
C
      DALF = DDOT(N,R,1,R(NQ(4)),1)
      DO 10 I=1,2
         CALL DAXPY(N,-DALF,R(NQ(1)),1,R,1)
         ALF = ALF + DALF
         CALL OPM(R,R(NQ(3)),N)
         RNM2 = DDOT(N,R,1,R(NQ(3)),1)
         ALPH = ALF
C
         IF (RNM2*ORTFAC .GT. (ALF**2 + BET2) .OR. I .EQ. 2) RETURN
C
C....    REPEAT LOCAL REORTHOGONALIZATION WHEN WARRANTED
C
         DALF = DDOT(N,R(NQ(1)),1,R(NQ(3)),1)
         DBET = DDOT(N,R(NQ(2)),1,R(NQ(3)),1)
         CALL DAXPY(N,-DBET,R(NQ(2)),1,R,1)
         BET = BET + DBET
         BET2 = BET**2
10    CONTINUE
C
      END
```

Subroutine ORTBND

The subroutine ORTBND updates the orthogonality bounds, h_j, of (10.6.36), and the τ recurrence of (10.6.40). These quantities are later examined in subroutine PURGE for possible loss of semiorthogonality. LANSEL is the only routine that calls ORTBND.

```
C
      SUBROUTINE ORTBND (ALF,BET,J,EPS1,ETA,OLDETA,TAU,OLDTAU,EIG,INFO,
     1                   RNM,NEIG,N)
C
C.... UPDATE THE ETA AND TAU RECURRENCES.
C
C.... INPUTS
C
C     ALF(J)         DIAGONAL OF THE TRIDIAGONAL T
C     BET(J)         OFF-DIAGONAL OF T
C     J              DIMENSION OF T
C     EPS1           ROUNDOFF ESTIMATE FOR DOT PRODUCT OF TWO UNIT VECTORS
C     ETA(J)         ORTHOGONALITY ESTIMATE OF LANCZOS VECTORS AT STEP J
C     OLDETA(J)      ORTHOGONALITY ESTIMATE OF LANCZOS VECTORS AT STEP J-1
C     TAU(NEIG)      ORTHOGONALITY ESTIMATE OF RITZ VECTORS AT STEP J
C     OLDTAU(NEIG)   ORTHOGONALITY ESTIMATE OF RITZ VECTORS AT STEP J-1
C     EIG(NEIG)      ARRAY OF CONVERGED EIGENVALUES
C     INFO(NEIG)     INFORMATION ARRAY ABOUT EIGENVECTORS OF T
C     RNM            NORM OF THE NEXT RESIDUAL VECTOR
C     NEIG           NUMBER OF CONVERGED EIGENVALUES
C     N              DIMENSION OF THE EIGENPROBLEM
C
C.... OUTPUTS
C
C     ETA(J)         ORTHOGONALITY ESTIMATE OF LANCZOS VECTORS AT STEP J+1
C     OLDETA(J)      ORTHOGONALITY ESTIMATE OF LANCZOS VECTORS AT STEP J
C     TAU(NEIG)      ORTHOGONALITY ESTIMATE OF RITZ VECTORS AT STEP J+1
C     OLDTAU(NEIG)   ORTHOGONALITY ESTIMATE OF RITZ VECTORS AT STEP J
C
      IMPLICIT DOUBLE PRECISION (A-H,O-Z)
      DIMENSION ALF(1),BET(1),ETA(1),OLDETA(1),TAU(1),OLDTAU(1)
      DIMENSION EIG(1),INFO(1)
      COMMON /LANCON/ ZERO, TENTH, EIGHTH, FOURTH, HALF, ONE, TWO,
     1                FOUR, TEN, ONE28, TWO56, FIVE12, ORTFAC, OVRFLW
C
```

```
      IF ( J .GT. 1 ) THEN
         OLDETA(1) = ( BET(2)*ETA(2) + (ALF(1) - ALF(J))*ETA(1)
     1               - BET(J)*OLDETA(1)) / RNM
         J1 = J - 1
         IF ( J .GT. 2 ) THEN
            DO 100 K=2,J1
               OLDETA(K) = (BET(K+1)*ETA(K+1) + (ALF(K) - ALF(J))*ETA(K)
     1                   + BET(K)*ETA(K-1) - BET(J)*OLDETA(K)) / RNM
100         CONTINUE
         END IF
         DO 200 K=1,J1
            T = OLDETA(K)
            OLDETA(K) = ETA(K)
            ETA(K) = T
200      CONTINUE
      END IF
      ETA(J) = EPS1*MAX(BET(2)/RNM,ONE)
C
C.... UPDATE THE TAU RECURRENCE.
C
      DO 300 I=1,NEIG
         IF ( INFO(I) .NE. 0 ) THEN
            T = TAU(I)
            TAU(I) = (EIG(I) - ALF(J))*TAU(I) - BET(J)*OLDTAU(I)
            OLDTAU(I) = T

         END IF
300   CONTINUE
C
      RETURN
      END
```

Subroutine PURGE

PURGE first examines the array ETA, which holds the vector $\boldsymbol{h}_{j+1}$. See Sec. 10.6.8. If the element of ETA with largest absolute value is less then REPS, $\sqrt{\epsilon}$, indicating no loss of semiorthogonality, then PURGE does nothing and returns. Otherwise, the elements of TAU are examined to determine if loss of orthogonality might be due to the return of a previously banished Ritz vector. If TAU(I) is greater than REPS in absolute value, it indicates loss of semiorthogonality against the Ritz vector with index I. The corresponding eigenvector of $\boldsymbol{T}_j$ is computed using GIVENS. ETA and OLDETA are orthogonalized against the computed eigenvector of $\boldsymbol{T}_j$. See Fig. 10.6.5.

The elements of ETA are examined for a second time. If ETA still holds an element with absolute value greater than REPS, then loss of orthogonality is also due to the convergence of a Ritz value. Then RITVEC is called and the contents of TAU, OLDTAU, ETA, and OLDETA are all set to EPS1 $= \sqrt{n\epsilon}$. Otherwise, $\boldsymbol{q}_j$ and $\boldsymbol{r}_j$, held in Q and R, are orthogonalized against those Ritz vectors in columns of Y with TAU value greater than REPS in absolute value. These elements of TAU and OLDTAU are then set to EPS1.

All orthogonalizations of the Lanczos vectors are performed with respect to the inertial inner product, and therefore we require two vectors, QA and RA, that hold $\boldsymbol{M}\boldsymbol{q}_j$ and $\boldsymbol{M}\boldsymbol{r}_j$, respectively. At the end of PURGE, $\boldsymbol{M}\boldsymbol{q}_j$ and $\boldsymbol{M}\boldsymbol{r}_j$ are recomputed if $\boldsymbol{q}_j$ and $\boldsymbol{r}_j$ are modified.

```
C
      SUBROUTINE PURGE(R,Q,RA,QA,T,Y,ALF,BET,S,EIG,ETA,OLDETA,TAU,
     1                 OLDTAU,WORK,INFO,N,J,NBUF,IERR)
C
C.... THIS ROUTINE EXAMINES ETA, OLDETA, TAU AND OLDTAU TO DECIDE
C     WHICH FORM OF REORTHOGONALIZATION IF ANY SHOULD BE PERFORMED.
C
C.... INPUT/OUTPUT
C
C     R       THE RESIDUAL VECTOR TO BECOME THE NEXT LANCZOS VECTOR
C     Q       THE CURRENT LANCZOS VECTOR
C     RA      THE PRODUCT OF THE MASS MATRIX AND R
C     QA      THE PRODUCT OF THE MASS MATRIX AND Q
```

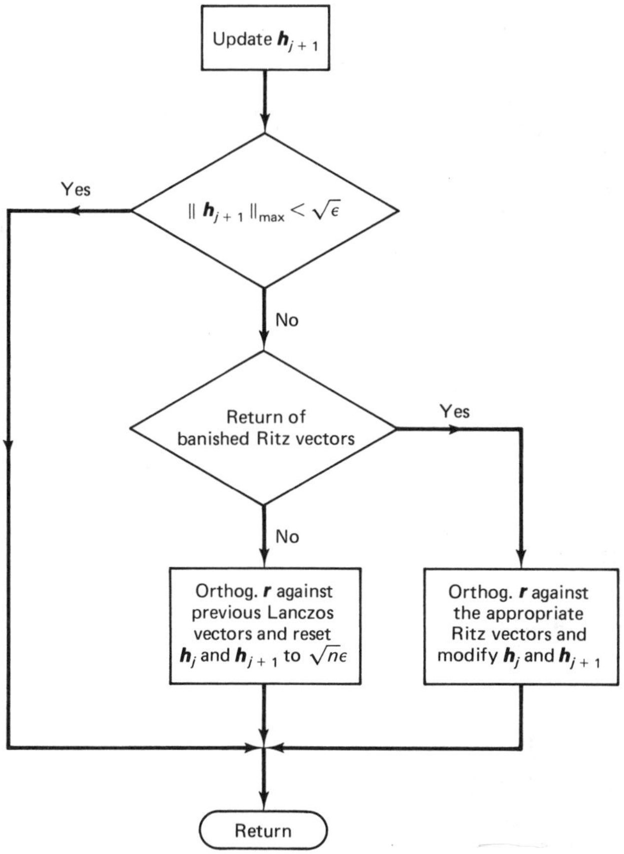

Figure 10.6.5 A flowchart for the selective orthogonalization method.

```
C       T       A TEMPORARY VECTOR TO HOLD THE PREVIOUS LANCZOS VECTORS
C       Y       CONTAINS THE COMPUTED RITZ VECTORS
C       ALF     THE NEW DIAGONAL OF T
C       BET     THE NEW OFF-DIAGONAL OF T
C       S       VECTOR FOR COMPUTING EIGENVECTORS OF T(J)
C       EIG     HOLDS THE CONVERGED RITZ VALUES
C       ETA     STATE OF ORTHOGONALITY BETWEEN R AND PREVIOUS LANCZOS VECTORS
C       OLDETA  STATE OF ORTHOGONALITY BETWEEN Q AND PREVIOUS LANCZOS VECTORS
C       TAU     STATE OF ORTHOGONALITY BETWEEN R AND COMPUTED RITZ VECTORS
C       OLDTAU  STATE OF ORTHOGONALITY BETWEEN Q AND COMPUTED RITZ VECTORS
C       WORK    WORKING ARRAY EXPLAINED IN LANSEL
C       INFO    INFORMATION ABOUT THE EIGENVECTORS OF T(J)
C       N       DIMENSION OF THE EIGENPROBLEM
C       J       CURRENT LANCZOS STEP
C       NEIG    NUMBER OF RITZ VALUES
C       NBUF    NUMBER OF VECTORS IN S
C
C....   SUBROUTINES: DAXPY,DZERO,DDOT,GIVENS,IDAMAX,OPM,RITVEC,SUBTJ
C
        IMPLICIT DOUBLE PRECISION(A-H,O-Z)
        DIMENSION R(1),Q(1),RA(1),QA(1),T(1),Y(N,1),ALF(1),BET(1),S(J,1)
        DIMENSION EIG(1),ETA(1),OLDETA(1),TAU(1),OLDTAU(1),INFO(1),WORK(1)
        LOGICAL ORTHO
C
        COMMON/IDATA/EIGL,EIGR,NEIG
        COMMON/RDATA/RNM,RNM2,SPREAD,DUMMY,EPS,EPS1,REPS
        COMMON /LANCON/ ZERO, TENTH, EIGHTH, FOURTH, HALF, ONE, TWO,
     1                  FOUR, TEN, ONE28, TWO56, FIVE12, ORTFAC, OVRFLW
C
        ORTHO  = .FALSE.
        TOL    = REPS*SPREAD
        REPSOJ = SQRT(EPS/FLOAT(J))
        K      = IDAMAX(J-2,ETA,1)
        IF (ABS(ETA(K)) .GT. REPSOJ) THEN
```

```
         DO 10 I=1,NEIG
            IF (INFO(I) .LT. 0 .AND. ABS(TAU(I)) .GT. REPSOJ) THEN
               CALL DZERO(J,S,1)
               CALL SUBTJ(ALF,BET,EIG,INFO,TOL,I,N,NEIG,J,L,K,M,
     1                    WORK,IERR)
               IF (IERR .GT. 0) RETURN
               CALL GIVENS(K-M+1,L-M+1,ALF(M),BET(M),EIG(I),EPS,
     1                     S,RESID,RAYCOR,IERR)
               IF (IERR .GT. 0) RETURN
               ZETA = -DDOT(J,ETA,1,S,1)
               CALL DAXPY(J,ZETA,S,1,ETA,1)
               ZETA = -DDOT(J,OLDETA,1,S,1)
               CALL DAXPY(J,ZETA,S,1,OLDETA,1)
            END IF
10       CONTINUE
C
         K = IDAMAX(J-2,ETA,1)
C
         IF (ABS(ETA(K)) .GT. REPSOJ) THEN
C
C....       GRAM-SCHMID NEEDED
C
            ORTHO = .TRUE.
            CALL RITVEC(R,Q,RA,QA,T,Y,ALF,BET,EIG,S,INFO,N,J,NEIG,
     1                  NBUF,.FALSE.,WORK,IERR)
            IF (IERR.GT.0) RETURN
            DO 20 I=1,NEIG
               TAU(I) = EPS1
               OLDTAU(I) = EPS1
20          CONTINUE
            DO 30 I=1,J-1
               ETA(I) = EPS1
               OLDETA(I) = EPS1
30          CONTINUE
         ELSE
C
C....       REMOVE COMPONENTS OF A RITZ VECTOR
C
            ORTHO = .TRUE.
            DO 40 I = 1,NEIG
               IF (ABS(TAU(I)).GT.REPSOJ)THEN
                  TAU(I) = EPS1
                  OLDTAU(I) = EPS1
                  ZETA = DDOT(N,RA,1,Y(1,I),1)
                  CALL DAXPY(N,-ZETA,Y(1,I),1,R,1)
                  ZETA = DDOT(N,QA,1,Y(1,I),1)
                  CALL DAXPY(N,-ZETA,Y(1,I),1,Q,1)
               END IF
40          CONTINUE
         END IF
      END IF
      IF (ORTHO) THEN
         CALL OPM(Q,QA,N)
         QNORM = SQRT(DDOT(N,Q,1,QA,1))
         CALL DSCAL(N,ONE/QNORM,Q,1)
         CALL DSCAL(N,ONE/QNORM,QA,1)
         BET(J) = BET(J) * QNORM
         ZETA   = DDOT(N,R,1,QA,1)
         ALF(J) = ALF(J) + ZETA
         CALL DAXPY(N,-ZETA,Q,1,R,1)
         CALL OPM(R,RA,N)
         RNM2 = DDOT(N,R,1,RA,1)
         RNM  = SQRT(RMN2)
      END IF
C
      RETURN
      END
```

Subroutine RITVEC

The purpose of this routine is twofold:

1. Compute the Ritz vector corresponding to a converged Ritz value.
2. Perform a full reorthogonalization step.

The first step of the program is to compute some eigenvectors of the tridiagonal $\boldsymbol{T}_j$. The eigenvectors of $\boldsymbol{T}_j$ are stored in an array S, and the number of computed vectors depends on the available storage indicated by NBUF. GIVENS is used to obtain the wanted eigenvectors. The next step is to recall all the Lanczos vectors from secondary storage (by calling STORE) and compute the Ritz vectors using (10.6.23), accumulating the result in columns of Y. The old Lanczos vector that is brought in is stored in the temporary vector T. In the same loop, when EVONLY is false, the two current Lanczos vectors, $\boldsymbol{q}_j$ and $\boldsymbol{r}_j$, held in Q and R, are orthogonalized against the previous Lanczos vectors held in T. In this way the number of I/O transfers is minimized. Arrays QA and RA that hold $\boldsymbol{M}\boldsymbol{q}_j$ and $\boldsymbol{M}\boldsymbol{r}_j$, respectively, are also needed by RITVEC to perform the orthogonalization with respect to the inertial inner product.

```
C
      SUBROUTINE RITVEC(R,Q,RA,QA,T,Y,ALF,BET,EIG,S,INFO,N,J,NEIG,NBUF,
     1                  EVONLY,WORK,IERR)
C
C.... THIS ROUTINE COMPUTES SOME RITZ VECTORS AND PERFORMS A
C     REORTHOGONALIZATION OF THE LANCZOS VECTORS.
C
C.... INPUT/OUTPUT
C
C     R       THE RESIDUAL VECTOR TO BECOME THE NEXT LANCZOS VECTOR
C     Q       THE CURRENT LANCZOS VECTOR
C     RA      THE PRODUCT OF THE MASS MATRIX AND R
C     QA      THE PRODUCT OF THE MASS MATRIX AND Q
C     T       A TEMPORARY VECTOR TO HOLD THE PREVIOUS LANCZOS VECTORS
C     Y       CONTAINS THE COMPUTED RITZ VECTORS
C     ALF     THE NEW DIAGONAL OF T
C     BET     THE NEW OFF-DIAGONAL OF T
C     EIG     HOLDS THE CONVERGED RITZ VALUES
C     S       VECTOR FOR COMPUTING EIGENVECTORS OF T(J)
C     INFO    INFORMATION ABOUT THE EIGENVECTORS OF T(J)
C     N       DIMENSION OF THE EIGENPROBLEM
C     J       CURRENT LANCZOS STEP
C     NEIG    NUMBER OF RITZ VALUES
C     NBUF    NUMBER OF VECTORS IN S
C     EVONLY  IF .TRUE. NO REORTHO. IS PERFORMED, COMPUTES ONLY RITZ VECTORS
C
C.... SUBROUTINES: DAXPY,DDOT,DZERO,GIVENS,IDAMAX,NUMLES,STORE
C
      IMPLICIT DOUBLE PRECISION (A-H,O-Z)
      DIMENSION R(1),Q(1),RA(1),QA(1),T(1),Y(N,1),ALF(1),BET(1),S(J,1)
      DIMENSION EIG(1),INFO(1),WORK(1)
      LOGICAL EVONLY
C
      COMMON /RDATA/ RNM,RNM2,SPREAD,DUMMY,EPS,EPS1,REPS
      COMMON /LANCON/ ZERO, TENTH, EIGHTH, FOURTH, HALF, ONE, TWO,
     1                FOUR, TEN, ONE28, TWO56, FIVE12, ORTFAC, OVRFLW
C
C     EPS14 = EPS**(1/4)
C     TOL14 = EPS**(1/4)*SPREAD
C     TOL34 = EPS**(3/4)*SPREAD
C
      EPS14 = SQRT(REPS)
      TOL14 = EPS14*SPREAD
      TOL34 = REPS*TOL14
C
      IBUF = 0
      TOL = REPS*SPREAD
      RNEPS = RNM*EPS
C
C.... COMPUTE EIGENVECTORS OF T AND PUT IN THE BUFFER.
C
      DO 30 I=NEIG,1,-1
         IF (INFO(I) .GE. 0 .AND. IBUF .LT. NBUF) THEN
            IF (EVONLY .AND. INFO(I) .NE. 0) GO TO 30
            IBUF = IBUF + 1
            CALL DZERO(N,Y(1,I),1)
C
C....       EIG(I) ISOLATED
C
            M = 1
            L = J
            K = M
            CALL DZERO(J,S(1,IBUF),1)
            CALL GIVENS(K-M+1,L-M+1,ALF(M),BET(M),EIG(I),EPS1,
     1           S(M,IBUF),RESID,RAYCOR,IERR)
            IF (IERR .GT. 0) GO TO 900
            IF (RESID .LE. TOL14) GO TO 15
C
```

```
C....       CHECK THAT EIG(I) IS AN EIGENVALUE OF T
C
            DO 5 IDUM = M, L
               WORK(IDUM) = BET(IDUM)*BET(IDUM)
 5          CONTINUE
            ZETA = EIG(I)*(ONE - EPS14)
            NUL = NUMLES(ALF(M),WORK(M),ZETA,L-M+1,1,EPS)
            ZETA = EIG(I)*(ONE + EPS14)
            NUR = NUMLES(ALF(M),WORK(M),ZETA,L-M+1,1,EPS)
C
C....       EIG(I) IS NOT AN EIGENVALUE OF T, GOODBYE.
C
            IF (NUR .EQ. NUL) THEN
               IERR = IERR + 1024
               GO TO 900
            ENDIF
C
C....       EIG(I) IS O.K., NOW FIND A GOOD K
C
            DO 10 K=M+1,L
               CALL DZERO(J,S(1,IBUF),1)
               CALL GIVENS(K-M+1,L-M+1,ALF(M),BET(M),EIG(I),EPS1,
     1                     S(M,IBUF),RESID,RAYCOR,IERR)
               IF (IERR.GT.0) GO TO 900
               IF (RESID .LE. TOL14) GO TO 15
10          CONTINUE
C
C....       NO SUITABLE K WAS FOUND, GOODBYE.
C
            IERR = IERR + 512
            GO TO 900
C
C....       NOW WE CAN REFINE THE EIGENVECTOR
C
15          DO 20 LOOP = 1,10
               K = IDAMAX(L-M+1,S(M,IBUF),1)+M-1
               CALL DZERO(J,S(1,IBUF),1)
               CALL GIVENS(K-M+1,L-M+1,ALF(M),BET(M),EIG(I),EPS1,
     1                     S(M,IBUF),RESID,RAYCOR,IERR)
               IF (RESID.LE.TOL) GO TO 25
               EIG(I) = EIG(I) + RAYCOR
20          CONTINUE
C
C....       RESID FAILED TO DIMINISH, GOODBYE.
C
            IF (RESID .GT. TOL) THEN
               IERR = IERR + 2048
               GO TO 900
            ENDIF
C
25          INFO(I) = N*IDAMAX(J,S(1,IBUF),1)+J-1
         END IF
30    CONTINUE
C
C.... COMPUTE THE RITZ VECTORS AND PERFORM G-S ORTHOGONALIZATION.
C
      DO 50 I=1,J-1
C
C.....   RETRIEVING THE LANCZOS VECTOR AND PUT IT IN T
C
         CALL STORE(T,N,I,2)
         KBUF = 0
         DO 40 K=NEIG,1,-1
            IF (INFO(K) .GE. 0 .AND. KBUF .LT. IBUF) THEN
               KBUF = KBUF + 1
               SI = S(I,KBUF)
               IF (ABS(SI) .GT. EPS) CALL DAXPY(N,SI,T,1,Y(1,K),1)
            END IF
 40      CONTINUE
         IF (.NOT.EVONLY) THEN
            ZETOLD = -DDOT(N,QA,1,T,1)
            IF ( ABS(ZETOLD) .GT. EPS ) CALL DAXPY(N,ZETOLD,T,1,Q,1)
            ZETA = -DDOT(N,RA,1,T,1)
            IF ( ABS(ZETA) .GT. RNEPS ) CALL DAXPY(N,ZETA,T,1,R,1)
         END IF
50    CONTINUE
C
C.... ADD IN CONTRIBUTION OF QJ TO Y
C
      KBUF = 0
      DO 60 I=NEIG,1,-1
         IF (INFO(I) .GE. 0 .AND. KBUF .LT. IBUF) THEN
            KBUF = KBUF + 1
            IF (ABS(S(J,KBUF)) .LT. EPS1) INFO(I) = -INFO(I)
            CALL DAXPY(N,S(J,KBUF),Q,1,Y(1,I),1)
         END IF
```

```
60    CONTINUE
C
900   RETURN
      END
```

Subroutine GIVENS

This routine computes the eigenvector of a tridiagonal corresponding to an eigenvalue stored in THET. It first assumes a value for the bottom element of the eigenvector and solves for the rest using the Givens recurrence [28] going backward. This phase is terminated once the Kth element of S has been obtained. Then all the terms computed thus far are scaled to make S(K) unity. A similar procedure is performed starting from the top element and running the recurrence in the forward direction. Then the computed vector is normalized, and finally the residual and the Rayleigh correction to THETA are computed and returned in RES and COR, respectively. The algorithm scales S to avoid overflow whenever it becomes necessary.

```
C
      SUBROUTINE GIVENS(K,J,ALF,BET,THET,EPS,S,RES,COR,IERR)
C
C.... GIVENS RECURRENCE FOR COMPUTING EIGENVECTORS OF A TRIDIAGONAL T.
C
C.... INPUTS
C
C     K       INDEX OF THE RIGHT HAND SIDE E(K)
C     J       DIMENSION OF THE TRIDIAGONAL MATRIX
C     ALF(J)  DIAGONALS OF T
C     BET(J)  OFF-DIAGONALS OF T
C     THET    EIGENVALUE OF T
C     EPS     COMPUTER PRECISION
C
C.... OUTPUTS
C
C     S(J)    COMPUTED EIGENVECTOR
C     RES     NORM OF THE RESIDUAL
C     COR     RAYLEIGH CORRECTION FOR THET
C
C.... SUBROUTINES : DSCAL
C
      IMPLICIT DOUBLE PRECISION (A-H,O-Z)
      DIMENSION ALF(J),BET(J),S(J)
C
C.... OVRFLW IS THE MACHINE OVERFLOW THRESHOLD
C
      COMMON /LANCON/ ZERO, TENTH, EIGHTH, FOURTH, HALF, ONE, TWO,
     1                FOUR, TEN, ONE28, TWO56, FIVE12, ORTFAC, OVRFLW
C
C.... BACKWARD RECURRENCE
C
      BIG = SQRT(OVRFLW/FLOAT(J+1))
      RES = ZERO
C
      S(J) = ONE
      SUM2 = ONE
      IF ( K .LT. J ) THEN
         S(J-1) = -(ALF(J) - THET)*S(J)/BET(J)
         SUM2 = SUM2 + S(J-1)**2
      END IF
      DO 10 I = J-1,K+1,-1
         S(I-1) = -((ALF(I) - THET)*S(I) + BET(I+1)*S(I+1))/BET(I)
C
C....    SCALE TO AVOID OVERFLOW
C
         IF ( ABS(S(I-1)) .GT. BIG ) THEN
            F = ONE/S(I-1)
            S(I-1) = ONE
            CALL DSCAL(J-I+1,F,S(I),1)
            SUM2 = (SUM2*F)*F
         END IF
         SUM2 = SUM2 + S(I-1)**2
10    CONTINUE
C
C.... STOP EXECUTION GRACEFULLY IF S(K) IS EXACTLY ZERO
C
      IF (S(K) .EQ. ZERO) THEN
         IERR = IERR + 256
```

```
            RETURN
         ENDIF
         F = ONE/S(K)
         S(K) = ONE
         SUM2 = (SUM2*F)*F
         CALL DSCAL(J-K,F,S(K+1),1)
         IF ( K .LE. 1 ) GO TO 30
C
C.... FORWARD RECURRENCE
C
            X = ZERO
            S(1) = ONE
            SUM1 = ONE
            IF ( K .GT. 2 ) THEN
               S(2) = -((ALF(1) - THET)*S(1))/BET(2)
               SUM1 = SUM1 + S(2)**2
               DO 20 I = 2,K-2
                  S(I+1) = -((ALF(I) - THET)*S(I) + BET(I)*S(I-1))/BET(I+1)
                  IF ( ABS(S(I+1)) .GT. BIG ) THEN
                     F = ONE/S(I+1)
                     S(I+1) = ONE
                     CALL DSCAL(I,F,S,1)
                     SUM1 = (SUM1*F)*F
                  END IF
                  SUM1 = SUM1 + S(I+1)**2
20             CONTINUE
               X = BET(K-1)*S(K-2)
            END IF
            X = -(X + (ALF(K-1) - THET)*S(K-1))/BET(K)
C
C....       MATCH X WITH S(K)
C
            IF ( X .EQ. ZERO ) THEN
               RES = ONE
               RETURN
            END IF
            F = S(K)/X
            CALL DSCAL(K-1,F,S,1)
            SUM2 = SUM2 + SUM1*F**2
            RES = BET(K)*S(K-1)
C
C.... NORMALIZE S
C
30       F = ONE/SQRT(SUM2)
         CALL DSCAL(J,F,S,1)
         RES = RES*F + (ALF(K) - THET)*S(K)
         IF ( K .LT. J ) RES = RES + BET(K+1)*S(K+1)
         COR = S(K)*RES
         RES = ABS(RES)
C
         RETURN
         END
```

Subroutine SUBTJ

In theory, unreduced tridiagonal matrices do not have equal eigenvalues, but in practice they can have two eigenvalues that are so close to one another that they are indistinguishable in finite precision. This creates certain difficulties when computing eigenvectors with close eigenvalues. The problem is solved by working with a submatrix $\boldsymbol{T}_{l,m}$ of the tridiagonal $\boldsymbol{T}_j$. This routine obtains an estimate of the indices l and m that defines a submatrix for which EIG(I) is simple. INFO(I) = N×K + J stores the two indices J and K. J is the step at which the eigenvector corresponding to EIG(I) was computed and K is the index of the element of this eigenvector with largest absolute value. When INFO(I) is negative it indicates that the Ritz vector in Y(I) is converged.

```
C
         SUBROUTINE SUBTJ(ALF,BET,EIG,INFO,TOL,I,N,NEIG,J,L,K,M,
     1                    U,IERR)
C
C.... THIS ROUTINE SCANS BACK THROUGH THE CONVERGED EIGENVALUES
C     FOR COPIES OF EIG(I) TO DETERMINE THE SUBMATRIX T(M,L) AND
C     THE RIGHT HAND SIDE E(K) FOR THE GIVENS RECURRENCE.
C
C.... INPUTS
```

```
C
C      EIG(NEIG)      LIST OF THE CONVERGED EIGENVALUES
C      INFO(NEIG)     INFORMATION ABOUT THE EIGENVECTORS
C      TOL            TOLERANCE FOR FINDING COPIES OF EIG(I)
C      I              INDEX OF THE EIGENVALUE CONSIDERED
C      N              DIMENSION OF THE EIGENPROBLEM
C      NEIG           NUMBER OF EIGENVALUES IN EIG
C
C....  OUTPUTS
C
C      K              INDEX OF THE RIGHT HAND SIDE FOR GIVENS
C      L              INDEX OF THE LAST ELEMENT OF THE SUBMATRIX
C      M              INDEX OF THE FIRST ELEMENT OF THE SUBMATRIX
C
       IMPLICIT DOUBLE PRECISION (A-H,O-Z)
       DIMENSION ALF(1),BET(1),EIG(1),INFO(1),U(1)
C
       COMMON /LANCON/ ZERO, TENTH, EIGHTH, FOURTH, HALF, ONE, TWO,
     1                 FOUR, TEN, ONE28, TWO56, FIVE12, ORTFAC, OVRFLW
C
       L = J
       EIGI = EIG(I)
C
C....  CHECK SPECIAL CASE OF EIGI = ALL ALTERNATE ALF'S
C
       DO 10 JM = J,1,-2
          IF (ABS(ALF(JM)-EIGI) .GE. EPS*SPREAD) GO TO 20
10     CONTINUE
C
C....  FALSE RITZ VALUE ACCEPTED, STOP EXECUTION GRACEFULLY.
C
       IERR = IERR + 2048
       RETURN
C
C....  RUN RECURRENCE UNTIL "PIVOT = DIAGONAL" TO FIND M
C
20     U(JM)  = (ALF(JM)-EIGI)*(ONE - TEN*EPS)
       UMIN   = ABS(U(JM))
       IUMIN  = JM
       DO 30 M = JM-1,1,-1
          U(M)  = ALF(M)-EIGI-BET(M+1)**2/U(M+1)
          IF (ABS(U(M)).LT.TOL*EIGHTH) GO TO 40
          IF (ABS(U(M)).LE.UMIN) THEN
             UMIN = ABS(U(M))
             IUMIN = M
          ENDIF
30     CONTINUE
C
C....  TEMPORARILY
C
       M = IUMIN
C
C....  NOW SET K
C
40     IF (M .EQ. 1) THEN
          K = M
       ELSE
          K = M+1
       ENDIF
C
C....  NOW CHECK FOR CLOSE EIGENVALUES
C
       IF (M .GT. 1) THEN
          DO 50 I1 = I-1,1,-1
             IF (ABS(EIGI-EIG(I1)).LT.ONE28*EPS*SPREAD)
     1          GO TO 60
50        CONTINUE
C
          RETURN
C
60        KPREV = MOD(ABS(INFO(I1)),N)
          IF (KPREV .LE. K) K = KPREV+1
       ENDIF
C
C....  FIND L
C
       DO 200 I2 = I+1,NEIG
          IF (ABS(EIGI - EIG(I2)) .LT. TOL) GO TO 210
200    CONTINUE
210    IF (I2 .LE. NEIG) THEN
          L = ABS(INFO(I2))/N - 1
          IF (INFO(I2) .EQ. 0) L = MOD(ABS(INFO(I)),N)
       END IF
C
       RETURN
       END
```

Subroutine ANALZT

The goal of this routine is to obtain the smallest possible interval that contains a single eigenvalue of $\boldsymbol{T}_j$ (see [24] for more details). It makes full use of the corresponding information that was obtained at the previous step. When J = 2 it simply computes the eigenvalues of the 2 × 2 tridiagonal matrix. Phase I of the routine updates the data structure (THETA, BJ) containing those eigenvalues of $\boldsymbol{T}_{j-1}$ that are about to converge to eigenvalues of the big problem. Their residual bounds go in BJ. Phase II of the routine appends some new items to this data structure. Any converged eigenvalue is removed from THETA, put into EIG, and then explicitly deflated from the tridiagonal. This routine is called by LANSEL. For a more detailed description see [24].

```
C
      SUBROUTINE ANALZT(J,ALF,BET2,EIG,TAU,OLDTAU,RHO,INFO)
C
C.... THIS ROUTINE UPDATES SOME EIGENVALUES OF A TRIDIAGONAL T(J) USING
C     THE EIGENVALUES OF T(J-1).
C
C.... INPUTS
C
C     J           ORDER OF THE TRIDIAGONAL T.
C     ALF         DIAGONAL OF T.
C     BET2        SQUARES OF THE OFFDIAGONAL TERMS, BET2(1) = ZERO
C     EIG         ARRAY OF CONVERGED EIGENVALUES
C     TAU         ORTHOGONALITY ESTIMATE OF RITZ VECTORS AT STEP J
C     OLDTAU      ORTHOGONALITY ESTIMATE OF RITZ VECTORS AT STEP J-1
C     RHO         WORKING ARRAY USED IN DEFLAT
C     INFO        INFORMATION ARRAY ABOUT EIGENVECTORS OF T
C
C.... INTERNAL VARIABLES
C
C     THET        EXTERIOR EIGENVALUES OF T, NEARLY CONVERGED
C                 RITZ VALUES. THET(1)=LEFTMOST, THET(NDST)=RIGHTMOST.
C     NDST        SIZE OF THET AND BJ
C     BJ          ERROR BOUND ON THET
C                 BJ(I) IS SET TO -1 IF THET(I) DISAPPEARS.
C     NBD         CONTAINS L AND R IN THE TEXT.
C     SPREAD      THET(NDST) - THET(1)
C     EPS         PRECISION OF ARITHMETIC OPERATIONS
C     IP          IP =1 FOR UPDATING LEFT END, IP = 2 FOR THE RIGHT END.
C     INC         INC=1 FOR UPDATING LEFT END, INC=-1 FOR THE RIGHT END.
C     IS          STARTING INDEX (EITHER 1 OR NDST)
C     START       LEFT BOUND ON EIGENVALUES (INC=1), RIGHT BOUND (INC=-1)
C     PROBE       THE OUTER END OF THE NEXT SUBINTERVAL TO BE UPDATED.
C     INDXOK      TRUE, IF THERE ARE I-INC RITZ VALUES EXTERIOR TO THE
C                 NEW THET(I).
C
C.... SUBROUTINES: DEFLAT,MOVE1,NEWCOR,NUMLES
C
      IMPLICIT DOUBLE PRECISION(A-H,O-Z)
      PARAMETER (NMAX = 128)
C
      DIMENSION ALF(1),BET2(1),EIG(1),TAU(1),OLDTAU(1),RHO(1),INFO(1)
      LOGICAL INSERT,INDXOK,APPEND
C
      COMMON /RDATA/ RNM,RNM2,SPREAD,TOL,EPS,EPS1,REPS
      COMMON /ATDATA/ THET(NMAX),BJ(NMAX),WINDOW,NBD(2),NDST
      COMMON /IDATA/ EIGL,EIGR,NEIG
      COMMON /LANCON/ ZERO, TENTH, EIGHTH, FOURTH, HALF, ONE, TWO,
     1                FOUR, TEN, ONE28, TWO56, FIVE12, ORTFAC, OVRFLW
C
      IF (J.LE.1) RETURN
      IF (J.EQ.2) THEN
         NDST = 16
         WINDOW = FOURTH/SQRT(REPS)
         THET(1) = (ALF(1) + ALF(2) - SQRT(FOUR*BET2(2) +
     1             (ALF(1) - ALF(2))**2))*HALF
         THET(NDST) = ALF(1) + ALF(2) - THET(1)
         BJ(1) = SQRT(RNM2/(ONE + BET2(2)/(THET(1) - ALF(1))**2))
         BJ(NDST) = SQRT(RNM2/(ONE + BET2(2)/(THET(NDST) - ALF(1))**2))
         NBD(1) = 1
         NBD(2) = NDST
         SPREAD = THET(NDST) - THET(1)
         RETURN
      END IF
      SPREAD = THET(NDST) - THET(1)
```

```
      TOL = TWO*REPS*SPREAD
      W = WINDOW*TOL
C
C.... BEGIN PHASE 1.
C
C.... LOOP FOR LEFT END, THEN RIGHT
C
      DO 100 IP = 1,2
         INC = 3 - 2*IP
         IS = (NDST-1)*IP - (NDST-2)
         I = IS
         INSERT = .FALSE.
         START = (THET(I) + ALF(J) - INC*SQRT(BET2(J)*FOUR +
     1           (ALF(J) - THET(I))**2))*HALF
         PROBE = THET(I) - INC*BJ(I)
         INDXOK = NUMLES(ALF,BET2,PROBE,J,INC,EPS) .EQ. 0
C
         DO 50 IDUMMY =1,NDST
C
            IF (I-NBD(IP).EQ.INC) GO TO 100
C
C....       EXAMINE I-TH SUBINTERVAL
C
            IF (INDXOK) THEN
               IF (INSERT) THEN
                  START = THET(I)
                  THET(I) = START + INC*MIN(B**2/
     1               ABS(START-THET(I-INC)),B)
               ELSE
                  IF (INT(SIGN(ONE,PROBE-START)).EQ.INC) START = PROBE
               END IF
C
C....          CHECK FOR DISJOINT SUBINTERVALS
C
               IF (I .EQ. NBD(IP)) THEN
                  PROBE=THET(I)+INC*(THET(NBD(2))-THET(NBD(1)))/(FOUR*J)
               ELSE
                  PROBE = THET(I+INC) - INC*BJ(I+INC)
               END IF
               IF (INT(SIGN(ONE,PROBE-THET(I))).EQ.INC) THEN
C
C....             CHECK FOR AN EXTRA RITZ VALUE
C
                  K = NUMLES(ALF,BET2,PROBE,J,INC,EPS)
C
                  IF (K.LT. ABS(I-IS+INC)) THEN
C
C....                THET(I) DISAPPEARS
C
                     BJ(I) = -ONE
                  ELSE
C
C....                RECORD INDXOK FOR NEXT LOOP. USE REFINED BOUNDS.
C
                     IF (.NOT.INSERT) THEN
                        B = BJ(I)
                        INDXOK = (K .LE. ABS(I-IS+INC))
                        BND = MIN(B**2/ABS(PROBE-THET(I)),B)
                        IF (INDXOK.AND.BND.LT.ABS(THET(I)-START)) THEN
                           START = THET(I) - INC*BND
                        END IF
                     END IF
                  END IF
               END IF
            ELSE
C
C....          PREPARE FOR AN INTRUDING RITZ VALUE
C
               IF ((IS.EQ.NBD(IP).OR.BJ(NBD(IP)-INC).LT.W).AND.
     1            NBD(2)-NBD(1).GT.1) NBD(IP) = NBD(IP) + INC
               CALL MOVE1(THET,I,NBD(IP),-INC,PROBE)
               CALL MOVE1(BJ,I,NBD(IP),-INC,TWO*TOL)
               INSERT = .TRUE.
               INDXOK = .TRUE.
            END IF
C
            IF (BJ(I).GT.TOL) THEN
C
C....       USE NEWTON ITERATION TO FIND NEW THET(I)
C
               CALL NEWCOR(ALF,BET2,START,THET,BJ,INC,I,J,NDST)
C
            END IF
C
            IF (BJ(I).LT.0) THEN
C
```

```
C....            THET(I) DISAPPEARS
C
                 CALL MOVE1(THET,NBD(IP),I,INC,ZERO)
                 CALL MOVE1(BJ,NBD(IP),I,INC,ZERO)
                 NBD(IP) = NBD(IP) - INC
                 INSERT = .FALSE.
                 INDXOK = .TRUE.
C
C
                 I = I - INC
               END IF
               I = I + INC
50         CONTINUE
C
C....      END OF PHASE 1
C
100    CONTINUE
C
C.... BEGIN PHASE 2.
C
C.... APPEND MORE RITZ VALUES AND CHECK FOR CONVERGED RITZ VALUES.
C
       DO 200 IP = 1,2
           INC = 3 - 2*IP
           IS  = (NDST-1)*IP - (NDST-2)
           I = IS
           DO 150 IDUMMY = 1,J
               NREM = J - NBD(1) - ((NDST+1) - NBD(2))
               IF ((I-NBD(IP))*INC.GT.0) GO TO 200
               APPEND = I.EQ.NBD(IP).AND.(BJ(I).LT.W.OR.(J.EQ.4
     1                .AND.NBD(IP).EQ.IS)).AND.NREM.GT.0
               IF (APPEND) THEN
                 START = THET(I) + FLOAT(INC)*BJ(I)
                 PROBE = INC*(THET(NBD(2)) - THET(NBD(1)))/NREM
               END IF
               IF (BJ(I) .LE. TOL) THEN
C
C....            APPLY QR ALGOR. TO DEFLATE
C
                 CALL DEFLAT(ALF,BET2,THET(I),J)
C
C....            INSERT THET(I) INTO EIG
C
                 NEIG = NEIG + 1
                 IF (IP.EQ.1) THEN
                     EIGL = MAX(EIGL,THET(I))
                 ELSE
                     EIGR = MIN(EIGR,THET(I))
                 END IF
                 EIG(NEIG) = THET(I)
                 INFO(NEIG) = 0
C
C....            REMOVE STABILIZED RITZ VALUES
C
                 CALL MOVE1(THET,NBD(IP),I,INC,ZERO)
                 CALL MOVE1(BJ,NBD(IP),I,INC,ZERO)
                 NBD(IP) = NBD(IP) - INC
                 I = I - INC
               END IF
               IF (APPEND.AND.NBD(2)-NBD(1).GT.1) THEN
                 T = START + PROBE
                 NBD(IP) = NBD(IP) + INC
                 IK = ABS(IS - NBD(IP))
                 DO 110 IDUM = 1,J
                     IF (NUMLES(ALF,BET2,T,J,INC,EPS).NE.IK) GO TO 120
                     T = T + PROBE
110              CONTINUE
120              THET(NBD(IP)) = T
                 START = T - PROBE
                 CALL NEWCOR(ALF,BET2,START,THET,BJ,INC,NBD(IP),J,NDST)
C
               END IF
               IF (J.GT.NDST.AND.I.EQ.NBD(IP).AND.I.NE.IS.AND.BJ(I).GT.
     1            BJ(I-INC).AND.BJ(I-INC).GT.W) NBD(IP) = NBD(IP) - INC
               I = I + INC
150        CONTINUE
200    CONTINUE
C
C.... RE-ESTABLISH AN END MARKER, IF NECESSARY, AT EARLY STAGE
C
       DO 300 IP = 1,2
           INC = 3 - 2*IP
           IS  = (NDST-1)*IP - (NDST-2)
           IF (NBD(IP).EQ.IS-INC) THEN
               THET(IS) = THET(NBD(3-IP))
               BJ(IS) = BJ(NBD(3-IP))
```

```
            NBD(IP) = IS
            NBD(3-IP) = NBD(3-IP) + INC
         END IF
300   CONTINUE
C
      RETURN
      END
```

Subroutine NEWCOR

The object of this routine is to find an eigenvalue of T_j in a given interval. It uses a combination of bisection and Newton's method. For a detailed description see [24]. INDEX points to the position in THETA that will eventually hold the computed eigenvalue. On entry THETA(INDEX) contains the eigenvalue of T_{j-1} that is also one end of the interval; ZETA is the other. ZETA is also the starting value for Newton's method. If this interval is too large then the algorithm performs a few steps of bisection to obtain a better estimate, ZETA, to the desired eigenvalue. The eigenvalue counts are obtained by calling NUMLES. Then Newton's iteration is used to compute the eigenvalue to the full computer precision. The Newton correction is computed using a recurrence that is given in [24]. The convergence of Newton's method is guaranteed by deflating the eigenvalues exterior to ZETA implicitly. NEWCOR is called only by ANALZT.

```
C
      SUBROUTINE NEWCOR(ALF,BET2,ZETA,THET,BJ,INC,INDX,J,NDST)
C
C.... COMPUTES EXTERIOR EIGENVALUES OF A TRIDIAGONAL USING A
C     COMBINATION OF BISECTIONS AND NEWTON'S METHOD
C
C.... INPUT
C
C     ALF         DIAGONAL OF T
C     BET2        SQUARES OF THE OFFDIAGONAL TERMS, BET2(1) = ZERO
C     SPREAD      THET(NDST) - THET(1)
C     INC         INC=1 FOR UPDATING LEFT END, INC=-1 FOR THE RIGHTEND.
C     INDX        INDEX OF TO-BE-UPDATED THET.
C     J           ORDER OF THE TRIDIAGONAL T.
C     NDST        SIZE OF THET AND BJ
C
C.... INPUT/OUTPUT
C
C     ZETA        EXTERIOR BOUND FOR EIGENVALUE OF T IN THET[INDX]
C     THET        EXTERIOR EIGENVALUES OF T, NEARLY CONVERGED RITZ VALUES;
C                 THET(1)=LEFTMOST, THET(NDST)=RIGHTMOST.
C     BJ          ERROR BOUND ON THET
C
C     SUBROUTINES: NUMLES,QLBOT
C
      IMPLICIT DOUBLE PRECISION(A-H,O-Z)
      PARAMETER (MAXBIS = 15, MAXNEW = 40)
      DIMENSION ALF(1),BET2(1),THET(1),BJ(1)
C
      COMMON /RDATA/ RNM,RNM2,SPREAD,TOL,EPS,EPS1,REPS
      COMMON /LANCON/ ZERO, TENTH, EIGHTH, FOURTH, HALF, ONE, TWO,
     1                FOUR, TEN, ONE28, TWO56, FIVE12, ORTFAC, OVRFLW
C
      ZOLD = ZETA
      IF (J .EQ. 1) THEN
         ZETA = THET(INDX)
         THET(INDX) = ALF(J)
         BJ(INDX) = SQRT(RNM2)
         RETURN
      END IF
C
C.... PERFORM BISECTION FOR AN IMPROVED ZETA
C
      IS = ((NDST+1) - (NDST-1)*INC)/2
      FACT = FIVE12*FLOAT(J)*LOG(FLOAT(J))
      WIDTH = (THET(INDX) - ZETA)*HALF
      IF (WIDTH .EQ. ZERO) THEN
         WIDTH = BJ(INDX)*HALF
      ENDIF
      IOLD = ABS(IS - INDX)
```

```
      DO 10 IDUMMY = 1,MAXBIS
         IF (ABS(WIDTH)*FACT .LE. ABS(THET((NDST+1)-IS)-THET(INDX)))
     1      GO TO 20
         ZNEW = ZETA + WIDTH
         INEW = NUMLES(ALF,BET2,ZNEW,J,INC,EPS)
         WIDTH = WIDTH*HALF
         IF (INEW .EQ. IOLD) THEN
            ZETA = ZNEW
         ENDIF
C
10    CONTINUE
20    CONTINUE
C
      DO 50 IDUMMY = 1,MAXNEW
         U = ALF(1) - ZETA
         IF ( U .EQ. ZERO ) U = TENTH*EPS*BET2(2)
         RAT = ONE/U
         SUM = RAT
         DO 30 I = 2,J
            H   = BET2(I)/U
            U = ALF(I) - ZETA - H
            IF ( U .EQ. ZERO ) U = TENTH*EPS*(H + BET2(I))
            RAT = ( ONE + H*RAT )/U
            SUM = SUM + RAT
30       CONTINUE
         BOT2 = U*SUM
C
C....    DEFLATION
C
         DO 40 I = IS,INDX-INC,INC
            DEL = ZETA - THET(I)
            IF (ABS(DEL).LT.EPS*ABS(ZETA)) THEN
               DEL = EPS*ABS(ZETA)
            ENDIF
            SUM = SUM + ONE/DEL
40       CONTINUE
C
C....    CHECK FOR CONVERGENCE
C
         ZNEW = ZETA + ONE/SUM
         ZETA = ZNEW
         IF (SPREAD+TENTH/SUM.EQ.SPREAD.OR.
     1      FLOAT(INC)/SUM.LT.ZERO) THEN
            GO TO 60
         ENDIF
50    CONTINUE
60    CONTINUE
C
      CALL QLBOT(ALF,BET2,ZETA,BOT2,J)
C
      ZNEW = THET(INDX)
      THET(INDX) = ZETA
      ZETA = ZNEW
      BJ(INDX) = SQRT(RNM2*BOT2)
C
      RETURN
      END
```

Subroutine DEFLAT

The subroutine DEFLAT deflates the tridiagonal matrix T_j using an eigenvalue THETA. One step of the QR algorithm is used to perform the deflation. Each of the last few QR rotations will be almost a row and column interchange. For a more detailed description of the algorithm see [24]. ANALZT is the only routine that calls DEFLAT.

```
C
      SUBROUTINE DEFLAT(ALF,BET2,THET,J)
C
C.... THIS ROUTINE PERFORMS DEFLATION OF T USING SHIFT THET.
C
C.... INPUTS
C
C     ALF(J)    DIAGONALS OF T
C     BET2(J)   SQUARES OF OFF-DIAGONALS OF T
C     THET      EIGENVALUE OF T TO BE DEFLATED EXPLICITLY
C     J         DIMENSION OF THE TRIDIAGONAL MATRIX
C
C.... OUTPUTS
C
C     ALF(J-1)  MODIFIED DIAGONALS OF T
```

```
C       BET2(J-1) MODIFIED OFF-DIAGONALS OF T
C
        IMPLICIT DOUBLE PRECISION(A-H,O-Z)
        PARAMETER (MAXITR = 3)
        DIMENSION ALF(J),BET2(J)
C
        COMMON /RDATA/ RNM,RNM2,SPREAD,TOL,EPS,EPS1,REPS
        COMMON /LANCON/ ZERO, TENTH, EIGHTH, FOURTH, HALF, ONE, TWO,
     1                  FOUR, TEN, ONE28, TWO56, FIVE12, ORTFAC, OVRFLW
C
C.... PWK ALGORITHM
C
        DO 200 LOOP=1,MAXITR
           C = ONE
           S = ZERO
           G = ALF(1) - THET
           P = G**2
           DO 100 I=1,J-1
              B = BET2(I+1)
              R = P + B
              BET2(I) = S*R
              OLDC = C
              C = P/R
              S = B/R
              OLDG = G
              A = ALF(I+1)
              G = C*(A - THET) - S*OLDG
              ALF(I) = OLDG + (A - G)
              IF ( C .EQ. ZERO ) THEN
                 P = OLDC*B
              ELSE
                 P = G*(G/C)
              END IF
100        CONTINUE
           BET2(J) = S*P
           ALF(J) =  G + THET
           IF (BET2(J) .LE. EPS*SPREAD) THEN
              GOTO 300
           ENDIF
200     CONTINUE
300     J = J - 1
C
        RETURN
        END
```

Function NUMLES

This routine performs the LDL^T factorization of the tridiagonal matrix T_j to obtain eigenvalue counts for $T_j - \zeta$. By Sylvester's inertia theorem [21], D has the same signature as $T_j - \zeta$. L is not needed and therefore is not computed. The diagonal elements of D are not preserved, but a record of the number of negative or positive terms is kept and returned in NUMLES. NUMLES is called by ANALZT and NEWCOR.

```
C
        INTEGER FUNCTION NUMLES(ALF,BET2,ZETA,N,INC,EPS)
C
C.... ROUTINE TO PERFOM THE SPECTRUM SLICING OF A TRIDIAGONAL MATRIX.
C
C       IF INC =  1, NUMLES RETURNS THE NUMBER OF EIGENVALUES BELOW ZETA.
C       IF INC = -1, NUMLES RETURNS THE NUMBER OF EIGENVALUES ABOVE ZETA.
C
C.... INPUTS
C
C       ALF(N)     DIAGONALS OF T
C       BET2(N)    SQUARE OF THE OFF-DIAGONALS OF T
C       ZETA       THE SHIFT TO BE APPLIED TO T
C       N          DIMENSION OF THE TRIDIAGONAL MATRIX
C       INC        INDEX TO INDICATE ABOVE OR BELOW
C       EPS        COMPUTER PRECISION
C
C.... OUTPUTS
C
C       NUMLES     THE NUMBER OF EIGENVALUES ABOVE/BELOW ZETA.
C
        IMPLICIT DOUBLE PRECISION (A-H,O-Z)
        DIMENSION ALF(1),BET2(1)
C
```

```
      COMMON /LANCON/ ZERO, TENTH, EIGHTH, FOURTH, HALF, ONE, TWO,
     1                FOUR, TEN, ONE28, TWO56, FIVE12, ORTFAC, OVRFLW
C
      SAVE = BET2(1)
      BET2(1) = ZERO
      DEL = ONE
      K = 0
      DO 10 J=1,N
         DEL = (ALF(J) - ZETA) - BET2(J)/DEL
         IF ( DEL .EQ. ZERO ) DEL = EPS*BET2(J+1)*INC
         IF ( DEL .LT. ZERO ) K = K + 1
10    CONTINUE
C
      NUMLES = K
      IF ( INC .LT. 0 ) NUMLES = N - K
      BET2(1) = SAVE
C
      RETURN
      END
```

Subroutine QLBOT

This routine computes the bottom element of the eigenvector of the tridiagonal matrix $\mathbf{T}_j$ corresponding to the eigenvalue contained in THET. The algorithm performs a step of the QL factorization. The product of the sines of the rotation angles is the desired bottom element and is returned in BOT. NEWCOR is the only routine that calls QLBOT.

```
C
      SUBROUTINE QLBOT(ALF,BET2,THET,BOT,J)
C
C.... COMPUTE THE BOTTOM ELEMENT OF THE NORMALIZED EIGENVECTOR OF A
C     TRIDIAGONAL MATRIX CORRESPONDING TO EIGENVALUE THET.
C
C.... INPUTS
C
C     ALF(J)    DIAGONALS OF T
C     BET2(J)   SQUARE OF THE OFF-DIAGONALS OF T
C     THET      EIGENVALUE OF T
C     J         DIMENSION OF THE TRIDIAGONAL MATRIX
C
C.... OUTPUTS
C
C     BOT       BOTTOM ELEMENT OF THE NORMALIZED EIGENVECTOR
C
      IMPLICIT DOUBLE PRECISION(A-H,O-Z)
      DIMENSION ALF(1),BET2(1)
C
      COMMON /LANCON/ ZERO, TENTH, EIGHTH, FOURTH, HALF, ONE, TWO,
     1                FOUR, TEN, ONE28, TWO56, FIVE12, ORTFAC, OVRFLW
C
      BOT = ONE
      C = ONE
      S = ZERO
      G = ALF(J) - THET
      P = G**2
C
      DO 100 I=J-1,1,-1
         B = BET2(I+1)
         R = P + B
         OLDC = C
         C = P/R
         S = B/R
         OLDG = G
         A = ALF(I)
         G = C*(A - THET) - S*OLDG
         IF ( C .EQ. ZERO ) THEN
            P = OLDC*B
         ELSE
            P = G**2/C
         END IF
         BOT = BOT*S
100   CONTINUE
C
      RETURN
      END
```

Subroutine MOVE1

This routine inserts the value of T into the Kth location of the array Y. Elements of Y are shifted up or down depending on the sign of MINC. MOVE1 is only called by ANALZT for data management.

```
C
      SUBROUTINE MOVE1(Y,K,L,MINC,T)
C
C.... MOVES THE CONTENT OF Y TO OPEN A SPACE FOR T.
C
C.... INPUT/OUTPUT
C
C     Y      THE ARRAY TO TO BE REORGANIZED
C     K      THE POSITION IN Y OF THE NEW ELEMENT T
C     L      END OF THE DATA IN Y
C     MINC   THE INCREMENT +1 OR -1
C     T      THE NEW ELEMENT TO BE INSERTED
C
      IMPLICIT DOUBLE PRECISION(A-H,O-Z)
      DIMENSION Y(1)
C
      DO 100 I=L,K-MINC,MINC
         Y(I) = Y(I+MINC)
100   CONTINUE
      Y(K) = T
C
      RETURN
      END
```

REFERENCES

Section 10.1

1. K. J. Bathe, *Finite Element Procedures in Engineering Analysis*. Englewood Cliffs, N. J.: Prentice-Hall, 1982.
2. B. Noble, *Applied Linear Algebra*. Englewood Cliffs, N.J.: Prentice-Hall, 1969.

Section 10.2

3. E. L. Wilson, "The Static Condensation Algorithm," *International Journal for Numerical Methods in Engineering*, 8 (1974), 199–203.

Section 10.4

4. B. M. Irons, "Eigenvalue Economisers in Vibration Problems," *Journal of the Royal Aeronautical Society*, 67 (1963), 526.
5. B. M. Irons, "Structural Eigenvalue Problems: Elimination of Unwanted Variables," *AIAA Journal,* 3 (1965), 961.
6. R. J. Guyan, "Reduction of Stiffness and Mass Matrices," *AIAA Journal*, 3 (1965), 380.
7. R. D. Henshell and J. H. Ong, "Automatic Masters for Eigenvalue Economization," *Earthquake Engineering and Structural Dynamics,* 3 (1975), 375–383.

Section 10.5

8. K. J. Bathe, *Finite Element Procedures in Engineering Analysis*. Englewood Cliffs, N. J.: Prentice-Hall, 1982.
9. E. L. Wilson, "Numerical Methods for Dynamic Analysis," *International Symposium on Numerical Methods in Offshore Engineering,* Swansea, January 11–15, 1977.

Section 10.6

10. J. J. Dongarra, C. B. Moler, J. R. Bunch, and G. W. Stewart, *LINPACK Users' Guide*, SIAM, Philadelphia, 1979.

11. T. Ericsson and A. Ruhe, "The Spectral Transformation Lanczos Method for the Numerical Solution of Large Sparse Generalized Symmetric Eigenvalue Problems," *Mathematics of Computation*, 35 (1980), 1251–1268.

12. G. Golub, R. Underwood, and J. H. Wilkinson, "The Lanczos Algorithm for the Symmetric $Ax = \lambda Bx$ Problem," Tech. Rep. STAN-CS-72-720, Computer Science Department, Stanford University, 1972.

13. J. Grcar, "Analyses of the Lanczos Algorithm and of the Approximation Problem in Richardson's Method," Ph.D. Thesis, University of Illinois at Urbana-Champaign, 1981.

14. C. Lanczos, "An Iteration Method for the Solution of the Eigenvalue Problem of Linear Differential and Integral Operators," *Journal of Research of the National Bureau of Standards*, 45 (1950), 255–281.

15. B. Nour-Omid, B. N. Parlett, and R. L. Taylor, "Lanczos versus Subspace Iteration for Solution of Eigenvalue Problems," *International Journal for Numerical Methods in Engineering*, 19 (1983), 859–871.

16. B. Nour-Omid and B. N. Parlett, "How to Implement the Spectral Transformation," Tech. Rep. PAM-224, Center for Pure and Applied Mathematics, University of California, Berkeley, 1984.

17. B. Nour-Omid and R. W. Clough, "Dynamic Analysis of Structures Using Lanczos Coordinates," *Earthquake Engineering and Structural Dynamics*, 12 (1984), 565–577.

18. C. C. Paige, "Computational Variants of the Lanczos Method for the Eigenproblem," *Journal of the Institute for Mathematics and its Applications*, 10 (1972), 373–381.

19. C. C. Paige, "Error Analysis of the Lanczos Algorithm for Tridiagonalizing a Symmetric Matrix," *Journal of the Institute for Mathematics and its Applications*, 18 (1976), 341–349.

20. C. C. Paige, "Accuracy and Effectiveness of the Lanczos Algorithm for Symmetric Eigenproblems," *Linear Algebra and its Applications*, 34 (1980), 235–258.

21. B. N. Parlett, *The Symmetric Eigenvalue Problem*. Englewood Cliffs, N.J.: Prentice-Hall, 1980.

22. B. N. Parlett and D. Scott, "The Lanczos Algorithm with Selective Orthogonalization," *Mathematics of Computation*, 33, no. 145 (1979), 217–238.

23. B. N. Parlett, H. D. Simon, and L. M. Stringer, "On Estimating the Largest Eigenvalue with the Lanczos Algorithm," *Mathematics of Computation*, 38, no. 157 (1982), 153–165.

24. B. N. Parlett and B. Nour-Omid, "The Use of Refined Error Bounds When Updating Eigenvalues of Tridiagonals," *Linear Algebra and its Applications*, 68 (1985), 179–219.

25. R. L. Taylor, private communication, 1978.

26. D. S. Scott, "Analysis of the Symmetric Lanczos Process," Tech. Rep. ERL-M78/40, Electronics Research Laboratory, University of California, Berkeley, 1978.

27. H. D. Simon, "The Lanczos Algorithm with Partial Reorthogonalization," *Mathematics of Computation*, 42, no. 165 (1984), 115–142.

28. H. D. Wilkinson, *The Algebraic Eigenvalue Problem*. Oxford: Clarendon, 1965.

29. E. L. Wilson, M. Yuan, and J. M. Dickens, "Dynamic Analysis by Direct Superposition of Ritz Vectors," *Earthquake Engineering and Structural Dynamics*, 10 (1982), 813–821.

11

DLEARN—A Linear Static and Dynamic Finite Element Analysis Program

Thomas J. R. Hughes, Robert M. Ferencz, and Arthur M. Raefsky

11.1. INTRODUCTION

This chapter describes DLEARN, a finite element code for linear static and dynamic analysis. For instructional purposes it may be considered a "first example" in that the use of auxiliary disk storage to reduce memory requirements is omitted, and blocking (segmenting large jobs via auxiliary storage) and overlay structures are avoided. Also, the class of capabilties—linear static and dynamic analysis—is relatively simple. Nevertheless, many sophisticated procedures are employed (e.g., compacted column solution and data structures, dynamic storage allocation, a memory manager, element routines, data generation routines, etc.) that are used in more advanced situations.

An important feature of DLEARN is that it has been written to be an integral constituent of this book. Methods and data-processing techniques developed throughout the first 10 chapters are employed in DLEARN. The names of variables and arrays are kept as close as possible to those used in the development of the theory. It is believed that this will expedite the learning process. Finite element methods find their ultimate manifestations in computer programs. Unless the student comprehends the intricacies of finite element programs, he or she will never fully appreciate the finite element method.

The major features and capabilities of DLEARN are listed below along with relevant subsections of the Input Instructions (these are denoted by "Sec. In." followed by a decimal number):

- Modular program structure to enhance readability and extendability.
- Memory pointer dictionary printed with each execution to provide storage information and illustrate the concept of dynamic storage allocation.
- Simple, but labor-saving, nodal generation capabilities in one, two, and three

dimensions (Sec. In.5.0) as well as element generation capabilities (Sec. In.10.0).

- Complex loadings can be synthesized by combination of multiple load vectors and load-time functions (Secs. In.7.0 and In.8.0).
- Static analysis with full stress recovery (Sec. In.2.0).
- Dynamic analysis with an α-method predictor/corrector algorithm allowing calculation of consistent initial accelerations (Sec. In.2.0). Includes Rayleigh viscous damping with mixed implicit-explicit element groups (Sec. In.10.0).
- Four-node quadrilaterial, elastic continuum element, with plane stress, plane strain, and torsionless axisymmetric capabilities. Includes a mean-dilatational $\bar{\boldsymbol{B}}$ capability for applications to nearly incompressible media. (Sec. In.10.1).
- Three-dimensional. elastic truss element (Sec. In.10.2).

DLEARN is not intended to be a "full-service" analysis package with a broad range of capabilities. For instance, the number of element types included has been purposely limited. However, major emphasis has been placed on providing a flexible and comprehensible program structure to which the student or researcher can profitably add new features of his or her own choosing. Moreover, the code has been structured around techniques having natural extensions to nonlinear analysis.

DLEARN may be used for problem solution in finite element courses and serves the following additional purposes:

1. It is felt that the first step in learning to implement finite element methods is to add a new element to an existing program. In this regard, DLEARN is written so that it is very easy to add new elements. (Instructions are given in Sec.11.4 entitled "Adding an Element to DLEARN.") Students may elect to perform projects of this kind. For this reason, only two sets of element routines are included with the listing so that many standard element types (e.g., three-dimensional elasticity; beam, plate, and shell elements; heat conduction elements; etc.) are available for project topics.

2. DLEARN makes available to students a good, first, finite element "software package." The subroutines are efficiently and simply coded and, as the student becomes more advanced, the subroutines may be used as a set of building blocks for more-sophisticated applications. With this idea in mind it was decided to include a very efficient equation solver in DLEARN, which employs a compacted column-storage mode.

The reader is forwarned that the level of documentation is not considered self-sufficient (nor is it intended to be) for fully understanding all aspects of the program but rather is supplementary to the remainder of the text.

11.2. DESCRIPTION OF CODING TECHNIQUES USED IN DLEARN

In keeping with contemporary programming practice, DLEARN has been written with the objectives of readability, maintainability and extendability. The 1000-line subroutines of an earlier generation of finite element codes have been eschewed in favor of compact subroutines with a limited number of tasks to perform. While large subroutines have a place in codes dedicated to efficient solution of large-scale problems, we believe smaller program units will be more easily comprehended by students first learning how to implement finite element methodologies. To this end, utility subroutines are widely used to remove "in-line" code, which, though faster, might obscure the algorithmic outline of a procedure. To understand DLEARN and ultimately extend its capabilities, the user must become familiar with the data structure.

Definition. The ***data structure*** is the mode of storage, location, and history of existence of the data employed.

In DLEARN, all the data are stored in central ("core") memory in the unnamed common block COMMON A(MTOT), where the size parameter MTOT is set in the main program. Many operating systems, such as those for Control Data Corporation (CDC) and Cray Research computers, allow greater run-time flexibility in memory allocation if the storage vector A is kept in an unnamed COMMON block. This fact is responsible for the commonly made remark that finite element programs store data in "blank common." On virtual memory machines such as the Digital Equipment Corporation (DEC) VAX-11/7xx series, however, no penalty is incurred by using a named common block. The basic features of the data structure are:

Compacted column architecture This feature is engendered by the method of equation solving employed. The solver is called variously a "profile," "skyline," or "active column" solver. Since, in large problems, the dominant portion of storage is usually that devoted to the effective mass matrix $\boldsymbol{M}^*$, the storage scheme for $\boldsymbol{M}^*$ is the most important feature of the storage.

Dynamic storage allocation Because of the variable sizes of the arrays needed in different problems, it would be very inefficient to "fix" the dimensioning. In the dynamic storage allocation concept the dimensions of arrays are set in the program at the time of execution.

Element groups Many features of finite element computer programs are the same even though the intended applications may be quite different. To some extent this is also true of the individual finite element subroutines. However, the data structure and options available may vary considerably from one element to another. To accommodate these differences, the elements are read in, stored and operated upon in groups. The data structure for the group is set up via dynamic storage allocation in individual element subroutines.

11.2.1 Compacted Column Storage Scheme

Assume the symmetric effective mass matrix $\boldsymbol{M}^*$ has dimension 8×8 and has zero and nonzero elements indicated by 0 and X, respectively, in the figure on page 634. The dashed line enveloping the nonzero terms is sometimes called the "profile" or "skyline."

$$
\boldsymbol{M}^* = [M^*_{ij}] = \begin{bmatrix}
X & X & 0 & 0 & 0 & 0 & 0 & X \\
 & X & X & 0 & X & 0 & 0 & 0 \\
 & & X & X & X & 0 & 0 & 0 \\
 & & & X & 0 & X & 0 & X \\
 & & & & X & X & 0 & 0 \\
 & & & & & X & X & 0 \\
 & & & & & & X & X \\
 & & & & & & & X
\end{bmatrix}
$$

$\boldsymbol{M}^*$ is stored in a one-dimensional array, ALHS, column-wise beginning with the first nonzero element in each column and ending with the diagonal term. The locations of the diagonal terms in ALHS are stored in a one-dimensional integer array IDIAG. The dimension of IDIAG is NEQ, the number of equations (e.g., for $\boldsymbol{M}^*$, NEQ = 8). The dimension of ALHS is NALHS, the sum of the "column heights." The column height is the number of terms in a column beginning with the first nonzero term and ending with the diagonal term (e.g., for the above matrix $\boldsymbol{M}^*$, column 1 has height 1, column 4 has height 2, column 6 has height 3, etc.). NALHS for $\boldsymbol{M}^*$ is 24. The arrays ALHS and IDIAG corresponding to $\boldsymbol{M}^*$ are given explicitly as follows:

$$
\begin{aligned}
&\text{ALHS}(\ 1) = M^*_{11} && \} \text{ column 1} \\
&\text{ALHS}(\ 2) = M^*_{12} && \multirow{2}{*}{\} \text{ column 2}} \\
&\text{ALHS}(\ 3) = M^*_{22} && \\
&\text{ALHS}(\ 4) = M^*_{23} && \} \text{ column 3} \\
&\text{ALHS}(\ 5) = M^*_{33} && \\
&\text{ALHS}(\ 6) = M^*_{34} && \} \text{ column 4} \\
&\text{ALHS}(\ 7) = M^*_{44} && \\
&\text{ALHS}(\ 8) = M^*_{25} && \\
&\text{ALHS}(\ 9) = M^*_{35} && \} \text{ column 5} \\
&\text{ALHS}(10) = M^*_{45} && \\
&\text{ALHS}(11) = M^*_{55} && \\
&\text{ALHS}(12) = M^*_{46} && \\
&\text{ALHS}(13) = M^*_{56} && \} \text{ column 6} \\
&\text{ALHS}(14) = M^*_{66} && \\
&\text{ALHS}(15) = M^*_{67} && \} \text{ column 7} \\
&\text{ALHS}(16) = M^*_{77} &&
\end{aligned}
$$

$$\left.\begin{aligned}
\text{ALHS}(17) &= M^*_{18} \\
\text{ALHS}(18) &= M^*_{28} \\
\text{ALHS}(19) &= M^*_{38} \\
\text{ALHS}(20) &= M^*_{48} \\
\text{ALHS}(21) &= M^*_{58} \\
\text{ALHS}(22) &= M^*_{68} \\
\text{ALHS}(23) &= M^*_{78} \\
\text{ALHS}(24) &= M^*_{88}
\end{aligned}\right\} \quad \text{column 8}$$

$$\begin{aligned}
\text{IDIAG}(1) &= 1 \\
\text{IDIAG}(2) &= 3 \\
\text{IDIAG}(3) &= 5 \\
\text{IDIAG}(4) &= 7 \\
\text{IDIAG}(5) &= 11 \\
\text{IDIAG}(6) &= 14 \\
\text{IDIAG}(7) &= 16 \\
\text{IDIAG}(8) &= 24
\end{aligned}$$

The zero terms beneath the skyline must be stored because they become nonzero during the factorization process. All information in the original $\boldsymbol{M}^*$ is now contained in the arrays ALHS and IDIAG in compact fashion. For large finite element effective mass matrices, the saving of storage is considerable.

The heights of the columns in $\boldsymbol{M}^*$ are determined by subroutine COLHT from the element group LM arrays. The column heights are initially stored in IDIAG. When the calculation of column heights is completed, subroutine DIAG determines the diagonal address and "overwrites" them into IDIAG.

The right-hand-side vector (residual, or out-of-balance, force) in

$$\boldsymbol{M}^* \, \Delta\boldsymbol{a} = \boldsymbol{R}$$

is stored in an array BRHS of dimension NEQ in the obvious way:

$$\begin{aligned}
\text{BRHS}(1) &= R_1 \\
\text{BRHS}(2) &= R_2 \\
&\vdots \\
\text{BRHS}(8) &= R_8
\end{aligned}$$

The elimination scheme employed is based upon the following theorem.

Theorem. Let $\boldsymbol{M}^*$ be symmetric and positive-definite. Then there exists a nonsingular upper triangular matrix $\boldsymbol{U}$, with unit diagonal entries, and a diagonal matrix $\boldsymbol{D}$ such that

$$\boldsymbol{M}^* = \boldsymbol{U}^T\boldsymbol{D}\boldsymbol{U}$$

Remarks

1. In the case when $\boldsymbol{M}^*$ is positive-definite, the diagonal entries of $\boldsymbol{D}$ are all strictly positive. Thus examining the diagonal entries of $\boldsymbol{D}$ enables us to determine the rank of $\boldsymbol{M}^*$. This is the purpose of the ***rank check*** option in the program.

2. The matrix $\boldsymbol{U}$ has the same profile as $\boldsymbol{M}^*$. The fact that $\boldsymbol{U}$ has diagonal entries equal to one allows us to save storage by placing the diagonal entries of $\boldsymbol{D}$ in the diagonal entries of the array ALHS. Thus after factorization the nonzero entries of both $\boldsymbol{U}$ and $\boldsymbol{D}$ are stored in array ALHS. During the factorization procedure *no* auxiliary storage is needed.

3. Upper triangular matrices with unit diagonal entries are sometimes referred to as ***unit upper triangular matrices***.

11.2.2 Crout Elimination

The procedure used for performing the factorization of the left-hand-side matrix is called ***Crout elimination***, a convenient variant of Gauss elimination. In the Crout algorithm, one column at a time is completely factorized, beginning with the first column and not involving subsequent columns. This feature proves convenient in nonlinear analysis in which only a small zone of the entire matrix is nonlinear. The linear portion can be located in the first columns, factorized once and retained for the duration of the calculation.

After factorization, the solution is carried out in three steps:

$$\boldsymbol{U}^T\boldsymbol{z} = \boldsymbol{R} \qquad \text{forward reduction}$$

$$\boldsymbol{D}\boldsymbol{y} = \boldsymbol{z} \qquad \text{diagonal scaling}$$

$$\boldsymbol{U}\Delta\boldsymbol{a} = \boldsymbol{y} \qquad \text{back substitution}$$

In each step the solution is an explicit process, the coefficient matrix being triangular or diagonal. The intermediate vectors $\boldsymbol{y}$ and $\boldsymbol{z}$ are in turn stored in the array BRHS. Thus again no additional storage besides that for ALHS and BRHS (and IDIAG) is needed to carry out the factorization and solution processes.

Derivation of Computational Formulas for Crout Elimination

To simplify the notation, let us write

$$\boldsymbol{A}\boldsymbol{x} = \boldsymbol{b}$$

and assume $\boldsymbol{x}$ is n-dimensional. We wish to derive algorithms for

$$\boldsymbol{A} = \boldsymbol{U}^T\boldsymbol{D}\boldsymbol{U}$$

$$\boldsymbol{U}^T\boldsymbol{z} = \boldsymbol{b}$$

$$\boldsymbol{D}\boldsymbol{y} = \boldsymbol{z}$$

and

$$\boldsymbol{U}\boldsymbol{x} = \boldsymbol{y}$$

Factorization

The index version of the Crout factorization is expressed as

$$A_{ij} = \sum_{k=1}^{j} U_{ki} D_{kk} U_{kj} \qquad (1 \leq i \leq j)$$

where

$$U_{ii} = 1,$$
$$U_{ij} = 0, \qquad \text{for } i > j$$

Useful formulas for programming may be derived by expanding the above for $j = 1, 2, \ldots,$ as follows:

$$A_{11} = U_{11}D_{11}U_{11} = D_{11}$$

$$A_{12} = U_{11}D_{11}U_{12} + U_{21}D_{22}U_{22}$$
$$= D_{11}U_{12}$$
$$A_{22} = U_{12}D_{11}U_{12} + U_{22}D_{22}U_{22}$$
$$= U_{12}D_{11}U_{12} + D_{22}$$

$$A_{13} = U_{11}D_{11}U_{13} + U_{21}D_{22}U_{23} + U_{31}D_{33}U_{33}$$
$$= D_{11}U_{13}$$
$$A_{23} = U_{12}D_{11}U_{13} + U_{22}D_{22}U_{23} + U_{32}D_{33}U_{33}$$
$$= U_{12}D_{11}U_{13} + D_{22}U_{23}$$
$$A_{33} = U_{13}D_{11}U_{13} + U_{23}D_{22}U_{23} + U_{33}D_{33}U_{33}$$
$$= U_{13}D_{11}U_{13} + U_{23}D_{22}U_{23} + D_{33}$$

$$A_{14} = U_{11}D_{11}U_{14} + U_{21}D_{22}U_{24} + U_{31}D_{33}U_{34} + U_{41}D_{44}U_{44}$$
$$= D_{11}U_{14}$$
$$A_{24} = U_{12}D_{11}U_{14} + U_{22}D_{22}U_{24} + U_{32}D_{33}U_{34} + U_{42}D_{44}U_{44}$$
$$= U_{12}D_{11}U_{14} + D_{22}U_{24}$$
$$A_{34} = U_{13}D_{11}U_{14} + U_{23}D_{22}U_{24} + U_{33}D_{33}U_{34} + U_{43}D_{44}U_{44}$$
$$= U_{13}D_{11}U_{14} + U_{23}D_{22}U_{24} + D_{33}U_{34}$$
$$A_{44} = U_{14}D_{11}U_{14} + U_{24}D_{22}U_{24} + U_{34}D_{33}U_{34} + U_{44}D_{44}U_{44}$$
$$= U_{14}D_{11}U_{14} + U_{24}D_{22}U_{24} + U_{34}D_{33}U_{34} + D_{44}$$

And so on. These formulas may be solved for the U_{ij}'s and D_{jj}'s:

$$D_{11} = A_{11}$$

$$U_{12} = \frac{A_{12}}{D_{11}}$$

$$D_{22} = A_{22} - D_{11}U_{12}^2$$

$$U_{13} = \frac{A_{13}}{D_{11}}$$

$$U_{23} = \frac{A_{23} - U_{12}D_{11}U_{13}}{D_{22}}$$

$$D_{33} = A_{33} - D_{11}U_{13}^2 - D_{22}U_{23}^2$$

$$U_{14} = \frac{A_{14}}{D_{11}}$$

$$U_{24} = \frac{A_{24} - U_{12}D_{11}U_{14}}{D_{22}}$$

$$U_{34} = \frac{A_{34} - U_{13}D_{11}U_{14} - U_{23}D_{22}U_{24}}{D_{33}}$$

$$D_{44} = A_{44} - D_{11}U_{14}^2 - D_{22}U_{24}^2 - D_{33}U_{34}^2$$

And so on. For purposes of efficient programming, it is worthwhile to introduce the auxiliary variable

$$L_{ji} \overset{\text{def}}{=} D_{ii}U_{ij}$$

Then, equivalent to the above, we have

$$D_{11} = A_{11}$$

$$L_{21} = A_{12}$$

$$U_{12} = L_{21}/D_{11}$$

$$D_{22} = A_{22} - L_{21}U_{12}$$

$$L_{31} = A_{13}$$

$$L_{32} = A_{23} - U_{12}L_{31}$$

$$U_{13} = L_{31}/D_{11}$$

$$U_{23} = L_{32}/D_{22}$$

$$D_{33} = A_{33} - L_{31}U_{13} - L_{32}U_{23}$$

$$L_{41} = A_{14}$$

$$L_{42} = A_{24} - U_{12}L_{41}$$

$$L_{43} = A_{34} - U_{13}L_{41} - U_{23}L_{42}$$
$$U_{14} = L_{41}/D_{11}$$
$$U_{24} = L_{42}/D_{22}$$
$$U_{34} = L_{43}/D_{33}$$
$$D_{44} = A_{44} - L_{41}U_{14} - L_{42}U_{24} - L_{43}U_{34}$$

And so on. Summarizing, for $j = 1, 2, \ldots, n$

$$L_{ji} = A_{ij} - \sum_{k=1}^{i-1} U_{ki}L_{jk}, \qquad 1 \leq i \leq j - 1$$
$$U_{ij} = L_{ji}/D_{ii}$$
$$D_{jj} = A_{jj} - \sum_{i=1}^{j-1} L_{ji}U_{ij}$$

Observe that no additional storage besides that needed for $\boldsymbol{A}$ is necessary in the factorization process if $\boldsymbol{A}$ is overwritten by $\boldsymbol{U}$ and $\boldsymbol{D}$, viz.,

For $i = 2, 3, \ldots, j - 1$

$$A_{ij} \leftarrow A_{ij} - \sum_{k=1}^{i-1} A_{ki}A_{kj}$$

For $i = 1, 2, \ldots, j - 1$

$$T \leftarrow A_{ij}$$
$$A_{ij} \leftarrow T/A_{ii}$$
$$A_{jj} \leftarrow A_{jj} - TA_{ij}$$

This is the procedure coded in subroutine FACTOR in DLEARN. Additionally, FACTOR takes account of the profile storage of the coefficient matrix. The indexing necessary to account for the profile somewhat complicates the procedure but is necessary to achieve optimal efficiency. Figures 11.2.1(a) and 11.2.1(b) are presented as an aid for understanding the indexing and one-dimensional storage of the coefficient matrix in FACTOR. Note that if $A_{ii} = 0$, subroutine FACTOR skips the operations in the second do loop.

Forward Reduction

The indicial form of the forward reduction is

$$\sum_{i=1}^{n} U_{ij} z_i = b_j$$

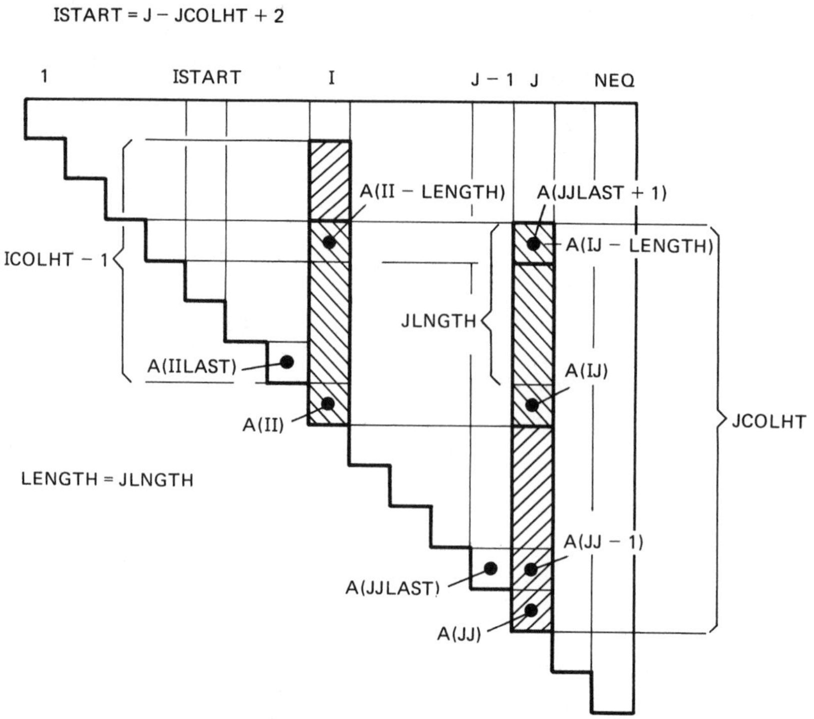

Figure 11.2.1(a) Indexing and storage for subroutine FACTOR; the case ICOLHT − 1 > JLNGTH.

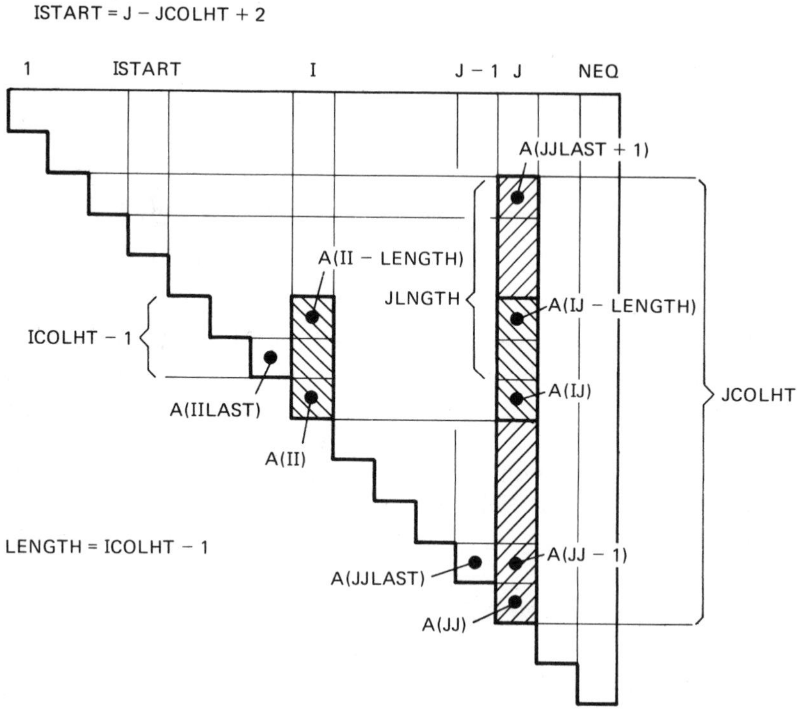

Figure 11.2.1(b) Indexing and storage for subroutine FACTOR; the case ICOLHT − 1 < JLNGTH.

Using the properties of the U_{ij}'s, we can immediately derive

$$\begin{aligned} z_1 &= b_1 \\ z_2 &= b_2 - U_{12} z_1 \\ z_3 &= b_3 - U_{13} z_1 - U_{23} z_2 \\ &\vdots \end{aligned}$$

from which we conclude that

$$z_j = b_j - \sum_{i=1}^{j-1} U_{ij} z_i, \qquad (2 \le j \le n)$$

It is clear from this formula that b_j may be overwritten by z_j, viz.,

$$\boxed{b_j \leftarrow b_j - \sum_{i=1}^{j-1} A_{ij} b_i \qquad (2 \le j \le n)}$$

where we have used the fact that A_{ij} has been overwritten by U_{ij}. This procedure is coded in subroutine BACK, which takes account of the profile storage of the coefficient matrix. The indexing is illustrated in Fig. 11.2.2(a).

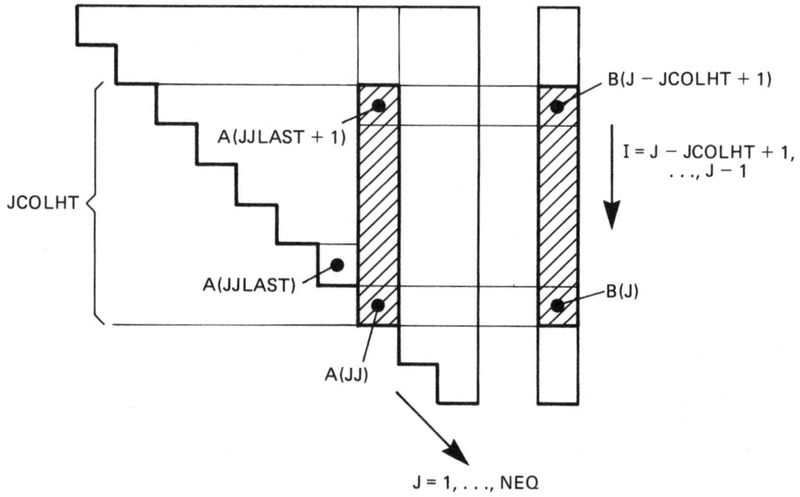

Figure 11.2.2(a) Indexing for forward reduction phase of subroutine BACK.

Diagonal Scaling

Diagonal scaling is also performed in subroutine BACK. Again b_j is overwritten:

$$
y_j = \frac{z_j}{D_{jj}} \quad \Leftrightarrow \quad \boxed{b_j \leftarrow b_j / A_{jj} \qquad (1 \leq j \leq n)}
$$

If $A_{jj} = 0$, diagonal scaling is skipped in subroutine BACK.

Back Substitution

The indicial form of the back substitution is

$$
\sum_{j=1}^{n} U_{ij} x_j = y_i
$$

Expanding this expression, we obtain

$$
\begin{array}{lclclclclclcl}
x_1 = y_1 & - & U_{1n} & x_n & - & U_{1,n-1} & x_{n-1} \cdots & - & U_{13} & x_3 & - & U_{12} & x_2 \\
x_2 = y_2 & - & U_{2n} & x_n & - & U_{2,n-1} & x_{n-1} \cdots & - & U_{23} & x_3 & & \text{2nd column} & \\
\vdots & & & & & & & & & & & & \\
x_{n-2} = y_{n-2} & - & U_{n-2,n} & x_n & - & U_{n-2,n-1} & x_{n-1} & & \text{3rd column} & & & & \\
x_{n-1} = y_{n-1} & - & U_{n-1,n} & x_n & & & & & & & & & \\
 & & & & & (n-1)\text{st column} & & & & & & & \\
x_n = y_n & & n\text{th column} & & & & & & & & & &
\end{array}
$$

There are several ways of organizing the calculations. The most computationally efficient proceeds from left to right, subtracting multiples of the columns of $\boldsymbol{U}$. Proceeding in this manner allows $\boldsymbol{b}$ to be overwritten with $\boldsymbol{x}$:

$$
\boxed{
\begin{array}{l}
\text{For } j = n, n-1, \ldots, 2 \\
\qquad \text{For } i = 1, 2, \ldots, j-1 \\
\qquad b_i \leftarrow b_i - A_{ij} b_j.
\end{array}
}
$$

The indexing used in subroutine BACK is illustrated in Fig. 11.2.2(b). The number of "flops" (floating-point operations [1]) involved in Crout factorization is approximately $m^2 n/2$, where m is the mean half-bandwidth (defined to be the sum of the column heights divided by the number of equations). In obtaining this result we assume $1 \ll m \ll n$, and thus lower-order terms may be ignored. Given the factorized matrix, a "solution" (which consists of a forward reduction, diagonal scaling, and back

[1] Roughly speaking, one flop consists of a floating-point multiply, an addition, and some subscripting. For example, X = X + Y(I)*Z(J) would amount to one flop.

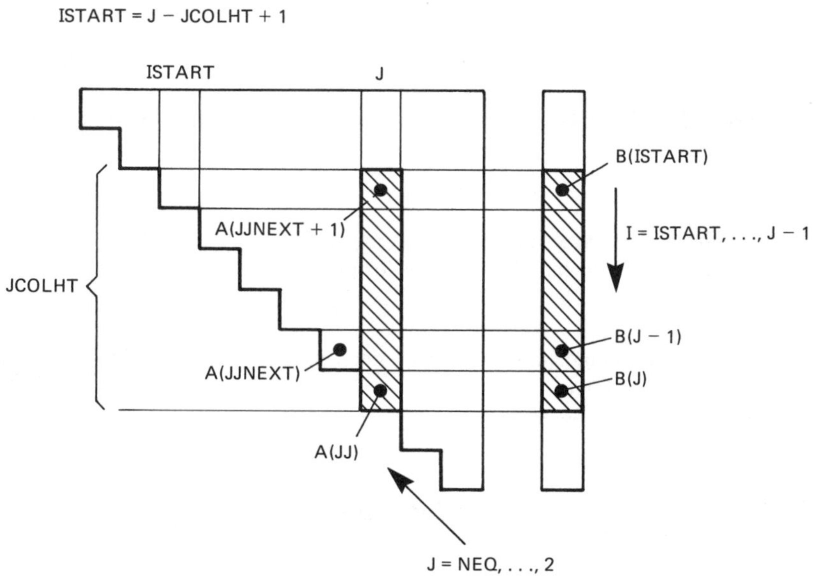

Figure 11.2.2(b) Indexing for back substitution phase of subroutine BACK. (Note: ISTART has a different significance in this routine than in subroutine FACTOR.)

substitution) involves approximately $2mn$ flops. Clearly, for large systems, factorization of the coefficient matrix is much more expensive than solution. Factorization costs vary quadratically with m and thus there is considerable interest in schemes for ordering the equations to minimize m. For additional information on this subject, see [1–4]. An excellent reference for equation solving in general is [5].

Exercise 1. Consider the system

$$\boldsymbol{Kd} = \boldsymbol{F}$$

where

$$\boldsymbol{K} = \begin{bmatrix} 1 & -1 & 0 & 0 \\ -1 & 2 & -1 & 0 \\ 0 & -1 & 2 & -1 \\ 0 & 0 & -1 & 2 \end{bmatrix}$$

$$\boldsymbol{d} = \begin{Bmatrix} d_1 \\ d_2 \\ d_3 \\ d_4 \end{Bmatrix} \qquad \boldsymbol{F} = \begin{Bmatrix} 1 \\ 0 \\ 0 \\ 0 \end{Bmatrix}$$

i. Obtain the Crout factorization of $\boldsymbol{K}$ (i.e., define the diagonal matrix $\boldsymbol{D}$ and the unit upper triangular matrix $\boldsymbol{U}$).

ii. Obtain the solution $\boldsymbol{d}$ by performing the following steps:

$$\boldsymbol{U}^T\boldsymbol{z} = \boldsymbol{F} \qquad \text{forward reduction}$$

$$\boldsymbol{Dy} = \boldsymbol{z} \qquad \text{diagonal scaling}$$

$$\boldsymbol{Ud} = \boldsymbol{y} \qquad \text{back substitution}$$

iii. Check parts (i) and (ii) by verifying $\boldsymbol{U}^T\boldsymbol{DU} = \boldsymbol{K}$ and $\boldsymbol{Kd} = \boldsymbol{F}$.

Exercise 2. The ***Cholesky decomposition*** of a symmetric positive-definite matrix $\boldsymbol{A}$ takes the form

$$\boldsymbol{A} = \boldsymbol{G}^T \boldsymbol{G}$$

where $\boldsymbol{G}$ is an upper triangular matrix with positive diagonal entries. The Cholesky decomposition may be obtained from the Crout factorization by way of

$$\boldsymbol{G} = \boldsymbol{D}^{1/2} \boldsymbol{U}$$

Develop an algorithm for directly determining $\boldsymbol{G}$ along the lines of the Crout factorization presented above.

11.2.3 Dynamic Storage Allocation

Examples

In the initialization subroutine STATIN, memory storage is allocated for the global arrays required for static analysis. On lines 24 and 25 the "pointers" are determined for the X and ID arrays:

```
MPX = MPOINT('X       ',NSD ,NUMNP,0,IPREC)
MPID = MPOINT('ID      ',NDOF,NUMNP,0,1)
```

FUNCTION MPOINT manages memory allocation within the unnamed or "blank" COMMON A(MTOT). The first argument is a 2A4 string, which serves as the array's label when the storage dictionary is printed. The next three arguments are array dimensions; in this case we see that both ID and X are two-dimensional arrays. The last argument determines the precision of the array (IPREC = 1 or 2 corresponding to single or double precision, respectively). Notice that as an integer array, ID is always defined to be single precision.

The calling statements:

```
CALL COORD(A(MPX),NSD,NUMNP,IPRTIN)
CALL BC(A(MPID),NDOF,NUMNP,NEQ,IPRTIN)
```

and the corresponding subroutine statements:

```
SUBROUTINE COORD(X,NSD,NUMNP, IPRTIN)
SUBROUTINE BC(ID,NDOF,NUMNP,NEQ,IPRTIN)
```

mean *array* X *begins at address* MPX *of blank common and array* ID *begins at address* MPID *of blank common,* respectively.

In COORD and BC, the arrays X and ID are treated as *two-dimensonal,* as evidenced by the DIMENSION statements:

```
DIMENSION X(NSD,1)
```

and

```
DIMENSION ID(NDOF,1)
```

respectively. It does not matter that X and ID correspond to portions of a one-dimensional array in the calling program STATIN. The significance of the dimensioning is as follows: *The arrays* X *and* ID *are stored in blank common in consecutively addressed cells by columns*. Consequently the addresses of X(M,N) and ID(M,N) in A are given by

$$\text{IPREC}*(\text{NSD}*(\text{N} - 1)+\text{M} - 1) + \text{MPX}$$

and

$$\text{NDOF}*(\text{N} - 1)+\text{M} - 1 + \text{MPID}$$

Thus the only information necessary to determine the appropriate addresses in A are the starting addresses (i.e., MPX and MPID, respectively), which are provided by the CALL statements, and the number of rows (i.e., NSD and NDOF, respectively), which are provided by the DIMENSION statements in the subroutines. This explains why it is not necessary to define the number of columns of two-dimensional arrays "passed" to subroutines. The 1's in the column slots could be replaced by any positive integer; they only serve notice that these arrays are to be regarded as two-dimensional in the subroutine—the integer is never used. For three-dimensional arrays, the first two dimensions must be precisely defined, whereas the third may be set to 1, etc.

The array names, pointers, and other dictionary entries are stored at the end of blank common. Thus for the first two pointers declared during execution:

$$\text{A}(\text{MTOT} - 4) = \text{MPSTEP} = 1$$

$$\text{A}(\text{MTOT} - 11) = \text{MPDPRT}, \ldots.$$

Note that seven words of storage are required for each entry in the dictionary. The memory FUNCTION MPOINT monitors the blank common allocation to ensure that data arrays starting from the beginning of blank common do not overwrite the array pointer information. If the user receives a memory error, he or she must reduce the size of the problem or increase MTOT. Given the availability of sufficient machine memory size, *the latter action requires only recompiling the main program* (*which amounts to only a few statements*), *followed by linking to create a new executable file*.

The following tables present the storage arrays and associated pointers for all global data used during static or dynamic analysis. Also, an example of local element data storage requirements is presented for the four-node quadrilateral, elastic continuum element QUADC.

TABLE 11.2.1 Storage in Blank Common—Time Sequence and Time History Data*

Array Pointer	Array
MPSTEP	NSTEP(NUMSEQ)
MPDPRT	NDPRT(NUMSEQ)
MPSPRT	NSPRT(NUMSEQ)
MPHPLT	NHPLT(NUMSEQ)
MPITER	NITER(NUMSEQ)
MPALPH	ALPHA(NUMSEQ)
MPBETA	BETA(NUMSEQ)

*Required for static and dynamic analysis, storage allocated in SUBROUTINE TSEQ.

TABLE 11.2.1 (cont.) Storage in Blank Common—Time Sequence and Time History Data

Array Pointer	Array
MPGAMM	GAMMA(NUMSEQ)
MPDT	DT(NUMSEQ)
MPIDHS	IDHIST(3,NDOUT)†
MPDOUT	DOUT(NDOUT+1,NPLPTS)†

where, for each time sequence I=1,NUMSEQ:

NSTEP(I)	=	number of time steps
NDPRT(I)	=	number of time steps between printing of kinematic output
NSPRT(I)	=	number of time steps between printing of element stress/strain output
NHPLT(I)	=	number of time steps between plotting of output time-history data
NITER(I)	=	number of iterations in corrector loop
ALPHA(I)	=	α coefficient for time-integration algorithm
BETA(I)	=	β coefficient for time-integration algorithm
GAMMA(I)	=	γ coefficient for time-integration algorithm
DT(I)	=	time step Δt

End time sequence data.

IDHIST(1,N)	=	node number for output history N
IDHIST(2,N)	=	degree of freedom number for output history N
IDHIST(3,N)	=	kinematic-variable-type number for output history N
DOUT(1,J)	=	value of time at plot step J (stored as single precision for IPREC = 1 or 2)
DOUT(N+1,J)	=	output history N at plot step J (stored as single precision for IPREC = 1 or 2)
NDOUT	=	number of nodal output time-histories
NPLPTS	=	number of points in output time-history plots
IPREC	=	precision flag: 1 ⇒ single, 2 ⇒ double

† If NDOUT=0, no storage is allocated for IDHIST and DOUT.

TABLE 11.2.2 Storage in Blank Common—Dynamic Analysis Data*

Array Pointer	Array
MPVPRD	VPRED(NDOF,NUMNP)
MPDPRD	DPRED(NDOF,NUMNP)
MPA	A(NDOF,NUMNP)
MPV	V(NDOF,NUMNP)

where:

VPRED(I,J)	=	Ith predictor velocity component of node J
DPRED(I,J)	=	Ith predictor displacement component of node J
A(I,J)	=	Ith acceleration component of node J
V(I,J)	=	Ith velocity component of node J
NDOF	=	number of degrees of freedom per node
NUMNP	=	number of nodal points

* Required only for dynamic analysis; storage allocated in SUBROUTINE DYNPTS.

TABLE 11.2.3 Storage in Blank Common—Static Analysis Data*

Array Pointer	Array
MPD	D(NDOF,NUMNP)
MPX	X(NSD,NUMNP)
MPID	ID(NDOF,NUMNP)
MPF	F(NDOF,NUMNP,NLVECT)†
MPG	G(NPTSLF,2,NLTFTN)‡
MPG1	G1(NLTFTN)‡
MPDIAG	IDIAG(NEQ)
MPNGRP	NGRP(NUMEG)

where:

D(I,J)	=	Ith displacement component of node J
X(I,J)	=	Ith coordinate of node J
ID(I,J)	=	equation number for degree of freedom I, node J
F(I,J,L)	=	Ith prescribed force/kinematic-variable§ component for node J, load vector L
G(K,1,L)	=	time t_k for load-time function L
G(K,2,L)	=	function value at time t_k, load-time function L
G1(L)	=	value of load-time function L at current time t
IDIAG(N)	=	location of Nth diagonal element in coefficient array ALHS
NGRP(N)	=	pointer to beginning address of element group N
NDOF	=	number of degrees of freedom per node
NUMNP	=	number of nodal points
NSD	=	number of space dimensions
NLVECT	=	number of load vectors
NPTSLF	=	number of points in a load-time function
NLTFTN	=	number of load-time functions
NEQ	=	number of equations
NUMEG	=	number of element groups

*Required for static and dynamic analysis; storage allocated in SUBROUTINE STATIN.

†If NLVECT = 0, no storage is allocated for F.

‡If NLTFTN = 0, no storage is allocated for G and G1.

§Displacements for static analysis; displacements, velocities or accelerations for dynamic analysis.

TABLE 11.2.4 Storage Requirements for Four-Node Quadrilateral, Elastic Continuum Element (NTYPE = 1)

Each element group has an associated array NPAR with pointer MPNPAR; group control data can be stored in positions 1 through 16. This is followed immediately in blank common by the MP array, which stores the memory pointers for the element group data. This allows each element type to have a unique data structure. For the present element, MP contains 35 words and stores the following pointers:

Array Pointer	Array
MP(MW)	W(NINT)
MP(MDET)	DET(NINT)
MP(MR)	R(NINT)
MP(MSHL)	SHL(NROWSH,NEN,NINT)

TABLE 11.2.4 (cont.) Storage Requirements for Four-Node Quadrilateral, Elastic Continuum Element (NTYPE = 1)

Array Pointer	Array
MP(MSHG)	SHG(NROWSH,NEN,NINT)
MP(MSHGBR)	SHGBAR(NROWSH,NEN,NRINT)
MP(MRHO)	RHO(NUMAT)
MP(MRDPM)	RDAMPM(NUMAT)
MP(MRDPK)	RDAMPK(NUMAT)
MP(MTH)	TH(NUMAT)
MP(MC)	C(NROWB,NROWB,NUMAT)
MP(MGRAV)	GRAV(NESD)
MP(MIEN)	IEN(NEN,NUMEL)
MP(MMAT)	MAT(NUMEL)
MP(MLM)	LM(NED,NEN,NUMEL)
MP(MIELNO)	IELNO(NSURF)
MP(MISIDE)	ISIDE(NSURF)
MP(MPRESS)	PRESS(2,NSURF)
MP(MSHEAR)	SHEAR(2,NSURF)
MP(MISHST)	ISHIST(3,NSOUT)*
MP(MSOUT)	SOUT(NSOUT+1,NPLPTS)*
MP(MELEFM)	ELEFFM(NEE,NEE)
MP(MXL)	XL(NESD,NEN)
MP(MWORK)	WORK(16)
MP(MB)	B(NROWB,NEE)
MP(MDMAT)	DMAT(NROWB,NROWB)
MP(MDB)	DB(NROWB,NEE)
MP(MVL)	VL(NED,NEN)
MP(MAL)	AL(NED,NEN)
MP(MELRES)	ELRESF(NEE)
MP(MDL)	DL(NED,NEN)
MP(MSTRN)	STRAIN(NROWB)
MP(MSTRS)	STRESS(NROWB)
MP(MPSTRN)	PSTRN(NROWB)
MP(MPSTRS)	PSTRS(NROWB)

where, for each quadrature point L=1,NINT:

W(L)	=	integration-rule weight
DET(L)	=	Jacobian determinant
R(L)	=	radius at quadrature point (axisymmetric option)
SHL(1,J,L)	=	local derivative $N_{j,\xi}$
SHL(2,J,L)	=	local derivative $N_{j,\eta}$
SHL(3,J,L)	=	N_j
SHG(1,J,L)	=	global derivative N_{j,x_1}
SHG(2,J,L)	=	global derivative N_{j,x_2}
SHG(3,J,L)	=	N_j

End quadrature-point data.

*If NSOUT = 0, no storage is allocated for ISHIST and SOUT.

TABLE 11.2.4 (cont.) Storage Requirements for Four-Node Quadrilateral, Elastic Continuum Element (NTYPE = 1)

For mean-dilatational $\bar{\boldsymbol{B}}$ formulation:

SHGBAR(1,J) = $\int_{\Omega^e} N_{j,x_1}\, d\Omega \Big/ \int_{\Omega^e} d\Omega$

SHGBAR(2,J) = $\int_{\Omega^e} N_{j,x_2}\, d\Omega \Big/ \int_{\Omega^e} d\Omega$

SHGBAR(3,J) = $\int_{\Omega^e} N_j\, d\Omega \Big/ \int_{\Omega^e} d\Omega$

For each material set M=1,NUMAT:

RHO(M)	=	mass density
RDAMPM(M)	=	Rayleigh mass-proportional damping coefficient
RDAMPK(M)	=	Rayleigh stiffness-proportional damping coefficient
TH(M)	=	thickness
C(I,J,M)	=	I,Jth component of material moduli matrix

End material data.

GRAV(I)	=	gravity-load multiplier in x_i direction
IEN(J,NEL)	=	global node number for local node J of element NEL
MAT(NEL)	=	material set number for element NEL
LM(I,J,NEL)	=	global equation number for Ith degree of freedom, Jth local node of element NEL
IELNO(N)	=	element number of Nth applied surface force
ISIDE(N)	=	element-side number of Nth applied surface force
PRESS(1,N)	=	pressure at local node I for Nth applied surface force†
PRESS(2,N)	=	pressure at local node J for Nth applied surface force†
SHEAR(1,N)	=	shear at local node I for Nth applied surface force†
SHEAR(2,N)	=	shear at local node J for Nth applied surface force†
ISHIST(1,N)	=	element number for output history N
ISHIST(2,N)	=	quadrature-point number for output history N
ISHIST(3,N)	=	stress/strain-component number for output history N
SOUT(1,J)	=	time at plot step J (stored as single precision for IPREC = 1 or 2)
SOUT(N+1,J)	=	output history N at plot step J (stored as single precision for IPREC = 1 or 2)
ELEFFM(I,J)	=	element effective mass matrix
XL(I,J)	=	Ith coordinate component of Jth element node
WORK(I)	=	work space used for temporary calculations
B(I,J)	=	strain-displacement matrix
DMAT(I,J)	=	scalar multiple of element material moduli matrix
DB(I,J)	=	matrix product of DMAT and B
VL(I,J)	=	Ith velocity component of Jth element node
AL(I,J)	=	Ith acceleration component of Jth element node
ELRESF(I)	=	Ith entry of element residual force vector
DL(I,J)	=	Ith displacement component of Jth element node
STRAIN(I)	=	strains
STRESS(I)	=	stresses
PSTRN(I)	=	principal strains

† For the four-node quadrilateral, $1 \le I \le 4$; if $I \le 3$, $J = I + 1$; if $I = 4$, $J = 1$.

TABLE 11.2.4 (cont.) Storage Requirements for Four-Node Quadrilateral, Elastic Continuum Element (NTYPE = 1)

PSTRS(I)	=	principal stresses
NINT	=	number of integration points
NUMAT	=	number of materials in this group
NROWSH	=	number of rows in shape-function arrays (=3)
NEN	=	number of element nodes (=4)
NRINT	=	number of reduced-integration points; for mean-dilatational $\bar{\boldsymbol{B}}$ formulation = 1
NROWB[‡]	=	number of rows in strain-displacement matrix (=4)
NUMEL	=	number of elements in this group
NSURF	=	number of element surface loads in this group
NSOUT	=	number of stress/strain time-histories in this group
NPLPTS	=	number of points in output time-histories.
NEE	=	number of element equations (=8)
NED	=	number of element degrees of freedom per node (=2)

[‡] Note that while the strain-displacement matrix $\boldsymbol{B}$ has a constant row dimension of 4, the parameter NSTR specified in SUBROUTINE QUADC denotes the number of nonzero strain-energy components in the element. For plane stress (IOPT = 0), and plane strain (IOPT = 1) without mean-dilatational $\bar{\boldsymbol{B}}$ formulation (IBBAR = 0), NSTR equals 3. However, for plane strain with $\bar{\boldsymbol{B}}$ (IBBAR = 1), and torsionless axisymmetry (IOPT = 2) with or without $\bar{\boldsymbol{B}}$, NSTR equals 4.

TABLE 11.2.5 Storage in Blank Common—Equation System Data*

Array Pointer	Array
MPALHS	ALHS(NALHS)
MPBRHS	BRHS(NEQ)

where:

ALHS(N)	=	Nth term of coefficient matrix stored in compacted column form
BRHS(N)	=	Nth active residual (dynamic analysis) or force (static analysis) component[†]
NALHS	=	number of terms in ALHS
NEQ	=	number of equations

* Required for static and dynamic analysis; storage allocated in SUBROUTINE EQSET.

[†] Upon back substitution BRHS contains acceleration increments (dynamic analysis) or displacements (static analysis) for active degrees of freedom.

11.3 PROGRAM STRUCTURE

In this section we present an organizational description of DLEARN, including algorithmic equations when they are deemed illustrative. All FORTRAN subroutines appear in boldface and variable names are capitalized. In the example element loops, the term "generic subroutine" identifies a fundamental program unit which could be accessed by any element type, e.g., **COLHT**, or by a class of elements, e.g., **PROP2D** for two-dimensional, elastic continuum elements.

11.3.1 Global Control

G1. MAIN PROGRAM

1. Set blank common capacity MTOT, and initialize addresses of first and last available words.
2. Specify numerical precision IPREC.
3. Specify input/output unit numbers and open files.
4. Call global driver **DLEARN.**
5. Close files and exit program.

G2. DLEARN: Global driver

1. **ECHO:** IF (IECHO.EQ.1) write copy of input data (uninterpreted) to output file.
2. Read TITLE: IF (TITLE(1).EQ.'*END') RETURN to G1.5.
3. Read execution control card. If static analysis LDYN = .FALSE., otherwise .TRUE. .
4. Call initialization drivers, GO TO I1.
5. IF (IEXEC.EQ.1) **DRIVER:** solution driver, GO TO S1.
6. IF (IWRITR.EQ.1) **RSOUT:** write restart file.
7. **PRTDC:** print memory pointer dictionary.
8. GO TO 2 (begin new problem).

11.3.2 Initialization Phase

I1. TSEQ: driver to input time sequence control data.

1. Set memory pointers for time sequence data.
2. **TSEQIN:** read time sequence data and compute storage for time histories. Set defaults for static analysis.
3. Set memory pointers for nodal time history data arrays.

I2. IF (LDYN) **DYNPTS:** set pointers for dynamic analysis data arrays.

I3. IF (NDOUT.GT.0) **DHIST:** read nodal output time-history data, store in IDHIST(3,NDOUT).

I4. STATIN: driver to set pointers for static analysis data arrays and read geometric and boundary condition input data. IF (.NOT.LDYN) MPDPRD = MPD, then element task 'FORM_RHS' accesses proper displacement array during static analysis. (cf. S10.1.a).

1. **COORD:** read/generate nodal coordinates, store in X(NSD,NUMNP).

2. **BC:** read/generate nodal boundary condition codes into ID(NDOF,NUMNP), then overwrite array with equation numbers.
3. IF (NLVECT.GT.0) **INPUT:** read/generate prescribed nodal force/kinematic boundary condition data and store in F(NDOF, NUMNP, NLVECT).
4. IF (NLTFTN.GT.0) **LTIMEF:** read load-time functions, store in G(NPTSLF,2,NLTFTN).

I5. IF (LDYN): input kinematic initial conditions.

1. IF (IREADR .EQ. 1) THEN
   RSIN: read initial conditions from restart file.
   ELSE
   INPUT: read initial conditions from input file.
   ENDIF

I6. IF (LDYN .AND. NDOUT.GT.0) **STORED:** store kinematic initial conditions for time histories in DOUT(NDOUT+1,NPLPTS).

I7. **ELEMNT**('INPUT__'): loop over element groups NEG=1,NUMEG and read/generate element definition data; element task 1.

Example

QUADC: four-node quadrilateral, elastic continuum element.

1. Set memory pointers.
2. **QDCT1:** element task 1.
   a. **QDCSHL:** compute bilinear shape functions N_a and their local derivatives $N_{a,\xi}$, $N_{a,\eta}$ at integration points, store in SHL(NROWSH,NEN,NINT); generic subroutine.
   b. **PROP2D:** read material property data for two-dimensional elastic continuum. Compute material moduli matrix, store in C(NROWB,NROWB, NUMAT); generic subroutine.
   c. Read body-force vector GRAV(NESD).
   d. **GENEL:** read/generate element data; store element node numbers in IEN(NEN,NUMEL) and material set numbers in MAT(NUMEL).
   e. **FORMLM:** form assembly mapping LM(NED,NEN,NUMEL); generic subroutine.
   f. If element group contributes to global left-hand-side matrix, then **COLHT:** determine column heights of each element; generic subroutine.
   g. IF (NSURF.GT.0) **QDCRSF:** read surface force data.
   h. IF (NSOUT.GT.0) **SHIST:** read stress/strain output time-history data, store in ISHIST(3,NSOUT); generic subroutine.

I8. EQSET: complete specification of data structure for global equation system.

1. **DIAG:** determine addresses of diagonal entries of global left-hand-side matrix ALHS, store in IDIAG(NEQ). Set NALHS equal to number of terms in ALHS.
2. Allocate memory for ALHS(NALHS) and global right-hand-side vector BRHS(NEQ).

I9. Resume global driver **DLEARN;** GO TO G2.5.

11.3.3 Solution Phase

Time sequence loop (DO S19 NSQ = 1,NUMSEQ)

S1. TIMCON: set current time sequence parameters: NSTEP1, NDPRT1, NSPRT1, NHPLT1, NITER1, ALPHA1, BETA1, GAMMA1, DT1 and compute coefficients for α-method transient algorithm:

$$\begin{aligned}
\text{COEFF1} &= (1 + \alpha), \\
\text{COEFF2} &= \gamma\Delta t, \\
\text{COEFF3} &= \beta\Delta t^2 \\
\text{COEFF4} &= (1 + \alpha)\gamma\Delta t, \\
\text{COEFF5} &= (1 + \alpha)\beta\Delta t^2, \\
\text{COEFF6} &= (1 + \alpha)\Delta t, \\
\text{COEFF7} &= \frac{1}{2}(1 + \alpha)(1 - 2\beta)\Delta t^2, \\
\text{COEFF8} &= (1 + \alpha)(1 - \gamma)\Delta t.
\end{aligned}$$

S2. ELEMNT('FORM_LHS'): loop over element groups NEG=1,NUMEG and form effective mass matrix $\boldsymbol{M}^*$; element task 2.

$$\boldsymbol{M}^* = \boldsymbol{M}^E + \boldsymbol{M}^I + (1 + \alpha)\gamma\Delta t\boldsymbol{C}^I + (1 + \alpha)\beta\Delta t^2\boldsymbol{K}^I \qquad \text{(dynamic analysis)}$$

$$\boldsymbol{M}^* = \boldsymbol{K} \qquad \text{(static analysis)}$$

where superscripts E and I denote explicit and implicit element group contributions, respectively.

Example

QUADC: four-node quadrilateral, elastic continuum element.

1. **QDCT2:** element task 2.

DO h NEL=1,NUMEL

a. **CLEAR:** zero element effective mass array ELEFFM(NEE,NEE).

b. **LOCAL:** localize nodal coordinates X, store in XL(NESD,NEN); generic subroutine.

c. **QDCSHG:** compute shape function global derivatives $N_{a,x}$, $N_{a,y}$ at integration points, store in SHG(NROWSH,NEN,NINT); generic subroutine.

d. IF (IOPT.EQ.2), compute radial coordinates of integration points for torsionless axisymmetry option, store in R(NINT). Include radial weighting with Jacobian determinants DET(NINT).

e. IF (LDYN . AND. IMASS.NE.2) THEN

   CONTM: form consistent or lumped mass matrix for continuum element and store in ELEFFM(NEE,NEE); generic subroutine.

$$\boldsymbol{m}^* = \boldsymbol{m}^* + [1 + \text{RDAMPM(MAT(NEL))}*(1 + \alpha)\gamma\Delta t]\boldsymbol{m}^e$$

ENDIF

f. IF ((NOT.LDYN) .OR. IMPEXP.EQ.0) THEN

   QDCK: form stiffness matrix $\boldsymbol{k}^e = \int \boldsymbol{B}^T\boldsymbol{D}\boldsymbol{B}\, d\Omega$ and sum to ELEFFM(NEE,NEE).

$$\boldsymbol{m}^* = \boldsymbol{m}^* + [\text{RDAMPK(MAT(NEL))}*(1 + \alpha)\gamma\Delta t + (1 + \alpha)\beta\Delta t^2]\boldsymbol{k}^e$$

ENDIF

g. **ADDLHS:** assemble ELEFFM into ALHS; generic subroutine.

h. CONTINUE to next element.

S3. FACTOR: perform Crout factorization of ALHS.

$$\boldsymbol{M}^* = \boldsymbol{U}^T\boldsymbol{D}\boldsymbol{U}$$

S4. Perform rank check if requested.

1. IF (IRANK.EQ.1) **PIVOTS:** print numbers of negative and zero pivots; RETURN to S19.
2. IF (IRANK.EQ.2) **PRINTP:** print all pivots; RETURN to S19.

Time-step loop (DO S18 N=1,NSTEP1)

S5. IF (LDYN) THEN

PREDCT: predictor update of all degrees of freedom.

$$\tilde{\boldsymbol{d}}^{(1)} = \boldsymbol{d}_n + (1 + \alpha)\Delta t\boldsymbol{v}_n + \frac{1}{2}(1 + \alpha)(1 - 2\beta)\Delta t^2\boldsymbol{a}_n \qquad \text{DPRED} = \text{D} + \text{COEFF6} * \text{V} + \text{COEFF7} * \text{A}$$

$\widetilde{\boldsymbol{v}}^{(1)} = \boldsymbol{v}_n + (1 + \alpha)(1 - \gamma)\Delta t \boldsymbol{a}_n$ VPRED = V + COEFF8 * A

$\widetilde{\boldsymbol{a}}^{(1)} = \boldsymbol{0}$ A = ZERO

ELSE

CLEAR: zero displacement array D.

ENDIF

S6. IF (NLVECT.GT.0) **LFAC:** evaluate load-time functions at time t_{n+1}, store values in G1(NLTFTN).

S7. IF (DT1.NE.ZERO) **COMPBC:** overwrite predictors to account for kinematic boundary conditions.

1. If d_{n+1} given:

IF (LDYN) THEN

$d^* = (1 + \alpha)d_{n+1} - \alpha d_n$ TEMP = COEFF1 * VAL − ALPHA1 * D

$\widetilde{a}^{(1)} = \dfrac{d^* - \widetilde{d}^{(1)}}{(1 + \alpha)\beta\Delta t^2}$ A = (TEMP − DPRED)/COEFF5

$\widetilde{d}^{(1)} = d^*$ DPRED = TEMP

$\widetilde{v}^{(1)} = \widetilde{v}^{(1)} + (1 + \alpha)\gamma\Delta t \widetilde{a}^{(1)}$ VPRED = VPRED + COEFF4 * A

ELSE

$\widetilde{d} = d_{n+1}$ DPRED = VAL

ENDIF

2. If v_{n+1} given:

$v^* = (1 + \alpha)v_{n+1} - \alpha v_n$ TEMP = COEFF1 * VAL − ALPHA1 * V

$\widetilde{a}^{(1)} = \dfrac{v^* - \widetilde{v}^{(1)}}{(1 + \alpha)\gamma\Delta t}$ A = (TEMP − VPRED)/COEFF4

$\widetilde{v}^{(1)} = v^*$ VPRED = TEMP

$\widetilde{d}^{(1)} = \widetilde{d}^{(1)} + (1 + \alpha)\beta\Delta t^2 \widetilde{a}^{(1)}$ DPRED = DPRED + COEFF5 * A

3. If a_{n+1} given:

$\widetilde{d}^{(1)} = \widetilde{d}^{(1)} + (1 + \alpha)\beta\Delta t^2 a_{n+1}$ DPRED = DPRED + COEFF5 * VAL

$\widetilde{v}^{(1)} = \widetilde{v}^{(1)} + (1 + \alpha)\gamma\Delta t a_{n+1}$ VPRED = VPRED + COEFF4 * VAL

$\widetilde{a}^{(1)} = a_{n+1}$ A = VAL

Multicorrector iteration loop (DO S13 I=1,NITER1)

S8. IF (NLTFTN.GT.0) **LFAC:** evaluate load-time functions at time $t_{n+1+\alpha}$, store values in G1(NLTFTN).

S9. IF (NLVECT.GT.0) **LOAD:** form nodal contribution to residual force, $\boldsymbol{F}_{n+1+\alpha}$, store in BRHS(NEQ).

S10. ELEMNT('FORM_RHS'): loop over element groups NEG=1,NUMEG and form element contributions to residual force $\boldsymbol{R}$; element task 3.

$$\boldsymbol{R} = \underbrace{\boldsymbol{F}_{n+1+\alpha}}_{\text{Nodal contribution}} \underbrace{-\boldsymbol{M}\widetilde{\boldsymbol{a}}^{(i)} - \boldsymbol{C}\widetilde{\boldsymbol{v}}^{(i)} - \boldsymbol{K}\widetilde{\boldsymbol{d}}^{(i)} + \mathop{\mathsf{A}}_{e=1}^{n_{el}} \boldsymbol{f}^{e}_{n+1+\alpha}}_{\text{Element contribution}} \qquad \text{(dynamic analysis)}$$

$$\boldsymbol{R} = \underbrace{\boldsymbol{F}}_{\text{Nodal contribution}} + \underbrace{\mathop{\mathsf{A}}_{e=1}^{n_{el}} \boldsymbol{f}^{e}}_{\text{Element contribution}} \qquad \text{(static analysis)}$$

Example

QUADC: four-node quadrilateral, elastic continuum element.

1. **QDCT3:** element task 3.

DO h NEL=1,NUMEL

a. **LOCAL:** localize displacements DPRED, store in DL(NED,NEN); generic subroutine. Note: *For static analysis* MPDPRD = MPD, *so reference to* DPRED *accesses the array* D.
b. IF (LDYN) THEN

   i. **LOCAL:** localize VPRED and A, store in VL(NED,NEN) and AL(NED,NEN), respectively; generic subroutine.
   ii. Compute effective displacement and acceleration accounting for Rayleigh damping:

$$\widetilde{\boldsymbol{d}}^{e} = \widetilde{\boldsymbol{d}}^{e} + \text{RDAMPK} * \widetilde{\boldsymbol{v}}^{e},$$

$$\widetilde{\boldsymbol{a}}^{e} = \widetilde{\boldsymbol{a}}^{e} + \text{RDAMPM} * \widetilde{\boldsymbol{v}}^{e}.$$

ENDIF

c. Determine if element makes inertial (FORMMA = .TRUE.) and/or stiffness (FORMKD = .TRUE.) contribution to right-hand-side vector.
d. IF (FORMMA .OR.FORMKD) THEN

   i. **LOCAL:** localize nodal coordinates X, store in XL(NESD,NEN); generic subroutine.
   ii. **QDCSHG:** compute shape function global derivatives $N_{a,x}$, $N_{a,y}$ at integration points, store in SHG(NROWSH,NEN,NINT); generic subroutine.

iii. IF (IOPT.EQ.2) compute radial coordinates of integration points for torsionless axisymmetry option, store in R(NINT). Include radial weighting with Jacobian determinants DET(NINT).

ENDIF

e. IF (FORMMA) **CONTMA:** compute $-\boldsymbol{m}^e(\widetilde{\boldsymbol{a}}^e - \boldsymbol{g}^e)$ for continuum element and store in ELRESF(NEE); generic subroutine.

f. IF (FORMKD) **QDCKD:** compute internal force $-\boldsymbol{k}^e\widetilde{\boldsymbol{d}}^e = -\int \boldsymbol{B}^T\boldsymbol{\sigma}\, d\Omega$, and add to ELRESF(NEE).

g. **ADDRHS:** assemble ELRESF into BLHS; generic subroutine.

h. CONTINUE to next element.

i. IF (NSURF.GT.0) **QDCSUF:** compute element surface forces and assemble into BRHS.

S11. BACK: perform forward reduction, diagonal scaling, and back substitution to find $\Delta\boldsymbol{a}^{(i)}$ (dynamic analysis) or $\boldsymbol{d}^{(1)}$ (static analysis).

S12. ITERUP: intermediate update of active degrees of freedom.

IF (LDYN) THEN

$$\widetilde{\boldsymbol{d}}^{(i+1)} = \widetilde{\boldsymbol{d}}^{(i)} + (1+\alpha)\beta\Delta t^2\Delta\boldsymbol{a}^{(i)} \qquad \text{DPRED = DPRED + COEFF5 * BRHS}$$

$$\widetilde{\boldsymbol{v}}^{(i+1)} = \widetilde{\boldsymbol{v}}^{(i)} + (1+\alpha)\gamma\Delta t\Delta\boldsymbol{a}^{(i)} \qquad \text{VPRED = VPRED + COEFF4 * BRHS}$$

$$\widetilde{\boldsymbol{a}}^{(i+1)} = \widetilde{\boldsymbol{a}}^{(i)} + \Delta\boldsymbol{a}^{(i)} \qquad \text{A = A + BRHS}$$

ELSE

$$\boldsymbol{d} = \boldsymbol{d}^{(1)} \qquad \text{D = BRHS}$$

ENDIF

S13. CONTINUE to next iteration (GO TO S8).

S14. IF (LDYN) **CORRCT:** corrector update of all degrees of freedom.

$$\boldsymbol{d}_{n+1} = \frac{\widetilde{\boldsymbol{d}}^{(i+1)} - \boldsymbol{d}_n}{1+\alpha} + \boldsymbol{d}_n, \qquad \text{D = (DPRED - D)/COEFF1 + D}$$

$$\boldsymbol{v}_{n+1} = \frac{\widetilde{\boldsymbol{v}}^{(i+1)} - \boldsymbol{v}_n}{1+\alpha} + \boldsymbol{v}_n \qquad \text{V = (VPRED - V)/COEFF1 + V}$$

$$\boldsymbol{a}_{n+1} = \widetilde{\boldsymbol{a}}^{(i+1)}$$

S15. IF (**LOUT**(N,NDPRT1)) **PRINTD:** write kinematic output.

S16. IF (**LOUT**(N,NSPRT1)) **ELEMNT**('STR_PRNT'): loop over element groups NEG=1,NUMEG and perform stress recovery for printing; element task 4.

Example

QUADC: four-node quadrilateral, elastic continuum element.

1. **QDCT4:** element task 4.

DO h NEL=1,NUMEL

a. **LOCAL:** localize coordinates X and displacements D, store in XL(NESD,NEN) and DL(NED,NEN), respectively; generic subroutine.
b. **QDCSHG:** compute shape function global derivatives $N_{a,x}$, $N_{a,y}$ at integration points, store in SHG(NROWSH,NEN,NINT); generic subroutine.
c. Compute coordinates of integration points, store in XINT(NESD,NINT). IF (IOPT.EQ.2) set R(L) = XINT(1,L) and include radial weighting in Jacobian determinants DET(NINT).
d. IF (IBBAR.EQ.1) **MEANSH:** compute volume average of global shape function derivatives for mean-dilatational $\bar{\boldsymbol{B}}$; IF (IOPT.EQ.2), compute volume average of shape functions; store in SHGBR(NROWSH,NEN).

DO g L=1,NINT

e. **QDCSTR:** compute stress and strain for two-dimensional continuum element, store in STRESS(NROWB) and STRAIN(NROWB), respectively.
f. **PRTS2D:** print stresses and strains for two-dimensional element; generic subroutine.
g. CONTINUE to next quadrature point.
h. CONTINUE to next element.

S17. IF (LDYN .AND. **LOUT**(N,NHPLT1)) THEN

1. **STORED:** store kinematic time history data in DOUT(NDOUT+1,NPLPTS).
2. **ELEMNT**('STR_STOR'): loop over element groups NEG=1,NUMEG and perform stress recovery for element time-histories; element task 5.

Example

QUADC: four-node quadrilateral, elastic continuum element.

1. **QDCT5:** element task 5.

a. Store current time value in SOUT(1,LOCPLT).

DO h I=1,NSOUT

b. For active element **LOCAL:** localize coordinates X and displacements D, store in XL(NESD,NEN) and DL(NED,NEN), respectively; generic subroutine.
c. **QDCSHG:** Compute shape function global derivatives $N_{a,x}$, $N_{a,y}$ at integration points, store in SHG(NROWSH,NEN,NINT); generic subroutine.
d. IF (IOPT.EQ.2) compute radial coordinates of integration points, store in R(NINT). Include radial weighting in Jacobian determinants DET(NINT).
e. IF (IBBAR.EQ.1) **MEANSH:** compute volume average of global shape function derivatives for mean-dilatational $\bar{\boldsymbol{B}}$; IF (IOPT.EQ.2), compute volume average of shape functions; store in SHGBR(NROWSH,NEN).
f. For active integration point **QDCSTR:** compute stress and strain for two-dimensional continuum element, store in STRESS(NROWB) and STRAIN(NROWB), respectively.
g. Store required stress/strain component in SOUT(I+1,LOCPLT).
h. CONTINUE to next active element/quadrature point pair.

ENDIF

S18. CONTINUE to next time step (GO TO S5).

S19. CONTINUE to next time sequence (GO TO S1).

S20. IF (LDYN) THEN

1. **HPLOT:** plot nodal time-histories.
2. **ELEMNT**('STR_PLOT'): loop over element groups NEG=1,NUMEG and plot element time-histories with **HPLOT**; element task 6.

ENDIF

S21. RETURN to **DLEARN** (GO TO G2.6).

11.4 ADDING AN ELEMENT TO DLEARN

Example

Suppose we wish to add a shell element to the program. Let us call the subroutine that sets element group parameters and allocates blank common storage SUBROUTINE SHELL. The steps are as follows:

1. Study and understand thoroughly the subroutines QUADC and TRUSS before beginning. Notice the great similarity in their structure. Although a two-node truss

could be programmed more explicitly, i.e, in the spirit of matrix structural analysis, we feel the present implementation emphasizes more generalized finite element methodologies.

2. When studying the elements QUADC and TRUSS, note that the elements are broken into six calls to different subroutines; the variable ITASK defines which routine will be called. In the global routines the call to ELEMNT defines the value of ITASK. For example, in DRIVER, the statement CALL ELEMNT ('FORM_LHS') results in ITASK having the value 2 (see routine ELEMNT). In the routine QUADC, if ITASK has the value 2, the routine QDCT2 is executed. Therefore, to make any new element compatible with the "global element functions," the new element should be designed with these six functions in mind. However, if the user is extending the capabilities of DLEARN, he or she is totally free to define additional element functions. The six current functions are defined as follows:

    1. 'INPUT_____' : Read/generate element input data.
    2. 'FORM_LHS' : Compute element contribution to global left-hand-side matrix.
    3. 'FORM_RHS' : Compute element contribution to global right-hand-side vector.
    4. 'STR_PRNT' : Compute and print element stress/strain output.
    5. 'STR_STOR' : Compute element stress/strain for output time histories.
    6. 'STR_PLOT' : Plot element time histories.

As an example, we present an outline of SUBROUTINE QUADC.

```
      SUBROUTINE QUADC(ITASK,NPAR,MP,NEG)
C
C.... PROGRAM TO SET STORAGE AND CALL TASKS FOR THE
C        FOUR-NODE QUADRILATERAL, ELASTIC CONTINUUM ELEMENT
C
              *
              *
C
C.... DEFINE STORAGE POINTER NAMES
C
              *
              *
C
C.... EQUIVALENCE GROUP PARAMETERS FROM NPAR ARRAY
C
              *
              *
C
C.... SET ELEMENT PARAMETERS
C
              *
              *
      IF (ITASK.EQ.1) THEN
C
C...... SET MEMORY POINTERS
C
              *
              *
      ENDIF
C
C.... TASK CALLS
C
      IF (ITASK.GT.6) RETURN
      GO TO (100,200,300,400,500,600),ITASK
```

```
C
  100 CONTINUE
C
C.... INPUT ELEMENT DATA ('INPUT___')
C
      CALL QDCT1(A(MP(MSHL  )),A(MP(MW     )),A(MP(MRHO  )),
     *
     *
C
      RETURN
C
  200 CONTINUE
C
C.... FORM ELEMENT EFFECTIVE MASS AND ASSEMBLE INTO GLOBAL
C        LEFT-HAND-SIDE MATRIX  ('FORM_LHS')
C
      CALL QDCT2(A(MP(MELEFM)),A(MP(MIEN  )),A(MPX        ),
     *
     *
C
      RETURN
C
  300 CONTINUE
C
C.... FORM ELEMENT RESIDUAL-FORCE VECTOR AND ASSEMBLE INTO GLOBAL
C        RIGHT-HAND-SIDE VECTOR ('FORM_RHS')
C
      CALL QDCT3(A(MP(MMAT  )),A(MP(MIEN  )),A(MPDPRD     ),
     *
     *
C
      RETURN
C
  400 CONTINUE
C
C.... CALCULATE AND PRINT ELEMENT STRESS/STRAIN OUTPUT ('STR_PRNT')
C
      IF (ISTPRT.EQ.0)
     &   CALL QDCT4(A(MP(MMAT  )),A(MP(MIEN  )),A(MPD         ),
     *
     *
C
      RETURN
C
  500 CONTINUE
C
C.... CALCULATE AND STORE ELEMENT TIME-HISTORIES ('STR_STOR')
C
      IF (NSOUT.GT.0)
     &   CALL QDCT5(A(MP(MISHST)),A(MP(MSOUT )),A(MP(MMAT  )),
     *
     *
C
      RETURN
C
  600 CONTINUE
C
C.... PLOT ELEMENT TIME-HISTORIES ('STR_PLOT')
C
      IF (NSOUT.GT.0)
     &   CALL HPLOT(A(MP(MISHST)),A(MP(MSOUT )),NSOUT ,3,NTYPE )
      RETURN
C
      END
```

3. When coding SHELL and its task routines, use existing utility routines, such as MULTAB, whenever possible.[2]
4. Code a new version of subroutine ELMLIB to include a call to SHELL. The necessary modifications are indicated:

[2] The experienced programmer will recognize that the use of utility routines such as MULTAB is highly inefficient. Nevertheless, this approach facilitates implementation and results in very readable coding. Significant gains in efficiency can be made by avoiding subroutine calls and explicitly writing out the matrix products line by line, taking account of zeros and repeated calculations. See Sec. 3.10.

```
      GO TO (100,200, 300 ),NTYPE
C
  100 CALL QUADC(ITASK,A(MPNPAR),A(MPNPAR+16),NEG)
      RETURN
C
  200 CALL TRUSS(ITASK,A(MPNPAR),A(MPNPAR+16),NEG)
      RETURN
```

```
C
  300 CALL SHELL(ITASK,A(MPNPAR),A(MPNPAR+16),NEG)
      RETURN
```

5. Define an array LABEL3 in BLOCK DATA and specify the desired labels for stress, strain, and any other time-histories. Include the appropriate WRITE statements in HPLOT and SHIST.
6. Write a description for the Input Instructions to be included in Sec.In.10, ELEMENT DATA. Mimic the style of Secs. In.10.1 and In.10.2, and use the same variable names as far as possible.

Instructions for Debugging[3] and Running the New Element

1. Compile all your new routines separately before including them in DLEARN.

2. Link all routines compiled and execute with simple, static, one-element problems (e.g., homogeneous deformation states), which are easy to check. Check all options (e.g., gravity loads, etc.) on the one-element problems.

3. Next, proceed to "patch tests," which also should be easy to check. Finally, after you are convinced your element is working properly, try some interesting static and dynamic test problems for which you have exact analytical, or reliable experimental data. Assess the accuracy of the element you have programmed.

11.5 DLEARN USER'S MANUAL

11.5.1 Remarks for the New User

This section presents the necessary input instructions for the user to execute DLEARN. The new user is advised to read *all* the instructions, paying particular

[3] Debugging is boring and frustrating, though the use of interactive debug utilities can reduce the tedium. When you are feeling particularly depressed by this experience, remember *everyone* who has ever programmed has experienced what you are feeling now. Misery loves company.

attention to the appended notes, to gain an initial overview of the data required to specify an analysis task to DLEARN. Readers without prior finite element computational experience will benefit from careful study of the example problems; they are intended to help clarify the use of the data generation features. The amount of data required to specify even a small problem can become burdensome if the user does not learn to exercise these capabilities.

Upon examining the input instructions, the reader should recognize that no system of units is assumed by the program. Rather, *it is the user's responsibility to specify data in the consistent units of his or her choice*. Overlooking this requirement is a common source of erroneous results, especially in dynamic analysis.

It is often the case that new users have difficulty in getting a program to successfully run. It is tempting to assume that there are "bugs" in the program and that the fault does not lie with oneself. DLEARN has undergone extensive testing to minimize the possibility of any "bugs" being present. Thus, the new user should normally avoid this alibi, for the vast majority of aborted runs are the result of input specification errors. Train yourself to look closely at generation sequences; a single slipped digit can lead to unbelievably chaotic input data and frequently the legendary "floating point overflow/divide by zero."

Unfortunately, input errors need not be this apparent, though nonetheless catastrophic. For example, an error in generating initial displacements for a dynamic analysis may lead INPUT to write beyond the storage allocated for the D array. The D array is followed in COMMON A(MTOT) by the nodal coordinates array X(NSD,NUMNP), thus INPUT would unknowingly overwrite the coordinate data. This may later lead to an error termination, e.g., calculation of a nonpositive Jacobian determinant in an element or the successful solution of the *wrong* initial/boundary-value problem. All this would occur after the coordinate data are written to the output file, making detection difficult. This can be a frustrating task for the new user. However, with a little experience, efficiency in recognizing errors will be gained. Take heart, even the most seasoned analyst occasionally wastes time (and money) searching for such anomalies.

11.5.2 Input Instructions

In.0.0 Input Data Echo Card[4] (I5)

Note	Columns	Variable	Description
(1)	1–5	IECHO	Input file data echo EQ.0, no data echo EQ.1, data echo

Notes

1. Data from input file are read and written, without interpretation, at beginning of output file before execution continues.

[4] Recognizing the preponderance of interactive computing today, the term "card" as used here refers to an 80-character line in the input file.

In.1.0 Title Card (20A4)

Note	Columns	Variable	Description
(1)	1–80	TITLE(20)	Job title for heading the output

Notes

1. Each data set begins with a title card. *To terminate execution, the last card in the input file, which will be read as a new title card, must contain the characters* *END *in the first four columns.*

In.2.0 Execution Control Card (15I5)

Note	Columns	Variable	Description
(1)	1–5	IEXEC	Execution code EQ.0, data check only EQ.1, execution
(2)	6–10	IACODE	Analysis code EQ.0, dynamic analysis EQ.1, static analysis
(3)	11–15	IREADR	Read restart file code EQ.0, do not read restart file EQ.1, read restart file
(4)	16–20	IWRITR	Write restart file code EQ.0, do not write restart file EQ.1, write restart file
(5)	21–25	IPRTIN	Input data print code EQ.0, print nodal and element input data EQ.1, do not print nodal and element input data
(6)	26–30	IRANK	Rank check code EQ.0, do not perform rank check EQ.1, print numbers of zero and negative pivots EQ.2, print all pivots
(7)	31–35	NUMSEQ	Number of time sequences, GE.1
(8)	36–40	NDOUT	Number of nodal output time histories
(9)	41–45	NSD	Number of space dimensions
	46–50	NUMNP	Number of nodal points
(10)	51–55	NDOF	Number of nodal degrees of freedom
(11)	56–60	NLVECT	Number of load vectors
(12)	61–65	NLTFTN	Number of load-time functions
(13)	65–70	NPTSLF	Number of points defining a load-time function
(14)	70–75	NUMEG	Number of element groups

Notes

1. In the data check mode, input data are printed and storage requirements are indicated. This mode should be employed prior to making expensive calculations.
2. A static analysis consists of solution of the equilibrium equations, resulting in displacement and stress output. A dynamic analysis consists of time-integration of the semi-discrete equations of motion, resulting in displacement, velocity, acceleration, and stress output.
3. If IREADR .EQ. 1, then initial displacement, velocity, and acceleration data will be read from the file associated with FORTRAN unit IRSIN at the beginning of a dynamic analysis (i.e. IACODE .EQ. 0).
4. If IWRITR .EQ. 1, then final displacement, velocity, and acceleration data will be written to the file associated with FORTRAN unit IRSOUT at the conclusion of a dynamic analysis (i.e., IACODE .EQ. 0).
5. If IPRTIN .EQ. 1, then only control input data, e.g., time-sequence data, is printed in the output file. This option is useful in reducing output for very large amounts of generated input data. However, it should only be employed after thorough checking of data has been performed prior to execution.
6. A rank check consists of factorization of the effective mass matrix, $\boldsymbol{M}^* = \boldsymbol{U}^T\boldsymbol{D}\boldsymbol{U}$, and either printing the numbers of zero and negative entries along the diagonal of $\boldsymbol{D}$ (IRANK .EQ. 1), or printing the entire diagonal of $\boldsymbol{D}$ (IRANK .EQ. 2), for each time sequence. The number of zero-energy modes can be deduced from the number of zeroes along the diagonal of $\boldsymbol{D}$. This option allows for the checking of elements and meshes suspected of containing spurious zero-energy modes, such as those arising in so-called "under-integrated" finite elements. If IRANK .GT. 0, then no other executable calculations are performed (i.e., dynamic or static analysis).
7. For static analysis, a single time sequence should be adequate, although the use of multiple sequences is not prohibited. A dynamic analysis can be subdivided into multiple contiguous time intervals in which different time-integration constants and output parameters may be specified. See Sec. In.3.0.
8. Output histories are ignored in static analysis as denoted by IACODE .EQ. 1. See Sec. In.4.0.
9. Consult individual element routines in Sec. In.10.0 for requirements.
10. In problems for which there are different numbers of degress of freedom at different nodes, NDOF should be set equal to the maximum number. Consult individual element routines in Sec. In.10.0 for requirements. Superfluous degrees of freedom can be eliminated by employing boundary condition codes. See Sec. In.6.0.
11. The prescribed nodal forces and kinematic boundary conditions are synthesized by NLVECT vectors and corresponding load-time functions. Details are presented in Secs. In.7.0 and In.8.0.
12. The user must define NLTFTN .GE. NLVECT. The first NLVECT load-time functions scale the corresponding load vectors. Additional load-time functions may be defined to scale element surface and body forces specified in the element group data.
13. The load-time functions are discretized by NPTSLF points. See Sec. In.8.0 for details.
14. The elements are read in groups.

In.3.0 Time Sequence Data Cards (6I5,4F10.0)

Note	Columns	Variable	Description
(1)	1–5	N	Sequence number
	6–10	NSTEP(N)	Number of time steps for sequence N
(2)	11–15	NDPRT(N)	Print kinematic output every NDPRT(N) time steps
(2)	16–20	NSPRT(N)	Print stress/strain output every NSPRT(N) time steps
(2)	21–25	NHPLT(N)	Plot time history data every NHPLT(N) time steps
	26–30	NITER(N)	Number of iterations in corrector loop; GE. 1
(3)	31–40	ALPHA(N)	α parameter for sequence N, LE. 0.0
(3)	41–50	BETA(N)	β parameter for sequence N, GT. 0.0
(3)	51–60	GAMMA(N)	γ parameter for sequence N, GE. 0.0
(3)	61–70	DT(N)	Time step Δt for sequence N

Notes

1. NUMSEQ time sequence data cards must be defined but need not be input in any particular order. NUMSEQ .GE. 1 is defined on the Execution Control Card; see Sec. In.2.0.
2. If set Eq. 0, no output. In each sequence, the final print/plot state is output at *sequence* time step NSTEP(N) − MOD(NSTEP(N), NDPRT(N)), etc., and the output logic will reinitialize at the beginning of the following sequence.
3. Many different transient algorithms can be realized by the proper choice of parameters α, β, and γ. The choice of $\alpha \in [-\frac{1}{3}, 0]$, $\beta = \frac{1}{4}(1 - \alpha)^2$, $\gamma = \frac{1}{2}(1 - 2\alpha)$ leads to a second-order accurate method, which is unconditionally stable *if all element groups are specified to be implicit.* The user should be cognizant of any stability restrictions upon the time step size Δt; *such limitations can be severe when using explicit element groups.*

   To compute the consistent initial acceleration for all *active* degrees of freedom, specify the first time sequence to consist of one time-step with $\alpha = 0$ and $\Delta t = 0$. The code will then solve $\boldsymbol{M}\Delta\boldsymbol{a} = \boldsymbol{F}_0 - \boldsymbol{M}\boldsymbol{a}_0 - \boldsymbol{C}\boldsymbol{v}_0 - \boldsymbol{K}\boldsymbol{d}_0$ and update: $\boldsymbol{a}_0 = \boldsymbol{a}_0 + \Delta\boldsymbol{a}$, $\boldsymbol{v}_0 = \boldsymbol{v}_0$, and $\boldsymbol{d}_0 = \boldsymbol{d}_0$. The values of β and γ are irrelevant, but the user should specify NDPRT = NSPRT = NHPLT = 1 if he or she wishes to examine the results of this calculation. Note that there is no rational way to *calculate* the initial acceleration for any node controlled by kinematic boundary conditions (see Sec. In.6.0): These data must be specified by the user (See Sec. In.9.0).

   For *static analysis*, the following defaults are set internally:

$$\text{NSTEP(N)} = \text{MAX0(1, NSTEP(N))}$$

$$\text{NDPRT(N)} = 1 \qquad \alpha = 0.0$$

$$\text{NSPRT(N)} = 1 \qquad \beta = 1.0$$

$$\text{NHPLT(N)} = 0 \qquad \gamma = 0.0$$

$$\text{NITER(N)} = 1 \qquad \Delta t = 1.0$$

In.4.0 Nodal History Data Cards (3I5)

If there are no nodal output time-histories (i.e., NDOUT .EQ. 0 on the Execution Control Card; see Sec. In.2.0), then no nodal history data cards should be input; proceed directly to Sec. In.5.0.

If the analysis is static (i.e., IACODE .EQ. 1 on the Execution Control Card), then output history cards are read, but subsequently ignored.

Each component requested is plotted versus time in the output file. Plots of this type are useful in providing quick information concerning the time histories of important components. Data for line plots of nodal variables are stored in array DOUT. The user may wish to modify DLEARN to write this information to a disk file for subsequent off-line plotting.

Note	Columns	Variable	Description
(1)	1–5	IDHIST(1,N)	Node number; GE. 1 and LE. NUMNP
	6–10	IDHIST(2,N)	Degree of freedom number; LE. NDOF
(2)	11–15	IDHIST(3,N)	Kinematic quantity specifier

Notes

1. History output data must be input for each node/degree of freedom combination at which a time history output is desired. The index N in the array IDHIST corresponds to the order in which the cards are read: 1 .LE. N .LE. NDOUT. The cards need not be specified in any particular order. *The total number of degrees of freedom to be plotted for all nodes must equal* NDOUT, *defined on the Execution Control Card* (see Sec. In.2.0).
2. IDHIST(3,N) = 1, displacement
   IDHIST(3,N) = 2, velocity
   IDHIST(3,N) = 3, acceleration

In.5.0 Coordinate Data

In.5.1 Nodal Coordinate Cards (2I5,NSD×F10.0)

Note	Columns	Variable	Description
(1)	1–5	N	Node number; GE. 1 and LE. NUMNP
(2)	6–10	NUMGP	Number of generation points; EQ.0, no generation GT.0, generate nodal data
(3)	11–20	X(1,N)	x_1-coordinate of node N
	21–30	X(2,N)	x_2-coordinate of node N
	31–40	X(3,N)	x_3-coordinate of node N

Notes

1. The coordinates of each node must be defined, but need not be input in any particular

order. If NSD is specified as 2 on the Execution Control Card (See Sec. In.2.0), any nonzero x_3-coordinates will be ignored. *Terminate with a blank card.*

2. If NUMGP is greater than zero, this card initiates an isoparametric data generation sequence. Cards 2 to NUMGP of the sequence define the coordinates of the additional generation points (see Sec. In.5.2). The final card of the sequence defines the nodal increment information (see Sec. In.5.3). After the generation sequence is completed, additional nodal coordinate cards, or generation sequences, may follow.

The generation may be performed along a line, over a surface, or within a volume. A description of each of these options follows.

Generation along a line

The line may be defined by two or three generation points (see Fig. 11.5.1), and the physical space may be one-, two-, or three-dimensional.

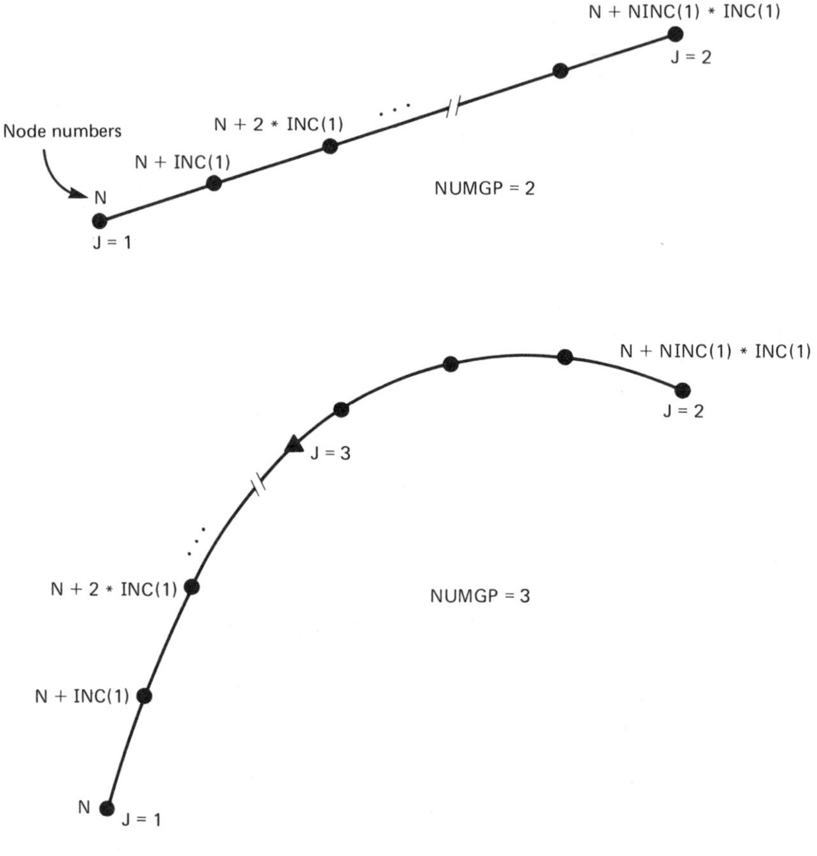

Figure 11.5.1 Nodal generation along a line.

In the case NUMGP = 2, linear interpolation takes place, resulting in equally spaced nodal points.

In the case NUMGP = 3, quadratic interpolation is employed and graded nodal spacing may be achieved by placing the third generation point (J = 3) off center. Note that the third generation point does not generally coincide with any nodal point. The spacing in this case may be determined from the following mapping:

$$\boldsymbol{x}_A = \boldsymbol{x}(\xi_A) = \frac{1}{2}\xi_A(\xi_A - 1)\boldsymbol{x}_1^g + \frac{1}{2}\xi_A(\xi_A + 1)\boldsymbol{x}_2^g + (1 - \xi_A^2)\boldsymbol{x}_3^g$$

where ξ_A is the location of node number A in ξ-space. The nodes are placed at equal intervals in ξ-space; $\boldsymbol{x}_1^g$, $\boldsymbol{x}_2^g$ and $\boldsymbol{x}_3^g$ are the coordinates of the three generation points in $\boldsymbol{x}$-space; and $\boldsymbol{x}_A$ denotes the coordinates of the Ath node in $\boldsymbol{x}$-space (see Fig. 11.5.2).

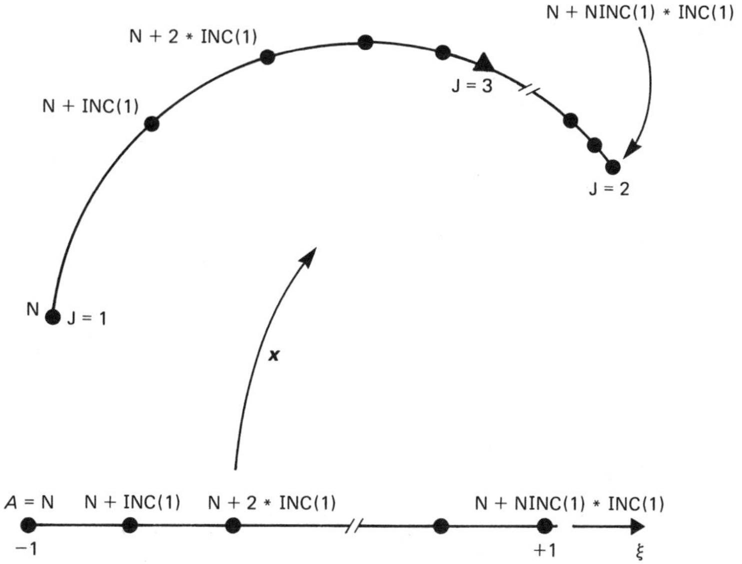

Figure 11.5.2 Nodal generation along a line: mapping from local interval to physical space.

Generation over a surface

The surface may be defined by four or eight generation points (see Fig. 11.5.3) and the physical space may be two- or three-dimensional. In the three-dimensional case, the surfaces may be curved.

In the case NUMGP = 4, bilinear interpolation is employed, resulting in equally spaced nodal points along generating lines.

In the case NUMGP = 8, biquadratic serendipity interpolation is employed, and graded nodal spacing may be achieved by placing generation points 5–8 off center. Note that generation points 5–8 do not generally coincide with any nodal points. The spacing of the nodal points may be determined from the serendipity mapping.

Generation within a volume

The volume is brick shaped and may be defined by 8 or 20 generation points (see Fig. 11.5.4). In this case the physical space must be three-dimensional.

If NUMGP = 8, trilinear interpolation is employed, resulting in equally spaced nodal points along generating lines.

If NUMGP = 20, triquadratic serendipity interpolation is employed and graded nodal spacing may be achieved by placing generation points 9–20 off center. Note that generation points 9–20 do not generally coincide with any nodal points. The spacing of the nodal points may be determined by the serendipity mapping.

3. If the coordinates of node N are input and/or generated more than one time, the last values take precedence.

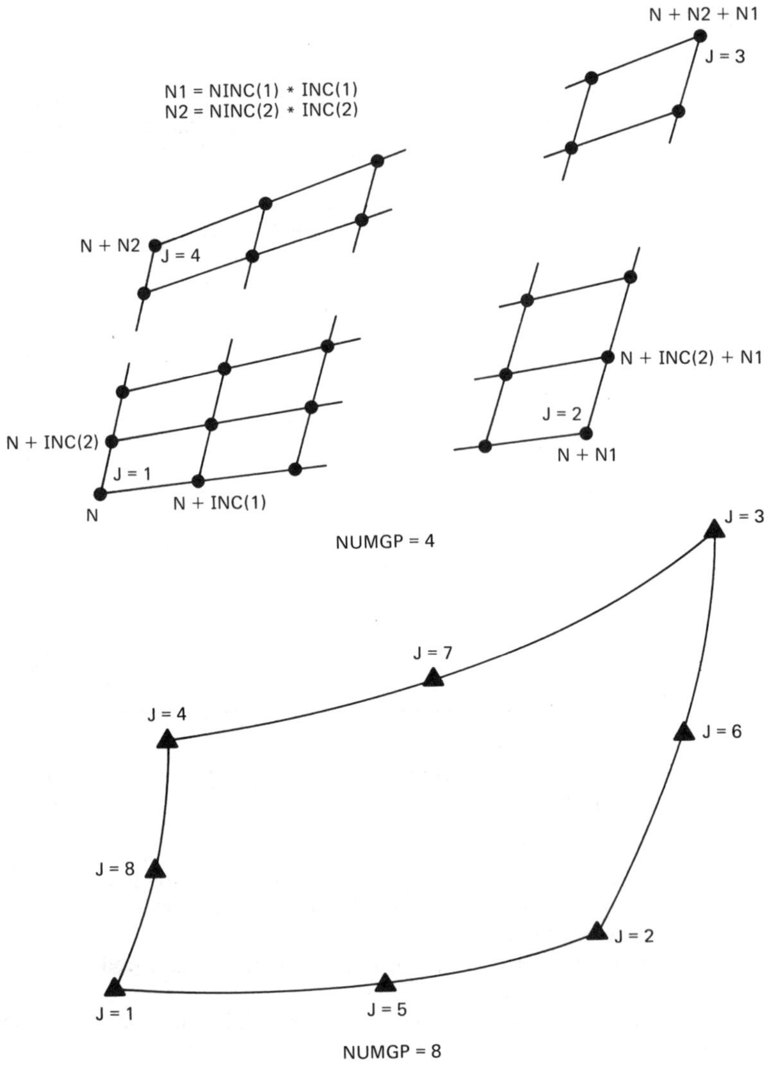

Figure 11.5.3 Nodal generation over a surface.

In.5.2 Generation Point Coordinate Cards (2I5,NSD×F10.0)

Note	Columns	Variable	Description
(1)	1–5	M	Node number
	6–10	MGEN	Generation parameter; EQ.0, coordinates of the Jth generation point are input on this card; M is ignored.

Note	Columns	Variable	Description
			EQ.1, coordinates of the Jth generation point are set equal to coordinates of the Mth node which was previously defined; coordinates on this card are ignored.
	11–20	TEMP(1,J)	x_1-coordinate of generation point J
	21–30	TEMP(2,J)	x_2-coordinate of generation point J
	31–40	TEMP(3,J)	x_3-coordinate of generation point J

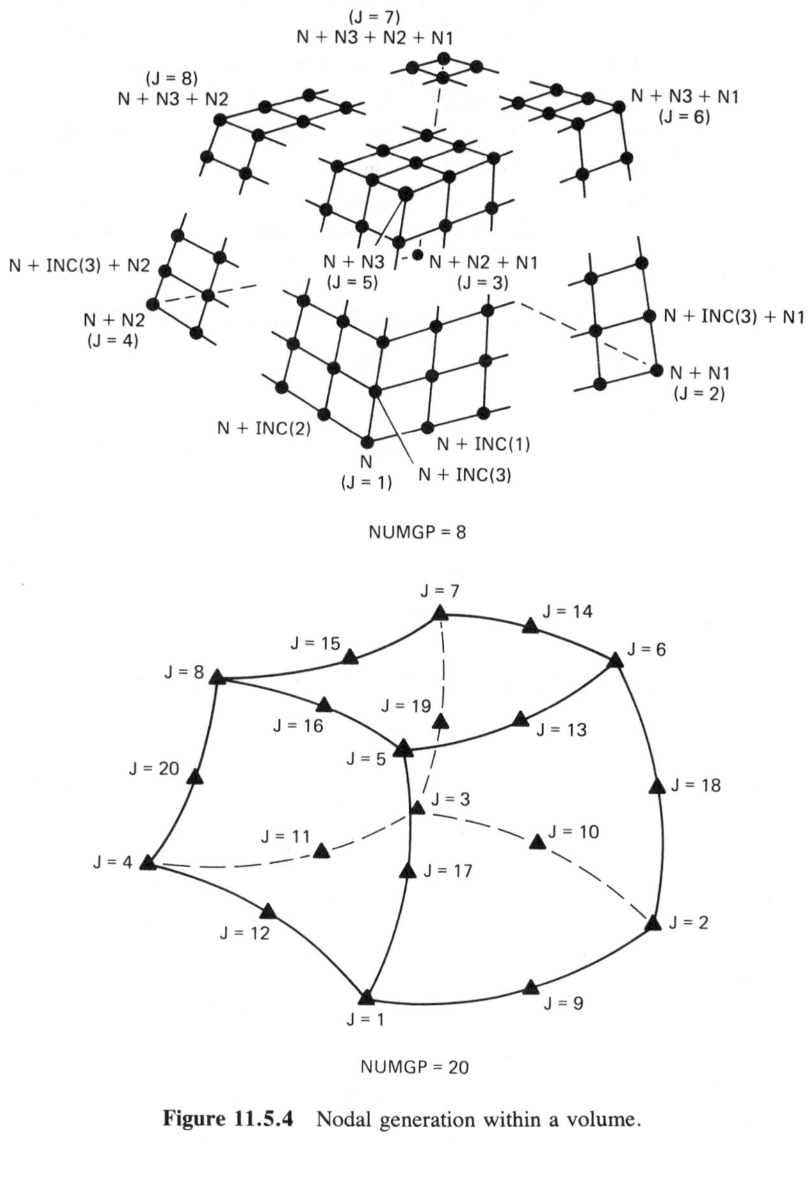

Figure 11.5.4 Nodal generation within a volume.

Notes

1. The coordinates of each generation point are defined by a generation point coordinate card. The cards must be read in order (J = 2, 3, . . . , NUMGP) following the nodal coordinate card which initiated the generation sequence (J = 1). A nodal increments card (see Sec. In.5.3), which completes the sequence, follows the last generation point card (J = NUMGP).

In.5.3 Nodal Increments Card (6I5)

Note	Columns	Variable	Description
	1–5	NINC(1)	Number of node increments for direction 1; GE.0
	6–10	INC(1)	Node number increment for direction 1
(1)	11–15	NINC(2)	Number of node increments for direction 2; GE.0
	16–20	INC(2)	Node number increment for direction 2
(1)	21–25	NINC(3)	Number of node increments for direction 3; GE.0
	26–30	INC(3)	Node number increment for direction 3

Notes

1. Each generation option is assigned an option code (IOPT) as follows:

IOPT	**Option**
1	generation along a line
2	generation over a surface
3	generation within a volume

IOPT is determined by the following logic:

IOPT = 3

IF (NINC(3).EQ.0) IOPT = 2

IF (NINC(2).EQ.0) IOPT = 1

In.6.0 Boundary Conditions Data Cards((3+NDOF) × I5)

Note	Columns	Variable	Description
(1)	1–5	N	Number of first node in sequence GE.1 and LE. NUMNP
(2)	6–10	NE	Number of last node in generation sequence LE. NUMNP
	11–15	NG	Generation increment
(3)	16–20	ID(1,N)	Degree of freedom 1 boundary code
	21–25	ID(2,N)	Degree of freedom 2 boundary code
	⋮	⋮	⋮
		ID(NDOF,N)	Degree of freedom NDOF boundary code

Notes

1. Boundary condition data must be input for each node which has one or more specified displacements, velocities or accelerations. Cards need not be specified in any particular order. *Terminate with a blank card.*
2. Each data card will generate identical boundary condition codes for the following sequence of nodes:

   $$\text{N, N+NG, N+2}\times\text{NG, } \ldots \text{, NE}-\text{MOD(NE}-\text{N,NG)}$$

   If NG is blank or zero, no generation takes place and only boundary conditions for node N are defined. In this case NE is ignored.
3. Boundary condition codes may be assigned the following values:

   ID(I,N) = 0, unspecified displacement (active degree of freedom)

   ID(I,N) = 1, specified displacement

   ID(I,N) = 2, specified velocity

   ID(I,N) = 3, specified acceleration

   Upon exiting subroutine BC, the ID array holds the following values:

   ID(I,N) = M (> 0), global equation number for active degree of freedom I of node N

   ID(I,N) = 0, displacement specified for degree of freedom I of node N

   ID(I,N) = −1, velocity specified for degree of freedom I of node N

   ID(I,N) = −2, acceleration specified for degree of freedom I of node N

In.7.0 Prescribed Nodal Forces and Kinematic Boundary Conditions

If NLVECT .EQ. 0 on the Execution Control Card (see Sec. In.2.0), no prescribed nodal forces and boundary condition cards should be input; proceed directly to Sec. In.8.0.

Prescribed nodal forces and kinematic boundary conditions, i.e., displacements, velocities or accelerations, are defined by expansions of the form:

$$\boldsymbol{F}(\boldsymbol{x}_A, t) = \sum_{i=1}^{\text{NLVECT}} G_i(t)\boldsymbol{F}_i(\boldsymbol{x}_A)$$

where $\boldsymbol{F}(\boldsymbol{x}_A, t)$ is the resultant force or kinematic boundary condition at node A at time t; G_i is the ith load-time function; $\boldsymbol{F}_i$ is the ith load vector; and NLVECT is the total number of load vectors as specified on the Execution Control Card. The data preparation for the load-time functions is described in Sec. In.8.0. In this section, the data specification for each $\boldsymbol{F}_i$ is described.

The load vectors should be read in the order $\boldsymbol{F}_1, \boldsymbol{F}_2, \ldots, \boldsymbol{F}_{\text{NLVECT}}$. Note that each entry of a single vector can be interpreted as a force or kinematic quantity depending on its corresponding boundary condition code defined in Sec. In.6.0.

Data cards for a typical load vector are described next.

In.7.1 Nodal Prescribed Forces/Kinematic Boundary Conditions Cards (2I5,NDOF×F10.0)

Note	Columns	Variable	Description
(1)	1–5	N	Node number; GE.1 and LE.NUMNP
(2)	6–10	NUMGP	Number of generation points EQ.0, no generation GT.1, generate data
(3)	11–20	F(1,N)	Degree of freedom 1 prescribed force/ kinematic boundary condition
	21–30	F(2,N)	Degree of freedom 2 prescribed force/ kinematic boundary condition
	⋮	⋮	⋮
		F(NDOF,N)	Degree of freedom NDOF prescribed force/ kinematic boundary condition

Notes

1. Data must be included for each node subjected to a nonzero prescribed force or kinematic boundary condition. Cards need not be specified in any particular order. *Terminate input for each load vector with a blank card.*
2. If NUMGP is greater than zero, this card initiates an isoparametric data generation sequence. The scheme used is the same as that employed for coordinate generation described previously in Sec. In.5. Cards 2 to NUMGP of the sequence define the prescribed forces/kinematic boundary condition data of the additional generation points (see Sec. In.7.2). The final card of the sequence defines the nodal increment information (see Sec. In.5.3). After the generation sequence is completed, additional nodal prescribed forces/kinematic boundary conditions cards, or generation sequences, may follow.

   The generation may be performed along a line, over a surface, or within a volume. A description of each of these options is given below.

   Generation along a line

   The line may be defined by two or three generation points (see Fig. 11.5.5), and the physical space ($\boldsymbol{x}$-space) may be one-, two-, or three-dimensional.

   In the case NUMGP = 2, linear interpolation takes place with respect to ξ-space. If the nodes are equally spaced in $\boldsymbol{x}$-space, then the variation will also be linear in $\boldsymbol{x}$-space. Otherwise, a nonlinear variation will be induced by the unequal nodal spacing.

   In the case NUMGP = 3, quadratic interpolation is performed with respect to ξ-space. Note that the third generation point does not generally coincide with any nodal point. The variation of the data may be determined from the following mapping:

$$\boldsymbol{d}_A = \boldsymbol{d}(\xi_A) = \tfrac{1}{2}\xi_A(\xi_A - 1)\boldsymbol{d}_1^g + \tfrac{1}{2}\xi_A(\xi_A + 1)\boldsymbol{d}_2^g + (1 - \xi_A^2)\boldsymbol{d}_3^g$$

   where ξ_A is the location of node number A in ξ-space (recall that nodes are assumed to be placed at equal intervals in ξ-space); $\boldsymbol{d}_1^g$, $\boldsymbol{d}_2^g$, and $\boldsymbol{d}_3^g$ are the data assigned to the three generation points in $\boldsymbol{x}$-space; and $\boldsymbol{d}_A$ denotes the prescribed force/kinematic boundary condition data of the Ath node in $\boldsymbol{x}$-space.

   The case in which NUMGP = 2 may be deduced from the case NUMGP = 3 by setting $\boldsymbol{d}_3^g = (\boldsymbol{d}_1^g + \boldsymbol{d}_2^g)/2$.

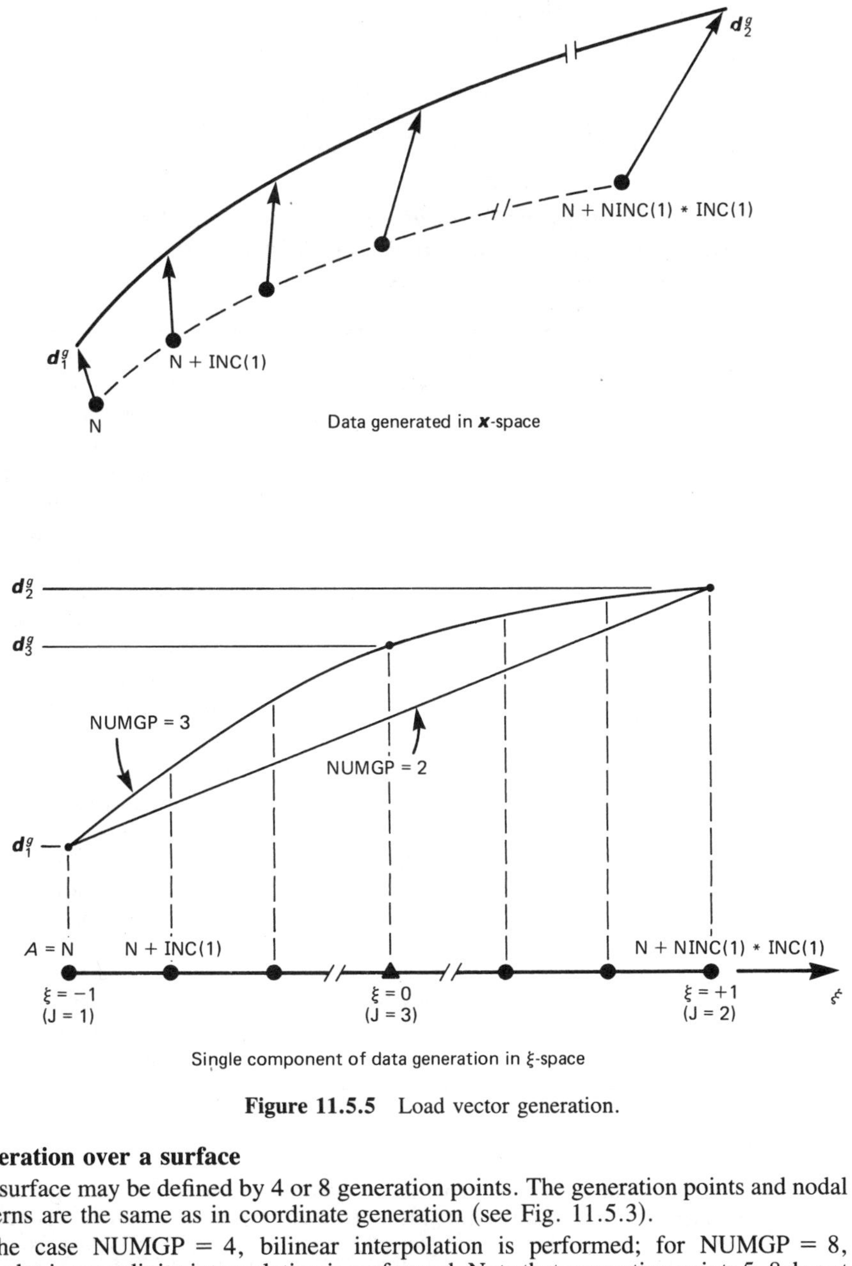

Figure 11.5.5 Load vector generation.

Generation over a surface

The surface may be defined by 4 or 8 generation points. The generation points and nodal patterns are the same as in coordinate generation (see Fig. 11.5.3).

In the case NUMGP = 4, bilinear interpolation is performed; for NUMGP = 8, biquadratic serendipity interpolation is performed. Note that generation points 5–8 do not generally coincide with any nodal points.

Generation within a volume

Generation of data within a brick-shaped volume may be performed using 8 or 20 generation points. The generation points and nodal patterns are the same as for coordinate generation (see Fig. 11.5.4).

If NUMGP = 8, trilinear interpolation is performed; if NUMGP = 20, triquadratic serendipity interpolation is performed. Note that generation points 9–20 do not generally coincide with any nodal points.

3. The elements of the array F(NDOF,NUMNP,NLVECT) are initialized to zero. Prescribed forces/kinematic boundary conditions for node N are written within this array. If values for node N are defined more than once, the last value will take precedence.

In.7.2 Generation Point Prescribed Forces/Kinematic Boundary Conditions Cards (2I5,NDOF×F10.0)

Note	Columns	Variable	Description
1	1–5	M	Node number
	6–10	MGEN	Generation parameter; EQ.0, prescribed forces/kinematic boundary conditions of the Jth generation point are input on this card; M is ignored. EQ.1, prescribed forces/kinematic boundary conditions of the Jth generation point are set equal to the prescribed forces/kinematic boundary conditions of the Mth node which were previously defined; prescribed forces/ kinematic boundary conditions on this card are ignored.
	11–20	TEMP(1,J)	Degree of freedom 1 prescribed force/ kinematic boundary condition for generation point J
	21–30	TEMP(2,J)	Degree of freedom 2 prescribed force/ kinematic boundary condition for generation point J
	⋮	⋮	⋮
		TEMP(NDOF,J)	Degree of freedom NDOF prescribed force/kinematic boundary condition for generation point J

Notes

1. The prescribed forces/kinematic boundary conditions of each generation point are defined by a generation point prescribed forces/kinematic boundary conditions card. The cards must be read in order (J = 2, 3, . . . , NUMGP) following the nodal prescribed forces/kinematic boundary conditions card which initiated the generation sequence (J = 1). A nodal increments card (see Sec. In.5.3) follows the last generation point card (J = NUMGP) and completes the sequence.

In.7.3 Nodal Increments Card (6I5)

See instructions in Sec. In.5.3.

In.8.0 LOAD-TIME FUNCTIONS

If NLTFTN .EQ. 0 on the Execution Control Card (see Sec. In.2.0), no load-time function cards should be input; proceed directly to Sec. In.9.0.

If NLTFTN .GT. 0, the user must input NLTFTN load-time functions, each defined by NPTSLF pairs of time instants and function values, where NLTFTN and NPTSLF are defined on the Execution Control Card. A schematic representation of a typical load-time function is shown in Fig. 11.5.6. The time instants must be in

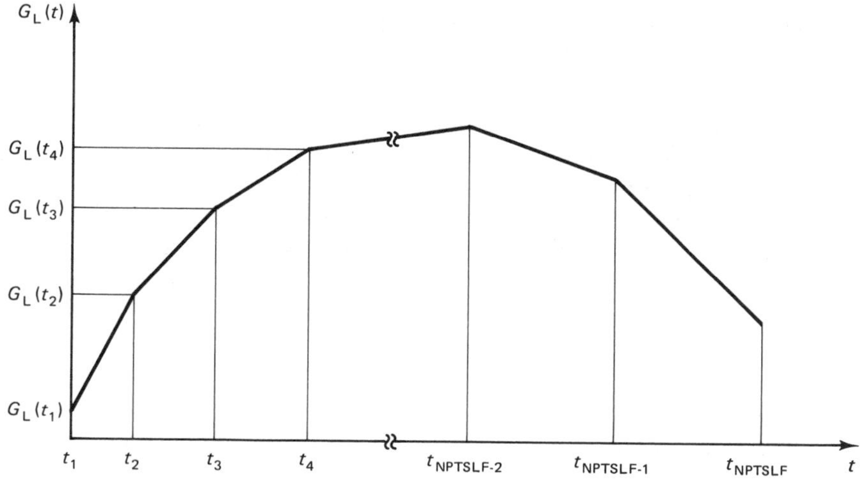

Figure 11.5.6 Schematic representation of load-time function L.

ascending order (i.e., $t_k < t_{k+1}$, $1 \leq k \leq \text{NPTSLF} - 1$). Load step intervals need not be equal and need not be the same from one load-time function to another. However, there must be the same number of load steps (i.e., NPTSLF) for each load-time function. As may be seen from Fig. 11.5.6, the load-time function is assumed to behave in a piecewise linear fashion between data points. For values of t outside the interval $[t_1, t_{\text{NPTSLF}}]$ we define G_i by constant extrapolation: $G_i(t) = G_i(t_1)$ for all $t \leq t_1$ and $G_i(t) = G_i(t_{\text{NPTSLF}})$ for all $t \geq t_{\text{NPTSLF}}$. For example, if the user wishes to define all load-time functions with a constant magnitude for all time, he may set NPTSLF = 1 and define a single point of $G_i(t)$ for any time t.

The load-time functions must be read in the order $G_1, G_2, \ldots, G_{\text{NLTFTN}}$, where the first NLVECT functions scale the corresponding load vectors specified in Sec. In.7.0. Element loads (e.g., pressure, gravity, etc.) are also multiplied by load-time functions. The applicable function number is defined in the element group data (see Sec. In.10).

Data cards for a typical load-time function are described next.

In.8.1 Load-time Function Cards (2F10.0)

Note	Columns	Variable	Description
(1)	1–10	G(K,1)	Time value K, i.e., t_k
	11–20	G(K,2)	Value of load-time function at time t_k

Notes

1. The local array G(K,I) is equivalent to global array G(K,I,L) for load-time function L.

In.9.0 KINEMATIC INITIAL CONDITION DATA

If the analysis is static (i.e., if IACODE .EQ. 1 on the Execution Control Card, see Sec. In.2.0) or if the program is instructed to read a restart file (i.e., if IREADR .EQ. 1 on the Execution Control Card), then no nodal initial condition data cards should be input; proceed directly to Sec. In.10.0.

Initial nodal displacements, velocities and accelerations are specified in separate data blocks having identical form. Each block is terminated with a blank card, which must be present even if zero initial conditions are desired at all nodes for one of the kinematic variables. For example, if the user wishes only to specify nonzero initial velocities, the data will have the form:

1. A blank card representing the absence of nonzero initial displacement data.
2. A block of initial velocity data input in the manner specified by Secs. In.9.1 to In.9.3 and terminating with a blank card.
3. A blank card representing the absence of nonzero initial acceleration data.

In.9.1 Nodal Initial Condition Cards (2I5,NSD×F10.0)

Note	Columns	Variable	Description
(1)	1–5	N	Node number; GE.1 and LE. NUMNP
(2)	6–10	NUMGP	Number of generation points EQ.0, no generation GT.0, generate initial conditions
(3)	11–20	D(1,N)*	Degree of freedom 1 initial condition
	21–30	D(2,N)	Degree of freedom 2 initial condition
	⋮	⋮	⋮
		D(NDOF,N)	Degree of freedom NDOF initial condition

* Here the array D is used generically to refer to either displacements (D), velocities (V), or accelerations (A).

Notes

1. Data must be included for each node subjected to nonzero initial displacements, velocities, or accelerations but need not be specified in any particular nodal order. *Terminate each data block with a blank card.*
2. If NUMGP is greater than zero, this card initiates an isoparametric data generation sequence. The scheme used is the same as that for prescribed nodal force/boundary condition data generation described in Sec. In.7.0. Cards 2 to NUMGP of the sequence define the initial condition data of the additional generation points and are described in Sec. In.9.2. The final card of the sequence defines the nodal increment information and is identical to the one used for coordinate generation (see Sec. In.5.3). After the generation sequence is completed, additional nodal initial condition cards, or generation sequences, may follow.

   The generation may be performed along a line, over a surface, or within a volume. For additional information concerning these options see Note (2) of Sec. In.7.1.
3. The elements of the arrays D(NDOF,NUMNP), V(NDOF,NUMNP) and A(NDOF, NUMNP) are initialized to zero. If the initial condition data of node N is input and/or generated more than one time, the last value takes precedence.

In.9.2 Generation Point Initial Condition Cards (2I5,NDOF×F10.0)

Note	Columns	Variable	Description
(1)	1–5	M	Node number
	6–10	MGEN	Generation parameter EQ.0, initial condition data of the Jth generation point are input on this card; M is ignored. EQ.1, initial condition data of the Jth generation point are set equal to the initial condition data of the Mth node which were previously defined; initial condition data on this card are ignored.
	11–20	TEMP(1,J)	Degree of freedom 1 initial condition for generation point J
	21–30	TEMP(2,J)	Degree of freedom 2 initial condition for generation point J
	. . .	. . .	. . .
		TEMP(NDOF,J)	Degree of freedom NDOF initial condition for generation point J

Notes

1. The initial condition data of each generation point are defined by a generation point initial condition card. The cards must be read in order (J = 2,3,...,NUMGP) following the nodal initial condition card that initiated the generation sequence (J = 1). A nodal increments card (see Sec. In.5.3), which completes the sequence, follows the last generation point card (J = NUMGP).

In.9.3 Nodal Increments Card (6I5)

See the discussion in Sec. In.5.3.

In.10.0 ELEMENT DATA

In.10.1 Two-Dimensional, Isotropic Elasticity Element

The present element may be used in triangular (three-node) or quadrilateral (four-node) form for plane stress, plane strain, or torsionless axisymmetric analysis. Two displacement degrees of freedom, in the x_1- and x_2-directions, are assigned to each node. Therefore, the user must specify NSD .GE. 2 and NDOF .GE. 2 on the Execution Control Card (see Sec. In.2.0).

Stress and strain output is computed at each of the integration points. Printed results include stresses and strains in the global coordinate system, principal stresses and strains, maximum shear stress or strain in the x_1, x_2-plane and angle of inclination, in degrees, of principal states (see Fig. 11.5.7). Any one of these quantities may also be output as a time history. All shear strains are reported according to the "engineering" convention (i.e., *twice* the value of the tensor components). See Sec. In.10.1.6 for precise definitions of the output quantities.

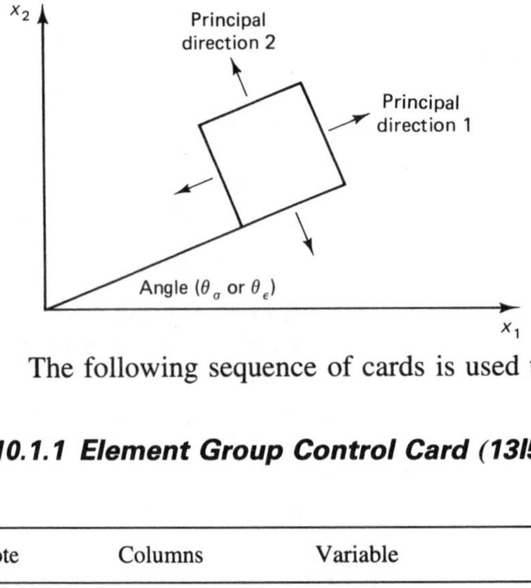

Figure 11.5.7 Orientation of principal axes.

The following sequence of cards is used to describe the elements in a group.

In.10.1.1 Element Group Control Card (13I5)

Note	Columns	Variable	Description
(1)	1–5	NTYPE	The number 1
	6–10	NUMEL	Number of elements in this group; GE. 1
	11–15	NUMAT	Number of material sets in this group; GE. 1
	16–20	NSURF	Number of surface force cards
(2)	21–25	NSOUT	Number of stress/strain time-history cards
	26–30	IOPT	Analysis option EQ.0, plane stress EQ.1, plane strain EQ.2, torsionless axisymmetric

Note	Columns	Variable	Description
(3)	31–35	ISTPRT	Stress output print code EQ.0, print stress output EQ.1, do not print stress output
(4)	36–40	LFSURF	Surface force load function number
(4)	41–45	LFBODY	Body force load function number
(5)	46–50	NICODE	Numerical integration code EQ.0, 2×2 Gaussian quadrature EQ.1, one-point Gaussian quadrature
(6)	51–55	IBBAR	Strain-displacement option EQ.0, standard formulation EQ.1, mean-dilatational $\overline{\boldsymbol{B}}$ formulation
(7)	56–60	IMASS	Mass type code EQ.0, consistent mass matrix EQ.1, lumped mass matrix EQ.2, no mass matrix
	61–65	IMPEXP	Implicit/explicit code EQ.0, implicit element group EQ.1, explicit element group

Notes

1. All data on this card are stored in an array (NPAR(I),I=1,16):

   NPAR(1)=NTYPE
   NPAR(2)=NUMEL
   ⋮
   NPAR(13)=IMPEXP

2. Output histories are ignored in static analysis as denoted by IACODE .EQ.1 on the Execution Control Card, Sec. In.2.0. See also Sec. In.10.1.6.
3. This option is useful, especially during dynamic analysis, for suppressing output from element groups in physically uninteresting regions of the mesh.
4. The user may specify any of the load-time functions defined in Sec. In.8.0.
5. The standard four-node quadrilateral employs 2 × 2 Gaussian quadrature. One-point Gaussian quadrature produces a dangerous element in that zero energy modes of deformation, so-called hourglass or keystone modes, may be present, resulting in a singular stiffness matrix. If enough displacement boundary conditions are present these modes may be eliminated. *However, one-point Gaussian quadrature should only be used if you know exactly what you are doing*. Internally, NINT denotes the number of integration points; NICODE .EQ. 0 ⇒ NINT .EQ. 4, and NICODE .EQ. 1 ⇒ NINT .EQ. 1.
6. The use of the mean-dilatational $\overline{\boldsymbol{B}}$ strain-displacement matrix is effective in the analysis of nearly incompressible materials, where standard elements are susceptible to mesh "locking." Use *only* if IOPT .EQ. 1 or 2 (i.e., plane strain or axisymmetry).
7. IMASS .EQ. 2 will cause inertial and body forces to be neglected for this element group in both static and dynamic analysis.

In.10.1.2 Material Properties Cards (I5,5X,6F10.0)

Note	Columns	Variable	Description
(1)	1–5	M	Material identification number; GE. 1 and LE. NUMAT
	11–20	E	Young's modulus
(2)	21–30	POIS	Poisson's ratio
	31–40	RHO(M)	Mass density
(3)	41–50	RDAMPM(M)	Mass matrix Rayleigh damping coeffcient
(3)	51–60	RDAMPK(M)	Stiffness matrix Rayleigh damping coefficient
(4)	61–70	TH(M)	Thickness

Notes

1. Material property sets may be specified in any order. Note that E and POIS are not stored directly; rather they are used to compute the entries of the material stiffness matrix $\boldsymbol{C}$.
2. Poisson's ratio cannot be set equal to 0.5 since it results in division by zero. A value close to 0.5—for instance, 0.4999—can be employed for incompressible applications.
3. The element damping matrix is computed as:

$$\boldsymbol{c}^e = \text{RDAMPM(MAT(NEL))}*\boldsymbol{m}^e + \text{RDAMPK(MAT(NEL))}*\boldsymbol{k}^e.$$

4. For IOPT .EQ. 1 or 2, the thickness will default to 1.0.

In.10.1.3 Gravity Vector Card (2F10.0)

Note	Columns	Variable	Description
(1)	1–10	GRAV(1)	Component of gravity vector in x_1-direction
	11–20	GRAV(2)	Component of gravity vector in x_2-direction

Notes

1. These components will be scaled by the load-time function LFBODY when computing the element body force.

In.10.1.4 Element Definition

In.10.1.4.1 Nodal data cards (7I5)

Note	Columns	Variable	Description
(1)	1–5	N	Element number
	6–10	M	Element material number
(2)	11–15	IEN(1,N)	Number of first node

Note	Columns	Variable	Description
	16–20	IEN(2,N)	Number of second node
	21–25	IEN(3,N)	Number of third node
(3)	26–30	IEN(4,N)	Number of fourth node
(4)	31–35	NG	Generation parameter EQ.0, no generation EQ.1, generate data

Notes

1. All elements must be input on a nodal data card or generated. Each group begins with an element number 1. *Terminate with a blank card.*
2. Element nodes must be listed in counterclockwise order. See Fig. 11.5.8.
3. For triangular elements, set node number IEN (4,N) equal to IEN(3,N).
4. If the generation parameter is set to 1, an Element Generation Data Card must be input next.

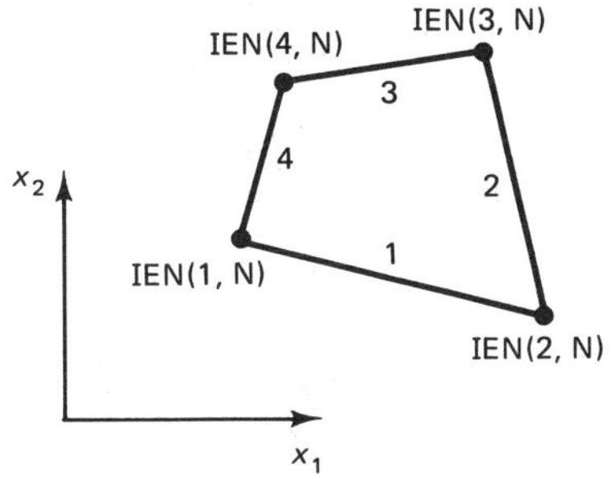

Figure 11.5.8 Nodal ordering for element number N.

In.10.1.4.2 Element generation data cards (6I5)

Note	Columns	Variable	Description
(1)	1–5	NEL(1)	Number of elements in direction 1; GE. 0 If EQ.0, set internally to 1.
	6–10	INCEL(1)	Element number increment for direction 1 If EQ.0, set internally to 1.
	11–15	INC(1)	Node number increment for direction 1 If EQ.0, set internally to 1.
	16–20	NEL(2)	Number of elements in direction 2; GE. 0 If EQ.0, set internally to 1.
	21–25	INCEL(2)	Element number increment for direction 2 If EQ.0, set internally to NEL(1).
	26–30	INC(2)	Node number increment for direction 2 If EQ.0, set internally to (NEL(1) + 1)*INC(1).

Notes

1. See Fig. 11.5.9 for a schematic representation of the generation algorithm.

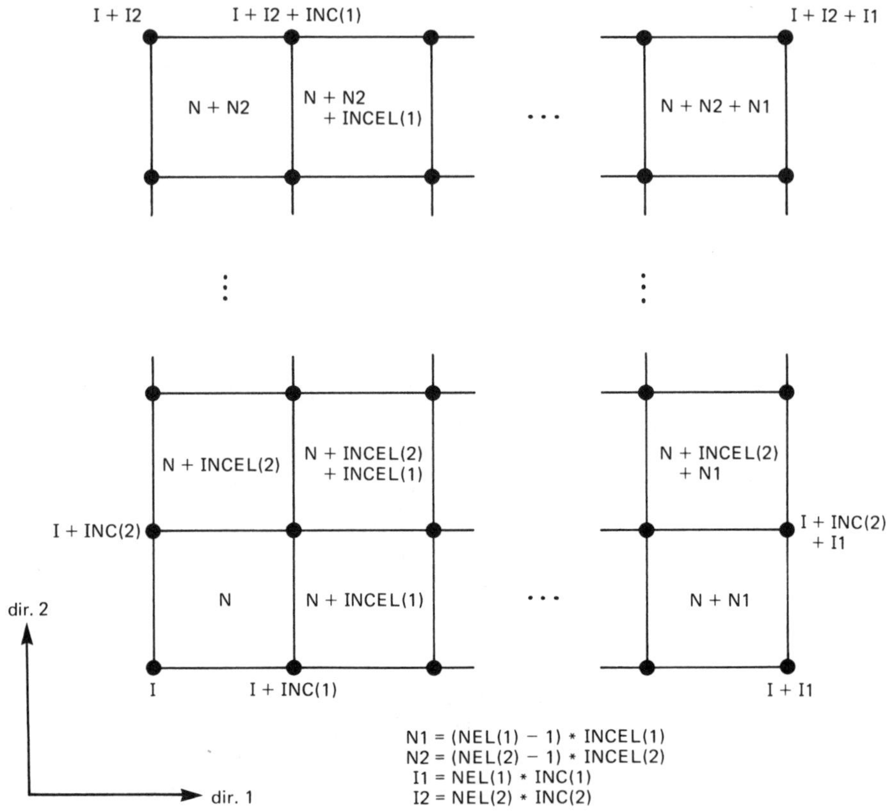

Figure 11.5.9 Element generation over a logically rectangular region.

In.10.1.5 Element Surface Load Cards (2I5,4F10.0)

Note	Columns	Variable	Description
(1)	1–5	IELNO(K)	Element number; GE. 1 and LE. NUMEL
(2)	6–10	ISIDE(K)	Element side number; GE. 1 and LE. 4
(3)	11–20	PRESS(1,K)	Pressure at node I
	21–30	PRESS(2,K)	Pressure at node J
(4)	31–40	SHEAR(1,K)	Shear stress at node I
	41–50	SHEAR(2,K)	Shear stress at node J

Notes

1. Each element side subjected to a surface load must be specified. If more than one side of the element is loaded, one card must be input for each loaded side. The index K in the arrays IELNO and ISIDE corresponds to the order in which the cards are read: 1 .LE. K .LE. NSURF. The cards need not be specified in any particular order. *The total number*

of Element Surface Load Cards must equal NSURF, *as specified on the Element Group Control Card* (see Sec. In.10.1.1).

2. The element side number is deduced as follows: Consider the element of Fig. 11.5.8. Let N = IELNO(K). Side 1 connects nodes IEN(1,N) and IEN(2,N), side 2 connects nodes IEN(2,N) and IEN(3,N), etc. For the purpose of element surface force specification, locally we refer to the node beginning the loaded side as I and the node ending the side as J.
3. The pressure is assumed to be positive pointing inward, and is linearly interpolated between nodal values; see Fig. 11.5.10. In computing the element surface forces, pressure is scaled by load-time function LFSURF.

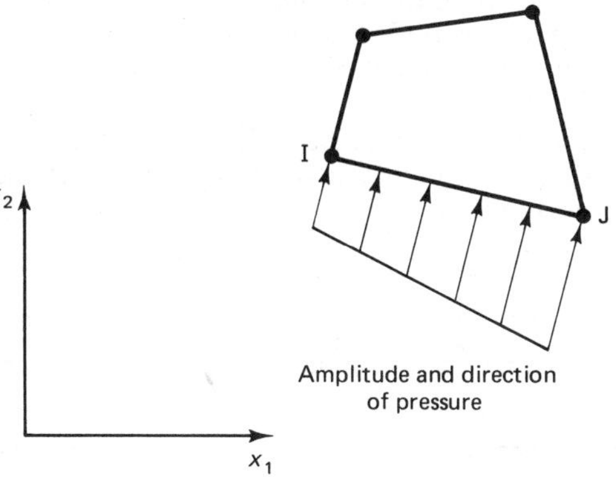

Figure 11.5.10

4. The shear stress is assumed to be positive pointing in a counterclockwise fashion and is linearly interpolated between nodal values; see Fig. 11.5.11. In computing the element surface forces, the shear stress is scaled by load-time function LFSURF.

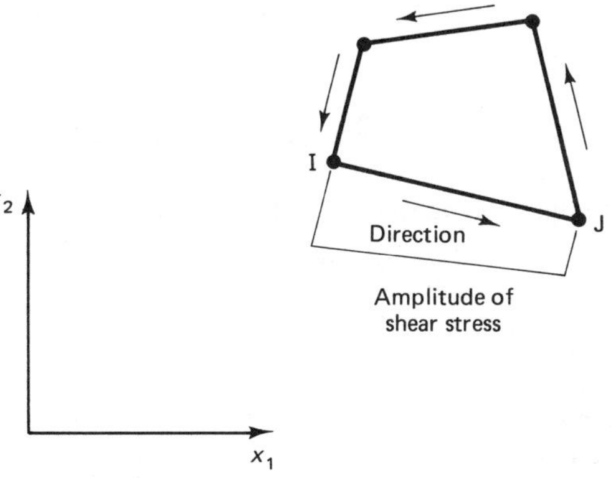

Figure 11.5.11

In.10.1.6 Element Time History Cards (3I5)

If the analysis is static (i.e., IACODE .EQ. 1 on the Execution Control Card; see Sec. In.2.0), then output history cards are read but subsequently ignored.

Note	Columns	Variable	Description
(1)	1–5	ISHIST(1,N)	Element number; GE. 1 and LE. NUMEL
(2)	6–10	ISHIST(2,N)	Quadrature point number; GE. 1 and LE. NINT
(3)	11–15	ISHIST(3,N)	Stress/strain component number, GE. 1 and LE. 16

Notes

1. Each desired element time history must be specified; thus an element for which the user requests multiple time histories will engender multiple data cards. The index N in the array ISHIST corresponds to the order in which the cards are read: 1 .LE. N .LE. NSOUT. The cards need not be specified in any particular order. *The total number of Element Time History Cards must equal* NSOUT *as specified on the Element Group Control Card* (see Sec. In.10.1.1).
2. NINT denotes the number of integration points; NICODE .EQ. 0 ⇒ NINT .EQ. 4, and NICODE .EQ. 1 ⇒ NINT .EQ. 1. In local coordinates (i.e., ξ, η), the quadrature point number defines the following locations:

NINT	l	$\tilde{\xi}_l$	$\tilde{\eta}_l$
1	1	0	0
4	1	$\frac{-1}{\sqrt{3}}$	$\frac{-1}{\sqrt{3}}$
	2	$\frac{1}{\sqrt{3}}$	$\frac{-1}{\sqrt{3}}$
	3	$\frac{1}{\sqrt{3}}$	$\frac{1}{\sqrt{3}}$
	4	$\frac{-1}{\sqrt{3}}$	$\frac{1}{\sqrt{3}}$

3. The following element components may be specified:

NCOMP	Description	Output Label
1	σ_{11}, stress 11	S 11
2	σ_{22}, stress 22	S 22
3	σ_{12}, stress 12	S 12
4	σ_{33}, stress 33	S 33
5	σ_1, principal stress 1	PS 1
6	σ_2, principal stress 2	PS 2
7	$\tau = \frac{1}{2}\lvert\sigma_1 - \sigma_2\rvert$, shear stress	TAU
8	θ_σ, stress angle (i.e., angle between x_1 and σ_1 axes)	SANG
9	ϵ_{11}, strain 11	E 11
10	ϵ_{22}, strain 22	E 22
11	$\gamma_{12} = 2\epsilon_{12}$ (i.e., engineering shear strain)	G 12
12	ϵ_{33}, strain 33	E 33

13	ϵ_1, principal strain 1	PE 1
14	ϵ_2, principal strain 2	PE 2
15	$\gamma = \|\epsilon_1 - \epsilon_2\|$, (i.e., engineering shear strain)	GAM
16	θ_ϵ, strain angle (i.e., angle between x_1 and ϵ_1 axes)	EANG

In.10.2 Three-Dimensional, Elastic Truss Element

The present element connects two or three nodes in space and transmits axial force only. There are three degrees of freedom at each node, i.e., the x_1, x_2, and x_3 translations. When employing truss elements, the user must specify NSD .EQ. 3 and NDOF .GE. 3 on the Execution Control Card (see Sec. In.2.0). The following sequence of cards is used to describe truss elements:

In.10.2.1 Element Group Control Card (10I5)

Note	Columns	Variable	Description
(1)	1–5	NTYPE	The number 2
	6–10	NUMEL	Number of elements in this group; GE. 1
	11–15	NUMAT	Number of material sets in this group; GE. 1
	16–20	NEN	Maximum number of nodes for all elements in this group EQ. 2 or 3
(2)	21–25	NSOUT	Number of stress/strain time-history cards
(3)	26–30	ISTPRT	Stress output print code EQ.0, print stress output EQ.1, do not print stress output
(4)	31–35	LFBODY	Body force load function number
(5)	36–40	NINT	Number of integration points EQ.1, one-point Gaussian quadrature EQ.2, two-point Gaussian quadrature EQ.3, three-point Gaussian quadrature
(6)	41–45	IMASS	Mass type code EQ.0, consistent mass matrix EQ.1, lumped mass matrix EQ.2, no mass matrix
	46–50	IMPEXP	Implicit/explicit code EQ.0, implicit element group EQ.1, explicit element group

Notes

1. All data on this card are stored in an array (NPAR(I),I = 1,16):

$$\begin{aligned} \text{NPAR}(1) &= \text{NTYPE} \\ \text{NPAR}(2) &= \text{NUMEL} \\ &\vdots \\ \text{NPAR}(10) &= \text{IMPEXP} \end{aligned}$$

2. Output histories are ignored in static analysis as denoted by IACODE .EQ. 1 on the Execution Control Card, Sec. In.2.0. See also Sec. In.10.2.5.
3. This option is useful, especially during dynamic analysis, for suppressing output from element groups in physically uninteresting regions of the mesh.
4. The user may specify any of the load-time functions defined in Sec. In.8.0.
5. The following facts may be used to guide the selection of an appropriate integration rule:
   For two-node truss elements, the stiffness is exactly integrated by the one-point Gauss rule and the mass is exactly integrated by the two-point Gauss rule.
   For three-node truss elements, if the location of the "middle node," $\boldsymbol{x}_3$, satisfies $\boldsymbol{x}_3 = (\boldsymbol{x}_1 + \boldsymbol{x}_2)/2$, then the stiffness is exactly integrated by the two-point Gauss rule and the mass is exactly integrated by the three-point Gauss rule.
6. IMASS .EQ. 2 will cause inertial and body forces to be neglected for this element group in both static and dynamic analysis.

In.10.2.2 Material Properties Cards (I5,5X,5F10.0)

Note	Columns	Variable	Description
(1)	1–5	M	Material identification number GE. 1 and LE. NUMAT
	11–20	E	Young's modulus
	21–30	RHO(M)	Mass density
(2)	31–40	RDAMPM(M)	Mass matrix Rayleigh damping coefficient
(2)	41–50	RDAMPK(M)	Stiffness matrix Rayleigh damping coefficient
	51–60	AREA(M)	Area

Notes

1. Material property sets may be specified in any order. Note that E is stored as the single entry in the material stiffness matrix $\boldsymbol{C}$.
2. The element damping matrix is computed as

$$\boldsymbol{c}^e = \text{RDAMPM(MAT(NEL))}*\boldsymbol{m}^e + \text{RDAMPK(MAT(NEL))}*\boldsymbol{k}^e.$$

In.10.2.3 Gravity Vector Card (3F10.0)

Note	Columns	Variable	Description
(1)	1–10	GRAV(1)	Component of gravity vector in x_1-direction
	11–20	GRAV(2)	Component of gravity vector in x_2-direction
	21–30	GRAV(3)	Component of gravity vector in x_3-direction

Notes

1. These components will be scaled by the load-time function LFBODY when computing the element body force.

In.10.2.4 Element Definition

In.10.2.4.1. Nodal Data Cards (6I5)

If NEN .EQ. 2, all nodal data cards in this group have the following form.

Note	Columns	Variable	Description
(1)	1–5	N	Element number
	6–10	M	Element material number
(2)	11–15	IEN(1,N)	Number of first node
	16–20	IEN(2,N)	Number of second node
(3)	21–25	NG	Generation parameter EQ.0, no generation EQ.1, generate data

If NEN .EQ. 3, all nodal data cards in this group have the following form.

Note	Columns	Variable	Description
(1)	1–5	N	Element number
	6–10	M	Element material number
(2)	11–15	IEN(1,N)	Number of first node
	16–20	IEN(2,N)	Number of second node
(4)	21–25	IEN(3,N)	Number of third node
(3)	26–30	NG	Generation parameter EQ.0, no generation EQ.1, generate data

Notes

1. All elements must be input on a nodal data card or generated. Each group begins with an element number 1. *Terminate with a blank card.*
2. See Fig. 11.5.12 for definition of nodal ordering.
3. If the generation parameter is set to 1, a generation data card must be input next.
4. Two-node elements may be defined within a group of three-node elements (i.e., NEN = 3) by setting IEN(3,N) equal to IEN(2,N).

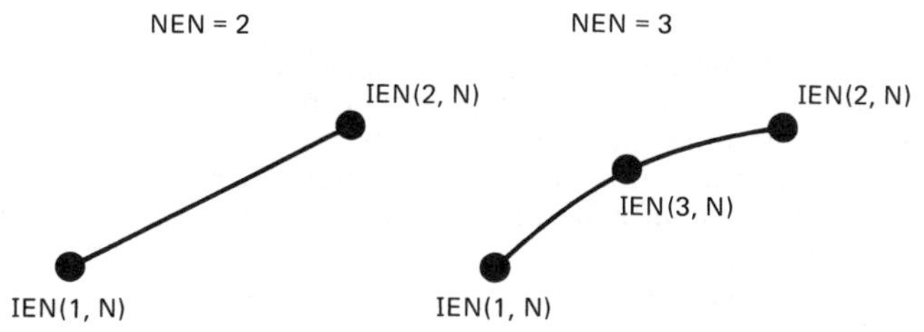

Figure 11.5.12 Nodal ordering for the truss element.

In.10.2.4.2 Element Generation Data Cards (3I5)

Note	Columns	Variable	Description
(1)	1–5	NEL(1)	Number of elements to be generated
	6–10	INCEL(1)	Element number increment If EQ.0, set internally to 1.
	11–15	INC(1)	Node number increment If EQ.0, set internally to 1.

Notes

1. See Fig. 11.5.13 for a schematic representation of the generation algorithm.

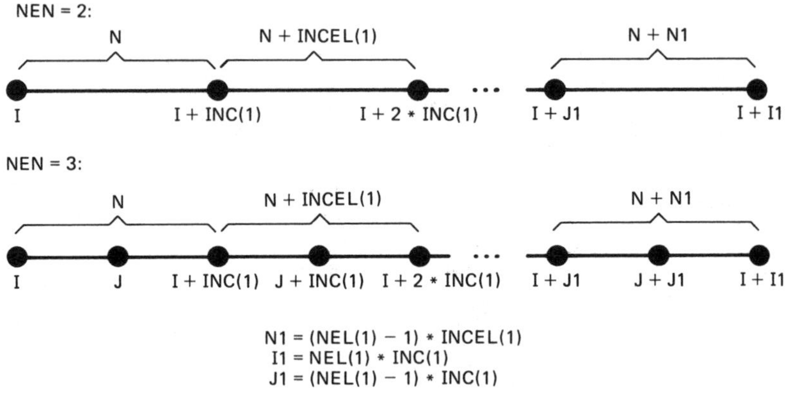

Figure 11.5.13 Generation of truss elements along a line.

In.10.2.5 Element Time History Cards (3I5)

If the analysis is static (i.e., IACODE .EQ. 1 on the Execution Control Card; see Sec. In.2.0), then output history cards are read, but subsequently ignored.

Note	Columns	Variable	Description
(1)	1–5	ISHIST(1,N)	Element number; GE. 1 and LE. NUMEL
(2)	6–10	ISHIST(2,N)	Quadrature point number; GE. 1 and LE. NINT
(3)	11–15	ISHIST(3,N)	Stress/strain component number, GE. 1 and LE. 3

Notes

1. Each desired element time history must be specified; thus an element for which the user requests multiple time histories will engender multiple data cards. The index N in the array ISHIST corresponds to the order in which the cards are read: 1 .LE. N .LE. NSOUT. The cards need not be specified in any particular order. *The total number of Element Time History Cards must equal* NSOUT, *as specified on the Element Group Control Card* (see Sec. In.10.2.1).
2. In the local coordinate (i.e., ξ), the quadrature point number corresponds to the following locations:

NINT	l	$\tilde{\xi}_l$
1	1	0
2	1	$\frac{-1}{\sqrt{3}}$
	2	$\frac{1}{\sqrt{3}}$
3	1	$\frac{-1}{\sqrt{3/5}}$
	2	$\frac{1}{\sqrt{3/5}}$
	3	0

3. The following element components may be specified:

NCOMP	Description	Output Label
1	Axial stress	STRS
2	Axial force	FORC
3	Axial strain	STRN

11.5.3 Examples

Example 1 (Planar Truss)

Here we consider a planar truss, shown schematically with node numbers in Fig. 11.5.14, subjected to multiple static loadings. Looking at the input file given in

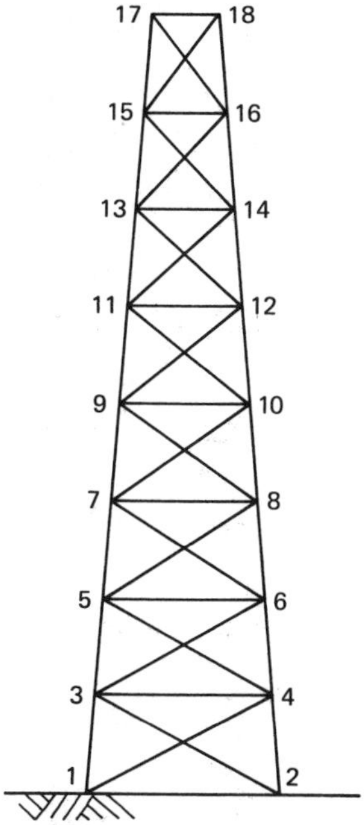

Figure 11.5.14 Planar truss.

Table 11.5.1, note the use of generation capabilities, including the definition of the truss elements. In the time-sequence data we request three time steps and have set $\Delta t = 1.0$, although the code would default to this value. Two load vectors and load-time functions have been specified. The first load vector corresponds to a linearly varying lateral load, as might be used for preliminary seismic design. The second defines a uniform vertical load acting at every node. The load-time functions, shown schematically in Fig. 11.5.15, produce three different load cases combining varying proportions of the two load vectors. The output on pages 693 through 704 can be used to verify DLEARN upon installation on a new computer system.

TABLE 11.5.1 Input File for Planar Truss Problem

```
ln 0.0          0
ln 1.0      TRUSS TOWER WITH STATIC VERTICAL AND LATERAL LOADS
ln 2.0          1    1    0    0    0    0    1    0    3   18    3    2    2    4    1
ln 3.0          1    3    1    1    0    1      0.00      1.00      0.00      1.00
ln 5.1          1    2         0.        0.       )
ln 5.2         17    0        48.      768.       )
ln 5.3          8    2                            )
ln 5.1          2    2       144.        0.       )  Coordinates
ln 5.2         18    0        96.      768.       )
ln 5.3          8    2                            )

              { 1   18    1    0    0    1        )
ln 6.0        { 1    0    0    1    1    1        )  Boundary conditions
              { 2    0    0    1    1    1        )

ln 7.1          3    2      500.0      0.00      0.00   )
ln 7.2         17    0     4000.0      0.00      0.00   )  Load vector 1
ln 7.3          7    2                                  )

ln 7.1          3    2       0.00   -3000.0      0.00   )
ln 7.2         18    0       0.00   -3000.0      0.00   )  Load vector 2
ln 7.3         15    1                                  )

              {     0.00      0.0                       )
              {     1.00      1.0                       )  Load-time function 1
              {     2.00      0.0                       )
ln 8.1.       {     3.00      1.7                       )
              {     0.00      0.0                       )
              {     1.00      0.0                       )  Load-time function 2
              {     2.00      1.0                       )
              {     3.00      1.3                       )
ln 10.2.1       2   40    2    2    0    0    0    1    1    0             )
ln 10.2.2     { 1      30000000.  7.30E-4      0.00      0.00      10.0    )
              { 2      30000000.  7.30E-4      0.00      0.00       6.0    )
ln 10.2.3           0.00   -386.4      0.00                                )
              { 1    1    1    3    1         )                            )
              { 4    1    2                   )                            )
              { 5    1    2    4    1         )                            )
              { 4    1    2                   )                            )
              { 9    1    3    4    1         )                            )
              { 4    1    2                   )                            )
              {13    2    1    4    1         )                            )
              { 4    1    2                   )                            )
              {17    2    2    3    1         )                            )  Element group
              { 4    1    2                   )                            )
ln 10.2.4.1   {21    2    9   11    1         )  Element nodes             )
ln 10.2.4.2   { 4    1    2                   )                            )
              {25    2   10   12    1         )                            )
              { 4    1    2                   )                            )
              {29    2   11   12    1         )                            )
              { 4    1    2                   )                            )
              {33    2    9   12    1         )                            )
              { 4    1    2                   )                            )
              {37    2   10   11    1         )                            )
              { 4    1    2                   )                            )
            *END
```

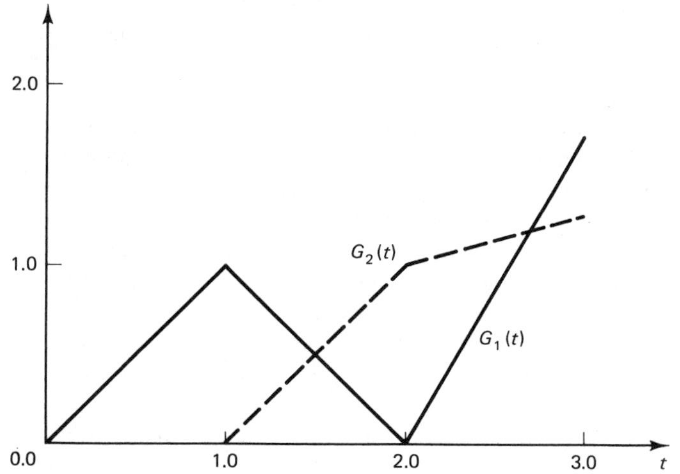

Figure 11.5.15 Load-time functions for planar truss problem.

```
TRUSS TOWER WITH STATIC VERTICAL AND LATERAL LOADS

E X E C U T I O N   C O N T R O L   I N F O R M A T I O N
     EXECUTION CODE . . . . . . . . . . . . . . . (IEXEC ) =      1

        EQ. 0, DATA CHECK
        EQ. 1, EXECUTION

     ANALYSIS CODE . . . . . . . . . . . . . . . (IACODE) =      1

        EQ. 0, DYNAMIC ANALYSIS
        EQ. 1, STATIC ANALYSIS

     READ RESTART FILE CODE . . . . . . . . . . . (IREADR) =      0

        EQ. 0, DO NOT READ RESTART FILE
        EQ. 1, READ RESTART FILE

     WRITE RESTART FILE CODE . . . . . . . . . . (IWRITR) =      0

        EQ. 0, DO NOT WRITE RESTART FILE
        EQ. 1, WRITE RESTART FILE

     INPUT DATA PRINT CODE . . . . . . . . . . . (IPRTIN) =      0

        EQ. 0, PRINT NODAL AND ELEMENT INPUT DATA
        EQ. 1, DO NOT PRINT NODAL AND ELEMENT INPUT DATA

     RANK CHECK CODE . . . . . . . . . . . . . . (IRANK ) =      0

        EQ. 0, DO NOT PERFORM RANK CHECK
        EQ. 1, PRINT NUMBERS OF ZERO AND NEGATIVE PIVOTS
        EQ. 2, PRINT ALL PIVOTS

     NUMBER OF TIME SEQUENCES . . . . . . . . . . (NUMSEQ) =      1

     NUMBER OF NODAL OUTPUT TIME-HISTORIES . . . (NDOUT ) =      0

     NUMBER OF SPACE DIMENSIONS . . . . . . . . . (NSD   ) =      3

     NUMBER OF NODAL POINTS . . . . . . . . . . . (NUMNP ) =     18

     NUMBER OF NODAL DEGREES-OF-FREEDOM . . . . . (NDOF  ) =      3

     NUMBER OF LOAD VECTORS . . . . . . . . . . . (NLVECT) =      2

     NUMBER OF LOAD-TIME FUNCTIONS . . . . . . . (NLTFTN) =      2

     NUMBER OF POINTS ON LOAD-TIME FUNCTIONS . . (NPTSLF) =      4

     NUMBER OF ELEMENT GROUPS . . . . . . . . . . (NUMEG ) =      1
```

T I M E   S E Q U E N C E   D A T A

NUMBER OF TIME SEQUENCES . . . . . . (NUMSEQ) = 1

TIME SEQUENCE NUMBER . . . . (N) = 1

NUMBER OF TIME STEPS . . . . (NSTEP(N)) = 3

KINEMATIC PRINT INCREMENT . . . . . (NDPRT(N)) = 1

STRESS/STRAIN PRINT INCREMENT . . . (NSPRT(N)) = 1

TIME HISTORY PLOT INCREMENT . . . . (NHPLT(N)) = 0

NUMBER OF ITERATIONS . . . . (NITER(N)) = 1

FIRST INTEGRATION PARAMETER . . . . (ALPHA(N)) = 0.00000E+00

SECOND INTEGRATION PARAMETER . . . . (BETA(N)) = 1.00000E+00

THIRD INTEGRATION PARAMETER . . . . (GAMMA(N)) = 0.00000E+00

TIME STEP . . . . (DT(N)) = 1.00000E+00

N O D A L   C O O R D I N A T E   D A T A

NODE NO.	X1	X2	X3
1	0.00000000E+00	0.00000000E+00	0.00000000E+00
2	1.44000000E+02	0.00000000E+00	0.00000000E+00
3	6.00000000E+00	9.60000000E+01	0.00000000E+00
4	1.38000000E+02	9.60000000E+01	0.00000000E+00
5	1.20000000E+01	1.92000000E+02	0.00000000E+00
6	1.32000000E+02	1.92000000E+02	0.00000000E+00
7	1.80000000E+01	2.88000000E+02	0.00000000E+00
8	1.26000000E+02	2.88000000E+02	0.00000000E+00
9	2.40000000E+01	3.84000000E+02	0.00000000E+00
10	1.20000000E+02	3.84000000E+02	0.00000000E+00
11	3.00000000E+01	4.80000000E+02	0.00000000E+00
12	1.14000000E+02	4.80000000E+02	0.00000000E+00
13	3.60000000E+01	5.76000000E+02	0.00000000E+00
14	1.08000000E+02	5.76000000E+02	0.00000000E+00
15	4.20000000E+01	6.72000000E+02	0.00000000E+00
16	1.02000000E+02	6.72000000E+02	0.00000000E+00
17	4.80000000E+01	7.68000000E+02	0.00000000E+00
18	9.60000000E+01	7.68000000E+02	0.00000000E+00

N O D A L   B O U N D A R Y   C O N D I T I O N   C O D E S

NODE NO.	DOF1	DOF2	DOF3
1	1	1	1
2	1	1	1
3	0	0	1
4	0	0	1
5	0	0	1
6	0	0	1
7	0	0	1
8	0	0	1
9	0	0	1
10	0	0	1
11	0	0	1
12	0	0	1
13	0	0	1
14	0	0	1
15	0	0	1
16	0	0	1
17	0	0	1
18	0	0	1

P R E S C R I B E D   F O R C E S   A N D   K I N E M A T I C
B O U N D A R Y   C O N D I T I O N S

LOAD VECTOR NUMBER = 1

NODE NO.	DOF1	DOF2	DOF3
3	5.00000000E+02	0.00000000E+00	0.00000000E+00
5	1.00000000E+03	0.00000000E+00	0.00000000E+00
7	1.50000000E+03	0.00000000E+00	0.00000000E+00
9	2.00000000E+03	0.00000000E+00	0.00000000E+00
11	2.50000000E+03	0.00000000E+00	0.00000000E+00
13	3.00000000E+03	0.00000000E+00	0.00000000E+00
15	3.50000000E+03	0.00000000E+00	0.00000000E+00
17	4.00000000E+03	0.00000000E+00	0.00000000E+00

```
 P R E S C R I B E D   F O R C E S   A N D   K I N E M A T I C
   B O U N D A R Y   C O N D I T I O N S

      LOAD VECTOR NUMBER =      2

      NODE NO.              DOF1               DOF2                DOF3
           3         0.00000000E+00    -3.00000000E+03    0.00000000E+00
           4         0.00000000E+00    -3.00000000E+03    0.00000000E+00
           5         0.00000000E+00    -3.00000000E+03    0.00000000E+00
           6         0.00000000E+00    -3.00000000E+03    0.00000000E+00
           7         0.00000000E+00    -3.00000000E+03    0.00000000E+00
           8         0.00000000E+00    -3.00000000E+03    0.00000000E+00
           9         0.00000000E+00    -3.00000000E+03    0.00000000E+00
          10         0.00000000E+00    -3.00000000E+03    0.00000000E+00
          11         0.00000000E+00    -3.00000000E+03    0.00000000E+00
          12         0.00000000E+00    -3.00000000E+03    0.00000000E+00
          13         0.00000000E+00    -3.00000000E+03    0.00000000E+00
          14         0.00000000E+00    -3.00000000E+03    0.00000000E+00
          15         0.00000000E+00    -3.00000000E+03    0.00000000E+00
          16         0.00000000E+00    -3.00000000E+03    0.00000000E+00
          17         0.00000000E+00    -3.00000000E+03    0.00000000E+00
          18         0.00000000E+00    -3.00000000E+03    0.00000000E+00

 L O A D - T I M E   F U N C T I O N   D A T A

      NUMBER OF LOAD-TIME FUNTIONS . . . . . (NLTFTN  ) =      2

      FUNCTION NUMBER      1

                TIME              LOAD FACTOR

          0.00000000E+00       0.00000000E+00
          1.00000000E+00       1.00000000E+00
          2.00000000E+00       0.00000000E+00
          3.00000000E+00       1.70000000E+00

      FUNCTION NUMBER      2

                TIME              LOAD FACTOR

          0.00000000E+00       0.00000000E+00
          1.00000000E+00       0.00000000E+00
          2.00000000E+00       1.00000000E+00
          3.00000000E+00       1.30000000E+00


   E L E M E N T   G R O U P   D A T A

        ELEMENT GROUP NUMBER . . . . . . . . . . . (NEG    ) =      1

   T W O / T H R E E - N O D E   T R U S S   E L E M E N T S

        ELEMENT TYPE NUMBER . . . . . . . . . . . . (NTYPE ) =      2

        NUMBER OF ELEMENTS  . . . . . . . . . . . . (NUMEL ) =     40

        NUMBER OF ELEMENT MATERIAL SETS . . . . . . (NUMAT ) =      2

        NUMBER OF ELEMENT NODES . . . . . . . . . . (NEN   ) =      2

        NUMBER OF STRESS/STRAIN TIME HISTORIES  . . (NSOUT ) =      0

        STRESS OUTPUT PRINT CODE  . . . . . . . . . (ISTPRT) =      0

           EQ.0, STRESS OUTPUT PRINTED
           EQ.1, STRESS OUTPUT NOT PRINTED

        BODY FORCE LOAD-TIME FUNCTION NUMBER  . . . (LFBODY) =      0

        INTEGRATION CODE  . . . . . . . . . . . . . (NINT  ) =      1

           EQ.1, 1-POINT GAUSSIAN QUADRATURE
           EQ.2, 2-POINT GAUSSIAN QUADRATURE
           EQ.3, 3-POINT GAUSSIAN QUADRATURE
```

M A T E R I A L   S E T   D A T A

NUMBER OF MATERIAL SETS . . . . (NUMAT) = 2

SET NUMBER	YOUNG'S MODULUS	MASS DENSITY	MASS DAMPING	STIFFNESS DAMPING	AREA
1	3.0000E+07	7.3000E-04	0.0000E+00	0.0000E+00	1.0000E+01
2	3.0000E+07	7.3000E-04	0.0000E+00	0.0000E+00	6.0000E+00

G R A V I T Y   V E C T O R   C O M P O N E N T S

X-1 DIRECTION . . . . = 0.00000000E+00

X-2 DIRECTION . . . . = -3.86400000E+02

X-3 DIRECTION . . . . = 0.00000000E+00

E L E M E N T   D A T A

ELEMENT	MATERIAL	NODE 1	NODE 2
1	1	1	3
2	1	3	5
3	1	5	7
4	1	7	9
5	1	2	4
6	1	4	6
7	1	6	8
8	1	8	10
9	1	3	4
10	1	5	6
11	1	7	8
12	1	9	10
13	2	1	4
14	2	3	6
15	2	5	8
16	2	7	10
17	2	2	3
18	2	4	5
19	2	6	7
20	2	8	9
21	2	9	11
22	2	11	13
23	2	13	15
24	2	15	17
25	2	10	12
26	2	12	14
27	2	14	16
28	2	16	18
29	2	11	12
30	2	13	14
31	2	15	16
32	2	17	18
33	2	9	12
34	2	11	14
35	2	13	16
36	2	15	18
37	2	10	11
38	2	12	13
39	2	14	15
40	2	16	17

```
TRUSS TOWER WITH STATIC VERTICAL AND LATERAL LOADS

E Q U A T I O N   S Y S T E M   D A T A
     NUMBER OF EQUATIONS . . . . . . . . . . . . . . (NEQ    ) =        32
     NUMBER OF TERMS IN LEFT-HAND-SIDE MATRIX  . (NALHS ) =       192
     MEAN HALF BANDWIDTH . . . . . . . . . . . . . . (MEANBW) =         6
     TOTAL LENGTH OF BLANK COMMON REQUIRED . . . (NWORDS) =      2012

 D I S P L A C E M E N T S
     STEP NUMBER =            1
     TIME        =  1.000E+00
```

NODE NO.	DOF1	DOF2	DOF3
3	-3.55458742E-02	-3.80776188E-02	0.00000000E+00
4	-3.39701332E-02	2.82149641E-02	0.00000000E+00
5	-1.18072704E-01	-6.85865605E-02	0.00000000E+00
6	-1.16727240E-01	4.99146888E-02	0.00000000E+00
7	-2.46087075E-01	-9.10080677E-02	0.00000000E+00
8	-2.44821303E-01	6.49165276E-02	0.00000000E+00
9	-4.15844342E-01	-1.05185369E-01	0.00000000E+00
10	-4.145476337E-01	7.29927551E-02	0.00000000E+00
11	-6.33329338E-01	-1.22704195E-01	0.00000000E+00
12	-6.31285639E-01	8.31843360E-02	0.00000000E+00
13	-9.00366267E-01	-1.27236953E-01	0.00000000E+00
14	-8.98780674E-01	8.21569828E-02	0.00000000E+00
15	-1.20224430E+00	-1.20066166E-01	0.00000000E+00
16	-1.20101352E+00	7.13913000E-02	0.00000000E+00
17	-1.52011008E+00	-1.04034181E-01	0.00000000E+00
18	-1.51917308E+00	5.35084016E-02	0.00000000E+00

E L E M E N T   S T R E S S E S   A N D   S T R A I N S

ELEMENT GROUP NUMBER . . . . (NEG) = 1

ELEMENT NUMBER	INT. PT. NUMBER	X1 COORD.	X2 COORD.	X3 COORD.	STRESS	FORCE	STRAIN
1	1	3.00E+00	4.80E+01	0.00E+00	-1.25E+04	-1.25E+05	-4.18E-04
2	1	9.00E+00	1.44E+02	0.00E+00	-1.11E+04	-1.11E+05	-3.70E-04
3	1	1.50E+01	2.40E+02	0.00E+00	-9.47E+03	-9.47E+04	-3.16E-04
4	1	2.10E+01	3.36E+02	0.00E+00	-7.72E+03	-7.72E+04	-2.57E-04
5	1	1.41E+02	4.80E+01	0.00E+00	9.44E+03	9.44E+04	3.15E-04
6	1	1.35E+02	1.44E+02	0.00E+00	8.36E+03	8.36E+04	2.79E-04
7	1	1.29E+02	2.40E+02	0.00E+00	7.16E+03	7.16E+04	2.39E-04
8	1	1.23E+02	3.36E+02	0.00E+00	5.82E+03	5.82E+04	1.94E-04
9	1	7.20E+01	9.60E+01	0.00E+00	3.58E+02	3.58E+03	1.19E-05
10	1	7.20E+01	1.92E+02	0.00E+00	3.36E+02	3.36E+03	1.12E-05
11	1	7.20E+01	2.88E+02	0.00E+00	3.52E+02	3.52E+03	1.17E-05
12	1	7.20E+01	3.84E+02	0.00E+00	4.05E+02	4.05E+03	1.35E-05
13	1	6.90E+01	4.80E+01	0.00E+00	-2.10E+03	-1.26E+04	-7.00E-05
14	1	6.90E+01	1.44E+02	0.00E+00	-2.13E+03	-1.28E+04	-7.10E-05
15	1	6.90E+01	2.40E+02	0.00E+00	-2.21E+03	-1.32E+04	-7.35E-05
16	1	6.90E+01	3.36E+02	0.00E+00	-2.20E+03	-1.32E+04	-7.33E-05
17	1	7.50E+01	4.80E+01	0.00E+00	1.33E+03	7.96E+03	4.42E-05
18	1	7.50E+01	1.44E+02	0.00E+00	1.56E+03	9.35E+03	5.20E-05
19	1	7.50E+01	2.40E+02	0.00E+00	1.65E+03	9.87E+03	5.49E-05
20	1	7.50E+01	3.36E+02	0.00E+00	1.70E+03	1.02E+04	5.68E-05
21	1	2.70E+01	4.32E+02	0.00E+00	-9.68E+03	-5.81E+04	-3.23E-04
22	1	3.30E+01	5.28E+02	0.00E+00	-6.61E+03	-3.96E+04	-2.20E-04
23	1	3.90E+01	6.24E+02	0.00E+00	-3.64E+03	-2.18E+04	-1.21E-04
24	1	4.50E+01	7.20E+02	0.00E+00	-1.19E+03	-7.16E+03	-3.98E-05

E L E M E N T   S T R E S S E S   A N D   S T R A I N S

ELEMENT GROUP NUMBER . . . . (NEG) = 1

ELEMENT NUMBER	INT. PT. NUMBER	X1 COORD.	X2 COORD.	X3 COORD.	STRESS	FORCE	STRAIN
25	1	1.17E+02	4.32E+02	0.00E+00	7.39E+03	4.43E+04	2.46E-04
26	1	1.11E+02	5.28E+02	0.00E+00	4.88E+03	2.93E+04	1.63E-04
27	1	1.05E+02	6.24E+02	0.00E+00	2.53E+03	1.52E+04	8.43E-05
28	1	9.90E+01	7.20E+02	0.00E+00	6.23E+02	3.74E+03	2.08E-05
29	1	7.20E+01	4.80E+02	0.00E+00	7.30E+02	4.38E+03	2.43E-05
30	1	7.20E+01	5.76E+02	0.00E+00	6.61E+02	3.96E+03	2.20E-05
31	1	7.20E+01	6.72E+02	0.00E+00	6.15E+02	3.69E+03	2.05E-05
32	1	7.20E+01	7.68E+02	0.00E+00	5.86E+02	3.51E+03	1.95E-05
33	1	6.90E+01	4.32E+02	0.00E+00	-2.26E+03	-1.36E+04	-7.54E-05
34	1	6.90E+01	5.28E+02	0.00E+00	-2.04E+03	-1.22E+04	-6.79E-05
35	1	6.90E+01	6.24E+02	0.00E+00	-1.71E+03	-1.03E+04	-5.71E-05
36	1	6.90E+01	7.20E+02	0.00E+00	-1.12E+03	-6.69E+03	-3.72E-05
37	1	7.50E+01	4.32E+02	0.00E+00	1.57E+03	9.39E+03	5.22E-05
38	1	7.50E+01	5.28E+02	0.00E+00	1.54E+03	9.27E+03	5.15E-05
39	1	7.50E+01	6.24E+02	0.00E+00	1.36E+03	8.16E+03	4.53E-05
40	1	7.50E+01	7.20E+02	0.00E+00	9.65E+02	5.79E+03	3.22E-05

D I S P L A C E M E N T S

STEP NUMBER = 2

TIME = 2.000E+00

NODE NO.	DOF1	DOF2	DOF3
3	-1.03904457E-03	-7.03705120E-03	0.00000000E+00
4	1.03904457E-03	-7.03705120E-03	0.00000000E+00
5	-7.97810718E-04	-1.32947618E-02	0.00000000E+00
6	7.97810718E-04	-1.32947618E-02	0.00000000E+00
7	-6.72223927E-04	-1.85330487E-02	0.00000000E+00
8	6.72223927E-04	-1.85330487E-02	0.00000000E+00
9	-6.43972712E-04	-2.28047225E-02	0.00000000E+00
10	6.43972712E-04	-2.28047225E-02	0.00000000E+00
11	-9.38067044E-04	-2.78375422E-02	0.00000000E+00
12	9.38067044E-04	-2.78375422E-02	0.00000000E+00
13	-5.64200278E-04	-3.15277555E-02	0.00000000E+00
14	5.64200278E-04	-3.15277555E-02	0.00000000E+00
15	-2.80376309E-04	-3.38077792E-02	0.00000000E+00
16	2.80376309E-04	-3.38077792E-02	0.00000000E+00
17	-6.37576302E-05	-3.48593216E-02	0.00000000E+00
18	6.37576302E-05	-3.48593216E-02	0.00000000E+00

```
E L E M E N T   S T R E S S E S   A N D   S T R A I N S
    ELEMENT GROUP NUMBER . . . . . . . . . . . (NEG    ) =      1
```

ELEMENT NUMBER	INT. PT. NUMBER	X1 COORD.	X2 COORD.	X3 COORD.	STRESS	FORCE	STRAIN
1	1	3.00E+00	4.80E+01	0.00E+00	-2.21E+03	-2.21E+04	-7.37E-05
2	1	9.00E+00	1.44E+02	0.00E+00	-1.94E+03	-1.94E+04	-6.48E-05
3	1	1.50E+01	2.40E+02	0.00E+00	-1.63E+03	-1.63E+04	-5.43E-05
4	1	2.10E+01	3.36E+02	0.00E+00	-1.33E+03	-1.33E+04	-4.43E-05
5	1	1.41E+02	4.80E+01	0.00E+00	-2.21E+03	-2.21E+04	-7.37E-05
6	1	1.35E+02	1.44E+02	0.00E+00	-1.94E+03	-1.94E+04	-6.48E-05
7	1	1.29E+02	2.40E+02	0.00E+00	-1.63E+03	-1.63E+04	-5.43E-05
8	1	1.23E+02	3.36E+02	0.00E+00	-1.33E+03	-1.33E+04	-4.43E-05
9	1	7.20E+01	9.60E+01	0.00E+00	4.72E+02	4.72E+03	1.57E-05
10	1	7.20E+01	1.92E+02	0.00E+00	3.99E+02	3.99E+03	1.33E-05
11	1	7.20E+01	2.88E+02	0.00E+00	3.73E+02	3.73E+03	1.24E-05
12	1	7.20E+01	3.84E+02	0.00E+00	4.02E+02	4.02E+03	1.34E-05
13	1	6.90E+01	4.80E+01	0.00E+00	-5.65E+02	-3.39E+03	-1.88E-05
14	1	6.90E+01	1.44E+02	0.00E+00	-4.42E+02	-2.65E+03	-1.47E-05
15	1	6.90E+01	2.40E+02	0.00E+00	-4.53E+02	-2.72E+03	-1.51E-05
16	1	6.90E+01	3.36E+02	0.00E+00	-4.22E+02	-2.53E+03	-1.41E-05
17	1	7.50E+01	4.80E+01	0.00E+00	-5.65E+02	-3.39E+03	-1.88E-05
18	1	7.50E+01	1.44E+02	0.00E+00	-4.42E+02	-2.65E+03	-1.47E-05
19	1	7.50E+01	2.40E+02	0.00E+00	-4.53E+02	-2.72E+03	-1.51E-05
20	1	7.50E+01	3.36E+02	0.00E+00	-4.22E+02	-2.53E+03	-1.41E-05
21	1	2.70E+01	4.32E+02	0.00E+00	-1.57E+03	-9.43E+03	-5.24E-05
22	1	3.30E+01	5.28E+02	0.00E+00	-1.14E+03	-6.85E+03	-3.80E-05
23	1	3.90E+01	6.24E+02	0.00E+00	-7.04E+02	-4.23E+03	-2.35E-05
24	1	4.50E+01	7.20E+02	0.00E+00	-3.23E+02	-1.94E+03	-1.08E-05

E L E M E N T   S T R E S S E S   A N D   S T R A I N S

ELEMENT GROUP NUMBER . . . . (NEG) = 1

ELEMENT NUMBER	INT. PT. NUMBER	X1 COORD.	X2 COORD.	X3 COORD.	STRESS	FORCE	STRAIN
25	1	1.17E+02	4.32E+02	0.00E+00	-1.57E+03	-9.43E+03	-5.24E-05
26	1	1.11E+02	5.28E+02	0.00E+00	-1.14E+03	-6.85E+03	-3.80E-05
27	1	1.05E+02	6.24E+02	0.00E+00	-7.04E+02	-4.23E+03	-2.35E-05
28	1	9.90E+01	7.20E+02	0.00E+00	-3.23E+02	-1.94E+03	-1.08E-05
29	1	7.20E+01	4.80E+02	0.00E+00	6.70E+02	4.02E+03	2.23E-05
30	1	7.20E+01	5.76E+02	0.00E+00	4.70E+02	2.82E+03	1.57E-05
31	1	7.20E+01	6.72E+02	0.00E+00	2.80E+02	1.68E+03	9.35E-06
32	1	7.20E+01	7.68E+02	0.00E+00	7.97E+01	4.78E+02	2.66E-06
33	1	6.90E+01	4.32E+02	0.00E+00	-5.90E+02	-3.54E+03	-1.97E-05
34	1	6.90E+01	5.28E+02	0.00E+00	-4.65E+02	-2.79E+03	-1.55E-05
35	1	6.90E+01	6.24E+02	0.00E+00	-3.61E+02	-2.16E+03	-1.20E-05
36	1	6.90E+01	7.20E+02	0.00E+00	-2.04E+02	-1.22E+03	-6.79E-06
37	1	7.50E+01	4.32E+02	0.00E+00	-5.90E+02	-3.54E+03	-1.97E-05
38	1	7.50E+01	5.28E+02	0.00E+00	-4.65E+02	-2.79E+03	-1.55E-05
39	1	7.50E+01	6.24E+02	0.00E+00	-3.61E+02	-2.16E+03	-1.20E-05
40	1	7.50E+01	7.20E+02	0.00E+00	-2.04E+02	-1.22E+03	-6.79E-06

D I S P L A C E M E N T S

STEP NUMBER = 3

TIME = 3.000E+00

NODE NO.	DOF1	DOF2	DOF3
3	3.34677851E-02	2.40035164E-02	0.00000000E+00
4	3.60482223E-02	-4.22890665E-02	0.00000000E+00
5	1.16477083E-01	4.19970369E-02	0.00000000E+00
6	1.18322861E-01	-7.65042123E-02	0.00000000E+00
7	2.44742627E-01	5.39419704E-02	0.00000000E+00
8	2.46165751E-01	-1.01982625E-01	0.00000000E+00
9	4.14556396E-01	5.95759242E-02	0.00000000E+00
10	4.15835582E-01	-1.18602200E-01	0.00000000E+00
11	6.31453204E-01	6.70291111E-02	0.00000000E+00
12	6.33161773E-01	-1.38859420E-01	0.00000000E+00
13	8.99237866E-01	6.41814424E-02	0.00000000E+00
14	8.99909074E-01	-1.45212494E-01	0.00000000E+00
15	1.20168354E+00	5.24506079E-02	0.00000000E+00
16	1.20157427E+00	-1.39006859E-01	0.00000000E+00
17	1.51998256E+00	3.43155377E-02	0.00000000E+00
18	1.51930059E+00	-1.23227045E-01	0.00000000E+00

```
E L E M E N T   S T R E S S E S   A N D   S T R A I N S
   ELEMENT GROUP NUMBER . . . . . . . . . . . . (NEG   ) =     1
```

ELEMENT NUMBER	INT. PT. NUMBER	X1 COORD.	X2 COORD.	X3 COORD.	STRESS	FORCE	STRAIN
1	1	3.00E+00	4.80E+01	0.00E+00	8.12E+03	8.12E+04	2.71E-04
2	1	9.00E+00	1.44E+02	0.00E+00	7.22E+03	7.22E+04	2.41E-04
3	1	1.50E+01	2.40E+02	0.00E+00	6.21E+03	6.21E+04	2.07E-04
4	1	2.10E+01	3.36E+02	0.00E+00	5.06E+03	5.06E+04	1.69E-04
5	1	1.41E+02	4.80E+01	0.00E+00	-1.39E+04	-1.39E+05	-4.62E-04
6	1	1.35E+02	1.44E+02	0.00E+00	-1.23E+04	-1.23E+05	-4.08E-04
7	1	1.29E+02	2.40E+02	0.00E+00	-1.04E+04	-1.04E+05	-3.47E-04
8	1	1.23E+02	3.36E+02	0.00E+00	-8.47E+03	-8.47E+04	-2.82E-04
9	1	7.20E+01	9.60E+01	0.00E+00	5.86E+02	5.86E+03	1.95E-05
10	1	7.20E+01	1.92E+02	0.00E+00	4.61E+02	4.61E+03	1.54E-05
11	1	7.20E+01	2.88E+02	0.00E+00	3.95E+02	3.95E+03	1.32E-05
12	1	7.20E+01	3.84E+02	0.00E+00	4.00E+02	4.00E+03	1.33E-05
13	1	6.90E+01	4.80E+01	0.00E+00	9.71E+02	5.83E+03	3.24E-05
14	1	6.90E+01	1.44E+02	0.00E+00	1.25E+03	7.48E+03	4.16E-05
15	1	6.90E+01	2.40E+02	0.00E+00	1.30E+03	7.80E+03	4.33E-05
16	1	6.90E+01	3.36E+02	0.00E+00	1.36E+03	8.14E+03	4.52E-05
17	1	7.50E+01	4.80E+01	0.00E+00	-2.46E+03	-1.47E+04	-8.19E-05
18	1	7.50E+01	1.44E+02	0.00E+00	-2.44E+03	-1.47E+04	-8.14E-05
19	1	7.50E+01	2.40E+02	0.00E+00	-2.55E+03	-1.53E+04	-8.50E-05
20	1	7.50E+01	3.36E+02	0.00E+00	-2.55E+03	-1.53E+04	-8.49E-05
21	1	2.70E+01	4.32E+02	0.00E+00	6.54E+03	3.92E+04	2.18E-04
22	1	3.30E+01	5.28E+02	0.00E+00	4.32E+03	2.59E+04	1.44E-04
23	1	3.90E+01	6.24E+02	0.00E+00	2.23E+03	1.34E+04	7.44E-05
24	1	4.50E+01	7.20E+02	0.00E+00	5.47E+02	3.28E+03	1.82E-05

```
E L E M E N T   S T R E S S E S   A N D   S T R A I N S
   ELEMENT GROUP NUMBER . . . . . . . . . . . . (NEG   ) =      1
```

ELEMENT NUMBER	INT. PT. NUMBER	X1 COORD.	X2 COORD.	X3 COORD.	STRESS	FORCE	STRAIN
25	1	1.17E+02	4.32E+02	0.00E+00	-1.05E+04	-6.32E+04	-3.51E-04
26	1	1.11E+02	5.28E+02	0.00E+00	-7.17E+03	-4.30E+04	-2.39E-04
27	1	1.05E+02	6.24E+02	0.00E+00	-3.94E+03	-2.36E+04	-1.31E-04
28	1	9.90E+01	7.20E+02	0.00E+00	-1.27E+03	-7.62E+03	-4.23E-05
29	1	7.20E+01	4.80E+02	0.00E+00	6.10E+02	3.66E+03	2.03E-05
30	1	7.20E+01	5.76E+02	0.00E+00	2.80E+02	1.68E+03	9.32E-06
31	1	7.20E+01	6.72E+02	0.00E+00	-5.46E+01	-3.28E+02	-1.82E-06
32	1	7.20E+01	7.68E+02	0.00E+00	-4.26E+02	-2.56E+03	-1.42E-05
33	1	6.90E+01	4.32E+02	0.00E+00	1.08E+03	6.49E+03	3.61E-05
34	1	6.90E+01	5.28E+02	0.00E+00	1.11E+03	6.64E+03	3.69E-05
35	1	6.90E+01	6.24E+02	0.00E+00	9.91E+02	5.94E+03	3.30E-05
36	1	6.90E+01	7.20E+02	0.00E+00	7.08E+02	4.25E+03	2.36E-05
37	1	7.50E+01	4.32E+02	0.00E+00	-2.75E+03	-1.65E+04	-9.15E-05
38	1	7.50E+01	5.28E+02	0.00E+00	-2.47E+03	-1.48E+04	-8.25E-05
39	1	7.50E+01	6.24E+02	0.00E+00	-2.08E+03	-1.25E+04	-6.94E-05
40	1	7.50E+01	7.20E+02	0.00E+00	-1.37E+03	-8.24E+03	-4.58E-05

D Y N A M I C   S T O R A G E   A L L O C A T I O N   I N F O R M A T I O N

ARRAY NO.	ARRAY	ADDRESS	DIM1	DIM2	DIM3	PREC.
1	NSTEP	1	1	0	0	1
2	NDPRT	3	1	0	0	1
3	NSPRT	5	1	0	0	1
4	NHPLT	7	1	0	0	1
5	NITER	9	1	0	0	1
6	ALPHA	11	1	0	0	2
7	BETA	13	1	0	0	2
8	GAMMA	15	1	0	0	2
9	DT	17	1	0	0	2
10	D	19	3	18	0	2
11	X	127	3	18	0	2
12	ID	235	3	18	0	1
13	F	289	3	18	2	2
14	G	505	4	2	2	2
15	G1	537	2	0	0	2
16	IDIAG	541	32	0	0	1
17	NGRP	573	1	0	0	1
*****	BEGIN ELEMENT GROUP NUMBER		1			
18	NPAR	575	16	0	0	1
19	MP	591	29	0	0	1
20	W	621	1	0	0	2
21	DET	623	1	0	0	2
22	SHL	625	2	2	1	2
23	SHG	633	2	2	1	2
24	XS	641	3	1	0	2
25	RHO	647	2	0	0	2
26	RDAMPM	651	2	0	0	2
27	RDAMPK	655	2	0	0	2
28	AREA	659	2	0	0	2
29	C	663	1	1	2	2
30	GRAV	667	3	0	0	2
31	IEN	673	2	40	0	1
32	MAT	753	40	0	0	1
33	LM	793	3	2	40	1
34	ELEFFM	1033	6	6	0	2
35	XL	1105	3	2	0	2
36	WORK	1117	16	0	0	2
37	B	1149	1	6	0	2
38	DMAT	1161	1	1	0	2
39	DB	1163	1	6	0	2
40	VL	1175	3	2	0	2
41	AL	1187	3	2	0	2
42	ELRESF	1199	6	0	0	2
43	DL	1211	3	2	0	2
44	STRAIN	1223	1	0	0	2
45	STRESS	1225	1	0	0	2
46	FORCE	1227	1	0	0	2
*****	END ELEMENT GROUP DATA					
47	ALHS	1229	192	0	0	2
48	BRHS	1613	32	0	0	2

TRUSS TOWER WITH STATIC VERTICAL AND LATERAL LOADS

E X E C U T I O N   T I M I N G   I N F O R M A T I O N

I N I T I A L I Z A T I O N P H A S E	=	1.578E+00
S O L U T I O N P H A S E	=	2.461E+00
FORMATION OF LEFT-HAND-SIDE MATRICES	=	1.797E-01
FACTORIZATIONS	=	1.953E-02
FORMATION OF RIGHT-HAND-SIDE VECTORS	=	5.859E-02
FORWARD REDUCTIONS/BACK SUBSTITUTIONS	=	1.953E-02
CALCULATION OF ELEMENT OUTPUT	=	1.781E+00

SUBTOTAL	=	2.059E+00

Example 2 (Static Analysis of a Plane Strain Cantilever Beam)

This problem illustrates the use of the four-node quadrilateral, elastic continuum element. The mathematical model and corresponding mesh are shown in Figs. 4.4.1 and 4.4.2. In Table 11.5.2 we present the input file needed for one of the static analyses discussed in Chapter 4. Notice that the nodes have been numbered in a manner that minimizes the bandwidth of the effective mass matrix $\boldsymbol{M}^*$ (i.e., the stiffness matrix $\boldsymbol{K}$). Observe that the boundary condition codes are set to allow warping of the root section of the cantilever, which is consistent with the analytic solution. The single load vector specified represents the consistent nodal loads $f_a^e = \int N_a \ell \, d\Omega$ computed from the surface tractions defined in Chapter 4. In this example, the nodal loads were computed by hand. An alternative procedure is to use the element surface loads described in Sec. In.10.1.5. However, note that the element surface loads are restricted to piecewise linear variation. The parabolic variation of end shear cannot be exactly represented by *continuous* piecewise linear variation, but it can be exactly represented by *discontinuous* piecewise linear interpolation (see Sec. 1.15, Exercise 1).

TABLE 11.5.2 Input File for Static Analysis of Cantilever Beam

```
In 0.0         0
In 1.0      PLANE STRAIN CANTILEVER BEAM: STATIC ANALYSIS WITH CONSISTENT NODAL FORCES
In 2.0         1    1    0    0    0    0    1    0    2   45    2    1    1    1    1
In 3.0         1    1    1    1    0    1   0.0       0.0       0.0       1.0
In 5.1         1    4        0.0       0.0   ⎫
            ⎧                16.0       0.0   ⎪
In 5.2      ⎨                16.0       2.0   ⎬ Coordinates
            ⎩                 0.0       2.0   ⎪
In 5.3         8    5    4    1               ⎭

            ⎧  1              1    1          ⎫
In 6.0      ⎨  5              1    0          ⎬ Boundary conditions
            ⎩  6   41    5    1    0          ⎭

            ⎧  2      -7.500E-01   0.17383    ⎫
            ⎪  3      -1.500E+00   0.13867    ⎪
            ⎪  4      -2.250E+00   0.08008    ⎪
            ⎪  5       0.000E+00   0.01465    ⎪
In 7.1      ⎨ 41       0.000E+00  -0.09280    ⎬ Load vector
            ⎪ 42       0.000E+00  -0.17383    ⎪
            ⎪ 43       0.000E+00  -0.13867    ⎪
            ⎪ 44       0.000E+00  -0.08008    ⎪
            ⎩ 45       0.000E+00  -0.01465    ⎭
In 8.1             0.0       1.0
In 10.1.1      1   32    1    0    0    1    0    0    0    0    0    0    0   ⎫
In 10.1.2      1        1.00     0.300    0.0       0.0       0.0            ⎪
In 10.1.3                                                                    ⎬ Element
In 10.1.4.1    1    1    1    6    7    2    1 ⎫                             ⎪ group
In 10.1.4.2    8    4    5    4    1    1      ⎭ Element nodes               ⎭
            *END
```

Example 3 (Dynamic Analysis of a Plane Strain Cantilever Beam)

This problem, like the previous one, employs the configuration illustrated in Figs. 4.4.1 and 4.4.2. However, here the beam is forced dynamically. The beam is assumed initially at rest and all loads, namely, those of the static linear elasticity solution, are applied instantaneously at $t = 0^+$. The following data were employed in this calculation:

$$P(t) = 1000H(t) \qquad \text{(load factor, see Sec. 4.4)}$$

$$\left.\begin{aligned} \lambda &= 1.2 \times 10^6 \\ \mu &= 0.8 \times 10^6 \end{aligned}\right\} \qquad \text{(Lamé parameters)}$$

$$\rho = 2.0 \times 10^{-3} \qquad \text{(density)}$$

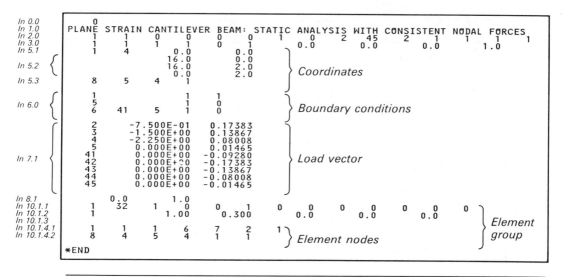

where $H(t)$ denotes the Heaviside step function (i.e., $H(t) = 1$ if $t > 0$; otherwise, $H(t) = 0$).

The dynamic response is calculated with the Hilber-Hughes-Taylor α-method. The parameters employed are

$$\alpha = -0.1, \qquad \beta = 0.3025, \qquad \gamma = 0.60$$

which results in an implicit, second-order accurate, unconditionally stable algorithm with some high-frequency numerical dissipation. One hundred equal time steps of length $\Delta t = 2.5 \times 10^{-4}$ were used in the analysis. This amounts to about 50 time steps for the fundamental period* and ensures an accurate integration of the response in the fundamental mode. By comparing the input files of Tables 11.5.2 and 11.5.3, it may be seen that little additional information is required to specify the dynamic problem. The same load vector is utilized as in the static case, but it is now scaled by a factor of 1000 by way of a constant load-time function. Three blank lines are inserted to indicate zero kinematic initial conditions. Notice that setting NDPRT and NSPRT equal to zero minimizes the printed output. Time-history plots of the vertical displacement of the tip (i.e., node 45) and the σ_{11}-component of stress in element 20 are requested. The output file is presented on pages 707 through 715.

TABLE 11.5.3 Input File for Dynamic Analysis of a Plane Strain Cantilever Beam

```
In 0.0         0
In 1.0     PLANE STRAIN CANTILEVER BEAM: DYNAMIC ANALYSIS WITH ALPHA METHOD
In 2.0         1     0     0     0     0     0     1     1     2    45     2     1     1     1     1
In 3.0         1   100     0     0     2     1      -0.10    0.3025        0.60   2.50E-4
In 4.0        45     2     1
In 5.1         1     4         0.0         0.0    }
In 5.2                        16.0         0.0    }
                              16.0         2.0    }  Coordinates
                               0.0         2.0    }
In 5.3         8     5     4     1                }

In 6.0         1                 1     1          }
               5                 1     0          }  Boundary conditions
               6    41     5     1     0          }

In 7.1         2      -7.500E-01   0.17383        }
               3      -1.500E+00   0.13867        }
               4      -2.250E+00   0.08008        }
               5       0.000E+00   0.01465        }
              41       0.000E+00  -0.09280        }  Load vector
              42       0.000E+00  -0.17383        }
              43       0.000E+00  -0.13867        }
              44       0.000E+00  -0.08008        }
              45       0.000E+00  -0.01465        }
In 8.1               0.0    1000.0                   Load-time function
In 9.1                                            }  Initial conditions

In 10.1.1      1    32     1     0     1     1     0     0     0     0     0     0     0  }
In 10.1.2      1       2.08E+6        0.30     0.0020      0.00        0.00             }
In 10.1.3                                                                               }  Element
In 10.1.4.1    1     1     1     6     7     2     1  }                                 }  group
In 10.1.4.2    8     4     5     4     1     1        }  Element nodes                  }
In 10.1.6     20     4     1                             Element time-history           }
           *END
```

* The fundamental period can be estimated from classical beam theory:

$$T_1 = \frac{2\pi}{\omega_1} \cong \frac{2\pi}{1.875^2}\sqrt{\frac{ml^4}{EI}} = \frac{2\pi}{1.875^2}\sqrt{\frac{(0.008)(16)^4}{(2.08 \times 10^6)(5.3333)}} = 1.23 \times 10^{-2}$$

where E is Young's modulus, I is the cross-sectional moment of inertia, l is the length and m denotes the mass per unit length. (See W. C. Hurty and M. F. Rubinstein, *Dynamics of Structures,* p. 203. Englewood Cliffs, N.J.: Prentice-Hall, 1964.)

```
PLANE STRAIN CANTILEVER BEAM: DYNAMIC ANALYSIS WITH ALPHA METHOD

E X E C U T I O N   C O N T R O L   I N F O R M A T I O N
     EXECUTION CODE . . . . . . . . . . . . . . . (IEXEC ) =      1
        EQ. 0, DATA CHECK
        EQ. 1, EXECUTION
     ANALYSIS CODE . . . . . . . . . . . . . . . . (IACODE) =      0
        EQ. 0, DYNAMIC ANALYSIS
        EQ. 1, STATIC ANALYSIS
     READ RESTART FILE CODE . . . . . . . . . . . (IREADR) =      0
        EQ. 0, DO NOT READ RESTART FILE
        EQ. 1, READ RESTART FILE
     WRITE RESTART FILE CODE . . . . . . . . . . . (IWRITR) =      0
        EQ. 0, DO NOT WRITE RESTART FILE
        EQ. 1, WRITE RESTART FILE
     INPUT DATA PRINT CODE . . . . . . . . . . . . (IPRTIN) =      0
        EQ. 0, PRINT NODAL AND ELEMENT INPUT DATA
        EQ. 1, DO NOT PRINT NODAL AND ELEMENT INPUT DATA
     RANK CHECK CODE . . . . . . . . . . . . . . . (IRANK ) =      0
        EQ. 0, DO NOT PERFORM RANK CHECK
        EQ. 1, PRINT NUMBERS OF ZERO AND NEGATIVE PIVOTS
        EQ. 2, PRINT ALL PIVOTS
     NUMBER OF TIME SEQUENCES . . . . . . . . . . (NUMSEQ) =      1
     NUMBER OF NODAL OUTPUT TIME-HISTORIES . . . (NDOUT ) =      1
     NUMBER OF SPACE DIMENSIONS . . . . . . . . . (NSD   ) =      2
     NUMBER OF NODAL POINTS . . . . . . . . . . . (NUMNP ) =     45
     NUMBER OF NODAL DEGREES-OF-FREEDOM . . . . . (NDOF  ) =      2
     NUMBER OF LOAD VECTORS . . . . . . . . . . . (NLVECT) =      1
     NUMBER OF LOAD-TIME FUNCTIONS . . . . . . . . (NLTFTN) =      1
     NUMBER OF POINTS ON LOAD-TIME FUNCTIONS . . (NPTSLF) =      1
     NUMBER OF ELEMENT GROUPS . . . . . . . . . . (NUMEG ) =      1

T I M E   S E Q U E N C E   D A T A
    NUMBER OF TIME SEQUENCES . . . . . . (NUMSEQ  ) =      1

    TIME SEQUENCE NUMBER . . . . . . . . (N       ) =      1
    NUMBER OF TIME STEPS . . . . . . . . (NSTEP(N)) =    100
    KINEMATIC PRINT INCREMENT . . . . . (NDPRT(N)) =      0
    STRESS/STRAIN PRINT INCREMENT . . . (NSPRT(N)) =      0
    TIME HISTORY PLOT INCREMENT . . . . (NHPLT(N)) =      2
    NUMBER OF ITERATIONS . . . . . . . . (NITER(N)) =      1
    FIRST INTEGRATION PARAMETER . . . . (ALPHA(N)) = -1.00000E-01
    SECOND INTEGRATION PARAMETER . . . . (BETA(N) ) =  3.02500E-01
    THIRD INTEGRATION PARAMETER . . . . (GAMMA(N)) =  6.00000E-01
    TIME STEP . . . . . . . . . . . . . . (DT(N)   ) =  2.50000E-04

N O D A L   T I M E - H I S T O R Y   I N F O R M A T I O N
    NUMBER OF NODAL TIME HISTORIES . . . . . . (NDOUT ) =      1

      NODE        DOF       KINEMATIC
     NUMBER      NUMBER       TYPE
       45           2         DISP
```

N O D A L   C O O R D I N A T E   D A T A

NODE NO.	X1	X2
1	0.00000000E+00	0.00000000E+00
2	0.00000000E+00	5.00000000E-01
3	0.00000000E+00	1.00000000E+00
4	0.00000000E+00	1.50000000E+00
5	0.00000000E+00	2.00000000E+00
6	2.00000000E+00	0.00000000E+00
7	2.00000000E+00	5.00000000E-01
8	2.00000000E+00	1.00000000E+00
9	2.00000000E+00	1.50000000E+00
10	2.00000000E+00	2.00000000E+00
11	4.00000000E+00	0.00000000E+00
12	4.00000000E+00	5.00000000E-01
13	4.00000000E+00	1.00000000E+00
14	4.00000000E+00	1.50000000E+00
15	4.00000000E+00	2.00000000E+00
16	6.00000000E+00	0.00000000E+00
17	6.00000000E+00	5.00000000E-01
18	6.00000000E+00	1.00000000E+00
19	6.00000000E+00	1.50000000E+00
20	6.00000000E+00	2.00000000E+00
21	8.00000000E+00	0.00000000E+00
22	8.00000000E+00	5.00000000E-01
23	8.00000000E+00	1.00000000E+00
24	8.00000000E+00	1.50000000E+00
25	8.00000000E+00	2.00000000E+00
26	1.00000000E+01	0.00000000E+00
27	1.00000000E+01	5.00000000E-01
28	1.00000000E+01	1.00000000E+00
29	1.00000000E+01	1.50000000E+00
30	1.00000000E+01	2.00000000E+00
31	1.20000000E+01	0.00000000E+00
32	1.20000000E+01	5.00000000E-01
33	1.20000000E+01	1.00000000E+00
34	1.20000000E+01	1.50000000E+00
35	1.20000000E+01	2.00000000E+00
36	1.40000000E+01	0.00000000E+00
37	1.40000000E+01	5.00000000E-01
38	1.40000000E+01	1.00000000E+00
39	1.40000000E+01	1.50000000E+00
40	1.40000000E+01	2.00000000E+00
41	1.60000000E+01	0.00000000E+00
42	1.60000000E+01	5.00000000E-01
43	1.60000000E+01	1.00000000E+00
44	1.60000000E+01	1.50000000E+00
45	1.60000000E+01	2.00000000E+00

```
N O D A L   B O U N D A R Y   C O N D I T I O N   C O D E S

     NODE NO.          DOF1      DOF2
         1               1         1
         5               1         0
         6               1         0
        11               1         0
        16               1         0
        21               1         0
        26               1         0
        31               1         0
        36               1         0
        41               1         0

P R E S C R I B E D   F O R C E S   A N D   K I N E M A T I C
  B O U N D A R Y   C O N D I T I O N S

     LOAD VECTOR NUMBER =       1

     NODE NO.              DOF1               DOF2
         2           -7.50000000E-01     1.73830000E-01
         3           -1.50000000E+00     1.38670000E-01
         4           -2.25000000E+00     8.00800000E-02
         5            0.00000000E+00     1.46500000E-02
        41            0.00000000E+00    -9.28000000E-02
        42            0.00000000E+00    -1.73830000E-01
        43            0.00000000E+00    -1.38670000E-01
        44            0.00000000E+00    -8.00800000E-02
        45            0.00000000E+00    -1.46500000E-02

L O A D - T I M E   F U N C T I O N   D A T A

     NUMBER OF LOAD-TIME FUNTIONS . . . . . (NLTFTN  ) =       1

     FUNCTION NUMBER      1
               TIME                 LOAD FACTOR

         0.00000000E+00        1.00000000E+03

THERE ARE NO NONZERO INITIAL DISPLACEMENTS

THERE ARE NO NONZERO INITIAL VELOCITIES

THERE ARE NO NONZERO INITIAL ACCELERATIONS
```

```
E L E M E N T   G R O U P   D A T A
    ELEMENT GROUP NUMBER . . . . . . . . . . . . (NEG    ) =      1

F O U R - N O D E   Q U A D R I L A T E R A L   E L E M E N T S
    ELEMENT TYPE NUMBER . . . . . . . . . . . . . (NTYPE ) =      1
    NUMBER OF ELEMENTS  . . . . . . . . . . . . . (NUMEL ) =     32
    NUMBER OF ELEMENT MATERIAL SETS . . . . . . . (NUMAT ) =      1
    NUMBER OF SURFACE FORCE CARDS . . . . . . . . (NSURF ) =      0
    NUMBER OF STRESS/STRAIN TIME HISTORIES  . . . (NSOUT ) =      1
    ANALYSIS OPTION . . . . . . . . . . . . . . . (IOPT  ) =      1
       EQ.0, PLANE STRESS
       EQ.1, PLANE STRAIN
       EQ.2, AXISYMMETRIC
    STRESS OUTPUT PRINT CODE  . . . . . . . . . . (ISTPRT) =      0
       EQ.0, STRESS OUTPUT PRINTED
       EQ.1, STRESS OUTPUT NOT PRINTED
    SURFACE FORCE LOAD-TIME FUNCTION NUMBER . . . (LFSURF) =      0
    BODY FORCE LOAD-TIME FUNCTION NUMBER  . . . . (LFBODY) =      0
    NUMERICAL INTEGRATION CODE  . . . . . . . . . (NICODE) =      0
       EQ.0, 2 X 2 GAUSSIAN QUADRATURE
       EQ.1, 1-POINT GAUSSIAN QUADRATURE
    STRAIN-DISPLACEMENT OPTION  . . . . . . . . . (IBBAR ) =      0
       EQ.0, STANDARD FORMULATION
       EQ.1, B-BAR FORMULATION
    MASS TYPE CODE  . . . . . . . . . . . . . . . (IMASS ) =      0
       EQ.0, CONSISTENT MASS MATRIX
       EQ.1, LUMPED MASS MATRIX
       EQ.2, NO MASS MATRIX
    IMPLICIT/EXPLICIT CODE  . . . . . . . . . . . (IMPEXP) =      0
       EQ.0, IMPLICIT ELEMENT GROUP
       EQ.1, EXPLICIT ELEMENT GROUP


M A T E R I A L   S E T   D A T A
    NUMBER OF MATERIAL SETS . . . . . . . . . . . (NUMAT ) =      1

     SET      YOUNG'S     POISSON'S      MASS         MASS       STIFFNESS    THICKNESS
    NUMBER    MODULUS       RATIO       DENSITY      DAMPING      DAMPING
      1     2.0800E+06   3.0000E-01   2.0000E-03   0.0000E+00   0.0000E+00   1.0000E+00


G R A V I T Y   V E C T O R   C O M P O N E N T S
    X-1 DIRECTION . . . . . . . . . . . . . . . = 0.00000000E+00
    X-2 DIRECTION . . . . . . . . . . . . . . . = 0.00000000E+00
```

```
E L E M E N T   D A T A

     ELEMENT    MATERIAL  NODE 1     NODE 2     NODE 3     NODE 4
        1           1        1          6          7          2
        2           1        2          7          8          3
        3           1        3          8          9          4
        4           1        4          9         10          5
        5           1        6         11         12          7
        6           1        7         12         13          8
        7           1        8         13         14          9
        8           1        9         14         15         10
        9           1       11         16         17         12
       10           1       12         17         18         13
       11           1       13         18         19         14
       12           1       14         19         20         15
       13           1       16         21         22         17
       14           1       17         22         23         18
       15           1       18         23         24         19
       16           1       19         24         25         20
       17           1       21         26         27         22
       18           1       22         27         28         23
       19           1       23         28         29         24
       20           1       24         29         30         25
       21           1       26         31         32         27
       22           1       27         32         33         28
       23           1       28         33         34         29
       24           1       29         34         35         30
       25           1       31         36         37         32
       26           1       32         37         38         33
       27           1       33         38         39         34
       28           1       34         39         40         35
       29           1       36         41         42         37
       30           1       37         42         43         38
       31           1       38         43         44         39
       32           1       39         44         45         40

E L E M E N T   T I M E   H I S T O R Y   I N F O R M A T I O N
     NUMBER OF STRESS/STRAIN TIME HISTORIES  . . (NSOUT ) =     1

       ELEMENT   INT PT   COMPONENT
       NUMBER    NUMBER

         20         4        S 11

PLANE STRAIN CANTILEVER BEAM: DYNAMIC ANALYSIS WITH ALPHA METHOD

E Q U A T I O N   S Y S T E M   D A T A
     NUMBER OF EQUATIONS . . . . . . . . . . . . . (NEQ    ) =      79
     NUMBER OF TERMS IN LEFT-HAND-SIDE MATRIX  . (NALHS ) =     872
     MEAN HALF BANDWIDTH . . . . . . . . . . . . . (MEANBW) =      11
     TOTAL LENGTH OF BLANK COMMON REQUIRED . . . (NWORDS) =    5196
```

PLANE STRAIN CANTILEVER BEAM: DYNAMIC ANALYSIS WITH ALPHA METHOD

NODE NUMBER = 45

DOF NUMBER = 2 OUTPUT: DISP

TIME	VALUE
0.0000E+00	0.0000E+00
5.0000E-04	-5.4078E-03
1.0000E-03	-1.8582E-02
1.5000E-03	-3.3184E-02
2.0000E-03	-4.6195E-02
2.5000E-03	-6.2725E-02
3.0000E-03	-8.8417E-02
3.5000E-03	-1.1853E-01
4.0000E-03	-1.4416E-01
4.5000E-03	-1.6146E-01
5.0000E-03	-1.7541E-01
5.5000E-03	-1.9146E-01
6.0000E-03	-2.0595E-01
6.5000E-03	-2.1102E-01
7.0000E-03	-2.0382E-01
7.5000E-03	-1.9046E-01
8.0000E-03	-1.7812E-01
8.5000E-03	-1.6520E-01
9.0000E-03	-1.4554E-01
9.5000E-03	-1.1748E-01
1.0000E-02	-8.8231E-02
1.0500E-02	-6.5911E-02
1.1000E-02	-4.9822E-02
1.1500E-02	-3.4154E-02
1.2000E-02	-1.6705E-02
1.2500E-02	-3.9318E-03
1.3000E-02	-2.8680E-03
1.3500E-02	-1.1958E-02
1.4000E-02	-2.3960E-02
1.4500E-02	-3.4801E-02
1.5000E-02	-4.9081E-02
1.5500E-02	-7.2129E-02
1.6000E-02	-1.0136E-01
1.6500E-02	-1.2838E-01
1.7000E-02	-1.4806E-01
1.7500E-02	-1.6424E-01
1.8000E-02	-1.8210E-01
1.8500E-02	-1.9972E-01
1.9000E-02	-2.0945E-01
1.9500E-02	-2.0710E-01
2.0000E-02	-1.9753E-01
2.0500E-02	-1.8743E-01
2.1000E-02	-1.7690E-01
2.1500E-02	-1.6005E-01
2.2000E-02	-1.3415E-01
2.2500E-02	-1.0517E-01
2.3000E-02	-8.0990E-02
2.3500E-02	-6.2825E-02
2.4000E-02	-4.5339E-02
2.4500E-02	-2.5752E-02
2.5000E-02	-9.3527E-03

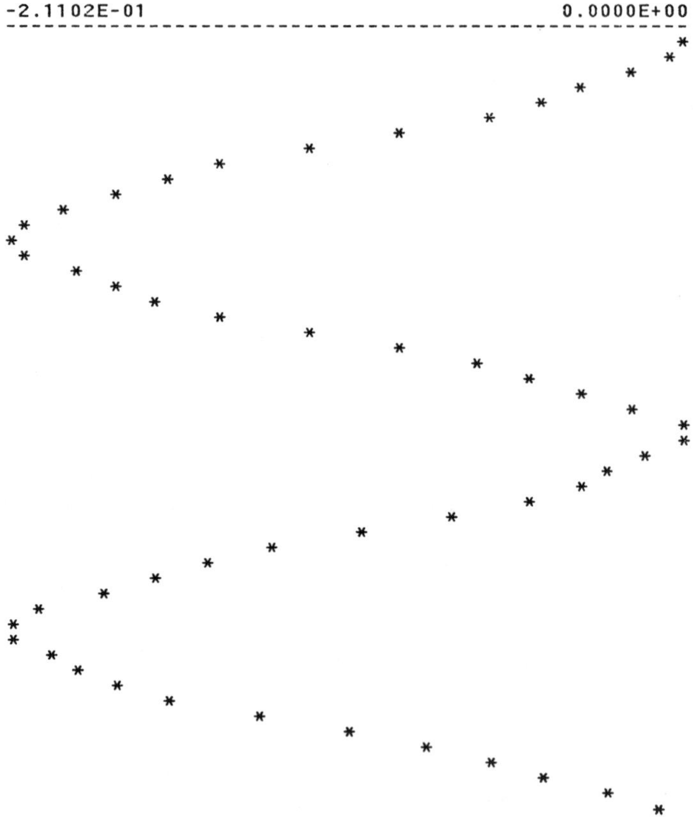

```
PLANE STRAIN CANTILEVER BEAM: DYNAMIC ANALYSIS WITH ALPHA METHOD

ELEMENT NUMBER             =    20
INTEGRATION POINT NUMBER   =     4      OUTPUT: S 11

    TIME         VALUE       0.0000E+00                              4.4685E+03
 ----------   ----------    --------------------------------------------------
 0.0000E+00   0.0000E+00    *
 5.0000E-04   4.6883E+02         *
 1.0000E-03   1.7634E+03                        *
 1.5000E-03   2.1571E+03                             *
 2.0000E-03   1.5223E+03                      *
 2.5000E-03   8.3195E+02             *
 3.0000E-03   1.5578E+03                      *
 3.5000E-03   2.9402E+03                                *
 4.0000E-03   3.6048E+03                                        *
 4.5000E-03   3.0491E+03                                  *
 5.0000E-03   2.2760E+03                         *
 5.5000E-03   2.8644E+03                               *
 6.0000E-03   4.0393E+03                                            *
 6.5000E-03   4.4685E+03                                                 *
 7.0000E-03   3.5778E+03                                        *
 7.5000E-03   2.5135E+03                           *
 8.0000E-03   2.6804E+03                             *
 8.5000E-03   3.4800E+03                                       *
 9.0000E-03   3.6342E+03                                         *
 9.5000E-03   2.4908E+03                           *
 1.0000E-02   1.2421E+03                 *
 1.0500E-02   1.2176E+03                 *
 1.1000E-02   1.9710E+03                      *
 1.1500E-02   2.2058E+03                         *
 1.2000E-02   1.2270E+03                 *
 1.2500E-02   1.9409E+02      *
 1.3000E-02   3.8807E+02        *
 1.3500E-02   1.4569E+03                    *
 1.4000E-02   2.0683E+03                       *
 1.4500E-02   1.4862E+03                    *
 1.5000E-02   8.0191E+02             *
 1.5500E-02   1.2554E+03                 *
 1.6000E-02   2.5711E+03                            *
 1.6500E-02   3.3970E+03                                      *
 1.7000E-02   2.9637E+03                                 *
 1.7500E-02   2.3049E+03                          *
 1.8000E-02   2.6465E+03                             *
 1.8500E-02   3.8000E+03                                          *
 1.9000E-02   4.4250E+03                                                 *
 1.9500E-02   3.7470E+03                                         *
 2.0000E-02   2.7592E+03                              *
 2.0500E-02   2.6999E+03                              *
 2.1000E-02   3.4782E+03                                       *
 2.1500E-02   3.7809E+03                                         *
 2.2000E-02   2.8405E+03                               *
 2.2500E-02   1.6044E+03                      *
 2.3000E-02   1.3247E+03                  *
 2.3500E-02   1.9881E+03                      *
 2.4000E-02   2.2995E+03                          *
 2.4500E-02   1.4800E+03                    *
 2.5000E-02   4.0438E+02        *
```

D Y N A M I C   S T O R A G E   A L L O C A T I O N   I N F O R M A T I O N

ARRAY NO.	ARRAY	ADDRESS	DIM1	DIM2	DIM3	PREC.
1	NSTEP	1	1	0	0	1
2	NDPRT	3	1	0	0	1
3	NSPRT	5	1	0	0	1
4	NHPLT	7	1	0	0	1
5	NITER	9	1	0	0	1
6	ALPHA	11	1	0	0	2
7	BETA	13	1	0	0	2
8	GAMMA	15	1	0	0	2
9	DT	17	1	0	0	2
10	IDHIST	19	3	1	0	1
11	DOUT	23	2	51	0	1
12	VPRED	125	2	45	0	2
13	DPRED	305	2	45	0	2
14	A	485	2	45	0	2
15	V	665	2	45	0	2
16	D	845	2	45	0	2
17	X	1025	2	45	0	2
18	ID	1205	2	45	0	1
19	F	1295	2	45	1	2
20	G	1475	1	2	1	2
21	G1	1479	1	0	0	2
22	IDIAG	1481	79	0	0	1
23	NGRP	1561	1	0	0	1
*****	BEGIN ELEMENT GROUP NUMBER 1					
24	NPAR	1563	16	0	0	1
25	MP	1579	35	0	0	1
26	W	1615	4	0	0	2
27	DET	1623	4	0	0	2
28	R	1631	4	0	0	2
29	SHL	1639	3	4	4	2
30	SHG	1735	3	4	4	2
31	SHGBAR	1831	3	4	1	2
32	RHO	1855	1	0	0	2
33	RDAMPM	1857	1	0	0	2
34	RDAMPK	1859	1	0	0	2
35	TH	1861	1	0	0	2
36	C	1863	4	4	1	2
37	GRAV	1895	2	0	0	2
38	IEN	1899	4	32	0	1
39	MAT	2027	32	0	0	1
40	LM	2059	2	4	32	1
41	IELNO	2315	0	0	0	1
42	ISIDE	2315	0	0	0	1
43	PRESS	2315	2	0	0	2
44	SHEAR	2319	2	0	0	2
45	ISHIST	2323	3	1	0	1
46	SOUT	2327	2	51	0	1
47	ELEFFM	2429	8	8	0	2
48	XL	2557	2	4	0	2
49	WORK	2573	16	0	0	2
50	B	2605	4	8	0	2

D Y N A M I C   S T O R A G E   A L L O C A T I O N   I N F O R M A T I O N

ARRAY NO.	ARRAY	ADDRESS	DIM1	DIM2	DIM3	PREC.
51	DMAT	2669	4	4	0	2
52	DB	2701	4	8	0	2
53	VL	2765	2	4	0	2
54	AL	2781	2	4	0	2
55	ELRESF	2797	8	0	0	2
56	DL	2813	2	4	0	2
57	STRAIN	2829	4	0	0	2
58	STRESS	2837	4	0	0	2
59	PSTRN	2845	4	0	0	2
60	PSTRS	2853	4	0	0	2
*****	END ELEMENT GROUP DATA					
61	ALHS	2861	872	0	0	2
62	BRHS	4605	79	0	0	2

PLANE STRAIN CANTILEVER BEAM: DYNAMIC ANALYSIS WITH ALPHA METHOD

E X E C U T I O N   T I M I N G   I N F O R M A T I O N

I N I T I A L I Z A T I O N P H A S E	=	1.340E+00
S O L U T I O N P H A S E	=	9.855E+01
FORMATION OF LEFT-HAND-SIDE MATRICES	=	1.891E+00
FACTORIZATIONS	=	1.289E-01
FORMATION OF RIGHT-HAND-SIDE VECTORS	=	8.767E+01
FORWARD REDUCTIONS/BACK SUBSTITUTIONS	=	3.773E+00
CALCULATION OF ELEMENT OUTPUT	=	1.006E+00

SUBTOTAL	=	9.447E+01

Example 4 (Implicit-Explicit Dynamic Analysis of a Rod)

The initial-value problem consists of a rod, modeled with two-node truss elements, fixed at the left end, and subjected to an initial displacement. The initial velocity and external forces are assumed equal to 0. The length of the rod is 10.5 and there are 21 equal-length elements. Young's modulus equals 100 in all elements except the last, in which it equals 10^7. The initial displacement varies linearly between 0 on the left end of the rod (i.e., $x = 0$) and 0.1 at $x = 10.0$ and is constant over the last element. The cross-sectional area of the rod is 1.0 and its density is 0.01. The implicit-explicit algorithm described in Sec. 9.4 is employed. The algorithmic parameters used are

$$\alpha = 0.0, \qquad \beta = 0.3025, \qquad \gamma = 0.60$$

The critical time step of explicit integration may be computed from (9.4.29), namely,

$$\omega^h \Delta t \leq \Omega_{\text{bif}} = \frac{2(1-\xi)}{\gamma + \frac{1}{2}}$$

There is no viscous damping, so $\xi = 0$. Assuming lumped mass, the maximum element frequency is (see (9.2.1)):

$$\omega^h_{\text{max}} = \frac{2c}{h} = \frac{2\sqrt{E/\rho}}{h}$$

Evaluating these formulas reveals that the stable critical time step for the last element is $\frac{1}{316}$ that of the first 20 elements. This suggests a mesh partition in which the last element is treated implicitly and the first 20 elements are treated explicitly, thus enabling the larger critical time step to be employed. Specifically, $\Delta t = 4.5 \times 10^{-3}$ was used, and the analysis was run for 100 steps. Output histories are requested for displacement at the end of the rod (i.e., node 22) and stress in element 10. The initial displacements require that the initial accelerations be computed. This is facilitated by a one-step, "zero Δt" time sequence as described in Sec. In.3.0. The data for the analysis is presented in Table 11.5.4 and the output follows on pages 717 through 729.

Exercise 1. The addition of Rayleigh damping can significantly reduce critical time steps in explicit integration. This can be seen from (9.4.28):

$$\Omega_{\text{crit}} = \frac{(\xi^2 + 2\gamma)^{1/2} - \xi}{\gamma}$$

In the physically undamped case,

$$\Omega_{\text{crit}} = \sqrt{\frac{2}{\gamma}}$$

Estimate the fundamental mode of the rod from

$$\omega_1 = \frac{c\pi}{2L}$$

Select the stiffness-proportional damping factor RDAMPK such that the damping ratio in the fundamental mode, namely, ξ_1, is 10%. Determine the maximum damping factor from

$$\xi_{\max} = \tfrac{1}{2}(\text{RDAMPM}/\omega^h_{\max} + \text{RDAMPK} * \omega^h_{\max})$$

Consider the setup of Example 4. Due to *implicit* treatment of the single stiff element, it may be ignored in calculating $\omega^h_{\max}$, i.e., employ $E = 100.0$ and $h = 0.5$. Estimate ω_1 using $E = 100.0$ and $L = 10.0$. Assuming $\gamma = \frac{1}{2}$, calculate the reduction in critical time step compared with the physically undamped case, namely,

$$\frac{[(\xi^2_{\max} + 2\gamma)^{1/2} - \xi_{\max}]/\gamma}{\sqrt{2/\gamma}}$$

[*Ans.* 0.189]

Repeat the calculation assuming the rod is discretized by 210 elements of length 0.05.

TABLE 11.5.4 Input File for Implicit-Explicit Dynamic Analysis of a Rod

```
ln 0.0          0
ln 1.0        IMPLICIT-EXPLICIT DYNAMIC ANALYSIS OF A ROD
ln 2.0          1    0    0    0    0    0    2    1    3   22    3    0    0    0    2
ln 3.0   {      1    1    1    1    1    1         0.0       0.0       0.0       0.0
                2  100    0    0    2    1         0.0    0.3025       0.6    0.0045
ln 4.0         22    1    1
ln 5.1          1    2          0.0       0.0      0.0  }
ln 5.2                         10.5       0.0      0.0  }  Coordinates
ln 5.3         21    1                                  }
ln 6.0   {      1    0    0    1    1    1              }  Boundary conditions
                2   22    1    0    1    1              }
ln 9.1          1    2          0.0                     }
ln 9.2               0          0.1                     }  Initial displacements
ln 9.3         20    1                                  }
         {     22    0          0.1                     }
ln 9.1   {                                              }  Initial velocities/accelerations
ln 10.2.1       2    1    1    2    0    0    0    1    1    0                }
ln 10.2.2       1         1.0E+7      0.01        0.0       0.0       1.0    }  El. group 1
ln 10.2.3                                                                    }
ln 10.2.4.1     1    1   21   22                        }  Element nodes     }
ln 10.2.1       2   20    1    2    1    0    0    1    1    1                }
ln 10.2.2       1          100.0      0.01        0.0       0.0       1.0    }
ln 10.2.3                                                                    }  El. group 2
ln 10.2.4.1     1    1    1    2    1                   }  Element nodes     }
ln 10.2.4.2    20    1    1                             }                    }
ln 10.2.5      10    1    1                                El. time-history  }
              *END
```

```
IMPLICIT-EXPLICIT DYNAMIC ANALYSIS OF A ROD

E X E C U T I O N   C O N T R O L   I N F O R M A T I O N
     EXECUTION CODE . . . . . . . . . . . . . . . (IEXEC ) =      1

        EQ. 0, DATA CHECK
        EQ. 1, EXECUTION

     ANALYSIS CODE . . . . . . . . . . . . . . . . (IACODE) =      0

        EQ. 0, DYNAMIC ANALYSIS
        EQ. 1, STATIC ANALYSIS

     READ RESTART FILE CODE  . . . . . . . . . . . (IREADR) =      0

        EQ. 0, DO NOT READ RESTART FILE
        EQ. 1, READ RESTART FILE

     WRITE RESTART FILE CODE . . . . . . . . . . . (IWRITR) =      0

        EQ. 0, DO NOT WRITE RESTART FILE
        EQ. 1, WRITE RESTART FILE

     INPUT DATA PRINT CODE . . . . . . . . . . . . (IPRTIN) =      0

        EQ. 0, PRINT NODAL AND ELEMENT INPUT DATA
        EQ. 1, DO NOT PRINT NODAL AND ELEMENT INPUT DATA

     RANK CHECK CODE . . . . . . . . . . . . . . . (IRANK ) =      0

        EQ. 0, DO NOT PERFORM RANK CHECK
        EQ. 1, PRINT NUMBERS OF ZERO AND NEGATIVE PIVOTS
        EQ. 2, PRINT ALL PIVOTS

     NUMBER OF TIME SEQUENCES  . . . . . . . . . . (NUMSEQ) =      2

     NUMBER OF NODAL OUTPUT TIME-HISTORIES . . . . (NDOUT ) =      1

     NUMBER OF SPACE DIMENSIONS  . . . . . . . . . (NSD   ) =      3

     NUMBER OF NODAL POINTS  . . . . . . . . . . . (NUMNP ) =     22

     NUMBER OF NODAL DEGREES-OF-FREEDOM  . . . . . (NDOF  ) =      3

     NUMBER OF LOAD VECTORS  . . . . . . . . . . . (NLVECT) =      0

     NUMBER OF LOAD-TIME FUNCTIONS . . . . . . . . (NLTFTN) =      0

     NUMBER OF POINTS ON LOAD-TIME FUNCTIONS . . . (NPTSLF) =      0

     NUMBER OF ELEMENT GROUPS  . . . . . . . . . . (NUMEG ) =      2
```

```
T I M E   S E Q U E N C E   D A T A
    NUMBER OF TIME SEQUENCES . . . . . . (NUMSEQ  ) =      2

    TIME SEQUENCE NUMBER . . . . . . . . (N       ) =      1
    NUMBER OF TIME STEPS . . . . . . . . (NSTEP(N)) =      1
    KINEMATIC PRINT INCREMENT  . . . . . (NDPRT(N)) =      1
    STRESS/STRAIN PRINT INCREMENT  . . . (NSPRT(N)) =      1
    TIME HISTORY PLOT INCREMENT  . . . . (NHPLT(N)) =      1
    NUMBER OF ITERATIONS . . . . . . . . (NITER(N)) =      1
    FIRST INTEGRATION PARAMETER  . . . . (ALPHA(N)) =  0.00000E+00
    SECOND INTEGRATION PARAMETER . . . . (BETA(N) ) =  0.00000E+00
    THIRD INTEGRATION PARAMETER  . . . . (GAMMA(N)) =  0.00000E+00
    TIME STEP  . . . . . . . . . . . . . (DT(N)   ) =  0.00000E+00

    TIME SEQUENCE NUMBER . . . . . . . . (N       ) =      2
    NUMBER OF TIME STEPS . . . . . . . . (NSTEP(N)) =    100
    KINEMATIC PRINT INCREMENT  . . . . . (NDPRT(N)) =      0
    STRESS/STRAIN PRINT INCREMENT  . . . (NSPRT(N)) =      0
    TIME HISTORY PLOT INCREMENT  . . . . (NHPLT(N)) =      2
    NUMBER OF ITERATIONS . . . . . . . . (NITER(N)) =      1
    FIRST INTEGRATION PARAMETER  . . . . (ALPHA(N)) =  0.00000E+00
    SECOND INTEGRATION PARAMETER . . . . (BETA(N) ) =  3.02500E-01
    THIRD INTEGRATION PARAMETER  . . . . (GAMMA(N)) =  6.00000E-01
    TIME STEP  . . . . . . . . . . . . . (DT(N)   ) =  4.50000E-03

N O D A L   T I M E - H I S T O R Y   I N F O R M A T I O N
    NUMBER OF NODAL TIME HISTORIES  . . . . . . (NDOUT ) =     1

      NODE       DOF      KINEMATIC
     NUMBER     NUMBER       TYPE
       22          1         DISP
```

N O D A L   C O O R D I N A T E   D A T A

NODE NO.	X1	X2	X3
1	0.00000000E+00	0.00000000E+00	0.00000000E+00
2	5.00000000E-01	0.00000000E+00	0.00000000E+00
3	1.00000000E+00	0.00000000E+00	0.00000000E+00
4	1.50000000E+00	0.00000000E+00	0.00000000E+00
5	2.00000000E+00	0.00000000E+00	0.00000000E+00
6	2.50000000E+00	0.00000000E+00	0.00000000E+00
7	3.00000000E+00	0.00000000E+00	0.00000000E+00
8	3.50000000E+00	0.00000000E+00	0.00000000E+00
9	4.00000000E+00	0.00000000E+00	0.00000000E+00
10	4.50000000E+00	0.00000000E+00	0.00000000E+00
11	5.00000000E+00	0.00000000E+00	0.00000000E+00
12	5.50000000E+00	0.00000000E+00	0.00000000E+00
13	6.00000000E+00	0.00000000E+00	0.00000000E+00
14	6.50000000E+00	0.00000000E+00	0.00000000E+00
15	7.00000000E+00	0.00000000E+00	0.00000000E+00
16	7.50000000E+00	0.00000000E+00	0.00000000E+00
17	8.00000000E+00	0.00000000E+00	0.00000000E+00
18	8.50000000E+00	0.00000000E+00	0.00000000E+00
19	9.00000000E+00	0.00000000E+00	0.00000000E+00
20	9.50000000E+00	0.00000000E+00	0.00000000E+00
21	1.00000000E+01	0.00000000E+00	0.00000000E+00
22	1.05000000E+01	0.00000000E+00	0.00000000E+00

N O D A L   B O U N D A R Y   C O N D I T I O N   C O D E S

NODE NO.	DOF1	DOF2	DOF3
1	1	1	1
2	0	1	1
3	0	1	1
4	0	1	1
5	0	1	1
6	0	1	1
7	0	1	1
8	0	1	1
9	0	1	1
10	0	1	1
11	0	1	1
12	0	1	1
13	0	1	1
14	0	1	1
15	0	1	1
16	0	1	1
17	0	1	1
18	0	1	1
19	0	1	1
20	0	1	1
21	0	1	1
22	0	1	1

I N I T I A L   D I S P L A C E M E N T S

STEP NUMBER = 0

TIME = 0.000E+00

NODE NO.	DOF1	DOF2	DOF3
2	5.00000000E-03	0.00000000E+00	0.00000000E+00
3	1.00000000E-02	0.00000000E+00	0.00000000E+00
4	1.50000000E-02	0.00000000E+00	0.00000000E+00
5	2.00000000E-02	0.00000000E+00	0.00000000E+00
6	2.50000000E-02	0.00000000E+00	0.00000000E+00
7	3.00000000E-02	0.00000000E+00	0.00000000E+00
8	3.50000000E-02	0.00000000E+00	0.00000000E+00
9	4.00000000E-02	0.00000000E+00	0.00000000E+00
10	4.50000000E-02	0.00000000E+00	0.00000000E+00
11	5.00000000E-02	0.00000000E+00	0.00000000E+00
12	5.50000000E-02	0.00000000E+00	0.00000000E+00
13	6.00000000E-02	0.00000000E+00	0.00000000E+00
14	6.50000000E-02	0.00000000E+00	0.00000000E+00
15	7.00000000E-02	0.00000000E+00	0.00000000E+00
16	7.50000000E-02	0.00000000E+00	0.00000000E+00
17	8.00000000E-02	0.00000000E+00	0.00000000E+00
18	8.50000000E-02	0.00000000E+00	0.00000000E+00
19	9.00000000E-02	0.00000000E+00	0.00000000E+00
20	9.50000000E-02	0.00000000E+00	0.00000000E+00
21	1.00000000E-01	0.00000000E+00	0.00000000E+00
22	1.00000000E-01	0.00000000E+00	0.00000000E+00

THERE ARE NO NONZERO INITIAL VELOCITIES

THERE ARE NO NONZERO INITIAL ACCELERATIONS

E L E M E N T   G R O U P   D A T A

ELEMENT GROUP NUMBER . . . . (NEG) = 1

T W O / T H R E E - N O D E   T R U S S   E L E M E N T S

ELEMENT TYPE NUMBER . . . . (NTYPE) = 2

NUMBER OF ELEMENTS . . . . (NUMEL) = 1

NUMBER OF ELEMENT MATERIAL SETS . . . . (NUMAT) = 1

NUMBER OF ELEMENT NODES . . . . (NEN) = 2

NUMBER OF STRESS/STRAIN TIME HISTORIES . . (NSOUT) = 0

```
     STRESS OUTPUT PRINT CODE  . . . . . . . . . (ISTPRT) =      0

        EQ.0, STRESS OUTPUT PRINTED
        EQ.1, STRESS OUTPUT NOT PRINTED

     BODY FORCE LOAD-TIME FUNCTION NUMBER  . . . (LFBODY) =      0

     INTEGRATION CODE  . . . . . . . . . . . . . (NINT  ) =      1

        EQ.1, 1-POINT GAUSSIAN QUADRATURE
        EQ.2, 2-POINT GAUSSIAN QUADRATURE
        EQ.3, 3-POINT GAUSSIAN QUADRATURE

     MASS TYPE CODE  . . . . . . . . . . . . . . (IMASS ) =      1

        EQ.0, CONSISTENT MASS MATRIX
        EQ.1, LUMPED MASS MATRIX
        EQ.2, NO MASS MATRIX

     IMPLICIT/EXPLICIT CODE  . . . . . . . . . . (IMPEXP) =      0

        EQ.0, IMPLICIT ELEMENT GROUP
        EQ.1, EXPLICIT ELEMENT GROUP


 M A T E R I A L   S E T   D A T A

     NUMBER OF MATERIAL SETS . . . . . . . . . . (NUMAT ) =      1

      SET      YOUNG'S        MASS         MASS       STIFFNESS      AREA
     NUMBER    MODULUS       DENSITY      DAMPING      DAMPING

       1     1.0000E+07  1.0000E-02  0.0000E+00  0.0000E+00  1.0000E+00


 G R A V I T Y   V E C T O R   C O M P O N E N T S

     X-1 DIRECTION . . . . . . . . . . . . . . =  0.00000000E+00

     X-2 DIRECTION . . . . . . . . . . . . . . =  0.00000000E+00

     X-3 DIRECTION . . . . . . . . . . . . . . =  0.00000000E+00


 E L E M E N T   D A T A

     ELEMENT   MATERIAL   NODE 1     NODE 2
         1         1        21         22
```

```
E L E M E N T   G R O U P   D A T A
    ELEMENT GROUP NUMBER . . . . . . . . . . . . (NEG   ) =     2

T W O / T H R E E - N O D E   T R U S S   E L E M E N T S
    ELEMENT TYPE NUMBER . . . . . . . . . . . . . (NTYPE ) =     2
    NUMBER OF ELEMENTS  . . . . . . . . . . . . . (NUMEL ) =    20
    NUMBER OF ELEMENT MATERIAL SETS . . . . . . . (NUMAT ) =     1
    NUMBER OF ELEMENT NODES . . . . . . . . . . . (NEN   ) =     2
    NUMBER OF STRESS/STRAIN TIME HISTORIES  . . . (NSOUT ) =     1
    STRESS OUTPUT PRINT CODE  . . . . . . . . . . (ISTPRT) =     0

       EQ.0, STRESS OUTPUT PRINTED
       EQ.1, STRESS OUTPUT NOT PRINTED

    BODY FORCE LOAD-TIME FUNCTION NUMBER  . . . . (LFBODY) =     0
    INTEGRATION CODE  . . . . . . . . . . . . . . (NINT  ) =     1

       EQ.1, 1-POINT GAUSSIAN QUADRATURE
       EQ.2, 2-POINT GAUSSIAN QUADRATURE
       EQ.3, 3-POINT GAUSSIAN QUADRATURE

    MASS TYPE CODE  . . . . . . . . . . . . . . . (IMASS ) =     1

       EQ.0, CONSISTENT MASS MATRIX
       EQ.1, LUMPED MASS MATRIX
       EQ.2, NO MASS MATRIX

    IMPLICIT/EXPLICIT CODE  . . . . . . . . . . . (IMPEXP) =     1

       EQ.0, IMPLICIT ELEMENT GROUP
       EQ.1, EXPLICIT ELEMENT GROUP


M A T E R I A L   S E T   D A T A
    NUMBER OF MATERIAL SETS . . . . . . . . . . . (NUMAT ) =     1

     SET      YOUNG'S       MASS        MASS      STIFFNESS       AREA
    NUMBER    MODULUS      DENSITY     DAMPING     DAMPING

      1     1.0000E+02  1.0000E-02  0.0000E+00  0.0000E+00  1.0000E+00
```

```
G R A V I T Y   V E C T O R   C O M P O N E N T S
    X-1 DIRECTION . . . . . . . . . . . . . . . =  0.00000000E+00
    X-2 DIRECTION . . . . . . . . . . . . . . . =  0.00000000E+00
    X-3 DIRECTION . . . . . . . . . . . . . . . =  0.00000000E+00

E L E M E N T   D A T A
    ELEMENT   MATERIAL  NODE 1    NODE 2
       1          1        1         2
       2          1        2         3
       3          1        3         4
       4          1        4         5
       5          1        5         6
       6          1        6         7
       7          1        7         8
       8          1        8         9
       9          1        9        10
      10          1       10        11
      11          1       11        12
      12          1       12        13
      13          1       13        14
      14          1       14        15
      15          1       15        16
      16          1       16        17
      17          1       17        18
      18          1       18        19
      19          1       19        20
      20          1       20        21

E L E M E N T   T I M E   H I S T O R Y   I N F O R M A T I O N
    NUMBER OF STRESS/STRAIN TIME HISTORIES  . . (NSOUT ) =      1

      ELEMENT    INT PT    COMPONENT
      NUMBER     NUMBER

        10         1         STRS

IMPLICIT-EXPLICIT DYNAMIC ANALYSIS OF A ROD

E Q U A T I O N   S Y S T E M   D A T A
    NUMBER OF EQUATIONS . . . . . . . . . . . . (NEQ    ) =        21
    NUMBER OF TERMS IN LEFT-HAND-SIDE MATRIX  . (NALHS  ) =        22
    MEAN HALF BANDWIDTH . . . . . . . . . . . . (MEANBW) =         1
    TOTAL LENGTH OF BLANK COMMON REQUIRED . . . (NWORDS) =      2542
```

D I S P L A C E M E N T S

STEP NUMBER = 1

TIME = 0.000E+00

NODE NO.	DOF1	DOF2	DOF3
2	5.00000000E-03	0.00000000E+00	0.00000000E+00
3	1.00000000E-02	0.00000000E+00	0.00000000E+00
4	1.50000000E-02	0.00000000E+00	0.00000000E+00
5	2.00000000E-02	0.00000000E+00	0.00000000E+00
6	2.50000000E-02	0.00000000E+00	0.00000000E+00
7	3.00000000E-02	0.00000000E+00	0.00000000E+00
8	3.50000000E-02	0.00000000E+00	0.00000000E+00
9	4.00000000E-02	0.00000000E+00	0.00000000E+00
10	4.50000000E-02	0.00000000E+00	0.00000000E+00
11	5.00000000E-02	0.00000000E+00	0.00000000E+00
12	5.50000000E-02	0.00000000E+00	0.00000000E+00
13	6.00000000E-02	0.00000000E+00	0.00000000E+00
14	6.50000000E-02	0.00000000E+00	0.00000000E+00
15	7.00000000E-02	0.00000000E+00	0.00000000E+00
16	7.50000000E-02	0.00000000E+00	0.00000000E+00
17	8.00000000E-02	0.00000000E+00	0.00000000E+00
18	8.50000000E-02	0.00000000E+00	0.00000000E+00
19	9.00000000E-02	0.00000000E+00	0.00000000E+00
20	9.50000000E-02	0.00000000E+00	0.00000000E+00
21	1.00000000E-01	0.00000000E+00	0.00000000E+00
22	1.00000000E-01	0.00000000E+00	0.00000000E+00

V E L O C I T I E S

STEP NUMBER = 1

TIME = 0.000E+00

NODE NO.	DOF1	DOF2	DOF3

THERE ARE NO NONZERO COMPONENTS

A C C E L E R A T I O N S

STEP NUMBER = 1

TIME = 0.000E+00

NODE NO.	DOF1	DOF2	DOF3
3	-5.55111512E-15	0.00000000E+00	0.00000000E+00
4	5.55111512E-15	0.00000000E+00	0.00000000E+00
5	-2.22044605E-14	0.00000000E+00	0.00000000E+00
6	-1.94289029E-14	0.00000000E+00	0.00000000E+00
7	3.60822483E-14	0.00000000E+00	0.00000000E+00
8	-1.66533454E-14	0.00000000E+00	0.00000000E+00
9	-3.60822483E-14	0.00000000E+00	0.00000000E+00
10	6.93889390E-14	0.00000000E+00	0.00000000E+00
11	-1.02695630E-13	0.00000000E+00	0.00000000E+00
12	1.02695630E-13	0.00000000E+00	0.00000000E+00
13	-7.21644966E-14	0.00000000E+00	0.00000000E+00
14	7.21644966E-14	0.00000000E+00	0.00000000E+00
16	-6.93889390E-14	0.00000000E+00	0.00000000E+00
17	1.41553436E-13	0.00000000E+00	0.00000000E+00
18	-1.41553436E-13	0.00000000E+00	0.00000000E+00
19	6.93889390E-14	0.00000000E+00	0.00000000E+00
21	-2.00000000E+02	0.00000000E+00	0.00000000E+00
22	2.77555756E-08	0.00000000E+00	0.00000000E+00

E L E M E N T   S T R E S S E S   A N D   S T R A I N S

  ELEMENT GRØUP NUMBER . . . . (NEG) =   1

ELEMENT NUMBER	INT. PT. NUMBER	X1 COORD.	X2 COORD.	X3 COORD.	STRESS	FØRCE	STRAIN
1	1	1.02E+01	0.00E+00	0.00E+00	-6.94E-11	-6.94E-11	-6.94E-18

E L E M E N T   S T R E S S E S   A N D   S T R A I N S

  ELEMENT GRØUP NUMBER . . . . (NEG) =   2

ELEMENT NUMBER	INT. PT. NUMBER	X1 COORD.	X2 COORD.	X3 COORD.	STRESS	FØRCE	STRAIN
1	1	2.50E-01	0.00E+00	0.00E+00	1.00E+00	1.00E+00	1.00E-02
2	1	7.50E-01	0.00E+00	0.00E+00	1.00E+00	1.00E+00	1.00E-02
3	1	1.25E+00	0.00E+00	0.00E+00	1.00E+00	1.00E+00	1.00E-02
4	1	1.75E+00	0.00E+00	0.00E+00	1.00E+00	1.00E+00	1.00E-02
5	1	2.25E+00	0.00E+00	0.00E+00	1.00E+00	1.00E+00	1.00E-02
6	1	2.75E+00	0.00E+00	0.00E+00	1.00E+00	1.00E+00	1.00E-02
7	1	3.25E+00	0.00E+00	0.00E+00	1.00E+00	1.00E+00	1.00E-02
8	1	3.75E+00	0.00E+00	0.00E+00	1.00E+00	1.00E+00	1.00E-02
9	1	4.25E+00	0.00E+00	0.00E+00	1.00E+00	1.00E+00	1.00E-02
10	1	4.75E+00	0.00E+00	0.00E+00	1.00E+00	1.00E+00	1.00E-02
11	1	5.25E+00	0.00E+00	0.00E+00	1.00E+00	1.00E+00	1.00E-02
12	1	5.75E+00	0.00E+00	0.00E+00	1.00E+00	1.00E+00	1.00E-02
13	1	6.25E+00	0.00E+00	0.00E+00	1.00E+00	1.00E+00	1.00E-02
14	1	6.75E+00	0.00E+00	0.00E+00	1.00E+00	1.00E+00	1.00E-02
15	1	7.25E+00	0.00E+00	0.00E+00	1.00E+00	1.00E+00	1.00E-02
16	1	7.75E+00	0.00E+00	0.00E+00	1.00E+00	1.00E+00	1.00E-02
17	1	8.25E+00	0.00E+00	0.00E+00	1.00E+00	1.00E+00	1.00E-02
18	1	8.75E+00	0.00E+00	0.00E+00	1.00E+00	1.00E+00	1.00E-02
19	1	9.25E+00	0.00E+00	0.00E+00	1.00E+00	1.00E+00	1.00E-02
20	1	9.75E+00	0.00E+00	0.00E+00	1.00E+00	1.00E+00	1.00E-02

IMPLICIT-EXPLICIT DYNAMIC ANALYSIS OF A ROD

NODE NUMBER = 22

DOF NUMBER = 1 OUTPUT: DISP

TIME	VALUE
0.0000E+00	1.0000E-01
0.0000E+00	1.0000E-01
9.0000E-03	9.5603E-02
1.8000E-02	8.7134E-02
2.7000E-02	7.8211E-02
3.6000E-02	6.9223E-02
4.5000E-02	6.0225E-02
5.4000E-02	5.1225E-02
6.3000E-02	4.2225E-02
7.2000E-02	3.3225E-02
8.1000E-02	2.4225E-02
9.0000E-02	1.5225E-02
9.9000E-02	6.2250E-03
1.0800E-01	-2.7750E-03
1.1700E-01	-1.1775E-02
1.2600E-01	-2.0775E-02
1.3500E-01	-2.9775E-02
1.4400E-01	-3.8775E-02
1.5300E-01	-4.7775E-02
1.6200E-01	-5.6775E-02
1.7100E-01	-6.5775E-02
1.8000E-01	-7.4774E-02
1.8900E-01	-8.3659E-02
1.9800E-01	-9.1395E-02
2.0700E-01	-9.5442E-02
2.1600E-01	-9.4162E-02
2.2500E-01	-8.8593E-02
2.3400E-01	-8.0782E-02
2.4300E-01	-7.2120E-02
2.5200E-01	-6.3202E-02
2.6100E-01	-5.4220E-02
2.7000E-01	-4.5224E-02
2.7900E-01	-3.6225E-02
2.8800E-01	-2.7225E-02
2.9700E-01	-1.8225E-02
3.0600E-01	-9.2250E-03
3.1500E-01	-2.2500E-04
3.2400E-01	8.7750E-03
3.3300E-01	1.7775E-02
3.4200E-01	2.6775E-02
3.5100E-01	3.5775E-02
3.6000E-01	4.4775E-02
3.6900E-01	5.3776E-02
3.7800E-01	6.2782E-02
3.8700E-01	7.1777E-02
3.9600E-01	8.0459E-02
4.0500E-01	8.7779E-02
4.1400E-01	9.2123E-02
4.2300E-01	9.2408E-02
4.3200E-01	8.8829E-02
4.4100E-01	8.2499E-02
4.5000E-01	7.4625E-02

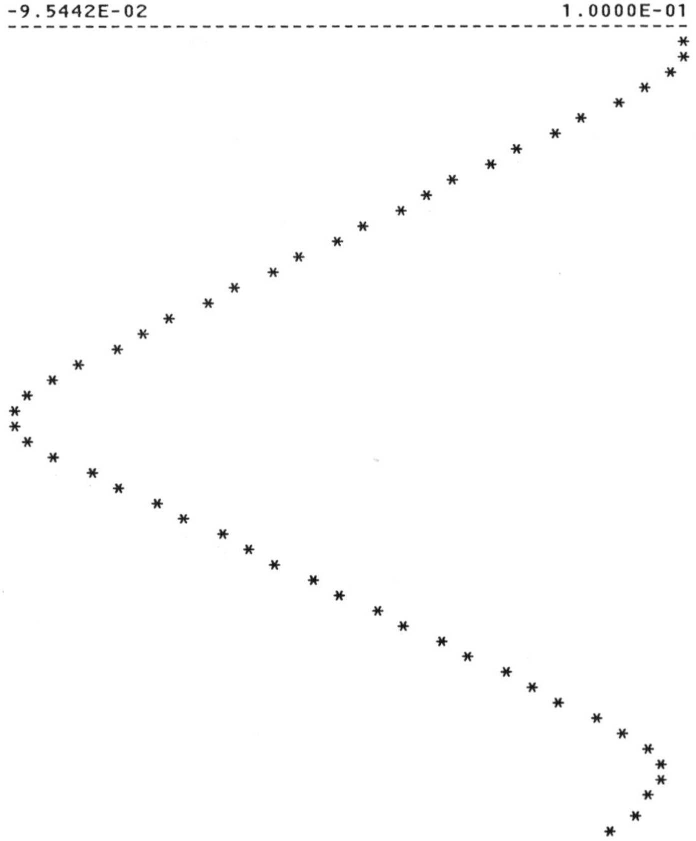

```
IMPLICIT-EXPLICIT DYNAMIC ANALYSIS OF A ROD

ELEMENT NUMBER                 =    10
INTEGRATION POINT NUMBER       =     1     OUTPUT: STRS

    TIME        VALUE       -1.0018E+00                                 1.0021E+00
----------    ----------    -------------------------------------------------------
0.0000E+00    1.0000E+00                                                          *
0.0000E+00    1.0000E+00                                                          *
9.0000E-03    1.0000E+00                                                          *
1.8000E-02    1.0000E+00                                                          *
2.7000E-02    1.0000E+00                                                          *
3.6000E-02    1.0000E+00                                                          *
4.5000E-02    9.8613E-01                                                          *
5.4000E-02    6.9567E-01                                                  *
6.3000E-02    2.1295E-01                                      *
7.2000E-02    3.7574E-02                                *
8.1000E-02    5.5897E-03                               *
9.0000E-02    8.4809E-04                               *
9.9000E-02    1.4168E-04                               *
1.0800E-01    2.8074E-05                               *
1.1700E-01    6.3156E-06                               *
1.2600E-01    8.1778E-07                               *
1.3500E-01   -1.4638E-02                               *
1.4400E-01   -1.7536E-01                          *
1.5300E-01   -5.3637E-01                 *
1.6200E-01   -8.3211E-01        *
1.7100E-01   -9.5730E-01     *
1.8000E-01   -9.9151E-01    *
1.8900E-01   -9.9855E-01    *
1.9800E-01   -9.9977E-01    *
2.0700E-01   -9.9996E-01    *
2.1600E-01   -9.9999E-01    *
2.2500E-01   -1.0000E+00    *
2.3400E-01   -1.0018E+00    *
2.4300E-01   -9.9973E-01    *
2.5200E-01   -9.2160E-01      *
2.6100E-01   -6.9174E-01            *
2.7000E-01   -3.9669E-01                     *
2.7900E-01   -1.7502E-01                          *
2.8800E-01   -6.1613E-02                             *
2.9700E-01   -1.8053E-02                               *
3.0600E-01   -4.5810E-03                               *
3.1500E-01   -1.1041E-03                               *
3.2400E-01   -1.2314E-03                               *
3.3300E-01    1.1990E-03                               *
3.4200E-01    4.6505E-02                                *
3.5100E-01    1.9879E-01                                    *
3.6000E-01    4.4648E-01                                            *
3.6900E-01    6.9268E-01                                                  *
3.7800E-01    8.6165E-01                                                       *
3.8700E-01    9.4836E-01                                                         *
3.9600E-01    9.8359E-01                                                          *
4.0500E-01    9.9545E-01                                                          *
4.1400E-01    9.9886E-01                                                          *
4.2300E-01    9.9975E-01                                                          *
4.3200E-01    1.0009E+00                                                          *
4.4100E-01    1.0021E+00                                                          *
4.5000E-01    9.8302E-01                                                          *
```

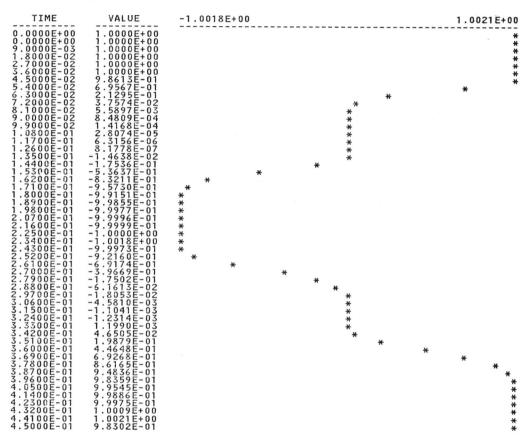

D Y N A M I C   S T Ø R A G E   A L L Ø C A T I Ø N   I N F Ø R M A T I Ø N

ARRAY NØ.	ARRAY	ADDRESS	DIM1	DIM2	DIM3	PREC.
1	NSTEP	1	2	0	0	1
2	NDPRT	3	2	0	0	1
3	NSPRT	5	2	0	0	1
4	NHPLT	7	2	0	0	1
5	NITER	9	2	0	0	1
6	ALPHA	11	2	0	0	2
7	BETA	15	2	0	0	2
8	GAMMA	19	2	0	0	2
9	DT	23	2	0	0	2
10	IDHIST	27	3	1	0	1
11	DØUT	31	2	52	0	1
12	VPRED	135	3	22	0	2
13	DPRED	267	3	22	0	2
14	A	399	3	22	0	2
15	V	531	3	22	0	2
16	D	663	3	22	0	2
17	X	795	3	22	0	2
18	ID	927	3	22	0	1
19	IDIAG	993	21	0	0	1
20	NGRP	1015	2	0	0	1
*****	BEGIN ELEMENT GRØUP NUMBER 1					
21	NPAR	1017	16	0	0	1
22	MP	1033	29	0	0	1
23	W	1063	1	0	0	2
24	DET	1065	1	0	0	2
25	SHL	1067	2	2	1	2
26	SHG	1075	2	2	1	2
27	XS	1083	3	1	0	2
28	RHØ	1089	1	0	0	2
29	RDAMPM	1091	1	0	0	2
30	RDAMPK	1093	1	0	0	2
31	AREA	1095	1	0	0	2
32	C	1097	1	1	1	2
33	GRAV	1099	3	0	0	2
34	IEN	1105	2	1	0	1
35	MAT	1107	1	0	0	1
36	LM	1109	3	2	1	1
37	ELEFFM	1115	6	6	0	2
38	XL	1187	3	2	0	2
39	WØRK	1199	16	0	0	2
40	B	1231	1	6	0	2
41	DMAT	1243	1	1	0	2
42	DB	1245	1	6	0	2
43	VL	1257	3	2	0	2
44	AL	1269	3	2	0	2
45	ELRESF	1281	6	0	0	2
46	DL	1293	3	2	0	2
47	STRAIN	1305	1	0	0	2
48	STRESS	1307	1	0	0	2
49	FØRCE	1309	1	0	0	2
*****	BEGIN ELEMENT GRØUP NUMBER 2					
50	NPAR	1311	16	0	0	1

D Y N A M I C   S T O R A G E   A L L O C A T I O N   I N F O R M A T I O N

ARRAY NO.	ARRAY	ADDRESS	DIM1	DIM2	DIM3	PREC.
51	MP	1327	29	0	0	1
52	W	1357	1	0	0	2
53	DET	1359	1	0	0	2
54	SHL	1361	2	2	1	2
55	SHG	1369	2	2	1	2
56	XS	1377	3	1	0	2
57	RHO	1383	1	0	0	2
58	RDAMPM	1385	1	0	0	2
59	RDAMPK	1387	1	0	0	2
60	AREA	1389	1	0	0	2
61	C	1391	1	1	1	2
62	GRAV	1393	3	0	0	2
63	IEN	1399	2	20	0	1
64	MAT	1439	20	0	0	1
65	LM	1459	3	2	20	1
66	ISHIST	1579	3	1	0	1
67	SOUT	1583	2	52	0	1
68	ELEFFM	1687	6	6	0	2
69	XL	1759	3	2	0	2
70	WORK	1771	16	0	0	2
71	B	1803	1	6	0	2
72	DMAT	1815	1	1	0	2
73	DB	1817	1	6	0	2
74	VL	1829	3	2	0	2
75	AL	1841	3	2	0	2
76	ELRESF	1853	6	0	0	2
77	DL	1865	3	2	0	2
78	STRAIN	1877	1	0	0	2
79	STRESS	1879	1	0	0	2
80	FORCE	1881	1	0	0	2
*****	END ELEMENT GROUP DATA					
81	ALHS	1883	22	0	0	2
82	BRHS	1927	21	0	0	2

```
IMPLICIT-EXPLICIT DYNAMIC ANALYSIS OF A ROD

     E X E C U T I O N   T I M I N G   I N F O R M A T I O N

     I N I T I A L I Z A T I O N   P H A S E      =  1.711E+00

     S O L U T I O N   P H A S E                  =  1.584E+01

        FORMATION OF LEFT-HAND-SIDE MATRICES      =  1.211E-01
        FACTORIZATIONS                            =  0.000E+00
        FORMATION OF RIGHT-HAND-SIDE VECTORS      =  1.122E+01
        FORWARD REDUCTIONS/BACK SUBSTITUTIONS     =  1.973E-01
        CALCULATION OF ELEMENT OUTPUT             =  4.473E-01
                                                     ---------
        SUBTOTAL                                  =  1.199E+01
```

11.5.4 Subroutine Index for Program Listing

11.5.4 Subroutine Index for Program Listing (cont.)

Routine	Description	Page
SUBROUTINE BC	Program to read, generate and write boundary condition data and establish equation numbers	741
BLOCK DATA	Program to define output labels and numerical constants	741
SUBROUTINE BTDB	Program to multiply $\boldsymbol{B}^T\boldsymbol{DB}$, taking account of symmetry, and accumulate into element stiffness matrix	742
SUBROUTINE CLEAR	Program to clear a floating-point array	742
FUNCTION COLDOT	Program to compute the dot product of vectors stored column-wise	742
SUBROUTINE COLHT	Program to compute column heights in global left-hand-side matrix	742
SUBROUTINE COMPBC	Program to compute displacement, velocity and acceleration boundary conditions	743
SUBROUTINE CONTM	Program to form mass matrix for a continuum element with NEN nodes	743
SUBROUTINE CONTMA	Program to calculate inertial and gravity/body force $(-\boldsymbol{m}^e * (\boldsymbol{a}^e - \boldsymbol{g}^e))$ for a continuum element with NEN nodes	745
SUBROUTINE COORD	Program to read, generate, and write coordinate data	745
SUBROUTINE CORRCT	Program to perform corrector update of displacements and velocities	746
SUBROUTINE DCTNRY	Program to store pointer information in dictionary	746
SUBROUTINE DHIST	Program to read, write, and store nodal time history input	746
SUBROUTINE DIAG	Program to compute diagonal addresses of left-hand-side matrix	747
SUBROUTINE DYNPTS	Program to set memory pointers for dynamic analysis data arrays	747
SUBROUTINE ECHO	Program to echo input data	747
SUBROUTINE ELEMNT	Program to calculate element task number	747
SUBROUTINE ELCARD	Program to read element group control card	748
SUBROUTINE ELMLIB	Program to call element routines	748
SUBROUTINE EQSET	Program to allocate storage for global equation system	748
SUBROUTINE FACTOR	Program to perform Crout factorization $\boldsymbol{A} = \boldsymbol{U}^T\boldsymbol{DU}$	749
SUBROUTINE FORMLM	Program to form LM array	750
SUBROUTINE GENEL	Program to read and generate element node and material numbers	750
SUBROUTINE GENEL1	Program to generate element node and material numbers	750
SUBROUTINE GENELD	Program to set defaults for element node and material number generation	751
SUBROUTINE GENELI	Program to increment element node numbers	751
SUBROUTINE GENFL	Program to read and generate floating-point nodal data	751

11.5.4 Subroutine Index for Program Listing (cont.)

11.5.4 Subroutine Index for Program Listing (cont.)

11.5.4 Subroutine Index for Program Listing (cont.)

11.5.4 Subroutine Index for Program Listing (cont.)

11.5.5 Program Listing

```
      PROGRAM MAIN                                                      MAIN   1
C                                                                       MAIN   2
C     ************************************************************      MAIN   3
C     *                                                          *      MAIN   4
C     *              * * *   D L E A R N   * * *                 *      MAIN   5
C     *                                                          *      MAIN   6
C     *                                                          *      MAIN   7
C     *            A LINEAR STATIC AND DYNAMIC                   *      MAIN   8
C     *                                                          *      MAIN   9
C     *         FINITE ELEMENT ANALYSIS PROGRAM                  *      MAIN  10
C     *                                                          *      MAIN  11
C     *                                                          *      MAIN  12
C     *            T. J. R.   H U G H E S                        *      MAIN  13
C     *                                                          *      MAIN  14
C     *               R. M.   F E R E N C Z                      *      MAIN  15
C     *                                                          *      MAIN  16
C     *               A. M.   R A E F S K Y                      *      MAIN  17
C     *                                                          *      MAIN  18
C     ************************************************************      MAIN  19
C                                                                       MAIN  20
C.... PROGRAM TO SET STORAGE CAPACITY, PRECISION AND INPUT/OUTPUT UNITS MAIN  21
C                                                                       MAIN  22
      COMMON /BPOINT/ MFIRST,MLAST,MTOT,IPREC                           MAIN  23
      COMMON /IOUNIT/ IIN,IOUT,IRSIN,IRSOUT                             MAIN  24
      COMMON A(50000)                                                   MAIN  25
C                                                                       MAIN  26
C        MFIRST = ADDRESS OF FIRST AVAILABLE WORD IN BLANK COMMON       MAIN  27
C        MLAST  = ADDRESS OF LAST AVAILABLE WORD IN BLANK COMMON        MAIN  28
C        MTOT   = TOTAL STORAGE ALLOCATED TO BLANK COMMON               MAIN  29
C        IPREC  = PRECISION FLAG; EQ.1, SINGLE PRECISION                MAIN  30
C                                 EQ.2, DOUBLE PRECISION                MAIN  31
C                                                                       MAIN  32
      MFIRST = 1                                                        MAIN  33
      MTOT   = 50000                                                    MAIN  34
      MLAST  = MTOT                                                     MAIN  35
      IPREC  = 2                                                        MAIN  36
C                                                                       MAIN  37
C        IIN    = INPUT UNIT NUMBER                                     MAIN  38
C        IOUT   = OUTPUT UNIT NUMBER                                    MAIN  39
C        IRSIN  = RESTART INPUT UNIT NUMBER                             MAIN  40
C        IRSOUT = RESTART OUTPUT UNIT NUMBER                            MAIN  41
```

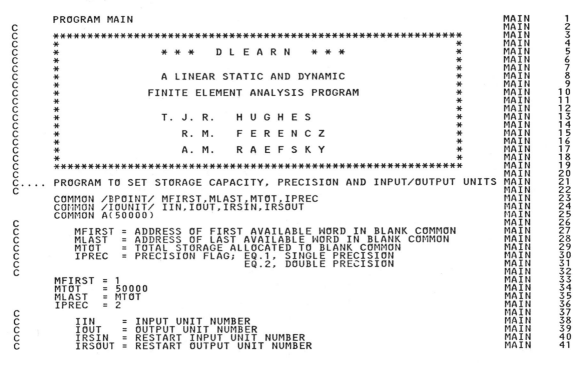

```
C                                                                    MAIN   42
         IIN   = 5                                                   MAIN   43
         IOUT  = 6                                                   MAIN   44
         IRSIN = 7                                                   MAIN   45
         IRSOUT = 8                                                  MAIN   46
C                                                                    MAIN   47
C.... SYSTEM-DEPENDENT UNIT/FILE SPECIFICATIONS                      MAIN   48
C                                                                    MAIN   49
C        THE FOLLOWING LINES ARE APPROPRIATE FOR THE VMS OPERATING   MAIN   50
C        SYSTEM -- CHANGE AS NECESSARY FOR OTHER OPERATING SYSTEMS   MAIN   51
C                                                                    MAIN   52
         CALL ASSIGN(   IIN,  'INPUT.DAT', 9)                        MAIN   53
         CALL ASSIGN(  IOUT,  'OUTPUT.DAT',10)                       MAIN   54
         CALL ASSIGN( IRSIN, 'RSINPUT.DAT',11)                       MAIN   55
         CALL ASSIGN(IRSOUT,'RSOUTPUT.DAT',12)                       MAIN   56
C                                                                    MAIN   57
         CALL DLEARN                                                 MAIN   58
C                                                                    MAIN   59
C.... SYSTEM-DEPENDENT UNIT/FILE SPECIFICATIONS                      MAIN   60
C                                                                    MAIN   61
C        THE FOLLOWING LINES ARE APPROPRIATE FOR THE VMS OPERATING   MAIN   62
C        SYSTEM -- CHANGE AS NECESSARY FOR OTHER OPERATING SYSTEMS   MAIN   63
C                                                                    MAIN   64
         CALL CLOSE(IIN)                                             MAIN   65
         CALL CLOSE(IOUT)                                            MAIN   66
         CALL CLOSE(IRSIN)                                           MAIN   67
         CALL CLOSE(IRSOUT)                                          MAIN   68
C                                                                    MAIN   69
         STOP                                                        MAIN   70
         END                                                         MAIN   71
C                                                                    MAIN   72
C********************************************************************************
         SUBROUTINE DLEARN                                           DLEARN  1
C                                                                    DLEARN  2
C.... DLEARN - A LINEAR STATIC AND DYNAMIC FINITE ELEMENT            DLEARN  3
C              ANALYSIS PROGRAM: GLOBAL DRIVER                       DLEARN  4
C                                                                    DLEARN  5
         DOUBLE PRECISION                                            DLEARN  6
     &          TIME,ZERO,PT1667,PT25,PT5,ONE,TWO,THREE,FOUR,FIVE,TEMPF,  DLEARN  7
     &          COEFF1,COEFF2,COEFF3,COEFF4,COEFF5,COEFF6,           DLEARN  8
     &          COEFF7,COEFF8,ALPHA1,BETA1,GAMMA1,DT1                DLEARN  9
C                                                                    DLEARN 10
C.... DEACTIVATE ABOVE CARD(S) FOR SINGLE-PRECISION OPERATION        DLEARN 11
C                                                                    DLEARN 12
         LOGICAL LDYN                                                DLEARN 13
         CHARACTER*4 TITLE,CIAO                                      DLEARN 14
C                                                                    DLEARN 15
C.... CATALOG OF COMMON STATEMENTS                                   DLEARN 16
C                                                                    DLEARN 17
         COMMON /BPOINT/ MFIRST,MLAST,MTOT,IPREC                     DLEARN 18
         COMMON /COEFFS/ COEFF1,COEFF2,COEFF3,COEFF4,COEFF5,COEFF6,  DLEARN 19
     &                   COEFF7,COEFF8,ALPHA1,BETA1 ,GAMMA1,DT1      DLEARN 20
         COMMON /COLHTC/ NEQ                                         DLEARN 21
         COMMON /CONSTS/ ZERO,PT1667,PT25,PT5,ONE,TWO,THREE,FOUR,FIVE  DLEARN 22
         COMMON /DPOINT/ MPSTEP,MPDPRT,MPSPRT,MPHPLT,MPITER,MPALPH,MPBETA,  DLEARN 23
     &                   MPGAMM,MPDT ,MPIDHS,MPDOUT,MPVPRD,MPDPRD,MPA,MPV  DLEARN 24
         COMMON /ETIMEC/ ETIME(7)                                    DLEARN 25
         COMMON /GENELC/ N,NEL(3),INCEL(3),INC(3)                    DLEARN 26
         COMMON /GENFLC/ TEMPF(6,20),NF,NUMGPF,NINCF(3),INCF(3)      DLEARN 27
         COMMON /HPLOTC/ NPLPTS,LOCPLT,TIME                          DLEARN 28
         COMMON /INFO  / IEXEC,IACODE,LDYN,IREADR,IWRITR,IPRTIN,IRANK,  DLEARN 29
     &                   NUMSEQ,NDOUT,NSD,NUMNP,NDOF,NLVECT,NLTFTN,NPTSLF,  DLEARN 30
     &                   NUMEG                                       DLEARN 31
         COMMON /IOUNIT/ IIN,IOUT,IRSIN,IRSOUT                       DLEARN 32
         COMMON /LABELS/ LABELD(3),LABEL1(16),LABEL2(3)              DLEARN 33
         COMMON /SPOINT/ MPD,MPX,MPID,MPF,MPG,MPG1,MPDIAG,MPNGRP,    DLEARN 34
     &                   MPALHS,MPBRHS                               DLEARN 35
         COMMON /TITLEC/ TITLE(20)                                   DLEARN 36
         COMMON A(1)                                                 DLEARN 37
         DATA CIAO/'*END'/                                           DLEARN 38
C                                                                    DLEARN 39
C.... INPUT PHASE                                                    DLEARN 40
C                                                                    DLEARN 41
         CALL ECHO                                                   DLEARN 42
  100 CONTINUE                                                       DLEARN 43
         DO 200 I=1,7                                                DLEARN 44
  200 ETIME(I) = 0.0                                                 DLEARN 45
         CALL TIMING(T1)                                             DLEARN 46
         READ(IIN,1000) TITLE                                        DLEARN 47
         IF (TITLE(1).EQ.CIAO) RETURN                                DLEARN 48
         READ(IIN,2000) IEXEC,IACODE,IREADR,IWRITR,IPRTIN,IRANK,NUMSEQ,  DLEARN 49
     &                  NDOUT,NSD,NUMNP,NDOF,NLVECT,NLTFTN,NPTSLF,NUMEG  DLEARN 50
         LDYN = .TRUE.                                               DLEARN 51
         IF (IACODE.EQ.1) LDYN = .FALSE.                             DLEARN 52
         WRITE(IOUT,3000) TITLE , IEXEC,IACODE,IREADR,IWRITR,IPRTIN  DLEARN 53
         WRITE(IOUT,4000) IRANK ,NUMSEQ, NDOUT,    NSD, NUMNP,  NDOF,  DLEARN 54
     &                    NLVECT,NLTFTN,NPTSLF, NUMEG               DLEARN 55
C                                                                    DLEARN 56
C.... INITIALIZATION PHASE                                           DLEARN 57
```

```
C                                                                      DLEARN   58
        CALL TSEQ                                                      DLEARN   59
        IF (LDYN) CALL DYNPTS                                          DLEARN   60
        IF (NDOUT.GT.0) CALL DHIST(A(MPIDHS),NDOUT)                    DLEARN   61
        CALL STATIN(NEQ)                                               DLEARN   62
C                                                                      DLEARN   63
        NTSTEP = 0                                                     DLEARN   64
        TIME    = ZERO                                                 DLEARN   65
        IF (LDYN) THEN                                                 DLEARN   66
C                                                                      DLEARN   67
           IF (IREADR.EQ.1) THEN                                       DLEARN   68
C                                                                      DLEARN   69
C......... READ INITIAL CONDITIONS FROM RESTART FILE                   DLEARN   70
C                                                                      DLEARN   71
              CALL RSIN(A(MPD),A(MPV),A(MPA),NDOF,NUMNP,NTSTEP,TIME)    DLEARN   72
              IF (IPRTIN.EQ.0) THEN                                    DLEARN   73
              CALL PRINTD(' R E S T A R T   D I S P L A C E M E N T S ',DLEARN   74
     &             A(MPD),NDOF,NUMNP,NTSTEP,TIME)                      DLEARN   75
              CALL PRINTD(' R E S T A R T   V E L O C I T I E S       ',DLEARN   76
     &             A(MPV),NDOF,NUMNP,NTSTEP,TIME)                      DLEARN   77
              CALL PRINTD(' R E S T A R T   A C C E L E R A T I O N S ',DLEARN   78
     &             A(MPA),NDOF,NUMNP,NTSTEP,TIME)                      DLEARN   79
              ENDIF                                                    DLEARN   80
C                                                                      DLEARN   81
           ELSE                                                        DLEARN   82
C                                                                      DLEARN   83
C......... READ INITIAL CONDITIONS FROM INPUT FILE                     DLEARN   84
C                                                                      DLEARN   85
              CALL INPUT(A(MPD),NDOF,NUMNP,1,1,IPRTIN,TIME)            DLEARN   86
              CALL INPUT(A(MPV),NDOF,NUMNP,2,1,IPRTIN,TIME)            DLEARN   87
              CALL INPUT(A(MPA),NDOF,NUMNP,3,1,IPRTIN,TIME)            DLEARN   88
C                                                                      DLEARN   89
           ENDIF                                                       DLEARN   90
C                                                                      DLEARN   91
        ENDIF                                                          DLEARN   92
C                                                                      DLEARN   93
C.... STORE INITIAL KINEMATIC DATA FOR TIME HISTORIES                  DLEARN   94
C                                                                      DLEARN   95
        IF ( LDYN .AND. NDOUT.GT.0 ) THEN                              DLEARN   96
           LOCPLT = 1                                                  DLEARN   97
           CALL STORED(A(MPIDHS),A(MPD),A(MPV),A(MPA),A(MPDOUT),        DLEARN   98
     &             NDOF,NDOUT)                                         DLEARN   99
        ENDIF                                                          DLEARN  100
C                                                                      DLEARN  101
C.... INPUT ELEMENT DATA                                               DLEARN  102
C                                                                      DLEARN  103
        CALL ELEMNT('INPUT___',A(MPNGRP))                             DLEARN  104
C                                                                      DLEARN  105
C.... STORE INITIAL STRESS/STRAIN DATA FOR ELEMENT TIME HISTORIES      DLEARN  106
C                                                                      DLEARN  107
        IF (LDYN) THEN                                                 DLEARN  108
           LOCPLT = 1                                                  DLEARN  109
           CALL ELEMNT('STR_STOR',A(MPNGRP))                          DLEARN  110
        ENDIF                                                          DLEARN  111
C                                                                      DLEARN  112
C.... ALLOCATE MEMORY FOR GLOBAL EQUATION SYSTEM                       DLEARN  113
C                                                                      DLEARN  114
        CALL EQSET(NEQ,NALHS)                                          DLEARN  115
        CALL TIMING(T2)                                                DLEARN  116
        ETIME(1) = T2 - T1                                             DLEARN  117
C                                                                      DLEARN  118
C.... SOLUTION PHASE                                                   DLEARN  119
C                                                                      DLEARN  120
        IF (IEXEC.EQ.1) CALL DRIVER(NTSTEP,NEQ,NALHS)                  DLEARN  121
C                                                                      DLEARN  122
C.... WRITE RESTART FILE                                               DLEARN  123
C                                                                      DLEARN  124
        IF ( LDYN .AND. (IWRITR.EQ.1) )                                DLEARN  125
     &     CALL RSOUT(A(MPD),A(MPV),A(MPA),NDOF,NUMNP,NTSTEP,TIME)     DLEARN  126
C                                                                      DLEARN  127
C.... PRINT MEMORY-POINTER DICTIONARY                                  DLEARN  128
C                                                                      DLEARN  129
        CALL PRTDC                                                     DLEARN  130
C                                                                      DLEARN  131
        CALL TIMING(T1)                                                DLEARN  132
        ETIME(2) = T1 - T2                                             DLEARN  133
C                                                                      DLEARN  134
C.... PRINT ELAPSED TIME SUMMARY                                       DLEARN  135
C                                                                      DLEARN  136
        CALL TIMLOG                                                    DLEARN  137
        GO TO 100                                                      DLEARN  138
```

```
C
 1000 FORMAT(20A4)
 2000 FORMAT(16I5)
 3000 FORMAT('1',20A4///
     &' E X E C U T I O N   C O N T R O L   I N F O R M A T I O N '//5X,
     &' EXECUTION CODE . . . . . . . . . . . . . . (IEXEC ) = ',I5//5X,
     &'     EQ. 0, DATA CHECK                                  ',   /5X,
     &'     EQ. 1, EXECUTION                                   ',  //5X,
     &' ANALYSIS CODE . . . . . . . . . . . . . . (IACODE) = ',I5//5X,
     &'     EQ. 0, DYNAMIC ANALYSIS                            ',   /5X,
     &'     EQ. 1, STATIC ANALYSIS                             ',  //5X,
     &' READ RESTART FILE CODE . . . . . . . . . . (IREADR) = ',I5//5X,
     &'     EQ. 0, DO NOT READ RESTART FILE                    ',   /5X,
     &'     EQ. 1, READ RESTART FILE                           ',  //5X,
     &' WRITE RESTART FILE CODE . . . . . . . . . (IWRITR) = ',I5//5X,
     &'     EQ. 0, DO NOT WRITE RESTART FILE                   ',   /5X,
     &'     EQ. 1, WRITE RESTART FILE                          ',  //5X,
     &' INPUT DATA PRINT CODE . . . . . . . . . . (IPRTIN) = ',I5//5X,
     &'     EQ. 0, PRINT NODAL AND ELEMENT INPUT DATA          ',   /5X,
     &'     EQ. 1, DO NOT PRINT NODAL AND ELEMENT INPUT DATA   ',   /5X)
 4000 FORMAT(5X,
     &' RANK CHECK CODE . . . . . . . . . . . . . (IRANK ) = ',I5//5X,
     &'     EQ. 0, DO NOT PERFORM RANK CHECK                   ',   /5X,
     &'     EQ. 1, PRINT NUMBERS OF ZERO AND NEGATIVE PIVOTS   ',   /5X,
     &'     EQ. 2, PRINT ALL PIVOTS                            ',  //5X,
     &' NUMBER OF TIME SEQUENCES  . . . . . . . . (NUMSEQ) = ',I5//5X,
     &' NUMBER OF NODAL OUTPUT TIME-HISTORIES . . . (NDOUT ) = ',I5//5X,
     &' NUMBER OF SPACE DIMENSIONS  . . . . . . . (NSD   ) = ',I5//5X,
     &' NUMBER OF NODAL POINTS  . . . . . . . . . (NUMNP ) = ',I5//5X,
     &' NUMBER OF NODAL DEGREES-OF-FREEDOM  . . . . (NDOF  ) = ',I5//5X,
     &' NUMBER OF LOAD VECTORS  . . . . . . . . . (NLVECT) = ',I5//5X,
     &' NUMBER OF LOAD-TIME FUNCTIONS . . . . . . (NLTFTN) = ',I5//5X,
     &' NUMBER OF POINTS ON LOAD-TIME FUNCTIONS . . (NPTSLF) = ',I5//5X,
     &' NUMBER OF ELEMENT GROUPS  . . . . . . . . (NUMEG ) = ',I5//5X)
C
      END
C**********************************************************************
      SUBROUTINE DRIVER(NTSTEP,NEQ,NALHS)
C
C.... SOLUTION DRIVER PROGRAM
C
      DOUBLE PRECISION
     &       TIME,COEFF1,COEFF2,COEFF3,COEFF4,COEFF5,COEFF6,
     &            COEFF7,COEFF8,ALPHA1,BETA1 ,GAMMA1,DT1
C
C.... DEACTIVATE ABOVE CARD(S) FOR SINGLE-PRECISION OPERATION
C
      LOGICAL LDYN,LOUT
      COMMON /COEFFS/ COEFF1,COEFF2,COEFF3,COEFF4,COEFF5,COEFF6,
     &                COEFF7,COEFF8,ALPHA1,BETA1 ,GAMMA1,DT1
      COMMON /DPOINT/ MPSTEP,MPDPRT,MPSPRT,MPHPLT,MPITER,MPALPH,MPBETA,
     &                MPGAMM,MPDT  ,MPIDHS,MPDOUT,MPVPRD,MPDPRD,MPA,MPV
      COMMON /ETIMEC/ ETIME(7)
      COMMON /HPLOTC/ NPLPTS,LOCPLT,TIME
      COMMON /INFO  / IEXEC,IACODE,LDYN,IREADR,IWRITR,IPRTIN,IRANK,
     &                NUMSEQ,NDOUT,NSD,NUMNP,NDOF,NLVECT,NLTFTN,NPTSLF,
     &                NUMEG
      COMMON /SPOINT/ MPD,MPX,MPID,MPF,MPG,MPG1,MPDIAG,MPNGRP,
     &                MPALHS,MPBRHS
      COMMON A(1)
C
C.... TIME SEQUENCE LOOP
C
      DO 300 NSQ=1,NUMSEQ
C
C.... SET CURRENT TIME SEQUENCE PARAMETERS
C
      CALL TIMCON(NSQ,A(MPSTEP),A(MPDPRT),A(MPSPRT),A(MPHPLT),A(MPITER),
     &                  NSTEP1,   NDPRT1,    NSPRT1,    NHPLT1,    NITER1,
     &                A(MPALPH),A(MPBETA),A(MPGAMM),A(MPDT   ))
C
C.... FORM EFFECTIVE MASS MATRIX
C
      CALL CLEAR(A(MPALHS),NALHS)
      CALL TIMING(T1)
      CALL ELEMNT('FORM_LHS',A(MPNGRP))
      CALL TIMING(T2)
      ETIME(3) = ETIME(3) + T2 - T1
C
C.... PERFORM FACTORIZATION OF EFFECTIVE MASS MATRIX
C
      CALL FACTOR(A(MPALHS),A(MPDIAG),NEQ)
      CALL TIMING(T1)
      ETIME(4) = ETIME(4) + T1 - T2
C
C.... RANK CHECK (NOTE: SUBROUTINES "PIVOTS" AND "PRINTP"
C        RETURN TO STATEMENT 300)
```

```
C                                                                      DRIVER  51
        IF (IRANK.EQ.1) CALL PIVOTS(A(MPALHS),A(MPDIAG),NEQ,NSQ,*300)  DRIVER  52
        IF (IRANK.EQ.2) CALL PRINTP(A(MPALHS),A(MPDIAG),NEQ,NSQ,*300)  DRIVER  53
C                                                                      DRIVER  54
C.... TIME STEP LOOP                                                   DRIVER  55
C                                                                      DRIVER  56
        DO 200 N=1,NSTEP1                                              DRIVER  57
C                                                                      DRIVER  58
        TIME = TIME + DT1                                              DRIVER  59
        NTSTEP = NTSTEP + 1                                            DRIVER  60
C                                                                      DRIVER  61
        IF (LDYN) THEN                                                 DRIVER  62
C                                                                      DRIVER  63
C...... PREDICTOR UPDATE OF ALL DEGREES-OF-FREEDOM                     DRIVER  64
C                                                                      DRIVER  65
           CALL PREDCT(A(MPD),A(MPV),A(MPA),A(MPDPRD),A(MPVPRD),       DRIVER  66
     &                 NDOF,NUMNP)                                     DRIVER  67
        ELSE                                                           DRIVER  68
           CALL CLEAR(A(MPD),NDOF*NUMNP)                               DRIVER  69
        ENDIF                                                          DRIVER  70
C                                                                      DRIVER  71
        IF (NLVECT.GT.0)                                               DRIVER  72
C                                                                      DRIVER  73
C.... EVALUATE LOAD-TIME FUNCTIONS AT TIME N+1                         DRIVER  74
C                                                                      DRIVER  75
     &     CALL LFAC(A(MPG),TIME,A(MPG1),NLTFTN,NPTSLF)                DRIVER  76
C                                                                      DRIVER  77
        IF (DT1.NE.0)                                                  DRIVER  78
C                                                                      DRIVER  79
C...... OVERWRITE PREDICTORS TO ACCOUNT FOR KINEMATIC                  DRIVER  80
C             BOUNDARY CONDITIONS                                      DRIVER  81
C                                                                      DRIVER  82
     &     CALL COMPBC(A(MPID),A(MPD),A(MPV),A(MPA),A(MPDPRD),A(MPVPRD),DRIVER 83
     &                 A(MPF),A(MPG1),NDOF,NUMNP,NLVECT,LDYN)          DRIVER  84
C                                                                      DRIVER  85
C.... MULTI-CORRECTOR ITERATION LOOP                                   DRIVER  86
C                                                                      DRIVER  87
        DO 100 I=1,NITER1                                              DRIVER  88
C                                                                      DRIVER  89
        CALL CLEAR(A(MPBRHS),NEQ)                                      DRIVER  90
C                                                                      DRIVER  91
        IF (NLTFTN.GT.0)                                               DRIVER  92
C                                                                      DRIVER  93
C...... EVALUATE LOAD-TIME FUNCTIONS AT TIME N+1+ALPHA                 DRIVER  94
C                                                                      DRIVER  95
     &     CALL LFAC(A(MPG),TIME+ALPHA1*DT1,A(MPG1),NLTFTN,NPTSLF)     DRIVER  96
C                                                                      DRIVER  97
        IF (NLVECT.GT.0)                                               DRIVER  98
C                                                                      DRIVER  99
C...... FORM NODAL CONTRIBUTION TO RESIDUAL FORCE VECTOR               DRIVER 100
C                                                                      DRIVER 101
     &     CALL LOAD(A(MPID),A(MPF),A(MPBRHS),A(MPG1),NDOF,NUMNP,NLVECT)DRIVER 102
C                                                                      DRIVER 103
C.... FORM ELEMENT CONTRIBUTION TO RESIDUAL FORCE VECTOR               DRIVER 104
C                                                                      DRIVER 105
        CALL TIMING(T1)                                                DRIVER 106
        CALL ELEMNT('FORM_RHS',A(MPNGRP))                             DRIVER 107
        CALL TIMING(T2)                                                DRIVER 108
        ETIME(5) = ETIME(5) + T2 - T1                                 DRIVER 109
C                                                                      DRIVER 110
C.... SOLVE EQUATION SYSTEM                                            DRIVER 111
C                                                                      DRIVER 112
        CALL BACK(A(MPALHS),A(MPBRHS),A(MPDIAG),NEQ)                   DRIVER 113
        CALL TIMING(T1)                                                DRIVER 114
        ETIME(6) = ETIME(6) + T1 - T2                                 DRIVER 115
C                                                                      DRIVER 116
C.... PERFORM INTERMEDIATE UPDATE OF ACTIVE DEGREES-OF-FREEDOM         DRIVER 117
C                                                                      DRIVER 118
        CALL ITERUP(A(MPID),A(MPD),A(MPDPRD),A(MPVPRD),A(MPA),A(MPBRHS),DRIVER 119
     &              NDOF,NUMNP,LDYN)                                   DRIVER 120
  100 CONTINUE                                                         DRIVER 121
C                                                                      DRIVER 122
        IF (LDYN)                                                      DRIVER 123
C                                                                      DRIVER 124
C.... PERFORM CORRECTOR UPDATE OF ALL DEGREES-OF-FREEDOM               DRIVER 125
C                                                                      DRIVER 126
     &     CALL CORRCT(A(MPID),A(MPD),A(MPV),A(MPDPRD),A(MPVPRD),      DRIVER 127
     &                 NDOF,NUMNP)                                     DRIVER 128
C                                                                      DRIVER 129
        IF (LOUT(N,NDPRT1)) THEN                                       DRIVER 130
C                                                                      DRIVER 131
C....... WRITE KINEMATIC OUTPUT                                        DRIVER 132
```

```
C                                                                        DRIVER 133
              CALL PRINTD(' D I S P L A C E M E N T S                    DRIVER 134
     &               A(MPD),NDOF,NUMNP,NTSTEP,TIME)                      DRIVER 135
           IF (LDYN) THEN                                               DRIVER 136
              CALL PRINTD(' V E L O C I T I E S                          DRIVER 137
     &               A(MPV),NDOF,NUMNP,NTSTEP,TIME)                      DRIVER 138
              CALL PRINTD(' A C C E L E R A T I O N S                    DRIVER 139
     &               A(MPA),NDOF,NUMNP,NTSTEP,TIME)                      DRIVER 140
           ENDIF                                                        DRIVER 141
        ENDIF                                                           DRIVER 142
C                                                                       DRIVER 143
      CALL TIMING(T1)                                                   DRIVER 144
C                                                                       DRIVER 145
C.... CALCULATE AND WRITE ELEMENT OUTPUT                                DRIVER 146
C                                                                       DRIVER 147
      IF (LOUT(N,NSPRT1)) CALL ELEMNT('STR_PRNT',A(MPNGRP))             DRIVER 148
C                                                                       DRIVER 149
      IF (LDYN.AND.LOUT(N,NHPLT1)) THEN                                 DRIVER 150
         LOCPLT = LOCPLT + 1                                            DRIVER 151
C                                                                       DRIVER 152
C...... NOTE: VARIABLES "LOCPLT" AND "TIME" ARE PASSED INTO SUB-        DRIVER 153
C             ROUTINE "STORED" AND ELEMENT ROUTINES PERFORMING          DRIVER 154
C             TASK5 ('STR_STOR') BY WAY OF COMMON /HPLOTC/.             DRIVER 155
C                                                                       DRIVER 156
C...... STORE KINEMATIC TIME HISTORY DATA                               DRIVER 157
C                                                                       DRIVER 158
         CALL STORED(A(MPIDHS),A(MPD),A(MPV),A(MPA),A(MPDOUT),          DRIVER 159
     &               NDOF,NDOUT)                                        DRIVER 160
C                                                                       DRIVER 161
C...... CALCULATE AND STORE ELEMENT TIME HISTORY DATA                   DRIVER 162
C                                                                       DRIVER 163
C                                                                       DRIVER 164
         CALL ELEMNT('STR_STOR',A(MPNGRP))                             DRIVER 165
      ENDIF                                                            DRIVER 166
      CALL TIMING(T2)                                                   DRIVER 167
      ETIME(7) = ETIME(7) + T2 - T1                                    DRIVER 168
C                                                                       DRIVER 169
  200 CONTINUE                                                          DRIVER 170
C                                                                       DRIVER 171
  300 CONTINUE                                                          DRIVER 172
      IF (LDYN) THEN                                                   DRIVER 173
C                                                                       DRIVER 174
C...... PLOT NODAL TIME-HISTORIES                                       DRIVER 175
C                                                                       DRIVER 176
      IF (NDOUT.GT.0) CALL HPLOT(A(MPIDHS),A(MPDOUT),NDOUT,3,0)         DRIVER 177
C                                                                       DRIVER 178
C...... PLOT ELEMENT TIME-HISTORIES                                     DRIVER 179
C                                                                       DRIVER 180
      CALL ELEMNT('STR_PLOT',A(MPNGRP))                                DRIVER 181
      ENDIF                                                            DRIVER 182
C                                                                       DRIVER 183
      RETURN                                                            DRIVER 184
      END                                                               DRIVER 185
C*********************************************************************************
      SUBROUTINE ADDLHS(ALHS,ELEFFM,IDIAG,LM,NEE,LDIAG)                ADDLHS   1
C                                                                       ADDLHS   2
C.... PROGRAM TO ADD ELEMENT LEFT-HAND-SIDE MATRIX TO                   ADDLHS   3
C        GLOBAL LEFT-HAND-SIDE MATRIX                                   ADDLHS   4
C                                                                       ADDLHS   5
C        LDIAG = .TRUE.,   ADD DIAGONAL ELEMENT MATRIX                  ADDLHS   6
C                                                                       ADDLHS   7
C        LDIAG = .FALSE., ADD UPPER TRIANGLE OF FULL ELEMENT MATRIX     ADDLHS   8
C                                                                       ADDLHS   9
      IMPLICIT DOUBLE PRECISION (A-H,O-Z)                              ADDLHS  10
C                                                                       ADDLHS  11
C.... DEACTIVATE ABOVE CARD(S) FOR SINGLE-PRECISION OPERATION           ADDLHS  12
C                                                                       ADDLHS  13
      LOGICAL LDIAG                                                    ADDLHS  14
      DIMENSION ALHS(1),ELEFFM(NEE,1),IDIAG(1),LM(1)                   ADDLHS  15
C                                                                       ADDLHS  16
      IF (LDIAG) THEN                                                 ADDLHS  17
C                                                                       ADDLHS  18
         DO 100 J=1,NEE                                                ADDLHS  19
         K = LM(J)                                                     ADDLHS  20
         IF (K.GT.0) THEN                                              ADDLHS  21
            L = IDIAG(K)                                               ADDLHS  22
            ALHS(L) = ALHS(L) + ELEFFM(J,J)                            ADDLHS  23
         ENDIF                                                         ADDLHS  24
  100    CONTINUE                                                      ADDLHS  25
C                                                                       ADDLHS  26
      ELSE                                                            ADDLHS  27
C                                                                       ADDLHS  28
         DO 300 J=1,NEE                                                ADDLHS  29
         K = LM(J)                                                     ADDLHS  30
         IF (K.GT.0) THEN                                              ADDLHS  31
```

```
C                                                                           ADDLHS  32
                   DO 200 I=1,J                                             ADDLHS  33
                   M = LM(I)                                                ADDLHS  34
                   IF (M.GT.0) THEN                                         ADDLHS  35
                      IF (K.GE.M) THEN                                      ADDLHS  36
                         L = IDIAG(K) - K + M                               ADDLHS  37
                      ELSE                                                  ADDLHS  38
                         L = IDIAG(M) - M + K                               ADDLHS  39
                      ENDIF                                                 ADDLHS  40
                      ALHS(L) = ALHS(L) + ELEFFM(I,J)                       ADDLHS  41
                   ENDIF                                                    ADDLHS  42
  200           CONTINUE                                                    ADDLHS  43
C                                                                           ADDLHS  44
             ENDIF                                                          ADDLHS  45
  300     CONTINUE                                                          ADDLHS  46
C                                                                           ADDLHS  47
      ENDIF                                                                 ADDLHS  48
C                                                                           ADDLHS  49
      RETURN                                                                ADDLHS  50
      END                                                                   ADDLHS  51
C****************************************************************************
      SUBROUTINE ADDRHS(BRHS,ELRESF,LM,NEE)                                 ADDRHS   1
C                                                                           ADDRHS   2
C.... PROGRAM TO ADD ELEMENT RESIDUAL-FORCE VECTOR TO                       ADDRHS   3
C        GLOBAL RIGHT-HAND-SIDE VECTOR                                      ADDRHS   4
C                                                                           ADDRHS   5
      IMPLICIT DOUBLE PRECISION (A-H,O-Z)                                   ADDRHS   6
C                                                                           ADDRHS   7
C.... DEACTIVATE ABOVE CARD(S) FOR SINGLE-PRECISION OPERATION               ADDRHS   8
C                                                                           ADDRHS   9
      DIMENSION BRHS(1),ELRESF(1),LM(1)                                     ADDRHS  10
C                                                                           ADDRHS  11
      DO 100 J=1,NEE                                                        ADDRHS  12
      K = LM(J)                                                             ADDRHS  13
      IF (K.GT.0) BRHS(K) = BRHS(K) + ELRESF(J)                             ADDRHS  14
  100 CONTINUE                                                              ADDRHS  15
C                                                                           ADDRHS  16
      RETURN                                                                ADDRHS  17
      END                                                                   ADDRHS  18
C****************************************************************************
      SUBROUTINE BACK(A,B,IDIAG,NEQ)                                        BACK     1
C                                                                           BACK     2
C.... PROGRAM TO PERFORM FORWARD REDUCTION AND BACK SUBSTITUTION            BACK     3
C                                                                           BACK     4
      IMPLICIT DOUBLE PRECISION (A-H,O-Z)                                   BACK     5
C                                                                           BACK     6
C.... DEACTIVATE ABOVE CARD(S) FOR SINGLE-PRECISION OPERATION               BACK     7
C                                                                           BACK     8
      DIMENSION A(1),B(1),IDIAG(1)                                          BACK     9
      COMMON /CONSTS/ ZERO,PT1667,PT25,PT5,ONE,TWO,THREE,FOUR,FIVE          BACK    10
C                                                                           BACK    11
C.... FORWARD REDUCTION                                                     BACK    12
C                                                                           BACK    13
      JJ = 0                                                                BACK    14
C                                                                           BACK    15
      DO 100 J=1,NEQ                                                        BACK    16
      JJLAST = JJ                                                           BACK    17
      JJ     = IDIAG(J)                                                     BACK    18
      JCOLHT = JJ - JJLAST                                                  BACK    19
      IF (JCOLHT.GT.1)                                                      BACK    20
     &   B(J) = B(J) - COLDOT(A(JJLAST+1),B(J-JCOLHT+1),JCOLHT-1)           BACK    21
  100 CONTINUE                                                              BACK    22
C                                                                           BACK    23
C.... DIAGONAL SCALING                                                      BACK    24
C                                                                           BACK    25
      DO 200 J=1,NEQ                                                        BACK    26
      AJJ = A(IDIAG(J))                                                     BACK    27
C                                                                           BACK    28
C.... WARNING: DIAGONAL SCALING IS NOT PERFORMED IF AJJ EQUALS ZERO         BACK    29
C                                                                           BACK    30
      IF (AJJ.NE.ZERO) B(J) = B(J)/AJJ                                      BACK    31
  200 CONTINUE                                                              BACK    32
C                                                                           BACK    33
C.... BACK SUBSTITUTION                                                     BACK    34
C                                                                           BACK    35
      IF (NEQ.EQ.1) RETURN                                                  BACK    36
      JJNEXT = IDIAG(NEQ)                                                   BACK    37
C                                                                           BACK    38
      DO 400 J=NEQ,2,-1                                                     BACK    39
      JJ     = JJNEXT                                                       BACK    40
      JJNEXT = IDIAG(J-1)                                                   BACK    41
      JCOLHT = JJ - JJNEXT                                                  BACK    42
      IF (JCOLHT.GT.1) THEN                                                 BACK    43
         BJ = B(J)                                                          BACK    44
         ISTART = J - JCOLHT + 1                                           BACK    45
         JTEMP  = JJNEXT - ISTART + 1                                       BACK    46
```

```
C                                                                        BACK    47
          DO 300 I=ISTART,J-1                                            BACK    48
          B(I) = B(I) - A(JTEMP+I)*BJ                                    BACK    49
  300     CONTINUE                                                       BACK    50
C                                                                        BACK    51
       ENDIF                                                             BACK    52
C                                                                        BACK    53
  400 CONTINUE                                                           BACK    54
C                                                                        BACK    55
      RETURN                                                             BACK    56
      END                                                                BACK    57
C*********************************************************************** 
      SUBROUTINE BC(ID,NDOF,NUMNP,NEQ,IPRTIN)                            BC       1
C                                                                        BC       2
C.... PROGRAM TO READ, GENERATE AND WRITE BOUNDARY CONDITION DATA        BC       3
C          AND ESTABLISH EQUATION NUMBERS                                BC       4
C                                                                        BC       5
      DIMENSION ID(NDOF,1)                                               BC       6
C                                                                        BC       7
      COMMON /IOUNIT/ IIN,IOUT,IRSIN,IRSOUT                              BC       8
      LOGICAL PFLAG                                                      BC       9
C                                                                        BC      10
      CALL ICLEAR(ID,NDOF*NUMNP)                                         BC      11
      CALL IGEN(ID,NDOF)                                                 BC      12
C                                                                        BC      13
      IF (IPRTIN.EQ.0) THEN                                              BC      14
         NN=0                                                            BC      15
         DO 200 N=1,NUMNP                                                BC      16
         PFLAG = .FALSE.                                                 BC      17
C                                                                        BC      18
         DO 100 I=1,NDOF                                                 BC      19
         IF (ID(I,N).NE.0) PFLAG = .TRUE.                                BC      20
  100    CONTINUE                                                        BC      21
C                                                                        BC      22
         IF (PFLAG) THEN                                                 BC      23
            NN = NN + 1                                                  BC      24
            IF (MOD(NN,50).EQ.1) WRITE(IOUT,1000) (I,I=1,NDOF)           BC      25
            WRITE(IOUT,2000) N,(ID(I,N),I=1,NDOF)                        BC      26
         ENDIF                                                           BC      27
  200    CONTINUE                                                        BC      28
      ENDIF                                                              BC      29
C                                                                        BC      30
C.... ESTABLISH EQUATION NUMBERS                                         BC      31
C                                                                        BC      32
      NEQ = 0                                                            BC      33
C                                                                        BC      34
      DO 400 N=1,NUMNP                                                   BC      35
C                                                                        BC      36
      DO 300 I=1,NDOF                                                    BC      37
      IF (ID(I,N).EQ.0) THEN                                             BC      38
         NEQ = NEQ + 1                                                   BC      39
         ID(I,N) = NEQ                                                   BC      40
      ELSE                                                               BC      41
         ID(I,N) = 1 - ID(I,N)                                          BC      42
      ENDIF                                                              BC      43
C                                                                        BC      44
  300 CONTINUE                                                           BC      45
C                                                                        BC      46
  400 CONTINUE                                                           BC      47
C                                                                        BC      48
      RETURN                                                             BC      49
C                                                                        BC      50
 1000 FORMAT('1',' N O D A L   B O U N D A R Y   C O N D I T I O N   C OBC      51
     & D E S'///                                                         BC      52
     & 5X,' NODE NO.',3X,6(6X,'DOF',I1:)//)                              BC      53
 2000 FORMAT(6X,I5,5X,6(5X,I5))                                          BC      54
C                                                                        BC      55
      END                                                                BC      56
C*********************************************************************** 
      BLOCK DATA                                                         BLOCK    1
C                                                                        BLOCK    2
C.... PROGRAM TO DEFINE OUTPUT LABELS AND NUMERICAL CONSTANTS            BLOCK    3
C                                                                        BLOCK    4
      IMPLICIT DOUBLE PRECISION (A-H,O-Z)                                BLOCK    5
C                                                                        BLOCK    6
C.... DEACTIVATE ABOVE CARD(S) FOR SINGLE-PRECISION OPERATION            BLOCK    7
C                                                                        BLOCK    8
C                                                                        BLOCK    9
      CHARACTER*4 LABELD,LABEL1,LABEL2                                   BLOCK   10
      COMMON /CONSTS/ ZERO,PT1667,PT25,PT5,ONE,TWO,THREE,FOUR,FIVE       BLOCK   11
      COMMON /LABELS/ LABELD(3),LABEL1(16),LABEL2(3)                     BLOCK   12
C                                                                        BLOCK   13
C         LABELD(3)  = DISPLACEMENT, VELOCITY AND ACCELERATION LABELS    BLOCK   14
C         LABEL1(16) = OUTPUT LABELS FOR ELEMENT-TYPE 1                  BLOCK   15
C         LABEL2(3)  = OUTPUT LABELS FOR ELEMENT-TYPE 2                  BLOCK   16
C                                                                        BLOCK   17
C.... NOTE: ADD LABEL ARRAYS FOR ANY ADDITIONAL ELEMENTS                 BLOCK   18
```

```
C                                                                              BLOCK   19
      DATA    ZERO,PT1667,PT25,PT5                                             BLOCK   20
     &        /0.00,0.1666666666666667,0.25,0.50/,                            BLOCK   21
     &        ONE,TWO,THREE,FOUR,FIVE                                          BLOCK   22
     &        /1.00,2.00,3.00,4.00,5.00/                                      BLOCK   23
C                                                                              BLOCK   24
      DATA LABELD/'DISP','VEL ','ACC '/                                        BLOCK   25
C                                                                              BLOCK   26
      DATA LABEL1/'S 11','S 22','S 12','S 33','PS 1','PS 2',                   BLOCK   27
     &           'TAU ','SANG','E 11','E 22','G 12','E 33',                   BLOCK   28
     &           'PE 1','PE 2','GAM ','EANG'/                                  BLOCK   29
C                                                                              BLOCK   30
      DATA LABEL2/'STRS','FORC','STRN'/                                        BLOCK   31
C                                                                              BLOCK   32
      END                                                                      BLOCK   33
C*****************************************************************************************
      SUBROUTINE BTDB(ELSTIF,B,DB,NEE,NROWB,NSTR)                              BTDB     1
C                                                                              BTDB     2
C.... PROGRAM TO MULTIPLY B(TRANSPOSE) * DB TAKING ACCOUNT OF SYMMETRY        BTDB     3
C        AND ACCUMULATE INTO ELEMENT STIFFNESS MATRIX                         BTDB     4
C                                                                              BTDB     5
      IMPLICIT DOUBLE PRECISION (A-H,O-Z)                                      BTDB     6
C                                                                              BTDB     7
C.... DEACTIVATE ABOVE CARD(S) FOR SINGLE-PRECISION OPERATION                 BTDB     8
C                                                                              BTDB     9
      DIMENSION ELSTIF(NEE,1),B(NROWB,1),DB(NROWB,1)                          BTDB    10
C                                                                              BTDB    11
      DO 200 J=1,NEE                                                           BTDB    12
C                                                                              BTDB    13
      DO 100 I=1,J                                                             BTDB    14
      ELSTIF(I,J) = ELSTIF(I,J) + COLDOT(B(1,I),DB(1,J),NSTR)                 BTDB    15
  100 CONTINUE                                                                 BTDB    16
C                                                                              BTDB    17
  200 CONTINUE                                                                 BTDB    18
C                                                                              BTDB    19
      RETURN                                                                   BTDB    20
      END                                                                      BTDB    21
C*****************************************************************************************
      SUBROUTINE CLEAR(A,M)                                                    CLEAR    1
C                                                                              CLEAR    2
C.... PROGRAM TO CLEAR A FLOATING-POINT ARRAY                                 CLEAR    3
C                                                                              CLEAR    4
      IMPLICIT DOUBLE PRECISION (A-H,O-Z)                                      CLEAR    5
C                                                                              CLEAR    6
C.... DEACTIVATE ABOVE CARD(S) FOR SINGLE-PRECISION OPERATION                 CLEAR    7
C                                                                              CLEAR    8
      DIMENSION A(1)                                                           CLEAR    9
      COMMON /CONSTS/ ZERO,PT1667,PT25,PT5,ONE,TWO,THREE,FOUR,FIVE            CLEAR   10
C                                                                              CLEAR   11
      DO 100 I=1,M                                                             CLEAR   12
      A(I) = ZERO                                                              CLEAR   13
  100 CONTINUE                                                                 CLEAR   14
C                                                                              CLEAR   15
      RETURN                                                                   CLEAR   16
      END                                                                      CLEAR   17
C*****************************************************************************************
      FUNCTION COLDOT(A,B,N)                                                   COLDOT   1
C                                                                              COLDOT   2
C.... PROGRAM TO COMPUTE THE DOT PRODUCT OF VECTORS STORED COLUMN-WISE        COLDOT   3
C                                                                              COLDOT   4
      IMPLICIT DOUBLE PRECISION (A-H,O-Z)                                      COLDOT   5
C                                                                              COLDOT   6
C.... DEACTIVATE ABOVE CARD(S) FOR SINGLE-PRECISION OPERATION                 COLDOT   7
C                                                                              COLDOT   8
      DIMENSION A(1),B(1)                                                      COLDOT   9
      COMMON /CONSTS/ ZERO,PT1667,PT25,PT5,ONE,TWO,THREE,FOUR,FIVE            COLDOT  10
C                                                                              COLDOT  11
      COLDOT = ZERO                                                            COLDOT  12
C                                                                              COLDOT  13
      DO 100 I=1,N                                                             COLDOT  14
      COLDOT = COLDOT + A(I)*B(I)                                              COLDOT  15
  100 CONTINUE                                                                 COLDOT  16
C                                                                              COLDOT  17
      RETURN                                                                   COLDOT  18
      END                                                                      COLDOT  19
C*****************************************************************************************
      SUBROUTINE COLHT(IDIAG,LM,NED,NEN,NUMEL)                                 COLHT    1
C                                                                              COLHT    2
C.... PROGRAM TO COMPUTE COLUMN HEIGHTS IN GLOBAL LEFT-HAND-SIDE MATRIX       COLHT    3
C                                                                              COLHT    4
      DIMENSION IDIAG(1),LM(NED,NEN,1)                                         COLHT    5
      COMMON /COLHTC/ NEQ                                                      COLHT    6
C                                                                              COLHT    7
      DO 500 K=1,NUMEL                                                         COLHT    8
      MIN = NEQ                                                                COLHT    9
C                                                                              COLHT   10
      DO 200 J=1,NEN                                                           COLHT   11
```

```
C                                                                        COLHT  12
      DO 100 I=1,NED                                                     COLHT  13
      NUM = LM(I,J,K)                                                    COLHT  14
      IF (NUM.GT.0) MIN = MINO(MIN,NUM)                                  COLHT  15
  100 CONTINUE                                                           COLHT  16
C                                                                        COLHT  17
  200 CONTINUE                                                           COLHT  18
C                                                                        COLHT  19
      DO 400 J=1,NEN                                                     COLHT  20
C                                                                        COLHT  21
      DO 300 I=1,NED                                                     COLHT  22
      NUM = LM(I,J,K)                                                    COLHT  23
      IF (NUM.GT.0) THEN                                                 COLHT  24
         M = NUM - MIN                                                   COLHT  25
         IF (M.GT.IDIAG(NUM)) IDIAG(NUM) = M                             COLHT  26
      ENDIF                                                              COLHT  27
C                                                                        COLHT  28
  300 CONTINUE                                                           COLHT  29
                                                                         COLHT  30
  400 CONTINUE                                                           COLHT  31
C                                                                        COLHT  32
  500 CONTINUE                                                           COLHT  33
C                                                                        COLHT  34
      RETURN                                                             COLHT  35
      END                                                                COLHT  36
C**********************************************************************************
      SUBROUTINE COMPBC(ID,D,V,A,DPRED,VPRED,F,G1,                       COMPBC   1
     &                  NDOF,NUMNP,NLVECT,LDYN)                          COMPBC   2
C                                                                        COMPBC   3
C.... PROGRAM TO COMPUTE DISPLACEMENT, VELOCITY AND                      COMPBC   4
C         ACCELERATION BOUNDARY CONDITIONS                               COMPBC   5
C                                                                        COMPBC   6
      IMPLICIT DOUBLE PRECISION (A-H,O-Z)                                COMPBC   7
C                                                                        COMPBC   8
C.... DEACTIVATE ABOVE CARD(S) FOR SINGLE-PRECISION OPERATION            COMPBC   9
C                                                                        COMPBC  10
      LOGICAL LDYN                                                       COMPBC  11
      DIMENSION ID(NDOF,1),D(NDOF,1),V(NDOF,1),A(NDOF,1),                COMPBC  12
     &          DPRED(NDOF,1),VPRED(NDOF,1),F(NDOF,NUMNP,1),G1(1)        COMPBC  13
C                                                                        COMPBC  14
      COMMON /COEFFS/ COEFF1,COEFF2,COEFF3,COEFF4,COEFF5,COEFF6,         COMPBC  15
     &                COEFF7,COEFF8,ALPHA1,BETA1 ,GAMMA1,DT1             COMPBC  16
      COMMON /CONSTS/ ZERO,PT1667,PT25,PT5,ONE,TWO,THREE,FOUR,FIVE       COMPBC  17
C                                                                        COMPBC  18
      DO 700 I=1,NDOF                                                    COMPBC  19
C                                                                        COMPBC  20
      DO 600 J=1,NUMNP                                                   COMPBC  21
C                                                                        COMPBC  22
      K = ID(I,J)                                                        COMPBC  23
      IF (K.GT.0) GO TO 500                                              COMPBC  24
      VAL = ZERO                                                         COMPBC  25
      DO 100 LV=1,NLVECT                                                 COMPBC  26
      VAL = VAL + F(I,J,LV)*G1(LV)                                       COMPBC  27
  100 CONTINUE                                                           COMPBC  28
C                                                                        COMPBC  29
      M = 1 - K                                                          COMPBC  30
      GO TO (200,300,400),M                                             COMPBC  31
C                                                                        COMPBC  32
  200 CONTINUE                                                           COMPBC  33
      IF (LDYN) THEN                                                     COMPBC  34
         TEMP = COEFF1*VAL - ALPHA1*D(I,J)                               COMPBC  35
         A(I,J) = (TEMP - DPRED(I,J))/COEFF5                             COMPBC  36
         DPRED(I,J) = TEMP                                               COMPBC  37
         VPRED(I,J) = VPRED(I,J) + COEFF4*A(I,J)                         COMPBC  38
      ELSE                                                               COMPBC  39
         D(I,J) = VAL                                                    COMPBC  40
      ENDIF                                                              COMPBC  41
      GO TO 500                                                          COMPBC  42
C                                                                        COMPBC  43
  300 TEMP = COEFF1*VAL - ALPHA1*V(I,J)                                  COMPBC  44
      A(I,J) = (TEMP - VPRED(I,J))/COEFF4                                COMPBC  45
      VPRED(I,J) = TEMP                                                  COMPBC  46
      DPRED(I,J) = DPRED(I,J) + COEFF5*A(I,J)                            COMPBC  47
      GO TO 500                                                          COMPBC  48
C                                                                        COMPBC  49
  400 DPRED(I,J) = DPRED(I,J) + COEFF5*VAL                               COMPBC  50
      VPRED(I,J) = VPRED(I,J) + COEFF4*VAL                               COMPBC  51
      A(I,J) = VAL                                                       COMPBC  52
C                                                                        COMPBC  53
  500 CONTINUE                                                           COMPBC  54
C                                                                        COMPBC  55
  600 CONTINUE                                                           COMPBC  56
  700 CONTINUE                                                           COMPBC  57
      RETURN                                                             COMPBC  58
      END                                                                COMPBC  59
C**********************************************************************************
      SUBROUTINE CONTM(SHG,XL,W,DET,ELMASS,WORK,CONSTM,IMASS,NINT,       CONTM    1
     &                 NROWSH,NESD,NEN,NED,NEE,COLUMN)                   CONTM    2
C                                                                        CONTM    3
```

```
C.... PROGRAM TO FORM MASS MATRIX FOR A CONTINUUM ELEMENT        CONTM    4
C          WITH "NEN" NODES                                      CONTM    5
C                                                                CONTM    6
C          IMASS = MASS CODE, EQ. 0, CONSISTENT MASS             CONTM    7
C                            EQ. 1, LUMPED MASS                  CONTM    8
C                            OTHERWISE RETURN                    CONTM    9
C                                                                CONTM   10
      IMPLICIT DOUBLE PRECISION (A-H,O-Z)                        CONTM   11
C                                                                CONTM   12
C.... DEACTIVATE ABOVE CARD(S) FOR SINGLE-PRECISION OPERATION    CONTM   13
C                                                                CONTM   14
      LOGICAL COLUMN                                             CONTM   15
      DIMENSION SHG(NROWSH,NEN,1),XL(NESD,1),W(1),DET(1),        CONTM   16
     &          ELMASS(NEE,1),WORK(1)                            CONTM   17
      COMMON /CONSTS/ ZERO,PT1667,PT25,PT5,ONE,TWO,THREE,FOUR,FIVE CONTM 18
C                                                                CONTM   19
      IF (IMASS.EQ.0) THEN                                       CONTM   20
C                                                                CONTM   21
C....... CONSISTENT MASS                                         CONTM   22
C                                                                CONTM   23
      DO 400 L=1,NINT                                            CONTM   24
      TEMP1 = CONSTM*W(L)*DET(L)                                 CONTM   25
C                                                                CONTM   26
      DO 300 J=1,NEN                                             CONTM   27
      N = (J - 1)*NED                                            CONTM   28
C                                                                CONTM   29
      DO 200 I=1,J                                               CONTM   30
      M = (I - 1)*NED                                            CONTM   31
      TEMP2 = TEMP1*SHG(NROWSH,I,L)*SHG(NROWSH,J,L)              CONTM   32
C                                                                CONTM   33
      DO 100 K=1,NED                                             CONTM   34
      ELMASS(M + K,N + K) = ELMASS(M + K,N + K) + TEMP2          CONTM   35
  100 CONTINUE                                                   CONTM   36
C                                                                CONTM   37
  200 CONTINUE                                                   CONTM   38
C                                                                CONTM   39
  300 CONTINUE                                                   CONTM   40
C                                                                CONTM   41
  400 CONTINUE                                                   CONTM   42
C                                                                CONTM   43
      ENDIF                                                      CONTM   44
C                                                                CONTM   45
      IF (IMASS.EQ.1) THEN                                       CONTM   46
C                                                                CONTM   47
C....... LUMPED MASS                                             CONTM   48
C                                                                CONTM   49
      DSUM   = ZERO                                              CONTM   50
      TOTMAS = ZERO                                              CONTM   51
      CALL CLEAR(WORK,NEN)                                       CONTM   52
C                                                                CONTM   53
      DO 600 L=1,NINT                                            CONTM   54
      TEMP1 = CONSTM*W(L)*DET(L)                                 CONTM   55
      TOTMAS = TOTMAS + TEMP1                                    CONTM   56
C                                                                CONTM   57
      DO 500 J=1,NEN                                             CONTM   58
      TEMP2 = TEMP1*SHG(NROWSH,J,L)**2                           CONTM   59
      DSUM = DSUM + TEMP2                                        CONTM   60
      WORK(J) = WORK(J) + TEMP2                                  CONTM   61
  500 CONTINUE                                                   CONTM   62
C                                                                CONTM   63
  600 CONTINUE                                                   CONTM   64
C                                                                CONTM   65
C....... SCALE DIAGONAL TO CONSERVE TOTAL MASS                   CONTM   66
C                                                                CONTM   67
      TEMP1 = TOTMAS/DSUM                                        CONTM   68
C                                                                CONTM   69
      IF (COLUMN) THEN                                           CONTM   70
C                                                                CONTM   71
C......... STORE TERMS IN FIRST COLUMN OF MATRIX                 CONTM   72
C                                                                CONTM   73
      DO 800 J=1,NEN                                             CONTM   74
      TEMP2 = TEMP1*WORK(J)                                      CONTM   75
      N = (J - 1)*NED                                            CONTM   76
C                                                                CONTM   77
      DO 700 K=1,NED                                             CONTM   78
      ELMASS(N + K,1) = ELMASS(N + K,1) + TEMP2                  CONTM   79
  700 CONTINUE                                                   CONTM   80
C                                                                CONTM   81
  800 CONTINUE                                                   CONTM   82
C                                                                CONTM   83
      ELSE                                                       CONTM   84
C                                                                CONTM   85
C......... STORE TERMS ALONG DIAGONAL OF MATRIX                  CONTM   86
C                                                                CONTM   87
      DO 1000 J=1,NEN                                            CONTM   88
      TEMP2 = TEMP1*WORK(J)                                      CONTM   89
      N = (J - 1)*NED                                            CONTM   90
```

```
C
            DO 900 K=1,NED
            ELMASS(N + K,N + K) = ELMASS(N + K,N + K) + TEMP2
  900       CONTINUE
C
 1000       CONTINUE
C
         ENDIF
C
      ENDIF
C
      RETURN
      END
C**********************************************************************
      SUBROUTINE CONTMA(SHG,XL,W,DET,AL,ELMASS,WORK,ELRESF,CONSTM,IMASS,
     &                  NINT,NROWSH,NESD,NEN,NED,NEE)
C
C.... PROGRAM TO CALCULATE INERTIAL AND GRAVITY/BODY FORCE ("-M*(A-G)")
C        FOR A CONTINUUM ELEMENT WITH "NEN" NODES
C
C        IMASS = MASS CODE, EQ. 0, CONSISTENT MASS
C                           EQ. 1, LUMPED MASS
C                           OTHERWISE RETURN
C
      IMPLICIT DOUBLE PRECISION (A-H,O-Z)
C
C.... DEACTIVATE ABOVE CARD(S) FOR SINGLE-PRECISION OPERATION
C
      DIMENSION SHG(NROWSH,NEN,1),XL(NESD,1),W(1),DET(1),
     &          AL(NED,1),ELMASS(NEE,1),WORK(1),ELRESF(NED,1)
C
      IF (IMASS.EQ.0) THEN
C
C....... CONSISTENT MASS
C
         DO 300 L=1,NINT
         TEMP = CONSTM*W(L)*DET(L)
C
         DO 200 I=1,NED
         ACC = ROWDOT(SHG(NROWSH,1,L),AL(I,1),NROWSH,NED,NEN)
C
         DO 100 J=1,NEN
         ELRESF(I,J) = ELRESF(I,J) + TEMP*ACC*SHG(NROWSH,J,L)
  100    CONTINUE
C
  200    CONTINUE
C
  300    CONTINUE
C
      ENDIF
C
      IF (IMASS.EQ.1) THEN
C
C....... LUMPED MASS
C
         CALL CLEAR(ELMASS,NEE)
         CALL CONTM(SHG,XL,W,DET,ELMASS,WORK,CONSTM,IMASS,NINT,
     &              NROWSH,NESD,NEN,NED,NEE,.TRUE.)
C
         DO 500 J=1,NEN
         K = (J - 1)*NED
C
         DO 400 I=1,NED
         ELRESF(I,J) = ELRESF(I,J) + AL(I,J)*ELMASS(K + I,1)
  400    CONTINUE
C
  500    CONTINUE
C
      ENDIF
C
      RETURN
      END
C**********************************************************************
      SUBROUTINE COORD(X,NSD,NUMNP,IPRTIN)
C
C.... PROGRAM TO READ, GENERATE AND WRITE COORDINATE DATA
C
C        X(NSD,NUMNP) = COORDINATE ARRAY
C
      IMPLICIT DOUBLE PRECISION (A-H,O-Z)
C
C.... DEACTIVATE ABOVE CARD(S) FOR SINGLE-PRECISION OPERATION
C
      DIMENSION X(NSD,1)
      COMMON /IOUNIT/ IIN,IOUT,IRSIN,IRSOUT
C
      CALL GENFL(X,NSD)
C
      IF (IPRTIN.EQ.1) RETURN
```

```
C                                                                    COORD   17
      DO 100 N=1,NUMNP                                               COORD   18
      IF (MOD(N,50).EQ.1) WRITE(IOUT,1000) (I,I=1,NSD)               COORD   19
      WRITE(IOUT,2000) N,(X(I,N),I=1,NSD)                            COORD   20
  100 CONTINUE                                                       COORD   21
C                                                                    COORD   22
      RETURN                                                         COORD   23
C                                                                    COORD   24
 1000 FORMAT('1',' N O D A L   C O O R D I N A T E   D A T A '///5X, COORD   25
     &' NODE NO.',3(13X,' X',I1,'   :')//)                           COORD   26
 2000 FORMAT(6X,I5,10X,3(1PE15.8,2X))                                COORD   27
      END                                                            COORD   28
C***********************************************************************
      SUBROUTINE CORRCT(ID,D,V,DPRED,VPRED,NDOF,NUMNP)               CORRCT   1
C                                                                    CORRCT   2
C.... PROGRAM TO PERFORM CORRECTOR UPDATE OF DISPLACEMENTS           CORRCT   3
C        AND VELOCITIES                                              CORRCT   4
C                                                                    CORRCT   5
      IMPLICIT DOUBLE PRECISION (A-H,O-Z)                            CORRCT   6
C                                                                    CORRCT   7
C.... DEACTIVATE ABOVE CARD(S) FOR SINGLE-PRECISION OPERATION        CORRCT   8
C                                                                    CORRCT   9
      DIMENSION ID(NDOF,1),D(NDOF,1),V(NDOF,1),DPRED(NDOF,1),        CORRCT  10
     &          VPRED(NDOF,1)                                        CORRCT  11
C                                                                    CORRCT  12
      COMMON /COEFFS/ COEFF1,COEFF2,COEFF3,COEFF4,COEFF5,COEFF6,     CORRCT  13
     &                COEFF7,COEFF8,ALPHA1,BETA1 ,GAMMA1,DT          CORRCT  14
      COMMON /CONSTS/ ZERO,PT1667,PT25,PT5,ONE,TWO,THREE,FOUR,FIVE   CORRCT  15
C                                                                    CORRCT  16
      TEMP = ONE/COEFF1                                              CORRCT  17
C                                                                    CORRCT  18
      DO 200 I=1,NDOF                                                CORRCT  19
C                                                                    CORRCT  20
      DO 100 J=1,NUMNP                                               CORRCT  21
      DN = D(I,J)                                                    CORRCT  22
      VN = V(I,J)                                                    CORRCT  23
      D(I,J) = (DPRED(I,J) - DN)*TEMP + DN                           CORRCT  24
      V(I,J) = (VPRED(I,J) - VN)*TEMP + VN                           CORRCT  25
  100 CONTINUE                                                       CORRCT  26
C                                                                    CORRCT  27
  200 CONTINUE                                                       CORRCT  28
C                                                                    CORRCT  29
      RETURN                                                         CORRCT  30
      END                                                            CORRCT  31
C***********************************************************************
      SUBROUTINE DCTNRY(NAME,NDIM1,NDIM2,NDIM3,MPOINT,IPR,MLAST)     DCTNRY   1
C                                                                    DCTNRY   2
C.... PROGRAM TO STORE POINTER INFORMATION IN DICTIONARY            DCTNRY   3
C                                                                    DCTNRY   4
      DIMENSION NAME(2)                                              DCTNRY   5
      COMMON IA(1)                                                   DCTNRY   6
C                                                                    DCTNRY   7
      MLAST = MLAST - 7                                              DCTNRY   8
      IA(MLAST+1) = NAME(1)                                          DCTNRY   9
      IA(MLAST+2) = NAME(2)                                          DCTNRY  10
      IA(MLAST+3) = MPOINT                                           DCTNRY  11
      IA(MLAST+4) = NDIM1                                            DCTNRY  12
      IA(MLAST+5) = NDIM2                                            DCTNRY  13
      IA(MLAST+6) = NDIM3                                            DCTNRY  14
      IA(MLAST+7) = IPR                                              DCTNRY  15
C                                                                    DCTNRY  16
      RETURN                                                         DCTNRY  17
      END                                                            DCTNRY  18
C***********************************************************************
      SUBROUTINE DHIST(IDHIST,NDOUT)                                 DHIST    1
C                                                                    DHIST    2
C.... PROGRAM TO READ, WRITE AND STORE NODAL TIME-HISTORY INPUT DATA DHIST    3
C                                                                    DHIST    4
      DIMENSION IDHIST(3,1)                                          DHIST    5
      COMMON /IOUNIT/ IIN,IOUT,IRSIN,IRSOUT                          DHIST    6
      COMMON /LABELS/ LABELD(3),LABEL1(16),LABEL2(3)                 DHIST    7
C                                                                    DHIST    8
      DO 100 N=1,NDOUT                                               DHIST    9
      READ(IIN,2000) NODE,IDOF,IDVA                                  DHIST   10
      IF (MOD(N,50).EQ.1) WRITE(IOUT,1000) NDOUT                     DHIST   11
      WRITE(IOUT,3000) NODE,IDOF,LABELD(IDVA)                        DHIST   12
      IDHIST(1,N) = NODE                                             DHIST   13
      IDHIST(2,N) = IDOF                                             DHIST   14
      IDHIST(3,N) = IDVA                                             DHIST   15
  100 CONTINUE                                                       DHIST   16
C                                                                    DHIST   17
      RETURN                                                         DHIST   18
```

```
C
 1000 FORMAT('1',' N O D A L   T I M E - H I S T O R Y  ',
     &' I N F O R M A T I O N'//5X,
     &' NUMBER OF NODAL TIME HISTORIES  . . . . . . (NDOUT ) = ',I5///
     &5X,'    NODE       DOF       KINEMATIC ',/
     &5X,'   NUMBER    NUMBER        TYPE     ',/)
 2000 FORMAT(3I5)
 3000 FORMAT(7X,I5,5X,I5,7X,A4)
C
      END
C**********************************************************************
      SUBROUTINE DIAG(IDIAG,NEQ,N)
C
C.... PROGRAM TO COMPUTE DIAGONAL ADDRESSES OF LEFT-HAND-SIDE MATRIX
C
      DIMENSION IDIAG(1)
C
      N = 1
      IDIAG(1) = 1
      IF (NEQ.EQ.1) RETURN
C
      DO 100 I=2,NEQ
      IDIAG(I) = IDIAG(I) + IDIAG(I-1) + 1
  100 CONTINUE
      N = IDIAG(NEQ)
C
      RETURN
      END
C**********************************************************************
      SUBROUTINE DYNPTS
C
C.... PROGRAM TO SET MEMORY POINTERS FOR DYNAMIC ANALYSIS DATA ARRAYS
C
      LOGICAL LDYN
      COMMON /BPOINT/ MFIRST,MLAST,MTOT,IPREC
      COMMON /DPOINT/ MPSTEP,MPDPRT,MPSPRT,MPHPLT,MPITER,MPALPH,MPBETA,
     &                MPGAMM,MPDT  ,MPIDHS,MPDOUT,MPVPRD,MPDPRD,MPA,MPV
      COMMON /INFO  / IEXEC,IACODE,LDYN,IREADR,IWRITR,IPRTIN,IRANK,
     &                NUMSEQ,NDOUT,NSD,NUMNP,NDOF,NLVECT,NLTFTN,NPTSLF,
     &                NUMEG
      COMMON A(1)
C
      MPVPRD = MPOINT('VPRED   ',NDOF   ,NUMNP ,0,IPREC)
      MPDPRD = MPOINT('DPRED   ',NDOF   ,NUMNP ,0,IPREC)
      MPA    = MPOINT('A       ',NDOF   ,NUMNP ,0,IPREC)
      MPV    = MPOINT('V       ',NDOF   ,NUMNP ,0,IPREC)
C
      RETURN
      END
C**********************************************************************
      SUBROUTINE ECHO
C
C.... PROGRAM TO ECHO INPUT DATA
C
      DIMENSION IA(20)
      COMMON /IOUNIT/ IIN,IOUT,IRSIN,IRSOUT
C
      READ(IIN,1000) IECHO
      IF (IECHO.EQ.0) RETURN
C
      WRITE(IOUT,2000) IECHO
      BACKSPACE IIN
C
      DO 100 I=1,100000
      READ(IIN,3000,END=200) IA
      IF (MOD(I,50).EQ.1) WRITE(IOUT,4000)
      WRITE(IOUT,5000) IA
  100 CONTINUE
C
  200 CONTINUE
      REWIND IIN
      READ(IIN,1000) IECHO
C
      RETURN
C
 1000 FORMAT(16I5)
 2000 FORMAT('1',' I N P U T   D A T A   F I L E              ',  //5X,
     &' ECHO PRINT CODE . . . . . . . . . . . . . (IECHO ) = ',I5//5X,
     &'    EQ. 0, NO ECHO OF INPUT DATA                        ',   /5X,
     &'    EQ. 1, ECHO INPUT DATA                              ',   ///)
 3000 FORMAT(20A4)
 4000 FORMAT(' ',8('123456789*'),//)
 5000 FORMAT(' ',20A4)
      END
C**********************************************************************
      SUBROUTINE ELEMNT(TASK,NGRP)
C
C.... PROGRAM TO CALCULATE ELEMENT TASK NUMBER
```

```
C                                                                      ELEMNT    4
      CHARACTER*8 TASK,ELTASK(6)                                       ELEMNT    5
      LOGICAL LDYN                                                     ELEMNT    6
      DIMENSION NGRP(1)                                                ELEMNT    7
      COMMON /INFO  / IEXEC,IACODE,LDYN,IREADR,IWRITR,IPRTIN,IRANK,     ELEMNT    8
     &                NUMSEQ,NDOUT,NSD,NUMNP,NDOF,NLVECT,NLTFTN,NPTSLF, ELEMNT    9
     &                NUMEG                                            ELEMNT   10
      COMMON IA(1)                                                     ELEMNT   11
      DATA NTASK,       ELTASK                                         ELEMNT   12
     &     /   6,'INPUT   ',                                           ELEMNT   13
     &            'FORM_LHS',                                          ELEMNT   14
     &            'FORM_RHS',                                          ELEMNT   15
     &            'STR_PRNT',                                          ELEMNT   16
     &            'STR_STOR',                                          ELEMNT   17
     &            'STR_PLOT'/                                          ELEMNT   18
C                                                                      ELEMNT   19
      DO 100 I=1,NTASK                                                 ELEMNT   20
      IF (TASK.EQ.ELTASK(I)) ITASK = I                                 ELEMNT   21
  100 CONTINUE                                                         ELEMNT   22
C                                                                      ELEMNT   23
      DO 200 NEG=1,NUMEG                                               ELEMNT   24
C                                                                      ELEMNT   25
      IF (ITASK.EQ.1) THEN                                             ELEMNT   26
         MPNPAR = MPOINT('NPAR    ',16    ,0,0,1)                      ELEMNT   27
         NGRP(NEG) = MPNPAR                                            ELEMNT   28
         CALL ELCARD(IA(MPNPAR),NEG)                                   ELEMNT   29
      ELSE                                                             ELEMNT   30
         MPNPAR = NGRP(NEG)                                            ELEMNT   31
      ENDIF                                                            ELEMNT   32
C                                                                      ELEMNT   33
      NTYPE  = IA(MPNPAR)                                              ELEMNT   34
      CALL ELMLIB(NTYPE,MPNPAR,ITASK,NEG)                              ELEMNT   35
  200 CONTINUE                                                         ELEMNT   36
C                                                                      ELEMNT   37
      RETURN                                                           ELEMNT   38
      END                                                              ELEMNT   39
C*******************************************************************************
      SUBROUTINE ELCARD(NPAR,NEG)                                      ELCARD    1
C                                                                      ELCARD    2
C.... PROGRAM TO READ ELEMENT GROUP CONTROL CARD                       ELCARD    3
C                                                                      ELCARD    4
      DIMENSION NPAR(1)                                                ELCARD    5
      COMMON /IOUNIT/ IIN,IOUT,IRSIN,IRSOUT                            ELCARD    6
C                                                                      ELCARD    7
      READ(IIN,1000) (NPAR(I),I=1,16)                                  ELCARD    8
      WRITE(IOUT,2000) NEG                                             ELCARD    9
C                                                                      ELCARD   10
      RETURN                                                           ELCARD   11
C                                                                      ELCARD   12
 1000 FORMAT(16I5)                                                     ELCARD   13
 2000 FORMAT('1',' E L E M E N T   G R O U P   D A T A       ',//5X,   ELCARD   14
     &' ELEMENT GROUP NUMBER . . . . . . . . . . (NEG   ) = ',I5///)   ELCARD   15
C                                                                      ELCARD   16
      END                                                             ELCARD   17
C*******************************************************************************
      SUBROUTINE ELMLIB(NTYPE,MPNPAR,ITASK,NEG)                        ELMLIB    1
C                                                                      ELMLIB    2
C.... PROGRAM TO CALL ELEMENT ROUTINES                                 ELMLIB    3
C                                                                      ELMLIB    4
      COMMON A(1)                                                      ELMLIB    5
C                                                                      ELMLIB    6
      GO TO (100,200),NTYPE                                            ELMLIB    7
C                                                                      ELMLIB    8
  100 CONTINUE                                                         ELMLIB    9
      CALL QUADC(ITASK,A(MPNPAR),A(MPNPAR+16),NEG)                     ELMLIB   10
      RETURN                                                           ELMLIB   11
C                                                                      ELMLIB   12
  200 CONTINUE                                                         ELMLIB   13
      CALL TRUSS(ITASK,A(MPNPAR),A(MPNPAR+16),NEG)                     ELMLIB   14
      RETURN                                                           ELMLIB   15
C                                                                      ELMLIB   16
C.... ADD ADDITIONAL ELEMENTS FOR FUN AND VALUABLE PRIZES              ELMLIB   17
C                                                                      ELMLIB   18
      END                                                             ELMLIB   19
C*******************************************************************************
      SUBROUTINE EQSET(NEQ,NALHS)                                      EQSET     1
C                                                                      EQSET     2
C.... PROGRAM TO ALLOCATE STORAGE FOR GLOBAL EQUATION SYSTEM           EQSET     3
C                                                                      EQSET     4
      CHARACTER*4 TITLE                                                EQSET     5
      COMMON /BPOINT/ MFIRST,MLAST,MTOT,IPREC                          EQSET     6
      COMMON /IOUNIT/ IIN,IOUT,IRSIN,IRSOUT                            EQSET     7
      COMMON /SPOINT/ MPD,MPX,MPID,MPF,MPG,MPG1,MPDIAG,MPNGRP,         EQSET     8
     &                MPALHS,MPBRHS                                    EQSET     9
      COMMON /TITLEC/ TITLE(20)                                        EQSET    10
      COMMON A(1)                                                      EQSET    11
C                                                                      EQSET    12
C.... DETERMINE ADDRESSES OF DIAGONALS IN LEFT-HAND-SIDE MATRIX        EQSET    13
```

```
C
      CALL DIAG(A(MPDIAG),NEQ,NALHS)
      MPALHS = MPOINT('ALHS    ',NALHS,0,0,IPREC)
      MPBRHS = MPOINT('BRHS    ',NEQ  ,0,0,IPREC)
      MEANBW = NALHS/NEQ
      NWORDS = MTOT - MLAST + MFIRST - 1
C
C.... WRITE EQUATION SYSTEM DATA
C
      WRITE(IOUT,1000) TITLE,NEQ,NALHS,MEANBW,NWORDS
C
      RETURN
 1000 FORMAT('1',20A4///
     &' E Q U A T I O N    S Y S T E M    D A T A              ',  //5X,
     &' NUMBER OF EQUATIONS . . . . . . . . . . . . (NEQ    ) = ',I8//5X,
     &' NUMBER OF TERMS IN LEFT-HAND-SIDE MATRIX  . (NALHS  ) = ',I8//5X,
     &' MEAN HALF BANDWIDTH . . . . . . . . . . . . (MEANBW) = ',I8//5X,
     &' TOTAL LENGTH OF BLANK COMMON REQUIRED . . . (NWORDS) = ',I8    )
C
      END
C**********************************************************************
      SUBROUTINE FACTOR(A,IDIAG,NEQ)
C
C.... PROGRAM TO PERFORM CROUT FACTORIZATION: A = U(TRANSPOSE) * D * U
C
C        A(I):  COEFFICIENT MATRIX STORED IN COMPACTED COLUMN FORM;
C               AFTER FACTORIZATION CONTAINS D AND U
C
      IMPLICIT DOUBLE PRECISION (A-H,O-Z)
C
C.... DEACTIVATE ABOVE CARD(S) FOR SINGLE-PRECISION OPERATION
C
      DIMENSION A(1),IDIAG(1)
      COMMON /CONSTS/ ZERO,PT1667,PT25,PT5,ONE,TWO,THREE,FOUR,FIVE
C
      JJ = 0
C
      DO 300 J=1,NEQ
C
      JJLAST = JJ
      JJ     = IDIAG(J)
      JCOLHT = JJ - JJLAST
C
      IF (JCOLHT.GT.2) THEN
C
C....... FOR COLUMN J AND I.LE.J-1, REPLACE A(I,J) WITH D(I,I)*U(I,J)
C
         ISTART = J - JCOLHT + 2
         JM1    = J - 1
         IJ     = JJLAST + 2
         II     = IDIAG(ISTART-1)
C
         DO 100 I=ISTART,JM1
C
         IILAST = II
         II     = IDIAG(I)
         ICOLHT = II - IILAST
         JLNGTH = I - ISTART + 1
         LENGTH = MIN0(ICOLHT-1,JLNGTH)
         IF (LENGTH.GT.0)
     &      A(IJ) = A(IJ) - COLDOT(A(II-LENGTH),A(IJ-LENGTH),LENGTH)
         IJ = IJ + 1
  100    CONTINUE
C
      ENDIF
C
      IF (JCOLHT.GE.2) THEN
C
C....... FOR COLUMN J AND I.LE.J-1, REPLACE A(I,J) WITH U(I,J);
C            REPLACE A(J,J) WITH D(J,J).
C
         JTEMP = J - JJ
C
         DO 200 IJ=JJLAST+1,JJ-1
C
         II = IDIAG(JTEMP + IJ)
C
C....... WARNING: THE FOLLOWING CALCULATIONS ARE SKIPPED
C                 IF A(II) EQUALS ZERO
C
         IF (A(II).NE.ZERO) THEN
            TEMP  = A(IJ)
            A(IJ) = TEMP/A(II)
            A(JJ) = A(JJ) - TEMP*A(IJ)
         ENDIF
  200    CONTINUE
C
      ENDIF
```

```
C                                                                    FACTOR  68
   300 CONTINUE                                                      FACTOR  69
C                                                                    FACTOR  70
       RETURN                                                        FACTOR  71
       END                                                           FACTOR  72
C*************************************************************************
       SUBROUTINE FORMLM (ID,IEN,LM,NDOF,NED,NEN,NUMEL)              FORMLM   1
C                                                                    FORMLM   2
C.... PROGRAM TO FORM LM ARRAY                                       FORMLM   3
C                                                                    FORMLM   4
       DIMENSION ID(NDOF,1),IEN(NEN,1),LM(NED,NEN,1)                 FORMLM   5
C                                                                    FORMLM   6
       DO 300 K=1,NUMEL                                              FORMLM   7
C                                                                    FORMLM   8
       DO 200 J=1,NEN                                                FORMLM   9
       NODE=IEN(J,K)                                                 FORMLM  10
C                                                                    FORMLM  11
       DO 100 I=1,NDOF                                               FORMLM  12
       LM(I,J,K) = ID(I,NODE)                                        FORMLM  13
   100 CONTINUE                                                      FORMLM  14
C                                                                    FORMLM  15
   200 CONTINUE                                                      FORMLM  16
C                                                                    FORMLM  17
   300 CONTINUE                                                      FORMLM  18
C                                                                    FORMLM  19
       RETURN                                                        FORMLM  20
       END                                                           FORMLM  21
C*************************************************************************
       SUBROUTINE GENEL(IEN,MAT,NEN)                                 GENEL    1
C                                                                    GENEL    2
C.... PROGRAM TO READ AND GENERATE ELEMENT NODE AND MATERIAL NUMBERS GENEL    3
C                                                                    GENEL    4
C          IEN(NEN,NUMEL) = ELEMENT NODE NUMBERS                     GENEL    5
C          MAT(NUMEL)     = ELEMENT MATERIAL NUMBERS                 GENEL    6
C          NEN            = NUMBER OF ELEMENT NODES (LE.27)          GENEL    7
C          N              = ELEMENT NUMBER                           GENEL    8
C          NG             = GENERATION PARAMETER                     GENEL    9
C          NEL(I)         = NUMBER OF ELEMENTS IN DIRECTION I        GENEL   10
C          INCEL(I)       = ELEMENT NUMBER INCREMENT FOR DIRECTION I GENEL   11
C          INC(I)         = NODE NUMBER INCREMENT FOR DIRECTION I    GENEL   12
C                                                                    GENEL   13
       DIMENSION IEN(NEN,1),MAT(1),ITEMP(27)                         GENEL   14
       COMMON /IOUNIT/ IIN,IOUT,IRSIN,IRSOUT                         GENEL   15
       COMMON /GENELC/ N,NEL(3),INCEL(3),INC(3)                      GENEL   16
C                                                                    GENEL   17
   100 CONTINUE                                                      GENEL   18
       READ(IIN,1000) N,M,(ITEMP(I),I=1,NEN),NG                      GENEL   19
       IF (N.EQ.0) RETURN                                            GENEL   20
       CALL IMOVE(IEN(1,N),ITEMP,NEN)                                GENEL   21
       MAT(N)=M                                                      GENEL   22
       IF (NG.NE.0) THEN                                             GENEL   23
C                                                                    GENEL   24
C...... GENERATE DATA                                                GENEL   25
C                                                                    GENEL   26
       READ(IIN,1000) (NEL(I),INCEL(I),INC(I),I=1,3)                 GENEL   27
       CALL GENEL1(IEN,MAT,NEN)                                      GENEL   28
       ENDIF                                                         GENEL   29
       GO TO 100                                                     GENEL   30
C                                                                    GENEL   31
  1000 FORMAT(16I5,10X,14I5)                                         GENEL   32
C                                                                    GENEL   33
       END                                                           GENEL   34
C*************************************************************************
       SUBROUTINE GENEL1(IEN,MAT,NEN)                                GENEL1   1
C                                                                    GENEL1   2
C.... PROGRAM TO GENERATE ELEMENT NODE AND MATERIAL NUMBERS          GENEL1   3
C                                                                    GENEL1   4
       DIMENSION IEN(NEN,1),MAT(1)                                   GENEL1   5
       COMMON /GENELC/ N,NEL(3),INCEL(3),INC(3)                      GENEL1   6
C                                                                    GENEL1   7
C.... SET DEFAULTS                                                   GENEL1   8
C                                                                    GENEL1   9
       CALL GENELD                                                   GENEL1  10
C                                                                    GENEL1  11
C.... GENERATION ALGORITHM                                           GENEL1  12
C                                                                    GENEL1  13
       IE = N                                                        GENEL1  14
       JE = N                                                        GENEL1  15
       KE = N                                                        GENEL1  16
C                                                                    GENEL1  17
       II = NEL(1)                                                   GENEL1  18
       JJ = NEL(2)                                                   GENEL1  19
       KK = NEL(3)                                                   GENEL1  20
C                                                                    GENEL1  21
       DO 300 K=1,KK                                                 GENEL1  22
C                                                                    GENEL1  23
       DO 200 J=1,JJ                                                 GENEL1  24
C                                                                    GENEL1  25
       DO 100 I=1,II                                                 GENEL1  26
```

```
C                                                                      GENEL1 27
      IF (I.NE.II) THEN                                                GENEL1 28
         LE = IE                                                       GENEL1 29
         IE = LE + INCEL(1)                                            GENEL1 30
         CALL GENELI(IEN(1,IE),IEN(1,LE),INC(1),NEN)                   GENEL1 31
         MAT(IE) = MAT(LE)                                             GENEL1 32
      ENDIF                                                            GENEL1 33
  100 CONTINUE                                                         GENEL1 34
C                                                                      GENEL1 35
      IF (J.NE.JJ) THEN                                                GENEL1 36
         LE = JE                                                       GENEL1 37
         JE = LE + INCEL(2)                                            GENEL1 38
         CALL GENELI(IEN(1,JE),IEN(1,LE),INC(2),NEN)                   GENEL1 39
         MAT(JE) = MAT(LE)                                             GENEL1 40
         IE = JE                                                       GENEL1 41
      ENDIF                                                            GENEL1 42
  200 CONTINUE                                                         GENEL1 43
C                                                                      GENEL1 44
      IF (K.NE.KK) THEN                                                GENEL1 45
         LE = KE                                                       GENEL1 46
         KE = LE + INCEL(3)                                            GENEL1 47
         CALL GENELI(IEN(1,KE),IEN(1,LE),INC(3),NEN)                   GENEL1 48
         MAT(KE) = MAT(LE)                                             GENEL1 49
         IE = KE                                                       GENEL1 50
         JE = KE                                                       GENEL1 51
      ENDIF                                                            GENEL1 52
  300 CONTINUE                                                         GENEL1 53
C                                                                      GENEL1 54
      RETURN                                                           GENEL1 55
      END                                                              GENEL1 56
C***********************************************************************************
      SUBROUTINE GENELD                                                GENELD  1
C                                                                      GENELD  2
C.... PROGRAM TO SET DEFAULTS FOR ELEMENT NODE                         GENELD  3
C        AND MATERIAL NUMBER GENERATION                                GENELD  4
C                                                                      GENELD  5
      COMMON /GENELC/ N,NEL(3),INCEL(3),INC(3)                         GENELD  6
C                                                                      GENELD  7
      IF (NEL(1).EQ.0) NEL(1) = 1                                      GENELD  8
      IF (NEL(2).EQ.0) NEL(2) = 1                                      GENELD  9
      IF (NEL(3).EQ.0) NEL(3) = 1                                      GENELD 10
C                                                                      GENELD 11
      IF (INCEL(1).EQ.0) INCEL(1) = 1                                  GENELD 12
      IF (INCEL(2).EQ.0) INCEL(2) = NEL(1)                             GENELD 13
      IF (INCEL(3).EQ.0) INCEL(3) = NEL(1)*NEL(2)                      GENELD 14
C                                                                      GENELD 15
      IF (INC(1).EQ.0) INC(1) = 1                                      GENELD 16
      IF (INC(2).EQ.0) INC(2) = (1+NEL(1))*INC(1)                      GENELD 17
      IF (INC(3).EQ.0) INC(3) = (1+NEL(2))*INC(2)                      GENELD 18
C                                                                      GENELD 19
      RETURN                                                           GENELD 20
      END                                                              GENELD 21
C***********************************************************************************
      SUBROUTINE GENELI(IEN2,IEN1,INC,NEN)                             GENELI  1
C                                                                      GENELI  2
C.... PROGRAM TO INCREMENT ELEMENT NODE NUMBERS                        GENELI  3
C                                                                      GENELI  4
      DIMENSION IEN1(1),IEN2(1)                                        GENELI  5
C                                                                      GENELI  6
      DO 100 I=1,NEN                                                   GENELI  7
      IF (IEN1(I).EQ.0) THEN                                           GENELI  8
         IEN2(I) = 0                                                   GENELI  9
      ELSE                                                             GENELI 10
         IEN2(I) = IEN1(I) + INC                                       GENELI 11
      ENDIF                                                            GENELI 12
  100 CONTINUE                                                         GENELI 13
C                                                                      GENELI 14
      RETURN                                                           GENELI 15
      END                                                              GENELI 16
C***********************************************************************************
      SUBROUTINE GENFL(A,NRA)                                          GENFL   1
C                                                                      GENFL   2
C.... PROGRAM TO READ AND GENERATE FLOATING-POINT NODAL DATA           GENFL   3
C                                                                      GENFL   4
C        A      = INPUT ARRAY                                          GENFL   5
C        NRA    = NUMBER OF ROWS IN A (LE.6)                           GENFL   6
C        N      = NODE NUMBER                                          GENFL   7
C        NUMGP  = NUMBER OF GENERATION POINTS                          GENFL   8
C        NINC(I) = NUMBER OF INCREMENTS FOR DIRECTION I                GENFL   9
C        INC(I) = INCREMENT FOR DIRECTION I                            GENFL  10
C                                                                      GENFL  11
      IMPLICIT DOUBLE PRECISION (A-H,O-Z)                              GENFL  12
C                                                                      GENFL  13
C.... DEACTIVATE ABOVE CARD(S) FOR SINGLE-PRECISION OPERATION          GENFL  14
C                                                                      GENFL  15
      DIMENSION A(NRA,1)                                               GENFL  16
      COMMON /IOUNIT/ IIN,IOUT,IRSIN,IRSOUT                            GENFL  17
      COMMON /GENFLC/ TEMP(6,20),N,NUMGP,NINC(3),INC(3)                GENFL  18
```

```
C                                                               GENFL    19
   100 CONTINUE                                                 GENFL    20
       READ(IIN,1000) N,NUMGP,(TEMP(I,1),I=1,NRA)               GENFL    21
       IF (N.EQ.0) RETURN                                       GENFL    22
       CALL MOVE(A(1,N),TEMP,NRA)                               GENFL    23
       IF (NUMGP.NE.0) THEN                                     GENFL    24
          DO 200 J=2,NUMGP                                      GENFL    25
C                                                               GENFL    26
          READ(IIN,1000) M,MGEN,(TEMP(I,J),I=1,NRA)             GENFL    27
          IF (MGEN.NE.0) CALL MOVE(TEMP(1,J),A(1,M),NRA)        GENFL    28
C                                                               GENFL    29
   200    CONTINUE                                              GENFL    30
          READ(IIN,2000) (NINC(I),INC(I),I=1,3)                 GENFL    31
          CALL GENFL1(A,NRA)                                    GENFL    32
       ENDIF                                                    GENFL    33
       GO TO 100                                                GENFL    34
C                                                               GENFL    35
 1000 FORMAT(2I5,6F10.0)                                        GENFL    36
 2000 FORMAT(16I5)                                              GENFL    37
C                                                               GENFL    38
       END                                                      GENFL    39
C****************************************************************************
       SUBROUTINE GENFL1(A,NRA)                                 GENFL1    1
C                                                               GENFL1    2
C.... PROGRAM TO GENERATE FLOATING-POINT NODAL DATA             GENFL1    3
C          VIA ISOPARAMETRIC INTERPOLATION                      GENFL1    4
C                                                               GENFL1    5
C          IOPT = 1, GENERATION ALONG A LINE                    GENFL1    6
C               = 2, GENERATION OVER A SURFACE                  GENFL1    7
C               = 3, GENERATION WITHIN A VOLUME                 GENFL1    8
C                                                               GENFL1    9
       IMPLICIT DOUBLE PRECISION (A-H,O-Z)                      GENFL1   10
C                                                               GENFL1   11
C.... DEACTIVATE ABOVE CARD(S) FOR SINGLE-PRECISION OPERATION   GENFL1   12
C                                                               GENFL1   13
       DIMENSION A(NRA,1),SH(20)                                GENFL1   14
       COMMON /GENFLC/ TEMP(6,20),N,NUMGP,NINC(3),INC(3)        GENFL1   15
       COMMON /CONSTS/ ZERO,PT1667,PT25,PT5,ONE,TWO,THREE,FOUR,FIVE GENFL1 16
C                                                               GENFL1   17
       IOPT = 3                                                 GENFL1   18
       IF (NINC(3).EQ.0) IOPT = 2                               GENFL1   19
       IF (NINC(2).EQ.0) IOPT = 1                               GENFL1   20
C                                                               GENFL1   21
       DR = ZERO                                                GENFL1   22
       DS = ZERO                                                GENFL1   23
       DT = ZERO                                                GENFL1   24
C                                                               GENFL1   25
       IF (NINC(1).NE.0) DR = TWO/NINC(1)                       GENFL1   26
       IF (NINC(2).NE.0) DS = TWO/NINC(2)                       GENFL1   27
       IF (NINC(3).NE.0) DT = TWO/NINC(3)                       GENFL1   28
C                                                               GENFL1   29
       II = NINC(1)+1                                           GENFL1   30
       JJ = NINC(2)+1                                           GENFL1   31
       KK = NINC(3)+1                                           GENFL1   32
C                                                               GENFL1   33
       NI = N                                                   GENFL1   34
       NJ = N                                                   GENFL1   35
       NK = N                                                   GENFL1   36
C                                                               GENFL1   37
       T = -ONE                                                 GENFL1   38
       DO 300 K=1,KK                                            GENFL1   39
C                                                               GENFL1   40
       S = -ONE                                                 GENFL1   41
       DO 200 J=1,JJ                                            GENFL1   42
C                                                               GENFL1   43
       R = -ONE                                                 GENFL1   44
       DO 100 I=1,II                                            GENFL1   45
C                                                               GENFL1   46
       CALL GENSH(R,S,T,SH,NUMGP,IOPT)                          GENFL1   47
       CALL MULTAB(TEMP,SH,A(1,NI),6,20,NRA,NUMGP,NRA,1,1)      GENFL1   48
       NI = NI + INC(1)                                         GENFL1   49
       R = R + DR                                               GENFL1   50
   100 CONTINUE                                                 GENFL1   51
C                                                               GENFL1   52
       NJ = NJ + INC(2)                                         GENFL1   53
       NI = NJ                                                  GENFL1   54
       S = S + DS                                               GENFL1   55
   200 CONTINUE                                                 GENFL1   56
C                                                               GENFL1   57
       NK = NK + INC(3)                                         GENFL1   58
       NI = NK                                                  GENFL1   59
       T = T + DT                                               GENFL1   60
   300 CONTINUE                                                 GENFL1   61
C                                                               GENFL1   62
       RETURN                                                   GENFL1   63
       END                                                      GENFL1   64
```

```
C************************************************************************
      SUBROUTINE GENSH(R,S,T,SH,NUMGP,IOPT)                          GENSH    1
C                                                                    GENSH    2
C.... PROGRAM TO CALL SHAPE FUNCTION ROUTINES                        GENSH    3
C          FOR ISOPARAMETRIC GENERATION                              GENSH    4
C                                                                    GENSH    5
      IMPLICIT DOUBLE PRECISION (A-H,O-Z)                            GENSH    6
C                                                                    GENSH    7
C.... MODIFY ABOVE CARD FOR SINGLE-PRECISION OPERATION               GENSH    8
C                                                                    GENSH    9
      DIMENSION SH(1)                                                GENSH   10
C                                                                    GENSH   11
      GO TO (100,200,300),IOPT                                       GENSH   12
C                                                                    GENSH   13
  100 CALL GENSH1(R,SH,NUMGP)                                        GENSH   14
      RETURN                                                         GENSH   15
C                                                                    GENSH   16
  200 CALL GENSH2(R,S,SH,NUMGP)                                      GENSH   17
      RETURN                                                         GENSH   18
C                                                                    GENSH   19
  300 CALL GENSH3(R,S,T,SH,NUMGP)                                    GENSH   20
      RETURN                                                         GENSH   21
C                                                                    GENSH   22
      END                                                            GENSH   23
C************************************************************************
      SUBROUTINE GENSH1(R,SH,N)                                      GENSH1   1
C                                                                    GENSH1   2
C.... PROGRAM TO COMPUTE 1D SHAPE FUNCTIONS                          GENSH1   3
C          FOR ISOPARAMETRIC GENERATION                              GENSH1   4
C                                                                    GENSH1   5
      IMPLICIT DOUBLE PRECISION (A-H,O-Z)                            GENSH1   6
C                                                                    GENSH1   7
C.... MODIFY ABOVE CARD(S) FOR SINGLE-PRECISION OPERATION            GENSH1   8
C                                                                    GENSH1   9
      DIMENSION SH(1)                                                GENSH1  10
      COMMON /CONSTS/ ZERO,PT1667,PT25,PT5,ONE,TWO,THREE,FOUR,FIVE   GENSH1  11
C                                                                    GENSH1  12
      SH(2) = PT5*R                                                  GENSH1  13
      SH(1) = PT5 - SH(2)                                            GENSH1  14
      SH(2) = PT5 + SH(2)                                            GENSH1  15
      IF (N.EQ.3) THEN                                               GENSH1  16
         SH(3) = ONE - R*R                                           GENSH1  17
         SH(1) = SH(1) - PT5*SH(3)                                   GENSH1  18
         SH(2) = SH(2) - PT5*SH(3)                                   GENSH1  19
      ENDIF                                                          GENSH1  20
C                                                                    GENSH1  21
      RETURN                                                         GENSH1  22
      END                                                            GENSH1  23
C************************************************************************
      SUBROUTINE GENSH2(R,S,SH,N)                                    GENSH2   1
C                                                                    GENSH2   2
C.... PROGRAM TO COMPUTE 2D SHAPE FUNCTIONS                          GENSH2   3
C          FOR ISOPARAMETRIC GENERATION                              GENSH2   4
C                                                                    GENSH2   5
      IMPLICIT DOUBLE PRECISION (A-H,O-Z)                            GENSH2   6
C                                                                    GENSH2   7
C.... MODIFY ABOVE CARD FOR SINGLE-PRECISION OPERATION               GENSH2   8
C                                                                    GENSH2   9
      DIMENSION SH(1)                                                GENSH2  10
      COMMON /CONSTS/ ZERO,PT1667,PT25,PT5,ONE,TWO,THREE,FOUR,FIVE   GENSH2  11
C                                                                    GENSH2  12
      R2 = PT5*R                                                     GENSH2  13
      R1 = PT5 - R2                                                  GENSH2  14
      R2 = PT5 + R2                                                  GENSH2  15
      S2 = PT5*S                                                     GENSH2  16
      S1 = PT5 - S2                                                  GENSH2  17
      S2 = PT5 + S2                                                  GENSH2  18
      SH(1) = R1*S1                                                  GENSH2  19
      SH(2) = R2*S1                                                  GENSH2  20
      SH(3) = R2*S2                                                  GENSH2  21
      SH(4) = R1*S2                                                  GENSH2  22
      IF (N.EQ.4) RETURN                                             GENSH2  23
C                                                                    GENSH2  24
      R3 = ONE - R*R                                                 GENSH2  25
      S3 = ONE - S*S                                                 GENSH2  26
      SH(5) = R3*S1                                                  GENSH2  27
      SH(6) = S3*R2                                                  GENSH2  28
      SH(7) = R3*S2                                                  GENSH2  29
      SH(8) = S3*R1                                                  GENSH2  30
      SH(1) = SH(1) - PT5*(SH(5) + SH(8))                            GENSH2  31
      SH(2) = SH(2) - PT5*(SH(6) + SH(5))                            GENSH2  32
      SH(3) = SH(3) - PT5*(SH(7) + SH(6))                            GENSH2  33
      SH(4) = SH(4) - PT5*(SH(8) + SH(7))                            GENSH2  34
C                                                                    GENSH2  35
      RETURN                                                         GENSH2  36
      END                                                            GENSH2  37
```

```
C*************************************************************************
      SUBROUTINE GENSH3(R,S,T,SH,N)                          GENSH3    1
C                                                            GENSH3    2
C.... PROGRAM TO COMPUTE 3D SHAPE FUNCTIONS                  GENSH3    3
C         FOR ISOPARAMETRIC GENERATION                       GENSH3    4
C                                                            GENSH3    5
      IMPLICIT DOUBLE PRECISION (A-H,O-Z)                    GENSH3    6
C                                                            GENSH3    7
C.... MODIFY ABOVE CARD FOR SINGLE-PRECISION OPERATION       GENSH3    8
C                                                            GENSH3    9
      DIMENSION SH(1)                                        GENSH3   10
      COMMON /CONSTS/ ZERO,PT1667,PT25,PT5,ONE,TWO,THREE,FOUR,FIVE  GENSH3  11
C                                                            GENSH3   12
      R2 = PT5*R                                             GENSH3   13
      R1 = PT5 - R2                                          GENSH3   14
      R2 = PT5 + R2                                          GENSH3   15
      S2 = PT5*S                                             GENSH3   16
      S1 = PT5 - S2                                          GENSH3   17
      S2 = PT5 + S2                                          GENSH3   18
      T2 = PT5*T                                             GENSH3   19
      T1 = PT5 - T2                                          GENSH3   20
      T2 = PT5 + T2                                          GENSH3   21
C                                                            GENSH3   22
      RS1 = R1*S1                                            GENSH3   23
      RS2 = R2*S1                                            GENSH3   24
      RS3 = R2*S2                                            GENSH3   25
      RS4 = R1*S2                                            GENSH3   26
      SH(1) = RS1*T1                                         GENSH3   27
      SH(2) = RS2*T1                                         GENSH3   28
      SH(3) = RS3*T1                                         GENSH3   29
      SH(4) = RS4*T1                                         GENSH3   30
      SH(5) = RS1*T2                                         GENSH3   31
      SH(6) = RS2*T2                                         GENSH3   32
      SH(7) = RS3*T2                                         GENSH3   33
      SH(8) = RS4*T2                                         GENSH3   34
      IF (N.EQ.8) RETURN                                     GENSH3   35
C                                                            GENSH3   36
      R3 = ONE - R*R                                         GENSH3   37
      S3 = ONE - S*S                                         GENSH3   38
      T3 = ONE - T*T                                         GENSH3   39
      SH(17) = T3*RS1                                        GENSH3   40
      SH(18) = T3*RS2                                        GENSH3   41
      SH(19) = T3*RS3                                        GENSH3   42
      SH(20) = T3*RS4                                        GENSH3   43
      RS1 = R3*S1                                            GENSH3   44
      RS2 = S3*R2                                            GENSH3   45
      RS3 = R3*S2                                            GENSH3   46
      RS4 = S3*R1                                            GENSH3   47
      SH( 9) = RS1*T1                                        GENSH3   48
      SH(10) = RS2*T1                                        GENSH3   49
      SH(11) = RS3*T1                                        GENSH3   50
      SH(12) = RS4*T1                                        GENSH3   51
      SH(13) = RS1*T2                                        GENSH3   52
      SH(14) = RS2*T2                                        GENSH3   53
      SH(15) = RS3*T2                                        GENSH3   54
      SH(16) = RS4*T2                                        GENSH3   55
C                                                            GENSH3   56
      SH(1) = SH(1) - PT5*(SH( 9) + SH(12) + SH(17))         GENSH3   57
      SH(2) = SH(2) - PT5*(SH( 9) + SH(10) + SH(18))         GENSH3   58
      SH(3) = SH(3) - PT5*(SH(10) + SH(11) + SH(19))         GENSH3   59
      SH(4) = SH(4) - PT5*(SH(11) + SH(12) + SH(20))         GENSH3   60
      SH(5) = SH(5) - PT5*(SH(13) + SH(16) + SH(17))         GENSH3   61
      SH(6) = SH(6) - PT5*(SH(13) + SH(14) + SH(18))         GENSH3   62
      SH(7) = SH(7) - PT5*(SH(14) + SH(15) + SH(19))         GENSH3   63
      SH(8) = SH(8) - PT5*(SH(15) + SH(16) + SH(20))         GENSH3   64
C                                                            GENSH3   65
      RETURN                                                 GENSH3   66
      END                                                    GENSH3   67
C*************************************************************************
      SUBROUTINE HPLOT(IH,XT,NPLOTS,NROWS,IO)                HPLOT     1
C                                                            HPLOT     2
C.... PROGRAM TO PLOT OUTPUT HISTORIES                       HPLOT     3
C                                                            HPLOT     4
C        IH(NROWS,NPLOTS) = DOF/COMPONENT INFORMATION        HPLOT     5
C        XT(NPLOTS+1,NPLPTS) = OUTPUT HISTORY DATA           HPLOT     6
C        XT(        1,NPLPTS) = TIME RECORD                  HPLOT     7
C              NPLOTS = NUMBER OF HISTORIES TO BE PLOTTED     HPLOT     8
C              NPLPTS = NUMBER OF TIME POINTS AT WHICH        HPLOT     9
C                       DATA IS TO BE PLOTTED                HPLOT    10
C              NROWS = NUMBER OF ROWS IN IH ARRAY            HPLOT    11
C              IO = OUTPUT CODE                              HPLOT    12
C                    EQ.0, NODAL OUTPUT HISTORIES            HPLOT    13
C                    EQ.N.GT.0, ELEMENT OUTPUT HISTORIES     HPLOT    14
C                    (N = NTYPE IN CALLING ROUTINE)          HPLOT    15
C                                                            HPLOT    16
      DOUBLE PRECISION TIME                                  HPLOT    17
C                                                            HPLOT    18
C.... DEACTIVATE ABOVE CARD(S) FOR SINGLE-PRECISION OPERATION  HPLOT   19
```

```
C                                                                     HPLOT  20
      CHARACTER*1 IBLANK,ISTAR,LINE(53)                              HPLOT  21
      CHARACTER*4 TITLE,LABELD,LABEL1,LABEL2                         HPLOT  22
      DIMENSION IH(NROWS,1),XT(NPLOTS+1,1)                           HPLOT  23
      COMMON /HPLOTC/ NPLPTS,LOCPLT,TIME                             HPLOT  24
      COMMON /IOUNIT/ IIN,IOUT,IRSIN,IRSOUT                          HPLOT  25
      COMMON /LABELS/ LABELD(3),LABEL1(16),LABEL2(3)                 HPLOT  26
      COMMON /TITLEC/ TITLE(20)                                      HPLOT  27
C                                                                     HPLOT  28
      DATA IBLANK,ISTAR/' ','*'/,NCHAR/53/                           HPLOT  29
C                                                                     HPLOT  30
      DO 300 I=1,NPLOTS                                              HPLOT  31
C                                                                     HPLOT  32
      I1 = IH(1,I)                                                   HPLOT  33
      I2 = IH(2,I)                                                   HPLOT  34
      I3 = IH(3,I)                                                   HPLOT  35
C                                                                     HPLOT  36
      IF (IO.EQ.0) WRITE(IOUT,1000) TITLE,I1,I2,LABELD(I3)           HPLOT  37
      IF (IO.EQ.1) WRITE(IOUT,2000) TITLE,I1,I2,LABEL1(I3)           HPLOT  38
      IF (IO.EQ.2) WRITE(IOUT,2000) TITLE,I1,I2,LABEL2(I3)           HPLOT  39
C                                                                     HPLOT  40
C.... ADD IF/WRITE STATEMENTS AS ABOVE FOR ADDITIONAL ELEMENT TYPES  HPLOT  41
C                                                                     HPLOT  42
      CALL MINMAX(XT,XMAX,XMIN,NPLOTS+1,NPLPTS,I+1)                  HPLOT  43
      IF (XMAX.EQ.XMIN) THEN                                         HPLOT  44
C                                                                     HPLOT  45
         WRITE(IOUT,3000) XMAX                                       HPLOT  46
C                                                                     HPLOT  47
      ELSE                                                           HPLOT  48
C                                                                     HPLOT  49
         SCALE = XMAX - XMIN                                         HPLOT  50
         WRITE(IOUT,4000) XMIN,XMAX                                  HPLOT  51
C                                                                     HPLOT  52
         DO 200 J=1,NPLPTS                                           HPLOT  53
         T = XT(1,J)                                                 HPLOT  54
C                                                                     HPLOT  55
         DO 100 K = 1,NCHAR                                          HPLOT  56
         LINE(K) = IBLANK                                            HPLOT  57
  100    CONTINUE                                                    HPLOT  58
C                                                                     HPLOT  59
         XK = ((XT(I+1,J) - XMIN)/SCALE)*NCHAR                       HPLOT  60
         K  = XK + 1                                                 HPLOT  61
         IF (K.GT.NCHAR) K = NCHAR                                   HPLOT  62
         LINE(K) = ISTAR                                             HPLOT  63
         WRITE(IOUT,5000) T,XT(I+1,J),(LINE(K),K=1,NCHAR)            HPLOT  64
  200    CONTINUE                                                    HPLOT  65
      ENDIF                                                          HPLOT  66
C                                                                     HPLOT  67
  300 CONTINUE                                                       HPLOT  68
C                                                                     HPLOT  69
      RETURN                                                         HPLOT  70
C                                                                     HPLOT  71
 1000 FORMAT('1',20A4///                                            HPLOT  72
     &' NODE NUMBER = ',I5//                                        HPLOT  73
     &' DOF NUMBER  = ',I5,5X,'OUTPUT: ',A4//5X)                    HPLOT  74
 2000 FORMAT('1',20A4///                                            HPLOT  75
     &' ELEMENT NUMBER           = ',I5//                           HPLOT  76
     &' INTEGRATION POINT NUMBER = ',I5,5X,'OUTPUT: ',A4//5X)       HPLOT  77
 3000 FORMAT(' ',                                                   HPLOT  78
     &' VALUE IS CONSTANT ( = ',1PE11.4,' ), PLOT OMITTED')         HPLOT  79
 4000 FORMAT(' ',4X,'TIME',8X,'VALUE',6X,1PE11.4,31X,1PE11.4/       HPLOT  80
     &2X,10('-'),3X,10('-'),3X,53('-'))                             HPLOT  81
 5000 FORMAT(' ',1PE11.4,2X,1PE11.4,3X,53A1)                        HPLOT  82
      END                                                           HPLOT  83
C*********************************************************************
      SUBROUTINE ICLEAR(IA,M)                                       ICLEAR  1
C                                                                     ICLEAR  2
C.... PROGRAM TO CLEAR AN INTEGER ARRAY                             ICLEAR  3
C                                                                     ICLEAR  4
      DIMENSION IA(1)                                               ICLEAR  5
C                                                                     ICLEAR  6
      DO 100 I=1,M                                                  ICLEAR  7
      IA(I) = 0                                                     ICLEAR  8
  100 CONTINUE                                                      ICLEAR  9
C                                                                     ICLEAR 10
      RETURN                                                        ICLEAR 11
      END                                                           ICLEAR 12
C*********************************************************************
      SUBROUTINE IGEN(IA,M)                                         IGEN    1
C                                                                     IGEN    2
C.... PROGRAM TO READ AND GENERATE INTEGER NODAL DATA               IGEN    3
C                                                                     IGEN    4
C        IA = INPUT ARRAY                                           IGEN    5
C         M = NUMBER OF ROWS IN IA                                  IGEN    6
C         N = NODE NUMBER                                           IGEN    7
C        NE = END NODE IN GENERATION SEQUENCE                       IGEN    8
C        NG = GENERATION INCREMENT                                  IGEN    9
```

```
C                                                                        IGEN    10
      DIMENSION IA(M,1),IB(13)                                           IGEN    11
      COMMON /IOUNIT/ IIN,IOUT,IRSIN,IRSOUT                              IGEN    12
C                                                                        IGEN    13
  100 CONTINUE                                                           IGEN    14
      READ(IIN,1000) N,NE,NG,(IB(I),I=1,M)                               IGEN    15
      IF (N.EQ.0) RETURN                                                 IGEN    16
      IF (NG.EQ.0) THEN                                                  IGEN    17
         NE = N                                                          IGEN    18
         NG = 1                                                          IGEN    19
      ELSE                                                               IGEN    20
         NE = NE - MOD(NE-N,NG)                                          IGEN    21
      ENDIF                                                              IGEN    22
C                                                                        IGEN    23
      DO 200 I=N,NE,NG                                                   IGEN    24
      CALL IMOVE(IA(1,I),IB,M)                                           IGEN    25
  200 CONTINUE                                                           IGEN    26
C                                                                        IGEN    27
      GO TO 100                                                          IGEN    28
C                                                                        IGEN    29
 1000 FORMAT(16I5)                                                       IGEN    30
      END                                                                IGEN    31
C************************************************************************
      SUBROUTINE IMOVE(IA,IB,N)                                          IMOVE    1
C                                                                        IMOVE    2
C.... PROGRAM TO MOVE AN INTEGER ARRAY                                   IMOVE    3
C                                                                        IMOVE    4
      DIMENSION IA(1),IB(1)                                              IMOVE    5
C                                                                        IMOVE    6
      DO 100 I=1,N                                                       IMOVE    7
      IA(I)=IB(I)                                                        IMOVE    8
  100 CONTINUE                                                           IMOVE    9
C                                                                        IMOVE   10
      RETURN                                                             IMOVE   11
      END                                                                IMOVE   12
C************************************************************************
      SUBROUTINE INPUT(F,NDOF,NUMNP,J,NLVECT,IPRTIN,TIME)                INPUT    1
C                                                                        INPUT    2
C.... PROGRAM TO READ, GENERATE AND WRITE NODAL INPUT DATA               INPUT    3
C                                                                        INPUT    4
C         F(NDOF,NUMNP,NLVECT) = PRESCRIBED FORCES/KINEMATIC DATA (J=0)  INPUT    5
C                              = INITIAL DISPLACEMENTS (J=1)             INPUT    6
C                              = INITIAL VELOCITIES(J=2)                 INPUT    7
C                              = INITIAL ACCELERATIONS (J=3)             INPUT    8
C                                                                        INPUT    9
      IMPLICIT DOUBLE PRECISION (A-H,O-Z)                                INPUT   10
C                                                                        INPUT   11
C.... DEACTIVATE ABOVE CARD(S) FOR SINGLE-PRECISION OPERATION            INPUT   12
C                                                                        INPUT   13
      LOGICAL LZERO                                                      INPUT   14
      DIMENSION F(NDOF,NUMNP,1)                                          INPUT   15
      COMMON /CONSTS/ ZERO,PT1667,PT25,PT5,ONE,TWO,THREE,FOUR,FIVE       INPUT   16
      COMMON /IOUNIT/ IIN,IOUT,IRSIN,IRSOUT                              INPUT   17
C                                                                        INPUT   18
      CALL CLEAR(F,NLVECT*NUMNP*NDOF)                                    INPUT   19
C                                                                        INPUT   20
      DO 100 NLV=1,NLVECT                                                INPUT   21
      CALL GENFL(F(1,1,NLV),NDOF)                                        INPUT   22
      CALL ZTEST(F(1,1,NLV),NDOF*NUMNP,LZERO)                            INPUT   23
C                                                                        INPUT   24
      IF (IPRTIN.EQ.0) THEN                                              INPUT   25
C                                                                        INPUT   26
         IF (LZERO) THEN                                                 INPUT   27
            IF (J.EQ.0) WRITE(IOUT,1000) NLV                            INPUT   28
            IF (J.EQ.1) WRITE(IOUT,2000)                                INPUT   29
            IF (J.EQ.2) WRITE(IOUT,3000)                                INPUT   30
            IF (J.EQ.3) WRITE(IOUT,4000)                                INPUT   31
         ELSE                                                            INPUT   32
            IF (J.EQ.0) CALL PRINTF(F,NDOF,NUMNP,NLV)                   INPUT   33
C                                                                        INPUT   34
            IF (J.EQ.1)                                                  INPUT   35
     &      CALL PRINTD(' I N I T I A L   D I S P L A C E M E N T S ',  INPUT   36
     &                  F,NDOF,NUMNP,0,TIME)                            INPUT   37
C                                                                        INPUT   38
            IF (J.EQ.2)                                                  INPUT   39
     &      CALL PRINTD(' I N I T I A L   V E L O C I T I E S       ',  INPUT   40
     &                  F,NDOF,NUMNP,0,TIME)                            INPUT   41
C                                                                        INPUT   42
            IF (J.EQ.3)                                                  INPUT   43
     &      CALL PRINTD(' I N I T I A L   A C C E L E R A T I O N S ',  INPUT   44
     &                  F,NDOF,NUMNP,0,TIME)                            INPUT   45
C                                                                        INPUT   46
         ENDIF                                                           INPUT   47
      ENDIF                                                              INPUT   48
C                                                                        INPUT   49
  100 CONTINUE                                                           INPUT   50
```

```
C                                                                       INPUT   51
      RETURN                                                            INPUT   52
 1000 FORMAT('1'//,' THERE ARE NO NONZERO PRESCRIBED FORCES AND ',      INPUT   53
     &       'KINEMATIC BOUNDARY CONDITIONS FOR LOAD VECTOR NUMBER ',I5) INPUT   54
 2000 FORMAT('1'//,' THERE ARE NO NONZERO INITIAL DISPLACEMENTS')       INPUT   55
 3000 FORMAT('1'//,' THERE ARE NO NONZERO INITIAL VELOCITIES   ')       INPUT   56
 4000 FORMAT('1'//,' THERE ARE NO NONZERO INITIAL ACCELERATIONS')       INPUT   57
      END                                                               INPUT   58
C************************************************************************
      SUBROUTINE INTERP(X,Y,XX,YY,N)                                    INTERP   1
C                                                                       INTERP   2
C.... PROGRAM TO PERFORM LINEAR INTERPOLATION                           INTERP   3
C                                                                       INTERP   4
C        X(I) = ABSCISSAS                                               INTERP   5
C        Y(I) = ORDINATES                                               INTERP   6
C          XX = INPUT ABSCISSA                                          INTERP   7
C          YY = OUTPUT ORDINATE                                         INTERP   8
C           N = TOTAL NUMBER OF DATA POINTS (1.LE.I.LE.N)               INTERP   9
C                                                                       INTERP  10
      IMPLICIT DOUBLE PRECISION (A-H,O-Z)                               INTERP  11
C                                                                       INTERP  12
C.... DEACTIVATE ABOVE CARD(S) FOR SINGLE-PRECISION OPERATION           INTERP  13
C                                                                       INTERP  14
      DIMENSION X(1),Y(1)                                               INTERP  15
C                                                                       INTERP  16
      IF (XX.LE.X(1)) THEN                                              INTERP  17
         YY = Y(1)                                                      INTERP  18
C                                                                       INTERP  19
      ELSE IF (XX.GE.X(N)) THEN                                         INTERP  20
         YY = Y(N)                                                      INTERP  21
C                                                                       INTERP  22
      ELSE                                                              INTERP  23
         DO 100 I=1,N                                                   INTERP  24
         IF (X(I).GE.XX)                                                INTERP  25
     &      YY = Y(I-1) + (XX - X(I-1))*(Y(I) - Y(I-1))/(X(I) - X(I-1)) INTERP  26
  100 CONTINUE                                                          INTERP  27
C                                                                       INTERP  28
      ENDIF                                                             INTERP  29
C                                                                       INTERP  30
      RETURN                                                            INTERP  31
      END                                                               INTERP  32
C************************************************************************
      SUBROUTINE ITERUP(ID,D,DPRED,VPRED,A,BRHS,NDOF,NUMNP,LDYN)        ITERUP   1
C                                                                       ITERUP   2
C.... PROGRAM TO PERFORM INTERMEDIATE UPDATE OF DISPLACEMENTS,          ITERUP   3
C        VELOCITIES AND ACCELERATIONS DURING ITERATIVE LOOP IN          ITERUP   4
C        PREDICTOR/CORRECTOR ALGORITHM                                  ITERUP   5
C                                                                       ITERUP   6
      IMPLICIT DOUBLE PRECISION (A-H,O-Z)                               ITERUP   7
C                                                                       ITERUP   8
C.... DEACTIVATE ABOVE CARD(S) FOR SINGLE-PRECISION OPERATION           ITERUP   9
C                                                                       ITERUP  10
      LOGICAL LDYN                                                      ITERUP  11
      DIMENSION ID(NDOF,1),D(NDOF,1),DPRED(NDOF,1),VPRED(NDOF,1),       ITERUP  12
     &          A(NDOF,1),BRHS(1)                                       ITERUP  13
      COMMON /COEFFS/ COEFF1,COEFF2,COEFF3,COEFF4,COEFF5,COEFF6,        ITERUP  14
     &          COEFF7,COEFF8,ALPHA1,BETA1 ,GAMMA1,DT1                  ITERUP  15
C                                                                       ITERUP  16
      IF (LDYN) THEN                                                    ITERUP  17
C                                                                       ITERUP  18
         DO 200 I=1,NDOF                                                ITERUP  19
C                                                                       ITERUP  20
         DO 100 J=1,NUMNP                                               ITERUP  21
         K = ID(I,J)                                                    ITERUP  22
         IF (K.GT.0) THEN                                               ITERUP  23
            DPRED(I,J) = DPRED(I,J) + COEFF5*BRHS(K)                    ITERUP  24
            VPRED(I,J) = VPRED(I,J) + COEFF4*BRHS(K)                    ITERUP  25
            A(I,J) = A(I,J) + BRHS(K)                                   ITERUP  26
         ENDIF                                                          ITERUP  27
  100    CONTINUE                                                       ITERUP  28
C                                                                       ITERUP  29
  200    CONTINUE                                                       ITERUP  30
C                                                                       ITERUP  31
      ELSE                                                              ITERUP  32
C                                                                       ITERUP  33
         DO 400 I=1,NDOF                                                ITERUP  34
C                                                                       ITERUP  35
         DO 300 J=1,NUMNP                                               ITERUP  36
         K = ID(I,J)                                                    ITERUP  37
         IF (K.GT.0) D(I,J) = BRHS(K)                                   ITERUP  38
  300    CONTINUE                                                       ITERUP  39
C                                                                       ITERUP  40
  400    CONTINUE                                                       ITERUP  41
C                                                                       ITERUP  42
      ENDIF                                                             ITERUP  43
C                                                                       ITERUP  44
      RETURN                                                            ITERUP  45
      END                                                               ITERUP  46
```

```
C********************************************************************************
      SUBROUTINE LFAC(G,T,G1,NLTFTN,NPTSLF)                      LFAC     1
C                                                                LFAC     2
C.... PROGRAM TO COMPUTE LOAD FACTORS AT TIME T                  LFAC     3
C                                                                LFAC     4
      IMPLICIT DOUBLE PRECISION (A-H,O-Z)                        LFAC     5
C                                                                LFAC     6
C.... DEACTIVATE ABOVE CARD(S) FOR SINGLE-PRECISION OPERATION    LFAC     7
C                                                                LFAC     8
      DIMENSION G(NPTSLF,2,1),G1(1)                              LFAC     9
C                                                                LFAC    10
      DO 100 NLF=1,NLTFTN                                        LFAC    11
      CALL INTERP(G(1,1,NLF),G(1,2,NLF),T,G1(NLF),NPTSLF)        LFAC    12
  100 CONTINUE                                                   LFAC    13
C                                                                LFAC    14
      RETURN                                                     LFAC    15
      END                                                        LFAC    16
C********************************************************************************
      SUBROUTINE LOAD(ID,F,BRHS,G1,NDOF,NUMNP,NLVECT)            LOAD     1
C                                                                LOAD     2
C.... PROGRAM TO ACCUMULATE NODAL FORCES AND TRANSFER INTO       LOAD     3
C          RIGHT-HAND-SIDE VECTOR                                LOAD     4
C                                                                LOAD     5
      IMPLICIT DOUBLE PRECISION (A-H,O-Z)                        LOAD     6
C                                                                LOAD     7
C.... DEACTIVATE ABOVE CARD(S) FOR SINGLE-PRECISION OPERATION    LOAD     8
C                                                                LOAD     9
      DIMENSION ID(NDOF,1),F(NDOF,NUMNP,1),BRHS(1),G1(1)         LOAD    10
C                                                                LOAD    11
      DO 300 I=1,NDOF                                            LOAD    12
C                                                                LOAD    13
      DO 200 J=1,NUMNP                                           LOAD    14
      K = ID(I,J)                                                LOAD    15
      IF (K.GT.0) THEN                                           LOAD    16
C                                                                LOAD    17
      DO 100 NLV=1,NLVECT                                        LOAD    18
      BRHS(K) = BRHS(K) + F(I,J,NLV)*G1(NLV)                     LOAD    19
  100    CONTINUE                                                LOAD    20
C                                                                LOAD    21
      ENDIF                                                      LOAD    22
C                                                                LOAD    23
  200 CONTINUE                                                   LOAD    24
C                                                                LOAD    25
  300 CONTINUE                                                   LOAD    26
C                                                                LOAD    27
      RETURN                                                     LOAD    28
      END                                                        LOAD    29
C********************************************************************************
      SUBROUTINE LOCAL(IEN,X,XL,NEN,NROWX,NROWXL)                LOCAL    1
C                                                                LOCAL    2
C.... PROGRAM TO LOCALIZE A GLOBAL ARRAY                         LOCAL    3
C                                                                LOCAL    4
C         NOTE: IT IS ASSUMED NROWXL.LE.NROWX                    LOCAL    5
C                                                                LOCAL    6
      IMPLICIT DOUBLE PRECISION (A-H,O-Z)                        LOCAL    7
C                                                                LOCAL    8
C.... DEACTIVATE ABOVE CARD(S) FOR SINGLE-PRECISION OPERATION    LOCAL    9
C                                                                LOCAL   10
      DIMENSION IEN(1),X(NROWX,1),XL(NROWXL,1)                   LOCAL   11
C                                                                LOCAL   12
      DO 200 J=1,NEN                                             LOCAL   13
      NODE = IEN(J)                                              LOCAL   14
C                                                                LOCAL   15
      DO 100 I=1,NROWXL                                          LOCAL   16
      XL(I,J)= X(I,NODE)                                         LOCAL   17
  100 CONTINUE                                                   LOCAL   18
C                                                                LOCAL   19
  200 CONTINUE                                                   LOCAL   20
C                                                                LOCAL   21
      RETURN                                                     LOCAL   22
      END                                                        LOCAL   23
C********************************************************************************
      FUNCTION LOUT(I,J)                                         LOUT     1
C                                                                LOUT     2
C.... PROGRAM TO DETERMINE LOGICAL SWITCH                        LOUT     3
C                                                                LOUT     4
      LOGICAL LOUT                                               LOUT     5
C                                                                LOUT     6
      LOUT = .FALSE.                                             LOUT     7
      IF (J.EQ.0) RETURN                                         LOUT     8
      IF (MOD(I,J).EQ.0) LOUT = .TRUE.                           LOUT     9
C                                                                LOUT    10
      RETURN                                                     LOUT    11
      END                                                        LOUT    12
C********************************************************************************
      SUBROUTINE LTIMEF(G,NPTSLF,NLTFTN,IPRTIN)                  LTIMEF   1
C                                                                LTIMEF   2
C.... PROGRAM TO READ, WRITE AND STORE LOAD-TIME FUNCTIONS       LTIMEF   3
C                                                                LTIMEF   4
```

```
C           G(I,1,L) = TIME I FOR LOAD-TIME FUNCTION L                  LTIMEF   5
C           G(I,2,L) = LOAD FACTOR AT TIME I FOR LOAD-TIME FUNCTION L   LTIMEF   6
C                                                                       LTIMEF   7
      IMPLICIT DOUBLE PRECISION (A-H,O-Z)                               LTIMEF   8
C                                                                       LTIMEF   9
C.... DEACTIVATE ABOVE CARD(S) FOR SINGLE-PRECISION OPERATION           LTIMEF  10
C                                                                       LTIMEF  11
      DIMENSION G(NPTSLF,2,NLTFTN)                                      LTIMEF  12
      COMMON /IOUNIT/ IIN,IOUT,IRSIN,IRSOUT                             LTIMEF  13
C                                                                       LTIMEF  14
      DO 200 L=1,NLTFTN                                                 LTIMEF  15
C                                                                       LTIMEF  16
      DO 100 I=1,NPTSLF                                                 LTIMEF  17
      READ(IIN,1000) G(I,1,L),G(I,2,L)                                  LTIMEF  18
  100 CONTINUE                                                          LTIMEF  19
C                                                                       LTIMEF  20
  200 CONTINUE                                                          LTIMEF  21
C                                                                       LTIMEF  22
      IF (IPRTIN.EQ.1) RETURN                                           LTIMEF  23
C                                                                       LTIMEF  24
      WRITE(IOUT,2000) NLTFTN                                           LTIMEF  25
      DO 400 L=1,NLTFTN                                                 LTIMEF  26
C                                                                       LTIMEF  27
      DO 300 I=1,NPTSLF                                                 LTIMEF  28
      IF (MOD(I,50).EQ.1) WRITE(IOUT,3000) L                            LTIMEF  29
      WRITE(IOUT,4000) G(I,1,L),G(I,2,L)                                LTIMEF  30
  300 CONTINUE                                                          LTIMEF  31
C                                                                       LTIMEF  32
  400 CONTINUE                                                          LTIMEF  33
C                                                                       LTIMEF  34
      RETURN                                                            LTIMEF  35
C                                                                       LTIMEF  36
 1000 FORMAT(2F10.0)                                                    LTIMEF  37
 2000 FORMAT('1',' L O A D - T I M E   F U N C T I O N   D A T A ',//5X,LTIMEF  38
     &' NUMBER OF LOAD-TIME FUNTIONS . . . . . (NLTFTN  ) = ',I5       )LTIMEF  39
 3000 FORMAT(///5X,' FUNCTION NUMBER ',I5,//                            LTIMEF  40
     &        16X,'TIME',13X,'LOAD FACTOR'/)                            LTIMEF  41
 4000 FORMAT(5X,2(1PE20.8))                                             LTIMEF  42
C                                                                       LTIMEF  43
      END                                                               LTIMEF  44
C***************************************************************************
      SUBROUTINE MATADD(A,B,C,MA,MB,MC,M,N,IOPT)                        MATADD   1
C                                                                       MATADD   2
C.... PROGRAM TO ADD RECTANGULAR MATRICES                               MATADD   3
C                                                                       MATADD   4
      IMPLICIT DOUBLE PRECISION (A-H,O-Z)                               MATADD   5
C                                                                       MATADD   6
C.... DEACTIVATE ABOVE CARD(S) FOR SINGLE-PRECISION OPERATION           MATADD   7
C                                                                       MATADD   8
      DIMENSION A(MA,1),B(MB,1),C(MC,1)                                 MATADD   9
C                                                                       MATADD  10
      GO TO (1000,2000,3000),IOPT                                       MATADD  11
C                                                                       MATADD  12
C.... IOPT = 1, ADD ENTIRE MATRICES                                     MATADD  13
C                                                                       MATADD  14
 1000 DO 1200 J=1,N                                                     MATADD  15
C                                                                       MATADD  16
      DO 1100 I=1,M                                                     MATADD  17
      C(I,J) = A(I,J) + B(I,J)                                          MATADD  18
 1100 CONTINUE                                                          MATADD  19
C                                                                       MATADD  20
 1200 CONTINUE                                                          MATADD  21
      RETURN                                                            MATADD  22
C                                                                       MATADD  23
C.... IOPT = 2, ADD LOWER TRIANGULAR AND DIAGONAL ELEMENTS              MATADD  24
C                                                                       MATADD  25
 2000 DO 2200 J=1,N                                                     MATADD  26
C                                                                       MATADD  27
      DO 2100 I=J,M                                                     MATADD  28
      C(I,J) = A(I,J) + B(I,J)                                          MATADD  29
 2100 CONTINUE                                                          MATADD  30
C                                                                       MATADD  31
 2200 CONTINUE                                                          MATADD  32
      RETURN                                                            MATADD  33
C                                                                       MATADD  34
C.... IOPT = 3, ADD UPPER TRIANGULAR AND DIAGONAL ELEMENTS              MATADD  35
C                                                                       MATADD  36
 3000 DO 3200 J=1,N                                                     MATADD  37
C                                                                       MATADD  38
      DO 3100 I=1,J                                                     MATADD  39
      C(I,J) = A(I,J) + B(I,J)                                          MATADD  40
 3100 CONTINUE                                                          MATADD  41
C                                                                       MATADD  42
 3200 CONTINUE                                                          MATADD  43
      RETURN                                                            MATADD  44
C                                                                       MATADD  45
      END                                                               MATADD  46
```

```
C*********************************************************************************
      SUBROUTINE MEANSH(SHGBAR,W,DET,R,SHG,NEN,NINT,IOPT,NESD,NROWSH)    MEANSH  1
C                                                                        MEANSH  2
C.... PROGRAM TO CALCULATE MEAN VALUES OF SHAPE FUNCTION                 MEANSH  3
C        GLOBAL DERIVATIVES FOR B-BAR METHOD                             MEANSH  4
C                                                                        MEANSH  5
C        NOTE: IF IOPT.EQ.2, DET(L) = DET(L)*R(L) UPON ENTRY             MEANSH  6
C                                                                        MEANSH  7
      IMPLICIT DOUBLE PRECISION (A-H,O-Z)                                MEANSH  8
C                                                                        MEANSH  9
C.... DEACTIVATE ABOVE CARD(S) FOR SINGLE-PRECISION OPERATION            MEANSH 10
C                                                                        MEANSH 11
      DIMENSION SHGBAR(3,1),W(1),DET(1),R(1),SHG(NROWSH,NEN,1)           MEANSH 12
      COMMON /CONSTS/ ZERO,PT1667,PT25,PT5,ONE,TWO,THREE,FOUR,FIVE       MEANSH 13
C                                                                        MEANSH 14
      CALL CLEAR(SHGBAR,3*NEN)                                           MEANSH 15
C                                                                        MEANSH 16
      VOLINV = ONE/COLDOT(W,DET,NINT)                                    MEANSH 17
C                                                                        MEANSH 18
      DO 300 L=1,NINT                                                    MEANSH 19
      TEMP1 = W(L)*DET(L)*VOLINV                                         MEANSH 20
      IF (IOPT.EQ.2) TEMP2 = TEMP1/R(L)                                  MEANSH 21
C                                                                        MEANSH 22
      DO 200 J=1,NEN                                                     MEANSH 23
C                                                                        MEANSH 24
      DO 100 I=1,NESD                                                    MEANSH 25
      SHGBAR(I,J) = SHGBAR(I,J) + TEMP1*SHG(I,J,L)                       MEANSH 26
  100 CONTINUE                                                           MEANSH 27
C                                                                        MEANSH 28
      IF (IOPT.EQ.2) SHGBAR(3,J) = SHGBAR(3,J) + TEMP2*SHG(3,J,L)        MEANSH 29
  200 CONTINUE                                                           MEANSH 30
C                                                                        MEANSH 31
  300 CONTINUE                                                           MEANSH 32
C                                                                        MEANSH 33
      RETURN                                                             MEANSH 34
      END                                                                MEANSH 35
C*********************************************************************************
      SUBROUTINE MINMAX(X,XMAX,XMIN,L,M,N)                               MINMAX  1
C                                                                        MINMAX  2
C.... PROGRAM TO COMPUTE THE MIN AND MAX IN THE ROW OF A MATRIX          MINMAX  3
C                                                                        MINMAX  4
C        X = MATRIX                                                      MINMAX  5
C        L = NUMBER OF ROWS IN X                                         MINMAX  6
C        M = NUMBER OF COLUMNS IN X                                      MINMAX  7
C        N = ROW NUMBER                                                  MINMAX  8
C                                                                        MINMAX  9
      DIMENSION X(L,1)                                                   MINMAX 10
C                                                                        MINMAX 11
      XMAX = X(N,1)                                                      MINMAX 12
      XMIN = X(N,1)                                                      MINMAX 13
C                                                                        MINMAX 14
      DO 100 I = 2,M                                                     MINMAX 15
         IF (X(N,I).GT.XMAX) XMAX = X(N,I)                               MINMAX 16
         IF (X(N,I).LT.XMIN) XMIN = X(N,I)                               MINMAX 17
  100 CONTINUE                                                           MINMAX 18
C                                                                        MINMAX 19
      RETURN                                                             MINMAX 20
      END                                                                MINMAX 21
C*********************************************************************************
      SUBROUTINE MOVE(A,B,N)                                             MOVE    1
C                                                                        MOVE    2
C.... PROGRAM TO MOVE A FLOATING-POINT ARRAY                             MOVE    3
C                                                                        MOVE    4
      IMPLICIT DOUBLE PRECISION (A-H,O-Z)                                MOVE    5
C                                                                        MOVE    6
C.... DEACTIVATE ABOVE CARD(S) FOR SINGLE-PRECISION OPERATION            MOVE    7
C                                                                        MOVE    8
      DIMENSION A(1),B(1)                                                MOVE    9
C                                                                        MOVE   10
      DO 100 I=1,N                                                       MOVE   11
      A(I) = B(I)                                                        MOVE   12
  100 CONTINUE                                                           MOVE   13
C                                                                        MOVE   14
      RETURN                                                             MOVE   15
      END                                                                MOVE   16
C*********************************************************************************
      FUNCTION MPOINT(NAME,NDIM1,NDIM2,NDIM3,IPR)                        MPOINT  1
C                                                                        MPOINT  2
C.... PROGRAM TO CALCULATE STORAGE POINTER                               MPOINT  3
C                                                                        MPOINT  4
      DIMENSION NAME(2)                                                  MPOINT  5
      COMMON /BPOINT/ MFIRST,MLAST,MTOT,IPREC                            MPOINT  6
C                                                                        MPOINT  7
      MPOINT = MFIRST                                                    MPOINT  8
      IF ( IPREC.EQ.2 .AND. MOD(MPOINT,2).EQ.0 ) MPOINT = MPOINT + 1     MPOINT  9
      CALL DCTNRY(NAME,NDIM1,NDIM2,NDIM3,MPOINT,IPR,MLAST)               MPOINT 10
      MFIRST = MPOINT + NDIM1*MAX0(1,NDIM2)*MAX0(1,NDIM3)*IPR            MPOINT 11
      IF (MFIRST.GE.MLAST) CALL SERROR(NAME,MFIRST-MLAST)                MPOINT 12
```

```
C                                                                      MPOINT   13
        RETURN                                                         MPOINT   14
        END                                                            MPOINT   15
C******************************************************************************
        SUBROUTINE MULTAB(A,B,C,MA,MB,MC,L,M,N,IOPT)                   MULTAB    1
C                                                                      MULTAB    2
C.... PROGRAM TO MULTIPLY TWO MATRICES                                 MULTAB    3
C                                                                      MULTAB    4
C           L = RANGE OF DOT-PRODUCT INDEX                             MULTAB    5
C           M = NUMBER OF ACTIVE ROWS IN C                             MULTAB    6
C           N = NUMBER OF ACTIVE COLUMNS IN C                          MULTAB    7
C                                                                      MULTAB    8
        IMPLICIT DOUBLE PRECISION (A-H,O-Z)                            MULTAB    9
C                                                                      MULTAB   10
C.... DEACTIVATE ABOVE CARD(S) FOR SINGLE-PRECISION OPERATION          MULTAB   11
C                                                                      MULTAB   12
        DIMENSION A(MA,1),B(MB,1),C(MC,1)                              MULTAB   13
C                                                                      MULTAB   14
        GO TO (1000,2000,3000,4000),IOPT                               MULTAB   15
C                                                                      MULTAB   16
C.... IOPT = 1, C(I,J) = A(I,K)*B(K,J) , (C = A * B)                   MULTAB   17
C                                                                      MULTAB   18
 1000 DO 1200 I=1,M                                                    MULTAB   19
C                                                                      MULTAB   20
        DO 1100 J=1,N                                                  MULTAB   21
        C(I,J) = RCDOT(A(I,1),B(1,J),MA,L)                             MULTAB   22
 1100 CONTINUE                                                         MULTAB   23
C                                                                      MULTAB   24
 1200 CONTINUE                                                         MULTAB   25
        RETURN                                                         MULTAB   26
C                                                                      MULTAB   27
C.... IOPT = 2, C(I,J) = A(K,I)*B(K,J) (C = A^T * B)                   MULTAB   28
C                                                                      MULTAB   29
 2000 DO 2200 I=1,M                                                    MULTAB   30
C                                                                      MULTAB   31
        DO 2100 J=1,N                                                  MULTAB   32
        C(I,J) = COLDOT(A(1,I),B(1,J),L)                               MULTAB   33
 2100 CONTINUE                                                         MULTAB   34
C                                                                      MULTAB   35
 2200 CONTINUE                                                         MULTAB   36
        RETURN                                                         MULTAB   37
C                                                                      MULTAB   38
C.... IOPT = 3, C(I,J) = A(I,K)*B(J,K) (C = A * B^T )                  MULTAB   39
C                                                                      MULTAB   40
 3000 DO 3200 I=1,M                                                    MULTAB   41
C                                                                      MULTAB   42
        DO 3100 J=1,N                                                  MULTAB   43
        C(I,J) = ROWDOT(A(I,1),B(J,1),MA,MB,L)                         MULTAB   44
 3100 CONTINUE                                                         MULTAB   45
C                                                                      MULTAB   46
 3200 CONTINUE                                                         MULTAB   47
        RETURN                                                         MULTAB   48
C                                                                      MULTAB   49
C.... IOPT = 4, C(I,J) = A(K,I)*B(J,K) (C = A^T * B^T )                MULTAB   50
C                                                                      MULTAB   51
 4000 DO 4200 I=1,M                                                    MULTAB   52
C                                                                      MULTAB   53
        DO 4100 J=1,N                                                  MULTAB   54
        C(I,J) = RCDOT(B(J,1),A(1,I),MB,L)                             MULTAB   55
 4100 CONTINUE                                                         MULTAB   56
C                                                                      MULTAB   57
 4200 CONTINUE                                                         MULTAB   58
C                                                                      MULTAB   59
        RETURN                                                         MULTAB   60
        END                                                            MULTAB   61
C******************************************************************************
        SUBROUTINE PIVOTS(A,IDIAG,NEQ,NSQ,*)                           PIVOTS    1
C                                                                      PIVOTS    2
C.... PROGRAM TO DETERMINE THE NUMBER OF ZERO AND NEGATIVE TERMS IN    PIVOTS    3
C          ARRAY D OF FACTORIZATION A = U(TRANSPOSE) * D * U           PIVOTS    4
C                                                                      PIVOTS    5
        IMPLICIT DOUBLE PRECISION (A-H,O-Z)                            PIVOTS    6
C                                                                      PIVOTS    7
C.... DEACTIVATE ABOVE CARD(S) FOR SINGLE-PRECISION OPERATION          PIVOTS    8
C                                                                      PIVOTS    9
        DIMENSION A(1),IDIAG(1)                                        PIVOTS   10
        COMMON /IOUNIT/ IIN,IOUT,IRSIN,IRSOUT                          PIVOTS   11
C                                                                      PIVOTS   12
        IZ = 0                                                         PIVOTS   13
        IN = 0                                                         PIVOTS   14
C                                                                      PIVOTS   15
        DO 100 N=1,NEQ                                                 PIVOTS   16
        I = IDIAG(N)                                                   PIVOTS   17
        IF (A(I).EQ.0.) IZ = IZ + 1                                    PIVOTS   18
        IF (A(I).LT.0.) IN = IN + 1                                    PIVOTS   19
  100 CONTINUE                                                         PIVOTS   20
C                                                                      PIVOTS   21
        WRITE(IOUT,1000) NSQ,IZ,IN                                     PIVOTS   22
```

```
C                                                            PIVOTS  23
      RETURN 1                                               PIVOTS  24
C                                                            PIVOTS  25
 1000 FORMAT(' ',                                            PIVOTS  26
     &' ZERO AND/OR NEGATIVE PIVOTS ENCOUNTERED           ',///5X,PIVOTS  27
     &' TIME SEQUENCE NUMBER   . . . . . . . . (NSQ  ) = ',I5//5X,PIVOTS  28
     &' NUMBER OF ZEROES  . . . . . . . . . . . . . . = ',I5//5X,PIVOTS  29
     &' NUMBER OF NEGATIVES . . . . . . . . . . . . . = ',I5//5X)PIVOTS  30
C                                                            PIVOTS  31
      END                                                    PIVOTS  32
C*************************************************************************
      SUBROUTINE PREDCT(D,V,A,DPRED,VPRED,NDOF,NUMNP)        PREDCT   1
C                                                            PREDCT   2
C.... PROGRAM TO CALCULATE PREDICTOR FOR DISPLACEMENTS,VELOCITIES PREDCT   3
C        AND ACCELERATIONS                                   PREDCT   4
C                                                            PREDCT   5
      IMPLICIT DOUBLE PRECISION (A-H,O-Z)                    PREDCT   6
C                                                            PREDCT   7
C.... DEACTIVATE ABOVE CARD(S) FOR SINGLE PRECISION OPERATION PREDCT   8
C                                                            PREDCT   9
      DIMENSION D(NDOF,1),V(NDOF,1),A(NDOF,1),               PREDCT  10
     &          DPRED(NDOF,1),VPRED(NDOF,1)                  PREDCT  11
      COMMON /COEFFS/ COEFF1,COEFF2,COEFF3,COEFF4,COEFF5,COEFF6, PREDCT  12
     &                COEFF7,COEFF8,ALPHA1,BETA1 ,GAMMA1,DT1 PREDCT  13
      COMMON /CONSTS/ ZERO,PT1667,PT25,PT5,ONE,TWO,THREE,FOUR,FIVE PREDCT  14
C                                                            PREDCT  15
      DO 200 I=1,NDOF                                        PREDCT  16
C                                                            PREDCT  17
      DO 100 J=1,NUMNP                                       PREDCT  18
      DPRED(I,J) = D(I,J) + COEFF6*V(I,J) + COEFF7*A(I,J)    PREDCT  19
      VPRED(I,J) = V(I,J) + COEFF8*A(I,J)                    PREDCT  20
      A(I,J) = ZERO                                          PREDCT  21
  100 CONTINUE                                               PREDCT  22
C                                                            PREDCT  23
  200 CONTINUE                                               PREDCT  24
C                                                            PREDCT  25
      RETURN                                                 PREDCT  26
      END                                                    PREDCT  27
C*************************************************************************
      SUBROUTINE PRINC(N,S,P)                                PRINC    1
C                                                            PRINC    2
C.... PROGRAM TO COMPUTE PRINCIPAL VALUES OF SYMMETRIC 2ND-RANK TENSOR PRINC    3
C                                                            PRINC    4
C        S = SYMMETRIC SECOND-RANK TENSOR STORED AS A VECTOR PRINC    5
C        N = NUMBER OF DIMENSIONS (2 OR 3)                   PRINC    6
C        P = VECTOR OF PRINCIPAL VALUES                      PRINC    7
C                                                            PRINC    8
C.... THE COMPONENTS OF S MUST BE STORED IN THE FOLLOWING ORDERS PRINC    9
C                                                            PRINC   10
C        2-D PROBLEMS: S11,S22,S12                           PRINC   11
C        3-D PROBLEMS: S11,S22,S33,S12,S23,S31               PRINC   12
C                                                            PRINC   13
      IMPLICIT DOUBLE PRECISION (A-H,O-Z)                    PRINC   14
C                                                            PRINC   15
C.... DEACTIVATE ABOVE CARD(S) FOR SINGLE PRECISION OPERATION PRINC   16
C                                                            PRINC   17
      DIMENSION S(1),P(1)                                    PRINC   18
      COMMON /CONSTS/ ZERO,PT1667,PT25,PT5,ONE,TWO,THREE,FOUR,FIVE PRINC   19
      DATA RT2/1.41421356237309/,PI23/2.09439510239321/,     PRINC   20
     &     TWO2/22.50/,FOUR5/45.0/                            PRINC   21
C                                                            PRINC   22
      IF (N.EQ.2) THEN                                       PRINC   23
C                                                            PRINC   24
C....... 2-D PROBLEM                                         PRINC   25
C                                                            PRINC   26
      A = TWO2/ATAN(ONE)                                     PRINC   27
      X = PT5*(S(1) + S(2))                                  PRINC   28
      Y = PT5*(S(1) - S(2))                                  PRINC   29
      R = SQRT(Y*Y + S(3)*S(3))                              PRINC   30
      P(1) = X + R                                           PRINC   31
      P(2) = X - R                                           PRINC   32
      P(3) = R                                               PRINC   33
      P(4) = FOUR5                                           PRINC   34
      IF (Y.NE.ZERO .OR. S(3).NE.ZERO) P(4) = A*ATAN2(S(3),Y) PRINC   35
      ENDIF                                                  PRINC   36
C                                                            PRINC   37
      IF (N.EQ.3) THEN                                       PRINC   38
C                                                            PRINC   39
C....... 3-D PROBLEM                                         PRINC   40
```

```
C
  100     R = ZERO
          X = (S(1) + S(2) + S(3))/THREE
          Y = S(1)*(S(2) + S(3)) + S(2)*S(3)
     &        - S(4)*S(4) - S(6)*S(6) - S(5)*S(5)
          Z = S(1)*S(2)*S(3) - TWO*S(4)*S(6)*S(5) - S(1)*S(5)*S(5)
     &        - S(2)*S(6)*S(6) - S(3)*S(4)*S(4)
          T = THREE*X*X - Y
          U = ZERO
          IF (T.NE.ZERO) THEN
             U = SQRT(TWO*T/THREE)
             UCUBED = U*U*U
             IF (UCUBED.NE.ZERO) THEN
                A = (Z + (T - X*X)*X)*RT2/UCUBED
                R = SQRT(ABS(ONE - A*A))
                IF ((R.NE.ZERO) .OR. (A.NE.ZERO)) THEN
                   R = ATAN2(R,A)/THREE
                ELSE
                   R = ZERO
                ENDIF
             ELSE
                U = ZERO
             ENDIF
          ENDIF
          P(1) = X + U*RT2*COS(R)
          P(2) = X + U*RT2*COS(R - PI23)
          P(3) = X + U*RT2*COS(R + PI23)
      ENDIF
C
      RETURN
      END
C**************************************************************************
      SUBROUTINE PRINTD(NAME,DVA,NDOF,NUMNP,NTSTEP,TIME)
C
C.... PROGRAM TO PRINT KINEMATIC DATA
C
      IMPLICIT DOUBLE PRECISION (A-H,O-Z)
C
C.... DEACTIVATE ABOVE CARD(S) FOR SINGLE PRECISION OPERATION
C
      LOGICAL LZERO,LSKIP
      DIMENSION NAME(11),DVA(NDOF,1)
      COMMON /IOUNIT/ IIN,IOUT,IRSIN,IRSOUT
C
      NN = 0
      LSKIP = .TRUE.
C
      DO 100 N=1,NUMNP
      CALL ZTEST(DVA(1,N),NDOF,LZERO)
      IF (.NOT.LZERO) THEN
         NN = NN + 1
         IF (MOD(NN,50).EQ.1)
     &      WRITE(IOUT,1000) NAME,NTSTEP,TIME,(I,I=1,NDOF)
         WRITE(IOUT,2000) N,(DVA(I,N),I=1,NDOF)
         LSKIP = .FALSE.
      ENDIF
  100 CONTINUE
C
      IF (LSKIP) THEN
         WRITE(IOUT,1000) NAME,NTSTEP,TIME,(I,I=1,NDOF)
         WRITE(IOUT,3000)
      ENDIF
C
      RETURN
C
 1000 FORMAT('1',11A4//5X,
     &' STEP NUMBER = ',I10//5X,
     &' TIME        = ',1PE10.3///5X,
     &' NODE NO.',6(13X,'DOF',I1,:)/)
 2000 FORMAT(6X,I5,10X,6(1PE15.8,2X))
 3000 FORMAT(' ',//,' THERE ARE NO NONZERO COMPONENTS')
      END
C**************************************************************************
      SUBROUTINE PRINTF(F,NDOF,NUMNP,NLV)
C
C.... PROGRAM TO PRINT PRESCRIBED FORCE AND BOUNDARY CONDITION DATA
C
      IMPLICIT DOUBLE PRECISION (A-H,O-Z)
C
C.... DEACTIVATE ABOVE CARD(S) FOR SINGLE PRECISION OPERATION
C
      LOGICAL LZERO
      DIMENSION F(NDOF,NUMNP,1)
      COMMON /IOUNIT/ IIN,IOUT,IRSIN,IRSOUT
C
      NN = 0
```

```
C                                                               PRINTF  14
      DO 100 N=1,NUMNP                                          PRINTF  15
      CALL ZTEST(F(1,N,NLV),NDOF,LZERO)                         PRINTF  16
      IF (.NOT.LZERO) THEN                                      PRINTF  17
         NN = NN + 1                                            PRINTF  18
         IF (MOD(NN,50).EQ.1)                                   PRINTF  19
     &      WRITE(IOUT,1000) NLV,(I,I=1,NDOF)                   PRINTF  20
         WRITE(IOUT,2000) N,(F(I,N,NLV),I=1,NDOF)               PRINTF  21
      ENDIF                                                     PRINTF  22
  100 CONTINUE                                                  PRINTF  23
C                                                               PRINTF  24
      RETURN                                                    PRINTF  25
C                                                               PRINTF  26
 1000 FORMAT('1',                                               PRINTF  27
     &' P R E S C R I B E D   F O R C E S   A N D   K I N E M A T I C ',PRINTF  28
     &' B O U N D A R Y   C O N D I T I O N S'//5X,             PRINTF  29
     &' LOAD VECTOR NUMBER = ',I5///5X,                         PRINTF  30
     &' NODE NO.',6(13X,'DOF',I1,:)/)                           PRINTF  31
 2000 FORMAT(6X,I5,10X,6(1PE15.8,2X))                           PRINTF  32
      END                                                       PRINTF  33
C*******************************************************************************
      SUBROUTINE PRINTP(A,IDIAG,NEQ,NSQ,*)                      PRINTP   1
C                                                               PRINTP   2
C.... PROGRAM TO PRINT ARRAY D AFTER CROUT FACTORIZATION        PRINTP   3
C         A = U(TRANSPOSE) * D * U                              PRINTP   4
C                                                               PRINTP   5
      IMPLICIT DOUBLE PRECISION (A-H,O-Z)                       PRINTP   6
C                                                               PRINTP   7
C.... DEACTIVATE ABOVE CARD(S) FOR SINGLE PRECISION OPERATION   PRINTP   8
C                                                               PRINTP   9
      DIMENSION A(1),IDIAG(1)                                   PRINTP  10
      COMMON /IOUNIT/ IIN,IOUT,IRSIN,IRSOUT                     PRINTP  11
C                                                               PRINTP  12
      DO 100 N=1,NEQ                                            PRINTP  13
      IF (MOD(N,50).EQ.1) WRITE(IOUT,1000) NSQ                  PRINTP  14
      I = IDIAG(N)                                              PRINTP  15
      WRITE(IOUT,2000) N,A(I)                                   PRINTP  16
  100 CONTINUE                                                  PRINTP  17
C                                                               PRINTP  18
      RETURN 1                                                  PRINTP  19
C                                                               PRINTP  20
 1000 FORMAT('1',' ARRAY D OF FACTORIZATION',/                  PRINTP  21
     &' A = U(TRANSPOSE) * D * U ',                    //5X,    PRINTP  22
     &' TIME SEQUENCE NUMBER  . . . . . . . . . (NSQ) = ',I5//5X)PRINTP  23
 2000 FORMAT(1X,I5,4X,1PE20.8)                                  PRINTP  24
      END                                                       PRINTP  25
C*******************************************************************************
      SUBROUTINE PRNTEL(MAT,IEN,NEN,NUMEL)                      PRNTEL   1
C                                                               PRNTEL   2
C.... PROGRAM TO PRINT DATA FOR ELEMENT WITH "NEN" NODES        PRNTEL   3
C                                                               PRNTEL   4
C        NOTE: PRESENTLY THE LABEL FORMATS ARE LIMITED TO       PRNTEL   5
C              ELEMENTS WITH ONE TO NINE NODES                  PRNTEL   6
C                                                               PRNTEL   7
      DIMENSION MAT(1),IEN(NEN,1)                               PRNTEL   8
      COMMON /IOUNIT/ IIN,IOUT,IRSIN,IRSOUT                     PRNTEL   9
C                                                               PRNTEL  10
      DO 100 N=1,NUMEL                                          PRNTEL  11
      IF (MOD(N,50).EQ.1) WRITE(IOUT,1000) (I,I=1,NEN)          PRNTEL  12
      WRITE(IOUT,2000) N,MAT(N),(IEN(I,N),I=1,NEN)              PRNTEL  13
  100 CONTINUE                                                  PRNTEL  14
C                                                               PRNTEL  15
      RETURN                                                    PRNTEL  16
C                                                               PRNTEL  17
 1000 FORMAT('1',                                               PRNTEL  18
     &' E L E M E N T   D A T A',//5X,                          PRNTEL  19
     &' ELEMENT   MATERIAL',9('  NODE ',I1,:,2X),/5X,           PRNTEL  20
     &'  NUMBER    NUMBER'//)                                   PRNTEL  21
 2000 FORMAT(6X,I5,9(5X,I5))                                    PRNTEL  22
      END                                                       PRNTEL  23
C*******************************************************************************
      SUBROUTINE PROP2D(RHO,RDAMPM,RDAMPK,TH,C,NUMAT,IOPT,NROWB)PROP2D   1
C                                                               PROP2D   2
C.... PROGRAM TO READ, WRITE AND STORE PROPERTIES FOR TWO-DIMENSIONALPROP2D   3
C         CONTINUUM ELEMENTS                                    PROP2D   4
C                                                               PROP2D   5
C        NOTE: THIS ROUTINE IS PRESENTLY RESTRICTED TO THE      PROP2D   6
C              ISOTROPIC LINEARLY-ELASTIC CASE                  PROP2D   7
C                                                               PROP2D   8
C              IOPT = 0; PLANE STRESS                           PROP2D   9
C                   = 1; PLANE STRAIN                           PROP2D  10
C                   = 2; TORSIONLESS AXISYMMETRIC               PROP2D  11
C                                                               PROP2D  12
      IMPLICIT DOUBLE PRECISION (A-H,O-Z)                       PROP2D  13
C                                                               PROP2D  14
C.... DEACTIVATE ABOVE CARD(S) FOR SINGLE PRECISION OPERATION   PROP2D  15
```

```
C                                                                    PROP2D  16
      DIMENSION RHO(1),RDAMPM(1),RDAMPK(1),TH(1),C(NROWB,NROWB,1)     PROP2D  17
      COMMON /CONSTS/ ZERO,PT1667,PT25,PT5,ONE,TWO,THREE,FOUR,FIVE    PROP2D  18
      COMMON /IOUNIT/ IIN,IOUT,IRSIN,IRSOUT                           PROP2D  19
C                                                                    PROP2D  20
      DO 100 N=1,NUMAT                                               PROP2D  21
      IF (MOD(N,50).EQ.1) WRITE(IOUT,1000) NUMAT                     PROP2D  22
      READ(IIN,2000) M,E,POIS,RHO(M),RDAMPM(M),RDAMPK(M),TH(M)       PROP2D  23
      IF (TH(M).EQ.ZERO) TH(M) = ONE                                 PROP2D  24
      WRITE(IOUT,3000) M,E,POIS,RHO(M),RDAMPM(M),RDAMPK(M),TH(M)     PROP2D  25
C                                                                    PROP2D  26
C.... SET MATERIAL CONSTANTS FOR OUT-OF-PLANE COMPONENTS             PROP2D  27
C                                                                    PROP2D  28
      AMU2 = E/(ONE + POIS)                                          PROP2D  29
      ALAM = AMU2*POIS/(ONE - TWO*POIS)                              PROP2D  30
C                                                                    PROP2D  31
      C(1,4,M) = ALAM                                                PROP2D  32
      C(2,4,M) = ALAM                                                PROP2D  33
      C(3,4,M) = ZERO                                                PROP2D  34
      C(4,4,M) = ALAM + AMU2                                         PROP2D  35
C                                                                    PROP2D  36
      C(4,1,M) = C(1,4,M)                                            PROP2D  37
      C(4,2,M) = C(2,4,M)                                            PROP2D  38
      C(4,3,M) = C(3,4,M)                                            PROP2D  39
C                                                                    PROP2D  40
C.... SET MATERIAL CONSTANTS FOR IN-PLANE COMPONENTS                 PROP2D  41
C                                                                    PROP2D  42
      IF (IOPT.EQ.0) ALAM = ALAM*AMU2/(ALAM + AMU2)                  PROP2D  43
C                                                                    PROP2D  44
      C(1,1,M) = ALAM + AMU2                                         PROP2D  45
      C(1,2,M) = ALAM                                                PROP2D  46
      C(2,2,M) = C(1,1,M)                                            PROP2D  47
      C(1,3,M) = ZERO                                                PROP2D  48
      C(2,3,M) = ZERO                                                PROP2D  49
      C(3,3,M) = PT5*AMU2                                            PROP2D  50
C                                                                    PROP2D  51
      C(2,1,M) = C(1,2,M)                                            PROP2D  52
      C(3,1,M) = C(1,3,M)                                            PROP2D  53
      C(3,2,M) = C(2,3,M)                                            PROP2D  54
C                                                                    PROP2D  55
  100 CONTINUE                                                       PROP2D  56
C                                                                    PROP2D  57
      RETURN                                                         PROP2D  58
C                                                                    PROP2D  59
 1000 FORMAT('1',                                                    PROP2D  60
     &' M A T E R I A L   S E T   D A T A          ',//5X,          PROP2D  61
     &' NUMBER OF MATERIAL SETS . . . . . . (NUMAT ) = ',I5///,     PROP2D  62
     & 7X,'SET',5X,'YOUNG''S',4X,'POISSON''S',5X,'MASS',8X,'MASS',  PROP2D  63
     & 6X,'STIFFNESS',3X,'THICKNESS',/6X,'NUMBER',3X,'MODULUS',     PROP2D  64
     & 6X,'RATIO',6X,'DENSITY',5X,'DAMPING',5X,'DAMPING',/)         PROP2D  65
 2000 FORMAT(I5,5X,7F10.0)                                           PROP2D  66
 3000 FORMAT(4X,I5,3X,6(2X,1PE10.4))                                 PROP2D  67
      END                                                            PROP2D  68
C***********************************************************************  PRTDC
      SUBROUTINE PRTDC                                               PRTDC    1
C                                                                    PRTDC    2
C.... PROGRAM TO PRINT MEMORY-POINTER DICTIONARY                     PRTDC    3
C                                                                    PRTDC    4
      COMMON /BPOINT/ MFIRST,MLAST,MTOT,IPREC                        PRTDC    5
      COMMON /IOUNIT/ IIN,IOUT,IRSIN,IRSOUT                          PRTDC    6
      COMMON IA(1)                                                   PRTDC    7
C                                                                    PRTDC    8
      N = (MTOT-MLAST)/7                                             PRTDC    9
      J = MTOT + 1                                                   PRTDC   10
C                                                                    PRTDC   11
      DO 100 I=1,N                                                   PRTDC   12
      IF (MOD(I,50).EQ.1) WRITE(IOUT,1000)                          PRTDC   13
      J = J - 7                                                      PRTDC   14
      CALL PRTDC1(I,IA(J),IA(J+2),IA(J+3),IA(J+4),IA(J+5),IA(J+6))   PRTDC   15
  100 CONTINUE                                                       PRTDC   16
C                                                                    PRTDC   17
      RETURN                                                         PRTDC   18
C                                                                    PRTDC   19
 1000 FORMAT('1',                                                    PRTDC   20
     &' D Y N A M I C   S T O R A G E   A L L O C A T I O N',        PRTDC   21
     &'   I N F O R M A T I O N'//                                   PRTDC   22
     & 12X,'ARRAY NO.',5X,'ARRAY',8X,'ADDRESS',6X,'DIM1',6X,'DIM2',  PRTDC   23
     & 6X,'DIM3',6X,'PREC.'/)                                        PRTDC   24
C                                                                    PRTDC   25
      END                                                            PRTDC   26
C***********************************************************************  PRTDC1
      SUBROUTINE PRTDC1(I,INAME,IADD,NDIM1,NDIM2,NDIM3,IPR)          PRTDC1   1
C                                                                    PRTDC1   2
C.... PROGRAM TO PRINT MEMORY-POINTER INFORMATION FOR AN ARRAY       PRTDC1   3
C                                                                    PRTDC1   4
      DIMENSION INAME(2)                                             PRTDC1   5
      COMMON /IOUNIT/ IIN,IOUT,IRSIN,IRSOUT                          PRTDC1   6
      SAVE NEG                                                       PRTDC1   7
      DATA NELPAR,LEFTHS/'NPAR','ALHS'/                              PRTDC1   8
```

```
C                                                                    PRTDC1   9
      IF (I.EQ.1) NEG = 1                                            PRTDC1  10
      IF (INAME(1).EQ.NELPAR) THEN                                   PRTDC1  11
        WRITE (IOUT,1000) NEG                                        PRTDC1  12
        NEG = NEG + 1                                                PRTDC1  13
      ENDIF                                                          PRTDC1  14
      IF (INAME(1).EQ.LEFTHS) WRITE (IOUT,2000)                      PRTDC1  15
      WRITE(IOUT,3000) I,INAME,IADD,NDIM1,NDIM2,NDIM3,IPR            PRTDC1  16
C                                                                    PRTDC1  17
      RETURN                                                         PRTDC1  18
C                                                                    PRTDC1  19
 1000 FORMAT(/14X,'*****',7X,'BEGIN ELEMENT GROUP NUMBER',I5/' ')    PRTDC1  20
 2000 FORMAT(/14X,'*****',7X,'END ELEMENT GROUP DATA',/' ')          PRTDC1  21
 3000 FORMAT(14X,I5,7X,2A4,1X,6I10)                                  PRTDC1  22
      END                                                            PRTDC1  23
C************************************************************************
      SUBROUTINE PRTS2D(XINT,STRESS,PSTRS,STRAIN,PSTRN,              PRTS2D   1
     &                  NN,NNTOT,NEG,NEL,LINT)                       PRTS2D   2
C                                                                    PRTS2D   3
C.... PROGRAM TO PRINT STRESS, STRAIN, AND PRINCIPAL VALUES          PRTS2D   4
C         FOR TWO-DIMENSIONAL CONTINUUM ELEMENTS                     PRTS2D   5
C                                                                    PRTS2D   6
      IMPLICIT DOUBLE PRECISION (A-H,O-Z)                            PRTS2D   7
C                                                                    PRTS2D   8
C.... DEACTIVATE ABOVE CARD(S) FOR SINGLE PRECISION OPERATION        PRTS2D   9
C                                                                    PRTS2D  10
      DIMENSION XINT(2),STRESS(4),PSTRS(4),STRAIN(4),PSTRN(4)        PRTS2D  11
      COMMON /IOUNIT/ IIN,IOUT,IRSIN,IRSOUT                          PRTS2D  12
C                                                                    PRTS2D  13
      NN = NN+1                                                      PRTS2D  14
      IF (MOD(NN,NNTOT).EQ.1) WRITE(IOUT,1000) NEG                   PRTS2D  15
      WRITE(IOUT,2000) NEL,LINT,XINT,STRESS,PSTRS,STRAIN,PSTRN       PRTS2D  16
C                                                                    PRTS2D  17
      RETURN                                                         PRTS2D  18
C                                                                    PRTS2D  19
 1000 FORMAT('1',                                                    PRTS2D  20
     &' E L E M E N T   S T R E S S E S   A N D   S T R A I N S ',  //5X,PRTS2D  21
     &' ELEMENT GROUP NUMBER  . . . . . . . . . .(NEG  ) = ',I5///  PRTS2D  22
     &'  ELEMENT  INT. PT.          X1          X2       ',5X,     PRTS2D  23
     &'   STRESS    STRESS      STRESS     STRESS    ',            PRTS2D  24
     &'    PRINC.    PRINC.      SHEAR     STRESS    ',  /,         PRTS2D  25
     &'   NUMBER    NUMBER                            ',5X,        PRTS2D  26
     &'      11        22         12         33      ',            PRTS2D  27
     &'  STRESS 1  STRESS 2    STRESS     ANGLE     ',//,49X,      PRTS2D  28
     &'    STRAIN    STRAIN      STRAIN     STRAIN    ',           PRTS2D  29
     &'    PRINC.    PRINC.      SHEAR     STRAIN    ',  /,49X,     PRTS2D  30
     &'      11        22         12         33      ',            PRTS2D  31
     &'  STRAIN 1  STRAIN 2    STRAIN     ANGLE     ')             PRTS2D  32
 2000 FORMAT(/2X,I5,6X,I2,8X,2(1PE10.2),5X,8(1PE10.2)/48X,8(1PE10.2))PRTS2D  33
      END                                                            PRTS2D  34
C************************************************************************
      FUNCTION RCDOT(A,B,MA,N)                                       RCDOT    1
C                                                                    RCDOT    2
C.... PROGRAM TO COMPUTE THE DOT PRODUCT OF A VECTOR STORED ROW-WISE RCDOT    3
C         WITH A VECTOR STORED COLUMN-WISE                           RCDOT    4
C                                                                    RCDOT    5
      IMPLICIT DOUBLE PRECISION (A-H,O-Z)                            RCDOT    6
C                                                                    RCDOT    7
C.... DEACTIVATE ABOVE CARD(S) FOR SINGLE-PRECISION OPERATION        RCDOT    8
C                                                                    RCDOT    9
      DIMENSION A(MA,1),B(1)                                         RCDOT   10
      COMMON /CONSTS/ ZERO,PT1667,PT25,PT5,ONE,TWO,THREE,FOUR,FIVE   RCDOT   11
C                                                                    RCDOT   12
      RCDOT = ZERO                                                   RCDOT   13
C                                                                    RCDOT   14
      DO 100 I=1,N                                                   RCDOT   15
      RCDOT = RCDOT + A(1,I)*B(I)                                    RCDOT   16
  100 CONTINUE                                                       RCDOT   17
C                                                                    RCDOT   18
      RETURN                                                         RCDOT   19
      END                                                            RCDOT   20
C************************************************************************
      FUNCTION ROWDOT(A,B,MA,MB,N)                                   ROWDOT   1
C                                                                    ROWDOT   2
C.... PROGRAM TO COMPUTE THE DOT PRODUCT OF VECTORS STORED ROW-WISE  ROWDOT   3
C                                                                    ROWDOT   4
      IMPLICIT DOUBLE PRECISION (A-H,O-Z)                            ROWDOT   5
C                                                                    ROWDOT   6
C.... DEACTIVATE ABOVE CARD(S) FOR SINGLE PRECISION OPERATION        ROWDOT   7
C                                                                    ROWDOT   8
      DIMENSION A(MA,1),B(MB,1)                                      ROWDOT   9
      COMMON /CONSTS/ ZERO,PT1667,PT25,PT5,ONE,TWO,THREE,FOUR,FIVE   ROWDOT  10
C                                                                    ROWDOT  11
      ROWDOT = ZERO                                                  ROWDOT  12
C                                                                    ROWDOT  13
      DO 100 I=1,N                                                   ROWDOT  14
      ROWDOT = ROWDOT + A(1,I)*B(1,I)                                ROWDOT  15
  100 CONTINUE                                                       ROWDOT  16
```

```
C                                                                    ROWDOT  17
      RETURN                                                         ROWDOT  18
      END                                                            ROWDOT  19
C*******************************************************************************
      SUBROUTINE RSIN(D,V,A,NDOF,NUMNP,NTSTEP,TIME)                  RSIN     1
C                                                                    RSIN     2
C.... PROGRAM TO READ RESTART FILE                                  RSIN     3
C                                                                    RSIN     4
      IMPLICIT DOUBLE PRECISION (A-H,O-Z)                           RSIN     5
C                                                                    RSIN     6
C.... DEACTIVATE ABOVE CARD(S) FOR SINGLE PRECISION OPERATION        RSIN     7
C                                                                    RSIN     8
      DIMENSION D(NDOF,1),V(NDOF,1),A(NDOF,1)                        RSIN     9
      COMMON /IOUNIT/ IIN,IOUT,IRSIN,IRSOUT                          RSIN    10
C                                                                    RSIN    11
      READ(IRSIN,1000) NTSTEP,TIME,IJUNK                             RSIN    12
C                                                                    RSIN    13
      DO 100 J=1,NUMNP                                               RSIN    14
      READ(IRSIN,2000) (D(I,J),V(I,J),A(I,J),I=1,NDOF)               RSIN    15
  100 CONTINUE                                                       RSIN    16
C                                                                    RSIN    17
      RETURN                                                         RSIN    18
C                                                                    RSIN    19
 1000 FORMAT(//,15X,I5,/,15X,E12.5/,I1)                              RSIN    20
 2000 FORMAT(3(6E16.8/))                                             RSIN    21
      END                                                            RSIN    22
C*******************************************************************************
      SUBROUTINE RSOUT(D,V,A,NDOF,NUMNP,NTSTEP,TIME)                 RSOUT    1
C                                                                    RSOUT    2
C.... PROGRAM TO WRITE RESTART FILE                                 RSOUT    3
C                                                                    RSOUT    4
      DOUBLE PRECISION A,D,V,TIME                                    RSOUT    5
C                                                                    RSOUT    6
C.... DEACTIVATE ABOVE CARD(S)S FOR SINGLE PRECISION OPERATION       RSOUT    7
C                                                                    RSOUT    8
      CHARACTER*4 TITLE                                              RSOUT    9
      DIMENSION D(NDOF,1),V(NDOF,1),A(NDOF,1)                        RSOUT   10
      COMMON /IOUNIT/ IIN,IOUT,IRSIN,IRSOUT                          RSOUT   11
      COMMON /TITLEC/ TITLE(20)                                      RSOUT   12
C                                                                    RSOUT   13
      WRITE(IRSOUT,1000) TITLE,NTSTEP,TIME                           RSOUT   14
C                                                                    RSOUT   15
      DO 100 J=1,NUMNP                                               RSOUT   16
      WRITE(IRSOUT,2000) (D(I,J),V(I,J),A(I,J),I=1,NDOF)             RSOUT   17
  100 CONTINUE                                                       RSOUT   18
C                                                                    RSOUT   19
      RETURN                                                         RSOUT   20
C                                                                    RSOUT   21
 1000 FORMAT(' ',20A4//,' STEP NUMBER = ',I5/,                       RSOUT   22
     &               '  TIME        = ',1PE12.5/' ')                 RSOUT   23
 2000 FORMAT(3(6E16.8/))                                             RSOUT   24
      END                                                            RSOUT   25
C*******************************************************************************
      SUBROUTINE SERROR(NAME,I)                                      SERROR   1
C                                                                    SERROR   2
C.... PROGRAM TO PRINT ERROR MESSAGE IF AVAILABLE STORAGE IS EXCEEDED SERROR  3
C                                                                    SERROR   4
      DIMENSION NAME(2)                                              SERROR   5
      COMMON /IOUNIT/ IIN,IOUT,IRSIN,IRSOUT                          SERROR   6
C                                                                    SERROR   7
      CALL PRTDC                                                     SERROR   8
      WRITE(IOUT,1000) I,NAME                                        SERROR   9
      STOP                                                           SERROR  10
C                                                                    SERROR  11
 1000 FORMAT(1X,5('*'),'STORAGE EXCEEDED BY ',I10,                   SERROR  12
     &' WORDS IN ATTEMPTING TO STORE ARRAY ',2A4)                    SERROR  13
      END                                                            SERROR  14
C*******************************************************************************
      SUBROUTINE SETUPD(C,DMAT,CONST,NSTR,NROWB)                     SETUPD   1
C                                                                    SETUPD   2
C.... PROGRAM TO CALCULATE THE D MATRIX                             SETUPD   3
C                                                                    SETUPD   4
      IMPLICIT DOUBLE PRECISION (A-H,O-Z)                           SETUPD   5
C                                                                    SETUPD   6
C.... DEACTIVATE ABOVE CARD(S) FOR SINGLE PRECISION OPERATION        SETUPD   7
C                                                                    SETUPD   8
      DIMENSION C(NROWB,1),DMAT(NROWB,1)                             SETUPD   9
C                                                                    SETUPD  10
      DO 200 J=1,NSTR                                                SETUPD  11
C                                                                    SETUPD  12
      DO 100 I=1,J                                                   SETUPD  13
      DMAT(I,J) = CONST*C(I,J)                                       SETUPD  14
      DMAT(J,I) = DMAT(I,J)                                          SETUPD  15
  100 CONTINUE                                                       SETUPD  16
C                                                                    SETUPD  17
  200 CONTINUE                                                       SETUPD  18
C                                                                    SETUPD  19
      RETURN                                                         SETUPD  20
      END                                                            SETUPD  21
```

```
C*****************************************************************************
      SUBROUTINE SHIST(ISHIST,NSOUT,NTYPE)                        SHIST    1
C                                                                 SHIST    2
C.... PROGRAM TO READ, WRITE AND STORE ELEMENT TIME-HISTORY INPUT DATA  SHIST  3
C                                                                 SHIST    4
      DIMENSION ISHIST(3,1)                                       SHIST    5
      COMMON /IOUNIT/ IIN,IOUT,IRSIN,IRSOUT                       SHIST    6
      COMMON /LABELS/ LABELD(3),LABEL1(16),LABEL2(3)              SHIST    7
C                                                                 SHIST    8
      DO 100 N=1,NSOUT                                            SHIST    9
      IF (MOD(N,50).EQ.1) WRITE(IOUT,1000) NSOUT                  SHIST   10
      READ(IIN,2000) NEL,INTPT,NCOMP                              SHIST   11
      IF (INTPT.EQ.0) INTPT = 1                                   SHIST   12
C                                                                 SHIST   13
      IF (NTYPE.EQ.1) WRITE(IOUT,3000) NEL,INTPT,LABEL1(NCOMP)    SHIST   14
      IF (NTYPE.EQ.2) WRITE(IOUT,3000) NEL,INTPT,LABEL2(NCOMP)    SHIST   15
C                                                                 SHIST   16
C.... ADD IF/WRITE STATEMENTS AS ABOVE FOR ADDITIONAL ELEMENT TYPES  SHIST  17
C                                                                 SHIST   18
      ISHIST(1,N) = NEL                                           SHIST   19
      ISHIST(2,N) = INTPT                                         SHIST   20
      ISHIST(3,N) = NCOMP                                         SHIST   21
  100 CONTINUE                                                    SHIST   22
C                                                                 SHIST   23
      RETURN                                                      SHIST   24
C                                                                 SHIST   25
 1000 FORMAT('1',                                                 SHIST   26
     &' E L E M E N T   T I M E   H I S T O R Y ',                SHIST   27
     &' I N F O R M A T I O N '//5X,                              SHIST   28
     &' NUMBER OF STRESS/STRAIN TIME HISTORIES  . . (NSOUT ) = ',I5///  SHIST  29
     &5X,'   ELEMENT    INT PT    COMPONENT',/,                   SHIST   30
     &5X,'   NUMBER     NUMBER              ',/)                  SHIST   31
 2000 FORMAT(3I5)                                                 SHIST   32
 3000 FORMAT(7X,I5,5X,I5,7X,1A4)                                  SHIST   33
      END                                                         SHIST   34
C*****************************************************************************
      SUBROUTINE SMULT(A,B,C,MB,MC,M,N,IOPT)                      SMULT    1
C                                                                 SMULT    2
C.... PROGRAM TO PERFORM SCALAR MULTIPLICATION OF A MATRIX        SMULT    3
C                                                                 SMULT    4
C         C(I,J) = A*B(I,J)                                       SMULT    5
C                                                                 SMULT    6
      IMPLICIT DOUBLE PRECISION (A-H,O-Z)                         SMULT    7
C                                                                 SMULT    8
C.... DEACTIVATE ABOVE CARD(S) FOR SINGLE PRECISION OPERATION     SMULT    9
C                                                                 SMULT   10
      DIMENSION B(MB,1),C(MC,1)                                   SMULT   11
C                                                                 SMULT   12
      GO TO (1000,2000,3000),IOPT                                 SMULT   13
C                                                                 SMULT   14
C.... IOPT = 1, MULTIPLY ENTIRE MATRIX                            SMULT   15
C                                                                 SMULT   16
 1000 DO 1200 J=1,N                                               SMULT   17
C                                                                 SMULT   18
      DO 1100 I=1,M                                               SMULT   19
      C(I,J) = A*B(I,J)                                           SMULT   20
 1100 CONTINUE                                                    SMULT   21
C                                                                 SMULT   22
 1200 CONTINUE                                                    SMULT   23
      RETURN                                                      SMULT   24
C                                                                 SMULT   25
C.... IOPT = 2, MULTIPLY LOWER TRIANGULAR AND DIAGONAL ELEMENTS   SMULT   26
C                                                                 SMULT   27
 2000 DO 2200 J=1,N                                               SMULT   28
C                                                                 SMULT   29
      DO 2100 I=J,M                                               SMULT   30
      C(I,J) = A*B(I,J)                                           SMULT   31
 2100 CONTINUE                                                    SMULT   32
C                                                                 SMULT   33
 2200 CONTINUE                                                    SMULT   34
      RETURN                                                      SMULT   35
C                                                                 SMULT   36
C.... IOPT = 3, MULTIPLY UPPER TRIANGULAR AND DIAGONAL ELEMENTS   SMULT   37
C                                                                 SMULT   38
 3000 DO 3200 J=1,N                                               SMULT   39
C                                                                 SMULT   40
      DO 3100 I=1,J                                               SMULT   41
      C(I,J) = A*B(I,J)                                           SMULT   42
 3100 CONTINUE                                                    SMULT   43
C                                                                 SMULT   44
 3200 CONTINUE                                                    SMULT   45
      RETURN                                                      SMULT   46
C                                                                 SMULT   47
      END                                                         SMULT   48
C*****************************************************************************
      SUBROUTINE STATIN(NEQ)                                      STATIN   1
C                                                                 STATIN   2
C.... PROGRAM TO SET MEMORY POINTERS FOR STATIC ANALYSIS DATA ARRAYS,  STATIN  3
C        AND CALL ASSOCIATED INPUT ROUTINES                       STATIN   4
```

```
C                                                                         STATIN     5
       DOUBLE PRECISION ZERO,PT1667,PT25,PT5,ONE,TWO,THREE,FOUR,FIVE      STATIN     6
C                                                                         STATIN     7
C.... DEACTIVATE ABOVE CARD(S) FOR SINGLE PRECISION OPERATION             STATIN     8
C                                                                         STATIN     9
       LOGICAL LDYN                                                       STATIN    10
       COMMON /BPOINT/ MFIRST,MLAST,MTOT,IPREC                            STATIN    11
       COMMON /CONSTS/ ZERO,PT1667,PT25,PT5,ONE,TWO,THREE,FOUR,FIVE       STATIN    12
       COMMON /DPOINT/ MPSTEP,MPDPRT,MPSPRT,MPHPLT,MPITER,MPALPH,MPBETA,  STATIN    13
      &                MPGAMM,MPDT  ,MPIDHS,MPDOUT,MPVPRD,MPDPRD,MPA,MPV   STATIN    14
       COMMON /INFO  / IEXEC,IACODE,LDYN,IREADR,IWRITR,IPRTIN,IRANK,      STATIN    15
      &                NUMSEQ,NDOUT,NSD,NUMNP,NDOF,NLVECT,NLTFTN,NPTSLF,  STATIN    16
      &                NUMEG                                              STATIN    17
       COMMON /SPOINT/ MPD,MPX,MPID,MPF,MPG,MPG1,MPDIAG,MPNGRP,           STATIN    18
      &                MPALHS,MPBRHS                                      STATIN    19
       COMMON A(1)                                                       STATIN    20
C                                                                         STATIN    21
       MPD    = MPOINT('D       ',NDOF   ,NUMNP ,0,IPREC)                STATIN    22
       IF (.NOT.LDYN) MPDPRD = MPD                                       STATIN    23
       MPX    = MPOINT('X       ',NSD    ,NUMNP ,0       ,IPREC)         STATIN    24
       MPID   = MPOINT('ID      ',NDOF   ,NUMNP ,0       ,1)             STATIN    25
C                                                                         STATIN    26
       IF (NLVECT.EQ.0) THEN                                            STATIN    27
          MPF = 1                                                       STATIN    28
       ELSE                                                             STATIN    29
          MPF = MPOINT('F       ',NDOF   ,NUMNP ,NLVECT,IPREC)          STATIN    30
       ENDIF                                                            STATIN    31
C                                                                         STATIN    32
       IF (NLTFTN.EQ.0) THEN                                            STATIN    33
          MPG  = 1                                                      STATIN    34
          MPG1 = 1                                                      STATIN    35
       ELSE                                                             STATIN    36
          MPG  = MPOINT('G       ',NPTSLF,2      ,NLTFTN,IPREC)         STATIN    37
          MPG1 = MPOINT('G1      ',NLTFTN,0      ,0       ,IPREC)       STATIN    38
       ENDIF                                                            STATIN    39
C                                                                         STATIN    40
C.... INPUT COORDINATE DATA                                               STATIN    41
C                                                                         STATIN    42
       CALL COORD(A(MPX),NSD,NUMNP,IPRTIN)                              STATIN    43
C                                                                         STATIN    44
C.... INPUT BOUNDARY CONDITION DATA AND ESTABLISH EQUATION NUMBERS        STATIN    45
C                                                                         STATIN    46
       CALL BC(A(MPID),NDOF,NUMNP,NEQ,IPRTIN)                           STATIN    47
C                                                                         STATIN    48
C.... INPUT NODAL FORCE AND PRESCRIBED KINEMATIC BOUNDARY-VALUE DATA      STATIN    49
C                                                                         STATIN    50
       IF (NLVECT.GT.0) CALL INPUT(A(MPF),NDOF,NUMNP,0,NLVECT,          STATIN    51
      &                            IPRTIN,ZERO)                         STATIN    52
C                                                                         STATIN    53
C.... INPUT LOAD-TIME FUNCTIONS                                           STATIN    54
C                                                                         STATIN    55
       IF (NLTFTN.GT.0) CALL LTIMEF(A(MPG),NPTSLF,NLTFTN,IPRTIN)        STATIN    56
C                                                                         STATIN    57
C.... ALLOCATE MEMORY FOR IDIAG ARRAY AND CLEAR                           STATIN    58
C                                                                         STATIN    59
       MPDIAG = MPOINT('IDIAG   ,NEQ    ,0       ,0       ,1)           STATIN    60
       CALL ICLEAR(A(MPDIAG),NEQ)                                       STATIN    61
C                                                                         STATIN    62
       MPNGRP = MPOINT('NGRP    ,NUMEG ,0       ,0       ,1)            STATIN    63
C                                                                         STATIN    64
       RETURN                                                           STATIN    65
       END                                                              STATIN    66
C*********************************************************************************
       SUBROUTINE STORED(IDHIST,D,V,A,DOUT,NDOF,NDOUT)                   STORED     1
C                                                                         STORED     2
C.... PROGRAM TO STORE NODAL TIME HISTORIES AS SINGLE-PRECISION DATA      STORED     3
C                                                                         STORED     4
       IMPLICIT DOUBLE PRECISION (A-H,O-Z)                              STORED     5
C                                                                         STORED     6
C.... DEACTIVATE ABOVE CARD(S) FOR SINGLE PRECISION OPERATION             STORED     7
C                                                                         STORED     8
       REAL DOUT(NDOUT+1,1)                                            STORED     9
       DIMENSION IDHIST(3,1),D(NDOF,1),V(NDOF,1),A(NDOF,1)             STORED    10
       COMMON /HPLOTC/ NPLPTS,LOCPLT,TIME                              STORED    11
C                                                                         STORED    12
       DOUT(1,LOCPLT) = REAL(TIME)                                     STORED    13
C                                                                         STORED    14
       DO 100 I=1,NDOUT                                                STORED    15
       NODE = IDHIST(1,I)                                              STORED    16
       IDOF = IDHIST(2,I)                                              STORED    17
       IDVA = IDHIST(3,I)                                              STORED    18
       IF (IDVA .EQ. 1) DOUT(I+1,LOCPLT) = REAL(D(IDOF,NODE))          STORED    19
       IF (IDVA .EQ. 2) DOUT(I+1,LOCPLT) = REAL(V(IDOF,NODE))          STORED    20
       IF (IDVA .EQ. 3) DOUT(I+1,LOCPLT) = REAL(A(IDOF,NODE))          STORED    21
  100 CONTINUE                                                         STORED    22
C                                                                         STORED    23
       RETURN                                                           STORED    24
       END                                                              STORED    25
```

```
C***********************************************************************
      SUBROUTINE TIMCON(NSQ,NSTEP ,NDPRT ,NSPRT ,NHPLT ,NITER ,     TIMCON  1
     &                      NSTEP1,NDPRT1,NSPRT1,NHPLT1,NITER1,      TIMCON  2
     &                      ALPHA ,BETA  ,GAMMA ,DT     )            TIMCON  3
C                                                                    TIMCON  4
C.... PROGRAM TO COMPUTE CURRENT TIME SEQUENCE PARAMETERS            TIMCON  5
C        AND TIME-INTEGRATION COEFFICIENTS                           TIMCON  6
C                                                                    TIMCON  7
      IMPLICIT DOUBLE PRECISION (A-H,O-Z)                            TIMCON  8
C                                                                    TIMCON  9
C.... DEACTIVATE ABOVE CARD(S) FOR SINGLE PRECISION OPERATION        TIMCON 10
C                                                                    TIMCON 11
      DIMENSION NSTEP(1),NDPRT(1),NSPRT(1),NHPLT(1),NITER(1),        TIMCON 12
     &          ALPHA(1),BETA(1) ,GAMMA(1) ,DT(1)                    TIMCON 13
      COMMON /COEFFS/ COEFF1,COEFF2,COEFF3,COEFF4,COEFF5,COEFF6,     TIMCON 14
     &                COEFF7,COEFF8,ALPHA1,BETA1 ,GAMMA1,DT1         TIMCON 15
      COMMON /CONSTS/ ZERO,PT1667,PT25,PT5,ONE,TWO,THREE,FOUR,FIVE   TIMCON 16
C                                                                    TIMCON 17
      NSTEP1 = NSTEP(NSQ)                                            TIMCON 18
      NDPRT1 = NDPRT(NSQ)                                            TIMCON 19
      NSPRT1 = NSPRT(NSQ)                                            TIMCON 20
      NHPLT1 = NHPLT(NSQ)                                            TIMCON 21
      NITER1 = NITER(NSQ)                                            TIMCON 22
      ALPHA1 = ALPHA(NSQ)                                            TIMCON 23
      BETA1  = BETA(NSQ)                                             TIMCON 24
      GAMMA1 = GAMMA(NSQ)                                            TIMCON 25
      DT1    = DT(NSQ)                                               TIMCON 26
C                                                                    TIMCON 27
      COEFF1 = ONE + ALPHA1                                          TIMCON 28
      COEFF2 = GAMMA1*DT1                                            TIMCON 29
      COEFF3 = BETA1*DT1*DT1                                         TIMCON 30
      COEFF4 = COEFF1*COEFF2                                         TIMCON 31
      COEFF5 = COEFF1*COEFF3                                         TIMCON 32
      COEFF6 = COEFF1*DT1                                            TIMCON 33
      COEFF7 = PT5*COEFF1*(ONE - TWO*BETA1)*DT1*DT1                  TIMCON 34
      COEFF8 = COEFF1*(ONE - GAMMA1)*DT1                             TIMCON 35
C                                                                    TIMCON 36
      RETURN                                                        TIMCON 37
      END                                                           TIMCON 38
C***********************************************************************
      SUBROUTINE TIMING(TIME)                                       TIMING  1
C                                                                    TIMING  2
C.... PROGRAM TO DETERMINE ELAPSED CPU TIME                          TIMING  3
C                                                                    TIMING  4
C.... **** THIS IS A SYSTEM-DEPENDENT ROUTINE ****                   TIMING  5
C                                                                    TIMING  6
C....     NOTE: CAN ONLY ACCESS CLOCK TIME ON VAX/VMS                TIMING  7
C                                                                    TIMING  8
      TIME = SECNDS(0.0)                                            TIMING  9
C                                                                    TIMING 10
      RETURN                                                        TIMING 11
      END                                                           TIMING 12
C***********************************************************************
      SUBROUTINE TIMLOG                                             TIMLOG  1
C                                                                    TIMLOG  2
C.... PROGRAM TO PRINT LOG OF EXECUTION TIMES                        TIMLOG  3
C                                                                    TIMLOG  4
      CHARACTER*4 TITLE                                             TIMLOG  5
      COMMON /ETIMEC/ ETIME(7)                                      TIMLOG  6
      COMMON /IOUNIT/ IIN,IOUT,IRSIN,IRSOUT                         TIMLOG  7
      COMMON /TITLEC/ TITLE(20)                                     TIMLOG  8
C                                                                    TIMLOG  9
      SUBTOT = 0.0                                                  TIMLOG 10
      DO 100 I=3,7                                                  TIMLOG 11
      SUBTOT = SUBTOT + ETIME(I)                                    TIMLOG 12
  100 CONTINUE                                                      TIMLOG 13
C                                                                    TIMLOG 14
      WRITE(IOUT,1000) TITLE,ETIME,SUBTOT                           TIMLOG 15
C                                                                    TIMLOG 16
      RETURN                                                        TIMLOG 17
C                                                                    TIMLOG 18
 1000 FORMAT('1',20A4///5X,                                         TIMLOG 19
     &' E X E C U T I O N   T I M I N G   I N F O R M A T I O N' ///5X,TIMLOG 20
     &' I N I T I A L I Z A T I O N   P H A S E       = ',1PE10.3///5X,TIMLOG 21
     &' S O L U T I O N   P H A S E                   = ',1PE10.3///5X,TIMLOG 22
     &'    FORMATION OF LEFT-HAND-SIDE MATRICES       = ',1PE10.3 //5X,TIMLOG 23
     &'    FACTORIZATIONS                             = ',1PE10.3 //5X,TIMLOG 24
     &'    FORMATION OF RIGHT-HAND-SIDE VECTORS       = ',1PE10.3 //5X,TIMLOG 25
     &'    FORWARD REDUCTIONS/BACK SUBSTITUTIONS      = ',1PE10.3 //5X,TIMLOG 26
     &'    CALCULATION OF ELEMENT OUTPUT              = ',1PE10.3  /5X,TIMLOG 27
     &51X,'-------',//5X,                                          TIMLOG 28
     &'            SUBTOTAL                           = ',1PE10.3 )TIMLOG 29
C                                                                    TIMLOG 30
      END                                                           TIMLOG 31
C***********************************************************************
      SUBROUTINE TSEQ                                               TSEQ    1
C                                                                    TSEQ    2
C.... PROGRAM TO SET MEMORY POINTERS FOR TIME SEQUENCE AND           TSEQ    3
C        NODAL TIME HISTORY DATA ARRAYS                              TSEQ    4
```

```
C                                                                         TSEQ      5
       LOGICAL LDYN                                                       TSEQ      6
       COMMON /BPOINT/ MFIRST,MLAST,MTOT,IPREC                            TSEQ      7
       COMMON /DPOINT/ MPSTEP,MPDPRT,MPSPRT,MPHPLT,MPITER,MPALPH,MPBETA,  TSEQ      8
      &                MPGAMM,MPDT  ,MPIDHS,MPDOUT,MPVPRD,MPDPRD,MPA,MPV   TSEQ      9
       COMMON /HPLOTC/ NPLPTS,LOCPLT,TIME                                 TSEQ     10
       COMMON /INFO  / IEXEC,IACODE,LDYN,IREADR,IWRITR,IPRTIN,IRANK,      TSEQ     11
      &                NUMSEQ,NDOUT,NSD,NUMNP,NDOF,NLVECT,NLTFTN,NPTSLF,  TSEQ     12
      &                NUMEG                                              TSEQ     13
       COMMON A(1)                                                        TSEQ     14
C                                                                         TSEQ     15
       MPSTEP = MPOINT('NSTEP   ',NUMSEQ,0,0,1)                           TSEQ     16
       MPDPRT = MPOINT('NDPRT   ',NUMSEQ,0,0,1)                           TSEQ     17
       MPSPRT = MPOINT('NSPRT   ',NUMSEQ,0,0,1)                           TSEQ     18
       MPHPLT = MPOINT('NHPLT   ',NUMSEQ,0,0,1)                           TSEQ     19
       MPITER = MPOINT('NITER   ',NUMSEQ,0,0,1)                           TSEQ     20
       MPALPH = MPOINT('ALPHA   ',NUMSEQ,0,0,IPREC)                       TSEQ     21
       MPBETA = MPOINT('BETA    ',NUMSEQ,0,0,IPREC)                       TSEQ     22
       MPGAMM = MPOINT('GAMMA   ',NUMSEQ,0,0,IPREC)                       TSEQ     23
       MPDT   = MPOINT('DT      ',NUMSEQ,0,0,IPREC)                       TSEQ     24
C                                                                         TSEQ     25
       CALL TSEQIN(A(MPSTEP),A(MPDPRT),A(MPSPRT),A(MPHPLT),               TSEQ     26
      &            A(MPITER),A(MPALPH),A(MPBETA),A(MPGAMM),               TSEQ     27
      &            A(MPDT  ),NUMSEQ,NPLPTS,LDYN)                          TSEQ     28
C                                                                         TSEQ     29
       IF (NDOUT.EQ.0) THEN                                              TSEQ     30
          MPIDHS = 1                                                      TSEQ     31
          MPDOUT = 1                                                      TSEQ     32
       ELSE                                                               TSEQ     33
          MPIDHS = MPOINT('IDHIST  ',3       ,NDOUT  ,0,1)                TSEQ     34
          MPDOUT = MPOINT('DOUT    ',NDOUT+1,NPLPTS,0,1)                  TSEQ     35
       ENDIF                                                              TSEQ     36
C                                                                         TSEQ     37
       RETURN                                                             TSEQ     38
       END                                                                TSEQ     39
C*************************************************************************************
       SUBROUTINE TSEQIN(NSTEP,NDPRT,NSPRT,NHPLT,NITER,ALPHA,BETA,        TSEQIN    1
      &                  GAMMA,DT,NUMSEQ,NPLPTS,LDYN)                     TSEQIN    2
C                                                                         TSEQIN    3
C.... PROGRAM TO READ, WRITE AND STORE TIME SEQUENCE DATA                 TSEQIN    4
C                                                                         TSEQIN    5
C        NOTE: "NPLPTS" IS PASSED TO SUBROUTINE HPLOT BY WAY OF           TSEQIN    6
C              COMMON /HPLOTC/                                            TSEQIN    7
C                                                                         TSEQIN    8
       IMPLICIT DOUBLE PRECISION (A-H,O-Z)                               TSEQIN    9
C                                                                         TSEQIN   10
C.... DEACTIVATE ABOVE CARD(S) FOR SINGLE PRECISION OPERATION             TSEQIN   11
C                                                                         TSEQIN   12
       LOGICAL LDYN                                                       TSEQIN   13
       DIMENSION NSTEP(1),NDPRT(1),NSPRT(1),NHPLT(1),NITER(1)             TSEQIN   14
      &         ,ALPHA(1),BETA(1),GAMMA(1),DT(1)                          TSEQIN   15
       COMMON /CONSTS/ ZERO,PT1667,PT25,PT5,ONE,TWO,THREE,FOUR,FIVE       TSEQIN   16
       COMMON /IOUNIT/ IIN,IOUT,IRSIN,IRSOUT                              TSEQIN   17
C                                                                         TSEQIN   18
       NPLPTS = 1                                                         TSEQIN   19
C                                                                         TSEQIN   20
       DO 100 I=1,NUMSEQ                                                  TSEQIN   21
       READ(IIN,1000) N,NSTEP(N),NDPRT(N),NSPRT(N),NHPLT(N),NITER(N),     TSEQIN   22
      &               ALPHA(N),BETA(N) ,GAMMA(N),DT(N)                    TSEQIN   23
       IF (NHPLT(N).GT.0) NPLPTS = NPLPTS + NSTEP(N)/NHPLT(N)             TSEQIN   24
  100 CONTINUE                                                            TSEQIN   25
C                                                                         TSEQIN   26
C.... SET DEFAULT SEQUENCE PARAMETERS FOR STATIC ANALYSIS                 TSEQIN   27
C                                                                         TSEQIN   28
       IF (.NOT.LDYN) THEN                                               TSEQIN   29
          DO 200 I=1,NUMSEQ                                               TSEQIN   30
          NSTEP(I) = MAX0(1,NSTEP(I))                                     TSEQIN   31
          NDPRT(I) = 1                                                    TSEQIN   32
          NSPRT(I) = 1                                                    TSEQIN   33
          NHPLT(I) = 0                                                    TSEQIN   34
          NITER(I) = 1                                                    TSEQIN   35
          ALPHA(I) = ZERO                                                 TSEQIN   36
          BETA(I)  = ONE                                                  TSEQIN   37
          GAMMA(I) = ZERO                                                 TSEQIN   38
          DT(I)    = ONE                                                  TSEQIN   39
  200     CONTINUE                                                        TSEQIN   40
       ENDIF                                                              TSEQIN   41
C                                                                         TSEQIN   42
       DO 300 N=1,NUMSEQ                                                  TSEQIN   43
       IF (MOD(N,2).EQ.1) WRITE(IOUT,2000) NUMSEQ                         TSEQIN   44
       WRITE(IOUT,3000) N,NSTEP(N),NDPRT(N),NSPRT(N),NHPLT(N),NITER(N),   TSEQIN   45
      &                 ALPHA(N),BETA(N) ,GAMMA(N),DT(N)                  TSEQIN   46
  300 CONTINUE                                                            TSEQIN   47
C                                                                         TSEQIN   48
       RETURN                                                             TSEQIN   49
```

```
C
 1000 FORMAT(6I5,4F10.0)
 2000 FORMAT('1',' T I M E   S E Q U E N C E   D A T A      ',     //5X,
     &' NUMBER OF TIME SEQUENCES . . . . . . (NUMSEQ  ) = ',   I5///  )
 3000 FORMAT(5X,
     &' TIME SEQUENCE NUMBER . . . . . . . . (N       ) = ',     I5//5X,
     &' NUMBER OF TIME STEPS . . . . . . . . (NSTEP(N)) = ',     I5//5X,
     &' KINEMATIC PRINT INCREMENT  . . . . . (NDPRT(N)) = ',     I5//5X,
     &' STRESS/STRAIN PRINT INCREMENT  . . . (NSPRT(N)) = ',     I5//5X,
     &' TIME HISTORY PLOT INCREMENT  . . . . (NHPLT(N)) = ',     I5//5X,
     &' NUMBER OF ITERATIONS . . . . . . . . (NITER(N)) = ',     I5//5X,
     &' FIRST INTEGRATION PARAMETER  . . . . (ALPHA(N)) = ',1PE12.5//5X,
     &' SECOND INTEGRATION PARAMETER . . . . (BETA(N) ) = ',1PE12.5//5X,
     &' THIRD INTEGRATION PARAMETER  . . . . (GAMMA(N)) = ',1PE12.5//5X,
     &' TIME STEP  . . . . . . . . . . . . . (DT(N)   ) = ',1PE12.5////)
C
      END
C**************************************************************************
      SUBROUTINE ZTEST(A,N,LZERO)
C
C.... PROGRAM TO DETERMINE IF AN ARRAY CONTAINS ONLY ZERO ENTRIES
C
      IMPLICIT DOUBLE PRECISION (A-H,O-Z)
C
C.... DEACTIVATE ABOVE CARD(S) FOR SINGLE PRECISION OPERATION
C
      DIMENSION A(1)
      COMMON /CONSTS/ ZERO,PT1667,PT25,PT5,ONE,TWO,THREE,FOUR,FIVE
      LOGICAL LZERO
C
      LZERO = .TRUE.
C
      DO 100 I=1,N
      IF (A(I).NE.ZERO) THEN
         LZERO = .FALSE.
         RETURN
      ENDIF
  100 CONTINUE
C
      RETURN
      END
C**************************************************************************
      SUBROUTINE QUADC(ITASK,NPAR,MP,NEG)
C
C.... PROGRAM TO SET STORAGE AND CALL TASKS FOR THE
C        FOUR-NODE QUADRILATERAL, ELASTIC CONTINUUM ELEMENT
C
      DOUBLE PRECISION TIME
C
C.... DEACTIVATE ABOVE CARD(S) FOR SINGLE-PRECISION OPERATION
C
      LOGICAL LDYN
      DIMENSION NPAR(1),MP(1)
      COMMON /BPOINT/ MFIRST,MLAST,MTOT,IPREC
      COMMON /DPOINT/ MPSTEP,MPDPRT,MPSPRT,MPHPLT,MPITER,MPALPH,MPBETA,
     &                MPGAMM,MPDT  ,MPIDHS,MPDOUT,MPVPRD,MPDPRD,MPA,MPV
      COMMON /HPLOTC/ NPLPTS,LOCPLT,TIME
      COMMON /INFO  / IEXEC,IACODE,LDYN,IREADR,IWRITR,IPRTIN,IRANK,
     &                NUMSEQ,NDOUT,NSD,NUMNP,NDOF,NLVECT,NLTFTN,NPTSLF,
     &                NUMEG
      COMMON /SPOINT/ MPD,MPX,MPID,MPF,MPG,MPG1,MPDIAG,MPNGRP,
     &                MPALHS,MPBRHS
      COMMON A(1)
```

```
C                                                                        QUADC  22
      MW       = 1                                                       QUADC  23
      MDET     = 2                                                       QUADC  24
      MR       = 3                                                       QUADC  25
      MSHL     = 4                                                       QUADC  26
      MSHG     = 5                                                       QUADC  27
      MSHGBR   = 6                                                       QUADC  28
      MRHO     = 7                                                       QUADC  29
      MRDPM    = 8                                                       QUADC  30
      MRDPK    = 9                                                       QUADC  31
      MTH      = 10                                                      QUADC  32
      MC       = 11                                                      QUADC  33
      MGRAV    = 12                                                      QUADC  34
      MIEN     = 13                                                      QUADC  35
      MMAT     = 14                                                      QUADC  36
      MLM      = 15                                                      QUADC  37
      MIELNO   = 16                                                      QUADC  38
      MISIDE   = 17                                                      QUADC  39
      MPRESS   = 18                                                      QUADC  40
      MSHEAR   = 19                                                      QUADC  41
      MISHST   = 20                                                      QUADC  42
      MSOUT    = 21                                                      QUADC  43
      MELEFM   = 22                                                      QUADC  44
      MXL      = 23                                                      QUADC  45
      MWORK    = 24                                                      QUADC  46
      MB       = 25                                                      QUADC  47
      MDMAT    = 26                                                      QUADC  48
      MDB      = 27                                                      QUADC  49
      MVL      = 28                                                      QUADC  50
      MAL      = 29                                                      QUADC  51
      MELRES   = 30                                                      QUADC  52
      MDL      = 31                                                      QUADC  53
      MSTRN    = 32                                                      QUADC  54
      MSTRS    = 33                                                      QUADC  55
      MPSTRN   = 34                                                      QUADC  56
      MPSTRS   = 35                                                      QUADC  57
C                                                                        QUADC  58
      NTYPE    = NPAR( 1)                                                QUADC  59
      NUMEL    = NPAR( 2)                                                QUADC  60
      NUMAT    = NPAR( 3)                                                QUADC  61
      NSURF    = NPAR( 4)                                                QUADC  62
      NSOUT    = NPAR( 5)                                                QUADC  63
      IOPT     = NPAR( 6)                                                QUADC  64
      ISTPRT   = NPAR( 7)                                                QUADC  65
      LFSURF   = NPAR( 8)                                                QUADC  66
      LFBODY   = NPAR( 9)                                                QUADC  67
      NICODE   = NPAR(10)                                                QUADC  68
      IBBAR    = NPAR(11)                                                QUADC  69
      IMASS    = NPAR(12)                                                QUADC  70
      IMPEXP   = NPAR(13)                                                QUADC  71
C                                                                        QUADC  72
C.... SET ELEMENT PARAMETERS                                             QUADC  73
C                                                                        QUADC  74
      NEN      = 4                                                       QUADC  75
      NED      = 2                                                       QUADC  76
      NEE      = NEN*NED                                                 QUADC  77
      NESD     = 2                                                       QUADC  78
      NROWSH   = 3                                                       QUADC  79
      NEESQ    = NEE*NEE                                                 QUADC  80
      NROWB    = 4                                                       QUADC  81
      NSTR     = 3                                                       QUADC  82
      IF ( (IOPT.EQ.2) .OR. (IBBAR.EQ.1) ) NSTR = 4                      QUADC  83
      NINT     = 1                                                       QUADC  84
      IF (NICODE.EQ.0) NINT = 4                                          QUADC  85
      NRINT    = 1                                                       QUADC  86
      IF (ITASK.EQ.1) THEN                                               QUADC  87
C                                                                        QUADC  88
C....... SET MEMORY POINTERS                                             QUADC  89
C                                                                        QUADC  90
C                                                                        QUADC  91
C         NOTE:   THE MP ARRAY IS STORED DIRECTLY AFTER THE NPAR ARRAY,  QUADC  92
C                 BEGINNING AT LOCATION MPNPAR + 16 OF BLANK COMMON.     QUADC  93
C                 THE VARIABLE "JUNK" IS NOT USED SUBSEQUENTLY.          QUADC  94
C                                                                        QUADC  95
          JUNK       = MPOINT('MP      ',35      ,0      ,0      ,1)     QUADC  96
```

```
C
          MP(MW     ) = MPOINT('W       ',NINT   ,0     ,0     ,IPREC)
          MP(MDET   ) = MPOINT('DET     ',NINT   ,0     ,0     ,IPREC)
          MP(MR     ) = MPOINT('R       ',NINT   ,0     ,0     ,IPREC)
          MP(MSHL   ) = MPOINT('SHL     ',NROWSH ,NEN   ,NINT  ,IPREC)
          MP(MSHG   ) = MPOINT('SHG     ',NROWSH ,NEN   ,NINT  ,IPREC)
          MP(MSHGBR) = MPOINT('SHGBAR  ',NROWSH ,NEN   ,NRINT ,IPREC)
          MP(MRHO   ) = MPOINT('RHO     ',NUMAT  ,0     ,0     ,IPREC)
          MP(MRDPM  ) = MPOINT('RDAMPM  ',NUMAT  ,0     ,0     ,IPREC)
          MP(MRDPK  ) = MPOINT('RDAMPK  ',NUMAT  ,0     ,0     ,IPREC)
          MP(MTH    ) = MPOINT('TH      ',NUMAT  ,0     ,0     ,IPREC)
          MP(MC     ) = MPOINT('C       ',NROWB  ,NROWB ,NUMAT ,IPREC)
          MP(MGRAV  ) = MPOINT('GRAV    ',NESD   ,0     ,0     ,IPREC)
          MP(MIEN   ) = MPOINT('IEN     ',NEN    ,NUMEL ,0     ,1)
          MP(MMAT   ) = MPOINT('MAT     ',NUMEL  ,0     ,0     ,1)
          MP(MLM    ) = MPOINT('LM      ',NED    ,NEN   ,NUMEL ,1)
          MP(MIELNO) = MPOINT('IELNO   ',NSURF  ,0     ,0     ,1)
          MP(MISIDE) = MPOINT('ISIDE   ',NSURF  ,0     ,0     ,1)
          MP(MPRESS) = MPOINT('PRESS   ',2      ,NSURF ,0     ,IPREC)
          MP(MSHEAR) = MPOINT('SHEAR   ',2      ,NSURF ,0     ,IPREC)
C
      IF (NSOUT.EQ.0) THEN
          MP(MISHST) = JUNK
          MP(MSOUT ) = JUNK
      ELSE
          MP(MISHST) = MPOINT('ISHIST  ',3      ,NSOUT ,0     ,1)
          MP(MSOUT ) = MPOINT('SOUT    ',NSOUT+1,NPLPTS,0     ,1)
      ENDIF
C
          MP(MELEFM) = MPOINT('ELEFFM  ',NEE    ,NEE   ,0     ,IPREC)
          MP(MXL    ) = MPOINT('XL      ',NESD   ,NEN   ,0     ,IPREC)
          MP(MWORK  ) = MPOINT('WORK    ',16     ,0     ,0     ,IPREC)
          MP(MB     ) = MPOINT('B       ',NROWB  ,NEE   ,0     ,IPREC)
          MP(MDMAT  ) = MPOINT('DMAT    ',NROWB  ,NROWB ,0     ,IPREC)
          MP(MDB    ) = MPOINT('DB      ',NROWB  ,NEE   ,0     ,IPREC)
          MP(MVL    ) = MPOINT('VL      ',NED    ,NEN   ,0     ,IPREC)
          MP(MAL    ) = MPOINT('AL      ',NED    ,NEN   ,0     ,IPREC)
          MP(MELRES) = MPOINT('ELRESF  ',NEE    ,0     ,0     ,IPREC)
          MP(MDL    ) = MPOINT('DL      ',NED    ,NEN   ,0     ,IPREC)
          MP(MSTRN  ) = MPOINT('STRAIN  ',NROWB  ,0     ,0     ,IPREC)
          MP(MSTRS  ) = MPOINT('STRESS  ',NROWB  ,0     ,0     ,IPREC)
          MP(MPSTRN) = MPOINT('PSTRN   ',NROWB  ,0     ,0     ,IPREC)
          MP(MPSTRS) = MPOINT('PSTRS   ',NROWB  ,0     ,0     ,IPREC)
      ENDIF
C
C.... TASK CALLS
C
      IF (ITASK.GT.6) RETURN
      GO TO (100,200,300,400,500,600),ITASK
C
  100 CONTINUE
C
C.... INPUT ELEMENT DATA ('INPUT___')
C
      CALL QDCT1(A(MP(MSHL  )),A(MP(MW    )),A(MP(MRHO  )),
     &           A(MP(MRDPM )),A(MP(MRDPK )),A(MP(MTH   )),
     &           A(MP(MC    )),A(MP(MGRAV )),A(MP(MIEN  )),
     &           A(MP(MMAT  )),A(MPID       ),A(MP(MLM   )),
     &           A(MPDIAG     ),A(MP(MIELNO)),A(MP(MISIDE)),
     &           A(MP(MPRESS)),A(MP(MSHEAR)),A(MP(MISHST)),
     &           NTYPE ,NUMEL ,NUMAT ,NSURF ,NSOUT ,IOPT  ,
     &           ISTPRT,LFSURF,LFBODY,NICODE,NINT  ,IBBAR ,
     &           IMASS ,IMPEXP,NROWSH,NROWB ,NESD  ,NEN   ,
     &           NDOF  ,NED   ,IPRTIN,LDYN  )
C
      RETURN
C
  200 CONTINUE
C
C.... FORM ELEMENT EFFECTIVE MASS AND ASSEMBLE INTO GLOBAL
C        LEFT-HAND-SIDE MATRIX  ('FORM_LHS')
C
      CALL QDCT2(A(MP(MELEFM)),A(MP(MIEN  )),A(MPX         ),
     &           A(MP(MXL   )),A(MP(MMAT  )),A(MP(MDET  )),
     &           A(MP(MSHL  )),A(MP(MSHG  )),A(MP(MR    )),
     &           A(MP(MRDPM )),A(MP(MRDPK )),A(MP(MTH   )),
     &           A(MP(MRHO  )),A(MP(MW    )),A(MP(MWORK )),
     &           A(MP(MSHGBR)),A(MP(MB    )),A(MP(MC    )),
     &           A(MP(MDMAT )),A(MP(MDB   )),A(MPALHS      ),
     &           A(MPDIAG     ),A(MP(MLM   )),
     &           IMPEXP,IMASS ,NUMEL ,NEESQ ,NEN   ,NSD   ,
     &           NESD  ,NINT  ,NEG   ,NROWSH,LDYN  ,NED   ,
     &           IOPT  ,IBBAR ,NROWB ,NSTR  ,NEE   )
C
      RETURN
C
  300 CONTINUE
C
C.... FORM ELEMENT RESIDUAL-FORCE VECTOR AND ASSEMBLE INTO GLOBAL
```

```
C        RIGHT-HAND-SIDE VECTOR ('FORM_RHS')
C
      CALL QDCT3(A(MP(MMAT  )),A(MP(MIEN  )),A(MPDPRD     ),
     &           A(MP(MDL   )),A(MPVPRD    ),A(MP(MVL   )),
     &           A(MPA         ),A(MP(MAL   )),A(MP(MRDPK )),
     &           A(MP(MRDPM )),A(MP(MRHO  )),A(MP(MGRAV )),
     &           A(MP(MELRES)),A(MPX       ),A(MP(MXL   )),
     &           A(MP(MDET  )),A(MP(MSHL  )),A(MP(MSHG  )),
     &           A(MP(MR    )),A(MPG1      ),A(MP(MWORK )),
     &           A(MP(MTH   )),A(MP(MW    )),A(MP(MELEFM)),
     &           A(MP(MSHGBR)),A(MP(MB    )),A(MP(MSTRN )),
     &           A(MP(MC    )),A(MP(MSTRS )),A(MPBRHS     ),
     &           A(MP(MLM   )),A(MP(MIELNO)),A(MP(MISIDE)),
     &           A(MP(MPRESS)),A(MP(MSHEAR)),
     &           NUMEL ,NED   ,NEN   ,NDOF  ,LDYN  ,NEE   ,
     &           IMASS ,NESD  ,LFBODY,NSD   ,NINT  ,NROWSH,
     &           NEG   ,IOPT  ,NROWB ,NSTR  ,IBBAR , NSURF,
     &           LFSURF)
C
      RETURN
C
  400 CONTINUE
C
C.... CALCULATE AND PRINT ELEMENT STRESS/STRAIN OUTPUT ('STR_PRNT')
C
      IF (ISTPRT.EQ.0)
     &   CALL QDCT4(A(MP(MMAT  )),A(MP(MIEN  )),A(MPD         ),
     &              A(MP(MDL   )),A(MPX        ),A(MP(MXL   )),
     &              A(MP(MDET  )),A(MP(MSHL  )),A(MP(MSHG  )),
     &              A(MP(MWORK )),A(MP(MR    )),A(MP(MSHGBR)),
     &              A(MP(MW    )),A(MP(MB    )),A(MP(MSTRN )),
     &              A(MP(MC    )),A(MP(MSTRS )),A(MP(MPSTRN)),
     &              A(MP(MPSTRS)),
     &              NINT  ,NUMEL ,NEN   ,NDOF  ,NED   ,NSD   ,
     &              NESD  ,NROWSH,NEG   ,IOPT  ,IBBAR ,NROWB ,
     &              NEE   ,NSTR  )
C
      RETURN
C
  500 CONTINUE
C
C.... CALCULATE AND STORE ELEMENT TIME-HISTORIES ('STR_STOR')
C
      IF (NSOUT.GT.0)
     &   CALL QDCT5(A(MP(MISHST)),A(MP(MSOUT )),A(MP(MMAT  )),
     &              A(MP(MIEN  )),A(MPD         ),A(MP(MDL   )),
     &              A(MPX         ),A(MP(MXL   )),A(MP(MDET  )),
     &              A(MP(MSHL  )),A(MP(MSHG  )),A(MP(MR    )),
     &              A(MP(MSHGBR)),A(MP(MW    )),A(MP(MB    )),
     &              A(MP(MSTRN )),A(MP(MC    )),A(MP(MSTRS )),
     &              A(MP(MPSTRN)),A(MP(MPSTRS)),A(MP(MWORK )),
     &              NSOUT ,NEN   ,NDOF  ,NED   ,NSD   ,NESD  ,
     &              NROWSH,NINT  ,NEG   ,IOPT  ,IBBAR ,NROWB ,
     &              NEE   ,NSTR  )
C
      RETURN
C
  600 CONTINUE
C
C.... PLOT ELEMENT TIME-HISTORIES ('STR_PLOT')
C
      IF (NSOUT.GT.0)
     &   CALL HPLOT(A(MP(MISHST)),A(MP(MSOUT )),NSOUT ,3,NTYPE )
      RETURN
C
      END
C**********************************************************************
      SUBROUTINE QDCT1(SHL   ,W     ,RHO   ,RDAMPM,RDAMPK,TH    ,
     &                 C     ,GRAV  ,IEN   ,MAT   ,ID    ,LM    ,
     &                 IDIAG ,IELNO ,ISIDE ,PRESS ,SHEAR ,ISHIST,
     &                 NTYPE ,NUMEL ,NUMAT ,NSURF ,NSOUT ,IOPT  ,
     &                 ISTPRT,LFSURF,LFBODY,NICODE,NINT  ,IBBAR ,
     &                 IMASS ,IMPEXP,NROWSH,NROWB ,NESD  ,NEN   ,
     &                 NDOF  ,NED   ,IPRTIN,LDYN  )
C
C.... PROGRAM TO READ, GENERATE AND WRITE DATA FOR THE
C        FOUR-NODE QUADRILATERAL, ELASTIC CONTINUUM ELEMENT
C
C
C
      IMPLICIT DOUBLE PRECISION (A-H,O-Z)
C
C.... DEACTIVATE ABOVE CARD(S) FOR SINGLE-PRECISION OPERATION
C
      LOGICAL LDYN
      DIMENSION SHL(NROWSH,NEN,1),W(1),RHO(1),RDAMPM(1),RDAMPK(1),
     &          TH(1),C(NROWB,NROWB,1),GRAV(NESD),IEN(NEN,1),MAT(1),
     &          ID(NDOF,1),LM(NED,NEN,1),IDIAG(1),IELNO(1),ISIDE(1),
     &          PRESS(2,1),SHEAR(2,1),ISHIST(3,1)
      COMMON /IOUNIT/ IIN,IOUT,IRSIN,IRSOUT
```

```
C
      WRITE(IOUT,1000) NTYPE,NUMEL,NUMAT,NSURF,NSOUT,IOPT,ISTPRT,
     &                 LFSURF,LFBODY
      WRITE(IOUT,2000) NICODE,IBBAR
      IF (LDYN) WRITE(IOUT,3000) IMASS,IMPEXP
C
      CALL QDCSHL(SHL,W,NINT)
C
      CALL PROP2D(RHO,RDAMPM,RDAMPK,TH,C,NUMAT,IOPT,NROWB)
C
      READ (IIN,4000) GRAV
      WRITE (IOUT,5000) GRAV
C
      CALL GENEL(IEN,MAT,NEN)
C
      IF (IPRTIN.EQ.0) CALL PRNTEL(MAT,IEN,NEN,NUMEL)
C
      CALL FORMLM(ID,IEN,LM,NDOF,NED,NEN,NUMEL)
C
      IF ( (.NOT.LDYN) .OR. (IMPEXP.EQ.0) .OR. (IMASS.EQ.0) )
     &   CALL COLHT(IDIAG,LM,NED,NEN,NUMEL)
C
      IF (NSURF.GT.0) CALL QDCRSF(IELNO,ISIDE,PRESS,SHEAR,NSURF)
C
      IF (NSOUT.GT.0) CALL SHIST(ISHIST,NSOUT,NTYPE)
C
      RETURN
C
 1000 FORMAT(' ',' F O U R - N O D E ',
     &' Q U A D R I L A T E R A L   E L E M E N T S',          //5X,
     &' ELEMENT TYPE NUMBER . . . . . . . . . . . (NTYPE ) = ',I5//5X,
     &' NUMBER OF ELEMENTS . . . . . . . . . . . (NUMEL ) = ',I5//5X,
     &' NUMBER OF ELEMENT MATERIAL SETS . . . . . (NUMAT ) = ',I5//5X,
     &' NUMBER OF SURFACE FORCE CARDS . . . . . . (NSURF ) = ',I5//5X,
     &' NUMBER OF STRESS/STRAIN TIME HISTORIES . . (NSOUT ) = ',I5//5X,
     &' ANALYSIS OPTION . . . . . . . . . . . . . (IOPT  ) = ',I5//5X,
     &'    EQ.0, PLANE STRESS                              ',    /5X,
     &'    EQ.1, PLANE STRAIN                              ',    /5X,
     &'    EQ.2, AXISYMMETRIC                              ',   //5X,
     &' STRESS OUTPUT PRINT CODE . . . . . . . . (ISTPRT) = ',I5//5X,
     &'    EQ.0, STRESS OUTPUT PRINTED                     ',    /5X,
     &'    EQ.1, STRESS OUTPUT NOT PRINTED                 ',   //5X,
     &' SURFACE FORCE LOAD-TIME FUNCTION NUMBER . . (LFSURF) = ',I5//5X,
     &' BODY FORCE LOAD-TIME FUNCTION NUMBER  . . . (LFBODY) = ',I5 /5X)
 2000 FORMAT(5X,
     &' NUMERICAL INTEGRATION CODE . . . . . . . (NICODE) = ',I5//5X,
     &'    EQ.0, 2 X 2 GAUSSIAN QUADRATURE                 ',    /5X,
     &'    EQ.1, 1-POINT GAUSSIAN QUADRATURE               ',   //5X,
     &' STRAIN-DISPLACEMENT OPTION . . . . . . . . (IBBAR ) = ',I5//5X,
     &'    EQ.0, STANDARD FORMULATION                      ',    /5X,
     &'    EQ.1, B-BAR FORMULATION                         ',    /5X)
 3000 FORMAT(5X,
     &' MASS TYPE CODE . . . . . . . . . . . . . (IMASS ) = ',I5//5X,
     &'    EQ.0, CONSISTENT MASS MATRIX                    ',    /5X,
     &'    EQ.1, LUMPED MASS MATRIX                        ',    /5X,
     &'    EQ.2, NO MASS MATRIX                            ',   //5X,
     &' IMPLICIT/EXPLICIT CODE . . . . . . . . . (IMPEXP) = ',I5//5X,
     &'    EQ.0, IMPLICIT ELEMENT GROUP                    ',    /5X,
     &'    EQ.1, EXPLICIT ELEMENT GROUP                    ',   //5X)
 4000 FORMAT(8F10.0)
 5000 FORMAT(////' ',
     &' G R A V I T Y   V E C T O R   C O M P O N E N T S     ',  //5X,
     &' X-1 DIRECTION . . . . . . . . . . . . . . = ',    1PE15.8//5X,
     &' X-2 DIRECTION . . . . . . . . . . . . . . = ',    1PE15.8//5X)
C
      END
C**********************************************************************
      SUBROUTINE QDCT2(ELEFFM,IEN   ,X    ,XL    ,MAT   ,DET   ,
     &                 SHL   ,SHG   ,R    ,RDAMPM,RDAMPK,TH    ,
     &                 RHO   ,W     ,WORK ,SHGBAR,B     ,C     ,
     &                 DMAT  ,DB    ,ALHS ,IDIAG ,LM    ,
     &                 IMPEXP,IMASS ,NUMEL,NEESQ ,NEN   ,NSD   ,
     &                 NESD  ,NINT  ,NEG  ,NROWSH,LDYN  ,NED   ,
     &                 IOPT  ,IBBAR ,NROWB,NSTR  ,NEE   )
C
C.... PROGRAM TO CALCULATE EFFECTIVE MASS MATRIX FOR THE
C        FOUR-NODE QUADRILATERAL, ELASTIC CONTINUUM ELEMENT AND
C        ASSEMBLE INTO THE GLOBAL LEFT-HAND-SIDE MATRIX
C
C         IMPEXP = 0, IMPLICIT TIME INTEGRATION
C                = 1, EXPLICIT TIME INTEGRATION
C
      IMPLICIT DOUBLE PRECISION (A-H,O-Z)
C
C.... DEACTIVATE ABOVE CARD(S) FOR SINGLE-PRECISION OPERATION
```

```
C                                                                        QDCT2  19
         LOGICAL LDYN,LDIAG,LQUAD                                        QDCT2  20
         DIMENSION ELEFFM(NEE,1),IEN(NEN,1),X(NSD,1),XL(NESD,1),MAT(1),  QDCT2  21
        &          DET(1),SHL(NROWSH,NEN,1),SHG(NROWSH,NEN,1),R(1),      QDCT2  22
        &          RDAMPM(1),RDAMPK(1),TH(1),RHO(1),W(1),WORK(1),        QDCT2  23
        &          SHGBAR(3,1),B(NROWB,1),C(NROWB,NROWB,1),DMAT(NROWB,1),QDCT2  24
        &          DB(NROWB,1),ALHS(1),IDIAG(1),LM(NED,NEN,1)            QDCT2  25
         COMMON /COEFFS/ COEFF1,COEFF2,COEFF3,COEFF4,COEFF5,COEFF6,      QDCT2  26
        &                COEFF7,COEFF8,ALPHA1,BETA1 ,GAMMA1,DT1          QDCT2  27
         COMMON /CONSTS/ ZERO,PT1667,PT25,PT5,ONE,TWO,THREE,FOUR,FIVE    QDCT2  28
C                                                                        QDCT2  29
         LDIAG = .FALSE.                                                 QDCT2  30
         IF ( (IMPEXP.EQ.1) .AND. (IMASS.EQ.1) ) LDIAG = .TRUE.          QDCT2  31
C                                                                        QDCT2  32
         DO 200 NEL=1,NUMEL                                              QDCT2  33
C                                                                        QDCT2  34
         CALL CLEAR(ELEFFM,NEESQ)                                        QDCT2  35
         CALL LOCAL(IEN(1,NEL),X,XL,NEN,NSD,NESD)                        QDCT2  36
         M = MAT(NEL)                                                    QDCT2  37
         LQUAD = .TRUE.                                                  QDCT2  38
         IF (IEN(3,NEL).EQ.IEN(4,NEL)) LQUAD = .FALSE.                   QDCT2  39
         CALL QDCSHG(XL,DET,SHL,SHG,NINT,NEL,NEG,LQUAD)                  QDCT2  40
C                                                                        QDCT2  41
         IF (IOPT.EQ.2) THEN                                             QDCT2  42
C                                                                        QDCT2  43
            DO 100 L=1,NINT                                              QDCT2  44
            R(L) = ROWDOT(SHG(NROWSH,1,L),XL,NROWSH,NESD,NEN)            QDCT2  45
            DET(L) = DET(L)*R(L)                                         QDCT2  46
  100       CONTINUE                                                     QDCT2  47
C                                                                        QDCT2  48
         ENDIF                                                           QDCT2  49
C                                                                        QDCT2  50
         IF ( LDYN .AND. (IMASS.NE.2) ) THEN                             QDCT2  51
C                                                                        QDCT2  52
C....... FORM MASS MATRIX                                                QDCT2  53
C                                                                        QDCT2  54
            CONSTM = (ONE + RDAMPM(M)*COEFF4)*TH(M)*RHO(M)               QDCT2  55
            IF (CONSTM.NE.ZERO) CALL CONTM(SHG,XL,W,DET,ELEFFM,WORK,     QDCT2  56
        &           CONSTM,IMASS,NINT,NROWSH,NESD,NEN,NED,NEE,.FALSE.)   QDCT2  57
C                                                                        QDCT2  58
         ENDIF                                                           QDCT2  59
C                                                                        QDCT2  60
         IF ( (.NOT.LDYN) .OR. (IMPEXP.EQ.0) ) THEN                      QDCT2  61
C                                                                        QDCT2  62
C....... FORM STIFFNESS MATRIX                                           QDCT2  63
C                                                                        QDCT2  64
            CONSTK = (COEFF4*RDAMPK(M) + COEFF5)*TH(M)                   QDCT2  65
            CALL QDCK(SHGBAR,W,DET,R,SHG,B,C(1,1,M),DMAT,DB,ELEFFM,CONSTK,QDCT2 66
        &          IBBAR,NEN,NINT,IOPT,NESD,NROWSH,NROWB,NSTR,NEE)       QDCT2  67
C                                                                        QDCT2  68
         ENDIF                                                           QDCT2  69
C                                                                        QDCT2  70
C.... ASSEMBLE ELEMENT EFFECTIVE MASS MATRIX INTO GLOBAL                 QDCT2  71
C           LEFT-HAND-SIDE MATRIX                                        QDCT2  72
C                                                                        QDCT2  73
         CALL ADDLHS(ALHS,ELEFFM,IDIAG,LM(1,1,NEL),NEE,LDIAG)            QDCT2  74
C                                                                        QDCT2  75
  200 CONTINUE                                                           QDCT2  76
C                                                                        QDCT2  77
         RETURN                                                          QDCT2  78
         END                                                             QDCT2  79
C*************************************************************************************
         SUBROUTINE QDCT3(MAT     ,IEN     ,DPRED   ,DL      ,VPRED   ,VL      , QDCT3  1
        &                 A       ,AL      ,RDAMPK,RDAMPM,RHO     ,GRAV    , QDCT3  2
        &                 ELRESF,X       ,XL      ,DET     ,SHL     ,SHG     , QDCT3  3
        &                 R       ,G1      ,WORK    ,TH      ,W       ,ELEFFM, QDCT3  4
        &                 SHGBAR,B       ,STRAIN,C       ,STRESS,BRHS    , QDCT3  5
        &                 LM      ,IELNO   ,ISIDE   ,PRESS   ,SHEAR   , QDCT3  6
        &                 NUMEL   ,NED     ,NEN     ,NDOF    ,LDYN    ,NEE , QDCT3  7
        &                 IMASS   ,NESD    ,LFBODY,NSD     ,NINT    ,NROWSH, QDCT3  8
        &                 NEG     ,IOPT    ,NROWB   ,NSTR    ,IBBAR , NSURF, QDCT3  9
        &                 LFSURF)                                        QDCT3  10
C                                                                        QDCT3  11
C.... PROGRAM TO CALCULATE RESIDUAL-FORCE VECTOR FOR THE                 QDCT3  12
C        FOUR-NODE QUADRILATERAL, ELASTIC CONTINUUM ELEMENT AND          QDCT3  13
C        ASSEMBLE INTO THE GLOBAL RIGHT-HAND-SIDE VECTOR                 QDCT3  14
C                                                                        QDCT3  15
         IMPLICIT DOUBLE PRECISION (A-H,O-Z)                             QDCT3  16
C                                                                        QDCT3  17
C.... DEACTIVATE ABOVE CARD(S) FOR SINGLE-PRECISION OPERATION            QDCT3  18
```

```
C
      LOGICAL LDYN,FORMMA,FORMKD,ZEROAL,ZERODL,ZEROG,LQUAD
      DIMENSION MAT(1),IEN(NEN,1),DPRED(NDOF,1),DL(NED,1),VPRED(NDOF,1),
     &          VL(NED,1),A(NDOF,1),AL(NED,1),RDAMPK(1),RDAMPM(1),
     &          RHO(1),GRAV(1),ELRESF(1),X(NSD,1),XL(NESD,1),DET(1),
     &          SHL(NROWSH,NEN,1),SHG(NROWSH,NEN,1),R(1),G1(1),WORK(1),
     &          TH(1),W(1),ELEFFM(NEE,1),SHGBAR(3,1),B(NROWB,1),
     &          STRAIN(1),C(NROWB,NROWB,1),STRESS(1),BRHS(1),
     &          LM(NED,NEN,1),IELNO(1),ISIDE(1),PRESS(2,1),SHEAR(2,1)
      COMMON /CONSTS/ ZERO,PT1667,PT25,PT5,ONE,TWO,THREE,FOUR,FIVE
C
      DO 600 NEL=1,NUMEL
C
      FORMMA = .FALSE.
      FORMKD = .FALSE.
      M = MAT(NEL)
C
C.... NOTE: FOR STATIC ANALYSIS MPDPRD = MPD, HENCE REFERENCE TO
C           ARRAY "DPRED" WILL ACCESS THE CONTENTS OF ARRAY "D".
C
      CALL LOCAL(IEN(1,NEL),DPRED,DL,NEN,NDOF,NED)
      IF (LDYN) THEN
C
         CALL LOCAL(IEN(1,NEL),VPRED,VL,NEN,NDOF,NED)
         CALL LOCAL(IEN(1,NEL),A,AL,NEN,NDOF,NED)
C
         DO 200 J=1,NEN
C
         DO 100 I=1,NED
         DL(I,J) = DL(I,J) + RDAMPK(M)*VL(I,J)
         AL(I,J) = AL(I,J) + RDAMPM(M)*VL(I,J)
  100    CONTINUE
C
  200    CONTINUE
C
         CALL ZTEST(AL,NEE,ZEROAL)
         IF ( (.NOT.ZEROAL) .AND. (IMASS.NE.2) .AND. (RHO(M).NE.ZERO) )
     &      FORMMA = .TRUE.
C
      ELSE
C
         CALL CLEAR(AL,NEE)
C
      ENDIF
C
      CALL ZTEST(DL,NEE,ZERODL)
      IF (.NOT.ZERODL) FORMKD = .TRUE.
      CALL ZTEST(GRAV,NESD,ZEROG)
C
      IF ((.NOT.ZEROG) .AND. (LFBODY.NE.0) .AND. (RHO(M).NE.ZERO)
     &   .AND. (IMASS.NE.2)) THEN
         FORMMA = .TRUE.
         DO 400 I=1,NED
         TEMP = GRAV(I)*G1(LFBODY)
C
         DO 300 J=1,NEN
         AL(I,J) = AL(I,J) - TEMP
  300    CONTINUE
C
  400    CONTINUE
C
      ENDIF
C
      IF (FORMMA.OR.FORMKD) THEN
C
         CALL CLEAR(ELRESF,NEE)
         CALL LOCAL(IEN(1,NEL),X,XL,NEN,NSD,NESD)
         LQUAD = .TRUE.
         IF (IEN(3,NEL).EQ.IEN(4,NEL)) LQUAD = .FALSE.
         CALL QDCSHG(XL,DET,SHL,SHG,NINT,NEL,NEG,LQUAD)
C
         IF (IOPT.EQ.2) THEN
            DO 500 L=1,NINT
            R(L) = ROWDOT(SHG(NROWSH,1,L),XL,NROWSH,NESD,NEN)
            DET(L) = DET(L)*R(L)
  500       CONTINUE
         ENDIF
C
         IF (FORMMA) THEN
C
C.......... FORM INERTIAL AND/OR BODY FORCE
C
            CONSTM = - TH(M)*RHO(M)
            CALL CONTMA(SHG,XL,W,DET,AL,ELEFFM,WORK,ELRESF,CONSTM,IMASS,
     &                  NINT,NROWSH,NESD,NEN,NED,NEE)
         ENDIF
C
         IF (FORMKD) THEN
C
```

```
C......... FORM INTERNAL FORCE                                    QDCT3 108
C                                                                 QDCT3 109
                                                                  QDCT3 110
             CONSTK = - TH(M)                                     QDCT3 111
             CALL QDCKD(SHGBAR,W,DET,R,SHG,B,DL,STRAIN,C(1,1,M),STRESS, QDCT3 112
     &            WORK,ELRESF,CONSTK,IBBAR,NEN,NINT,IOPT,NROWSH,  QDCT3 113
     &            NESD,NROWB,NEE,NSTR)                            QDCT3 114
          ENDIF                                                   QDCT3 115
C                                                                 QDCT3 116
          CALL ADDRHS(BRHS,ELRESF,LM(1,1,NEL),NEE)                QDCT3 117
C                                                                 QDCT3 118
       ENDIF                                                      QDCT3 119
C                                                                 QDCT3 120
  600 CONTINUE                                                    QDCT3 121
C                                                                 QDCT3 122
C.... FORM SURFACE FORCE                                          QDCT3 123
C                                                                 QDCT3 124
C        NOTE: ASSEMBLY OF SURFACE LOADS IS PERFORMED INSIDE QDCSUF QDCT3 125
C                                                                 QDCT3 126
       IF ( (NSURF.GT.0) .AND. (LFSURF.GT.0) )                    QDCT3 127
     &    CALL QDCSUF(IELNO,IEN,X,XL,ISIDE,MAT,TH,PRESS,SHEAR,ELRESF, QDCT3 128
     &          BRHS,LM,G1(LFSURF),NSURF,NEN,NSD,NESD,NED,NEE,IOPT)QDCT3 129
C                                                                 QDCT3 130
       RETURN                                                     QDCT3 131
       END                                                        QDCT3
C***************************************************************************
       SUBROUTINE QDCT4(MAT    ,IEN    ,D     ,DL    ,X      ,XL    , QDCT4   1
     &            DET   ,SHL    ,SHG   ,XINT  ,R      ,SHGBAR, QDCT4   2
     &            W     ,B      ,STRAIN,C     ,STRESS,PSTRN , QDCT4   3
     &            PSTRS ,                                     QDCT4   4
     &            NINT  ,NUMEL  ,NEN   ,NDOF  ,NED    ,NSD   , QDCT4   5
     &            NESD  ,NROWSH ,NEG   ,IOPT  ,IBBAR  ,NROWB , QDCT4   6
     &            NEE   ,NSTR   )                              QDCT4   7
C                                                                 QDCT4   8
C.... PROGRAM TO CALCULATE AND PRINT STRESS, STRAIN AND           QDCT4   9
C        PRINCIPAL VALUES FOR THE FOUR-NODE QUADRILATERAL,        QDCT4  10
C        ELASTIC CONTINUUM ELEMENT                                QDCT4  11
C                                                                 QDCT4  12
       IMPLICIT DOUBLE PRECISION (A-H,O-Z)                        QDCT4  13
C                                                                 QDCT4  14
C.... DEACTIVATE ABOVE CARD(S) FOR SINGLE-PRECISION OPERATION     QDCT4  15
C                                                                 QDCT4  16
       LOGICAL LQUAD                                              QDCT4  17
       DIMENSION MAT(1),IEN(NEN,1),D(NDOF,1),DL(NED,1),X(NSD,1),  QDCT4  18
     &         XL(NESD,1),DET(1),SHL(NROWSH,NEN,1),SHG(NROWSH,NEN,1), QDCT4  19
     &         XINT(NESD,1),R(1),SHGBAR(3,1),W(1),B(NROWB,1),STRAIN(1), QDCT4  20
     &         C(NROWB,NROWB,1),STRESS(1),PSTRN(1),PSTRS(1)       QDCT4  21
       COMMON /CONSTS/ ZERO,PT1667,PT25,PT5,ONE,TWO,THREE,FOUR,FIVE QDCT4  22
C                                                                 QDCT4  23
       NNTOT = 16                                                 QDCT4  24
       NN = 0                                                     QDCT4  25
C                                                                 QDCT4  26
       DO 300 NEL=1,NUMEL                                         QDCT4  27
C                                                                 QDCT4  28
       M = MAT(NEL)                                               QDCT4  29
       CALL LOCAL(IEN(1,NEL),D,DL,NEN,NDOF,NED)                   QDCT4  30
       CALL LOCAL(IEN(1,NEL),X,XL,NEN,NSD,NESD)                   QDCT4  31
       LQUAD = .TRUE.                                             QDCT4  32
       IF (IEN(3,NEL).EQ.IEN(4,NEL)) LQUAD = .FALSE.              QDCT4  33
       CALL QDCSHG(XL,DET,SHL,SHG,NINT,NEL,NEG,LQUAD)             QDCT4  34
C                                                                 QDCT4  35
C.... CALCULATE COORDINATES OF INTEGRATION POINTS                QDCT4  36
C                                                                 QDCT4  37
       DO 100 L=1,NINT                                            QDCT4  38
       XINT(1,L) = ROWDOT(SHG(NROWSH,1,L),XL(1,1),NROWSH,NESD,NEN) QDCT4  39
       XINT(2,L) = ROWDOT(SHG(NROWSH,1,L),XL(2,1),NROWSH,NESD,NEN) QDCT4  40
       IF (IOPT.EQ.2) THEN                                        QDCT4  41
          R(L) = XINT(1,L)                                        QDCT4  42
          DET(L) = DET(L)*R(L)                                    QDCT4  43
       ENDIF                                                      QDCT4  44
  100 CONTINUE                                                    QDCT4  45
C                                                                 QDCT4  46
       IF (IBBAR.EQ.1)                                            QDCT4  47
     &    CALL MEANSH(SHGBAR,W,DET,R,SHG,NEN,NINT,IOPT,NESD,NROWSH) QDCT4  48
C                                                                 QDCT4  49
C.... LOOP OVER INTEGRATION POINTS                               QDCT4  50
C                                                                 QDCT4  51
       DO 200 L=1,NINT                                            QDCT4  52
C                                                                 QDCT4  53
C.... CALCULATE STRESS, STRAIN AND PRINCIPAL VALUES              QDCT4  54
C                                                                 QDCT4  55
       CALL QDCSTR(SHG(1,1,L),SHGBAR,B,R(L),DL,STRAIN,C(1,1,M),STRESS, QDCT4  56
     &        PSTRN,PSTRS,NROWSH,NESD,NROWB,IBBAR,NEN,NED,NEE,NSTR,IOPT) QDCT4  57
C                                                                 QDCT4  58
C.... PRINT STRESS, STRAIN AND PRINCIPAL VALUES                  QDCT4  59
C                                                                 QDCT4  60
       CALL PRTS2D(XINT(1,L),STRESS,PSTRS,STRAIN,PSTRN,           QDCT4  61
     &         NN,NNTOT,NEG,NEL,L)                                QDCT4  62
  200 CONTINUE                                                    QDCT4  63
```

```
C
  300 CONTINUE
C
      RETURN
      END
C**************************************************************************
      SUBROUTINE QDCT5(ISHIST,SOUT  ,MAT   ,IEN   ,D     ,DL    ,
     &                 X     ,XL    ,DET   ,SHL   ,SHG   ,R     ,
     &                 SHGBAR,W     ,B     ,STRAIN,C     ,STRESS,
     &                 PSTRN ,PSTRS ,WORK  ,
     &                 NSOUT ,NEN   ,NDOF  ,NED   ,NSD   ,NESD  ,
     &                 NROWSH,NINT  ,NEG   ,IOPT  ,IBBAR ,NROWB ,
     &                 NEE   ,NSTR  )
C
C.... PROGRAM TO CALCULATE AND STORE ELEMENT TIME-HISTORIES FOR THE
C        FOUR-NODE QUADRILATERAL, ELASTIC CONTINUUM ELEMENT
C
      IMPLICIT DOUBLE PRECISION (A-H,O-Z)
C
C.... DEACTIVATE ABOVE CARD(S) FOR SINGLE-PRECISION OPERATION
C
      REAL SOUT
      LOGICAL LQUAD
      DIMENSION ISHIST(3,1),SOUT(NSOUT+1,1),MAT(1),IEN(NEN,1),D(NDOF,1),
     &          DL(NED,1),X(NSD,1),XL(NESD,1),DET(1),SHL(NROWSH,NEN,1),
     &          SHG(NROWSH,NEN,1),R(1),SHGBAR(3,NEN,1),W(1),B(NROWB,1),
     &          STRAIN(1),C(NROWB,NROWB,1),STRESS(1),PSTRN(1),PSTRS(1),
     &          WORK(1)
      COMMON /HPLOTC/ NPLPTS,LOCPLT,TIME
C
      SOUT(1,LOCPLT) = REAL(TIME)
C
      DO 300 I=1,NSOUT
C
      NEL   = ISHIST(1,I)
      INTPT = ISHIST(2,I)
      NCOMP = ISHIST(3,I)
C
      M = MAT(NEL)
      CALL LOCAL(IEN(1,NEL),D,DL,NEN,NDOF,NED)
      CALL LOCAL(IEN(1,NEL),X,XL,NEN,NSD,NESD)
      LQUAD = .TRUE.
      IF (IEN(3,NEL).EQ.IEN(4,NEL)) LQUAD = .FALSE.
      CALL QDCSHG(XL,DET,SHL,SHG,NINT,NEL,NEG,LQUAD)
C
      IF (IOPT.EQ.2) THEN
C
         DO 100 L=1,NINT
         R(L) = ROWDOT(SHG(NROWSH,1,L),XL,NROWSH,NESD,NEN)
         DET(L) = DET(L)*R(L)
  100    CONTINUE
C
      ENDIF
C
      IF (IBBAR.EQ.1)
     &   CALL MEANSH(SHGBAR,W,DET,R,SHG,NEN,NINT,IOPT,NESD,NROWSH)
C
C.... CALCULATE STRESS, STRAIN AND PRINCIPAL VALUES
C
      CALL QDCSTR(SHG(1,1,INTPT),SHGBAR,B,R(INTPT),DL,STRAIN,C(1,1,M),
     &            STRESS,PSTRN,PSTRS,NROWSH,NESD,NROWB,IBBAR,NEN,NED,
     &            NEE,NSTR,IOPT)
C
      DO 200 J=1,4
      WORK(J     ) = STRESS(J)
      WORK(J +  4) = PSTRS(J)
      WORK(J +  8) = STRAIN(J)
      WORK(J + 12) = PSTRN(J)
  200 CONTINUE
C
      SOUT(I+1,LOCPLT) = REAL(WORK(NCOMP))
C
  300 CONTINUE
C
      RETURN
      END
C**************************************************************************
      SUBROUTINE QDCB(SHG,SHGBAR,B,R,IOPT,NROWSH,NROWB,NEN,IBBAR)
C
C.... PROGRAM TO SET UP THE STRAIN-DISPLACEMENT MATRIX "B" FOR
C        TWO-DIMENSIONAL CONTINUUM ELEMENTS
C
C        IBBAR = 0, STANDARD B-MATRIX
C
C        IBBAR = 1, MEAN-DILATATIONAL B-MATRIX
C
C
      IMPLICIT DOUBLE PRECISION (A-H,O-Z)
C
```

```
C.... DEACTIVATE ABOVE CARD(S) FOR SINGLE-PRECISION OPERATION         QDCB   13
C                                                                     QDCB   14
      DIMENSION SHG(NROWSH,1),SHGBAR(3,1),B(NROWB,1)                   QDCB   15
      COMMON /CONSTS/ ZERO,PT1667,PT25,PT5,ONE,TWO,THREE,FOUR,FIVE     QDCB   16
C                                                                     QDCB   17
      DO 100 J=1,NEN                                                  QDCB   18
C                                                                     QDCB   19
      J2   = 2*J                                                      QDCB   20
      J2M1 = J2 - 1                                                   QDCB   21
C                                                                     QDCB   22
      B(1,J2M1) = SHG(1,J)                                            QDCB   23
      B(1,J2  ) = ZERO                                                QDCB   24
      B(2,J2M1) = ZERO                                                QDCB   25
      B(2,J2  ) = SHG(2,J)                                            QDCB   26
      B(3,J2M1) = SHG(2,J)                                            QDCB   27
      B(3,J2  ) = SHG(1,J)                                            QDCB   28
C                                                                     QDCB   29
      IF (IOPT.EQ.2) THEN                                             QDCB   30
         B(4,J2M1) = SHG(3,J)/R                                       QDCB   31
         B(4,J2  ) = ZERO                                             QDCB   32
      ENDIF                                                           QDCB   33
C                                                                     QDCB   34
  100 CONTINUE                                                        QDCB   35
C                                                                     QDCB   36
      IF (IBBAR.EQ.0) RETURN                                          QDCB   37
C                                                                     QDCB   38
C.... ADD CONTRIBUTIONS TO FORM B-BAR                                 QDCB   39
C                                                                     QDCB   40
      CONSTB = ONE/THREE                                              QDCB   41
C                                                                     QDCB   42
      DO 200 J=1,NEN                                                  QDCB   43
C                                                                     QDCB   44
      J2   = 2*J                                                      QDCB   45
      J2M1 = J2 - 1                                                   QDCB   46
C                                                                     QDCB   47
      IF (IOPT.EQ.2) THEN                                             QDCB   48
         TEMP3 = CONSTB*(SHGBAR(3,J) - SHG(3,J)/R)                    QDCB   49
         B(1,J2M1) = B(1,J2M1) + TEMP3                               QDCB   50
         B(2,J2M1) = B(2,J2M1) + TEMP3                               QDCB   51
         B(4,J2M1) = B(4,J2M1) + TEMP3                               QDCB   52
      ELSE                                                            QDCB   53
         B(4,J2M1) = ZERO                                             QDCB   54
         B(4,J2  ) = ZERO                                             QDCB   55
      ENDIF                                                           QDCB   56
C                                                                     QDCB   57
      TEMP1 = CONSTB*(SHGBAR(1,J) - SHG(1,J))                         QDCB   58
      TEMP2 = CONSTB*(SHGBAR(2,J) - SHG(2,J))                         QDCB   59
C                                                                     QDCB   60
      B(1,J2M1) = B(1,J2M1) + TEMP1                                  QDCB   61
      B(1,J2  ) = B(1,J2  ) + TEMP2                                  QDCB   62
      B(2,J2M1) = B(2,J2M1) + TEMP1                                  QDCB   63
      B(2,J2  ) = B(2,J2  ) + TEMP2                                  QDCB   64
      B(4,J2M1) = B(4,J2M1) + TEMP1                                  QDCB   65
      B(4,J2  ) = B(4,J2  ) + TEMP2                                  QDCB   66
C                                                                     QDCB   67
  200 CONTINUE                                                        QDCB   68
C                                                                     QDCB   69
      RETURN                                                          QDCB   70
      END                                                             QDCB   71
C*********************************************************************
      SUBROUTINE QDCK(SHGBAR,W,DET,R,SHG,B,C,DMAT,DB,ELSTIF,CONSTK,    QDCK    1
     &                IBBAR,NEN,NINT,IOPT,NESD,NROWSH,NROWB,NSTR,NEE)  QDCK    2
C                                                                     QDCK    3
C.... PROGRAM TO FORM STIFFNESS MATRIX FOR A CONTINUUM ELEMENT        QDCK    4
C        WITH "NEN" NODES                                             QDCK    5
C                                                                     QDCK    6
C        NOTE: THE B-BAR OPTION IS RESTRICTED TO THE MEAN-DILATATION  QDCK    7
C              FORMULATION. TO GENERALIZE TO OTHER FORMULATIONS,       QDCK    8
C              REDIMENSION ARRAY "SHGBAR", AND REPLACE ROUTINES        QDCK    9
C              "MEANSH" AND "QDCB".                                    QDCK   10
C                                                                     QDCK   11
      IMPLICIT DOUBLE PRECISION (A-H,O-Z)                             QDCK   12
C                                                                     QDCK   13
C.... DEACTIVATE ABOVE CARD(S) FOR SINGLE-PRECISION OPERATION         QDCK   14
C                                                                     QDCK   15
      DIMENSION SHGBAR(3,NEN,1),W(1),DET(1),R(1),SHG(NROWSH,NEN,1),    QDCK   16
     &          B(NROWB,1),C(NROWB,1),DMAT(NROWB,1),DB(NROWB,1),       QDCK   17
     &          ELSTIF(NEE,1)                                         QDCK   18
C                                                                     QDCK   19
C.... CALCULATE MEAN VALUES OF SHAPE FUNCTION GLOBAL DERIVATIVES      QDCK   20
C        FOR MEAN-DILATATIONAL B-BAR FORMULATION                      QDCK   21
C                                                                     QDCK   22
      IF (IBBAR.EQ.1)                                                 QDCK   23
     &   CALL MEANSH(SHGBAR,W,DET,R,SHG,NEN,NINT,IOPT,NESD,NROWSH)    QDCK   24
C                                                                     QDCK   25
C.... LOOP ON INTEGRATION POINTS                                      QDCK   26
C                                                                     QDCK   27
      DO 100 L=1,NINT                                                 QDCK   28
      TEMP = CONSTK*W(L)*DET(L)                                       QDCK   29
```

```
C                                                                       QDCK   30
C.... SET UP THE STRAIN-DISPLACEMENT MATRIX                             QDCK   31
C                                                                       QDCK   32
      CALL QDCB(SHG(1,1,L),SHGBAR,B,R(L),IOPT,NROWSH,NROWB,NEN,IBBAR)    QDCK   33
C                                                                       QDCK   34
C.... SET UP THE CONSTITUTIVE MATRIX                                    QDCK   35
C                                                                       QDCK   36
      CALL SETUPD(C,DMAT,TEMP,NSTR,NROWB)                               QDCK   37
C                                                                       QDCK   38
C.... MULTIPLY D*B                                                      QDCK   39
C                                                                       QDCK   40
      CALL MULTAB(DMAT,B,DB,NROWB,NROWB,NROWB,NSTR,NSTR,NEE,1)          QDCK   41
C                                                                       QDCK   42
C.... MULTIPLY B(TRANSPOSE) * DB, TAKING ACCOUNT OF SYMMETRY,           QDCK   43
C        AND ACCUMULATE IN ELSTIF                                       QDCK   44
C                                                                       QDCK   45
      CALL BTDB(ELSTIF,B,DB,NEE,NROWB,NSTR)                             QDCK   46
C                                                                       QDCK   47
  100 CONTINUE                                                          QDCK   48
C                                                                       QDCK   49
      RETURN                                                            QDCK   50
      END                                                               QDCK   51
C************************************************************************
      SUBROUTINE QDCKD(SHGBAR,W,DET,R,SHG,B,DL,STRAIN,C,STRESS,WORK,     QDCKD   1
     &               ELRESF,CONSTK,IBBAR,NEN,NINT,IOPT,NROWSH,          QDCKD   2
     &               NESD,NROWB,NEE,NSTR)                               QDCKD   3
C                                                                       QDCKD   4
C.... PROGRAM TO FORM INTERNAL FORCE ("-K*D") FOR A CONTINUUM ELEMENT   QDCKD   5
C        WITH "NEN" NODES                                               QDCKD   6
C                                                                       QDCKD   7
C        NOTE: THE B-BAR OPTION IS RESTRICTED TO THE MEAN-DILATATION    QDCKD   8
C              FORMULATION. TO GENERALIZE TO OTHER FORMULATIONS,        QDCKD   9
C              REDIMENSION ARRAY "SHGBAR", AND REPLACE ROUTINES         QDCKD  10
C              "MEANSH" AND "QDCB".                                     QDCKD  11
C                                                                       QDCKD  12
      IMPLICIT DOUBLE PRECISION (A-H,O-Z)                               QDCKD  13
C                                                                       QDCKD  14
C.... DEACTIVATE ABOVE CARD(S) FOR SINGLE-PRECISION OPERATION           QDCKD  15
C                                                                       QDCKD  16
      DIMENSION SHGBAR(3,NEN,1),W(1),DET(1),R(1),SHG(NROWSH,NEN,1),      QDCKD  17
     &          B(NROWB,1),DL(1),STRAIN(1),C(NROWB,1),STRESS(1),        QDCKD  18
     &          WORK(1),ELRESF(1)                                       QDCKD  19
C                                                                       QDCKD  20
      IF (IBBAR.EQ.1)                                                   QDCKD  21
     &   CALL MEANSH(SHGBAR,W,DET,R,SHG,NEN,NINT,IOPT,NESD,NROWSH)       QDCKD  22
C                                                                       QDCKD  23
C.... LOOP ON INTEGRATION POINTS                                        QDCKD  24
C                                                                       QDCKD  25
      DO 100 L=1,NINT                                                   QDCKD  26
      TEMP = CONSTK*W(L)*DET(L)                                         QDCKD  27
C                                                                       QDCKD  28
C.... SET UP THE STRAIN-DISPLACEMENT MATRIX                             QDCKD  29
C                                                                       QDCKD  30
      CALL QDCB(SHG(1,1,L),SHGBAR,B,R(L),IOPT,NROWSH,NROWB,NEN,IBBAR)    QDCKD  31
C                                                                       QDCKD  32
C.... CALCULATE STRAINS                                                 QDCKD  33
C                                                                       QDCKD  34
      CALL MULTAB(B,DL,STRAIN,NROWB,NEE,NSTR,NEE,NSTR,1,1)              QDCKD  35
C                                                                       QDCKD  36
C.... CALCULATE STRESSES                                                QDCKD  37
C                                                                       QDCKD  38
      CALL MULTAB(C,STRAIN,STRESS,NROWB,NSTR,NSTR,NSTR,NSTR,1,1)        QDCKD  39
C                                                                       QDCKD  40
C.... CALCULATE ELEMENT INTERNAL FORCE                                  QDCKD  41
C                                                                       QDCKD  42
      CALL SMULT(TEMP,STRESS,STRESS,NSTR,NSTR,NSTR,1,1)                 QDCKD  43
      CALL MULTAB(B,STRESS,WORK,NROWB,NSTR,NEE,NSTR,NEE,1,2)            QDCKD  44
      CALL MATADD(ELRESF,WORK,ELRESF,NEE,NEE,NEE,NEE,1,1)               QDCKD  45
C                                                                       QDCKD  46
  100 CONTINUE                                                          QDCKD  47
C                                                                       QDCKD  48
      RETURN                                                            QDCKD  49
      END                                                               QDCKD  50
C************************************************************************
      SUBROUTINE QDCRSF(IELNO,ISIDE,PRESS,SHEAR,NSURF)                  QDCRSF  1
C                                                                       QDCRSF  2
C.... PROGRAM TO READ, WRITE AND STORE SURFACE FORCE DATA FOR THE       QDCRSF  3
C        FOUR-NODE QUADRILATERAL, ELASTIC CONTINUUM ELEMENT             QDCRSF  4
C                                                                       QDCRSF  5
      IMPLICIT DOUBLE PRECISION (A-H,O-Z)                               QDCRSF  6
C                                                                       QDCRSF  7
C.... DEACTIVATE ABOVE CARD(S) FOR SINGLE PRECISION OPERATION           QDCRSF  8
C                                                                       QDCRSF  9
      DIMENSION IELNO(1),ISIDE(1),PRESS(2,1),SHEAR(2,1)                 QDCRSF 10
      COMMON /IOUNIT/ IIN,IOUT,IRSIN,IRSOUT                             QDCRSF 11
```

```
C                                                                      QDCRSF 12
      DO 100 N=1,NSURF                                                 QDCRSF 13
      IF (MOD(N,50).EQ.1) WRITE(IOUT,1000) NSURF                       QDCRSF 14
      READ(IIN,2000) IELNO(N),ISIDE(N),PRESS(1,N),PRESS(2,N),          QDCRSF 15
     &                SHEAR(1,N),SHEAR(2,N)                            QDCRSF 16
      WRITE(IOUT,3000) IELNO(N),ISIDE(N),PRESS(1,N),PRESS(2,N),        QDCRSF 17
     &                SHEAR(1,N),SHEAR(2,N)                            QDCRSF 18
  100 CONTINUE                                                         QDCRSF 19
C                                                                      QDCRSF 20
      RETURN                                                           QDCRSF 21
C                                                                      QDCRSF 22
 1000 FORMAT('1',                                                      QDCRSF 23
     &' E L E M E N T   S U R F A C E   F O R C E   D A T A ',//5X,    QDCRSF 24
     &' NUMBER OF SURFACE FORCE CARDS . . . . . . (NSURF ) = ',I5///   QDCRSF 25
     &5X,'  ELEMENT     SIDE    ',2('   PRESSURE  '),                  QDCRSF 26
     &        2('   SHEAR    '),/                                      QDCRSF 27
     &5X,2('  NUMBER   '),2('   NODE I        NODE J   '),/)           QDCRSF 28
 2000 FORMAT(2I5,4F10.0)                                               QDCRSF 29
 3000 FORMAT(6X,I5,7X,I2,3X,4(2X,E12.4))                               QDCRSF 30
      END                                                              QDCRSF 31
C*********************************************************************  QDCSHG  1
      SUBROUTINE QDCSHG(XL,DET,SHL,SHG,NINT,NEL,NEG,LQUAD)             QDCSHG  1
C                                                                      QDCSHG  2
C.... PROGRAM TO CALCULATE GLOBAL DERIVATIVES OF SHAPE FUNCTIONS AND   QDCSHG  3
C         JACOBIAN DETERMINANTS FOR A FOUR-NODE QUADRILATERAL ELEMENT  QDCSHG  4
C                                                                      QDCSHG  5
C         XL(J,I)    = GLOBAL COORDINATES                              QDCSHG  6
C         DET(L)     = JACOBIAN DETERMINANT                            QDCSHG  7
C         SHL(1,I,L) = LOCAL ("XI") DERIVATIVE OF SHAPE FUNCTION       QDCSHG  8
C         SHL(2,I,L) = LOCAL ("ETA") DERIVATIVE OF SHAPE FUNCTION      QDCSHG  9
C         SHL(3,I,L) = LOCAL   SHAPE FUNCTION                          QDCSHG 10
C         SHG(1,I,L) = X-DERIVATIVE OF SHAPE FUNCTION                  QDCSHG 11
C         SHG(2,I,L) = Y-DERIVATIVE OF SHAPE FUNCTION                  QDCSHG 12
C         SHG(3,I,L) = SHL(3,I,L)                                      QDCSHG 13
C         XS(I,J)    = JACOBIAN MATRIX                                 QDCSHG 14
C              I     = LOCAL NODE NUMBER OR GLOBAL COORDINATE NUMBER   QDCSHG 15
C              J     = GLOBAL COORDINATE NUMBER                        QDCSHG 16
C              L     = INTEGRATION-POINT NUMBER                        QDCSHG 17
C           NINT     = NUMBER OF INTEGRATION POINTS, EQ. 1 OR 4        QDCSHG 18
C                                                                      QDCSHG 19
      IMPLICIT DOUBLE PRECISION (A-H,O-Z)                              QDCSHG 20
C                                                                      QDCSHG 21
C.... DEACTIVATE ABOVE CARD(S) FOR SINGLE PRECISION OPERATION          QDCSHG 22
C                                                                      QDCSHG 23
      LOGICAL LQUAD                                                    QDCSHG 24
      DIMENSION XL(2,1),DET(1),SHL(3,4,1),SHG(3,4,1),XS(2,2)           QDCSHG 25
      COMMON /CONSTS/ ZERO,PT1667,PT25,PT5,ONE,TWO,THREE,FOUR,FIVE     QDCSHG 26
      COMMON /IOUNIT/ IIN,IOUT,IRSIN,IRSOUT                            QDCSHG 27
C                                                                      QDCSHG 28
      CALL MOVE(SHG,SHL,12*NINT)                                       QDCSHG 29
C                                                                      QDCSHG 30
      DO 700 L=1,NINT                                                  QDCSHG 31
C                                                                      QDCSHG 32
      IF (.NOT.LQUAD) THEN                                             QDCSHG 33
         DO 100 I=1,3                                                  QDCSHG 34
         SHG(I,3,L) = SHL(I,3,L) + SHL(I,4,L)                          QDCSHG 35
         SHG(I,4,L) = ZERO                                             QDCSHG 36
  100    CONTINUE                                                      QDCSHG 37
      ENDIF                                                            QDCSHG 38
C                                                                      QDCSHG 39
      DO 300 J=1,2                                                     QDCSHG 40
      DO 200 I=1,2                                                     QDCSHG 41
      XS(I,J) = ROWDOT(SHG(I,1,L),XL(J,1),3,2,4)                       QDCSHG 42
  200 CONTINUE                                                         QDCSHG 43
  300 CONTINUE                                                         QDCSHG 44
C                                                                      QDCSHG 45
      DET(L) = XS(1,1)*XS(2,2)-XS(1,2)*XS(2,1)                         QDCSHG 46
      IF (DET(L).LE.ZERO) THEN                                         QDCSHG 47
         WRITE(IOUT,1000) NEL,NEG                                      QDCSHG 48
         STOP                                                          QDCSHG 49
      ENDIF                                                            QDCSHG 50
C                                                                      QDCSHG 51
      DO 500 J=1,2                                                     QDCSHG 52
      DO 400 I=1,2                                                     QDCSHG 53
      XS(I,J) = XS(I,J)/DET(L)                                         QDCSHG 54
  400 CONTINUE                                                         QDCSHG 55
  500 CONTINUE                                                         QDCSHG 56
C                                                                      QDCSHG 57
      DO 600 I=1,4                                                     QDCSHG 58
      TEMP = XS(2,2)*SHG(1,I,L) - XS(1,2)*SHG(2,I,L)                   QDCSHG 59
      SHG(2,I,L) = - XS(2,1)*SHG(1,I,L) + XS(1,1)*SHG(2,I,L)           QDCSHG 60
      SHG(1,I,L) = TEMP                                                QDCSHG 61
  600 CONTINUE                                                         QDCSHG 62
C                                                                      QDCSHG 63
  700 CONTINUE                                                         QDCSHG 64
C                                                                      QDCSHG 65
      RETURN                                                           QDCSHG 66
```

```
C                                                                       QDCSHG 67
 1000 FORMAT('1','NON-POSITIVE DETERMINANT IN ELEMENT NUMBER   ',I5,    QDCSHG 68
      &           ' IN ELEMENT GROUP   ',I5)                            QDCSHG 69
      END                                                               QDCSHG 70
C***********************************************************************QDCSHL  1
      SUBROUTINE QDCSHL(SHL,W,NINT)                                     QDCSHL  1
C                                                                       QDCSHL  2
C.... PROGRAM TO CALCULATE INTEGRATION-RULE WEIGHTS, SHAPE FUNCTIONS    QDCSHL  3
C        AND LOCAL DERIVATIVES FOR A FOUR-NODE QUADRILATERAL ELEMENT    QDCSHL  4
C                                                                       QDCSHL  5
C            R,S = LOCAL ELEMENT COORDINATES ("XI", "ETA", RESP.)       QDCSHL  6
C      SHL(1,I,L) = LOCAL ("XI") DERIVATIVE OF SHAPE FUNCTION           QDCSHL  7
C      SHL(2,I,L) = LOCAL ("ETA") DERIVATIVE OF SHAPE FUNCTION          QDCSHL  8
C      SHL(3,I,L) = LOCAL   SHAPE FUNCTION                              QDCSHL  9
C           W(L) = INTEGRATION-RULE WEIGHT                              QDCSHL 10
C              I = LOCAL NODE NUMBER                                    QDCSHL 11
C              L = INTEGRATION POINT NUMBER                             QDCSHL 12
C           NINT = NUMBER OF INTEGRATION POINTS, EQ. 1 OR 4             QDCSHL 13
C                                                                       QDCSHL 14
      IMPLICIT DOUBLE PRECISION (A-H,O-Z)                               QDCSHL 15
C                                                                       QDCSHL 16
C.... DEACTIVATE ABOVE CARD(S) FOR SINGLE PRECISION OPERATION           QDCSHL 17
C                                                                       QDCSHL 18
      DIMENSION SHL(3,4,1),W(1),RA(4),SA(4)                             QDCSHL 19
      COMMON /CONSTS/ ZERO,PT1667,PT25,PT5,ONE,TWO,THREE,FOUR,FIVE      QDCSHL 20
      DATA RA/-0.50,0.50,0.50,-0.50/,SA/-0.50,-0.50,0.50,0.50/          QDCSHL 21
C                                                                       QDCSHL 22
      G = ZERO                                                          QDCSHL 23
      W(1) = FOUR                                                       QDCSHL 24
      IF (NINT.EQ.4) THEN                                               QDCSHL 25
         G = TWO/SQRT(THREE)                                            QDCSHL 26
         W(1) = ONE                                                     QDCSHL 27
         W(2) = ONE                                                     QDCSHL 28
         W(3) = ONE                                                     QDCSHL 29
         W(4) = ONE                                                     QDCSHL 30
      ENDIF                                                             QDCSHL 31
C                                                                       QDCSHL 32
      DO 200 L=1,NINT                                                   QDCSHL 33
      R = G*RA(L)                                                       QDCSHL 34
      S = G*SA(L)                                                       QDCSHL 35
C                                                                       QDCSHL 36
      DO 100 I=1,4                                                      QDCSHL 37
      TEMPR = PT5 + RA(I)*R                                             QDCSHL 38
      TEMPS = PT5 + SA(I)*S                                             QDCSHL 39
      SHL(1,I,L) = RA(I)*TEMPS                                          QDCSHL 40
      SHL(2,I,L) = TEMPR*SA(I)                                          QDCSHL 41
      SHL(3,I,L) = TEMPR*TEMPS                                          QDCSHL 42
  100 CONTINUE                                                          QDCSHL 43
C                                                                       QDCSHL 44
  200 CONTINUE                                                          QDCSHL 45
C                                                                       QDCSHL 46
      RETURN                                                            QDCSHL 47
      END                                                               QDCSHL 48
C***********************************************************************QDCSTR
      SUBROUTINE QDCSTR(SHG,SHGBAR,B,R,DL,STRAIN,C,STRESS,PSTRN,PSTRS,  QDCSTR  1
      &                 NROWSH,NESD,NROWB,IBBAR,NEN,NED,NEE,NSTR,IOPT)  QDCSTR  2
C                                                                       QDCSTR  3
C.... PROGRAM TO CALCULATE STRESS, STRAIN AND PRINCIPAL VALUES AT AN    QDCSTR  4
C        INTEGRATION POINT FOR A TWO-DIMENSIONAL CONTINUUM ELEMENT      QDCSTR  5
C                                                                       QDCSTR  6
      IMPLICIT DOUBLE PRECISION (A-H,O-Z)                               QDCSTR  7
C                                                                       QDCSTR  8
C.... DEACTIVATE ABOVE CARD(S) FOR SINGLE PRECISION OPERATION           QDCSTR  9
C                                                                       QDCSTR 10
      DIMENSION SHG(NROWSH,1),SHGBAR(3,1),B(NROWB,1),DL(NED,1),         QDCSTR 11
      &          STRAIN(1),C(NROWB,1),STRESS(1),PSTRN(1),PSTRS(1)       QDCSTR 12
      COMMON /CONSTS/ ZERO,PT1667,PT25,PT5,ONE,TWO,THREE,FOUR,FIVE      QDCSTR 13
C                                                                       QDCSTR 14
C.... SET UP STRAIN-DISPLACEMENT MATRIX                                 QDCSTR 15
C                                                                       QDCSTR 16
      CALL QDCB(SHG,SHGBAR,B,R,IOPT,NROWSH,NROWB,NEN,IBBAR)             QDCSTR 17
C                                                                       QDCSTR 18
C.... CALCULATE STRAINS                                                 QDCSTR 19
C                                                                       QDCSTR 20
      CALL MULTAB(B,DL,STRAIN,NROWB,NEE,NSTR,NEE,NSTR,1,1)              QDCSTR 21
C                                                                       QDCSTR 22
C.... CALCULATE STRESSES                                                QDCSTR 23
C                                                                       QDCSTR 24
      CALL MULTAB(C,STRAIN,STRESS,NROWB,NSTR,NSTR,NSTR,NSTR,1,1)        QDCSTR 25
C                                                                       QDCSTR 26
C.... CALCULATE PRINCIPAL STRAINS; ACCOUNT FOR ENGINEERING SHEAR STRAIN QDCSTR 27
C                                                                       QDCSTR 28
      STRAIN(3) = PT5*STRAIN(3)                                         QDCSTR 29
      CALL PRINC(NESD,STRAIN,PSTRN)                                     QDCSTR 30
      STRAIN(3) = TWO*STRAIN(3)                                         QDCSTR 31
      PSTRN(3) = TWO*PSTRN(3)                                           QDCSTR 32
C                                                                       QDCSTR 33
C.... CALCULATE PRINCIPAL STRESS                                        QDCSTR 34
```

```
C
      CALL PRINC(NESD,STRESS,PSTRS)
C
      IF (IOPT.EQ.0) THEN
         STRESS(4) = ZERO
         STRAIN(4) = - ( C(4,1)*STRAIN(1) + C(4,2)*STRAIN(2)
     &                 + C(4,3)*STRAIN(3) )/C(4,4)
      ENDIF
C
      IF ( (IOPT.EQ.1) .AND. (IBBAR.EQ.0) ) THEN
         STRAIN(4) = ZERO
         STRESS(4) = C(4,1)*STRAIN(1) + C(4,2)*STRAIN(2)
     &             + C(4,3)*STRAIN(3)
      ENDIF
C
      RETURN
      END
C**********************************************************************
      SUBROUTINE QDCSUF(IELNO,IEN,X,XL,ISIDE,MAT,TH,PRESS,SHEAR,ELRESF,
     &                  BRHS,LM,FAC,NSURF,NEN,NSD,NESD,NED,NEE,IOPT)
C
C.... PROGRAM TO COMPUTE CONSISTENT SURFACE LOADS FOR THE
C        FOUR-NODE QUADRILATERAL, ELASTIC CONTINUUM ELEMENT
C
C        NOTE: TWO-POINT GAUSSIAN QUADRATURE IS EMPLOYED
C
      IMPLICIT DOUBLE PRECISION (A-H,O-Z)
C
C.... DEACTIVATE ABOVE CARD(S) FOR SINGLE-PRECISION OPERATION
C
      DIMENSION Z(2),WORK(2),IELNO(1),IEN(NEN,1),X(NSD,1),XL(NESD,1),
     &          ISIDE(1),MAT(1),TH(1),PRESS(2,1),SHEAR(2,1),
     &          ELRESF(NED,1),BRHS(1),LM(NED,NEN,1)
      COMMON /CONSTS/ ZERO,PT1667,PT25,PT5,ONE,TWO,THREE,FOUR,FIVE
C
      Z(2) = PT5/SQRT(THREE)
      Z(1) = - Z(2)
C
      DO 300 K=1,NSURF
      NEL = IELNO(K)
      CALL LOCAL(IEN(1,NEL),X,XL,NEN,NSD,NESD)
      CALL CLEAR(ELRESF,NEE)
      I = ISIDE(K)
      J = I + 1
      IF (J.EQ.5) J = 1
      DX = XL(1,J) - XL(1,I)
      DY = XL(2,J) - XL(2,I)
      M = MAT(NEL)
      TEMP = PT5*FAC*TH(M)
C
      DO 200 L=1,2
      SHI = PT5 - Z(L)
      SHJ = PT5 + Z(L)
      P = SHI*PRESS(1,K) + SHJ*PRESS(2,K)
      S = SHI*SHEAR(1,K) + SHJ*SHEAR(2,K)
C
      IF (IOPT.EQ.2) THEN
         R = SHI*XL(1,I) + SHJ*XL(1,J)
         P = P*R
         S = S*R
      ENDIF
C
      WORK(1) = TEMP*( - P*DY + S*DX)
      WORK(2) = TEMP*(   P*DX + S*DY)
C
      DO 100 N=1,2
      ELRESF(N,I) = ELRESF(N,I) + SHI*WORK(N)
      ELRESF(N,J) = ELRESF(N,J) + SHJ*WORK(N)
  100 CONTINUE
C
  200 CONTINUE
C
      CALL ADDRHS(BRHS,ELRESF,LM(1,1,NEL),NEE)
C
  300 CONTINUE
C
      RETURN
      END
C**********************************************************************
      SUBROUTINE TRUSS(ITASK,NPAR,MP,NEG)
C
C.... PROGRAM TO SET STORAGE AND CALL TASKS FOR THE
C        THREE-DIMENSIONAL, ELASTIC TRUSS ELEMENT
C
      DOUBLE PRECISION TIME
C
C.... DEACTIVATE ABOVE CARD(S) FOR SINGLE PRECISION OPERATION
```

```
C                                                          TRUSS    9
      LOGICAL LDYN                                         TRUSS   10
      DIMENSION NPAR(1),MP(1)                              TRUSS   11
      COMMON /BPOINT/ MFIRST,MLAST,MTOT,IPREC              TRUSS   12
      COMMON /DPOINT/ MPSTEP,MPDPRT,MPSPRT,MPHPLT,MPITER,MPALPH,MPBETA, TRUSS 13
     &               MPGAMM,MPDT ,MPDOUT,MPIDHS,MPVPRD,MPDPRD,MPA,MPV  TRUSS 14
      COMMON /HPLOTC/ NPLPTS,LOCPLT,TIME                   TRUSS   15
      COMMON /INFO  / IEXEC,IACODE,LDYN,IREADR,IWRITR,IPRTIN,IRANK,  TRUSS 16
     &               NUMSEQ,NDOUT,NSD,NUMNP,NDOF,NLVECT,NLTFTN,NPTSLF, TRUSS 17
     &               NUMEG                                 TRUSS   18
      COMMON /SPOINT/ MPD,MPX,MPID,MPF,MPG,MPG1,MPDIAG,MPNGRP,  TRUSS 19
     &               MPALHS,MPBRHS                         TRUSS   20
      COMMON A(1)                                          TRUSS   21
C                                                          TRUSS   22
      MW     = 1                                           TRUSS   23
      MDET   = 2                                           TRUSS   24
      MSHL   = 3                                           TRUSS   25
      MSHG   = 4                                           TRUSS   26
      MXS    = 5                                           TRUSS   27
      MRHO   = 6                                           TRUSS   28
      MRDPM  = 7                                           TRUSS   29
      MRDPK  = 8                                           TRUSS   30
      MAREA  = 9                                           TRUSS   31
      MC     = 10                                          TRUSS   32
      MGRAV  = 11                                          TRUSS   33
      MIEN   = 12                                          TRUSS   34
      MMAT   = 13                                          TRUSS   35
      MLM    = 14                                          TRUSS   36
      MISHST = 15                                          TRUSS   37
      MSOUT  = 16                                          TRUSS   38
      MELEFM = 17                                          TRUSS   39
      MXL    = 18                                          TRUSS   40
      MWORK  = 19                                          TRUSS   41
      MB     = 20                                          TRUSS   42
      MDMAT  = 21                                          TRUSS   43
      MDB    = 22                                          TRUSS   44
      MVL    = 23                                          TRUSS   45
      MAL    = 24                                          TRUSS   46
      MELRES = 25                                          TRUSS   47
      MDL    = 26                                          TRUSS   48
      MSTRN  = 27                                          TRUSS   49
      MSTRS  = 28                                          TRUSS   50
      MFORCE = 29                                          TRUSS   51
C                                                          TRUSS   52
      NTYPE  = NPAR( 1)                                    TRUSS   53
      NUMEL  = NPAR( 2)                                    TRUSS   54
      NUMAT  = NPAR( 3)                                    TRUSS   55
      NEN    = NPAR( 4)                                    TRUSS   56
      NSOUT  = NPAR( 5)                                    TRUSS   57
      ISTPRT = NPAR( 6)                                    TRUSS   58
      LFBODY = NPAR( 7)                                    TRUSS   59
C                                                          TRUSS   60
      IF (NPAR(8).EQ.0) NPAR(8) = 2                        TRUSS   61
C                                                          TRUSS   62
      NINT   = NPAR( 8)                                    TRUSS   63
      IMASS  = NPAR( 9)                                    TRUSS   64
      IMPEXP = NPAR(10)                                    TRUSS   65
C                                                          TRUSS   66
C.... SET ELEMENT PARAMETERS                               TRUSS   67
C                                                          TRUSS   68
      NED    = 3                                           TRUSS   69
      NEE    = NEN*NED                                     TRUSS   70
      NESD   = 3                                           TRUSS   71
      NROWSH = 2                                           TRUSS   72
      NEESQ  = NEE*NEE                                     TRUSS   73
      NROWB  = 1                                           TRUSS   74
      NSTR   = 1                                           TRUSS   75
C                                                          TRUSS   76
      IF (ITASK.EQ.1) THEN                                 TRUSS   77
C                                                          TRUSS   78
C....... SET MEMORY POINTERS                               TRUSS   79
C                                                          TRUSS   80
C         NOTE:  THE MP ARRAY IS STORED DIRECTLY AFTER THE NPAR ARRAY, TRUSS 81
C                BEGINNING AT LOCATION MPNPAR + 16 OF BLANK COMMON.  TRUSS 82
C                THE VARIABLE "JUNK" IS NOT USED SUBSEQUETLY.  TRUSS 83
C                                                          TRUSS   84
         JUNK       = MPOINT('MP      ',29     ,0     ,0     ,1)  TRUSS 85
```

```
C
         MP(MW     ) = MPOINT('W        ',NINT   ,0      ,0      ,IPREC)
         MP(MDET   ) = MPOINT('DET      ',NINT   ,0      ,0      ,IPREC)
         MP(MSHL   ) = MPOINT('SHL      ',NROWSH ,NEN    ,NINT   ,IPREC)
         MP(MSHG   ) = MPOINT('SHG      ',NROWSH ,NEN    ,NINT   ,IPREC)
         MP(MXS    ) = MPOINT('XS       ',NESD   ,NINT   ,0      ,IPREC)
         MP(MRHO   ) = MPOINT('RHO      ',NUMAT  ,0      ,0      ,IPREC)
         MP(MRDPM  ) = MPOINT('RDAMPM   ',NUMAT  ,0      ,0      ,IPREC)
         MP(MRDPK  ) = MPOINT('RDAMPK   ',NUMAT  ,0      ,0      ,IPREC)
         MP(MAREA  ) = MPOINT('AREA     ',NUMAT  ,0      ,0      ,IPREC)
         MP(MC     ) = MPOINT('C        ',NROWB  ,NROWB  ,NUMAT  ,IPREC)
         MP(MGRAV  ) = MPOINT('GRAV     ',NESD   ,0      ,0      ,IPREC)
         MP(MIEN   ) = MPOINT('IEN      ',NEN    ,NUMEL  ,0      ,1)
         MP(MMAT   ) = MPOINT('MAT      ',NUMEL  ,0      ,0      ,1)
         MP(MLM    ) = MPOINT('LM       ',NED    ,NEN    ,NUMEL  ,1)
C
      IF (NSOUT.EQ.0) THEN
         MP(MISHST) = JUNK
         MP(MSOUT ) = JUNK
      ELSE
         MP(MISHST) = MPOINT('ISHIST   ',3       ,NSOUT ,0       ,1)
         MP(MSOUT ) = MPOINT('SOUT     ',NSOUT+1,NPLPTS,0       ,1)
      ENDIF
C
         MP(MELEFM) = MPOINT('ELEFFM   ',NEE    ,NEE    ,0      ,IPREC)
         MP(MXL    ) = MPOINT('XL       ',NESD   ,NEN    ,0      ,IPREC)
         MP(MWORK  ) = MPOINT('WORK     ',16     ,0      ,0      ,IPREC)
         MP(MB     ) = MPOINT('B        ',NROWB  ,NEE    ,0      ,IPREC)
         MP(MDMAT  ) = MPOINT('DMAT     ',NROWB  ,NROWB  ,0      ,IPREC)
         MP(MDB    ) = MPOINT('DB       ',NROWB  ,NEE    ,0      ,IPREC)
         MP(MVL    ) = MPOINT('VL       ',NED    ,NEN    ,0      ,IPREC)
         MP(MAL    ) = MPOINT('AL       ',NED    ,NEN    ,0      ,IPREC)
         MP(MELRES) = MPOINT('ELRESF   ',NEE    ,0      ,0      ,IPREC)
         MP(MDL    ) = MPOINT('DL       ',NED    ,NEN    ,0      ,IPREC)
         MP(MSTRN  ) = MPOINT('STRAIN   ',NROWB  ,0      ,0      ,IPREC)
         MP(MSTRS  ) = MPOINT('STRESS   ',NROWB  ,0      ,0      ,IPREC)
         MP(MFORCE) = MPOINT('FORCE    ',NROWB  ,0      ,0      ,IPREC)
      ENDIF
C
C.... TASK CALLS
C
      IF (ITASK.GT.6) RETURN
      GO TO (100,200,300,400,500,600     ),ITASK
C
  100 CONTINUE
C
C.... INPUT ELEMENT DATA ('INPUT___')
C
      CALL TRUST1(A(MP(MSHL  )),A(MP(MW    )),A(MP(MRHO  )),
     &            A(MP(MRDPM )),A(MP(MRDPK )),A(MP(MAREA )),
     &            A(MP(MC    )),A(MP(MGRAV )),A(MP(MIEN  )),
     &            A(MP(MMAT  )),A(MPID        ),A(MP(MLM   )),
     &            A(MPDIAG     ),A(MP(MISHST)),
     &            NTYPE ,NUMEL ,NUMAT ,NEN   ,NSOUT ,ISTPRT,
     &            LFBODY,NINT  ,IMASS ,IMPEXP,NROWSH,NROWB ,
     &            NESD  ,NDOF  ,NED   ,IPRTIN,LDYN  )
C
      RETURN
C
  200 CONTINUE
C
C.... FORM ELEMENT EFFECTIVE MASS AND ASSEMBLE INTO GLOBAL
C        LEFT-HAND-SIDE MATRIX  ('FORM_LHS')
C
      CALL TRUST2(A(MP(MELEFM)),A(MP(MIEN  )),A(MPX         ),
     &            A(MP(MXL   )),A(MP(MMAT  )),A(MP(MDET  )),
     &            A(MP(MSHL  )),A(MP(MSHG  )),A(MP(MRDPM )),
     &            A(MP(MRDPK )),A(MP(MAREA )),A(MP(MRHO  )),
     &            A(MP(MW    )),A(MP(MWORK )),A(MP(MB    )),
     &            A(MP(MC    )),A(MP(MDMAT )),A(MP(MDB   )),
     &            A(MPALHS     ),A(MPDIAG     ),A(MP(MLM   )),
     &            A(MP(MXS   )),
     &            IMPEXP,IMASS ,NUMEL ,NEESQ ,NEN   ,NSD   ,
     &            NESD  ,NINT  ,NEG   ,NROWSH,LDYN  ,NED   ,
     &            NROWB ,NSTR  ,NEE   )
C
      RETURN
C
  300 CONTINUE
C
C.... FORM ELEMENT RESIDUAL-FORCE VECTOR AND ASSEMBLE INTO GLOBAL
C        RIGHT-HAND-SIDE VECTOR ('FORM_RHS')
```

```
C
      CALL TRUST3(A(MP(MMAT   )),A(MP(MIEN   )),A(MPDPRD      ),
     &            A(MP(MDL    )),A(MPVPRD      ),A(MP(MVL    )),
     &            A(MPA        ),A(MP(MAL    )),A(MP(MRDPK  )),
     &            A(MP(MRDPM  )),A(MP(MRHO   )),A(MP(MGRAV  )),
     &            A(MP(MELRES)),A(MPX          ),A(MP(MXL    )),
     &            A(MP(MDET   )),A(MP(MSHL   )),A(MP(MSHG   )),
     &            A(MPG1       ),A(MP(MWORK  )),A(MP(MAREA  )),
     &            A(MP(MW     )),A(MP(MELEFM)),A(MP(MB     )),
     &            A(MP(MSTRN  )),A(MP(MC     )),A(MP(MDMAT  )),
     &            A(MP(MSTRS  )),A(MPBRHS      ),A(MP(MLM    )),
     &            A(MP(MXS    )),
     &            NUMEL ,NED   ,NEN   ,NDOF  ,LDYN  ,NEE   ,
     &            IMASS ,NESD  ,LFBODY,NSD   ,NINT  ,NROWSH,
     &            NEG   ,NROWB )
C
      RETURN
C
  400 CONTINUE
C
C.... CALCULATE AND PRINT ELEMENT STRESS/STRAIN OUTPUT ('STR_PRNT')
C
      IF (ISTPRT.EQ.0)
     &   CALL TRUST4(A(MP(MMAT   )),A(MP(MIEN   )),A(MPD          ),
     &               A(MP(MDL    )),A(MPX          ),A(MP(MXL    )),
     &               A(MP(MDET   )),A(MP(MSHL   )),A(MP(MSHG   )),
     &               A(MP(MXS    )),A(MP(MWORK  )),A(MP(MB     )),
     &               A(MP(MSTRN  )),A(MP(MC     )),A(MP(MSTRS  )),
     &               A(MP(MFORCE)),A(MP(MAREA  )),
     &               NINT  ,NUMEL ,NEN   ,NDOF  ,NED   ,NSD   ,
     &               NESD  ,NROWSH,NEG   ,NROWB ,NEE   )
C
      RETURN
C
  500 CONTINUE
C
C.... CALCULATE AND STORE ELEMENT TIME-HISTORIES ('STR_STOR')
C
      IF (NSOUT.GT.0)
     &   CALL TRUST5(A(MP(MISHST)),A(MP(MSOUT  )),A(MP(MMAT   )),
     &               A(MP(MIEN   )),A(MPD          ),A(MP(MDL    )),
     &               A(MPX          ),A(MP(MXL    )),A(MP(MDET   )),
     &               A(MP(MSHL   )),A(MP(MSHG   )),A(MP(MXS    )),
     &               A(MP(MB     )),A(MP(MSTRN  )),A(MP(MC     )),
     &               A(MP(MSTRS  )),A(MP(MFORCE)),A(MP(MAREA  )),
     &               A(MP(MWORK  )),
     &               NSOUT ,NEN   ,NDOF  ,NED   ,NSD   ,NESD  ,
     &               NROWSH,NINT  ,NEG   ,NROWB ,NEE   )
C
      RETURN
C
  600 CONTINUE
C
C.... PLOT ELEMENT TIME-HISTORIES ('STR_PLOT')
C
      IF (NSOUT.GT.0)
     &   CALL HPLOT(A(MP(MISHST)),A(MP(MSOUT  )),NSOUT ,3,NTYPE )
      RETURN
C
      END
C**********************************************************************
      SUBROUTINE TRUST1(SHL   ,W     ,RHO   ,RDAMPM,RDAMPK,AREA  ,
     &                  C     ,GRAV  ,IEN   ,MAT   ,ID    ,LM    ,
     &                  IDIAG ,ISHIST,
     &                  NTYPE ,NUMEL ,NUMAT ,NEN   ,NSOUT ,ISTPRT,
     &                  LFBODY,NINT  ,IMASS ,IMPEXP,NROWSH,NROWB ,
     &                  NESD  ,NDOF  ,NED   ,IPRTIN,LDYN  )
C
C.... PROGRAM TO READ, GENERATE AND WRITE ELEMENT DATA FOR THE
C        THREE-DIMENSIONAL, ELASTIC TRUSS ELEMENT
C
      IMPLICIT DOUBLE PRECISION (A-H,O-Z)
C
C.... DEACTIVATE ABOVE CARD(S) FOR SINGLE PRECISION OPERATION
C
      LOGICAL LDYN
      DIMENSION SHL(NROWSH,NEN,1),W(1),RHO(1),RDAMPM(1),RDAMPK(1),
     &          AREA(1),C(NROWB,NROWB,1),GRAV(NESD),IEN(NEN,1),MAT(1),
     &          ID(NDOF,1),LM(NED,NEN,1),IDIAG(1),ISHIST(3,1)
      COMMON /IOUNIT/ IIN,IOUT,IRSIN,IRSOUT
C
      WRITE(IOUT,1000) NTYPE,NUMEL,NUMAT,NEN,NSOUT,ISTPRT,LFBODY,NINT
      IF (LDYN) WRITE(IOUT,2000) IMASS,IMPEXP
C
      CALL TRUSHL(SHL,W,NINT,NEN)
C
      CALL TRUSPR(RHO,RDAMPM,RDAMPK,AREA,C,NUMAT)
```

```
C
      READ(IIN,3000) GRAV
      WRITE(IOUT,4000) GRAV
C
      CALL GENEL(IEN,MAT,NEN)
C
      IF (IPRTIN.EQ.0) CALL PRNTEL(MAT,IEN,NEN,NUMEL)
C
      CALL FORMLM(ID,IEN,LM,NDOF,NED,NEN,NUMEL)
C
      IF ( (.NOT.LDYN) .OR. (IMPEXP.EQ.0) .OR. (IMASS.EQ.0) )
     &   CALL COLHT(IDIAG,LM,NED,NEN,NUMEL)
C
      IF (NSOUT.GT.0) CALL SHIST(ISHIST,NSOUT,NTYPE)
C
      RETURN
C
 1000 FORMAT(' ',
     &' T W O / T H R E E - N O D E   T R U S S   E L E M E N T S',//5X,
     &' ELEMENT TYPE NUMBER . . . . . . . . . . . . (NTYPE ) = ',I5//5X,
     &' NUMBER OF ELEMENTS . . . . . . . . . . . . (NUMEL ) = ',I5//5X,
     &' NUMBER OF ELEMENT MATERIAL SETS . . . . . . (NUMAT ) = ',I5//5X,
     &' NUMBER OF ELEMENT NODES . . . . . . . . . . (NEN   ) = ',I5//5X,
     &' NUMBER OF STRESS/STRAIN TIME HISTORIES  . . (NSOUT ) = ',I5//5X,
     &' STRESS OUTPUT PRINT CODE . . . . . . . . . (ISTPRT) = ',I5//5X,
     &'      EQ.0, STRESS OUTPUT PRINTED                    ',/5X,
     &'      EQ.1, STRESS OUTPUT NOT PRINTED                ',//5X,
     &' BODY FORCE LOAD-TIME FUNCTION NUMBER  . . . (LFBODY) = ',I5//5X,
     &' INTEGRATION CODE . . . . . . . . . . . . . (NINT  ) = ',I5//5X,
     &'      EQ.1, 1-POINT GAUSSIAN QUADRATURE              ',/5X,
     &'      EQ.2, 2-POINT GAUSSIAN QUADRATURE              ',/5X,
     &'      EQ.3, 3-POINT GAUSSIAN QUADRATURE              ',/5X)
 2000 FORMAT(' ',/5X,
     &' MASS TYPE CODE . . . . . . . . . . . . . . (IMASS ) = ',I5//5X,
     &'      EQ.0, CONSISTENT MASS MATRIX                   ',/5X,
     &'      EQ.1, LUMPED MASS MATRIX                       ',/5X,
     &'      EQ.2, NO MASS MATRIX                           ',//5X,
     &' IMPLICIT/EXPLICIT CODE . . . . . . . . . . (IMPEXP) = ',I5//5X,
     &'      EQ.0, IMPLICIT ELEMENT GROUP                   ',/5X,
     &'      EQ.1, EXPLICIT ELEMENT GROUP                   ',//5X)
 3000 FORMAT(8F10.0)
 4000 FORMAT(////' ',
     &' G R A V I T Y   V E C T O R   C O M P O N E N T S     ',//5X,
     &' X-1 DIRECTION . . . . . . . . . . . . . . = ',  1PE15.8//5X,
     &' X-2 DIRECTION . . . . . . . . . . . . . . = ',  1PE15.8//5X,
     &' X-3 DIRECTION . . . . . . . . . . . . . . = ',  1PE15.8//5X)
C
      END
C**********************************************************************
      SUBROUTINE TRUST2(ELEFFM,IEN   ,X     ,XL    ,MAT   ,DET   ,
     &                  SHL   ,SHG   ,RDAMPM,RDAMPK,AREA  ,RHO   ,
     &                  W     ,WORK  ,B     ,C     ,DMAT  ,DB    ,
     &                  ALHS  ,IDIAG ,LM    ,XS    ,
     &                  IMPEXP,IMASS ,NUMEL ,NEESQ ,NEN   ,NSD   ,
     &                  NESD  ,NINT  ,NEG   ,NROWSH,LDYN  ,NED   ,
     &                  NROWB ,NSTR  ,NEE   )
C
C.... PROGRAM TO CALCULATE EFFECTIVE MASS MATRIX FOR THE
C        THREE-DIMENSIONAL, ELASTIC TRUSS ELEMENT AND
C        ASSEMBLE INTO THE GLOBAL LEFT-HAND-SIDE MATRIX
C
C         IMPEXP = 0, IMPLICIT TIME INTEGRATION
C                = 1, EXPLICIT TIME INTEGRATION
C
      IMPLICIT DOUBLE PRECISION (A-H,O-Z)
C
C.... DEACTIVATE ABOVE CARD(S) FOR SINGLE-PRECISION OPERATION
C
      LOGICAL LDYN,LDIAG,LNODE3
      DIMENSION ELEFFM(NEE,1),IEN(NEN,1),X(NSD,1),XL(NESD,1),MAT(1),
     &          DET(1),SHL(NROWSH,NEN,1),SHG(NROWSH,NEN,1),RDAMPM(1),
     &          RDAMPK(1),AREA(1),RHO(1),W(1),WORK(1),B(NROWB,1),
     &          C(NROWB,NROWB,1),DMAT(NROWB,1),DB(NROWB,1),ALHS(1),
     &          IDIAG(1),LM(NED,NEN,1),XS(NESD,1)
      COMMON /COEFFS/ COEFF1,COEFF2,COEFF3,COEFF4,COEFF5,COEFF6,
     &                COEFF7,COEFF8,ALPHA1,BETA1 ,GAMMA1,DT1
      COMMON /CONSTS/ ZERO,PT1667,PT25,PT5,ONE,TWO,THREE,FOUR,FIVE
C
      LDIAG = .FALSE.
      IF ( (IMPEXP.EQ.1) .AND. (IMASS.EQ.1) ) LDIAG = .TRUE.
C
      DO 100 NEL=1,NUMEL
C
      CALL CLEAR(ELEFFM,NEESQ)
      CALL LOCAL(IEN(1,NEL),X,XL,NEN,NSD,NESD)
      M = MAT(NEL)
      LNODE3 = .TRUE.
      IF ( NEN .EQ. 3 .AND. IEN(2,NEL).EQ.IEN(3,NEL) ) LNODE3 = .FALSE.
      CALL TRUSHG(XL,DET,SHL,SHG,XS,NEN,NINT,NEL,NEG,LNODE3)
```

```
C                                                                      TRUST2 41
      IF ( LDYN .AND. (IMASS.NE.2) ) THEN                             TRUST2 42
C                                                                      TRUST2 43
C....... FORM MASS MATRIX                                              TRUST2 44
C                                                                      TRUST2 45
         CONSTM = (ONE + RDAMPM(M)*COEFF4)*AREA(M)*RHO(M)              TRUST2 46
         IF (CONSTM.NE.ZERO) CALL CONTM(SHG,XL,W,DET,ELEFFM,WORK,      TRUST2 47
     &              CONSTM,IMASS,NINT,NROWSH,NESD,NEN,NED,NEE,.FALSE.) TRUST2 48
C                                                                      TRUST2 49
      ENDIF                                                            TRUST2 50
C                                                                      TRUST2 51
      IF ( (.NOT.LDYN) .OR. (IMPEXP.EQ.0) ) THEN                      TRUST2 52
C                                                                      TRUST2 53
C....... FORM STIFFNESS MATRIX                                         TRUST2 54
C                                                                      TRUST2 55
         CONSTK = (COEFF4*RDAMPK(M) + COEFF5)*AREA(M)                 TRUST2 56
         CALL TRUSK(W,DET,SHG,XS,XL,B,C(1,1,M),DMAT,DB,ELEFFM,         TRUST2 57
     &              CONSTK,NEN,NINT,NESD,NROWSH,NROWB,NSTR,NEE)        TRUST2 58
C                                                                      TRUST2 59
      ENDIF                                                            TRUST2 60
C                                                                      TRUST2 61
C.... ASSEMBLE ELEMENT EFFECTIVE MASS MATRIX INTO GLOBAL               TRUST2 62
C         LEFT-HAND-SIDE MATRIX                                        TRUST2 63
C                                                                      TRUST2 64
      CALL ADDLHS(ALHS,ELEFFM,IDIAG,LM(1,1,NEL),NEE,LDIAG)             TRUST2 65
C                                                                      TRUST2 66
  100 CONTINUE                                                         TRUST2 67
C                                                                      TRUST2 68
      RETURN                                                           TRUST2 69
      END                                                              TRUST2 70
C*******************************************************************************
      SUBROUTINE TRUST3(MAT     ,IEN     ,DPRED ,DL     ,VPRED ,VL    , TRUST3  1
     &              A       ,AL     ,RDAMPK,RDAMPM,RHO    ,GRAV  ,      TRUST3  2
     &              ELRESF,X       ,XL     ,DET   ,SHL    ,SHG   ,      TRUST3  3
     &              G1      ,WORK   ,AREA   ,W     ,ELEFFM,B     ,      TRUST3  4
     &              STRAIN,C       ,DMAT    ,STRESS,BRHS  ,LM    ,      TRUST3  5
     &              XS      ,                                          TRUST3  6
     &              NUMEL  ,NED    ,NEN    ,NDOF  ,LDYN  ,NEE   ,       TRUST3  7
     &              IMASS  ,NESD   ,LFBODY,NSD    ,NINT  ,NROWSH,       TRUST3  8
     &              NEG    ,NROWB )                                     TRUST3  9
C                                                                      TRUST3 10
C.... PROGRAM TO CALCULATE RESIDUAL-FORCE VECTOR FOR THE               TRUST3 11
C         THREE-DIMENSIONAL, ELASTIC TRUSS ELEMENT AND                 TRUST3 12
C         ASSEMBLE INTO THE GLOBAL RIGHT-HAND-SIDE VECTOR              TRUST3 13
C                                                                      TRUST3 14
      IMPLICIT DOUBLE PRECISION (A-H,O-Z)                              TRUST3 15
C                                                                      TRUST3 16
C.... DEACTIVATE ABOVE CARD(S) FOR SINGLE-PRECISION OPERATION          TRUST3 17
C                                                                      TRUST3 18
      LOGICAL LDYN,FORMMA,FORMKD,ZEROAL,ZERODL,ZEROG,LNODE3            TRUST3 19
      DIMENSION MAT(1),IEN(NEN,1),DPRED(NDOF,1),DL(NED,1),VPRED(NDOF,1),TRUST3 20
     &              VL(NED,1),A(NDOF,1),AL(NED,1),RDAMPK(1),RDAMPM(1), TRUST3 21
     &              RHO(1),GRAV(1),ELRESF(1),X(NSD,1),XL(NESD,1),DET(1),TRUST3 22
     &              SHL(NROWSH,NEN,1),SHG(NROWSH,NEN,1),G1(1),WORK(1),  TRUST3 23
     &              AREA(1),W(1),ELEFFM(NEE,1),B(NROWB,1),STRAIN(1),    TRUST3 24
     &              C(NROWB,NROWB,1),DMAT(NROWB,1),STRESS(1),BRHS(1),   TRUST3 25
     &              LM(NED,NEN,1),XS(NESD,1)                            TRUST3 26
      COMMON /CONSTS/ ZERO,PT1667,PT25,PT5,ONE,TWO,THREE,FOUR,FIVE      TRUST3 27
C                                                                      TRUST3 28
      DO 500 NEL=1,NUMEL                                               TRUST3 29
C                                                                      TRUST3 30
      FORMMA = .FALSE.                                                 TRUST3 31
      FORMKD = .FALSE.                                                 TRUST3 32
      M = MAT(NEL)                                                     TRUST3 33
C                                                                      TRUST3 34
C.... NOTE: FOR STATIC ANALYSIS MPDPRD = MPD, HENCE REFERENCE          TRUST3 35
C         TO ARRAY "DPRED" WILL ACCESS THE CONTENTS OF ARRAY "D".      TRUST3 36
C                                                                      TRUST3 37
      CALL LOCAL(IEN(1,NEL),DPRED,DL,NEN,NDOF,NED)                     TRUST3 38
      IF (LDYN) THEN                                                   TRUST3 39
C                                                                      TRUST3 40
         CALL LOCAL(IEN(1,NEL),VPRED,VL,NEN,NDOF,NED)                 TRUST3 41
         CALL LOCAL(IEN(1,NEL),A,AL,NEN,NDOF,NED)                     TRUST3 42
C                                                                      TRUST3 43
         DO 200 J=1,NEN                                                TRUST3 44
C                                                                      TRUST3 45
         DO 100 I=1,NED                                                TRUST3 46
         DL(I,J) = DL(I,J) + RDAMPK(M)*VL(I,J)                         TRUST3 47
         AL(I,J) = AL(I,J) + RDAMPM(M)*VL(I,J)                         TRUST3 48
  100    CONTINUE                                                      TRUST3 49
C                                                                      TRUST3 50
  200    CONTINUE                                                      TRUST3 51
C                                                                      TRUST3 52
         CALL ZTEST(AL,NEE,ZEROAL)                                     TRUST3 53
         IF ( (.NOT.ZEROAL) .AND. (IMASS.NE.2) .AND. (RHO(M).NE.ZERO) ) TRUST3 54
     &        FORMMA = .TRUE.                                          TRUST3 55
C                                                                      TRUST3 56
      ELSE                                                             TRUST3 57
```

```
C                                                                   TRUST3    58
            CALL CLEAR(AL,NEE)                                      TRUST3    59
C                                                                   TRUST3    60
         ENDIF                                                      TRUST3    61
C                                                                   TRUST3    62
         CALL ZTEST(DL,NEE,ZERODL)                                  TRUST3    63
         IF (.NOT.ZERODL) FORMKD = .TRUE.                           TRUST3    64
         CALL ZTEST(GRAV,NESD,ZEROG)                                TRUST3    65
C                                                                   TRUST3    66
         IF ((.NOT.ZEROG) .AND. (LFBODY.NE.0) .AND. (RHO(M).NE.ZERO) TRUST3   67
      &       .AND. (IMASS.NE.2)) THEN                              TRUST3    68
            FORMMA = .TRUE.                                         TRUST3    69
            DO 400 I=1,NED                                          TRUST3    70
            WORK(I) = GRAV(I)*G1(LFBODY)                            TRUST3    71
C                                                                   TRUST3    72
            DO 300 J=1,NEN                                          TRUST3    73
            AL(I,J) = AL(I,J) - WORK(I)                             TRUST3    74
  300       CONTINUE                                                TRUST3    75
C                                                                   TRUST3    76
  400       CONTINUE                                                TRUST3    77
C                                                                   TRUST3    78
         ENDIF                                                      TRUST3    79
C                                                                   TRUST3    80
         IF (FORMMA.OR.FORMKD) THEN                                 TRUST3    81
C                                                                   TRUST3    82
            CALL CLEAR(ELRESF,NEE)                                  TRUST3    83
            CALL LOCAL(IEN(1,NEL),X,XL,NEN,NSD,NESD)                TRUST3    84
            LNODE3 = .TRUE.                                         TRUST3    85
            IF (NEN.EQ.3 .AND. IEN(2,NEL).EQ.IEN(3,NEL)) LNODE3 = .FALSE. TRUST3 86
            CALL TRUSHG(XL,DET,SHL,SHG,XS,NEN,NINT,NEL,NEG,LNODE3)  TRUST3    87
C                                                                   TRUST3    88
            IF (FORMMA) THEN                                        TRUST3    89
C                                                                   TRUST3    90
C......... FORM INERTIAL AND/OR BODY FORCE                          TRUST3    91
C                                                                   TRUST3    92
               CONSTM = - AREA(M)*RHO(M)                            TRUST3    93
               CALL CONTMA(SHG,XL,W,DET,AL,ELEFFM,WORK,ELRESF,CONSTM,IMASS, TRUST3 94
      &                NINT,NROWSH,NESD,NEN,NED,NEE)                TRUST3    95
            ENDIF                                                   TRUST3    96
C                                                                   TRUST3    97
            IF (FORMKD) THEN                                        TRUST3    98
C                                                                   TRUST3    99
C......... FORM INTERNAL FORCE                                      TRUST3   100
C                                                                   TRUST3   101
               CONSTK = - AREA(M)                                  TRUST3   102
               CALL TRUSKD(W,DET,SHG,XS,XL,B,DL,STRAIN,C(1,1,M),DMAT, TRUST3  103
      &                STRESS,WORK,ELRESF,CONSTK,NEN,NINT,NROWSH,   TRUST3   104
      &                NESD,NROWB,NEE)                              TRUST3   105
            ENDIF                                                   TRUST3   106
C                                                                   TRUST3   107
            CALL ADDRHS(BRHS,ELRESF,LM(1,1,NEL),NEE)               TRUST3   108
C                                                                   TRUST3   109
         ENDIF                                                      TRUST3   110
C                                                                   TRUST3   111
  500 CONTINUE                                                      TRUST3   112
C                                                                   TRUST3   113
      RETURN                                                        TRUST3   114
      END                                                           TRUST3   115
C*****************************************************************************
      SUBROUTINE TRUST4(MAT    ,IEN    ,D      ,DL     ,X      ,XL    , TRUST4  1
      &                 DET    ,SHL    ,SHG    ,XS     ,XINT   ,B     , TRUST4  2
      &                 STRAIN,C      ,STRESS,FORCE  ,AREA   ,        TRUST4  3
      &                 NINT   ,NUMEL ,NEN    ,NDOF   ,NED    ,NSD   , TRUST4  4
      &                 NESD   ,NROWSH,NEG    ,NROWB ,NEE    )        TRUST4  5
C                                                                   TRUST4    6
C.... PROGRAM TO CALCULATE AND PRINT STRESS, STRAIN AND FORCE FOR THE TRUST4  7
C         THREE-DIMENSIONAL, ELASTIC TRUSS ELEMENT                  TRUST4    8
C                                                                   TRUST4    9
      IMPLICIT DOUBLE PRECISION (A-H,O-Z)                           TRUST4   10
C                                                                   TRUST4   11
C.... DEACTIVATE ABOVE CARD(S) FOR SINGLE-PRECISION OPERATION       TRUST4   12
C                                                                   TRUST4   13
      LOGICAL LNODE3                                                TRUST4   14
      DIMENSION MAT(1),IEN(NEN,1),D(NDOF,1),DL(NED,1),X(NSD,1),     TRUST4   15
      &          XL(NESD,1),DET(1),SHL(NROWSH,NEN,1),SHG(NROWSH,NEN,1), TRUST4 16
      &          XS(NESD,1),XINT(NESD,1),B(NROWB,1),STRAIN(1),      TRUST4   17
      &          C(NROWB,NROWB,1),STRESS(1),FORCE(1),AREA(1)        TRUST4   18
C                                                                   TRUST4   19
      NNTOT = 24                                                    TRUST4   20
      NN = 0                                                        TRUST4   21
C                                                                   TRUST4   22
      DO 300 NEL=1,NUMEL                                            TRUST4   23
C                                                                   TRUST4   24
      M = MAT(NEL)                                                  TRUST4   25
      CALL LOCAL(IEN(1,NEL),D,DL,NEN,NDOF,NED)                      TRUST4   26
      CALL LOCAL(IEN(1,NEL),X,XL,NEN,NSD,NESD)                      TRUST4   27
      LNODE3 = .TRUE.                                               TRUST4   28
      IF ( NEN .EQ. 3 .AND. IEN(2,NEL).EQ.IEN(3,NEL) ) LNODE3 = .FALSE. TRUST4 29
      CALL TRUSHG(XL,DET,SHL,SHG,XS,NEN,NINT,NEL,NEG,LNODE3)        TRUST4   30
```

```
C                                                                       TRUST4  31
C.... LOOP OVER INTEGRATION POINTS                                      TRUST4  32
C                                                                       TRUST4  33
      DO 200 L=1,NINT                                                   TRUST4  34
C                                                                       TRUST4  35
C.... CALCULATE COORDINATES OF INTEGRATION POINTS                       TRUST4  36
C                                                                       TRUST4  37
      DO 100 I=1,NESD                                                   TRUST4  38
      XINT(I,L) = ROWDOT(SHG(NROWSH,1,L),XL(I,1),NROWSH,NESD,NEN)       TRUST4  39
  100 CONTINUE                                                          TRUST4  40
C                                                                       TRUST4  41
C.... CALCULATE STRESS, STRAIN AND FORCE                                TRUST4  42
C                                                                       TRUST4  43
      CALL TRUSTR(SHG(1,1,L),XS(1,L),B,DL,STRAIN,C(1,1,M),STRESS,       TRUST4  44
     &            FORCE,AREA(M),NROWSH,NESD,NROWB,NEN,NEE)              TRUST4  45
C                                                                       TRUST4  46
C.... PRINT STRESS, STRAIN AND FORCE                                    TRUST4  47
C                                                                       TRUST4  48
      CALL TRUSPT(XINT(1,L),STRESS,FORCE,STRAIN,NN,NNTOT,NEG,NEL,L)     TRUST4  49
C                                                                       TRUST4  50
  200 CONTINUE                                                          TRUST4  51
C                                                                       TRUST4  52
  300 CONTINUE                                                          TRUST4  53
C                                                                       TRUST4  54
      RETURN                                                            TRUST4  55
      END                                                               TRUST4  56
C**********************************************************************  TRUST5   1
      SUBROUTINE TRUST5(ISHIST,SOUT   ,MAT    ,IEN    ,D     ,DL     ,  TRUST5   1
     &                  X      ,XL     ,DET    ,SHL    ,SHG   ,XS     ,  TRUST5   2
     &                  B      ,STRAIN,C      ,STRESS,FORCE ,AREA    ,  TRUST5   3
     &                  WORK   ,                                        TRUST5   4
     &                  NSOUT  ,NEN    ,NDOF   ,NED    ,NSD   ,NESD   ,  TRUST5   5
     &                  NROWSH,NINT    ,NEG    ,NROWB  ,NEE   )          TRUST5   6
C                                                                       TRUST5   7
C.... PROGRAM TO CALCULATE AND STORE ELEMENT TIME-HISTORIES FOR THE     TRUST5   8
C        THREE-DIMENSIONAL, ELASTIC TRUSS ELEMENT                       TRUST5   9
C                                                                       TRUST5  10
      IMPLICIT DOUBLE PRECISION (A-H,O-Z)                               TRUST5  11
C                                                                       TRUST5  12
C.... DEACTIVATE ABOVE CARD(S) FOR SINGLE-PRECISION OPERATION           TRUST5  13
C                                                                       TRUST5  14
      REAL SOUT                                                         TRUST5  15
      LOGICAL LNODE3                                                    TRUST5  16
      DIMENSION ISHIST(3,1),SOUT(NSOUT+1,1),MAT(1),IEN(NEN,1),D(NDOF,1),TRUST5  17
     &          DL(NED,1),X(NSD,1),XL(NESD,1),DET(1),SHL(NROWSH,NEN,1), TRUST5  18
     &          SHG(NROWSH,NEN,1),XS(NESD,1),B(NROWB,1),STRAIN(1),       TRUST5  19
     &          C(NROWB,NROWB,1),STRESS(1),FORCE(1),AREA(1),WORK(1)      TRUST5  20
      COMMON /HPLOTC/ NPLPTS,LOCPLT,TIME                                TRUST5  21
C                                                                       TRUST5  22
      SOUT(1,LOCPLT) = REAL(TIME)                                       TRUST5  23
C                                                                       TRUST5  24
      DO 100 I=1,NSOUT                                                  TRUST5  25
C                                                                       TRUST5  26
      NEL   = ISHIST(1,I)                                               TRUST5  27
      INTPT = ISHIST(2,I)                                               TRUST5  28
      NCOMP = ISHIST(3,I)                                               TRUST5  29
C                                                                       TRUST5  30
      M = MAT(NEL)                                                      TRUST5  31
      CALL LOCAL(IEN(1,NEL),D,DL,NEN,NDOF,NED)                          TRUST5  32
      CALL LOCAL(IEN(1,NEL),X,XL,NEN,NSD,NESD)                          TRUST5  33
      LNODE3 = .TRUE.                                                   TRUST5  34
      IF ( NEN .EQ. 3 .AND. IEN(2,NEL).EQ.IEN(3,NEL) ) LNODE3 = .FALSE. TRUST5  35
      CALL TRUSHG(XL,DET,SHL,SHG,XS,NEN,NINT,NEL,NEG,LNODE3)            TRUST5  36
C                                                                       TRUST5  37
      CALL TRUSTR(SHG(1,1,INTPT),XS(1,INTPT),B,DL,STRAIN,C(1,1,M),       TRUST5  38
     &            STRESS,FORCE,AREA(M),NROWSH,NESD,NROWB,NEN,NEE)        TRUST5  39
C                                                                       TRUST5  40
      WORK(1) = STRESS(1)                                               TRUST5  41
      WORK(2) = FORCE(1)                                                TRUST5  42
      WORK(3) = STRAIN(1)                                               TRUST5  43
      SOUT(I+1,LOCPLT) = REAL(WORK(NCOMP))                              TRUST5  44
C                                                                       TRUST5  45
  100 CONTINUE                                                          TRUST5  46
C                                                                       TRUST5  47
      RETURN                                                            TRUST5  48
      END                                                               TRUST5  49
C**********************************************************************  TRUSB
      SUBROUTINE TRUSB(B,SHG,XS,NEN,NESD,NROWB,NROWSH)                   TRUSB    1
C                                                                       TRUSB    2
C.... PROGRAM TO SET UP THE STRAIN-DISPLACEMENT MATRIX FOR THE          TRUSB    3
C        THREE-DIMENSIONAL, ELASTIC TRUSS ELEMENT                       TRUSB    4
C                                                                       TRUSB    5
      IMPLICIT DOUBLE PRECISION (A-H,O-Z)                               TRUSB    6
C                                                                       TRUSB    7
C.... DEACTIVATE ABOVE CARD(S) FOR SINGLE PRECISION OPERATION           TRUSB    8
C                                                                       TRUSB    9
      DIMENSION B(NROWB,1),SHG(NROWSH,1),XS(1)                          TRUSB   10
```

```
C                                                                    TRUSB  11
       DO 200 J=1,NEN                                                TRUSB  12
       K = (J - 1)*NESD                                              TRUSB  13
C                                                                    TRUSB  14
       DO 100 I=1,NESD                                               TRUSB  15
       B(1,K+I) = SHG(1,J)*XS(I)                                     TRUSB  16
  100 CONTINUE                                                       TRUSB  17
C                                                                    TRUSB  18
  200 CONTINUE                                                       TRUSB  19
C                                                                    TRUSB  20
       RETURN                                                        TRUSB  21
       END                                                           TRUSB  22
C********************************************************************************
       SUBROUTINE TRUSHG(XL,DET,SHL,SHG,XS,NEN,NINT,NEL,NEG,LNODE3)  TRUSHG  1
C                                                                    TRUSHG  2
C.... PROGRAM TO CALCULATE GLOBAL DERIVATIVES OF SHAPE FUNCTIONS     TRUSHG  3
C        AND JACOBIAN DETERMINANTS FOR THE THREE-DIMENSIONAL,        TRUSHG  4
C        ELASTIC TRUSS ELEMENT                                       TRUSHG  5
C                                                                    TRUSHG  6
C          XL(J,L) = GLOBAL COORDINATES OF NODAL POINTS              TRUSHG  7
C         SHL(1,I,L) = LOCAL ("XI") DERIVATIVE OF SHAPE FUNCTION     TRUSHG  8
C         SHL(2,I,L) = SHAPE FUNCTION                                TRUSHG  9
C         SHG(1,I,L) = GLOBAL ("ARC-LENGTH") DERIVATIVE OF SHAPE FTN TRUSHG  10
C         SHG(2,I,L) = SHL(2,I,L)                                    TRUSHG  11
C          XS(J,L) = JTH COMPONENT OF THE LOCAL DERIVATIVE           TRUSHG  12
C                    OF THE POSITION VECTOR; THEN SCALED TO          TRUSHG  13
C                    DIRECTION COSINE                                TRUSHG  14
C          DET(L) = EUCLIDEAN LENGTH OF XS                           TRUSHG  15
C               I = LOCAL NODE NUMBER                                TRUSHG  16
C               J = GLOBAL COORDINATE NUMBER                         TRUSHG  17
C               L = INTEGRATION-POINT NUMBER                         TRUSHG  18
C            NINT = NUMBER OF INTEGRATION POINTS                     TRUSHG  19
C                                                                    TRUSHG  20
       IMPLICIT DOUBLE PRECISION (A-H,O-Z)                           TRUSHG  21
C                                                                    TRUSHG  22
C.... DEACTIVATE ABOVE CARD(S) FOR SINGLE PRECISION OPERATION        TRUSHG  23
C                                                                    TRUSHG  24
       LOGICAL LNODE3                                                TRUSHG  25
       DIMENSION XL(3,1),DET(1),SHL(2,NEN,1),SHG(2,NEN,1),XS(3,1)    TRUSHG  26
       COMMON /CONSTS/ ZERO,PT1667,PT25,PT5,ONE,TWO,THREE,FOUR,FIVE  TRUSHG  27
       COMMON /IOUNIT/ IIN,IOUT,IRSIN,IRSOUT                         TRUSHG  28
C                                                                    TRUSHG  29
       CALL MOVE(SHG,SHL,2*NEN*NINT)                                 TRUSHG  30
C                                                                    TRUSHG  31
       DO 400 L=1,NINT                                               TRUSHG  32
       IF (.NOT.LNODE3) THEN                                         TRUSHG  33
          TEMP = PT5*SHG(1,3,L)                                      TRUSHG  34
          SHG(1,1,L) = SHG(1,1,L) + TEMP                            TRUSHG  35
          SHG(1,2,L) = SHG(1,2,L) + TEMP                            TRUSHG  36
          TEMP = PT5*SHG(2,3,L)                                      TRUSHG  37
          SHG(2,1,L) = SHG(2,1,L) + TEMP                            TRUSHG  38
          SHG(2,2,L) = SHG(2,2,L) + TEMP                            TRUSHG  39
       ENDIF                                                         TRUSHG  40
       DET(L) = ZERO                                                 TRUSHG  41
C                                                                    TRUSHG  42
       DO 100 J=1,3                                                  TRUSHG  43
       XS(J,L) = ROWDOT(SHL(1,1,L),XL(J,1),2,3,NEN)                  TRUSHG  44
       DET(L) = DET(L) + XS(J,L)**2                                  TRUSHG  45
  100 CONTINUE                                                       TRUSHG  46
C                                                                    TRUSHG  47
       DET(L) = SQRT(DET(L))                                         TRUSHG  48
C                                                                    TRUSHG  49
       IF (DET(L).LE.ZERO) THEN                                      TRUSHG  50
          WRITE(IOUT,1000) NEL,NEG                                   TRUSHG  51
          STOP                                                       TRUSHG  52
       ENDIF                                                         TRUSHG  53
C                                                                    TRUSHG  54
       DO 200 J=1,3                                                  TRUSHG  55
       XS(J,L) = XS(J,L)/DET(L)                                      TRUSHG  56
  200 CONTINUE                                                       TRUSHG  57
C                                                                    TRUSHG  58
       DO 300 I=1,NEN                                                TRUSHG  59
       SHG(1,I,L) = SHL(1,I,L)/DET(L)                                TRUSHG  60
  300 CONTINUE                                                       TRUSHG  61
C                                                                    TRUSHG  62
  400 CONTINUE                                                       TRUSHG  63
C                                                                    TRUSHG  64
       RETURN                                                        TRUSHG  65
C                                                                    TRUSHG  66
 1000 FORMAT('1','NON-POSITIVE DETERMINANT IN ELEMENT NUMBER  ',I5,  TRUSHG  67
      &            ' IN ELEMENT GROUP  ',I5)                         TRUSHG  68
       END                                                           TRUSHG  69
C********************************************************************************
       SUBROUTINE TRUSHL(SHL,W,NINT,NEN)                             TRUSHL  1
C                                                                    TRUSHL  2
C.... PROGRAM TO CALCULATE INTEGRATION-RULE WEIGHTS, SHAPE FUNCTIONS TRUSHL  3
C        AND LOCAL DERIVATIVES FOR A TWO OR THREE NODE,              TRUSHL  4
C        ONE-DIMENSIONAL ELEMENT                                     TRUSHL  5
```

```
C                                                                        TRUSHL    6
C                  R = LOCAL ELEMENT COORDINATE ("XI")                   TRUSHL    7
C         SHL(1,I,L) = LOCAL ("XI") DERIVATIVE OF SHAPE FUNCTION         TRUSHL    8
C         SHL(2,I,L) = SHAPE FUNCTION                                    TRUSHL    9
C               W(L) = INTEGRATION-RULE WEIGHT                           TRUSHL   10
C                  I = LOCAL NODE NUMBER                                 TRUSHL   11
C                  L = INTEGRATION-POINT NUMBER                          TRUSHL   12
C               NINT = NUMBER OF INTEGRATION POINTS, EQ. 1, 2 OR 3       TRUSHL   13
C                                                                        TRUSHL   14
      IMPLICIT DOUBLE PRECISION (A-H,O-Z)                                TRUSHL   15
C                                                                        TRUSHL   16
C.... DEACTIVATE ABOVE CARD(S) FOR SINGLE PRECISION OPERATION            TRUSHL   17
C                                                                        TRUSHL   18
      DIMENSION SHL(2,NEN,1),W(1),RA(3)                                  TRUSHL   19
      COMMON /CONSTS/ ZERO,PT1667,PT25,PT5,ONE,TWO,THREE,FOUR,FIVE       TRUSHL   20
      DATA RA/-1.00,1.00,0.00/,                                         TRUSHL   21
     &     FIVE9/0.5555555555555555/,EIGHT9/0.8888888888888888/         TRUSHL   22
C                                                                        TRUSHL   23
      IF (NINT.EQ.1) THEN                                                TRUSHL   24
         W(1) = TWO                                                      TRUSHL   25
         G = ZERO                                                        TRUSHL   26
      ENDIF                                                              TRUSHL   27
C                                                                        TRUSHL   28
      IF (NINT.EQ.2) THEN                                                TRUSHL   29
         W(1) = ONE                                                      TRUSHL   30
         W(2) = ONE                                                      TRUSHL   31
         G = ONE/SQRT(THREE)                                            TRUSHL   32
      ENDIF                                                              TRUSHL   33
C                                                                        TRUSHL   34
      IF (NINT.EQ.3) THEN                                                TRUSHL   35
         W(1) = FIVE9                                                    TRUSHL   36
         W(2) = FIVE9                                                    TRUSHL   37
         W(3) = EIGHT9                                                   TRUSHL   38
         G = SQRT(THREE/FIVE)                                           TRUSHL   39
      ENDIF                                                              TRUSHL   40
C                                                                        TRUSHL   41
      DO 100 L=1,NINT                                                    TRUSHL   42
      R = G*RA(L)                                                        TRUSHL   43
C                                                                        TRUSHL   44
      SHL(1,1,L) = - PT5                                                 TRUSHL   45
      SHL(1,2,L) =   PT5                                                 TRUSHL   46
      SHL(2,1,L) =   PT5*(ONE - R)                                       TRUSHL   47
      SHL(2,2,L) =   PT5*(ONE + R)                                       TRUSHL   48
C                                                                        TRUSHL   49
      IF (NEN.EQ.3) THEN                                                 TRUSHL   50
         SHL(1,3,L) = - TWO*R                                            TRUSHL   51
         SHL(2,3,L) = ONE - R**2                                         TRUSHL   52
C                                                                        TRUSHL   53
         TEMP = - PT5*SHL(2,3,L)                                         TRUSHL   54
         SHL(1,1,L) = SHL(1,1,L) + R                                     TRUSHL   55
         SHL(1,2,L) = SHL(1,2,L) + R                                     TRUSHL   56
         SHL(2,1,L) = SHL(2,1,L) + TEMP                                  TRUSHL   57
         SHL(2,2,L) = SHL(2,2,L) + TEMP                                  TRUSHL   58
C                                                                        TRUSHL   59
      ENDIF                                                              TRUSHL   60
C                                                                        TRUSHL   61
  100 CONTINUE                                                           TRUSHL   62
C                                                                        TRUSHL   63
      RETURN                                                             TRUSHL   64
      END                                                                TRUSHL   65
C************************************************************************
      SUBROUTINE TRUSK(W,DET,SHG,XS,XL,B,C,DMAT,DB,ELSTIF,CONSTK,        TRUSK     1
     &                 NEN,NINT,NESD,NROWSH,NROWB,NSTR,NEE)              TRUSK     2
C                                                                        TRUSK     3
C.... PROGRAM TO FORM STIFFNESS MATRIX FOR THE                           TRUSK     4
C        THREE-DIMENSIONAL, ELASTIC TRUSS ELEMENT                        TRUSK     5
C                                                                        TRUSK     6
      IMPLICIT DOUBLE PRECISION (A-H,O-Z)                                TRUSK     7
C                                                                        TRUSK     8
C.... DEACTIVATE ABOVE CARD(S) FOR SINGLE PRECISION OPERATION            TRUSK     9
C                                                                        TRUSK    10
      DIMENSION W(1),DET(1),SHG(NROWSH,NEN,1),XS(NESD,1),XL(NESD,1),     TRUSK    11
     &          B(NROWB,1),DB(NROWB,1),ELSTIF(NEE,1)                     TRUSK    12
C                                                                        TRUSK    13
C.... LOOP ON INTEGRATION POINTS                                         TRUSK    14
C                                                                        TRUSK    15
      DO 100 L=1,NINT                                                    TRUSK    16
      TEMP1 = CONSTK*W(L)*DET(L)                                         TRUSK    17
C                                                                        TRUSK    18
C.... SET UP THE STRAIN-DISPLACEMENT MATRIX                              TRUSK    19
C                                                                        TRUSK    20
      CALL TRUSB(B,SHG(1,1,L),XS(1,L),NEN,NESD,NROWB,NROWSH)             TRUSK    21
C                                                                        TRUSK    22
C.... SET UP THE CONSTITUTIVE "MATRIX"                                   TRUSK    23
C                                                                        TRUSK    24
      DMAT = C*TEMP1                                                     TRUSK    25
C                                                                        TRUSK    26
C.... MULTIPLY DMAT * B                                                  TRUSK    27
```

```
C                                                                         TRUSK  28
        CALL SMULT(DMAT,B,DB,NROWB,NROWB,NSTR,NEE,1)                       TRUSK  29
C                                                                         TRUSK  30
C.... MULTIPLY B(TRANSPOSE) * DB, TAKING ACCOUNT OF SYMMETRY,             TRUSK  31
C        AND ACCUMULATE IN ELSTIF                                         TRUSK  32
C                                                                         TRUSK  33
        CALL BTDB(ELSTIF,B,DB,NEE,NROWB,NSTR)                             TRUSK  34
C                                                                         TRUSK  35
  100 CONTINUE                                                            TRUSK  36
C                                                                         TRUSK  37
        RETURN                                                            TRUSK  38
        END                                                               TRUSK  39
C*********************************************************************     *
        SUBROUTINE TRUSKD(W,DET,SHG,XS,XL,B,DL,STRAIN,C,DMAT,STRESS,WORK,  TRUSKD  1
     &              ELRESF,CONSTK,NEN,NINT,NROWSH,NESD,NROWB,NEE)          TRUSKD  2
C                                                                         TRUSKD  3
C.... PROGRAM TO FORM INTERNAL FORCE ("-K*D") FOR THE                     TRUSKD  4
C        THREE-DIMENSIONAL, ELASTIC TRUSS ELEMENT                         TRUSKD  5
C                                                                         TRUSKD  6
        IMPLICIT DOUBLE PRECISION (A-H,O-Z)                               TRUSKD  7
C                                                                         TRUSKD  8
C.... DEACTIVATE ABOVE CARD(S) FOR SINGLE-PRECISION OPERATION             TRUSKD  9
C                                                                         TRUSKD 10
        DIMENSION W(1),DET(1),SHG(NROWSH,NEN,1),XS(NESD,1),XL(NESD,1),     TRUSKD 11
     &            B(NROWB,1),DL(1),WORK(1),ELRESF(1)                       TRUSKD 12
C                                                                         TRUSKD 13
C.... LOOP ON INTEGRATION POINTS                                          TRUSKD 14
C                                                                         TRUSKD 15
        DO 100 L=1,NINT                                                   TRUSKD 16
        TEMP = CONSTK*W(L)*DET(L)                                         TRUSKD 17
C                                                                         TRUSKD 18
C.... SET UP THE STRAIN-DISPLACEMENT MATRIX                               TRUSKD 19
C                                                                         TRUSKD 20
        CALL TRUSB(B,SHG(1,1,L),XS,NEN,NESD,NROWB,NROWSH)                  TRUSKD 21
C                                                                         TRUSKD 22
C.... CALCULATE STRAIN                                                    TRUSKD 23
C                                                                         TRUSKD 24
        STRAIN = RCDOT(B,DL,NROWB,NEE)                                    TRUSKD 25
C                                                                         TRUSKD 26
C.... CALCULATE STRESS                                                    TRUSKD 27
C                                                                         TRUSKD 28
        STRESS = C*STRAIN                                                 TRUSKD 29
C                                                                         TRUSKD 30
C.... CALCULATE ELEMENT INTERNAL FORCE                                    TRUSKD 31
C                                                                         TRUSKD 32
        STRESS = TEMP*STRESS                                              TRUSKD 33
        CALL SMULT(STRESS,B,WORK,NROWB,1,1,NEE,1)                         TRUSKD 34
        CALL MATADD(ELRESF,WORK,ELRESF,NEE,NEE,NEE,NEE,1,1)               TRUSKD 35
C                                                                         TRUSKD 36
  100 CONTINUE                                                            TRUSKD 37
C                                                                         TRUSKD 38
        RETURN                                                            TRUSKD 39
        END                                                               TRUSKD 40
C*********************************************************************     *
        SUBROUTINE TRUSPR(RHO,RDAMPM,RDAMPK,AREA,C,NUMAT)                 TRUSPR  1
C                                                                         TRUSPR  2
C.... PROGRAM TO READ, WRITE AND STORE PROPERTIES FOR                     TRUSPR  3
C        THREE-DIMENSIONAL, ELASTIC TRUSS ELEMENT                         TRUSPR  4
C                                                                         TRUSPR  5
        IMPLICIT DOUBLE PRECISION (A-H,O-Z)                               TRUSPR  6
C                                                                         TRUSPR  7
C.... DEACTIVATE ABOVE CARD(S) FOR SINGLE-PRECISION OPERATION             TRUSPR  8
C                                                                         TRUSPR  9
        DIMENSION RHO(1),RDAMPM(1),RDAMPK(1),AREA(1),C(1)                 TRUSPR 10
        COMMON /CONSTS/ ZERO,PT1667,PT25,PT5,ONE,TWO,THREE,FOUR,FIVE      TRUSPR 11
        COMMON /IOUNIT/ IIN,IOUT,IRSIN,IRSOUT                             TRUSPR 12
C                                                                         TRUSPR 13
        DO 100 N=1,NUMAT                                                  TRUSPR 14
        IF (MOD(N,50).EQ.1) WRITE(IOUT,1000) NUMAT                        TRUSPR 15
        READ(IIN,2000) M,E,RHO(M),RDAMPM(M),RDAMPK(M),AREA(M)             TRUSPR 16
        WRITE(IOUT,3000) M,E,RHO(M),RDAMPM(M),RDAMPK(M),AREA(M)           TRUSPR 17
        C(M) = E                                                          TRUSPR 18
  100 CONTINUE                                                            TRUSPR 19
C                                                                         TRUSPR 20
        RETURN                                                            TRUSPR 21
C                                                                         TRUSPR 22
 1000 FORMAT('1',                                                         TRUSPR 23
     &' M A T E R I A L   S E T   D A T A                       //5X,     TRUSPR 24
     &' NUMBER OF MATERIAL SETS . . . . . . . . (NUMAT ) = ',I5///,       TRUSPR 25
     & 7X,'SET',5X,'YOUNG''S',6X,'MASS',8X,'MASS',                        TRUSPR 26
     & 6X,'STIFFNESS',6X,'AREA'/6X,'NUMBER',3X,'MODULUS',                 TRUSPR 27
     & 5X,'DENSITY',5X,'DAMPING',5X,'DAMPING',/)                          TRUSPR 28
 2000 FORMAT(I5,5X,7F10.0)                                                TRUSPR 29
 3000 FORMAT(4X,I5,3X,5(2X,1PE10.4))                                      TRUSPR 30
        END                                                               TRUSPR 31
C*********************************************************************     *
        SUBROUTINE TRUSPT(XINT,STRESS,FORCE,STRAIN,NN,NNTOT,NEG,NEL,LINT) TRUSPT  1
C                                                                         TRUSPT  2
C.... PROGRAM TO PRINT STRESS, STRAIN AND FORCE FOR THE                   TRUSPT  3
```

```
C          THREE-DIMENSIONAL, ELASTIC TRUSS ELEMENT            TRUSPT    4
C                                                              TRUSPT    5
      IMPLICIT DOUBLE PRECISION (A-H,O-Z)                      TRUSPT    6
C                                                              TRUSPT    7
C.... DEACTIVATE ABOVE CARD(S) FOR SINGLE-PRECISION OPERATION  TRUSPT    8
C                                                              TRUSPT    9
      DIMENSION XINT(3)                                        TRUSPT   10
      COMMON /IOUNIT/ IIN,IOUT,IRSIN,IRSOUT                    TRUSPT   11
C                                                              TRUSPT   12
      NN = NN + 1                                              TRUSPT   13
C                                                              TRUSPT   14
      IF (MOD(NN,NNTOT).EQ.1) THEN                             TRUSPT   15
         WRITE(IOUT,1000) NEG                                  TRUSPT   16
         NN = 1                                                TRUSPT   17
      ENDIF                                                    TRUSPT   18
C                                                              TRUSPT   19
      WRITE(IOUT,2000) NEL,LINT,XINT,STRESS,FORCE,STRAIN       TRUSPT   20
C                                                              TRUSPT   21
      RETURN                                                   TRUSPT   22
C                                                              TRUSPT   23
 1000 FORMAT('1',                                              TRUSPT   24
     &' E L E M E N T   S T R E S S E S   A N D   S T R A I N S',//5X,TRUSPT   25
     &' ELEMENT GROUP NUMBER  . . . . . . . (NEG  ) = ',I5///, TRUSPT   26
     &'   ELEMENT   INT. PT.           X1            X2            X3    ',5X,TRUSPT   27
     &'   STRESS    FORCE     STRAIN    ',/                    TRUSPT   28
     &'   NUMBER    NUMBER            COORD.        COORD.        COORD.   ')TRUSPT   29
 2000 FORMAT(/2X,I5,7X,I2,8X,3(1PE10.2),5X,3(1PE10.2))         TRUSPT   30
      END                                                      TRUSPT   31
C****************************************************************************
      SUBROUTINE TRUSTR(SHG,XS,B,DL,STRAIN,C,STRESS,FORCE,AREA,TRUSTR    1
     &                  NROWSH,NESD,NROWB,NEN,NEE)             TRUSTR    2
C                                                              TRUSTR    3
C.... PROGRAM TO CALCULATE STRESS, STRAIN AND FORCE AT AN INTEGRATION TRUSTR    4
C          POINT FOR THE THREE-DIMENSIONAL, ELASTIC TRUSS ELEMENT TRUSTR    5
C                                                              TRUSTR    6
      IMPLICIT DOUBLE PRECISION (A-H,O-Z)                      TRUSTR    7
C                                                              TRUSTR    8
C.... DEACTIVATE ABOVE CARD(S) FOR SINGLE PRECISION OPERATION  TRUSTR    9
C                                                              TRUSTR   10
      DIMENSION SHG(NROWSH,1),XS(NESD,1),B(NROWB,1),DL(1)      TRUSTR   11
      COMMON /CONSTS/ ZERO,PT1667,PT25,PT5,ONE,TWO,THREE,FOUR,FIVE TRUSTR   12
C                                                              TRUSTR   13
C.... SET UP STRAIN-DISPLACEMENT MATRIX                        TRUSTR   14
C                                                              TRUSTR   15
      CALL TRUSB(B,SHG,XS,NEN,NESD,NROWB,NROWSH)               TRUSTR   16
C                                                              TRUSTR   17
C.... CALCULATE STRAIN                                         TRUSTR   18
C                                                              TRUSTR   19
      STRAIN = RCDOT(B,DL,NROWB,NEE)                           TRUSTR   20
C                                                              TRUSTR   21
C.... CALCULATE STRESS                                         TRUSTR   22
C                                                              TRUSTR   23
      STRESS = C*STRAIN                                        TRUSTR   24
C                                                              TRUSTR   25
C.... CALCULATE FORCES                                         TRUSTR   26
C                                                              TRUSTR   27
      FORCE = AREA*STRESS                                      TRUSTR   28
C                                                              TRUSTR   29
      RETURN                                                   TRUSTR   30
      END                                                      TRUSTR   31
```

References   Section 11.2

1. E. Cuthill, "Several Strategies for Reducing the Bandwidth of Matrices," in *Sparse Matrices and Their Applications,* eds. D. J. Rose and R. A. Willoughby. New York: Plenum Press, 1972.

2. W.-H. Liu and A. H. Sherman, "Comparitive Analysis of the Cuthill-McKee and the Reversed Cuthill-McKee Ordering Algorithms for Sparse Matrices," *SIAM Journal on Numerical Analysis,* 13 (1976), 198–213.

3. N. E. Gibbs, W. G. Poole, Jr., and P. K. Stockmeyer, "An Algorithm for Reducing the Bandwidth and Profile of a Sparse Matrix," *SIAM Journal on Numerical Analysis,* 13 (1976), 236–250.

4. N. E. Gibbs, W. G. Poole, and P. K. Stockmeyer, "A Comparison of Several Bandwidth and Profile Reduction Algorithms," *Association for Computing Machinery Transactions of Mathematical Software*, 2 (1976), 322–330.

5. G. H. Golub and C. F. Van Loan, *Matrix Computations*. Baltimore: The Johns Hopkins University Press, 1983.

Index